Lecture Notes in Electrical Engineering

Volume 792

The book series *Lecture Notes in Electrical Engineering* (LNEE) publishes the latest developments in Electrical Engineering - quickly, informally and in high quality. While original research reported in proceedings and monographs has traditionally formed the core of LNEE, we also encourage authors to submit books devoted to supporting student education and professional training in the various fields and applications areas of electrical engineering. The series cover classical and emerging topics concerning:

- Communication Engineering, Information Theory and Networks
- Electronics Engineering and Microelectronics
- Signal, Image and Speech Processing
- Wireless and Mobile Communication
- Circuits and Systems
- Energy Systems, Power Electronics and Electrical Machines
- Electro-optical Engineering
- Instrumentation Engineering
- Avionics Engineering
- Control Systems
- Internet-of-Things and Cybersecurity
- Biomedical Devices, MEMS and NEMS

For general information about this book series, comments or suggestions, please contact leontina.dicecco@springer.com.

To submit a proposal or request further information, please contact the Publishing Editor in your country:

China

Jasmine Dou, Editor (jasmine.dou@springer.com)

India, Japan, Rest of Asia

Swati Meherishi, Editorial Director (Swati.Meherishi@springer.com)

Southeast Asia, Australia, New Zealand

Ramesh Nath Premnath, Editor (ramesh.premnath@springernature.com)

USA, Canada:

Michael Luby, Senior Editor (michael.luby@springer.com)

All other Countries:

Leontina Di Cecco, Senior Editor (leontina.dicecco@springer.com)

**** This series is indexed by EI Compendex and Scopus databases. ****

More information about this series at http://www.springer.com/series/7818

A. Sivasubramanian · Prasad N. Shastry ·
Pua Chang Hong
Editors

Futuristic Communication and Network Technologies

Select Proceedings of VICFCNT 2020

Springer

Editors
A. Sivasubramanian
School of Electronics Engineering
Vellore Institute of Technology
Chennai, Tamil Nadu, India

Pua Chang Hong
Department of Electrical and Electronic Engineering
Universiti Tunku Abdul Rahman
Bandar Sungai Long, Malaysia

Prasad N. Shastry
Department of Electrical and Computer Engineering
Bradley University
Peoria, IL, USA

ISSN 1876-1100 ISSN 1876-1119 (electronic)
Lecture Notes in Electrical Engineering
ISBN 978-981-16-4624-9 ISBN 978-981-16-4625-6 (eBook)
https://doi.org/10.1007/978-981-16-4625-6

This Springer imprint is published by the registered company Springer Nature Singapore Pte Ltd.
The registered company address is: 152 Beach Road, #21-01/04 Gateway East, Singapore 189721, Singapore

Preface

The rapid development in communication technologies is happening day by day. The new challenges posed and the ideas generated to solve problems need to be shared and discussed to bring awareness about necessary research in this field. This book is a collated work of researchers at the Virtual International Conference on Futuristic Communication and Network Technologies 2020 (VICFCNT 2020) conducted from November 6 to 7, 2020, at Vellore Institute of Technology (VIT) Chennai. The book aims to discuss issues, challenges, opportunities, and findings of research in communication and network technologies. The book provides research findings on all aspects in the emerging technology trends in the field of optical communication and networks, MEMS and sensors, active and passive RF components, and devices, antenna systems and applications, RF devices and antennas for microwave emerging technologies, wireless communication for future networks, signal and image processing, machine learning/AI for networks, Internet of Intelligent Things, network security, and blockchain technologies.

We are pleased to share the proceedings of VICFCNT 2020. We believe that the book would enrich the knowledge of the intellectuals and would lead to further exploration and innovative research in the various domains.

Chennai, India — A. Sivasubramanian
Peoria, USA — Prasad N. Shastry
Bandar Sungai Long, Malaysia — Pua Chang Hong

Acknowledgements

Our heartfelt gratitude to all the authors invited and keynote speakers, symposium co-chairs, committee members, and the organizing secretary of VICFCNT 2020 for their untiring efforts to make this conference a reality.

Our sincere thanks to the respected Chancellor Dr. G. Viswanathan for his outstanding support and blessings. Our deep sense of appreciation to Dr. Sekar Viswanathan, Vice President, VIT University, Chennai, for his tremendous support, enthusiasm, and encouragement to organize this great event in VIT University, Chennai, India.

Our sincere thanks to the Vice-chancellor and Pro-Vice-chancellor of VIT University for their support and encouragement.

Editors wish to acknowledge the essential contributions of publication committee members in coordinating and assembling the contributions for the book.

We want to thank and appreciate our publication partners, Springer LNEE (Mrs. Swati Meherishi Springer India), for bringing out this special VICFCNT 2020 proceeding as a book series.

The editorial team would like to record its acknowledgment of the effort and time devoted by the advisory editors, the members of the editorial board, and the researchers from various universities in reviewing the research papers submitted for the symposium, which ensured high quality in the review process.

Contents

About the Editors

Dr. A. Sivasubramanian has graduated in Electronics and Communication Engineering from University of Madras in the year 1990 and completed doctorate in Optical Communication from Anna University in the year 2008. His working experience in teaching and research includes 28 years. He is presently working as a Professor in the School of Electronics Engineering, VIT University, Chennai Campus. He has held various administrative positions in his work experience including as a Principal in Tagore Institute of Engineering and Technology and Head of the Department of ECE in St. Joseph's College of Engineering for a long span of around 18 years. He is also a reviewer in various peer-reviewed journals. He has acted as chairperson in various international conferences and had been the technical program committee member. He has delivered invited talks in various universities both at national and international level. He is guiding many research scholars in the area of optical communication and networks. His research contribution has led to around 45 publications in peer-reviewed journals and 32 conference publications, both at the national and international level. He has also served as a syllabus committee member in various universities. He had been to Taiwan to present his research work and established a joint collaboration between National Taiwan University of Science and Technology, Taipei and VIT University, Chennai. His research interests include optical communication and networks, non-linear optics, optical amplifiers, WDM Networks, free space optics, silicon photonics and energy harvesting for optical applications.

Dr. Prasad N. Shastry is a Professor of microwave and wireless engineering in the Department of Electrical and Computer Engineering at Bradley University, Peoria, Illinois. Since joining Bradley University in 1991, he has spearheaded the development of the curriculum, laboratories, and an active research program in the area of microwave and wireless engineering. He has established a National Science Foundation-supported advanced microwave engineering laboratory for testing MMICs, and an anechoic chamber for antenna measurements. His areas of research are wideband distributed architecture circuits, reconfigurable circuits and antennas, wireless power transfer, and ambient RF energy-harvesting. During the past thirty years, he has made significant contributions in his area of expertise in Industry-University collaborative projects sponsored by several leading microwave companies. Dr. Shastry was awarded a Ph.D. degree in microwave engineering by IIT-Bombay in 1980. He earned an M.Tech. degree in microwave and radar engineering from IIT-Kharagpur in 1974, and a Bachelor of Engineering (Electronics) degree from Bangalore University in 1972. He was awarded a Post-Doctoral Fellowship in the Department of Electrical and Computer Engineering at University of Wisconsin, Madison, in 1982. He is a fellow of the IETE, India, a senior member of IEEE, a member of IEEE Microwave Theory and Techniques Society, and a member of Sigma Xi.

Dr. Pua Chang Hong graduated in Applied Physics from University of Malaya with a B.Sc. in 2006, followed by a M.Sc. in 2009 and Ph.D. in 2012 in Photonics. His Master research involved the fabrication of optical waveguide for FTTh application and doctoral research in nonlinear optics field for acoustic sensing using fiber optic sensors. After graduation, he became a Research Fellow in University of Malaya to work on the development of new optical waveguide and as well as fiber sensors. Following University of Malaya, in 2014 he joined Universiti Tunku Abdul Rahman (UTAR) as Assistant Professor and he was appointed as the Chairperson of Centre for Photonics and Advanced Materials. His current focus in research is developing different types of optical sensors including fabricate tapering based Mach Zender interferometer for sensing applications, studying the nonlinear interaction of acousto-optic in fiber laser for different applications such as building structure health monitoring, pipeline leakage monitoring and localization, refrigerant leakage detection, and etc.

Deep Learning-Based Image Preprocessing Techniques for Crop Disease Identification

Nakhatha Arun Kumar and S. Sathish Kumar

Abstract Plants image preprocessing is a challenge as it is emerging area in the research for various disease detections and diagnoses which aids for the agriculturist for easy diagnosis of disease in percentage and applies appropriate fertilizers to get the disease-free crop. Preprocessing is being major operation with respect to image data at lowest level of abstraction. The aim of the preprocessing is to enhance the image features to avoid the distortion. Image preprocessing is very necessary aspect as the image should not have any impurities, and it is accomplished to be better for the further process such as segmentation, feature extraction, and classification. Collection of data is challenging because of the type of device the image captured and factors like climatic condition, etc. The factors like noise and other aspects need to be eliminated. It is necessary for the preprocessing to remove noise. Preprocessing is the process which is used to enhance the evaluation and interpretation of an image. Image preprocessing is executed systematically; then, the further detection of disease in plants will be successful.

Keywords Preprocessing · Convolutional neural network · Deep learning

1 Introduction

In India, 61.5% of population is dependent on agriculture, as the vegetables are the main source of the vitamins. In large scale, tomato and potato are the crops are grown and produced due to demand, and the advances in agriculture have made tomato crop to gain enormous commercial importance. The environmental condition and pollution affect the tomato, and more than hundreds of diseases exist in tomato. This impact on the quality of tomato is detrimental. The deep learning is the sub field of machine learning. The machine learning has achieved certain rate of accuracy. Later, deep learning is meant to be one of the efficient among the various trends.

N. A. Kumar (✉)
RNSIT Research Center, VTU, Bangaluru, India

S. Sathish Kumar
ISE Dept, RNSIT, VTU, Bengaluru, Karnataka, India

A. Sivasubramanian et al. (eds.), *Futuristic Communication and Network Technologies*, Lecture Notes in Electrical Engineering 792,
https://doi.org/10.1007/978-981-16-4625-6_1

Deep learning turned faster because of scalability when applied on large data set. In deep learning, various libraries exists, and among various libraries, tensor flow is one of the most popular deep learning libraries. Tensor flow is one of the open source libraries. Implementing machine learning techniques such as deep learning neural network on agriculture data has immense attention in recent times. Thus, the neural network is prominent strategy among various deep learning techniques. As in the agriculture crops are affected by the disease, prediction of the disease at early stage plays an important role. To identify the diseased plants and to apply the appropriate fertilizers are tedious. Hence, deep learning techniques are suitable for multiple classes preprocessing.

2 Related Results

The paper [1] have proposed a novel approach based on principle component analysis and artificial neural network. We compared the optimality of bands selected by both algorithms. To overcome the drawback, parallel convolutional neural network has been proposed. In the paper [2], they identified how much percentage of fruit is affected by disease, and it has been done using the deep learning techniques. They have also specified the work can be extended for the detecting disease in vegetables. In paper [3], they have applied the convolutional neural network with experimenting using three-layer model, four-layer model, five-layer model, and the result is obtained with 15 epochs. The study of paper [4] is that the applications of deep convolutional neural network has been formulated to classify and identify disease in plants. Hence, the work can be extended to classify the images captured in non-controlled environment and with multiple orientations. The paper [5] feature learning and data training occupies dominant place; image classification is tedious as the data is huge. The time required for the training the data is more; it is required to conduct a more comprehensive and in-depth study on image classification algorithm based on deep learning. In the paper [6], it is mentioned that the accuracy rate is 70% using transfer learning in identification and classification of floral images paper [7]. Image classification is done only on two types of images; the accuracy got is 88.31%. The enhancement can be done by gathering additional training data or streamlining the design including more layers and hyper-parameters of the system; the test accuracy can be significantly improved. In paper [8], the training is longer and batch size is smaller with decreased accuracy. As per paper [9], they have not chosen the pixel instead sub-statistical region merging has been chosen; various conventional methods have been compared, and more images operations can be tested, and identification accuracy can be increased. In paper 10, various techniques are employed in detection of diseases like back propagation neural network, support vector machine, K-means, color co-occurrence method, spatial gray-level dependencies, Otsu's algorithm. These techniques are used to detect healthy or diseased. The challenges specified in the system is that complex images captured in outdoor lightening and intense environment conditions need to be considered. In paper [10], the rate of accuracy is 90.9% for both prediction of backward and healthy segmentation. The accuracy obtained in segmentation is only 57%. So, in average, 79.65 was obtained from the algorithm proposed. In paper [11], the cause of the disease bacterial blight is Xanthomas axonopodis pv, punicae. The formation of disease takes place over the leaves, stem as well as on fruits. It reduces nearly 75% of yield, and the chemicals applied over the plants will not affect positively in curing the disease. It needs a periodic day-to-day observation. The canny edge detection technique is used for the detection of the disease. The grab cut segmentation is used to segment affected areas in the fruit. Pixel-based analysis takes place as pixel counts are considered; the infection is evaluated in fruits; the preventive measures will be taken, and chemical solutions are provided. In the paper [12], disease in plants detected by applying shape descriptors. Precision recall and retrieval efficiency has been improved. In the comparative study, one of the algorithm is K-means algorithm and its accuracy is high, but it is not for multiclass database. The process of feature extraction and disease classification is been done; feature extraction is being processed using statistic-based gray-level co-occurrence matrix and texture feature equations.

3 Data Collection and Disease Classes

The collection and maintaining the data in efficient manner plays an important role in the research. The reason of this section is to understand the data sets, diseases, and classes. The crop is affected by many diseases; tomato crop is affected by diseases like fungal, bacterial, viral, and nematodes. The data set consists of various diseases images. The diseases are bacterial spot, healthy, late blight, leaf mold, early blight, septoria_leaf_spot, spider mite, two spider mite, target spot, tomato mosaic virus, yellow curls. There are samples of images.

3.1 Diseases in Tomato Plants

The above-shown images (Fig. 1: Sample of images from data set) are from plant village data set extracted from git hub, and it consists of various classes like bacterial spot, healthy, late blight, leaf mold, early blight, septoria_leaf_spot, spider mite, two spider_mite, target_spot, tomato mosaic virus, yellow. The data set consists of 30000 images approximately related to tomato plant. The various diseases are affected to the plants, caused by a virus and bacteria. The variation of effect on plants is that the leaves can be affected from discoloration or it can lead to death of plants. In the paper [16], the fungi, microbes, viruses, nematodes, etc., are the aspects which affect the plants and disease. The various diseases are as follows:

Rust: The lower surface of the leaves is affected; matured plants are major victims of this rust disease. Undersides of leaves, spot is initially raised. After certain days, these spots turn into reddish-orange spore masses. Leaf pustules later change to

Fig. 1 Sample of images from data set

yellow-green and gradually turn black. Finally, severe infestations will bend and yellow leaves cause leaf drop.

Yellow leaf disease: Pathogen phytoplasma is the cause for this disease where green leaves turn into yellow and gradually yield is declined.

Leaf rot: Leaf spots will be raised over leaf, and these spots vary in size, shape, and colors.

Leaf curl: The various aspects are there with respect to leaf curl. The causes are fungus, genus Taphrina, or virus.

Leaf Spot: It appears on the leaves by the formation of black spot and small hole on the leaves.

Late Blight: This type of diseases spreads the crops at faster rate. The diseases are caused due to the cool climate and wet weather.

Figure 2 is image of tomato leaf extracted from plant village data set, as image indicates that the tomato plant is healthy without lesions or yellow spots. Thus, the comparison can be made and healthy leaf is considered as one of the classes in data set.

The survey carried out on various aspects that cause the diseases in plants. The identification and specification of key issues and challenges on disease and techniques are as follows.

Fig. 2 Healthy image of tomato leaf

3.2 Quality Image

- Data collected and framed into large amount.
- Acquired images need to be free from noises.
- Training and testing samples from data set are carried.
- Classify the disease-based certain parameters like size, shape, color by applying segmentation techniques. Color of plant leaf, size, and texture are varying when climate changed.
- Manual intervention and frequent observation are needed for plants.

4 Methodology

The discussion is about the acquisition, image preprocessing and classification is as in Fig. 3.

Image Acquisition: The images are acquired from the plant village dataset. The plant village dataset consists of images of various diseases in plants. It is classified

Fig. 3 Basic steps of acquisition, image preprocessing, and classification

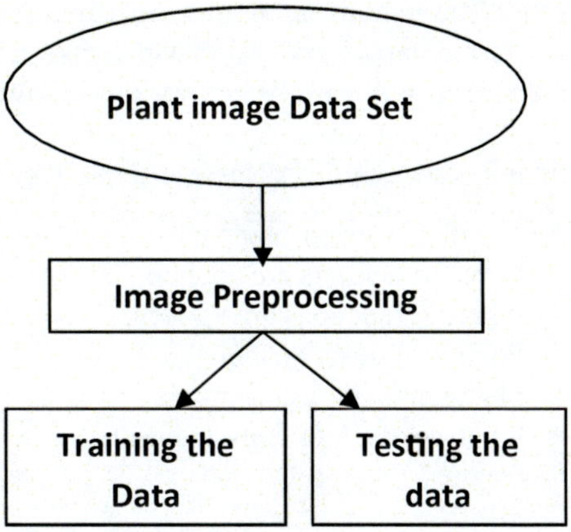

into various classes of diseases along with the healthy images. The data set consists of around approximately 30,000 images of plant disease.

Image Preprocessing: The images collected need to be clear with respect to the colors RGB. The image preprocessing consists of image resizing, image de-blurring, noise removal, and compression to maintain the well-defined data set for the research. The snipping and smoothing are performed on the image to increase the quality of the image, and unwanted noise is removed to focus on the particular object in an image. The step before model creation is to preprocess our data. This simply means applying some prior transformations to our data before feeding it the model.

Training and testing data: The model is built in a ratio of 8:2, 80% for training and 20% for testing. Images will be chosen from the plant village datasets. Dataset consists of both healthy and diseased leaf images and is categorized into various classes based on diseases of plants.

Deep learning and its libraries: The model deep learning is a part of machine learning; software like Torch, Caffe, Theano, tensor flow, and deep learning4J along with graphical processing unit technology and new training methods provides efficient training models in neural networks easy. The sub-branch of machine learning is deep learning, the results that are derived in this field are pattern matching and recognition, construction, or machine learning system using convolution neural network. Before the data been sent for the training data set, the raw data needs to preprocessed into suitable internal representation from which the system could classify and detect the patterns in the data given as input (Table 1).

5 Convolutional Neural Network

The computer program is built as a technique that learns from data. It is based very loosely on how we think the human brain works. First, a collection of software "neurons" is created and connected together, allowing them to send messages to each other. The neurons are the units on neural network. The neural network are framed and organized into several layers with connections like either full or partial connections. Capturing and learning the various features from the data at distinguished levels similar to human vision are the advantages of the neural network. The convolutional neural network process has got steps. The various steps are as follows (Fig. 4).

- The data is given as input.
- Level-wise filters are applied.
- Filtering functions are applied.
- Pooling strategy is applied.
- Flattening.
- Input vector into fully connected network processing the data by considering features.

Table 1 Various libraries of deep learning

Libraries Parameter	Thea n o	Deep learni ng4j	Caffe	Torch	TensorFlow	Keras
Creator	University de Montral	Sky mind engineering team	Berkeley vision and learning center	Renan collobert koray Kavukcuoglu clement Farabet	Google train team	ONEIROS project
Platform	Cross-platform	Cross-platform	Ubuntu, OS, X AWS, Unofficial port for windows and android	Linux Android, iOS, Windows, MAC OS	Linux MAC OS	Cross-platform or Linux MAC OS
Interface	Python	Java scala clojure	C++	C, Lua, C++	C++ , Pyhton	Python
Written	Python	C, C++	C++	C, Lua	C++ , Python	Python
Parallelizing technique support	Open CUDA	Open C uda	CUDA Open MP open CL	CUDA Open MP.Open CL	CUDA	Python
Modeling capability	RNN, CNN, RB M, DBN	RNN, CNN , RBM, DBN	RNN, CNN, RBM, DBN	RNN, CNN, RB M , DBN	RNN, CNN, R BM, DBN	RNN CNN RBM DBN
Parallel execution	Multiple GPUs	Yes performs parallel execution	Yes performs parallel execution	Yes performs parallel execution	Yes performs parallel execution	YES

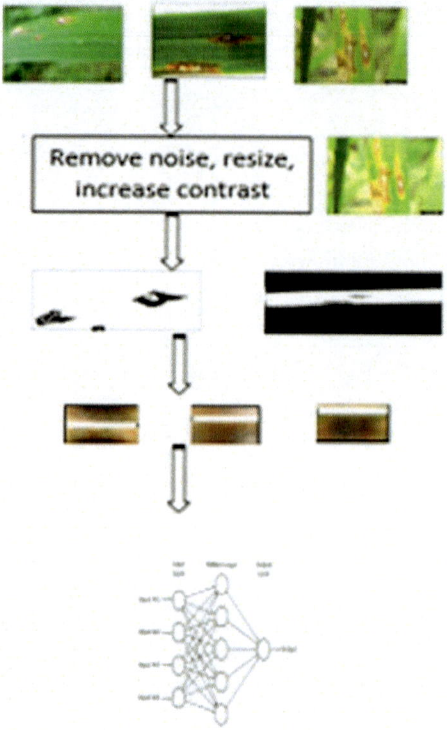

Fig. 4 Applying the convolutional neural network

5.1 Critical Issue

The three process used here are: convolution, max pooling, and neural network.

Convolution: Initially, split the original image into many overlapping small images (tiles) using the sliding window technique (can use any size and can use any operators). Each tile is saved as a separate image. The original image is divided into small number of tiles. In the convolutional neural network, connections of all units in the network are organized and each layer unto feature maps. Within small window, the weighted sums are performed. The same filter has been applied to each color channel. The convolutional neural network linearly combines the activations from input channel in previous layer and different channels for the output. Some of the models in neural network do not perform convolutions but rather combine various channels on a per-pixel basis, often with the goal of reducing dimensionality of feature space because weights for all the pixel are uniform. The additional parameters have been incorporated in the observations of convolutions neural networks. These includes padding, strides, dilation, grouping.

Max Pooling: The maximum response within a square window to compute is widely used variant, which is called max pooling. Common strides and window sizes for max pooling are a stride of 2 and 2 × 2 non-overlapping windows or 3 × 3

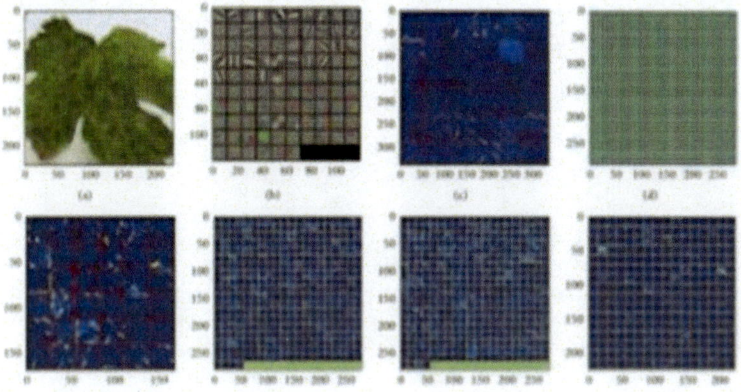

Fig. 5 Convolutional and pooling using plant data set

overlapping windows. Assuming max pooling layers can be as a "logical or," since they only require one of the units in the pooling region to be turned on. They are also supposed to provide some shift invariance over the inputs. However, most deep network is not all that shift invariant, which degrades their performance.

Neural Network: It is supervised model where we need both outputs target. Error value is determined using the outputs and the targets (MSE function normally used).

5.2 *Features and Outputs*

The color, shape, and disease are features of the plants with ranges. The system is fed with millions of odd diseases of plant data which model characterize and put in the system. The layered output is next step after categorizing various images which are decoded to execute and get results. Detection of disease at different layers of layered outputs takes place. These can be gray scale or HSV images. This aspect helps in differentiating by using features of plants, thus categorization of the images data carried. The black white images are converted numerically into 1 and 0, which help in locating the diseased region in plants and are helpful in understanding disease in plants (Fig. 5).

6 Conclusion

Tensorflow is one of the advantageous and flexible framework for researchers and through git hub and online forum to progress its efficiency. Tensor flow has many pertained models, for example MobileNets, etc. This method is very appropriate for preprocessing and detection of disease in real time so as to maintain the quality for vegetables using TensorFlow CNN implementation method for multi-disease detection using tensor flow platform.

References

1. Roopa Sri M, Uma S et al "Fruit recognition and grade of disease detection using inception v3 model". Proceedings of the 3rd ICECA 978–1–7281–0167- 5/19/ 2019 IEEE
2. Kosamkar PK et al "Leaf disease detection and recommendation of pesticides using convolution neural network", Fourth international conference on computing communication control and automation (ICCUBEA)
3. Kulkarni O et al "Crop disease detection using deep learning fourth international conference" ICCUBEA 978–1- 5386–5257/18
4. Dong YN et.al (2019) "Research and discussion on image recognition and classification algorithm based on deep learning". International conference on machine learning, Big, data and business intelligence MLBDBI 978–1–7281–5094/19 IEEE
5. Dharwadkar S et al (2017) "Floriculture classification using simple neural network and deep learning". International conference on recent trends in Electronics Information and Communication technology 978–1–5090–3704 IEEE
6. panigrahi S et al "Deep learning Approach for image classification". 2018 2nd international conference on data science and business analytics 978–1–5386–8431–3 IEEE
7. Durmus H et al "Disease detection on the leaves of the tomato plants by using deep learning", I.T.U TARBIL environment agriculture informatics applied research center
8. Huang T "Detection and classification of various image operations using deep learning technology Proceedings" of the 2018 International Conference on Machine Learning and Cybernetics,978–1–5386–5214/18 IEEE.
9. Sandhu GK et.al "Plant disease detection techniques: a review international conference on automation"
10. Wahab AHBA et al "Detection diseases in chilli plants using k means segemented support vector machine", 3rd international conference on ICCS ,978–1–7281–3663–9/19 IEEE
11. Sharath DM et al "Image based plant disease detection in pomegranate plant for bacterial blight". International conference on communication and signal Processing 978- 1–5386–7595–3/2019
12. Meenakshi K et al "Grading of quality in tomatoes using multi-class". 3rd international conference on computing methodologies and communication (ICCMC 2019) ISBN :978–1–5386–7808–4 2019 IEEE

Performance Evaluation and Comparison Analysis of AODV and RPL Using NetSim in Low Power, Lossy Networks

B. Priyeash and Jayavignesh Thyagarajan

Abstract Low-power and Lossy Networks (LLNs) are a type of wireless sensor networks (WSNs). These types of networks have constraints like memory, power, size, etc. The devices deployed in such networks need to be optimized to carefully use the available minimum resources for a long time. At times, these networks are deployed in places where frequent human intervention is impossible. Thus, the management of resources is a vital task while considering the deployment of LLNs. Since the devices like sensors are used in such networks for continuous monitoring, the data collected by them has to be transmitted to the user end. The selection of routing protocols for these low-power devices is very important. Routing Protocol for Low-power and Lossy Networks (RPL) and ad hoc on-demand distance vector (AODV) are two important protocols used in them. The merits and demerits of these two protocols are analyzed in detail using the Tetcos NettSim network simulator. The results showed that RPL consumes more energy, but the RPL performs better in terms of throughput. Due to the frequent transmission of control packets, the data packets take more time to reach the destination than AODV.

Keywords Wireless sensor networks (WSNs) · Internet of Things (IoT) · RPL · AODV · Routing protocol · 6LoWPAN · IPv6 · Sensors · Energy · NetSim

1 Introduction

Sensors are devices that detect events or changes in its environment and transmit the information to an electronic device usually a processor or a controller unit. Sensors are one of the key aspects in the field of Internet of Things (IoT). The sensors and a transceiver circuit as a combination are known as nodes or motes. Wireless sensor networks (WSNs) [1] refer to a group of sensors that are used for monitoring and

B. Priyeash · J. Thyagarajan (✉)
School of Electronics Engineering, Vellore Institute of Technology, Chennai, India
e-mail: jayavignesh.t@vit.ac.in

B. Priyeash
e-mail: bpriyeash.2017@vitstudent.ac.in

A. Sivasubramanian et al. (eds.), *Futuristic Communication and Network Technologies*, Lecture Notes in Electrical Engineering 792,
https://doi.org/10.1007/978-981-16-4625-6_2

collecting the information regarding the environment and transmitting it to a central hub known as gateway. The gateway is the end device for the WSN. It is further connected to the router for transmitting the collected data over the Internet to the destined location. At present, most of the motes have non-IP network layer protocol such as ZigBee [2] that does not support TCP/IP protocol suite [3]. However, there is a strong requirement for the motes to be connected to the Internet for the storage, processing and/or analysis of data sent by the sensors. Thus, for using the TCP/IP stack for the low power devices like the sensor motes in WSN, a new standard has been proposed by the Internet Engineering Task Force (IETF) known as the IPv6 over Low-power Wireless Personal Area Networks (6LoWPAN) [4]. The 6LoWPAN uses IEEE 802.15.4 for carrying data packets using IPv6. This standard enables different low-power devices to send and receive data packets. The 6LoWPAN introduces an adaptation layer above the data link layer (IEEE 802.15.4) for the compression of headers. Since the motes used in LLNs are low powered, thus short ranged, requires multiple hops for reaching the gateway [5]. Therefore, routing protocols are required for the transmission of data in WSN. The data from a mote might take different routes to reach the gateway, thereby having many hops but the final hop to the gateway is made from the motes that are closest to the gateway, known as rank 1 motes. The motes in rank 1 that acts as a bridge between the sensor network and the gateway are known as hotspot motes.

There are two types of ad hoc routing protocols, topology-based protocols, and geographical protocols [12]. The classification of routing protocols for ad hoc network is shown in Fig. 1.

Under topology based, there are two subtypes, proactive, and reactive [6]. Proactive routing protocols are the ones in which each node of the network frequently updates their routing table by constantly broadcasting control packets to all the neighboring nodes in its range to check for any update in the network. On the other hand, reactive routing protocols determine the route only when prompted. Both of them have their merits as well as demerits. Proactive routing protocols show higher throughput than reactive routing protocols, but there is a huge control overhead compared to reactive routing protocols [13]. Examples of proactive routing protocols are Routing Protocol for Low-Power and Lossy Networks (RPL), Destination Sequence Distance Vector (DSDV), etc. Examples of reactive routing protocols are

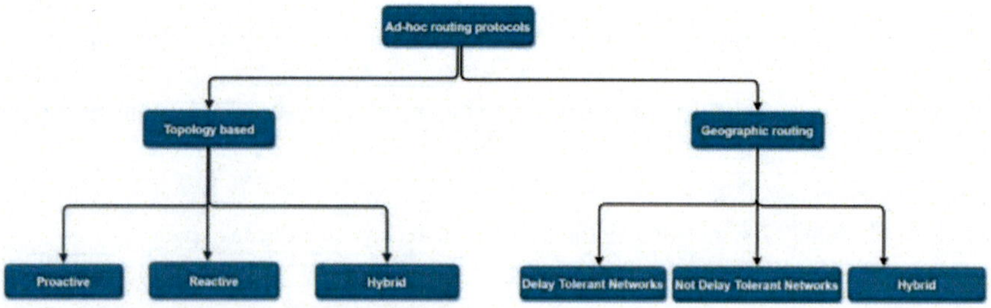

Fig. 1 Ad hoc routing protocols classification

ad hoc on-demand distance vector (AODV), dynamic source routing (DSR), etc. Few developments were made on the 6LoWPAN routing protocols. In this paper, the performance analysis was done between the routing protocols RPL and AODV using NetSim v10 simulator.

The remainder of the paper is organized as follows. Section 2 describes the working mechanism of RPL and AODV routing protocols followed by the related works in Section 3. Section 4 describes the simulation parameters followed by the results obtained in Sect. 5. Finally, Sect. 6 concludes the paper.

2 RPL Versus AODV

Routing Protocol for Low power and Lossy Networks (RPL) is a 6LoWPAN routing protocol for Low-power and Lossy Networks (LLNs). The RPL forms Destination-Oriented Directed Acyclic Graph (DODAG) for reaching the sink node (gateway). The formation of DODAG is based on rank. The rank of a node in RPL defines the node's individual position with respect to its DODAG root. The rank of the DODAG root is always 0. The rank of a node keeps decreasing from the root that is the gateway. The metrics used for calculating the rank are known as objective functions (OF). RPL uses three different types of control packets as shown in Fig. 2 [11]. In RPL, DIO is a control packet that is used to maintain the routes in the network. The DODAG construction process begins from the gateway [7]. DIO is frequently broadcasted to all the neighboring nodes. When a node received DIO packet from many nodes, it adds them all in its table but chooses the parent node by considering certain metrics, OFs and lowest rank. Then, the rank of this node is calculated based on the OFs, and DIO packets are generated by this node and broadcasted to other nodes. This process continues until all the nodes are discovered in the network. DIS is a control packet that is used to solicitate a DODAG Information Object from an RPL node. DAO is a control packet that is used to forward the DODAG information of a node to its parent/DODAG root.

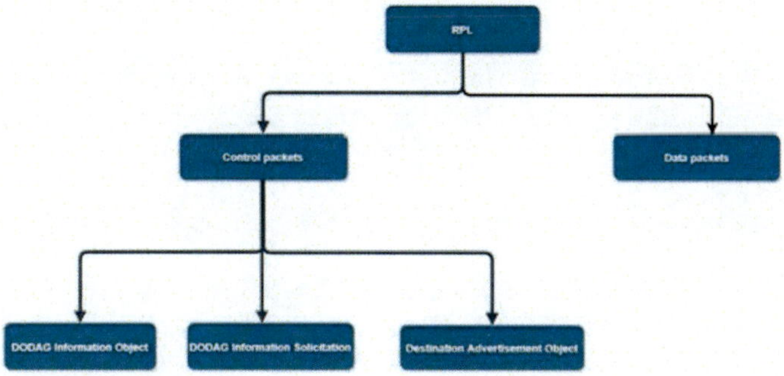

Fig. 2 RPL packets distribution

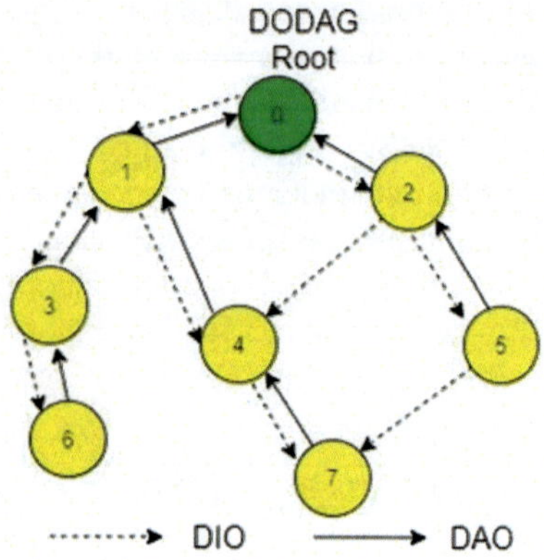

Fig. 3 Route discovery in RPL using DIO and DAO

The route discovery process in RPL is shown in Fig. 3.

AODV is an enhancement of the DSDV routing protocol. It is a reactive routing protocol that acts only upon the route request by any other node. AODV mainly uses three control packets, route request (RREQ), route reply (RREP) and route error (RERR). Whenever a node has a data to be transmitted and it does not have the route to the destination in its routing table, it uses the RREQ control packet to find the route. The node broadcasts the RREQ packet to its neighboring nodes and forwards it further until the packet reaches the destination node. The destination node sends the RREP message to source node that is a unicast message. The RREP message is transmitted back to the initial node through the same route as the RREQ packet with the information of number of hops required to reach the destination. The RREQ packets usually contains source ID, destination ID, source sequence number, destination sequence number, and Time To Live (TTL) [10]. The RERR packet is used in case there is an error discovered in the existing route. The route discovery process in AODV is shown in Figs. 4 and 5.

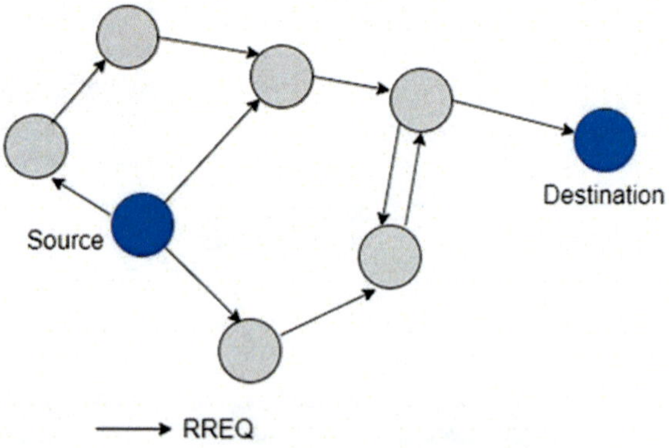

Fig. 4 Route request in AODV

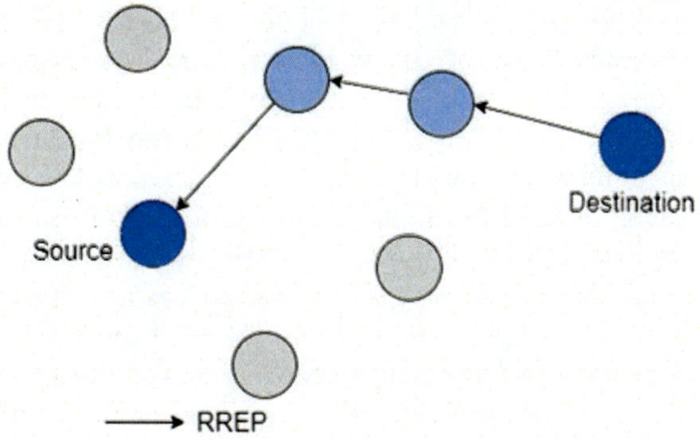

Fig. 5 Route reply in AODV

3 Related Work

In [8], the power and energy analysis of RPL routing protocol was performed under various topologies. Cooja network simulator with sky motes were used for the simulation of different RPL-based network topologies. Initially, the energy consumed in each state of operation was found. The four different states were ON state (EON), transmission state (ETX), receiving state (ERX), and IDLE state (EI). On analysis, the order of energy consuming states was found to be ERX > ETX > EON > EI. Then, the power consumption was found by calculating the quotient of energy in that state by total time spent in that state. The above analysis is carried out for both linear and random topologies. In linear topology, it was found that the nodes that are close to the border router consume more energy in the transmission mode since they are involved in frequent transmissions due to forwarding of packets from the network. Similarly, the nodes that are out of range of the border router consume the least energy since they are not involved in the transmission process. In receiving state, the border router is the highest energy consumer since it receives all the signals. The energy consumption in ON and IDLE state is very minimal as the current requirement in those states is in micro-Ampere (μA). The total energy consumed by border router was found to be the highest in all topologies. In [9], the energy analysis was performed according to the different node state and different network protocols. The different node states discussed are central processing unit (CPU), interrupt request (IRQ), low-power mode (LPM), transmitting mode (Tx), and receiving mode (Rx). The different network protocols considered in the 6LoWPAN stack are CoAP, RPL, 6LoWPAN, and ContikiMAC. Among the protocols, the low-power listening state protocol is the most energy consuming. For a CCR of 4 Hz and observe period of 6 s, the energy required was maximum among the others. Thus, modifying the CCR has a great impact in the Rx and Tx states average power consumption. Reducing the CCR decreases the Rx power but increases the Tx power. In [10], the comparison of Destination Sequenced Distance Vector (DSDV) and ad hoc on-demand distance vector (AODV) routing protocols was performed. A 500mx500m network was used for the simulation with 20 mobile sensor nodes. The network was simulated for three different velocities of the sensor nodes, namely 0 m/s, 5 m/s, and 10 m/s with an initial power of 100 W in each node. The network simulation time was 100 s. On analysis, it was found that DSDV takes more energy to update its routing table. Therefore, the ratio of energy spent for the transmission of data packets to energy spent for the transmission of control packets is low in DSDV. More than 50% of the energy is used for the transmission of control packets. Therefore, packet delivery ratio (PDR) of DSDV is very low. However, AODV is energy efficient. Since AODV is on demand, the mobility of nodes does not affect the routing in AODV routing protocol in a great manner. In [11], the energy analysis of source node, intermediate node, and destination node in a network using AODV protocol was performed. The results showed that the energy consumption of intermediate node was the most followed by the source node and finally the destination node. This is due to the fact that the intermediate node has the dual responsibility of receiving the data packet as well as forwarding it to the next node.

4 Simulation Parameters

The simulation parameters and the network configuration for the comparison of RPL and AODV routing protocols and to analyze the energy dissipation of hotspot nodes are discussed in this section. The power configurations used for the wireless sensor nodes are given in Table 1. All the nodes were uniformly configured with the given below settings.

The path loss model for the simulation environment was set as log distance with path loss exponent (n) as 3.5. The simulation was performed on different size of networks. The physical and datalink layer properties of each node were set with default values as given in Table 2.

The nodes in each network were placed in a uniform distance of 75 m between them. On verifying the maximum distance of transmission with the above set properties, each node has a transceiver range of approximately 100 m. Thus, 75 m was chosen to be the optimum distance for the simulation to reduce the overlapping of

Table 1 Power configurations

Parameters	Values
Power source	Battery
Initial energy (mAh)	0.5
Voltage (V)	3.6
Transmitting current (mA)	8.8
Receiving current (mA)	9.6
Idle mode current (mA)	3.3
Sleep mode current (mA)	0.237

Table 2 Physical configurations

Parameters	Values
Protocol	IEEE 802.15.4
Frequency band (MHz)	2400
Data rate (kbps)	250
Modulation technique	O-QPSK
Receiver sensitivity (dBm)	−85
Transmitted power (dBm)	1
Antenna gain	1

nodes. The gateway for the network was set at a perpendicular distance of 75 m from the middle of the rank 1 node(s). The gateway is further connected to a router and a PC using a wired link for the transmission of message. The dynamic routing protocol used by the router was open shortest path first (OSPF). In each network, four applications are used from four different sources to the same destination as mentioned above. Each packet from each application is generated at an interval of 1 s with 50 bytes. The networks were simulated for a time period of 100 s. Therefore, 100 packets would have been generated from each source. In total, 400 application packets were generated in each network. Since all the packets were destined toward the same node, they tend to choose different paths among each other to avoid collision. In Fig. 6, the layout of a 5 × 5 network is shown along with the imaginary lines marking the source to destination of each application is shown.

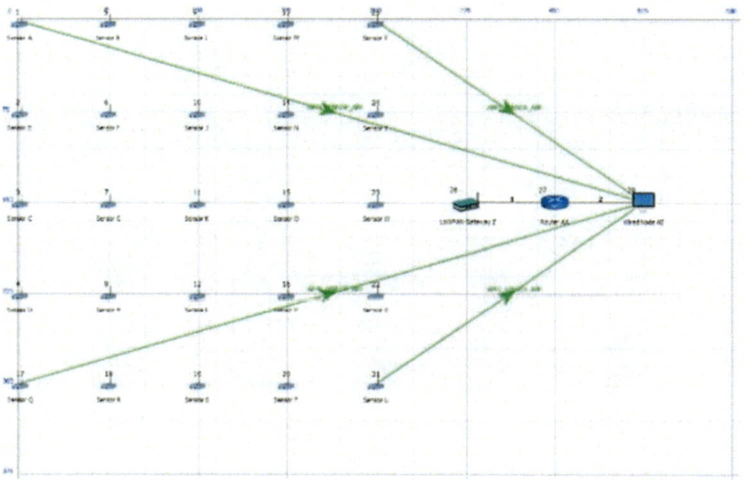

Fig. 6 5 × 5 network topology

Similarly, other networks were designed and simulated.

On having a fixed number of application traffic, the aim was to observe the change in the ratio of control packets to total number of packets transmitted by each node. Also, the energy dissipation model of all the nodes according to their ranks was expected to be observed. Keeping a standard uniform distance between the nodes and the gateway helps in observing the selection of hotspot node in each network.

5 Conclusion

On simulating the network with above given specifications, the following results were obtained. The comparison between RPL and AODV based on the number of hops required to reach the destination versus the grid size is given in Fig. 7. On analysis, it was noted that the number of hops required by AODV and RPL remains almost same in all the grid sizes. In Fig. 8, grid size versus average throughput is plotted for both AODV and RPL routing protocols. The results show that the throughput of RPL is better than AODV in almost all the cases. Also, another pattern is noted in this plot. The throughput of even-sized grid is always higher than the throughput of odd sized grid. This is due to the traffic overloading on a single hotspot node in odd grids. In Figs. 9 and 10, the comparison between number of control packets and number of app data packets was performed between AODV and RPL protocols, respectively. The number of control packets is more than the number of data packets in both the cases. The ratio between number of data packets to number of control packets is greater in AODV compared to RPL. This is due to the large control overhead in RPL compared to AODV.

As stated above RPL being a proactive protocol, the control packets are sent even if there is no data transmission. Frequent transmission of DIO packets by the nodes in RPL network mainly attributes to this poor data to control packet ratio. Therefore,

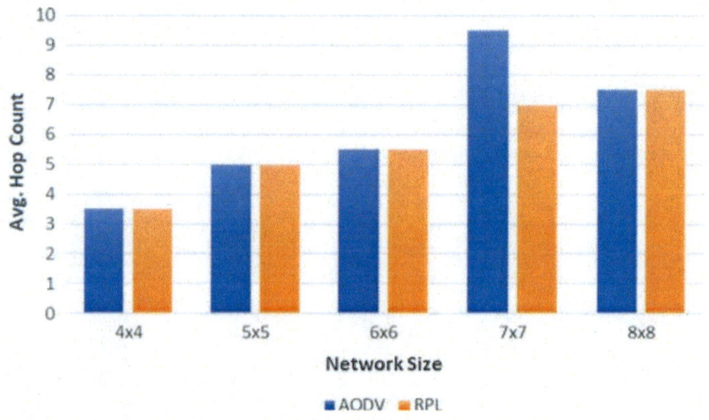

Fig. 7 Network size versus average hop count

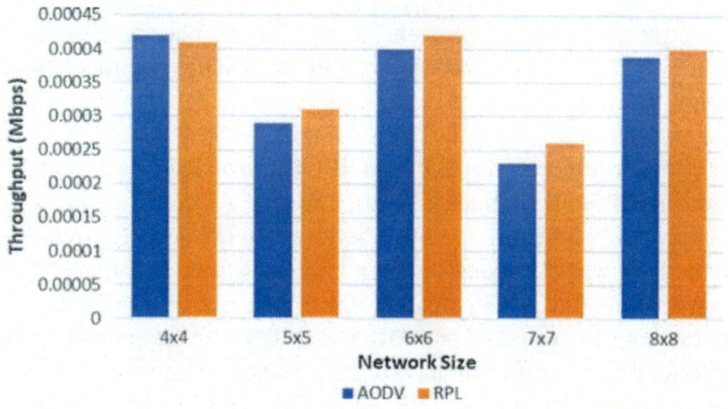

Fig. 8 Network size versus average throughput

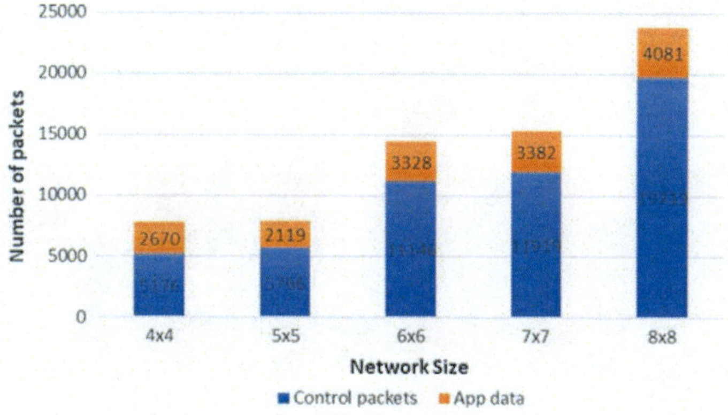

Fig. 9 Network size versus distribution of Tx packets in AODV

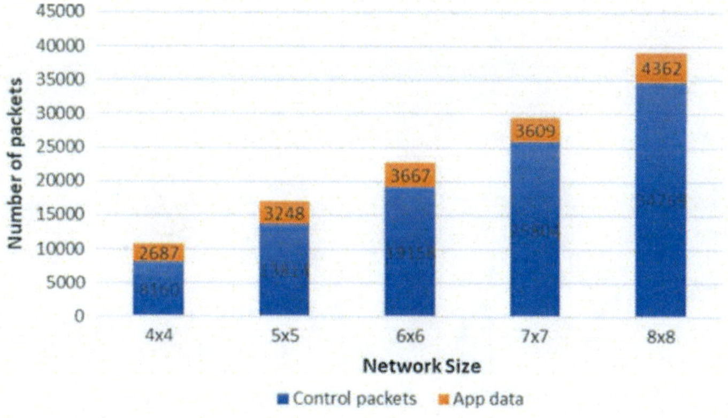

Fig. 10 Network size versus distribution of Tx packets in RPL

this large control overhead decreases the number of data packets in the case of RPL. Figure 11 shows the comparison between RPL and AODV protocols based on the average delay. The RPL routing protocol exhibits more delay in the transmission of packets compared to AODV. Similarly, the comparison based on energy consumption between the networks based on AODV and RPL is shown in Fig. 12. In both the cases, RPL exhibits higher values compared to AODV.

Finally, the node-wise energy analysis between AODV and RPL routing protocols in a 5 × 5 network with 25 nodes is shown in Fig. 13. It is clear that RPL consumes more energy than AODV. This is mainly due to the control overhead. RPL being a proactive protocol keeps sending control packets like DIO for maintaining the routes in the network. Thus, it is no wonder that RPL consumes more energy than AODV. The average delay and energy consumption of the nodes in RPL network are higher compared to AODV due to the fact that RPL is proactive. The frequent transmission of DIO control packet is also one of the reasons for the increase in delay of the

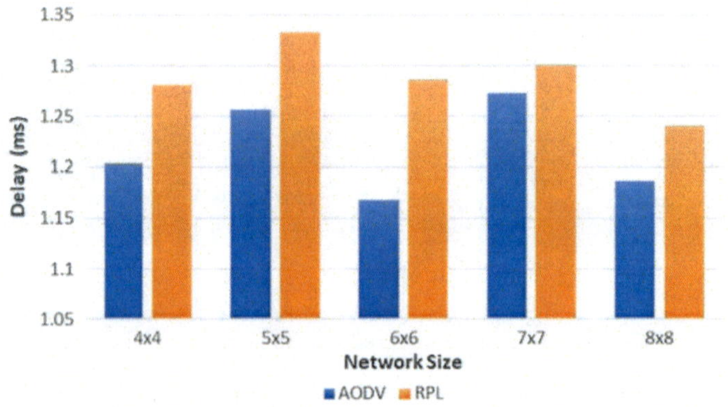

Fig. 11 Network size versus average delay

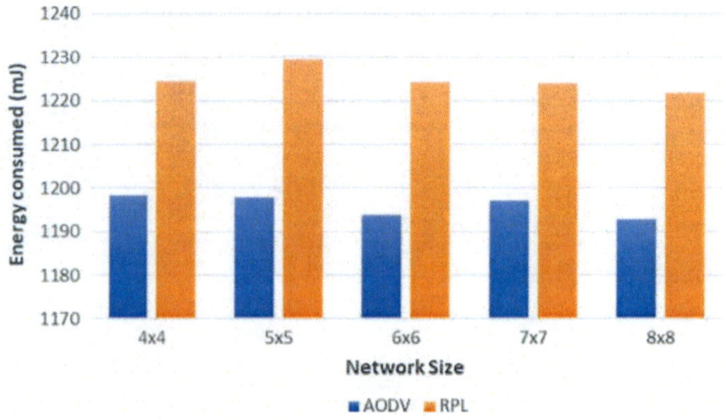

Fig. 12 Network size versus average energy consumption

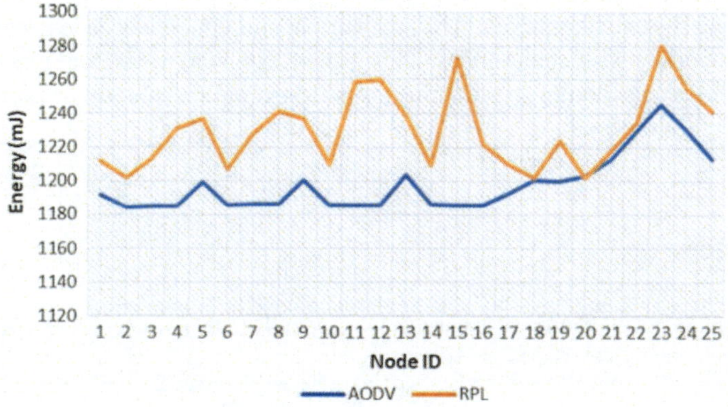

Fig. 13 Node-wise energy analysis in 5 × 5 network

transmission of data packet as well as the energy consumption of the nodes in RPL networks. The average energy consumption can also be evidently observed from the individual energy consumption of each node in the RPL as well as AODV network.

6 Conclusion and Future Scope

Thus, on analysis, it is clear that RPL consumes more energy compared to AODV. On calculations, RPL uses 2.643% more energy than AODV protocol in a 5 × 5 network with linear topology. This is due to the mechanism of RPL routing protocol. RPL broadcasts control packets frequently for the maintenance of the network. But AODV issues them only when deemed necessary for the transmission of data packets. The high average delay of RPL can also be mapped to this reason. RPL has its own advantages like higher throughput compared to AODV. The throughput in RPL is 6.9% more than the throughput in AODV. The average delay in RPL is 6.05% more compared to AODV. The advantage of higher throughput compared to delay and energy consumption is very meagre in RPL. Thus, AODV proves better in Low-power and Lossy Networks (LLNs) due to its advantages. The current work is limited to the comparison of the RPL and AODV for low power, lossy static IoT-based wireless sensor network. Since RPL is the standardized protocol as per RFC 6550 march 2012 [14], the comparative evaluation is limited to comparison of RPL with other standardized AODV protocol (Zigbee). The recent works are mainly focused on RPL enhancements such as evaluation of various routing metrics for RPL objective function and adaptive timer feature, etc., and AODV enhancements such as light weight on-demand ad hoc distance-vector routing protocol (LOADng). WSN applications such as environmental monitoring, precision agriculture, etc., are the potential applications where routing protocol plays an important role. A more detailed analysis can be done on optimizing the performance of AODV in terms of throughput while maintaining the existing merits. Also, the current simulation has been performed on a linear topology with static nodes. This can be extended to random placement of nodes and by considering the mobility of nodes in the calculation of performance metrics in future works.

References

1. Akyildiz IF, Su W, Sankarasubramaniam Y, Cayirci E (2002) Wireless sensor networks: a survey. Comput Netw 38(4):393–422
2. Safaric S, Malaric K (2006) "ZigBee wireless standard," Proceedings ELMAR 2006, Zadar, pp. 259–262, https://doi.org/10.1109/ELMAR.2006.329562
3. Ee GK, Ng CK, Noordin NK, Ali BM (2010) A Review of 6LoWPAN Routing Protocols. Proceedings of the Asia-Pacific Advanced Network 30:71
4. Mulligan G (2007) The 6LoWPAN architecture. Proceedings of the 4th workshop on embedded networked sensors emnets '07
5. Singh SK, Singh M, Singh DK (2010) Routing protocols in wireless sensor networks—a survey. Int J Comput Sci Eng Surv 1(2):63–83
6. Pirzada AA, Mcdonald C, Datta A (2006) Performance comparison of trust-based reactive routing protocols. IEEE Trans Mob Comput 5(6):695–710
7. Alvi SA, Hassan FU, Mian AN (2017) "On the energy efficiency and stability of RPL routing protocol," 2017 13th international wireless communications and mobile computing conference (IWCMC), Valencia, pp 1927–1932
8. Gokilapriya V, Bhuvaneswari PTV, SaraswathiPriyadharshini A (Apr 2018) "Energy and power analysis of rpl under topology variations". Int J Adv Res Trends Eng Technol (IJARTET) 5(4)
9. Schandy J, SteinfeldL, Silveira F (2015) "Average power consumption breakdown of wireless sensor network nodes using ipv6 over llns," 2015 international conference on distributed computing in sensor systems, Fortaleza, 242–247. https://doi.org/10.1109/DCOSS.2015.37
10. Khokhar K, Pandya V, Prof. Mehta ND (April 2013) Performance analysis based on energy consumption of aodv and dsdv routing protocol. Int J Eng Res Technol (IJERT), 02(04)
11. Wang ZM, Li W, Dong HL (2018) Analysis of energy consumption and topology of routing protocol for low-power and lossy networks. J Phys: Conf Series, 1087:052004
12. Smiri S, Boushaba A, Ben Abbou R, Zahi A (2018) Geographic and topology-based routing protocols in vehicular ad-hoc networks: performance evaluation and qos analysis. 2018 international conference on intelligent systems and computer vision (ISCV)
13. Istikmal V, Leanna, Y, Rahmat B (2013) "Comparison of proactive and reactive routing protocol in mobile adhoc network based on "Ant-algorithm", 2013 international conference on computer, control, informatics and its applications (IC3INA), Jakarta, pp 153–158
14. https://tools.ietf.org/html/rfc6550

Three Notched Bands Modified Hexagonal Patch Monopole Antenna

Manish Sharma, Shiva Sharma, and Rakesh Ahuja

Abstract This research paper concentrates on design aspects of three band-stop filters super-wideband monopole antenna. Modified hexagon shape radiation patch provides the required super-wideband bandwidth (2.85–21.36 GHz). Inverted T-type stub, T-type parasitic placed backed plane, and pair of C-type stub adjacent to microstrip feed line provide the function of band-stop filter, thereby mitigating WiMAX, WLAN, and DSS interfering bands. Super-wideband characteristics which include applications for UWB, X, Ku band applications, and excellent far-near field results suggest the proposed designed is well suited for wireless applications.

Keywords Super-wideband · WiMAX · WLAN · DSS · Parasitic elements

1 Introduction

A super-wideband antenna reported in the literature covers larger bandwidth of interest which not only is useful for UWB applications but is also useful for higher microwave band applications. Super-wideband bandwidth encounters interference at lower existing wireless communication systems such as WiMAX (3.30–3.70 GHz), WLAN (5.150–5.825 GHz), and satellite communication systems in X-band (7.25–7.75 GHz). Various technologies have been reported to remove interfering bands which are discussed in the literature. A compact CPW-fed UWB antenna eliminates interfering bands by use of quarter wavelength stubs/slots [1]. Also, a multiple-input–multiple-output (MIMO) resonates for three operating bands and also includes notched interfering ultra-wideband signals [2]. A defected Minkowski fractal defected in the ground plane provides better isolation in the MIMO system and rotated C-shaped slots rectangular slots (identical in shape) in-ground plane removes WiMAX and C-band [3]. Inverted L-shaped stubs that are embedded within

M. Sharma (✉) · S. Sharma · R. Ahuja
Chitkara University Institute of Engineering and Technology, Chitkara University, Punjab, India
e-mail: manish.sharma@chitkara.edu.in

R. Ahuja
e-mail: rakesh.ahuja@chitkara.edu.in

A. Sivasubramanian et al. (eds.), *Futuristic Communication and Network Technologies*, Lecture Notes in Electrical Engineering 792,
https://doi.org/10.1007/978-981-16-4625-6_3

monopole (polygon shape) suppress WiMAX, WLAN, and X-band [4]. Fibonacci-type GPS-ultra-wideband signals suppress potential interference of WiMAX and WLAN bands by introducing rotated C-shaped slots in microstrip and ground plane [5]. Asymmetric coplanar-fed monopole antenna fabricated with dimensions 14×16 mm^2 uses a rectangular patch and folded type slit with stubs (rejecting elements) [6]. U-slots etched on radiating patch with beveled ground plane removes WiMAX/WLAN/X-band interfering bands which also cover operating ultra-wideband signals [7]. Dual radiating strips fed by microstrip including inverted T-shaped parasitic element at back and etched inverted C-shaped slot on rectangular ground plane cover 2.80–12.50 GHz excluding WiMAX/WLAN/X-band signals [8]. Two elliptical slots on a semi-elliptical patch with pair of C-shaped slots near microstrip or dual pairs of complimentary capacitive-loaded loop slots in uniplanar bow-tie antenna also suppresses interfering bands [9, 10]. Split ring resonators on the patch, complementary split ring resonator on the ground, or two different type slits also notches interfering bands (WiMAX, WLAN, and ITU bands) [11, 12]. Triple notched bands are also achieved using a couple of modified capacitive loaded loops (MCLL), quadra-pair of meander lines, realizing notched bands by various length elliptical single CSR on the patch and stepped impedance resonator in the defected ground plane with fork-shaped stubs which also eliminates interfering bands [13–16]. Higher permittivity microwave dielectrics encountering lower losses provide good filter characteristics. The band-notched characteristics are obtained by the insertion of quarter wavelength vertical or horizontal stubs [17].

In current reported research, a smaller-sized super-wideband antenna with dimensions of 20×25 mm^2 including suppression of potential interfering bands (WiMAX, WLAN, and downlink satellite system) is presented. To verify results, antenna parameters are simulated on two different analytical platform simulators which almost agree with each other. Complete design, analysis, and results are discussed.

2 Antenna Configuration

Figure 1 shows the total plan examination of the triple notched band antenna. Figure 1a shows a perspective view of the designed antenna. The proposed aerial is imprinted on microwave substrate Rogers RTDuroid5870 substrate with volume measurements $W_1 \times L_1 \times h_1$ mm^3. As watched, the emanating patch is imprinted on one plane of the substrate, while ground and π-formed electromagnetic coupling parasitic component on the opposite side of the substrate. The proposed antenna is taken care of by 50Ω microstrip feed of measurement $W_f \times L_f$ mm^2 which is soldered by a 50Ω SMA connector for signal transmission. Figure 1b shows the front perspective of the designed antenna demonstrating emanating patch and ground without interfering bands. It comprises of three different patches embedded together with rectangular, hexagon, and half-circular.

As observed from Fig. 1b, the rectangular ground is chamfered at edges by an angle δ which ensures matched impedance all through the working band. Figure 1c

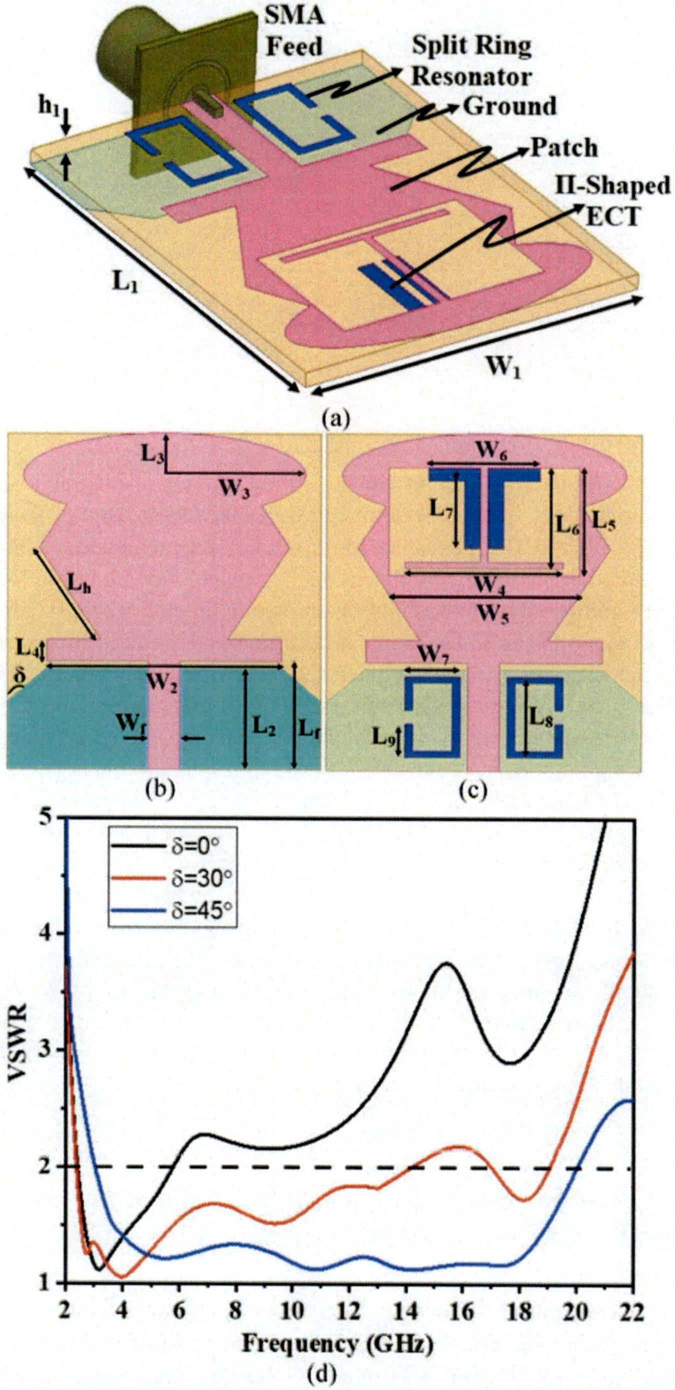

Fig. 1 Proposed antenna **a** acute view, **b** front view with ground, **c** front view with ground and triple notch function, **d** optimization of the ground plane

Table 1 Optimized parameters

Parameter	mm	Parameter	mm
W_1	20.0	L_4	1.65
W_2	15.0	L_5	8.00
W_3	9.00	L_6	7.50
W_4	10.00	$L_7 = L_8$	6.00
W_5	12.00	L_9	2.50
W_6	7.00	h_l	0.787
W_7	3.50	W_f	2.00
L_1	25.0	L_f	8.00
L_2	7.50	δ	45°
L_3	3.15	L_h	8.00

shows the front perspective antenna with mitigation of potential obstruction WiMAX, WLAN, and downlink satellite system. WiMAX interfering band is accomplished by methods for rotated T-formed stub, while WLAN indented band is accomplished by using π-molded parasitic component backed plane. Downlink satellite system interference is mitigated by embedding a pair of complimentary split ring resonators. Also, Fig. 1d shows the optimized value for the modified ground which is chamfered at edges by angle δ. For $\delta = 0°$, the measured bandwidth is 2.42–5.78 GHz, and as the chamfered angle is increased, better matching of impedance is observed. For $\delta = 45°$, wider impedance bandwidth is obtained (3.04–20.02 GHz). Key parameters involved in designing the proposed antenna are enhanced utilizing EM test simulator Ansys HFSSv13 which are organized in Table 1.

Figure 2 represents the proposed antenna design without notched bands which are explained in four iterations. Figure 2a shows Iteration 1 which is a combination of a hexagonal patch and a rectangular ground plane covering bandwidth 2.25–7.85 GHz. Iteration 2 is a modification of Iteration 1 which is the addition of rectangular slot $W_2 \times L_4$ mm^2 which covers new bandwidth of 3.69–10.59 GHz as shown by Fig. 2b. Figure 2c shows the addition of the ellipse of the major radius of W_3 mm which results in new bandwidth of 3.98–16.50 GHz. Iteration 4 is the final version of the proposed antenna shown by Fig. 2d where the ground plane is chamfered at edges by an angle δ leading to required bandwidth of 2.81–21.85 GHz.

3 Parametric Variation (Notched Bands) and Analysis of Surface Current Density Distribution

Figure 3 shows the optimization of key parameters of notched band characteristics of the proposed antenna. For the WiMAX band, the physical length of inverted T-shaped stub W_4 is varied from 8.00 mm to 10.00 mm, resulting in shifting of the band from 3.78 GHz–4.25 GHz to 3.11 GHz–3.81 GHz. For the optimized value of

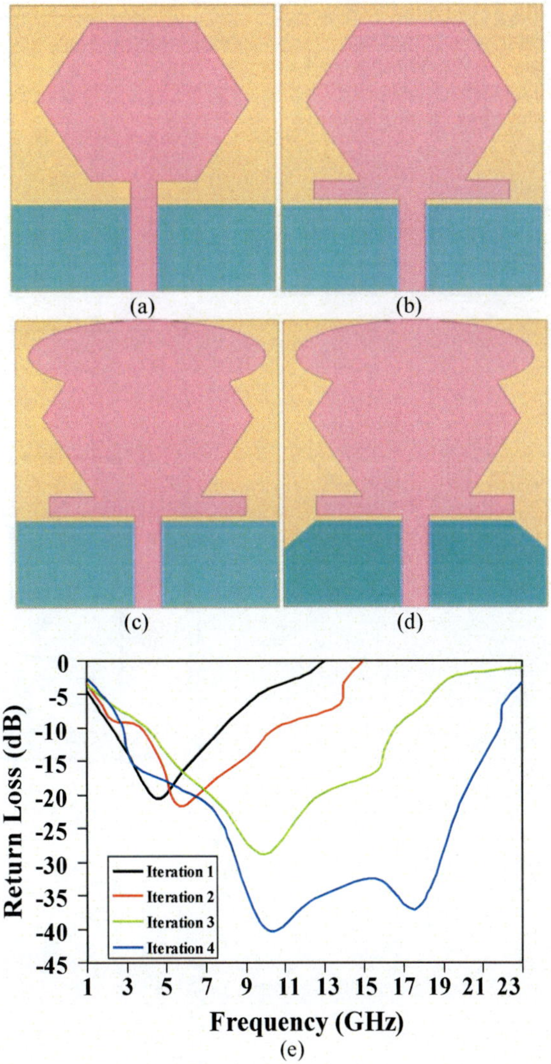

Fig. 2 Iteration used in the design of antenna **a** iteration-1, **b** iteration-2, **c** iteration-3, **d** iteration-4, **e** return loss curve of antenna for Iteration 1-iteration 4

$W_4 = 10.00$ mm, intended notched bandwidth for WiMAX is obtained. Figure 3b represents the optimization of the WLAN band. Physical length L_7 is changed from 4.0 mm to 6.00 mm with step size of 1.00 mm for which notched bandwidth also varies from 4.15 GHz–4.65 GHz to 5.11 GHz–5.75 GHz. For the optimized value of $L_7 = 6.00$ mm, the intended notched bandwidth for WLAN (5.11 GHz–5.78 GHz) is obtained. For the third interfering band, downlink satellite system, physical length L_9 is varied from 2.00 mm to 2.50 mm with a step size of 0.25 mm, changes of corresponding notched bandwidth are seen from 6.18 GHz–6.97 GHz to 7.21 GHz–7.72 GHz. Summary of optimization for notched band characteristics is shown in

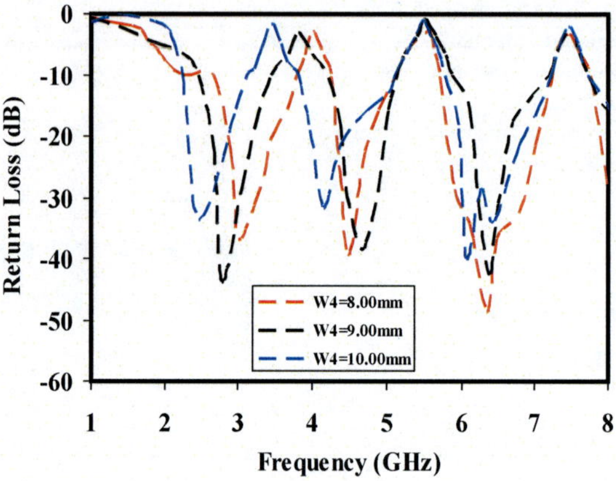

(a)

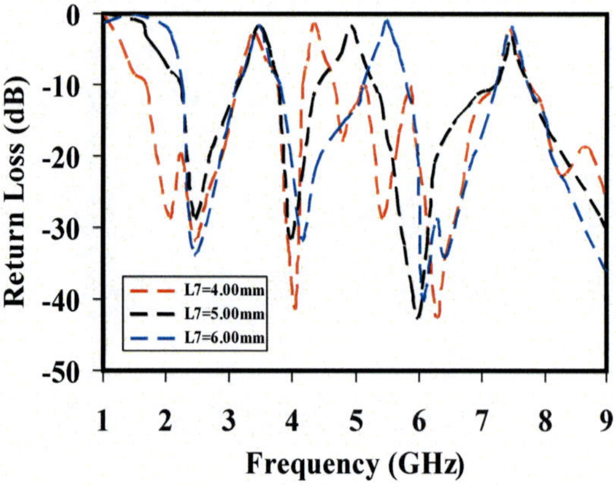

(b)

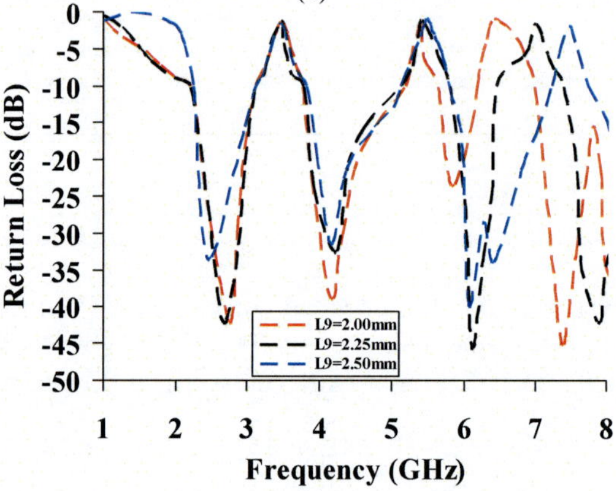

(c)

Fig. 3 Optimization of key parameters for notched bands **a** W_4 (WiMAX) and **b** L_7 (WLAN) **c** L_9 (downlink satellite system)

Table 2 Parametric variation of optimized parameters

WiMAX	Parameter W_4 (mm)	Bandwidth (GHz)
	8.00	3.78–4.25
	9.00	3.44–4.22
	10.00	3.11–3.81
WLAN	Parameter L_7 (mm)	Bandwidth (GHz)
	4.00	4.15–4.65
	5.00	4.41–5.28
	6.00	5.11–5.78
Satellite downlink system	Parameter L_9 (mm)	Bandwidth (GHz)
	2.00	6.18–6.97
	2.25	6.45–7.38
	2.50	7.21–7.72

Table 2, and respective equations are given below:

$$L_{\text{Notch Band}}=c/(2 \times f_{\text{Notch}}\sqrt{\varepsilon_{\text{reff}}}) \quad (1)$$

$$\varepsilon_{\text{reff}} = \frac{\varepsilon_r + 1}{2} + \frac{\varepsilon_r - 1}{2}\left[1 + 12\frac{h_1}{W_f}\right]^{\frac{-1}{2}} \quad (2)$$

Further, to study the characteristics of interfering bands, the distribution of current density on the surface of the radiator and ground is analyzed. Figure 4a shows surface current density at a designed frequency of 6.85 GHz where low surface current density distribution is observed all over the surface indicating the radiation. Figure 4b–d shows surface current density distribution at 3.48 GHz, 5.52 GHz, and 7.39 GHz, respectively. From Fig. 4b, the maximum density of surface current is concentrated inside the rotated T-shaped stub which correspondingly notches the WiMAX band. The maximum current indicates that the antenna offers a high mismatch condition for 3.48 GHz center frequency of WiMAX notched band, and the antenna does not radiate. For Fig. 4c–d, the maximum surface current distribution is observed inside a π-shaped parasitic element backed plane and single complementary split ring resonator placed near microstrip feed. The same condition prevails for both WLAN and DSS notched band where high mismatch condition leads to concentration of surface current leading to the antenna in the non-radiating mode for 5.52 GHz and 7.39 GHz.

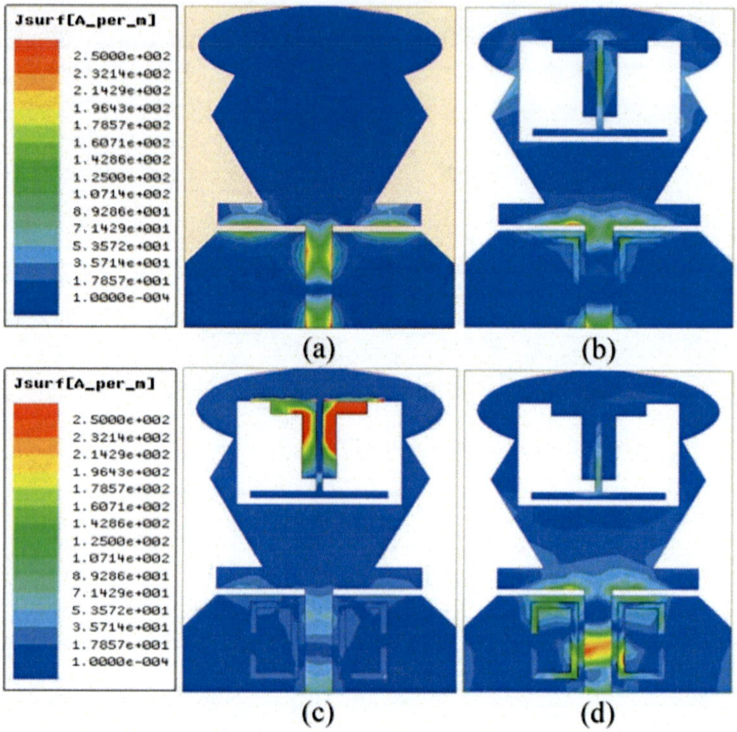

Fig. 4 Simulated surface current density distribution of proposed triple notched band antenna at **a** 6.85 GHz, **b** 3.48 GHz, **c** 5.52 GHz, and **d** 7.39 GHz

4 Results and Discussions

Figure 5a shows verification of results by two different RF simulators, Ansys HFSSv13 and CST Microwave Studio. There is almost agreement between the two results, and compared values are shown in Table 3.

Figure 5b shows two different curves, one gain and the other radiation efficiency. For the antenna without notched bands, gain varies between 3.52 dBi and 5.27 dBi. For notched band antenna, gain falls to −24.52 at 3.52 GHz, −21.89 at 5.48 GHz and −28.96 at 7.54 GHz, respectively, and remains between 3.11 dBi and 5.96 dBi for remaining radiating signals. Figure 5b also shows the radiation curve for notched band characteristics which remains between 73.69% and 88.58% in the operating band but falls to 12.59% at 3.44 GHz, 18.96% at 5.44 GHz, and 14.25% at 7.48 GHz.

Figure 6 shows the radiation design plot for various operating frequencies. Standardized radiation design for frequencies 4.50 GHz. 7.00 GHz, 11.00 GHz, and 16 GHz appear in Fig. 6 for E-plane and H-plane. It tends to be seen that there is an acceptable agreement between two test systems which shows an omnidirectional example in H-plane and dipole like the example in E-plane.

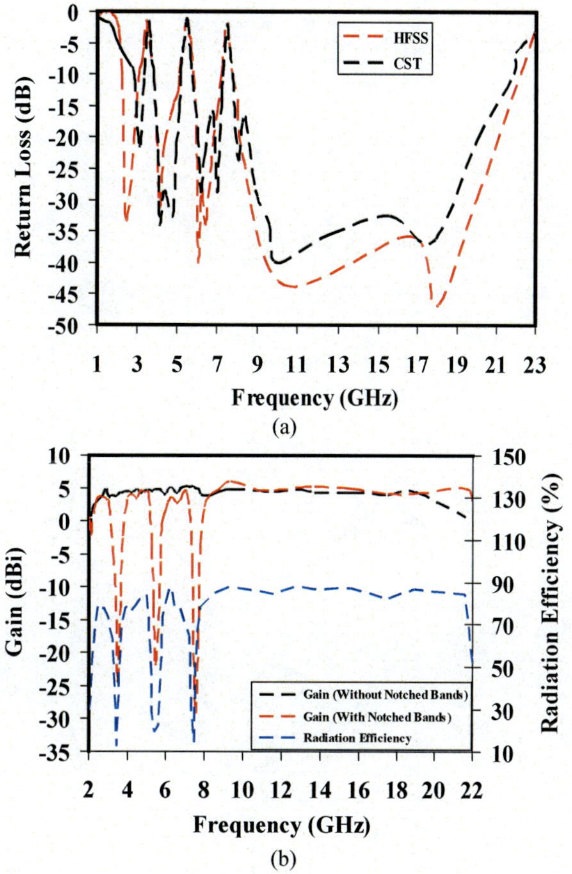

Fig. 5 Fairfield result of proposed antenna **a** comparison of triple notched band antenna and **b** gain of the proposed antenna without/with notched bands and radiation efficiency

Table 3 Return loss comparison

Simulator	Bandwidth (GHz)	The bandwidth of notched bands (GHz)
HFSS	2.83–21.85	3.32–3.78 5.16–5.77 7.23–7.73
CST Microwave Studio	2.24–22.25	3.11–3.81 5.11–5.78 7.21–7.72

5 Conclusions

Super-wideband antenna with triple band-notched characteristics is proposed with compact dimensions of 20 $\times$ 25 mm^2 printed on Rogers RTDuroid5870. Notched

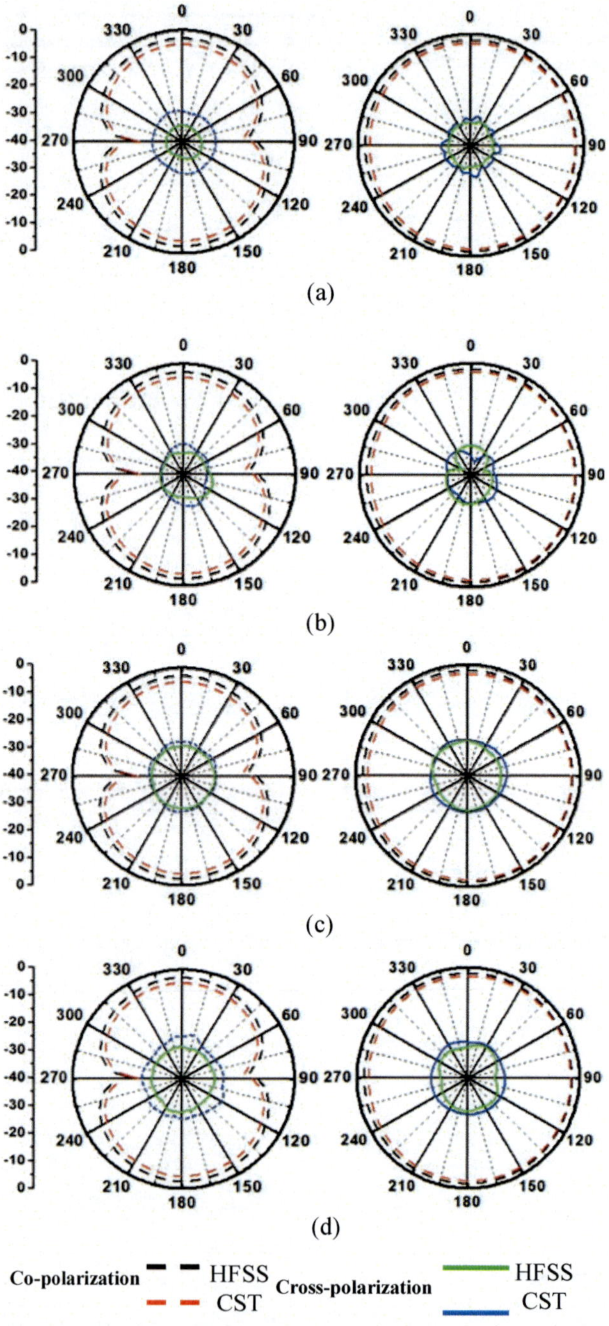

Fig. 6 Simulated normalized radiation patterns in E- and H-plane (dB) at **a** 4.50 GHz, **b** 7.00.0 GHz, **c** 11.00 GHz, and **d** 16.00 GHz

bands are achieved using C-shaped slots, inverted T-shaped electromagnetic coupling element, and C-shaped parasitic elements near the feed line. Notched bands gain correspond to −24.52, −21.89, and −28.96 dBi at 3.52 GHz, 5.48 GHz, and 7.54 GHz, while gain remains 3.11–5.96 dBi in the remaining operating bands. The simulated radiation pattern shows omnidirectional with radiation efficiency ranges from 73.69% to 88.58% except for notched bands (12.59% at 3.44 GHz, 18.96% at 5.44 GHz, and 14.25% at 7.48 GHz). All the discussed parameters of the designed antenna make it a promising candidate for UWB, X, and Ku band applications.

References

1. Dang Trang N, Dong Hyun L, Hyun Chang P (2012) Very compact printed triple band-notched uwb antenna with quarter-wavelength slots. IEEE Antennas Wireless Propag Lett 11:411–414
2. Jiang ZH et al (2018) A compact triple-band antenna with a notched ultra-wideband and its mimo array. IEEE Trans Antennas Propag 66(12):7021–7031
3. Banerjee J, Karmakar A, Ghatak R, Poddar DR (2017) Compact CPW-fed UWB MIMO antenna with a novel modified Minkowski fractal defected ground structure (DGS) for high isolation and triple band-notch characteristic. J Electromagnetic Waves Appl 31(15):1550–1565
4. Hosseini H, Hassani HR, Amini MH (2018) Miniaturised multiple notched omnidirectional UWB monopole antenna. Electron Lett 54(8):472–474
5. Srivastava K et al (2018) Integrated GSM-UWB fibonacci-type antennas with single, dual, and triple notched bands. IET Microwaves Antennas Propag 12(6):1004–1012
6. Lotfi P, Azarmanesh M, Soltani S (2013) Rotatable dual band-notched uwb/triple-band wlan reconfigurable antenna. IEEE Antennas Wirel Propag Lett 12:104–107
7. Murugan NA, Balasubramanian R, Patnam HR (2016) Printed ultra-wideband monopole U-slotted antenna for triple band-rejection. J Electromagnetic Waves Appl 30(12):1532–1544
8. Rojhani N, Sebak A, Akbari M (2015) Controllable triple band-notched monopole antenna for ultra-wideband applications. IET Microwaves Antennas Propag 9(4):336–342
9. Sarkar D, Srivastava KV, Saurav K (2014) A compact microstrip-fed triple band-notched uwb monopole antenna. IEEE Antennas Wirel Propag Lett 13:396–399
10. Si L-M et al (2015) A uniplanar triple-band dipole antenna using complementary capacitively loaded loop. IEEE Antennas Wirel Propag Lett 14:743–746
11. Tang Z, Lian R, Yin Y (2015) Differential-fed UWB patch antenna with triple band-notched characteristics. Electron Lett 51(22):1728–1730
12. Tang Z, Wu X, Zhan J, Xi Z, Hu S (2018) A novel miniaturized antenna with multiple band-notched characteristics for UWB communication applications. J Electromagnetic Waves Appl 32(15):1961–1972
13. Wang J, Yin Y, Liu X (2014) Triple band-notched ultra-wideband antenna using a pair of novel symmetrical resonators. IET Microwaves Antennas Propag 8(14):1154–1160
14. Weng YF, Cheung SW, Yuk TI (2012) "Design of multiple band-notch using meander lines for compact ultra-wide band antennas." IET Microwaves Antennas Propag 6(8)
15. Xiao W, Mei T, Lan Y, Wu Y, Xu R, Xu Y (2017) Triple band-notched uwb monopole antenna on ultra-thin liquid crystal polymer based on escsrr. Electron Lett 53(2):57–58
16. Zhang C, Zhang J, Li L (2014) Triple band-notched uwb antenna based on sir-dgs and fork-shaped stubs. Electron Lett 50(2):67–69
17. Zhu F et al (2013) Multiple band-notched uwb antenna with band-rejected elements integrated in the feed line. IEEE Trans Antennas Propag 61(8):3952–3960

An Internet of Things (IoT)-based Approach for Real-Time Kitchen Monitoring Using NodeMCU 1.0

N. Umapathi and Saiteja Sabbani

Abstract Monitoring the kitchen is one thing everyone should consider as it is the most probable place in the house where accidents tend to happen. This paper explores the idea of building an automated kitchen monitoring system using Internet of Things (IoT) and microcontroller-based sensors. An IoT approach for monitoring the kitchen is encouraged due to the fact that the number of Internet users has increased exponentially in recent years by technological advancements in the field of telecommunication and also increased the standard of living of people. This automated kitchen monitoring system has sensor unit (to monitor the kitchen parameters) which is interfaced to a microcontroller development board (Arduino UNO), and the transmission of this parameter values through Internet is carried out by NodeMCU 1.0. The system comes with a user-friendly mobile application to monitor kitchen parameters like temperature (in degree Celsius), percentage of humidity, potential LPG gas leakage, accidental kitchen fires and weight scale for groceries. The mobile application is developed using an online tool MIT APP inventor and the database which stores the kitchen parameters temporarily is Google's Firebase Database. Additionally, the mobile application also has a feature of controlling two electronic appliances (connected to a two-channel relay) from anywhere around the world.

Keywords IoT · Kitchen monitoring · LPG · NodeMCU

1 Introduction

The vast majority of reported kitchen accidents people come across are significantly small. These mishaps when identified early can reduce the chance of accidents, thereby reducing the loss of life, property, etc. Hence, a need for a proper automated kitchen monitoring system is clearly seen, which drives the motivation for designing the project.

N. Umapathi (✉) · S. Sabbani
Department of ECE, Jyothismathi Institute of Technology & Science, Karimnagar, Telangana, India

A. Sivasubramanian et al. (eds.), *Futuristic Communication and Network Technologies*, Lecture Notes in Electrical Engineering 792,
https://doi.org/10.1007/978-981-16-4625-6_4

NPFA's report in reference [6] suggests cooking caused an average of 173,200 reported home structure fires per year. These fires resulted in an average of 550 deaths (22% of all home fire deaths) and 5,020 home fire injuries (45% of all reported home fire injuries) annually.

2 Literature Survey

The work referenced in [1] developed a system which uses sensors like LDR, weight cell along with microcontroller-based boards Arduino Mega and NodeMCU to monitor the grocery weight placed in a contraption and light intensity in the kitchen. The system updates this information in a web server/database and notifies if there is a shortage. System referenced in [2] uses DHT 11, IR flame and MQ-3 sensor which are interfaced to Arduino UNO to measure the corresponding parameters and display on a 16 × 2 LCD display. Subsequently, it uses Thingspeak web server to display information to user through Internet. System in [3] used an android application/Web site to operate the kitchen. The system gives the quantity of ingredients and sends the information about weight for cooking side node by means of developed application to avoid the discomfort of shortage beforehand. Reference [4] is a design of smart kitchen cabinet which helps in grocery identification and inventory management. The system has weighing section with RFID tag defining the description of product and ultrasonic sensor for measuring the level of items in cabinet. The system continuously monitors and sends this data to database and thus helps in generating automatic shopping list. IoT system in [5] is focused on NB-IOT safety system. The system uses STM32F103 as main microcontroller and Alibaba cloud servers for communicating with NB-IOT node about the safety parameters which are under observation by the system. As reviewed in [7], a load cell is to measure grocery weight and NodeMCU 1.0 to transmit the data by tethering it to a local Wi-Fi network in order to implement an active measurement system for smart kitchen. Reference [8] is an idea of system based around development of IoT prototype for controlling the AC appliances in kitchen using a DC relay of 5 V and microcontroller. The work in [9] focuses on developing a monitoring system using microcontroller and web server. The work is solely centered on the development of web server architecture. System stated in reference [10] is an example prototype which is built keeping in mind to be more power efficient and usability. The system is less power hunger and can be used with DC battery. The URL links [11] and [12] are for MIT App Inventor and Google's Firebase Database Web site, respectively, which are used in the development of the application.

3 Design and Implementation

The proposed system uses Arduino UNO as the primary microcontroller for controlling the sensor unit and NodeMCU. Sensors like DHT11, MQ-2, IR flame, HX711 and load cell are used to monitor temperature, humidity, LPG gas, flames groceries weight, respectively. NodeMCU is just there to receive the values from Arduino UNO and send to the app through using Wi-Fi network. A two-channel relay is used to control two kitchen appliances using the developed application.

The IoT communication technology uses Wi-Fi network at the sensor end of the system. However, on the user end, Internet can be provided using either Wi-Fi or cellular network based on user's convenience. Figure 1 shows the block diagram of the system with kitchen end (Wi-Fi for NodeMCU) and user end mobile (Wi-Fi or cellular network for mobile).

3.1 Explanation of Circuit Connections

The connections for the prototype are shown in Fig. 2. Arduino UNO uses the built-in function like analogRead() connected to pin A0, to read the data from DHT11 sensor. As the other sensors like HX711, IR flame and MQ-2 give digital output, they are connected through pins 4,5,6,7 and interfaced used digitalRead() function. Arduino and NodeMCU are serially connected by using pins 9, 10 of Arduino UNO (as Tx and Rx, respectively) to NodeMCU pins D6, D7 (as Tx and Rx, respectively). A 220VAC is connected to relay's COM and NO terminals to switch the appliances using NodeMCU pins D3 and D2.

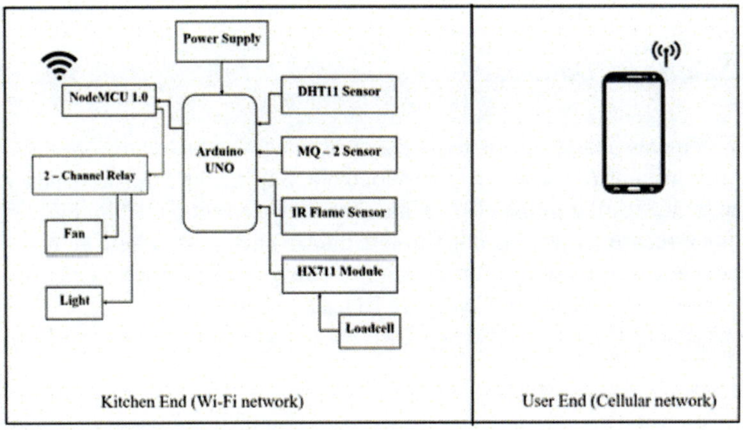

Fig. 1 Block diagram of proposed system

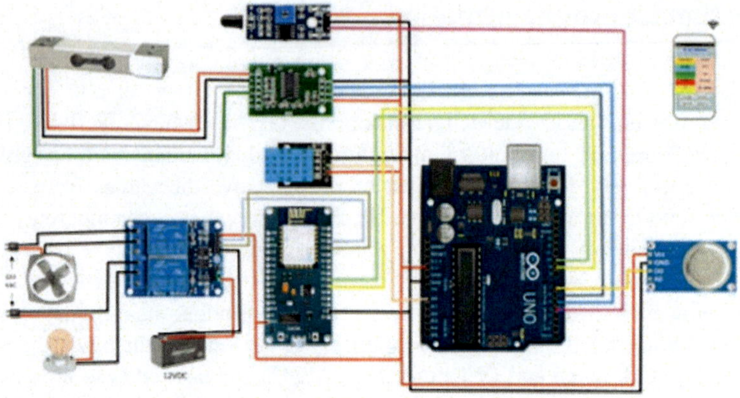

Fig. 2 Schematic of proposed system

3.2 Design of Mobile Application

The design of the application to display the parameters from Arduino is done using an online free tool, namely MIT App Inventor. This tool has two main sections: They are design window and blocks window. Initially, the user interface of application is done using the design window of the tool. We place required labels and buttons and adjust their properties like visibility, size and color in this section.

Next in the blocks window, we have building blocks for which our components from the user interface has to be connected. There is no need for any coding knowledge while using this MIT App Inventor, since we have to place all the actions in form of building blocks.

3.3 Setting up Database

Google's Firebase Database is a free high-speed database which can be used alongside MIT APP Inventor for mobile app development. Setting the Firebase Database can be done by navigating to the URL. Under the console section, add project. Stating mode for Firebase security is done in test mode. This mode offers getting started with mobile and web clients. But in order to make the data inaccessible by others, change the Firebase Real-time Database Rules accordingly.

4 Output and Analysis

4.1 Output in App

Figure 3 demonstrates the user interface of app along with the columns for displaying parameters which are monitored. The system monitors the temperature, humidity, groceries weight, flames, LPG, and subsequently four buttons are used to control the

Fig. 3 Screenshot of output in mobile app

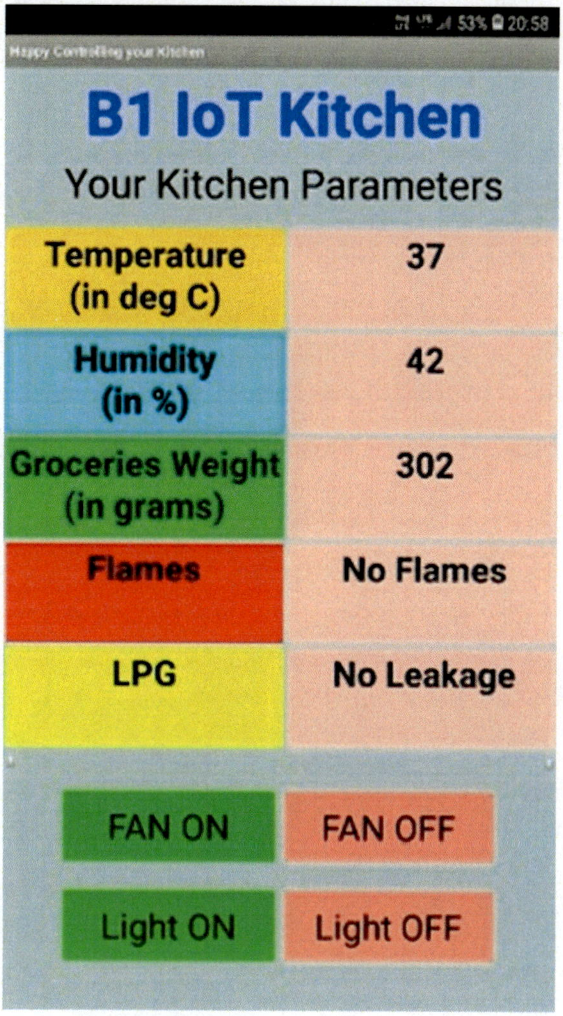

two appliances which are connected to two-channel relay. The background color of the buttons changes to green when they are in ON state and to red when they are in OFF state. Now for the flames and LPG, the text is turned into red when there is potential presence of LPG and flames.

4.2 Alert Message in Mobile

The app is designed in a way that when there is a leakage of LPG gases or in case of a fire accident it sends an alert SMS to the registered mobile number which is given during app design through MIT Web site. The alert message is shown in Figs. 3 and 4.

Fig. 4 Screenshot of alert message

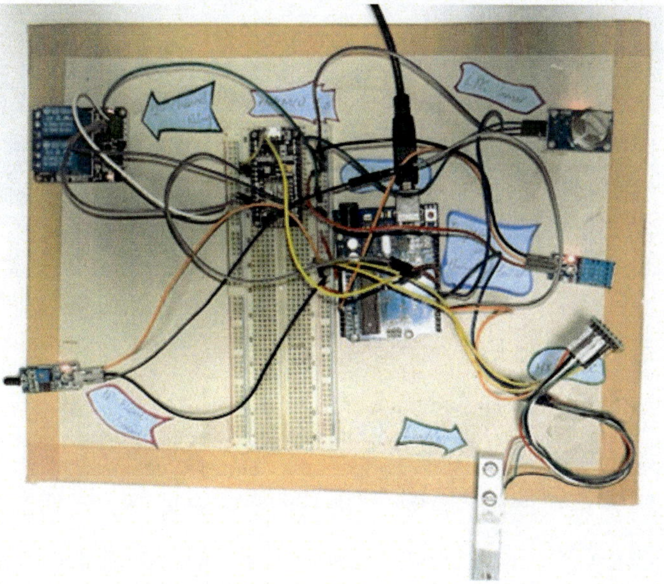

Fig. 5 Hardware demonstration

4.3 Analysis of the Output

In order to test, the reliability of the prototype analysis of the output must be done. This analysis includes the time of response, which means the time taken for the prototype to send its data from sensors to the application and the time taken by the app to send data to prototype's relay.

The experimental analysis is done by tethering the NodeMCU with a local Wi-Fi network giving the user name and password through programming. A standard 4G cellular network is provided to mobile. The experiment is done three times, and every sensor's time response is tabulated in Tables 1 and 2.

Table 1 Time of response (in seconds) from sensors to app

	Temperature	Humidity	Flames	LPG	Groceries weight
Trail–1	14	14	32	30	9
Trail–2	21	21	19	28	8
Trail–3	18	18	26	33	10
Average time	17.6	17.6	25.6	30.3	9

Table 2 Time of response (in seconds) from app to relay

	Channel–1	Channel–2
Trail–1	5	8
Trail–1	8	7
Trail–1	7	6
Average time	6.6	7

5 Challenges in Implementation of System

The sensors like LPG and flame used in developing the system since having least ranges typically in order of 20–30 cm have to be placed as near as possible to the gas cylinder to make the system more reliable.

Weight sensor HX711 has to be calibrated accordingly since it shows different values depending upon the room temperature. This problem can be addressed by using the zero error value of weight scale and calculating the weight based on this zero error.

Using a wireless network (Wi-Fi/cellular), add some unreliability for the system since the problems on network signal impact the performance of system. Network congestion, signal strength, interference during transmission, antenna bandwidth, etc., all impact the alert message. Latency, i.e., the time taken by the signal to reach to its intended target plays a crucial role as well since the issue we are addressing here is related to safety concerns.

All these problems are tackled to an extent by employing an optimal design by using components which are picked keeping these issues under consideration, and an extra layer of optimization is done in code as well.

6 Conclusion

The built prototype is able to deliver the expected results with acceptable accuracy. The developed prototype is portable, less expensive and is also worth affording keeping the safety it can provide and accidents which can potentially occur, in mind. The time taken by the system to respond to the accident and send the alert message is also less as the Google's Firebase Real-Time Database can update the information sent from sensors very quickly. This is an important feature when it comes to handling fire and LPG leakage. The developed application has a user-friendly interface and can even work on any low-end devices with Internet access. As the system is built using an IoT approach, it adds a strength to the system as the range IoT can provide is not limited to a state or country it can reach up to continents overseas.

7 Future Scope

The overall project is very efficient and worth building for its price, but like always there is a lot of room for new things which can be added to this system to increase its scope. We can build the same system using a Raspberry Pi and include a camera module and a (WSN) wireless sensor network for monitoring the kitchen. It can also be improvised to order groceries when there is a shortage and even call fire department when flames are detected.

This is more useful in large restaurants where groceries consumption is more and helps to keep a track to monthly/daily usage and also apartments where there are many kitchens, hence more chance for accident.

References

1. Rezwan S, Ahmed W, Mahima MA, Islam MR (2018) IoT based smart inventory management system for kitchen using weight sensors, LDR, LED, Arduino Mega and NodeMCU (ESP8266) WiFi Module with Website and App. 978–1–5386–7167–2/18/$31.00
2. Pudugosula H (2019) Automatic smart and safety monitoring system kitchen using internet of things. International conference on intelligent computing and control systems, 978–1–5386–8113–8
3. Arya A, Taliyan A, Chauhan P, Gautam A (2019) smart kitchen with new measurment, web and application based with affordable design. 978–1–7281–1253–4/19/$31.00
4. Thakare A, Gandhe PR (Feb 2017) A review paper on kitchen monitoring system using embedded web server. Int Res J Eng Technol 04(02): 2395–0056
5. Chang Yun K, Chang Wen A (2019) Design of safety for kitchen based on nb-iot. 3rd IEEE international conference on robotics and automation sciences, 978–1–7281–0855–1/19/$31.00
6. NFPA's REPORT 2019, https://www.nfpa.org/News-and-Research/Data-research-and-tools/US-Fire-Problem/Home-Cooking-Fires, Last accessed 17 Aug 2020
7. Azran MAM, Zaid MFM, Sakeeb MSM, Althaff MMM, Fernando SGS (Dec 2017) Smart kitchen—a measurment system. Int J Comput Appl 179(2):0975–8887
8. Sinha A, Sharma S, Mahboob M (2017) An internet of things based prototype for smart appliance control. International conference on computing, communication and automation ICCCA
9. Damodhar J, Swathi S (Jun–Jul 2016) Internet based monitoring system for smart kitchen using embedded web server architecture. Int J Innovative Technol Res 4(4):3306–3311
10. Ceccacci S, Menghi R, Germani M (2015) Example of a new smart kitchen for energy efficiency and usability. The fourth international conference on smart systems, devices and technology
11. MIT App Inventor Website, https://appinventor.mit.edu. Last accessed 17 Aug 2020
12. Firebase link, https://firebase.google.com. Last accessed 14 Aug 2020

IoT-Based Smart Irrigation and Monitoring System in Smart Agriculture

D. Manikandan, Sadhish Prabhu, Parnasree Chakraborty, T. Manthra, and M. Kumaravel

Abstract Developing countries such as India regards agriculture as the backbone. It plays a vital role in the growth of the Nation's economy. Despite the fact that the world is moving into new technology, agriculture must also be allowed to drift away. One such innovation is the Internet of Things (IoT). The Internet of Things (IoT) is defined as interconnecting different devices like sensors via the Internet. This paper proposes and tests a cloud-based wireless communication framework for tracking and managing a series of sensors and actuators for determining plant water needs and environmental parameters that impact plants. An ESP8266 Wi-Fi module-based automatic irrigation system and environment monitoring systems are implemented in this project. A smart irrigation system is implemented using soil moisture sensor which senses the moisture present in soil and actuates motor accordingly. For healthy growth of plants, it needs fresh and clean air; hence, it is monitored by MQ135 sensor called air quality sensor. The DHT11 sensor also known as temperature and humidity sensor is used for low-cost and more efficient monitoring of temperature and humidity. Blynk application in mobile is used to analyze all wireless sensor data through the cloud. These implementations will maximize the yield of the crops.

Keywords IoT · Soil moisture · Air quality · Cloud computing · Wireless sensors

1 Introduction

Agriculture is known to be one of the most momentous economic activities. It is crucial for the sustainability and development of the Indian economy. In average, 70% of households and 10% of the urban population depend on farming as their

D. Manikandan (✉) · S. Prabhu · P. Chakraborty · T. Manthra · M. Kumaravel
Department of Electronics and Communication Engineering, B. S. Abdur Rahman Crescent Institute of Science and Technology, Chennai, Tamil Nadu 600048, India

S. Prabhu
e-mail: sadhishprabhu@crescent.education

P. Chakraborty
e-mail: prernasree@crescent.education

A. Sivasubramanian et al. (eds.), *Futuristic Communication and Network Technologies*, Lecture Notes in Electrical Engineering 792,
https://doi.org/10.1007/978-981-16-4625-6_5

livelihood [1]. Also one of the world's biggest employers, particularly in developing and underdeveloped nations, is the agriculture sector [2]. Millions of people around the world, directly or indirectly depend on the agriculture sector for their livelihood. Water and nutrients are normally taken up by the roots from the soil [3]; this is why it is important to water plants whenever the soil is dry.

We should switch the manual cultivation and irrigation methods toward automation in order to improve plant productivity. This automation can be done using the Internet of Things (IoT). In the near future, IoT contains some advanced technologies for use, such as intelligent homes, smarter towns, smart transport and intelligent agriculture. The engineering can be used in the system to improve productivity and quality of specific amounts of fertilizer, food, pesticide, etc. [4]. IoT sensors provide agricultural field data. This paper presents an IoT and smart agriculture system using automation.

This IoT-based system for smart agriculture tracks wireless sensor systems which gather data from various sensors at specific nodes that the device processes and analyzes. This IoT system is powered by the ESP8266 Wi-Fi module; it consists of a DHT11-sensor, soil moisture sensor, MQ135-sensor, DC motor, and relay module. The IoT-based monitoring system senses the moisture content present in the soil, and if it is low, then it automatically actuates the water pump. The cloud server contains soil humidity, temperature, humidity, and the sensor data to analyze and control actions. The user may always monitor data and also control irrigation on the mobile phone via the Blynk android app.

2 Related Work

Rabadiya Kinjal et al. [5] have given an automatic irrigation system because the most critical and pain consuming job is watering to the entire crop field. One of the active parameters of the greenhouse effect on crops is the amount of water needed by them. To facilitate the irrigation done by the farmers, some theoretical and analytical data are required. This research is based on an ESP8266 microcontroller with an inbuilt Wi-Fi Module. The data collected from this system would help farmers to identify the suitable soil for the crops.

Ji-Chun Zhao et al. [6] researched the applications of IoT in agriculture. The researchers proposed a remote detector or Internet-based monitoring system. In order to provide information for smart agriculture research, a data management framework has been created. The software developed offers detailed control of the greenhouse.

Rajalakshmi P. et al. [7] described plant field control with various sensors that are used for the automatic irrigation system. Wireless information has been sent to the webserver, and JSON format is used to encrypt the information to manage the repository of the client. The farmland's moisture and also temperature dropped below its borders, and the irrigation system will be actuated.

Balasaraswathy et al. [8] have investigated that sunlight pitfall because the origin of UVA is unique from the data, location, and environmental conditions of phototherapy. The peak exposure of UVB ranges from 12mid—1.00 p.m. to 12midnight—1.15 a.m. The early and late evenings, the UVA rate was 1:50 and 1:60 ratios. The UVA was peak from 11.30 a.m. to 1.30 p.m., and the giveaway time was 1 June of UVA. Such study findings suggest that UV radiation is high during the peak hour (11.30 in the morning to 1.30 in the afternoon), and therefore, plants need to be watered every hour to prevent drying up.

Nakutis et al. [9] proposed an automation system for remote farming with IoT-based OPC UA servers connected sensors and actuators. Cloud services (install and upgrade system controllers) are used to adjust control rules without modifying remote sensor / actuator software.

3 Proposed System Model

ESP8266 (Node MCU) is the Wi-Fi module that is responsible for all the control, data analyzing, and cloud storing actions. It is a microcontroller with high processing power. It is cheap, low-power consuming, and portable. It is connected to the Internet via Wi-Fi or hotspots.

Figure 1 represents the block diagram of proposed system model. The overall yield of the plant will be affected by environmental changes. For optimal growth and health, plants require appropriate conditions. It is very important to track the conditions of the crop so that different sensors are used. DHT11 sensor is used to

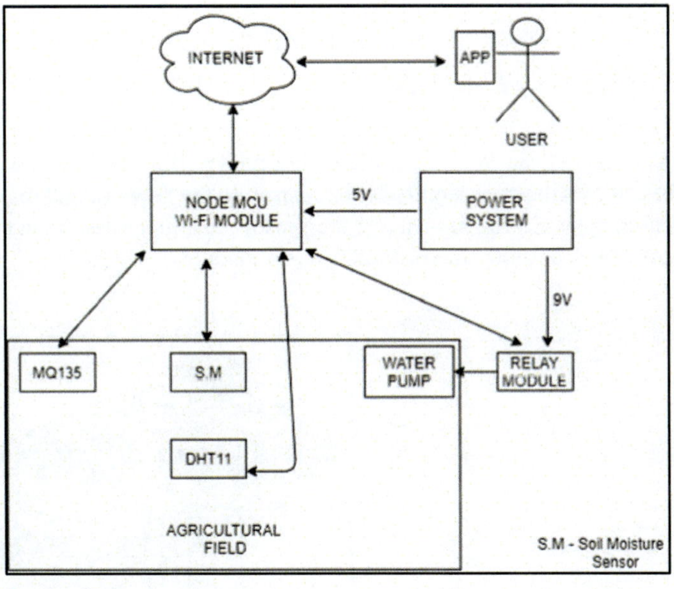

Fig. 1 Proposed system block diagram

sense the humidity and temperature present in the environment. By sensing these parameters, we can decide when to water the plants and when not to water them. This can save the amount of water being used. DHT11 sensor is easily available in the market. Its technology ensures high accuracy and excellent long-term uniformity. The MQ135—sensor is used to sense the quality of air in Parts per Million (PPM); accordingly, we can determine the presence of CO2 and ammonia. It is stable with long life and has a simple driving circuit in air quality control devices that is ideal for the sensing of NH3, NOx, alcohol, benzene, smoke, CO2, –etc. The sensor is designed with a high detection range, quick reactivity, and high sensitivity. Soil moisture sensor is used for accurate measurement of moisture level present in the soil. There are two nickel rods inserted into the field in the soil moisture detector. They are used to flow current through the ground [10]. The microcontroller analyzes the humidity information and allows the water pump to be operated and disconnected accordingly. Node MCU is powered with 5 V by mobile adapter, and water pump is powered by a 9 V battery. These sensors are low cost and offer good accuracy.

4 Description of the Hardware Used

The proposed system model consists of the various module such as ESP8266 Wi-Fi module, DHT11—temperature and humidity sensor, MQ135—air quality sensor, soil moisture sensor, submersible water pump, relay module, etc., which are discussed in details in the following section.

4.1 ESP8266 Wi-Fi Module

The ESP8266 is a cheap Wi-Fi module with a high performance microcontroller. This module allows the microcontroller to connect to the network and makes TCP / IP connections easy. ESP8266 acquires the sensor data, processes it, and pushes it to the cloud. Figure 2 shows the ESP8266 Wi-Fi module.

Fig. 2 ESP8266 Wi-Fi module

Fig. 3 DHT11—Temperature and humidity sensor

4.2 DHT11—Temperature and Humidity Sensor

The DHT11 is a primitive digital temperature and humidity sensor. It uses a powerful moisture sensor and a thermistor for calculating environmental parameters and sends a digital signal to the data pin. Figure 3 shows the DHT11 sensor module.

4.3 MQ135—Air Quality Sensor

The sensitive material for the gas sensor MQ 135 is the SnO2 (Tin Oxide), which has reduced clean air conductivity and where the target gases are present the conductivity of the sensor increases when the gas concentration increases. The sensor can find the presence of gases such as CO2, ammonia, and smoke. Figure 4 shows the MQ135 sensor module.

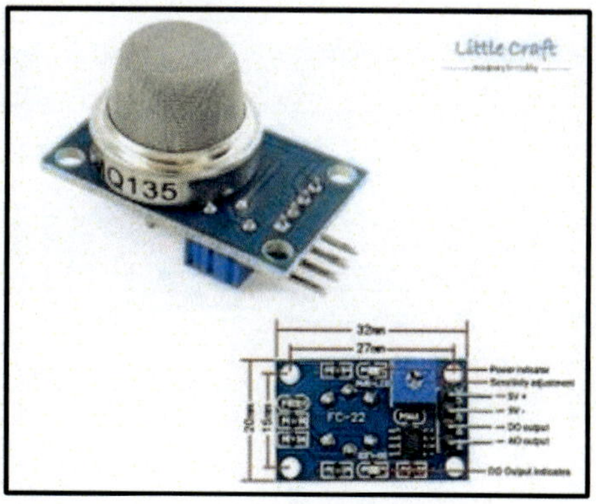

Fig. 4 MQ135—air quality sensor

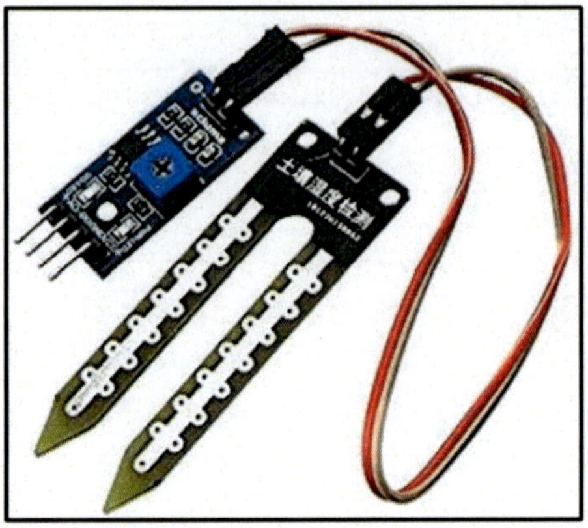

Fig. 5 Soil moisture sensor

4.4 *Soil Moisture Sensor*

The probes of the soil moisture sensor are inserted in the soil. The moisture soil has less resistance and hence allows more current through the soil, whereas the dry soils have high resistance and allow less current to flow through the soil. The resistance value helps to detect the soil moistures. Figure 5 shows the soil moisture sensor.

4.5 *Submersible Water Pump*

To irrigate the field when there is less water content in the soil, the submersible water pump is turned ON which is being controlled by the Node MCU and soil moisture sensor. This is a small and cheap submersible pump which can be supplied with 6~12V. Figure 6 shows the Submersible water pump.

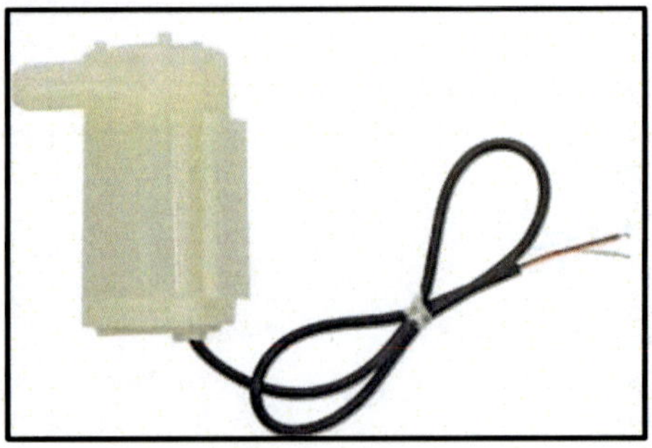

Fig. 6 Submersible water pump

Fig. 7 Relay module

4.6 Relay Module

Relays are fundamental switches that open and close circuits electromechanically or electronically. One circuit is operated by the relay by another circuit control. The lower current can control the higher in a power and high voltage system. It is widely used in power protection, automation and communication devices [11] (Fig. 7).

5 Software Implementation Using Arduino IDE

The Arduino Integrated Development Environment (IDE) is an open-source cross-platform framework. It is used to write and upload programs to boards that are adaptive for Arduino, but also for other developers ' boards supported by the cores of third parties. The Arduino IDE supports languages C and C + + . It has huge community and libraries support, which allows the user to code the boards in ease. It is very much reliable and user-friendly. Arduino is a famous tool for IoT product prototyping and development. It is used to program all the sensors that are integrated to the ESP8266 Wi-Fi module and connect with Wi-Fi to enhance IoT communication. Figure 8 shows the Arduino IDE.

6 Blynk Mobile Application

For the Internet of Things, the software Blynk is built. The hardware can be remotely controlled; sensor data displayed, and data can be saved, viewed and done many more. The Blynk app enables users to create a dashboard for the project and to arrange on screen buttons, sliders, illustrations, and other widgets. Using the widgets, users can turn pins ON and OFF or display data from sensors in the mobile phone. In smart agriculture, farmers can view the data such as temperature, soil moisture, and air quality from a remote place. Figure 9 shows the Blynk mobile application.

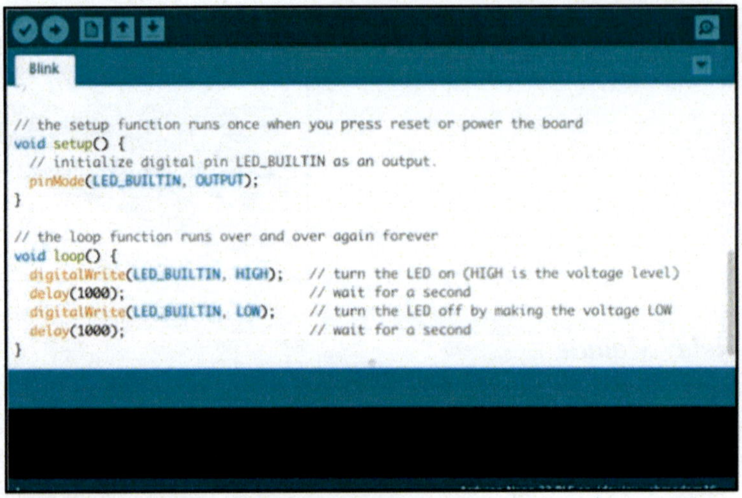

Fig. 8 Arduino IDE

7 System Implementation

The data received from the field is being processed by the ESP8266 Wi-Fi module. The data that has been received is sent to the Blynk cloud server using the Wi-Fi network. The task of data processing in a field with already established threshold values is to test the values of each sensor received. Depending upon the crops planted, thresholds vary. Different plants require different water amounts. In the same way, different plants require different temperatures and moisture. The sensor values also differ depending on the weather [12]. In summer and winter, soil moisture will be different. During spring, winter or rainy season, the temperature and humidity also differ. When the soil moisture value drops under the threshold and vice versa, the motor is switched ON and OFF automatically. The farmer can even switch on the motor from the mobile using the mobile application.

Figure 10 shows that the soil moisture sensor is placed in the high moisture soil; hence, ESP8266 sends a low signal to the relay module which keeps the water pump in OFF state. It continuously acquires all the sensor data and pushes them to the cloud.

Figure 11 shows that the sensor is not placed in the soil; hence, moisture will be zero, and ESP8266 sends high signal to relay module which turns ON the water pump. Various sensor data such as temperature, humidity, air quality, and soil moisture can be viewed by the farmer in the mobile phone application.

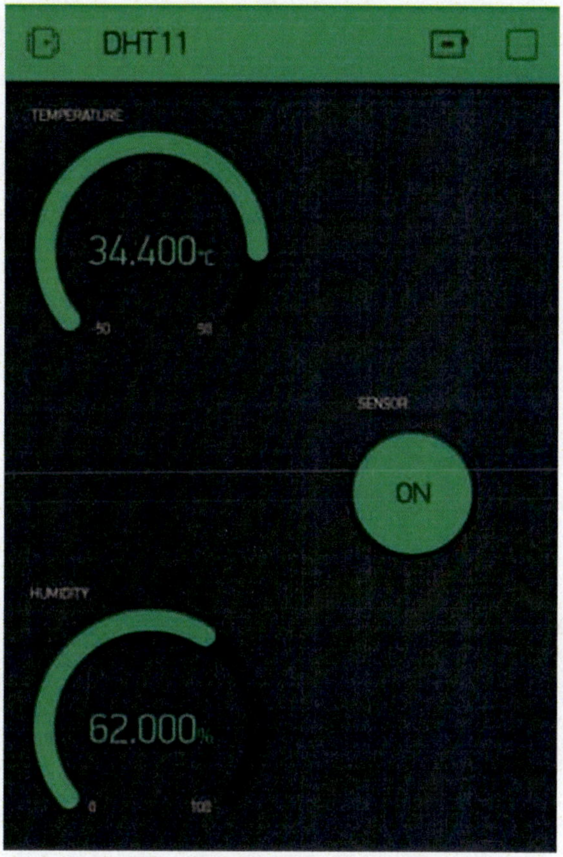

Fig. 9 Blynk mobile application

8 Results

The Blynk mobile app provides users with textual representations of information collected from the network of wireless sensors that are stored on the cloud platform. This representation includes the readings of each sensor. The application user interface is created by inserting widgets such as buttons and display meters which both enables crop control and monitoring of the device. To track and control the field, the Internet connection should be given. Figure 12 shows DHT11 sensor data, MQ135 sensor data, and soil moisture sensor data in the Blynk mobile application. This helps the farmer to know various parameters and can do further actions. The water pump motor turns ON automatically that is being controlled by ESP8266 controller when the moisture is below the threshold level.

Dirty air created by fire, pollutants, and other chemicals may be detrimental to crops, limiting their ability to absorb carbon dioxide from the atmosphere to produce food (photosynthesis). It can also block sunlight, important for healthy growth of plants. In addition to water and nutrients, clean air and good soil help plants grow

Fig. 10 Hardware implementation (Motor is OFF at High Moisture)

Fig. 11 Hardware Implementation (Motor is ON at Low Moisture)

very well. The table below illustrates the different values of the air quality index and its consequences [13]. The MQ135 sensor reading shows the air quality index (AQI) as 49 PPM, and the condition is good [15]. The DHT11 sensor shows the humidity values and temperature values of the environment. The farmer can not only view the sensor data but can turn ON and OFF the sensors if required.

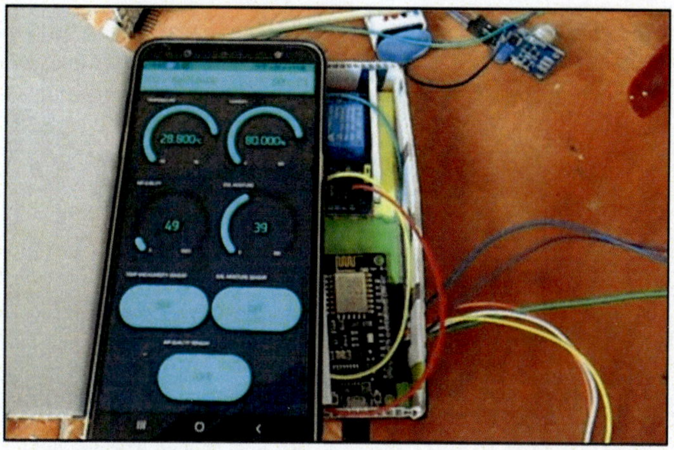

Fig. 12 Sensor data in Blynk app

9 Conclusion

In the field of digital tools and data selection, the Internet of Things is wide-ranging. This farm surveillance method acts as a reliable, effective process, and corrective measures can be taken. Wireless field tracks reduce human power and also allow users to see precise changes in crop yield. Prices are cheaper because power levels are better. The process of intelligent agriculture was created and synthesized [14]. This solution also offers a fully automated irrigation system allowing farmers to determine in real time regarding land and crops. Cloud computing is "a new computer design, which provides the web with dynamically dispersed and often virtualized assets."

All results and preliminary trials indicate that the solution proposed would certainly lead to increasing crop production and overall production in the region of agriculture, irrigation problems, and application of this system in the sector. Data shall be made available to farmers by means of a mobile application on soil moisture, temperature, air humidity, and air quality in the agricultural field. The device can be used in greenhouses and temperature-dependent crops. In areas where water scarcity is a major problem, the method is very useful. The system that has been built is easier, and tests are more realistic.

10 Future Works

In the future, this process can be improved by adding a number of modern techniques such as advanced irrigation methods and the use of renewable resources such as solar power. A thorough understanding of industrial functionality and standards on considerations such as cost, security, confidentiality, and danger needs to be discussed before the IoT is widely accepted and implemented in all fields. In addition, the varieties of plants and their water requirement can be analyzed. By this, an algorithm can be framed and implemented in the future. The most important feature is to make cost-effective to the farmers.

References

1. Limbore, Nilesh,. "An Analytical Study Of Indian Agriculture Crop Production And Export With Reference To Wheat", Review of Research 2249–894X Impact points 3.1402(UIF). Volume 4. 11 (2015).
2. https://www.indiacelebrating.com/essay/agriculture- essay/
3. https://www.gardeningknowhow.com/special/children/ho w-plants-grow.htm
4. Saraf, S. B., & Gawali, D. H., "IoT based smart irrigation monitoring and controlling system", 2017 2nd IEEE International Conference on Recent Trends in Electronics, Information & Communication Technology (RTEICT).
5. Kinjal, A. Rabadiya, B. Shivangi Patel, and C. Chintan Bhatt, "Smart irrigation: Towards next generation agriculture", In Internet of Things and Big Data Analytics Toward Next-Generation Intelligence, pp. 265–282. Springer, Cham, 2018.
6. Zhao, Ji-chun, Jun-feng Zhang, Yu Feng, and Jian-xin Guo.,"The study and application of the IOT technology in agriculture", In Computer Science and Information Technology ICCSIT, 2010 3rd IEEE International Conference on, vol. 2, pp. 462–465. IEEE, 2010.
7. Rajalakshmi P., Mrs.S.Devi Mahalakshmi. "IOT Based Crop-Field Monitoring And Irrigation Automation", 10th International conference on Intelligent systems and control (ISCO), 7–8 Jan 2016 published in IEEE Xplore Nov 2016.
8. Balasaraswathy P, Kumar U, Srinivas C R, Nair S. ,"UVA and UVB in sunlight, Optimal Utilization of UV rays in Sunlight for phototherapy", Indian Journal of Dermatology Venereology Leprology, 2002;Vol 68:Pg 198- 201.
9. Nakutis et al., "Remote Agriculture Automation Using Wireless Link and IoT Gateway Infrastructure", 2015 26th International Workshop on Database and Expert Systems Applications (DEXA), Valencia, 2015, pp.99–103.
10. Rajalakshmi, P., & Devi Mahalakshmi, S. ,"IOT based crop-field monitoring and irrigation automation",. 10th International Conference on Intelligent Systems and Control (ISCO) (2016).
11. Rao, R. N., & Sridhar, B., "IoT based smart crop- field monitoring and automation irrigation system", 2nd International Conference on Inventive Systems and Control (ICISC) (2018).
12. Ashmi V K, Sahaya Edal Queen G, Anisha V, Sree Sankar.J, Dr.D. Ramalingam, "Fire Detection and Irrigation Automation based on IOT", 2018 International Journal of Electronics, Electrical and Computational System, ISSN 2348–117X, Volume 7, Issue 3, March 2018.
13. Prathibha, S. R., Hongal, A., & Jyothi, M. P, "IOT Based Monitoring System in Smart Agriculture", International Conference on Recent Advances in Electronics and Communication Technology (ICRAECT) IEEE 2017.
14. Mekala, M. S., & Viswanathan, P, "A Survey: Smart agriculture IoT with cloud computing", International Conference on Microelectronic Devices, Circuits and Systems (ICMDCS) IEEE 2017.
15. https://airnow.gov/index.cfm?action=aqibasics.aqi

Design of Rover with Live Video Transmission

T. Muhamed Sharrukh, V. R. Rajapriya, P. Paviethra, and S. Sadhishprabhu

Abstract The paper deals with the design and control of vehicle-type robot with live video transmission, and it can move in different directions according to the user's commands. The design of rover is highly reliable, cost efficient and controlled using a universal android mobile application. The Arduino is programmed for speed control and motion control of rover. Arduino is inexpensive compared to other controllers. It runs on Windows and Linux working frameworks. It has an easy and clear programming environment, open source and applications that can be generalized. The Arduino is connected to Bluetooth module through UART protocol. Depending on the commands from the Arduino through Bluetooth device, the rover motion is controlled. With the help of FTDI programmer kit, camera module is programmed for live video streaming. The camera module is fixed to the rover. The camera module has two megapixels, resolution of 1600 × 1200 and refresh rate is 15fps. Thus, the robotic vehicle can be used for number of applications especially in surveillance, military and security.

Keywords Rover · Bluetooth module · Arduino · Wireless camera

1 Introduction

The rover is an electromechanical device that is controlled by system program to perform multiple operations. Nowadays, mobile phones have become relatively powerful with high technology processors, massive storage capacities and more communication methods. In this process, the Arduino board is used as a controller which handles the overall activity of the rover. The motor driver is used to move the rover to other places provided by the directions in the mobile application. The Bluetooth device is connected to the Arduino board and merged with the mobile application for controlling the movements (left, right, forward, backward) of the rover. Wireless-controlled spy robots are useful if they are controlled remotely for a

T. M. Sharrukh · V. R. Rajapriya · P. Paviethra · S. Sadhishprabhu (✉)
Department of Electrical Communication Engineering, B.S. Abdur Rahman Crescent Institute of Science and Technology, Chennai, India
e-mail: sadhishprabhu@crescent.education

A. Sivasubramanian et al. (eds.), *Futuristic Communication and Network Technologies*, Lecture Notes in Electrical Engineering 792,
https://doi.org/10.1007/978-981-16-4625-6_6

large communication range. The Bluetooth technology used is serial communication with the rover. It is used to transfer data between two devices and be contained in the range within two devices. The Bluetooth device is connected with the rover, and the commands are provided over android application.

For live video transmission, FTDI programmer is used with ESP-32. By using FTDI programmer, the camera module is programmed, and live video streaming is monitored using laptop and mobile phones.

2 Literature Review

The robot is controlled by Bluetooth, and video streaming is done by camera module using mobile application. These are the following literatures that we have gone through for our project [1]. The new technologies lead to complete modification in the area of robotics and automation. The robot will serve as an appropriate machine for the defense sector to reduce the loss of human life and will also prevent illegal activities [2]. A mobile phone is built on a mobile computing platform, with more additional computing ability and connectivity [3]. Mobile phone restrained spy devices that performs complex and multiple tasks by the commands which is stored in the programmable device. [4]. Android platform has revolutionized the new dimensions for technical exploration and innovations [5, 6]. According to the gesture commands received from Android, the movements of the robot are controlled. Gestures are captured through the Android smartphone [7]. Finding and tracking people is the core of any vision based gesture recognition system. After all, the robot must know where in the image the person is [8]. It has camera with wireless transmission and robot is controlled with wireless links like Bluetooth, RF frequencies [9]. LOBOT identifies the local relative movement through a set of integrated inexpensive sensors and well corrects the localization drift by infrequent GPS-augmentation [10].

3 Hardware Description

3.1 Bluetooth Module

Wireless Bluetooth RF transceiver module HC-05 is a Bluetooth serial transmitter and receiver module master slave for Arduino. It operates on every Bluetooth USB adapter. Compared to the smart Bluetooth module, this includes the main interface and state interface. The Bluetooth module is lightweight and compatible with master, slave and master–slave mode modes. It is user-friendly, and it is simple to allow Bluetooth connectivity to communicate with the module.

3.2 Arduino UNO

The Arduino Uno is an open-source microcontroller board designed by Arduino.CC based on the MicrochipATmega328P microcontroller. It is primarily composed of digital and analog input/output (I/O) pin sets that are attached to separate expansion boards (shields) and other circuits. The board consists of 14 digital pins and six analog pins, which are programmed with a USB cable with the Arduino integrated development environment (IDE) app.

3.3 Motor Driver

This L298 dual H-Bridge chip, which is connected to a board along with all the essential required peripherals, operates on a single two-phase bipolar stepper motor or two DC motors. In both directives, the circuit gives power to two DC motors with a minimum current of 2A each, or it can be used to power a stepper motor. Standard switches, TTL logic gates and relays can also be attached to turn the driver on or to allow access.

3.4 Camera Module

The camera module (ESP32) is a budget-oriented, low-power consumption device with built-in Wi-Fi and dual-mode Bluetooth on a chip microcontroller. In both dual-core and single-core variations, it is the LX6 microchip. It has built-in antenna switches, power amplifiers, amplifiers for low noise detectors and modules for power control. The memory space is SRAM 520kib. The peripheral interfaces provide up to 18 channels of 12-bit SAR ADC, 2 × 8-bit DACs, 10 × touch sensors, etc.

4 Schematic Circuit Diagram

The motor driver module in Fig. 1 is connected with the motors in the chassis. Linked to the Arduino are the control pins. For interaction with the rover, the Bluetooth sensor is embedded to the Arduino. Two separate 9 V batteries are installed for the power supply for the motor driver and the Arduino. The power supply from the Arduino is supplied to the Bluetooth system.

The relation between the ESP32 module and the FTDI programmer is represented in Fig. 2. For live video streaming, this is fixed on the rover. The battery power supply for this module is 9 V.

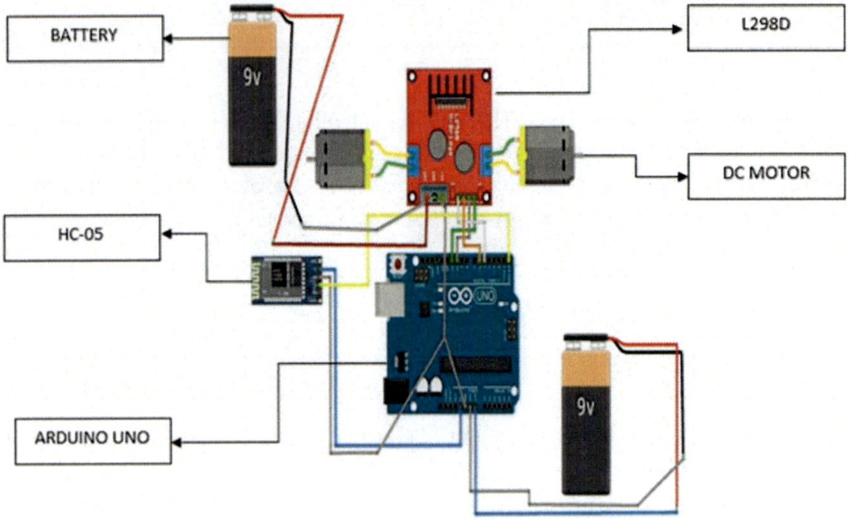

Fig. 1 Rover schematic diagram

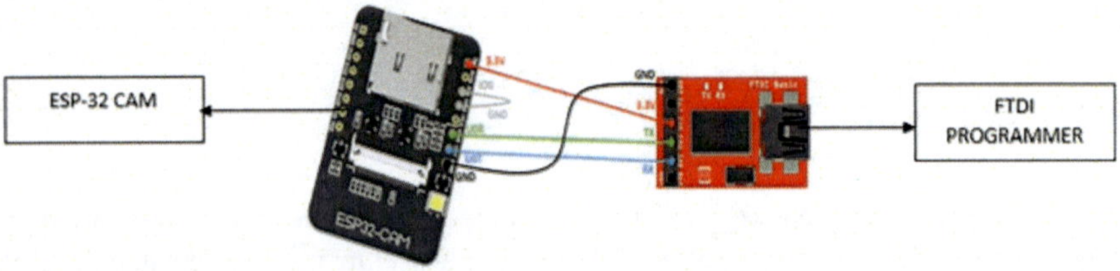

Fig. 2 Live streaming schematic diagram

5 Software Description

5.1 Direction Controlled Application

The android application used is Bluetooth RC control application which is used for direction control of rover through mobile phone. The mobile phone and the Bluetooth module are paired. In Fig. 3, the rover goes forward if the up arrow is pressed, the rover goes backward if the back arrow is pressed, and for other actions accordingly.

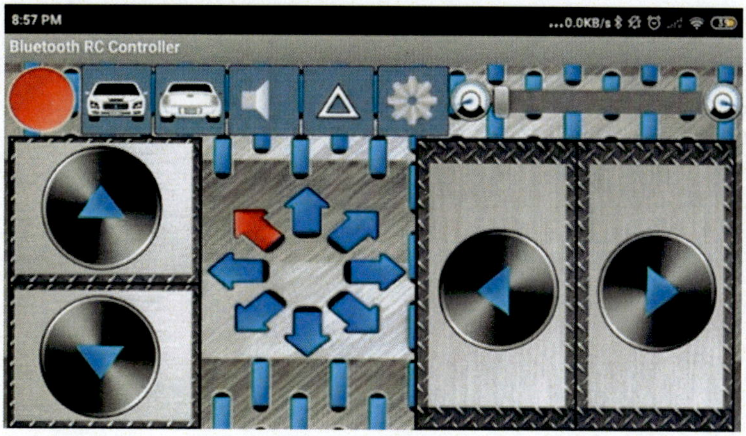

Fig. 3 Direction control

Fig. 4 Rover with live video transmission

6 Experimental Implementation

The hardware design consists of two sections: They are implementation of rover and interface of live video streaming.

6.1 Implementation of Rover

This section consists of interfacing Bluetooth module (HC-05) with Arduino and interfacing DC motors with motor driver (L298D). The Bluetooth module's VCC and GND pins are wired to the Arduino 5 V and GND pins. The Bluetooth module's TXD and RXD pins are attached to the Arduino TXD (1) and RXD (0).

In interfacing of DC motors with motor driver, the control pins IN1, IN2, IN3 and IN4 are connected to pins 5, 6, 10 and 11 of Arduino. The output A and output B are connected to motors in the chassis for clockwise and anticlockwise rotation.

6.2 Interface of Live Video Streaming

The ESP-32 module is programmed using FTDI programmer for live video streaming. The 3.3 V and GND pins of ESP-32 module and FTDI programmer are connected. The UOR and UOT pins are connected to Tx and Rx pins of FTDI programmer. In ESP-32 module the IO0 pin is grounded (Fig. 4).

7 Result and Discussion

Figures 5 and 6 are the obtained results during the live streaming using the ESP32 module. In our proposed system, minimum power (5 V-2A) is needed to run the entire rover and the live video transmission run in 2.4 GHz frequency. The android device is linked to the Bluetooth module which is coupled to the Arduino. Once the input signal is received from the Bluetooth module, according to the signal received by the Arduino, the rover moves in the respective direction. In the ESP32 module, we can change the resolution for the better clarity. We can record or capture during our process and can be stored locally. When we burn the ESP32, the module will provide the IP address. Once the IP address is obtained, we can use the IP address in the browser for the live video streaming, and it can also be used for image capturing and video recording. It is used for military applications such as spying; with some minor adjustment, it can also be used for bomb diffusing. By linking multiple devices to the same channel, we can watch live feeds on multiple screens. Some of the images that were captured are given below:

8 Future Works

The future implementations of the rover are by replacing myRIO instead of Arduino board, and the Bluetooth module can be replaced by Zigbee, Wi-Fi or LoRa module

Fig. 5 Output1

for communicating through wider range. The quality of the camera can be adjusted or increased.

9 Conclusion

In this paper, the model is described to build a rover with wireless camera which runs on android application and streamed using IP address in browser. It has two approaches of controlling the rover using a universal remote-controlled android application which has button control and gesture control. It is adaptable to any kind of Arduino, and no additional software is required. The android mobile phones are handy and reasonable price in the market. There is a huge development in the number of people using mobile phones, and the probability of many applications using the

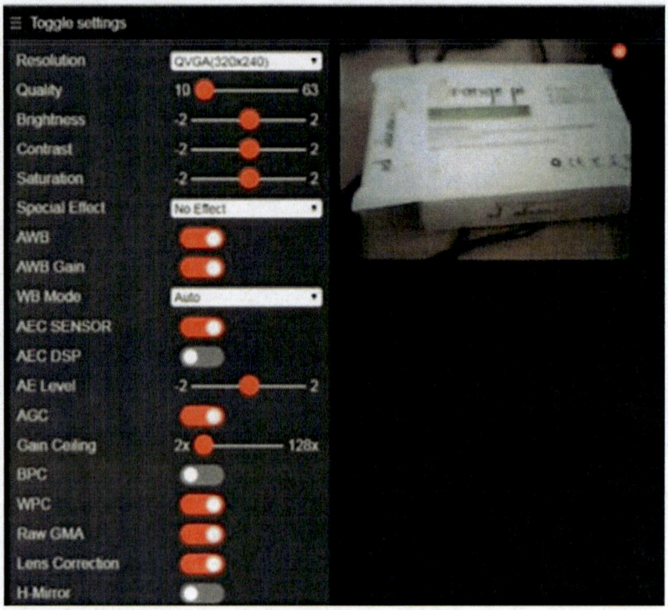

Fig. 6 Output2

mobile phones also increased. The android program uses sensors on android devices. The connection is then fixed between the rover and the android device with the aid of the Bluetooth module, which is operated by Arduino when the user sends the input. The direction is transmitted from a smartphone to a microchip. On the basis of the instructions, specific signals are sent to the motor driver, which drives the motors which are attached to the chassis. The suggested method also demonstrates how the rover can be used for live broadcasting. For this purpose, wireless camera module (ESP32) has been used through which we can see, click photos and record videos on the laptop. In the future, the size of the rover can be minimized, and camera can be rotated. The Bluetooth module is for limited range. For modification in the range, other modules like Wi-Fi and ZigBee can be used.

References

1. Maheshwari T, Mittal VK (2015) "Capturing the spied image-video data using a flexi-controlled spy-robot," (ICIIP)
2. Patoliya J, Mehta H, Patel H (2015) "Arduino controlled war field spy robot using night vision wireless camera and Android application," (NUiCONE)
3. Singh A, Gupta T, Korde M (2017) "Bluetooth controlled spy robot," (ICICIC)
4. Jewel MKH, Mostakim MN (2017) "Smartphone controlled spy robot with video transmission and object collector". Int J Eng Manuf

5. Dey A, Roy L (2015) "Three way controlled android smartphone based robotic vehicle via Bluetooth". Int J Adv Res Comput Commun Eng
6. Harun-Ar-Rashid M, Islam MT (2017) "Implementation of android control bluetooth spy car with live video streaming". Int J Comput Appl
7. Goel V, Riya, Kumari P (2018) "Design of smartphone controlled robot using bluetooth". Lecture Notes Elect Eng
8. Waldherr S, Romero R, Thrun S (2000) "A gesture based interface for human-robot interaction"
9. Balakrishnan M, Gowthaman S, Jaya Kumaran SP, Sabhapathy GR (2015) "A smart spy robot charged and controlled by wireless systems, (ICIIECS)
10. Zhan G, Shi W (2013) "LOBOT: low-cost, self-contained localization of small-sized ground robotic vehicles", IEEE Trans Parallel Distrib Syst

Stock Price Prediction Based on Deep Learning Using Long Short-Term Memory

R. Shanmuga Priya and C. Sruthi

Abstract Predicting the stock market is either the easiest or the toughest task in the field of computations. There are many factors related to prediction—physical factors versus physiological, rational and irrational behaviour, capitalist sentiment, market rumours, etc. All these aspects combine to make stock costs volatile and are extremely tough to predict with high accuracy. The prices of a stock market depend very much on demand and supply. High-demand stocks will increase in price, while heavy selling stocks will decrease. Fluctuations in stock prices affect investor perception, and thus, there is a need to predict future share price and to predict stock market prices to make more acquaint and precise investment decisions. We examine data analysis in this domain as a game-changer. This paper proposes that historical value bears the impact of all other market events and can be used to predict future movement. Machine learning techniques can detect paradigms and insights that can be used to construct surprisingly correct predictions. We propose the long short-term memory (LSTM) model to examine the future price of a stock. This paper is to predict stock market prices to make more acquaint and precise investment decisions.

Keywords Machine learning · Deep learning · RNN · LSTM

1 Introduction

Financial markets are one of the most significant inventions of our time. They have a notable impact on business, education, employment, technology and thus the economy in many areas. Stock market forecasting has always attracted researchers and analysts. It acts like a voting machine in the short term, but it is actually weighing machine in the long term, and therefore has the scope to predict market movements over a longer period. The stock market fluctuations are incredibly violent which leads to more complex financial indicators. Also, the greatest advantage of price prediction is to maximize the benefits of purchase options simultaneously minimizing its risk. [1, 2] The "Efficient Market Hypothesis" (EMH) suggests it is not possible to

R. S. Priya (✉) · C. Sruthi
Madras Institute of Technology, Anna University, Chennai, Tamil Nadu, India

A. Sivasubramanian et al. (eds.), *Futuristic Communication and Network Technologies*, Lecture Notes in Electrical Engineering 792,
https://doi.org/10.1007/978-981-16-4625-6_7

make the market low with the usage of the same information because the current overall stock price reflects all available information about a firm. Despite this, in the proposed model historical data is used to predict future stock price. In recent years, the increasing prominence of machine learning has produced quite promising results in various fields and industries that have enlightened many analysts, researchers and traders to use machine learning techniques. LSTM is the widely preferred RNNs architecture. LSTM has a memory cell, which acts as a unit for computation. This method is well suited for classifying, processing and predicting statistical data. The paper intends to predict stocks using LSTM. The main objective of forecasting is to reverse the uncertainty associated with investment determination.

2 Existing Approaches

Many existing systems are used to predict stock prices such as linear regression, k-nearest neighbour, Auto Arima and prophet. To estimate the stock market, two-way back-propositional neural networks, auto-regressive moving average models as well as multi-layer perception have been used [3]. In existing systems, the performance of a company, in terms of its stock price movement, is foreseen by internal communication patterns. In these systems, accuracy is unreliable when determining multiple levels of stock movements. For example, using the decision tree as a classifier, the average prediction accuracy was 63.7, 31.92 and 12% for "two levels", "three levels" and "five levels", respectively [4]. For each input knowledge, it performs the same function with its output of the current state of input that depends on the previous calculations; from this, it shows recurrent neural networks (RNN) have recurring properties. There exist a vanishing gradient and exploding gradient problem. This result clearly shows that stock prices are not predictable using these methods [5].

3 Proposed Method

Accuracy is vital in market forecasting. Although many algorithms are available for this purpose, making precise selections continues to be the authentic work to achieve the most constructive results. In this work, an effective and reliable stock market forecasting system is hypothesized by combining informative input variables along with RNN. Primarily, the problem of processing the concatenation of data, such as audio, video and text can be solved by RNN. Recurrent neural networks can be discernible of feed-forward neural networks that pose an attached memory. For each input knowledge, it performs the same function with its output of the current state of input that depends on the previous calculations; from this, it shows RNN has recurring properties. Once the output is produced, it is cloned and forwarded to the recurrent networks. From the previous inputs, it can learn contemporaneous input and output which is in turn used to create a call. To process a sequence of inputs, RNNs

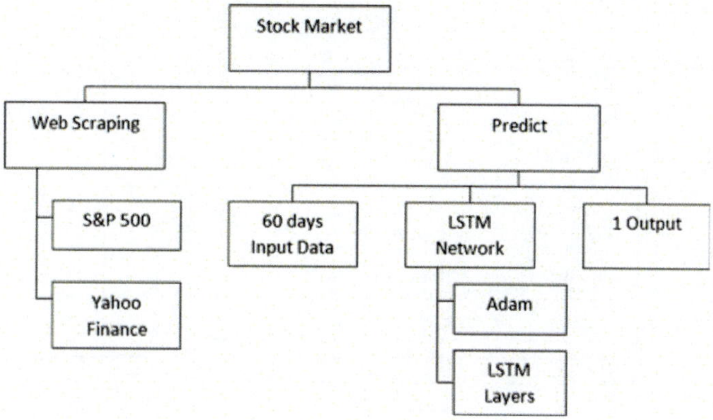

Fig. 1 Architecture diagram

uses its internal state memory which is disparate to feed-forward neural networks. All inputs are independent of every instance input in other neural networks. Whereas, in RNN, whole inputs are interconnected to each other.

In addition, a novel stock market price prediction model based on LSTM using deep learning and stock market basic data is being proposed, and further validation is done. Figure 1. Architecture Diagram. The justification on which they work is that the LSTM is ready to store previous information that is important and neglect information that is not [6, 7].

3.1 Architecture Diagram

See Fig. 1.

3.2 LSTM Approach

Long short-term memory (LSTM) networks are distinctively different from traditional RNNs and in neuron formations. In traditional RNN and normal neural networks, it does not pose any difference between its neurons. But, LSTM connects previous information to the current errand by its apiece neuron which is also a "memory cell". In LSTM, a unit of computation is being introduced by the memory cell that supplants conventionally in the hidden layer of the network by artificial neurons. Networks can effectively able to connect to its memory and input gradually using these memory cells, therefore with a higher predictive capacity to suit the structure of knowledge over time. RNN poses a vanishing gradient problem that is solved here. Figure 2. LSTM architecture.

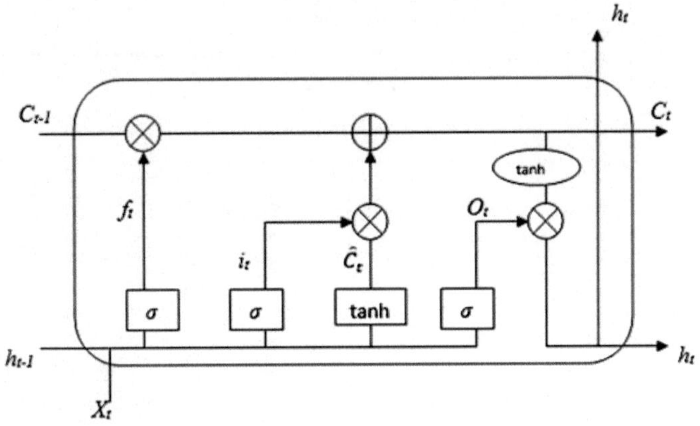

Fig. 2 LSTM architecture

3.3 LSTM Architecture

Functions in LSTM:

tanh: Function's second derivative can be maintained for a prolonged distance before zero is required to overcome the vanishing gradient problem. The suitable function for the above property can be *tanh*.

Sigmoid: Since the sigmoid output can either be 0 or 1, it is accustomed to obliterate or recall its knowledge. By using such LSTM, unit's information can be passed.

LSTM unit has three main components:

1. *Forget gate*—Search for details to be dismissed from the block. Here sigmoid layer determines input as *X(t)* and $h_{(t\text{-}1)}$ and elects the parts that can be removed from the previous output (giving an output of 0). This is known as forget gate *f(t)*. The output is *f(t) * c(t-1)*.

$$f_t = \sigma\left(W_f.\left[h_{t-1}, x_t\right] + b_f\right) \quad (1)$$

2. *Input gate*—To modify its memory which value is to be used from the input can be identified. A value that passes over 0 and 1 is decided by the sigmoid function. The weightage for the values that are passed is given by tanh function, setting its significance levels from -1 to 1. Along with old memory *c(t-1)*, the new memory is computed to offer *c(t)*.

$$i_t = \sigma\left(W_i.\left[h_{t-1}, x_t\right] + b_i\right) \quad (2)$$

$$\tilde{C}_t = \tanh\left(W_C.\left[h_{t-1}, x_t\right] + b_C\right) \quad (3)$$

3. *Output gate*—To determine the output, input and block memory is being used. A value that passes over 0 and 1 is decided by the sigmoid function. The weightage for the values that are passed is given by tanh function, setting its significance levels from -1 to 1, which is multiplied by the sigmoid function's output.

$$O_t = \sigma(W_\circ\left[h_{t-1}, x_t\right] + b_\circ \quad (4)$$

$$h_t = O_t * \tanh(C_t) \quad (5)$$

It is observed that RNN and LSTM have a major architectural difference between them. During it's long-term memory basis these prototype pedagogies which information is to be stocked and eliminated in the LSTM.

4 Methodologies

The data used here is from Yahoo Finance, S&P 500 component shares. This includes open, high, low, close, volume values of 500 companies.

This paper uses a close price to estimate the future close price. This section discusses the functioning of our system. There are several stages in this system.

They are:

- *Stage I*: *Raw data:* For forecasting stock prices in future, historical stock data is collected and employed in this phase.
- *Stage II*: *Data preprocessing*: A portion of data is reduced but with specific criteria, especially for numerical data this involves the preprocessing stages and knowledge file integration. After the transformation of the dataset into a proper data fields, it has been segregated into training data and testing data for evaluation. The test data is retained for about 5 to 10% of the entire datasets.
- *Stage III: Feature extraction*: During this stage, only the selected features are fed into the neural network. Here, the close price is selected.

In the recurrent network, the neural network is being used, where the sequence data is formed by the transformation of input data. By using 60 width sliding windows, segmentation of input data is done. This process is described in Fig. 3. Input Sequence Diagram.

- *Stage IV: Training neural network*: Here, our model assigns random biases and weights by feeding information to the current neural network, and predictions are also done by training the network. In our model, the neural network has four layers including the input layer, two LSTM layers (hidden layer) and the output layer (refer Fig. 4. Prediction Model. All units in adjacent layers are connected by each unit in a layer. The LSTM layer consists of 46 LSTM nerve entities. The fully connected layer is formed by the output layer which has only one unit.

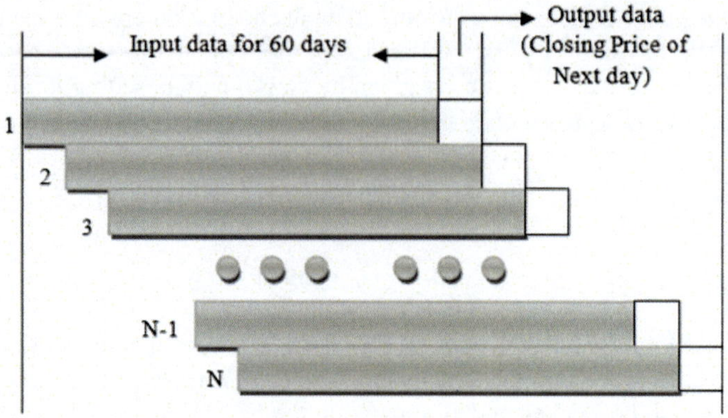

Fig. 3 Input sequence diagram

Fig. 4 Prediction model

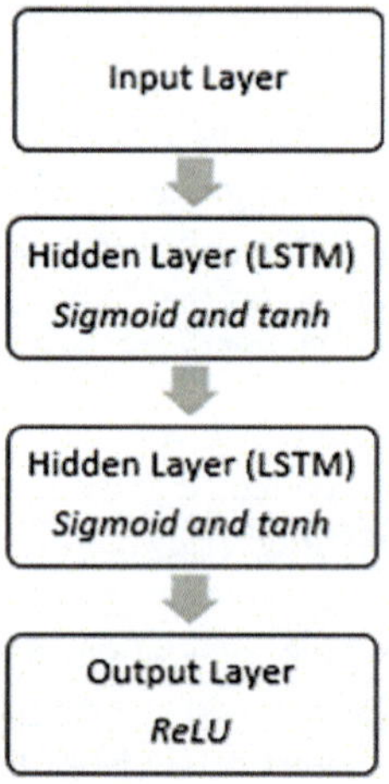

In the stock price prediction prototype, it poses three types of activation functions: rectified linear unit (ReLU), hard sigmoid and *tanh*. The ReLU is present in the output layer; σ(hard sigmoid) and *tanh* are shown in Sect. 3.3.

- *Stage V: Output generation*: To the target value, the output value is being compared which was in turn generated by the output layer of our model. A back-propagation algorithm is used to find inaccuracy and dissimilarities that ally the targeted and received output value which are being minimized by adjusting its weights; therefore, it is the bias of a network.

5 Analysis

For predicting stock index, a predictive model requires less error that can be taken into account of processing the input data. In our model, rectified linear unit (ReLU) activation is used. The activation function *f(x) = max (0, x)* is applied in ReLU to remove the vanishing gradient and uses element activation when it limits to 0. However, ReLU has linear properties so it will speed up the sigmoid function / tanh being compared to a stochastic gradient. Moreover, it creates an activation function matrix when its threshold bound to 0 and exponential operations like *sigmoid / tanh* are not being used. Neuron becomes re-indolent in data points because the major flow of gradients in ReLU occurs to weight gain where it becomes fragile during its training. The flow of gradient will bound to zero perpetually if this proceeds over that point. From this, ReLU can be terminated irreparably while its training and it can debilitate numerous data units. Using LSTM, unfamiliar values for the next day can be found, and also mediocre values can be estimated on the next day. Investors, analysts or anyone can invest in the stock market in terms of their own interest. This paves the way for analysts to be aware of the stock market in the future.

Also, a series of tests and results are being conducted by measuring a few criteria to evaluate the neural network model's performance. MSE, MAE, RMSE and R-squared metrics are primarily used to evaluate regression error rates and model performance in regression analysis.

- The AAE (average absolute error): represents to extract the absolute difference of the dataset on an average basis by calculating its difference from the original and predicted values (Eq. 6).
- The MSE (mean squared error): shows the squaring mean difference of the dataset to extract values by finding its difference from the original and predicted values (Eq. 7).
- RMSE (root mean square error): is an error rate obtained by the square root of MSE (Eq. 8).
- The R-squared (coefficient of determination): represents the value that fits perfectly comparing to original values (Eq. 9). Values from 0 to 1 are interpreted as a percentage. The higher the value, the better the model.

$$\mathrm{MAE} = \frac{1}{n}\sum_{i=1}^{n}\left|Y_i - \hat{Y}\right| \tag{6}$$

$$\mathrm{MSE} = \frac{1}{n}\sum_{i=1}^{n}\left(Y_i - \hat{Y}\right)^2 \tag{7}$$

$$\mathrm{RMSE} = \sqrt{\frac{1}{n}\sum_{i=1}^{n}\left(Y_i - \hat{Y}\right)^2} \tag{8}$$

Table 1 Prediction performance of LSTM for S&P 500

Company	MAE	MSE	RMSE	R^2
AAPL	1.8533	3.8756	1.9686	0.9745
GOOGL	0.5524	0.0679	0.26	0.9927
AMZN	0.8278	0.975	0.987	0.9836

$$R^2 = 1 - \frac{\sum(Y_i - \widehat{Y})^2}{\sum(Y_i - \overline{Y})^2} \tag{9}$$

where Y_t—actual value, $\hat{Y}$—predicted value and $\overline{Y}$—mean of Y.

Table 1 Prediction performance of LSTM for S&P 500. Figure 5, 6, 7 Close predictions of AAPL (Apple), GOOGL (Google) and AMZN (Amazon), respectively. From the table prediction performance of three companies AAPL, GOOGL and AMZN, it is clear that the accuracy of our model is more than 97%.

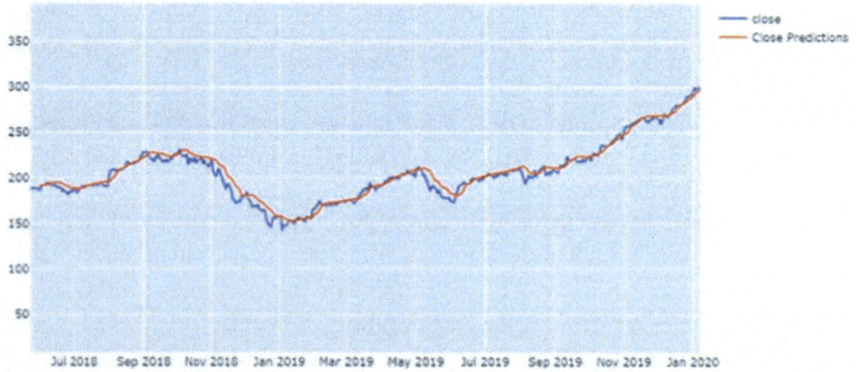

Fig. 5 Close prediction of AAPL

Fig. 6 Close prediction of GOOGL

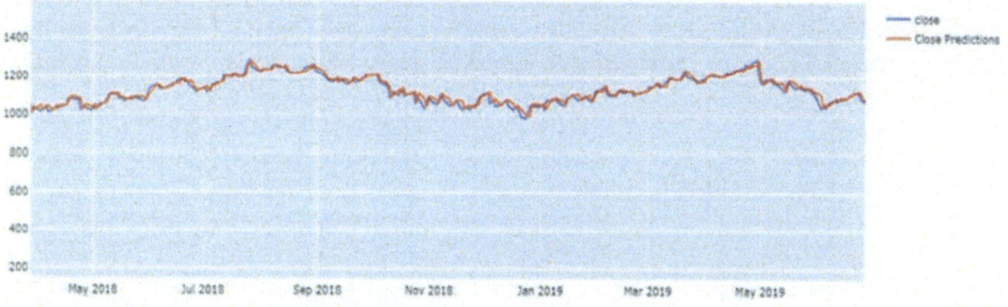

Fig. 7 Close prediction of AMZN

6 Conclusion

The stock market's trading popularity is increasing swiftly, and it encourages scholars to discover intact methods in forecasting using new terms and techniques. Also, the researchers, investors and individuals are being helped by this forecasting technology in exchange. A forecasting method with ethical accuracy is used to predict stock indices. LSTM outperforms other models because this model learns by using long-term dependency. LSTM is a step forth to RNN; its ability to update, obliterate and recollect information is more effective. In this work, one of the most accurate precision forecasting techniques is used that helps investors, analysts and individuals eagerly to invest by granting good knowledge for the future state of exchange in the stock market exchanges. This can be further improved by an LSTM model with multiple inputs that can extract relevant information by employing additional input gates with low correlated factors to discard their noise controlled by convincing factors called mainstream.

References

1. Silva TR, Li AW, Pamplona EO (19–24 Jul 2020) "Automated trading system for stock index using LSTM neural networks and risk management". International joint conference on neural networks (IJCNN), United Kingdom
2. Sarkar A, Sahoo AK, Sah S, Pradhan C (13–14 Mar 2020) "LSTMSA: a novel approach for stock market prediction using lstm and sentiment analysis". International conference on computer science, engineering and applications (ICCSEA), India
3. Pahwa K, Agarwal N (14–16 Feb 2019) "Stock market analysis using supervised machine learning". International conference on machine learning, big data, cloud and parallel computing (Com-IT-Con), India
4. Wei D (16–18 Oct 2019) "Prediction of stock price based on LSTM neural network. International conference on artificial intelligence and advanced manufacturing (AIAM)
5. Shin HG, Ra I, Choi YH (16–18 Oct 2019) "A deep multimodal reinforcement learning system combined with CNN and LSTM for stock trading". International conference on information and communication technology convergence (ICTC), Korea

6. Lai CY, Chen RC, Caraka RE (7–10 Jul 2019) "Prediction stock price based on different index factors using LSTM". International conference on machine learning and cybernetics (ICMLC), Japan
7. Zhang Y, Yang S (18–20 Oct 2019) "Prediction on the highest price of the stock based on PSO-LSTM neural network". 3rd international conference on electronic information technology and computer engineering (EITCE), China

Design of Surface Quality Identification Device

Vangala Indhrani, K. Mounika, and Srinivasan Ashok Kumar

Abstract A novel surface quality identification device for vehicles is a low-cost sensor platform mounted on vehicles to identify the quality of the streets. Design and development of street quality identification device have been proposed, and it is mounted on public vehicles. The proposed device designing and development are made with microcomputers, accelerometer, GPS, camera to build an image database and Internet of things (IOT). By combining of all the above, we can collect data and represent a low-cost device to identify the street surface quality. And the concept of surface quality identification device has been proposed for real-time applications.

Keywords Raspberry Pi · Accelerometer · GPS · LINUX · Potholes · Internet of things · Street defects · Database

1 Introduction

The maintenance of city streets is one of the most visible indicators of operational effectiveness of a city government. The main initiative exists that involve in the design and development of this low-cost device to passively collect the data about the surface quality in a city continuously and in a reusable manner [1]. Previous devices have been largely experimental, and the vehicle is fitted with large, expensive, military grade sensor equipment. But data collection in a scalable manner is necessary to resurfacing decisions [2]. In [3], recognize that the problem is complex and includes a large property loss for the public as we explained before. Our design philosophy is to overcome this complex environment. From the last 2 years, a report by "TIMES OF INDIA", approximately 11,106 crashes and 3,597 occurred across India and 8% is the rate at which the road fatality in India is increasing every year [4, 5]. These problems are not unique to a particular city but any city that depends on road transportation.

V. Indhrani (✉) · K. Mounika
Jyothishmathi Institute of Technological Sciences, Karimnagar, India

S. Ashok Kumar
Jyothishmathi Institute of Technology & Science, Karimnagar, India

A. Sivasubramanian et al. (eds.), *Futuristic Communication and Network Technologies*, Lecture Notes in Electrical Engineering 792,
https://doi.org/10.1007/978-981-16-4625-6_8

Several transportation research reports highlighted the enormous costs of poor street quality on individual drivers [6, 7].

This paper mainly highlighted a particular problem, i.e. when we are driving a vehicle (e.g. car) in new streets/in an area, there are many defects on the surface with those defects we suffer for unexpected losses such as crash of our vehicle or stuck in that. So, to avoid small accidents and property losses and to drive easily in uncomfortable roads, we proposed this design. To overcome this problem, we have prototyped a device using cost-effective components, collected data from the fields and demonstrated results to inform our discussion around achieving better situational awareness for surface quality in a city.

This paper is organized as follows: Sect. 2 explains about literature review of the device. Section 3 describes about methodology and core components. Section 4 explains about result and discussion of the device. In Sect. 5, conclusion has been done.

2 Literature Survey

This paper [1] investigated on the maintenance of street surfaces in cities and the defects on surface (i.e. manholes, potholes, cracks, etc.) and the problems occur with them. Previously, we have devices with largely experimental and more expensive.

In this paper [2], we discuss the overview of the device and core components. Raspberry Pi combines with sensors in the device and passively collects the data. The data is transmitted by a Wi-Fi module attached to a USB for remote login and the need of Internet of Things (IOT) in this device.

This paper [3] proposed a new thing that the collected data can display on the user's screen in real time as well as it stores the data and shows in web app user's dashboard in different types of graphs (potholes, manholes, etc.), and this data can also use for the city government to maintain the street surface defects.

This paper [4] concludes the total prototype of the device. The entire device is installed inside the vehicle with only camera and GPS antenna exposed to the elements. And the result on mobile screen after the device is inbuilt in vehicle, and we were able to get consistent image position.

3 Methodology

3.1 Surface Quality Identification Device

This idea, surface quality identification device, is a device which can display the quality of street surface, and it identifies the defects (e.g. manholes, cracks) by using an accelerometer, GPS, camera, microcomputers and sensors. The combination of a

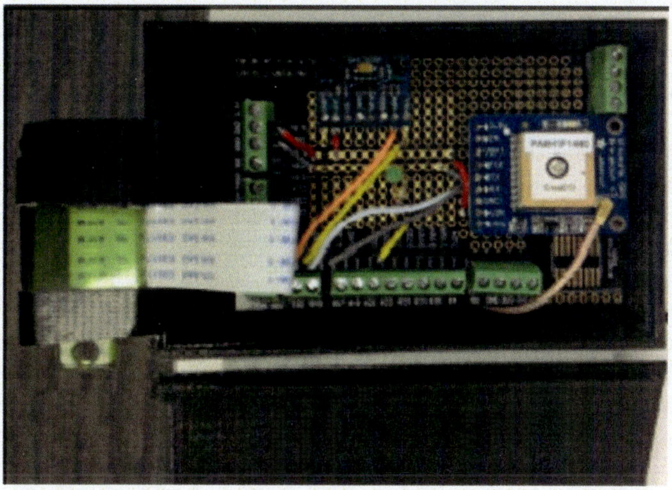

Fig. 1 SQUID prototype with accelerometer, GPS and camera

microcomputer and sensors installed in it is used to identify the defects and passively collect the data in a scalable manner towards achieving a citywide map for decision-making. In this device, Raspberry Pi is a core component and is based on LINUX. We can be able to get done image position and accelerometer to differentiate one street from the next in terms of overall quality as well as identify the specific defects that caused spikes in the accelerometer readings, and the collected data stores in the Pi processor attached along with the sensors and peripherals. At last, a Wi-Fi module is attached through the USB for remote login and data transmission of the device (Fig. 1).

3.2 *About the Device*

This device is the combination of a microcomputer and associated sensors installed in it and is used to identify the surface defects and passively collects the data in a scalable manner towards achieving an overall city map of surface defects.

3.3 *Analysis of Core Components*

Raspberry Pi is used as a single-board computers, and the Raspberry Pi link is one of the popular microcomputers. And the accelerometer is used to measure acceleration forces. These both are combining together to measure overall ride quality. The accelerometer which was inbuilt in the SQUID measures acceleration in such directions at nearly 1500 Hz it means, and it measures 1500 readings per second. In each second, the maximum variations from a standard reading in all directions are recorded. It also records the latitude, longitude; vehicle speed and timestamp of reading as well as it take a photo for each second to present an image to driver. There are two main reasons for collecting image data, the primary thing is to provide literal surface quality to accelerometer data, and the secondary reason is every cracks and potholes are not identified by accelerometer in some cases, so image will recover the data, which is missed by the accelerometer.

Internet of things (IOT) is the emerging technology that promotes new levels of connectivity. The need for IOT implementation is the network of interconnected things embedded with sensors, software, network connectivity and necessary electronics enable to collect data easily. The real-time data will be automatically updated in the server. The sensor attached to this device collects the data, and they can translate to operational decisions. And the collected data is stored in an inbuilt SD card. We developed a web app to visualize the data which is collected by the device. The data is continuously updated in the server through the concept of IOT, and each server page consists of its own IP address that allows the user to access the output page. With this device, we can identify the street defects in near real-time conditions to leading better situational awareness. By this project, we can be able to predict the problems that occur in our streets and near our locations. Then we can be able to solve the problem, and we can make our streets and locations free from problems. This results in the less cost and smart performance (Fig. 2).

The maintenance of the city surface quality is one of the most visible indicators of a city government's and municipal performance. In earlier days, it is very difficult to find out all the defects in each and every street, but with the help of this device, they can easily view the number of potholes, cracks and manholes in a particular street in the web app dashboard, and the overall collect data is visualized in the dashboard in different types of graphs.

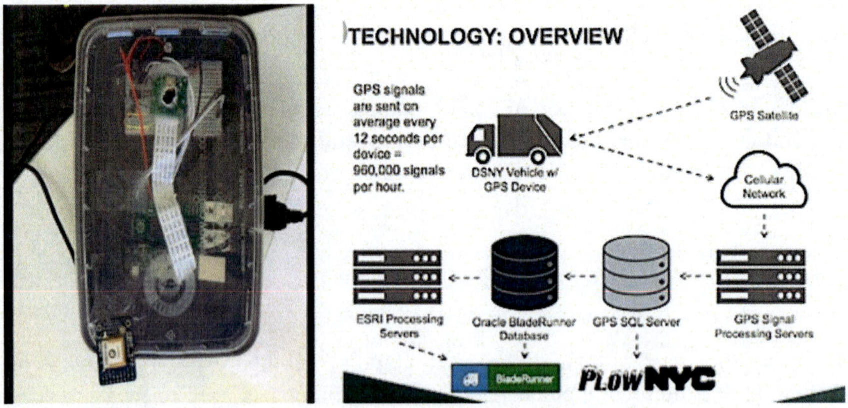

Fig. 2 Device and the block diagram of technology used in SQUID

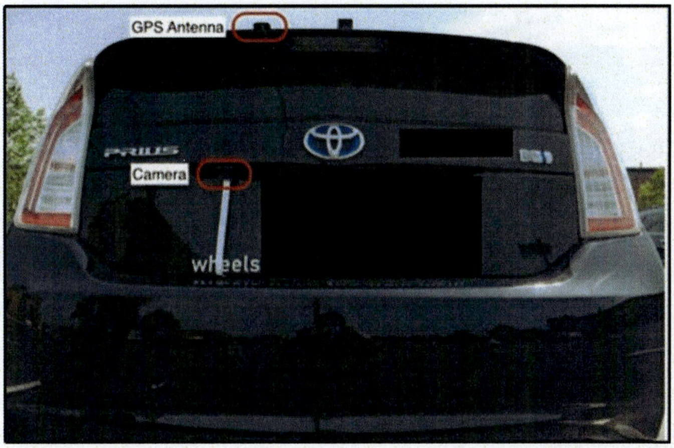

Fig. 3 Entire device is housed inside the vehicle with only the camera and GPS antenna exposed to the elements

4 Results and Discussion

In earlier days, the roads are not properly maintained, and roads contain a lot of manholes, potholes and pooling water. This may lead to the cause of accidents. So to overcome this problem this project is very much useful. This device gives us an alert mark in the GPS where there is any problem in the travelling area. So as a result, the people can move to another way. This may reduce the cause of accidents due to poor maintenance of the road.

Recently, many of the automobile industries design vehicles along with built camera in them. Due to that, the person who drives the vehicles can get to know what is there along his path. Due to that, some of the accidents may reduce. But in that case, the person just knows the defects or problems on the road only when the person reached there. But he cannot get to know about that place until he reached there. But with the help of this device, one can get to know where the defects are there on the road even he is not there in that area. This device gives us an alert symbol in the GPS; with the help of that GPS, one can easily find where the problem will be and then he can slide to another area. So as a result, one can save their time to reach their destination and also prevent the causing of accidents (Fig. 3).

5 Conclusion

Surface quality identification device is based on Pi processor and LINUX operating system to collect and store the data about the surface quality of the streets and the Wi-Fi module attached along with a USB to remote login and transferring the data

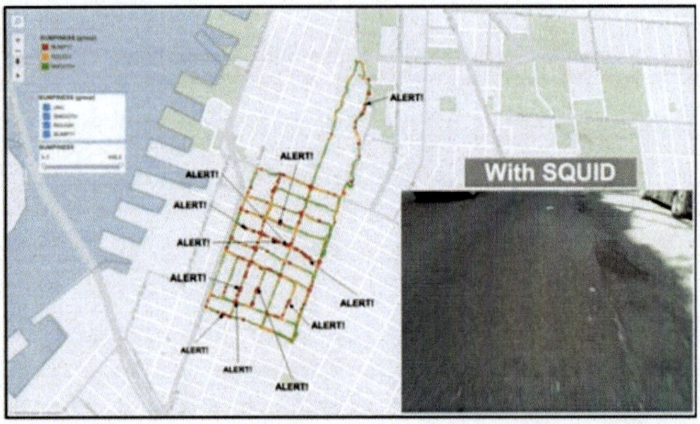

Fig. 4 Result on mobile screen after the device is inbuilt in vehicle and we were able to get consistent image position and accelerometer to differentiate one street from the next in terms of overall quality

and determined to improve reporting the defects around the street and aware the person before the problem occurs. This overall device designing and development are done at a low-cost baseline (Fig. 4).

References

1. Surajit B, Somashekhar C (2017) "Artificial intelligence to impart surveillance, tracking, & actuation on suspicious activities". IEEE Int Conf Advance Computing Conference (IACC)
2. Ekasit V, Nick M, Havlicek JP, Ronald DB, Alan RS (2011) "A low-cost distributed instrumentation system for monitoring, identifying and diagnosing irregular patterns of behavior in critical ITS components". IEEE Int Conf Intell Transp Syst (ITSC)
3. Kumar A, Gupta S (2000) "Real time DSP based identification of surface defects using content-based imaging technique" IEEE Int. Conf Ind Technol 2
4. Usman TV, Sinh Le HN, Katsuyuki H, Jean-Frederic W (2018) "On-site permittivity estimation at 60 ghz through reflecting surface identification". IEEE Trans Antennas Propag 66
5. Toshishige S, Hiroki M, Satoshi S, Mamoru N, Katsuyuki M (2010) "Capacitive-sensing circuit technique for image quality improvement on fingerprint sensor LSIs". IEEE J Solid-State Circuits 45
6. Kun Z, Shiquan Q, Kai G (2015) "A new point cloud reconstruction algorithm based-on geometrical features". Int Conf Mod Identification Control (ICMIC)
7. Ino S, Shimizu S, Odagawa T, Sato M, Takahashi M, Izumi T, Ifukube T (1993) "A tactile display for presenting quality of materials by changing the temperature of skin surface". IEEE Int Workshop on Robot and Human Communication

HVIATC: Ontology-Based Efficient Query Optimization for Declarative Crowdsourcing System Using OAF Measures

N. Bhaskar, P. Mohan Kumar, and J. Arokia Renjit

Abstract The problem of query optimization with declarative crowdsourcing systems [1] has been well studied. There exist some approaches discussed toward the problem of query optimization, but still, the methods suffer to achieve higher performance in query optimization. To improve the performance of query optimization, object access frequency (OAF) measure-based horizontal–vertical-information-access-time–cost (HVIATC)-based approach is presented in this paper. The HVIATC method receives the crowdsource on the specific query submitted. The query evaluator receives the crowdsource, and for each source is received, the query evaluator extracts the fields and verifies the completeness of the query toward the requirement. Then, the optimizer computes the access frequency of each source on different objects of the data set. Similarly, the method computes the horizontal information cost, horizontal access cost. Similarly, the cost of vertical features is estimated. The cost on time feature has been estimated with each query recommended by the crowdsourcing. In conclusion, the method computes the query sequence weight for each recommendation. Based on computed weight measure, a single sequence has been selected. The proposed algorithm improves the performance of optimization in relational databases. The method also reduces the time complexity of optimization and execution.

Keywords Query optimization · Relational databases · Cost · OAF · HVIATC · Ontology

1 Introduction

The modern information world has various scenarios in presenting the information to the end users. The user has a great expectation in the results where they would not know much about the way of asking. As the data representation and the amount of

N. Bhaskar (✉)
Department of IT, KCG College of Technology, Chennai, TamilNadu, India

P. M. Kumar · J. A. Renjit
Department of CSE, Jeppiaar SRR Engineering College, Chennai, TamilNadu, India

A. Sivasubramanian et al. (eds.), *Futuristic Communication and Network Technologies*, Lecture Notes in Electrical Engineering 792,
https://doi.org/10.1007/978-981-16-4625-6_9

Table 1 Student

Regno	Marks
101	98
102	100
103	39

information are growing each day, the database management systems cannot provide the required results in time. In earlier days, the data has been managed in a data table when the relational data set has emerged; the problem of claiming the data enters. In this scheme, the user has to submit the query as input where the DBMS has to produce the result of the user.

The problem here is, how to raise the query. For example, in a student database, if you want to know the meritorious student then you can generate a different form of queries as follows (Table 1):

```
SELECT*
FROM student
WHERE regno=101
OR regno=102
OR regno=103
```

On the other way, you can write the same query as follows:

```
SELECT *
FROM student
WHERE regno IN (101,102,103)
```

However, the queries produce the same result; their cost and time will be different. Similarly to extract a result from the large relational data set, the way of producing the query should be efficient.

The query optimization is the process of producing the exact query from a list of queries. If the user issues a query then the optimization process has to find the sequence of execution in which the query has to be executed. However, the query would be a simple, multiple, and nested one. If it is a nested one, you have to identify that which part of the query has to be executed first and which has to be executed as second. Also in a nested query [2] the result of a query part would become an input for another part. So it is necessary to identify the impacting query part and if there is any query part that does not make any difference that has to be identified before executing. The fewer impact queries can be executed at the same time to reduce the cost. However, the efficiency of query optimization depends on the way you are writing the query.

In general, not all the users have the efficiency and knowledge of DBMS and they would not know about generating an efficient query. To generate an efficient query, the scheme of the database must be familiar. There are some automatic systems available for generating an efficient query and to perform query optimization. However, the system cannot think intelligently like a human. So utilizing the human brain in generating the query has been emerged. The crowdsourcing [3] is the concept of collecting the user recommendations to construct an efficient query [4] to the system. The recommendation from the crowd can be collected for the same requirement. Based on the crowdsource result, the optimization process can select an efficient one.

The usage of the crowd has been used in different problems like tagging in the encyclopedia, where the people would specify the information to the entity considered. On the other side, the social network collects information about a particular entity through crowdsourcing [5] to identify the popularity. Similarly, there is some application available for the problem of crowdsourcing. For a query optimization problem, the same can be used by collecting queries from different units or persons. From the queries being collected, the method can identify a single one which will be more efficient. The domain ontology could be used in query optimization. The ontology of the relational database can be used in verifying the presence of the data objects in the query sequences. The same can be used in producing higher results in query optimization.

To identify an efficient query from crowdsourcing [6], that must be analyzed. In query optimization [7], crowdsourcing [3] can be used to identify an efficient query. Through the crowd, you can collect a different way of generating the query to the system. From the list of queries received, an efficient query from optimization has been selected. There exist several approaches around these terms. Some methods identify the query based on their time complexity, but some methods select the query based on their cost. However, the methods are not efficient in identifying an efficient query. To overcome this, an object access frequency-based approach is presented in this paper.

The access frequency of each object must be considered in selecting the query. Because each access increases the cost of query, they increase the time complexity. Similarly, the amount of information produced and the amount of information being used must be considered. By considering all these aspects, the method is proposed. The detailed method has been presented in the next section.

2 Object Access Frequency-Based Query Optimization

The proposed optimization algorithm considers the object access frequency as the major key in optimizing and analyzing the quality of the query sequence produced by the crowdsourcing. The method accepts the query sequences from the crowdsource. Then for each result of the source, the method computes different factors and costs. The method computes cost on access with both horizontal and vertical. Similarly, the methods compute the cost of execution time and information generated by the query. Using all these methods computes the sequence weight for the query sequence produced by the crowdsourcing [8]. Finally, a single one is selected based on the value of the sequence weight. The detailed approach is described in this section.

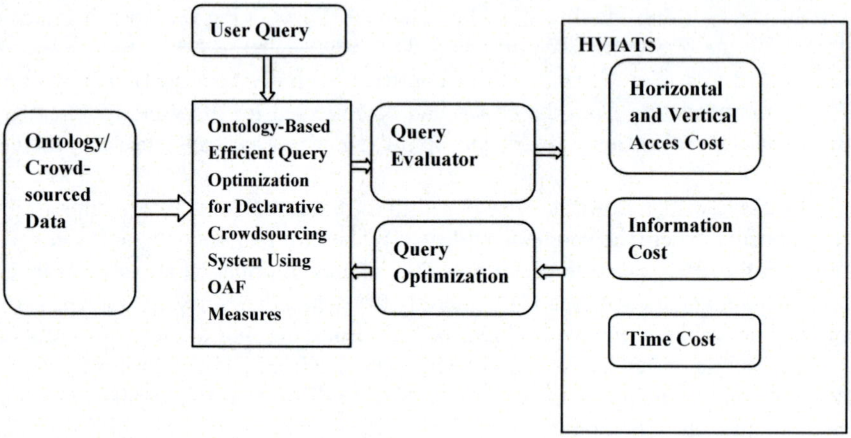

Fig. 1 Architecture of proposed algorithm

Figure 1 shows the architecture of the proposed query optimization algorithm and shows various stages involved.

2.1 *Query Evaluator*

The query evaluation [9] is the process of identifying the fitness [4] of the query for optimization. In this stage, the method receives the query sequences from the crowdsource. Then for each query, the method identifies the keywords and the entities of the data object. Then for each term or keyword identified, the method verifies the presence of the entities in concern data objects. Also, the key words are verified for their presence in the schema. If everything passes the check then it will be selected for optimization. Otherwise, the sequence generated by a particular crowdsource will be neglected. The evaluated query sequences are used to perform optimization.

Algorithm:

```
Input: Crowd-Source-Data Csd, Ontology O, Sql Set sqs
Output: query sequence set Qss
Start
    Read crowdsourcedata csd
    Read ontology O
    For each source si of Csd
        Sequence set ses = ∫_{i=1}^{size(Csd)} ∑ Csd(i).source == si
        For each query sequence qs
            Identify Key set ks = ∑_{i=1}^{size(sqs)} Sqs(i) ∈ sq
            Identify data attributes
            Da = ∑_{i=1}^{size(O)} O(i) ∈ sq
            Generate term set Ts = Split(qs,' ')
            For each term Ti
                If Ti∈ ks || Ti ∈ Da then
                Else
                    Flag==1
                    Break
                End
            End
        End
            If flag==1 then
                Remove the query sequence from the set
            Csd =∑ Seq(Csd) ∩ seq(si)
            End
    End
Stop.
```

The above-discussed algorithm identifies the incomplete sequence or false sequence and removes them. The evaluated query set has been used to perform optimization later.

2.2 *Horizontal Access Cost*

Any query can access the data objects in the number of rows. But not all the rows get to participate in the final result. Based on the number of rows being accessed for any query like SELECT, DELETE, JOIN [10] the method estimates the horizontal access cost. The horizontal access cost represents the cost being incurred by horizontally accessing the data object. It has been estimated dependent on the number of tuples being accessed and the number of tuples being produced in the result.

Algorithm:

Input: Sequence S, Ontology O
Output: HAC
Start
Read ontology O
Read sequence s
Compute no of rows being accessed on each entity.
Nor = $\sum_{i=1}^{size(s)} Da(k) \in s \ \&\& \sum S \rightarrow Da(k)$
Number of tuples in result Nt
Nt = $\sum Tuples \in\leftarrow s$
Compute HAC = $Nt/Nor \times \mu$ //here μ minimum cost of single tuple access
Stop

The above-discussed algorithm estimates the cost of the query sequence in horizontal access and will be used to perform query optimization.

2.3 *Vertical Access Cost*

The vertical access cost represents the cost being produced by accessing the data object in properties. Not all the properties of a data object need to be accessed. The sequence may access only a few properties of the object. It has been estimated dependent on the number of properties being accessed and the number of rows being selected. It has been measured as follows:

Algorithm:

Input: Ontology O, Sequence S
Output: VAC
Start
Read ontology O
Read sequence s
Compute no of properties being accessed on each entity.
Nop = $\sum_{i=1}^{size(s)} Da(k).Properties \in s \ \&\& \sum S \rightarrow Da(k).Properties$
Number of tuples in result Nt
Nt = $\sum Tuples \in\leftarrow s$
Compute VAC=Nt/Nop×μ //here μ minimum cost of single tuple access
Stop

The above-discussed algorithm computes the vertical access cost of the query sequence of each crowdsource. The estimated measure will be used to perform query optimization.

2.4 *Information Cost*

The information cost represents the amount of information being produced by the query sequence and how much it cost in producing the result. It has been measured based on the number of data objects being accessed and the amount of information is obtained as result from each object considered. Using these two metrics, the method computes the information cost.

Algorithm:

Input: Sequence s, Ontology O
Output: IC
Start
  Read ontology O
  Read sequence s
  Compute no of data objects being accessed
    Nodoa = $\sum_{i=1}^{size(s)} Da(k) \in s$
  Total amount of information obtained TI
    TI = $\sum_{i=1}^{Nodoa} \sum Tuples \leftarrow s(k)$
  Compute average information produced
  Aip = TI/Nodoa
  Compute IC = $Aip \times \mu$ //here μ minimum cost of single tuple access
Stop

The above-discussed algorithm computes the information cost produced by the sequence being considered. An estimated measure has been used to perform query optimization.

2.5 *OAF Query Optimization [3]*

The query optimization algorithm receives the input query and submits it to the crowd. The source returned from the crowd has been obtained. Then, the method performs the query evaluation to identify an unsuitable query sequence. With the evaluated result, the method computes the different costs for each sequence of queries submitted by different crowdsource. Third, the method computes the object access frequency

OAF for each sequence. Using the cost measures and the OAF value, the method computes the sequence weight. Based on the sequence weight, a single sequence has been selected.

Algorithm:

Input: query q, Ontology O
Output: selected query sequence Qs
Start
Read query q
Read ontology O
Crowd source Cs = Submit query q to Crowd
Preprocessed query set Pqs = Query-Evaluation(cs)
For each query qi
Hc = HorizontalAccessCost(qi)
Vc = VerticalAccessCost(qi)
Ic = InfomationCost(qi)
Compute OAF = $\sum_{i=1}^{size(s)} \sum Da(k) \in s(i) / No.of\ data\ objects$
Compute Time Cost Tc = Hc×Vc×β // β-time in accessing a row/col
Computesequenc weight Sw = $Hc/Tc \times Vc/Tc \times Ic/Tc \times OAF$

End
Choose the sequence with a higher sequence weight.
Stop

The above-discussed algorithm computes the object access frequency and different cost measures. Using the measures estimated, a single sequence weight is estimated. Based on the sequence weight, the method selects the most weighted one as the final result.

3 Results and Discussion

The proposed algorithm has been implemented and evaluated for its efficiency. The method has been evaluated for its performance with different relational databases. The data from crowdsourcing has been obtained and based on that the method has produced efficient results. The method has produced the following results.

Table 2 shows the details of the data set being used to assess the presence of the proposed algorithm. The UCI data set has been taken for evaluation and it contains 10,546 numbers of tuples, and each tuple has 29 attributes of different types. The data set has been used for evaluation, and the method has produced the following results.

Table 2 Details of the data set

Parameter	Value
Data set	UCI, HKU emotion, HKU web search
No of instances	10,546, 70 million, 2665 million
No of attributes	29, 1,1
No of types	Multivariate, Text, Text

The HKU Sentiment data set contains more than a 70 million instances, to rate the emotion of a given text. There are 7 emotions (anger, disgust, fear, joy, sadness, surprise, valence), and a user gives a value from -100 to 100 for each emotion about the text.

The HKU web search data set contains more than 2665 million data instances that have been used to judge the significance of query-URL pairs with a 5-level rating scale (from 1 to 5).

Table 3 shows the comparison of latency and completion being produced by various techniques on executing the queries. The proposed HVIATC method has produced higher performance in completion as well as latency.

Table 4 shows the assessment on cost of executing different complex queries. The proposed algorithm reduces the cost on each case than other methods.

Figure 2 shows the correlation result on query optimization, and the result produced has been compared with other methods. The proposed strategy has created a higher result than other methods.

Figure 3 shows the correlation result in cost reduction. The proposed strategy has produced less cost than other methods.

Figure 4 shows the correlation on time complexity delivered by various strategies. The result shows that the proposed algorithm produces less time complexity than other methods.

Table 3 Comparison percentage of completion and latency

Method	Task numbers	% of completion	Latency in seconds
Crowd DB	220	78.6	12
Crowd OP	220	82.3	10
Greedy	220	89.4	8
HVIATC	220	98.6	4

Table 4 Comparison of cost on various complex queries

Method	CSelect	CJOIN [11]	CFill	Total
Crowd DB	84.15	307.8	0	391.96
Crowd OP	76.37	7.6	31.02	114.95
Greedy	71.43	7.3	49.4	102.6
HVIATC	32.46	5.2	62.3	78.6

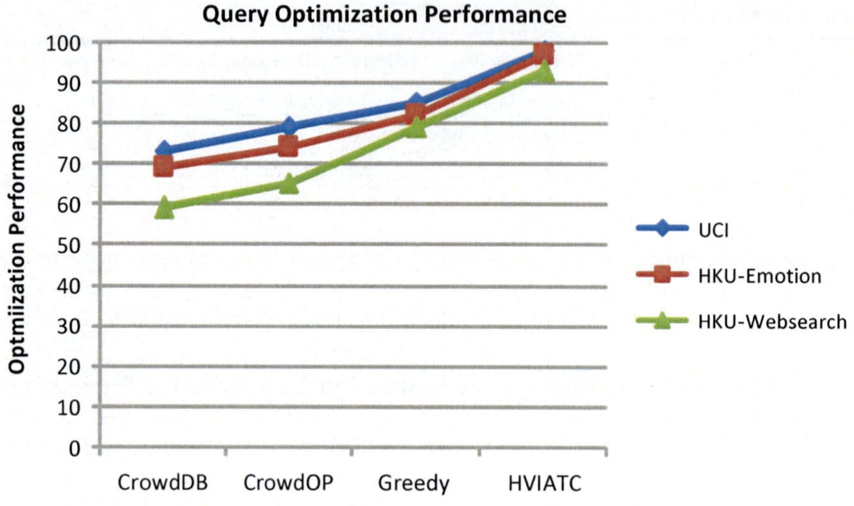

Fig. 2 Comparison on query optimization performance

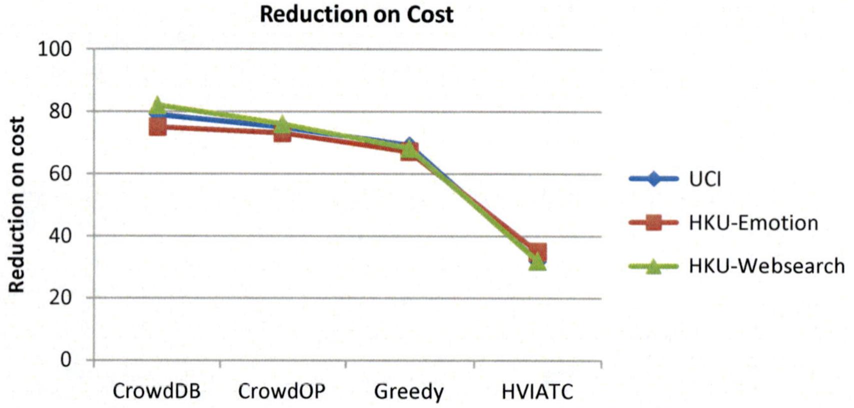

Fig. 3 Comparison of cost reduction

In this paper, an efficient query optimization algorithm with OAF has been presented. The method obtains the sequences from the crowd and identifies incomplete and false queries. Then with the remaining, it computes various cost measures and access frequency of objects. Using these measures, a sequence weight is computed for each sequence. Finally, a single one is selected based on the sequence weight estimated. The method produces a higher result in query optimization and decreases the time complexity as well.

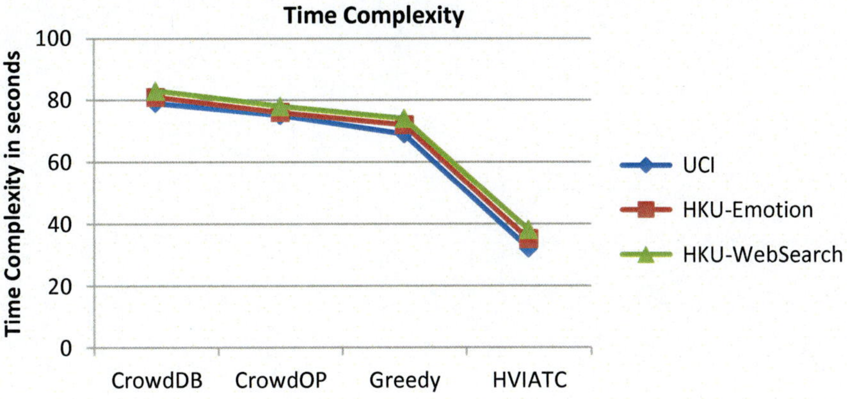

Fig. 4 Comparison of time complexity

References

1. Fan J, Zhang M, Kok S, Lu M, Ooi BC (Aug 2015) Crowdop: query optimization for declarative crowdsourcing systems. IEEE Trans Knowl Data Eng 27(8)
2. Davidson SB, Khanna S, Milo T, Roy S (2013) Using the crowd for top-k and group-by queries, ICDT, pp 225–236
3. Guo S, Parameswaran A, Garcia-Molina H (2012) So who won? Dynamic Max Discovery with the Crowd, SIGMOD, pp 385–396
4. Bhaskar N, Mohan Kumar P (2020) Optimal processing of nearest-neighbor user queries in crowdsourcing based on the whale optimization algorithm. Soft Comput 24:13037–13050. https://doi.org/10.1007/s00500-020-04722-0
5. Parameswaran A, Park H, Garcia-Molina H, Poluzotis N, Ramesh A, Widom J (2012) CrowdScreen: algorithms for filtering data with humans. SIGMOD, pp 361–372
6. Marcus A, Wu E, Karger DR, Madden S, Miller RC (2011) Crowdsourced databases: query processing with people. CIDR, pp 211–214
7. Parameswaran AG, Boyd S, Garcia-Molina H, Gupta A, Polyzotis N, Widom J (2014) Optimal crowd-powered rating and filtering algorithms. PVLDB 7(9):685–696
8. Michael J.Franklin, Donald Kossmann, Tim Kraska, Sukriti Ramesh, *Crowd DB: Answering Queries with Crowdsourcing,SIGMOD*, pages 61–72, 2011.
9. Parameswaran A, Park H, Garcia-Molina H, Poluzotis N, Widom J (2012) Deco: declarative crowdsourcing, CIKM, ACM, pp 1203–1212
10. Wang S, Xiao X, Lee C (2015) Crowd-based deduplication: an adaptive approach. SIGMOD, pp 1263–1277
11. Wang, Li G, Kraska T, Franklin MJ, Feng J (2013) "Leveraging transitive relations for crowdsourced joins". In SIGMOD

Fire Escape Route Tracing Using Sensors, Artificial Intelligence, and Azure Cloud

D. Geetha, A. Logarathinam, and R. Parvathi

Abstract Nowadays, indoor fire accidents are common. This, in turn, makes us think of a smart solution for this modern era. Many methods have been developed and implemented to control fire accidents. However, most of the preventing measures do not focus on how to escape the building during the initial stage of the fire accident. In this paper, a strategy has been developed to help people inside the building to find an efficient and safest route to exit from their current location. This model is developed based on A* algorithm which is a popular AI technique in finding the shortest path between the source and the destination. Temperature sensors have been used to detect the room temperature, and based on the collected data from the sensors, the safest and the shortest path is decided. The Azure Cloud platform is used to process the collected data and to execute the algorithm.

Keywords A* algorithm · Sensors · Artificial intelligence · Fire escape route · Azure cloud · Fire accidents solutions

1 Introduction

Technology improvement is emerging in all sectors nowadays. However, some important fields in which the lives of the people have to be saved are still lacking technical improvements. One such life-critical field is fire accidents, which is the focus of this paper. During the initial stage of fire, people have higher chances of escaping the building before the fire spreads rapidly. However, due to the lack of path guidance on how to exit the building, people get stuck and lose their chances of escaping the danger. Currently, there are software systems for controlling and monitoring the smoke and maintaining room temperature [1]. In this paper, a model has been developed, which finds an efficient and safest route to exit from the current location of the user during the initial stages of the fire accident. The first thing that comes to people's mind during tough times is their smartphones. The cloud platform [2] incorporated in this model will bring the solution to their smartphones through an application. The

D. Geetha (✉) · A. Logarathinam · R. Parvathi
School of Computer Science and Engineering, VIT Chennai, Chennai, Tamil Nadu, India

A. Sivasubramanian et al. (eds.), *Futuristic Communication and Network Technologies*, Lecture Notes in Electrical Engineering 792,
https://doi.org/10.1007/978-981-16-4625-6_10

application gets the current position of the user which is considered as the source and uses the A* algorithm [3] to find the path from the source to the safest and nearest exit which is considered as the destination. The A* algorithm, unlike other conventional algorithms, works a little smarter. It is the combination of uniform cost search (UCS) [4] and best first search [5]. In this model, the A* algorithm uses the sensor data collected in the cloud database to compute the path. The cloud database fetches the data from the temperature sensors that are already fixed in the building through Raspberry Pi. The flow diagram of the system is shown in Fig. 1.The rest of the paper is organized as follows. Section 2 talks about hardware and software requirements. Section 3 talks about the hardware implementation, and Sect. 4 talks about software implementation. Section 5 shows the results and analysis. The conclusion is made in Sect. 6 (Fig. 1).

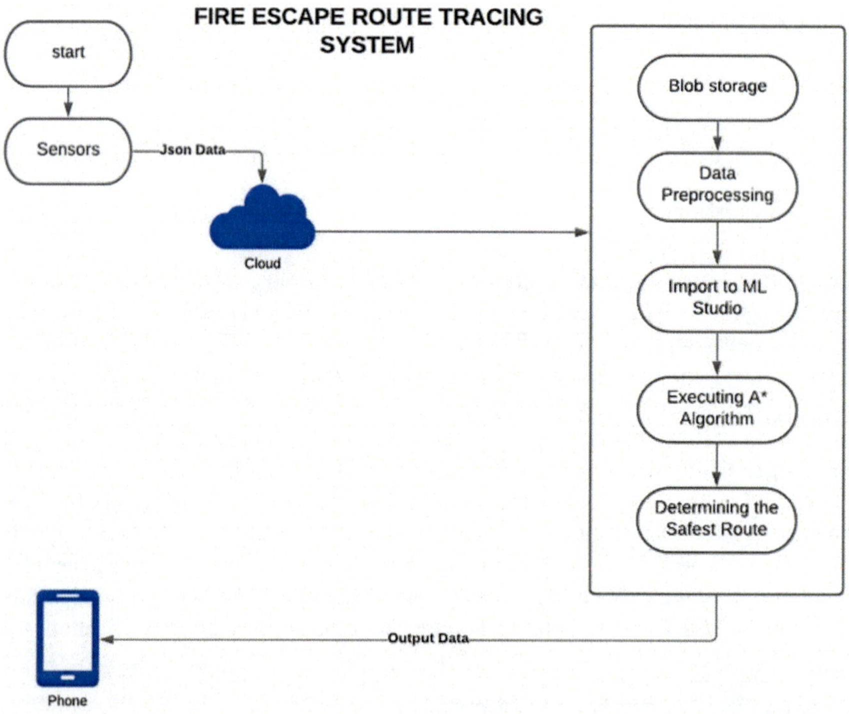

Fig. 1 Flow diagram of the system

2 Hardware and Software Requirements

The application that we have designed works using the cloud, sensor, Raspberry Pi, and AI algorithm.

- The Sensor used: Temperature Sensor—DS18B20 sensor [6]
- The Raspberry Pi used: Raspberry Pi 2 model B Rev 1.1 [7]
- The AI Algorithm used is A* algorithm
- The Cloud Platform used: Microsoft Azure [8].

The resources that have been created in the cloud platform are:

IoT Hub—It allows bidirectional communication between IoT application and the devices it manages.

Stream Analytics Job—It ingests data from Azure IoT Hub, Azure Events Hubs or Azure Blob storage.

Storage Container—The Azure blob storage is used in this application. It stores a large amount of unstructured data.

ML Studio—It is a GUI-based IDE for constructing and operationalizing machine learning workflow in Azure.

3 Hardware Implementation

3.1 Deploying Temperature Sensor

The temperature sensor (DS18B20) will detect the temperature within the range of 20–30 m for indoor and 75–100 m for outdoor, respectively. For example, to cover an indoor region of 100 m, a minimum of three sensors and a maximum of five sensors are required. Based on this calculation, the temperature sensors are placed in any buildings in such a way that all regions are completely covered. The temperature sensors will send the temperature data to the Raspberry Pi through the connections.

3.2 Setting Up the Raspberry Pi

The following are the steps to setup a Raspberry Pi:

The OS of the Raspberry Pi should be etched in an 8 GB or 16 GB micro SD card.

- The components needed for giving input to Raspberry Pi such as keyboard and mouse should be connected.
- The Raspberry Pi is connected to an LCD monitor using HDMI cable.
- Then the SSH and the other initial settings should be done.
- Then the Internet connection is established using Ethernet cable or dongle.

Fig. 2 Sensor connection with Raspberry Pi

3.3 Connecting Temperature Sensor to Raspberry Pi

The temperature sensor is connected to the Raspberry Pi using jumper wires [9]. The connections are made as follows:

- The VDD pin of the sensor is connected to the pin 1 of Raspberry Pi.
- The GND pin of the sensor is connected to the pin 6 of Raspberry Pi.
- The SDI pin of the sensor is connected to the pin 3 of Raspberry Pi.

The Fig. 2 shows the connection between a single ds18b20 temperature sensor and the Raspberry Pi.

4 Software Implementation

4.1 Designing the Graph

A graph is designed based on the internal structure of the building. Each node in the graph depicts a particular position in the building. The temperature sensors are efficiently placed all over the building. The nodes are divided into different groups based on their location such as rooms and corridors. Each group of nodes will have a dedicated sensor, which will update the value of its respective nodes with its detected temperature data. The nodes, which represents the walls or any other obstacles, are marked as 0, whereas the other remaining nodes will have the temperature data as their value. If a sensor is damaged during a fire accident, then its group of nodes' value is set to 0 to indicate that place has caught fire and is not accessible. The exits of the buildings are also stored in the model as end nodes from which the model will choose the safest exit.

4.2 Setting Up the Microsoft Azure Cloud

Four main resources are created in this model. The IoT Hub [10] is created to register the Raspberry Pi and sensors [11]. One new device id is created in the IoT hub. The primary connection string is generated which will be used to connect the device to the cloud securely.

The stream analytics job [12] is created to process the query. It consists of input query and output. It ingests data from IoT hub to aggregate and joins the streaming data. The data from the Raspberry Pi is collected in JSON format, which is stored in the blob storage. The storage container resource is created to organize the blobs. The machine learning studio [13] is created and is linked to the storage account to access the blob in the storage container. It is then used to execute the A* algorithm on the data.

4.3 Setting Up the Microsoft Azure Cloud

A python script is created in the Raspberry Pi to live stream the data from the sensor to the cloud in JSON format. There are a few primary parameters in the script to connect the pi to the cloud. They are the sensors device ID, uniform resource identifier (URI) of the IoT Hub, the primary string of the IoT hub, and the device id that is created in the IoT hub. The sensor device ID is identified by executing the following commands in the terminal.

```
sudo modprobe w1-gpio
sudo modprobe w1-therm
```

The remaining parameters are collected while creating the IoT Hub resource. The stream analytics job is used to sample the live stream data using SQL queries which is then automatically stored inside the blob storage in JSON format. Figure 3 shows the overall data flow in the system.

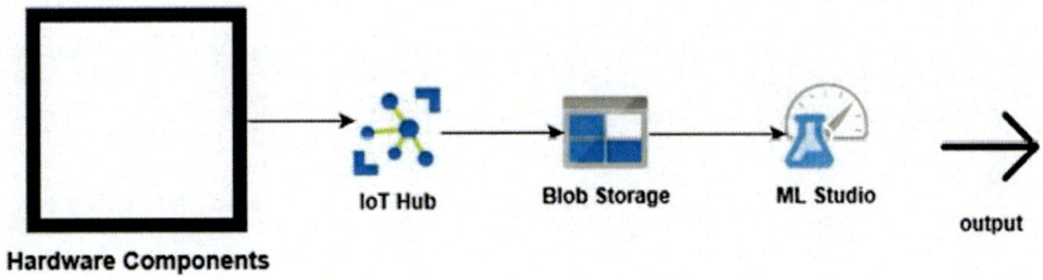

Fig. 3 Data flow in the system

4.4 Data Preprocessing

An experiment is created in the machine learning (ML) studio. A python module is added to import the data from blob storage using the access key of the storage account. The imported data is then cleaned and modified in this module. The data is converted from JSON format to CSV for further processing [14]. Each node value is assigned by picking the average of the data that has been sampled from its respective sensor.

4.5 Implementing the Algorithm

Another python module is added to the experiment to implement the A* algorithm. The output port of the previously mentioned data-preprocessing module is connected to the input port of this module. The files necessary to execute the A* algorithm are downloaded and are given as an input to the zip bundle port [15]. The python script in this module reads the CSV file from the previous data preprocessing module.

Procedure

The CSV file contains the temperature data from all the sensors inserted in the building. Based on the column-id, the region to which this sensor's temperature value belongs is found. The graph is converted into a two-dimensional array. The values of the array is fetched from the CSV file, and it is updated. To implement the A* algorithm in PyPi library, the 2D array is converted into a grid object. All the temperature values are taken as the weights of the nodes. The user's current position is taken as the start node. All the safest exit nodes are found from the 2D array based on whether the temperature value is bearable by humans or not. The A* algorithm finds the shortest and the safest path to all the safest nodes which were found previously. The path with the minimum cost is selected as the best path and is given as the output.

A* Algorithm

- The processed data, which is in the form of a grid, is given as the input parameter for the A* algorithm [16].
- The A* algorithm combines best first search (BFS) and uniform cost search (UCS) to compute the shortest path between the source node and the destination node.
- The A*algorithm is executed using the following function in the pathfinding library.

  Model = pathfinder.finder.a_star.AStarFinder()

- The algorithm computes the cost of the path in every iteration using the below equation and selects the minimum cost path to process further.

  $F(n) = G(n) + H(n)$

Where

F (n) is the estimated cost of the cheapest solution, G (n) is the cost to reach node n from the start node, and H (n) is the cost to reach from node n to end node.

- Once the goal node is reached, the algorithm terminates its iteration and outputs the cost and the shortest path.
- The time complexity of this algorithm is O (b^d).

5 Results and Discussion

The prototype is tested with three sensors and 50 test cases which covered the combinations for all three sensors. The prototype has managed to pass the test cases if at least one exit has a human bearable temperature. Thus, the scope of the model is limited to find a path only during the initial ignition of fire in the building.

Table 1 shows three-sample test cases that have been used to test the prototype. The values were set by increasing the temperature manually. The model calculates the safest end node by finding the reachable minimum temperature node from the exit nodes which we have declared previously in the model.

Explanation of Test Case 1

The node values are the input temperatures given to the model where 0 represents obstacles as mentioned in the Sect. 4.1. The border nodes of the graph excluding the start node are considered as the exits of the buildings in this example and are stored in the model while designing the graph. Therefore in this example, the exit nodes of

Table 1 Sample test cases

Test case number	1	2	3
Node 0 (1,1)	30	30	0
Node 1 (1,2)	65	73	150
Node 2 (1,3)	115	100	55
Node 3 (1,4)	200	75	32
Node 4 (2,1)	33	0	0
Node 5 (2,2)	35	43	55
Node 6 (2,3)	0	0	0
Node 7 (2,4)	90	300	41
Node 8 (3,1)	0	0	51
Node 9 (3,2)	30.5	32	45
Node 10 (3,3)	30.2	35	52
Node 11 (3,4)	32	50	53
Start node	(1,1)	(1,2)	(2,2)
Calculated end node	(3,3)	(3,2)	(1,4)

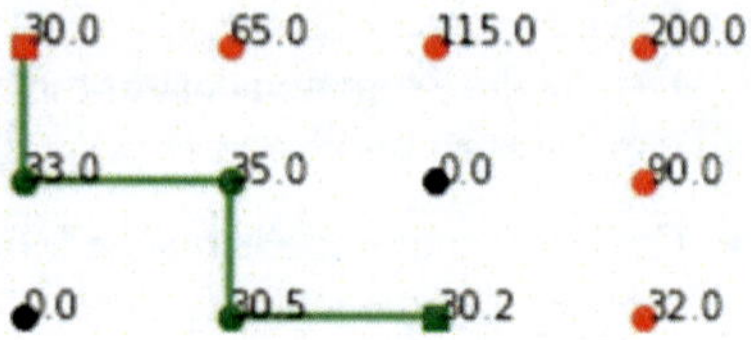

Fig. 4 Output of test case 1

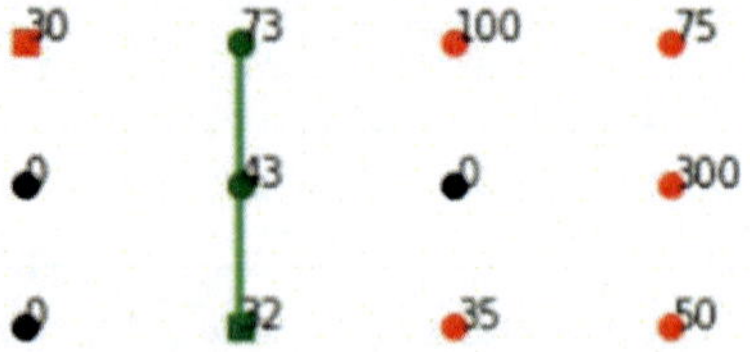

Fig. 5 Output of test case 2

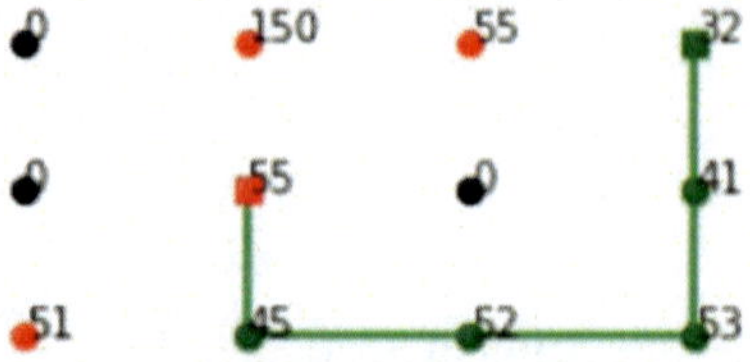

Fig. 6 Output of test case 3

the buildings are Node 1, Node 2, Node 3, Node 4, Node 7, Node 8, Node 9, Node 10, and Node 11. From this list, the model computes the reachable safest node with the minimum node value. Thus, the end nodes are calculated.

These tests have given the expected results. This model can provide the safest path possible to exit the building. However, the model fails when there is a safe place present within the building. Therefore, this model is limited to always find the path to exit the building. These observations prove that the important part of this model is the development and designing of the graph of the internal structure of the building. This method can further be improved by calculating the safest place inside the building by designing the graph to have multiple subgraphs where each subgraph should have at least one safe node to go.

The output of all the three test cases is given in the Figs. 4, 5, and 6

6 Conclusion

Fire accidents occur frequently which results in severe loss of life. This paper aims in helping people to escape the building during the initial ignition of fire by finding the shortest and the safest path to the exit. However, if there is heavy smoke, the opacity increases, and the subject may not be able to find the path given by the application. This application can be used in all types of buildings if the graph for the building is designed appropriately. The requirements would be designing the graph for the building and then installing the temperature sensors inside the buildings, which is the nodes in the graph. The preliminary results are promising. Moreover, the application can be upgraded by including the position of the firefighter to calculate the safest path and integrating its working with the smoke suction machine. This is a small initiative to march toward fire safety.

References

1. National Institute of Standards and Technology, U.S Department of Commerce, Fire modeling programs, https://www.nist.gov/el/fire-research-division-73300/fire-modeling-programs, last accessed 2019/11/20
2. Pratiksha P, Nikam Ranjeetsingh S, Suryawanshi (2013) Microsoft windows azure: developing applications for highly available storage of cloud service
3. Belwariar R A* Search Algorithm. https://www.geeksforgeeks.org/a-search-algorithm, last accessed 2019/11/20
4. Kundu A Uniform-cost search (Dijkstra for large Graphs). https://www.geeksforgeeks.org/uniform-cost-search-dijkstra-for-large-graphs, last accessed 2019/11/20
5. Sharma S Best first search (Informed Search). https://www.tutorialspoint.com/best-first-search-informed-search, last accessed 2019/11/22
6. DS18B20 Temperature Sensor, https://wiki.eprolabs.com/index.php?title=DS18B20_Temperature_Sensor, last accessed 2019/11/22
7. Introducing the Raspberry Pi 2 - Model B, https://learn.adafruit.com/introducing-the-raspberry-pi-2-model-b?view=all, last accessed 2019/11/22
8. Microsoft Azure, https://azure.microsoft.com/en-in/, last accessed 2019/11/22
9. Raspberry Pi DS18B20 Temperature Sensor Tutorial, https://www.circuitbasics.com/raspberry-pi-ds18b20-temperature-sensor-tutorial, last accessed 2019/11/22
10. Microsoft Azure, Azure IoT Hub, https://azure.microsoft.com/en-in/services/iot-hub, last accessed 2019/11/22
11. Microsoft Docs, Connect Raspberry Pi to Azure IoT Hub (Node.js). https://docs.microsoft.com/en-us/azure/iot-hub/iot-hub-raspberry-pi-kit-node-get-started, last accessed 2019/11/22
12. Microsoft Azure, Azure Stream Analytics, https://azure.microsoft.com/en-in/services/stream-analytics, last accessed 2019/11/22
13. Microsoft Docs, Azure Machine Learning Studio (classic) documentation. https://docs.microsoft.com/en-us/azure/machine-learning/studio, last accessed 2019/11/22
14. Google, Data Preparation and Feature Engineering in ML, https://developers.google.com/machine-learning/data-prep, last accessed 2019/11/22
15. Pypi, https://pypi.org/project/pathfinding/, last accessed 2019/11/22
16. Talan K, Bamnote GR (2015) Shortest path finding using a star algorithm and minimum weight node first principle. Int J Innov Res Comput Commun Eng 3:4290–4294

A Compact Penta-Band Printed Monopole Antenna for Multiple Wireless Communication Systems

Sudipta Das, Apurba Chowdhury, Bikram Lala, Ravi Prakash Dwivedi, K. Vasu Babu, and Mohammed EL Ghzaoui

Abstract A compact offset microstrip line-fed printed multiband monopole antenna is proposed. In this work, the suggested antenna operates at multiple resonant frequencies due to the incorporation of narrow parallel slits and a hammer-shaped slot on the radiating patch with a modified ground plane. The proposed antenna has been configured on a 26 × 25 mm^2 FR-4 substrate ($\varepsilon_r = 4.4$ and thickness = 1.6 mm). The simulated return loss results confirm the penta-band resonances from 2.9 to 3.2 GHz, 3.22 to 3.65 GHz, 5.6 to 6.5 GHz, 7.8–8.4 GHz, and 9.4–11.1 GHz to support multi-standard wireless systems such as maritime radio navigation, Wi-MAX, WLAN, X-band satellite communication uplink, ITU band, and aeronautical radio navigation bands. The simulated antenna parameters such as VSWR, radiation pattern, gain and directivity along with surface current distribution have been analyzed in detail and presented in this article. The designed structure maintains VSWR < 2 within the operating penta-bands which indicates very low mismatch losses in the design consideration. Furthermore, almost stable radiation patterns are obtained at all the operating frequencies. Simulated maximum peak gain of about 3.43 dBi at 5.8 GHz is reported for the proposed design.

Keywords Printed antenna · Multiband antenna · Slot · Modified ground plane · Wireless communication

---

S. Das (✉) · B. Lala
Department of Electronics and Communication Engineering, IMPS College of Engineering and Technology, Malda, West Bengal, India

A. Chowdhury
Mallabhum Institute of Polytechnic, Bishnupur, Bankura, India

R. P. Dwivedi
SENSE, VIT University Chennai Campus, Chennai, Tamilnadu, India

K. Vasu Babu
Department of Electronics and Communication Engineering, Vasireddy Venkatadri Institute of Technology, Guntur, India

M. EL Ghzaoui
Intelligent Systems Networks and Telecommunications Team, IMAGE Laboratory, ENS-Moulay Ismail University of Meknes, Meknes, Morocco

A. Sivasubramanian et al. (eds.), *Futuristic Communication and Network Technologies*, Lecture Notes in Electrical Engineering 792,
https://doi.org/10.1007/978-981-16-4625-6_11

1 Introduction

In the last few years, the expeditious progressive expansion of wireless communication systems necessitates an unbelievable demand of multiband antennas. The lightweight, low-cost and compact microstrip antennas are preferred to be used in modern wireless communication applications. The portability, cost-effectiveness and better performance are the major key concerns for the primary components that are used in wireless devices. As the microstrip antenna is one of the vital components in wireless devices, the design and implementation of miniaturized microstrip antenna [1, 2] is on high demand to make the wireless devices portable. Other than compactness, there is an increasing demand for a single antenna that can support many wireless systems operating at distinct frequency bands. Usually, the employment of many single-band antennas will increase the size and cost of implementation. These requirements lead to the usage of a compact multiband antenna to execute several applications on a single device. Nowadays, researchers have put great efforts to design multiband antennas through different design techniques such as defected microstrip structure [3], fractal [4], and defected ground structure [5]. In addition to these, different monopole configurations have been adopted for achieving multiband functionally, such as loading circular split ring resonator [6], incorporating pair of L-shaped slots [7], using an F-shaped radiator [8], dual-coupled strips of C-shape [9], E- and C-shaped radiators [10], modified patch loaded with multiple strips [11], step-shaped feed line connected with folded slots loaded patch [12], complementary split ring resonator [13], tuning stub[14], meta-material [15], L-loaded IFA [16], modified feed line and parallel strips [17].

The objective of this suggested work is to develop a small multiband antenna with suitable characteristic parameters to support multiple wireless communication systems In this work, a very compact, low profile printed penta-band antenna with an area of 25×26 mm^2 is structured by introducing a slot of hammer shape and three simple parallel slits on the patch along with a modified partial ground plane. The suggested compact antenna radiates at five distinct resonant frequencies with acceptable fractional bandwidth and radiation characteristics to support the multiple numbers of wireless communication applications. The structural design and detailed analysis of the characteristics parameters are performed using MOM-based IE3D software. The proposed antenna offers many advantages such as miniaturized dimension, penta-band operation, low VSWR, good impedance matching and thus low mismatch loss, acceptable gain, directivity, and stable radiation pattern characteristics.

2 Structure of the Proposed Antenna

The geometrical dimensions of the patch can be given as follows [18].

$$W_P = \frac{c}{2f_r}\sqrt{\frac{2}{1+\varepsilon_r}} \tag{1}$$

$$L_P = L_{\text{eff}} - 2\Delta L \tag{2}$$

$$\Delta L = 0.412h \times \frac{(\varepsilon_{\text{re}} + 0.3)\left(\frac{W_P}{h} + 0.264\right)}{(\varepsilon_{\text{re}} - 0.258)\left(\frac{W_P}{h} + 0.8\right)} \tag{3}$$

$$\varepsilon_{\text{re}} = \frac{\varepsilon_r + 1}{2} + \frac{\varepsilon_r - 1}{2}\left[1 + 12\frac{h}{w_P}\right]^{-1/2} \tag{4}$$

$$L_{\text{eff}} = \frac{c}{2f_r\sqrt{\varepsilon_{\text{re}}}} \tag{5}$$

where f_r is the operating frequency of the antenna, W is the width of the radiating patch, L_{eff} is the effective length of the radiator, and ΔL represents the extended normalized lengths of the patch.

The width of the microstrip line is calculated using the following equations [19]

$$\frac{W}{h} = \frac{2}{\pi}\left[B - 1 - \ln(2B - 1) + \frac{\varepsilon_r - 1}{2\varepsilon_r}\left\{\ln(B - 1) + 0.39 - \frac{0.61}{\varepsilon_r}\right\}\right] \tag{6}$$

where $B = \frac{377\pi}{2Z_0\sqrt{\varepsilon_r}}$.

The length of 50-Ω microstrip line is calculated as

$$L = \frac{\lambda_{\text{ge}}}{2.348} \tag{7}$$

where Effective wavelength (λ_{ge}) is calculated from

$$\lambda_{\text{ge}} = \frac{\lambda}{\sqrt{\varepsilon_{\text{re}}}} \text{ Where, } \varepsilon_{\text{re}} = \text{Effective dielectric constant.} \tag{8}$$

The length of the partial ground plane can be calculated using the following empirical equation:

$$L_g = L_p + 6.25h \tag{9}$$

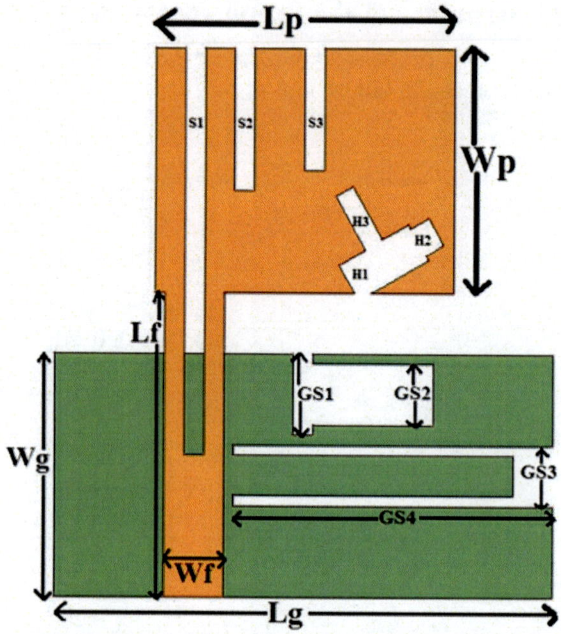

Fig. 1 Layout of the designed multiband monopole antenna

The suggested multiband antenna is shown in Fig. 1. The easily available cost-effective 1.6 mm thick FR-4 substrate with a loss tangent of 0.02 is applied for the design purpose. The designed structure occupies a total area of only 25 × 26 mm^2. The top side of the suggested antenna has been configured by introducing a hammer-shaped slot along with triple parallel slits into the microstrip line fed 15 × 12 mm^2 patch. The back plane of the antenna is comprised of a modified ground structure of 25 × 12mm^2. The optimal structural parameters are summarized as $L_P = 15$ mm, $W_P = 12$ mm, $W_g = 12$ mm, $L_g = 25$ mm,$W_f = 3$ mm, $L_f = 14$ mm, S1 = 20 mm, S2 = 8 mm, S3 = 6 mm, H1 = 4 mm, H2 = 1.5 mm, H3 = 3 mm, GS1 = 4 mm, GS2 = 3 mm, GS3 = 3 mm, GS4 = 14 mm.

3 Structural Development of the Proposed Antenna

The final geometry of the suggested penta-band antenna is constructed after some modifications in the initial design. The structural deployment is depicted in Fig. 2. The desired configuration is obtained after changing the feeding position of the reference antenna and then by introducing various slots on the patch and the ground plane. In case-I, single band performance is observed from 9.286 GHz to 10.403 GHz for −10 dB S_{11} parameter with a resonant frequency of 9.863 GHz. In case II, the feeding position of the antenna is changed. After changing the feeding position, the same structure as of case I show triple band characteristics. The antenna structure shown

Fig. 2 S_{11} parameters for different stages of structural development

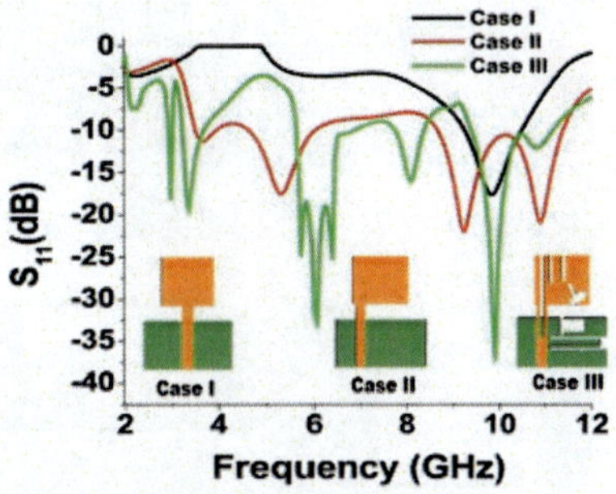

in case II provides −10 dB impedance bandwidths from 3.53 to 3.90 GHz, 4.70 GHz to 6.02 GHz, and 8.698 GHz to 11.4 GHz. But the intended applications such as WiMAX (3.3–3.6 GHz), X-band satellite communication uplink (7.9–8.4 GHz), and ITU (8–8.4 GHz) bands are not covered in the design case II. In the final design step (case III), the structure has been modified through slot perturbations on the patch as well as the ground plane. The suggested modifications in case III lead to achieving penta-band operation and the antenna structure covers frequency bands of 30 MHz (2.9–3.2 GHz), 430 MHz (3.22–3.65 GHz), 900 MHz (5.6–6.5 GHz), 600 MHz (7.8–8.4 GHz), and 1700 MHz (9.4–11.1 GHz). Hence, the layout (Case III) shown in Fig. 2 is considered as the proposed antenna configuration for the intended wireless applications such as maritime radio navigation (3 GHz), Wi-MAX (3.3–3.6 GHz), WLAN (5.725–5.825 GHz), X-band satellite communication uplink (7.9–8.4 GHz), ITU band (8–8.4 GHz), and aeronautical radio navigation (10–10.7 GHz) bands.

4 Explanation of the Surface Current Distribution

The variations of the surface current in the proposed penta-band antenna are presented in this section. Surface current distributions at different operating frequencies (3.05, 3.435, 6.05, 8.1, and 10.25 GHz) are investigated and displayed in Figs. 3a–e). An explanation of the variations in surface current distribution on the designed antenna for different operating frequencies will add some key insights regarding the structural validation of the proposed configuration and its excitation behavior to understand the working of the suggested antenna. For the 3.05 GHz operation, it can be observed from Fig. 3a that the surface currents strongly circulate around the vertically placed parallel slits and the vertical arm of the hammer-shaped slot which controls the resonance excitation of the first operating band (2.9–3.2 GHz). However, the second operating band is mainly generated and controlled by the slots incorporated in the partial ground plane as shown in Fig. 3b for 3.435 GHz. As observed from Fig. 3c, the center frequency (6.05 GHz) of the third resonance band is dependent on the slots loaded on both the patch and ground plane. A very dense surface current is noticed around the hammer-shaped slot along with the shortest vertical slit loaded

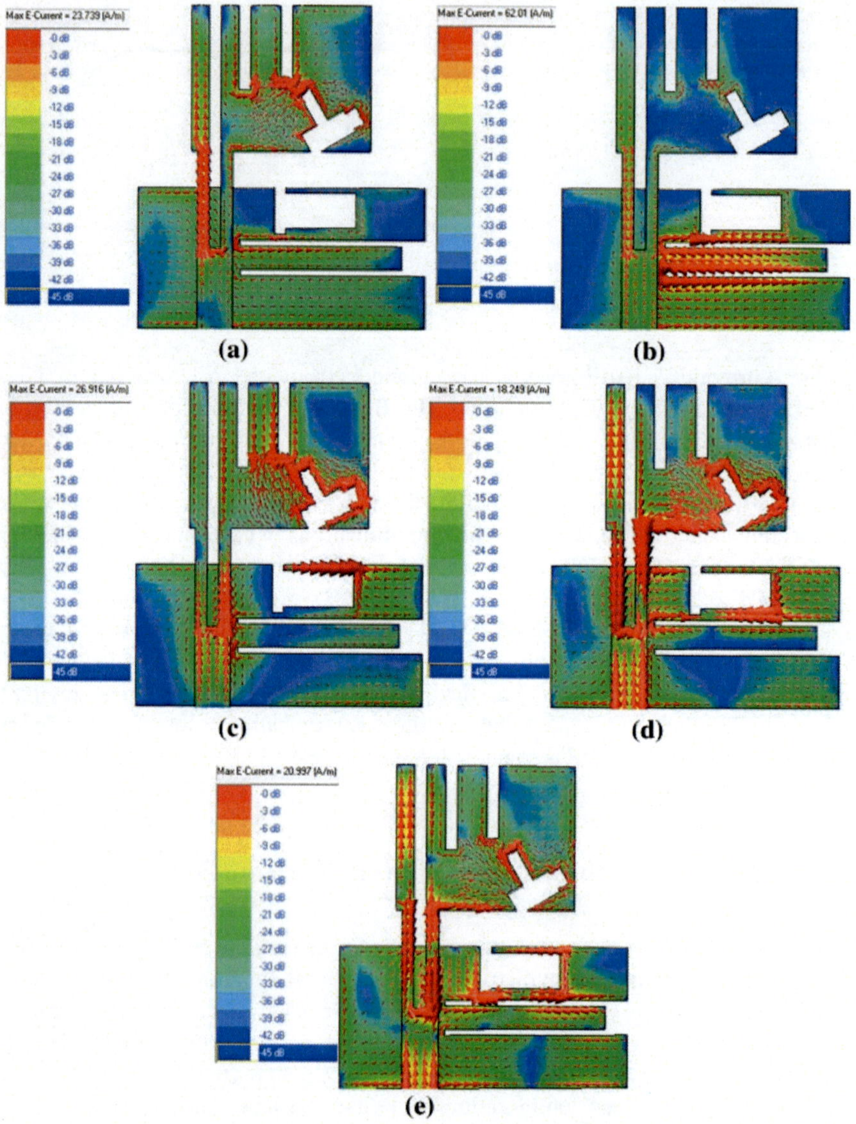

Fig. 3 Presentation of surface current variations at **a** 3.05 GHz, **b** 3.435 GHz, **c** 6.05 GHz, **d** 8.1 GHz, and **e** 10.25 GHz

on the patch and the horizontal arm of the open-ended slot in the ground plane. For the fourth resonance band with center frequency 8.1 GHz, the originated surface currents strongly circulate around the longest vertical slit and hammer-shaped slot on the patch along with a wide open-ended slot in the partial ground plane [see Fig. 3d]. Finally, for 10.25 GHz, it is observed from Fig. 3e that the maximum surface current is distributed along the defected ground plane structure, longest slit incorporated on the patch, and vertical arm of the hammer-shaped slot for which the fifth operating band is generated. The comprehensive analysis of the surface current distribution helps to understand the working mechanism of the designed antenna. The incorporated slots on the patch and ground plane have changed the electric and magnetic field distributions which further varies the mean current path of any resonant mode and as a result number of resonant frequency increases. Furthermore, the shifting of resonant frequencies to lower values may be explained by the fact that the concentration of the surface currents is enhanced around the sides of the slots and the elongation of the distributed surface current increases the current path thus making the antenna electrically larger thereby the excited resonant frequency decreases.

5 Results and Discussion

The simulation results of the presented antenna in terms of S_{11} parameter and VSWR are shown in Fig. 4. A clear observation shows that the suggested structure resonates at penta-band for VSWR < 2 and $S_{11} < -10$ dB. The results depicts five distinct frequency bands with impedance bandwidths (IBW) of 30 MHz (2.9–3.2 GHz, fractional bandwidth = 9.836%), 430 MHz (3.22–3.65 GHz, fractional bandwidth = 12.518%), 900 MHz (5.6–6.5 GHz, fractional bandwidth = 14.876%), 600 MHz (7.8–8.4 GHz, fractional bandwidth = 7.407%), and 1700 MHz (9.4–11.1 GHz, fractional bandwidth = 16.583%). The VSWR lies within 2:1 signifies good impedance matching and acceptable mismatch loss across the penta-band of operation. The variations of gain and directivity for the designed penta-band antenna are depicted in Fig. 5. A peak gain of about 3.46dBi is obtained at 5.8 GHz while the directivity reaches to a maximum value of 6.15dBi at 10.40 GHz. Figure 6a, b presents the radiation patterns of the designed penta-band antenna. The E and H plane radiation patterns are described for the center operating frequencies of each band, 3.05, 3.435, 6.05, 8.1, and 10.25 GHz, respectively. The proposed antenna offers almost stable monopole-like radiation responses at both the planes (E and H) but slight deterioration

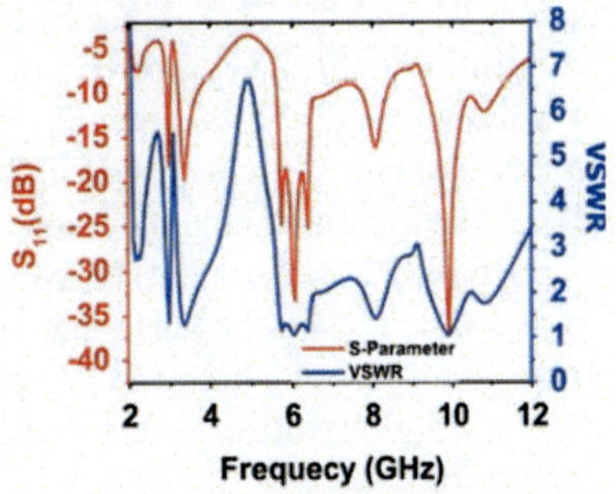

Fig. 4 S_{11} and VSWR of the proposed antenna

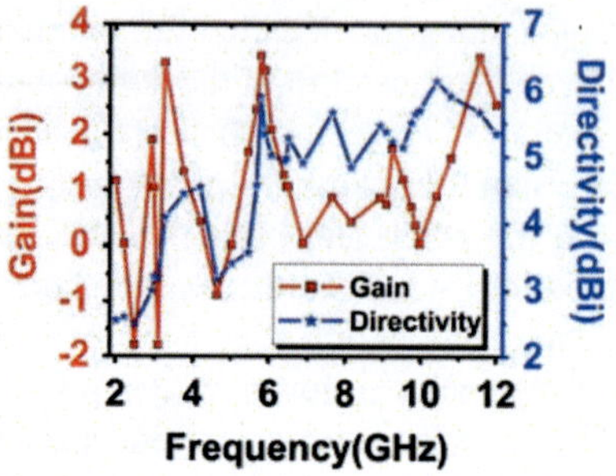

Fig. 5 Gain and Directivity versus frequency

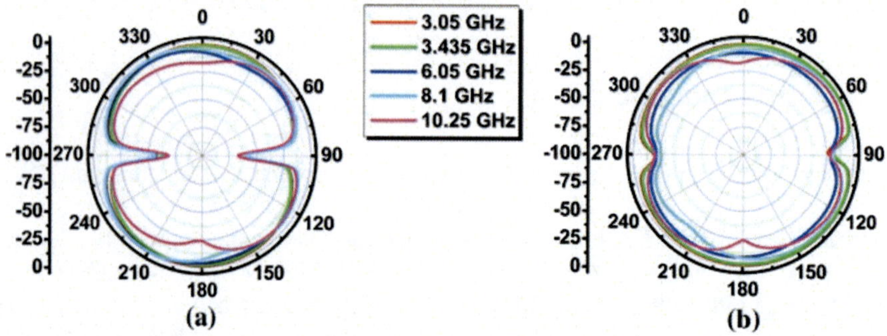

Fig. 6 Radiation pattern characteristics **a** E-plane **b** H-plane

is noted with an increase in frequency mainly at 10.25 GHz. The maximum radiation intensity is observed towards 0° with linear polarization. The E-plane patterns almost resemble a figure of "8" while the H-plane patterns resemble an oval-shaped with slight distortions at the upper band frequency.

6 Conclusion

A small printed antenna has been proposed in this article for maritime radio navigation, Wi-MAX, WLAN, X-band satellite communication uplink, ITU band, and aeronautical radio navigation bands. The presented antenna is suitable for penta-band operation. The proposed antenna has been configured by incorporating a hammer-shaped slot and vertical slits of different lengths on the patch and reformed ground plane. The presented radiator resonates at five distinct frequency bands at about 3.05, 3.435, 6.05, 8.1, and 10.25 GHz with desired stable radiation patterns and acceptable gains. Also, good impedance matching and thus low mismatch losses are obtained across the operating frequency bands. So, the proposed antenna is compact, low profile, easy to realize, offers penta-band operation and supports several wireless applications.

References

1. Das S, Sarkar PP, Chowdhury SK (2014) Design and analysis of a compact monitor-shaped multifrequency microstrip patch antenna. J Electromagn Waves Appl 28(7):827–837
2. Das S, Sarkar PP, Chowdhury SK (2015) Modified π-shaped slot loaded multifrequency microstrip antenna. Progress Electromagn Res B 64(1):103–117
3. Salamin MA, Das S, Zugari A (2018) Design and realization of low profile dual-wideband monopole antenna incorporating a novel ohm (Ω) shaped DMS and semi-circular DGS for wireless applications. AEU-Int J Electron Commun 97:45–53
4. Ali T, Aw MS, Biradar RC (2018) A fractal quad-band antenna loaded with L-shaped slot and metamaterial for wireless applications. Int J Microw Wirel Technol 10:826–834
5. Kunwar A, Gautam AK, Kanaujia BK (2017) Inverted L-slot triple-band antenna with defected ground structure for WLAN and WiMAX applications. Int J Microw Wirel Technol 9:191–196
6. Baba MA, Rahim MKA, Zubir F, Fairus M, Yusoff M (2018) Design of miniaturized multiband patch antenna using CSRR for WLAN/WiMAX applications. Telkomnika 16(4):1838–1845
7. Chen S, Fang M, Dong D, Han M, Liu G (2015) Compact multiband antenna for GPS/Wi-MAX/WLAN applications. Micowave Opt Technol Lett 57(8):1769–1773
8. Ren W, Hu SW, Jiang C (2017) An ACS-fed F-shaped monopole antenna for GPS/WLAN/WiMAX applications. Int J Microw Wirel Technol 9:1123–1129
9. Han Y, Yin YZ, Wei YQ, Zhao Y, Li B, Li XN (2011) A novel triple-band monopole antenna with double coupled c-shaped strips for wlan/wimax applications. J Electromagn Waves Appl 25:1308–1316
10. Honarvar MA, Hamidi N, Virdee BS (2015) Multiband antenna for portable device applications. Microw Opt Technol Lett 57:956–959
11. Mondal K, Samanta A, Sarkar PP (2017) Compact multiband monopole antenna for ISMband 2.4 GHz, Bluetooth, WiMAX Wi-Fi applications. Wirel Pers Commun 97:181–195
12. Jalali AR, Ahamdi JS, Emadian SR (2016) Compact multiband monopole antenna for UMTS, WiMAX, and WLAN applications. Microw Opt Technol Lett 58:844–847
13. Daniel RS, Pandeeswari R, Raghavan S (2017) Multiband monopole antenna loaded with-complementary split ring resonator and C-shaped slots. AEU-Int J Electron Commun 75:8–14
14. Brar RS, Saurav K, Sarkar D, Srivastava KV (2018) A quad-band dual-polarized monopole antenna for GNSS/UMTS/WLAN/WiMAX applications. Microw Opt Technol Lett 60:538–545
15. Ali T, Khaleeq MM, Pathan S, Biradar RC (2017) A multiband antenna loaded with metamaterial and slots for GPS/WLAN/WiMAX applications. Microw Opt Technol Lett 60:79–85
16. Al-Khaldi M (2017) A highly compact multiband antenna for Bluetooth/WLAN, W-MAX, and Wi-Fi Applications. Microw Opt Technol Lett 59:77–80
17. Bag B, Biswas P, Sarkar S, Sakar PP, Ghosal D (2020) Novel monopole microstrip antennas for GPS, WiMAX and WLAN applications. J Circuits Syst Comput 29(3):2050050
18. Balanis CA (2005) Antenna theory analysis and design, 3rd edn. Hoboken, NJ, USA, Wiley
19. David M (2005) Pozar: microwave engineering, 3rd edn. Wiley, USA

QoS-Driven AODV Algorithm for WSN

R. Avudaiammal, M. S. Sowmyaa Vathsan, and S. O. Sivashanmugam

Abstract Wireless sensor networks (WSNs) are useful in various applications like environmental analysis, industrial monitoring and surveillance, health care and in post-disaster relief operations. It requires collecting and directing data to a control station called sink or base station are to be performed reliably and in a timely manner. The variations in the delay, capacity and loss ratio of wireless links are directing to the highly unreliable route. Moreover, due to the mobility of nodes, provisioning of Quality of Service (QoS) is a challenging problem. In this work, QoS-driven ad hoc on-demand distance vector (QAODV) routing algorithm has been proposed that aims to establish a reliable and stable route by adding the power constraint along with the existing distance constraint to achieve QoS. The QoS metrics of QAODV are analysed using NS-2 simulator. The simulation outcomes proved that QAODV provides improvement of 14.28% throughput and reduction in delay and packet loss of about 27.83% and 11.6%, respectively, than AODV.

Keywords QoS · WSN · Energy-aware routing · NS-2 · AODV

1 Introduction

Routing in WSN is varied from formal routing in various means because of their infrastructure less architecture, the unreliability of wireless links and failure of sensor nodes [1]. Routing protocols for WSNs are classified into seven types such as location-based protocols, data-centric protocols, hierarchical protocols, mobility-based protocols, multipath-based protocols, heterogeneity-based protocols and QoS-based protocols. An enhance AODV routing protocol uses *Hello* packets to intermediate nodes to find the most stable paths for data transmission [2].

R. Avudaiammal
Professor, St. Joseph's College of Engineering, Chennai 600119, India

M. S. Sowmyaa Vathsan (✉)
Anna University, MIT Campus, Chennai 600025, India

S. O. Sivashanmugam
State University of New York At Buffalo, 14261 Buffalo, New York, US

A. Sivasubramanian et al. (eds.), *Futuristic Communication and Network Technologies*, Lecture Notes in Electrical Engineering 792,
https://doi.org/10.1007/978-981-16-4625-6_12

The location-based protocols address the sensor nodes by means of the location of the sensor nodes. Protocols that come under this type are MECN, GAF, GEAR, etc. [3] The data-centric protocols have lesser energy requirements as the source node sends data directly to sink which is different from the address centric protocol. Notable protocols that come under this type are SPIN, COUGAR, EAD, etc. The hierarchical protocols are essential in managing clump using a cluster head and coordinate the data transmission activities. Protocols in this type are LEECH, TEEN, HEED, etc. The multipath-based protocols find n number of paths with the shortest distance and then distribute its transmission load evenly across the multiple paths and thereby improving reliable transmission. Sensor-disjoint multipath and braided multipath are some of the multipath routing protocols. Heterogeneity-based protocols minimize the potential of data communication and computation by efficiently using the available energy. IDSQ and CADR are some of the heterogeneity-based routing protocols.

The QoS-based routing protocols help balance the consumption of energy and QoS, some of which are SAR, SPEED [4] and energy-aware routing [5–11].

The mobility-based protocols provide a solution to the challenges faced during routing mobile nodes as they generally have no pre-defined topology. Some of the protocols in this type are AODV, SEAD, TTDD, etc. [12].

The security-based routing algorithms provide the mechanism to overcome malicious attacks caused by internal nodes in WSN. Many security-based routing algorithms have been proposed by many researchers using a distributed trust evaluation model to find and to separate malicious node [13–15].

The development of wireless multimedia sensor networks (WMSNs) has stirred up the progressing of research from customary WSN to WMSN. QoS routing is a keystone of WMSN for fulfilling QoS requirements. An optimal path choice is found based on cost function estimation using several metrics of routing [16, 17].

Ad hoc on-demand distance vector (AODV) is the mobility-based routing protocol that aids both unicast and multicast services over wireless and mobile ad hoc networks [18, 19]. This protocol sets up routes to the destinations on demand, i.e. it avoids the formation of any excess traffic along with links. Sources are allowed to access the preserved routes and to check route freshness by using sequence numbers. Multicast group members are connected in trees kind structure which is loop-free and can be scaled to several mobile nodes. The source node broadcasts a connection request message. The other nodes in the path forward the message containing a sequence of momentary routes along with the source node id. A node that has a route to the destination node transmits a message through temporary routes to the source node. The source node chooses a route with the smallest hop count and discards the unused entries in routing tables later. During the occurrence of a link failure, the routing error is intimated to the source node, and the procedure is repeated.

From the literature, it is inferred that AODV has several advantages such as simplicity, high efficiency and potentially high reliability because the packets may be delivered through multiple paths [20]. It has some drawbacks such as very high overhead and may also potentially low reliability of data delivery because of the simultaneous reception of the same data through two individual paths may lead to packet loss at the destination. It is also inferred that in all the routing protocol, each source node directs its data to the destination through the shortest path [21]. Hence in this work, a modified version of the original AODV algorithm, termed as QoS-driven ad hoc on-demand distance vector (QAODV) algorithm, has been proposed that aims to establish a stable and reliable route by adding the power constraint along with the existing distance constraint to achieve QoS.

2 QAODV Algorithm

The proposed QoS-driven ad hoc on-demand distance vector routing algorithm aims to establish a stable and reliable by adding the power constraint along with the existing distance constraint. The proposed algorithm works in two stages, namely the discovery of the route stage and maintenance of route stage. The selection of the route is based on both the distance as well as on the initial power level of the intermediate nodes. It is aimed to achieve better QoS parameters such as packet loss, time delay and throughput per node.

Step 1: For route discovery, the source node sends an RREQ message via flooding to all its neighbouring nodes
Step 2: Those nodes which receive the RREQ update the hop count & power and then forward the same to its neighbouring nodes.
Step 3: Flooding continues until the destination node is reached.
Step 4: Destination node sends back RREP to the source through the same path of reception.
Step 5: Source node checks the power of the neighbouring nodes after receiving RREP.
Step 6: Source node verifies the power of the neighbouring nodes, and then the route is selected by neglecting the nodes with low power.
Step 7: If the power is high enough, i.e. above the already set value, then the routing takes place through that defined route.
Step 8: The intermediate nodes along the defined path also consider the power of the successive node before forwarding the data packet; if the power requirement is not met, then that node initiates a route discovery again.

3 Simulation Environment

The simulation of the proposed QoS-driven ad hoc on-demand distance vector routing algorithm (QAODV) has been carried out using Network Simulator-2 (NS-2).

Table 1 Simulation parameters

Simulator	NS-2.35
Simulation area	1000 m × 1000 m
Total no. of nodes	19
Queue length	240 Bytes
Packet size	100 Bytes
txPower	0.36 W
rxPower	0.31 W
Simulation time	30 cycles
Initial energy	1.0 J

3.1 Simulation Parameters

TCP traffic is generated randomly between mobile nodes using NS-2. Other parameters are shown in Table 1.

3.2 Energy Model

The energy level of the nodes is set by using the class energy model in NS-2 with attributes (**txPower** (mw), **rxPower** (mw) and **initialEnergy** (Joules)

3.3 Node Initialization

Step 1: Initialize the nodes with the parameters that are listed in Table 2,
Step 2: Set the number of nodes to 60 and the simulation environment size to 1000mx1000m.
Step 3: Initiate a new simulator window.

Table 2 Node initialization

Channel	Wireless Channel
Propagation	2-Ray Ground
Network interface	Wireless
Queue	Droptail & Priqueue
Antenna	Omnidirectional
Packet length	100 Bytes
Simulation period	30 cycles
Energy model	Present
Initial energy value	Watt

Step 4: Create a General Operations Director (GOD) which is used to store global information about the state of the simulation environment.
Step 5: Configure the nodes according to the requirements.
Step 6: Setup a TCP connection, i.e. set the source node by attaching a TCP agent and set the destination node by attaching a TCP sink.
Step 7: Set the packet size that is to be transmitted by the nodes.
Step 8: Create the FTP over TCP connection at the source.
Step 9: Generate the required trace files.
Step 10: After execution, the simulation takes place in nam window, and the QoS graphs are plotted in XGraph.

4 Network Configuration

In this work, the proposed QAODV is a refined version of original AODV with the aim to get a stable and reliable route by adding the power constraint along with the existing distance constraint. Therefore, during the route discovery stage, a route choice is built on the present power node along the path of transmission. Also, link failures due to degradation of power in nodes are considered for route maintenance, and then, the requirement for a rediscovery of the node is analysed.

Scenario 1: Figure 1 shows the example topology created for simulation and testing. It consists of 19 nodes, and their positions are randomly allocated. During the node initialization, the power level of all nodes is set as 1 J which reduces gradually while involving in data transmission. At 1.4th simulation cycle, the nodes

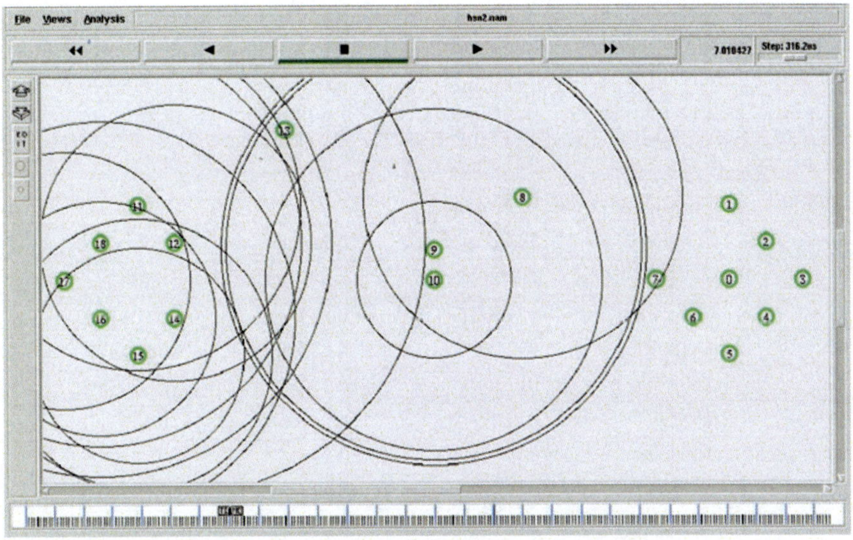

Fig. 1 Screenshot of re-initiated route discovery at the 7th simulation cycle

16 & 5 are set as source and destination, respectively. And at 2nd simulation cycle, packet transfer is done over the path 16–13-10–5, and the acknowledgements are transferred over the path 5–9-13–16.

The usage of separate routes for packet and acknowledgement delivery is due to the fact that node 10 at flat grid coordinates (300, 500) is relatively farther away from node 13 at coordinates (200, 600) when compared to the position of node 9 at position (300, 520) in the flat grid simulation surface. Since power is an important constraint to ensure reliable data transfer among the nodes, the node with the higher power is taken for data transmission, while the node with comparatively lesser power is taken for acknowledgement transfer. As it is using two different paths, one for data and one for acknowledgement transfer, the power of all the nodes is evenly consumed. Table 3 shows the routing paths for both AODV and QAODV algorithms.

Scenario 2: At 7th simulation cycle, the route discovery is initiated by node 13 when it experiences link failure which is intentionally created by forcing node 10 to drop its power down to zero. This forces the simulator to initiate alternative measures to create a path to the destination, i.e. route discovery. Once the node 10's power is realized to be zero, then the preceding node sends a RERR (Route Error) message to the source node and then initiates a route discovery to find a suitable alternate path. Here the intermediate node being 13 becomes responsible for the initiation of route discovery. Once the operation is done, an alternate path to reach destination node 5 is discovered as 16—13—8—4—5.

Table 4 shows the power of the active nodes that take part in the transmission of packets which are analysed over the entire simulation period. But nodes 16 and 5

Table 3 Routing paths for AODV and QAODV

	AODV		QAODV	
	data transmission	Acknowledgement	data transmission	Acknowledgement
Scenario 1	16–12-13–9-10–6-5	5–9-14–15	16–13-10–5	5–9-13–16
Scenario 2	16–12-13–9-8–6-5	5–9-14–15	16–13-8–4-5	5–9-13–16

Table 4 Power of nodes at various simulation time

Time	Power at 9	Power at 10	Power at 8	Power at 4
1	0.973413	0.978968	0.973782	0.973782
5	0.857098	0.743425	0.859746	0.859746
6.8	**0.848987**	**0**	**0.856889**	**0.857877**
10	0.697785	0	0.790875	0.790359
15	0.537994	0	0.541568	0.536311
20	0.375653	0	0.392157	0.385601
25	0.227605	0	0.223833	0.219803
30	0.15894	0	0.150942	0.147734

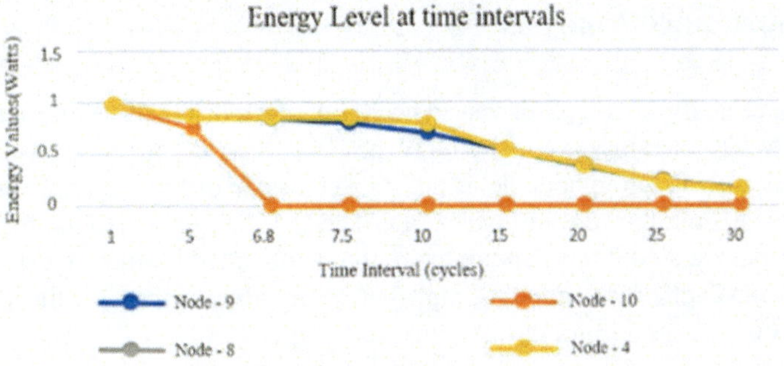

Fig. 2 Energy level at different time intervals

are not considered as they are always involved in transmission irrespective of their power value.

The simulation period 6.8 is highlighted in the table to show that it is at that moment when the node 10 drops its power to zero, and route discovery is initiated.

The graph is shown in Fig. 2, which shows the variation in the power of the nodes over the 30 simulation periods. Initially, power of all the nodes is set as 1 J, and some power is used up by all the nodes during the first route discovery initiated by the source node 16.

It is observed that the nodes that are far away from the source with the highest power will cause delay. In our proposed work, the path is selected is based on the power level of the nodes present in the path. The distance between the source and the next node may be greater, but the energy level of both the nodes is considered for the data transmission which ultimately leaves the nodes that have lesser energy.

In the first scenario, the path through node 10 is selected by node 13 even if it is comparatively farther than node 9 because the power of node 10 is higher at 0.978968 J, whereas the power of node 9 is 0.973413 J. So, node 10 is used for data packet delivery, whereas node 9 is used for acknowledgement.

In the second scenario, after the route discovery is completed by node 13, node 13 compares the viability of using node 8 and 9, and then node 8 is selected since it has 0.856889 J at 6.8th simulation cycle. So, node 8 is selected for data transmission, and node 9 is used for acknowledgement as before because the node 9's power is 0.848987 J which are lower than that of node 8's.

At the 7.5th simulation cycle node, 8 cannot directly connect with node 5 as it is not located within the coverage area of node 8. And so, node 8 finds an alternate route taking into consideration the power and the distance of the nodes. Node 4 is, thus, selected as an intermediate node for forwarding the packets to node 5. Therefore, the new path of transmission is 16–13-8–4-5, and the path of acknowledgement transmission is the same as scenario 1, i.e. 5–9-13–16. When the path is stable and the QoS parameters are stabilized, then the nodes change colour to indicate the stability of transmission in the topology using QAODV protocol.

5 Performance Analysis

Performance analysis of QoS-driven ad hoc on-demand distance vector routing algorithm (QAODV) against original AODV has been carried out by measuring QoS parameters such as throughput, delay and packet loss. In order to analyse, the routing algorithm, i.e. both the conventional AODV and QAODV are set to run for 30 simulation periods (at step 2 ms). The graph output is obtained by linking the generated trace file to an application called XGraph 4.30 (uses values in the trace file to generate the standard X–Y graph) for easier comparison and analysis.

5.1 *Throughput*

Throughput shows a successful transfer rate of packets from the source to the destination. It is measured in *packets per second*. The throughput of AODV & QAODV is shown in Fig. 3a.

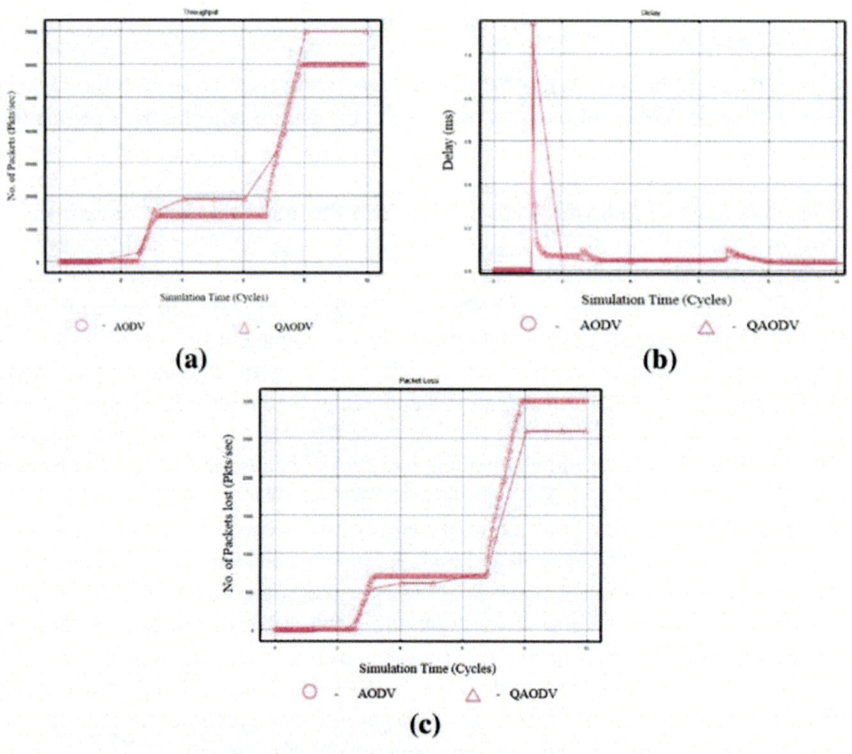

Fig. 3 a Throughput—AODV versus QAODV. **b** Delay—AODV versus QAODV. **c** Packet loss—AODV versus QAODV

Table 5 Performance measures AODV versus QAODV

	Throughput pkts/sec	Mean delay ms	Loss pkts/sec
AODV	60,000	0.265	3000
QAODV	70,000	0.19125	

Figure 3a shows that the power of the nodes considered in routing leads to a more reliable route. Therefore, the throughput of QAODV is improved by 14.28% compared to AODV.

5.2 *Mean Delay*

Mean delay of a network is a measure of average time taken by the packets to transit across the network from source to destination. Mean delay is determined by calculating the sum of individual delay samples to the total number of samples considered. The mean delay experienced by the packets in AODV & QAODV is shown in Fig. 3b. Figure 3b shows a reduction in delay experienced by packets in QAODV by 27.83% compared with AODV.

5.3 *Packet Loss*

Packet loss is the discarding of data or control packets in a network when a node is overloaded and cannot accept additional packets at a given moment. It is also measured in *packets per second*. The packet loss analysis of the AODV & QAODV algorithms shown in Fig. 3c depicts that the loss of packets in QAODV is reduced by 11.6% compared with AODV.

The simulation results are summarized in the following Table 5. The results of simulation proved that proposed QAODV provides throughput improvement of 14.28% reduction in delay, and packet loss of about 27.83% and 11.6%, respectively, than AODV.

6 Conclusion

In this work, the proposed QAODV protocol is an energy-based algorithm. It finds out the nodes that have almost exhausted their energy level during the transmission of data. A node with adequate energy for the transmission is chosen for providing better routing. The performance of QAODV has been analysed using QoS parameters such as throughput, delay and packet loss. It is inferred from the simulation results that throughput is improved by 14.28%, the delay is reduced by 27.83%, and the packet loss is reduced by 11.6% compared to AODV. The QAODV algorithm may be extended for reliable transmission of packets by incorporating virtual energy scheme. Our future work will focus on implementing enhanced QAODV algorithm for the mobile nodes.

References

1. Salhieh A et al (2001) "Power efficient topologies for wireless sensor networks." International conference on parallel processing, IEEE
2. Tajik S, Ghazal F, Sadan Z (2010) "Performance of modified AODV (waiting AODV) protocol in mobile ad-hoc networks." 2010 second international conference on ubiquitous and future networks (ICUFN). IEEE
3. Peterson LL, Davie BS (2007) Computer networks: a systems approach. Elsevier
4. Akyildiz IF et al (2002) Wireless sensor networks: a survey. Comput Netw 38(4):393–422
5. Hamrioui S, Lorenz P (2016) "EQ-AODV: energy and qos supported aodv for better performance in wmsns." 2016 IEEE international conference on communications (ICC). IEEE
6. Lee S-K, Koh J-G, Jung C-R (2014) An energy-efficient QoS-aware routing algorithm for wireless multimedia sensor networks. Int J Multi Ubiquitous Eng 9(2):245–252
7. Singh, Meeta, and Jigyasa Sharma. "Performance analysis of secure & efficient AODV (SE-AODV) with AODV routing protocol using NS2." Proceedings of 3rd International Conference on Reliability, Infocom Technologies and Optimization. IEEE, 2014.
8. Maurya PK et al (2012) An overview of AODV routing protocol. Int J Mod Eng Res 2(3):728–732
9. Ismail, Zahian, and Rosilah Hassan. "Performance of AODV routing protocol in mobile ad hoc network." 2010 International Symposium on Information Technology. Vol. 1. IEEE, 2010.
10. Singh SK, Singh MP, Singh DK (2010) Routing protocols in wireless sensor networks-A survey. Int J Comput Sci Eng Survey 1(2):63–83
11. He, Tian, et al. "SPEED: A stateless protocol for real-time communication in sensor networks." 23rd International Conference on Distributed Computing Systems, 2003. Proceedings.. IEEE, 2003.
12. Akkaya, Kemal, and Mohamed Younis. "An energy-aware QoS routing protocol for wireless sensor networks." 23rd International Conference on Distributed Computing Systems Workshops, 2003. Proceedings.. IEEE, 2003.
13. Karthick, S., E. Sree Devi, and R. V. Nagarajan. "Trust-distrust protocol for the secure routing in wireless sensor networks." 2017 International Conference on Algorithms, Methodology, Models and Applications in Emerging Technologies (ICAMMAET). IEEE, 2017.
14. Ahmed, Adnan, et al. "A secure and QoS aware routing protocol for Wireless Sensor Network." 2016 11th International Conference for Internet Technology and Secured Transactions (ICITST). IEEE, 2016.
15. Bhadauria HS, Singh A (2013) Performance Analysis of Ad hoc On-Demand Distance Vector Protocol for MANET. International Journal of Advanced Research in Computer Science and Software Engineering 3(3):1–6
16. Hussein, Wael Ali, et al. "Design and performance analysis of high reliability-optimal routing protocol for mobile wireless multimedia sensor networks." 2017 IEEE 13th Malaysia International Conference on Communications (MICC). IEEE, 2017.
17. Logambigai, R., and A. Kannan. "QEER: QoS aware energy efficient routing protocol for wireless sensor networks." 2014 Sixth international conference on advanced computing (ICoAC). IEEE, 2014.

18. Soni, Honey, Priyanka Tripathi, and Robin Singh Bhadoria. "An Investigation on Energy Efficient Routing Protocol for Wireless Sensor Network." 2013 5th International Conference and Computational Intelligence and Communication Networks. IEEE, 2013.
19. Barati, Mehdi, et al. "Performance evaluation of energy consumption for AODV and DSR routing protocols in MANET." 2012 International Conference on Computer & Information Science (ICCIS). Vol. 2. IEEE, 2012.
20. Xu, Zhen, and Gao Lu. "Energy-efficient and QoS-aware routing protocol for wireless sensor networks." ISWPC 2012 proceedings. IEEE, 2012.
21. Syarif, Abdusy, et al. "Adding gateway mode for r-aodv routing protocol in hybrid ad hoc network." TENCON 2011–2011 IEEE Region 10 Conference. IEEE, 2011.

Remote Sensing Image Fusion via Hybrid Image Decomposition with Spatial Frequency Motivated PCNN

K. Vanitha and K. Vijay Kumar

Abstract Pan sharpening is pivotal for getting a composite image which consists of the information related to spatial and spectral. In this paper, pan-sharpening method based on co-occurrence filtering with spatial frequency motivated PCNN is proposed. The three parts of panchromatic (PAN) image, i.e., small- and large-scale images, a base image have been obtained through hybrid of co-occurrence and Gaussian filtering (CoF-GF) decomposition. Next, intensity, saturation and hue components of multispectral (MS) image have been obtained through HSI transform. Thirdly, PCNN modulated with spatial frequency has been used to merge the base images and MS image intensity component. Finally, fused output has been reconstructed by an inverse HSI applied on addition of small-scale, large-scale and fused base image. Experiments in three datasets have been validated that the proposed outperforms most of the recently suggested methods.

Keywords Remote sensing image fusion · Co-occurrence filtering · Spatial frequency · PCNN

1 Introduction

Multispectral (MS) images are widely used for remote sensing field consisting of majorly spectral information have been applied in the fields of object recognition, disaster monitoring, marine research etc. [1]. MS image with spatial resolution not sufficient, more spectral resolution looks blurry. Panchromatic (PAN) image has spectral resolution not sufficient, more spatial resolution. Remote sensing images from only one sensor always lacks from complete details. Preserving both spectral and spatial data from MS and PAN images is possible with a technique called pan

K. Vanitha (✉)
Department of Electronics & Communication Engineering, Rayalaseema University College of Engineering, Kurnool, Andhra Pradesh, India

K. Vijay Kumar
Department of Electronics & Communication Engineering, RCEW, Kurnool, Andhra Pradesh, India

A. Sivasubramanian et al. (eds.), *Futuristic Communication and Network Technologies*, Lecture Notes in Electrical Engineering 792,
https://doi.org/10.1007/978-981-16-4625-6_13

sharpening [2, 3]. Mostly familiar pan sharpening with decomposition in multi-scale approaches (MSD) schemes such as wavelet, curvelet, NSCT and NSST [4]. The spatial resolution of fused image obtained by MSD is lower, but spectral data is good. In the recent, edge-preserving (EPF) filters such as Guided based, bilateral based, co-occurrence based have been good improvement in this field. The hybrid of EPF–Gaussian filtering include Rolling guidance Gaussian filtering decomposition (CGGFD), Bilateral Gaussian filtering decomposition (BGFD) and Curvature Gaussian filtering decomposition (CGFD) are also showing good results. Because of these EPF advantages, hybrid of co-occurrence filtering (CoF) and Gaussian filtering (CoF-GF) has been used, which preserves local region sharp details and smoothes all fine textures [10]. As we know that, PCNN easily provide gradient and edges information, but alone cannot give all details. However, all these results in fused image have blurredness and local block effect due to inappropriate weighted coefficients [10]. To overcome all these unwanted effects, an effective remote sensing image fusion via CoF-GF and PCNN modulated by spatial frequency has been proposed in this research article.

2 Preliminaries

In this section, the proposed PS algorithm requiring CoF-GF decomposition process has been introduced. The spatial frequency and PCNN model descriptions are available in [4, 10].

Co-occurrence filtering (CoF-GF) based decomposition

CoF-GF image decomposition model based on CoF and GF has three steps. Firstly, an input I is applied with the CoF and GF, to produce filtered image I_c and base image (BI), denoted as I_g. Secondly, the difference of I and I_C and difference of I and I_g gives small-scale image (SSI), large-scale image (LSI) [9].

$$\mathrm{BI} = \mathrm{Ig} = \mathrm{GF}(I, \sigma, \rho) \tag{1}$$

$$\mathrm{Ic} = \mathrm{CoF}(I, p) \tag{2}$$

$$\mathrm{SSI} = I - \mathrm{Ic} \tag{3}$$

$$\mathrm{LSI} = I - \mathrm{Ig} \tag{4}$$

where ρ = radium size of GF = 4.
δ = variance of the GF = 10.
p = filter window size = 10.
GF(·) = Gaussian filtering operation.

CoF(.) = co-occurrence filtering operation.

3 Proposed Method

The proposed pan-sharpening block diagram and algorithm are given below: (Fig. 1)

(1) Read the MS and PAN images for pan-sharpening process, which have been aligned, with size 256 × 256.
(2) At first, the three parts of panchromatic (PAN) image, i.e., small- and large-scale images (SSI, LSI), a base image (BI) has been obtained through co-occurrence filtering (CoF-GF) decomposition.
(3) Next, intensity (IN), saturation (S) and hue (H) components of multispectral (MS) image have been obtained through HSI transform.
(4) Thirdly, PCNN modulated with spatial frequency has been used to merge the base image (BI) and intensity (IN) component to give fused base image (FB).

$$\text{FB} = \text{SF - PCNN}(\text{BI}, \text{IN}) \tag{5}$$

(5) Finally, fused output has been reconstructed by an inverse HSI applied on addition of small scale, large scale and fused base image.

$$\text{Final Fused} = \text{inverse HSI}(\text{SSI} + \text{LSI} + \text{FB}) \tag{6}$$

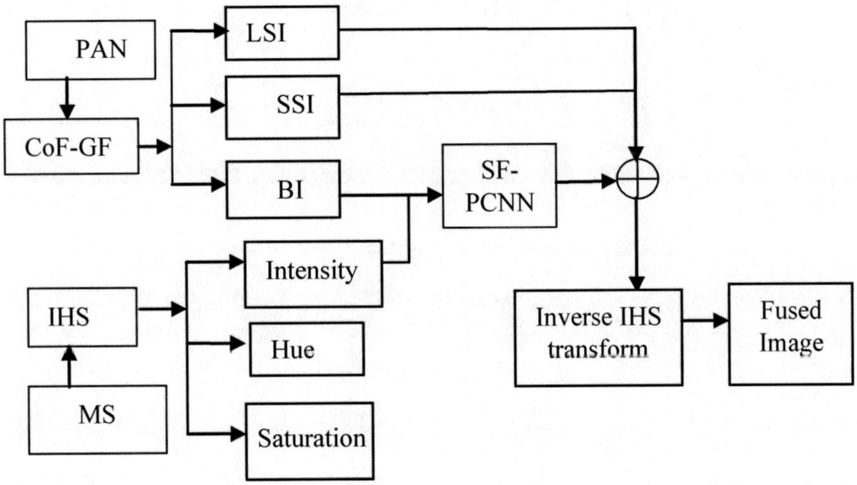

Fig. 1 Proposed pan-sharpening block diagram

4 Experimental Results

The experiments have been conducted on three sets of MS and PAN images as seen in Fig. 2. Methods like Brovey [5], HSI [6], PCA [7], Guided [8] and P + XS [9] have been taken for analysis purpose and fusion results are seen in Fig. 3a–f, 4 and 5a–f.

Subjective evaluation of results based on visual effects alone is not correct. Moreover, above visual effects of the guide filtering and P + XS methods are looking similar to our method. Therefore, it still needs enough quantitative evaluation to demonstrate the proposed method is the best. The objective assessment is carried out by estimating metrics like deviation (STD), quality indices (UIQI), correlation coefficient (CC), ERGAS and SAM. With respect to reference image and ERGAS tells us how much spectral distortion is contained by the fused image, UIQI evaluates distortion, contrast and correlation, spectral angle mapper (SAM) computes the spectral angle between the pixel, vector. CC computes the spectral features similarity. The method with minimum ERGAS value and maximum STD, UIQI, CC and SAM is best one. These values are tabulated, bolded and for further equations, see reference [11].

Figure 3 shows the fusion outputs of comparative methods, Brovey output has high contrast. HSI output is not clear as it lost some spectral data. PCA output has

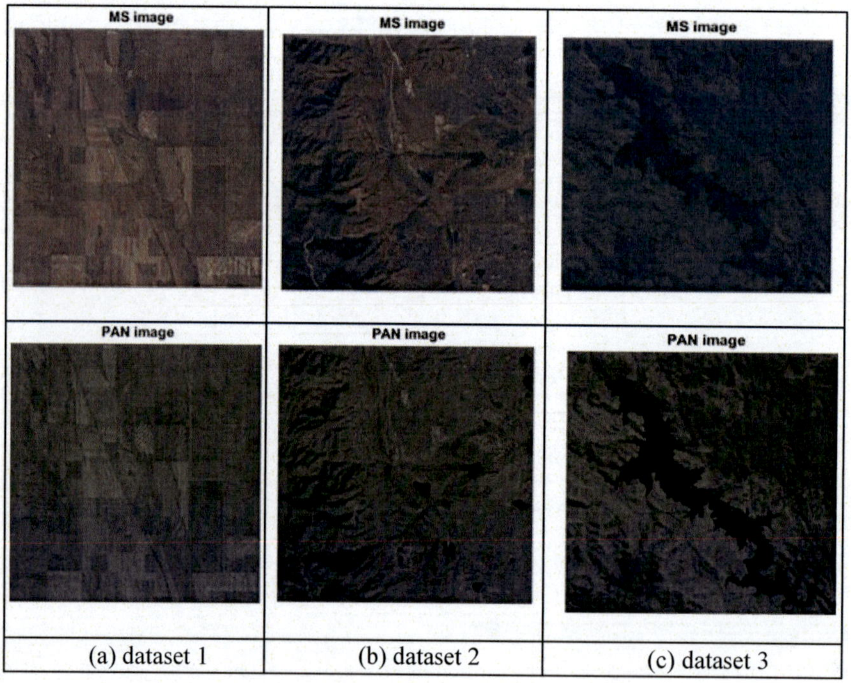

Fig. 2 Three sets of MS and PAN images

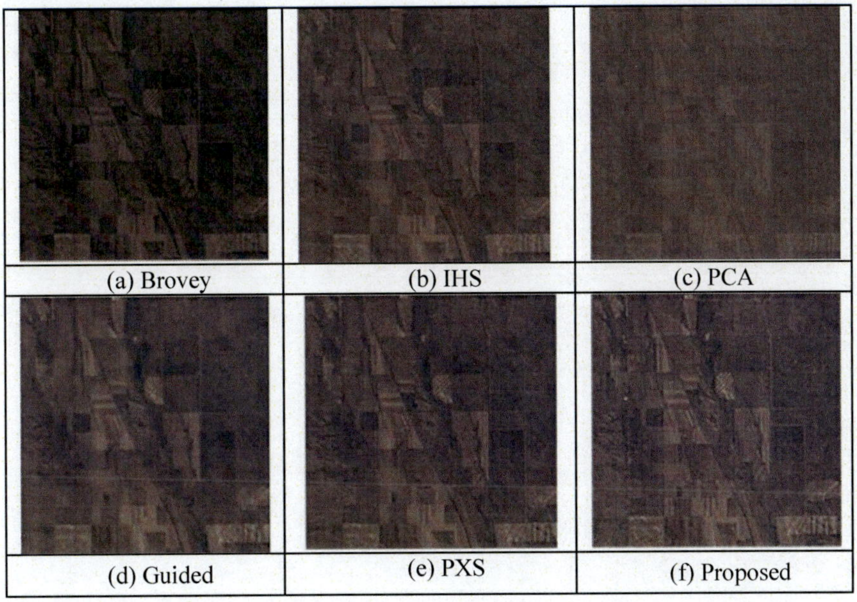

Fig. 3 Fusion results of proposed and various pan-sharpening methods

blurred spectral data compared to first two methods, so its visual performance is poor. Guide filtering output looks darker with reference to MS image and edges are not preserved. P + XS output is better than guided and PCA but still lacks of spectral details. As our method uses co-occurrence filtering, produces a good visual quality image with high spectral information (Table 1).

Our method has STD: 15.119, UIQI: 0.951, CC: 0.932, ERGAS: 1.739 and SAM: 0.071. Out of these, first three are highest, so they are bolded. This indicates that output image has strong correlation, spectral details with respect to MS and PAN images.

Figure 4 shows the fusion outputs of comparative methods, Brovey output has high spectral distortion among all methods. HSI output is somewhat clear than Brovey, but still, edges are not sharpened. PCA output is blurred compared to first two methods, so its visual performance is poor. Guide filtering output is completely blurred as it

Table 1 Quantitative evaluation of various pan-sharpening methods

Metrics	[5]	[6]	[7]	[8]	[9]	Proposed
STD	0.042	0.051	0.036	0.053	0.053	**15.119**
UIQI	0.838	0.925	0.820	0.943	0.951	**0.951**
CC	0.865	0.851	0.898	0.917	0.922	**0.932**
ERGAS	7.565	1.694	2.195	1.460	**1.395**	1.739
SAM	0.288	0.375	**1.189**	1.18	1.073	0.071

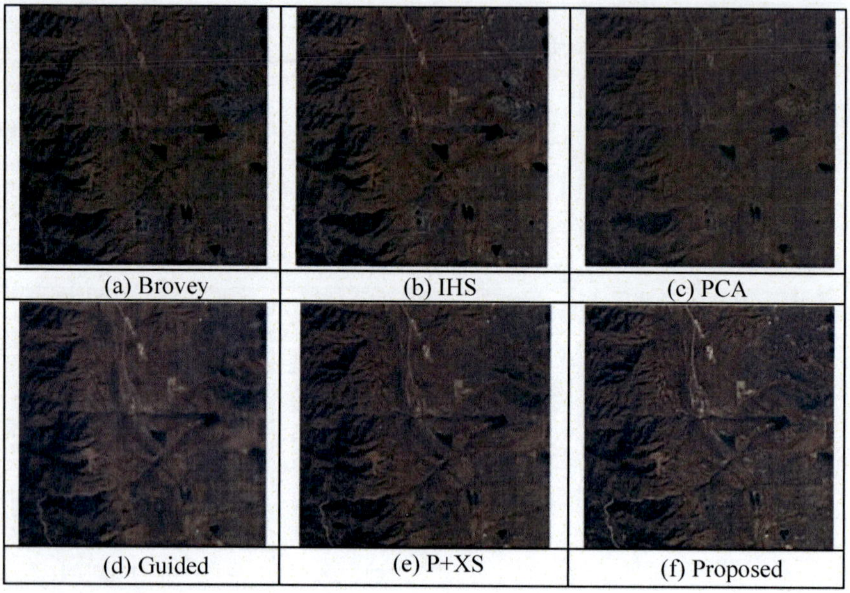

Fig. 4 Fusion results of proposed and various pan-sharpening methods

loses almost all details of MS and PAN images. P + XS output is better than guided and PCA but still looks somewhat darker. Our method output gives all spectral details, looks good by smoothened details and preserved edges (Table 2).

Our method has STD: 15.119, UIQI: 0.951, CC: 0.932, ERGAS: 1.739 and SAM: 0.071. Out of these, STD, UIQI and CC are the highest which reveals that visual performance and objective assessment of proposed method are good than others. ERGAS value of our method is observed as second minimum among other pan-sharpening methods.

Figure 5 shows the fusion outputs of comparative methods; Brovey output has high spectral distortion among all methods. HSI output is somewhat clear than Brovey, but still, edges are not sharpened. The output of PCA and guided methods is completely blurred as it loses almost all details while smoothening. P + XS output is better than

Table 2 Quantitative evaluation of various pan-sharpening methods

Metrics	[5]	[6]	[7]	[8]	[9]	Proposed
STD	0.0467	0.054	0.0347	0.059	0.0573	**16.8801**
UIQI	0.7202	0.769	0.6495	0.8951	0.91	**0.9354**
CC	0.6986	0.697	0.7393	0.8852	0.8957	**0.9304**
ERGAS	6.7721	4.006	4.2284	2.8012	**2.5759**	2.773
SAM	0.4355	0.854	**2.395**	1.99	1.6673	0.1064

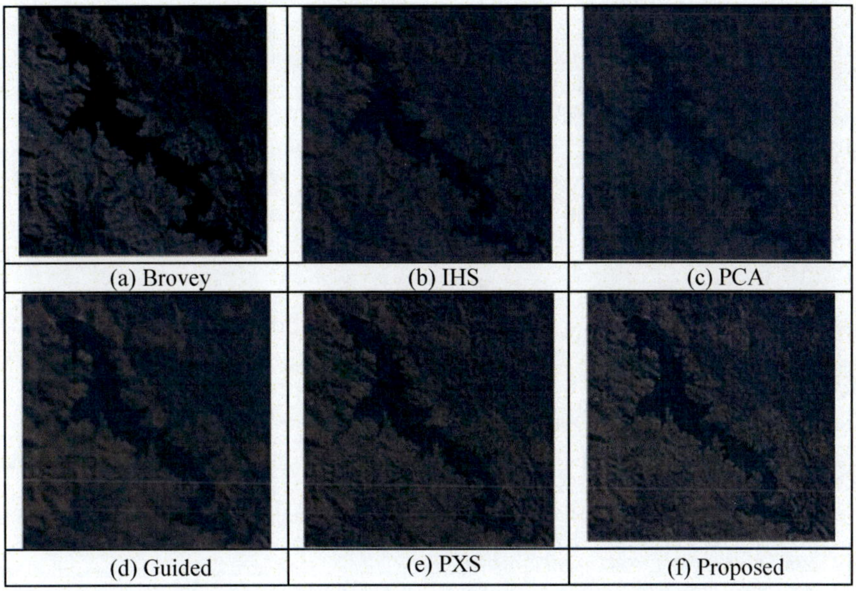

Fig. 5 Fusion results of proposed and various pan-sharpening methods

Table 3 Quantitative evaluation of various pan-sharpening methods

Metrics	[5]	[6]	[7]	[8]	[9]	Proposed
STD	0.0652	0.045	0.0371	0.0461	0.048	**13.6839**
UIQI	0.7346	0.924	0.8851	0.9573	0.9318	**0.9257**
CC	0.8093	0.786	0.8376	0.9059	0.8552	**0.8819**
ERGAS	6.2243	2.329	2.8177	**1.866**	2.3378	2.8964
SAM	0.4283	0.758	**2.0905**	1.8776	1.6544	0.1233

guided and PCA, but still edges are not retained. Our method output gives output with great spectal and spatial details without artifacts, distortions (Table 3).

Our method has STD: 13.6839, UIQI: 0.9257, CC: 0.8819, ERGAS: 2.8964 and SAM: 0.1233. ERGAS of guide filtering is high minimum, and SAM of PCA is high. However, the proposed gives high robustness with reference to all other pan-sharpening methods.

5 Conclusion

A novel pan-sharpening method based on CoF-GF decomposition with spatial frequency motivated PCNN is proposed. At first, the three parts of panchromatic

(PAN) image, i.e., small- and large-scale images, a base image have been obtained through decomposition using co-occurrence with Gaussian filtering (CoF-GF). Next, intensity, saturation and hue components of multispectral (MS) image have been obtained through HSI transform. Thirdly, PCNN modulated with spatial frequency has been used to merge the base images and MS image intensity component. Finally, fused output has been reconstructed by an inverse HIS applied on addition of small-scale, large-scale and fused base image. Experiments in three datasets have been validated that the proposed outperforms most of the recently pan-sharpening methods.

References

1. Ghassemian H (2016) A review of remote sensing image fusion methods. Inf Fusion 32:75–89
2. Metwalli MR, Nasr AH, Faragallah OS, El-Rabaie ESM, Abbas AM, Alshebeili SA, El-Samie FEA (2014) Efficient pansharpening of satellite images with the contourlet transform. Int J Remote Sens 35(5):1979–2003
3. Zhou Z, Wang B, Li S, Dong M (2016) Perceptual fusion of infrared and visible images through a hybrid multi-scale decomposition with Gaussian and bilateral filters. Inf Fusion 30:15–26
4. Kong W, Zhang L, Lei Y (2014) Novel fusion method for visible light and infrared images based on NSST–SF–PCNN. Infr Phys Technol 65:103–112
5. Vrabel J (1996) Multispectral imagery band sharpening study. Photogramm Eng Remote Sens 62(9):1075–1084
6. Masoudi R, KabiriP (2014) New intensity-hue-saturation pan-sharpening method based on texture analysis and genetic algorithm-adaption. J Appl Remote Sens 8(1)
7. Jelének J, Kopačková V, Koucká L, Mišurec J (2016) Testing a modified PCA-based sharpening approach for image fusion. Remote Sens 8(10):794
8. Li Q, Yang X, Wu W, Liu K, Jeon G (2018) Pansharpening multispectral remote-sensing images with guided filter for monitoring impact of human behavior on environment. Concurrency Comput Pract Exper p 5074
9. Ballester C, Caselles V, Igual L, Verdera J, Rougé B (2006) A variational model for P+XS image fusion. Int J Comput Vis 69(1):43–58
10. Tan W, Xiang P, Zhang J, Zhou H, Qin H (2020) Remote sensing image fusion via boundary measured dual-channel pcnn in multi-scale morphological gradient domain. IEEE Access 8:42540–42549
11. Jagalingam P, Hegde AV (2015) A review of quality metrics for fused image. Aquatic Procedia 4(Icwrcoe):133–142

Parametric Optimization and FWM Mitigations in 64-Channel DWDM System

K. Venkatesan, A. Chandrasekar, and P. G. V. Ramesh

Abstract Growth made over the decades in dense wavelength division multiplexing (DWDM) systems has been providing an opportunity to enhance channel capacity, wider bandwidth, and higher data rates in high-speed optical communication. This causes nonlinear effects, which include FWM leading to increase signal distortion and more cross talk, thereby degrading the performance of the DWDM system. The quality of data transmission can be improved by investigating optimal parameters of DWDM system. This paper proposes an influential parametric analysis to suppress FWM in DWDM system. Here simulations are performed to analyze and understand the influence of system parameters such as data rate, input power, modulation format, and channel spacing. The performance of the proposed 64-channel DWDM is evaluated by analyzing FWM mitigating factors such as Q-factor, minimum BER, eye-opening, and OSNR. The proposed system provides better performance under low input power (-10dBm), narrow channel spacing (50 GHz), and higher data rate (10Gbps/channel). Furthermore, this paper also presents the real trade-offs and estimates of bit rate, OSNR, signal power, and duty cycle in the presence of amplified spontaneous emission (ASE) noise to achieve min. BER and high Q-factor, which is required in a practical implementation of a DWDM transmission model.

Keywords DWDM · FWM · Q-factor

1 Introduction

There is an enormous increase in data rates being used nowadays, due to the prevailing pandemic situation across the globe due to COVID-19. This has forced many people to choose work from home, which drastically increased the utilization of Internet and other multimedia applications. These applications require wider bandwidths, higher data rates with more number of WDM channels for long-distance and ultra-speed data transmissions. DWDM and UDWDM techniques were developed to meet the

K. Venkatesan (✉) · A. Chandrasekar · P. G. V. Ramesh
Department of Electronics and Communication Engineering, St. Joseph's College of Engineering, Chennai, India

A. Sivasubramanian et al. (eds.), *Futuristic Communication and Network Technologies*, Lecture Notes in Electrical Engineering 792,
https://doi.org/10.1007/978-981-16-4625-6_14

demands of increasing number of users and to make the system more flexible [1]. Nonlinear impairments such as four-wave mixing (FWM), self- and cross-channel interferences, which are very hard to solve and also cause disturbances in ultra-speed DWDM data transmission should be eliminated [2]. In DWDM system, high data rate, lower channel spacing due to the increased number of channels also introduce pulse broadening with fiber nonlinearities, especially FWM which degrades system performance [3]. FWM in DWDM system causes more intersymbol interferences (ISI), which may affect transmission efficiency, increase network latency, and system throughput [4].

In an optical DWDM system, compared to any other nonlinearities FWM which is a predominant effect that totally disturbs the system's efficiency and hence suppressing FWM is a major objective [4] under consideration during the design of DWDN systems. Several techniques are proposed to reduce the FWM deteriorations in DWDM systems [5]. Dispersion compensation technique is an influential practice for improving the quality of signal transmission in a superchannel DWDM transmission system [6]. In this technique, single-mode fiber (SMF) is employed with dispersion compensation fiber (DCF), which has negative dispersion characteristics and tends to neutralize the positive dispersion, which occurred in SMF [7, 8]. Individual and combined effects of high dispersion due to FWM are observed using hybrid WDM/ TDM techniques [9]. Unequal channel spacing is a prominent method to overcome FWM but it fails due to its inefficient utilization of bandwidth and complex design structure [10]. Furthermore, many contributions have proved that polarization techniques and channel spacing methods reduce FWM to some extent [11–13].

From the literature survey, it is concluded that FWM is an important issue that causes high signal distortion and cross talk, thereby degrading the DWDM system's performance. Furthermore, in a DWDM system FWM characteristics are closely linked with the system parameters such as input power, channel spacing, data rate, and modulation format. At high data rates, FWM crosstalk influences DWDM system and exhibits dependency on modulation format. In this paper, an investigation is performed to analyze and understand the influence of various system parameters and different modulation formats to suppress FWM. In addition, a 64-channel DWDM system is proposed to meet the real trade-offs such as low input power, narrow channel spacing, and higher data rates. Thus, the proposed system when implemented over long distances would achieve a higher quality of data transmission, through optimizing influential parameters of FWM.

The rest of paper is organized as follows, Sect. 2 presents the proposed DWDM setup. Results and discussions are presented in Sect. 3. Section 4 presents overall conclusion and future scope of this research work.

2 Proposed DWDM Setup

Figure 1 shows the proposed DWDM system setup. In a long-haul DWDM data transmission, the transmitted signal requires high input power to travel longer distance and reach the receiver end. However, high power transmitter signal shows more FWM influence in DWDM system. In this proposed DWDM system setup, the input power is considered as -10 dBm. The choice of low input power forces the FWM signal to lose its crosstalk power generation, which additionally lowers FWM influence in DWDM system setup.

The proposed DWDM optical network configuration is designed at variable data rates of upto 15Gbps and variable channel spacing upto 100 GHz. DWDM transmitter, which has PRPS data source and electric pulse generator are combined with Mach–Zehnder modulator, which performs basic modulation technique using NRZ and RZ modulation format. The modulated signal from different DWDM transmitters is passed on to the DWDM multiplexer, which transmits the modulated signal over a nonlinear optical medium through single-mode optical fiber (SMOF) whose dispersion characteristic is 16.75 ps/nm/km and has 0.2 dB/km as attenuation coefficient. Optical signal from SMOF enters into the optical amplifier, which performs amplification and introduces amplified spontaneous emission noise. Furthermore, the amplified signal is transmitted into the dispersion compensating fiber (DCF), which has negative dispersion characteristics with attenuation and dispersion parameters of SMOF.

At the receiver end, a DWDM demultiplexer separates the received signals and transmits them into the individual channels with assigned frequency. The signal from demultiplexer is passed through a PIN photodetector, which performs optical to electrical conversion and then passed into a low pass (Bessel) filter. From BER analyzer, output spectral characteristics are observed as eye diagram characteristics. The mitigating factors such as Q-factor, minimum BER and eye height are measured, and then the performance of DWDM system is evaluated. Typically, the required Min.BER for optical DWDM system is around 10^{-9} or lower. As Q-factor increases

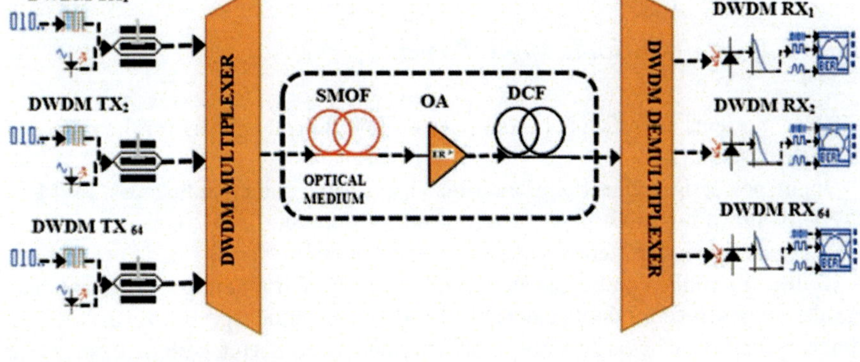

Fig. 1 Proposed DWDM setup

will lead to improve Min.BER and reaches 10^{-10} for the Q-factor value above 6 [14]. When BER above 10^{-10} will lead to the output signal infinitesimal.

Experiments are performed for various FWM influential parameters such as data rate, channel spacing, input power, and modulation format. Results are observed from WDM analyzer, optical spectrum analyzer, and BER analyzer. From the observed results, FWM mitigating factors such as OSNR, signal power, noise power, eye height, Q-factor, and BER are analyzed to choose optimized levels of parameters in order to suppress FWM. In results and discussion section, we discuss and evaluate the performance of DWDM system at 640Gbps data rate. This proposed DWDM system is designed with several FWM influencing parameters, and investigations are processed to suppress the effects of FWM nonlinearity and thus improve the quality of data transmission. Furthermore, a comparative study of various influencing parameters along with different modulation formats and variable duty cycle is also presented.

3 Results and Discussion

DWDM system achieves high channel capacity for high-speed optical communication. Erstwhile it also introduces linear and nonlinear effects, which degrades the system performance.

In this section, simulation for a DWDM system with various parameters is developed to analyze and understand the characteristics of FWM. Intensive study is carried out to identify and characterize the relationship of FWM issues with multiple factors. Influence of FWM is clearly understood by analyzing other factors such as maximum Q-factor, minimum BER, output signal power, noise power, and OSNR. It provides their correlation and shows how the factors such as output signal power, noise power, OSNR, Q-factor, eye diagrams characteristics are related to the arbitrary parameters such as input power, channel spacing, bit rate, and modulation format with duty cycle.

3.1 Impact of Variable Input Power

Figure 2 presents the effects of input power on the various factors for the proposed DWDM system.

Input power is a linearly proportional parameter, which influences FWM in DWDM. Lowering input power leads to decrease output signal power and hence OSNR gets decreased. Iterative analysis is carried out by varying input power from −10 dBm to 10dBm and considers all the other system parameters constant. The results are observed for 64-channels of DWDM on an output spectrum analyzer. Eye diagram characteristics are observed for factors such as Q-factor, BER, and eye height that lead to understand the significant changes in their dispersion. These are listed in

Influence of input power in DWDM system		
-10dBm	**0dBm**	**10dBm**
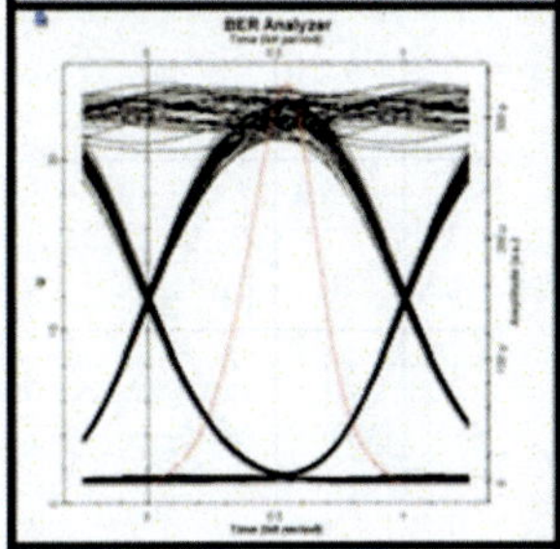	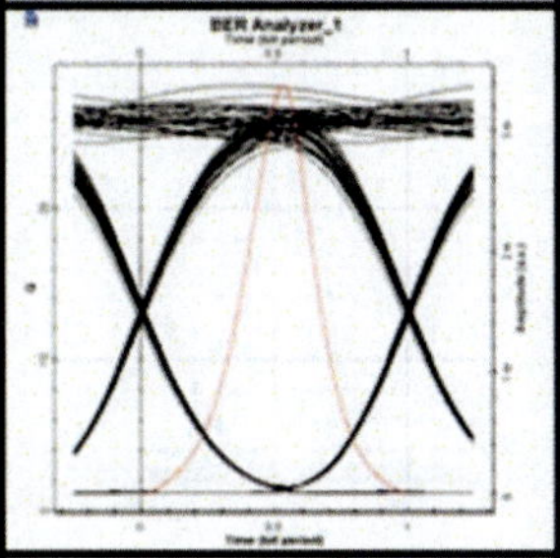	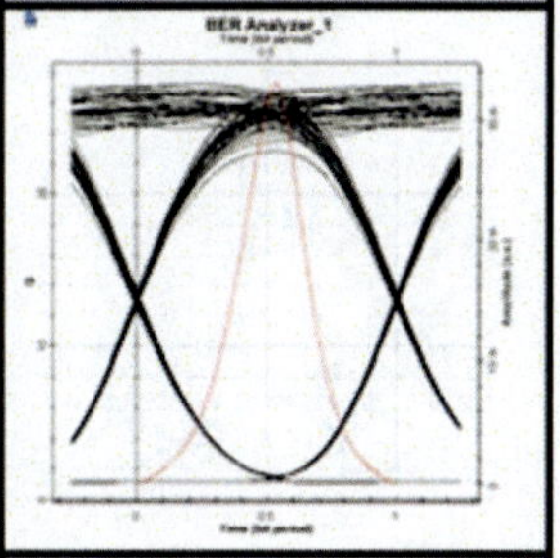
Max. Q Factor 24.4785 Min. BER 9.23027e-133 Eye Height0.000262887 Threshold 4.32252e-005 Decision Inst. 0.539063	Max. Q Factor 28.2577 Min. BER 4.25315e-176 Eye Height 0.00269426 Threshold 0.00044442 Decision Inst. 0.53125	Max. Q Factor 27.2973 Min. BER 1.68741e-164 Eye Height 0.0268266 Threshold 0.00434784 Decision Inst. 0.533203

Fig. 2 Impact of input power in DWDM system

Fig. 2. From this analysis, it can be concluded that lowering input power decreases FWM and it also happens to be an observation of Garg [15] too, when input power goes low, it results in high BER, an improper eye-opening and lowered Q-factor. By investigating the eye diagram characteristics, it shows that a linear relationship exists between the input power in a system and output signal power, noise power, and eye height. Relationship is inversely proportional for factors such as Q-factor and OSNR this is due to the established fact mentioned in Ahmed [16].

3.2 Impact of Variable Channel Spacing

Figure 3 represents the effects of channel spacing on the various factors understudy for the proposed DWDM system.

In this setup, channel spacing is varied from 25 to 100 GHz and iterative analysis is performed for the proposed 64-channel DWDM system. From the simulation results, it is observed that channel spacing provides inverse proportionality relations with the FWM parameters. Increasing the channel spacing has low FWM effect and produces Min.BER and maximum Q-factor [17]. Higher number of channels produces higher interference, resulting in higher FWM issues and increased BER. Narrow channel spacing in DWDM systems results in increased noise power, decreased Q-factor and overly affects OSNR. High FWM effects are observed with narrow channel spacing and hence leading to degradation of the DWDM system. Granada Torres [18] observes that lower channel spacing with high input power can accommodate more number of channels but at the cost of decreased eye height, which is observed here too, through

Influence of Channel Spacing in DWDM system		
25GHz	**50GHz**	**100GHz**
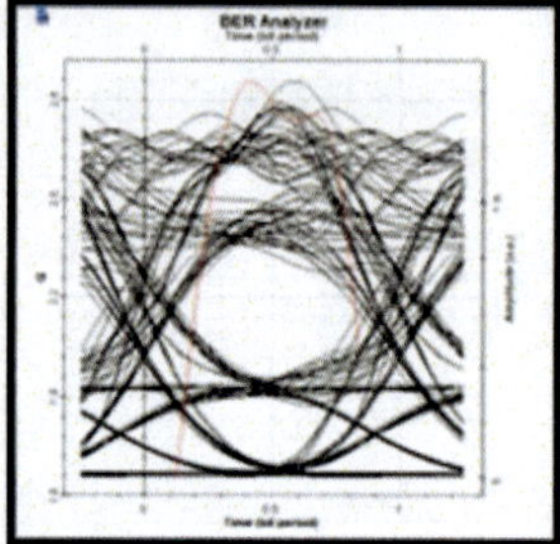	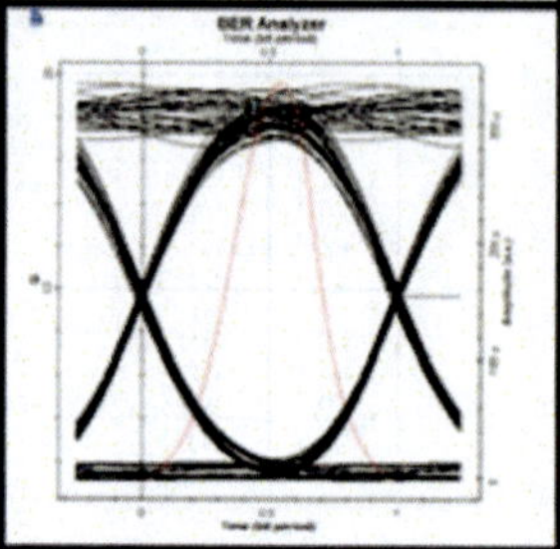	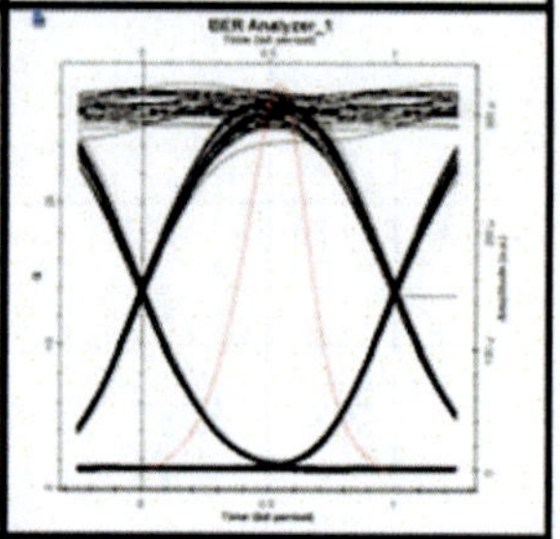
Max. Q Factor 2.85634 Min. BER 0.0021418 Eye Height -4.40669e-005 Threshold 0.000616898 Decision Inst. 0.414063	Max. Q Factor 19.595 Min. BER 7.94123e-086 Eye Height 0.000254206 Threshold 0.000103342 Decision Inst. 0.541016	Max. Q Factor 28.3641 Min. BER 2.19624e-177 Eye Height 0.00026881 Threshold 5.52912e-005 Decision Inst. 0.537109

Fig. 3 Impact of channel spacing in DWDM system

various iterations. From eye characteristics, it is concluded that increasing channel spacing has more correlations to factors such as Q-factor, output signal power, and eye spectral characteristics that show better changes in FWM. Increasing channel spacing will reduce noise power and gives improvement over OSNR.

3.3 *Impact of Duty Cycle in RZ Modulation Format*

Figure 4 represents the effects of RZ modulation format with duty cycle on the various factors for 64-channel DWDM system. In this section, the proposed DWDM system is designed with RZ format of modulation with duty cycle varying from 0.25 to 0.5. The iterative analysis shows linear relationship for FWM mitigating factors such as output signal power, BER, Q-factor, and eye height characteristics. Noise power shows low impact over RZ format with duty cycle variations. RZ modulation format with 0.4% duty cycle format gives Min.BER, and Max Q-factor of 40.7035. Spectral characteristics present an eye-opening with remarkable changes that make BER minimum and lead to high-quality data transmission, which achieves maximum Q-factor and high eye height characteristics. This comparative study shows system design with RZ modulation with low duty cycle results in narrow pulse and performs better with the reduction of FWM. It is also noticed that increasing duty cycle will lead to higher OSNR that will affect the Q-factor of the DWDM system introducing pulse broadening, thus exhibiting dispersion phenomenon which is also an observation from Secondini [19].

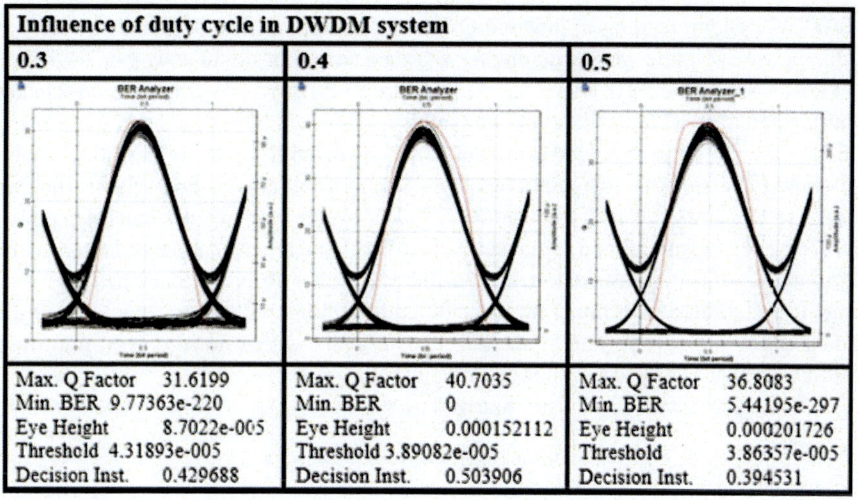

Fig. 4 Impact of duty cycle in DWDM system

3.4 *Impact of Varying Bit Rate*

Figure 5 presents the effects of bit rate on the various factors for 64-channel DWDM system.

In this section, data rate is varied from 5 to 15Gbps. Higher data rates induce more dispersion in DWDM system that affects BER. FWM mitigating factors such as noise power and OSNR are increased due to the presence of higher dispersion in

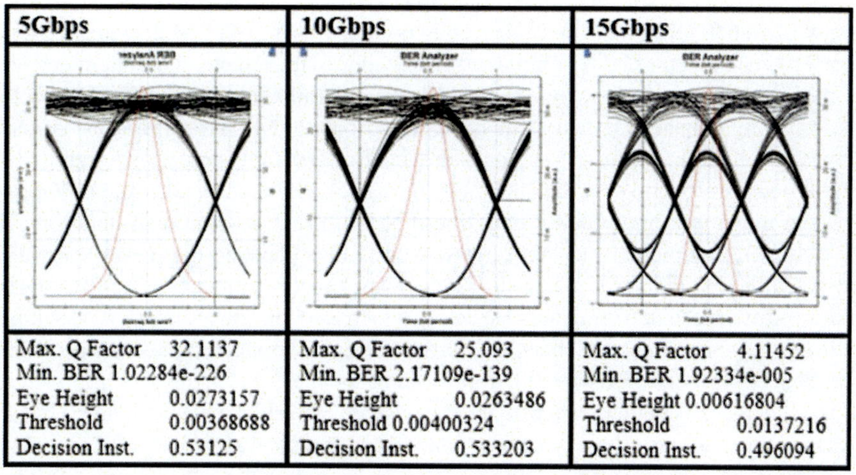

Fig. 5 Impact of bit rate in DWDM system

DWDM system, leading to lowered Q-factor. Simulation results represent relationship between bit rate and mitigating factors such as output signal power, noise power, OSNR, Q-factor, and eye characteristic. The proposed DWDM system provides better performance at 10 Gbps data rate and achieves appreciable Q-factor of 25.093 and BER of 2.17109 e^{-139}. However, the same setup with 15 Gbps data rate shows distorted eye spectral characteristics for -10dBm input power. From this result, we observed the importance of input power, DWDM system with insufficient input power will degrade the modulated signal and fails to reach the receiver and thus lowering the throughput of data transmission. Furthermore, it is also observed that when high data rate is transmitted with more number of channels along with good channel spacing, (100 GHz) and appropriate input power results in lowered noise power, improved OSNR and also achieves proper eye-opening,

The results obtain can be summarized as,

- FWM diminishes with decreasing input power.
- Increasing the channel spacing reduces the effects of FWM.
- RZ modulation with low duty cycle lowers pulse broadening and hence reduces the influence of FWM.
- Increasing data rate with low input power and narrow channel spacing introduces more FWM nonlinearity in the proposed DWDM system.

4 Conclusion

In this paper, a detailed investigation is performed to suppress FWM effects such that it improves the quality of data transmission for DWDM system by analyzing the influential parameters such as bit rate, modulation format, transmission input power levels, and channel spacing. The work emphasized on impacts of FWM in a DWDM network at 640 Gbps data rate. Fiber optic nonlinearity in DWDM degrades system performance and increases complexity in network management and system design. FWM should be minimized by investigating these parameters in order to achieve reliable and high-capacity data transmission. The proposed 64-channel DWDM system investigated these parameters and provides an effective implementation for allowing input power (-10dBm), narrow channel spacing (50 GHz), high data rate (10Gbps) system to achieve high-quality data transmission. From conventional iterations, it is observed that increase in input power results in broadened the power spectrum along with rise of FWM, which degrades quality factor. Effect of FWM becomes noteworthy, due to narrow channel spacing, which also introduces interference along with increase in number of channels. Furthermore, RZ modulation scheme provides better performance in FWM for all transmission distances. From the obtained results, it is noted that FWM increases with increasing data rate and duty cycle. Thus, a linear relationship exists for these parameters and FWM and shows inverse proportionality with duty cycle. Future scope of this work can be extending the analysis with different combinations of fiber optic amplifiers with varying doping ion concentration and their arbitrary parameters to achieve more gain and reduced FWM issues.

References

1. Sabapathi T, Poovitha R (2019) "Mitigation of nonlinearities in fiber optic DWDM system". Optik 185.https://doi.org/10.1016/j.ijleo.2019.02.073
2. Manzoor HU, Manzoor T, Hussain A, Aly M (2020) FWM mitigation in DWDM optical networks. J Phys: Conf Series 1447:012033. https://doi.org/10.1088/1742-6596/1447/1/012033
3. El-Deeb WS, Mohamed MT, Abdelnaiem AE (2016) Analysis of the nonlinear impairments on the DWDM optical communication systems. Egypt Int J Eng Sci Technol 22(January):19–26
4. Bhatia R, Sharma AK, Saxena J (2016) Improved analysis of four wave mixing with sub-plank higher-order dispersion parameters in optical communication systems. Optik (Stuttg) 127(20):9474–9478
5. Abed HJ, Din NM, Al-Mansoori MH, Fadhil HA, Abdullah F (2013) Recent four-wave mixing suppression methods. Optik (Stuttg) 124(15):2214–2218
6. Kaur G, Patterh MS (2014) "Suppression of four wave mixing in wavelength division multiplexed system with hybrid modules," Optik (Stuttg). Pp 3894–3896
7. Singh S, Kaler RS (2014) Novel optical flat-gain hybrid amplifier for dense wavelength division multiplexed system. IEEE Photonics Technol Lett 26(2):173–176
8. Agrawal GP (2001) Nonlinear fiber optics. Academic Press, New York
9. Agrawal GP (2001) Applications of nonlinear fiber optics. Academic Press, New York
10. Abd HJ et al (2014) Priority-based parameter optimization strategy for reducing the effects of four-wave mixing on WDM system. Optik 125:25
11. Abd H et al (2014) Four-wave Mixing crosstalk suppression based on the pairing combinations of differently linear-polarized optical signals. Sci World J 2014:1
12. Sandhu H, Kaler RS, Kaur G, Randhawa R (2018) "Transmission performance of 112 Gb/s POLMUX QPSK signal in OOK/DPSK WDM system". J Opt Commun 41. https://doi.org/10.1515/joc-2017-0158
13. Rani A, Singh M (2017) "Impact of different modulation data formats on DWDM system using SOA with narrow-channel spacing". J Opt Commun. https://doi.org/10.1515/joc-2017-0093
14. Kathpal N, Garg A (2019) "Analysis of radio over fiber system for mitigating four-wave mixing effect". Dig Commun Netw 6. https://doi.org/10.1016/j.dcan.2019.01.003
15. Neheeda, P. & Pradeep, M. & Shaija, P.J. "Analysis of WDM System with Dispersion Compensation Schemes". Procedia Computer Science. 93. 647–654. https://doi.org/10.1016/j.procs.2016.07.254. (2016)
16. Garg S, Kumar H (2015) Effect of fiber length on four wave mixing in WDM optical fiber systems. Commun. Appl. Electron. 3(5):50–54
17. Ahmed J, Hussain A, Siyal MY, Manzoor H, Masood A (2014) Parametric analysis of four wave mixing in DWDM systems. Optik (Stuttg) 125(7):1853–1859
18. Malhotra JS, Kumar M, Sharma AK (2013) Estimation and mitigation of FWM penalties in dispersion managed 32 channel long haul DWDM soliton link. Optik (Stuttg) 124(17):3029–3032
19. Granada Torres JJ et al (2017) Mitigation of time-varying distortions in Nyquist-WDM systems using machine learning. Opt Fiber Technol 38(September):130–135
20. Secondini M, Forestieri E (2014) On XPM mitigation in WDM fiber-optic systems. IEEE Photonics Technol Lett 26(22):2252–2255

Encryption and Decryption of Image for Secured Data Transmission

C. Thirumarai Selvi, R. S. Sankarasubramanian, and M. MuthuKrishnan

Abstract This work combines Arnold transform method with the spatial domain to encrypt a covered image and also recovers the covered image by the application of inverse Arnold transform. First, transform a cover image into subparts which consists of eight binary images by decimal value to eight-digit binary operation. Then, transform eight binary images into sub-blocks of eight binary scrambled images by the Arnold transform, respectively. Further, recombine the sequence of the eight binary scrambled matrices into a scrambled matrix with 256 Gy levels according to the specific club. Discrete wavelet transform (DWT) is used to perform image compression on the input image and secretly hidden image which is done using alpha blending. Finally, derive an encrypted image from the scrambled image by the Hartley transform. Second, decode the encrypted image using inverse Arnold Transform. Inverse DWT is performed to regain the compressed images. Simulations indicate that the hybrid new technique has more security, more robustness counter to occlusion and attacks about with high degree of image scrambling.

Keywords Arnold transform · Image encryption · DWT · Image decryption · Inverse Arnold transform

C. Thirumarai Selvi (✉)
Sri Krishna College of Engineering and Techology, Coimbatore, Tamil Nadu, India
e-mail: thirumaraiselvi@skcet.ac.in

R. S. Sankarasubramanian
PSG Institute of Technology and Applied Research, Coimbatore, Tamil Nadu, India
e-mail: rss@psgitech.ac.in

M. MuthuKrishnan
KIT-Kalaigarkarunanidhi Institute of Engineering and Technology, Coimbatore, Tamil Nadu, India

A. Sivasubramanian et al. (eds.), *Futuristic Communication and Network Technologies*, Lecture Notes in Electrical Engineering 792,
https://doi.org/10.1007/978-981-16-4625-6_15

1 Introduction

Image encryption for secured transmission applied with two stages. Encryption of image for the covered image has been applied in both time and frequency domain. In the existing methodology, intruders extract the covert image easily. Hence, it is necessary to protect the transmitted image from the intruders. The hybrid space-frequency domain enhances the security with its parallel operation. This hybrid methodology includes the benefit of both time and frequency domain. Image scrambling shows significant role, while encrypting the image in time domain. The applied improves the robustness nature of occlusion as well as noise attacks. Encrypting methods are classified divided into two groups. Among them, one of the class is matrix transformations named as Arnold transform. This method computes the inverse of the transformation matrix. It brings highly protected image. Then during the decoding stage, translation of the scrambled images uses the key. The second class is the cellular automata which operates with movement co-ordinates. The co-ordinating movements choose precise parameters on the scrambling route which decodes the image. Always the frequency domain encryption techniques provides high robustness, but it suffers due to complex nature in amplitude and phase. However, Hartley transformation is real for the applied image which also possesses properties of frequency. The problem statement can be explained briefly by considering the following example. The transmitter device sends the encrypted image with a hidden message or secured message over a secured channel to the receiver. Transmitter and receiver have the common secret key file. Transmitter hides the secret message within the decoded image by using the secret key file. The secret message which is inside the decoded image will be minimum than the decoded image. Intermediator is an observer who has the permission to access the data which is sent by the transmitter. But, the intermediator does not know about the secret key file. The process detail is public but that does not allow the intermediator to read the secret or hidden message within the decoded image or recover the image partial manner. The most common steganographic technique is the Least Significant Bit Replacement (LSBR) [4]. The significant bit is used to hide the secret image but converting the bits of pixels into smaller value. If the contrast and the intensity of the pixel tend to have 144 value range and the bit of the transmitted message is 1 then the value of the pixels in an image will be either 145 or 143 which the LSB will be same as the message bit numbers. When the changes takes place in the LSB bits, there would be small distortion in the input cover image.

Due to the changes in the LSB, the input image pixel values may change, but they cannot be identified by the human with the observing skill. There are several difficult methods to analyse which known as stegnalysis which can differentiate stegno images from the normal input or cover images. However, it is not applicable to match the input image through the given database. The main scrambling technique which is used to encrypt and decrypt the input image is Arnold transform. The scrambling technique is used to preserve the data without losing any information. This technique is called as Arnold cat map. After the recognition is one, the image can be completely hided using the certain mathematical functions. The transmitter holds the information and data about the transformation. It contain details of the transform applied, and it decoding of original image. The main purpose of the scrambling helps to encrypt the image. The transform applied depends upon the number of times the secret image expressed in higher base. It is used to combine the corrected number of transform used, and LSBs are appended. Thus, the proposed consists of the Arnold cat map of encryption and inverse Arnold cat transform using decryption process.

2 Literature Survey

There are various algorithm to implement the encryption and decryption of image with the data hiding. A symmetric color image encryption method using the intrinsic bit distributions was discussed in [1]. Ji Won et al. [2] had implemented the changes in the pixel as a reversible process to hide the large quantities of data. This reversible process can increase the capacity of data to hide the large quantities of pixel value. Histogram transferring in which the histogram equalized values and the error weights of histogram which were found in [3]. Wang [4] has discussed about logistic map based chaotic encryption. The image pixel values are changed from one bit to another and they were limited from zero ranges. This would allow the bits to transmit and the system had been proposed by Chen et al. [5] performs the encryption-decryption process over the colored image based on the Arnold interference method. The cover image which is encrypted and the secret message has been hided by extracting the channels. Enayatifar et al. [6] worked on chaotic encryption based on de-oxyriboneuclic acid Exor operated with cellular automata rules. This discussed algorithm was a hybrid encrypting method. Ma et al. Po-Yueh Chen et al. [7] had expressed that the hiding the secret image or information into the high-frequency layer using the coefficient of the wavelet transform by eliminating the low-frequency band and the values of the low-frequency band does not change. Arivazhagan1 et al. [8] suggested that the various methods and the various input image database which has the varying statistical and co-occurrence features could be changed by the discrete wavelet transform (DWT). But the classifier used in the existing system need to have the very large database but the output is used to have the larger spatial area. The input cover image can be encoded and decoded using the DWT and inverse DWT method, and it can be done using alpha blending. The main error in the existing work is the increment in the accuracy and the MSE value. Thus, the image have the lesser resolution value. Talwar et al. [9] carried out a strong blind watermarking set of rules primarily based on the algorithm which is used to discrete wavelet transform is the Arnold-Chaos encryption and decryption method. The above-mentioned two encryption and decryption techniques are used to employ the pixel value of the image during the encryption process. Thus, the algorithm used here is to give the greater value and the easier way of conversion because of the values generated by the Arnold and the chaos transform. These algorithms will have the many features with the excessive randomness in the pixel. Zhenjun Tang et al. [10] proposed that the encoding process takes place by the Arnold transform and the three different methods. The algorithm which is explained above is very efficient and there is higher accuracy. This proposed algorithm is independent of size where it can be used for encryption of any input images [10]. Pratibha Sharma et al. [11] discussed about the various three methods of DWT transform based on image watermarking methods which will have the lesser watermarking effect and will be invisible and ineffective. Thus, it can be followed by the alpha blending to extract the watermarked images and to decrypt the image. The main aim of this approach is taking the various number of bits and converting the number of bits into low-frequency sub band by the alpha blending. Watermark image is disperse or relying upon the scaling thing of alpha blending approach. The first stage of the watermarked image is the scaling process which is main algorithm used in embedding. This framing concept has the various comprehensive functions and eliminating the encrypted image without loss of data. Thus, the transferring data without loss of information is called lossless compression. Once the blue channel has been separated then it is encrypted. A coloration picture encoding scheme the use of discrete fractional random rework (DFRT) and Arnold transform. After the transformation process is done, the image can be subjected to the IHS correlation space. Arnold transformation helps to scramble the cover image and hide the secret data by changing pixel position and thus the DFRT[11] changes would tends to change the perimeter value of the image. Akram et al. [12] had examined the method of scrambling based on the chaos mapping which would continuously change the statistical characteristic of the cover image pixels. Liu et al. in [13] made the major designing in the cover and the secret images using the Arnold transform and the discrete cosine transform have been used for image decryption. Tang et al. [14] have been used to remove losage of data during the encryption process. Li et al. [15] carried out the transform which is done the safer encoding procedure. Arnold mapping gives the maximum dimensions values for the encrypted image which is proposed by the Xingyuan et al. [16]. Chang'e Dong [17] has worked on color image encryption using haotic process. They created one time key for both initial and hash functions and created three state variable to encrypt one pixel. Seyed Morteza et al. [18] have generated pseudo random numbers by combining two systems. They combined cellular automata and ant algorithm to solve complex behavior system. Reya et al. [19] have developed hybrid encryption algorithm by fusing chaotic method with cellular automata for encrypting color images. This hybrid process takes longer time for encryption because of block-by-block conversion. Siva Shankar and Rengarajan [20] in the proposed work of data hiding. They had represented 1340 images and the intensity of the image can be identified and valued. Thus, they would create the values in the pixel of the cover image.

3 Arnold Transform Techniques for Image Encryption and Decryption Process

The importance of the proposed work depends on the usage of Arnold transform in the image scrambling technique which is helpful in hiding the secret information

$$A^M : \begin{bmatrix} u_i \\ v_i \end{bmatrix} = \text{mod}\left(\begin{bmatrix} 1 & 1 \\ 1 & 2 \end{bmatrix}\begin{bmatrix} x_i \\ y_i \end{bmatrix}, M\right),$$

From the given matrix, the value of positions (x_i, y_i) which has the coordinate values before Arnold transform has been transferred to (u_i, v_i) by using the Arnold transform. Thus, the modulus operation which is represented 'mod' is used for the conversion of negative integers toward positive values after division operation is carried out. The size of the image which is represented as M influence the period value p. The main algorithm of the proposed work is the scrambling process which is carried out by the Arnold transform. Then, the scrambled image can be decoded using inverse Arnold transform with the matrix of m and n. To carry out the de-scrambling process, key value will be generated to decode the scrabble image. The main process of Arnold transform is c transform is carried out using the following algorithm. Consider the RGB images with the pixel value ranges from 0 to 255 at the gray scale image.

The input cover image is consider to be encoded by the Arnold transform and the secret message will be hided within the scrambled image using Arnold transform. The input image is indicated as C, and the encoded input image is considered as A. The process can be carried by the following steps. Initially, the input image pixels are transformed into binary images which have been divided into sub-blocks by the decimal to binary conversion operation. The conversion process can be carried out by using 'dec2bin'indication. Thus, the operator used here transforms the pixels value which is represented as decimal numbers into eight-digit binary numbers. Then, the converted binary values are scrambled by using the Arnold transform which is signified as $A7$, $A6$, $A5$, $A4$, $A3$, $A2$, $A1$, and $A0$. Then at the final stage, these converted binary values are scrambled into encoded image with the pixel range of 256 Gy level. Thus, the Arnold transform can be explained with the small illustration by considering the cover image of 3*3 matrix and secret image of 3*3 matrix size. This is shown in Fig. 1.

4 Proposed Work

The flow diagram shown in Figs. 2 and 3 represents the encryption and decryption process. The encryption process is carried out by Arnold transform, whereas the decryption process is carried out by the inverse Arnold transform.

$$C=\begin{bmatrix}63 & 64 & 65\\ 127 & 125 & 123\\ 224 & 226 & 228\end{bmatrix} \qquad A=\begin{bmatrix}252 & 71 & 222\\ 39 & 254 & 2\\ 190 & 130 & 7\end{bmatrix}$$

(a) (d)

$$B7=\begin{bmatrix}0&0&0\\0&0&0\\1&1&1\end{bmatrix}\; B6=\begin{bmatrix}0&1&1\\1&1&1\\1&1&1\end{bmatrix}\; B5=\begin{bmatrix}1&0&0\\1&1&1\\1&1&1\end{bmatrix}\; B4=\begin{bmatrix}1&0&0\\1&1&1\\0&0&0\end{bmatrix}$$

$$B3=\begin{bmatrix}1&0&0\\1&1&1\\0&0&0\end{bmatrix}\; B2=\begin{bmatrix}1&0&0\\1&1&0\\0&0&1\end{bmatrix}\; B1=\begin{bmatrix}1&0&0\\1&0&1\\0&0&0\end{bmatrix}\; B0=\begin{bmatrix}1&0&1\\1&1&1\\0&0&0\end{bmatrix}$$

(b)

$$A7=\begin{bmatrix}0&1&0\\1&0&0\\0&0&1\end{bmatrix}\; A6=\begin{bmatrix}0&1&1\\1&1&1\\1&1&1\end{bmatrix}\; A5=\begin{bmatrix}1&1&1\\1&1&0\\1&0&1\end{bmatrix}\; A4=\begin{bmatrix}1&0&1\\0&1&0\\1&0&0\end{bmatrix}$$

$$A3=\begin{bmatrix}1&0&1\\0&1&0\\1&0&0\end{bmatrix}\; A2=\begin{bmatrix}1&0&0\\1&1&0\\1&0&0\end{bmatrix}\; A1=\begin{bmatrix}1&1&1\\0&1&0\\0&0&0\end{bmatrix}\; A0=\begin{bmatrix}1&0&1\\0&1&0\\1&1&0\end{bmatrix}$$

(c)

Fig. 1 Matrix representation of the cover image, encrypted image, decrypted image

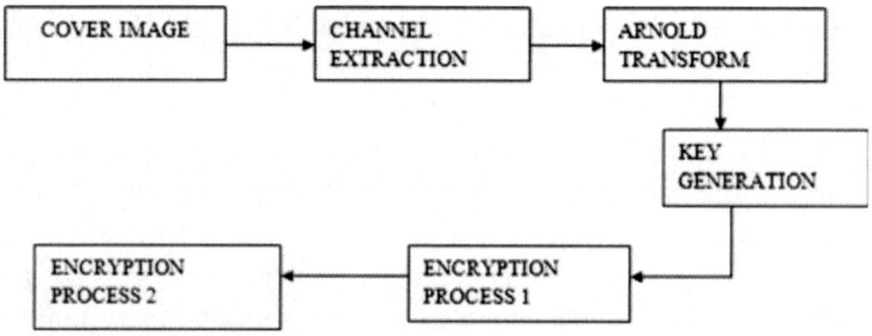

Fig. 2 Block diagram of encryption process

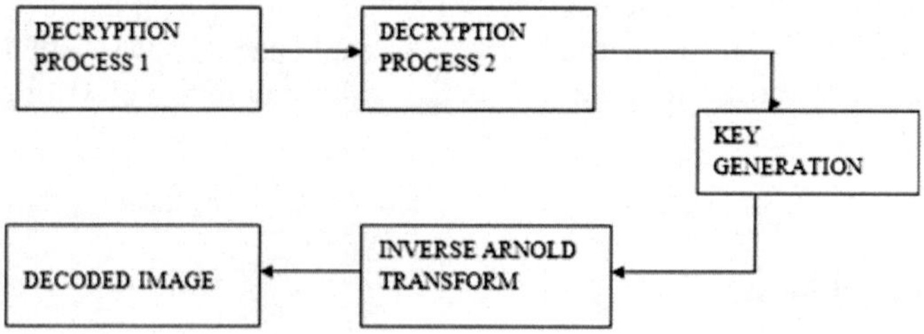

Fig. 3 Block diagram of decryption process

Encryption Process

First stage of the proposed method is encryption process which consists of various steps.

Step 1: The cover image is considered as an input image and then manually loaded in the transmitter side
Step 2: Channel extraction is done to find out the speckle noise and the compression ratio of the compressed image using DWT transform which is used in encryption process
Step 3: The Arnold transform process can be carried out to hide the message or to hide the given cover image
Step 4: Key values are generated using non-uniform mapping and the encryption process is carried out using the generated key values
Step 5: In the encryption process 1, there would 50% encoding will be done using chaotic sequence
Step 6: In the encryption process 2, 100% encoding has been implemented.

Decryption process

Second stage of the proposed method is encryption process which consists of various steps as it follows:

Step 1: Decryption process is carried out using the inverse Arnold transform process following by the chaos mapping
Step 2: In the decryption process 2, 100% encrypted image can be decoded
Step 3: Key values are generated using non-uniform mapping and the decryption process is carried out using the generated key values
Step 4: The Inverse Arnold transform process can be carried out to remove the message or to unhide the given cover image
Step 5: Finally, decoded image can be similar toward the original image.

5 Result

The proposed algorithm has been implemented for different images with various sizes. The mean square error is computed for various channels of the color images. Table1 presents the mean square error and peak signal to noise ratio for different 512 × 512 images. Table 2 lists the measurement of information entropy for color image Lena (512 × 512). Correlation coefficients for plain and cipher image for three planes were listed in Table 3. Figure 4 shows a low MSE as compared with the

Table 1 MSE & PSNR for Lena and Monolisa Images for size 512 × 512

Sample Images	MSE	PSNR
Lena(512 × 512)	0.2013	54.1722
Monolisa(512 × 512)	0.2368	54.3872

Table 2 Information entropy for color images

Color (Lena 512 × 512) Planes	Information entropy
Red plane	7.9773
Green plane	7.9772
Blue plane	7.9773

Table 3 Measurement of correlation coefficient for color images

Plain image			
	R	G	B
H	0.9655	0.9803	0.9686
V	0.9649	0.9799	0.9683
D	0.9568	0.9690	0.9467
Cipher image			
	R	G	B
H	−0.0142	−0.0061	0.0052
V	0.0017	0.0063	−0.0026
D	0.0134	−0.0085	0.0070

previous algorithm. Figure 4 gives the results of encrypted and decrypted images of Lena and Monolisa with two stages of encryption and recovery.

6 Conclusion

The proposed algorithm computes the encryption in the space domain using Arnold transform, and the inverse process decryption is done in the frequency domain. Two stages of encryption were done, and it improves the data security process during transmission. In future, wavelet transform can be embedded with Arnold transform to have lesser details for transmission which results faster transmission. The proposed work outperforms the PSNR rate with existing non hybrid method.

Fig. 4 Encrypted-Decrypted results for Lena and Monolisa images using proposed method

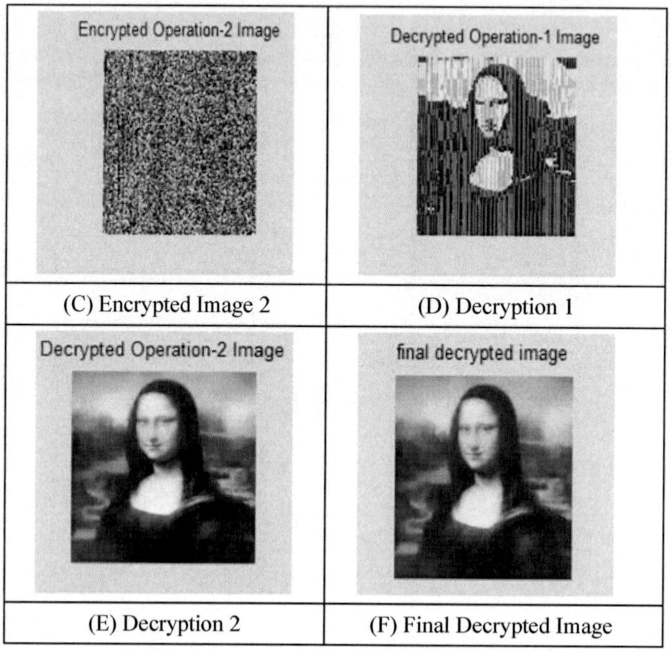

(C) Encrypted Image 2

(D) Decryption 1

(E) Decryption 2

(F) Final Decrypted Image

Fig. 4 (continued)

References

1. Wei Z, Hai Y, Zhiliang Z, Wong K (2013) A symmetric color tionalgorithm using the intrinsic features of bit distributions. Commun Nonlinear Sci Numer Simul 18(3):584–600
2. Yoon JW, Kim H (2010) An image encryption scheme with a pseudorandom permutationbased on chaotic maps. Commun Nonlinear Sci Numer Simul 15(12):3998–4006
3. Zhang G, Liu Q (2011) A novel image encryption method based on total shuffling scheme. Opt Commun 284(12):2775–2780
4. Wang XY, Gu SX, Zhang YQ (2015) Novel image encryption algorithm based oncycle shift and chaotic system. Opt Lasers Eng 68(Supplemrnt C):126–134
5. Wang X, Dapeng (2013) A novel image encryption algorithm using chaos and reversible cellular automata. Commun Nonlinear Sci Numer Simul 18(11):3075–3085
6. Enayatifar R, Abdullah AH, Sadaei HJ, Lee M, Isnin IF (2015) A novel chaotic based image encryption using a hybrid model of deoxyribo nucleic acid and cellular automata. Opt Lasers Eng 71:33–41
7. Wang L, Song H, Ping (2016) A novel hybrid color imageencryption algorithm using twocomplex chaotic systems. Opt Lasers Eng 77:118–125
8. Liu H, Kadir A (2015) A symmetric color image encryption scheme using 2d discrete-time map. Signal Process 113(Supplement C):104–112
9. Boora M, Gambhir M (2013) Arnold Transform Based Steganography. Int J Soft Comput Eng (IJSCE) 3(4):2231–2307
10. Zhou N, Zhang A, Zheng F, Gong L (2014) Novel image compressionen cryption hybridalgorithm based on key-controlled measurement matrix in compressive sensing. Optics Laser Technol 62:152–160

11. Lu Xu, Li Z, Li J, Hua W (2016) A novel bit level image encryption algorithm based on chaotic maps. Opt Lasers Eng 78:17–25
12. Belazi A, Abd El-Latif AA, Diaconu AV, Rhouma R, Belghith S (2017) Chaos-based partial image encryption scheme based onlinear fractional and lifting wavelet transforms. Opt Lasers Eng 88:37–50
13. Chai X, Chen Y, Broyde L (2017) A novel chaos-based image encryption algorithm using dnasequence operations. Opt Lasers Eng 88:197–213
14. Abdo AA, Lian S, Ismail IA, Amin M, Diab H (2013) A cryptosystem based on elementary cellular automata. Commun Nonlinear Sci Numer Simul 18(1):136–147
15. Gu G, Ling J (2014) A fast imageencryption method by using chaotic 3D cat maps. Optik-International Journal for Light and Electron Optics. Int J Pure Appl Math Special Issue 125(17):4700–4705
16. Wang X, Zhao Y, Zhang H, Guo K (2016) A novel color image encryption scheme using alternate chaotic mapping structure. Opt Lasers Eng 82(Supplement C):79–86
17. Dong C (2014) Color image encryption using one time keys and coupled chaotic systems. Sig Process: Image Commun 29(5):628–640
18. Hosseini SM, Karimi H, Jahan MV (2014) Generating pseudo-random numbersby combining two systems with complex behaviors. J Inf Security Appl 19(2):149–162
19. Reya K, Monika R (2018) Hybrid hyper chaotic system and cellular automata-based color image block wise encryption and decryption. Int J Pure Appl Math 118(22):33–40
20. Rengarajan A (2016) Datar hiding in encrypted images using arnold transform. ICTACT J Image Video Process 7(1)

Detection and Diagnosis of Fault Using Light-Weighted Midori Blocks

C. Thirumarai Selvi, R. S. Sankarasubramanian, and M. MuthuKrishnan

Abstract Modified block ciphers increase the performance and lower the time duration for high-speed applications. High-speed applications are mainly required for wearable medical devices, and compact devices which are implantable in the human body. Various security algorithm suffers due to inclusion of malicious and natural faults. The proposed model modifies the Midori block as light-weighted nature. This proposed algorithm introduces lower latency and less hardware complexity. The proposed model adds diagnosis algorithm in different stages of the work flow. Diagnosis algorithm reduces the fault value to the minimum. The fault diagnosis systems implemented with nonlinear S-box layer and beat structures. The beat structure includes 64-bit and 128-bit Midori symmetric key ciphers. The designed systems are benchmarked on a field programmable gate array (FPGA) with injected faults. The proposed light-weighted block cipher blocks yields, good energy efficiency rate and implemented using VHDL code.

Keywords Fault correction · Lower block cipher · Field programmable logic array gate (FPGA) · Midori blocks

1 Introduction

The main algorithm used for obtaining higher security protection with lower area consumption and to lower the energy consumption is light-weighted cryptography. These techniques have been used in many applications which have the following domain. They can be embedding system, wireless communication [1], RFID (Radio frequency identification tags) and other medical applicable devices. But obtaining

C. Thirumarai Selvi (✉)
Sri Krishna College of Engineering and Techology, Coimbatore, Tamil Nadu, India
e-mail: thirumaraiselvi@skcet.ac.in

R. S. Sankarasubramanian
PSG Institute of Technology and Applied Research, Coimbatore, Tamil Nadu, India

M. MuthuKrishnan
KIT-Kalaigarkarunanidhi Institute of Engineering and Technology, Coimbatore, Tamil Nadu, India

A. Sivasubramanian et al. (eds.), *Futuristic Communication and Network Technologies*, Lecture Notes in Electrical Engineering 792,
https://doi.org/10.1007/978-981-16-4625-6_16

lower energy consumption is a difficult process which is the drawback of the system. These technology have been used in implantable and mobile medical devices. The efficiency can be increased by the use of dischargeable batteries. The dischargeable batteries are difficult to remove the surgical operations. However, in the wireless technology by using the RFID tags and the sensor is the major reason for usage of light-weighted cryptography. By using these algorithm in the wireless technology, it will decrease the area consumption which would increase the efficiency of the system and decrease the source power. Thus, the requirements are given by the light-weighted ciphers [2]. By using the light-weighted ciphers, security level can be increase with the decrement in the optimal energy level. This light-weighted blocks lower the hardware difficulties. The main step noted in the light-weighted block [3] is the advance encryption standard (AES) that have limited the area and the power dissipation. In the proposed method, Midori Algorithm is used that has the required security level with the lower power optimization. The Midori which consists of S number of boxes that are obtained from the AES and other light-weighted block ciphers. Midori [4, 5] have been classified as two types. It has 4-bit S-boxes that are classified as ones variable. Midori which is the light-weighted block ciphers that can have the acceptable range of cells. The permutations of each cells are separated by using the certain distance calculation. These distance calculation method has classified as the MDS which is known as the maximum distance separable. It will arise due to the lower implementation of the values and the higher immunity against the several causes. The fault diagnosis in the cryptography is based on the crypto-architectures that have the center of values. As the existing work have implemented, the time and hardware redundant variation approaches. These variations which consists of variations in the different platforms due to the different light-weighted architectures [6–13]. In the lower weighted Midori domain, there doesn't exist any existing work and the main advantage of the proposed work is the higher value of accuracy of fault diagnosis as compared with the different existing methods. In the previous work, there exists the two-folded algorithms. These algorithms consist of two stages namely fault detection and fault diagnosis. During the first stage, there exists the comparison of logic-gate based and look up table (LUT). These two existing algorithm consists of the error detection techniques. The error detection depends upon the S number of boxes in the Midori algorithm. The key advantage of employing the Midori block is its low energy consumption, which allows it to meet other constraints such as area, power, or latency. The methods of implementation of Midori algorithm are that the implementation and the performance measuring schemes are different from the other algorithm. Second stage of the proposed algorithm is that it can implement the lower head detection and the performance rate can also increases. This is achieved by changing the blocks and the error containing values. Thus, to obtain the fault diagnosis, careful observation of the implementation is needed to achieve the required data value. The proposed algorithm will define the simulation results as the higher error coverage. The higher error coverage can be stated as the ratio of the number of error which is detected as a input and the number of injected error faults. From the proposed algorithm, the error detection can be achieved without any fault error [14]. These error diagnosis can be achieved by the detection of faults with the high-coverage value. From the proposed work, the permanent faults such as stuck-at faults, fault error can be occurred. These causes are the main reason for the VLSI manufacturing defects. The manufacturing defects [15] are due to the break in the intension during the run time. During the test generation, pattern will be generated to identify the faults within the system. The long transient faults can be identified by the leakage in the information and the changes in the output occurs due to the computation techniques. The faults attack are saved in the register for the further references. This action will discontinue the counter operation. The proposed algorithm can be implemented using the Field Programmable Logic Array (FPGA) in the Xilinx Vertex software. By using the proposed software, the work can be expected toward the constrained level. The proposed work consists of various steps which will give the detailed explanation about the fault detection and diagnosis using the Midori light-weighted blocks.

2 Proposed Method

The proposed work block diagram is shown in Fig. 1. The variants such as Midori-128 and Midori-64 have 128-bit secret key (K) and use two 64-bit sub keys K_0 and K_1 and the W_K are derived through modulo-2 addition of these 64-bit sub keys, respectively. In Midori64, the derivation of the round key is reiki = K (imod2) EX-ORαi, where $0 \leq i \leq 14$. In Midori128, $W_K = K$ and Reiki = K EX-orbetai, $0 \leq I \leq 18$, in which

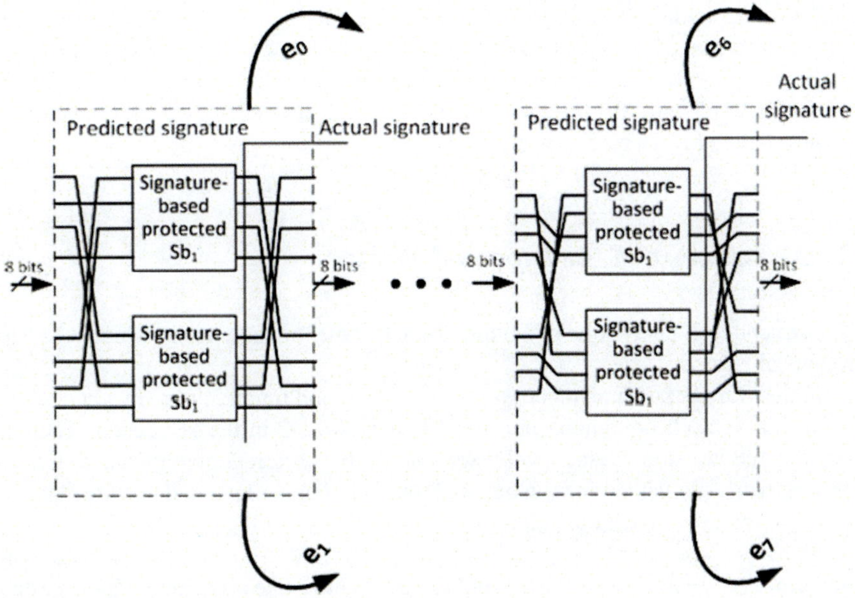

Fig. 1 Proposed method S-box transmission

Fig. 2 Three instances scheme transmission

$$\alpha_0 = \beta_0 = \begin{pmatrix} 0 & 0 & 1 & 0 \\ 0 & 1 & 0 & 0 \\ 0 & 0 & 1 & 1 \\ 1 & 1 & 1 & 1 \end{pmatrix}$$

$$\alpha_{14} = \beta_{14} = \begin{pmatrix} 1 & 1 & 1 & 0 \\ 1 & 1 & 0 & 0 \\ 0 & 1 & 0 & 0 \\ 1 & 1 & 1 & 0 \end{pmatrix}$$

$$\beta_{18} = \begin{pmatrix} 0 & 0 & 1 & 1 \\ 1 & 0 & 0 & 0 \\ 1 & 1 & 0 & 1 \\ 0 & 0 & 0 & 0 \end{pmatrix}.$$

betai = αi for $0 \leq I \leq 14$ (see Fig. 2). Hence, the round key matrix is either "1"or "0" as ^Signature.(Reiki) = Signature. (K) EX-OR Signature. (betai/αi).

To analyze matrices of constant values, same approaches are followed. Let us consider two examples for detections of errors by round key operation. In the first example, the input key by K and the output round key by rookie. To derive 16 signatures, each one is processed underage = k*j EX-orbeta j. The nibble of k3 jk2 jk1 jk0 j for Midori64 or a byte as k_7 jk_6 jk_5 jk_4 jk_3 jk_2 jk_1 jk_0 j for Midori128. Then, signatures are Midori128, ^Signature.(r*k*j) = Signature.(k_7 jk_6 jk_5 jk_4 jk_3 jk_2 jk_1 j(k_0 j EX-orbeta j)), where $0 \leq j \leq 15$. The received "1"or "0"elements of the round key matrix. In the second example, to create just one signature nature for error detection the modulo-2 addition of the entire elements in the state matrix is processed by one signature nature. For example, (r*k) = 15 j = 0 k*j + 15 j = 0 beta* j, and α_0 = $beta_0$,α_{14} = $beta_{14}$, and $beta_{18}$ elements of each matrix in modulo-2 addition drives "0," "1," and "0,", respectively. Hence, RK_0 and RK_{18}, the input signature natures, are intact, and RK_{14} is inverted.

3 Architecture of the Proposed Work

The structures of encryption of Midori128 detect mentioned error in which it has 20 rounds of 8bits cell size.

In this, variant has functioned to encrypt with round function and key generation. It has the SubCell operation, and the WK is modulo-2 in the last round. Then in the first and last steps, respectively (see Fig. 3). For different operations, they have derived with error detection schemes, and round function, illustrated by the predicted signature-nature of each operation.

Signature-based error detection architectures in Midori's encryption and decryption process (see Fig. 4) are carried out through a loop-like architecture by inserting a multiplexer (MUX) and a register. This will derive the signatures of input and the key. The similar functions operated in encryption and decryption, but with variation

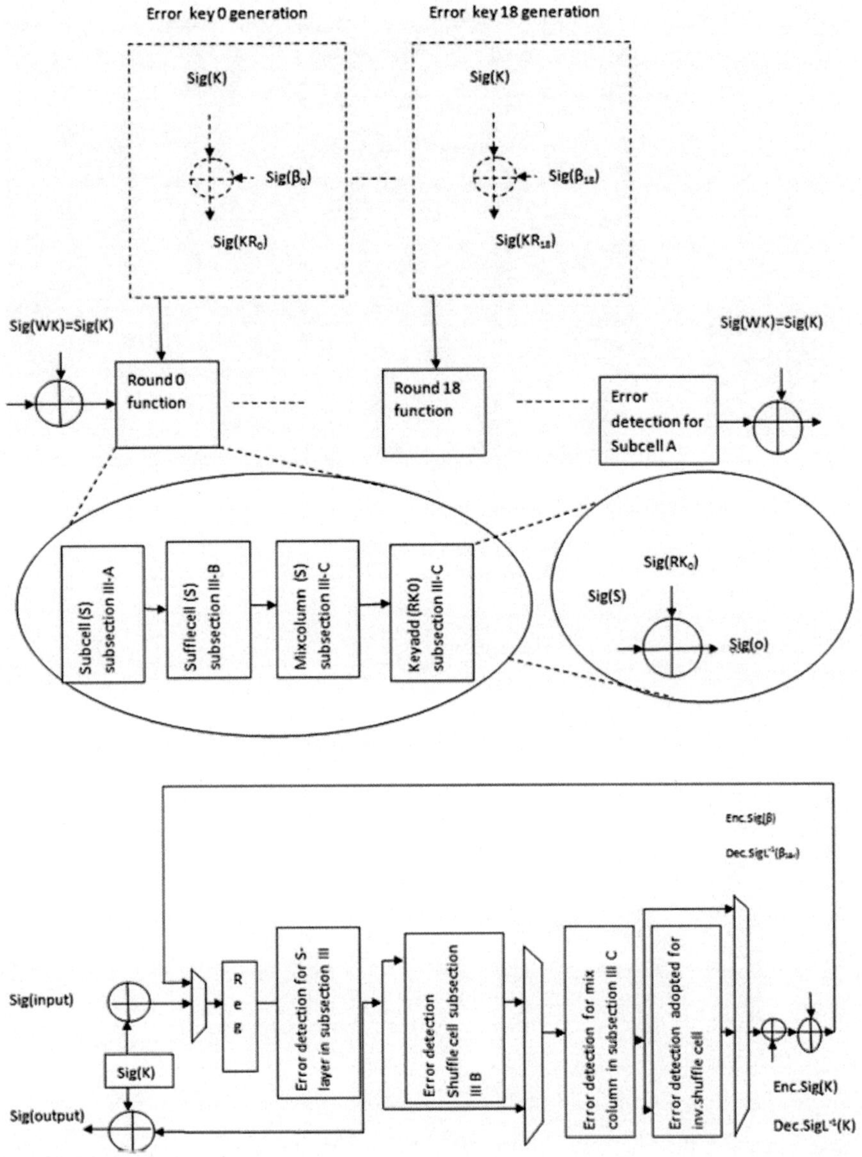

Fig. 3 Architecture of proposed work

as ShuffleCell (Sh) operation in encryption and InvShuffleCell (Sh−1) operation for decryption is adopted. In decryption key, generation process is altered, i.e., L−1 (K) instead of K and the corresponding signature natures "Signaturen."(L−1 (K)) and "Signaturen."(L). Similarly, the itch round constant is replaced by L−1 (beta*18−i), instead of betai (see Fig. 5). Rather than in other cases, in case of deriving a "black-

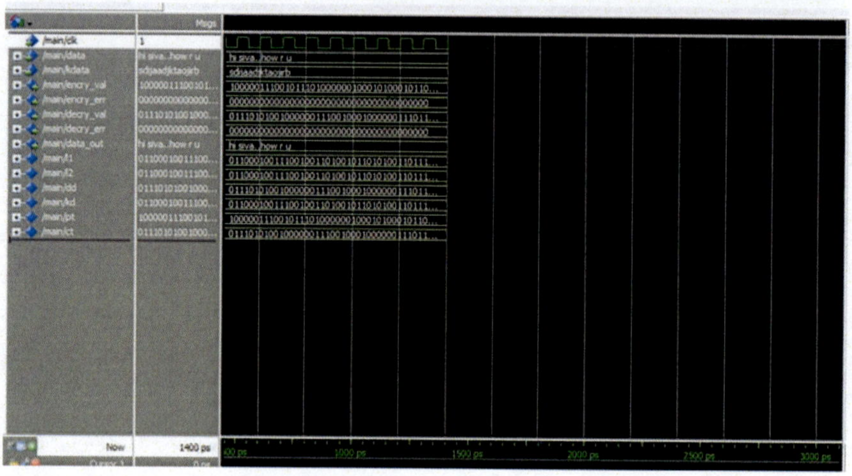

Fig. 4 Simulation of decrypted Midori block

```
Device utilization summary:
---------------------------

Selected Device : 7vx330tffg1157-3


Slice Logic Utilization:
 Number of Slice LUTs:                    816  out of  204000     0%
    Number used as Logic:                 816  out of  204000     0%

Slice Logic Distribution:
 Number of LUT Flip Flop pairs used:      816
   Number with an unused Flip Flop:       816  out of     816   100%
   Number with an unused LUT:               0  out of     816     0%
   Number of fully used LUT-FF pairs:       0  out of     816     0%
   Number of unique control sets:           0

IO Utilization:
 Number of IOs:                           389
 Number of bonded IOBs:                   388  out of     600    64%

Specific Feature Utilization:

---------------------------
```

Fig. 5 Reduced area using Midori block

box" signature, they have to choose the check arguments for the entire encryption/decryption. At the time, it had a huge probability for encountering error indication flags. In case of the wrong alarm ratio, fine-tuned multiple signaturenatures are avoided, and it uses the error indication flags with separate transformations. The formulation as per paper is used in each check point is higher for expanse of more overhead and wrong alarms probability. Atlast, best granularity is fixed with separate 4-bit S-boxes of Midori's 8-bit S-box checking. It leads to higher wrong alarms ratios at higher overhead and higher error coverage. This provides three models of different signaturenatures to detect odd faults in addition to single stuck-at faults. Here, it is possible to detect only VLSI defects such as burst faults and unable to find adjacent faults. Adjacent faults are diagnosed on interleaved parties at the expense of more overhead. In signaturenature-based diagnosis approach, it allows to linear codes to detect random errors of small multiplicity based on robust codes it detects any errors. The outcome of the methods to reach, reliability and hardware security.

3.1 Error Detection for Threshold Implementation of Midori

Threshold implementation (TI) for first order [16, 17] is widely used to protect light-weight block-ciphers. It also analysis the power consumption during the performance and complexity overheads. The S-box approach [5]: $A_{out}Q_{12}A_mQ_{12}$ A_{in}, where A_{out}:{0 A1B82934E 5FC6D7}$_{16}$, A_m: {84B70C3F 95 A61D2E$_{16}$, A_{in}:{8 A02DF57CE469B13}$_{16}$, and Q_{12}: {0123456789CDE FAB}$_{16.}$ Register sets have three instances of f_{Q12} over by an A_{out}. S-box combined with TI plots helps to find error, while duplicating native approach. $A_{out}Q_{12}A_mQ_{12}A_{in}$ reviles to high error coverage, no false alarms and low overhead. The performance, of signature, leads to lower overhead, while expanding the error coverage.

4 Proposed Technique Implementation in Fault Attacks

Analysis of differential fault intensity is the combinations of fault injection principles and differential power analysis derives based fault models in practical times the fault in original and redundant computations injections are rare. The insertion of few faults may possible in reduction of efforts, i.e., injecting of single faults with different intensities. The variation of a intensity simulation is used for biasing fault models by bits within higher intensities rather than using with low fault intensity. Fault categories such as single/two/three/four-bit modes. The approach to detecting error codes, signaturenatures on column are capable to throw more fault. Moreover, using parities SBUs and SBTBUs are detected fully. In addition, with burst faults, it is able to detect SBTBUs, SBQBUs and OSBs. Similar type of fault is introduce in both the original and redundant computation. This developed RESI architecture brings the null biasing effect in a fault model. Also, it injects f_0 and f_1 fault to the output registers. By the parting, both these attack schemes similarity are marked as in 23.

5 Simulating Errors by Injection with FPGA Benchmark

This section describes assessment of error coverage and benchmark overhead in the error detection structures. By transient random faults, the internal faults are modeled.

Table 1 Proposed MIDORI-128 method error coverage details

Type of faults	Injected faults	Detected faults	Error coverage (%)
Stuck-at zero	10,000	9909	99.09
	100,000	99,870	99.87
Stuck-at one	10,000	9908	99.08
	100,000	99,762	99.76

The errors are detected through multiple stuck-at fault injections. Then, the transient and permanent internal faults are identified after simulation. The natural and malicious faults are known as single and multiple stuck-at faults. In reality, technological constraints have led to multiple faults. So, single stuck-at faults are found to be ideal in nature. In the linear and nonlinear components of Midori, single-bit errors are detected in nibbles (Midori-64) and bytes (Midori-128). The proposed single stuck-at faults concludes to 100% error coverage. Hence, simulation is neglected. The proposed approaches give the detection of odd faults with the property of inheritance signaturenature is used with subsets as single stuck-at faults. It is an irregular system to detect all potential fault attacks, but the difficulty in mounting is achieved by proposing architecture. The architecture in different subparts alarms for the errors. The transient faults and permanent internal permanent faults and transient faults were simulated by injecting faults. This is developed with the help of Midori128 encryption. By injecting, 10,000 or 100,000 faults for the encryption process and monitored. Table 1 plots the error coverage which shows the increase in injection points and thus higher error prediction is calculated. Two experiments are implemented as follows: (a) Linear-feedback shift registers (LFSRs) used for detecting random multiple faults, with maximum tap polynomials. These polynomial projects the type, location and the total faults (b) Increased number of fault injections are used to get closer and realistic error coverage. Hence, 10,000 faults are injected to improve the performance.

Parity-interleaving or parity-swapping yields reduced are and power consumption, while operating with S-boxes. Reduction of error coverage results due to inputs swapping and column mixing in the S-boxes.

6 Simulation Results

Modified light-weighted blocks will lead to the changes in the accuracy rate and the delay time. The simulation results of the proposed work are presented from Figs. 6 and 7. These results represent the reduced area and increased speed. Also, the encrypted message using the modified light-weighted block is given in the results.

```
Timing Summary:
---------------
Speed Grade: -3

   Minimum period: No path found
   Minimum input arrival time before clock: No path found
   Maximum output required time after clock: No path found
   Maximum combinational path delay: 2.565ns

Timing Details:
---------------
All values displayed in nanoseconds (ns)
```

Fig. 6 Reduced time using Midori block

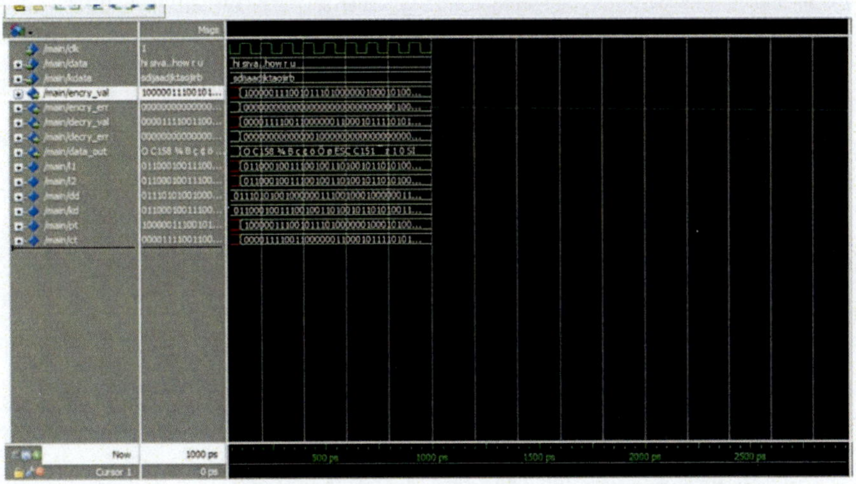

Fig. 7 Simulation of encrypted Midori block

7 Conclusion

Thus, the proposed method has the lesser area coverage, and there is the reduction in the time consumption. This reduction will have the better accuracy as comparison with the existing algorithm. The light-weighted Midori block will have the lesser values of the error. The fault detection can be shown in the encryption and decryption process of the proposed work. In future scope, the Midori blocks can be increased for achieving the more accuracy.

References

1. Kermani MM, Zhang M, Raghunathan A, Jha NK (2013) Emerging frontiers in embedded security. In: 2013 26th international conference on VLSI design,12th international conference

on embedded systems. IEEE, pp 203–208

2. Eisenbarth T, Kumar S, Paar C, Uhsadel L (2007) A survey of lightweight-cryptography implementations, IEEE Des Test Comput 24(6):522–533
3. Moradi A, Poschmann S, Ling C, Paar, Wang H (2011) Pushing the limits: a very compact and a threshold implementation of AES. In: Proceedings of EUROCRYPT, pp 69–88
4. Banik S, Bogdanov A, Regazzoni F (2015) A block cipher for low energy (extended version), Proc Cryptol ePrint Arch, pp 411–436
5. Moradi S (2016) Side-channel analysis protection andlow-latencyinaction-case study of prince and midori. [online]. Available: https://eprint.iacr.org/2016/481.pdf
6. Mozaffari-Kermani M, Azarderakhsh R, Lee CY, Bayat-Sarmadi S (2013) Reliable concurrent error detection architectures for extended Euclidean-based division OverGF(2^m). In: IEEE Trans Very Large Scale Integr (VLSI) Syst 22(5):995–1003
7. Jing Z, Zengrong L, Lei C, Shuo W, Zhiping W, Xun C, Chang Q (2012) An accurate fault location method based on configuration bitstream analysis. In: NORCHIP 2012. IEEE, pp 1–5
8. Sarkar S, Hembram PK, Purkait P, Das S (2016) Acquisition and pre-processing of three phase induction motor stator current signal for fault diagnosis using FPGA, NI Compact-RIO real time controller. In: 2016 IEEE Uttar Pradesh section international conference on electrical, computer and electronics engineering (UPCON). IEEE, pp 110–114
9. Hsu CL, Chen TH (2009) Built-in self-test design for fault detection and fault diagnosis in SRAM-based FPGA. IEEE Trans Instrument Meas 58(7):2300–2315
10. Jamshidpour E, Shahbazi M, Poure P, Gholipour E, Saadate S (2013) Fault tolerant operation of single-ended non-isolated DC-DC converters under open and short-circuit switch faults. In: 2013 15th European conference on power electronics and applications (EPE). IEEE, pp 1–7
11. Guo C, Zhang Y, Chen L, Zhou T, Li X, Wang M, Wen Z (2012) A novel application of FPGA-based partial dynamic reconfiguration system with CBSC. In: 2012 VIII Southern conference on programmable logic. IEEE, pp 1–4
12. Ghalaty NF, Yuce B, Taha M, Schaumont P (2014) Differential fault intensity analysis. In: 2014 Workshop on fault diagnosis and tolerance in cryptography. IEEE pp 49–58
13. Patranabis S, Chakraborty A, Nguyen PH Mukhopadhyay D (2015) A biased fault attack on the time redundancy countermeasure for AES. In: International workshop on constructive side-channel analysis and secure design. Springer, Cham, pp 189–203
14. Wang G, Wang S (2010) Differential fault analysis on PRESENT key schedule. In: 2010 international conference on computational intelligence and security. IEEE, pp 362–366
15. Biham E, Shamir A (1997) Differential fault analysis of secret key cryptosystems. In: Annual international cryptology conference. Springer, Berlin, Heidelberg, pp 513–525
16. Tunstall M, Mukhopadhyay D, Ali S (2011) Differential fault analysis of the advanced encryption standard using a single fault. In: IFIP international workshop on information security theory and practices. Springer, Berlin, Heidelberg, pp 224–233
17. Ali SS, Mukhopadhyay D (2013) Improved differential fault analysis of CLEFIA. In: 2013 workshop on fault diagnosis and tolerance in cryptography. IEEE, pp 60–70

Designing of Microstrip Patch Antenna for Satellite Communication

S. Nishandhi, Duvvuru Yugandhar Reddy, and R. Ajith Kumar

Abstract In modern era devices which are used in military, weather forecasting, government organizations, defense tracking not only perform single operation but also perform many operations at a time, and these can be achieved by designing microstrip patch antenna with many operating frequencies by creating slots in the patch of the antenna. These antenna design consists of sapphire as dielectric substrate which is suitable for X-band applications. The satisfactory impedance matching of octagonal microstrip patch antenna is achieved by providing feed through microstrip feeding technique. Due to the presence of *S*-slots in the patch, it get radiates at three different frequency. Current distribution and magnetic field distribution are provided from the basic working principles of antenna. These antenna operates at three different frequency such as 8.3, 9.2 and 9.8 GHz. It is concluded that the designed antenna was simulated and analyzed for various characteristics such as voltage standing wave ratio, axial ratio, field distributions, gain, radiation pattern and return loss. Results which are obtained by simulation are matched very well. The patch antenna designed with above parameters based on operating frequency and size. It is used in satellite communication.

Keywords SISO antenna X-band · Microstrip patch · Satellite communication

1 Introduction

Wireless telegraphy and existence of electromagnetic waves can be broadly described by the wireless communication in the modern society. Each and every individual life of citizen get enhanced by the wireless communication by transmitting and receiving huge amount of data and making quality and comfortable life through by wireless communication system such as LTE, 3G, GSM, Wi-Fi and Bluetooth. A part of conventional communication system, wireless communication system is also used for various applications, antennas are the key transmitting elements in

S. Nishandhi (✉) · D. Y. Reddy · R. Ajith Kumar
Department of Electronics and Communication Engineering, IFET College of Engineering, Villupuram, Tamil Nadu, India

A. Sivasubramanian et al. (eds.), *Futuristic Communication and Network Technologies*, Lecture Notes in Electrical Engineering 792,
https://doi.org/10.1007/978-981-16-4625-6_17

wireless communication, and these are used in medical applications such as hospitals, transmitting information from remote sensing areas, transmitting information for radars applications, wall imaging, and geographical location identifying applications [1].

By operating at particular resonant frequency i.e., 8.3, 9.2 and 9.8, the antenna converts guided energy into radiated energy at transmitting side and converting the radiated energy to guided energy at the receiver side at same particular operating frequency is the most important zone of the wireless communication system. The overall radio frequency link budget get increased by making designed antenna operating at multiple frequency because of the slotting technique in the patch or by using materials based on their orientation and size, and these increase the channel capacity by allowing more number of users should communicate through single interface by wireless communication system [1–5].

Antenna parameters like satisfactory return loss, bandwidth and gain, radiation pattern with desired physical size are the critical specifications that must be satisfied by antenna in modern wireless systems. Over the past decades, various wireless networks have been prosperously developed for different applications. Generally, they operate at different frequencies. Due to limited space [6], it is difficult to install numerous antennas operating at different frequencies in a compact wireless terminal.

To design a microstrip patch antenna which operates in X-band at a frequency range of (8–12) GHz with a wavelength of (2.5–3.75) cm for satellite communication and to analyze various Antenna parameters like *S*-parameter, radiation pattern, gain and VSWR.

The main goal of these work is to design a sapphire substrate with a patch which operates in X-band for desired bandwidth and to compare with the former antennas. The satisfactory impedance matching is achieved by feeding the antenna using microstrip feed line. Due to the presence of *S*-slots, triple resonant frequency is achieved [7, 8]. Triple resonant mechanism is provided by the current distribution. It is concluded that the desired antenna is analyzed and simulated for various characteristics such as return loss, radiation pattern, VSWR, current distribution and magnetic field distribution. The designed values and outcomes are matched well at desired frequency. Hence, the design antenna is suitable for satellite communication with triple resonant frequency.

And need for multiple resonance is antenna which is in resonance helps to operate a single device at that operating frequency. Antenna with multiple resonance frequency helps to operate multi devices with multiple operating frequency. Here, the proposed antenna is a multi-resonance antenna with three operating frequencies.

2 Analysis of Antenna Design

The three band octagonal microstrip patch antenna which is proposed is represented in Fig. 1. The dielectric material of antenna is designed with a sapphire substrate ($\varepsilon_r = 10$, tan $\delta = 0.02$) of 1.6-mm thickness. The antenna is excited by a 50 Ω stepped

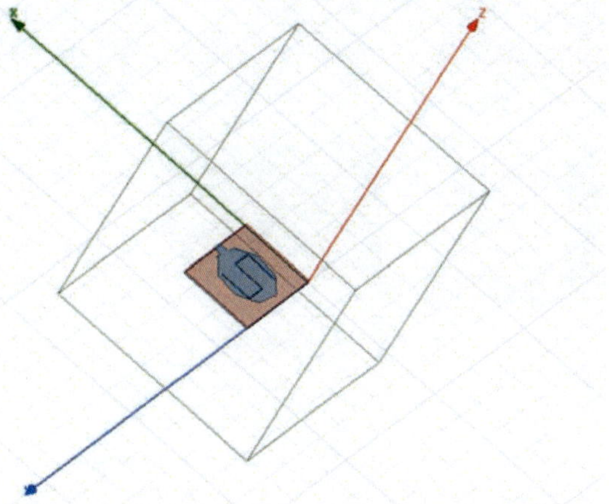

Fig. 1 Represents octagonal patch with radiation box

microstrip line. The *S*-shaped slots, etched at the patch plane, of approximately half-wavelength are used as radiator. While the *S*-shaped slot radiates at 8.3 GHz, the second *S*-slot creates another resonant frequency at 9.2 GHz, and the third *S*-slot creates another resonant frequency at 9.8 GHz [9, 10] (Fig. 2).

The antenna consists of octagonal patch with *s*-slot which mainly used in X-band applications for satellite communication. The energy radiated by octagonal the patch is more than the energy radiated by the rectangular patch with same features because by increasing more number of edges the radiated is also more. The antenna is designed with ground 50 * 70 mm size and octagonal patch with position (25, 35, 1.6) contains a dielectric substrate with a size 50 * 70 mm and patch contains slots

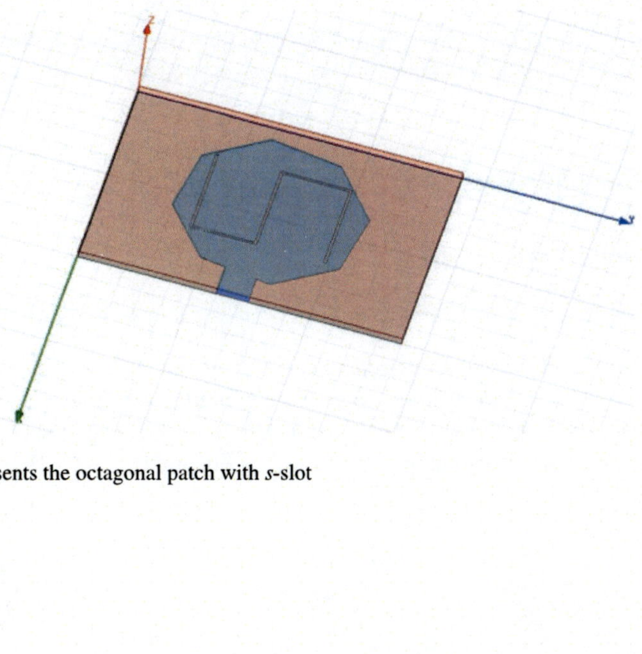

Fig. 2 Represents the octagonal patch with *s*-slot

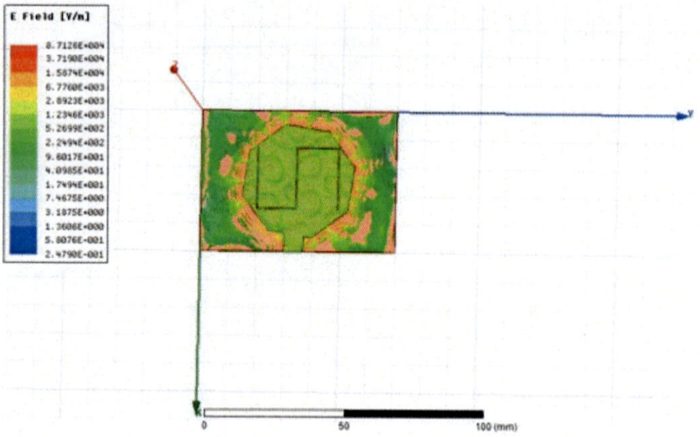

Fig. 3 E-field distribution in patch

with 22 * 0.5 mm parallel to *x* direction and having slots with size 0.5 * 14 mm parallel to *y* direction.

2.1 *Current Distribution*

Current distribution of microstrip antenna depends upon the type of antenna, which has maximum value at the center and zero at the edges. The radiation of octagonal patch antenna explained by the fringing fields the electric fields on the edge of the octagonal patch which contains *S*-slot of antenna are added up in phase and produce the radiation.

Figures 3 and 5 explain the electric field distribution of the patch and ground. Where Figs. 4 and 6 explain the magnetic distribution of octagonal microstrip patch and ground of the antenna, respectively.

3 Results and Discussion

3.1 *Return Loss Analysis*

The main aim of the desired antenna which is design is to achieve good return loss below −10 dB. The *S*-parameter characteristics of the desired antenna are represented in Fig. 7. It shows absolute impedance matching for all the desired resonant frequencies. The graph shows less return loss at multiple frequencies such as 8.3, 9.2, 9.8 GHZ. The corresponding return loss is obtained as −11 dB, −15 dB and

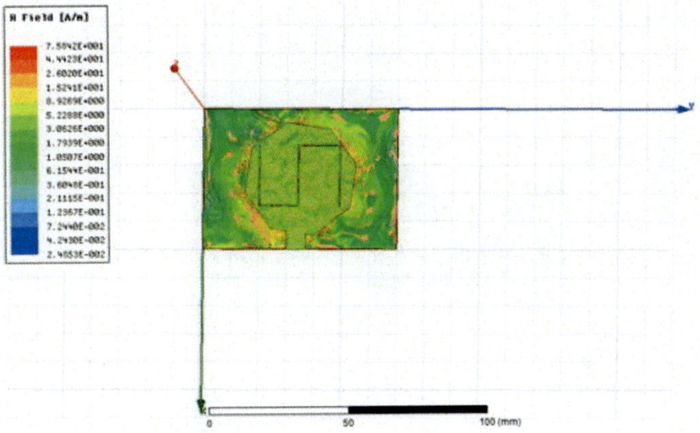

Fig. 4 H-field distribution in patch

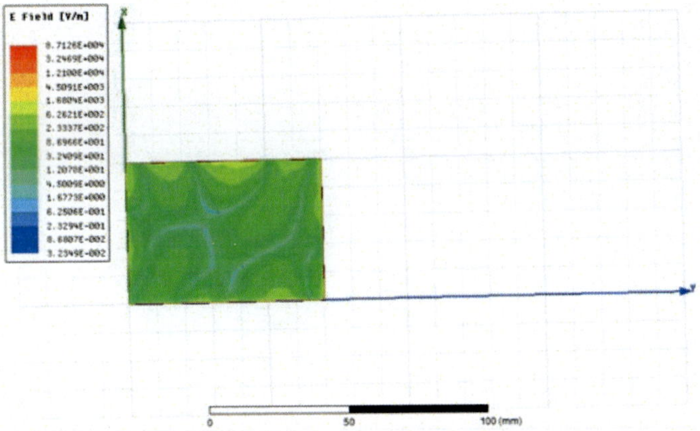

Fig. 5 E-field distribution in ground

−20 dB, respectively, at these frequencies. The large bandwidth is achieved with return loss −10dB which shows high performance at desired frequency.

3.2 VSWR Analysis

The parameter used to describe the impedance equalization between the constructed antenna and transmission line is voltage standing wave ratio, overall entire operating frequency of the desired antenna. The VSWR characteristics of the antenna which is designed for multiband applications is shown in Fig. 8. The ideal value of VSWR should be less than 2 which is achieved between the frequency of 8.3 and

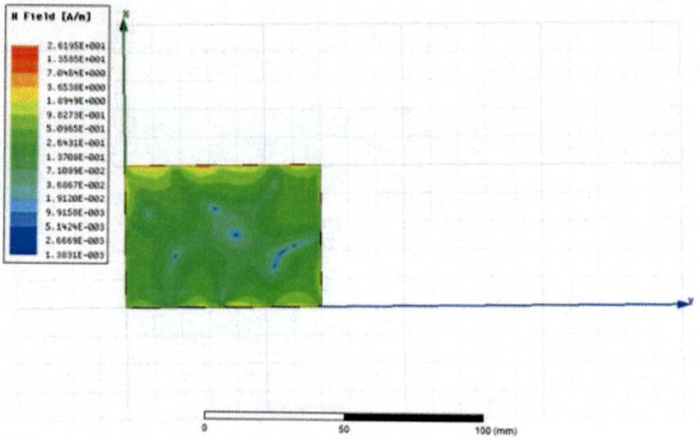

Fig. 6 H-field distribution in ground

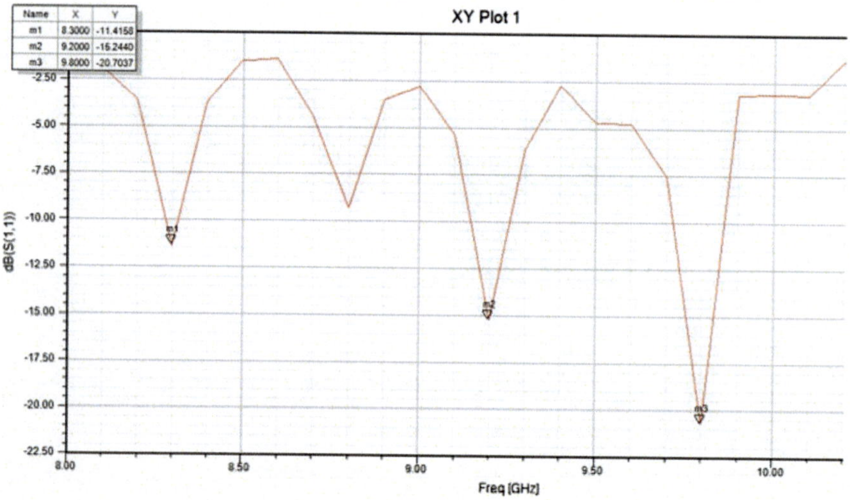

Fig. 7 Return loss of octagonal patch antenna

9.8 GHz which indicates that less amount of energy get reflected from the octagonal patch throughout the entire bandwidth these indicates there exists better matching in between the antenna and the transmission line. The graph illustrates the frequencies such as 8.3, 9.2, 9.8 GHz. The VSWR is obtained as 1.7 dB, 1.4 dB and 1.2 dB, respectively, at these frequencies.

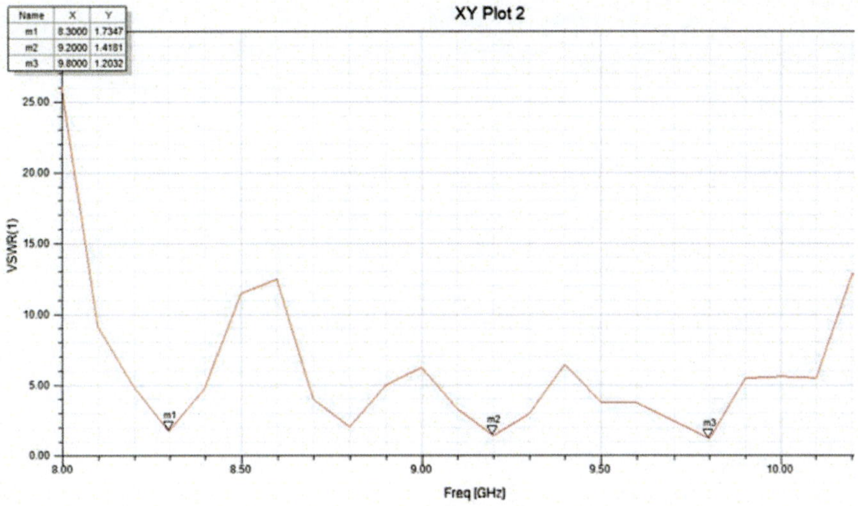

Fig. 8 VSWR of octagonal patch antenna

3.3 Radiation Pattern Analysis

The radiation pattern analysis for the proposed antenna is shown in Figs. 9 and 10. The obtained 8.3–9.8 GHz values depict the directional characteristics of the radiation pattern. The desired antenna shows better radiation performance, and these

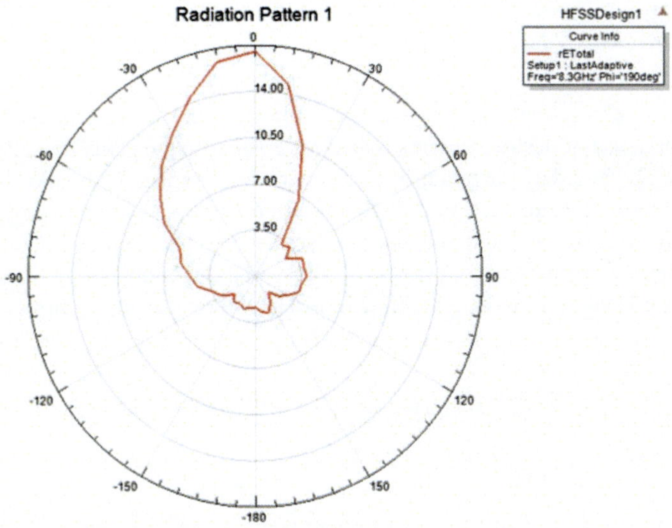

Fig. 9 Radiation pattern of octagonal patch antenna

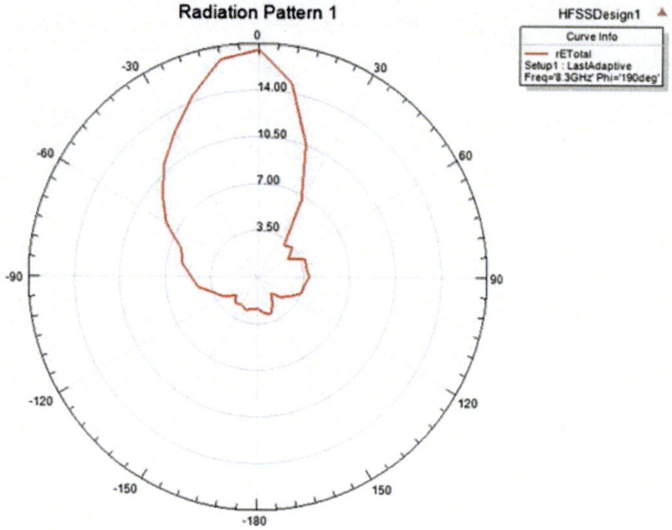

Fig. 10 Radiation pattern of microstrip patch antenna at phi 190°

can be analyzed based on radiation lobes at desired frequency. The antenna which has minimum side lobes at desired frequency increases the performance of the antenna. It also has excellent circular polarization features over the full operating frequency band.

3.4 *Gain Analysis*

Figures 11 and 12 show the proposed antenna gain in different orientation. The overall efficiency of desired antenna is analyzed based on the gain. The antenna gain can be decided based on the radiation of the antenna in a specific direction, and these antennas are highly operated in that direction the gain of antenna mainly depends upon edges, size of the patch and electric permittivity. The gain of a patch radiator describes how much power get generates in a desired direction among whole energy get obtained in comparing with all directions. Here, the maximum gain achieves 4.8 dB.

3.5 *Directivity Analysis*

The directivity of an antenna decides based upon the radiation pattern in a specific direction Figs. 13 and 14 represents the directivity of microstrip patch antenna in specific direction.

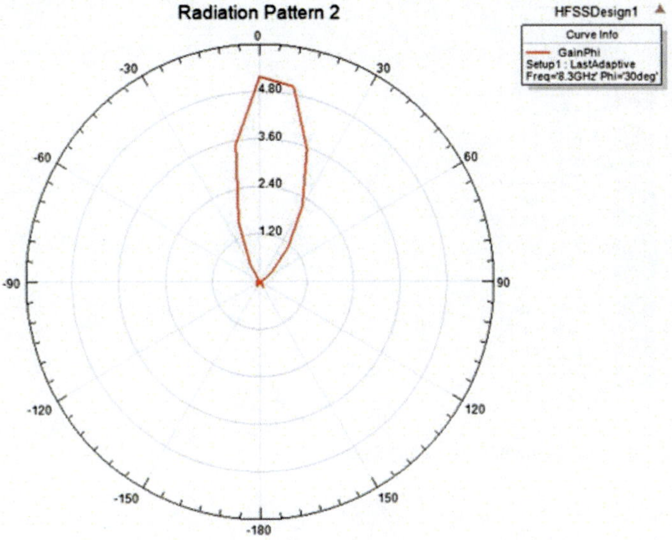

Fig. 11 Gain pattern of octagonal patch antenna at phi 30°

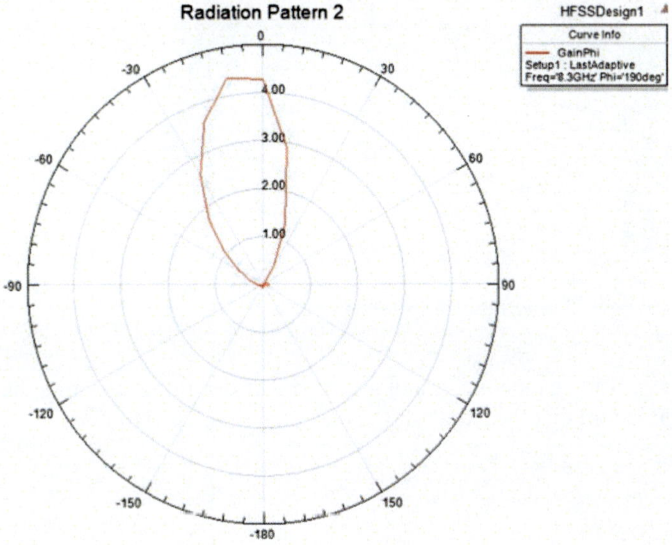

Fig. 12 Gain pattern of octagonal patch antenna at phi 30°

3.6 Axial Ratio

The parameter used to calculate the ratio between major and minor axis of polarization ellipse is known as axial ratio. If the octagonal patch is circularly polarized,

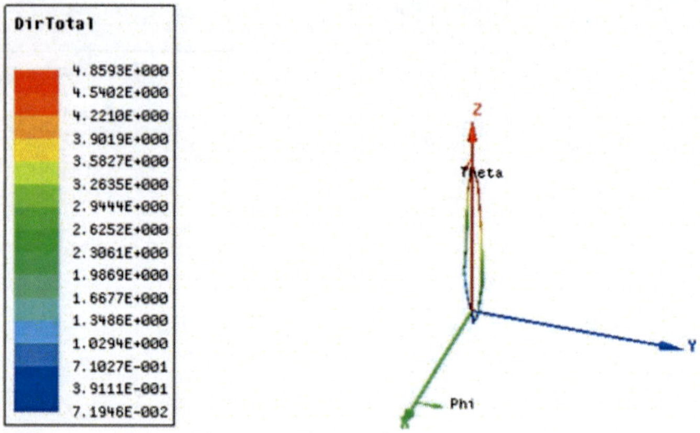

Fig. 13 Directivity of octagonal patch antenna

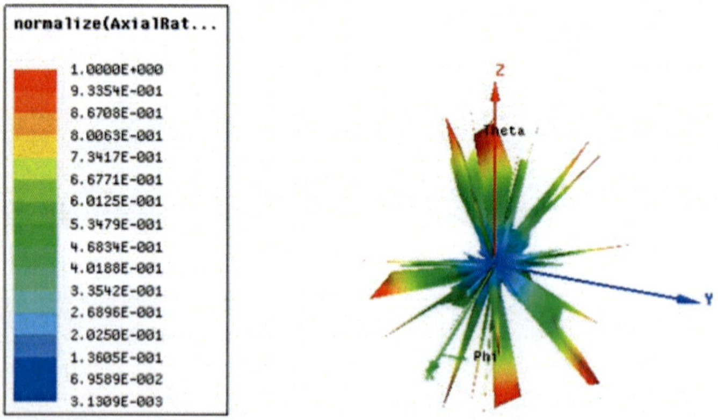

Fig. 14 Axial Ratio of octagonal patch antenna

then the ratio between the major and minor axis of the ellipse is said to be unity. In that case, it is said that the axial ratio of the antenna is equal to unity.

4 Conclusion

In this communication, the simple octagonal patch antenna is proposed for multi-band applications. The antenna uses an *S*-shaped slot radiator along with a octagonal patch structure which operates at multiple operating frequency and for performance improvement of an antenna to have minimum reflection loss, high gain, and better

Table 1 Comparison of existing and proposed work

Design parameters	Existing work	Proposed work
Operating frequency	(1–4) GHz	(8–10) GHz
Operating wavelength	(30–1.5) cm	(2.5–3.75) cm
Energy (%)	25	64
Dielectric material	Sapphire	Rogers
Width of patch	1.6	1.6
Permittivity	10	10.2
Slot	Stacked patch (four layers)	*S*-slot
Feeding technique	Microstrip feedline	Microstrip feedline

VSWR, stable directional radiation pattern and much more compact size and bandwidth. By utilizing the *S*-shape slot, a microstrip patch antenna is successfully developed. The *S*-shaped slot achieves excellent performance for multiband applications like X-band. The antenna is simulated using Ansoft HFSS with its characteristics parameters and the corresponding responses are obtained. The proposed antenna radiates from 8.3 to 9.8 GHz. The single antenna can be used for many X-band applications. The design and simulation are done. The antenna parameters such as return loss, gain, radiation pattern and VSWR have been improved and compared with existing antenna system. The proposed antenna achieves the return loss −21 dB and gain as 4.8 dB. It is used for short distance communications (Table 1).

References

1. Ivanchenko I, Khruslov M, Popenko N, Rönnow D (2019) L-S band combined antenna module. IET Microw. Antennas Propag 13(5):541–545
2. Ghosh A, Kumar V, Sen G, Das S (2018) Gain enhancement of triple-band patch antenna by using triple-band artificial magnetic conductor. IET Microw, Antennas Propoga 12(8):1400–1406
3. Boccia L, Amendola G, Di Massa G (2017) A dual frequency microstrip patch antenna for high-precision GPS application IEEE Antennas Wirel Propag Lett 3:157–160
4. Ivanchenko I, Popenko N, Khruslov M et al (2016) Combined system of the microstrip antennas with different frequencies. In: Proceedings 22nd international conference on applied electromagnetics and communications (ICECom), Dubrovnik, Chroatia, September 2016
5. Mingming F (2015) The microstrip antenna design for multiple frequency small broadband. In: Proceedings international conference on intelligent systems research and mechatronics engineering (ISRME), Zhengzhou, China, April 2015, pp 2159–2162
6. Dikmen CM (2014) Octagonal–shaped UWB antenna with reduced radar cross section. IEEE Trans Antennas propag 62:2946–2953

7. Pazin L, Leviathan Y (2013) Reconfigurable slot antenna for switchable multiband operation in a wide frequency range. IEEE Antenna Wirel Propag Lett 12:329–332
8. Aravindraj E, Nagarajan G, Kumaran RS (2020) Design and analysis of recursive square fractal antenna for WLAN applications. 2020 International Conference on Emerging Trends in Information Technology and Engineering (ic-ETITE), 2020, pp 1–5. https://doi.org/10.1109/ic-ETITE47903.2020.293
9. Menaka R, Nishandhi S, Sivaranjani S (2017) Designing of S shaped microstrip patch antenna for broadband application using slotting technique. IJSR 6(3):2284–2287
10. Nishandhi S, Nagarajan G, Prakash VC (2019) A compact microstrip-fed antenna for 2.4 GHz applications. In: IEEE-proceeding of international conference on systems, computation, automation and networking, March 2019

Skin Cancer Detection Using Deep Learning

Pradhumn Agrahari, Archit Agrawal, and N. Subhashini

Abstract Skin cancer is form of cancer which is a major health concern with 125,000 new melanoma cases every year. Even around 3 million non-melanoma cases can be seen worldwide each year. Initially, diagnosis of skin lesion was done visually by employing dermoscopic analysis, clinical screening, and several other methods. It is seen that inexperienced dermatologists can subsequently reduce the diagnostic accuracy of skin lesion. Early detection of skin cancer has the ability to decrease mortality rate. Earlier studies have demonstrated that deep learning has given a better performance than human experts in various computer vision tasks. The paper proposes an effective multiclass skin cancer detection system having high-performance comparable to that of a dermatology expert. Pre-trained MobileNet model is employed for model building. For training HAM10000 ISIC dataset is used. The model detects skin lesion with a categorical accuracy as high as 80.81%, top-2 accuracy of 91.25% and top-3 accuracy of 96.26%. This is fast and expansible method that can hold the ability for impeccable clinical advancement, including widening the scope and scale of primary healthcare practice.

Keywords Neural networks · Deep neural networks · Machine learning

1 Introduction

Skin malignant growth, a significant type of disease, is a basic general medical issue with 123,000 recently analyzed melanoma cases and around 300 thousands non-melanoma cases worldwide each year. Melanoma (Mel), basal cell carcinoma (Bcc), and squamous cell carcinoma are the prime skin cancer malignancy. Around 90% of non-melanoma and 86% of melanomas skin cancers are caused due to the exposure of harmful ultraviolet (UV) radiation. It has been surveyed that "a further 10% decline in ozone levels leads to an extra 300,000 non-melanoma and 4500 melanoma cases of skin malignant growth cases" [1]. Early detection and screening of skin malignancy can possibly lessen the death pace of skin cancer. Deep learning algorithms have

P. Agrahari (✉) · A. Agrawal · N. Subhashini
Vellore Institute of Technology, Chennai, India

A. Sivasubramanian et al. (eds.), *Futuristic Communication and Network Technologies*, Lecture Notes in Electrical Engineering 792,
https://doi.org/10.1007/978-981-16-4625-6_18

given unprecedented results in computer vision, and moreover, outclassed humans in the field of gaming, e.g., object recognition, which drives our research on automatic screening and analysis of skin cancers. We used the Classical CNN to develop a model for classification with 2–5 layers. Then, we also worked with ResNet50 Architecture and also MobileNet model for better accuracy and performance. MobileNet architecture is best suitable for mobile applications with less computing power. It essentially helps to reduce the number of training parameters by using depth-wise separable convolutions, which makes the model more lightweight.

1.1 Objectives

This paper aims to create an online web app that can tell user the three highest probability diagnoses for given skin cancer. We have also chosen and developed a suitable algorithm/architecture to develop machine learning model to detect skin cancer with appreciable accuracy. For this to be possible, a suitable dataset [2] with a large number of skin cancer dermoscopic images has to be selected. We have also focused on developing a preprocessing technique for image tested real time.

2 Literature Survey

In [3], the author presented a "completely automated system of dermatological disease recognition through lesion images, a machine intervention in contrast to conventional medical personnel-based detection. They have used the backpropagation algorithm (neural networks), support vector machine (SVM), and CNN for the detection of skin cancer. The model comprises of three phases. It starts with data collection, and then data augmentation is done. Finally, model designing and prediction are done. AI algorithms like convolutional neural network and support vector machine are combined to form a better framework with image processing tools leading to an accuracy of 85%."

In [4], the author focused on developing a method using deep convolution neural network for predicting skin cancer lesion. They have used GoogleNet (Inception v3) for the detection of skin cancer. Skin cancer image has been taken as input, and inception V3 is applied in which convolution, pooling, softmax, and fully connected operations are performed. They achieve 75–85% accuracy.

In [5], author gives a brief introduction to machine learning, deep learning, and neural networks. Deep learning can detect features through self-learning and training models and can provide better accuracy compared to AI. The paper uses gradient descent, ReLU, and also optimizers. They also use several pooling layers during training to get the important features. In case, in the absence of high computational power and hardware during the preparation of data, then the Google Colaboratory framework can be used. Its average accuracy was found to be around 75%.

3 Methodology

A dataset [6] which consists of more than 10,000 dermoscopy images of skin cancer is selected. Then preprocessing, model training and testing, and saving it for the next module is done. The project also consists of several modules that are needed to develop parallelly. The blocks used in the process of skin cancer detection application are shown (see Fig. 1). The first block is selection of dataset. Next step has eight sub-steps for model building and evaluation which is done in jupyter, which are as follows:

Step 1: Importing of essential libraries
Step 2: Loading pictures and making dictionary of images and labels
Step 3: Categorical labeling
Step 4: Normalization
Step 5: Train and test split
Step 6: Model building
Step 7: Cross-validation of model
Step 8: Testing of model.

The next step is the development of web application. The application is developed using Nodejs, HTML5, Nodejs, CSS8, etc. Front-end is developed using HTML, CSS, Nodejs, Jquery. The trained model is saved in a JSON format. Tensorflowjs is used for uploading model in server. Once the user uploads the picture in a web application, it automatically preprocesses the image and shows probability that a mole/discolored skin being cancerous within a fraction of a second. The block diagram is shown in Fig. 1.

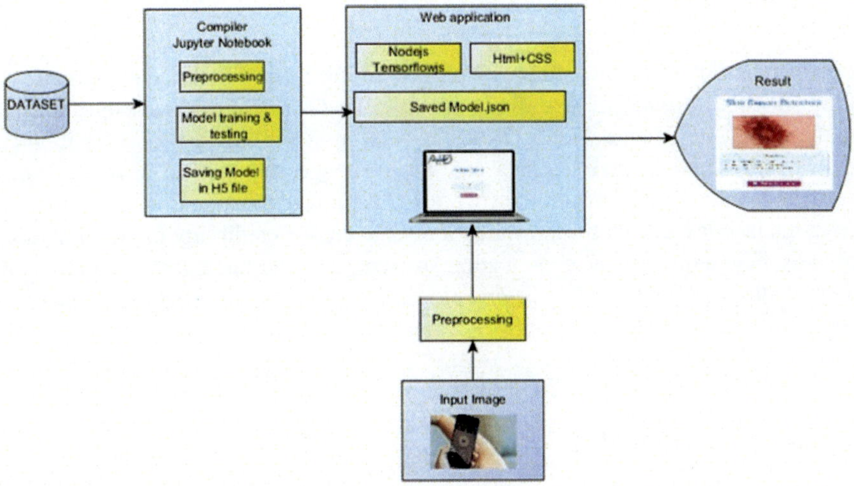

Fig. 1 Block diagram of skin cancer detection

Table 1 Categories of skin lesions

Categories of malignant skin cancer	Categories of benign skin cancer
Melanoma (mel)	Melanocytic Nevi (nv)
Basal cell carcinoma (bcc)	Benign keratosis (bkl)
Actinic keratoses and intraepithelial carcinoma (akiec)	Vascular lesions (vasc)
	Dermatofibroma (df)

3.1 Dataset

HAM10000 is a huge collection of dermoscopic skin cancer images collected by different sources, which are used for training and validation of the model. HAM10000 stands for human against machine with 10,000 training images. HAM10000 is a standard dataset with over 50% of lesions that are verified by experts and pathologists, and the rest of the cases have been approved by in-vivo confocal microscopy or expert consensus.

The dataset contains a total of 10,010 skin lesion which includes 6700 Melanocytic nevi images (nv), 1113 Melanoma images (mel), 1099 Benign keratosis images (bkl), 514 Basal cell carcinoma images (bcc), 327 Actinic keratosis images (akiec), 142 Vascular images (vasc), and 115 Dermatofibroma images (df) with 600 × 450 pixels. Table 1 shows the different categories of skin lesions.

3.2 Dermoscopy Image Preprocessing

Keras ImageDataGenerator [7] is used for preprocessing of skin lesion images. Mean filling method is used to fill 48 null age entries in the dataset. The dermoscopic images in the dataset are downscaled to 224 × 224 resolution to bring images compatibility with the MobileNet model. The dataset of 10,000 images was divided into the training set of 907 images and a validation set of around 940 images. Duplicates were extracted from the training dataset for the validation set to ensure credibility in the validation process. Few steps such as segmentation, overlapping, finding region of interest, and downscaling are shown in Fig. 2. The processes also involved is data augmentation for expanding dataset.

4 Training Algorithm

For this paper, we started our training with basic CNN with two layers. We implemented the two-layered CNN as it is basic building block of deep learning models.

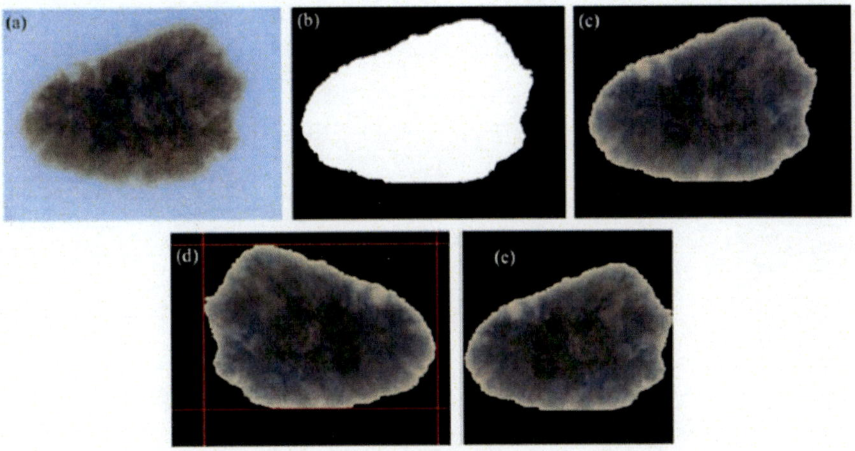

Fig. 2 Images after each preprocessing steps: **a** the original image obtained from ISIC 2018 challenge, **b** the segmentation mask of the image, **c** the overlap of (**a**) and (**b**), **d** Region of Interest, and **e** cropped and resized to $n \times n$

It helps in understanding complex architecture like ResNet. Further, we moved to complex algorithm namely ResNet and MobileNet for better accuracy. Details of the proposed algorithm are given in the following sections.

4.1 Convolutional Neural Network (CNN)

CNN [8] was incorporated first with the help of Keras. The convolutional (Conv2D) layer (see Fig. 3) is the first layer. We chose to set 64 filters for the first two conv2D layers. The next important layer in CNN is the MaxPool2D pooling layer. Combining convolutional and pooling layers, CNN can merge local features and learn important and global features of the image. Then dropout layer, flatten layer and Rectifier "ReLU" constitute the last few layers.

4.2 ResNet50

ResNet [9], short for residual networks, is a classic neural network used intensely in the field of object recognition. Skip connection is the strength of ResNet as shown (see Fig. 4). They mitigate the vanishing gradient's problem by allowing this alternate shortcut path for the gradient to flow through. It allows the model to learn an identity function that assures that the higher layer will give results at least as good as the lower layer.

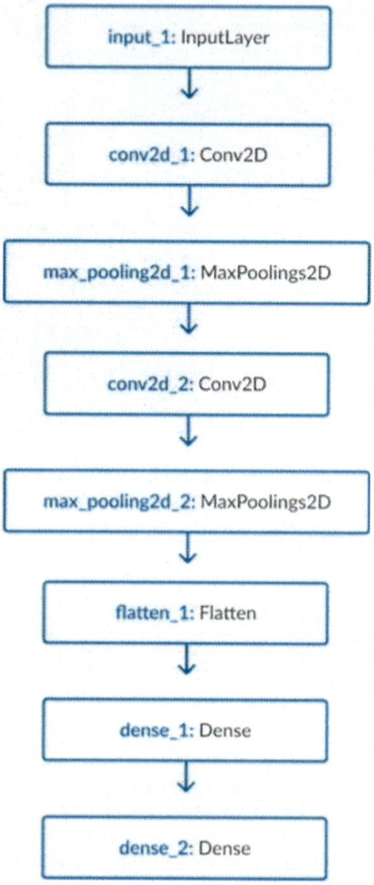

Fig. 3 Block diagram of CNN

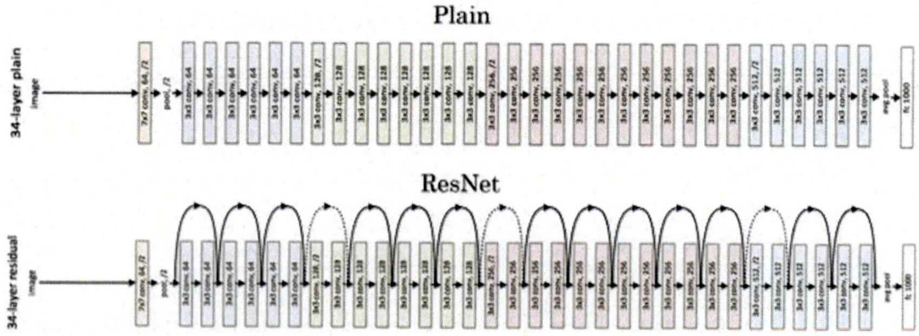

Fig. 4 ResNet50

4.3 MobileNet

MobileNet [8, 10] is ideal for mobile and embedded systems containing computer-aided vision applications as they have lightweight deep neural network architecture. We employed MobileNet CNN pre-trained on 1.28 million skin lesion image comprising 1000 object categories from ISIC dataset. The MobileNet [8] architecture was constructed which has four Conv2D layers, seven ReLU layers, three ZeroPadding2D layers, seven batch normalization layers, and single DepthwiseConv2D, and Global Average pooling along with dense layer and dropout. Transfer learning is used for training our model on a dataset of around 39,000 images with batch size and epochs of 10 and 50, respectively.

4.4 Deployment

As part of this project, we have developed a web application using TensorFlow and Keras libraries. We have built the model and convert it from Keras to Tensorflow.js. The basic layout of the page is developed through HTML5 and CSS8. We have used the Jquery and javascript to make web pages more interactive, simple, and user-friendly. Later on, with the help of Node.js which is an open-source server environment and allows us to run JavaScript on the server. Finally, we have hosted our web application on GitHub.com. GitHub Pages are public webpages hosted and easily published through GitHub. We have created an aid app repository in GitHub and pushed code with the help of git. GitHub pages were incorporated to host our web application.

4.4.1 Advantages of TensorFlow

Tensorflow.js is a new library that allows machine learning models to run in the browser—without having to download or install any additional software. Because the model is running locally, any data that a user submits never leaves his or her pc or mobile phone. We imagine that privacy is especially important when it comes to medical data.

5 Result

After model training and testing of the proposed algorithm, namely CNN, ResNet and MobileNet, give the following results.

Fig. 5 Model accuracy for classical convolutional neural network

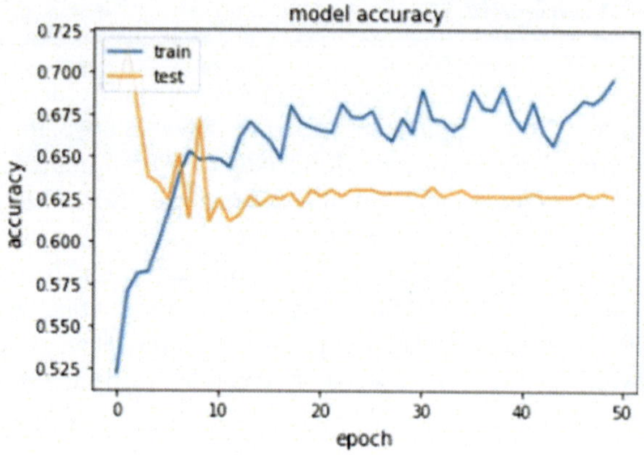

5.1 CNN

The model accuracy for training was around 70%, and testing was around 65%, as shown (see Fig. 5). It has been cleared after 40 epoch test accuracy is constant.

5.2 ResNet50

The model accuracy for training was around 95%, and testing was around 82%, as shown in see Fig. 6.

Fig. 6 Model accuracy for ResNet50

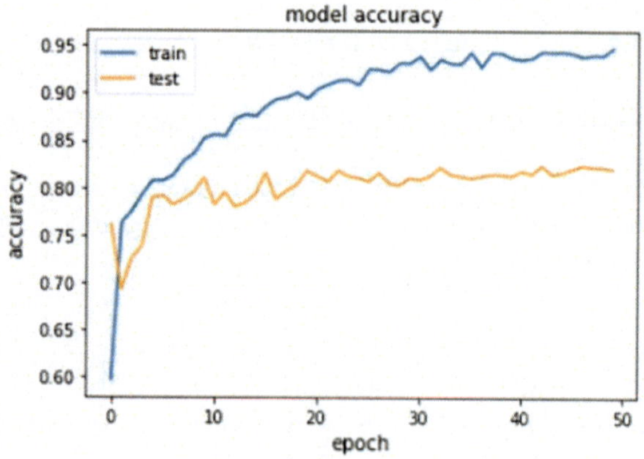

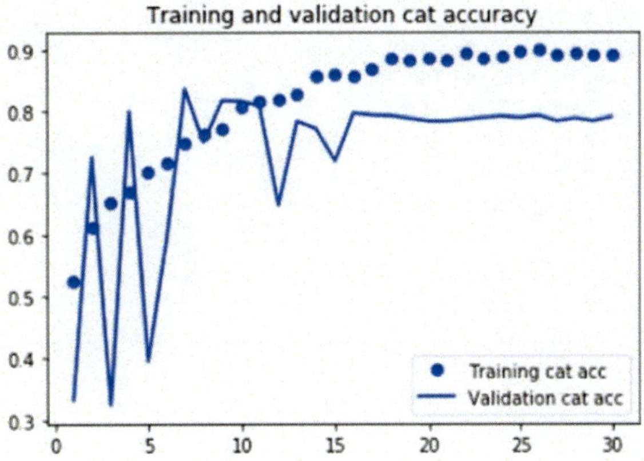

Fig. 7 Model accuracy for MobileNet architecture. Figure shows accuracy on X-axis and epoch on Y-axis

5.3 *MobileNet*

The MobileNet model shows a categorical accuracy of 80.81%, top2 accuracy of 91.2%, and top3 accuracy of 96.26% as shown (see Figs. 7 and 8). Since MobileNet gave the best result as shown in (Table 2) comparison chart, so we chose it for our web application.

5.4 *Web Application*

The homepage of our application named as "AID," where users can upload images of skin lesions is shown (see Fig. 9). It can be uploaded from the gallery or on-spot clicked with the camera. Now our web application is live at http://archit1998.github.io/Aid_app.

The sample output when we uploaded a Basal cell carcinoma (bcc) is shown (see Fig. 10). When we uploaded a bcc class skin lesion, we got accurate results with a probability of 0.9999995.

6 Conclusion

Using MobileNet, we trained on a total of around 39,000 dermoscopic skin lesions from the HAM10000, we matched the performance of dermatology experts among seven skin lesion class. The MobileNet model shows a categorical accuracy of 80.81%, top-2 accuracy of 91.25%, and top-3 accuracy of 96.26%. The weighted avg. of precision, recall, and F1-score was found to be 0.89, 0.73, and 0.76, respectively. The MobileNet model is optimized to develop efficient real-time computer-aided

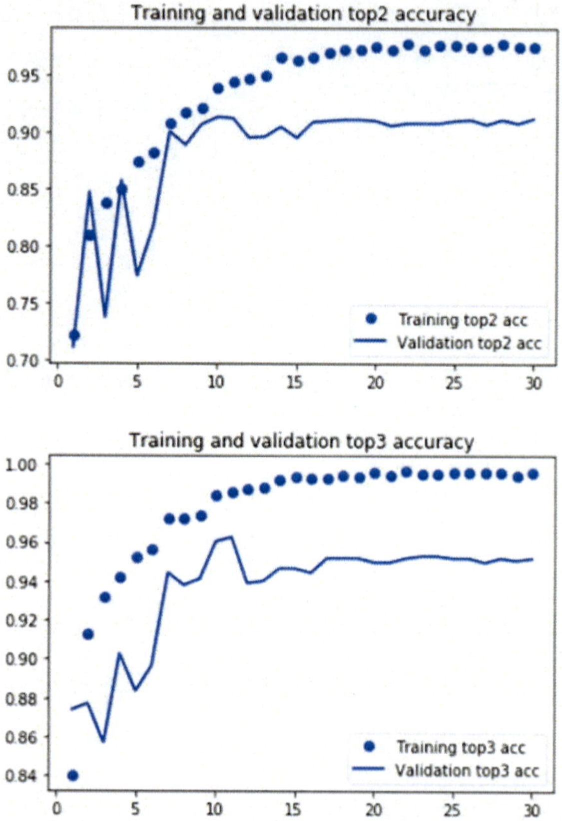

Fig. 8 Training and validation result for top-2 and top-3 accuracy for MobileNet. Figure shows accuracy on *X*-axis and epoch on *Y*-axis

Table 2 Comparison between different architecture

S. No.	Model	Training accuracy (%)	Testing accuracy (%)
1	Classical CNN	70	65
2	ResNet50	95	82
3	MobileNet	93 (cat) 98 (top2) 99.8 (top3)	80.81 91.2 (top2) 96.26 (top3)

systems to diagnose skin lesions. The MobileNet model has given significant accuracy along with robustness as compared to previously proposed models, in addition to its faster and lightweight architecture. We had deployed our deep learning model as a web application.

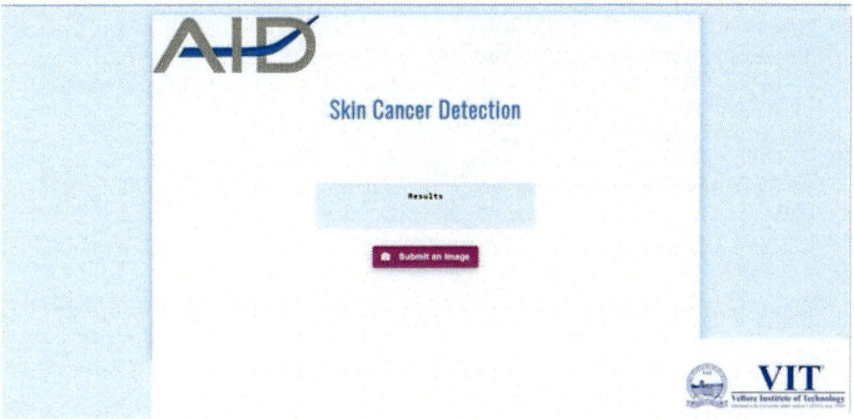

Fig. 9 This is the homepage of the application

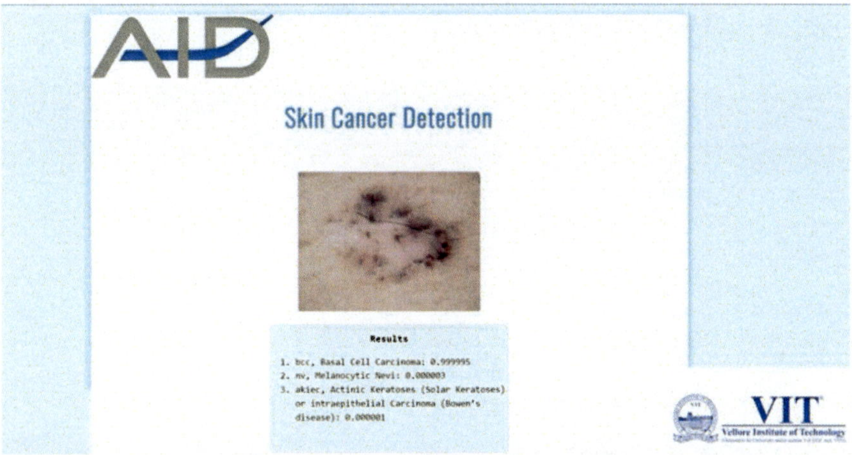

Fig. 10 Accurate result for BCC class skin lesion

References

1. https://www.who.int/docstore/peh-uv/pub/whoehg95-17.htm
2. ISIC Homepage (https://isic-archive.com/) [Dataset]
3. Vijayalakshmi MM (2019) Melanoma skin cancer detection using image processing and machine learning. Int J Trend Sci Res Dev (IJTSRD) 3(4):780–784
4. Bhavya V, Narasimha G (2018) Classification of skin cancer images using TensorFlow and inception v3. Int J Eng Technol 7(2.7):717–721
5. Kanani P, Padole M (2019) Deep learning to detect skin cancer using Google Colab. Int J Eng Adv Technol (IJEAT) 8(6)
6. https://challenge.kitware.com/#challenge/560d7856cad3a57cfde481ba
7. Keras Homepage https://keras.io/ last accessed 2020/06/30

8. Howard AG, Zhu M (2017) MobileNets: efficient convolutional neural networks for mobile vision applications. arXiv: 1704.04861v1
9. Szegedy C, Ioffe S, Vanhoucke V, Alemi, Alemi A (2017) Inception-v4, inception-resnet and the impact of residual connections on learning. In: Proceedings of the thirty-first AAAI conference on artificial intelligence, vol 4, p 12
10. Ramlakhan K, Shang Y A mobile automated skin lesion classification system. In: IEEE 23rd international conference on tools with artificial intelligence, Boca Raton, FL, pp 138–141 (2011)

Denoising X-Ray Image Using Discrete Wavelet Transform and Thresholding

P. Rajeswari, K. K. Thyagharajan, V. S. Prabhu, and G. Shree Devi

Abstract Medical imaging helps in acquiring structure of internal organs for diagnosing and analysis of the diseases. There is wide variety of medical image acquiring technologies. Despite the advancements in medical image acquiring technologies, the possibility of the presence of noise in the image is inevitable. Image denoising plays crucial role in the removal of such noise. This paper gives a clear insight into removal of noises present in X-ray images using discrete wavelet transformation with thresholding. The image quality assessment metrics PSNR and SNR are used to ascertain the performance of the denoising techniques.

Keywords Image denoising · Discrete wavelet transform · Image quality

1 Introduction

Diagnosis of a disease or degree of malfunctioning of an organ highly depends on the clarity of medical image acquired for diagnostics [1]. X-ray images are generally demanded by doctors when they want to have a clear idea on the problems present in lungs like pneumonia, breast cancer, bone fractures, dysphasia and tooth complications. X-rays are popular as they are less time consuming, economical and painless procedure used to look at the internal part of the body which cannot be seen with the naked eye [1]. The quality of the X-ray image depends on its sharpness and contrast. X-rays are electromagnetic radiations that pass through body to generate image of the internal structure of the body being exposed to it.

X-ray images are produced; the body part that needs to be imaged is placed between the X-ray source and X-ray detector. Tissues' radiological density helps in the formation of the X-ray images. Bones have high radiological density and hence produce high contrast on the X-ray detectors [2]. Whereas muscles and lungs have

P. Rajeswari (✉) · K. K. Thyagharajan · V. S. Prabhu
RMD Engineering College, Chennai, Tamil Nadu, India

G. Shree Devi
B. S. Abdur Rahman Crescent Institute of Science and Technology, Chennai, Tamil Nadu, India
e-mail: shreedevi@crescent.education

A. Sivasubramanian et al. (eds.), *Futuristic Communication and Network Technologies*, Lecture Notes in Electrical Engineering 792,
https://doi.org/10.1007/978-981-16-4625-6_19

less radiological density and thus produce low contrast (shades of gray) on the X-ray detectors. However, clarity in the X-rays is predominantly important for diagnosing the diseases [1]. X-rays are prone to noise. Noise present in X-ray images occurs due to the variation in the detection of the amount photons in every pixel. Not every pixel in a digital detector detects same number of photons. The pixels that accept more X-rays appear darker, and the pixels that accept less X-rays appear lighter. The distribution of lighter and darker pixel is random in nature. This results in a noisy image.

Noise is a disagreeable signal that gets attached to the required signal while X-ray acquisition [3]. Noise modifies the value of a single pixel or a group of pixels in an image, which reduces the image quality. Due to the randomness of the reception of photons in the X-ray detectors, the noise is characterized to follow Poisson distribution [4, 5]. Poisson distribution is represented by the equation

$$P(X = x) = \frac{\lambda^x e^{-\lambda}}{x!} \tag{1}$$

Poisson noise is otherwise called as quantum noise. The X-ray dose may be amplified to minimize Poisson noise. But, it needs to be limited due to the medical term known as maximum permissible dose (MPD) [4]. The X-ray dose must not surpass the MPD limit in view of safety of the patient. So, quality of the image can be enhanced by applying image denoising algorithms [6]. Subsequently, the various denoising algorithms are discussed to understand and ascertain what positive and negative impact these algorithms have on X-ray images.

2 Related Work

Image enhancement can be applied to images directly in the spatial domain or in frequency domain. In spatial domain, the image pixel values are manipulated for infusing change in the image. The manipulation of the pixels happens based on mathematical calculations. Frequency domain represents the rate of change of pixels in the spatial domain. So any change made in the frequency domain reflects on the whole image. Also, the main purpose to convert image from spatial to frequency domain is filtering. Image is a combination of smooth regions, edges and noise which depict abrupt changes in contradictions. Smooth regions form a part of low-frequency components, whereas edges and noise form a part of high-frequency components. These edges give us most valuable information about the image [3].

Discrete Fourier and cosine transformation is good tool for image analysis. Discrete Fourier transformation (DFT) represents image data as a sum of sine waves. Discrete cosine transformation (DCT) represents image data with fewer frequency components which leads to energy compaction [3]. They fail to represent the abrupt changes efficiently. Neither DFT nor DCT is localized in terms of time and space.

Discrete wavelet transformation (DWT) has emerged to be more significant than the discrete cosine and Fourier transformations [7]. With the help of wavelet transformation (WT), decomposing of the signal into a group of component signals, known as wavelets each with a clearly defined, dominant frequency. The problem with the Fourier transformation (FT) is when converting from time domain to the frequency domain, the information with respect to time or space is missing. The WT splits a signal into many scales demonstrating different frequency bands were in each scale the position of the WT can be obtained at the significant time characteristic subsequently the noise can be removed after identification [7, 8]. Short-time wavelets allow information to be extracted from high-frequency components [9]. This information is essential to get rid of noise as noise mostly exhibit high-frequency fluctuations. Wavelet analysis provides more accurate information about the image than DFT and DCT techniques [9].

The DWT is applied in two directions (n1, n2), namely horizontal direction and the vertical direction. To do this, we define four filters given in Eqs. 1–4. Change the equation numbers according to previous value, and the format $\varphi(.)$ denotes low-pass filter and $\psi(.)$ denotes high-pass filter.

$$\varphi(n1, n2) = \phi(n1)\phi(n2) \tag{2}$$

$$\psi(n1, n2) = \psi(n1)\phi(n2) \tag{3}$$

$$\psi(n1, n2) = \phi(n1)\psi(n2) \tag{4}$$

$$\psi(n1, n2) = \psi(n1)\psi(n2) \tag{5}$$

Equation 2 denotes low-pass filter in horizontal and vertical directions. Equation 3 denotes high-pass filter in horizontal direction and low-pass filter in the vertical direction. Equation 4 denotes low-pass filter in horizontal direction and high-pass filter in the vertical directions. Lastly, Eq. 5 denotes high-pass filter in both vertical and horizontal directions.

Hence, any given image is split into four sub-band images as given in Fig. 1.

LL^1 sub-band represents the low-pass filtered image both in the horizontal and vertical directions, HL^1 sub-band represents the input image that is high-pass filtered image horizontally and low-pass filtered vertically meaning, the horizontal edges are

LL^1	HL^1
LH^1	HH^1

(a)

<table>
<tr><td>LL^2</td><td>HL^2</td><td rowspan="2">HL^1</td></tr>
<tr><td>LH^2</td><td>HH^2</td></tr>
<tr><td colspan="2">LH^1</td><td>HH^1</td></tr>
</table>

(b)

Fig. 1 Two dimensional DWT on an image

highlighted, LH^1 sub-band represents the input image that is horizontally low-pass filtered and vertically high-pass filtered meaning, the vertical edges are highlighted. Finally in HH^1 sub-band, the diagonal edges are highlighted. All the sub band images are one-fourth size of the original image. Further when LL^1 sub-band is decomposed using discrete wavelet transformation, LL^2 sub-band represents the second level of decomposition of LL1 sub-band which is low-pass filtered horizontally and vertically. Similarly, we have HL2, LH2, HH2, which represent second-level decomposition of LL1 sub-band where horizontal edges are highlighted, vertical edges are highlighted and diagonal edges are highlighted, respectively [9].

3 Implementation

Let us consider an X-ray image of size n x n pixels. During acquisition, the image is populated with Poisson noise. Figure 2 shows the output after applying DWT on the X-ray image. After decomposing the image using discrete wavelet transformation, thresholding is applied for noise removal. Thresholding is categorized into two types, namely hard thresholding and soft thresholding [10]. This research work has been carried out using hard thresholding. Literature indicates that better edge preservation with hard thresholding in comparison with the soft thresholding. Soft thresholding gives rise to a smoother image compared to hard thresholding [11]. As the preservation of edges and identification of discontinuities are essential for abnormalities detection, hard thresholding is performed for noise removal. The hard thresholding is the process of setting the value of the pixels to zero if it is lower than the threshold [10]. Let the pixel value be x.

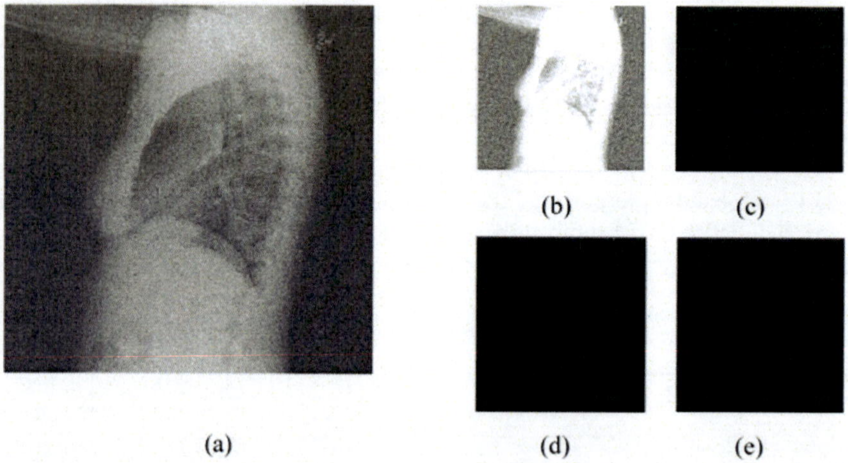

Fig. 2 **a** X-ray image with Poisson noise, **b** LL subband, **c** HL subband, **d** LH subband and **e** HH subband

if x > thr, then pixel value is set to x.
if x ≤ thr, then the pixel value is set to 0.

Thresholding is decided by using any one of the following three functions given in MATLAB. They are multithresh(), graythresh() and adaptthresh().

The MATLAB function multithresh(A) is a thresholding function to compute the threshold of an image which is given as input parameter. This function computes the threshold using Otsu's method. The threshold of given input image is returned as output. The MATLAB function graythresh(I) calculates the global threshold for the input image I using Otsu's method. In Otsu's method, threshold that minimizes the intra-class variance of the black and white pixels is chosen. The threshold value returned by multithresh() lies between [0 255]. The threshold value returned by graythresh() lies between [0 1]. The adaptthresh(I) calculates a locally adaptive threshold for 2-D grayscale images. The threshold in adaptthresh() is chosen assumed on the local mean intensity (first-order statistics) in each pixel's neighborhood.

The algorithm for applying DWT with hard thresholding using multithresh() is given below:

Step 1: Read the input image
Step 2: Add Poisson noise to the image
Step 3: Convert RGB image to a grayscale image
Step 4: Apply first level of DWT using Haar wavelet on gray image with Poisson noise.
Step 5: Apply multithresh() MATLAB function on LL subband, HLsubband, LH subband, HH subband images to find the threshold.
Step 6: Using the determined threshold, perform hard thresholding using wthresh() on HLsubband, LH subband, HH subband images and save the resultant images.
Step 7: Perform IDWT to reconstruct the image from LL subband and hard threshold HL, HLsubband, LH subband, HH subband images.
Step 8: Display the IDWT performed image.
Step 9: Estimate the peak signal-to-noise ratio.

In the above algorithm, multithresh() is replaced using graythresh() and adaptthresh() to apply and experiment with gray thresholding and adaptive thresholding, respectively, and this thresholding value is applied at level1 of the DWT obtained by the images which require noise removal. Compare the PSNR value of the images obtained after the application of various thresholding function to determine their efficiency. The methods multithresh(I), graythresh(I) and adaptthresh(I) return only a single threshold value.

4 Results

DWT is applied on the noisy X-ray image. Figure 3a, b show the X-ray image with noise and without noise, respectively. The results after applying DWT on the

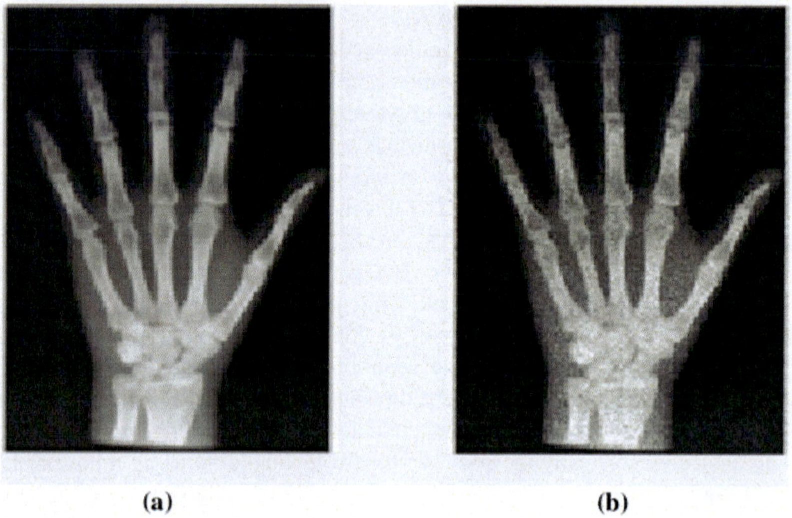

Fig. 3 **a** X-ray image without noise and **b** X-ray image with Poisson noise

noisy X-ray image is shown in Fig. 4. The LL subband represents the approximated image, and the HL subband shows the image that represents high-frequency information along the horizontal direction. The LH subband represents high-frequency information along the vertical direction, and finally, the HH subband represents the high-frequency information along the diagonal.

The thresholding function multithresh(.) is applied on LL, HL, LH, HH subband images. The MATLAB function wthresh(.) is used to apply the thresholding on the image using the value obtained by the multithresh(.) function. The wthresh(.) applies thresholding on LL, HL, LH and HH subband images. Inverse discrete wavelet transformation using the resultant images of LL, HL, LH, HH subband image after applying thresholding is used to reconstruct the X-ray image (Fig. 4).

After reconstruction of the image using inverse discrete wavelet transformation as shown in Fig. 5, to test the performance of the various thresholding techniques, peak signal-to-noise ratio (PSNR) and signal-to-noise ratio (SNR) are calculated, and results are listed in Table 1.

5 Conclusion

In this paper, DWT is applied on noisy X-ray images. On the resultant images of DWT, thresholding using various MATLAB functions is applied for denoising. The thresholding functions used are graythresh(), multithresh() and adapthresh(). It is evident that PSNR value is highest for the image reconstructed after applying thresholding using the adaptthresh(.). The percentage of increase in the PSNR value is

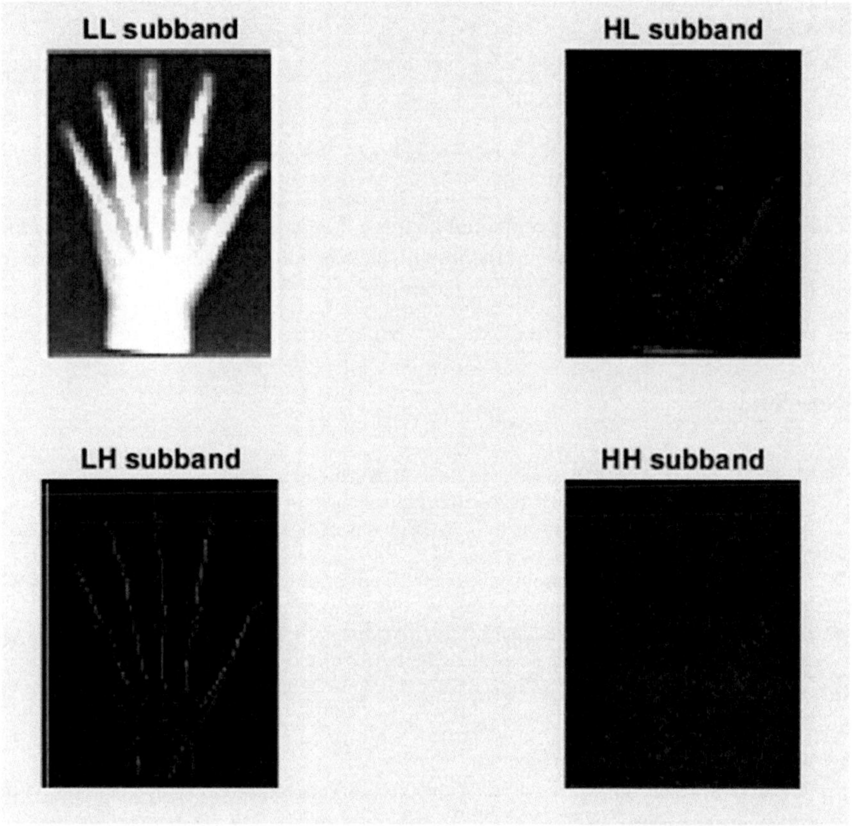

Fig. 4 Results after applying DWT on the X-ray image

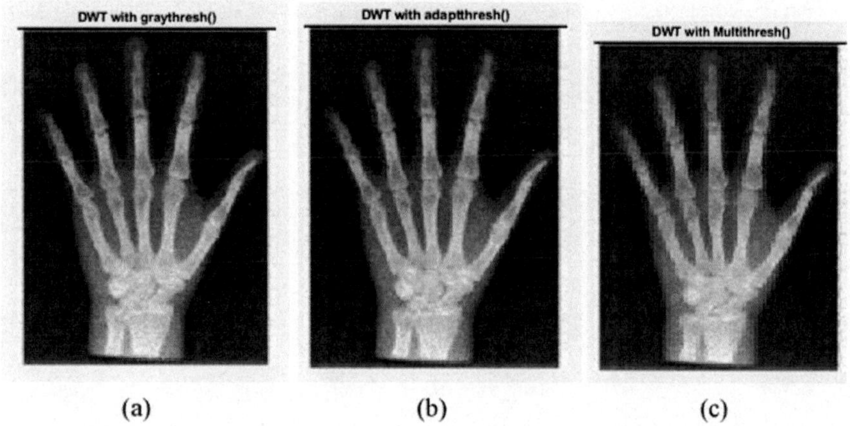

Fig. 5 **a** Reconstructed image after applying graythresh(.), **b** reconstructed image after applying adaptthresh(.) and **c** reconstructed image after applying multithresh(.)

Table 1 Peak signal-to-noise ratio and signal-to-noise ratio

	DWT with multithresh(.)	DWT with graythresh(.)	DWT with adaptthresh(.)
PSNR	21.7367	22.2889	22.9000
SNR	14.0286	14.4978	14.5989

2.74%. It is clear from the experimental analysis that an appropriate threshold value will help to minimize the noise in the image. Hence, further research can be carried out to devise an efficient thresholding technique.

References

1. Mandić I, Peić H, Lerg J, Štajduhar I (2018) Denoising of X-ray images using the adaptive algorithm based on the LPA-RICI algorithm. J Imaging 4:34
2. Ramamurthy PS (1995) Factors controlling the quality of radiograpy and the quality assurance. NTI Bull 31(3 & 4):37–41
3. Joharah F, Nisha S (2017) Noise removal based on discrete wavelet transform and filters. Int J Innov Res Sci Eng Technol 6(6)
4. Kirti T, Jitendra K, Ashok S, Sadhana (2017) Poisson noise reduction from X-ray images by region classification and response median filtering 42(6):855–863
5. Goyal B, Dogra A, Agrawal S, Sohi BS (2018) Noise issues prevailing in various types of medical images. Biomed Pharmacol J 11(3):1227–1237
6. Thanh VB, Surya Prasath Dang NH (2019) A review on CT and X-ray images denoising methods. Informatica 43:151–159
7. Shruthi G, Krishna R (2013) Image reconstruction using discrete wavelet transform. A.N IOSR J VLSI Signal Process (IOSR-JVSP) 2(4):14–20
8. Ameen M, Ahmed SA (2016) An extensive review of medical image denoising techniques. Glob J Med Res 16(2)
9. Kother Mohideen S, Arumuga Perumal S, Mohamed Sathik M (2008) Image de-noising using discrete wavelet transform. Int J Comput Sci Netw Secur 8(1):213
10. Anutam, Rajni R (2014) Performance analysis of image denoising with wavelet thresholding methods for different levels of decomposition. Int J Multim Its Appl 6(3)
11. German-Sallo Z (2016) Nonlinear wavelet denoising of data signals. Ubiquitous Comput Commun J 6(3)

Photobook Creation Using Face Recognition and Machine Learning

N. Aishwarya, N. G. Praveena, B. S. Akash Arumugam, and J. Pramod

Abstract Nowadays, mobile photography plays a pivotal role in major parts of human lives. Every individual captures a lot of photographs in their personal device. However, management of these photographs by person becomes very crucial. Hence, photobook provides an attractive solution for personal storing and printing. In fact, today many commercial products are available to support the above point. But all of them require human annotation which is a time-consuming process and also leads to unreliable authentication. Hence, in this work, face recognition and deep learning techniques are utilized for photobook creation, wherein the user can create albums for a particular group of people or individuals and annotate them. Also, for every new photograph captured, the photobook will update itself automatically. Further, user interaction is introduced in the proposed work which makes it more reliable and results in good performance.

Keywords Face recognition · Convolutional neural networks (CNNs) · Rectifying linear unit (ReLU) · Computer vision (CV)

1 Introduction

With the impetuous advancement in the digital world, people capture zillions of photographs every single day, and managing and organizing photographs manually is a laborious task and time-consuming process [1–3]. Photobook is found to be an attractive solution to overcome this limitation, which can be automated by albuming. Today, many commercial products are readily available to serve this purpose. But they involve human annotations, which will be a herculean process that very few will prefer. Therefore, it is extremely recommendable to automate this cataloging and organizing process. Organizing photographs into albums is one way to keep similar photographs together and makes it easily accessible to the user. Automatic

N. Aishwarya (✉)
Amrita School of Engineering, Amrita Vishwa Vidyapeetam, Chennai, India

N. G. Praveena · B. S. Akash Arumugam · J. Pramod
R.M.K College of Engineering and Technology, Anna University, Chennai, India

A. Sivasubramanian et al. (eds.), *Futuristic Communication and Network Technologies*, Lecture Notes in Electrical Engineering 792,
https://doi.org/10.1007/978-981-16-4625-6_20

annotation of faces in photographs can be done using face detection and face recognition algorithms. In the past decades, extensive studies have been done in the field of computer vision and machine learning for the above techniques [4–6]. With the advent of many vigorous face detection algorithms, the face annotation process does not appear to be an intricate process. However, the challenge still lies in the process of face recognition. With the advent of many robust face detection algorithms, the process of face annotation does not seem to be an intricate process. However, the challenge still lies in the process of face recognition. On the one hand, due to high variance in illumination, facial expressions, cues, poses, etc., in real-time pictures, it is often tough to extract precise facial features and develop an efficient face model. On the contrary, the existing face recognition algorithms fail to annotate all the faces in group photographs which greatly reduce their efficiency for real-time implementation [7]. Hence, deep learning technology using convolutional neural networks (CNNs) is proposed in this paper to address the above shortcomings.

CNN is a special type of neural network that transforms the given input images into corresponding feature maps using input layer, hidden layer which has convolutional layer, pooling layer, and output fully connected (FC) layer [8–10].

In this work, firstly, CNN is applied to an image to extract the set of unique complex facial features that include length and breadth of the face, width of the nose, eyes, lips, space between the eyes, skin tone, texture, etc. Then, these facial features are compared with the features obtained from the training data set. If the facial features are highly correlated with the training data set, then the image will be added to an existing album. Otherwise, a new album will be created for the annotated image.

The remaining paper is organized as follows. Related works are briefly described in Sect. 2. Section 3 explains the proposed system. Section 4 concludes the paper and future work.

2 Related Work

There is an extensive research on photobook creation using face recognition and machine learning. In this subsection, a brief description of the work done previously in the automatic photobook management was explained.

In [11, 12], two clustering methods were proposed in which the features of the representative images are taken from the data set. Platt [13] adopted a photograph clustering approach in which the temporal information was used for creating auto album. Platt et al. [14] extended their work proposed in [13] in which new interface was developed for assisting the users to get the photographs more efficiently.

In [15], clustering-based approach was proposed for the photograph selection from the set of photographs for creating an album. Papadopoulos et al. [16] developed a clustering approach for photograph representation related to landmarks and events in a particular place. In [17, 18], the face tagging was widely explored in photograph clustering technology based on the features extracted from the face and landmarks.

In this, photographs of same faces were grouped and formed a cluster and hence tagging cost was reduced. Suh et al. [19] grouped the faces into clusters and then labeled them which results in lack of supervision.

Chen et al. [20] proposed a semiautomatic family photo album framework which could categorize the faces in different angles of the same person into a small number of clusters very accurately and effectively. However, for better predication and retrieval of the photographs, time information for face annotation is not carried out. Yuansheng et al. [21] proposed a face album management system composed of two pools: certain and uncertain for organizing the photographs by identity. When more faces have to be recognized, both the pools work together for album organization. If new face is detected, it is initially stored in uncertain pool and later it is recognized with the new faces and then stored in the pool. However, recognizing such faces of conventional identities and discovering new identities in album is difficult.

Yan et al. [22] introduced CNN for face recognition in which Caffe was used for feature extraction during the training and validation process. The facial features were classified automatically on the nine-layered network. In [23], a subset of images was selected from an initially labeled data set and added to the training data set. Synthetic images were generated by feeding noise to every training sample. Once the training data set is created, these were given to the CNN to generate the trained model. The testing set is then recognized with the above trained network. This type of learning comes under supervised learning which involves human-annotated data and demands huge memory capacity and additional computational power due to deep learning models.

Coskun et al. [24] proposed face recognition process consisting of three stages: Preprocessing stage, then color space conversion and resizing of images were developed which extracts the facial features of the images using trained data set. Further, Softmax classifier is used to develop the classified resultant stage. In [25], Google Photos has built-in facial recognition similar to that of Facebook and Apple Photos. This software organizes photographs based on the faces by using an algorithm that identifies and groups photographs of people to help users easily find them. Google Photos provides unlimited online storage of digital images and videos. The original images are compressed after uploading, but the difference is hardly visible. It offers online backup as the images and videos are available in all the devices owned by the user. But, facial recognition feature is prohibited in few countries due to certain privacy laws. It also moves all the images and videos from device to the cloud.

In [26], iPhone has search option to find the pictures of a person, places, object, or event. Recognized faces can be labeled with a name. All the photographs of a person can be found under his/her label. iPhone groups the library photographs into Moments and Collections arranged by time and location. Album feature in iPhone consists of Created albums, Shared albums, and automatic collections of various types of photographs and videos like panoramas, selfies, slow motion, and time lapse. Photographs can also be manually added to the albums by the user. iPhone has made it easier to store the photographs online and access them from any device. The simulation results of the proposed method are found to have better results, but high computation time and supervised learning are essential to train the data set with respect to input images.

3 Proposed System

In this section, the proposed method is presented in detail that shows how CNN helps to extract the crucial features of an image to create a photobook.

3.1 *Overview of Convolutional Neural Network*

In the recent past, deployment of convolutional neural network (CNN) has led to prodigious success in many pattern recognition tasks. This is mainly due to the ability of CNN to work in a visual system similar to that of a human brain. One of the most exciting applications of pattern recognition which is the focal point of the proposed work is face recognition. A convolutional neural network basically has an input and output layer, along with multiple hidden layers. The hidden layers of the CNN consist of convolutional layers, ReLU layers, pooling layers, fully connected layers, and normalization layers. They are driving major growth in the field of computer vision (CV), which has noticeable applications for self-driving cars, drones, robotics, medical diagnoses, security, and treatments for the visually impaired persons.

3.2 *Feature Extraction*

The main motive of this work is to create a photobook which groups the images into appropriate album based on human facial features. Figure 1 shows the overview of the proposed method.

Initially, the training data set is formed randomly from a collection of images captured by multiple persons. The image is captured using digital camera. Then, the face detection technique using Haar cascade classifier [27] is used to detect the presence of a human face or a group of human faces. Then, the detected faces are resized to lower dimension and convoluted with different filters like horizontal edge, vertical edge, blur filter, sharpening filter to extract the features of the face. This forms the convolutional layer. Figure 2 shows the convolutional layer for an image segment. For example, a segment of input image of size 7×7 is taken and four convolutional filters are applied to it. The resultant convoluted matrix and the convoluted image segment are shown in Fig. 2.

After obtaining the convoluted layers, they are given to rectifying linear unit (ReLU) to increase the nonlinearity. The convolution operation and the activation function of ReLU in CNNs are given in Eq. 1.

$$y^{i} = \max\left(0, \sum_{i} k^{ij} \otimes x^{i} + b^{j}\right) \tag{1}$$

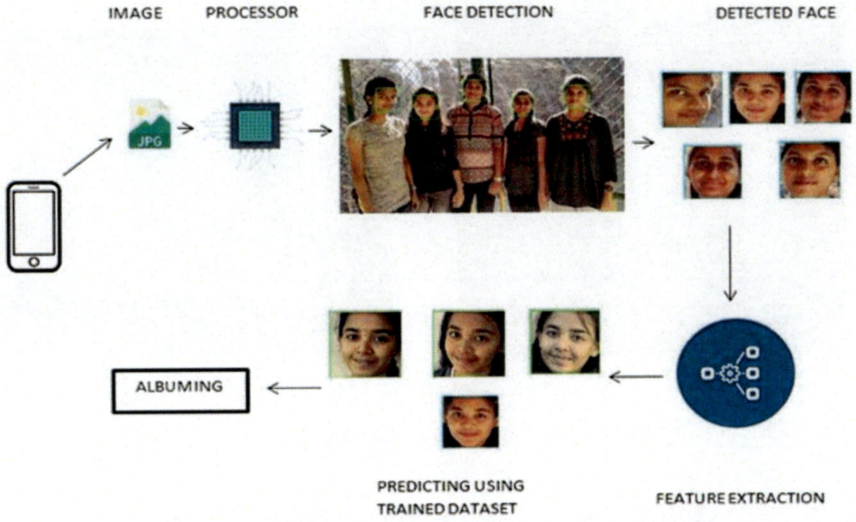

Fig. 1 Overview of the proposed system

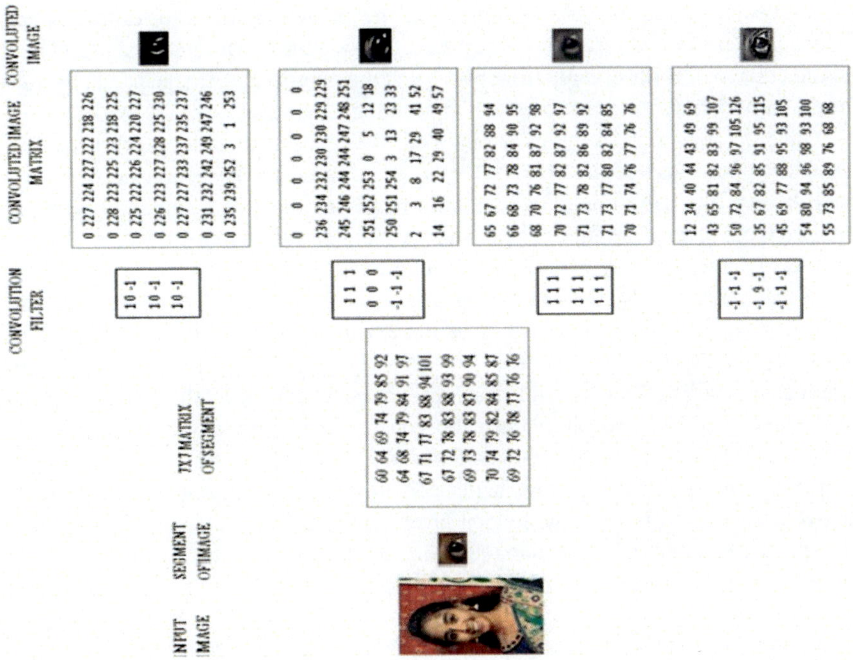

Fig. 2 Example of convolutional layer

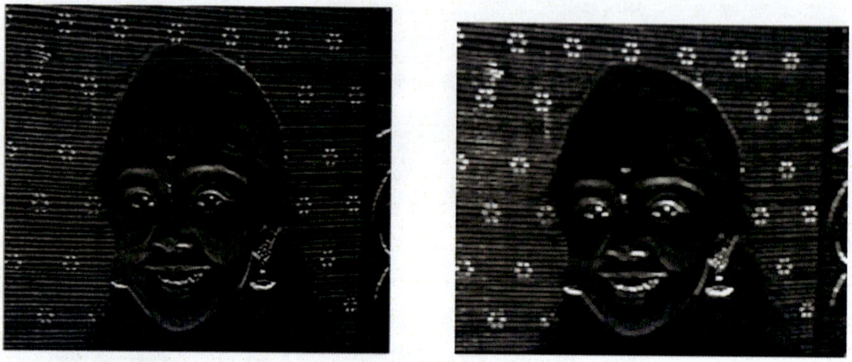

(a) (b)

Fig. 3 CNN outputs **a** Convoluted image and **b** max-pooled image

As a next step, max pooling is done to extract only the important features of the image. To accomplish this task, the network acquires a property known as spatial variance. This property is capable to detect the object in the image of the network without being confused by the differences in the image's textures, the distances from where they are shot, their angles, etc. Figure 3 shows the convoluted image and max-pooled image. It is observable from Fig. 3 that the max-pooled image preserves most of the important features of the given input image.

Once the max-pooled image is obtained, flattening is done to convert the max-pooled output into vector. This vector is fed as input to the neural network. It is well known that neural networks are built of perceptrons. In a neural network, the first layer of perceptron makes decisions by weighing the input evidence. The face features that are extracted from the convolutional layer of CNN are fed as input to first layer of perceptron. Each of the perceptrons in the second layer makes a decision by weighing the outputs of the first layer. For example, the eye extracted from the first layer is examined for iris, eyebrows, eye lashes, etc. In this way, the perceptron in the second layer will make a complex decision at more abstract level than the perceptrons in the first layer. And even more intricate decisions can be made by perceptrons in the third layer. Thus, a more sophisticated decision-making can be done. The weights of perceptron can be changed for better performance by backpropagation [28]. The above multiple layers form the hidden layer, which contains classes of training set. The processed input image vector is compared with all the classes, and the probability of matching is obtained. The input image is annotated by using the class with highest probability.

Figure 4 shows the formation of fully connected layer for the given input image. In Fig. 4, the input image is passed through input layer, hidden layer, and output layer of fully connected layer. The input layer forms a processed image vector of the input image. The fully connected layer contains many hidden layers, and these layers multiply the flattened vector with random weights and bias. The output layer

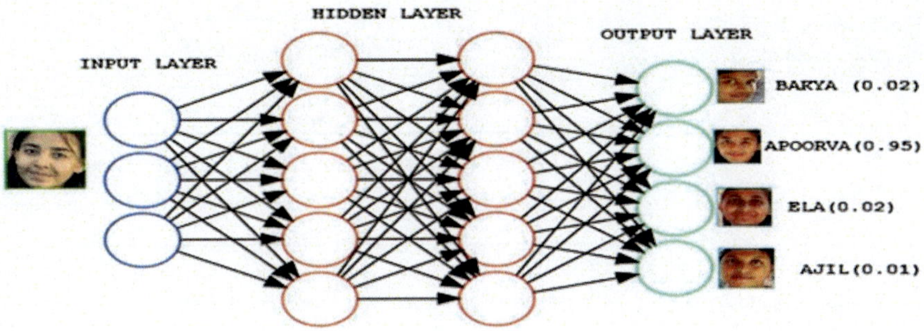

Fig. 4 Formation of fully connected layer

displays the array of face with their respective probabilities. Algorithm 1 explains the detailed process of fully connected layer.

Algorithm 1: Fully Connected Layer

Step 1: Let I be the processed image vector.

Step 2: Form hidden layer h1 by multiplying the image vector I with weight matrix w and add bias b.

$$h_1 = I * w + b$$

Step 3: Compare h1 with the pre-trained classes in the hidden layer.

Step 4: The class with highest probability is the matched class and the image is annotated with corresponding label

Step 5: If the prediction deviates from actual output, revise the weights using back propagation and go to step 2.

3.3 *Photo Albuming*

Once the human facial features are obtained, the features are compared with the trained data set. If there is only one person and the person already has an album, then the image is stored in their corresponding album. If the person does not have an album, then a dialog box will appear to indicate that the user has to create a new album to store the image. If a group of human faces are detected and the image does not has an existing album, then a dialog box will appear with two options "friends" and "family," and the user has to choose the appropriate album. Whenever an image is taken with the same group of people or some of the people in that group, the image is automatically stored in the appropriate album. Figure 5 shows the work flow of the proposed method.

Fig. 5 Work flow of the proposed method

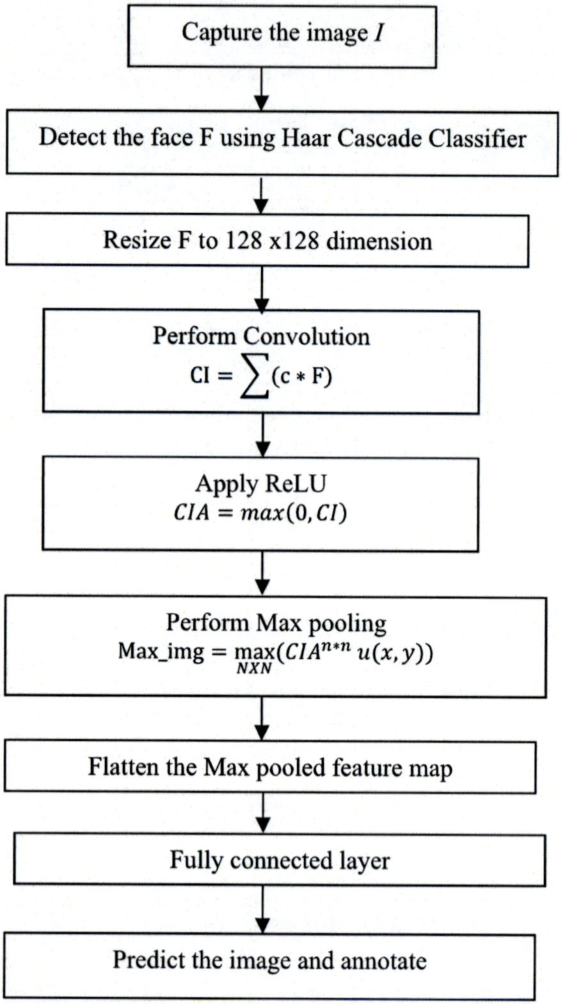

3.4 Results and Discussion

In general, the pooling layer in the neural network reduces the variance and computation effort of the algorithm. There are two types of pooling. One is max pooling which selects the maximum pixel value in the group, and the other one is average pooling which finds the mean of all the pixel values in the specified group. In this paper, max pooling operation is chosen as the accuracy results are better for this operation. To support this statement, Fig. 6 shows the comparison graph between max pooling and average pooling in terms of accuracy for various epochs. From Fig. 6, it is apparent that the proposed technique achieves better results for max pooling operation which

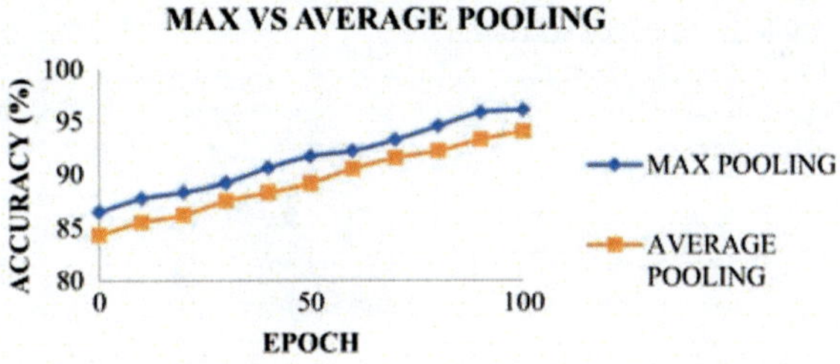

Fig. 6 Comparison between max pooling and average pooling

selects the brighter pixel values of an image and hence the sharp features of the image can be clearly distinguished.

We evaluate our technique with 100 images downloaded from Kaggle, and for each image, augmentation (shifting, rotation, resizing, cropping, etc.) has been done to create 10,000 samples for training the neural network. To show the efficiency of the proposed algorithm, accuracy results of two group of test images are given in Table 1. From , it can be verified that the proposed technique recognizes the faces correctly achieving greater accuracy values. Also, for these test images, only two albums are created in the database which further verifies the effectiveness of the algorithm. To validate the chosen classification model, two parameters precision and recall [20] are considered. Precision indicates the measure of the relevant data points, whereas recall indicates how the model correctly identifies the relevant data points. Figure 7 shows the plot between precision and recall. It is clearly visible that from Fig. 7 the algorithm achieves better classification rate with lesser false positives.

4 Conclusion

In this paper, we present the creation of photograph management system named "photobook" which groups photographs automatically into albums based on human facial features. Initially, once the image is captured, face detection technique using CNN is utilized to detect the presence of a human face or a group of human faces. Then, to enhance the face recognition accuracy, both the low-level and high-level features of the human face are extracted. The extracted features are compared with the training data set. Finally, the albums are created based on the compared facial features. The proposed algorithm provides a faster and easier browsing of photographs of a particular person or a group of people. As a practical system, user interaction is also introduced which makes it more reliable and results in good performance. In the future, the proposed work will be extended to categorize images with non-human faces and objects as well. Furthermore, it will be useful to group videos by recognizing the human faces in them.

Table 1 Accuracy scores for two test images with the application of image augmentation

Test image		Accuracy	Album
	Left shifted	0.9753	Folder 1
	Right shifted	0.9801	
	Shifted down	0.9868	
	Shifted up	0.9789	
	Mirror image	0.9413	
	Left shifted	0.9814	Folder 2
	Right shifted	0.9884	

(continued)

Table 1 (continued)

Test image		Accuracy	Album
	Shifted down	0.9868	
	Shifted up	0.9801	
	Mirror image	0.9445	

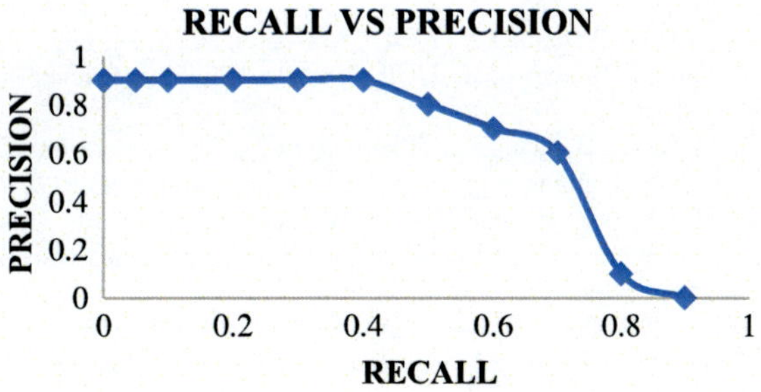

Fig. 7 Precision–recall curve

References

1. Tian Y, Liu W, Xiao R, Wen F, Tang X (2007) A face annotation framework with partial clustering and interactive labeling. In: IEEE conference on computer vision and pattern recognition, 2007. CVPR'07. IEEE, 2007, pp 1–8
2. Jheng H-W, Chen B-C, Chen Y-Y, Hsu W (2014) Automatic facial image annotation and retrieval by integrating voice label and visual appearance. In: Proceedings of the ACM international conference on multimedia. ACM, 2014, pp 1001–1004
3. Cui J, Wen F, Xiao R, Tian Y, Tang X (2007) Easyalbum: an interactive photo annotation system based on face clustering and re-ranking. In: Proceedings of the SIGCHI conference on human factors in computing systems. ACM, 2007, pp 367–376
4. Chellappa R, Wilson CL, Sirohey S (1995) Human and machine recognition of faces: a survey. Proc IEEE 83(5):705–740
5. Zhao W, Chellappa R, Phillips PJ, Rosenfeld A (2002) Face recognition: a literature survey, Technical Report CAR-TR-948, Center for Automation Research, University of Maryland
6. Canedo D, António JR (2019) Facial expression recognition using computer vision: a systematic review. Appl Sci 1–31
7. Wang M, Deng W (2019) Deep face recognition: a survey. arXiv, 1–26

8. LeCun Y, Bottou L, Bengio Y, Haffner P (1998) Gradient-based learning applied to document recognition. Proc IEEE 86(11):2278–324
9. [Online]. Available: https://en.wikipedia.org/wiki/Deep_learning (accessed 2 August 2018)
10. [Online]. Available: http://cs231n.github.io/convolutional-networks/ (accessed 2 August 2018)
11. Kennedy L, Naaman M (2008) Generating diverse and representative image search results for landmarks. In: International world wide web conference
12. Simon I, Snavely N, Seitz, SM (2007) Scene summarization for online image collections. In: International conference on computer vision
13. Platt JC (2000) AutoAlbum: clustering digital photographs using probabilistic model merging. In: Proceedings of IEEE workshop on content-based access of image and video libraries, pp 96–100
14. Platt JC, Czerwinski M, Field B (2003) PhotoTOC: automatic clustering for browsing personal photographs. In: Fourth IEEE Pacific Rim conference on multimedia
15. Sinha P, Pirsiavash H, Jain R (2009) Personal photo album summarization. In: Proceedings of ACM multimedia
16. Papadopoulos S, Zigkolis C, Kapiris S, Kompatsiaris Y, Vakali A (2010) ClustTour: City exploration by use of hybrid photo clustering. In: Proceedings of ACM multimedia
17. Cooper M, Foote J, Girgensohn A, Wilcox L (2003) Temporal event clustering for digital photo collections. In: Proceedings of ACM multimedia
18. Mei T, Wang B, Hua X-S, Zhou H-Q, Li S (2006) Probabilistic multimodality fusion for event based home photo clustering. In: Proceedings of international conference on multimedia and expo
19. Suh B, Bederson BB (2004) Semi-automatic image annotation using event and torso identification. Technical report, HCIL-2004-15, Computer Science Department, University of Maryland, MD
20. Chen L, Baogang Hu, Zhang L, Li M, Zhang HongJiang (2003) Face annotation for family photo album management. Int J Image Graph 3(01):81–94
21. Xu Y, Peng F, Yuan Y, Wang Y Face album: towards automatic photo management based on person identity on mobile phones. In: International conference on acoustics, speech and signal processing. IEEE, 2017
22. Yan K, Huang S, Song Y, Liu W, Fan N (2017) Face recognition based on convolution neural network. Published in IEEE conference 2017. Chinese control conference
23. Aiman U, Vishwakarma VP (2017) Face recognition using modified deep learning neural network. In: 2017 8th international conference on computing, communication and networking technologies (ICCCNT), Delhi, 2017, pp 1–5
24. Coşkun M, Uçar A, Yildirim Ö, Demir Y (2017) Face recognition based on convolutional neural network. In: 2017 International conference on modern electrical and energy systems (MEES), Kremenchuk, 2017, pp 376–379
25. Google Photos—https://www.google.com/photos/about/
26. Apple, macOS—Photos, http://www.apple.com/mac/iphoto/
27. Mustafa R, Min Y, Zhu D (2014) Obscenity detection using haar-like features and gentle Adaboost classifier. The Sci World J, 2014:1–6. Article ID 753860
28. Alsmadi M, Omar K, Noah M, Azman S (2009) Back propagation algorithm: the best algorithm among the multi-layer perceptron algorithm. Int J Comput Sci Netw Secur 9:378–383

Person Location Tracking Using Global Positioning System and ESP8266 with Internet of Things

N. Umapathi, Saiteja Sabbani, and S. Poovarasan

Abstract This paper explores the idea of locating the student's location within the organization/institute. This is done by using a GPS-based receiver modem attached to ID card of student and using a SQL coded database. Initially, the system is provided with all the coordinates of the places within organization/institute, and these coordinates (latitude, longitude and altitude) are stored in a lookup table (LUT) in the database. When the student with the GPS receiver roams inside the campus, his coordinates are recorded by the GPS modem and sent to the database via Internet (cellular/Wi-Fi network). This coordinates are then compared with the coordinates in the LUT; if the coordinates are matched with any of the contents in LUT, then the system identifies this and knows the location of student; otherwise, the student is considered as out of campus. Any authorized person can access this data, i.e., students' location at any time by either Web site or the application provided to the person by entering the unique ID number of the person who is to be tracked. Programming the NodeMCU has been developed in C environment with a self-made SQL database and person location data transfer to Internet of things using ThingSpeak software. This ThingSpeak software is used for the development of application.

Keywords GPS · SQL · NodeMCU ESP8266 · IoT · ThingSpeak

1 Introduction

The main aim of the project is to track a student position who is roaming inside the school or institute campus by using Global Positioning System (GPS) technology. Nowadays, parents face lot of problems while sending their children to the school and returning to home as they can get lost in the way. The idea is to overcome this problem by placing a GPS module in the identity card of the student.

N. Umapathi (✉) · S. Sabbani
Department of ECE, Jyothismathi Institute of Technology & Science, Karimnagar, Telangana, India

S. Poovarasan
R&D Engineer, Error Free Solutions, Erode, Tamil Nadu, India

A. Sivasubramanian et al. (eds.), *Futuristic Communication and Network Technologies*, Lecture Notes in Electrical Engineering 792,
https://doi.org/10.1007/978-981-16-4625-6_21

GPS module is used to track the student's current location, and later, this data is sent to the database. The NodeMCU ESP8266 is interfaced between the GPS module and database. The authorized user can access the data from database with valid unique login ID and can also do same thing using mobile application. Using Internet of things (IOT) technology with GPS module-based system helps access to database remotely and provides additional flexibility to the system. Programming has been developed in C environment with a self-made SQL database.

2 Related Works and Studies

System referenced in [1] has a tracking system based of GPS, GSM and RFID tags. The location of the buses is found by using a GPS module which is placed along with the RFID tags inside the buses, and this live location is monitored by authorized person using their mobile through IoT. Work referenced in [2] is a real-time attendance monitoring system based on GPS and fingerprint scanner. GPS is used to get the live location of the student and send this information to the smartphone. Work in [3] is developed based on widespread usage of Wi-Fi-enabled mobile devices. The system utilizes Wi-Fi-based MD and classifies these MD into static device SD and mobile phone MP using the nonintrusive method. Later, their location variations mapping between SD and its use are achieved. Work in [4] is an indoor positioning technology based on WSN research. Improvement in indoor positioning of user is done using Wi-Fi location of fingerprint. The system applies Gauss filter on the RSSI in fingerprint to remove unwanted multipath and noises. System with reference to [5] tracks the location of people within a bounded environment using RFID transponders and WPS. This location acquisition is achieved by data collection by RFID transponders and next position identification by using Wi-Fi positioning system. Reference [6] is a system based on IoT for monitoring school bus using RFID and GPS using the ESP8266 microcontroller's Wi-Fi server. This location of bus can be accessible by parents using a mobile application. Reference [7] is a system for child safety and security. The system has a combination of various sensors like MQ-3, IR proximity GPS and RFID to ensure the safety and get the live location of the children. Work referenced in [8] is built on ARM7 LPC2148 microcontroller. It uses GPS to get location of the child and GSM to send this information as an SMS of Google Maps link to parents. Work in [9] uses RFID satellite tag to detect the latitude and longitude of the user entity. This is achieved by generating a pseudocode and establishes connection to obtain the reference points on earth. LabVIEW is used for simulation and development of software. Reference [10] is location identification of Indian Railway using a GPS receiver and to alert the passengers about the arrival of trains.

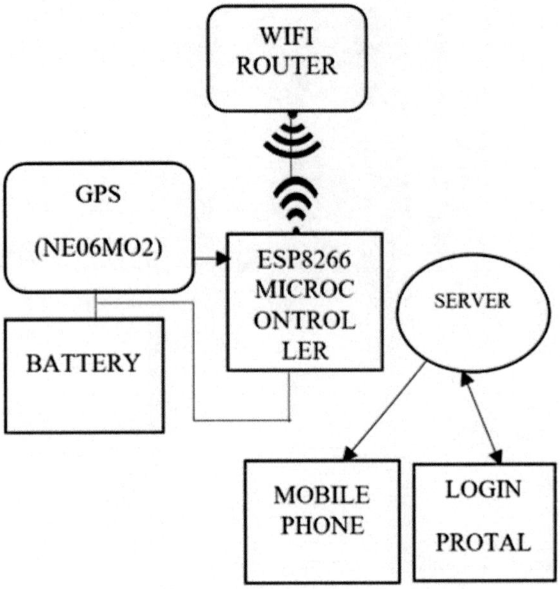

Fig. 1 Block diagram of proposed method of person location tracking using GPS and ESP8266 with IOT

3 Implementation Details

The students will have to wear the identity card which has NodeMCU and GPS module on it. As soon as the supply is turned on, the NodeMCU is automatically connected with the specified hotspot defined in the program. When student moves from one place to another inside the college campus, the GPS present in the ID card will track the latitude and longitude position of the student and sends the data to the database.

In database, we have lookup table (LUT) and tracking table. When the data arises from the GPS is stored in tracking table for few seconds and it is compared with the LUT for finding the location of the student. If the data is matched with the LUT, the current location of the student is identified and is displayed in the application provided (Fig. 1).

4 Components Used for System Design

In this project, system design and architecture comprise two parts:

- Hardware and Internet of things (IoT).

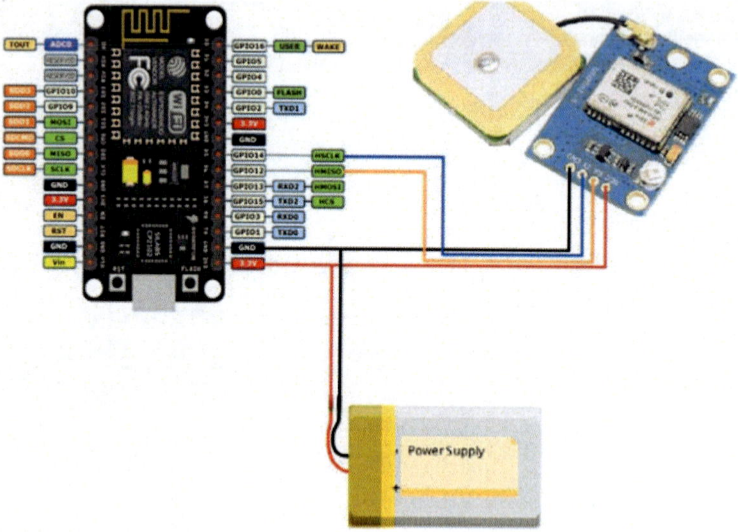

Fig. 2 Circuit diagram of proposed method

4.1 Hardware Architecture

NodeMCU ESP8266

NodeMCU ESP8266 is a microcontroller with a built-in Wi-Fi capability. It is a 32-bit MCU with 16 digital I/O's and 1 analog I/O (with 10-bit resolution). It has a whooping 4 Mb of flash memory for storing programs and 64 Kb of SRAM. It can communicate with sensors/other devices using serial, SPI and I2C protocols at 80 MHz clock speed. To program this microcontroller, we can use Arduino IDE by including required libraries and installing the drivers. One feature which separates this MCU from other is the Wi-Fi function it offers with a 2.4 GHz PCB antenna.

GPS 6MV2 Module

The NEO-6MV2 is a GPS module used for keeping track of location of students in campus. The functionality of the module is to provide the latitude and longitude data. It offers the data update rate of 5 Hz and works with 3.3 V. The module has 4 pins Vcc, GND, Tx and Rx that means it uses UART interface to communicate with the microcontroller. This module is suitable for this project as it has small size with dimensions 16 × 12.2 × 1.4 mm and has a microstrip antenna (Figs. 2 and 3).

5 Results and Experimental Analysis

The NodeMCU when provided with the power supply is connected to the Wi-Fi network which is given in programming. Next GPS data/coordinates from the GPS

Fig. 3 Person location tracking using GPS and ESP8266

module is automatically updated into database, and location of the student status inside the campus is recorded. Student location is then compared with lookup table, exact location is shown to the authorized user by their login, and it is also shown in mobile application and Internet of things.

At first, all the location coordinates of places within the campus (where a student can present) are entered into the LUT in the database. Data entry into LUT is done by giving the latitude and longitude values of particular place and saving them subsequently. Figure 4 shows the model LUT with coordinates of the places within the campus. To search for a student position within campus, user must enter the ID

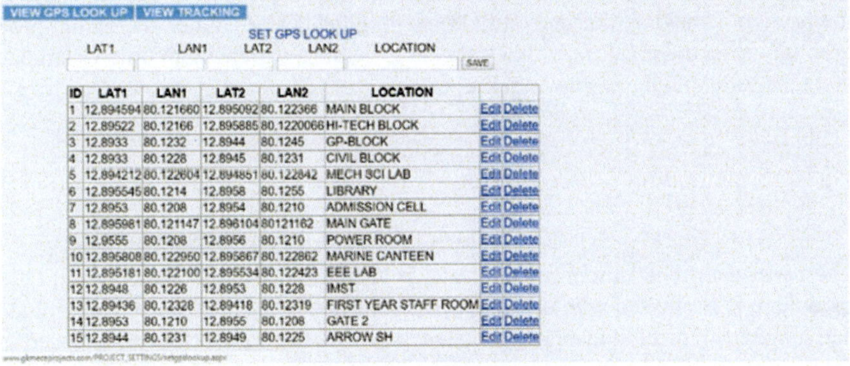

VIEW GPS LOOK UP | VIEW TRACKING

SET GPS LOOK UP

LAT1 LAN1 LAT2 LAN2 LOCATION SAVE

ID	LAT1	LAN1	LAT2	LAN2	LOCATION		
1	12.894594	80.121660	12.895092	80.122366	MAIN BLOCK	Edit	Delete
2	12.89522	80.12166	12.895885	80.1220066	HI-TECH BLOCK	Edit	Delete
3	12.8933	80.1232	12.8944	80.1245	GP-BLOCK	Edit	Delete
4	12.8933	80.1228	12.8945	80.1231	CIVIL BLOCK	Edit	Delete
5	12.894212	80.122604	12.894851	80.122842	MECH SCI LAB	Edit	Delete
6	12.895545	80.1214	12.8958	80.1255	LIBRARY	Edit	Delete
7	12.8953	80.1208	12.8954	80.1210	ADMISSION CELL	Edit	Delete
8	12.895981	80.121147	12.896104	80121162	MAIN GATE	Edit	Delete
9	12.9555	80.1208	12.8956	80.1210	POWER ROOM	Edit	Delete
10	12.895808	80.122950	12.895867	80.122862	MARINE CANTEEN	Edit	Delete
11	12.895181	80.122100	12.895534	80.122423	EEE LAB	Edit	Delete
12	12.8948	80.1226	12.8953	80.1228	IMST	Edit	Delete
13	12.89436	80.12328	12.89418	80.12319	FIRST YEAR STAFF ROOM	Edit	Delete
14	12.8953	80.1210	12.8955	80.1208	GATE 2	Edit	Delete
15	12.8944	80.1231	12.8949	80.1225	ARROW SH	Edit	Delete

Fig. 4 LUT with latitude and longitude values of places within campus

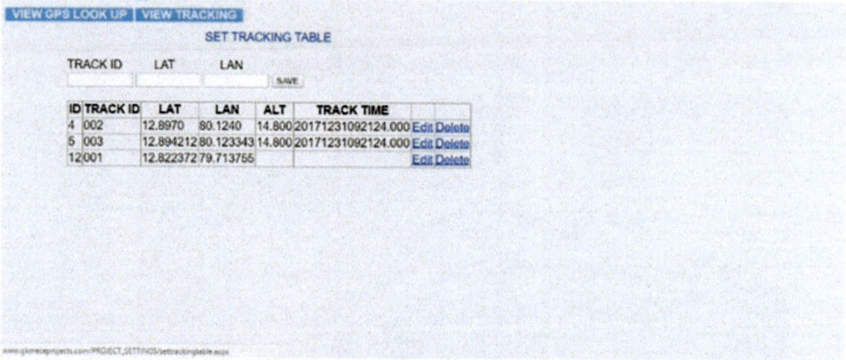

Fig. 5 Tracking of person with ID within campus

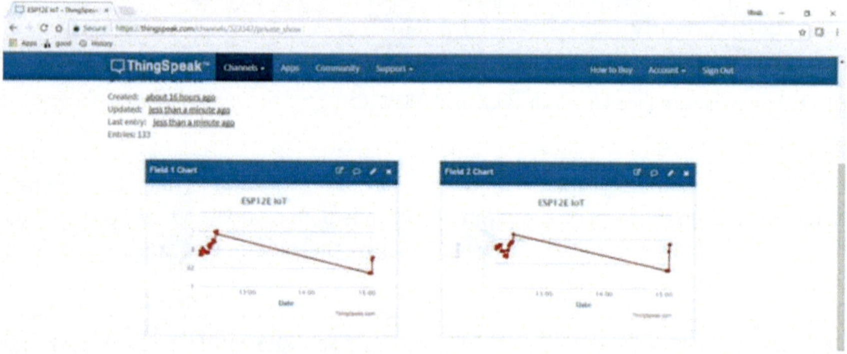

Fig. 6 Graphical representation of person's location

number in the respective column. The NodeMCU then compares these values with the values in the LUT, and if the values are matched, it provides a tracking table as in Fig. 5 with tracking time and latitude and longitude values. Later, the ThingSpeak Web site using these values within tracking table plots a graph with time vs latitude and time vs longitude as shown in Fig. 6.

6 Conclusion

The paper describes an identity card which operates based on ESP8266 and GPS module. It is controlled with the help of an html webpage created, and live latitude and longitude data of students are obtained with the help of GPS module placed in the identity card. The GPS data is received over a wireless technology Wi-Fi by using a particular Internet protocol. It is advantageous as it offers reliability and privacy on both sides and the data is transferred to Internet of things using ThingSpeak. It is authenticated only by the authorized user. Hence, it offers only the person concerned to view the details.

7 Future Work

Necessary action can be taken in short span in case of class bunking and roaming in and around the campus. In the future, latitude and longitude value of student is compared with attendance sheet whether the student is present or not inside the campus. Further, the identity card is implemented as a chip and the chip is placed inside the body to monitor continuously and to save the person in and anywhere in the world. This system is further implemented in areas like a hospital, industry, schools and colleges, and so far, it is used to find the soldiers in the military areas and festival occasions.

References

1. Deebika Shree A, Anusuya J, Malathy S (2019) Real time bus tracking and location updation system. In: 5th international conference on advanced computing & communication systems (ICACCS), 978-1-5386-9533-3/19/$31.00
2. Kamelia L, Darmalaksana W, Hamidi EAD, Nugraha A (2018) Real-time online attendance system based on fingerprint and GPS in the smartphone. IEEE. 978-1-5386-6163-5/18/$31.00
3. Zou H, Zhou Y, Yang J, Spanos CJ (2019) Unsupervised Wi-Fi IoT devices-user association for personalized location-based service. IEEE Internet Things J 6(1)
4. Zeng C, Zhao S, Zhong Y, Yuan Z, Luo X (2018) An improved method for indoor positioning of WIFI based on location fingerprint. In: 7th Internationa conference on digital home. https://doi.org/10.1109/ICDH.2018.00056
5. Deepika K, Usha J (2017) Design & development of location identification using RFID with Wi-Fi positioning systems. ICUFN. 978-1-5090-4749-9/17/$31.00
6. Raj JT, Sankar J (2017) IoT based smart school bus monitoring and notification system. In: IEEE region 10 humanitarian technology conference (R10-HTC) 21–23 Dec
7. Arun Kumar S, Kumaresan A (2017) Towards building intelligent systems to enhance the child safety and security. In: International conference on intelligent computing and control (I2C2)
8. Sunehra D, Priya PL, Bano A (2016) Children location monitoring on Google Maps using GPS and GSM Technologies. In: 6th International conference on advanced computing, IEEE
9. Deepika K, Usha J (2016) Investigations & implications on location tracking using RFID with global positioning systems. In: 3rd International conference on computer and information sciences 978-1-5090-2549-7/16/$31.00
10. Anish NK, Moorthi S (2013) Intelligent location identification and passenger-alert system in Indian Railways using GPS receiver. 978-1-4673-5943-6/13/$31.00

Oxygen Cylinders Monitoring System in Hospital Warehouse Using CNN

Pranav Suryadevara and S. Muthulakshmi

Abstract In recent years, a shortage of oxygen cylinders was observed in hospitals causing numerous deaths. Thus, a need for efficient monitoring of the quantity of the oxygen cylinders in a hospital storage facility becomes the need of the hour. To address this issue, we have come up with a solution of a surveillance system to keep track of the quantity of the oxygen cylinders. A surveillance system typically uses a camera to spy and observe a specific target. The process of monitoring can be done in both real time (live video-based) and in offline, where the surveillance can happen at a later time on a recorded video. This project employs the real time tracking of the oxygen cylinders in hospital store house using convolution neural network.

Keywords Oxygen cylinders · Surveillance · Real time tracking · Convolution neural network

1 Introduction

Currently, there is not an automated system in hospitals to check the number of oxygen cylinders in a hospital storage room. The process of monitoring the number of oxygen cylinders in a hospital is done manually, and the count is entered into spreadsheets electronically to maintain record. This process is prone to human error and may prove to be catastrophic when unrecognized in times of shortage. This is what happened in Gorakhpur, India in the year 2017 which led to the death of 60 infants in a hospital [1]. Also, considering the current covid-19 pandemic situation, it is very much essential to maintain enough stock of oxygen cylinders to serve the need of severely affected patients who are in need of oxygen [2]. Hence, we propose an oxygen cylinder monitoring system which can be considered as a solution for the said problem.

P. Suryadevara · S. Muthulakshmi (✉)
School of Electronics Engineering (SENSE), Chennai, India
e-mail: Muthulakshmi.s@vit.ac.in

P. Suryadevara
e-mail: Suryadevara.pranav2017@vitstudent.ac.in

A. Sivasubramanian et al. (eds.), *Futuristic Communication and Network Technologies*, Lecture Notes in Electrical Engineering 792,
https://doi.org/10.1007/978-981-16-4625-6_22

Oxygen cylinder monitoring system is a low-cost solution for alarming issues such as deaths due to shortage of essential necessities like oxygen cylinders. This can be implemented on a commercial scale in an economic and simple to use manner in hospital storage facilities [3]. It can also be extended to monitor other important medical gases like Carbon Dioxide, Nitrogen, Nitrous Oxide and commercially used gases like Medical Air (which would be indicating to the clean supply of compressed air used for driving the ventilators, as a carrier for anesthesia agents and also incubators in hospitals).

Our project mainly focusses on providing the count of the oxygen cylinders present in a storage at any point of time. The device is semi-autonomous which requires the human user to input the number of oxygen cylinders in case of newly added cylinders. It is proposed to use a communication module so that the oxygen cylinder supplier can directly receive the message which reduces the notification delay due to human fallacies. Also, a mobile or a web application would have an even more effective impact where it could be monitored by Health Officials. Convolutional neural networks (CNNs) are used in designing oxygen monitoring system (OMS), since it can be used in analyzing image data.

This paper is organized as follows. Section 2 discusses the hardware components and software tools used in implementing this work. This chapter further discusses the working principle of the hardware prototype and also the work flow of the software in detail. Result of this work is explained in Sect. 3.

2 Methods and Materials

2.1 *Hardware*

2.1.1 Components

The main components for this oxygen cylinders monitoring system are Raspberry Pi and a web camera (Fig. 1).

Raspberry Pi 4B

The Raspberry Pi 4B is a very compact, cost-effective, a powerful, and a miniaturized computer that is capable of being integrated with a standard computer. It comes in different models offering a wide range of specifications, namely, such as the Raspberry Pi 2, Raspberry Pi 3, Raspberry Pi 3B, Raspberry Pi 3B+, Raspberry Pi 4, Raspberry Pi 4B launched by the Raspberry Pi foundation [4].

USB Webcam

A webcam is typically a small video camera capable of recording a video that can feed or streams a video in a real time to a computer or through a computer to a computer network such as the Internet [5]. They are usually small and can be attached in any

Fig. 1 Hardware setup of the proposed system

desired location. It is interfaced to the Raspberry Pi's USB port through a USB hence making it easy to use.

2.2 *Software*

2.2.1 Programming Language and Platforms

Python: This is a general-purpose, high level, interpreted programming language used to program the Raspberry Pi. The execution command is written using this language is interpreted by the processor which in turn produces the desired results.

2.2.2 Operation System

The operating systems which was used in this prototype of Raspberry Pi is Raspbian (now Raspberry Pi OS). It is the operating system officially supported by the Raspberry Pi foundation for the series of Raspberry Pi. It comes in hand with ample amounts of software already installed for educational, programming and general-purpose uses. It has Scratch, Python, Java, Sonic pi and many more installed.

2.2.3 Machine Learning Algorithms

The machine learning algorithm used in this prototype is a convolution neural network (CNN) [6]. Typically, a feed forward neural network or a multiple hidden layer perceptron model with many layers in an artificial neural network (ANN) is usually known as deep neural networks (DNNs). CNN is an efficient algorithm for recognition which is extensively employed in the fields like recognition of patterns and processing of images. It has many advantageous aspects which include simplicity of the structure, requirement of less training parameters as well as examples and finally adaptability. It improves the model by reducing the complexity of the network by taking only a few features which are predominant for the identification and reducing associated weights [7].

Typically, the structure of a convolution neural network consists of two types of layers. They are a feature extraction layer (layers where filters are used to extract the data) and a feature mapping layer (a pooling layer where the image size is reduced by keep the features intact) [8]. The feature extraction layer is responsible for extracting local features from the input of every neuron connected to the local receiving fields of the previous layers. Once the features are extracted, the relationship between the extracted feature and other features would be determined. The other layer is responsible for the mapping of the features where every computing layer of the neural network composes a plurality of the feature map. Each feature map is a plane, the weight of the neurons in the plane is equal which helps the network learn in parallel. This is a major advantage attributed to a convolution neural network. The activation function used in a neural network is a sigmoid function which makes a feature map have full shift invariance. Every convolution neural network has a convolution layer which would be followed by a pooling layer (also known as computing layer). The pooling layer is used to calculate the local average and the second is to extract, and this distinct dual feature extraction structure reduces the size of the data (here resolution since it is an image). The process of the CNN is depicted in Fig. 2 which depicts the process the input being subjected to convolution filters and max-pooling layers and finally the output from the fully connected neurons.

A convolution neural network algorithm consists of two very essential processes which would be convolution and pooling. The process of convolution uses a set of standard filters in start so that it can derive features which can be used to develop the trainable filters [9]. This convoluted image is now further subjected to a process called pooling which reduces the size of the image keeping the features intact. This process helps a lot as the model can now be trained faster and would more effective. The convolutions used here were of the size 3X3 filters which make the input of size

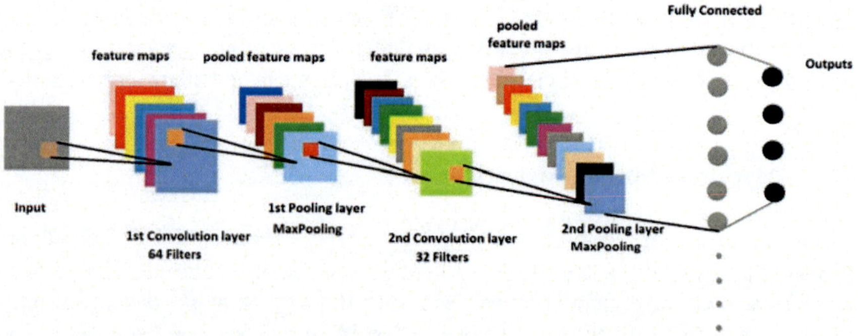

Fig. 2 Shows a representation of a multi-class classification convolution neural network

MXN pixels to result in an image of size M-2XN-2 (since the edge pixels could not have a convolution output thus canceling layers on all the 4 sides). The pooling layer helps us in reducing the size by grouping pixels and returning a pixel that represents the group of selected pixels (the pooling layer used here was a max-pooling layer which takes a set of 4 pixels and returns the max value as the pixel represented.) thereby reducing the size of the image (data) which also has the features enhanced [10].

2.2.4 Python Libraries

TensorFlow is a free and open-source software library for dataflow and differentiable programming useful across a range of tasks. It is a symbolic math library used for machine learning applications such as neural networks [11].

OpenCV (Open-Source Vision Library) is an open-source computer vision and machine learning software library. It was built to provide a common infrastructure for computer vision applications [12].

H5py package is a python interface to the HDF5 binary data format. It lets you store amounts of numerical data and easily manipulate that data from Numpy. A HDF5 (Hierarchical Data Format) file is a container for two kinds of objects: datasets which are array-like collections of data and groups which are folder-like containers that hold the datasets and other groups [13].

Scipy is a python based ecosystem of open-source software for mathematics, science and engineering used for scientific and technical computing. It contains modules for optimization, linear algebra, integration, interpolation, special functions, FFT, signal and image processing, ODE solvers and other tasks common in science and engineering. It has core packages that include Numpy, Pandas, Matplotlib [11].

Numpy is a library for the python programming language which is predominantly used in the data science community for handling data for machine learning implementations. It helps extensively by adding support for large, multi-dimensional arrays and matrices (as mainly most data is stored in these data structures), along with a large collection of high-level mathematical functions to operate on these arrays. From a company point of view, this library finds its main uses in cleaning the data before it is further processed and refined [11].

Pillow is a python imaging library is a free and open-source library for the python programming language that provides support for opening, manipulating and saving many different image file formats [11].

2.3 Working Methodology

2.3.1 Circuit and Setup

The proposed OMS is designed by interfacing USB web camera with a Raspberry Pi. The working model is shown in Fig. 1 The task of powering the raspberry pi can be done by using power bank (as done in this prototype, also as a backup) but it is suggested to use it via an electrical socket so that it could be switched on at all times leaving no room for fallacies. Adhesives are used to secure the components in the desired locations, and care is taken so that the USB camera is properly fixed.

2.3.2 Working

This system setup has a camera placed near the exit door of a facility which helps in detecting the oxygen cylinders when being taken for use. The main focus of the project is that the camera sends a live image taken every 3 min. This 3-min time delay is considered because the model takes this specified time to execute and return a result. During this time interval, the image is processed and check for the oxygen cylinders in the image. If the system detects the oxygen cylinders, then the number of oxygen cylinders in the image are counted. If the total number of cylinders drops below, a threshold value then a message is displayed to notify the issue.

This implementation assumes that oxygen cylinders as shown in Figs. 3 and 7 are distinguishable on the basis of a color (which is Green in this use case) apart from other medical supply cylinders in the storage facilities. When an image consisting

Fig. 3 Green color marked oxygen cylinder on a stand support

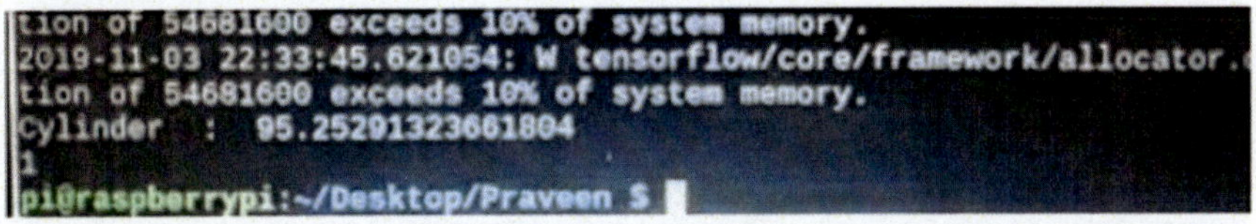

Fig. 4 Output of the machine leaning model in the console when input as in Fig. 3 is given

of a cylinder(s) with the appropriate color is obtained as an input then this process of detection, as in Figs. 4, and 8 would take place.

Code Summary

The flow diagram of the proposed work is given in Fig. 5. At first, the data has been collected, i.e., in this case, the input data is the images of the cylinders in a stock house. The entire dataset has been divided into training, validation and test sets. Now after dividing the data into the training, validation and test sets, the data has been reshaped and was normalized (to make all the features to be of the same scale which makes it easier for processing). Now using the keras library, we use a 2-D convolutional neural network to process the data [14]. We use max-pooling, followed by flattening and finally a dense layer to calculate the weights thus training the machine learning model. We can use different parameters to define the model. We have used an Adam optimizer, loss calculation using a binary categorical cross-entropy with the metric being used is accuracy [15]. The model learns the features from the training images to detect the desired objects from the set of images given as input [16]. We finally fit the model using these metrics and the data. We define a function to stop the model evaluation when the accuracy is reached to a desired threshold level (95% in this case) on the test data. Once the desired accuracy is obtained, we stop the model and it is capable of detecting a new test image on an unseen images thereby detecting the required object by using the feature learnt from the training data [17]. The code also draws a contour around the required object to be detected thus making it easier to cross verify the results as in the case of Figs. 6 and 9. Since a CNN used was a 2-D network, it can only detect the cylinders that are shown (clearly depicted) completely. When the multiple cylinders with slight colored to differentiate, Fig. 7, as oxygen cylinders are given as input [18], then we see a slight decrease in probability as the training had a few images of that sort hence the result obtained as in Fig. 8.

3 Results and Discussions

The prototype proposed detects all the oxygen cylinders present in an input image. It also gives a probability of the model predicting that the object would be an oxygen cylinder. The oxygen cylinders are detected based on the color which is green as in the use case presented here. It was also tested in various shades of lighting to observe the effects of lighting with the model always predicting the right output. It

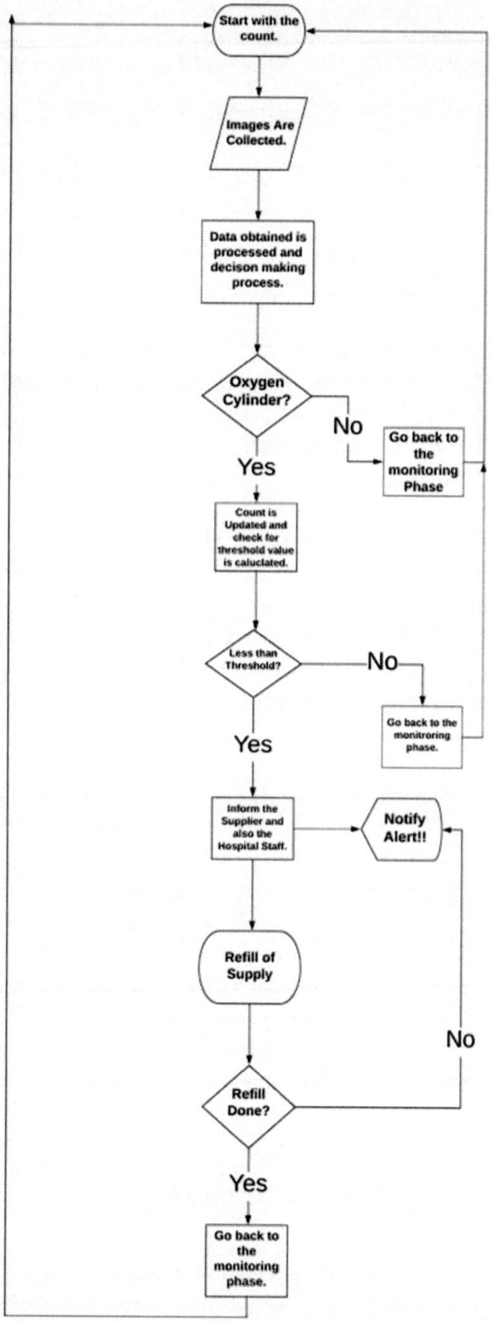

Fig. 5 Flowchart for the oxygen monitoring system using convolution neural networks (CNNs)

Fig. 6 The contour around the oxygen cylinder to represent the detected object

Fig. 7 Multiple oxygen cylinders in a plane surface

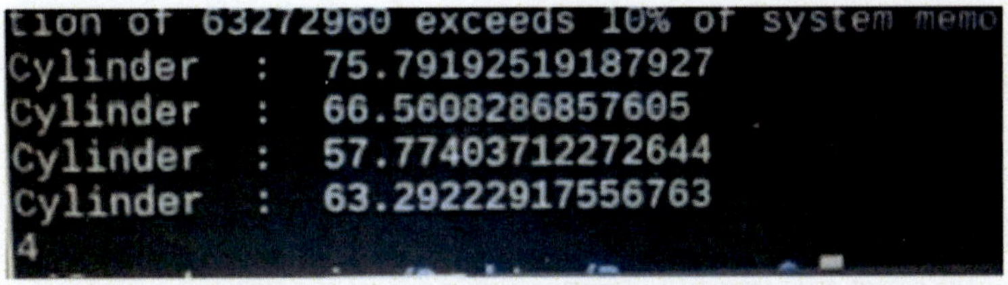

Fig. 8 Output of the machine leaning model in the console when input is a single-oxygen cylinder as in Fig. 3

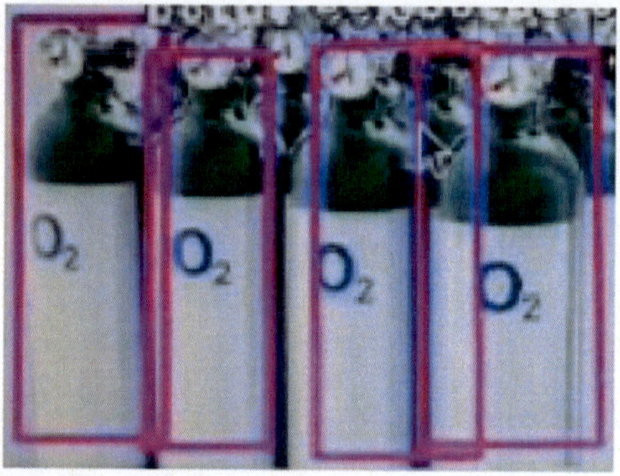

Fig. 9 The contour across each and every cylinder detected to represent detected objects

```
Threshold Set is 3 Cylinders.
Current Number of Oxygen Cylinders: 4.
Cylinder  :  95.25291323661804
1
1 Cylinders are detected.
Cylinders are at Threshold value!!!
Please notify the Supplier.
```

Fig. 10 Output in the console when the threshold limit is exceeded

also maintains the count and updates it as soon as an oxygen cylinders are detected. The count when decreases beyond a previously set threshold it notifies the vendor (Fig. 10) once the count falls beyond the threshold. The prototype is robust and would have the capabilities to operate in incase of a power cut as it uses a portable power bank as a backup supply.

4 Conclusion

This project can be further developed by employing the use of a communication module like a GSM (Global System for Mobile Communications) that would be helpful for direct communication with oxygen cylinder supplier so as to reduce any human fallacy. Also, a website with a database which could see all the entries being logged in would be very helpful for health officials monitoring the quantity of all cylinders in all hospitals.

Acknowledgements The authors wish to thank everyone who has contributed for the project study.

References

1. https://en.wikipedia.org/wiki/2017_Gorakhpur_hospital_deaths#:~:text=Acute%20encephalitis%20syndrome%20(AES)%20was,major%20cause%20for%20avoidable%20deaths
2. Cheng RZ, Kogan M, Davis D (2020) Ascorbate as prophylaxis and therapy for COVID-19—update from Shanghai and U.S. Medical Institutions. Glob Adv Health Med. https://doi.org/10.1177/2164956120934768
3. Rammohana Reddy E, Sankara K (2018) Internet of things based home automation control system using Raspberry Pi. Int J Sci Res Comput Sci, Eng Inf Technol, 3(4):34–38
4. https://en.wikipedia.org/wiki/Raspberry_Pi
5. Patrick A, Seidel, Böhnke K (2009) Camera-based control for industrial robots using OpenCV libraries. In: International conference on research and education in robotics, pp 144–157
6. Liu T, Fang S, Zhao Y, Wang P, Zhang J Implementation of training convolutional neural networks. Arxiv.org (in press)
7. Bosse S, Maniry D, Müller K, Wiegand T, Samek W (2018) Deep neural networks for no-reference and full-reference image quality assessment. IEEE Trans Image Process 27(1):206–219. https://doi.org/10.1109/TIP.2017.2760518
8. He K, Zhang X, Ren S, Sun J Deep residual learning for image recognition. Arxiv.org (in press)
9. Simonyan K, Zisserman A Very deep convolutional networks for large-scale image recognition. Arxiv.org (in press)
10. https://machinelearningmastery.com/object-recognition-with-deep-learnin
11. https://www.coursera.org/learn/introduction-tensorflow/home/welcome
12. Budiharto W (2015) Intelligent surveillance robot with obstacle avoidance capabilities using neural network. Hindawi Publishing Corporation Computational Intelligence and Neuroscience, vol 2015
13. Ziganshin E, Numerov M, Vyglov S (2010) UWB baby monitor. In: 2010 5th international conference on ultrawideband and ultrashort impulse signals, pp 159–161
14. https://www.tensorflow.org/guide/keras/sequential_model
15. Taqi AM, Awad A, Al-Azzo F, Milanova M (2018) The impact of multi-optimizers and data augmentation on tensorflow convolutional neural network performance. In: 2018 IEEE conference on multimedia information processing and retrieval (MIPR), Miami, FL, 2018, pp 140–145. https://doi.org/10.1109/MIPR.2018.00032.
16. https://www.coursera.org/specializations/deep-learning
17. Singh P, Deepak BBVL, Sethi T, Murthy MDP (2015) Real-time object detection and tracking using color feature and motion. In: International conference on communication and signal processing, pp 1252–1257
18. Oh SW, Kim SJ (2017) Approaching the computational color constancy as a classification problem through deep learning. Pattern Recogn 61:405–416

Fabric Defect Detection Using Modified Local Neighborhood Analysis

Maheshwari S. Biradar, P. M. Patil, and B. G. Sheeparamatti

Abstract Fabric defect detection plays a crucial role in the textile industry to improve the quality of service of the fabric texture. Automatic fault detection in fabric is challenging because of the variety of texture patterns, manufacturing defects, defects due to dyeing, and defects due to external environmental conditions. **Existing local neighborhood analysis (LNA) for defect detection has given poor performance for smaller and light color variation defects.** To deal with such conditions, this paper presents the unsupervised modified local neighborhood analysis (MLNA) for finding defect in non-patterned fabric. **The threshold value used for detection of defect depends upon mean, standard deviation, and entropy of local homogeneity measure.** The performance of the system is evaluated on the in-house database based on the percentage defect detection rate. The results of the proposed method are compared with previous methods such as wavelet transform and Gabor transform, **and it is observed that the proposed method detects 97.33% of defects and this is much better than the detection rates of LNAs and other existing methods.**

Keywords Modified local neighborhood analysis · Local homogeneity measure · Fabric defect detection · Non-patterned fabric

1 Introduction

In recent years, the textile industry is booming due to enormous growth in the fashion industry. The fabric material is made up of natural or synthetic threads by weaving, spreading, looping, crocheting, bonding, or knitting. Fabric materials are categorized into patterned and non patterned fabric images. Fabric is majorly made up of four sources such as a plant (cotton, bamboo, jute, flax), animal (silk and wool), mineral (glass fiber and asbestos), and synthetic (nylon, acrylic, polyester, rayon). Plant,

M. S. Biradar (✉) · P. M. Patil
Savitibai Phule Pune University, Pune, Maharashtra, India

B. G. Sheeparamatti
VTU, Belgaum, Karnataka, India

A. Sivasubramanian et al. (eds.), *Futuristic Communication and Network Technologies*, Lecture Notes in Electrical Engineering 792,
https://doi.org/10.1007/978-981-16-4625-6_23

animal, and minerals are considered as the natural resources of the fabric raw material. Cotton, bamboo, jute, or flax is obtained from plants [1, 2].

Defect in the fabric causes the irregularity in the pattern and textures of the fabric material which significantly reduces its cost by 45% to 75% in the market. Fabric defect can occur due to fault in the production machine, improper weaving, dyeing, oil, rust, finishing, high yarn tension, missing stitches, and contamination due to external agents such as dead fibers or husk. Common types of defects occurring in the fabrics are double ends, floats, holes, missing ends, thick bar, thin bar, broken pattern, broken picks, cut weft, double pick, gout, snarl, stain, tear, knots, etc. Some of the major fabric defects are shown in Fig. 1. Manual fabric defection is widely carried out for fabric defect detection. But manual fabric defect detection task is inefficient and unreliable because of boredom, tiredness, vision problem, and inattentiveness of the person. This leads to the automatic fabric defect detection system to improve the quality of service of fabric defect detection [3, 4].

In the past, various approaches have been carried out on computer vision-based automatic fabric defection for patterned as well as non-patterned fabrics. Computer vision and image processing play a vital role in the fabric defect detection which captures the fabric images, preprocesses it to remove the artifacts and noises present in the image, and employs computational algorithms to detect fabric defect. Fabric defect detection is categorized into statistical, spectral, model-based, structural, learning-based, and hybrid approaches [5, 6]. Statistical approaches such as auto-correlation models [7] and co-occurrence matrix [8] are simple to implement but

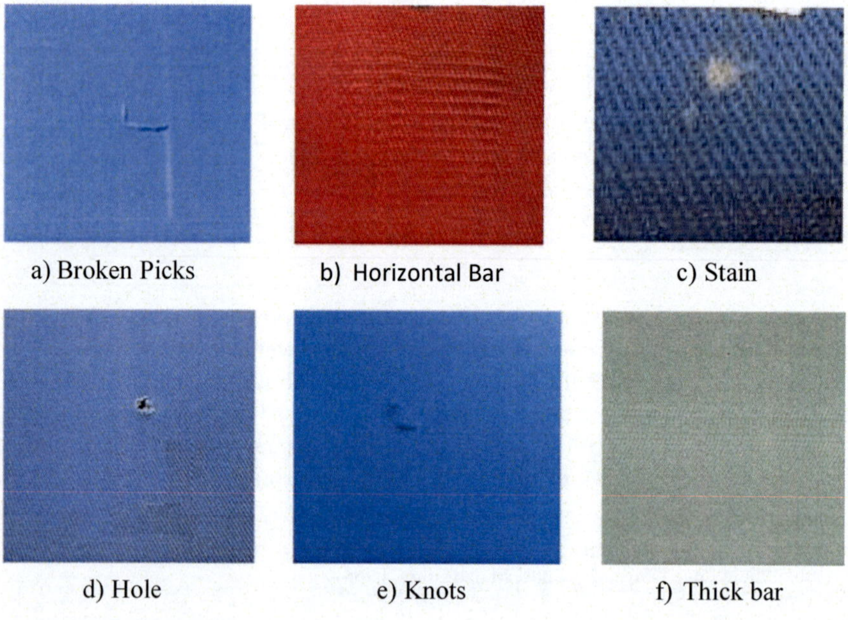

Fig. 1 Fabric defect samples

weak at describing the fine texture of the fabric. Spectral approaches such as Fourier transform [8], wavelet transforms [9], Gabor transforms [10], and contour-let transform [11] can detect and localize the defects effectively but suffering from lower detection rate and higher computation cost. Model-based approaches such as Markov random fields are weak in detecting smaller and light color variation defects in fabric [12]. Neural network-based learning approaches resulted in better detection accuracy for both online and offline modes but lack in reliability and highly complex for parameter tuning during training [13]. Structural approaches are more suitable for defect detection in complex patterns but perform poorly for smaller defects [6]. Hybrid approaches such as Bollinger Band (BB) [14], regular band (RB) [15, 16], and wavelet golden image subtraction (WGIS) [17] are efficient and combine the advantages of various defect detection approaches. Hybrid approaches generally give better results for patterned fabrics and are sensitive to the pattern period and illumination changes. Chengfei Li and Xinhua Chen [18] presented local neighborhood analysis for surface defect detection. They have given more focus on plain surface detection but not given much concentration on various types of texture defects so that cause of defect can be identified. In our previous approach [19], we further implemented local neighborhood analysis (LNA) for fabric defect detection which can deal with distinct fabric defects. In that, threshold calculation which is used to detect the defect depends upon the mean of local homogeneity measure only, due to which it gave a poor performance for the defect caused due to light color variation and smaller defects.

In the proposed method, we have presented modified local neighborhood analysis for the defect detection which can detect the light color variation and smaller defects. In this, the threshold value is dependent on mean, standard deviation, and entropy of local homogeneity measure.

Rest of the structure of paper is described as below: Section 2 depicts the details of proposed methodology. Section 3 describes the brief details about database, experimental results, and discussion. Finally, Sect. 4 concludes the paper.

2 Proposed Modified Local Neighborhood Analysis (MLNA)

The flow diagram of the proposed methodology is shown in Fig. 2 which consists of image preprocessing, local neighborhood analysis, and modified thresholding for the defect detection.

Input color image is converted to a grayscale image to minimize the computation efforts, and grayscale is enough to capture the texture information of the image. Local homogeneity of a grayscale image is measured using the coefficient of variation. The local neighborhood is the measure of regularity of the fabric texture and gives information about the distribution of gray scale. Fabric defect brings the abnormalities in the homogeneity of the defected region. The local homogeneity is computed over

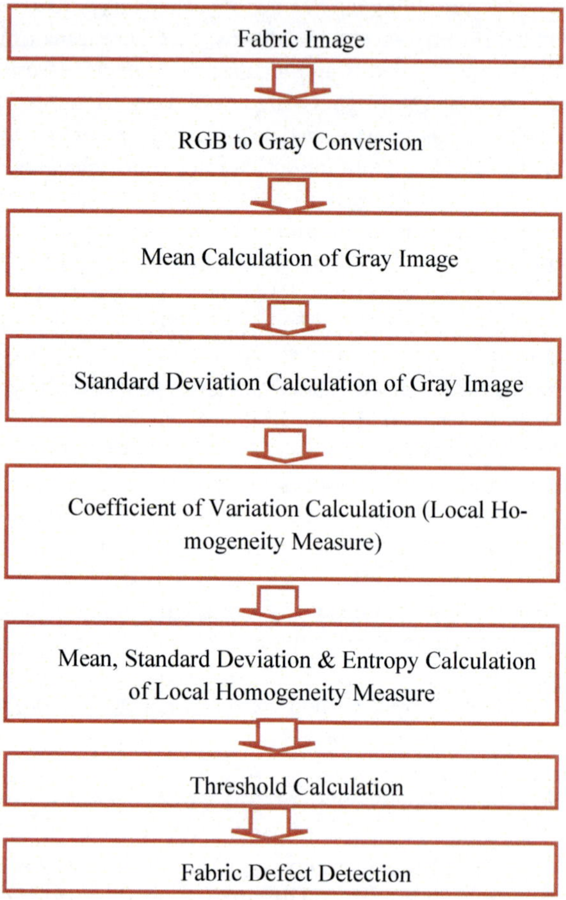

Fig. 2 Proposed system flow diagram

the local region of W x W, where W is given by W = 2w + 1. The coefficient of variation is the ratio of the mean of the grayscale image to the standard deviation of the grayscale image. The mean input grayscale image is computed over the local region using Eq. 1.

$$\mu_{x,y} = \frac{\sum_{i=-w}^{w} \sum_{j=-w}^{w} I(x+i, y+j)}{W \times W} \tag{1}$$

The standard deviation for the grayscale image gives the deviation of grayscale intensity from the mean over the local window and is computed using Eq. 2.

$$\delta_{x,y} = \sqrt{\frac{\sum_{i=-w}^{w} \sum_{j=-w}^{w} \left(I(x, y) - \mu_{x,y}\right)^2}{W \times W}} \tag{2}$$

The coefficient of variation is also called a local homogeneity measure (LHM). The percentage coefficient of variation for the local region is computed using Eq. 3.

$$C_v(x, y) = \frac{\delta_{x,y}}{\mu_{x,y}} \times 100 \tag{3}$$

where M is a total number of rows, whereas N represents a total number of columns, W is the local window, w is the factor that decided the local window, $\mu x,y$ is mean of a grayscale image, $\delta x,y$ is the standard deviation of grayscale image, and Cvx,y is coefficient of variation. In LHA, the threshold value is computed using the mean of local homogeneity measure. In the proposed method, the threshold value encompasses the mean, standard deviation, and entropy of the local homogeneity measure. Entropy gives the measure of randomness of the distribution of local homogeneity measure which helps to characterize the defected and defect-free region. Additional standard deviation and entropy in the calculation of threshold value make it compatible to detect fine and light color variation defects also. The threshold value ($\propto$) for MLHA is calculated using Eqs. 4–7.

$$\propto = \frac{(\mu_{cv} + \delta_{cv} + \varepsilon_{cv})}{3} + w \tag{4}$$

where

$$\mu_{cv} = \frac{\sum_{i=1}^{M} \sum_{j=1}^{N} C_v(i, j)}{M \times N} \tag{5}$$

$$\delta_{cv} = \sqrt{\frac{\sum_{i=1}^{M} \sum_{j=1}^{N} (C_v(i, j) - \mu_{cv})^2}{M \times N}} \tag{6}$$

$$\varepsilon = -\sum P \times \log_2(P) \tag{7}$$

where μ_{cv} is the mean of LHM, δ_{cv} is the standard deviation of LHM, and ε is the entropy of LHM. The defect in the image is detected by applying the threshold to the LHM. The performance of the thresholding is highly dependent upon the control variable w. Rather than selecting random values as a control variable, w is selected as a control variable which is also used as a factor for deciding the local window. If the LHM value is greater than the threshold value, then the region is considered as defected region; otherwise, it is considered as the defect-free region.

2.1 Experimental Results and Discussion

The proposed system is implemented using MATLAB software using Computer Vision Toolbox. The system specification used for the implementation has specifications such as a personal computer with a Core i3 processor, 2.64 GHz processor speed, 8 GB RAM, and Windows Operating Environment. The performance of the system is evaluated on the in-house database of non-patterned fabric textures based on % defect detection rate (DDR) as given in Eq. 8. Our database consists of a total of 450 images which consist of 75 images of the hole, thin bar, thick bar, broken picks, knot, and stain defect each.

$$\% DDR = \frac{DefectDetectedSamples}{TotalNumberofSamples} \times 100 \tag{8}$$

Table 1 presents the DDR for various types of defects for distinct window size. Very small window size (W = 3, W = 5) is unable to capture the fine texture of the local region of the image which tends to lower DDR. Larger window size (W = 9, W = 11) loses the fine texture and subsequently resulted in lower DDR. Moderate local window (W = 7) is well suitable for defining the texture of the local region, thus resulting in better DDR.

The performance of MLHA is compared with the simple LHA in , and it is observed that the addition of or standard deviation and entropy in the threshold value calculation significantly improves the performance of MLHA over LHA.

Table 1 Defect detection rate (%DDR) for various defects

Defect type	% Defect detection rate (DDR)				
	$W = 3$	$W = 5$	$W = 7$	$W = 9$	$W = 11$
Hole	85.33	96.00	98.67	93.33	81.33
Thin bar	84.00	92.00	97.33	89.33	70.67
Thick bar	86.67	90.67	97.33	92.00	77.33
Broken picks	82.67	94.67	98.67	88.00	74.67
Knots	88.00	92.00	97.33	92.00	78.67
Stain	74.67	90.67	94.67	85.33	69.33
Average DDR (%)	83.56	92.67	**97.33**	90.00	75.33

Table 2 Comparison of overall % DDR of MLNA with the previous methods

Method	% DDR
Wavelet transform [20]	94.00
Gabor transform [10]	95.00
LNA	96.40
MLNA	97.33

The experimental results for the proposed system for different kinds of defects are shown in Table 3. It shows that MLNA gives a precise defect outline than the simple LNA. The proposed method resulted in average Jaccard index (JI) of 0.87 for all types of fabric defects.

Table 3 Experimental results for various defects

Defected Image	Gray Scale Image	Defect Detection (LHA)	Defect Detection (MLNA)
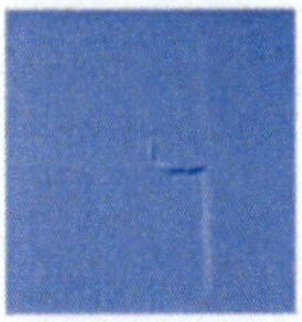			

			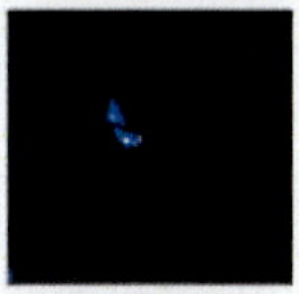
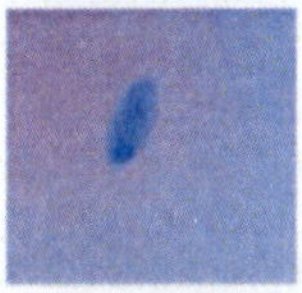		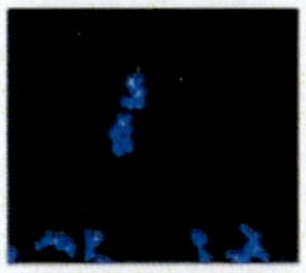	
		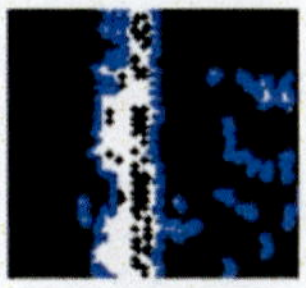	
		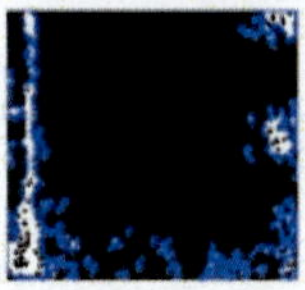	

3 Conclusion

Thus, in this work modified local neighborhood analysis is proposed for the fabric defect detection. The threshold value is dependent on mean, standard deviation, and entropy of local homogeneity measure. The proposed algorithm is simple to implement and faster and can be used for online fabric defection. It can detect the smaller, light color variation and stain defects. The proposed system resulted in a 97.33% defect detection rate which is better than the simple local neighborhood analysis. In the future, the proposed system is planned to use for patterned fabric defect detection.

References

1. An Introduction to Textile Terms. Contributor, Textile Museum (Washington, D.C.). Publisher, Textile Museum, (1997)
2. Hong Kong Productivity Council (2000) Textile Handbook 2000, The Hong Kong Cotton Spinners Association
3. Zhang W, Zhao Q, Liao L (2010) Development of a real-time machine vision system for detecting defeats of cord fabrics. In: 2010 International conference on computer application and system modeling (ICCASM 2010), vol 12, pp V12–539, IEEE
4. Xie X (2008) A review of recent advances in surface defect detection using texture analysis techniques. ELCVIA. In: Electronic letters on computer vision and image analysis, pp 1–22
5. Ngan HY, Pang GK, Yung NH (2011) Automated fabric defect detection—a review. Image Vis Comput 29(7):442–458
6. Hanbay K, Talu MF, Özgüven ÖF (2016) Fabric defect detection systems and methods—A systematic literature review. Optik 127(24):11960–11973
7. İzbudak H, Alkan A (2010) Denim fabric defect detection by using correlation method. In: National conference on electrical, electronics and computer engineering. IEEE, pp 638–642
8. Arnia F, Munadi K (2015) Real time textile defect detection using GLCM in DCT-based compressed images. In: 2015 6th International conference on modeling, simulation, and applied optimization (ICMSAO). IEEE, pp 1–6
9. Karlekar VV, Biradar MS, Bhangale KB (2015) Fabric defect detection using wavelet filter. In: 2015 International conference on computing communication control and automation. IEEE, pp 712–715
10. Lal Raheja J, Kumar S, Chaudhary A (2013) Fabric defect detection based on GLCM and Gabor filter: a comparison. Optik 124(23):6469–6474
11. Yapi D, Allili MS, Baaziz N (2017) Automatic fabric defect detection using learning-based local textural distributions in the contourlet domain. IEEE Trans Autom Sci Eng 15(3):1014–1026
12. Cohen FS, Fan Z, Attali S (1991) Automated inspection of textile fabrics using textural models. IEEE Trans Pattern Anal Mach Intell 8:803–808
13. Jmali M, Zitouni B, Sakli F (2014) Fabrics defects detecting using image processing and neural networks. In: 2014 Information and communication technologies innovation and application (ICTIA). IEEE, pp 1–6
14. Ngan HY, Pang GK (2006) Novel method for patterned fabric inspection using Bollinger bands. Opt Eng 45(8):087202
15. Ngan HY, Pang GK (2008) Regularity analysis for patterned texture inspection. IEEE Trans Autom Sci Eng 6(1):131–144
16. Biradar MS, Sheeparmatti BG, Patil PM, Naik SG (2017) Patterned fabric defect detection using regular band and distance matching function. In: 2017 International conference on computing, communication, control and automation (ICCUBEA). IEEE, pp 1–6

17. Naik SG, Biradar MS, Bhangale KB (2017) Patterned Fabric Defect Detection Using Wavelet Golden Image Subtraction Method. International Journal of Advance Research, Ideas and Innovations in Technology 3(3):767–771
18. Li C, Chen X (2017) Surface defect detection algorithm based on local neighborhood analysis. In: Information technology and intelligent transportation systems, Springer, Cham, pp 307–315
19. Kure N, Biradar MS, Bhangale KB (2017) Local neighborhood analysis for fabric defect detection. In: 2017 International conference on information, communication, instrumentation and control (ICICIC). IEEE, pp 1–5
20. Karayiannis YA, Stojanovic R, Mitropoulos P, Koulamas C, Stouraitis T, Koubias S, Papadopoulos G (1999) Defect detection and classification on web textile fabric using multiresolution decomposition and neural networks. In: ICECS'99, Proceedings of ICECS'99, 6th IEEE international conference on electronics, circuits and systems (Cat. No. 99EX357). IEEE, vol 2, pp 765–768

Speech Emotion Recognition Using Mel Frequency Log Spectrogram and Deep Convolutional Neural Network

Kishor Bhangale and K. Mohanaprasad

Abstract In recent years, speech emotion recognition (SER) has engrossed more attention in speech processing because of its potential in various speech-based intelligent systems. In deep learning algorithms to capture discriminative features of the audio emotion samples, a large number of features are required, which increases the computational complexity of the network. This paper presents a three-layered sequential deep convolutional neural network (DCNN) based on mel frequency log spectrogram (MFLS) for emotion recognition. Mel frequency log spectrogram that confines the salient information from the emotion speech corpus and two-dimensional DCNN. Exploratory outcomes on the Berlin Emo-DB dataset show that the proposed method gives 95.68 and 96.07% accuracy for the speaker-dependent and speaker-independent approaches. The performance of the proposed method is compared with CNN and CNN-LSTM on the Berlin Emo-DB dataset and results in improved accuracy.

Keywords Speech emotion recognition · Deep convolutional neural network · Mel frequency log spectrogram

1 Introduction

Speech emotion recognition (SER) is a crucial part of human–computer interaction (HCI) as speech is an efficient, fast and essential way of human interaction. This system eases the natural communication with the machine utilizing voice instructions rather than traditional input devices [1]. SER has widespread applications such as audio conferencing, interactive robot, call centre dialogue, aboard vehicle driving system, interactive game designing, medical psychological analysis, online learning and tutoring system [2, 3].

Determination of human speech emotion is an idiosyncratic task and can be used for any standard for any SER system. Human speech signal consists of verbal and para verbal information. The verbal information describes the meaning and context of

K. Bhangale · K. Mohanaprasad (✉)
SENSE, VIT, Chennai, India
e-mail: kmohanaprasad@vit.ac.in

A. Sivasubramanian et al. (eds.), *Futuristic Communication and Network Technologies*, Lecture Notes in Electrical Engineering 792,
https://doi.org/10.1007/978-981-16-4625-6_24

the speech, whereas the para verbal information describes the tacit information such as the emotion expressed in the speech signal. The speech signal consists of different emotions like happiness, sadness, anger, fear, surprise, joy, disgust, boredom and neutral. The paralinguistic information is usually independent of the lexical content, language and speaker [4, 5].

Different emotion has an immense effect on the various characteristics of the speech signal such as short-term features like energy, pitch and format [6], long-term features like mean and standard deviation [7]; and prosodic features like pitch, speaking rate, intensity, voice variation and quality [8].

Traditional, machine learning (ML)-based SER systems have two major phases, such as feature extraction and classification. The performance of ML-based approaches is highly dependent on handcrafted features. In the past, many feature extraction techniques for SER system are implemented such as mel frequency cepstrum coefficients (MFCC) [9], principal component analysis (PCA), linear predictor coefficients (LPC), Gaussian mixture model (GMM), perceptual linear prediction coefficients (PLP) and hidden Markov model (HMM). The classification phase learns the extracted features and depicts the correct emotion. SER system used various classifications algorithms such as support vector machine (SVM), K-nearest neighbour classifier (KNN) and artificial neural network (ANN). The performance of the ML classifiers depends upon raw features, database size, professional knowledge and manual tuning of features which are labour expensive [10].

In recent years, deep learning (DL) emerges as the advanced field for SER, which represents the low-level speech features into the high-level hierarchical features. Jianwei Niu et al. presented DNN for the modelling of complex and nonlinear features of emotion speech training data. It resulted in 92.1% accuracy for the five-layered DNN with MFCC features [11]. Huang et al. presented CNN for the representation of salient hierarchical features in two stages. It achieved better accuracy for speaker-dependent approach [12]. Zheng et al. presented deep CNN (DCNN) for SER, which uses PCA for dimension reduction and interference suppression of the input log spectrogram. It resulted in an accuracy of 40% for IEMOCAP database [13]. Abdul Malik Badshah et al. presented an SER system based on DCNN with three fully connected layers that used spectrogram. It resulted in 84.3% accuracy of the Berlin Emotion database [14]. Jianfeng Zhao et al. presented 1D CNN-long short-term memory (1D-CNN-LSTM) and 2D CNN-LSTM (2D-CNN-LSTM) to discover local and global emotion-specific features. It resulted in 95.33 and 95.89% accuracy on the Berlin Emo-DB database for speaker-dependent (SD) and speaker-independent (SID) approaches [15].

DL algorithms have several advantages such as the capability to deal with complex speech structure and features; ability to deal with un-labelled data; no need for feature tuning; and ability to handle more massive datasets. Though SER has made tremendous progress still, it faces many challenges such as variability in individual, variability in environmental conditions and effect of noise, generalizing the model for the distinct dataset and recognition of subtle expression. Therefore, there is a need for the design of a robust SER system that can overcome these limitations.

This paper presents three-layered sequential deep convolutional neural network (DCNN) for the speech emotion recognition that accepts two-dimensional mel frequency log spectrogram (MFLS) as input. For the performance evaluation Emo-DB database, this consists of seven acted emotions samples such as anger, boredom, disgust, fear, neutral, happiness and sadness.

This paper is systematized as follows: Sect. 2 illustrates the proposed methodology, along with implementation details of MFLS and DCNN implementation in detail. Section 3 focuses on the discussion of simulation results. Finally, the conclusion and future perspectives are given in Sect. 4.

2 Proposed Methodology

The proposed SER based on a three-layered sequential DCNN consists of three CNN layers for emotion recognition (see Fig. 1). For the two-dimensional CNN mel frequency log spectrogram is given as the input which helps to capture salient features of the speech signal and minimization of the random noise present in the signal. Each CNN layer comprises three essential layers, such as the convolution layer (CL), exponential linear layer (ELU) and max pooling layer (MP). The fully connected (FC) layer followed after the MP layer of the third CNN and softmax classifier followed by a fully connected layer classifies the emotion speech signal.

2.1 Mel Frequency Log Spectrogram (MFLS)

The human emotion speech signal is one-dimensional. Thus to avail, the simplicity and advantages of the two-dimensional CNN, input emotion speech signal are converted into two-dimensional mel frequency logarithmic spectrum (see Fig. 2). Mel frequency gives the relation between the human ear and sound perception frequency [16]. The mel frequency scale (Mel) can be obtained from the linear frequency (f) using Eq. 1.

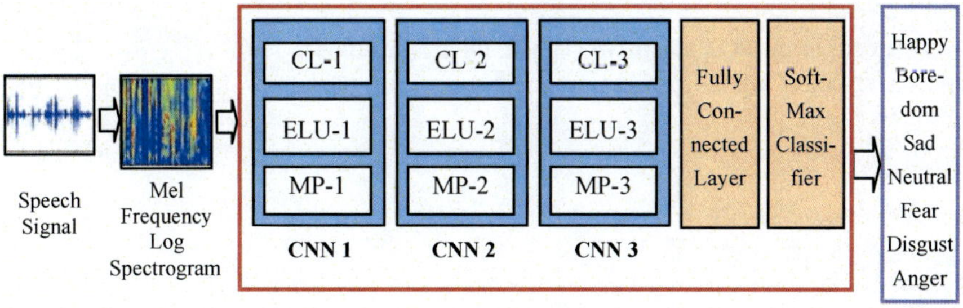

Fig. 1 Detailed process architecture of the proposed system

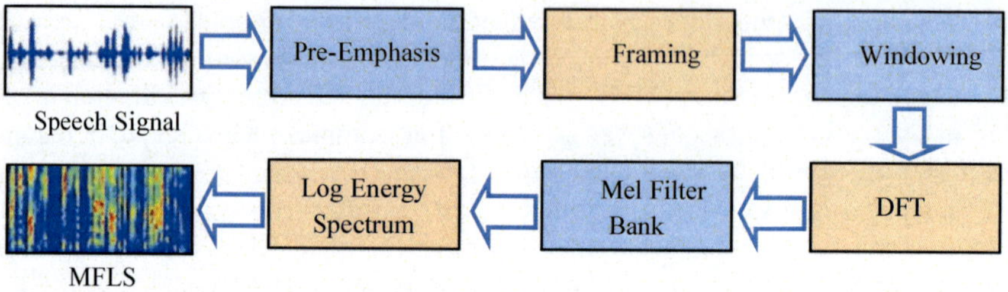

Fig. 2 Flow diagram of mel frequency log spectrogram (MFLS) process

$$\mathrm{Mel}(f) = 2595 \log\left(1 + \frac{f}{700}\right) \tag{1}$$

Complete processing of MFCC is time-consuming, and the application of discrete cosine transform (DCT) provides a higher frequency resolution but a lower spatial relationship. Thus, we formulated part of MFCC that can be input to two-dimensional DCNN can maintain better frequency and spatial relationship [17].

Pre-emphasis. The pre-emphasis filter suppresses the random noise and amplifies the high-frequency components of the speech emotion signal. The equation for the pre-emphasis filter H(z) is given by Eq. 2.

$$H(z) = 1 - \beta.z \tag{2}$$

where β is a pre-emphasis coefficient that lies between 0 and 1.

Framing and Windowing. The speech emotion signal is non-stationary; therefore, to process stable speech components, it is split into frames of the 40 ms. To obtain a smooth changeover between frames, 50% overlapping of frames is used. Further, Hamming window is used to collect the closest frequency components together and avoid the leakage phenomenon. The Hamming window $W(n)$ for $\alpha = 0.46$ and N samples can be expressed by using Eq. 3.

$$W(n) = (1 - \alpha) - \alpha \cdot \cos\left(\frac{2\pi n}{N - 1}\right), \quad 0 \le n \le N - 1 \tag{3}$$

Discrete Fourier Transform (DFT). DFT is used to transform the time-domain speech emotion signal into the frequency domain. The DFT $X(k)$ of the speech emotion signal $x(n)$ given is using Eq. 4.

$$X(k) = \sum_{n=0}^{N-1} x(n) \cdot e^{-j2\pi nk/N}, \quad 0 \le n, k \le N - 1 \tag{4}$$

The emotion power spectrum is by taking the square of the modulus of $X(k)$ as given in Eq. 5.

$$P(k) = \frac{1}{N}|X(k)|^2 \tag{5}$$

Mel Filter Bank. The mel spectrum can be obtained by passing the emotion power spectrum $P(k)$ through the mel-scale triangular filter bank. The product of $P(k)$ and $H_m(k)H_m\ (k)$ is computed at each frequency. Triangular filter bank frequency response $H_m(k)$ for $M = 32$ filters is computed using Eq. 6. We have considered $M = 32$, which can cover the frequency components between 133 and 3954 Hz.

$$H_m(k) = \begin{cases} 0, & k < f(m-1) \\ \dfrac{k - f(m-1)}{f(m) - f(m-1)}, & f(m-1) \le k \le f(m) \\ \dfrac{f(m+1) - k}{f(m+1) - f(m)}, & f(m) \le k \le f(m+1) \\ 0, & k > f(m+1) \end{cases} \tag{6}$$

where $f(m)$ stands for the centre frequency of the mel frequency filter.

Mel Frequency Logarithmic Spectrum. The logarithmic energy spectrum $S(m)$ for each frame is computed using Eq. 7.

$$S(m) = \log_e\left(\sum_{k=0}^{N-1} P(k).H_m(k)\right), \quad 0 \le m \le M \tag{7}$$

where $P(k)$ represents the power spectrum, $H_m(k)$ is a triangular filter bank response, and M is a number of filters.

2.2 *Deep Convolutional Neural Network*

Two-dimensional CNN is popular for images processing ability which represents the internal correlation and saliency information of the local region of two- or three-dimensional image. It also helps to describe the spatial, temporal and frequency domain representation of two-dimensional data [18]. Mini batch gradient descent method is employed for the learning of the DCNN. In this architecture, each feature map is convolved with each kernel filter at every layer. Various layers of CNN are as follows:

Convolution Layer. In the convolution layer, the mel frequency spectrogram convolved with the convolution filter bank. Convolution layer describes the spatial local connectivity and correlation of the local region of the mel frequency spectrogram. In the convolution layer, two-dimensional mel frequency spectrogram $S(m)$ convolved with the convolution kernel $w(i, j)$ having a size $(p \times q)$ as given in Eq. 8. The weights of kernel $w(i, j)$ are initialized randomly.

$$C(i,j) = S(i,j) * w(i,j) = \sum_{m=0}^{p}\sum_{n=0}^{q} S(m,n).w(i-m, j-n) \quad (8)$$

Exponential Linear Unit (ELU). ELU removes the negative weights from the convolution layer output and normalizes the convolution layer output using Eq. 9.

$$E(i,j) = \left\{ \begin{array}{ll} C(i,j), & \text{if } C(i,j) > 0 \\ e^{C(i,j)}, & \text{if } C(i,j) \leq 0 \end{array} \right. \quad (9)$$

Maximum Pooling Layer (MP). The maximum pooling layer acts as a nonlinear function, and it only considers the salient information of non-overlapping local sub-region. It increases the robustness of features against distortions and noise. Max pooling also helps to reduce the feature dimension. In this implementation, the maximum pooling window of 2 × 2 pixels is used with a stride of 2 × 2.

Fully Connected Layer (FC). The FC layer is similar to the multilayer perceptron network (MLP) that combines each neuron of a single layer to every neuron of other layers. The flattened output of the pooling layer is given to the FC layer as input.

Softmax Classifier. Softmax classifier is used for the multiclass emotion classification, which is the generalized framework of the logistic regression. Softmax function provides the probability of the predicted class (P_i), and the output class label (Y) is decided based on the maximum of P_i as given in Eq. 10–12.

$$P_i = \text{soft}\max(z)_i = \frac{e^{z_i}}{\sum_{j=1}^{n} e^{z_i}} \quad (10)$$

$$Y = \max_i(P_i) \quad (11)$$

$$z_i = \sum_{j}^{i} h_j.W_{ji} \quad (12)$$

where z_i is the output of an FC layer and input to the soft-max classifier, h_j is the activation function of the penultimate layer, and W_{ji} weight connecting penultimate and softmax layer.

Learning algorithm. For the learning of proposed DCNN, mini-batch gradient leaning method, which is a combination of stochastic gradient descent (SGD) and batch gradient descent (BGD) is used. It is robust and computationally cheaper [19]. In this, n samples are break up in smaller batches b. The error function for updating the weights (w) of DCNN is given in Eq. 13.

$$E_t[f(w)] = \frac{1}{b}\sum_{i=(t-1)b+1}^{tb} f(w, x_i) \quad (13)$$

where x_i is an ith sample of the training data, the weights are revised using mini-batch gradient update rule considering learning rate μ as given in Eq. 14.

$$w^{t+1} = w^t - \mu \nabla_w E[f(w^t)] \quad (14)$$

3 Experimental Results and Discussion

The proposed method is simulated using MATLAB software on a personal computer with a Core i5 CPU with 4 GB RAM on Windows environment. For the experimentation, Berlin Emo-DB speech emotion public database is used which consists of 535 utterances of 10 actors for seven emotions such as anger, boredom, fear, disgust, neutral, happiness and sadness [20]. The sampling rate used for data collecting is 16,000 Hz. We have considered the 5-s long speech samples to keep the uniformity in the mel frequency log spectrogram. If the sample length is less than 5 s, then it is padded to 5 s long using original signal. Otherwise, the samples are cropped to 5 s. The MFLS has the dimension of $\mathrm{M} \times F = 32 \times 249$ where M is number of triangular filter banks, and F is the number of frames for the 5-s signal for 40 ms duration and 50% frame-shift.

In the first CNN, input MFLS (32 × 249) convolved with the six kernels of 3 × 3 filter with the stride of one pixel and without zero-padding followed by ELU and max pooling. Each feature map of every layer is convolved with each filter kernel. The feature dimensions for various layers of DCNN are given in Table 1. The output layer has seven neurons that correspond to the output labels of seven emotions.

Table 2 shows the % accuracy for SER based on the speaker-dependent and speaker-independent approaches is considering the direct audio clip and MFLS as

Table 1 Details of feature map of various layers of proposed work

Layer	Sub-layer	Kernel size	Stride	Feature map
Input layer	Mel frequency log spectrogram	–	–	32 × 249
CNN layer 1	Convolution Layer 1	3 × 3 × 6	1	29 × 247 × 6
	ELU Layer 1	–	–	29 × 247 × 6
	Max Pooling Layer 1	2 × 2	2	14 × 123 × 6
CNN layer 2	Convolution Layer 2	3 × 3 × 6	1	12 × 121 × 36
	ELU Layer 2	–	–	12 × 121 × 36
	Max Pooling Layer 2	2 × 2	2	6 × 60 × 36
CNN layer 3	Convolution layer 3	3 × 3 × 6	1	4 × 58 × 216
	ELU layer 3	–	–	4 × 58 × 216
	Max pooling layer 3	2 × 2	2	2 × 29 × 216
FC Layer	–	–	–	12,528 × 1

Table 2 % accuracy for SER based on Emo-DB database

Emo-DB emotion	Speaker-dependent approach		Speaker-independent approach	
	Audio spectrogram as input	MFLS as input	Audio spectrogram as input	MFLS as input
Anger	94.49	100	96.06	100
Boredom	88.88	93.83	97.54	98.76
Disgust	84.78	97.82	78.19	91.31
Fear	89.85	94.2	94.2	97.1
Happy	89.43	91.55	73.83	93.66
Neutral	100	92.4	77.21	94.93
Sad	93.55	100	93.04	96.78
Average accuracy	91.56	95.68	87.15	96.07

Table 3 Comparison of the proposed method with the previous implementation based on % accuracy (Emo-DB)

Research work	Method	Speaker-dependent approach	Speaker-independent approach
Huang et al. [12]	CNN	88.30	85.20
Zhao et al. [15]	CNN-LSTM	95.33	95.89
Proposed work	MFLS + DCNN	95.68	96.07

the input. When simple speech spectrogram is provided to the system, it resulted in 91.56% and 87.15% accuracy for speaker-dependent and speaker-independent modes, respectively. While when MFLS is provided as an input to the system, it gives a better improvement in performance and results in 95.68 and 96.08% accuracy for speaker-dependent and speaker-independent modes, respectively.

When the proposed method performance is assimilated with other implementations on the Berlin Emo-DB database for speech emotion recognition based on % accuracy, it is noticed proposed method has given satisfactory results as shown in Table 3.

4 Conclusion and Future Scope

This paper has presented the SER system based on the mel frequency log spectrogram (MFLS) and three-layered sequential deep convolutional neural network (DCNN). MFLS extracts the salient information from the raw speech emotion signal, which further boosts the discriminative feature extraction ability of two-dimensional DCNN. The performance of the proposed system is estimated on the Emo-DB database considering speaker-dependent (SD) and speaker-independent

(SID) approaches. The proposed method has resulted in 95.68 and 96.07% accuracy for the speaker-dependent and speaker-independent approaches when MFLS is provided as the input to the proposed DCNN. Our future work consists of an investigation of a proposed method for noisy database and spontaneous emotion speech database.

References

1. Schuller BW (2018) Speech emotion recognition: two decades in a nutshell, benchmarks, and ongoing trends. Commun ACM 61(5):90–99
2. Khalil RA, Jones E, Babar MI et al (2019) Speech emotion recognition using deep learning techniques: a review. IEEE Access 7:17327–117345
3. Gunawan TS et al (2018) A review on emotion recognition algorithms using speech analysis. Indonesian J Elect Eng Inf (IJEEI) 6(1):12–20
4. Anagnostopoulos C, Iliou T, Giannoukos I (2012) Features and classifiers for emotion recognition from speech: a survey from 2000 to 2011. Artif Intell Rev 43(2):155–177
5. Guidi A, Vanello N, Bertschy G, Gentili C, Landini L, Scilingo E (2015) Automatic analysis of speech F0 contour for the characterization of mood changes in bipolar patients. Biomed Signal Process Control 17:29–37
6. Dellaert F, Polzin T, Waibel A (1996) Recognizing emotion in speech. In: Proceeding of fourth international conference on spoken language processing. ICSLP'96, vol 3. IEEE, pp 1970–1973
7. Zhou Y, Sun Y, Zhang J, Yan Y (2009) Speech emotion recognition using both spectral and prosodic features. In: Information engineering and computer science, pp 1–4
8. Haq S, Jackson P, Edge J (2008) Audio-visual feature selection and reduction for emotion classification. In: Proceedings of international conference on auditory-visual speech processing (AVSP'08), Tangalooma, Australia
9. Sonawane A, Inamdar M, Kishor B (2017) Sound based human emotion recognition using MFCC & multiple SVM. In: 2017 international conference on information, communication, instrumentation and control (ICICIC), pp 1–4. IEEE, Indore, India
10. El Ayadi M, Kamel MS, Karray F (2011) Survey on speech emotion recognition: features, classification schemes, and databases. Pattern Recogn 44(3):572–587
11. Jianwei N, Yanmin Q, Kai Y (2014) Acoustic emotion recognition using deep neural network. In: 9th international symposium on Chinese spoken language processing, pp 128–132. IEEE
12. Huang Z, Ming D, Qirong M, Yongzhao Z (2014) Speech emotion recognition using CNN. In: Proceedings of the 22nd ACM international conference on multimedia, pp 801–804
13. Zheng Q, Yu J, Zou Y (2015) An experimental study of speech emotion recognition based on deep convolutional neural networks. In 2015 international conference on affective computing and intelligent interaction (ACII), pp 827–831. IEEE)
14. Badshah A, Jamil M, A., Nasir, R., Sung, W.: Speech emotion recognition from spectrograms with deep convolutional neural network. In: 2017 international conference on platform technology and service (PlatCon), pp 1–5. IEEE (2017)
15. Zhao J, Xia M, Lijiang C (2019) Speech emotion recognition using deep 1D & 2D CNN LSTM networks. Biomed Signal Process Control 47:312–323
16. Bhangale K, Titare P, Pawar R, Bhavsar S (2018) Synthetic speech spoofing detection using MFCC and radial basis function SVM. IOSR J Eng 8(6):55–62
17. Zheng F, Guoliang Z, Zhanjiang S (2001) Comparison of different implementations of MFCC. J Comput Sci Technol 16(6):582–589
18. Goodfellow I, Bengio Y, Courville A (2016) Deep learning. 1st edn, MIT Press, Cambridge, MA

19. Shende P, Dandwate Y (2020) Convolutional neural network based multimodal biometric human authentication using face, Palm Veins and Fingerprint. Int J Innov Technol Explor Eng (IJITEE) 9(3):771–777
20. Burkhardt F, Paescke A, Rolfes M, Sendlmeirer WF, Weiss B (2005) A database of German emotional speech. In: 9th European conference of speech commincation and Technology

Voice Activity Detection for Monaural Speech Enhancement Using Visual Cues

S. Balasubramanian, R. Rajavel, and S. Shoba

Abstract Voice activity detection (VAD) is a vital module in various applications like speech recognition, speech enhancement, and dominant speaker identification. The performance of voice activity detectors using audio cues considerably declines under low SNR conditions. One method to improve the performance of VAD is to use the video signal, which is independent to acoustic background. At the present time, video calls have become a more popular way of communication, and recent products like laptops and smartphones, have cameras and microphones inbuilt. The accessibility of a video signal, along with the audio signal, can be used for voice activity detection particularly in a noisy environment, as the video signal is independent to background acoustic noise. This paper aims to develop a binary mask for voice activity detection to separate the target speech from the noisy speech mixture using the visual cues. The visual cues will be extracted by the mouth detection using Viola–Jones algorithm and the lip movement tracking using Kanade–Lucas–Tomasi (KLT) algorithm. Finally, the extracted mask using visual cues is compared with the mask obtained using audio cues under low SNR conditions. The performance is evaluated for the proposed system using PESQ. The experimental result shows that the proposed system performs well under low SNR conditions and improves the average PESQ score of 0.57 as compared to the other existing systems which use only auditory cues for voice activity detection.

Keywords Voice activity detector · Visual cues · Binary mask · Visual-VAD (V-VAD) · Audio-VAD (A-VAD)

S. Balasubramanian (✉)
Meenakshi Sundararajan Engineering College, Chennai, India

R. Rajavel
SSN Engineering College, Chennai, India
e-mail: rajavelr@ssn.edu.in

S. Shoba
VIT University (Chennai Campus), Chennai, India
e-mail: shoba.s@vit.ac.in

A. Sivasubramanian et al. (eds.), *Futuristic Communication and Network Technologies*, Lecture Notes in Electrical Engineering 792,
https://doi.org/10.1007/978-981-16-4625-6_25

1 Introduction

In noisy environment, the speech spoken by a person will be affected by the noise. For example, the noise can be other person speaking, noise due to a passing car, and many other sounds. Many applications need a system that separates the speech from the noisy speech and enhancement of the target speaker's speech. In cellular phone communication, the voice will be affected by the surrounding noise at the transmitter. A speech enhancement system can be used at the receiving end to improve the quality of the speech.

In speech separation process, the target speech signal is separated from acoustic mixture. The acoustic mixture may be another speech or environmental noise or both. Speech separation can be used in speech/speaker recognition, voice communication, air–ground communication, hearing aids, etc. Spectral subtraction, subspace analysis, hidden Markov modeling, and sinusoidal modeling are some of the methods propose earlier proposed for monaural speech separation. These approaches usually require a prior knowledge of noise signal. In the last few decades, many researchers have developed monaural speech separation using adaptive energy threshold [1], and image analysis techniques as in [2, 3]. The performance of the speech is improved using genetic algorithm-based fusion scheme in [4], and speech intangibility is improved by fusing voiced and unvoiced speech segments as in [5]. All these methods used audio as the only input to improve the speech quality.

Visual cues refer to the voice activity of the target speaker which is used in this paper to produce an enhanced output of the target speaker's speech. Normal speech enhancement techniques which are used to enhance a single (monaural) person speech do not provide the expected result.

Lip-reading depends on number of factors in human-to-human communication [6]. The quality of visual information plays a vital role. For example, in poor lighting condition it is hard to detect the shape of the mouth. Additionally, it is difficult to detect visual cues as the listener and the speaker are moving apart.

Hence simulating the target person's speech activity by tracking the lip movement of the person using Viola–Jones algorithm [7, 8] and Kanade–Lucas–Tomasi (KLT) [9] and comparing that with the original audio sample to ensure that the presence of voice activity from the lip moment matches with the original audio onsets and offsets.

The presence of speech activity and the voice activity can be detected from both audio and video stream. If the SNR is high, speech is dominant compared to noise; thus, it is more reliable to detect the presence of speech activity from audio stream itself. But when the SNR reduces, and as the noise dominates speech, it is not reliable to detect onsets and offsets from audio stream as it may treat some noisy parts as speech or vice-versa. In such cases, it is advisable to detect onsets and offsets from video stream as it is independent of SNR of the signal. The onset and offset times of video stream can be detected by tracking the mouth of the target speaker. The detected onset and offset time of the video stream is then plotted and compared with the plot of audio stream, and checked for one-to-one correspondence.

The reminder of this paper is organized as follows. Section 2 describes the proposed system about speech onset and offset detection using video stream. The existing system of speech onset and offset detection using audio stream is explained in Sect. 3. The experimental results are given in Sect. 4. Conclusion is outlined in Sect. 5.

2 Speech Onset and Offset Detection Using Visual Stream

In this paper, the main focus is to obtain a voice activity detection mask using visual cues (V-VAD) for target speech detection. The mask here is a binary mask which specifies the absence or presence of voice activity in each frame. First the video stream is splitted in sequence of image frames. Second, the face is detected in the given frame and the mouth region is encompassed with the bounding box in the detected face. The first frame is processed using the Viola–Jones algorithm [7, 8, 10] to get the bounding box for the mouth region. Viola–Jones algorithm uses Haar features to detect the face. The four stages of Viola–Jones algorithm are: (a) choice of Haar feature, (b) creation of integral image, (c) Adaboost training (d) cascading classifiers.

2.1 Binary Mask Detection

The mouth region in the image is identified using Viola–Jones algorithm. It uses a cascade of classifiers to detect the presence of target object in the image frame. Every stage in the cascade rejects the region which has no target object. As the sliding window travels over the image it is possible to produce multiple object detection near the target object. This multiple identification is combined to produce one bounding box for one target object. The bounding box is represented using the top left corner coordinates and the height and width of the box. Figure 1 highlights the bounding box of the mouth region detected.

2.2 KLT Feature Tracking

Kanade–Lucas–Tomasi (KLT) algorithm [9, 11, 12] is used for tracking the two points in the video frames. KLT tracking algorithm works with two simple steps: in initial frame tracking features are identified, and then tracks all the detected features in the remaining frames. Assume the first and the next images were taken at time t and $t + \tau$, respectively. Number of frames per second captured by the video camera determines the time τ. Let an image be represented as a function of two variables x and y. Now the variable t is added to represent the time at which the image was

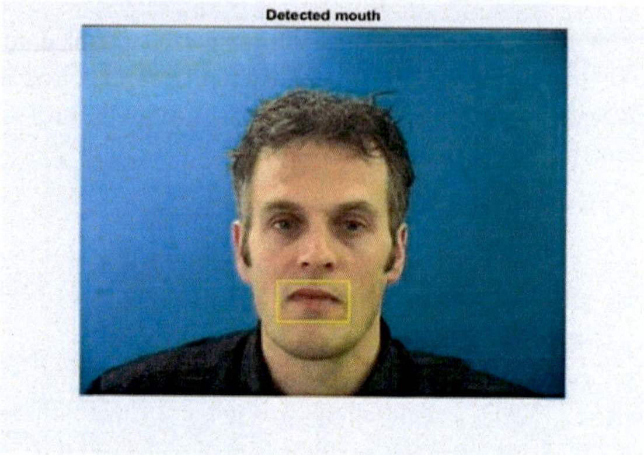

Fig. 1 Mouth detection of first frame

captured by the camera. Now any point in an image is defined by the function $f(x, y, t + \tau)$. The assumption made the KLT tracking algorithm is

$$f(x, y, t + \tau) = f(x - x, y - y, t) \tag{1}$$

From Eq. (1), it is understood that each point in the first frame is shifted by an amount (x, y) to obtain the second frame. This shifting amount is represented by displacement $d = (x, y)$, and the objective of tracking is to compute d. The two feature points taken from the first frame in our system for tracking throughout the frames are given as $[a + (w/2), b]$ and $[a + (w/2), b + h]$ where (a, b) are top left corner coordinates and w is the width and h is the height of the bounding box respectively.

Displacement between the feature points is calculated from one frame to other fame. Lip movement is computed using the displacement. This displacement calculation is used to find the onset and offset of speech using video. This decision will form the binary mask for speech enhancement. The steps involved in getting the binary mask using visual stream are also shown in Fig. 2.

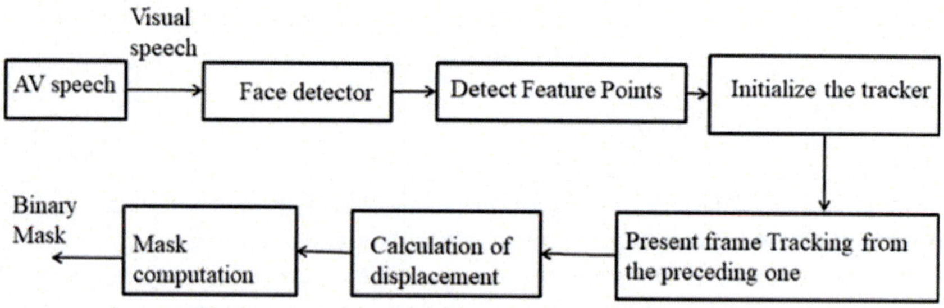

Fig. 2 Binary mask detection using video stream

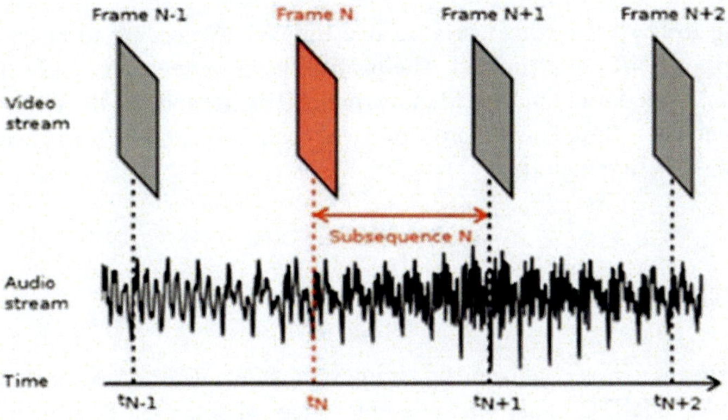

Fig. 3 Audio subsequence corresponding to a frame

2.3 Linear Interpolation

The number of video frames is significantly smaller than the number of determined audio frames. But the voice activity detection in audio stream and video stream is compared frame by frame for one-to-one correspondence. The original visual frames are interpolated to equate the number of audio stream frames. The onset/offset decision between two visual frames N and $N - 1$ is the onset/offset decision of Nth frame as in Fig. 3. The general formula to determine the number of frames in a given audio stream is that

$$\mathrm{NbFr} = (\mathrm{length}(x[n] - \mathrm{FL} + \mathrm{FS})/\mathrm{FS} \quad (2)$$

where NbFr = Number of audio frames, $x[n]$ = Audio samples, FL = Frame Length, FS = Frame Shift.

3 Speech Onset and Offset Detection Using Audio Stream

Varieties of information's are present in acoustic speech signal. In existing system, the voice activity detection using only audio is implemented as in [13]. The first step involved in conversion is framing. The input audio signal is segmented into 30 ms frames with 10 ms overlap between successive frames, and then Fourier transform is applied to obtain the frequency domain according to the Window, sidelobe attenuation, and FFT length properties. Using rectangular window will introduce high-frequency noise at the beginning and end of every frame. Hamming window is used to reduce this edge effect. The signal is then represented in power domain.

According to [14], the noise variance is estimated. According to [15], the posterior and prior SNR are estimated. The probability of speech present in the current frame is calculated using hidden Markov model (HMM) and log-likelihood ratio test according to [13]. Based on the probability of each frame, onset/offset decisions are made in the existing method.

4 Experimental Results

The experiment is first conducted by taking sample test videos from an audio–visual database like ‘GRID Corpus,’ where sample videos of both male and female speakers are available of the same duration. The noise samples with which the clean speech is mixed are taken from the ‘Noisex92’ database.

The frontview videos from the GRID corpus show the speakers uttering the same sentence consisting of voiced, unvoiced and silent segments to carry out the experiment. The mask using visual cues for voice activity detection is obtained using Viola–Jones and KLT algorithm. The target speech is separated using the mask obtained using visual cues. We have tested our system using three short videos from GRID corpus consisting of a total of (297 * 3) frames by mixing noise at − 5 dB to + 5 dB SNR conditions. The target speech separated using visual cues mask is compared with the target speech separated using mask obtained by the audio stream.

The experiment result is displayed in Table 1. The performance is evaluated using perceptual evaluation of speech quality (PESQ). PESQ score varies from 4.5 to − 0.5, higher scores better is the quality.

5 Conclusion

We have presented a mask using visual cues for voice activity detection. The proposed system detects the face and the mouth region to effectively distinguish the speaking from the silent frames in low SNR conditions. The binary mask presented here is independent of noise and effective for speech enhancement.

On comparing, the proposed VAD using visual cues to VAD using audio cues for three videos each of 297 frames it is evident that at low SNR conditions the target speech enhanced using visual cues are having higher PESQ score than target speech enhanced using audio cue.

Extremely low lighting situations and faces that move away considerably from a frontal pose or are too far from the camera to provide enough information for the mouth would cause the system to perform poorly.

Table 1 PESQ improvement of the proposed system

Video	Type of noise	SNR (dB)	PESQ score visual-VAD (proposed)	PESQ score audio-VAD (existing)	PESQ score improvement
Video 1	Babble	5	1.96	1.38	0.58
		0	1.85	0.96	0.89
		− 5	1.79	1.48	0.31
	Factory	5	2.05	1.5	0.55
		0	1.72	0.91	0.81
		− 5	1.47	0.67	0.8
Video 2	Babble	5	1.67	1.23	0.44
		0	1.45	0.98	0.47
		− 5	1.28	0.93	0.35
	Factory	5	1.67	1.4	0.27
		0	1.5	0.95	0.55
		− 5	1.26	0.92	0.34
Video 3	BABBLE	5	1.87	1.17	0.7
		0	1.59	0.96	0.63
		− 5	1.53	0.82	0.71
	Factory	5	1.92	1.37	0.55
		0	1.64	0.69	0.95
		− 5	1.39	0.88	0.51
	Average		1.64	1.06	0.57

References

1. Shoba S, Rajavel R (2017, April) Adaptive energy threshold for monaural speech separation. In 2017 international conference on communication and signal processing (ICCSP), pp 0905–0908. IEEE
2. Shoba S, Rajavel R (2018) Image processing techniques for segments grouping in monaural speech separation. Circ Syst Signal Process 37(8):3651–3670
3. Sivapatham S, Ramadoss R (2018) Performance improvement of monaural speech separation system using image analysis techniques. IET Signal Proc 12(7):896–906
4. Shoba S, Rajavel R (2020) A new Genetic Algorithm based fusion scheme in monaural CASA system to improve the performance of the speech. J Ambient Intell Humaniz Comput 11(1):433–446
5. Shoba S, Rajavel R (2019) Improving speech intelligibility in monaural segregation system by fusing voiced and unvoiced speech segments. Circ Syst Signal Process 38(8):3573–3590
6. Chen T, Rao RR (1998) Audio-visual integration in multimodal communication. Proc IEEE 86(5):837–852
7. Viola P, Jones M (2004) Robust real-time face detection. Int J Comput Vision 57:137–154
8. Viola P, Jones M (2001) Rapid object detection using a boosted cascade of simple features. In Proceedings of IEEE CVPR, Kauai, HI, USA, December 9–14, vol 1, pp 511–518
9. Tomasi C, Kanade T (1991, April) Detection and tracking of point features, Computer Science Department, Carnegie Mellon University

10. Lienhart R, Maydt J (2002) An extended set of Haar like features for rapid object detection. In: IEEE ICIP, Rochester, NY, USA, September 22–25. vol 1, pp 900–903
11. Shi J, Tomasi C (1994) Good features to track. In: IEEE conference on computer vision and pattern recognition, pp 593–600
12. Kalal Z, Mikolajczyk K, Matas J (2010) Forward-backward error: automatic detection of tracking failures. In: Proceedings of the 20th International conference on pattern recognition, pp 2756–2759
13. Sohn J, Kim NS, Sung W (1999) A statistical model-based voice activity detection. Signal Process Lett IEEE 6(1)
14. Martin R (2001) Noise power spectral density estimation based on optimal smoothing and minimum statistics. IEEE Trans Speech Audio Processing 9(5):504–512
15. Ephraim Y, Malah D (1984) Speech Enhancement using a minimum mean-square error short-time spectral amplitude estimator. IEEE Trans Acoust Speech Signal Process 32(6):1109–1121

Cognitive Vehicle Behavior Detector with Real-Time Analytics and Implementation

Mummareddy Yogendra Sai, Suri Kavya, Sravya Bhimavarapu, Mona Mudaliar, and Om Prakash Sahu

Abstract In this contemporary world, with the surge in automobiles, traffic has been a major issue. This project is going to deal with the violation of the vehicles in the traffic signals. By considering the surveillance footage, frames are achieved by using the open-source platform computer vision. Further, by applying image processing tools, details of the violated vehicles are generated in the form of tickets. By considering these tickets, analysis of data is performed so that we can visualize the traffic violations. The whole programming is performed using Python language, ELK stack includes Elastic search, Logstash, and Kibana are used for data analytics. Tickets generated are stored in a single file and loaded through Logstash. Then it is uploaded into the Kibana and visualization is performed.

Keywords Violation · Data analytics · Image processing

1 Introduction

The expanding number of vehicles in urban areas can cause high volume of traffic and suggests that petty criminal offenses become progressively basic these days around the globe. This causes serious annihilation of property and more mishaps that may imperil the lives of the individuals. Indeed the driving task is a very complicated one, involving numerous simultaneous actions [1].

In order to assist this task, several driver assistant systems have been suggested in the past years using either database information (i.e., learned geographic information systems) or on-vehicle sensors (i.e., laser, camera, etc.) to provide various environment information such as traffic signs, speed limits, traffic lights, crosswalks, or any other information like pedestrian or obstacles [3]. To take care of the disturbing issue

M. Y. Sai · S. Kavya · S. Bhimavarapu · M. Mudaliar
Electronics and Communication Department, Koneru Lakshmaiah University, Guntur, India

O. P. Sahu (✉)
SENSE, Vellore Institute of Technology, Chennai, India
e-mail: omprakash.sahu@vit.ac.in

A. Sivasubramanian et al. (eds.), *Futuristic Communication and Network Technologies*, Lecture Notes in Electrical Engineering 792,
https://doi.org/10.1007/978-981-16-4625-6_26

and forestall such inconceivable outcomes, petty criminal offense recognition frameworks are required. .This framework can distinguish most regular three sorts of petty criminal offense continuously which are signal infringement, stopping infringement and misguided course infringement. An easy-to-use graphical interface is related with the framework to make it basic for the client to work the framework, screen traffic and make a move against the infringement of traffic rules. To solve the alarming problem and prevent such unfathomable consequences, traffic violation detection systems are needed [3].

The objective of the project is to automate the traffic rules violation identification framework and make it simple for the traffic police office to screen the traffic and make a move against the disregarded vehicle proprietor in a quick and effective manner. Identifying and following the vehicle and their exercises precisely is the principle need of the framework.

2 Literature Survey

The literature survey of this project started with the announcement of the government to enhance the traffic violations which is available in the Web site [4]. The main concept of the idea for this project is taken from the Web site where the constraints of the project along with the dataset are provided.

Further, looking at the constraints for the project, it is clear that machine learning model along with Python would fit for this [5]. After having glance at the types of machine learning, reinforcement type of learning will be suited for this project.

Python libraries such as TensorFlow, Computervision and other libraries related to the video surveillance are used [6]. For extracting the video into frames and converting the video into respective frames, Computervision is used along with the TensorFlow. All the basic concepts of Computervision and TensorFlow are stated clearly in this Web site.

Further idea was to implement the data analytics by using the Elastic cloud where the analytics is performed [7]. Elastic is a specialized cloud where ingestion of the data and visualization is performed in a single platform called Elastic stack where working of Elasticsearch, Logstash and Kibana are the clearly stated in this Web site.

Image processing techniques such as gray scaling and blurring, binary thresholding, dilution, and finding the counter are the predefined libraries [8]. All the libraries for image processing are clearly stated in this Web site.

3 Components of the System

The framework comprises of two parts which incorporates:

- Vehicle detection model and Graphical UI.

First the CCTV camera footage from the street side is sent to the framework. Vehicles are recognized from the recording. Following the action of vehicles decides whether there is any infringement or not. Various kinds of violations have various calculations to decide the violation. A flowchart as shown in Fig. 2 shows how the framework functions. The Graphical UI (GUI) makes the framework intelligent for client to utilize. User can screen the traffic film and get the alert of violation with the caught vehicle picture. Client can make further move utilizing the GUI.

3.1 Database Structure

Making use of the Python programming language and SQL, whole data of the application is managed as we can observe in Fig. 1. Moreover, relational database such as Boyce–Codd normal form is also used for the main tables which includes (Fig. 2):

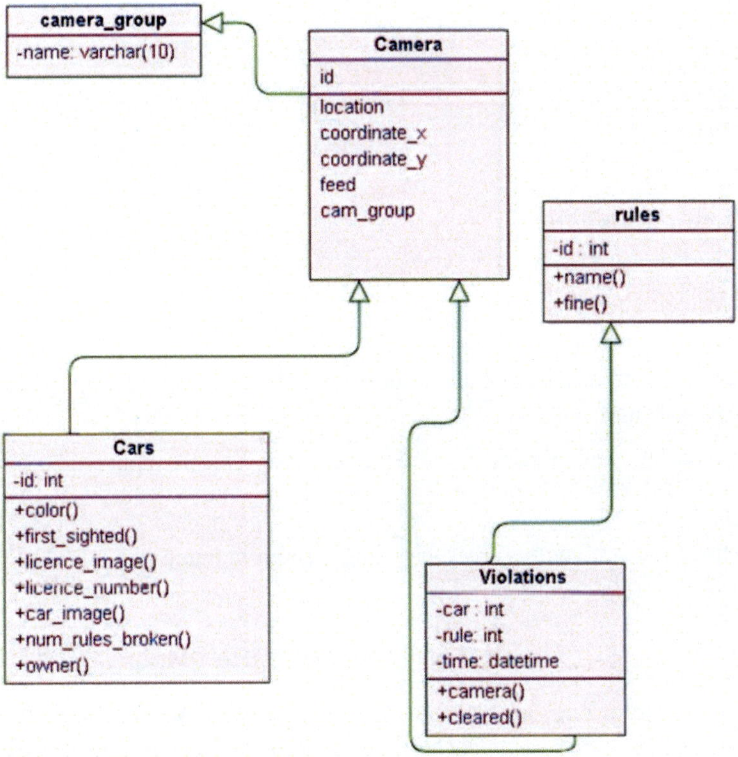

Fig. 1 Class diagram

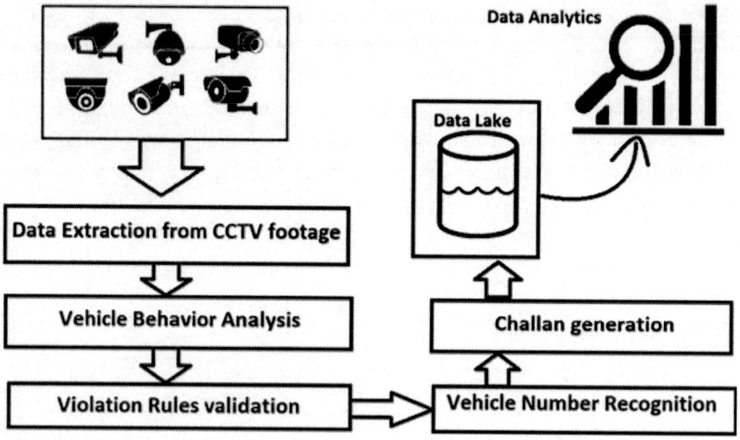

Fig. 2 Flow diagram of the project

- Cars
- Rules
- Cameras
- Violations
- Groups.

4 Methodology

4.1 Image Processing

The video consists of frames which can be extracted by performing image processing. Many steps such as

- **Gray scaling and blurring**: In the preprocessing stage, input frame which is received from CCTV footage is grayscaled and blurred using Gaussian blur method.
- **Background subtraction:** By subtracting the current frame with reference frame, objects area can be determined.

$$\text{dst}(I) = \text{saturate}(|\text{scr1}(I) - \text{scr2}(I)|)$$

- **Binary thresholding:** To obtain the accurate object area by removing all the noise and holes frame binarization method is used by providing threshold to the image.

$$\text{dst}(x, y) = \text{maxVal is scr}(x, y) > \text{thresh else } 0$$

Table 1 MobileNet body architecture

Type/stride	Filter shape	Input size
Conv/s2	3 × 3 × 3 × 32	224 × 224 × 3
Conv dw/s1	3 × 3 × 32 dw	112 × 112 × 32
Conv/s1	1 × 1 × 32 × 64	112 × 112 × 32
Conv dw/s2	3 × 3 × 64 dw	112 × 112 × 64
Conv/s1	1 × 1 × 64 × 128	56 × 56 × 64
Conv dw/s1	3 × 3 × 128 dw	56 × 56 × 128
Conv/s1	1 × 1 × 128 × 128	56 × 56 × 128
Conv dw/s2	3 × 3 × 128 dw	56 × 56 × 128
Conv/s1	1 × 1 × 128 × 128	28 × 28 × 128
Conv dw/s1	3 × 3 × 256 dw	28 × 28 × 256
Conv/s1	1 × 1 × 256 × 256	28 × 28 × 256
Conv dw/s2	3 × 3 × 256 dw	28 × 28 × 256
Conv/s1	1 × 1 × 256 × 512	14 × 14 × 256
Conv dw/s1	3 × 3 × 512 dw	14 × 14 × 512
Conv/s1	1 × 1 × 512 × 512	14 × 14 × 512
Conv dw/s2	3 × 3 × 512 dw	14 × 14 × 512
Conv/s1	1 × 1 × 512 × 1024	7 × 7 × 512
Conv dw/s2	3 × 3 × 1024 dw	7 × 7 × 1024
Conv/s1	1 × 1 × 1024 × 1024	7 × 7 × 1024
Avg Pool/s1	Pool 7 × 7	7 × 7 × 1024
FC/s1	1000 × 1024	1 × 1 × 1024
Softmax/s1	Classifier	1 × 1 × 1024

- **Dilation and finding the contour:** Image is diluted to fill the holes and contour is found from the image by drawing the rectangle box over desired object (Table 1).

Transfer learning approach is accustomed to prepare the model with our dataset. The dataset comprises of 500 pictures for every class. The preparation boundaries are referenced in table (Table 2).

Table 2 Training hyperparameters

Parameter Name	Value
Learning rate	0.01
Training steps	100

5 Implementation

OpenCV is the library for the picture handling where inbuilt libraries are utilized. Tensor stream AI structure is utilized for actualizing vehicle classifier. User can include menu things, for example, camera, area, the feed record for the camera. Here the feed record is introduced by the camera module over the Web. We have utilized Linux record sharing example for getting the video from the camera, where the camera will take care of the offered document to the server, and the server will take the feed record to process and identify violation. X and Y directions as shown in Fig. 3 of the camera area are utilized for possible future reason. JSON record is likewise required by determining a few standards for camera reason.

The client has numerous different articles to embed into the database. The administrator can include the accompanying elements as shown in Fig. 4 in the graphical UI:

- Camera
- Car
- Rule
- Violation.

A GUI is made for the purpose of gathering of cameras by the administrator consistently. He can see the rundown of rule violations and can see details of the vehicles that abused the standards. In the event that he taps on the detail button, another window will show up where the client will have the option to document the report or send/print ticket for the vehicle proprietor.

Likewise the administrator/client can erase the records for a false positive. However, there will be never a record erased. The database has a marker of which

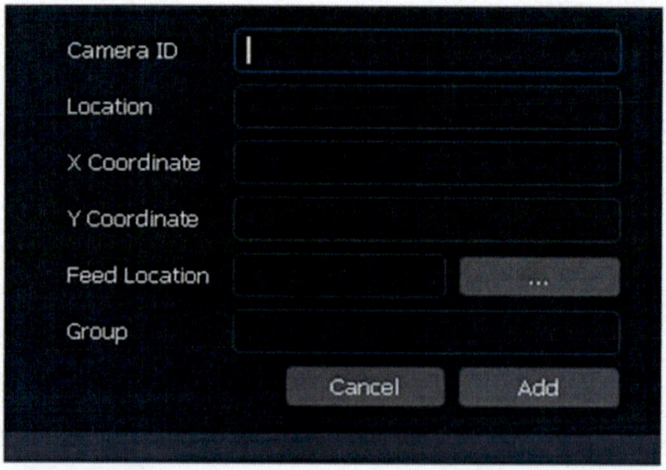

Fig. 3 Details of rule violation

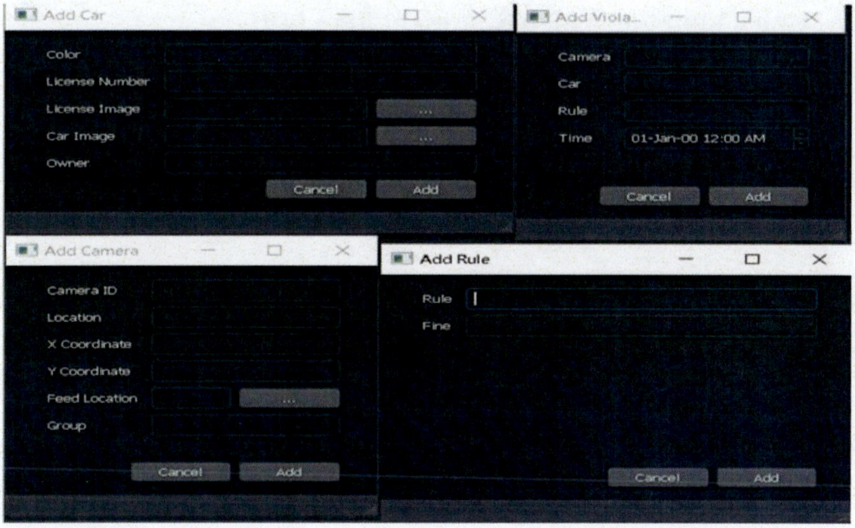

Fig. 4 Entity for adding windows

document have been chronicled. On the off chance if we need to recover a record from the erased once, at that point the administrator needs to go to the file window. There he can re-establish any record he needs.

The user can also look for a vehicle, with its license number, its color, or date of a rule violation as shown in Fig. 5. The permit number has content forecast so the client will be certain while composing a license number that it exists.

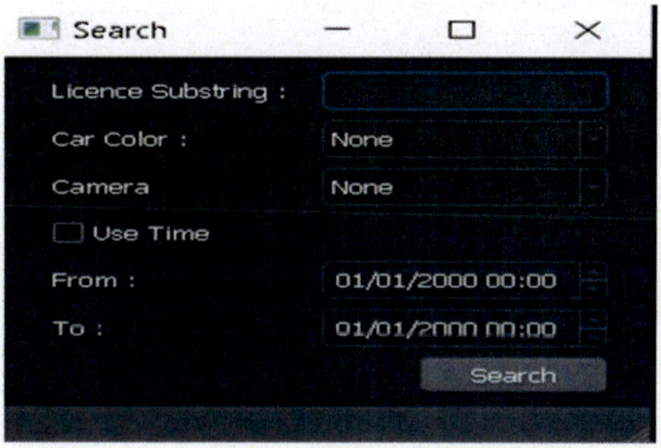

Fig. 5 Searching a car or rule violation

6 Rules Violation Video Representation in UI

There are as of now three guidelines we are worried about.

- Signal violation
- Parking violation
- Direction violation.

6.1 *Signal Violation*

We have utilized a straight line in the image. At the point when the traffic light is red and a vehicle is crossing the straight line, an image of that vehicle is enrolled in the database alongside some natural qualities. The client can find in the live review which vehicle is being identified continuous and tried in the event that they are disregarding.

6.2 *Parking Violation*

For parking violation, we have prefigured a square shape, which is the confined region for vehicle leaving. On the off chance that there is a vehicle in the square shape for more than a predefined time, at that point a picture with other ecological qualities is being enlisted to the database.

6.3 *Direction Violation*

For direction violation location as shown in Fig. 6, a few lines are attracted to isolate into districts. At that point when a vehicle moves starting with one area then onto the next, its heading is estimated. On the off chance that the course is not right, at that point it is enlisted as past.

7 Data Analytics

Now by considering each violation as a ticket, analytics is performed so that areas with high violations can be known .Data analytics is performed using Elastic stack which includes Elasticsearch, Logstash and Kibana [9].

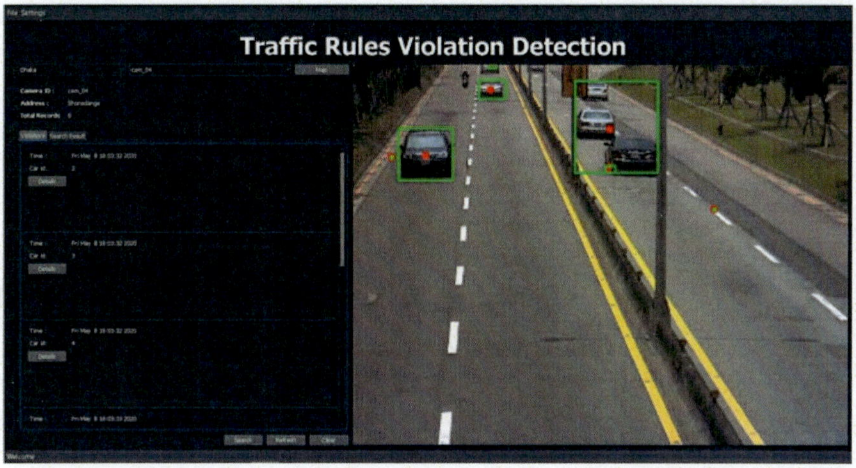

Fig. 6 Direction violation

7.1 Elasticsearch

Elasticsearch is a dispersed, open-source search and analytics engine for a wide range of information, including printed, numerical, geospatial, organized, and unstructured. Elasticsearch is the focal point of the Elastic Stack, a lot of open-source instruments, enemy information ingestion, improvement, stockpiling, investigation, and perception. To begin Elasticsearch, open the order brief and type elasticsearch

7.2 Logstash

Logstash is an open-source, server-side information handling pipeline that ingests information from a huge number of sources, changes it, and then sends it to our necessary reserve. In this project, tickets are transformed into the config file where every details of the violations lies. In the conf file, details of the violations are present and to start the logstash, open the command prompt by name of file.

7.3 Kibana

Kibana lets you to visualize your elastic search data and navigate the elastic stack so that you can do anything from tracking query load to understand the way requests flow through required applications. In this project, visualization is performed for violated vehicles. Kibana can be accessed by typing kibana.bat in the command prompt.

8 Results

Violated vehicle details are considered as tickets and stored in a file using the programming in Python. Subsequently, analytics is performed by using the ELK stack. Storing, searching and analyzing of the data will be done by elasticsearch. By using the Logstash, ingestion of data from the respective file to the cloud is done. Final output of the project is visualized in Kibana where the overall details of the violations can be known.

According to the violations of the vehicles, in Fig. 7, by using various visualization metrics, counts of the tickets are shown with the help display meter in the first sub-parts of figure. In the following sub-figure, numerical value of count of the tickets are depicted. The registration number of the vehicle with highest violations are visualized by means of heat map which enlarges the license number with more number of violations and types of violations are depicted in the fourth sub-figure.

In Fig. 8, as the wrong route violations are higher compared to parking violation, it gets enlarged by heat map. In the following figure, count of the violation type is shown.

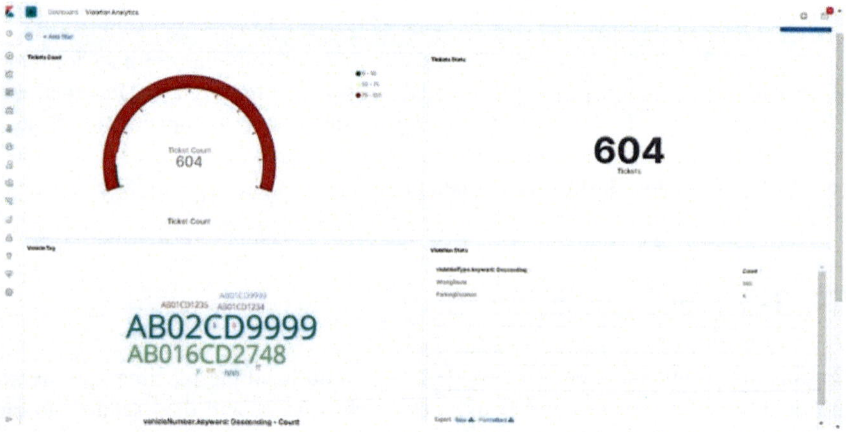

Fig. 7 Visualization in Kibana

Fig. 8 Violation tag and violation metrics

9 Conclusion

Vehicle challan can be generated by policeman by cross verifying the details generated by programming. Data visualization of the violated vehicles is done. In future, man power can be reduced at the traffic signals. Challan generations can be made fully automated with enhanced AI/ML models.

10 Discussion

The primary purpose of this study was to analyze the traffic and violations so that minimal man power can be used at the traffic signals. Initially, the video frames will be extracted and by using TensorFlow and image processing tools are applied where violations can be generated as a ticket.

Our results clearly proved the vehicles violation and analytics depict the type of violation from the specified location. From the analytics, while generating challan it will be clear for the person about the violation.

References

1. Rouzier B, Murakami T (2016) Hazard detection and cognition for an active driving assistance. In: 2016 IEEE 14th international workshop on advanced motion control (AMC)
2. de Charette R, Nashasihibi F Real time visual traffic lights recognition. In: World congress final paper template
3. Wang X, Meng L-M, Zhang B, Lu J, Du K-L (2013) A video-based traffic violation detection system. In: MEC, Shenyang, China, 2013
4. https://data.gov.in/ogpl_apis
5. https://www.analyticsvidhya.com/blog/2018/12/key-breakthroughs-ai-ml-2018-trends-2019/
6. https://www.geeksforgeeks.org/introduction-to-opencv/?ref=lbp
7. https://www.elastic.co/
8. https://www.geeksforgeeks.org/python-image-blurring-using-opencv/?ref=lbp
9. Elastic Stack and Product Documentation retrieved from https://www.elastic.co/guide

Overview of Fronthaul Technologies and the DBA Algorithms in XGPON-Based FH Technology in CRAN Architecture in 5G Network

Theresal Thangappan and Brintha Therese

Abstract The cloud radio access network or C-RAN is the vital technology involved in the fifth-generation (5G) radio access network. It is a centralized, cloud computing architecture for radio access network which is allowing many remote radio heads (RRHs) to connect to a centralized baseband unit (BBU) pool. The fronthaul (FH) approach has evolved due to significant difficulties in 5G transportation technology. Different fronthaul techniques were discussed in this paper, as well as the motivation to achieve the optimal FH solution through a time division multiplexing passive optical network (TDM-PON), and also an overview of the dynamic bandwidth allocation (DBA) algorithms in fronthaul technologies based on TDM-PON and their 5G network CRAN architecture characteristics.

Keywords Fronthaul · CRAN · TDM-PON · DBA

1 Introduction

High bandwidth user applications, such as the Internet of things (IoT), social networking sites, and video conferencing, have substantially evolved cellular communication paradigms. As the present technology (4G) is similar to the Shannon limit [1], the development of next-generation technology (5G) is essential to provide various services. The 5G technology aims to provide a data rate that is five times (i.e., >20 Gbps) more than the wireless system, and the latency is less than 1 ms [2]. For reaching the milestone of 5G technology, different techniques are in use. They are heterogeneous small cells, multiple inputs, multiple outputs (MIMO), energy-efficient antennas, non-orthogonal multiplexing scheme, sparse code multiple access schemes, and 3D beamforming and millimeter waves (mm). These are being researched upon as all the research organizations have extended their support for the 5G standards. The following methods need to be employed by mobile network operators (MNO) to develop the network capacity. The deployment of maximum cells

T. Thangappan · B. Therese (✉)
School of Electronics and Communication Engineering, VIT, Chennai Campus, Chennai 632014, India

A. Sivasubramanian et al. (eds.), *Futuristic Communication and Network Technologies*, Lecture Notes in Electrical Engineering 792,
https://doi.org/10.1007/978-981-16-4625-6_27

[3–6] and the use of advanced radio access technologies [7, 8] will achieve the use of additional spectrum. In the end, the most suitable solution for network capacity enhancement is the small cell method [6, 9, 10]. CRAN [11–15] used for the 5G network because of the benefits such as site installation process, cost-efficient, and need of the low power [16–18]. Hence, CRAN is an eminent method when immense numbers of remote radio heads (RRHs) are required to increase the network capacity [19]. CRAN introduces some changes in the 5G layer known as the fronthaul (FH) solution. The FH solution links the RRH to the BBU directly.

2 Centralized RAN Architecture

In the abbreviation CRAN, "C" refers to either cloud or centralized. CRAN divides the functions of traditional base transceiver station (BTS). The radio still functions through the RRH and remains either in the BTS capsule or at the radio site. During the baseband process's function, the BBU has been partially or fully moved from the cell to the COs (central office) at a common shared site and becomes the reason for developing a new RAN interface called FH. Therefore, CRAN is implemented through three network elements, i.e., FH interface, BBU, and RRH.

The sole purpose of RRH is to transmit and receive digitized radio signals. Since the functions of RRH are simple, the hardware that needed to carry out these tasks is cheap and makes the overall deployment cost of RRH less in a widespread area. The baseband unit performs three complex functions: resource allocation, resource scheduling, and baseband (radio) signal processing function. The tasks performed by BBU are complex, and the hardware used to perform those functions is a bit expensive. So, to reduce the overall maintenance cost of the network and CAPEX/OPEX, BBUs are implemented in a centralized manner.

The Metro Ethernet Forum (MEF) firstly describes the term FH as a network interface between centralized BBU and RRH. BBU and RRH will be connected via optical fiber using radio over fiber (RoF) technology. The RoF here falls into two categories, digital and analog RoFs. In these, digital RoF (DRoF) is more prominent and also a standardized one. It can also transmit signals (wireless) with less degradation [20], whereas analog RoF (ARoF) is not standardized [21]. Common Public Radio Interface (CPRI) is a widely used digital RoF interface and several telecommunications vendors [5, 12, 22–24]. Studies show that the most complex part of CRAN architecture is fronthaul since it ruins CRAN's overall performance and capacity. FH also makes the deployment cost of CRAN worse. Therefore, it should be built strategically [25, 26]. Copper, microwave, optical fiber, optical free-space contact, and mm waves are used to achieve the FH interface [12]. Out of all these, the optical fiber is considered the most preferred physical medium for FH because of its effective handling against low latency in gigabit connections [5, 27–30]. Figure 1 shows the CRAN architecture.

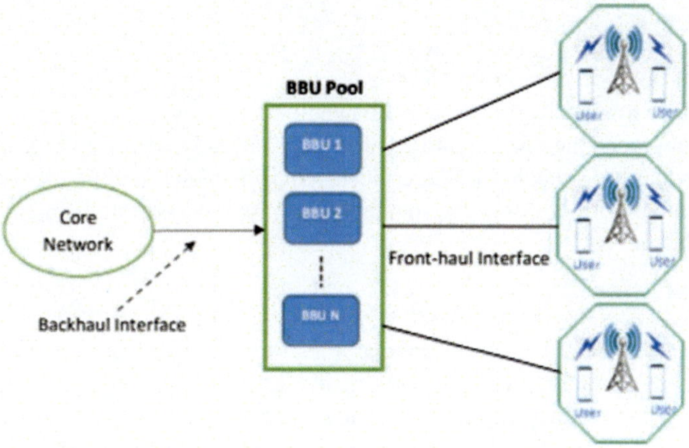

Fig. 1 CRAN architecture

3 Fronthaul Network Technology

Each fronthaul network will have its mix of challenges depending on deployment strategy, geographical area, number of sites, and RAN technology. It is essential to evaluate each deployment scenario and apply the right mix of technologies and solutions. Operators have several technologies that can be applied for the fronthaul use case, each comprising its advantages and disadvantages.

1. Wireless fronthaul solution:

Wireless technology is a feasible fronthaul transport, resulting in remarkable evolutions in telecommunication networks. It inherits the following advantages: simplicity, easy deployment, scalability, support roaming, efficient collaboration and cost-saving. It is a complement solution for fiber-based fronthaul solution, and because of its susceptibility to channel condition, it is suitable to implement for short range. The current wireless technology can adopt a few CPRI interface options, and that limits the bandwidth. Some of the promising wireless solutions like millimeter wave (MM-W) and Wireless Fidelity (Wi-Fi) are engaged within the fronthaul [31–33]. Wi-Fi network faces the limitation in coverage and mobility even though it provides sufficient data rates, and this problem might be reduced by Wi-Fi mesh network [31, 34].

2. Wired fronthaul solution:

Wired technologies' merits are less interference, maximum coverage, less latency, reliability, and security. Due to these advantages, the wired system can stand like a wireless system. Dark fiber, WDM-based solution, WDM-PON, WDM/optical transport network (OTN), and Ethernet are examples of fronthaul technology in wired systems. The superior wired fronthaul technology which is suitable for the solution discussed in the subsection.

a. Dark fiber solution

Dark fiber solution is considered an attractive and most comfortable deployment solution with low latency, as there is no need for transmission components between the BBU pools and the RRUs. It is a point-to-point solution; therefore, it lacks network security and is not suitable for 5G services, requiring high reliability. It also requires a high amount of fiber resources, making the deployment cost high and is a primary limiting factor.

b. WDM-based solution

WDM-based fronthaul solutions are classified into active and passive methods. It is beneficial for fronthaul solutions which need low latency and high data transmission rate. The active techniques provide a robust network with suitable flexibility, and with an optical amplifier, the system extends significantly, and it needs a power supply for the operation. The passive method depends only on CPRI multiplexing and demultiplexing since it uses passive components with no need for power supply and battery backup [31–35]. In passive methods, the switching granularity depends on the spectrum or time slot. The active method depends on packet or frame switching. Even though active methods provide good configuration flexibility, its power consumption is complicated. [36].

c. WDM/OTN

In WDM/OTN, the fiber capacity is boosted by adding multiple channels in a shared fiber to achieve multiplexed and transparent transmission of a signal through the fronthaul to link numerous sites [31, 33, 35]. The in-phase and quadrature component (I/Q) data encapsulated by the OTN frame and then multiplexed with WDM wavelength. This wavelength is used for routing the frame to the destination [36]. This method has features like low latency, high bandwidth, security, scalability, and reliability despite fiber resources. WDM transport network does not require a power supply, but it is necessary for wavelength translation and active management [31, 33, 35].

d. Packet/Ethernet-based Fronthaul

Packet technology in Ethernet-based technology emphasizes the statistical multiplexing features and helps achieve traffic convergence and enhance bandwidth usage. The most efficient Ethernet-based fronthaul option is to convert CPRI into eCPRI in the RAN domain. This approach leverages their CPRI data streams' baseband processing and converts the time domain signal to the frequency domain. The benefit of the CPRI to eCPRI conversion is that it drastically reduces the capacity required for Ethernet fronthaul by scaling traffic with used antenna bandwidth and removing the constant bit rate of the CPRI traffic. The conversion process will not add any additional latency in the fronthaul transmission, enabling more extended macrosites to CRAN hub site distance compared to RoE and greater flexibility in building the fronthaul network. Combining CPRI to eCPRI conversion with packet aggregation at the antenna site is the most efficient approach. The fronthaul capacity demands are reduced by 60–80% depending on radio configuration compared to other technologies such as traditional CPRI or mapping of CPRI utilizing RoE.

e. PON-Based Solutions

The intensive bandwidth traffic required for 5G and beyond network is not satisfied by the fiber-based and P2P Ethernet because of ultra-dense network deployment and highly network resources in the 5G network. The PON system reduces the number of interface and site space, and system power can also be saved [37]. PON technology has both WDM and TDM techniques used to improve the capacity and efficiency, and according to these techniques, the PON system is classified into WDM-PON and TDM-PON. TDM-PON provides higher bandwidth for different services even though available resources are limitedly delivered to the end user. But this is overcome by assigning one wavelength to the per user in WDM-PON high data rate and well secure because of the P2P channel and useful for the long-reach application. The WDM-PON does not recommend for fronthaul and backhaul technology.

Numerous optical solutions for 5G fronthaul technology have been proposed in a lot of literature. Depending on the following factor-like bandwidth requirements, latency, resource availability, business type, and deployment method, the deployment choice has been chosen from the following FH technology in the 5G network, i.e., P2P fiber access, OTN, WDM, PON, and carrier Ethernet [38, 39]. PON performs other solutions because of its high bandwidth, low installation and operational cost, and coverage. The PON system also uses the existing fiber cables or optical distribution networks (ODNs), optical network units (ONUs), and optical line terminal (OLT) for FH solution, even if it provides both wired and wireless services to the customers [40, 41]. Finally, PON is solemnly considered as a solution for FH technology as it effectively uses the fiber sources of existing fiber to the home (FTTH) [42, 43].

4 TDM-PON-Based Fronthaul Technology in CRAN

OLT, ONUs, and ODN have composed TDM-PON-based fronthaul technology. The OLT, which is located at the CO, is placed near BBU. The ONU, which is located beside the customer, is placed near to RRU at base stations. One OLT is capable of linking more than one ONUs through the passive splitter. TDM-PON needs a single wavelength, so it is a simple and cost-efficient solution, but the latency issue remains persistent in the FH solution. The following are the two factors conventional DBA and window size will be used to improve the latency of TDM-PON [44], for monitoring ONU registration and its activation time by using window size. In TDM-PON, many researchers are taking initiatives to provide the low-latency TDM-PON-based FH technology in the 5G network to prevent the collision between the ONU's simultaneous transmission by using DBA and providing the time slots to ONUs in each cycle to increase the overall latency performance. Figure 2 shows the TDM-PON-based FH technology for CRAN architecture.

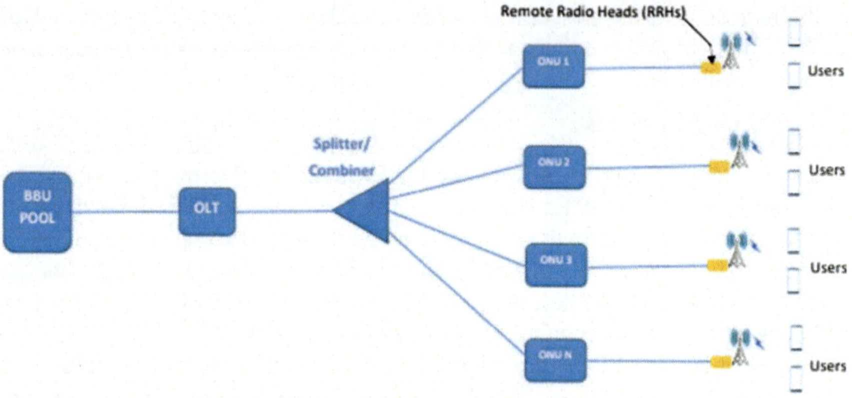

Fig. 2 TDM-PON-based FH technology for CRAN architecture

5 DBA Algorithm in XGPON-Based FH Technology

The latency concerns about the TDM-PON-based fronthaul technology have been proposed and studied widely among many papers. The author [45] of the article has depicted the optimized TDM-PON concept, which can be implemented for transparency optical transport of lower latency with constant bit rate data. The mathematical models analyzed that it was based on queuing theory and has achieved one-way latency, which is less than 250 μs. In paper [46], the author has presented the CPRI concept over the Ethernet-based TDM-PON. The delay jitter and the latency could be reduced using the IQ compression technique concept around 100 μs for a particular compression ratio. Paper [20] introduces the mobile DBA algorithms used to reduce the front haul latency by using mobile scheduling information from BBUs for the bandwidth allocation. It has achieved less than 50 μs by utilizing a fiber length of 10–20 kms. This approach needs a connection for the transfer of information from BBUs to OLT. Paper [47] introduces simple statistics functions in coordination with the fixed bandwidth allocation (FBA) to dynamically provide the data rate for the fronthaul transmission over G_EPON. This method relies on the periodical estimation of RRH traffic; therefore, it cannot capture low-scale variation in fronthaul traffic at each RRH.

In paper [48], the author reduced the DBA cycle length by lowering the DBA grant time to minimize the upstream latency of 10G EPON. This technique is not suitable for ITU TDM_PON as it is neither synchronous not has a fixed frame length as that of ITU TDM PONs. ITU PON [49] has fixed frame length and has frame support fragmentation that helps reduce the latency lower than IEEE PONs [50]. The above reasons require a lot of research work motivated based on fronthaul using XGPON.

Round robin DBA/RR DBA algorithm is the most straightforward algorithm that uses DBA compatible with the XG-PON model, and it is the first algorithm for using the fronthaul by XGPON [51]. RR-DBA uniformly allocates the bandwidth, which is less than or equal to a predetermined value circularly to all T-Conts. Good delay performance, but the utilization of upstream bandwidth under busty traffic conditions, is very poor and has not satisfied the 300 μs latency needed for the front haul technology. In group assured Giant DBA [52], which helps optimize XG-PON to transport backhaul traffic, and it is concluded that either RR-DBA or the gGAINT fulfills the latency requirement for the fronthaul. The gGAINT algorithm can assign group-assured bandwidth by sharing the unutilized data from individual assured bandwidth with other T_Conts of the same group and failed to fulfill the latency requirement for fronthaul technology. In optimized RR DBA, a modified RR DBA version. In this algorithm, firstly, it will find the total extra bandwidth from each upstream cycle, and then it will redistribute it to the heavily loaded T-Conts in the next transmission cycle. The lightly loaded T-Conts excess bandwidth will be utilized by heavily loaded T-Conts to eliminate the problems faced in the RR algorithm and satisfy the latency requirement for fronthaul technology.

The prediction method using deep learning algorithm has been widely used in PON. Adaptive learning approach DBA [53], to predict the future ONU buffer occupancy using the previous ONU buffer occupancy reports, is recorded in BBU processing. It eliminates the process of reporting ONU buffer occupancy to the OLT and eliminates the DBA process delay. In paper [54], the author proposed a novel DBA based on long short-term memory (LSTM) deep learning method. Recurrent neural network (RNN) introduces a recurrent structure for implementing a memory mechanism to track past information. Through this recurrent structure, the RNN has better performance than the FNN in predicting data with time series. The number of packets that will arrive at the ONU in the next cycle is expected by the LSTM recurrent neural network because of its accurate results the OLT grant bandwidth without waiting for ONU buffer occupancy reposts to the OLT. Some of the TDM-PON-based DBA algorithms for fronthaul technology and its characteristics are tabulated in Table 1.

6 Conclusion

This paper discussed the different solution available for fronthaul technology for CRAN in 5G. It concluded that TDM-PON-based fronthaul is well acknowledged and cost-efficient technology for the CRAN FH network due to inexpensive passive optical components. Using PON infrastructure, the cost of the FH network is estimated at around 60%. The latency issues in conventional TDM-PON are eliminated by using the DBA algorithm. This paper described many DBA algorithms used in TDM-PON-based fronthaul technology and its merit and demerits of traditional DBA and the prediction-based DBA algorithm using deep learning and its advantages.

Table 1 TDM-PON-based DBA algorithm for fronthaul technology and its characteristics

Algorithm	Year of publication	Characteristics
Mobile DBA	2015	Using mobile scheduling information from BBUs for the bandwidth allocation. Need a connection to transfer data from BBU to OLT
Round robin DBA	2014	Uniformly circularly allocates the bandwidth to all T-Conts. Good delay performance. utilization of upstream bandwidth under busty traffic condition is very poor
gGAINT algorithm	2014	Sharing the unutilized data from individual assured bandwidth with other T_Conts of the same group
Optimized RR DBA	2017	Total extra bandwidth from each upstream cycle redistributes to the heavily loaded T-Conts in the next transmission cycle
Adaptive learning approach DBA	2018	Using FANN, predict the future ONU buffer occupancy using the previous ONU buffer occupancy reports recorded in BBU processing
DBA-based on long short-term memory	2019	The number of packets that arrive at the ONU buffer from RRHs is predicted using LSTM RNN because of its accurate results, the OLT grant bandwidth without waiting for ONU buffer occupancy reposts to the OLT

References

1. Agiwal M, Roy A, Saxena N (2016) "Next generation 5G wireless networks": a comprehensive survey. IEEE Commun Surv Tutorials 18(3):1617–1655
2. Jiang D, Liu G (2017) An overview of 5G requirements. Springer, Berlin, pp 3–26
3. Akpakwu GA, Silva BJ, Hancke GP, Abu-Mahfouz AM (2018) A survey on 5G networks for the internet of things: communication technologies and challenges. IEEE Access 6:3619–3647
4. Etemad K, Baker M (2013) Evolution of 3GPP LTE in release 11 and beyond [Guest Editorial]. IEEE Commun Mag 51(2):73–73
5. Pizzinat A, Chanclou P, Saliou F, Diallo T (2015) Things you should know about fronthaul. J Lightwave Technol 33(5):1077–1083
6. Nakamura T, Nagata S, Benjebbour A, Kishiyama Y, Hai T, Xiaodong S et al (2013) Trends in small cell enhancements in LTE advanced. IEEE Commun Mag 51(2):98–105
7. Gesbert D, Kountouris M, Heath RW, Chae C-B, Salzer T (2007) From single user to multiuser communications: shifting the MIMO paradigm. IEEE Signal Process Mag 24(5):36–46
8. Hoydis J, Ten Brink S, Debbah M (2011) Massive MIMO: how many antennas do we need? In: 2011 49th Annual Allerton conference on communication, control, and computing (Allerton). IEEE, pp 545–550

9. Okumura Y, Terada J (2014) Optical network technologies and architectures for backhaul/fronthaul of future radio access supporting big mobile data. In: Optical fiber communication conference (pp. Tu3F-1). Optical Society of America
10. Xu J, Wang J, Zhu Y, Yang Y, Zheng X, Wang S et al (2014) Cooperative distributed optimization for the hyper-dense small cell deployment. IEEE Commun Mag 52(5):61–67
11. Mobile C (2011) C-RAN: The road towards green RAN. White Paper 2:1–10
12. Checko A, Christiansen HL, Yan Y, Scolari L, Kardaras G, Berger MS et al (2015) Cloud RAN for mobile networks—a technology overview. IEEE Commun Surv Tutorials 17(1):405–426
13. Lin Y, Shao L, Zhu Z, Wang Q, Sabhikhi RK (2010) Wireless network cloud: architecture and system requirements. IBM J Res Dev 54(1):4:1–4:12
14. Wu J, Rangan S, Zhang H (2016) Green communications: theoretical fundamentals, algorithms, and applications. Boca Raton: CRC Press
15. Chanclou P, Suzuki H, Wang J, Ma Y, Boldi MR, Tanaka K et al (2017) How does passiveoptical network tackle radio access network evolution? J Opt Commun Netw 9(11):1030–1040
16. Zhang J, Xiao Y, Li H, Ji Y (2017) Performance analysis of optical mobile fronthaul forcloud radio access networks. J Phys: Conf Ser, 910:012053
17. Boccardi F, Heath RW Jr, Lozano A, Marzetta TL, Popovski P (2014) Five disruptive technology directions for 5G. IEEE Commun Mag 52(2):74–80
18. Alliance N (2015) Further study on critical C-RAN technologies. Next Generation Mobile Networksv1.0
19. Morant M, Llorente R (2019) Performance analysis of multiple radio-access provision in a multicore-fibre optical fronthaul. Opt Commun 436:161–167
20. Shibata N, Tashiro T, Kuwano S, Yuki N, Fukada Y et al (2015) Performance evaluationof mobile front-haul employing Ethernet-based TDM-PON with IQ data compression. J Opt Commun Netw 7(11):B16–B22
21. Haddad A, Gagnaire M (2014) Radio-over-fiber (RoF) for mobile backhauling: a technical andeconomic comparison between analog and digitized RoF. In: 2014 International conference on optical network design and modeling. IEEE, pp 132–137
22. Gomes NJ, Chanclou P, Turnbull P, Magee A, Jungnickel V (2015) Fronthaul evolution: from CPRI to Ethernet. Opt Fiber Technol 26:50–58
23. de la Oliva A, Hernandez JA, Larrabeiti D, Azcorra A (2016) An overview of the CPRI specification and its application to C-RAN-based LTE scenarios. IEEE Commun Mag 54(2):152–159
24. Specification C (2014) V6. 1 (2014-07-01) Interface specification common public radio interface (CPRI)
25. Ranaweera C, Wong E, Nirmalathas A, Jayasundara C, Lim C (2017) 5G C-RAN architecture: a comparison of multiple optical fronthaul networks. In: 2017 International conference on optical network design and modeling (ONDM). IEEE, pp 1–6
26. Ranaweera C, Wong E, Nirmalathas A, Jayasundara C, Lim C (2018) 5G C-RAN with optical fronthaul: an analysis from a deployment perspective. J Lightwave Technol 36(11):2059–2068
27. Agrawal GP (2012) Fiber-optic communication systems, vol 222. Wiley, New York
28. Bernardos CJ, De Domenico A, Ortin J, Rost P, Wübben D (2013) Challenges of designing jointly the backhaul and radio access network in a cloud-based mobile network. In: 2013 Future network and mobile summit. IEEE, pp 1–10
29. Tashiro T, Kuwano S, Terada J, Kawamura T, Tanaka N et al (2014) A novel DBA scheme for TDM-PON based mobile fronthaul. In: OFC 2014. IEEE, pp 1–3
30. Wake D, Nkansah A, Gomes NJ (2010) Radio over fiber link design for next generation wireless systems. J Lightwave Technol 28(16):2456–2464
31. Alimi IA, Teixeira AL, Monteiro PP (2018) Toward an efficient c-ran opticalfronthaul for the future networks: A tutorial on technologies, requirements, challenges, and solutions. IEEE Commun Surv Tutorials 20(1):708–769
32. Guiomar FP, Alimi IA, Monteiro PP, Gameiro A (2018) Flexible infrastructure for the development and integration of access/fronthauling solutions in future wireless systems. In: 2018 IEEE 19th international workshop on signal processing advances in wireless communications (SPAWC), June 2018, pp 1–5

33. "Nokia Optical Anyhaul as an enabler of C-RAN: accelerating the delivery of 5G networks". White paper, documentcode: Sr1803022985en, Nokia, March 2018
34. Aijaz A, Aghvami H, Amani M (2013) A survey on mobile data offloading: technical and business perspectives. IEEE Wirel Commun 20(2):104–112
35. "5G-oriented Optical Transport Network Solution". White paper, ZTE Technologies, June 2017
36. Yu H, Zhang J, Ji Y, Tornatore M (2018) Energy-efficient dynamic lightpathadjustment in a decomposed awgrbasedpassive wdm fronthaul. IEEE/OSA J Opt Commun Netw 10(9):749–759
37. Otaka A (2017) "Flexible access system architecture": Fasa. In: NTT Tsukuba Forum 2016 Workshop Lectures. Tokyo, Japan: NTT, vol 15, pp 1–7
38. Guizani Z, Hamdi N (2017) CRAN, H-CRAN, and F-RAN for 5G systems: key capabilities and recent advances. Int J Netw Manag 27(5):e1973
39. Bhaumik P (2016) Next-generation broadband access network architectures and services. University of California, Davis
40. Abbas HS, Gregory MA (2016) "The next generation of passive optical networks": a review. J Netw Comput Appl 67:53–74
41. Diallo T, Le Guyader B, Pizzinat A, Gosselin S, Chanclou P et al (2015) A complete fronthaul CWDM single fiber solution including improved monitoring scheme. In: 2015 European conference on networks and communications (EuCNC). IEEE, pp 325–329
42. Kani J-I, Terada J, Suzuki K-I, Otaka A (2017) Solutions for future mobile fronthaul and access-network convergence. J Lightwave Technol 35(3):527–534
43. Pfeiffer T (2015) Next generation mobile fronthaul and midhaul architectures. J Opt Commun Netw 7(11):B38–B45
44. Wey JS, Zhang J (2018) Passive optical networks for 5G transport: technology and standards. J Lightwave Technol 37(12):2830–2837
45. Anthapadmanabhan NP, Walid A, Pfeiffer T (2015) Mobile fronthaul over latency-optimized time division multiplexed passive optical networks. In: IEEE international conference communication workshop (ICCW), pp 62–67
46. Shibata N, Tashiro T, Kuwano S, Yuki N, Terada J, Otaka A (2015) Mobile front-haul employing Ethernet-based TDM-PON system for small cells. In: Optical fiber communications conference and exhibition (OFC), pp 1–3
47. Kobayashi T, Hisano D, Shimada T, Terada J, Otaka A Bandwidth allocation scheme based on simple statistical traffic analysis for TDM-PON based mobile fronthaul. In: Optical fiber communication conference, 2016, paper W3C–7
48. Hatta S, Tanaka N, Sakamoto T (2017) Feasibility demonstration of low latency DBA method with high bandwidth-efficiency for TDM-PON. In: Optical fiber communication conference, paper M3I–2
49. Butt RA, Idrus SM, Qureshi KN, Zulkifli N, Mohammad SH (2017) Improved dynamic bandwidth allocation algorithm for XGPON. J Opt Commun Netw 9(1):87–97
50. Orphanoudakis T, Kosmatos E, Angelopoulos J, Stavdas A (2013) Exploiting PONs for mobile backhaul. IEEE Commun Mag 51(2):S27–S34
51. Arokkiam JA, Wu X, Brown KN, Sreenan CJ (2014) Experimental evaluation of TCP performance over 10 Gb/s passive optical networks (XG-PON). In: Global communications conference (GLOBECOM), pp 2223–2228
52. Alvarez P, Marchetti N, Payne D, Ruffini M (2014) "Backhauling mobile systems with XG-PON using grouped assured bandwidth. In: 19th European conference networks and optical communications (NOC), pp 91–96
53. Mikaeil AM, Hu W, Hussain SB, Sultan A (2018) Traffic-estimation-based low-latency XGS-PON mobile front-haul for small-cell C-RAN based on an adaptive learning neural network. Appl Sci 8(7):1097.
54. Zhang M, Xu B, Li X, Cai Y, Wu B, Qiu K (2019) Traffic estimation based on long short-term memory neural network for mobile front-haul with XG-PON. Chin Opt Lett 17(7):070603

Dynamic Behavior of a Pump-Modulated Erbium-Doped Fiber Linear Laser with Single Fiber Bragg Grating

Nneka Obianuju Onubogu and Chang Hong Pua

Abstract An experimental analysis of the dynamical behavior of a pump-modulated Erbium-doped fiber laser (EDFL) with single fiber Bragg grating has been carried out and reported in this paper. The linear laser cavity here resembles a Fabry Perot interferometer. While varying the modulation frequency with a consistent signal amplitude and pump power, a wide range of features were observed which include: bifurcation, regions of optical bi-stability, period doubling, intermittently chaotic paths to chaos and chaos. Based on the region of chaos observed around the resonant frequency, the propensity of the laser to be used as a sensor to detect acoustic waves is briefly discussed, as the EDFL displayed tunable sensitive frequency capability.

Keywords EDFL · Pump modulation · Bifurcation

1 Introduction

Erbium-doped fiber lasers (EDFL) are known to have very outstanding features such as single-mode operation, high amplification, self-pulsations, high sensitivity to external disturbances, etc. These characteristics have made them a good laser source for many applications including optical communication, medicine, science and technology, reflectometry, pipeline sensing, intrusion zone identification, airborne acoustic sensing, etc. [1–6]. The characteristic of "high sensitivity to external disturbances" has gained more focus over the years as it leads to nonlinear dynamics during pump or loss modulation [7], which has an immense range of applications. Experimental pump modulation has been carried out on fiber lasers doped with Erbium by various authors [8–20]; however, our configuration is different from theirs and is quite simple and affordable. More so, it is customizable as other optical devices or

N. O. Onubogu · C. H. Pua (✉)
Lee Kong Chian Faculty of Engineering and Science, Universiti Tunku Abdul Rahman, Bandar Sungai Long, Jalan Sungai Long, 43000 Kajang, Malaysia
e-mail: puach@utar.edu.my

N. O. Onubogu
e-mail: nneka.onubogu@1utar.my

A. Sivasubramanian et al. (eds.), *Futuristic Communication and Network Technologies*, Lecture Notes in Electrical Engineering 792,
https://doi.org/10.1007/978-981-16-4625-6_28

fiber can be attached to the laser set-up to alter the length of the cavity, reflectivity or loss in the cavity [6].

In this paper, a study of the dynamical behavior of a single-mode Erbium-doped fiber laser (EDFL) is carried out and reported. The dynamic behavior of the EDFL is triggered by the laser diode pump modulation. The EDFL having a linear cavity is pump-modulated experimentally and its dynamical behavior is examined. Compared to the linear cavity set-up by other researchers where two reflectors are employed, only one fiber Bragg grating (FBG) is attached to the Erbium-doped fiber sensing arm to serve as a mirror to reflect the photons back to the laser cavity. During pump modulation, bifurcation, optical bi-stability, including period doubling and the intermittently chaotic paths to chaos are observed at various pump modulation frequencies with a constant signal amplitude. When the pump modulation frequency is modulated within a certain range, chaos occurs. Pump modulation is actually an easy and straightforward approach to modulate the laser output power in order to examine the behavior exhibited by regulating the laser diode pump power using a function generator and an oscilloscope. Similar studies have been implemented on the dynamics of Erbium-doped fiber ring lasers in which these behaviors were also exhibited [8–18, 20, 21]. Nevertheless, research on the nonlinear dynamics of fiber lasers doped with Erbium is inexhaustible due to its many evolving potential applications. Based on the behaviors observed, the EDFL has a high propensity to be used as a sensor as it demonstrated high sensitivity at its resonance frequency.

This paper is organized such that: the introduction is presented in Sect. 1 and Sect. 2 gives details of the EDFL experimental set-up. In Sect. 3, the results of the experiments are reported graphically and explained in details; Sect. 4 is the conclusion.

2 Experimental Set-Up of the Pump-Modulated EDFL

The experimental set-up of the EDFL is displayed in Fig. 1 which resembles a Fabry Perot interferometer. The linear laser cavity consists mainly of a 980/1550 nm high-performance wavelength division multiplexer (WDM) coupler and an Erbium-doped fiber (EDF) as an active or gain medium (3 m length, core of 8 μm diameter, clad of 125 μm diameter) all in one box. As part of the laser cavity, a 3 m long bare fiber is attached to the EDF which is called a single-mode fiber (SMF) acting as a sensing arm to detect acoustic waves and vibration. Attached to the other side of the SMF is a single side fiber Bragg grating (FBG), which has a center wavelength of 1550 nm ± 0.3 nm, > 99% reflectivity, > 5 dB SLSR and bandwidth of < 0.3 nm @ 3 dB. The entire laser cavity length is about 10 m from the photodetector end to the FBG end. This cavity design takes less space and is affordable as no expensive devices are required. It should be noted that no polarizer is used in this set-up.

The working principle of our EDFL under pump modulation is very simple. A function generator (GWINSTEK GFG-8020H) is used to modulate the frequency and optical power of the 980 nm pump laser and a digital oscilloscope (MEGURO

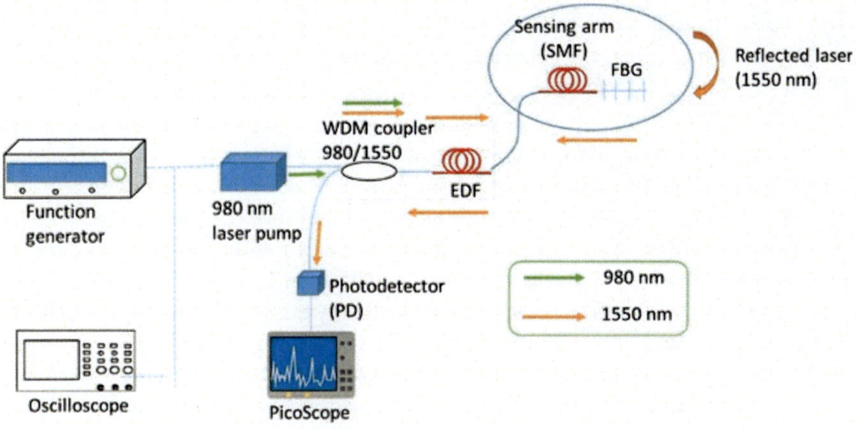

Fig. 1 Experimental set-up for pump modulation

MO-1020, 20 MHz) is used to monitor the modulation wave pattern. The modulation waveform of the pump laser is sinusoidal. A single wavelength 980 nm pump laser diode (Gooch & Housego EM595, with maximum pump power of 245 mW) emits 980 nm photons into the WDM and the WDM couples the photons into the EDF. The Erbium ions in the EDF absorb the photons of 980 nm wavelength and spontaneously emit photons with wavelength ranging from 1520 to 1560 nm. However, only photons of 1550 nm wavelength are reflected by the FBG back to the laser cavity and received by a high-speed photodetector (PD) made by THORLABS (DET08CFC/M). The PD changes the optical output to electrical output, so as to monitor the output light in waveforms on a PC via an oscilloscope (PicoScope 6 (3000 series)). The green and orange arrows in Fig. 1 symbolize the optical path of 980 and 1550 nm wavelengths.

3 Experimental Results

The optical spectrum of the EDFL was viewed using an optical spectrum analyzer (OSA) (Anritsu MS9740A 0.6 – 1.75 μm). With good launched pump power, the laser emission is obtained at 1549.7 nm ≈ 1550 nm, which matches the wavelength of the FBG. With continuous pumping of the laser, the lasing light remains uninterrupted.

After analyzing the optical spectrum of the lasing light, pump modulation experiments were carried out to observe the dynamic behaviors exhibited by the EDFL. During the experiments, optical output of the pump laser diode is modulated by the harmonic signal ($A \sin (2\pi f t)$) of the external modulation applied from the function generator to the laser diode driver, where A and f are the modulation amplitude and frequency, respectively [8]. The resonant peak displayed in the electrical frequency spectrum when the pump power is adjusted is the resonant frequency of the laser cavity which increases as the pump power is increased. The EDFL can actually have many resonant frequencies due to the long resonator cavity. Hence, for our pump modulation experiment, the pump power was fixed at 52.6 mW, corresponding to 19 kHz resonant frequency of the laser cavity, and the amplitude (A) of the signal was fixed at 0.4 V (corresponding to 33 mW pump power). Direct modulation of the laser diode was then carried out by modulating the frequency from 0 to 45 kHz via the function generator. PicoScope 6 was used in observing the laser optical output features in time and frequency domains. The pump laser output power linearly depends on the current from the laser diode. Maximum peak-to-peak amplitude of the laser intensity was recorded for each modulation frequency variation from 0 to 45 kHz and vice versa to get the graph in Fig. 2b which shows the dynamic behavior of the EDFL.

During frequency modulation, resonant peaks, bifurcation and three regions of optical bi-stability (where three resonant transmission states exist and are unwavering) were attained. On increasing the modulation frequency from 0 to 45 kHz (indicated by a blue line in Fig. 2b), the lasing light exhibited its first resonant peak at 5 kHz and repetitive resonant peaks at 9, 12, 17 and at 29 kHz. When the modulation frequency was decreased from 45 to 0 kHz (indicated by an orange line in Fig. 2b), resonant peaks were also observed but at 6, 9, 12 kHz and 20 kHz causing a delay (hysteresis) between the laser systems' input and output as a result of the alteration in route. The hysteresis curve created three optical bi-stability regions. The second bi-stable region (10–17 kHz) is wider than the first (5–7 kHz) and the third (19–29 kHz) is wider than the second.

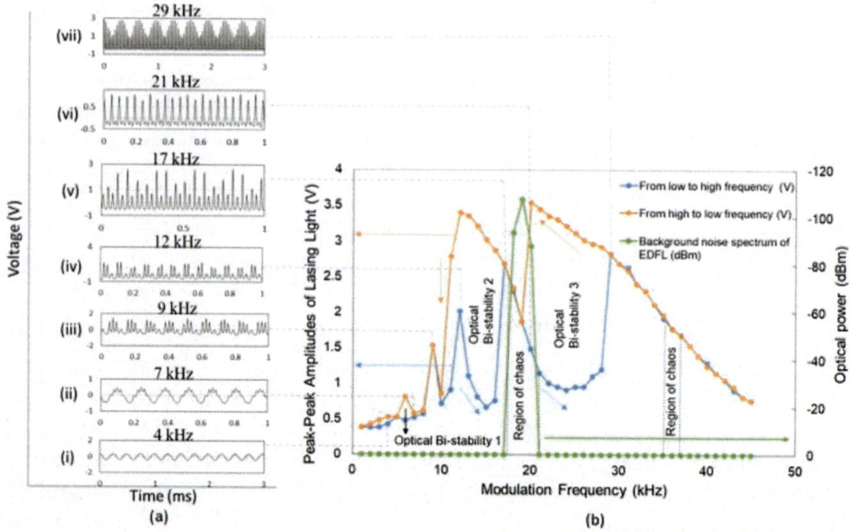

Fig. 2 **a** Time domain plots for the EDFL spectrum at different modulation frequencies; **b** graph showing the dynamical behaviors of the EDFL obtained by sweeping the modulation frequency from 0 to 45 kHz and vice versa

The time domain of the EDFL optical output was captured at different modulation frequencies and plotted in Fig. 2a with dashed lines showing the respective frequencies for each plot from (i) to (vii). When the modulation frequency is below 6 kHz, a lasing waveform with the resemblance of an amplitude modulation is obtained as shown in Fig. 2a (i) for 4 kHz. At 7 kHz (Fig. 2a (ii)), peaks are seen on the crest of each sinusoidal wave which reduces to 2 peaks in each sinusoidal wave at 8 kHz modulation frequency. At 9 kHz to 10 kHz, a period-doubling bifurcation repeating after every 3 long pulses and 1 short pulse is seen (Fig. 2a (iii)). The behavior goes back to 3 peaks on the crest of a sinusoidal waveform at 11 kHz modulation frequency, just as in 7 and 8 kHz. From 12 to 13 kHz, three stable lasing peaks are generated repetitively as shown in Fig. 2a (iv); two long peaks and a small peak. On increasing the pump modulation frequency to 14, 15 and 16 kHz, 2 tiny peaks appear on the crest of a single sinusoidal wave repeatedly. From 17 to 20 kHz, a period-doubling chaotic route observed as resonantly enhanced pulses repeating after a series of pulses exist as shown in Fig. 2a (v) leading to chaos at 21 kHz modulation frequency (Fig. 2a (vi)). Thus, the first region of chaos is marked from 17 to 21 kHz in Fig. 2b. At the region of chaos, the EDFL is most sensitive to external perturbations and forms some high intensity pulses. Chaos is indicated as irregular lasing peaks appearing at a particular frequency or range of frequencies and arises from the interaction of the population inversion with the intensity of the laser cavity [17]. From 22 to 34 kHz modulation frequency, the dynamic behavior changed again to period-doubling bifurcation with a higher order repetition of pulses. Chaos is seen to occur again at 35 kHz to 37 kHz modulation frequency. After the chaotic region, resonantly enhanced higher order repetition pulses leading to stable pulsation are obtained from 38 to 45 kHz. The existence of these routes to chaos and the occurrence of an irregularity after a series of period doubling shows that there is always a tendency for chaos to arise as expected in a pump-modulated EDFL for some variations of the modulation frequency and a constant pump power.

Resonant frequency aids in the characterization of the sensitivity of the EDFL as a sensor to acoustic wave modulation, although this is not the main focus of this paper. At the resonant frequencies of the EDFL, the system tends to oscillate at a maximum amplitude because of the periodic disturbance occurring at the same period of one of its natural frequencies. As said earlier, the resonant frequencies of the laser cavity can be regulated by varying the laser pump power from 0 to 245 mW (maximum limit of the laser diode). For the experiment carried out in this paper, the optical laser pump power was set at 52.6 mW to obtain a resonance frequency of 19 kHz. The background noise spectrum of the EDFL indicating the resonant peak at 19 kHz modulation frequency has been matched into Fig. 2b (indicated by the green line). It is clearly seen that irregular lasing peaks commonly called chaos occurs between modulation frequencies of 17–21 kHz, wherein 19 kHz is the resonance frequency. It can be deduced that at the resonance frequency of 19 kHz, the EDFL has a high propensity to be used as a sensor to detect ultrasonic waves at a low frequency and can thus be used for either gas, water or oil pipeline monitoring and leakage detection which is the proposed use of this sensor. Other possible applications include biomimicking and multi-stable switching. Also, one of the advantages of the EDFL sensor is its tunable sensitive frequency capability for acoustic wave sensing as the distribution of the resonant frequency at various pump power levels indicates a shift to a higher frequency with respect to the pump power increase [6]. Furthermore, theoretical simulation using the rate equations in MATLAB has been carried out and a comparison with the experiments showed similar results and would be published in another article. Similar comparison carried out by other authors actually showed a good agreement between their theoretical and experimental results [7–9, 12, 14].

4 Conclusion

In this paper, pump modulation has been carried out in the EDFL system to aid in the understanding of its behavior when exposed to external disturbances. Pump modulation on the EDFL performed from low to high frequency and from high to low frequency showed the existence of period-doubling bifurcation, intermittently chaotic paths to chaos, chaos and optical bi-stability. The existence of these routes to chaos shows that there is a tendency for chaos to obviously occur in an EDFL. Results from the experiments have also shown that chaos occurs at the resonance frequency of the EDFL which shows that it has a high propensity to be used as a sensor to detect ultrasonic waves at a low frequency and can thus be used for pipeline monitoring and leakage detection.

References

1. Luo L, Chu PL (1998) Optical secure communications with chaotic erbium-doped fiber lasers. J Opt Soc Am B 15:2425–2530
2. Digonnet MJF (2001) Rare Earth Doped Fiber Lasers and Amplifiers, Revised and Expanded, 2nd edn. Marcel Dekker, New York
3. Pua CH, Norizan SF, Harun SW, Ahmad H (2011) Non-membrane optical microphone based on longitudinal modes competition. Sens Actuators A Phys 168(2):281–285. https://doi.org/10.1016/j.sna.2011.04.034
4. Pua CH, Chong WY, Ahmad H (2013) Instantaneous response of wide area intrusion sensor with long haul monitoring capability. IEEE Photonics Tech. Lett 25(23):2255–2258. https://doi.org/10.1109/LPT.2013.2284608
5. Woon SL, Kwan KM, Chong WY, Lin HS, Pua CH (2017) Cascaded acoustic wave sensors based on erbium-doped fibre laser dynamics for intrusion zone identification. IEEE Sensors 17(6):1893–1898. https://doi.org/10.1109/JSEN.2017.2655019
6. Pua CH, Ahmad H, Harun S, De La Rue R (2012) Direct airborne acoustic wave modulation of Fabry-Perot fiber laser (FPFL) over 100 kHz of operating bandwidth. Appl Opt 51(15):2772–2777. https://doi.org/10.1364/AO.51.002772
7. Pua CH, Ahmad H, Harun S (2012) Study of Dual-wavelength mode competition in an Erbium-Doped Fiber Laser (EDFL) produced by acoustic waves. IEEE J Quantum Electron 48(12):1499–1504
8. Pisarchik, A.N., Barmenkov, Y.O., Kir'yanov, A.V.: Experimental characterization of the bifurcation structure in an Erbium-doped fiber laser with pump modulation. IEEE J. Quantum Electron 39(12), 1567–1571 (2003).

9. Pisarchik, A.N., Kir'yanov, A.V., Barmenkov, Y.O.: Dynamics of an erbium-doped fiber laser with pump modulation: theory and experiment. J. Opt. Soc. Am. B 22(10), 2107–2114 (2005). https://doi.org/10.1364/JOSAB.22.002107.
10. Reategui, R.J., Kir'yanov, A.V., Pisarchik, A.N., Barmenkov, Y.O., Il'ichev, N.N.: Experimental study and modelling of coexisting attractors and bifurcations in an Erbium-doped fiber laser with diode-pump modulation. Laser Phys 14(9), 1–5 (2004).
11. Pisarchik AN, Barmenkov YO (2005) Locking of self-oscillation frequency by pump modulation in an erbium-doped fiber laser. Opt Commun 254:128–137. https://doi.org/10.1016/j.optcom.2005.05.028
12. Lacot E, Stoeckel F, Chenevier M (1994) Dynamics of an erbium doped fiber laser. Phy. Rev. A. 49:3997–4008. https://doi.org/10.1364/JOSAB.22.002107
13. Luo L, Tee TJ, Chu PL (1998) Bistability of erbium-doped fiber laser. Opt Commun 146:151–157. https://doi.org/10.1016/S0030-4018(97)00502-6
14. Luo L, Tee TJ, Chu PL (1998) Chaotic behaviour in erbium-doped fiber-ring lasers. J Opt Soc Am B 15(3):972–978. https://doi.org/10.1364/JOSAB.15.000972
15. Kruger TS, Rech PC (2012) Dynamics of an erbium-doped fiber dual-ring laser. Eur Phys J D 66(12):3997–4008. https://doi.org/10.1140/epjd/e2011-20396-4
16. Arellano-Sotelo, H., Kir'yanov, A.V., Barmenkov, Y.O., Aboites, V.: The use of nonlinear dynamics of erbium-doped fiber laser at pump modulation for intra-cavity sensing. Opt Laser Technol 43 132–137 (2011). https://doi.org/10.1016/j.optlastec.2010.05.017.
17. Kumar G, Vijaya R (2015) Dynamical features of loss and pump modulation in an erbium-doped fiber ring laser. J Opt 17(125402):1–9. https://doi.org/10.1088/2040-8978/17/12/125402
18. Sola IJ, Martin JC, Alvarez JM (2002) Non-linear response of a unidirectional erbium-doped fiber ring laser to a sinusoidally modulated pump power. Opt Commun 212:359–369
19. Sola IJ, Martin JC, Alvarez JM (2006) Analytical treatment for the study of an erbium-doped fiber ring laser fed by a sinusoidally modulated pump power. Opt Commun 258:59–66
20. Onubogu NO, Pua CH, Lin HS, Faidz A-R (2020) The dynamic behavior of non-polarized erbium-doped fiber ring. Optik 207(164442):1–8
21. Zhang L, Yang R, Qin JM, Yang LZ (2013) An experimental study of high frequency chaotic dynamics in an erbium-doped fiber ring laser. Laser Phys 23(12):125104. https://doi.org/10.1088/1054-660X/23/12/125104

Design and Simulation of Dual-Band MIMO Antenna for Radar and Sub-6-GHz 5G Applications

K. Vasu Babu, Suneetha Kokkirigadda, and Sudipta Das

Abstract The MIMO antenna proposed in this article presents dual-band operations for radar in C-band and sub-6-GHz 5G applications. The simulated results show that the suggested dual-band antenna offers impedance bandwidths of 1300 MHz (2.8–4.1 GHz) for the lower band of operation and 500 MHz (7–7.5 GHz) for the higher band. The minimum isolation achieved for the first band is more than 40 dB and its noted more than 45 dB for the second band of operation. The structure is designed with Minkowski shape structure and dumb-bell shape is inserted between the two patches with a separation distance of 3 mm to reduce the interference between the identical patches. The designed antenna offers good radiation characteristics in terms of better gain, better efficiency, and stable radiation patterns. The proposed MIMO structure shows good diversity performance offering desired values of envelope correlation coefficient (ECC) and diversity gain (DG).

Keywords Dumb-bell shape structure · Mutual coupling · Diversity gain · Dual band · MIMO antenna

1 Introduction

MIMO array antennas are essential for 5G, LTE-A, Wi-Fi, and LTE wireless systems. In recent years, MIMO technology plays a major role to draw attention for the researchers working on wireless communications. It has various features like spectral efficiency enhancement, reliability improvement and channel capacity enhancement without using of any additional bandwidth or power. In [1], an eight-element MIMO

K. Vasu Babu (✉)
Departmentof ECE, Vasireddy Venkatadri Institute of Technology, Nambur, Andhra Pradesh, India

S. Kokkirigadda
Department of Electrical and Computer Engineering, Bule Hora University, Oromia, Ethiopia

S. Das
Department of Electronics & Communication Engineering, IMPS College of Engineering & Technology, Malda, West Bengal, India

A. Sivasubramanian et al. (eds.), *Futuristic Communication and Network Technologies*, Lecture Notes in Electrical Engineering 792,
https://doi.org/10.1007/978-981-16-4625-6_29

Table 1 Current system with other literature works

Literature works	Size (mm^2)	Resonant frequency (GHz)	Edge-to-edge distance	Isolation (dB)	Efficiency (%)
[1]	4200	3.5	0.08 λ_0	15	72
[3]	5400	2.5	0.04 λ_0	17	75
[4]	6400	2.5	0.25 λ_0	25	69
[5]	3600	3.2	0.33 λ_0	35	65
[7]	4200	5.2	0.22 λ_0	27	70
[9]	4800	5.8	0.14 λ_0	37	68
[9]	6800	3.5	0.35 λ_0	26	63
Proposed structure	2500	3.5	0.04 λ_0	55	74
		7.2	0.03 λ_0	49	69

dual-polarized-type slot antenna is proposed. For the band-rejection capability, dual-polarized compact UWB self-complimentary diversity/ MIMO has been designed [2]. In [3], and four-element MIMO system has been implemented using closely space elements with the help of discrete mushrooms to reduce the mutual coupling. A uniplanar compact slot antenna is proposed to achieve polarization diversity in UWB communication [4]. In [5], A compact MIMO system has been designed with band-notched characteristics for the applications in portable UWB. A dual-port low-profile UWB-MIMO system with better isolation, ECC, and band rejected capability is presented in Ref. [6]. In [7], mutual coupling reduction technique is explained using an F-shaped stub for a UWB-MIMO system. In Ref. [8], researchers have employed quasi-self-complementary structures for the design of the UWB-MIMO antenna with higher efficiency and gain at the operating band of required operation. To enhance the isolation of current system, a tree-like structure is inserted in the design for the UWB diversity/MIMO structure proposed in [9]. The novelty of the suggested MIMO structure is that the insertion of a dumb-bell shaped structure between the two Minkowski patches reduces the isolation by a greater amount compared to the previous works mentioned in the literature. The antenna parameters like ECC, DG, and peak gain are also higher compared to the other reported structures. Table 1 shows advantages of current system with literature works have better isolation and compact size.

2 Analysis of the Current Geometry

The dumb-bell shaped incorporated MIMO structure is designed using an FR-4 substrate having 1.6 mm thickness and relative permittivity of 4.4 as depict in Fig. 1. The presented structure dimensions listed in Table 2. From Fig. 1a represents the front plane of the current design with rectangular slots at the end of the patches.

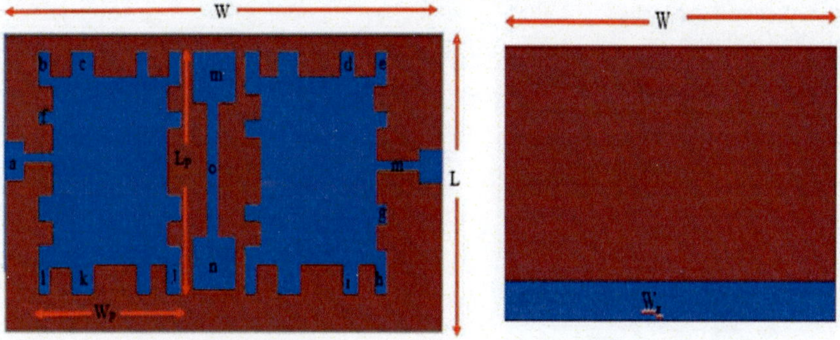

(a) Front-plane (b) Back Plane

Fig. 1 Dual-band MIMO antenna design

Table 2 Proposed MIMO design parameters

Parameters	L	W	L_p	W_p	a	b	c	d	e	f	g	h	i	J	K	l	m	n	o	w_g
(mm)	50	50	20	20	4	2	3	3	2	3	2	4	4	4	2	2	3	3	2	4

Figure 1b shows that the monopole ground is designed to improve the bandwidth of the system.

3 Results and Discussion

Figure 2 shows the surface current distribution of the designed structure. Proposed MIMO structure *S*-parameters shown in Fig. 3. The suggested MIMO structure antenna resonates at dual operating bands with the resonant frequencies of 3.5 and 7.2 GHz. The proposed MIMO aerial offers impedance bandwidths (IBW) of 1300 MHz (2.8–4.1 GHz) for the lower band of operation which completely covers

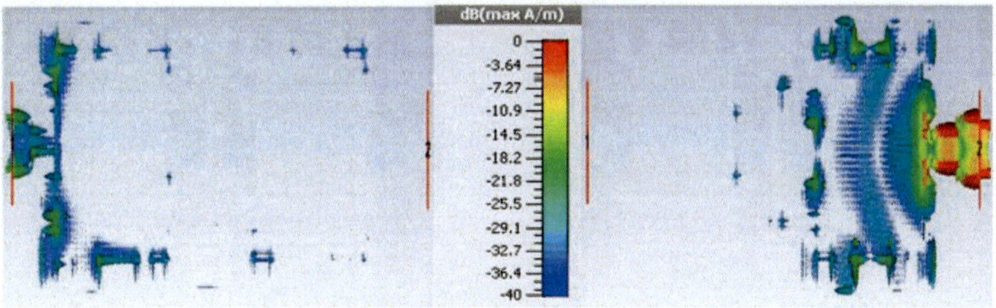

Fig. 2 Distribution of surface currents in the MIMO structure

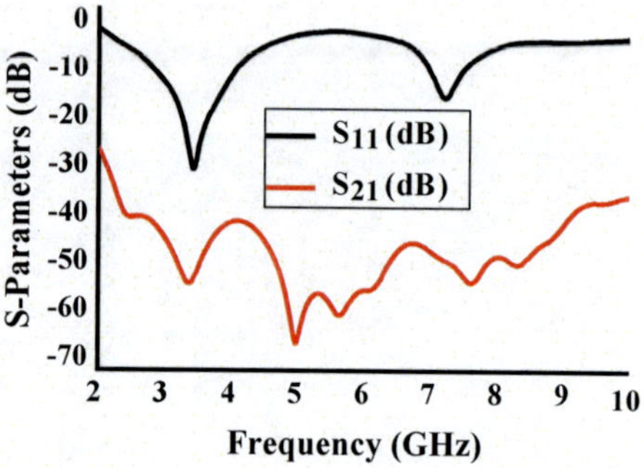

Fig. 3 *S* parameters of the MC MIMO antenna

the bandwidth requirement of sub-6-GHz 5 g application. An IBW of about 500 MHz (7–7.5 GHz) is obtained for the upper band for the applications of radar in C-band. For the resonant frequency of the lower band, S_{11} and S_{21} are of −32 dB and −55 dB, respectively. However, at the resonant frequency of the upper band, the *S*-parameters (S_{11} and S_{21}) reaches to −19 dB and −49 dB, respectively. So, the results indicate that the suggested MIMO structure produces greater isolation at the dual resonant frequencies and good impedance bandwidths. Figure 4 shows the variations of efficiency and peak gain about the current design antenna. From Fig. 4, noticed that current antenna structure efficiencies are 74 and 69% and peak gains are around 6.7 and 4.8 dBi at the dual resonant frequencies. Figure 5 identified that voltage standing wave ratios (VSWRs) are well maintained within ≤2 for the dual the resonant frequencies, which indicates better considerable impedance matching and hence lower mismatch losses. Figure 6a, b represents the E and H plane radiation patterns including co and cross-polarization at the resonant frequencies of 3.5 and 7.2 GHz. The designed antenna produces good stable radiation patterns which are useful for the intended applications. The diversity performance of the dumb-bell shaped inserted MIMO structure indicated using envelope correlation coefficient (ECC) and diversity gain (DG) as evaluated using Eqs. (1)–(3). [8]

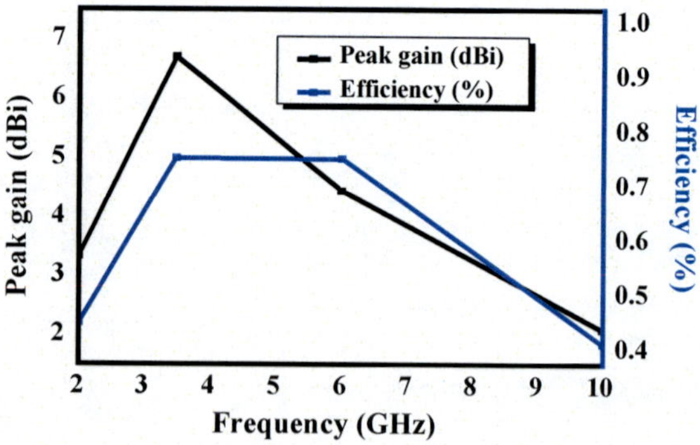

Fig. 4 Variations of efficiency and peak gain versus frequency

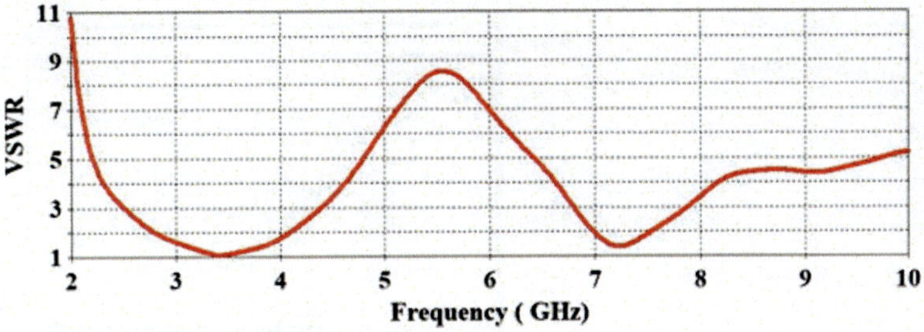

Fig. 5 VSWR graph

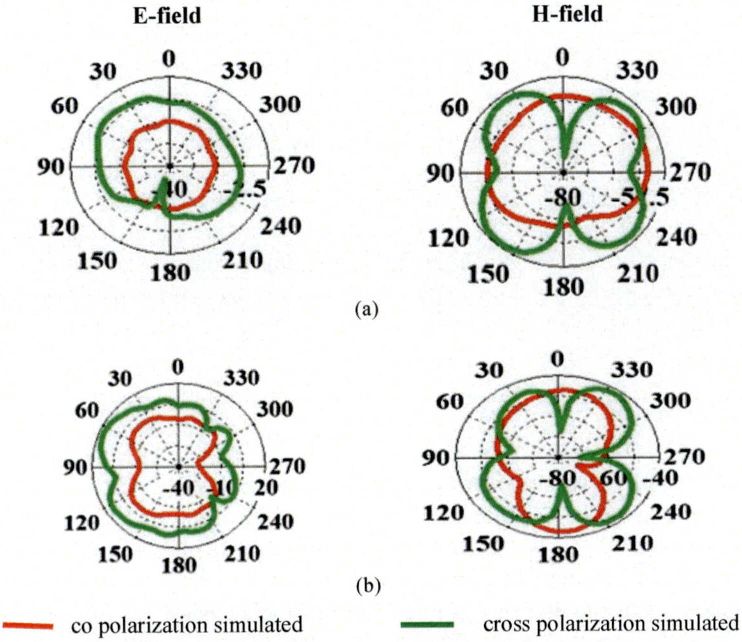

Fig. 6 **a** 3.5 GHz, **b** 7.2 GHz radiation patterns of E & H planes

$$\mathrm{ECC} = \frac{\left|\iint_{4\pi}[E_1(\theta,\phi) * E_2(\theta,\phi)]\mathrm{d}\Omega\right|^2}{\iint_{4\pi}|E_1(\theta,\phi)|^2\mathrm{d}\Omega \iint_{4\pi}|E_2(\theta,\phi)|^2\mathrm{d}\Omega} \quad (1)$$

where $E_i(\theta, \phi)$ Complex 3D far-field radiated pattern.
The ECC expression using S-parameters

$$\mathrm{ECC} = \left|\frac{\iint_{4\pi}\left|S_{11}^* S_{21} + S_{21}^* S_{22}\right|}{\left[\left(1-\left(|S_{11}|^2+|S_{21}|^2\right)\right)\left(1-\left(|S_{22}|^2+|S_{12}|^2\right)\right)\right]^{1/2}}\right| \quad (2)$$

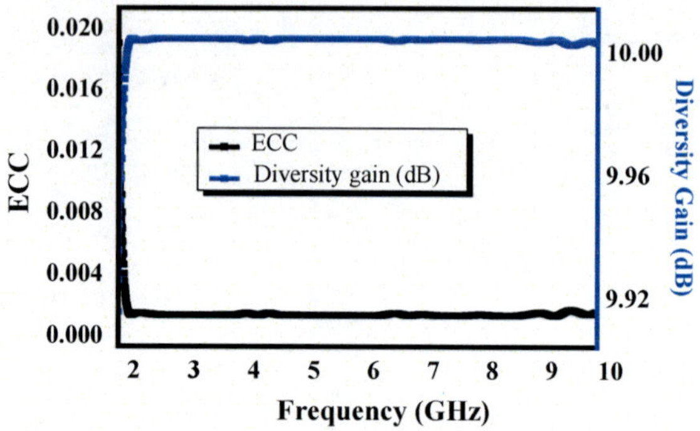

Fig. 7 Variations of ECC and diversity gain versus frequency

The diversity gain is evaluated with the relation is

$$\mathrm{DG} = 10\sqrt{1 - |\rho_e|^2} = 10\sqrt{1 - |\mathrm{ECC}|^2} \tag{3}$$

The ECC and DG about the current dual-band Minkowski MIMO aerial are depicted in Fig. 7. The MIMO parameter of ECC values are within 0.002. The diversity gains are more than 9.99 dBi at the dual operating bands. The values are well accepted and confirms good diversity performance of the proposed MIMO antenna.

4 Conclusion

In this MIMO structure, a dumb-bell shaped slit is placed between the two Minkowski patches to improve the isolation. In this design, the distance among two patches is considered as 0.04 λ_0 to downsize the mutual coupling. This structure occupied an overall area of 2500 mm^2 to design the MIMO dumb-bell inserted patch antenna. The proposed antenna offers good impedance matching as well as isolation characteristics. Furthermore, good MIMO parameters like ECC, diversity gain are obtained in addition to stable radiation patterns representations, realized gain of the system and efficiency of the radiator. The proposed dual-band MIMO structure antenna is suitable for sub-6-GHz 5G applications and Radar application in C-Band.

References

1. Parchin, NO et al (2019) Eight-element dual-polarized MIMO slot antenna system for 5G smartphone applications. IEEE Access 7:15612–15622
2. Zhu J et al (2015) Compact dual-polarized UWB quasi-self-complementary MIMO/diversity antenna with band-rejection capability. IEEE Antennas Wirel Propag Lett 15:905–908 (2015)

3. Zhai G, Chen ZN, Qing X (2016) Mutual coupling reduction of a closely spaced four-element MIMO antenna system using discrete mushrooms. IEEE Trans Microw Theory Techniques 64(10):3060–3067
4. Chacko BP, Augustin G, Denidni TA (2013) Uniplanar slot antenna for ultrawideband polarization-diversity applications. IEEE Antennas Wirel Propag Lett 12:88–91
5. Liu L, Cheung SW, Yuk TI (2015) Compact MIMO antenna for portable UWB applications with band-notched characteristic. IEEE Trans Antennas Propag 63(5):1917–1924
6. Biswal, SP, Das S (2018) A low-profile dual port UWB-MIMO/diversity antenna with band rejection ability. Int J RF Microw Comput-Aided Eng 28(1):e21159
7. Iqbal A et al (2017) Mutual coupling reduction using F-shaped stubs in UWB-MIMO antenna. IEEE Access 6:2755–2759
8. Liu L, Cheung SW, Yuk TI (2014) Compact multiple-input–multiple-output antenna using quasi-self-complementary antenna structures for ultrawideband applications. IET Microw, Antennas Propag 8(13):1021–1029
9. Zhang S et al (2009) Ultrawideband MIMO/diversity antennas with a tree-like structure to enhance wideband isolation. IEEE Antennas Wirel Propag Lett 8:1279–1282

Contact-Free Interaction with Computer Systems Using Computer Vision Techniques

Andrew Moses, E. Sarah Rajkumar, and R. Parvathi

Abstract Interaction with computer systems has evolved a long way. It started with buttons, and now, we have touch screens the most commonly used interface using which we can feed inputs to the systems. But the ongoing pandemic pushed humanity to find ways to minimize contact with anything and everything. Hence, this research is about finding an alternate way of interacting with systems with zero contact. Hand gesture recognition is a field which has been evolving a lot over the years but now is the immediate need of the hour for such techniques to be implemented in everyday life. In this paper, we have proposed to utilize and evaluate popular computer vision techniques to solve the above said objective. One among the popular object detection deep learning model is residual neural networks (ResNet). The model with transfer learning approach is utilized and evaluated using a standardized labeled dataset. Further comparisons are made to bench mark the solution with different models. Thus, the proposal is to find a suitable model to accomplish a simple contact-free inputs to a computer system. It is believed that this research would be a stepping stone to build more complex contact-free interface with digital systems.

Keywords Hand gesture · ResNet · Computer vision · Transfer learning · Contact free

1 Introduction

1.1 Motivation and Challenges

Technology has been taking over our world in almost all aspects of life. The human and computer interaction has been getting better and easier with all the new innovations and technology. COVID-19 has alarmed our entire planet about the importance of keeping one's environment hygienic. The ultimate goal of this project is to build

A. Moses (✉) · E. Sarah Rajkumar · R. Parvathi
School of Computer Science, Vellore Institute of Technology, Chennai, India

R. Parvathi
e-mail: parvathi.r@vit.ac.in

A. Sivasubramanian et al. (eds.), *Futuristic Communication and Network Technologies*, Lecture Notes in Electrical Engineering 792,
https://doi.org/10.1007/978-981-16-4625-6_30

a user-friendly model to aid humans to maintain distance while they do their day-to-day activities such as going to the ATM, check-in and check-out in a corporate environment, etc. People perform their daily work which is integrated with many electronic equipment. This pandemic has led people to give such importance to social distancing that people find it hard even to stand nearby someone they know. Finding a way to interact with these digital products without touching them would be a great contribution in fighting against the spread of this notorious virus.

This could be achieved by a few techniques which are built purely for zero contact purposes. The gestures recognition has been introduced long before this pandemic; we could use them in this case to build a creative model to interact with the computer systems.

Microsoft's famous console interface, the Kinect, is a brilliant product. Also, Samsung had its class apart device with gesture recognition capabilities. The only setback with these innovative products was their outcomes did not satisfy the requirement or were not an best alternate to the existing interfaces. Thus, it lost the place it had in consumer segment. But now is the time to rethink and shed light on these hand gesture recognition techniques to rescue mankind. In fact, this technology of recognizing hand gestures has been there for a very time which exhibits unique of visual pattern analysis study with a broad range of application, including interaction between people and machines, sign language interaction and virtual communication.

The hand gesture serves as an interface for the recognition models for tracking and detecting moving objects.

In this project, we have taken an open dataset of hand symbols. A subset of the largely available dataset was taken such that three classes are present. Then recent advances in computer vision helped us to choose certain models for understanding their capabilities. Among the recent advances in computer vision, residual neural networks (ResNet) with transfer learning approach outperformed the other explored models. The research utilized around 8 k raw images of hand symbols shot during various sessions to induce variance in the dataset. Thus, the dataset plays a major role in this research findings. Still, we focus on the model's performance to promote various studies to accomplish this challenging task.

All means needs to be identified to minimize the risk of spread of the ongoing pandemic due to corona virus. It is true that this spread of virus cannot be contained, but humanity does not give up that easily. We could try all means and methods to contain the spread of this virus as much as possible by inventing new touch-free technology. Findings show that the virus spreads through droplets in the atmosphere. Hence, it is recommended to not touch anyone or any device unnecessarily and to wash hands frequently. Still, it remains a challenging task; thus, various methods need to be explored to reduce physical contact with the equipment.

2 Literature Survey

Deep learning field has found its way into all fields such as computer vision, speech and audio recognition, image recognition, natural language processing and many more fields. In simple words, if one has to break down deep learning to a lay man, deep learning is all about various neural network architectures that are convoluted together predict a particular outcome. The most famous deep learning architectures include deep neural networks (DNN), deep belief networks (DBN), recurrent neural network (RNN) and convolution neural network (CNN). Among them, CNN has been extensively studied. Even though CNN has found its place in computer vision extensively, still its applications spread across multiple domains like natural language processing and speech recognitions. The advances in CNN [1] have been studied in depth by the folks at Singapore University. As our focus is on image recognition, deeper networks were explored; however, it had its conflicts due to the many layers involved in the network. The deeper the network, the difficult it is to train it. Residual neural network (ResNet) and its advantages over deeper VGGNet with lowering complexity give a different clarity over the underlying capabilities of residual network as explained at deeper residual learning for images [2]. Fusion of CNN layers was stacked and built to identify hand gestures; the activity in [3] paper talks about reducing the response time as the gestures need to be identified on the fly with minimum resources to put it into production. Thus, these stages separate the complexity and approach the problem using the divide and conquer approach. Another paper [4] also talks in same lines of bringing the gestures recognition real time. The American sign language (ASL) [5] being educationally taught at universities, the paper talks about utilizing 3D ResNet to identify grammar connections in a continuous video of gestures posed. Fuzzy inference systems [6] have also been utilized in hand symbol recognition activities to assist the differently challenged people. Even paper [7] talks of aiding the hearing-impaired people with a system to recognize sign language using segmented CNN layers. Ultrasonic waves were utilized to identify hang gestures, and using support vector machines (SVM), a considerable accuracy of 88% was recorded in [8]. Human interactions with smart phones with touchless inputs [9, 10] researchers have used LSTM and RNN algorithms to bench mark their models. Challenges like running the heavy models in mobile devices were addressed by utilizing various variants of MobileNets [11]. Long-term recurrent convolutional network (LRCN) is spatial and adaptable enough to be applied to computer vision tasks [12]. 3D CNN model for action recognition is implemented and captured motion information that is encoded in multiple adjacent frames [13]. The survey made by Rautaray and Agrawal helps in understanding the present vision based gesture capabilities and facilitates human-computer interactions [14]. A review on semantic segmentation using deep learning techniques gives deeper insight on analyzing a portion of image [15].

3 Experimentation

After much research, an ASL public dataset which consists of finger spelling images, that is all the English alphabets, its equivalent in American sign language, was captured as images in various sessions and was posted publicly for research purpose. The dataset is huge; it has been divided into two subsets where one dataset is about the static images of all letters except *j* and *z* has it has motion involved with it. Whereas, the other subset is about variety in environment setting like lighting and distinct positioning of the hand in the same image. With regard to this research, due to resource constrains, the previously discussed subset was taken and sliced into an even smaller set such that only 3 distinct classes of images were taken.

A total of 8 K images were taken containing 3 classes. Each class had images taken from 5 different sessions to induce diversity in the classification process. The raw images were taken and resized into a standardized size of 130 × 130 dimension as the images were of different sessions and different sizes. In order to keep the model more versatile, no further explicit preprocessing steps were performed. The resized images were put into data loaders and were split into three sets, i.e., training dataset of 5 K images and testing dataset of 2 K images and remaining 1 K images for validation. The split ratio was done to enhance the effectiveness of the performance of the model.

Figure 1 shows the workflow of our model where the raw dataset of 8 K images

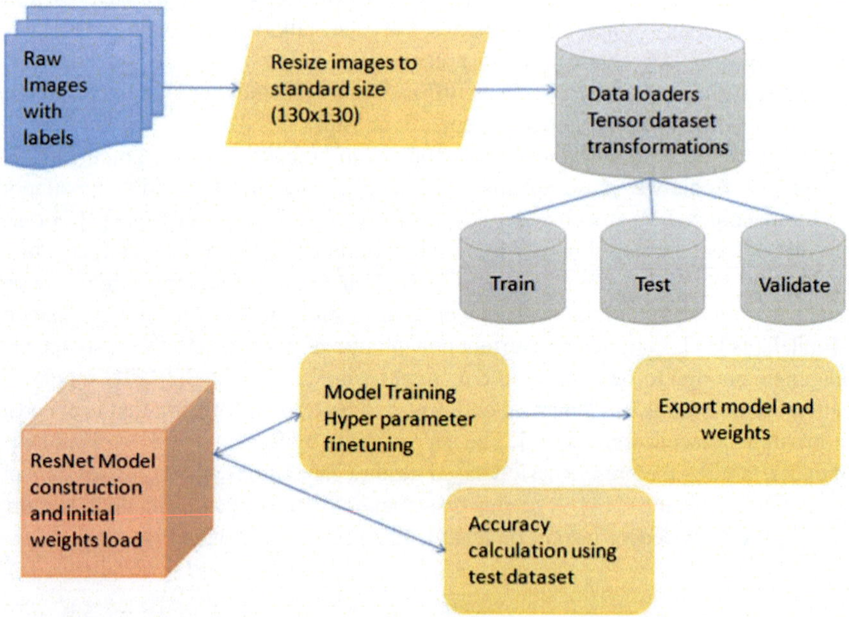

Fig. 1 Workflow of the proposed model and its construction

is taken and sliced, which is further resized and fed into the training, testing and validation sets, respectively.

The model was constructed using Pytorch framework. The data loaders were of Pytorch tensor datasets. The final proposed model had ResNet 34 with initialized weights. The raw model was pretrained with coco dataset where there are 80 object categories. Thus, this powerful model has the capability to filter the noise and extract the important features from a given images. This is another reason why dataset-specific image preprocessing was not performed in this research. The training dataset is fed into the model with varying hyper parameters. The training stage was experimented with different parameters multiple times until the desired validation loss was able to reach the global minimum. Later, the trained model was fed with the test set. The accuracy calculation was performed such that each image is being correctly identified with their respective original label. With the 2 K test images, the model was able to predict more than 1 K images correctly with misclassification of around 100–150 images. Further steps were taken to feed in live streams of images for the model to predict the signs real time. The integration was made with OpenCV packages. It was noted that the model's performance was not as matched with its test accuracy marks. On further analysis, it was found that the model did not perform up to mark due to the varying environmental conditions of the real stream images. So, it can be said the model needs to undergo more training with varying hyper parameters to generalize it.

The model which was finally built in this research can be said that it is been little overly fit with the dataset. To further understand its performance with other models, the steps were taken to build a simple CNN model; this model had 2 convolutional layers with drop outs and a fully connect layer in the end. It had more than 50 filter to capture the varying features of the images. After a set of training also keeping in mind the generalization of the model, the final model was bench marked with 88% of accuracy, and the accuracy calculation was followed same as the ResNet experiment.

The last model comparison was made with support vector machine from the sci-kit learn package. The sci-kit learn package when it was first applied caused the system to crash many times because of the huge image dataset. The dataset could not be held in memory which is one of the requirements while using a this package to perform SVM classifier. Hence, the dataset was sliced, and fewer sets of images were used to build the classifier; it was noted that SVM outperformed CNN model. Hence after multiple reduction of training dataset, it was found that with just 100 images as training set, 900 images were used as test dataset; the SVM model was able to bench mark 94% as accuracy with the 900 odd test images, which seems to be very surprising. Dataset characteristics needs to be considered while making conclusions about the performance of the proposed model. The 100 images were of different sessions; hence, the hyperplanes were able to distinctly segregate the classes as the size of the classes were also minimal with our research. Even if the SVM model seems so promising, it cannot be benchmarked with real-time streams of images. However, it still can be seen as a go-to model if the use case involves same background with very minimal changing environmental conditions.

4 Result and Analysis

The experiment had utilized Adam optimizer. Adam optimizer is an optimization technique used in stochastic gradient descent algorithm. It utilizes best of both Ada and RMS property techniques. Thus, the optimization helps in our dataset. Even if the dataset has sparse and noise data points, the optimizer helps in training the weights appropriately. In Fig. 2, we can see the ResNet model's accuracy against the CNN model. The ResNet algorithm gives us an accuracy of 98%, whereas the CNN algorithm gives us an accuracy of 88%. Both the models were experimented with same dataset. In the case of SVM algorithm, the dataset taken originally was too huge for the algorithm to process so we sliced the dataset down to a smaller amount. By doing do, our SVM model's hyperplane provided a better and accurate prediction on the dataset. Therefore, the SVM algorithm yields 94% accuracy. The proposed ResNet model has been able to increase the accuracy with 10% when compared again the conventional CNN model. Figure 3 is another graphic showing the dominance of ResNet yielding an accuracy of 98.87% when compared with CNN which gives us an accuracy of 88.89%. As mentioned above, the SVM model performed with better accuracy with a smaller dataset yielding an accuracy of 94.54%

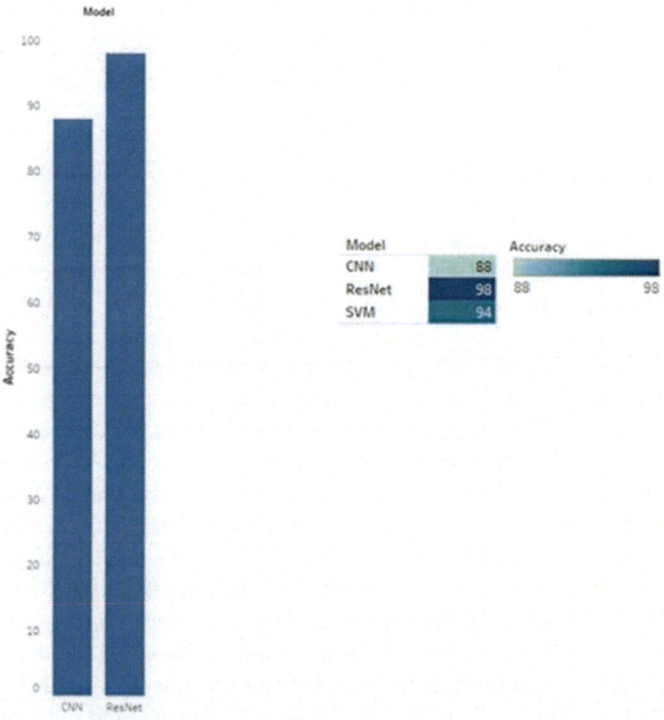

Fig. 2 Accuracy metrics of proposed model with other models

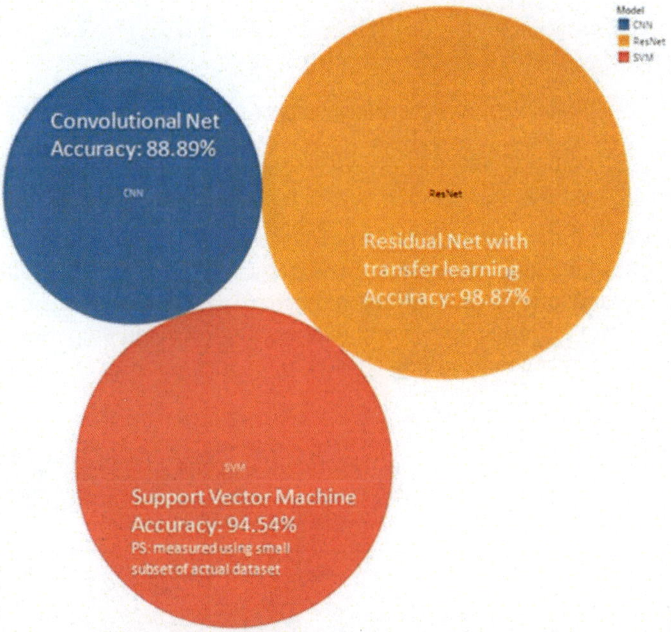

Fig. 3 Accuracy metrics of proposed model with other models

Eqs. (1) and (2) are the underlying equations for the Adam optimizer where most of the hyperparameters are left constant. The resulting moving averages are derived by the gradient along with the hyperparameter.

$$m_t = \beta_1 m_{t-1} + (1 - \beta_1) g_t \quad (1)$$

$$v_t = \beta_2 v_{t-1} + (1 - \beta_2) g_t^2 \quad (2)$$

m,v—moving averages.
g—gradient.
β·—hyperparameter.

Table 1 shows each model's predicted class from the given two sample images. As noted, all the models have a confidence score more than 50% in class 0; sample 1 is a person showing zero hand symbol; hence, all the three models have predicted

Table 1 Models' prediction and its confidence score for sample 1

Model name	Class: 0	Class: 5	Class: 3
CNN	0.905	0.003	0.09
ResNet (with transfer learning)	0.725	0.047	0.226
SVM	0.976	0.012	0.011

Table 2 Models' prediction and its confidence score for sample 2

Model name	Class: 0	Class: 5	Class: 3
CNN	0.009	0.515	0.475
ResNet (with transfer learning)	0.046	0.606	0.347
SVM	0.018	0.926	0.054

correctly, but on further analysis, we understand the performance of the model by seeing its corresponding confidence score over other classes. Similarly, Table 2 shows the results for sample image 2. ResNet does not have the highest confidence score, but on the other hand of unseen real-world images, ResNet could outperform, because the model has been generalized (Figs. 4 and 5).

Fig. 4 Sample 1

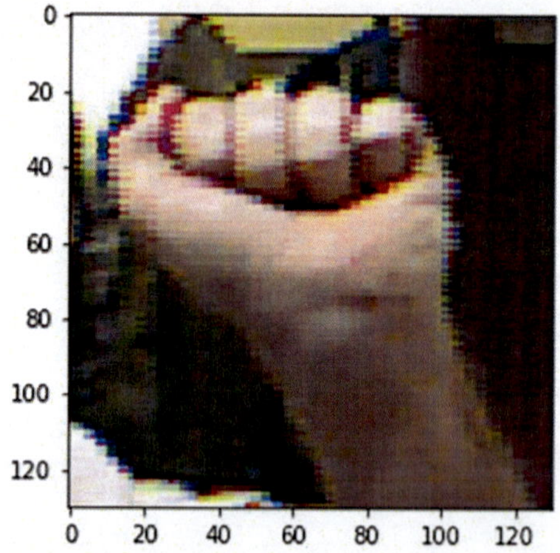

Fig. 5 Sample 2

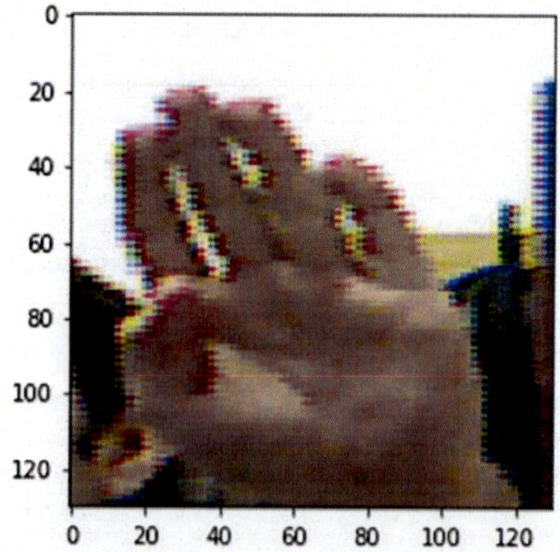

5 Conclusion

As discussed in this paper, a contact-free interaction with the computer systems using computer vision techniques can be achieved by applying deep learning algorithms to various gesture images. This enables us to identify inputs to computer systems in an effective manner. Further steps could be taken to improve this model to suit humanity according to their daily needs. This can be done by training the model with different kinds of gesture datasets which contain complex variables.

References

1. Gu J, Wang Z, Kuen J, Ma L, Amir S, Bing S, Liu T, Wang X, Li W, Gang W, Cai J, Chen T (2017) Recent advances in convolutional neural networks. arXiv:1512.07108v6 [cs.CV] Oct 2017
2. He K, Zhang X, Ren S, Sun J (2015) Deep residual learning for image recognition. arXiv:1512.03385v1 [cs.CV] Dec 2015
3. Dadashzadeh A, Targhi AT, Tahmasbi M, Mirmehdi M (2019) HGR-Net: a fusion network for hand gesture segmentation and recognition. arXiv:1806.05653v3 [cs.CV]
4. Chevtchenkoa SF, Valeb RF, Macarioa V, Cordeiroa FR (2018) A convolutional neural network with feature fusion for real-time hand posture recognition. Appl Soft Comput 73:748–766
5. Vahdani E, Jing L, Tian Y, Huenerfauth M (2020) Recognizing American sign language nonmanual signal grammar errors in continuous videos. arXiv:2005.00253v1 [cs.CV]
6. Mufarroha FA, Utaminingrum F, Mahmudy WF (2017) Segmentation algorithm to determine group for hand gesture recognition. J Telecommun Electron Comput Eng 9
7. Huang J, Zhou W, Zhang Q, Li H, Li W (2018) Video-based sign language recognition without temporal segmentation. arXiv:1801.10111v1 [cs.CV]
8. AlSharif MH, Saad M, Al-Naffouri TY (2017) Hand gesture recognition using ultrasonic waves. arXiv:1710.08623v1 [eess.SP]
9. Haseeb MAA, Parasuraman R (2017) Wisture: RNN-based learning of wireless signals for gesture recognition in unmodified smartphones. arXiv:1707.08569v2 [cs.HC]
10. Deng X, Ye Y, Zhang Y, Tan P, Chang L, Yang S, Wang H (2016) Joint hand detection and rotation estimation by using CNN. arXiv:1612.02742v1 [cs.CV]
11. Zhang C-L, Liu X-X, Wu J (2019) Towards real-time action recognition on mobile devices using deep models. arXiv:1906.07052v1 [cs.CV]
12. Donahue J, Anne Hendricks L, Rohrbach M, Venugopalan S, Guadarrama S, Saenko L, Darrell T (2016) Long-term recurrent convolutional networks for visual recognition and description. arXiv:1411.4389v4 [cs.CV]
13. Ji S, Xu W, Yang M, Yu K (2010) 3D convolutional neural networks for human action recognition. In: Proceedings of the 27th international conference on machine learning, Haifa, Israel
14. Rautaray SS, Agrawal A (2015) Vision based hand gesture recognition for human computer interaction: a survey. Artif Intell Rev 43(1):1–54
15. Guo Y, Liu Y, Georgiou T, Lew MS (2018) A review of semantic segmentation using deep neural networks. Int J Multimedia Inf Retrieval 7(2):87–93

SRL Video Recommender for Syllabus Driven E-Learning Platforms

Liya Laiju, N. S. Saurav, P. Rishad, S. Krishna Bhat, and G. Pankaj Kumar

Abstract "Acharya: community learning platform" is an intelligent online community-contributed learning platform with an integrated set of interactive online services that provides the teachers, learners, and others involved in education with information, tools, and resources to support and enhance educational delivery and management. Acharya includes study materials, including multiple videos on the same topic, with multiple languages so that students can choose their behavioral content to study. Educators can post their videos on related topics through the portal. They can also improve existing videos. Acharya automatically rates each content quality and recommends to each student according to their preference and taste. The paper describes the SRL algorithm used to address the challenges in recommending content in the Acharya platform.

Keywords SRL · Acharya

1 Introduction

Acharya is a free and open-source contemporary e-learning platform developed with the prime motive to promote free high-quality education. An e-learning platform is an integrated set of interactive online services that provide trainers, learners, and others involved in education seeking personalized content in high quality, with information, tools, and resources to support and enhance education delivery. The letter "e" in e-learning stands for "electronic", e-learning in-corporates all educational activities that are carried out by individuals or groups working online or offline, and synchronously or asynchronously via networked or standalone computers and other electronic devices [1]. Acharya is an innovative idea to provide free education by collaborating knowledge seekers and knowledge providers under one roof. It encourages peer-learning through multiple tutorial videos, which makes their learning process more effective. Acharya is a free and open-source contemporary

L. Laiju · N. S. Saurav · P. Rishad · S. Krishna Bhat · G. Pankaj Kumar (✉)
Department of Computer Science and Engineering, Federal Institute of Science and Technology, (FISAT), Angamaly, Kerala 683577, India

A. Sivasubramanian et al. (eds.), *Futuristic Communication and Network Technologies*, Lecture Notes in Electrical Engineering 792,
https://doi.org/10.1007/978-981-16-4625-6_31

e-learning platform developed with the prime motive to promote free high-quality education. Video tutorials are one of the best methods to grab a variety of content at a time. One of the major goals of using video multimedia tutorials is to transform a passive viewing experience into an active learning experience. So learners get a clear cut picture of their topic, which helps them to perform well in their future. Allowing students to submit online assessment forms can help teachers identify and apply many strategies to motivate them [2]. Acharya provides multiple collaborator's videos on a single topic so that learners can choose the better one. We provide learners with a choice. If a learner is not interested in the style in which a collaborator explains a topic, learners can switch to another presentation. Acharya is a community learning platform, so every interested one can share their knowledge in the topics that are specified in the syllabus of Acharya.

2 Related Works

Recommender systems are used to predict the rating that the user will give to an item if this is going to be the first time to be seen on the site [3], by using information filtering systems from the user's rating list history on the site [4] or by finding the item's similarity specification [5] or by finding the common interests using demographic information from their profile [6].

In [7], suggests a system which creates a comparison metric for related videos on YouTube. The metric is created by extracting the available comments. The comments are analyzed for the keywords. The analysis leads to the rejection of videos which does not satisfy certain constraints.

Rishad et al. [8] presents a syllabus driven E-learning platform which uses the application of fuzzy logic in video recommendation for its system. A recommendation technique is used to rank the videos by considering factors like comments and likes, which helps to recommend the best videos to the user.

3 Design and Challenges

Acharya recommends the videos to the user based on their behavior. It is an intelligent learning platform that studies a user's behavior and recommends videos that the user will likely prefer.

The system provides an equal chance for contributors to increase the reach of their content [9]. The content is analyzed based on various factors and is ranked accordingly. Tutor/collaborators can upload videos and take part in the discussion panel. Learners who choose a domain can move to its syllabus and take down its contents. The contents are available as multilingual videos and virtual labs. The virtual lab contains a compiler with hints and debugging functions.

The multilingual videos are ranked as per their unique SRL value. A separate algorithm was outlined for each problem statement. The user's past behavior was the essential criteria taken into consideration. Comments and ratings for each video also contribute to the recommendation. A value called SRL is calculated, taking the average of all these factors.

The recommendation is challenging, mainly due to the following factors. The old videos may lose their significance as time progresses. The system should not purely rely on old ratings to recommend the content. The user may update existing videos. These videos should not be rated from zero. Instead, the improved content should be given a boost in the current rankings. The boosting of the video should only be done for a period until the video ages. The ranking of the videos should be done independently for all users.

4 SRL Algorithms

A novel algorithm was designed to deal with the challenges in the e-learning platform. The algorithm evaluates the user data, user history, user similarity, video ratings, video comments, video age, video status, etc., to determine the SRL value. The value will be different depending on the user, thus offering a unique experience to the user. The algorithm also supports the user to upload improved versions of the video, rather than considering it as a new video.

Algorithm 1. SRL().
Input: new_list, main_list, threshold, user, video.
Output: new_list, main_list.

```
1. If user.age < threshold.user_age:
        1.1 If video.age < threshold.video_age:
                1.1.1 SRL = get_score(video, preference)
                1.1.2 new_list.append(video, SRL)
        1.2 Else If video.age > threshold.video_age:
                1.2.1 avg = get_avg(rating + comments, preference)
                1.2.2 SRL = get_score(video, avg)
                1.2.3 main_list.append(video, SRL)
        1.3 end if
2. Else If user.age > threshold.user_age:
        2.1 If video.age < threshold.video_age:
                2.1.1 avg = get_avg(matrix_factorization_output, preference)
```

2.1.2 SRL = get_score(video, avg)
2.1.3 new_list.append(video, SRL)
2.2 else If video_age > threshold.video_age:
2.2.1 avg = get_avg(matrix_factorization_output, avg(in case 1))
2.2.2 SRL = get_score(video, avg)
2.2.3 main_list.append(video, SRL)
2.3 end if
3. Else If video.improved == True:
3.1 avg = get_avg(matrix_factorization_output, video_boost)
3.2 SRL = get_score(video, avg)
3.3 new_list.append(video, SRL)
4. End if
5. Return new_list, main_list

new_list: list of newly added videos.
main_list: list of videos.
threshold: constraints.
avg: average value.
preference: the user preference.
get_score: returns the score of the video, which maps with the passed argument.
get_avg: returns the average value of the passed arguments.
video_boost: boosting value.

Scenario 1 (new user, new video): Here, the SRL recommender will initialize the video's SRL values based on the user preference and add to the new_list.

Scenario 2 (new user, Old video): In this case, the sentiment score from the top n comments are calculated. The average of the last n ratings, comments, and user preference is calculated to initialize the SRL value and used for appending the video to the main_list.

Scenario 3 (old user, new video): Here, the SRL is calculated with the average of the matrix factorization output (U, L) and user preference and used to append the video to the new_list.

Scenario 4 (old user, old video): In this case, the average of matrix factorization output and the average calculated for the video in the new_list is used for SRL calculation to be used for adding it in the main_list.

Scenario 5 (improved video): If the video is an improved version, the SRL value is boosted by a threshold. Then the video is added to the new_list. The boosting of the SRL score will stop when the video ages.

Then the recommender will sort the videos as per the SRL values obtained.

5 Experimental Results

There are numerous videos uploaded on the Acharya platform. For this experiment, we considered video lectures on C programming done by students and faculty

Table 1 SRL calculation for new user

Video	Date	Total views	Total rating	Rating in past n views	Total Comments	Number of + ve comments	Expected rating	SRL
V1	18-10-2019	256	1024	232	129	64	Nil	5.136
V2	22-10-2019	80	280	84	39	13	Nil	2.013
V3	02-11-2019	200	600	239	93	59	Nil	5.414
V4	08-11-2019	180	450	189	80	43	Nil	4.318
V5	15-12-2019	110	410	183	40	33	Nil	4.485

members. The videos were uploaded in June 2019 on the platform and shared among students of our college. The server was run on a Debian 10 machine with the Django-React framework and PostgreSQL.

Here we consider two different scenarios:

1. **Recommending videos to a new user**

For this experiment, we logged in as a new user. Table 1 represents the data related to the calculation of SRL. Since the user is new and we do not have any user's past data to calculate the expected rating he would give to a video. So we calculate SRL using the average rating acquired in n views and comment score. The comments are first analyzed and classified into +ve and −ve comments. Then SRL is calculated as:

$$\begin{aligned} \text{SRL} = & \left(\text{Total rating in } `n' \text{ views}/n\right) \\ & + (\text{Number of positive comments}/\text{Number of positive comments and negative comments}) \\ & + \text{Expected Rating} \end{aligned}$$

Video—name of the video considered.

Date—Date on which the video was uploaded.

Total views—Total viewed obtained so far.

Rating in past '*n*' views—For this experiment, the n is 50.

Comments—Only + ve and −ve comments are considered (whether the video is good/bad).

Expected Rating—The rating that a user is expected to give to a video.

(Here expected rating is NIL for every entry, a user is new, and it cannot be determined).

2. **Recommending videos to an existing user**

To test how the system recommends videos to an existing user, we logged in as a student who regularly uses this platform. Since the user is an existing user, we compute the expected rating that he would give to each video using the matrix factorization algorithm. The following table shows how the SRL value is calculated in this case (Table 2).

Table 2 SRL calculation for existing user

Video	Date	Total views	Total rating	Rating in past n views	Total Comments	Number of +ve comments	Expected rating	SRL
V1	18-10-2019	256	1024	232	129	64	4.674	9.810
V2	22-10-2019	80	280	84	39	13	2.735	4.748
V3	02-11-2019	200	600	239	93	59	4.290	9.704
V4	08-11-2019	180	450	189	80	43	3.568	7.886
V5	15-12-2019	110	410	183	40	33	3.338	7.823

Video—the name of the video considered.

Date—Date on which the video was uploaded.

Total views—Total viewed obtained so far.

Rating in past 'n' views—For this experiment, the n is 50.

Comments—Only +ve and −ve comments are considered (whether the video is good/bad).

Expected Rating—The rating that a user is expected to give to a video.

(Here, expected rating is obtained by matrix factorization, i.e., the result of U X L).

6 Conclusion

Recommending video contents in a community-contributed syllabus driven platform is challenging. The presence of multiple and multilingual contents, individual preferences, etc. increases the recommendation process's complexity. The paper described a novel algorithm—SRL, which addresses these challenges, and the experimental results were verified.

References

1. Khribi MK, Jemni M, Nasraoui O (2008) Automatic recommendations for e-learning personalization based on web usage mining techniques and information retrieval. In: 2008 Eighth IEEE international conference on advanced learning technologies. IEEE, pp 241–245
2. Manning Publications, Scarfone K, Mell P (2012) Guide to intrusion detection and prevention systems (idps) (No. NIST Special Publication (SP) 800–94 Rev. 1 (Draft)). National Institute of Standards and Technology
3. Aggarwal CC (2016) Recommender systems, vol 1. Springer International Publishing, Cham
4. Liu H, Hu Z, Mian A, Tian H, Zhu X (2014) A new user similarity model to improve the accuracy of collaborative filtering. Knowl-Based Syst 56:156–166
5. Falk K (2019) Practical recommender systems

6. Abou El-Seoud M, Taj-Eddin I, Seddiek N, El-Khouly M, Nosseir A (2014) E-learning and students' motivation: a research study on the effect of e-learning on higher education. Int J Emerg Technol Learn (iJET) 9(4):20–26
7. Jerrin J, Paul A, Anto S, Eswar NV, Pankaj Kumar G (2019) YouTube comment based video ranking system. In: International conference on digital technologies and transformation in academic libraries. NIT Surathkal
8. Rishad P, Saurav NS, Liya L, Jerrin J, Pankaj Kumar G, Sheela C (2020) Application of fuzzy logic in video recommendation system for syllabus driven E-learning platform. In: International digital conference CSMCS 2020
9. Fee K (2013) Delivering e-learning. A complete strategy for design, application and assessment. Dev Learn Organ Int J

IMD Signaling-Based Automated Safety Aid System for Fishermen

R. Avudaiammal, K. Jasmine Mystica, K. P. Raveendran, and Renjith George

Abstract Fishing industry is one of the most dangerous industries to work. Though fishing is a dangerous profession, many fishermen staunchly defend their independence. Indian Meteorological Department (IMD) periodically sends information regarding the weather conditions to ports through signals, ranging from 1 to 11. When the weather conditions are not normal, the flags will be raised by the port authorities to indicate various signal levels to alert the fishermen and public. Most of the fishermen ignore these signaling flags and take life risk to do fishing. It leads to motivation of the project "IMD Signaling-Based Automated Safety Aid System for Fishermen." The main objective of this work is to ensure safety for fishermen by monitoring the warning flags issued by IMD. The warning flags are monitored by the Raspberry Pi which acts as the decision maker in the proposed system to take the necessary actions like sending warning notifications and locking of boat's motor. The proposed system works in three modes, namely lock mode, notification mode, and rescue mode. If the warning flags are from signal 3 to signal 12 and if the ship is in the port, then the motor of the ship will be locked. If the warning flag is signal 1 or signal 2, the motor will be allowed to start and automatically a notification of the weather conditions will be sent to the registered fishermen using the cloud server with protocols like HTTP, MQTT, WebSocket, and SMTP.

Keywords Raspberry Pi · Color sensor · Proximity sensor · Cloud server

1 Introduction

One of the most dangerous industries to work is the fishing industry. Though fishing is a dangerous profession, many fishermen staunchly defend their independence. The main reasons behind the disasters occurring in fishing are poor emergency facilities, bad vessel maintenance, improper safety equipment, lack of awareness, and ignoring

R. Avudaiammal · K. Jasmine Mystica (✉) · K. P. Raveendran · R. George
Department of ECE, St. Joseph's College of Engineering, Chennai, India

R. Avudaiammal
e-mail: avudaiammalr@stjosephs.ac.in

A. Sivasubramanian et al. (eds.), *Futuristic Communication and Network Technologies*, Lecture Notes in Electrical Engineering 792,
https://doi.org/10.1007/978-981-16-4625-6_32

stability issues. Indian Meteorological Department (IMD) is a part of the Ministry of Earth Sciences of the Government of India. It collaborated with the Indian Space Research Organization and uses the IRS series and the Indian National Satellite System (INSAT) for weather monitoring of the Indian subcontinent by maintaining its own satellite system. IMD observes many ports across India. In 1864, Kolkata suffered from back to back storms and the government agreed to establish a cyclone alert system. The next year, a storm warning system was set up in Kolkata. India uses a signaling system ranging from Signal 1 to Signal 11, which is raised at the ports to alert vessels of a probable cyclonic storm. The IMD periodically delivers data to ports generally four moments a day and once in every three hours if there is a cyclone [1, 2].

Signal 1: There is a low-pressure zone developed far away in the sea. The expected wind speed is 33 knots (about 60 kmph).

Signal 2: There is a depression formed far in the sea, and wind speeds may reach between 34 and 47 knots (60–90 kmph).

Signal 3: There is a depression formed, and it can affect the port. Surface wind speed will be in the range of 22 knots to 27 knots (40–50 kmph).

Signal 4: There is a deep depression developed in the sea, which may affect the port later with wind speeds in the range of 28 knots to 33 knots (50–60 kmph).

Signal 5: There is a cyclonic storm which will cross the coast with the port to the left. Surface wind speed can be in the range of 34 knots to 47 knots (about 60–80 kmph).

Signal 6: Similar to Signal 5, but the cyclonic storm will cross the coast with the port to its right.

Signal 7: This means that the cyclone will cross very close to the port. Signal 5, Signal 6, and Signal 7 show and warn a potential risk to the port.

Signal 8: This is a "very danger" acute cyclone warning, and the cyclone will make landfall with port to its left. Wind speed will range from 48 to 63 knots (90–120 kmph).

Signal 9: Similar to Signal 8, but the cyclone will make landfall with port to its right.

Signal 10: This is a "very danger" warning, with wind speed 64–119 knots (120–220 kmph), and we call it a super-cyclone if wind speed is above 120 knots.

Signal 11: This means that there is no communication intact between the port and the warning office.

At present, the systems used for fishermen safety include the portable life protection system, border alert system, and the marine wireless networks for communication. The work presented in [3] focused on helping the fishermen by providing them with the details of their location in the sea with the help of GPS. The work also focused on establishing communication between the fishermen boat and the nearby ships. The work [4] focused on ensuring safety of fishermen by alerting them on the border area, thus preventing them from crossing the border of the country. The author made use of GPS and GSM techniques for border monitoring. The work [5] helped in providing a solution to the hardships faced by fishermen due to lack of communication. The author made use of GPS for real-time location and Zigbee for wireless communication. It is inferred from the literature that the existing systems to safeguard the fishermen make use of GSM technology and GPS technology. Moreover, there is no strict mechanism to restrict the movement of boats under critical climatic conditions.

The proposed **IMD Signaling-Based Automated Safety Aid System for Fishermen** focuses on replacing the GSM technology and Zigbee, with IOT and cloud services for saving the fishermen and by monitoring the flag raised by the Indian Meteorological Department (IMD) to take the necessary safety measures like locking the boat motor and sending notifications to authorities.

The outline of the paper is as follows: Sect. 2 gives the objective and explains the methodology of the proposed **IMD Signaling-Based Automated Safety Aid System for Fishermen**. Section 3 describes working prototype model with screenshots, and finally, Sect. 4 concludes the proposed work and future scope related to it.

2 IMD Signaling-Based Automated Safety Aid System

Indian Meteorological Department (IMD) periodically sends information to ports through signals ranging from 1 to 11 ports will raise warning flags according to signal levels. Most of the fishermen ignore those signaling flags and take life risk to do fishing. It leads to motivation of the work IMD Signaling-Based Automated Safety Aid for Fishermen. The objectives of our proposed **IMD Signaling-Based Automated Safety Aid System for Fishermen** are:

- To ensure safety for fishermen by monitoring the flag raised by the Indian Meteorological Department (IMD) and allow to access data to take the necessary safety measures like locking the boat motor and sending notifications to authorities.
- To use the cloud platform for establishing communication between the transmitter and receiver.

The proposed **"IMD Signaling-Based Automated Safety Aid System for Fishermen"** consists of transmitter module and receiver module. The transmitter module monitors flags and the proximity [6] and stores the information in the cloud. The receiver module monitors the stored information in the cloud and takes the necessary actions like locking the motor [7] and sending notifications. These modules are interfaced using a cloud server. There are several cloud providers who provide a pay-as-you-use model where customers pay for the specific resources used. In our system, we have used Heroku [8] as our cloud platform. Heroku is a service based on a managed container system with integrated data services, and it is a powerful ecosystem, for deploying and running modern apps. Nginx a free Web server of the platform is used. The information from the Raspberry Pi using the onboard Wi-Fi is stored in the Web server. GitLab is used to access the Heroku platform. The program in GitLab is written using Node.js. The cloud server is used to store the information from the transmitter module and to establish communication between the transmitter module and receiver module. The data is transferred using the protocols like HTTP

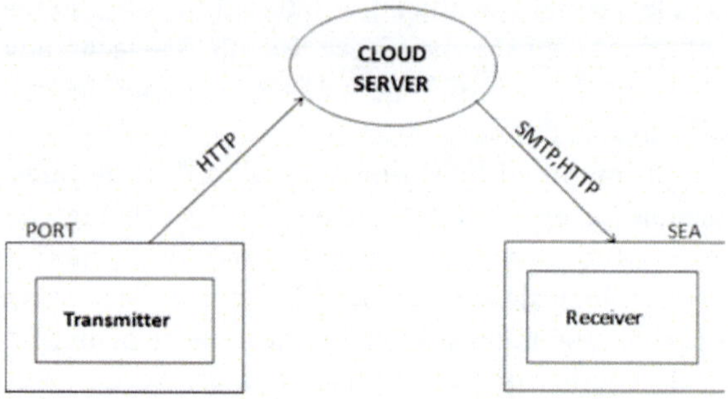

Fig. 1 Modules of the Proposed System

[9], MQTT [10], WebSocket [11], and SMTP [12] between the modules and the cloud (Fig. 1).

The transmitter section as shown in Fig. 2 comprises a color sensor, proximity sensor, and a Raspberry Pi. The Raspberry Pi functions as a decision maker. The color sensor is used to identify the color of the warning flag based on the signal level received from IMD ranging from 1 to 12. A proximity sensor is used to identify the location of the ship, whether the ship is at port or in the sea based on the proximity range. The information such as warning signal level and proximity range which are gathered from these color sensors and proximity sensors is analyzed by the Raspberry Pi to find out the modes of operation, and then, the identified mode is sent to the cloud server. When the weather conditions are not normal, the flags will indicate high signal levels. In our system, we represent the bad weather conditions with a red flag and the intermediate conditions with a green flag. The monitoring system monitors the warning flags, and it takes the necessary actions. If the warning flag is red in color, the motor of the ship will be locked. If the warning flag is green in color, the motor will be allowed to start and automatically a notification of the weather conditions will be sent to the fishermen using the cloud server [13, 14].

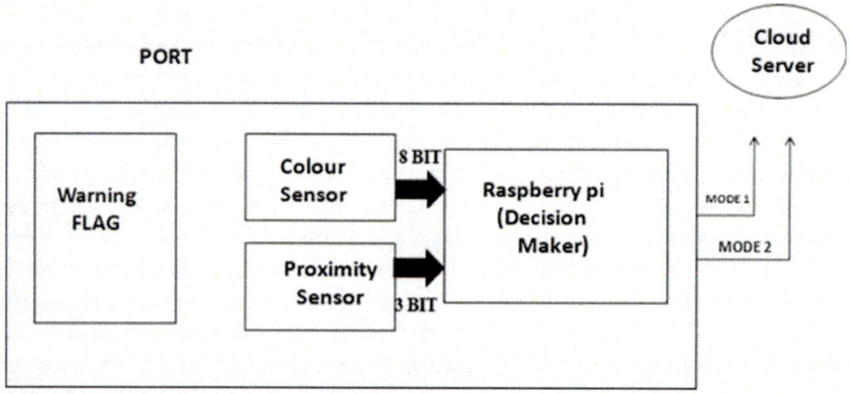

Fig. 2 Transmitter module

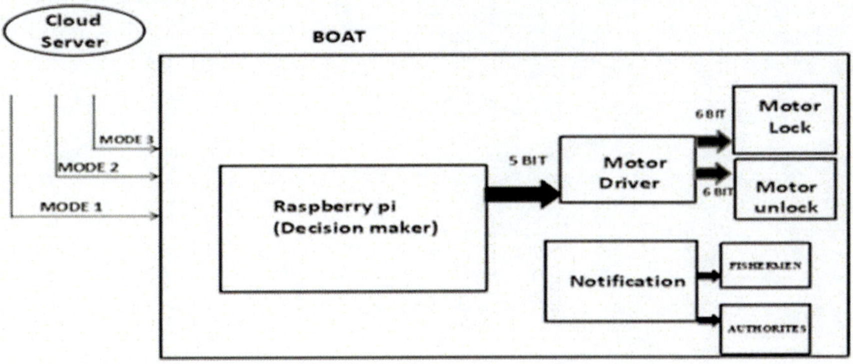

Fig. 3 Receiver module

Receiver section as shown in Fig. 3 includes a Raspberry Pi, a DC motor, and a motor driver [14, 14]. The Raspberry Pi with a LTE interface at the receiver is used to monitor the data in the cloud at regular time intervals and takes the necessary actions according to the respective modes. The proposed system works in three modes, namely lock mode, notification mode, and rescue mode.

1. **Lock mode**: If the proximity sensor indicates that the ship is at the port and if the warning flag level is above signal level 3 (red in color), then the proposed system automatically locks the engine. Thus, it aids fishermen to save their lives even if they ignore the warning flags.
2. **Notification mode**: If the proximity sensor indicates that the ship is at the port and if the warning flag lies between signal levels 1 and 2 (green in color), the proposed system sends a notification "This is a gentle warning that the weather conditions are risky for fishing" to the fishermen and the motor is allowed to start. The notification is sent via the cloud server.
3. **Rescue mode**: If the proximity sensor indicates that the ship is in the sea and if the warning flag level is above signal level 3 (red in color), the system sends a notification "The fishermen are stranded in the sea due to bad weather conditions" to the Fishermen Rescue Team and other authorities.

Figure 4 shows the flow diagram of the proposed system.

3 Results and Discussion

In this section, the operating modes of the proposed system such as lock mode, notification mode, and rescue mode are explained with screenshots. Figure 5 shows the output sent to the cloud by the transmitter module which includes the color sensed by the color sensor, and Fig. 6 shows the output received from the cloud by the receiver module.

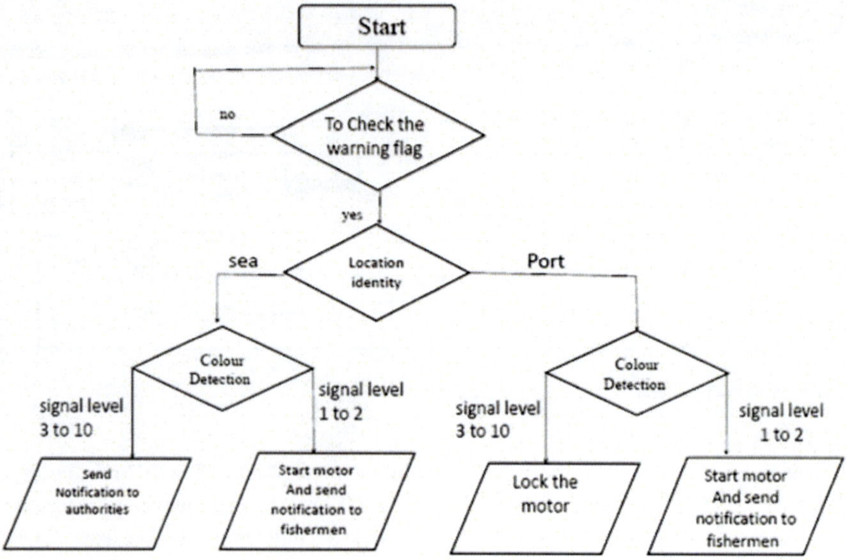

Fig. 4 Flow diagram

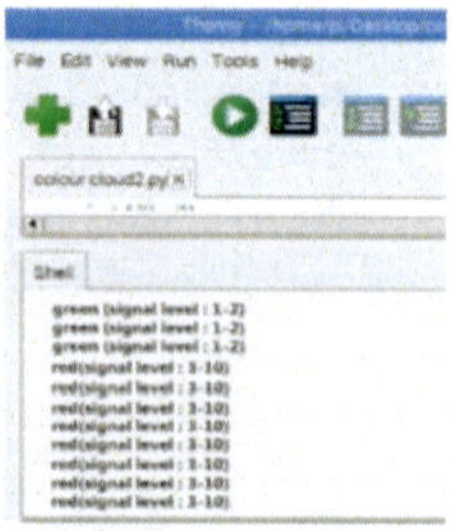

Fig. 5 Transmitter module output

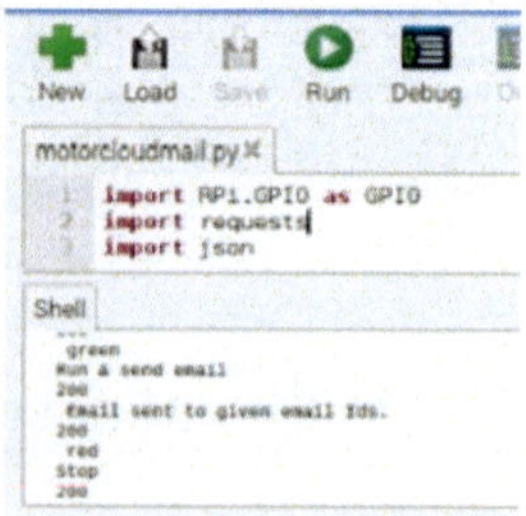

Fig. 6 Receiver module output

Figure 7 shows warning notification message sent by the fishermen safety system created in our work to give a gentle warning that the weather conditions are risky for fishing to the fishermen. When the proximity sensor indicates that the ship is at the port and if the warning flag lies between signal levels 1 and 2 (green in color), the

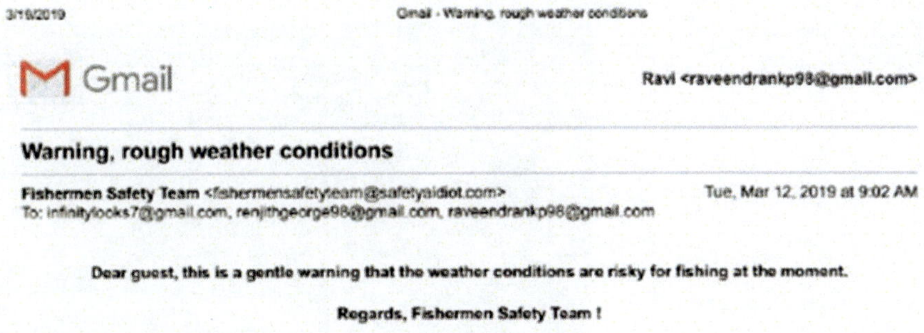
3/19/2019 Gmail - Warning, rough weather conditions

Gmail Ravi <raveendrankp98@gmail.com>

Warning, rough weather conditions

Fishermen Safety Team <fishermensafetyteam@safetyaidiot.com> Tue, Mar 12, 2019 at 9:02 AM
To: infinitylooks7@gmail.com, renjithgeorge98@gmail.com, raveendrankp98@gmail.com

Dear guest, this is a gentle warning that the weather conditions are risky for fishing at the moment.

Regards, Fishermen Safety Team !

Fig. 7 Warning notification

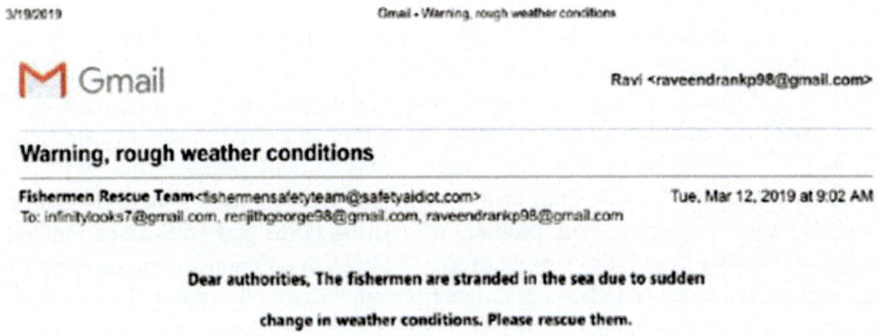
3/19/2019 Gmail - Warning, rough weather conditions

Gmail Ravi <raveendrankp98@gmail.com>

Warning, rough weather conditions

Fishermen Rescue Team<fishermensafetyteam@safetyaidiot.com> Tue, Mar 12, 2019 at 9:02 AM
To: infinitylooks7@gmail.com, renjithgeorge98@gmail.com, raveendrankp98@gmail.com

Dear authorities, The fishermen are stranded in the sea due to sudden change in weather conditions. Please rescue them.

Fig. 8 Rescue notification

proposed system sends this notification message and allows the motor to start. The notification is sent via the cloud server.

Figure 8 shows rescue notification message sent by the fishermen to the Fishermen Rescue Team and other authorities stating "The fishermen are stranded in the sea due to bad weather conditions." If the proximity sensor indicates that the ship is in the sea and if the warning flag level is above signal level 3 (red in color), the system sends this notification message.

The proposed IMD Signaling-Based Automated Safety Aid System for Fishermen is shown in Fig. 9.

4 Conclusion

Fishermen safety is becoming a very big concern in our day-to-day life. The proposed work is attempted to save lives of many fishermen who will not adhere to the IMD's protocol of entering the sea and alert message to save fishermen's vessels in the port

Fig. 9 Signaling system hardware

automatically according to the warning signal from Indian Meteorological Department (IMD). The proposed system "IMD Signaling-Based Automated Safety Aid System for Fishermen" chooses cloud technology to transmit and receive data which overcomes the existing problem. It also takes the necessary safety measures like locking the boat motor and sending notifications to authorities. In this work, the proximity sensor is used to monitor the distance. The proposed system overcomes limited range and connectivity problem of existing GSM and GPS-based methodology by making use of the cloud platform to establish communication between the transmitter and receiver and to access data from anywhere and anytime. The proposed system can be further enhanced by deploying multiple wireless sensor nodes having GPS modules to monitor the location of the boat and warn them of the anticipated weather condition.

References

1. https://www.thehindu.com/news/national/Cyclone-warning-signals-What-they mean/article16797019.ece
2. https://mausam.imd.gov.in/
3. Senthilkumar A (2013) Portable life protection system for fishermen using global positioning system. Int J Emerg Technol Adv Eng (2013)
4. Ranjith S (2017) Automatic border alert system for fishermen using GPS and GSM techniques. Indo J Electr Eng Comput Sci
5. Rajaparthiban M (2013) Multi-purpose marine wireless networks for fishermen aid and other applications. Int J Eng Res Technol (2013)
6. Kim YS (2014) Single chip dual plate capacitive proximity sensor with high noise immunity. IEEE Sens J 14(2)
7. Bubnov AV, Digital automatic control system of phase-lock motor drive. IEEE dynamics of systems, mechanisms and machines (Dynamics) (Omsk, Russia)
8. https://www.heroku.com/what
9. https://tools.ietf.org/html/rfc2616
10. https://tools.ietf.org/html/draft-sengul-ace-mqtt-tls-profile-00
11. https://tools.ietf.org/html/rfc6455
12. https://tools.ietf.org/html/rfc5321

13. Bellini P (2015) A knowledge base driven solution for smart cloud management. In: IEEE 8th international conference on cloud computing
14. Birgit S-W (2011) Cloud user roles establishing standards for describing core tasks of cloud creators, providers, and consumers. In: IEEE 4th international conference on cloud computing
15. Singh B (2014) A unity power factor NI-BIBRED converter fed brushless DC motor drive. In: IEEE conference
16. Lam Z-Y (2012) The design of DC motor driver for solar tracking applications. In: IEEE-ICSE2012 proceedings, Kuala Lumpur Malaysia

Comparative Study of Low-Energy Adaptive Clustering Hierarchy Protocols

Savita Jadhav, Ishwar Panpaliya, and Sangeeta Jadhav

Abstract The wireless sensor networks are composed of miniature power sensors that reach in remote regions. Sensors are alienated into diverse clusters. Among the randomly deployed cluster, one node is elected as cluster head (CH) and all other nodes act as member nodes (MNs) of that cluster. The foremost purpose of cluster head is to aggregate the sensed data from the member nodes to the sink node. Energy expenditure is a vital challenge in WSN as the sensor nodes are equipped with the batteries that are not replaceable. This paper put forward a relative revision of the LEACH protocols for wireless sensor network. The study starts with the review of preceding surveys of LEACH-based protocols. The assessment is carried out on the basis of use of location information, energy efficiency, hop count, base station centralized control, work distributed, self-organization and scalability. Further advantages and disadvantages of these protocols are also mentioned.

Keywords LEACH · EE-LEACH · C-LEACH · WSN

1 Introduction

The wireless sensor network (WSN) is mounting like most impressive and attractive technologies for communicating the information in the current world of Internet of things-based technology. To gather round useful information from the environment, numerous minuscule sensor nodes are arranged in the environment. For sensing and monitoring the factual environment, these sensor nodes are outfitted with signal sensing, processing and trans-receiving units. WSN is swiftly escalated in nearly every field like industries, health care, structural health monitoring, home automation, target tracking, military, agriculture, etc. But the major issue in WSN is the

S. Jadhav (✉) · I. Panpaliya
Electronics and Telecommunication Department, Dr. D. Y. Patil Institute of Technology, Pune, Maharashtra, India

S. Jadhav
IT Department, Army Institute of Technology, Pune, Maharashtra, India

A. Sivasubramanian et al. (eds.), *Futuristic Communication and Network Technologies*, Lecture Notes in Electrical Engineering 792,
https://doi.org/10.1007/978-981-16-4625-6_33

energy scarce of sensors. The nodes perform sensing, processing and trans-receiving of data based on their battery power. These tiny sensor nodes consist of very small size of the battery that provisions fewer sum of energy. Basically, it becomes difficult to transform the drain battery of the sensor nodes from inaccessible environment to get information. The total energy of sensor nodes is dissipated at the time of actual communication; the network lifetime reduces as the sensor nodes become dead. By considering limitations of WSN such as low power, low processing speed, less bandwidth and limited memory capacity, it is essential to utilize this sparse energy reserve competently to enhance the life span of WSN. The large amount of energy is degenerate in small period of time, and the network dies as the data packets are send from the member nodes to the base station (BS). To overcome these predicaments, LEACH routing protocol is a solution in which based on the probability theory every node can become cluster head (CH) and member node transmits there data in each iteration. The selection of cluster head randomly makes LEACH energy economy protocol. The probability approach is used in design of LEACH protocol to select any node as a cluster head. This procedure takes place in random manner. The tasks performed by member node are sensing the data from environment and sending this data to cluster head. The CH performs some advanced tasks like data reception from all member nodes which is the most energy-consuming processes, aggregating data and transmitting it to the sink node. Hence, cluster head guzzles huge quantity of energy compared to all other member nodes. After completion of iteration, one of the member nodes becomes cluster head and processes continue till all nodes in network become cluster head. By using this approach, none of sensor node gets energy discharge rapidly and there is impartial and consistent power eating in the sensor network; this step directs to improvement in lifetime of network. Several surveys [1–6] are scrutinized based on diverse approaches like node mobility, routing protocols, node localization techniques, optimization strategies and intelligent schemes. This comparative study proposed structured and inclusive review of the literature containing enrichment in LEACH clustering and routing protocols. The innovative categorization is based on diverse parameters like distance between nodes, selection of route, network connectivity and coverage. In [1], authors present a survey consisting of advancements in LEACH routing protocol. The taxonomy of protocols is based on data communication. Sixty LEACH alternatives are offered in this review with merits and demerits of each protocol. Also, nine diverse techniques are used for comparison of parameters. For classification of protocols, two different approaches are used as single-hop and multi-hop communication. In [2], authors presented an outline of advancements in LEACH. The comparison of protocols is based on node arrangement knowledge, node mobility and data communication process. The limitation of this survey is that it does not present any categorization. In [3], authors pay attention on the improvements in the LEACH-based clustering protocol. This survey presents six different types of LEACH protocol which is lacking in classification technique. The paper [4] surveyed twenty-one routing protocols based on LEACH. This paper also mentioned qualitative comparison with other existing surveys. The paper ends up with future guidelines. In [5], authors presented improvement over conventional LEACH. The limitation of [5] is only five protocols are used for survey without any classification

Table 1 Comparison of preceding reviews based on LEACH protocols

Ref	Key idea	Limitations
[1]	To discuss and compare several LEACH protocols with different performance parameters	Classification is based on two communication categories as single hop and multi-hop
[2]	To describe various LEACH variants	Three parameters compare the performance exclusive of further classification
[3]	To conquer uneven cluster head allotment in LEACH	This paper discusses merely six LEACH versions, not mentioned classification technique
[4]	Twenty-one types of improvements in LEACH are surveyed. Qualitative comparison with other existing surveys	Six types of parameters are used for giving comparison of different protocols in this survey. For selection of CH's selection and cluster formation, future guidelines are given
[5]	Improvement over conventional LEACH is presented	Only five protocols are used for survey without any classification and parameter comparison
[6]	Pros and cons of LEACH along with its descendants are discussed	Ten protocols are discussed without any comparison and classification

and parameter comparison. Table 1 indicates comparison of most accessible surveys available for reference.

The organization of this survey follows the following sequence: The overview of fundamental LEACH protocol architecture in detail with two phases is presented in Sect. 2; the comprehensive study of advancements taken place in basic LEACH protocol is conferred in Sect. 3; relative analyses of LEACH have been presented in Sect. 4; at the end, Sect. 5 concludes the paper.

2 LEACH: Protocol Architecture

This is a state-of-the-art protocol working on random revolution of cluster heads for uniformly distributing energy between cluster members and entire sensor network. The main working of LEACH [7] consists of the following two vital suppositions:

1. The base station situated at fixed position distant from the sensing devices,
2. The network type is homogeneous.

The cluster configuration and to employ cluster heads as routers for communication is the major initiative of LEACH. The construction of clusters is based on localized coordination. The pattern of clusters manages data transmitted to sink, reduces data congestion on communication link and creates network more scalable and robust. Figure 1 illustrates cluster arrangement and routing in LEACH protocol.

(i) Setup stage: A random number is generated by each node among 0 to 1 after that selected number is judged against $T(n)$, the threshold value. If chosen numeral is

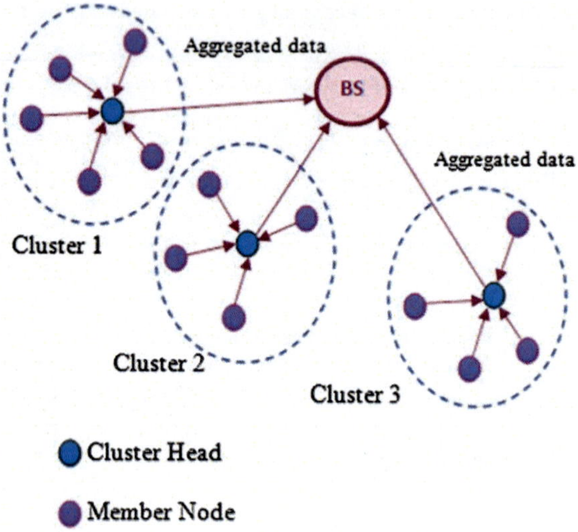

Fig. 1 LEACH cluster configuration and routing

fewer than $T(n)$, that node acts as a cluster head. Equation (1) expresses the threshold $T(n)$.

$$T(n) = \frac{P}{1 - P \times \left(r \bmod \frac{1}{P}\right)} \quad \text{and } T(n) = 0 \quad \text{if } n \neq G \tag{1}$$

where n indicates the node identification, p shows the cluster head prospect, r gives the recent encircling number and G indicates non-cluster head nodes in the first $1/P$ rounds. Following to $1/P$ rounds, all nodes will act as cluster head using this threshold value with probability P. The node which becomes cluster head in this round will not be cluster head in next $1/P$ rounds. In this way, the number of nodes which are capable to become cluster head goes on reducing and the probability of becoming cluster head among the remaining member nodes increases.

(ii) Steady-state stage: The key function of this stage is data processing, compressing and transmitting to the sink node. This data transmission is in frame format with combination of multiple frames. The structure is revealed in Fig. 2.

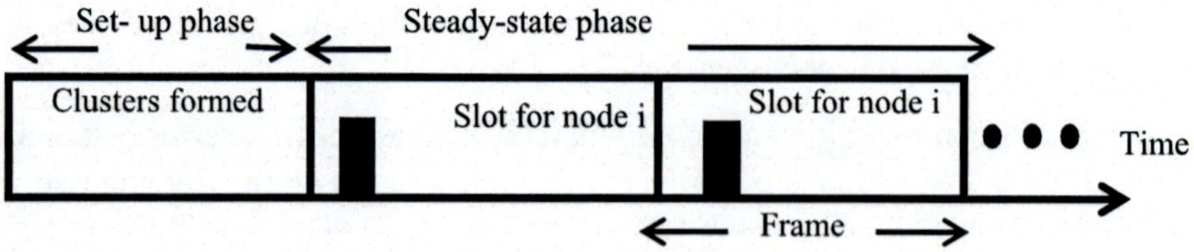

Fig. 2 LEACH protocol process

3 Advancements in Low-Energy Adaptive Clustering Hierarchy (LEACH)

3.1 Advanced LEACH (A-LEACH) (2008)

Ali et al. [8] present advanced low-energy adaptive clustering hierarchy (A-LEACH). In this procedure, nodes take independent assessment with no central involvement. In A-LEACH best cluster head selection, rotating cluster head position, adaptive clustering are the important processes takes place which gives even allocation of energy among all nodes of the network. Similar to LEACH, the A-LEACH works on two phases. In setup phase cluster formation and in steady-state phase data broadcasting to sink node. The formula for threshold is given in Eqs. (2) and (3). The advantages of A-LEACH are it increases network lifetime and energy efficiency and reduces communication distance with the sink node and cluster formation takes place without node position information.

$$T(n) = G_p + CS_p \tag{2}$$

Here,

$$G_p = \frac{K}{N - k(r \mod N/k)} \tag{3}$$

3.2 LEACH-B (Balanced) (2010)

Tong and Tang [9] present a LEACH-balanced (LEACH-B) protocol. The operation is divided in two rounds. Cluster formation and cluster head selection in first round are followed by steady-state phase in which nodes' residual energy is measured for cluster head node selection. The main theme of this protocol is that it has balanced and uniform cluster formation to save the energy consumption. LEACH-B has improved energy efficiency and longer network lifetime than LEACH.

3.3 Centralized LEACH (LEACH-C) (2013)

Tripathi et al. [10] propose LEACH-C which is advancement over basic LEACH protocol. For cluster formation, node's leftover energy and distance of all the member nodes to cluster head are considered. The cluster formation is centrally coordinated by the sink node. In the first round, all nodes of cluster drive their location information and energy level to sink node. The sink node using Eq. (4) estimates the average energy of the network. The higher energy nodes are marked eligible for becoming cluster head. The selected cluster head (CH) and member nodes are broadcasted to the network by base station. Next matching of node's own ID with cluster head ID takes place. If it matches, it acts as cluster head or else to transmit data to cluster head it will find TDMA slot. The LEACH-C has same data transmission phase as LEACH. The advantage of LEACH-C is that as cluster head node is positioned at the center of cluster energy expenditure is uniform, and the network lifetime is improved. But a GPS receiver is placed on every node which increases energy cost and LEACH-C is not compatible for large area of network is the limitations.

$$T(n)_{\text{new}} = \frac{P}{1 - P \times (r \mod \frac{1}{P})} * E_{n\_\text{current}} / E_{n\_\max} \tag{4}$$

where

E_{n_current} is the present sum energy and $E_{n\_\max}$ is the preliminary quantity of energy.

3.4 *E-LEACH (2014)*

Patel and Jinwala [11] propose E-LEACH (end-to-end secure LEACH) as an improvement of LEACH. E-LEACH applies homomorphic encryption for safe data processing in WSN. By using algebraic characteristics in homomorphic encryption, data processed algebraically devoid of decryption. This results in less energy with data privacy. It works on three types of keys. The first is a pairwise shared key. This key is to guarantee integrity and authentication at the time of communication. The second key is session key which works for each sitting of CH. Last is group key to authenticate nodes for CH. To achieve homomorphic encryption, it uses Paillier encryption scheme.

3.5 *Fuzzy LEACH (F-LEACH) (2017)*

Balakrishnan et al. [12] present fuzzy logic-based energy-efficient clustering hierarchy (F-LEACH) for non-uniform WSN. For increasing the network lifetime and selecting optimal cluster head node, F-LEACH uses residual energy, centralized nodes and distance between nodes. Again for appropriate cluster head selection, the probabilistic and weight-based techniques are combined together. The performance of F-LEACH may be increased in the future by introducing coverage redundancy and number of hop counts.

3.6 LEACH-PSO (2013)

Natarajan et al. [13] propose LEACH-PSO based on swarm particle algorithm. In particle swarm optimization (PSO) approach, objective function is formulated based on the outstanding energy, distance of member node to cluster head and node density. The combination of LEACH and PSO results in optimized cluster head selection, reduction in distance between cluster head and member node and uniform energy distribution in entire network. By using this approach, network can balance load among all the nodes and improves network life span. Where nodes distributed unevenly in WSN, LEACH-PSO is not suitable there. In addition to this, every isolated node should join the nearest cluster. More energy can be consumed while transferring information from the cluster head node to sink node.

3.7 LEACH-K (2014)

Nigam and Dabas [14] propose LEACH-K which makes changes in LEACH protocol by using K-means algorithm. In LEACH for selecting cluster head, residual energy of particular node is not considered and routing is of single-hop type. LEACH-K takes these two issues for further improvement. The equation for calculating threshold value is given by (5). As the selection of cluster head has the major impact on the advancement process of LEACH, the cluster head number K must be considered to improve result. The limitation of LEACH-K is that more isolation in nodes reduces node utilization, and increased energy consumption as LEACH-K is not responsive in remote nodes.

$$T(n) = \frac{P}{1 - P \times (r \mod \frac{1}{P})} * E_{n\_\text{current}}/E_{n\_\max} \times K_{\text{optimals}} \quad \text{if} \quad n \notin G$$
$$T(n) = 0 \quad \text{Otherwise} \tag{5}$$

3.8 LEACH-R (2012)

Wang and Zhu [15] present the cluster head selection stage which is improved in LEACH-R. In comparison with LEACH, cluster head and relay node selection are improved in LEACH-R. Residual energy is used in threshold calculation according to which small energy nodes do not act as cluster head. Again selection of relay node is based on lingering energy and distance to base station. The relay nodes are placed for communication in between cluster head to base station and cluster head to another cluster head in network. LEACH-R algorithm works on setup phase and steady-state phase. In comparison with LEACH, the LEACH-R saves almost 20% energy. Incomplete information transmission is the limitation of LEACH-R.

3.9 TL-LEACH (Two-Level Hierarchy) (2005)

Loscri et al. [16] propose multi-level communication. A new level of hierarchy is used in TL-LEACH for information transmission in between cluster head and base station. This technique increases number of cluster heads and decreases distance of data transmission to base station which results in decrement in node density for data transmission. The hierarchical network design is based on two levels as primary level in which cluster head is situated in the hierarchy at the topmost position after base station denoted as CHi and secondary level is a lower level denoted as CHij. Some end nodes are positioned at the end of hierarchy. At the second level, TL-LEACH initiates fractional limited working out in each cluster head. The top level performs the local computation for data transmission to base station. This technique shares the energy consumption equally among all the sensors of large density networks. TL-LEACH uses node localization to facilitate scalability and robustness.

3.10 Cluster Head LEACH (CH-LEACH) (2017)

Abushiba et al. [17] proposed cluster head selection (CH-LEACH) build on cluster count in the network gird area. To check the network coverage, the maximum number of the cluster head is elected in diverse situation. CH-LEACH selects cluster heads randomly, and in every round, each cluster head is assigned to nearest central location. The CH-LEACH minimizes energy consumption. The network lifetime is improved 91% than LEACH.

3.11 Multi-hop LEACH (MH-LEACH) (2007)

Xiangning and Yulin [18] presented MH-LEACH protocol in which every cluster head directly communicates with base station without considering distance between them. This consumes a lot of energy if distance is large. This drawback of LEACH is overcome in multi-hop LEACH (MH-LEACH). The MH-LEACH presents best possible pathway. It takes multiple hops from cluster head to base station. It performs first multi-hop communication between different cluster heads and second in between cluster heads to base station. In second stage, it uses optimal path for information transfer from cluster head to base station. The remaining things are same as the state-of-the-art LEACH protocol. In MH-LEACH, communication changes from single hop to multi-hop which results in improved energy consumption.

3.12 Enhanced Energy-Efficient LEACH (EEE-LEACH) (2015)

Bharti et al. [19] present an enhanced energy-efficient LEACH (EEE-LEACH) routing algorithm. In EEE-LEACH, member node uses shortest path for information transmission to cluster head via multi-hop communication technique. For improvement in basic LEACH routing protocol, EEE-LEACH uses another technique that is formation of master node for information transmission. The master head node takes the information from cluster head. Here, both master head and cooperative master head use cooperative MIMO approach to transmit the processed data to base station which is located far away in the network. For selection of cluster head, cooperative master head node and master head neighborhood distance among sensor nodes and residual energy of the nodes are considered.

3.13 Improved LEACH (I-LEACH) (2014)

Kehar and Singh [20] propose an integrated approach in which cluster head selection is based on three-level decision tree. The first important parameter of decision tree is residual energy. For balancing the clusters, protection scheme is planned in I-LEACH. Last option is the position of the base station which is placed in the network such that it covers high density of nodes. The cluster head selection depends on residual energy and distance between adjacent cluster head nodes. Routing technique is of reactive type. Nodes will send information as on if sensed information is above threshold value.

3.14 ETL-LEACH (Enhanced Two-Level Hierarchy) (2019)

Manzoor et al. [21] present ETL-LEACH which is advancement over the TL-LEACH. The TL-LEACH has two major limitations as it is not compatible for large-scale networks and node communication. The main task of ETL-LEACH is to decrease energy utilization. Similar to TL-LEACH, secondary cluster head nodes are treated as relay nodes which are placed between the primary cluster heads and the end nodes. ETL-LEACH prepares an energy table at the primary and the secondary cluster head nodes for cluster formation. This table gives the information of residual energy of end nodes to secondary cluster head nodes. This process is useful in switching the role of cluster head. The advantage of ETL-LEACH is that it is applicable to large-scale network. Use of mobile sink will be future advancement in ETL-LEACH with added mobility.

3.15 LEACH-T (2016)

Al Sibahee et al. [22] propose three layers (LEACH-T). The function of layer formation is to decrease distance between cluster head and base station. A separate cluster head is allotted to each layer. The LEACH-T operation is divided into three layers. The function of first layer is to collect the data from all member nodes of cluster. To collect the data from cluster head is the function of second layer. At last, the third layer functions if there is large distance between the second-layer cluster head and sink nodes. LEACH-T minimizes power consumption and solves the distance problem between cluster head and sink nodes.

4 Comparative Analysis

The comparison is carried out on the basis of use of location information, energy efficiency, hop count, base station centralized control, work distributed, self-organization and scalability as shown in Table 2. Advantages and disadvantages of recent clustering-based protocols are presented in Table 3. The conclusions are drawn on the basis of this review, and comparison tables are summed up below.

1. In WSN, sensor nodes are energy constrained, so vital design criteria for the majority of the LEACH protocols are to reduce energy consumption.
2. In WSN, enhanced security can be attained but it results in high energy consumption which increases cost. By applying innovative cryptographic techniques, researchers can accomplish improved protection through least energy consumption.
3. In large-scale wireless sensor networks, LEACH experiences communication problems as cluster heads directly communicate to the sink node. This reduces network lifetime. To solve this problem, researchers can use multi-hop protocols as they are accomplished with the optimal and shortest path techniques.
4. The optimization techniques are helpful in configuring best possible numbers of clusters and to decide most favorable numbers of cluster heads. Optimization can also utilize for most favorable deployment of nodes in network.
5. In WSN, one of the most talented research areas is energy harvesting. Harvesting techniques can be properly utilized in WSN as it increases hardware cost. The cosmological power, storm energy, kinetic energy and wireless charging are the diverse sources for energy harvesting.
6. In WSN, mobility is a popular research area. In comparison with static sensor networks, mobility of sensor nodes is flexible, and mobile nodes can be positioned in any scenario and supervised with quick topology transform. This area is less explored in modifications of LEACH. Researchers can recommend novel mobility patterns for cluster heads and sink nodes.

Table 2 Comparison of recent clustering-based protocols

Routing protocol	Location information	Energy efficiency	Hop count	Centralized control	Work distributed	Self-organization	Scalability	Improvements over LEACH
A-LEACH	Yes	Very high	Single hop	No	Yes	Yes	Good	All nodes are distributed in clusters. Member nodes send data directly to sink in LEACH
LEACH-B	No	High	Single hop	No	Yes	Yes	Good	The communication restricted between non-cluster head nodes. Distance between all nodes is known in LEACH
LEACH-C	Yes	High	Single hop	Yes	No	Yes	Good	Selection of cluster heads is done by base station. In LEACH, all nodes in cluster decide cluster head
E-LEACH	Yes	Very high	Single hop	No	No	Yes	Good	Predetermined number of clusters in E-LEACH
LEACH-F	Yes	Very high	Single hop	Yes	No	No	Limited	Fixed number of clusters and number of nodes in network
I-LEACH	Yes	Very high	Single hop	No	Yes	Yes	Very good	Twin node detection and assignment of sub-CH (SCH) nodes

(continued)

Table 2 (continued)

Routing protocol	Location information	Energy efficiency	Hop count	Centralized control	Work distributed	Self-organization	Scalability	Improvements over LEACH
K-LEACH	Yes	High	Single hop	No	No	No	Limited	Cluster formation is based on K-mediod technique
M-LEACH	Yes	Very high	Multi-hop	No	No	Yes	Very good	Multi-hop between CH and BS
T-LEACH	Yes	Very high	Single hop	No	Yes	Yes	Good	Role of cluster head nodes continues in next round if that node has more energy than threshold value
TL-LEACH	Yes	Very high	Multi-hop	No	Yes	Yes	Very good	Use of multi-hop communication

Table 3 Advantage and disadvantages of recent clustering-based protocols

Algorithm	Advantages	Disadvantages	Technique used
LEACH	Direct communication between cluster head and base station. The first state-of-the-art clustering protocol	More power consumption, addition or removal of dead nodes is not possible	Based on two phases, setup phase and steady-state phase
LEACH-C	Improvement in performance	Mobile nodes give poor performance, not suitable for large-scale networks; GPS installed on every node increases overhead	For cluster formation, apply a central control algorithm
A-LEACH	Increases network lifetime and throughput. Decreases node failure	Increased energy consumption, not appropriate for large-scale networks	Use of distributed algorithm for cluster selection. Cluster head mobility reduces distance; the member nodes need to transmit data
Multi-hop LEACH	For large size networks consumes less energy	Hotspot created at sink as multiple cluster heads transmit data to sink simultaneously	Implement multi-hop communication among cluster heads and sink and select optimal path for data routing
E-LEACH	Network lifetime improved	Constant round time consumes more energy	Leftover energy of nodes is used for cluster head selection
LEACH-R	Inter-cluster multi-hop routing improves network performance	Only suitable for small-scale networks	To enhance cluster structure uses node localization and node density
TL-LEACH	Power consumption reduction	End-to-end delay increases	Employ two-level transmission approaches

7. In WSN, network coverage is one of the important research areas which is not extended in depth in LEACH advancements, hence requiring more attention in this field.
8. The optimization techniques need to be developed for node localization as major LEACH variants use GPS for getting location coordinates of randomly deployed sensor nodes which results in increased cost and energy consumption.
9. In large area, WSN fault tolerance put forth a main impact on the network lifetime. Hence, resourceful algorithms must be designed that deal with faults.
10. In WSNs, network complexity is a vital factor in protocol design, as it affects overall network performance. Hence, low-complexity LEACH-based protocol design is essential.

11. Congestion control is one of the important areas of WSN, as it affects quality of service parameters like packet delivery rate, latency and energy consumption. Due to this, additional research attention is required for design of congestion-aware clustering and routing protocols.

5 Conclusion

The paper presents inclusive review of LEACH and its variants. This paper discusses and compares 15 improvements in LEACH. Localization, hop count, overheads, scalability and energy efficiency are the performance comparison parameters in this paper. Most of the discussed protocols are of distributed type and use node localization techniques. Using GPS for getting position coordinates is costly. Only few protocols have considered the energy expenditure factor in cluster head selection process and in cluster formation. Many other parameters such as node position, hop count, number of nodes, neighboring distance and mobility pattern are concentrated by researchers other than energy consumption which is essential factor of network design. Some future research points are highlighted in this paper which can be considered in designing improved and advanced versions of LEACH-based protocol. In multimedia and real-time WSN applications, LEACH-based clustering and routing techniques should be addressed profoundly.

References

1. Singh S, Kumar P, Singh J (2017) A survey on successors of LEACH protocol. IEEE Access 5:4298–4328
2. Maurya P, Kaur A (2016) A survey on descendants of LEACH protocol. Int J Inf Eng Electron Bus 8:46–58
3. Arora V, Sharma V, Sachdeva MA (2016) Survey on LEACH and other's routing protocols in wireless sensor network. Int J Light Electron Opt 127:6590–6600
4. Al-Shalabi M, Anbar M, Wan T-C, Khasawneh A (2018) Variants of the low-energy adaptive clustering hierarchy protocol: survey, issues and challenges. Electronics 7:136
5. Vyas IA, Panchal KJ (2017) A survey on LEACH protocol and its successors. Int J Adv Res Innov Ideas Educ 3(6)
6. Hitesh M, Shivarkar SA, Muzumdar AA, A survey on LEACH in WSN. https://www.researchgate.net/publication/315728088
7. Heinzelman WB, Chandrakasan AP, Balakrishnan H, An application-specific protocol architecture for wireless microsensor networks. IEEE Trans Wirel Commun 1(4):660–670
8. Ali MS, Dey T, Biswas R (2008) ALEACH: advanced LEACH routing protocol for wireless microsensor networks. In: Proceedings of international conference electrical computer engineering, pp 909–914
9. Tong M, Tang M (2010) LEACH-B: an improved LEACH protocol for wireless sensor network. In: 6th international conference on wireless communications networking and mobile computing (WiCOM), Chengdu, pp 1–4
10. Tripathi M, Gaur MS, Laxmi V, Battula RB (2013) Energy efficient leach-C protocol for wireless sensor network. In: Third international conference on computational intelligence and information technology (CIIT 2013). IET, pp 402–405

11. Patel HB, Jinwala DC (2014) E-LEACH: Improving the LEACH protocol for privacy preservation in secure data aggregation in wireless sensor networks. In: 2014 9th international conference on industrial and information systems (ICIIS). Gwalior, pp 1–5
12. Balakrishnan B, Balachandran S (2017) FLECH: fuzzy logic based energy efficient clustering hierarchy for nonuniform wireless sensor networks. Hindawi, Wirel Commun Mobile Comput 2017, Article ID 1214720
13. Natarajan M, Arthi R, Murugan K (2013) Energy aware optimal cluster head selection in wireless sensor networks. In: 2013 fourth international conference on computing, communications and networking technologies (ICCCNT), Tiruchengode, pp 1–4
14. Nigam GK, Dabas C (2018) ESO-LEACH: PSO based energy efficient clustering in LEACH. J King Saud Univ Comput Inform Sci
15. Wang N, Zhu H, An energy efficient algorithm based on LEACH protocol. In: 2012 international conference on computer science and electronics engineering
16. Loscri V, Morabito G, Marano S (2005) A two-levels hierarchy for lowenergy adaptive clustering hierarchy (TL-LEACH). In: Proceedings IEEE vehicle technology conference, vol. 62(3), 1809–1813
17. Abushiba W, Johnson P, Alharthi S, Wright C (2017) An energy efficient and adaptive clustering for wireless sensor network (CH-leach) using leach protocol. In: 2017 13th international computer engineering conference (ICENCO) Cairo, 50–54
18. Xiangning F, Yulin S (2007) Improvement on LEACH protocol of wireless sensor network. In: Proceedings of international conference sensor technology applied, pp 260–264
19. Bharti A, Devi C, Bhatia V (2015) Enhanced energy efficient LEACH (EEE-LEACH) algorithm using MIMO for wireless sensor network. In: 2015 IEEE international conference on computational intelligence and computing research (ICCIC), Madurai, pp 1–4
20. Kehar V, Singh R (2014) Evaluating the performance of reactive I-LEACH. In: 2014 international conference on advances in computing, communications and informatics (ICACCI), New Delhi, pp 2105–2109
21. Manzoor K, Jokhio SH, Khanzada TJS, Jokhio IA (2019) Enhanced TL-LEACH routing protocol for large-scale WSN applications. 2019 cybersecurity and cyber forensics conference (CCC) Melbourne, Australia, 35–39
22. Al Sibahee MA, Lu S et al (2016) LEACH-T: LEACH clustering protocol based on three layers. In: 2016 international conference on network and information systems for computers (ICNISC), Wuhan, pp 36–40

A Method of Hand Gestures Recognition using Convolutional Neural Network

Ragapriya Saravanan, Sindhu Retnaswamy, and Shirley Selvan

Abstract The progress in the realm of machine learning has caused the field of image processing to undergo a significant advancement. Image processing is a vast subject due to the diverse algorithms and techniques that can be implemented. Sign Language is the hearing disabled community's way of conveying information and this dissertation was drafted to understand and provide them with what little assistance we can. Visual cues and signs are used to convey messages and intentions of the speaker. It is a well-developed language that has its own vocabulary and grammar. In this paper, we analyze the various methods used in converting Sign Language into text which can be read or to audio that can be heard. This treatise also includes our very own methodology, which makes use of a CNN architecture called AlexNet and discusses the results of the same.

Keywords Sign Language · ASL · Fingerspelling · Pre-processing · Feature extraction · Recognition · Methodology · Algorithm · Dataset · Neural network

1 Introduction

Sign Language is a way of communication, rather a way of life practiced by the hearing impaired. It has existed among the hearing disabled community since ancient times. The spoken language that uses the reverberations from the mouth and is comprehended by the ear is difficult for them to understand. Sign Languages were established to exploit the unique features of visual modality (sight), along with tactile features (touch). Spoken Language is linear in nature as in, only one sound can be made or received at a time, whereas Sign Language, on other the hand, is visual and can therefore use a simultaneous expression through visual perception. Sign Language varies from each hearing disabled community and the place they live

R. Saravanan (✉) · S. Retnaswamy · S. Selvan
Department of Electronics and Communication Engineering, St. Joseph's College of Engineering, Chennai, India

A. Sivasubramanian et al. (eds.), *Futuristic Communication and Network Technologies*, Lecture Notes in Electrical Engineering 792,
https://doi.org/10.1007/978-981-16-4625-6_34

in; hence, Sign Languages are not universal. Linguists have perceived and distinguished 137 Sign Languages to date including American Sign Language, British Sign Language, Indo—Pakistani Sign Language, etc.

American Sign Language (ASL) is a language that is completely different and separate from the English language (see Fig. 1). ASL contains all the fundamental features of the language, with its own rules for pronunciation, word formation, and order. Parents are the main source of a child's early acquisition of language, but for children who are hearing impaired and dumb, additional people are necessary as models for language acquisition. A child who has a hearing impairment, born to parents with a similar impairment, naturally picks up Sign Language as a hearing abled child acquires spoken language from hearing abled parents.

A few methods of hand sign recognition from different authors over the years have been elucidated in the literature. A simple rule classifier to predict gestures is proposed [1]. Background subtraction is applied to the captured images to separate the hand region, which is the region of interest in this case from the background. After the hand is detected, the fingers and the palm region are segmented separately with the help of 3 parameters, which include the palm point [2], inner circle of

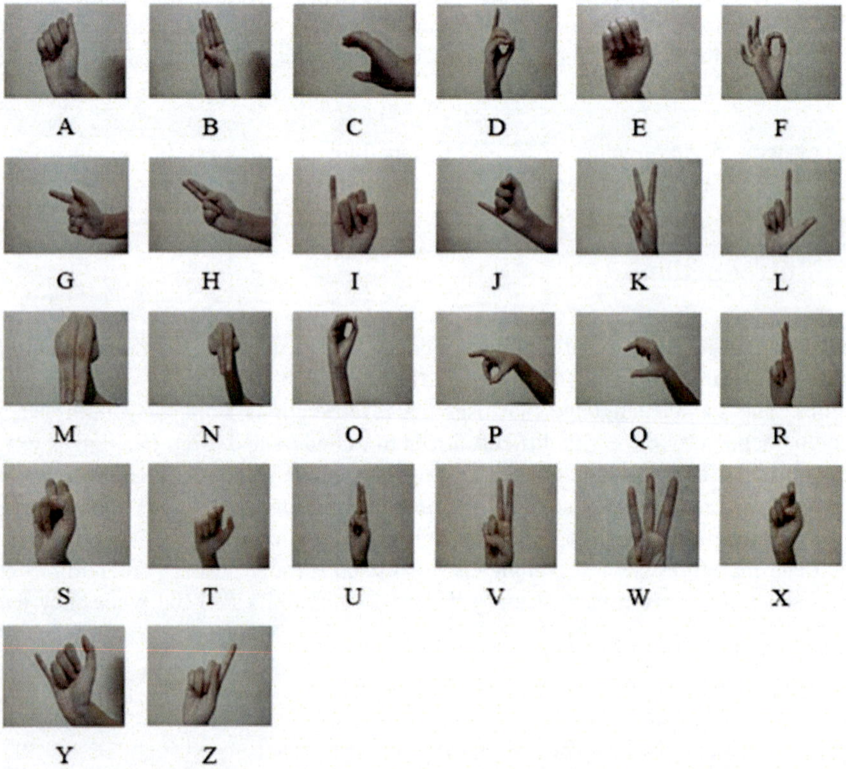

Fig. 1 Alphabets in ASL

maximal radius, and the wrist points and palm mask. When the fingers are detected and recognized, a simple rule classifier can be used to classify the hand gesture. In the rule classifier, the hand gesture is predicted according to the number and type of fingers detected. This classifier has high accuracy and is quick in detecting the distance transform to identify the sections of the hand such as the palm and the fingers. Despite having high accuracy, this method does not predict the gesture as well as other methods [3].

The speed up robust features (SURF) algorithm operates on still images of hand gestures shown in front of the webcam created with the help of frame capture function from real-time video capture. The points of interest of the grayscale images are detected using the SURF algorithm [4]. These points of interest from the captured images as well as the reference images are stored in the database. Following this, features are extracted and selected by SURF. The final step would be the recognition of the hand gestures by matching the reference features and reference points of the reference image with that of the input image. The use of box filters in the SURF algorithm provides the benefit of real-time applications such as tracking and object recognition. It provides robustness to rotation and also performs well with scale illumination changes. But the problems associated with SURF include its difficulty to track edges, dependency on view, and fragility [5].

An approach that uses the scale invariant feature transform (SIFT) algorithm was put forward in which a video of the hand is captured through a live camera using OpenCV. The SIFT algorithm detects key points or interest points of an object to provide a unique feature descriptor for the object, which are grouped to form numerous feature vectors for each image. The successive step would be orientation detection which will consider the movement of the hand. When the entire algorithm has completed compiling, the set of images fed as input are then translated into a one-dimensional array containing the characters corresponding to the alphabet shown in the image. When the alphabet is recognized, using speech conversion, the recognized text is converted to speech, and audio output is executed. The major interest in using the SIFT algorithm is that the feature vectors are not affected by object scaling or rotation. Also, the usage of multiple feature vectors would provide reasonable accuracy. The demerits of SIFT are that it is slow, mathematically complicated, and not good at illumination changes [11].

Another method of gesture recognition using a few segmentation techniques and convolutional neural network (CNN) is proposed. Once the hand is segmented, a morphological operation [12] is applied to the images to remove holes and noise present in both the object and the background. An arbitrary color model is then provided as input for the CNN architecture. Finally, based on the feature maps computed by the previous layers, the multilayer perceptron neural network performs classification. The architecture of the neural network and the provided training set influence the performance of the model greatly. There may be a wrong assignment of labels due to overfitting which is a major hurdle [13].

The paper is framed in the following way: Sect. 2 highlights the course of action which was developed for training a model using AlexNet [14] to recognize the signs of the ASL alphabets in a detailed manner. The trained model is tested with live frames captured through the webcam of the laptop and the results obtained for the same are discussed in Sect. 3. Finally, the conclusion is drafted and the means to improve the outcome along with the future scope of the project are enunciated in Sect. 4.

2 Methodology

We propose a method that uses AlexNet to recognize hand gestures. AlexNet is one of the architectures of CNN, designed by Alex Krizhevsky, published along with Ilya Sutskever, and Geoffrey Hinton. It is an incredibly powerful architecture that is adept at achieving high accuracies on very difficult datasets.

AlexNet has around 60 million parameters and 650,000 neurons making the network more accurate and intricate. AlexNet is suitable for object detection and many other applications in computer vision and image processing [15]. AlexNet architecture is 8 layers deep, comprising five convolutional layers and three fully connected layers.

2.1 Support Package

The pre-trained version of the network can be downloaded in the Add-On Explorer of the MATLAB software under the name "Deep Learning Toolbox Model for AlexNet Network". The pre-trained network is capable of classifying images into 1000 object categories, such as the keyboard, mouse, pencil, etc., as it was trained for 6 days using 2 powerful GPUs (see Fig. 2).

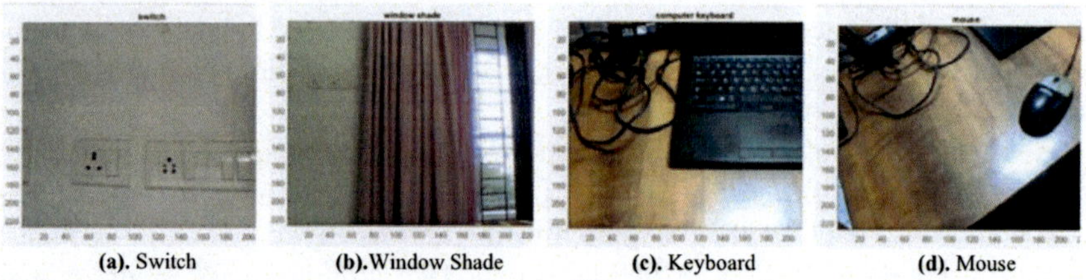

Fig. 2 Examples of classification by pre-trained network

2.2 *Dataset*

The dataset consists of a directory of 26 folders, each folder representing an alphabet of ASL, taken from Kaggle. Every folder contains a collection of 110 RGB images of dimensions 200 × 200 pixels, captured with a white background.

2.3 *Steps for Recognizing Hand Gestures Using AlexNet*

The steps for identifying hand gestures using images and videos with regard to AlexNet are presented below (see Fig. 3.).

Modify dataset. The input images fed to the AlexNet network should be RGB images of size 227 × 227 pixels. The dataset is modified and resized to fit this criterion.

Load dataset. The path to the dataset is fixed and the resized images are loaded onto the MATLAB workspace. The dataset is then split in such a way that 80% of the

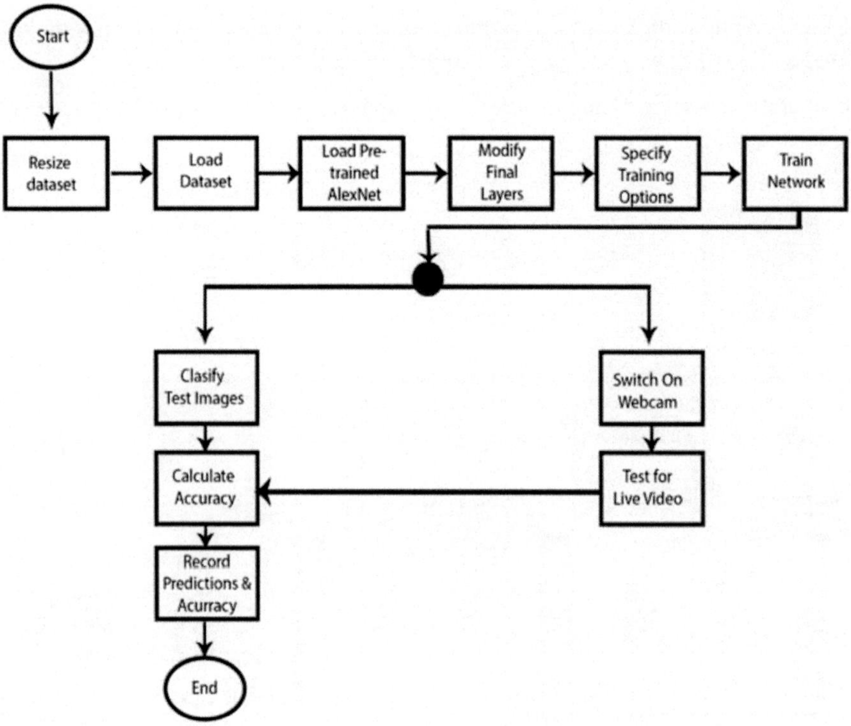

Fig. 3 Process flow diagram

Table 1 Dataset division

Dataset (for each alphabet)	Training images (for each alphabet)	Testing images (for each alphabet)
110	88	22

dataset is grouped into training images and the remaining 20% into testing images as indicated in Table 1.

Load AlexNet. The pre-trained AlexNet neural network model from the support package is then loaded onto the file.

Modify Final Layers. In order to fit the AlexNet to our custom specifications, the final layers are to be modified accordingly, which includes the classification layer and softmax. The basic architecture of AlexNet with 25 layers is used (see Fig. 4.).

Train Network. The network model is now trained with the input images from the dataset, the network architecture of AlexNet, and the training options. The training options specified include the optimizer: stochastic gradient descent with momentum (*sgdm*), initial learning rate: 0.001, maximum number of epochs: 10, and the mini batch size: 64.

Classify Validation Images. The trained model is then allowed to classify the test images. It assigns each image, a label representing the predicted output.

Calculate accuracy. Using the predicted output from the test images, the accuracy is calculated.

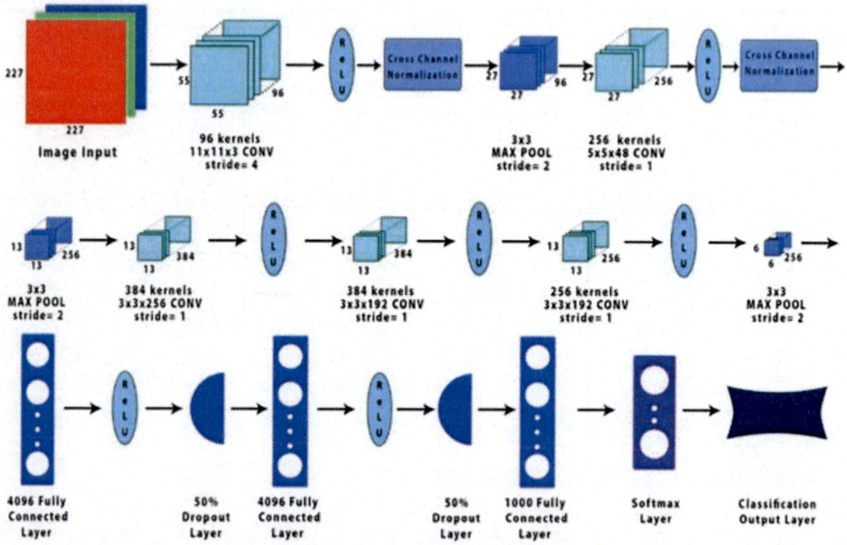

Fig. 4 Architecture of AlexNet to be modified for our methodology

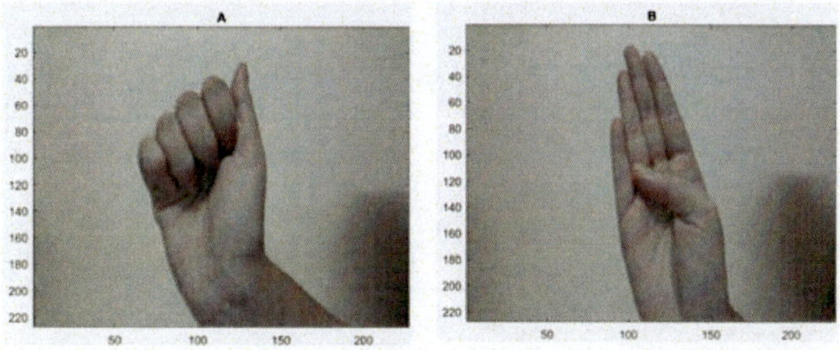

(a). Output for signs A and B

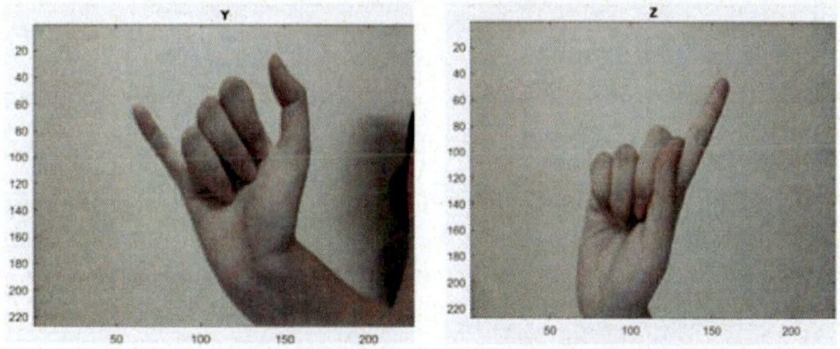

(b). Output for signs Y and Z

Fig. 5 Examples of correct prediction from live video

Live Testing. The webcam of the laptop is used to capture RGB images of the hand signs by taking individual frames from the video at particular intervals. The captured images are resized to 227 × 227 pixels, which is then classified using the trained network. The predicted output is displayed as the title of the image (see Fig. 5).

Problems Encountered During Training. This approach was executed on a machine that accommodated a GPU that proved to be incapable of training the AlexNet model for more than 10 epochs. In the case of using a sizeable dataset or increasing the number of iterations, the machine was unable to cope and halted before the completion of the training. Hence, it is recommended to use a fairly powerful GPU while training this model to get better results and accuracy.

3 Results and Discussion

The particulars of the results acquired are specified in Table 2. The accuracy obtained during the training of the AlexNet model using an *sgdm* optimizer for 10 epochs was 100% while testing using the test images. The same was tested and an accuracy of 76.92% was obtained for live images of hand signs captured by the laptop's webcam.

Table 2 Performance analysis of proposed method

Architecture used	Training time (min)	Test images		Live video	
		Correct prediction (alphabets)	Accuracy	Correct prediction (alphabets)	Accuracy
AlexNet	15	26/26	100%	20/26	76.92%

This model was unable to recognize alphabets R, S, T, U, V, and X for live images from the video. This variation in accuracy for test images and live video can be attested to any of the following reasons below:

- The similarity of the descriptor vectors of the signs.
- Inadequate time given for the model to train due to machine restrictions.

Limitations of Proposed Method. The high accuracy of any system that uses AlexNet can be attributed to the multiple layers present in its architecture but removing even a single layer leads to a 2% loss. Thus, the depth is important for the accuracy of AlexNet. It is also computationally expensive since it contains 3 fully connected layers.

4 Conclusion

This work explains how the ASL alphabet hand signs are captured using a webcam and converted into text. For the conversion of gesture to text, a CNN architecture called AlexNet is used. AlexNet is an architecture of CNN that has good accuracy and training speed. It avoids issues such as overfitting, making the network robust to errors and misrecognition. The proposed architecture can be used to determine ASL alphabet hand signs easily by using the trained model in a classification program.

Further work can be done on this approach by increasing the total dataset, as the model was trained only on 110 images for each alphabet for 10 epochs. It can also be developed to recognize words of ASL and non-stationary signs.

References

1. Patel R, Dhakad J, Desai K, Gupta T, Correia S (2018) Hand gesture recognition system using convolutional neural networks. In: 2018 4th international conference on computing communication and automation (ICCCA), pp 1–6
2. Gautam AK, Department of Electronics and Communication Engineering, Delhi Technological University Delhi, India, Kaushik A, Department of Computer Science and Engineering, Kurukshetra University, Haryana, India (2017) American Sign language recognition system using image processing method. Int J Comput Sci Eng (IJCSE)

3. Zhao Shan Chen Z-H, Kim J-T, Liang J, Zhang J, Yuan Y-B (2014) Real-time hand gesture 168 recognition using finger segmentation. The Scientific World Journal, Hindawi Publishing 169 Corporation. https://doi.org/10.1155/2014/267872
4. Soniya M, Sarah Suhasini P (2019) Integrated SURF and spatial augmented color feature based Bovw model with Svm for image classification. Int J Eng Adv Technol (IJEAT) 8(6). ISSN: 2249–8958
5. Kour K, Mathew L (2017) Sign language recognition using image processing. Int J Adv Res Comput Sci Softw Eng 7(142). https://doi.org/10.23956/ijarcsse.v7i8.41
6. Gaikwad S, Shetty A, Satam A, Rathod M, Shah P, Department of computer engineering, MCT "s rajiv gandhi institute of technology, Mumbai, India (2019) Recognition of American Sign language using image processing and machine learning. Int J Comput Sci Mobile Comput (IJCSMC)
7. Kumar RS, Srivastava M, Computer science and engineering department Jaypee University Anoopshahr, Patna, India (2019) Hand Gesture recognition using image analysis and neural network. Int J Res Advent Technol Special Issue
8. Pinto RF, Borges CDB, Almeida AMA, Paula IC, Universidade Federal do Ceará, Sobral, Ceará 62010–560, Brazil (2019) Static hand gesture recognition based on convolutional neural networks. J Electr Comput Eng
9. Krizhevsky A, Sutskever I, Hinton GE (2012) ImageNet classification with deep convolutional neural networks. In: Pereira F, Burges CJC, Bottou L, Weinberger KQ (eds) Advances in neural information processing systems 25. Curran Associates, Inc., pp 1097–1105
10. Sudha KK, Sujatha P (2019) A Qualitative analysis of googlenet and alexnet for fabric defect detection. Int J Recent Technol Eng (IJRTE) ISSN: 2277-3878

5G Modulation Techniques—A Systematic Literature Survey

J. Merin Joshiba, D. Judson, and A. Albert Raj

Abstract In the recent trend, data consumption goes on increasing day by day. In this way, the present 3G and 4G advancements cannot bolster those expansions in data usage, and the speed should be improved to accomplish better experience while at the same time accessing the data services. 5G technologies have a higher information rate and a better coverage area. It expends less power and has greater security, better spectral efficiency and energy efficiency. The speed of 5G technology reaches from roughly 50 mbps to 2G and even to 1000 Gbps which is much quicker than the 4G technology. Modulation is a process of influencing the data to a signal transmitted over radio carrier, which is the backbone of wireless communication system. Most remote transmissions today are computerized, and with the restricted range accessible, the modulation is more critical than it has ever been. We live in a digital era where wires are not needed to connect with loved ones. Messages, information and signals are sent across the globe within minutes. The modulation process is an important technique in the fast transmission of signals. The fundamental objective of modulation process is squeezing as much of data into a smaller possible spectrum is known as the spectral efficiency. It is used to measures how rapidly the information is transmitted in a single bandwidth. Its unit is b/s/Hz (bits per second per Hz). Various methods have risen to accomplish high spectral efficiency in different modulation techniques. Thus, in this paper, the various methods for modulation techniques are utilized for the development of the next generation communication system are contemplated.

Keywords 5G · MIMO · NOMA · OFDM · Modulation · Efficiency

J. Merin Joshiba (✉)
Lourdes Mount College of Engineering & Technology, Nagercoil, Tamil Nadu, India

D. Judson
St. Xavier's Catholic College of Engineering, Nagercoil, Tamil Nadu, India
e-mail: judson@sxcce.edu.in

A. Albert Raj
Sri Krishna College of Engineering & Technology, Coimbatore, Tamil Nadu, India
e-mail: albert@skcet.ac.in

A. Sivasubramanian et al. (eds.), *Futuristic Communication and Network Technologies*, Lecture Notes in Electrical Engineering 792,
https://doi.org/10.1007/978-981-16-4625-6_35

1 Introduction

Every decade, a new generation of wireless network has been implemented. In the wake of the new decade, the number of wireless devices is increasing and are becoming more diverse than ever, and this demand more robust resilient network system that caters these needs. For implementing this, many ideas had been discussed in the past few years. The accepted technologies will make into the 5G (Fifth-generation) network, which will succeed 4G (Fourth-generation).

5G communication systems ensure that it fulfil the experience of gigabit to all the mobile users, with an increase in capacity up to three times the magnitude concerning with the present LTE (long-term evolution) systems. It is highly acknowledged that a goal of achieving this technology is done by combining all the innovative technologies comprising of various network layers. In physical layer, the OFDM (orthogonal frequency division multiplexing), together with the multiple-access strategy called OFDMA (orthogonal frequency division multiple access), cannot be taken as it is, and many substitutions having promised larger spectral efficiency values are examined. In this paper, a literature survey of various modulation techniques regarding 5G modulation techniques are described.

In this paper, we will discuss the various modulation techniques proposed by various authors and the systematic literature survey of these papers.

2 Research Methodology

2.1 Eligibility and Exclusion Criteria

1. Identifying Keywords
   The keywords of the research topic are carefully selected to include the indented subject.
2. Screening
   The screening is done by reading titles and abstract. The first round of elimination which does not comply with our criteria.
3. Confirming the Eligibility Criteria
   After accessing the selected papers, affirming the eligibility of the journal papers by going through all the content.
4. The Actual Review
   Then the actual review is done and compared and presented in this paper.

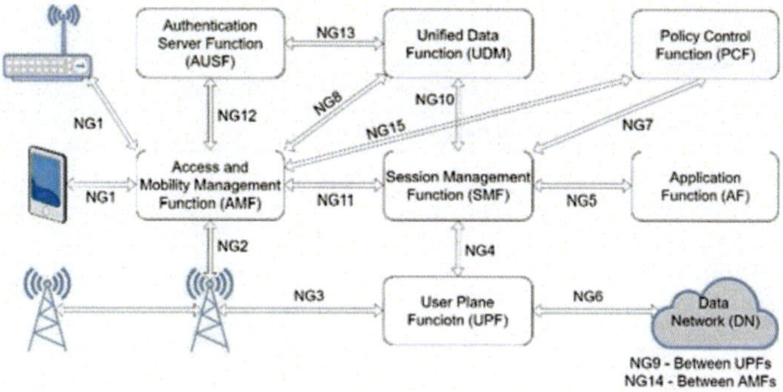

Fig. 1 5G network diagram

2.2 *Systematic Review Process*

The review resources have collected by electronic search in the various reputable online journals, which has more relevant databases which provide complete information on the topic "Modulation Techniques in Wireless Communication".

The journals include IEEE Explore, Springer, Elsevier and Google Scholar.

2.3 *Data Extraction and Synthesis*

After selecting appropriate papers, the next big step is to extract data and synthesis. Data extraction is a method of extracting information from selected research papers and to visualize the summary of that data. By using a spreadsheet, we create the data extraction form.

The network diagram of 5G in Fig. 1 shows the function of network process that are interconnected which activates the mobile communication. The 5G network gives better efficiency due to the selection of modulation techniques which satisfies the key performance indicator as in Table 1.

2.4 *Research Objectives*

The objective of the systematic literature survey (SLS) is to understand and present the existing empirical proofs and identify the areas for further research. To achieve this, we formulated our research objectives for our topic as below:

RO1: Key Performance Indicators (KPI).

RO2: Requirements for New Modulation Techniques.

Table 1 Comparison of KPIs of 4G and 5G network

KPI	4G	5G
User experienced data rate (Mbits/s)	10	100
Peak data rate (Gbits/s)	1	20
Spectrum efficiency	1x	3x
Latency (ms)	10	1
Mobility (km/h)	400	500
Network energy efficiency	10x	100x
Connection density (devices/sq.km)	10^5	10^6
Area traffic capacity (Mbits/s/m)	1	10

RO3: Potential Modulation Techniques for 5G Network and Their Advantages.
RO4: Disadvantages of Reviewed Modulation Techniques.

3 Research Exploration

3.1 RO1: Key Performance Indicators (KPI)

3.2 RO2: Requirements for New Modulation Techniques

1. Increased Throughput
   Bandwidth is the maximum possible data transfer per unit time, and the actual amount of the data transferred per unit time through a network is known as throughput. The challenge in 5G is to obtain a data rate of 100 Mbit/s everywhere and ~20 Gbit/s at peak.
2. Network Densification and New SPECTRUM
   Increasing the density of the network and including a new spectrum will increase the throughput. But this demands a new modulation and waveform design.
3. Coverage Extension
   Coverage is a crucial factor for wireless networks, but because of proposed is to include a high-frequency spectrum in communication made the coverage range smaller, and it leads to smaller cell size with a smaller antenna which leads to intercellular interference of cell edges.
4. Short Latency and High Reliability
   Latency is the time taken to get the response after the request sent. Reduced latency is a must. The proposed latency for the 5G network is 1 ms. Less latency network will provide an ultra-reliable communication system.
5. Support for A Wide Range of Frequencies

5G systems will operate in a wide range of frequencies in mm and cm range which makes traditional modulation techniques obsolete and demands for a more robust modulation technique.

6. Reducing Signalling Overhead
   Signalling overhead is the time taken by the 5G for signalling purpose. In 5G, machine type communication (MTC) is intrinsically done.
7. Multi-service Support
   Today, different services have different and diverging requirements. So 5G must have multi-service, multi-connectivity and wide range of frequencies support.
8. Energy Efficient
   5G must be flexible and energy-efficient to support sensors and other low-cost devices.

3.3 RO3: Potential Modulation Techniques for 5G Network and Their Advantages

Basar et al. [1] explain the whole index modulation (IM) concept and covers spatial modulation (SM), OFDM-IM schemes and CM in detail. This paper is one of the most comprehensive in IM field.

For last 50 years, the traditional AM/PM/FM systems used a sinusoidal signal as a carrier for transmission, unlike this, IM systems can map the information bits by switching the status of their transmission entities on/off [2]. That is, a new dimension for data transfer is created by IM.

In [3] and [4], it explores the potential of IM-based systems by introducing spatial modulation and OFDM with index modulation (OFDM-IM), which agitated the chase for alternative digital modulation schemes.

IM schemes can transfer stored energy for transmission from offline transmitting entity to live entity, and this leads to enhanced error performance. IM schemes delivers the information by a way which is more energy-efficient. This is made by deactivating idle parts of the system but using them for data transfers. Without increasing the complexity of hardware, we can increase spectral efficiency.

In backscatter communication, a tag is used which reflects the radio signal that is transmitted from reader. It is then modulated by controlling the coefficient of reflection.

In [5] and [6], 900 MHz and 2.45 GHz QPSK modulation were discussed, respectively. 16-QAM at 900 MHz and 32-QAM at 5.8 GHz to improves the range of backscattering communication systems in [7] and [8], respectively.

In this paper, authors designed a 16-QAM backscatter modulator with a 960 Mb/s of data rate and 59 μW of power consumption and have a wireless power transfer (WPT) capability.

The main achievement consists of WPT link, provides communication with 2.45 GHz, to improve the efficiency of conversion at 1.7 GHz a tuned matching network is used. The existing standards, ZigBee (50 mW), Bluetooth (100 mW) and Wi-Fi (800 mW) consume more power than this modulator (59 μW).

Neelamegam et al. [9] compare the cooperative WSN performance with performance of SM (CoSM) and enhanced SM (CoESM). It describes the index of the active antenna as well as constellations transmitted from each of them [10]. The spectral efficiency of the system is improved by SM. Also, the inter-carrier interference (ICI) is avoided by using the one radio-frequency chain active during symbols transmission [11].

CoESM outperforms the CoSM scheme. A data with good spectral-efficiency is given to the receiver even without knowing the channel state information and without any complex design. The authors show that we can increase minimum Euclidean distance even without increasing the transmitted power.

Sun et al. [12] describe an improved technique of phase modulation on the basis of 4-D array of antenna. The enhanced control unit consists of four delay lines namely the 0, 0.5π, π and 1.5π and two SPDT switches. This connects the elements in the array and feeding network. Without using conventional attenuators and phase shifters, at carrier frequency, this structure produces a non-uniform phase and amplitude excitations.

Arbitrary angle beam-forming and beam-steering is possible. This is done by changing the time scheme of the switches. By using two pairs of delay lines with two SPDT switches, we can increase radiation efficiency by placing all the switches in the "ON" state. Using time schemes design, we can obtain complex excitations.

Ozturk et al. [13] proposed a method of integration of a concept called GFDM with the space and the frequency index modulation schemes for providing a flexible and an advanced novel RAT for the wireless networks. Spatial multiplexing single carrier FDMA also gives low complex detection [14].

This framework provides excellent flexibility, and we can easily tune it to the required criterion.

Kim et al. [15] propose a system called LP-FBMC (linearly processed FBMC). It is a new method of FBMC system which is an interference-free system.

This paper proposes an FBMC system that transforms into a domain of interference-free from overlapped FBMC data by utilizing linear transformation based on singular value decomposition (SVD). The overlapped symbols will cause intrinsic interferences. To handle intrinsic interference, the authors uses matrix form to represent the FBMC transceiver. These will remove any residual intrinsic interference in MIMO techniques.

Also, to increase the capacity of the system, authors used power allocation across symbols. The LP-FBMC system overlays a new way of allowing traditional MIMO method in the system of FBMC.

Baig et al. [16] propose a technique called DHT-P-UFMC (Discrete-Hartley Transform Precoding based UFMC) which has low peak to average power ratio (PAPR) and is analysed through a simulation technique named Monte-Carlo simulation.

Universal filtered multi-carriers (UFMC) have less interference between users and have short burst support. But the UFMC transmitter efficiency is reduced due to higher value of PAPR.

Unlike selected-mapping UFMC and conventional UFMC systems, DHT-P-UFMC system has a lower value of PAPR. This proposed scheme does not require side information or complex optimizations.

Designing better RAN is vital for enhancing the machine type communication, which is an essential demand of 5G technology. RAN has a straight impact on many aspects, like energy efficiency, spectral efficiency, system capacity, latency, throughput and hardware complexity. OFDM is reliable along with spectral efficiency, but OFDM faces PAPR problem, have highly sensitive to time–frequency offset. To overcome these issues, new techniques like FBMC and UFMC have emerged.

In UFMC, fixed number of sub-carriers are formed by dividing full-band of subcarriers. Frequency to time-domain conversion is done by using the N point IFFT, by passing symbols of QAM from each sub-band over it. By using a filtering process, the spectral emission out of band is minimized. Then, the transmitter sums all the filtered values before transmitting.

Still, due to multi-carrier (MC) nature of the conventional UFMC system, it has high PAPR. Power constrains at the UE makes uplink mobile communications difficult if it has high PAPR.

Clipping and filtering scheme was proposed to be addressed to overcome these difficulties. These are easy to implement, but during clipping, information lost leads to high in-band distortion and also have higher BER.

To address these new set of problems, selected mapping UFMC and tone reservation UFMC are introduced. They have low PAPR and do not have any out-of-band distortion. But they require the side information and are computationally very complex.

These difficulties overcame in the proposed DHT-P-UFMC.

Roy et al. [17] proposed an innovative method called filtered OFDM on the basis of narrow band FIR filter operating on interpolated band pass technique.

OFDM is used to face the effects caused due to multipath reception. This can be done by the splitting of entire assigned bandwidth into a number of narrow sub-bands which leads to high spectral efficiency and minimizes the inter-symbol interferences (ISI) effect. Also, OFDM has a high bit rate and bandwidth flexibility.

Still, it fails to attain required data rate for the purpose of future communication networks. Also, OFDM is not suitable for the 5th Generation communication, due to the presence of larger side lobes larger PAPR and large out-of-band emission (OOBE).

To overcome these limitations, F-OFDM is proposed that separates the complete bandwidth into various flat sub-channels. With suitable filters, all these sub-channels are then filtered for setting up of a parameter of miscellaneous waveform compatible with various measures.

IBM-based narrow-band FIR filter built in F-OFDM achieves less OOBE when compared with F-OFDM having a technique of narrow transition band filtering.

The OTFS in Doppler fading channels are highly robust and have high spectral efficiency and high energy efficiency MIMO.

Future wireless systems need to operate in high mobility situations and millimetre wave (mmWave) bands are envisioned. Double-dispersion scenarios can occur, in which time dispersion occurs due to multipath propagation and frequency dispersion occurs due to frequency dispersion.

OFDM is used to alleviate the inter-symbol interference effect caused due to time dispersion. Still, in OFDM, Doppler shifts cause inter-carrier interference (ICI) and becomes inefficient.

OTFS modulation is a multiplexing technique suited for high-Doppler fading channels. A doubly dispersive channel is turned into non-fading channel by OTFS using the series of two-dimensional transformation, in delay-Doppler domain. Pulse shaped OFDM systems can overcome some of the problems, but cannot meet the needs for efficiently supporting high Doppler shifts.

The symbol coupling of OTFS channel and impulse response of sparse delay-Doppler channel facilitates the effective estimation of MIMO in high-Doppler circumstances.

An iterative algorithm is proposed for channel estimation in the domain of delay-Doppler system that suits the MIMO-OTFS and for signal detection based on message passing. An impulse in the domain of delay-Doppler system, used as a pilot for estimation, is used in this channel estimation scheme.

The OTFS modulation, with added blocks of pre-processing and post-processing, could be used for any type of multi-carrier modulation. From an implementation viewpoint, this is impressive than MIMO CC-CDMA systems [18, 19].

Zeng et al. [20] concentrate on the relation, from basic principles to enabling schemes, between SC-NOMA and MC-NOMA.

In [21], several key characteristics, important principles, merits and demerits, opportunities, challenges and future research trends and a brief comparison between various primal NOMA technologies are proposed. In [22], various basic downlink and uplink techniques for non-orthogonal transmission is discussed.

Liu et al. [23] propose NOMA transmission schemes assisted by two relays for 5G vehicle to everything (V2X) communication, i.e. full-Duplex relay-assisted NOMA and half-Duplex relay-assisted NOMA, and discussed the problems of optimum power allocation.

The V2X services should have high reliability and low latency. An LTE method based on orthogonal multiple access have limited spectrum and resources. Data congestion and low access efficiency will occur in dense networks. NOMA will provide a novel solution for 5G V2X services to reduce latency and decrease congestion.

Baig et al. [24] present Hadamard Transform (HT) based on F.I.R. filter and to decrease the PAPR, pre-coded uplink MC-NOMA scheme.

The advantages of higher data rate, spectral efficiency, massive connectivity and better cell coverage capability are achieved on MC-NOMA. But a disadvantage of PAPR is available in it. To reduce PAPR, excess bandwidth is limited by multiplying FIR filter coefficient with HT pre-coded modulated data.

Al Rabee et al. [25] concentrate on the technique of power-domain and analytically determines that the optimal uplink power received from SIC detector with any of the quantity of transmitters.

The number of users or devices supported is independent of the available resources in NOMA. Therefore, in NOMA, there can be more users fitted when compared with OMA [26]. No scheduling is used in uplink in most of the NOMA schemes, which reduces latency and signalling overhead [27]. Another attractiveness is, due to limited throughput, in IOT applications, to serve a huge number of sensors, few channels are enough [28, 29]. The error probability and interference are reduced when compared with complementary coded CDMA [30–32].

Hiari et al. [33] proposed implementation of different space modulation techniques using a new single Software Defined Radio (SDR) platform architecture.

Spatial multiplexing MIMO provides a promising linear gain in capacity with the quantity of antennas. But this linear gain suffers a practical shortcoming like spatial correlation, mutual coupling, and it requires complex transceivers.

To overcome these problems, many authors proposed space modulation techniques (SMTs) [2, 34–36]. At an increased data rate, SMTs enables the MIMO transceiver implementation with efficient hardware, low component cost and low power. This dynamically reconfigurable scheme can realise various techniques of space modulation with a minimum hardware up to no extra hardware.

Rajabi et al. [37] introduced the ratio phase modulation technique and merged the ratio amplitude phase modulation technique. The major accomplishment of this modulation scheme is the degree of freedom, even with symbol variations so that it can deliver constant dc power to sensors, which cannot be possible while using the modulation techniques like ON–OFF or M-ASK keying. For narrow band channel, demodulation of this proposed method is possible even without knowing the attenuation of the channel. This is due to the higher non-linearity of IRR. This feature of non-linearity is enough to achieve the goal of receiving the information and the power. It is then characterized that the IRR displays the symbol diagram influencing circuit's non-linearity and its limitations for modulation.

Song et al. [38] proposed and experimentally showed a 6-port direct modulator for downlinks having suppressed carrier leakage in future wireless communication systems with high carrier frequency and high speed.

Due to the growth of mobile devices and mobile data traffic rapidly, a wide variety of newly proposed communication systems are developed for various application systems [39–41]. In recent years, a 6-port communication technique is designed and used due to the low complexity in circuit, high linearity and high frequencies with smaller footprint [42, 43]. By using different methods like mechanical switch and Schottky diode, different types of 6-port modulators are attained [44–46]. But the mechanical switch can only be used for transmission with low data rate so that it restricts the application of high frequency and high speed. The Schottky diode has three characteristics that are unfavourable. They are: the T-structure is indispensable and is needed to be manufactured carefully, then the performance depth of modulation is likely to be linked with driving voltage and also with the performance, especially for higher carrier frequencies and finally due to non-linearity in coefficient of reflection, the performance of modulation is restrained. Thus, by using the MESFET, the RF switches realises the modulation with high speed [47]. Merging the circulator, this switch provides an excellent performance for modulation with high speed at a high carrier frequency.

This 6-port modulator can be easily integrated to optical link. The suppression of carrier leakage is realized to improve the modulation performance and alleviate the dc components that are harmful for wireless communication.

Kumaravelu et al. [48] proposed an improved model of spatial modulation for an 8×8 MIMO that utilizes transmitter antennas of one or two on the basis of incoming patterns of bits.

An energy-efficient and spectrum digital modulation method called spatial modulation uses the indices of transmitter antenna for conveying extra bits. The conventional spatial modulation method has a complex arrangement of complex receivers and has a poor performance in line of sight conditions and channels with spatial correlation.

To minimize the computational complexity, an adaptive mapping technique is used in antenna selection in transmitter side and the maximum likelihood detection is modified in receiver side. To minimize receiver complexity while resisting the effect of spatial correlation, mapping of each and every possible bit per channel use to transmit symbol vector. The average BER of MSM-ML is better than the average BER of conventional SM-ML, generalized spatial modulation-based ML and redesigned spatial modulation-based ML methods in line of sight conditions and channels with spatial correlation. This method results in providing a better alternative of spatial multiplexing in compact battery-based handheld devices.

Mesleh et al. [6] propose and analyse GQSSK and GQSM schemes.

In the past few years, space modulation techniques (SMTs) attracted research interest [2, 35, 36, 49–53]. A new spatial constellation is produced in SMTs that is composed of the spatially separated indices of transmitter antenna. To transmit a modulated RF carrier signal or unmodulated RF carrier signal at a specific time, the incoming data bits are used to modulate a spatial symbol [52, 54 and 55]. Various SMTs are examined in this paper and is expected to have a single chain of RF transmitter with lower computational complexity in receiver with minimum performance error when compared to MIMO systems. There are two families that are used to compare the various types of SMTs. They are the family which transmits an unmodulated RF carrier signal like SSK [54] and QSSK [52]. Here, the data bits modulate only the spatial constellation symbol and another only family is used to transmit a modulated RF carrier signal, like spatial modulation [49] and quadrature spatial modulation [52]. Here, the data bits modulate both the spatial symbol and signal to convey the information.

The major drawback of SMT is that the quantity of transmitting antennas used is equal to the power two. A generalized SMT called generalized SSK [56] and generalized SM is realized to mitigate this necessity. A group of transmitting antennas in GSMTs are activated at a particular time for the carrier signal transmission. The overall power transmitted is split among the activated antennas. In [57], the GSM performance analysis is described, and in [58–60] a sphere decoder with low complexity for GSM is described. A description for various SMTs and GSMTs are given in [50, 61, 62].

Here, a GQSSK scheme is used to activate each and every transmitting antenna block for transmitting unmodulated RF carrier signal cosine part and for transmitting the unmodulated carrier signal sine part. The GQSM uses two groups of transmitting antennas, as that in GQSSK, for transmitting the modulated RF carrier signals via the complex symbols.

Bashar et al. [63] investigate NOMA-based cell-free MIMO, in which the users get grouped into multiple forms of clusters.

By exploiting successive interference cancellation (SIC), for downlink bandwidth efficiency (DL-BE), a closed form of expression was derived without downlink training and the max–min DL BE problem is solved optimally. For several clustering schemes, numerical results were derived. Moreover, this paper investigates the effect of coherence time of the system. Also, for maximizing the minimum value of DL BE, a from the mode set mode = {OMA/NOMA} a switching point has been suggested.

Love et al. [64] consider an environment with multiple cells and explained the characteristics of non-orthogonal multiple access (NOMA) in it and explains its possibilities and difficulties.

NOMA is a potential candidate for 5G communication. The number of users served, user fairness enhancement, is scaled up by NOMA, enhance user-fairness, and improve spectral efficiency contrasted to orthogonal multiple access schemes by enabling multiple users for sharing same time and same frequency. While single-cell NOMA has attracted notable recognition recently, multi-cell NOMA has got limited attention.

When base station and device density increases, in multi cell networks, the intercellular interference is displayed. Thus, it is required to identify the methods for the approach of managing interference with NOMA.

Marcano et al. [65] provide NOMA's performance analysis in mmWave cells, using OMA benchmark.

Transmission of millimetre wave and NOMA are identified as key technologies among the capacity boosters for 5G mobile network. A huge number of spectra are present in mm Wave frequencies. If we can effectively use available resources, we can improve the overall capacity. NOMA allows to share resources of same frequency at same time, but by changing the signal power.

In recent approaches, it is assessed the mmWave performance [3, 55] and NOMA [4, 7] independently. However, when both the techniques are combined, it is not completely assessed. Further, the previously existing researches on NOMA mainly concentrates on the offered gain in capacity without the analysis of the shortcomings of intentional interference. In this paper, the NOMA-based system performance is assessed instead of combination OMA and mmWave transmission.

The results show that by combining NOMA and mmWave, up to 6.7 GBps channel capacity is obtained, corresponding to a gain of 70% which is in contrast with OMA. Also, the analysis of the mandatory signal to interference plus noise ratio (SINR) in NOMA is done, which leads to the achievement of target block error rate. 12 dB Average SINR penalty is available while choosing NOMA instead of OMA.

To compensate the penalty of SINR, a solution of performing MCS adjustment at BS is suggested with additional bandwidth that compensate the data rates.

Xiao et al. [66] focus on 5G communication with multicast service thereby developing a massive scheme MIMO-NOMA.

The users of NOMA signal are superimposed in to multicast signal. To separate the users of NOMA signal from multicast signal, successive interference cancellation is done in proposed scheme. It utilizes interference cancellation method-based null space to cancel leakage of signal caused due to other groups. Linear precoding of multi-user MIMO is used in unicast group.

With the help of broadcast characters of wireless medium, an efficient technology called multicast technology which transmits data from single source to multiple targets in the massive MIMO systems. This system is widely used in applications of point to multipoint systems like media broadcast, distance learning, resource synchronization and multi-players. It is standardized in 3GPP protocols in the form of evolved multicast services and evolved multimedia broadcast since there is an important and efficient technology for communication. The multicast communication system significantly improves the spectrum efficiency and system capacity as a physical layer beamforming, system delay and simultaneously decreasing system overhead [55].

Also, in NOMA simultaneous transmissions are supported for multiple users. This captured the attention for higher efficiency of spectrum in 5G. Also, NOMA enhances user fairness through the exploitation of a scheme for capacity achieving in downlink [4]. Also, to cancel intra-cluster interference, NOMA uses a methodology of successive interference cancellation in receiver side [7].

To achieve high peak spectral efficiency, MIMO and NOMA technologies are combined [8].

A massive MIMO-NOMA system based on hybrid unicast or multicast precoding scheme is developed to define the impact of the performance of massive MIMO-NOMA model using user selecting or grouping criteria and interference cancellation.

The numerical results of massive MIMO-NOMA system reveal that it significantly meets the necessity of peak value of spectral efficiency in typical scenarios.

It is also found that results of simulation justify this system can also significantly improves the throughput of the network in various situations.

Moltafet et al. [67] propose an approach in 5G communications networks of multiple access called PSMA (power domain sparse code multiple access).

In PSMA, both code domain and power domain are used simultaneously for transmitting signals of multiple users over subcarrier. In this kind of model, multiple users can use same sparse code multiple access (SCMA) codebook, the power domain non-orthogonal multiple access (PD-NOMA) techniques are used to send non orthogonal signals. Using the same codebook for multiple users will produce interference. In coverage area of each base station with PSMA, a codebook can be reused, in which the spectral efficiency is improved than multicarrier CC-CDMA [68].

To examine the PSMA performance, a heterogeneous cellular network is considered. The object of this design is maximizing the sum-rate of a network to some system levels and QoS constraints like transmitter power constraint.

This study shows that spectral efficiency of a system can be improved by PSMA around 50% when compared to PD-NOMA and SCMA with an increase in complexity.

3.4 RO4: Disadvantages of Reviewed Modulation Techniques

MC-NOMA has many disadvantages. First, in different propagation environments, the achievable rates are evaluated always by simulation, since theoretical capacity bounds of MC-NOMA are still inexplicit. Second, when combined with FDMA, the resource allocation and scheduling expanded to a greater order of dimension. Third, to design efficient constellations of multiple dimensions are more complicated than single-dimensional power splitting. Fourth, with lacking feasible applications and verifications, the large-scale deployments of MC-NOMA are seemed to be more urgent.

In **CP-OFDM**, every symbol requires cyclic prefix. The cyclic prefix insertion prevents obtaining a low latency and reduces the spectral efficiency by shortening the symbols. Also, it has emission of high out of band signal due to rectangular pulse shaping, and susceptible to time and frequency errors.

In **space modulation**, the implementation approaches and software-defined radio aspects are limited.

In **space modulation techniques (SMX)**, complex transceivers are required for linear gain requires and experiences practical imperfections like mutual coupling and spatial correlation.

In **GQSSK** and **GQSM**, the spatial modulation mapping tables are needed to be modified for some schemes for enabling transmission of single RF-chain, and in some of the schemes, additional number of antennas are needed for these designs.

4 Discussion and Conclusion

Modulation technique is the primary factor for solving the problems of the wireless network. There are many roadblocks to implement 5G networks. Every obstacle is dependent and can be resolved by improving the modulation technique. We here considered major ideas put forward by the scientific community and discussed the benefits of different methods, also challenges that are needed to solve in future.

We discussed some modulation schemes assumed to be suitable for implementing air interface for future 5G technology. In cellular environment, a comparison of those modulation schemes is carried out. The result shows that these schemes are allowing increased values of spectral efficiency than conventional techniques. Being there is no definite winner, the preferable model of modulation technique depends on some cases in terms of channel delay spread, channel Doppler spread and other parameters. Here, the air interface virtualization and cloud radio access network implementation provide the way towards adoption of tuneable and adaptive modulation techniques, where the parameters of waveform are chosen based on particular scenario.

The modulation techniques suitable for 5G can be extracted from Table 2.

5 Appendix

See Table 2.

Table 2 The summary of papers reviewed

Sl. No.	References	Title	Aim	Technology	Publisher
1	Basar et al. [1]	Index modulation techniques for next-generation wireless networks	This paper describes the concept of IM and covers spatial modulation (SM), OFDM-IM schemes and CM in detail. This paper is one of the most comprehensive in the field of IM	Index modulation	IEEE
2	Correia et al. [55]	Ultrafast backscatter modulator with low power consumption and wireless power transmission capabilities	This paper discusses backscatter modulator. Here, the tag reflects the radio signal which is transmitted from reader and by controlling the coefficient of reflection, this reflection is modulated	Backscatter modulation	IEEE
3	Neelamegam et al. [9]	Performance analysis of cooperative wireless sensor network with index-based modulation	This paper compares the performance of cooperative WSN associated with CoSM and CoESM	Index-based modulation	IEEE
4	Sun et al. [12]	An improved phase modulation technique based on four-dimensional arrays	This paper describes an improved technique of phase modulation based on 4D arrays of antenna	Phase modulation	IEEE
5	Ozturk et al. [13]	Generalized frequency division multiplexing with flexible index modulation	The integration of GFDM in space and the frequency IM is used to provide a flexible and an advanced RAT for WSNs	Frequency division multiplexing, index modulation	IEEE

(continued)

Table 2 (continued)

Sl. No.	References	Title	Aim	Technology	Publisher
6	Kim et al. [15]	A new filter-bank multicarrier system: the linearly processed FBMC system	This paper proposes a FBMC system without interference, named as LP-FBMC (linearly processed FBMC) system	Filter-bank multicarrier	IEEE
7	Baig et al. [16]	A low PAPR DHT precoding-based UFMC scheme for 5G communication systems	In this paper, a low PAPR DHT-P-UFMC (Discrete-Hartley Transform Precoding-based UFMC) and analysis through Monte-Carlo simulations	UFMC	IEEE
8	Roy et al. [17]	Interpolated band-pass method-based narrow-band FIR Filter: a prospective candidate in filtered-OFDM technique for the 5G cellular network	In this paper, a new innovative technique called filtered-OFDM (F-OFDM) based on narrow-band FIR filter operated in IBM (interpolated band-pass method) is proposed	Filtered-OFDM	IEEE
9	Zeng et al. [20]	Investigation on evolving single-carrier NOMA into multi-carrier NOMA in 5G	This paper concentrates on the relation, from basic principles to enabling schemes, between the SC-NOMA and MC-NOMA	Multi-carrier NOMA	IEEE
10	Liu et al. [23]	Cooperative NOMA broadcasting/multicasting for low-latency and high-reliability 5G cellular V2X communications	This paper proposes 2 relay assisted NOMA transmission systems for 5G-V2X i.e. vehicle to everything communication. It means the HDR4-NOMA (Half Duplex Relay Assisted NOMA) and FDR-NOMA (Full Duplex Relay Assisted NOMA) are used to investigate the optimum allocated power problems	NOMA	IEEE

(continued)

Table 2 (continued)

Sl. No.	References	Title	Aim	Technology	Publisher
11	Baig et al. [24]	On the PAPR reduction: a novel filtering based Hadamard transform pre-coded uplink MC-NOMA scheme for 5G cellular networks	This paper presents a Hadamard Transform (HT) pre-coded uplink MC-NOMA method based on F.I.R. filter to decrease the PAPR	MC-NOMA	IEEE
12	Al Rabee et al. [25]	The optimum received power levels of uplink non-orthogonal multiple access (NOMA) signals	In this paper, a power-domain technique is proposed and analytically determines the optimum received uplink level of power by using SIC detector for whatever number of the transmitters	NOMA	IEEE
13	Hiari et al. [33]	A reconfigurable SDR transmitter platform architecture for space modulation MIMO techniques	This paper proposes a novel single SDR platform architecture that implements different space modulation techniques	Space modulation, MIMO	IEEE
14	Rajabi et al. [37]	Modulation techniques for simultaneous wireless information and power transfer with an integrated rectifier–receiver	This paper introduces the ratio of phase modulation and merging with the ratio of amplitude phase modulation	Amplitude-Phase Modulation,	IEEE

(continued)

Table 2 (continued)

Sl. No.	References	Title	Aim	Technology	Publisher
15	Song et al. [38]	Six-port direct modulator with carrier suppression technology for high-speed high-frequency wireless communications	This paper experimentally shows a 6-port direct modulator having suppression of carrier leakage for downlink samples in future wireless communication systems with high speed and high carrier frequency	Six-Port Modulator	IEEE
16	Kumaravelu et al. [48]	Modified spatial modulation: an alternate to spatial multiplexing for 5G-based compact wireless devices	This paper proposes an improved spatial modulation called MSM for 8 × 8 configuration of MIMO, utilizing one or two transmitter antennas grounded on the incoming pattern of bits	Spatial Modulation	Springer
17	Mesleh et al. [6]	Generalized space modulation techniques: hardware design and considerations	This paper proposes and analyses generalized QSSK (GQSSK) and generalized QSM (GQSM)	Space Modulation	Elsevier
18	Bashar et al. [63]	NOMA/OMA mode selection-based cell-free massive MIMO	This paper investigates NOMA (Non-Orthogonal Multiple Access) based cell-free massive MIMO, in which the users are grouped in the form of multiple clusters	NOMA, OMA, MIMO	IEEE
19	Love et al. [64]	Non-orthogonal multiple access in multi-cell networks: theory, performance, and practical challenges	In this paper, NOMA in a multicell environment is studies and explains its possibilities and difficulties	Multi-Cell NOMA	IEEE

(continued)

Table 2 (continued)

Sl. No.	References	Title	Aim	Technology	Publisher
20	Marcano et al. [65]	Performance of non-orthogonal multiple access (NOMA) in mmWave wireless communications for 5G networks	The analysis of NOMA performance in mmWave cells is described in this paper, using a benchmark of OMA	mmWave, NOMA	IEEE
21	Xiao et al. [66]	A novel opportunistic NOMA scheme for 5G massive MIMO multicast communications	In this paper, a multicast service of 5G communication is described and a massive MIMO-NOMA system is developed	MIMO, NOMA	IEEE
22	Moltafet et al. [67]	A new multiple access technique for 5G: power domain sparse code multiple access (PSMA)	In this paper, an approach of multi access 5G communication network named PSMA (Power domain sparse code multiple access) is proposed	Power domain sparse code multiple access (PSMA)	IEEE

References

1. Basar E, Wen M, Mesleh R, Di Renzo M, Xiao Y, Haas H (2017) Index modulation techniques for next-generation wireless networks. IEEE Access 5:16693–16746
2. Basar E (2016) Index modulation techniques for 5G wireless networks. IEEE Commun Mag 54(7):168–175
3. Mesleh RY, Haas H, Sinanovic S, Ahn CW, Yun S (2008) Spatial modulation. IEEE Trans Veh Technol 57(4):2228–2241
4. Basar E, Aygölü Ü, Panayırcı E, Poor HV (2013) Orthogonal frequency division multiplexing with index modulation. IEEE Trans Sig Process 61(22):5536–5549
5. Thomas SJ, Wheeler E, Teizer J, Reynolds MS (2012) Quadrature amplitude modulated backscatter in passive and semi-passive UHF RFID systems. IEEE Trans Microw Theory Techn 60(4):1175–1182
6. Mesleh R, Hiari O, Younis A, Alouneh S (2017) Generalized space modulation techniques: hardware design and considerations. Elsevier
7. Thomas SJ, Reynolds MS (2012) A 96 Mbit/sec, 15.5 pJ/bit16-QAM modulator for UHF backscatter communication. In: Proceedings of IEEE international conference RFID (RFID), Apr 2012, pp 185–190
8. Shirane A et al (2015) A 5.8 GHz RF-powered transceiver with a 113μW32-QAM transmitter employing the IF-based quadrature backscattering technique. In: IEEE international solid-state circuits conference (ISSCC) Digital Technical Papers, pp 1–3
9. Neelamegam et al (2019) Performance analysis of cooperative wireless sensor network with index - based modulation. IET Digital Libr 2019(5):3438–3441
10. Cheng Q, Zhu J (2017) Generalised transmit-receive joint spatial modulation. Electron Lett 53(24):1613–1615
11. Murthy GR, Sankhe K (2016) Spatial modulation-spatial multiplexing in massive MIMO. arXiv: 1605.02969v1 [cs.IT]
12. Sun et al (2017) An improved phase modulation technique based on fourdimensional arrays, IEEE antenna and wave propagation letters, vol 16
13. Ozturk et al (2017) Generalized frequency division multiplexing with flexible index modulation. IEEE Access 5:24727–24746
14. Selvaraj K, Judson D, Ganeshkumar P, Anandaraj M (2020) Low complexity linear detection for multiuser MIMO SC-FDMA uplink transmission system. Springer-Wireless Personal Communication
15. Kim et al (2018) A new filter-bank multicarrier system: the linearly processed FBMC system. IEEE Trans Wireless Commun 17(7):4888–4898
16. Baig et al (2019) A low PAPR DHT precoding-based UFMC scheme for 5G communication systems. IEEE
17. Roy et al (2019) Interpolated band-pass method based narrow-band FIR Filter: a prospective candidate in filtered - OFDM technique for the 5G cellular network. IEEE
18. Judson D, Bhaskar V, Selvaraj K (2018) Pre-equalization schemes for MIMO CC-CDMA systems over frequency-selective fading channels. Springer-Wirel Personal Commun 98(1):1587–1603
19. Judson D, Bhaskar V, Arun S (2019) Space time regularized zero forcing in downlink code division multiple access systems with complementary codes. Springer-Wireless Personal Communication
20. Zeng et al (2018) Investigation on evolving singlecarrier NOMA into multi-carrier NOMA in 5G. IEEE Access
21. Dai L, Wang B, Yuan Y, Han S, Li C, Wang Z (2015) Non-orthogonal multiple access for 5G: solutions, challenges, opportunities, and future research trends. IEEE Commun Mag 53(9):74–81
22. Yuan Y et al (2016) Non-orthogonal transmission technology in LTE evolution. IEEE Commun Mag 54(7):68–74

23. Liu et al (2019) Cooperative NOMA broadcasting/multicasting for lowlatency and high-reliability 5G cellular V2X communications. IEEE
24. Baig et al (2018) On the PAPR reduction: a novel filtering based Hadamard transform pre-coded uplink MCNOMA scheme for 5G cellular networks. IEEE
25. Al Rabee et al (2017) The optimum received power levels of uplink non-orthogonal multiple access (NOMA) signals, IEEE
26. El-Sayed MM, Ibrahim AS, Khairy MM (2016) Power allocation strategies for non-orthogonal multiple access. In: 2016 international conference on selected topics in mobile & wireless networking (MoWNeT), pp 1–6
27. Saito Y, Benjebbour A, Kishiyama Y, Nakamura T (2013) System-level performance evaluation of downlink non-orthogonal multiple access (NOMA). In: Proceedings of IEEE personal indoor mobile radio communication (PIMRC), London, pp 611–615
28. Takeda T, Higuchi K (2011) Enhanced user fairness using non-orthogonal access with SIC in cellular uplink. In: Proceedings of IEEE Vehicular technology conference (VTC Fall), San Francisco, pp 1–5
29. Li Q, Niu H, Papathanassiou A, Wu G (2014) 5G network capacity: key elements and technologies. IEEE Veh Technol Mag 9(1):71–78
30. Judson D, Bhaskar V (2017) Error rate analysis of SIMO-CDMA with complementary codes under multipath fading channels. Springer-Wirel Personal Commun. https://doi.org/doi.org/10.1007/s11277-017-4938-0
31. Judson D, Bhaskar V (2018) Interference cancellation in complementary coded CDMA in Rician fading channels. Springer-Wireless Personal Communication. https://doi.org/10.1007/s11277-018-5732-3
32. Ascar Davix X, Judson D (2019) Successive interference cancellation in asynchronous CC-CDMA systems under Rician fading channels. Springer-Telecommunication systems
33. Hiari et al (2017) A reconfigurable SDR transmitter platform architecture for space modulation MIMO techniques, IEEE
34. Di Renzo M, Haas H, Grant PM (2011) Spatial modulation for multiple-antenna wireless systems: a survey. IEEE Commun Mag 49(12):182–191
35. Mesleh R, Ikki SS, Tumar I, Alouneh S (2017) Decode-and-forward with quadrature spatial modulation in the presence of imperfect channel estimation. Phys Commun Online" https://doi.org/10.1016/j.phycom.2017.06.005", Jun. 2017. [Online]. Available: http://www.sciencedirect.com/science/article/pii/S1874490717300071
36. Mesleh R, Hiari O, Younis A, Alouneh S (2017) Transmitter design and hardware considerations for different space modulation techniques. IEEE Trans Wirel Commun, pp 1–11 (to appear)
37. Rajabi et al (2018) Modulation techniques for simultaneous wireless information and power transfer with an integrated rectifier–receiver. IEEE
38. Song et al (2017) Six-port direct modulator with carrier suppression technology for high-speed high-frequency wireless communications. IEEE
39. Komine T, Nakagawa M (2004) Fundamental analysis for visible-light communication system using LED lights. IEEE Trans Consum Electron 50(1):100–107
40. Ghannouchi FM, Mohammadi A (2009) The six-port technique with microwave and wireless applications. Artech House, Norwood, pp 161–198
41. Koonen T, Oh J, Mekonnen K, Cao Z, Tangdiongga E (2016) Ultra-high capacity indoor optical wireless communication using 2D-steered pencil beams. J Lightw Technol 34(20): 4802–4809
42. Zhao Y, Viereck C, Frigon JF, Bosisio RG, Wu K (2005) Directquadrature phase shift keying modulator using six-port technology. Electron Lett 41(21):1180–1181
43. Luo B, Chia MYW (2008) Performance analysis of serial and parallel six-port modulators. IEEE Trans Microw Theory Techn 56(9):2062–2068
44. Malyshev SA, Galwas BA, Piotrowski J, Chizh AL, Szczepaniak ZR (2002) Photovaractor for remote optical control of microwave circuits. IEEE Microw Wirel Compon Lett 12(6):201–203
45. Östh J, Owais O, Karlsson M, Serban A, Gong S, Karlsson P (2011) Direct carrier six-port modulator using a technique to suppress carrier leakage. IEEE Trans Microw Theory Techn 59(3):741–747

46. Arshad NSA, Cheor WL, Ibrahim SZ, Razalli MS (2014) QPSK modulation using multi-port device. In: Proceedings of IEEE symposium wireless technology applied (ISWTA), Kota Kinabalu, Malaysia, Sep 2014, pp 58–63
47. Song X et al (2016) Integrating baseband-over-fibre and six-port direct modulation for high-speed high-frequency wireless communications. In: IEEE MTT-S international microwave symposium digital. San Francisco, pp 1–4
48. Kumaravelu et al (2018) Modified spatial modulation: an alternate to spatial multiplexing for 5G-based compact wireless devices. Springer
49. Mesleh R, Haas H, Sinanovi´c S, Ahn CW, Yun S (2008) Spatial modulation. IEEE Trans Veh Tech 57(4): 2228–2241
50. Di Renzo M, Haas H, Grant PM (2011) Spatial modulation for multiple-antenna wireless systems: a survey. IEEE Commun Mag 49(11):182–191
51. Basnayaka DA, Renzo MD, Haas H (2014) Massive but few active MIMO. IEEE Trans Commun (submitted)
52. Mesleh R, Ikki SS, Aggoune HM (2015) Quadrature spatial modulation. IEEE Trans Veh Technol 64(6):2738–2742
53. Badarneh OS, Mesleh R (2016) A comprehensive framework for quadrature spatial modulation in generalized fading scenarios. IEEE Trans Commun 64(7):2961–2970
54. Jeganathan J, Ghrayeb A, Szczecinski L, Ceron A (2009) Space shift keying modulation for MIMO channels. IEEE Trans Wirel Commun 8(7):3692–3703
55. Correia R, Boaventura A, Carvalho NB (2017) Quadrature amplitude backscatter modulator for passive wireless sensors in IoT applications. IEEE Trans Microw Theory Techn 65(4):1103–1110
56. Jeganathan J, Ghrayeb A, Szczecinski L (2008) Generalized space shift keying modulation for MIMO channels. In: Proceedings of IEEE 19th international symposium on personal, indoor and mobile radio communication. PIMRC 2008, Cannes, France, pp 1–5, 15–18 Sep. 2008
57. Younis A, Basnayaka DA, Haas H (2014) Performance analysis for generalised spatial modulation. In: Proceedings of European wireless conference (EW 2014), Barcelona, Spain, 14–16 May 2014, pp 207–212
58. Younis A, Di Renzo M, Mesleh R, Haas H (2011) Sphere decoding for spatial modulation. In: Proceedings of IEEE international conference on communication (ICC), Kyoto, Japan, 5–9 Jun 2011, pp 1–6
59. Younis A, Sinanovic S, Di Renzo M, Mesleh R, Haas H (2013) Generalised sphere decoding for spatial modulation. IEEE Trans Commun 61(7):2805–2815
60. Al-Nahhal I, Dobre O, Ikki S (2017) Quadrature spatial modulation decoding complexity: study and reduction. IEEE Wirel Commun Lett
61. Di Renzo M, Haas H, Ghrayeb A, Sugiura S, Hanzo L (2014) Spatial modulation for generalized MIMO: challenges, opportunities, and implementation. Proc IEEE 102(1):56–103
62. Basar E, Wen M, Mesleh R, Renzo MD, Xiao Y, Haas H (2017) Index modulation techniques for next-generation wireless networks. IEEE Access 5(99): 1–52 (invited)
63. Bashar et al (2019) NOMA/OMA mode selection-based cell-free massive MIMO. IEEE
64. Love et al (2017) Non-orthogonal multiple access in multi-cell networks: theory, performance, and practical challenges. IEEE
65. Marcano et al (2017) Performance of non-orthogonal multiple access (NOMA) in mmWave wireless communications for 5G networks. IEEE
66. Xiao et al (2017) A novel opportunistic NOMA scheme for 5G massive MIMO multicast communications. IEEE
67. Moltafet et al (2017) A new multiple access technique for 5G: power domain sparse code multiple access (PSMA). IEEE
68. Judson D, Albert Raj A (2016) Performance of multicarrier complementary coded CDMA under frequency-selective Nakagami-m fading channels. Springer-Eurasip J Wirel Commun Network 2016(67):1–9

A Tapered Slot Rectangular Ultra-wideband Microstrip Patch Antenna for Radio Frequency Energy Harvesting

P. S. Chindhi, H. P. Rajani, and G. B. Kalkhambkar

Abstract A tapered slot rectangular microstrip ultra-wideband (UWB) patch antenna is proposed in this chapter for radio frequency (RF) energy harvesting. The RF energy harvesting technologies are intensifying steadily because of constraints in energy storage and wired power supply. Hence, harvesting energy from RF sources is a promising technique for fulfilling the inimitable power prerequisites for powering smart sensors. Here, to achieve ultra-wideband, a rectangular structure is modified in a tapered slot. The antenna structure is optimized and simulated by using electromagnetic high-frequency structural simulator (HFSS) software, and the measured bandwidth defined by return loss <-10 dB is from 3.76 to 19.52 GHz with <2 voltage wave standing ratio (VSWR) throughout the band.

Keywords Radio frequency · Return loss · Wideband · Tapered slot · Notch

1 Introduction

As the demand for power is growing, the need for alternative energy sources has become essential; at the same time, the RF power density is increasing since there is an increase in frequency bands such as WLAN, GSM, DCS, 3G, 4G LTE, and 5G. Self-powered electronic devices have gained momentous attention in a wide range of applications such as permanently inserted sensors inside the human body,

P. S. Chindhi (✉)
Department of Electrical Engineering, Sant Gajanan Maharaj College of Engineering, Mahagaon, India

P. S. Chindhi · G. B. Kalkhambkar
Shivaji University Kolhapur, Kolhapur, India

H. P. Rajani
Department of Electronics and Communication Engineering, KLE's Dr. MSSCT, Visvesvaraya Technological University, Belagavi, India

G. B. Kalkhambkar
Department of Electronics and Telecommunication Engineering, Sant Gajanan Maharaj College of Engineering Mahagaon, Mahagaon, India

A. Sivasubramanian et al. (eds.), *Futuristic Communication and Network Technologies*, Lecture Notes in Electrical Engineering 792,
https://doi.org/10.1007/978-981-16-4625-6_36

Table1 Incremental development of RF energy sources in India

Years	Number of mobile towers in lakh	Number of mobile subscribers in million	Number of base transceiver stations (BTS) (^as on 12.07.2019)
2017	412,387	1170.59	1384737
2018	–	1188.99	2035488
2019	560162^	1161.71	2180224^

building walls, and textile wearable antenna. The RF energy harvesting technology is expected to rise with the Internet of Things (IoT). According to Gartner, in a few years, there will be 25 billion perpetually connected things and some year's later, 10 trillion connected sensors [1–3]. Devices that work within the limits of harvesting capability are Bluetooth transceiver, wireless sensor/hearing aid, remote sensors, RFID tags, watches/calculators, and processor clock [4]. According to the Economics Research Unit, Department of Telecommunications Ministry of Communications, Government of India, New Delhi, 2019, Table 1 indicates the growth in the RF energy surrounding us [5]. A triple-band differential antenna and reflector plate with a spacing of 28 mm are presented; the overall size of the rectenna is higher [6]. In Ref. [7], for higher bandwidth, CPW-feed fractal antenna is simulated and tested' to achieve higher bandwidth, slotted equilateral triangles are introduced in the CPW-feed circular patch. In Ref. [8], the E-shaped patch antenna with two parallel slots and the partial ground is incorporated to enhance the bandwidth. At 936 MHz, the current distribution density around the slot1, slot2, and at the edge of the slot is uniform. Circularly polarized shorted square-ring slot antenna for the GSM 900 band is presented in Ref. [9]. The circular patch antenna and parametric study on various parameters are taken into consideration to enhance the bandwidth of the antenna [10]. A multilayer stacked antenna loaded with an octagon radiator with tapered slits and an unequal slotted circular patch improves the bandwidth and polarization characteristics [11]. The various methods to enhance bandwidth are compressively reviewed in [12]. Different impedance matching techniques that are important to build an efficient energy harvesting circuit are studied in Ref. [13]. The hexagonal fractal antenna helps in improving bandwidth [14].

In the planned work, UWB rectangular microstrip patch antenna with a tapered slot is simulated in HFSS software, which is useful for the RF energy harvesting. The impact of stepwise change in the radius on the bandwidth and gain of an antenna is observed. In the first iteration, with an increase in the radius of the circular slot, the bandwidth is degraded; whereas in the second iteration where we decreased the radius of the circular slot, it is observed that the bandwidth is improved, but the gain is not satisfactory. To compensate for the loss in the gain, a notch is inserted at the periphery of the circular slot. The parametric study of the notch dimension is performed to arrive at good impedance matching. A good combination of circular slot and the notch gave the enhanced bandwidth and positive gain which make this antenna a good candidate for RF energy harvesting. The comparison of Ref. [6–11] in terms of overall dimensions and bandwidth with the proposed geometry is given in Table 7. The detail conceptual block diagram of a RF power harvesting system is presented in [15].

The chapter is organized as follows: Sects. 2 and 3 describe the antenna design and its parametric study and observations, respectively. Section 4 depicts results and discussion along with a comparison of other related works. Section 5 describes the conclusion, and in Sect. 6, future scope is specified.

2 Antenna Design

The proposed UWB rectangular microstrip patch antenna with a tapered slot is shown in Fig. 1. A low-cost double-sided printed circuit board (PCB) of flame retardant-4 (FR4) substrate with a thickness of 1.5 mm, dielectric constant 4.4, and loss tangent (tanδ) 0.02 is selected. The patch length and width are 110 mm and 80 mm, respectively. The detailed design parameters of the proposed antenna are given in Table 2

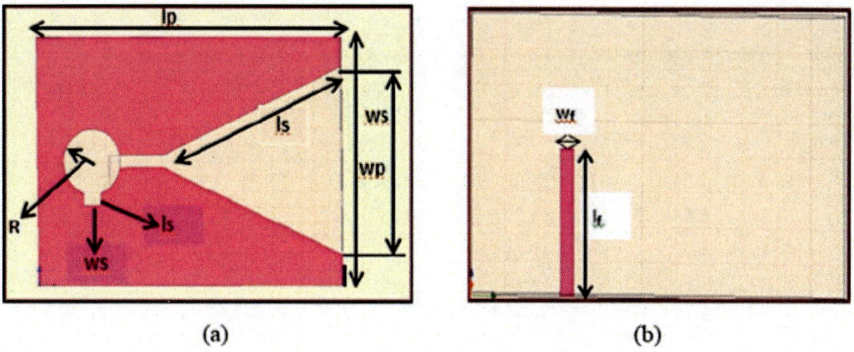

Fig. 1 Proposed UWB microstrip antenna. **a** Top view. **b** Bottom view

Table 2 The specification of the design parameters

S. No.	Parameters	Value
1	Design frequency (f_{r})	5 GHz
2	Dielectric constant (ε_r)	4.4
3	Height of the substrate (h)	1.5 mm
4	Loss tangent (tanδ)	0.02
5	Width of the patch (w_{p})	80 mm
6	Length of the patch (l_{p})	110 mm
7	Width of the ground (w_{p})	4 mm
8	Length of the ground (L_{p})	42 mm

3 Parametric Study and Observations

1. **Setp#1 Increase in radius of circular slot without notch.**

In the initial iteration, a circular slot with a radius $R_1 = 12$ mm is made; the bandwidth of 4.52 GHz is obtained. In the subsequent iterations, radius of the circular slot is increased with 2 mm. At $R_5 = 12.8$ mm, the bandwidth is reduced from 4.52 to 2.95 GHz. The various antenna parameters with an increasing radius of the circular slot are tabulated in Table 3. The return loss versus frequency plot is shown in Figs. 2 and 3.

2. **Step#2 Decrease in radius of circular slot without notch**

In the next iterations, the radius of the circular slot is decreased from $R_1 = 12$ mm to $R_6 = 11.8$ mm; the bandwidth is increased from 4.52 to 4.62 GHZ. In the next iteration, the radius of the circular slot is decreased at the steps of 2 mm. By decreasing the radius of the circular slot, there is a steady increase in the bandwidth from 4.52 to 4.93 GHz. The various antenna parameters with a decrease in the radius of the circular slot are tabulated in Table 4. The return loss versus frequency plot is shown in Fig. 4.

Table 3 Antenna performance comparison at R_1, R_2, R_3, R_4, and R_5 without notch

Variable parameter (mm)	Resonant frequency (GHz)	Return loss (dB)	Peak directivity	Peak gain (dBi)	Bandwidth (GHz)
$R_1 = 12.0$	4.54	−27.83	0.89	0.76	4.52
$R_2 = 12.2$	4.54	−34.77	0.93	0.97	4.53
$R_3 = 12.4$	4.72	−31.61	0.91	0.78	4.55
$R_4 = 12.6$	5.00	−32.80	0.88	0.75	4.55
$\mathbf{R_5 = 12.8}$	5.18	−41.35	0.82	0.70	2.95

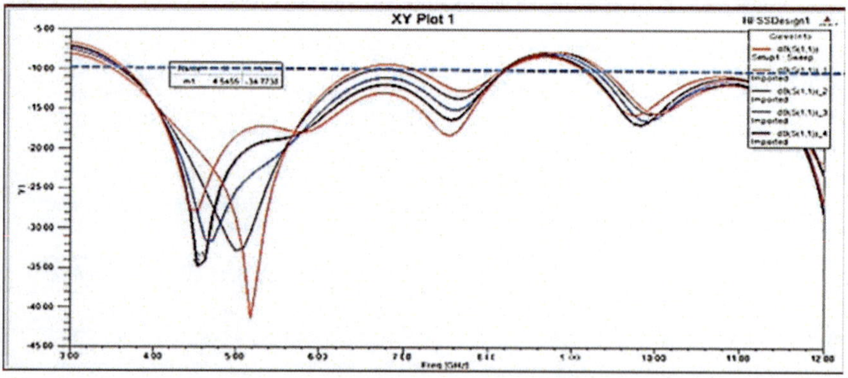

Fig. 2 Graph of the return loss versus frequency without notch at R_1, R_2, R_3, R_4, and R_5

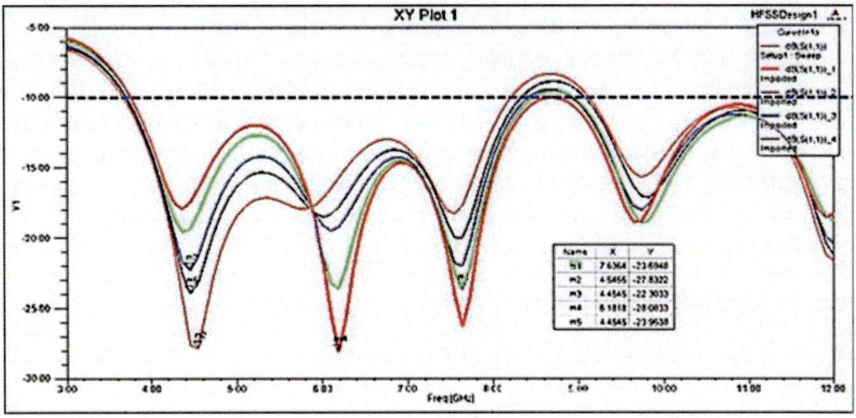

Fig. 3 Graph of the return loss versus frequency without notch at R_1, R_{10}, R_{11}, R_{12}, R_{13} and R_{14}

Table 4 Antenna performance comparison at R_1, R_6, R_7, R_8, and R_9 without notch

Variable parameter (mm)	Resonant frequency (GHz)	Return loss (dB)	Peak directivity	Peak gain (dBi)	Bandwidth (GHz)
$R_1 = 12.0$	4.54	−27.83	0.89	0.76	4.52
$R_6 = 11.8$	4.45	−23.95	0.95	0.80	4.62
$R_7 = 11.6$	4.45	−22.30	0.98	0.83	4.70
$R_8 = 11.4$	7.63	−23.69	0.94	0.78	4.73
$\mathbf{R_9 = 11.2}$	6.18	−28.08	0.96	0.80	4.93

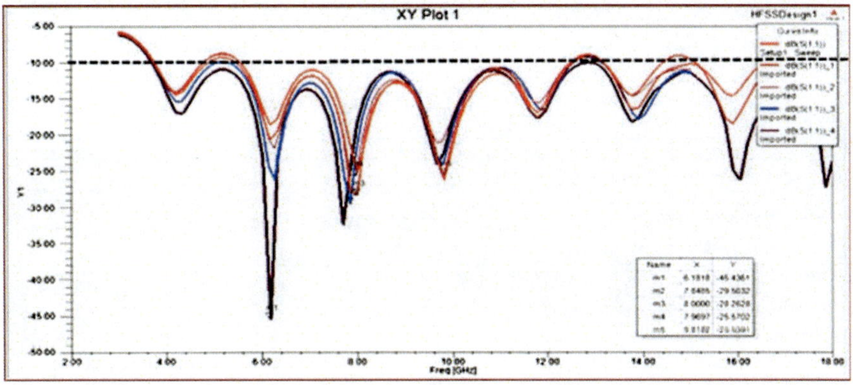

Fig. 4 Graph of the return loss versus frequency without notch at R_{10}, R_{11}, R_{12}, R_{13}, and R_{14}

3. **Step#3 Decrease in radius of circular slot without notch**
   As in step #2, there is a gradual increase in the bandwidth, so to get higher bandwidth in step #3, the radius of the circular slot is decreased by 10 that is 12–10 mm and in subsequent iterations, it decreased at a step of 2 mm. At R_{10} = 10 mm, the band width is 6.90 GHz, and at R_{14} = 10.8 mm, the bandwidth is 9.07 GHz. In Table 5, the bandwidth and other antenna parameters are tabulated with a decrease in the radius of the circular slot; Fig. 5 shows the return loss versus frequency plot for R_{10} to R_{14}.

4. **Step#4 Decrease in radius of circular slot with notch**

In step#4, the radius of the circular slot is maintained constant as in step#3, and a notch is inserted at the lower side of the circular slot as shown in Fig. 1. When the radius of the circular slot is 10 mm (R_{15}) with a notch, the bandwidth is 13.96 GHz, but at R16, R18 and R19, there is a decrease in bandwidth. At R = 17 and radius R = 10.4 with a notch, the bandwidth is 15.74 GHz. The parametric variation in the antenna results at R = 15 to R=19 is tabulated in Table 6. Fig. 6 shows the return loss versus frequency plot for R_{15} to R_{19}. In Fig. 6, green color indicates the higher

Table 5 Antenna performance comparison at R_{10}, R_{11}, R_{12}, R_{13}, and R_{14} without notch

Variable parameter (mm)	Resonant frequency (GHz)	Return loss (dB)	Peak directivity	Peak gain (dBi)	Bandwidth (GHz)
R_{10} = 10.0	9.81	−25.53	1.03	0.85	6.90
R_{11} = 10.2	8.00	−28.26	1.03	0.84	8.75
R_{12} = 10.4	7.96	−25.57	0.97	0.79	8.73
R_{13} = 10.6	7.84	−29.50	1.00	0.82	8.77
R_{14} = 10.8	6.18	−45.43	1.03	0.85	9.07

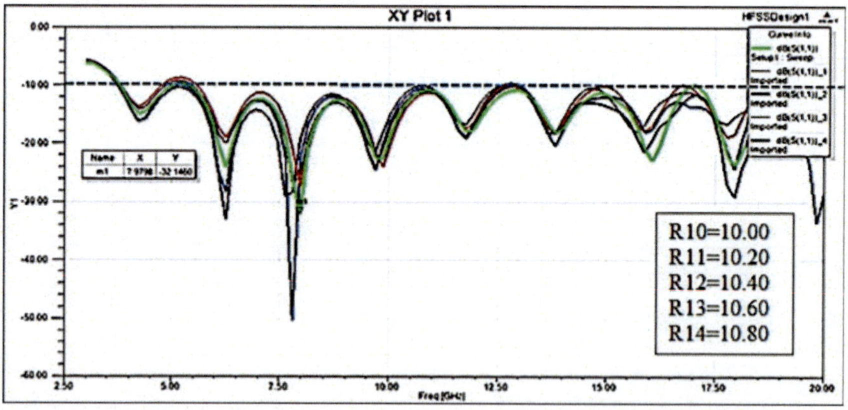

Fig. 5 Graph of the return loss versus frequency with notch at R_{15}, R_{16}, R_{17}, R_{18}, and R_{19}

Table 6 Antenna performance comparison at R_{15}, R_{16}, R_{17}, R_{18}, and R_{19} with notch

Variable parameter (mm)	Resonant frequency (GHz)	Return loss (dB)	Peak directivity	Peak gain (dBi)	Bandwidth (GHz)
$R_{15} = 10.0$	7.97	−27.34	1.01	0.82	13.96
$R_{16} = 10.2$	7.80	−26.84	1.00	0.82	7.03
$R_{17} = 10.4$	*7.97*	*−32.16*	*0.99*	*0.82*	*15.74*
$R_{18} = 10.6$	7.80	−50.24	1.01	0.83	7.29
$R_{19} = 10.8$	6.26	−33.14	0.96	0.78	7.03

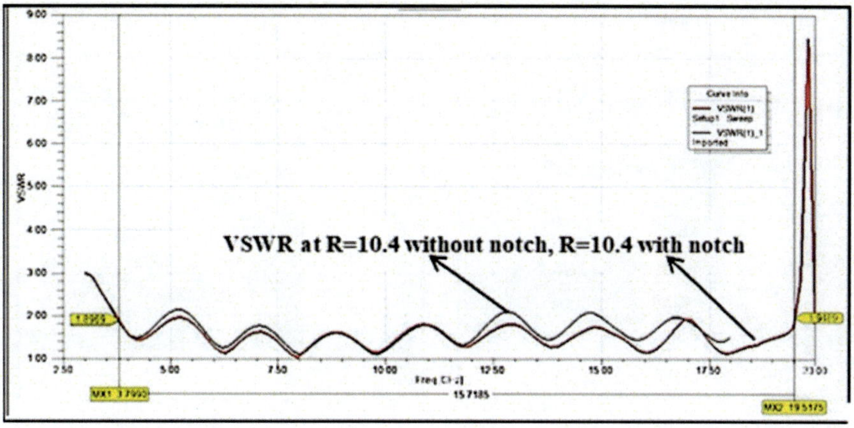

Fig. 6 Simulated VSWR of the proposed antenna with and without notch at $R = 10.4$

bandwidth; the range of bandwidth is 3.76–19.52 GHz. Fig. 7 shows the simulated VSWR of the proposed antenna with and without a notch at $R = 10.4$. The VSWR is <2 throughout the band. The VSWR at $R = 10.4$ mm without notch is slightly high, and at $R = 10.4$ with notch, the VSWR is within the standard value (<2).

4 Results and Discussion

In Fig. 7, the current distribution at 5 GHz is shown for the radius $R = 12$ mm (Fig. 7c), for $R = 10$ mm (Fig. 7d), for $R = 10.2$ mm with notch (Fig. 7e), and for $R = 10.4$ mm with notch (Fig. 7f). In the final iteration, the bandwidth is increased ranging from 3.76 to 19.52 GHz due to improved current distribution.

The proposed antenna performance is compared with the recent literature; it is observed that the present work shows the ultra-wideband performance which is suitable for the RF energy harvesting in the band ranging from 3.76 to 19.52 GHz.

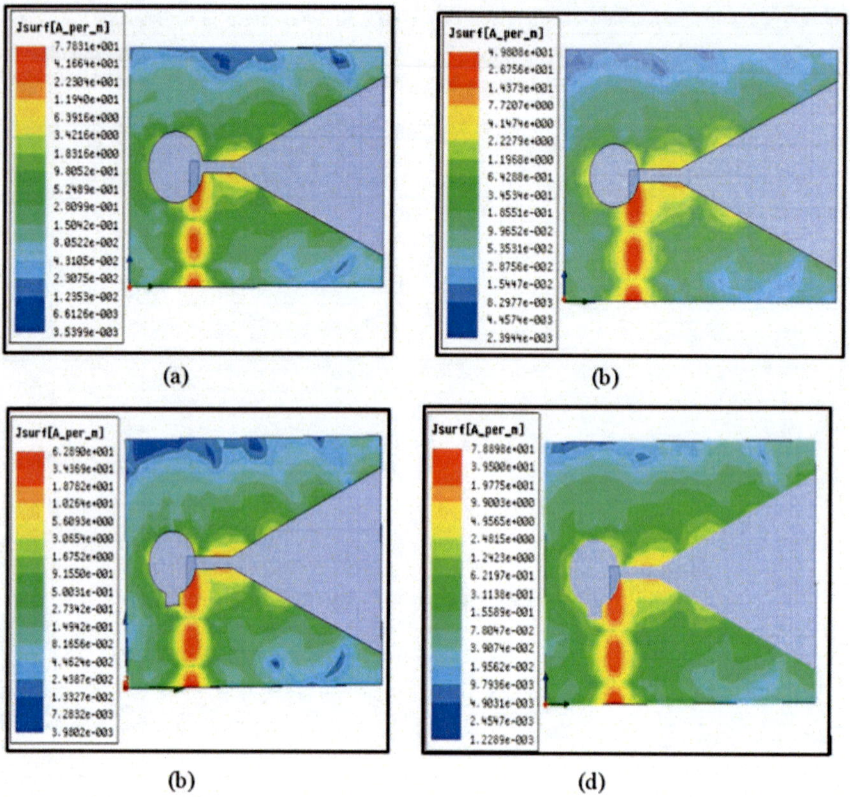

Fig. 7 Current distribution. **a** At $R = 12.00$ mm without notch. **b** At $R = 10.00$ mm without notch. **c** At $R = 10.2$ mm with notch. **d** At $R = 10.4$ mm with notch

Compared to the literature mentioned in Table 7 which works in the specific bands only, the proposed antenna is a better candidate in terms of bandwidth as well as size.

Table 7 Comparison of other related works

Sr. No	References	Frequency (GHz)	Size (mm^2)
1	Devi et al. [8]	GSM 900 band	106 × 89
2	Ghosh [9]	GSM 900 band	140 × 140
3	Agrawal et al. [10]	0.89–5.53 GHz	120 × 100
4	Chandravanshi et al. [6]	2.1 GHz, 2.4–2.48 GHz, 3.3–3.8 GHz	120 × 120
5	Bai et al. [7]	0.88–8.45 GHz	100 × 100
6	Jie et al. [11]	0.908–0.922 GHz, 2.35–2.50 GHz	120 × 120
7	Proposed work	3.76–19.52 GHz	110 × 80

5 Governing Equations for RF Energy Harvesting

Figure 8 shows the conceptual representation of RF energy harvesting, which consists of a transmission antenna and receiving antenna with a power management circuit. The received power is calculated by the Friis transmission equation (FTE),

$$P_{\mathrm{R}} = \frac{P_{\mathrm{T}} G_{\mathrm{T}} G_{\mathrm{R}} \lambda^2}{(4\pi R)^2} \tag{1}$$

where

P_{R} is power at the receiver antenna.

G_{R} is receiver antenna gain relative to the isotropic source (dBi).

λ is the wavelength of the electromagnetic signal, which is equal to the speed of light in vacuum divided by the signal frequency = c/f.

The loss of power in space can be characterized by free-space path loss (FSPL), which is the loss of signal power during propagation in free space. Calculating FSPL requires information about the antenna gain, frequency of transmitting wave, and distance between the transmitter and receiver. The FSPL for far field is,

$$P_{\mathrm{L}} = \frac{P_{\mathrm{T}}}{P_{\mathrm{R}}} = \frac{(4\pi R)^2}{G_{\mathrm{T}} G_{\mathrm{R}} \lambda^2} = \frac{(4\pi f R)^2}{G_{\mathrm{T}} G_{\mathrm{R}} c^2} = \frac{4}{G_{\mathrm{T}} G_{\mathrm{R}}} (kR)^2 \tag{2}$$

$$P_{\mathrm{L}}(\mathrm{dB}) = 20 \log_{10}(f) + 20 \log_{10}(R) + 20 \log_{10}\left(\frac{4\pi}{c}\right) - G_{\mathrm{T}} - G_{\mathrm{R}} \tag{3}$$

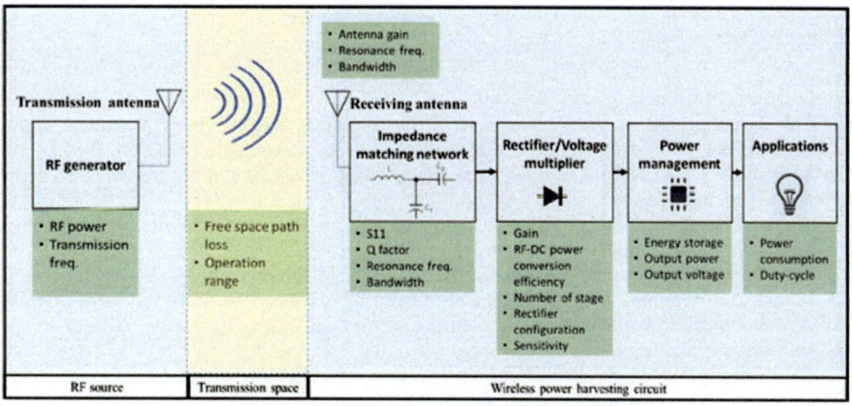

Fig. 8 Conceptual representation of RF energy harvesting system, Ref. [15]

In case f is measured in MHz, distance R is measured in km, and gain G_T and G_R are measured in dBi, the above function becomes

$$P_L(\text{dB}) = 20\log_{10}(f) + 20\log_{10}(R) + 32.44 - G_T - G_R \quad (4)$$

6 Conclusion

The bandwidth of the proposed tapered slot UWB rectangular microstrip patch antenna was enhanced by the optimization of circular slot dimensions. At initial iteration $R = 10$ mm without notch, the frequency band of the antenna lies between 5.57 and 6.89 GHz. At final iteration, the frequency band of the proposed rectangular tapered slot microstrip patch antenna with notch length 4.04 mm and width 6 mm antenna bandwidth lies between 3.76 and 19.52 GHz. The VSWR is <2 throughout the band. This antenna has a wide bandwidth, so it is suitable for RF energy harvesting and WLAN applications.

7 Future Work

The work given in this chapter shows the trade-off between bandwidth and gain due to inverse relationship between bandwidth and quality factor $Q = 1/\text{B.W}$, further attempts can be extended for improving the gain by using metamaterials, slits, parasitic patch, differential patches, folded ground, vertical embedded ground plane, etc.

References

1. Gartner (2015) Predicts 2015: the internet of things
2. Lim EG et al (2014) Compact size of textile wearable antenna. In: Proceedings of the international multi configuration of engineering and computer science (IMECS), vol 2
3. Shawalil S et al (2019) 2.45 GHz wearable rectenna array design for microwave energy harvesting. Ind J Electr Eng Comput Sci 14(2)
4. Charalampidisa G, Papadakisb A, Samarakoua M (2019) Power estimation of RF energy harvesters 157:892–900. https://doi.org/10.1016/j.egypro.2018.11.255
5. Telecom Statist India (2019) Economics Research Unit Department of Telecommunications Ministry of Communications Government of India, New Delhi
6. Chandravanshi S, Akhtar MA (2018) Design of triple band differential rectenna for RF energy harvesting. IEEE Trans Antennas Propagt 66(6). https://doi.org/10.1109/TAP.2018.2819699
7. Bai X, Zhang J-w, Xu L-j, Zhao B-h (2018) A broadband CPW fractal antenna for RF energy harvesting. Aces J 33(5):1054–4887 © ACES
8. Devi KKA et al (2012) Design of a wideband 377Ω E-shaped patch antenna for RF energy harvesting. Microwave Opt Technol Lett 54(3)

9. Ghosh S (2016) Design and testing of rectifying antenna for RF energy scavenging in GSM 900 band, 36–44, received 28 Jan 2016, Accepted 30 Oct 2016, published online: 23 Nov 2016. https://doi.org/10.1080/1206212X.2016.1259801
10. Agrawal S, Parihar MS, Kondekar PN (2017) Broadband Rectenna for radio frequency energy harvesting application, pp 347–353, published online: 29 Aug 2017. https://doi.org/10.1080/03772063.2017.1356755
11. Jie AM et al (2019) A dual-band efficient circularly polarized rectenna for RF energy harvesting systems. Int J RF Microw Comput Aided Eng 29: e21665
12. Geeta K, Rajashri K, Pradeep C (2018) A survey on microstrip patch antenna parameters enhancement techniques: a progress in last decade. JASC J Appl Sci Computat 5(7). ISSN NO: 0076-5131
13. Chindhi PS, Rajani HP, Kalkhambkar GB (2018) A review on radio frequency [RF] energy harvesting systems. JASC J Appl Sci Computat 5(8). ISSN NO: 1076-5131
14. Geeta K, Khanai RN, Chindhi P, Fractals: a novel method in the miniaturization of a patch antenna with bandwidth improvement. In: Information and communication technology for intelligent systems, pp 629–63. https://doi.org/10.1007/978-981-13-1742-2_63
15. Tran L, Cha H, Park W (2017) RF power harvesting: a review on designing methodologies and applications. Micro and Nano Syst Lett 5:14. https://doi.org/10.1186/s40486-017-0051-0

A Deep Learning Approach for Automatic Classification of Cognitive Task Using the Scalp Electroencephalogram Signals

Suchetha, Madhumitha, M. Sorna Meena, Sruthi, and Radha

Abstract Cognitive abilities are the vital skills that the human brain requires to perform any type of task, from simple to complex. These skills have a huge impact on the day-to-day lives of people. The electroencephalogram (EEG) is an efficacious technique to study the brain activity while it executes these different cognitive tasks. With low-cost EEG hardware available these days, measuring accurate EEG data has become an effortless task, thereby leading to new innovations like the brain–computer interfaces. In this study, four cognitive tests: Hopkins verbal learning test, Stroop test, Symbol digit modality test, and Benton's visual retention test were considered to stimulate the cognitive abilities. After performing the frequency domain analysis on the EEG signals from a predefined dataset, we obtain the spectral topographical images as classification data, and a deep learning model with convolutional neural network was used to classify these images based on the cognitive tasks. We have obtained a classification accuracy of 84% using our proposed model.

Keywords Electroencephalography · Cognitive tasks · Fast Fourier transform · Deep learning

1 Introduction

Cognitive abilities are a set of natural skills that the brain uses to carry out any task. It has been used to measure the productivity of human brain. They are based on the thinking mechanism of the brain, information processing and learning ability rather than the actual knowledge. They are a collection of skills such as attention, memory, reason, auditory and visual processing. Incompetence in any of these abilities will give rise to difficulties in comprehending a subject, solving problems, brain

Suchetha (✉)
Centre for Healthcare Advancement, Innovation and Research, Vellore Institute of Technology, Chennai, India
e-mail: suchetha.m@vit.ac.in

Madhumitha · M. Sorna Meena · Sruthi · Radha
School of Electronics Engineering, Vellore Institute of Technology, Chennai, India

A. Sivasubramanian et al. (eds.), *Futuristic Communication and Network Technologies*, Lecture Notes in Electrical Engineering 792,
https://doi.org/10.1007/978-981-16-4625-6_37

storming, analyzing and focusing. Hence, identifying these short comings becomes very important in developing efficient methods to cope with such problems. The idea of humans interacting with computers has gained importance recently. The brain–computer interface (BCI) is being seen as the future solution to cognitive impairment. Not only do BCIs have their applications in medicine, for instance, brain tumor diagnosis and neuronal rehabilitation, they are also being used for entertainment, neuromarketing and advertising [1].

Electroencephalography (EEG) is a convenient non-destructive instrument for examining brain activities with extraordinary temporal resolution that is on the scale of milliseconds. There has been a drastic improvement in the size and expense of EEG devices, which has made the process of procuring EEG signals much easier. EEG is effective in recording and analyzing brain functions for a plethora of that encompasses gaming, dealing with traumatic and sleep disorders, cognitive impairment, etc. The use of EEG has expanded from its conventional applications like neurological diagnosis [2] to neurotherapy and brain–computer interfacing [3]. With low-cost EEG hardware available these days, measuring accurate EEG data has become an effortless task. EEG records data from the scalp when electrical variations are detected due to synchronized activity of millions of neurons. BCI is a device that interprets the EEG signals and further translates it for communication purposes or for controlling appliances without manual commands.

Deep learning has caused radical breakthroughs in the field of data science. Convolutional neural networks (CNNs) are considered as the building blocks of deep learning architectures. They are incredible at capturing and learning the right set of features from any kind of input data at different levels similar to a human brain. The main task of BCI is to recognize brain signals. Since brain signals are prone to corruption by various biological artifacts and concentration levels, deep learning models with CNN are best suited for classifying such signals. In [4], the authors have used several complex CNN architectures like CifarNet, AlexNet and VGGNet to segregate the levels of consciousness of patients who have been under anesthetics. The capsule network [5] has been designed for the classification of motor imagery signals.

Although numerous types of EEG signals are used in the study of BCI, the signals of interest depend upon the brain states that generate the distinct cognitive abilities [6]. The various brain states are responsible for producing different brain waves at different frequency ranges. The classic names of these EEG bands are delta, theta, alpha, beta, and gamma. For our research, we have considered the three most predominant brain waves, i.e., theta, alpha, and beta bands.

EEG classification has been performed using many feature extraction techniques and with both deep and machine learning algorithms. Neural synchrony measures such as power analysis, phase amplitude coupling, and phase locking have been applied to EEG signals to assess which mathematical method provides the best discrimination between cognitive tasks using Support Vector Machines (SVM) for classification [6]. For classifying the Raven's advance progressive metric test and a baseline task, wavelet-based feature extraction was proposed by Amin et al. [7] along with the use K-nearest neighbors, SVM and naive Bayes classifiers. Radial basis function SVM has been suggested by Hosni et al. [8] for the classification of mental tasks, by using a new independent component analysis technique for EEG pre-processing. Letter composition and visual counting tasks can be distinguished using empirical wavelet transform [9].

The BCI dealing with EEG signals associated with cognitive tasks has features obtained from wavelet transform, fast Fourier transform, and other methods, with multilayer perceptron and SVM being used for classification [10]. An automated computer platform has been proposed for clustering EEG signals corresponding with left-hand and right-hand gestures, with the independent component analysis being applied on associated channels for noise alleviation of generated EEG sources. The independent components of the feature datasets were given as input to multilayer perceptron and SVM. This research shows that the categorization of diverse pairs of motor cognitive tasks can be utilized in a BCI to manage plethora of instruments [11]. Application of neural networks for identifying and grouping EEG patterns related to motor imagery, i.e., real and imaginary movement of left/right leg in untrained subjects is discussed [12]. CNN with stacked autoencoders, which are dense feed forward connections, has been utilized by Tabar and Halici [13] for the segregation of EEG-based motor imagery signals.

In this paper, we have used fast Fourier transform (FFT) for feature extraction and have transformed the intermediate results into three-channel images before applying deep learning CNN model for classification.

2 Data Acquisition

2.1 Subjects and Experimental Design

The EEG was recorded, while the test subjects performed different neurophysiological tests. The four cognitive tasks conducted were Hopkins Verbal Learning Test (HVLT), Stroop Test (ST), Symbol Digit Modality Test (SDMT), and Benton's Visual Retention Test (MODBENT). The tests were performed by each subject in the same order. The cognitive tests chosen focused on a variety of skill set like memory, attention, processing speed, visual retention, etc.

EEG was recorded from 14 electrodes placed in a standard 10–20 electrode system. The International 10–20 electrode placement system is recognized as a standard procedure for describing the location of electrodes on top of the human scalp. The system is based on the position of an electrode and the underlying area of the cerebral cortex. Each region in the cerebral cortex has a letter corresponding to the lobe and a number which identifies the segment of hemisphere. Frontal (F), Temporal (T), Central (C), Parietal (P), and Occipital (O) are the letters designated for the lobes [14].

EEG readings for the four cognitive tests along with the baseline readings were obtained from the dataset collected by Dvorak et al. [6] and were sampled at 250 Hz.

The main presumption is that, as the dissimilar cognitive tasks require distinct brain functions and different regions of brain to activate, the EEG characteristics during the variety of tasks would diverge. The extent to which attributes in the EEG can be used for mapping EEG signal analogous to its task provides insight into feature extraction for task discrimination.

2.2 Neuropsychological Tests

HVLT has been corroborated within brain disordered populations (e.g., Alzheimer's disease, amnesic disorders) as an estimate of verbal learning and memory. In this test, a list of words was read out to the subjects and they were asked to recall after a short period of time. ST is considered to measure selective attention and processing speed. Subjects had to skim through three different lists of words. The three lists contained names of colors printed in black ink, same colored ink, jumbled color ink (e.g., "red" printed in "blue" ink), respectively [6]. In SDMT, to assess the divided visual attention, the subjects had to promptly comprehend and coincide symbols with corresponding numbers. MODBENT assesses visual perception and memory. The subjects were given four figure designs and had to select the best match from the previously seen figures [9]. The tests were conducted in the same duration of time for each test (HVLT: 149 ± 96 s, ST: 207 ± 93 s, SDMT: 167 ± 85 s, MODBENT 950 ± 108 s) with baseline being conducted for 660 ± 350 s. Fifty-four subjects who participated in four cognitive activities and a baseline recording contributed to a total of 270 samples of data.

3 Design Methodology

3.1 Proposed Classification Model

The EEG time series data were down sampled from 250 to 125 Hz. From the down-sampled signals, 15,000 data points were extracted. This was done to exclude segments of the signal with extremely low relevance. The data points were converted into csv files with EEG signals from the 14 electrodes placed around the scalp. This marks the end of the preprocessing stage.

The spectral topography maps were generated with the preprocessed data. These maps are the three-channel 2D images [15]. These images were then classified using our CNN model. The process of generating the images and the CNN architecture are explained in the following sections. Figure 1 shows the classification model.

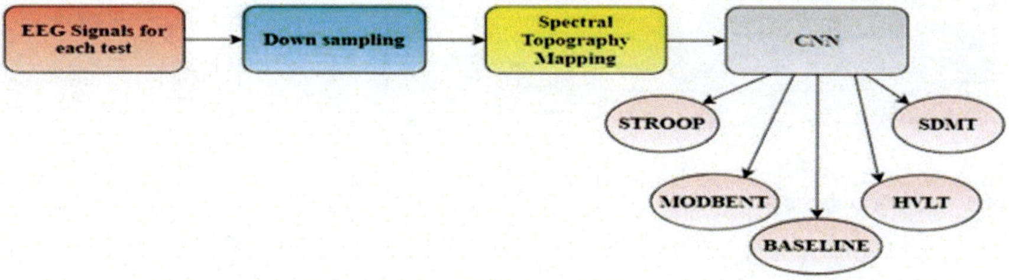

Fig. 1 Proposed classification model

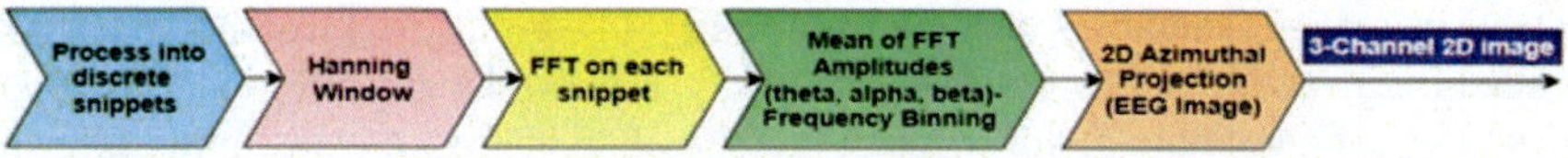

Fig. 2 Spectral topography mapping block

3.2 *Spectral Topography Mapping*

The EEG data were chopped into overlapping discrete one second "snippets". Hanning window was applied to each of these snippets. In order to obtain mean FFT amplitude, which is required for the succeeding steps, we must calculate the total number of intervals per electrode. The following specifications were considered to estimate the number of intervals. The sampling frequency is 125 Hz, and the frame duration is one second. The overlap percentage between snippets is taken as 75% with 15,000 data points. As a result, for one electrode we get 464 intervals. Since there are 14 electrodes, there are 6496 intervals for each subject.

Fast Fourier transform was applied for each interval and converted from time to frequency domain. The mean of the FFT amplitudes is calculated for each frequency band: beta (12–40 Hz), theta (4–8 Hz), alpha (8–12 Hz) for all intervals, giving three scalar values for each interval. This process is called as frequency binning, and it yields a three-dimensional array of size (464 × 14 × 3). These values were deciphered as RGB color channels and extrapolated onto a 2D map of the head [16]. This process of projecting theta, alpha, and beta values onto the map of electrode locations is known as Azimuthal projection. Thus, the entire procedure of spectral topography mapping results in three-channel 2D images of size 28 × 28 × 3 which are sent to the CNN network. Figure 2 shows the spectral mapping procedure.

3.3 *Convolutional Neural Network*

In the last decade, the CNN has become a very popular deep learning algorithm and has been applied to a variety of pattern recognition problems. In terms of accuracy and speed, it has been proven over time that CNNs outperform other classification

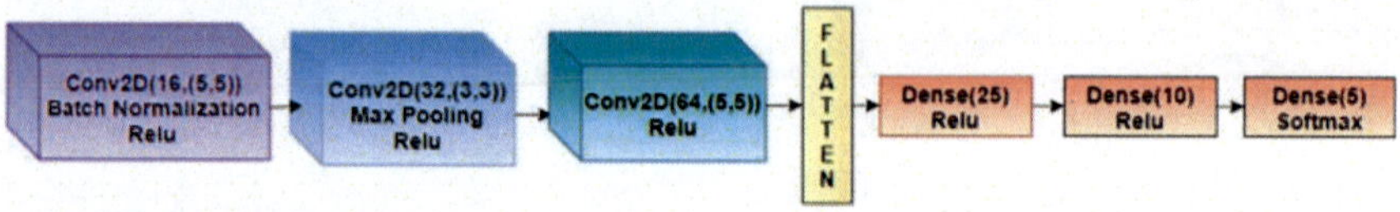

Fig. 3 CNN architecture model

algorithms. The hidden layers of a CNN typically consist of convolution and max pooling layers, fully connected layers, and activation function applied in the end [17]. For a completely new task/problem, CNNs prove to be very good feature extractors. Depth-wise and separable convolutions are used to extract EEG features in a CNN called EEGNet devised by Lawhern et al. [18]. The five-layer CNN model has been designed for automatic motor imagery classification. The first convolution layer in the five layer model aims at filtering EEG in the space domain, whereas the second convolution layer does subsampling on the EEG signal and transforms it into the time domain [19]. Baseline architecture was designed in [20] with the convolution block containing one-dimensional convolution layer, batch normalization layer [21], and ReLU activation. Three such blocks are stacked with global average pooling and softmax activation applied in the end.

The CNN architecture illustrated in Fig. 3 is used to classify the previously obtained EEG images. The first block consists of 16 two-dimensional convolution filters each of 5 × 5 kernel size, batch normalization layer and ReLU activation. The second block is constructed with a convolution layer having 32 filters each with 3 × 3 kernel size, followed by maxpooling operation and ReLU activation. The third block is made only with convolution layer (64 filters with 5 × 5 kernel size) and ReLU function. The resulting feature maps are flattened and given as input to the dense layers. Softmax activation is applied to the final dense layer. Scores for each data are then obtained from the network's output based on the interpolated images.

4 Results and Discussion

We obtained a training accuracy of 87.15% and testing accuracy of 84.33% on our proposed CNN. Figure 4 shows the accuracy plot. The number of EEG images generated from 270 data samples and 464 frames generated per data sample was 125,280. Training and testing data are split according to the ratio of 3:1. The model was trained for 30 epochs with a learning rate of 0.01 and a batch size of 256. Adadelta optimizer along with categorical cross entropy loss function was used to train the networks. The training was done on a Windows machine with 4 GB RAM. The preprocessing was done using MATLAB and Python with Keras API used to construct the CNN.

Figure 5 shows the testing accuracy along with a set of predicated and actual values for few images. Figure 6 shows the heat map obtained with the testing images, for the

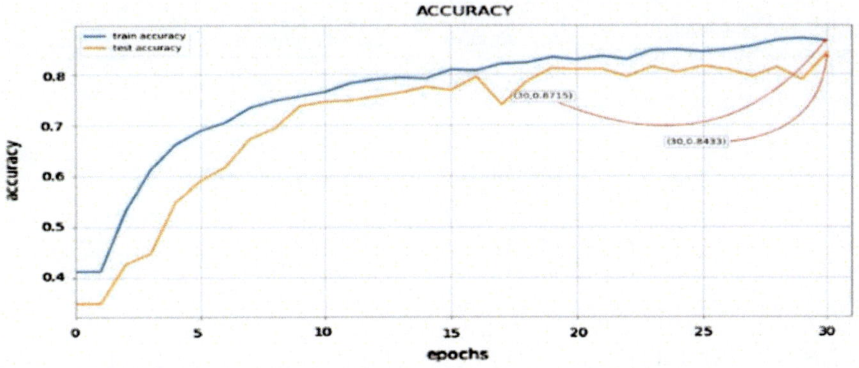

Fig. 4 Accuracy plot for the proposed CNN

Fig. 5 Predicted and actual values

```
[INFO] accuracy: 84.33%
[INFO] Predicted: 3, Actual: 3
[INFO] Predicted: 1, Actual: 1
[INFO] Predicted: 0, Actual: 0
[INFO] Predicted: 1, Actual: 1
[INFO] Predicted: 0, Actual: 0
[INFO] Predicted: 4, Actual: 4
[INFO] Predicted: 1, Actual: 1
[INFO] Predicted: 4, Actual: 2
[INFO] Predicted: 2, Actual: 2
[INFO] Predicted: 3, Actual: 3
[INFO] Predicted: 0, Actual: 0
[INFO] Predicted: 3, Actual: 3
[INFO] Predicted: 4, Actual: 4
[INFO] Predicted: 4, Actual: 1
[INFO] Predicted: 0, Actual: 0
[INFO] Predicted: 2, Actual: 2
[INFO] Predicted: 0, Actual: 4
[INFO] Predicted: 1, Actual: 1
[INFO] Predicted: 0, Actual: 0
[INFO] Predicted: 2, Actual: 2
[INFO] Predicted: 0, Actual: 0
[INFO] Predicted: 0, Actual: 0
[INFO] Predicted: 1, Actual: 3
```

Fig. 6 Heat map for cognitive task classification

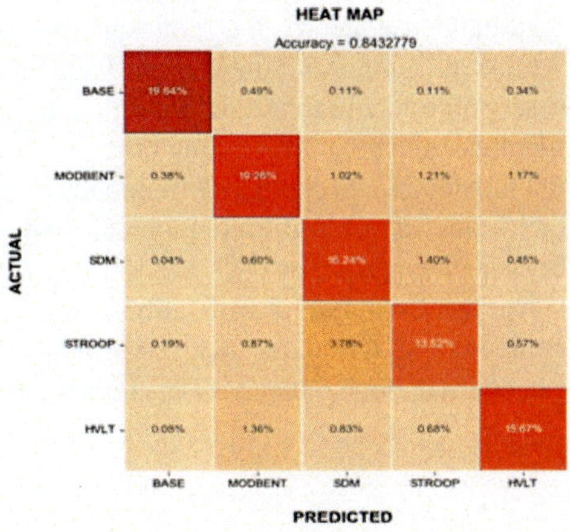

Table 1 Comparing accuracies with other methods

Architectures	Classification methods	Accuracy (%)
Machine learning	Radial basis function SVM [8]	70.00
	SVM [6]	78.80
Deep learning	CNN with autoencoders [13]	75.10
	CNN [15]	79.56
	Proposed CNN	84.33

CNN. We infer that the prediction percent is the highest for baseline task, followed by MODBENT, HVLT, SDMT and ST.

Radial basis function SVM was used as the classifier in [8], for distinguishing three cognitive tasks, and they obtained an accuracy of 70.00%. SVM was used by [6] for the classification of cognitive tasks, and the accuracy score was 78.80%. CNN perform better than machine learning methods since they deal with end-to-end classification. The deep learning model proposed by [13] has an accuracy of 75.10%, for classifying motor imagery signals. The CNN model designed by [15] consists of four convolution layers followed by three dense layers, and it classifies known skills from unknown skills. The accuracy obtained is 79.56%. Our CNN architecture can segregate more than two classes and achieves a higher accuracy score of 84.33%, performing better than all other methods listed in Table 1.

5 Conclusion

We transform EEG data into spectral 2D images that are topologically preserved, as opposed to traditional EEG analysis techniques that ignore vital spatial information. Consequently, the proposed approach is designed to conserve the spectral, spatial and temporal framework of EEG which aids in finding properties that are less susceptible to irregularities within each dimension. The process of spectral topographical mapping takes a long time; hence, it increases the computational time. The unique combination of EEG frequency band (alpha, beta and theta) components in the projected spectral topographical maps proved to be the distinguishing feature between the tasks. The CNN takes the spectral topography maps as features for classification. Cognitive classification can be used to facilitate interactions between a computer and a human, such as the brain–computer interface. The number of tests can be increased so that we can classify other cognitive behaviors.

References

1. Abdulkader SN, Atia A, Mostafa MSM (2015) Brain computer interfacing: applications and challenges. Egypt Inform J. https://doi.org/10.1016/j.eij.2015.06.002
2. Mcloughlin G, Makeig S, Tsuang MT (2014) In search of biomarkers in psychiatry: EEG-based measures of brain function. Am J Med Genet Part B Neuropsychiatr Genet. https://doi.org/10.1002/ajmg.b.32208
3. Nash JK (2000) Treatment of attention deficit hyperactivity disorder with neurotherapy. Clin EEG Neurosci. https://doi.org/10.1177/155005940003100109
4. Liu Q et al (2019) Spectrum analysis of EEG signals using CNN to model patient's consciousness level based on anesthesiologists' experience. IEEE Access. https://doi.org/10.1109/ACCESS.2019.2912273
5. Ha KW, Jeong JW (2019) Motor imagery EEG classification using capsule networks. Sensors (Switzerland). https://doi.org/10.3390/s19132854
6. Dvorak D, Shang A, Abdel-Baki S, Suzuki W, Fenton AA (2018) Cognitive behavior classification from scalp EEG signals. IEEE Trans Neural SystRehabil Eng. https://doi.org/10.1109/TNSRE.2018.2797547
7. Amin HU, Mumtaz W, Subhani AR, Saad MNM, Malik AS (2017) Classification of EEG signals based on pattern recognition approach. Front Comput Neurosci. https://doi.org/10.3389/fncom.2017.00103
8. Hosni SM, Gadallah ME, Bahgat SF, AbdelWahab MS (2007) Classification of EEG signals using different feature extraction techniques for mental-task BCI. In: ICCES'07—2007 international conference on computer engineering and systems. https://doi.org/10.1109/ICCES.2007.4447052
9. Tanveer M, Gupta A, Kumar D, Priyadarshini S, Chakraborti A, Mallipeddi R (2019) Cognitive task classification using fuzzy based empirical wavelet transform. In: Proceedings—2018 IEEE international conference on systems, man, and cybernetics, SMC 2018. https://doi.org/10.1109/SMC.2018.00304
10. El Bahy MM, Hosny M, Mohamed WA, Ibrahim S (2017) EEG signal classification using neural network and support vector machine in brain computer interface. Adv Intell Syst Comput. https://doi.org/10.1007/978-3-319-48308-5_24
11. Samaha MHA, AlKamha K (2013) Automated classification of L/R hand movement EEG signals using advanced feature extraction and machine learning. Int J Adv Comput Sci Appl. https://doi.org/10.14569/ijacsa.2013.040628
12. Maksimenko VA et al (2018) Artificial neural network classification of motor-related EEG: an increase in classification accuracy by reducing signal complexity. Complexity. https://doi.org/10.1155/2018/9385947
13. Tabar YR, Halici U (2017) A novel deep learning approach for classification of EEG motor imagery signals. J Neural Eng. https://doi.org/10.1088/1741-2560/14/1/016003
14. Kumar JS, Bhuvaneswari P (2012) Analysis of electroencephalography (EEG) signals and its categorization—a study. Procedia Eng. https://doi.org/10.1016/j.proeng.2012.06.298
15. Zulkifley M, Abdani SR (2019) EEG signals classification by using convolutional neural networks
16. Bashivan P, Rish I, Yeasin M, Codella N (2016) Learning representations from EEG with deep recurrent- convolutional neural networks. In: 4th international conference on learning representations, ICLR 2016—conference track proceedings
17. Al-Bander B, Al-Nuaimy W, Williams BM, Zheng Y (2018) Multiscale sequential convolutional neural networks for simultaneous detection of fovea and optic disc. Biomed Sig Process Control. https://doi.org/10.1016/j.bspc.2017.09.008
18. Lawhern VJ, Solon AJ, Waytowich NR, Gordon SM, Hung CP, Lance BJ (2018) EEGNet: a compact convolutional neural network for EEG-based brain-computer interfaces. J Neural Eng. https://doi.org/10.1088/1741-2552/aace8c
19. Tang Z, Li C, Sun S (2017) Single-trial EEG classification of motor imagery using deep convolutional neural networks. Optik (Stuttg). https://doi.org/10.1016/j.ijleo.2016.10.117

20. Wang Z, Yan W, Oates T (2017) Time series classification from scratch with deep neural networks: a strong baseline. In: Proceedings of the international joint conference on neural networks, vol 2017, pp 1578–1585. https://doi.org/10.1109/IJCNN.2017.7966039
21. Ioffe S, Szegedy C (2015) Batch normalization: accelerating deep network training by reducing internal covariate shift. In: 32nd international conference on machine learning, ICML 2015

Comparative Analysis on Frequent Itemset Mining Algorithms in Vertically Partitioned Cloud Data

M. Yogasini and B. N. Prathibha

Abstract Frequent itemset mining and Association Rule Mining are the extensively utilized data analysis techniques for a transactional database concerned with a trade-off. Data owners wish to acquire knowledge in these data analysis techniques to protect their sensitive data from additional data proprietor and third parties. This work emphasizes on privacy-preserving frequent itemset mining on a vertically partitioned database. An efficient homomorphic encryption scheme is designed to assure data privacy. Cryptography is a part of encryption that is used to guard information against third parties. In this work, frequent itemset mining algorithms such as Eclat, apriori, and FP-Growth are taken for analysis in terms of computation time and scalability of data. The analysis result shows that apriori algorithm is less time-consuming to generate rule in the cloud, irrespective of the number of transactions.

Keywords Frequent itemset · Encryption · Eclat · Apriori · FP-Growth · Association Rule Mining

1 Introduction

A transaction database consists of many transactions. Each transaction in a transaction database is categorized by an ordered pair as (TID, list of items), where TID is a transaction identifier. Original database from various data owners is stored as a joint database in the cloud and is crumbled as horizontal partition and vertical partition. Vaidya et al. [1] and Kantarcioglu et al. [2] address the privacy issues in the horizontal and vertical partitioned database. This work emphasizes vertically partitioned data in the cloud. Frequent itemset mining algorithms like apriori, Eclat, and FP-Growth are intended for the centralized database. Let "I" will be an itemset. It is considered

M. Yogasini (✉)
Department of Computer Science and Engineering, Manonmaniam Sundaranar University, Abishekapatti, Tamil Nadu 627012, India

B. N. Prathibha
Department of Computer Science, G. Venkataswamy Naidu College, Kovilpatti, Thoothukudi 628502, India

A. Sivasubramanian et al. (eds.), *Futuristic Communication and Network Technologies*, Lecture Notes in Electrical Engineering 792,
https://doi.org/10.1007/978-981-16-4625-6_38

as a frequent item when Supp(I) ≥ T_s, where T_s is a threshold value. When Supp (AUB) ≥ T_s and Conf (A B) ≥ T_c, association rule is generated in the frequent items of a transaction. T_s and T_c are the threshold value specified by the data owner and are constructed based on the type of transactions and size of the database. Based on the frequent itemset mining algorithms (FIM), the Association Rule Mining algorithms (ARM) are generated. In this work, cloud-based privacy-preserving frequent itemset mining is implemented. Fake transactions are added in the form of cipher contrary to frequency analysis attacks in the transaction. Apriori, Eclat, and FP-Growth itemset mining algorithms are adopted in the transactions with different *k*-values, where *k* denotes the items in a transaction. The results depict that the apriori algorithm takes less computation time to find frequent items in the cloud, compromising transaction count over Eclat and FP-Growth.

2 Related Work

Enormous research work had been carried in the last decades. Various strategies had been proposed for the privacy preserving in the distributed database. Some of the referred works here are in brief.

Vaidya and Clifton [1] developed a privacy preserving on frequent itemset, secure scalar product protocol, and two-party algorithm was implemented by them. Algebraic solution was also applied on vertically partitioned data to hide data. Li et al. [3] proposed efficient homomorphic encryption scheme and a secure comparison scheme for privacy preserving on vertically partitioned databases. Several data holders can exchange their information with more privacy by adopting this scheme. Keshavamurthy and Khan [4] adopted genetic algorithm for developing a global association rule model to find frequent items. Accurate or nearby solution for search problems is obtained by utilizing the robustness feature of Muthu Lakshmi and Sandhya Rani [5] adopted concept of scalar product with diverse cryptography to protect individual's site information. This model was designed for vertically partitioned database with 'n' number of sites to find global association rules. But this method is not suitable for large database. Kaosar and Paulet [6] proposed secure two-party Association Rule Mining protocol. The support and confidence of association rules are computed homomorphically, and the result was returned in a single bit. This model fails to remove unnecessary communication in a transaction. Abaril and Dehkordi [7] adopted DCR (Dual Clustering Rules) algorithm to hide sensitive association rules. It lacks to improve efficiency and security for the complex data. The server can identify the exact collection of items in the proprietor's data as well as the specific support of each item in the real information, by adopting the conservative frequency-based attack model proposed by Giannotti, et al. [8]. A novel FP array technique [9] was carried out by Gosta Grahne and Jianfei Zhu. This technique is an enhanced version of the FP tree algorithm and adopted for sparse data sets. Even though the data sets are dense, this algorithm works fast and consumes less memory when compared to other methods.

Tai et al. [10] proposed k-support anonymity to protect the data with the help of Pseudo-taxonomy tree technique. This method results in beneficial security assurance with adequate stockpiling overhead. Modi and Rao [11] proposed DSSRC (Decrease Support of R.H.S. item of Rule clusters) Algorithm. The privacy of the data is maintained without affecting the quality of the data in the sanitized databases like as original database and hide several rules at a time. Nomura et al. [12] proposed a secure Association Rule Mining (ARM) scheme for vertically partitioned data. Scalar product algorithm can be replaced by this algorithm by adopting flexible information sharing with less computation time. Efficiency and security of the data sharing are yet to be developed. Du-Atallah's efficient multiplication product for PPDFIM (Privacy-Preserving Distributed Frequent Itemset Mining) and SBDP (Secure Binary Dot Product) to mine frequency itemsets for a vertically partitioned data was proposed by Nirali and Devesh [13]. The proposed scheme's implementation cost is 38 times quicker than the seminal public key-based cryptographic scheme. Even though the computational cost is faster, this algorithm is not suitable for a large number of databases.

3 Methodology

Every data owner possesses a personal database and is stored in the cloud as a joint database. The joint database's association rules are cooperatively mine by the data owners with the assistance of the cloud. In this work, ciphertext substitution is adopted at the pre-processing stage for mining. Each data owner adds a fictitious transaction with the real data. These data are encrypted and stored in the cloud as a joint database. In a joint database, frequent items are discovered based on the support threshold value, and encrypted association rules are engendered in the database. The data owner decrypts the encrypted association rules which are returned by the cloud to recapture the real association rules. The dataset used for the complexity calculation is an open-source retail dataset of a Belgian retail store. The retail dataset consists of 88,162 transactions. The schematic representation for outsourced data mining on the joint database is depicted in Fig. 1.

3.1 Pre-processing

3.1.1 Substitution of Cipher Text

Substitution cipher is an eminent technique used in the encryption of the plain text. Data owner's data were encrypted by them with substitution cipher anterior to outsourcing. Attackers can understand some plaintext through frequency analysis and may also be aware of the cipher data with their knowledge. To avoid these, fake transactions are added in the original DB. The fake transactions are inserted into the

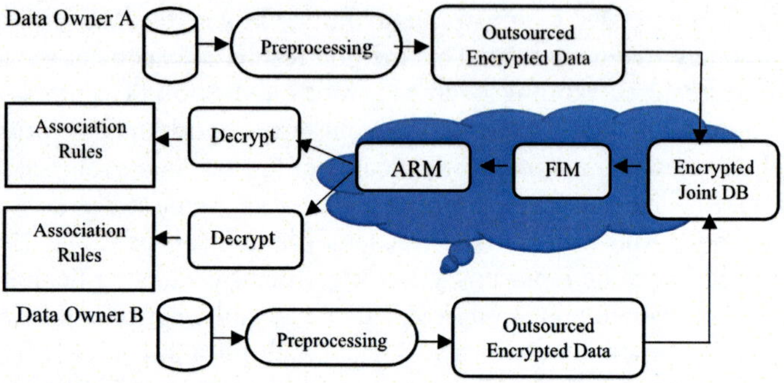

Fig. 1 Schematic representation of outsourced itemset mining on joint database

original database, and it may be equal to or exceeds the novel data. X forged transactions are embedded between every two factual transactions where X is a random variable. The data owner tags the transaction with Encrypted Realness Value (ERV). ERV is also known as Realness Value (RV), and the RV value is 0 and 1 for fake and real transactions, respectively.

3.2 Homomorphic Encryption

Homomorphic encryption is a cryptography technique that performs special mathematics operations on the ciphertext and allows computation on encrypted data. It is used in the privacy preservation of outsourced storage data. A secret key is used by the data owner for encryption. The size of the ciphertext should be polynomial for efficient homomorphic encryption. Paillier encryption [14] is the most common homographic encryption scheme adopted in a transaction. The public key and "●" modular multiplication are adapted to encrypt the data in Paillier encryption. EPK (), EPK(m_1), EPK(m_2), and the public key are utilized in the encryption. The data owner can compute EPK ($m_1 + m_2$) by executing a modular multiplication of EPK(m_1) and EPK(m_2), and likewise, the data owner can calculate EPK ($m_1 \times m_2$) by performing modular exponentiation of EPK(m_1) m_2 [3].

3.3 Frequent Itemset Mining

Frequent sets perform a dynamic role in data mining tasks that are used to discover interesting patterns from databases, such as association rules, correlations, sequences, classifiers, and clusters. Frequent itemset generation generates all itemsets whose support $\geq$ minsup. A set is known to be frequent if its support is not less than a given complete minimal support threshold min_sup with $0 \leq$ min_supabs $\leq |D|$.

The data owner initially decrypts the ESVR to fix whether the itemset is frequent or not. If the itemset is an actual frequent one, at that point it will be decrypted by the data owner. Cloud-aided privacy-preserving frequent itemset mining solution for vertically partitioned databases is proposed in this work by which data proprietors have a significant level of protection prerequisite, and the mining algorithms such as Eclat, apriori, and FP-Growth were compared.

3.4 *Association Rule Mining*

Association Rule Mining is a rule-based technique to find relations between items in a transaction and to forecast the manifestation of an item based on different items in the transaction. Rule mining is an inference expression of form, $X \rightarrow Y$, where X and Y are two items that exist in a transaction. Antecedent and consequent are the two parts of association rules. An antecedent is an item found in a transaction, and consequent is an item established in amalgamation with the antecedent. These are represented as if and then, respectively. Support, confidence, and lift are the measures of rules. If an item is in both A and B rows in a transaction, then the joint probability of A and B is known as support in a rule. Support $= \varepsilon(X + Y) -$ total. Confidence is the conditional probability of B given A. $\text{Conf}(X \rightarrow Y) = \text{Supp}(XUY) - \text{Supp}(X)$. Lift is the ration of confidence to support. Frequent itemsets are resolute through successive scans of the database based on a support and confidence values set by the user to construct a strong association rule among the items.

4 Performance Analysis

This work emphasizes on comparing the execution time of three mining algorithms such as apriori, Eclat, and FP-Growth on a vertically partitioned database. The execution time of different transactions like 5000, 10,000, 15,000. 20,000 and the running time for the same transactions with different k-values are evaluated, and the result is tabulated. The performance of these three algorithms is interpreted in the transaction set with the help of Python. The execution time for different transactions is shown in Table 2 and the corresponding chart in Fig 2.

From Table 1 and Fig. 2, it is found that the performance of apriori algorithm's execution time is less than Eclat and FP-Growth for the retail dataset with various transactions.

From Table 2, we can view that there is a change in running time with different k-values where k denotes the number of items in a transaction. From the result, we can perceive that the execution time of apriori is less. Even though the apriori algorithm scans database several times to find frequent items in a transaction, the computation time for apriori is less for different sizes of transactions with various k-values when

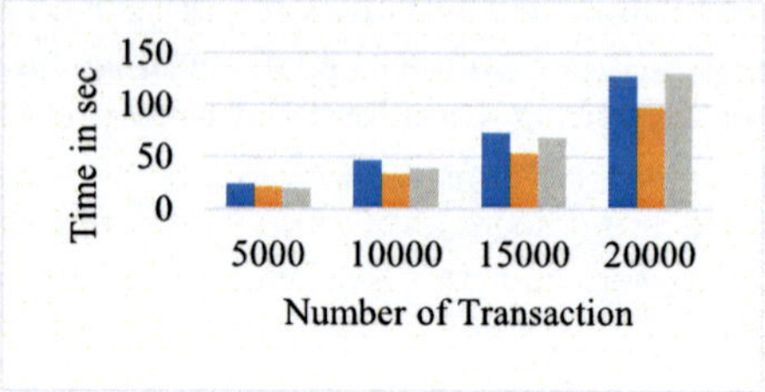

Fig. 2 Running time for different transactions

Table 1 Execution time in sec

Transactions	Eclat	Apriori	FP-growth
5000	24.9	22.4	20.5
10,000	47.6	34.9	39.8
15,000	73.5	54.5	68.8
20,000	128.9	98.4	130.7

compared to Eclat and FP-Growth. This algorithm is comprehensive to find frequent items in a transaction with specified support and confidence value.

5 Conclusion

To protect the data proprietor's data, one-to-one substitution of ciphertext and homomorphic encryption were adopted in a joint database. In this paper, a privacy-preserving outline for vertically partitioned data in a cloud environment using frequent itemset mining is implemented. Privacy-preserving frequent itemset mining with the help of the apriori algorithm was built to mine the data. This approach produces the result with less computation time and sustains constant privacy and performance of itemset mining when the number of transactions is increased with diverse k-values, when compared to Eclat and FP-Growth. With this system, the data proprietors can redistribute their data with a significant level of protection without negotiating the performance.

Table 2 Computational time for different *k*-values in sec

Transaction	5000			10,000			15,000			20,000		
	Eclat	Apriori	Fp-Growth	Eclat	Apriori	Fp-Growth	Eclat	Apriori	Fp-Growth	Eclat	Apriori	Fp-Growth
$k = 10$	17.6	13.4	15.2	31.7	24.6	32	52.7	40.2	49	74.3	52.3	75.2
$k = 20$	18.1	14.9	15.7	33.3	26.5	33	55.4	42.3	53	76.2	53.2	77.6
$k = 30$	18.1	15.2	15.9	33.9	26.5	37	56.7	43.5	55	76.8	54.4	78.7
$k = 40$	18.3	15.1	16.2	35.3	27.1	42	56.7	43.9	57	77.2	54.9	79.3

References

1. Vaidya J, Clifton C (2002) Privacy preserving association rule mining in vertically partitioned data. In: Proceedings of SIGKDD, pp 639–644
2. Kantarcioglu M, Clifton C (2004) Privacy-preserving distributed mining of association rules on horizontally partitioned data. IEEE Trans Knowl Data Eng 16(9):1026–1037
3. Li L, Lu R (2016) Privacy-preserving outsourced association rule mining on vertically partitioned databases. In: 2016, IEEE transactions on information forensics and security, vol. 11(8), 1847–1861
4. Keshavamurthy BN, Khan AM (2013) Privacy preserving association rule mining over distributed databases using genetic algorithm. Springer. Neural computing & application, pp S351–S364
5. Muthu Lakshmi NV, Sandhya Rani K (2012) Privacy preserving association rule mining in vertically partitioned databases. Int J Comput Appl 39:29–35
6. Kaosar MG, Paulet R, Yi X (2011) Secure two-party association rule mining. In: 2011 in Australian information security conference, vol 116
7. Abaril ZK, Dehkordi MN (2015) Privacy preserving in association rule mining. ACSIJ Adv Comput Sci Int J 4(1):13. ISSN: 2322-5157
8. Giannotto F, Lakshman LVS, Monreale A, Pedreschi D, Wang H (2013) Privacy-preserving mining of association rules from outsourced transaction databases. IEEE Syst J 7
9. Grahne G, Zhu J (2005) Fast algorithms for frequent itemset mining using FP-trees. IEEE Trans Knowl Data Eng 17(10)
10. Tai C-H, Yu PS, Chen M-S (2010) K-support anonymity based on pseudo taxonomy for outsourcing of frequent itemset mining. In: Proceedings of international knowledge discovery data mining, pp 473–482
11. Modi CN, Rao UP (2010) Maintaining privacy and data quality in privacy preserving association rule mining. In: IEEE second international conference on computing, communication and networking technologies
12. Kenta, Yoshiaki, Masami (2020) Secure association rule mining on vertically partitioned data using private-set intersection. IEEE Access 8:144458–144467. Electronic ISSN 2169-3536
13. Devesh N (2015) A novel privacy-preserving scheme for collaborative frequent itemset mining across vertically partitioned data. In: 2015, security and communication networks, pp 4407–4420. https://doi.org/10.1002/sec.1377
14. Paillier P (1999) Public-key cryptosystems based on composite degree residuosity classes. In: Eurocrypt

A Compact Complementary Split Ring Resonator-Based Notch Design for Wireless Access for Vehicular Environment/WLAN/ITU Band Application

Sanmugasundaram Ravichandran, Natarajan Somasundaram, and Rajkumar Rengasamy

Abstract A compact ultra-wideband (UWB) dual notch band antenna is presented. The proposed antenna possesses staircase typed radiating patch with half a ground, rectangular- and square-shaped complementary split ring resonators (CSRR) which are used for realizing ultra-wideband performance and filtering effects in the antenna structure, respectively. Filtering structure has a small size of $32 \times 20 \times 1.6\ \text{mm}^3$, and notching frequencies are obtained at 5.8/5.9 GHz and 8.3 GHz in the ultra-wideband region. Antenna has better reflection coefficient performance from 3.1 to 14 GHz and with filtering effect. Further, it has better radiation characteristics in the desired frequency bands.

Keywords Notch · Ultra-wideband · Compact · Complementary split ring resonator · WLAN · WAVE · ITU

1 Introduction

In modern wireless communication scenario, ultra-wideband communication has become famous one because of its unlicensed wideband frequency from 3.1 to 10.6 GHz allotted by FCC [1]. UWB technique meant for high bandwidth and it covers large wireless communication spectrum. It has average speed of 100 Mbps to 1 Gbps speed; due to this, many researchers have interest on this working area [2, 3]. Though UWB has quit advantages, it has issue of interference with short communication wireless applications such as 3.5 GHz IEEE 802.16 (WiMAX), 5.2/5.8 GHz IEEE 802.11 a (WLAN) and 8 GHz ITU band. Here, intention to design a compact notch band antenna and it address avoiding the interference in short wireless communications. The stop band filtering effect is realized using various methods such as

S. Ravichandran (✉) · N. Somasundaram · R. Rengasamy
VelTech Rangarajan Dr. Sagunthala R&D Institute of Science and Technology, Chennai 600062, India

N. Somasundaram
e-mail: drnatarajans@veltech.edu.in

A. Sivasubramanian et al. (eds.), *Futuristic Communication and Network Technologies*, Lecture Notes in Electrical Engineering 792,
https://doi.org/10.1007/978-981-16-4625-6_39

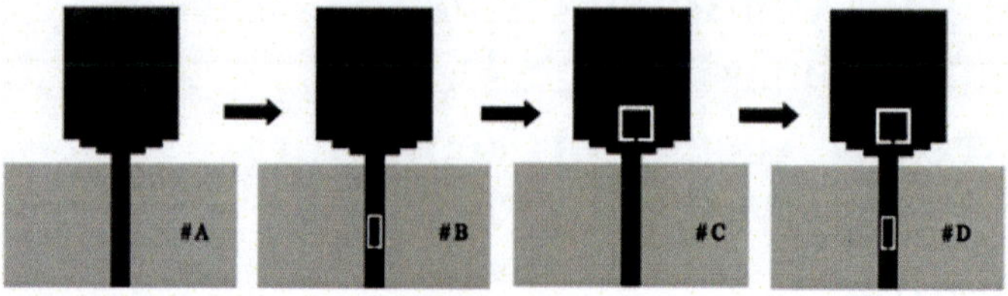

Fig. 1 Development stages of proposed multiband antenna

employing slots in radiating element, ground plane of antenna, introducing slits, embedding the resonators are discussed [4–16]. The single band notch antennas are achieved in [4–7]. The U-shaped slot was embedded to realize notch band performance but only one notch band was achieved [4]. In [5], antenna was designed to suppress unwanted WiMAX application using DGS structures. A compact notch antenna was designed using W-shaped slot in the ground plane with hooked shaped resonator [6]; though antenna has compact structure, it notches only one frequency. In order to cover various wireless applications in a single antenna, dual notch band antennas are used [7–10]. A UWB-based dual notch band antenna was achieved by using DGS structures in [8] but it requires multi-layer structures. A dual notch band antenna was proposed using electromagnetic band gap structures [9, 10]. To achieve more compact-based antennas metamaterial designs can be used by using we can realize gain enhancement, compact size, harmonic reduction and multiband performance [11–13]. Similarly, to realize notch band performance, CSRR/SRR structures were used [14–16]. The above antenna designs were achieved dual notch band performance using metamaterial technique. Here, SRR/CSRR was embedded top/bottom of the substrate of the antenna design.

A coplanar waveguide fed complementary split ring resonator-based notch band antenna is proposed. Antenna consists of two single CSRR structures, one is embedded in feed line and another one in radiating element to achieve the dual notch band antenna performance. It is effectively filtering out 5.8/5.9 GHz (WAVE/WLAN/ITU) and 8 GHz ITU band, respectively. Further, antenna design has decent radiation characteristics in the preferred frequencies expect notch band regions (Figs. 1 and 4).

2 Proposed Antenna Configuration

The proposed ultra-wideband antenna has a small dimension of $32 \times 20 \times 1.6\ \text{mm}^3$ which is designed commercially available FR4 substrate of thickness 1.6 mm (Fig. 2).

The antenna is fed with 50-Ω conventional microstrip feeding with a semi grounded plane. The microstrip has feed length (F_L) and width (F_w) of 14.5 mm and 2.2 mm, respectively. The model of proposed antenna is shown in Fig. 3.

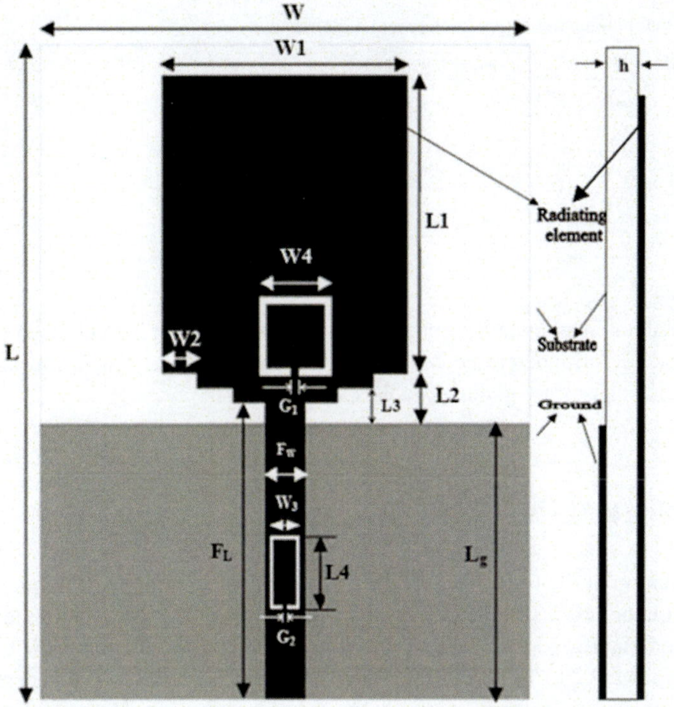

Fig. 2 Geometry of the proposed antenna

Fig. 3 Prototype of dual notch band antenna

Table 1 Dual notch antenna parameters and dimensions

L	W	L_g	F_L	F_w	W_3	L_4
32	20	13.5	14.5	2.2	1.6	3.6
L_1	L_3	L_2	W_1	W_2	L_3	L_4
14.6	1.7	2.4	13	1.8		3.6
G_1	G_2	W_3	W_4			
0.4	0.2	1.6	3.9			

The dual notch antenna has compact structure by which it has achieved number of notch band performance effectively at 5.8/5.9 and 8.3 GHz. The proposed filtering antenna dimensions are given in Table 1.

3 Results and Discussion

The simulation of dual notch band antenna is completed using high frequency structural simulator (HFSS) V.16.0. The microstrip fed half a grounded antenna (#A) is designed such a way to achieve the ultra-wideband performance. It covers the frequency from 3.2 to 14 GHz effectively. Antenna has achieved UWB performance due to stair cased designed in the main radiating element. Here, each stair case elements used to resonate separate frequencies. The nearer frequencies are merged together to produce to ultra-wide band performance from 3.2 to 14 GHz. To avoid the interferences, notch bands are very useful and achieved the desired notches in the ultra-wide band region, metamaterial-based complementary split ring resonator is introduced at microstrip feeding element of antenna (A) as shown in Fig. 1 (#B) and its reflection coefficient performance is exhibited in Fig. 4 (#B). Here, CSRR is introducing notch band effect at 5.8 GHz. In antenna (#C), square-shaped CSRR structure is introduced at radiating patch of microstrip-based antenna in order to achieve the notch band performance to the application of ITU band (8.05–8.5 GHz). Antenna (#C) notches 8–10 GHz effectively. In antenna (#D), both (#B) and (#C) are clubbed together to produce the dual notch band performance without affecting existing antenna performance. Antenna (#B) and (#C) notch band performances are retained as exhibited in Fig. 4 (#D).

Similarly, voltage standing wave ratio (VSWR) of the evolution of the antennas is depicted in Fig. 5. It depicts that the antennas are having <2 in VSWR scale at the desired frequencies except notch frequencies.

The notch antenna performance can be confirmed through another analysis of antennas impedance measurement. Figure 6 shows that antenna dual notch band antenna performance in which antenna's real and imaginary impedances are presented. Thus, the desired antenna notch frequency (5.8/5.9 and 8.3 GHz) impedances are varied much compared to other frequencies in UWB region. Through the above analysis, it is confirmed that antennas are having better notch effects. Here,

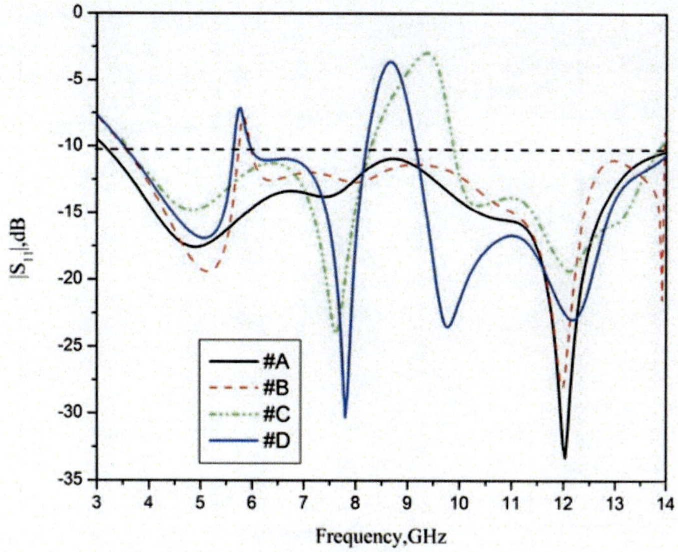

Fig. 4 Reflection coefficient of development stages of the antenna

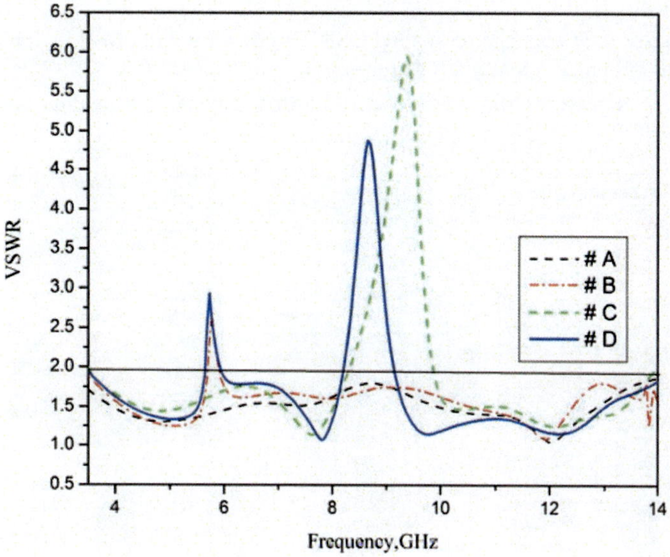

Fig. 5 VSWR of evolution of stages of antenna

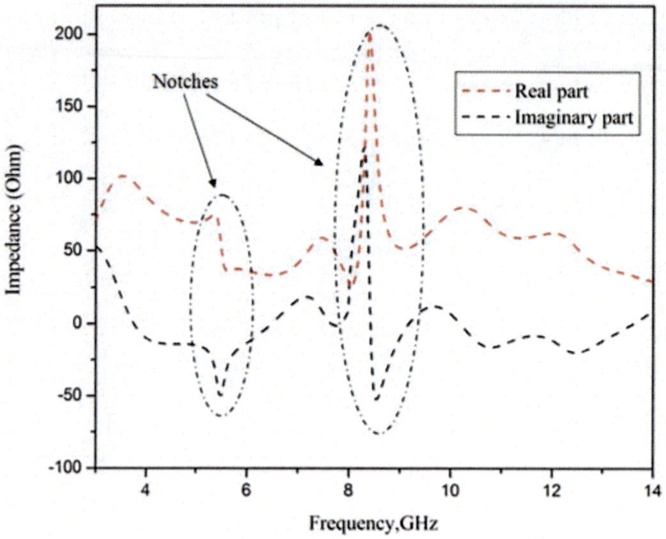

Fig. 6 Impedance performance of the proposed antenna

rectangular-shaped and square-shaped CSRR are having quit effect in achieving the notch band effects at 5.8 and 8.3 GHz.

Figure 7 represents the surface current distribution of the dual notch band

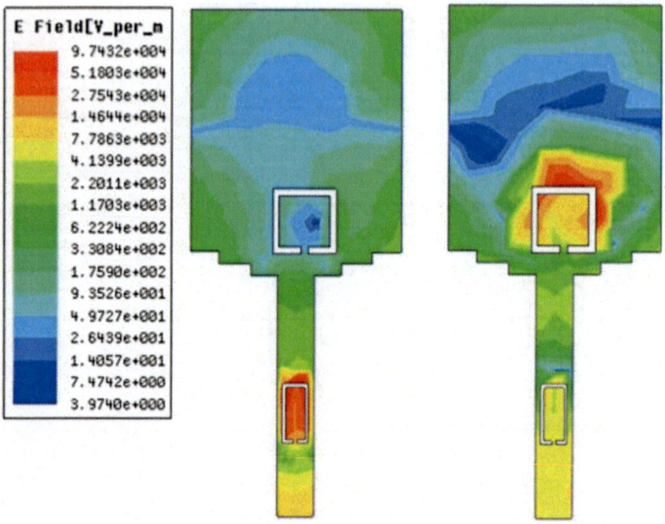

Fig. 7 Surface current distribution of the proposed antenna at **a** 5.8 GHz and **b** 8.3 GHz

antenna. Rectangular-shaped CSRR structure has an impact in the lower frequencies (5.8 GHz), and square-shaped CSRR has an impact at 8.3 GHz which can be confirmed through high current concentration at place of CSRR structure.

The dual band notch antenna radiation characteristics are exhibited in Figs. 8 and 9. It depicts that antennas power distribution in elevation and azimuth planes of the

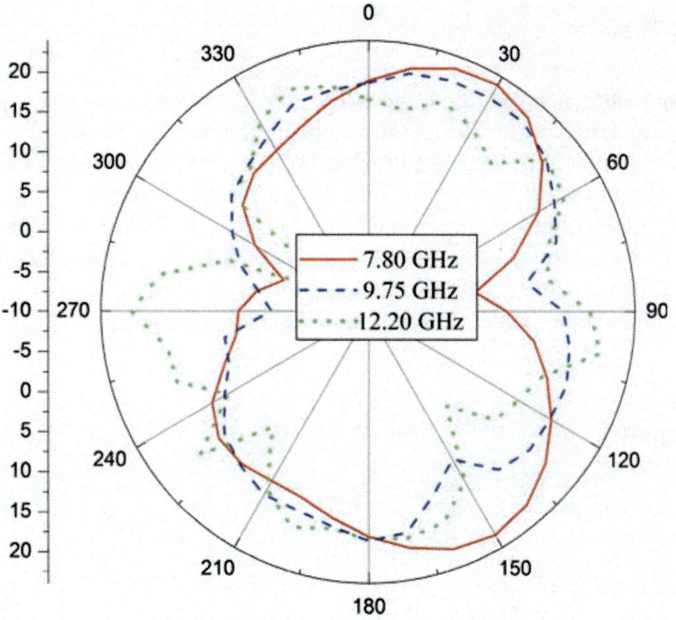

Fig. 8 E-plane radiation patterns for the proposed antenna

Fig. 9 H-plane radiation patterns for the proposed antenna

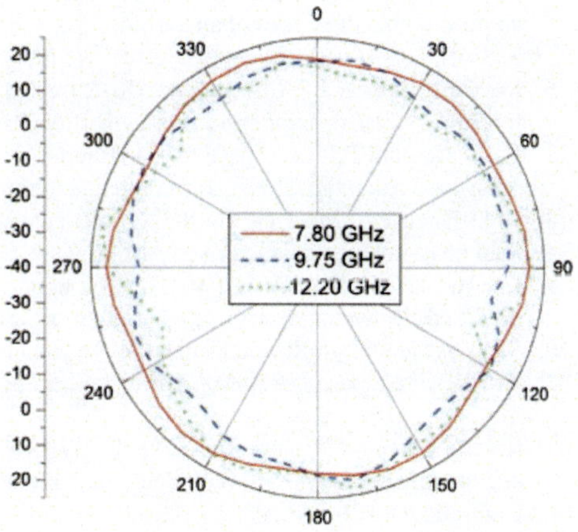

antenna. Omnidirectional and dipole-shaped radiation pattern is observed in E-plane and H-plane of the antenna, respectively.

The proposed antenna after embedding two CSRR structure has positive gain except the notch band frequencies.

4 Conclusion

A compact dual notched ultra-wideband (UWB) antenna has been presented. Filtering structure has a small size, and notching frequencies are obtained at 5.8/5.9 and 8.3 GHz in the ultra-wideband region. Antenna had cover from 3.1 to 14 GHz and with filtering effect. Further, antenna design has decent radiation characteristics in the preferred frequencies expect notch band regions. Similarly, notch band performance can be improved by introducing various CSRR and SRR structures in the antenna designs. Further, CSRR antenna can be used to cover various frequencies by embedding diodes and it can act as reconfigurable antennas.

References

1. Federal Communication Commission (2002) First report and order, revision of part 15 of the commission's rule regarding ultra-wideband transmission system FCC 02-48
2. Roy S, Foerster JR, Somayazulu VS, Leeper DG (2004) Ultrawideband radio design: the promise of high-speed, short-range wireless connectivity. Proc IEEE 92(2):295–311
3. Rajeshkumar V, Rajkumar R (2019) Corrugated fractal monopole antenna with enhanced bandwidth for ultrawideband applications. Int J Wirel Mobile Comput 16(1):66–71
4. Chung K, Kim J, Choi J (2005) Wideband microstrip-fed monopole antenna having frequency band-notch function. IEEE Microwave Wirel Compon Lett 15:766–768
5. Jaglan N, Gupta SD, Kanaujia BK, Srivastava S (2018) Band notched UWB circular monopole antenna with inductance enhanced modified mushroom EBG structures. Wirel Netw 24(2):383–393
6. Sanyal R, Sarkar PP, Chowdhury SK (2018) Miniaturized band notched UWB antenna with improved fidelity factor and pattern stability. Radioengineering 27(1): 39–46
7. Kelly JR, Hall PS, Gardner P (2011) Band-notched UWB antenna incorporating a microstrip open-loop resonator. IEEE Trans Antennas Propag 59(8):3045–3048
8. Bi D-H, Yu Z-Y, Mo S-G, Yin X-C (2009) Two new ultra-wideband antennas with 3.4/5.5 GHz dual band-notched characteristics. Microwave Opt Tech Lett 51:2942–2945
9. Liu H, Xu Z (2013) Design of UWB monopole antenna with dual notched bands using one modified electromagnetic-bandgap structure. Scient World J (Hindawi) 2013, ID 917975, pp1–9
10. Sanmugasundaram R, Somasundaram N, Rengasamy R (2020) Ultrawideband notch antenna with EBG structures for WiMAX and satellite application. Progr Electromagnet Res Lett 91:25–32
11. Rajesh Kumar V, Rajkumar R, Naidu PV, Kumar A (2019) A compact meta-atom loaded asymmetric coplanar strip-fed monopole antenna for multiband operation AEÜ. Int J Electron Commun 98:241–247
12. Rajkumar R, Usha Kiran K (2017) Gain enhancement of compact multiband antenna with metamaterial Superstrate. Lecture notes in Electrical Engineering, Springer LNEE, 468, 91–98

13. Rajkumar R, Usha Kiran K (2016) Study of harmonic suppression using defected ground structure for OSRR antenna. Microw Opt Technol Lett 58:1353–1358
14. Siddiqui JY, Saha C, Antar YM (2015) Compact dual-SRR-loaded UWB monopole antenna with dual frequency and wideband notch characteristics. IEEE Antennas Wirel Propag Lett 14:100–103
15. Li L, Zhou Z-L, Hong J-S, Wang B-Z (2011) Compact dual-band-notched UWB planar monopole antenna with modified SRR. Electron Lett 47(17): 950
16. Saha C, Siddiqui JY, Antar YMM (2014) Multilayered stacked square SRR coupled UWB monopole antenna with dual notch function. In: Proceedings of IEEE antennas and propagation society international symposium, 6–11 Jul 2014, Memphis. IEEE, pp 787–788

Decennial and Lifetime CV Risk Estimations Using JBS3 and ACC/AHA ASCVD Calculators in Kerala Based Indian Sub-Ethnics

Paulin Paul, B. Priestly Shan, and O. Jeba Shiney

Abstract The relative precision of JBS3 and ACC/AHA ASCVD cardiovascular risk assessment tools is unverified in Kerala based sub-ethnic population. In this context, a cross-sectional study with statistical comparison and validation of 10-year and lifetime estimations were done using JBS3 and ACC/AHA ASCVD tool. The proposed study was conducted in 579 Kerala based subjects aged between 40 and 60 years, with mean age (49.9 $\pm$ 5.7) years. Subject details were collected from the medical records of clinical sites at Ernakulum district, Kerala. The 10-year CV stratifications include 'Low 10-year' (<7.5%), 'Intermediate 10-year' ($\geq$7.5% and <20%), 'High 10-year' ($\geq$20%). The lifetime risk scores are 'Low lifetime' ($\leq$39%), 'High lifetime' ($\geq$40%) risk. JBS3 has identified 26.3% 'High 10-year' and 53% 'High Lifetime' risk. ACC/AHA ASCVD was able to predict 23% in 'High 10-year' and 58.4% in 'High lifetime' CV risk. The non-parametric Kruskal–Wallis H test was used to compare the models. A statistically significant difference was obtained only for the 10-year risk prediction models. The ACC/AHA ASCVD results demonstrate an increased proportion of reclassifications from the 'Intermediate' and 'Low' CV risk categories. With age being the strongest determinant for the risk in ACC/AHA ASCVD, it has reclassified several subjects from 'Low 10-year' to 'High lifetime' risk. Confirming the available literature, this study also identifies JBS3 as a better model for CV risk assessment in Kerala based Indian sub-ethnic population.

Keywords JBS3 · ACC/AHA ASCVD · CV risk prediction · Kerala CVD · 10-year CV risk · Lifetime CV risk

P. Paul (✉)
Sathyabama Institute of Science and Technology, Chennai, India

B. Priestly Shan
Galgotias University, Greater Noida, India

O. Jeba Shiney
Electronics and Communication Engineering, Galgotias University, Greater Noida, India

A. Sivasubramanian et al. (eds.), *Futuristic Communication and Network Technologies*, Lecture Notes in Electrical Engineering 792,
https://doi.org/10.1007/978-981-16-4625-6_40

1 Introduction

The intensity of the cardiovascular (CV) epidemic is high in the Kerala population compared to the North Indian and the Western populations. The high mortality and prevalence of CV events are rigorously connected with the increased metabolic risk factors. In conventional settings, according to the traditional routine, the CV risk factors are treated only when they become severe or with event occurrence. The prolonged exposure and impact of the multiple simultaneous severe and non-severe risk factors often ignored and lead to the development of acute cardiac manifestations. Studies highlight the role of lifestyle-related risk factors in the progression of CV disease [1–11]. The concept of CV risk prediction estimates the future possibility of CV risk, to ensure appropriate preventive and therapeutic intervention. Risk prediction is an evidence-based approach that can appropriately help the clinicians to administer suitable therapies with an added perspective based on lifestyle improvement. The risk prediction models are constructive tools for healthcare professionals. Several multivariable CV risk prediction calculators are available to predict the severity of CV risk both on a short-term and lifetime basis. Majority of these models are based on the traditional risk factors (TRFs) like age, gender, smoking, blood pressure, and lipid levels. In the absence of a customized Indian-based CV risk prediction model, we rely on western based tools. These models either tend to underestimate or overestimate the risk prediction when used in the Indian population. Different tools give varying results when used in Indian subjects. India with its co-existence of diverse ethnicities, the CV risk predictions may vary for each of the specific sub-ethnic populations and the accuracy of these calculators varies. Study based on northern population may be different from the southern population and cannot be generalized to be used for designing prevention strategies. The CV risk assessment tools available include Framingham risk score (FRS) [12], Prospective Cardiovascular Munster Score (PROCAM) [13], QRISK [14], Systemic Coronary Risk Evaluation (SCORE) [15], World Health Organization/International Society of Hypertension CVD risk prediction charts (WHO) [16], and the recently developed American College of Cardiology/American Heart Association (ACC/AHA) pooled cohort equations (ACC/AHA) [17] and the 3rd Joint British Societies' risk calculator (JBS3) [18]. The FRS tool developed in 1998 is the first CV risk prediction calculator [12], and the American College of Cardiology/American Heart Association and Atherosclerotic Cardiovascular Disease (ACC/AHA ASCVD) risk calculator [17] were also based on the basic TRFs. Some of the existing models have considered few additional risk markers to define the total CV risk prediction. The north Indian-based CV risk assessment studies available lack comprehensive risk assessment [19–21]. However, JBS3 and ACC/AHA ASCVD have shown comparatively superior performance in Indian-based studies with both short-term (10-years) and lifetime CV risk prediction available. With limited Indian-based studies available, no Kerala based studies have verified the accuracy of JBS3 and ACC/AHA ASCVD models. This study aims to compare these two clinically relevant models using subjects aged between 40 and 60 years. The objective is to statistically evaluate and compare the JBS3 and ACC/AHA ASCVD scores in Kerala ethnic population estimating the 10-year and lifetime CV risk.

2 Methodology

In the presence of several existing CV risk prediction models, two most promising models such as JBS3 and ACC/AHA ASCVD studied to be suitable in Indian ethnics were compared in the local settings of Indian-based Kerala sub-ethnics, to identify the most suitable model for Kerala population data, to further facilitate the investigation of extension with newer risk markers. The overall description of the study is shown in Fig. 1

2.1 Study Population and Risk Variables Studied

This is a cross-sectional population-based study using subjects aged between 40 and 60 years. The mean age is 49.9 ± 5.7 years, with 58.03% males. The study data collected from 579 subjects from the medical archives of clinical locations at Ernakulum district in Kerala, tracking the long-term historical records for over 1 year, from March 2017 to February 2018. Institutional Ethics committee approval was obtained, and all the records used were anonymized before the study analysis. 10-year and lifetime CVD risk assessment of both the models is done. The ACC/AHA

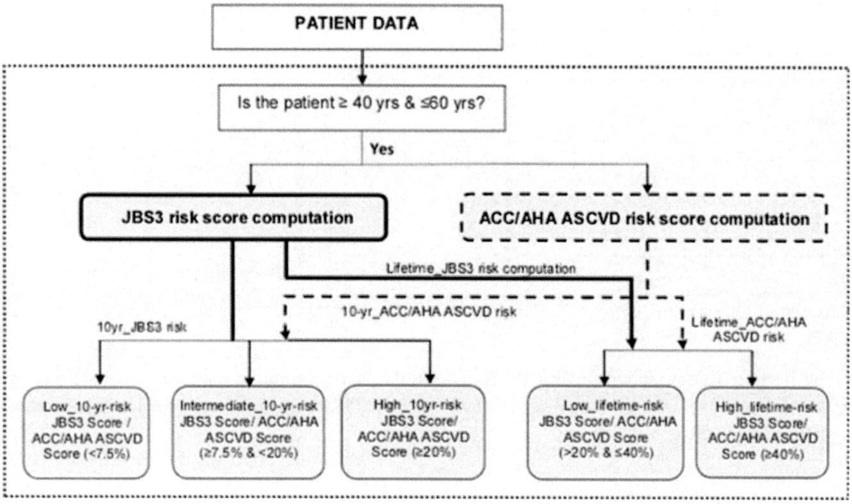

Fig. 1 Flow description of the study

ASCVD are primarily based on the traditional risk factors (TRFs), whereas JBS3 is more comprehensive and considers the family history of CVD, renal and peripheral vascular diseases that progressively leads to CVD events like chronic kidney disease (CKD), rheumatoid arthritis (RA), arterial fibrillation (AF) in addition to TRFs. The parameters used in the respective CV risk prediction tools are as shown in Table 1. The ACC/AHA ASCVD calculator has mentioned 'White', 'American-African' ethnicities mentioned, whereas JBS3 specifically has the 'Indian' ethnic data available in the model in addition to the other ethnicities.

Table 1 CVD risk variables in JBS3 and ACC/AHA ASCVD risk calculators

Variables	JBS3[a] model	ACC/AHA ASCVD[a] model
Type of risk predictions	10-year CVD risk	10-year CVD risk
	Lifetime risk prediction	Lifetime risk prediction
Age range	Yes, for 10 years (30–80 years)	Yes, for 10 years (20–79 years)
Gender	Yes	Yes
Ethnicity	Yes (Indian mentioned)	Yes (White/American/Others)
Presence of diabetics	Yes	Yes
Body mass index (BMI)	Yes (Calculated) ($\geq$20–50 kg/m^2)	No
Demography type	Yes	No
Smoking history	Yes	Yes
Total cholesterol	Yes (<290 mg/dL)	Yes (130–320 mg/dL)
HDL cholesterol	Yes (>18 mg/dL)	Yes (20–100 mg/dL)
Non-HDL cholesterol	Yes (Calculated)	Not used
Systolic blood pressure (SBP)	Yes (70–210 mmHg)	Yes (90–200 mmHg)
History of anti-hypertension treatment	Yes	Yes
History of premature CVD events	Yes	No
Chronic kidney disease (CKD)	Yes	No
History of rheumatoid arthritis (RA)	Yes	No
History of arterial fibrillation (AF)	Yes	No

[a] The lifetime risk prediction using JBS3 is possible only upto the maximum age of 95 year. In ACC/AHA ASCVD, the lifetime risk calculation is possible for 30 years, only for subjects between ages (40–59 years)

2.2 *Statistical Analysis*

We have compared the JBS3 and ACC/AHA ASCVD models to identify prospective statistical significance of the risk categories and the ability of the respective models to stratify the subjects with the presence of higher risk markers into high-risk. The data are managed using the Microsoft Excel spreadsheet (2007), and the statistical analysis was done with IBM SPSS (version 25) package for windows. There were no missing or incomplete data. Standard descriptive analysis was done to present the baseline characteristics of the study data. Data are presented as counts (and proportions) for categorical variables in the summary statistics. The non-parametric Kruskal–Wallis H test is used to compare the statistical significance of the categorized 10-year (low, intermediate, high) and lifetime (low, high) risk predictions derived across the JBS3 (available at www.jbs3risk.com as http://www.jbs3risk.com/JBS3Risk.swf) and ACC/AHAASCVD (https://tools.acc.org/ldl/ascvd_risk_estimator/index.html#!/calulate/estimator/) models. A Chi-square test is used to present the comparison between the dependent and categorical CVD risk variables. A p value of <0.05 is taken statistically significant. The distribution of subjects across the 10-year and lifetime risk categories and the reclassifications across the models is assessed and evaluated.

3 Results and Discussion

The mean age of the study population selected was (49.9 ± 5.7) years. Males were predominant with 58.03%. The intermediate age group is further sub-sectioned as 40–49 and 50–59 with 47.67 and 52.33% of subjects, respectively. About 1.73% of the study population had pre-existing CV events. About 13.64% had a family history of premature CV disease. The prevalence of BMI is very high in 68.57% and with an average BMI of (26.1 ± 3.10) kg/m^2. The prevalence of hypertension subjects with SBP > 140 mmHg is 31.43%, and average SBP of (26.4 ± 17.6) kg/m^2. About 25.91% of the study population is diabetic. Low HDL is prevalent in 57.5% of the population. Around 6.91% of the total population suffered from AF. Chronic inflammatory diseases like RA were present in 16.75% of subjects, and 7.77% has CKD.

3.1 *Ten year and Lifetime JBS3 and ACC/AHA ASCVD Risk Score*

JBS3 and ACC/AHA ASCVD models have both 10-year and lifetime CV risk estimations. In our study, JBS3 model has identified 26.3% in high 10-year risk (≥20%) and 53.0% as the high lifetime CVD risk, whereas the ACC/AHA ASCVD model has

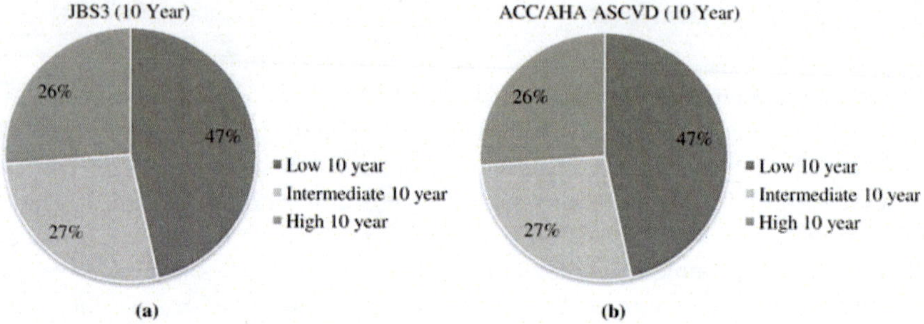

Fig. 2 **a** Distribution of JBS3 10-year risk stratification, **b** distribution of ACC/AHA ASCVD 10-year risk stratification

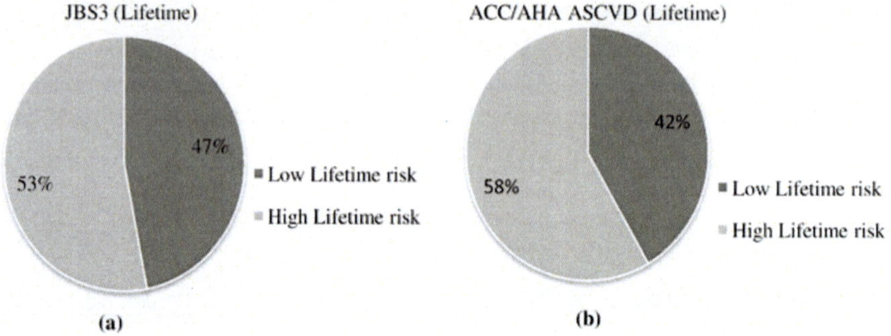

Fig. 3 **a** Distribution of JBS3 lifetime risk stratification, **b** distribution of ACC/AHA ASCVD lifetime risk stratification

stratified 23.0% in the high-10 year and 58.4% in the high lifetime CV risk categories. The 10-year JBS3 and ACC/AHA ASCVD distribution is shown in Fig. 2a, b, respectively. Similarly, the lifetime JBS3 and ACC/AHA ASCVD distributions are shown in Fig. 3a, b. When JBS3 identified 27.3% intermediate-risk subjects, ACC/AHA ASCVD model identified only 12.4%. In this study, the JBS3 model could classify a higher number of subjects with elevated risk profiles into the 'High' and 'Intermediate' CV risk categories. These are the subjects eligible to seek further clinical investigations of subclinical atherosclerotic status for appropriate therapeutic recommendations to avoid near-time event or avoid reclassification to irreversible lifetime CV risk. The ACC/AHA ASCVD model has a higher number of reclassifications, with a higher proportion of subjects with non-optimal risk factors reclassified to 'Intermediate' and 'Low' categories. The significance of the 10-year and lifetime CV risk distribution across the JBS3 and ACC/AHA CV risk prediction models was tested using the non-parametric Kruskal–Wallis H test. A statistically significant (p value <0.05) difference was noticed only for the 10-year CV risk prediction of the models. Compared to ACC/AHA ASCVD, the JBS3 model presents a significant variation in the 'Intermediate' risk category, which has contributed the higher number of reclassifications. Comparing the results of our study with the previously published literature; present evidences with varying accuracies for various Indian sub-ethnical regions. The 'High' 10-year risk (≥20%) classifications referred in the contemporary studies by Garg et al. have (41.4%) with JBS3 and (28.3%) with ACC/AHA ASCVD [19], whereas the study by Bansal et al. [20] has shown (55.9%) 'High' 10-year risk with JBS3, and (30.2%) with ACC/AHA ASCVD, similarly in an another study by Bansal et al. [21] JBS3 has shown (58.6%) and ACC/AHA ASCVD has (21.4%) 'High' 10-year risk. This evident variations in the 'High' and 'Intermediate' risk proportions in the Kerala based population data suggest that the generalizations based on any or few sub-populations would not fit for developing prevention strategies to whole of India. This suggests for the importance of local epidemiological data-based studies.

In this study, JBS3 has classified higher number of subjects with elevated risk profile to 'High' 10-year risk category. Also, JBS3 has shown improved classification in the 'high-risk' for female subjects with an elevated risk profile and has better accuracy for female subjects aged above 50 years with elevated risk profiles. Whereas, in the lifetime risk predicted, the ACC/AHA ASCVD model was able to classify higher high lifetime risk. Age is a strong determinant in ACC/AHA ASCVD model that classifies many of the low 10-year CV risk to high lifetime risk. In JBS3, about 53.3% of the subjects are stratified in the 'Intermediate' and 'High' 10-year category, eligible to be considered for therapeutic and lifestyle modifications, whereas the ACC/AHA ASCVD model could stratify only 35.4% of the subjects in spite of high prevalence of non-optimal risk factors. The results of the study confirm JBS3 to perform the best on the Kerala population. The age, gender-based distribution of both 10-year JBS3 and ACC/AHA ASCVD models are shown in Fig. 4a, b. The age, gender-based distribution of lifetime JBS3 and ACC/AHA ASCVD models are shown in Fig. 5a, b. For subjects aged above 50 years, the JBS3 model has presented a more reasonable proportion in the high CV risk with 22.2–33.3% in males and 20–27.5% in females, whereas ACC/AHA ASCVD model shows a very nominal

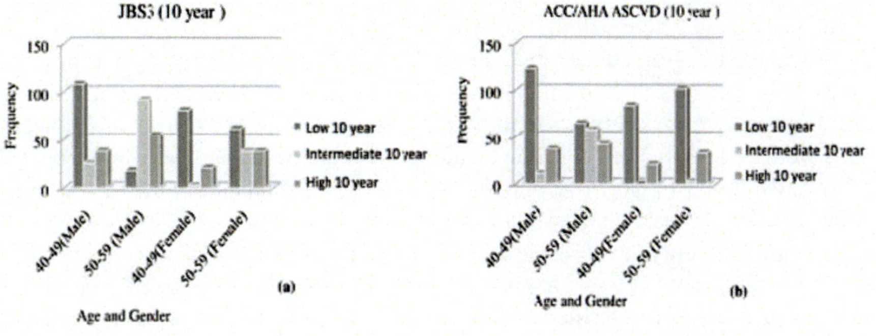

Fig. 4 **a** Age and gender distribution of 10-year JBS3 subjects, **b** age and gender distribution of 10-year ACC/AHA ASCVD subjects

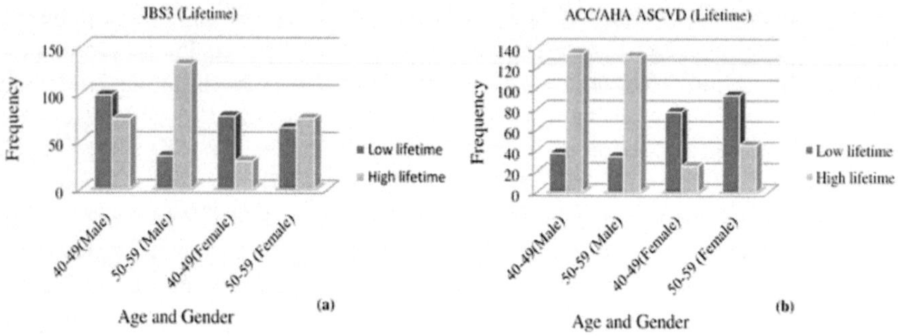

Fig. 5 **a** Age and gender distribution in lifetime JBS3, **b** age and gender distribution of lifetime ACC/AHA ASCVD

increase with 21.6–25.5% in males and 20% to 23.9% in females in spite of the presence of many high-risk variables.

In the JBS3 lifetime model, the proportion of subjects aged above 50 nearly doubles and has identified a higher proportion of subjects with elevated SBP levels (above 140 mmHg) and hypertension treatment (41.1%) classified into the 'high' CV risk category, compared to the ACC/AHA ASCVD model. JBS3 has categorized a higher proportion of subjects with diabetics, HDL-C and high TC in the 'intermediate' and 'High' risk category. Contradictory to this, the ACC/AHA ASCVD model has categorized comparatively more subjects with a 'High' risk profile into 'Low' risk category. In the lifetime risk predicted, ACC/AHA ASCVD model has outperformed the JBS3 model with a higher proportion of subjects with elevated risk profile reclassified to 'High Lifetime' CV risk category. With only minor difference in the 'High' 10-year risk classifications, but the 'Low' and 'Intermediate' strata of the models show a comparatively reasonable variation. The ACC/AHA ASCVD model shows a reasonably high percentage of low risk (64.5%) classifications. It has reduced percentage in 'Intermediate' with only 12.4% in spite of the presence of multiple 'High' CV risk variables. JBS3 shows a reasonably elevated percentage with 27.3% at the 'Intermediate' level.

The smaller 'Intermediate' strata of the ACC/AHA ASCVD risk model show the underestimation of risk in this population and have resulted in the reclassifications of the 'intermediate' risk subjects to the 'Low' risk, increasing the 'Low' risk subjects. This would result in clinical neglect leading to an increased number of unexpected CV events. In the 10-year risk classification, JBS3 has higher number in 'Intermediate' and 'High'-risk categories, in line with the risk profile. ACC/AHA ASCVD model considers only the basic parameters to compute the CV risk. This has resulted in a large number of reclassifications of subjects across the categories. The higher number of subjects reclassified to the 'Low' risk group in spite of the presence of multiple CV risk factors has resulted in a high number of underestimations, which may lead to a faulty prognosis and treatment plan. JBS3 model uses additional variables like BMI, history of premature CVD, AF, RA, and CKD in addition to the TRFs for improved absolute risk prediction, capable to classify the subjects with a

Table 2 Reclassification of subjects across the 10-year JBS3 risk category to ACC/AHA ACVD model

	Low ASCVD	Intermediate ASCVD	High ASCVD
Low JBS3	265 (45.8%)	4 (0.7%)	0 (0.0%)
Intermediate JBS3	104 (18%)	53 (9.2%)	1 (0.2%)
High JBS3	5 (0.9%)	15 (2.6%)	132 (22.8%)

Table 3 Reclassification of subjects across the lifetime risk strata's in JBS3 and ACC/AHA ACVD models

	Low lifetime ASCVD	High lifetime ASCVD
Low lifetime JBS3	189 (32.6%)	83 (14.3%)
High lifetime JBS3	52 (9%)	255 (44%)

higher number of CV risk factors more precisely to intermediate and high-risk categories. The reclassifications of subjects across the 10-year and lifetime risk strata are consolidated in Tables 2 and 3, respectively. The reclassification is highest in the 'Intermediate'-risk strata. Evaluating the reclassifications, only 1.5% of the JBS3 low risk is up-classified to the Intermediate 10-year ACC/AHA ASCVD risk strata; however, a reasonably high proportion of 65% of the 'Intermediate' 10-year JBS3 subjects got down-classified into the 'Low' 10-year ACC/AHA ASCVD.

The 'Intermediate' 10-year JBS3 to high 10-year ACC/AHA ASCVD up-classification is negligible (0.63%). Similarly from 'High' 10-year JBS3, the down-classifications to 'Intermediate' ACC/AHA ASCVD are 9.9% and 'Low' ACC/AHA ASCVD is 3.3%. In the lifetime risk computed, 30.5% of the JBS3 low Lifetime risk computed was reclassified to high ACC/AHA ASCVD and 16.9% of the JBS3 'High' lifetime risk computed got down-classified to 'Low' lifetime ACC/AHA ASCVD. With higher number of non-optimal risk variables seen in the intermediate age, the coexistence of two or more major risk variables in a subject has a significant impact on classifying the high 10-year and high lifetime categories. Periodic assessments and timely intervention of risk factors would avoid irreversible high lifetime reclassifications and the associated health consequences.

4 Conclusion

JBS3 tool performs better in Indian sub-ethnic Kerala population. JBS3 demonstrates a prominent variation in the 'Intermediate' category than the high-risk category. The subjects in the 'Intermediate' risk category deserve importance and appropriate follow up, else would gradually lead to the reclassification to a high-risk category. The difference between the 'High' risk predictions in the present study compared to the north Indian-based studies, and the difference in results points the importance of studies based on local data for population-based interventions. The

ACC/AHA ASCVD model has higher proportion of underestimations in the 'Low' and 'Intermediate' 10-year CV risk categories. The JBS3 tool with only small overall improvement in risk prediction compared to ACC/AHA ASCVD indicates the need to extend the risk prediction by incorporating more suitable risk markers. Quantified subclinical atherosclerotic CV risk markers such as carotid intima-media thickness (cIMT) and carotid plaque (cP) measurements can be considered for a secondary-level ASCV screening specifically for the 'Intermediate' and 'High' level risk classified subjects to improve the risk prediction. The main limitation of this study is that it lacks the advantages of a large-scale prospective epidemiologic research. Therefore, lacking the strength to represent a state-based generalized remark about the tool used, which requires broader studies. Moreover, only 'Intermediate' category subjects are included in the analysis and observations, which is not sufficient to make an overall generalized observation or suggestion for the entire Kerala population.

References

1. Soman CR (2011) All-cause mortality and cardiovascular mortality in Kerala State of India: results from a 5-year follow-up of 161 942 rural community dwelling adults. Asia-Pac J Public Heal 23(6):896–903
2. Prabhakaran AO (2016) Burden of non-communicable disease in Kerala. Presented in. Kerala Padana Congress, Thiruvananthapuram, Kerala 2016 1–12
3. Srinivasan S (2016) Metabolic syndrome in rural Kerala: a hospital based study. Int J Adv Med 3(4):898–904
4. Areekal B (2015) Prevalence of risk factors for cardiovascular disease among adults older than 30 years in a rural area in central Kerala, India. Int J Med Sci Public Heal 4(12):1655–1660
5. Ismail I (2016) Metabolic syndrome and its associated factors among the adult population residing in Kannavam tribal area of Kannur District. Kerala. Trop J Med Res 19(1):36–41
6. Kovil A (2013) Cardiovascular risk levels in patients with type 2 diabetes in rural and urban areas of Kerala: a cross sectional study. Indian J Basic Appl Med Res 2(6):658–663
7. Kumar SP (2017) A study on the glycaemic, lipid and blood pressure control among the type 2 diabetes patients of north Kerala, India. Indian Heart J 7–10
8. Thankappan KR (2010) Risk factor profile for chronic non-communicable diseases: results of a community-based study in Kerala. India. Indian J Med Res 131(1):53–63
9. Harikrishnan S (2018) Prevalence of metabolic syndrome and its risk factors in Kerala, South India: analysis of a community based cross-sectional study. PLOS-one 51(14):1–16
10. Zachariah G (2013) Prevalence of coronary artery disease and coronary risk factors in Kerala, South India: a population survey-design and methods. Indian Heart J 65(3):243–249
11. Paulin P (2020) Cardiovascular risk prediction using jbs3 tool: a Kerala based study. Curr Med Imaging 16. https://doi.org/10.2174/1573405616666200103144559
12. Wilson PWF (1998) Prediction of coronary heart disease using risk factor categories. Circ 97:1837–1847
13. Assmann G (2002) Simple scoring scheme for calculating the risk of acute coronary events based on the 10-year follow-up of the prospective cardiovascular Munster (PROCAM) study. Circulation 105:310–315
14. Hippisley-Cox J (2007) Derivation and validation of QRISK, a new cardiovascular disease risk score for the United Kingdom: prospective open cohort study. Br Med J 335:1–12
15. Conroy RM (2003) Estimation of ten-year risk of fatal cardiovascular disease in Europe: the SCORE project. Eur Heart J 24:987–1003

16. Hippisley-Cox J (2007) Derivation and validation of QRISK, a new cardiovascular disease risk score for the United Kingdom: Prospective open cohort study. Br Med J, 1–12
17. Goff DC (2013) ACC/AHA guideline on the assessment of cardiovascular risk: a report of the American college of cardiology/American heart association task force on practice guidelines. Circulation. 129
18. JBS3 Board: Joint British Societies' consensus recommendations for the prevention of cardiovascular disease (JBS3). Heart J 100 (2014)
19. Garg N (2007) Comparison of different cardiovascular risk score calculators for cardiovascular risk prediction and guideline recommended statin uses. Indian Heart J, 1–6
20. Bansal M (2014) Comparative accuracy of different risk scores in assessing cardiovascular risk in Indians: a study in patients with first myocardial infarction. Indian Hear J Sci Direct 66(6):580–586
21. Bansal M (2015) Relationship between different cardiovascular risk scores and measures of subclinical atherosclerosis in an Indian population. Indian Heart J 67(4):332–340

A Brief Survey on Metamaterial Antennas: Its Importance and Challenges

P. S. Chindhi, G. B. Kalkhambkar, H. P. Rajani, and Rajashree Khanai

Abstract Metamaterials are opening a new way of refining the material science and related areas. Metamaterials broke the limitation of naturally occurring substances by artificially creating the desired material properties. This chapter presents a brief survey on metamaterials and their importance in the area of antenna design. Metamaterials in fractal antenna geometries, their applications in different areas are discussed. A brief discussion on metamaterial parameter extracting methods, different metamaterial geometries, challenges, and the way forward is given.

Keywords Metamaterials · Negative refractive index · Permittivity · Permeability · Fractal antennas

1 Introduction

META means beyond; in the past two decades, metamaterial has been the popular and debated topic among researchers. Naturally occurring substances have some limit at ions in terms of material properties such as permittivity and permeability. The invent ion of metamaterials created a new breakthrough in material science which has the capability to alter the material properties artificially with the aid of specifically engineered structures. The elementary property that caused the difference is the negative refractive index of the material. Negative refractive index occurs if the real part of relative permittivity and relative permeability is concurrently negative,

P. S. Chindhi (✉)
Department of Electrical Engineering, Sant Gajanan Maharaj College of Engineering Mahagaon, Mahagaon, India

G. B. Kalkhambkar
Department of Electronics and Telecommunication Engineering, Shivaji University, Kolhapur, India

H. P. Rajani
Department of Electronics and Communication Engineering, KLE's Dr. MSSCT, Belgaum, India

R. Khanai
Visvesvaraya Technological University, Belagavi, India

A. Sivasubramanian et al. (eds.), *Futuristic Communication and Network Technologies*, Lecture Notes in Electrical Engineering 792,
https://doi.org/10.1007/978-981-16-4625-6_41

this condition is necessary but not sufficient to consider the material as a metamaterial. In the field of antennas, it is now a huge demand to enhance the antenna output parameters such as gain, bandwidth, directivity. Veselago a Russian scientist introduced the idea of metamaterials in the year 1968; after this, the researchers tried to explore naturally available materials for possible negative refraction phenomena but could not succeed, and the research in metamaterials was diluted for 30 years. After the year 2001, scientists experimentally demonstrated the negative refraction of light rays and the research in metamaterial got a new life. Various shapes of metamaterials and their parameters extraction techniques are given in many papers. An I-shaped left-handed material is used to enhance the resonance and improve the gain of patch antenna, the material gives the negative permittivity and permeability, and it is realized that the gain, bandwidth, and VSWR are improved with a reduction in the return loss [1]. Nicolson-Ross-Weir (NRW) a method is explained in [2]. An inclusion of metamaterial Split Ring Resonators (SRR) in the used to retrieve the dielectric parameters, the details of the NRW method are substrate of a slotted chopped patch antenna improves the bandwidth and helps in the miniaturization of the patch antenna. The size reduction of up to 37% is obtained with the aid of the metamaterial substrate [3]. Electromagnetic Band Gap (EBG) structures are employed for better return loss and efficiency with increased frequency bands [4]. Metamaterial as a superstrate provides tunability in patch antennas with the enhanced performance [5]. An active device such as a varactor diode with metamaterial structures is used to improve its performance. Nicolson RossWeir (NRW) method is used to measure effective permittivity and permeability in many papers [6]. Usually, it is very difficult to get high gain and high bandwidth simultaneously in patch antennas. With the aid of metamaterial, we can achieve this up to some extent [7]. Another very important fact about metamaterials is, it greatly helps in antenna miniaturization. The permittivity of the substrates is another important parameter. Substrates with high permittivity cause reduced bandwidth and efficiency, and also, it is very difficult to achieve impedance matching in such substrates, here comes the need for negative permittivity substrates, that is metamaterials [8]. Metamaterial structures in fractal microstrip patch antennas play a very important role. Due to the fabrication limitations, the field of fractal antennas has not been explored to the extraordinary level, but the inclusion of metamaterial geometries can give a different dimension to the fractal antenna research [9–17]. Different types of metamaterials are explored in [18–24]. This chapter explores the technicalities of metamaterials by combining them on fractal antennas. Various studies have been made on such designs, but the metamaterial loaded fractal antennas need the exclusive attention today. Various metamaterial unit cells are also explained, and the metamaterial parameters extraction methods are also briefed in the upcoming sections. Discussion on fractal metamaterials cannot be concluded in single literature as there is a huge part of it still need to be explored. The attempt is made to cover the fractal metamaterial in case of beam forming, wearable antennas, notch filter, wideband antennas, and wavefront shaping applications. Fundamentals of metamaterials.

Every metamaterial consists of metamolecules like atom in a substance, and the incident energy on the metamolecules is reradiated to form a reflected and transmitted wave. The resonance in metamolecule depends on its inductive and capacitive response given by Eq. (1). Metamolecules conduct oscillatory current to form standing waves. The wavelength is given by Eq. (2).

$$\omega_0 = 2\pi f_o = \frac{1}{\sqrt{LC}} \tag{1}$$

$$\lambda_o = \sqrt{\in}\frac{2l_e}{m} \tag{2}$$

where

'*l*e' is electrical length of the material.

'*m*' is an integer usually $m < 4$.

The rudimentary classification of metamaterials is given in Fig. 1. Apart from this classification nonlinear, amplifying, quantum, sensing tunable metamaterials also exists [18–24]. There are different metamaterial structures available, and basic ones are given in Fig. 2. The characterization of materials based on permittivity and permeability is shown in Fig. 3.

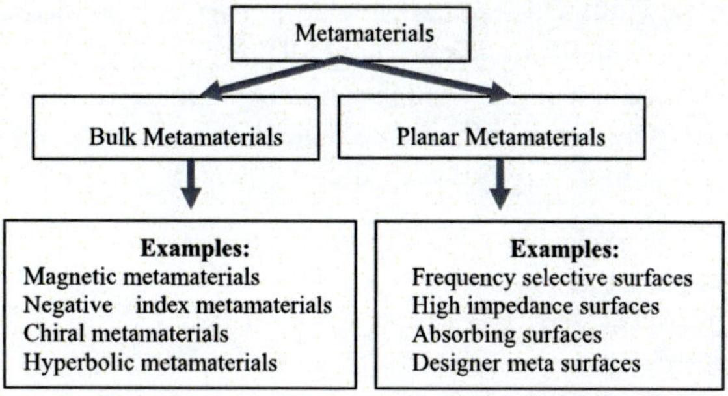

Fig. 1 General classification of metamaterials

Fig. 2 Different metamaterial shapes, Ref. [18]

Fig. 3 Material properties Ref. [12]

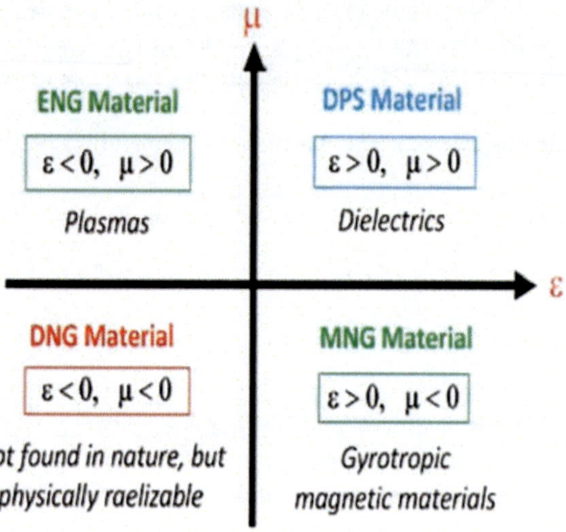

2 Metamaterials with Fractal Antennas

The inclusion of square SRR and micro-split SRR on fractal antennas improves the bandwidth with a remarkable difference, and multiband performance is obtained [9]. A unit cell SRR and micro-slit SRR is shown in Fig. 4a, b. A metamaterial unit cell is constructed in [10], and the structure is made tunable with the aid of two-pin diodes resulting in two different configurations as shown in Fig. 4c, with this structure high μ and high ε regions are detected which are useful in beamforming applications.

Fig. 4 a Unit cell of SRR, b Micro-slit SRR, c Tunable antenna

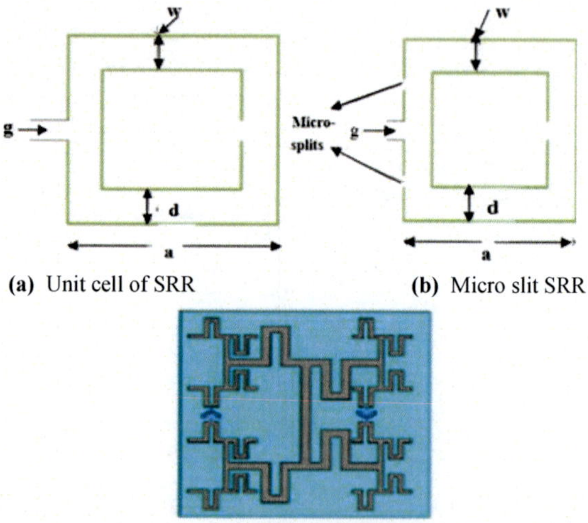

(a) Unit cell of SRR (b) Micro slit SRR

(c) Tunable antenna

Near zero performance is observed which when placed on positive media results in an enhanced gain. In the case of wearable antennas, metamaterials are included in the ground plane to minimize the Specific Absorption Rate (SAR) [11] since relative permittivity and permeability in the case of metamaterial leads to Negative Refractive Index (NRI) and it is related to tissue properties as $n = \sqrt{\mu_r \varepsilon_r}$, according to basics of optics the electromagnetic wave bent to the same side of normal as incident waves and absorption of wave becomes minimum, and thus, SAR reduces automatically.

$$\mathrm{SAR} = \frac{\sigma |E|^2}{2\rho} \tag{3}$$

The effective dielectric constant of metamaterials is derived from its polarizabilities as shown in Eqs. (4–6). N is a volume density. In Hilbert curve, it is realized that the permittivity curve becomes smoother as number of iterations increase and hence, add an extra benefit to metamaterial antennas in bandwidth improvement [12].

$$\boldsymbol{P} = \boldsymbol{\varepsilon}\left(\overline{\overline{a}}_{ee}\boldsymbol{E}\right) + \overline{\overline{a}}_{ee}\eta \boldsymbol{H} \tag{4}$$

$$\boldsymbol{m} = \frac{\overline{\overline{a}}_{me}}{\eta}\boldsymbol{E} + \overline{\overline{a}}_{mm}\boldsymbol{H} \tag{5}$$

$$\varepsilon_{\mathrm{eff}} = \varepsilon \frac{(1 + 0.66 N a_{ee})(1 - 0.33 N a_{mm}) - 0.22 N^2 a_{em}^2}{(1 - 0.33 N a_{ee})(1 - 0.33 N a_{mm}) * 0.11 N^2 a_{em}^2} \tag{6}$$

Fractal inspired metamaterial cell is utilized in the applications of notch filters [13]. Fractals with metamaterial structure are used as a absorber to suppress unwanted radiations in the application like mobile communication [14]. The equivalent circuit of absorber acts as a parallel RLC circuit, and it can be directly related to the permittivity and permeability of the materials and hence to the metamaterials using following Eqs. (7–10).

$$Z_I = Z_S \| Z_{\mathrm{TML}} \tag{7}$$

$$Z_S = R_S + j\omega L_S + \frac{1}{j\omega C_S} \tag{8}$$

$$Z_{\mathrm{TML}} = Z_d + Z_c \tag{9}$$

$$= j\sqrt{\frac{(\mu_0 \mu_r)}{(\varepsilon_0 \varepsilon_r)}} \tan(kd) + R_d + \frac{1}{j\omega C_c} \tag{10}$$

A fractal porous metal exhibits superior plasmonic properties; therefore, fractal like metamaterials allow greater coupling between incoming energy, and the metal thus increases the skin depth. When fractal dimensions reduce, it decreases the dielectric function and thus reduces the ohmic losses [15] space-coiling Mie-Resonance-based Metamaterials (MRMMs) are used as a slow wave structures with multiple band gaps [16], this is widely used in acoustics for sound waves but can also help in wavefront shaping applications in low frequency. Seismic metamaterials are good at attenuating surface waves within 1–20 Hz by controlling the bandgap properties of the fractal shapes [17].

3 Metamaterial Parameter Extraction Techniques

Simplest way to the extraction of parameters of metamaterials is by using transmission and reflection data, based on reflection coefficient expressed in Eq. (10), but in this method the real part of refractive index undergoes some change and the issue remains unhandled. It is expressed in Eqs. (11–14).

$$n = \frac{1}{kd}\cos^{-1}\left[\frac{1}{2S_{21}}\left(1 - S_{11}^2 + S_{21}^2\right)\right] \tag{11}$$

$$z = \sqrt{\frac{\left(1 + S_{11}^2\right)^2 - S_{21}^2}{\left(1 - S_{11}^2\right)^2 - S_{21}^2}} \tag{12}$$

$$\varepsilon = \frac{n}{z} \tag{13}$$

$$\mu = nz \tag{14}$$

where ω = Radian frequency, d = Thickness of unit cell.

Another way to determine the reflection and transmission data based on impedance is presented in [25], and it involves the complex algorithm. The imaginary part of the refractive index does not change, but for higher logarithmic branches, the real part undergoes some change this is called as branching problem. It is necessary to include a correction factor which can be done with the help of Kramers–Kronig relation Eq. (15) as given below. In [25], the author observed that the Kramers–Kronig relation does not hold good in case of increased thickness substrates.

$$n^{KK}(\omega') = 1 + \frac{2}{\pi\mathcal{P}}\int_0^\infty \frac{\omega\mathcal{K}_{eff(\omega)}}{\omega^2 - \omega'^2}d\omega \tag{15}$$

where $\mathcal{P}$ is principle value of improper integral.

Parameter extraction method with branching problem is mathematically discussed in [26]. The improvised method which solves the sigh ambiguity problem and the branch selection in extracting metamaterial parameter is presented in [27].

4 Conclusion

The metamaterial antennas provide the better options in the applications demanding high gain and bandwidth. The metamaterials serve the good candidate in the applications of wearable antennas where specific absorption rate of electromagnetic waves needs to be controlled. Higher order iterations of fractal antenna design impose a challenge in terms of space and fabrication if metamaterial unit cells are added. Effective parameter extraction technique imposes a challenge on metamaterial antenna research.

References

1. Song X-H, Wu W-Y, Shen T-G, Zhou Y-Q, Investigation of a patch antenna based on I-shaped left-handed material. https://doi.org/10.1016/j.ijleo.2010.09.02
2. Sharma R, Singh H (2015) Design of 8 - shaped DNG metamaterial for GSM 1.8 GHz applications: IJISET Int J Innov Sci Eng Technol 2(8)
3. Dwivedi S, Mishra V, Kosta YP (2012) Metamaterial inspired patch antenna miniaturization technique for satellite. In: 1st international conference on emerging technology trends in electronics, communication and networking
4. Kumar H, Kanth RK, Liljeberg P, Tenhunen H (2011) Metamaterial based slotted patch antenna. In: 10th international conference on telecommunication in modern satellite cable and broadcasting services (TELSIKS), 5–8 Oct. 2011
5. Jacob SB, Khot UP (2014) Frequency-tunable metamaterial for microstrip patch antenna. In: 2014 international conference on circuits systems communication and information technology applications
6. Varma R, Sharma SK (2014) Improvement in rectangular microstrip patch antenna parameters using metamaterial with active devices. 978-1-4799-6052-1/14/$31.00 ©2014
7. Das S, Sahu S, Design of high gain, broadband resonant cavity antenna with meta-material inspired superstrate. Int J Electron Commun https://doi.org/ https://doi.org/10.1016/j.aeue.2018.12.021
8. Akçelik H, Yücedağ OM, Torun E, Koçer H, A metamaterial based broadband microstrip. IEEE
9. Choudhury B, Manickam S, Jha RM (2013) Particle swarm optimization for multiband metamaterial fractal antenna. Hindawi Publ Corp J Optim 2013, Article ID 989135, 8 pages. https://doi.org/10.1155/2013/989135
10. Zarghooni B, Denidni TA (2014) New fractal metamaterial unit-cell for microwave applications. In: The 8th European conference on antennas and propagation (EuCAP 2014)
11. Ahmed MI, Ahmed MF, Shaalan AA, A novel wearable metamaterial fractal antenna for wireless applications. IEEE
12. Krzysztofik WJ, Fractals in antennas and metamaterials applications. Fractal Anal Appl Phys Eng Technol. https://doi.org/10.5772/intechopen.68188
13. Smith K, Adams R (2018) A broadband negative epsilon fractal metamaterial unit cell for coaxial notch filter applications. Progr Electromagnet Res C 86:257–267

14. Fan S, Song Y (2018) UHF metamaterial absorber with small-size unit cell by combining fractal and coupling lines. Hindawi Int J Antennas Propag 2018, Article ID 9409152, 9 pages https://doi.org/10.1155/2018/9409152
15. Garoli D, Calandrini E, Bozzola A, Toma A, Cattarin S, Ortolani M, De Angelis F, Fractal-like plasmonic metamaterial with a tailorable plasma frequency in the near-infrared. https://doi.org/10.1021/acsphotonics.8b00676 ACS Photonics
16. Man X, Liu T, Xia B, Luo Z, Xie L, Liu J (2018) Space-coiling fractal metamaterial with multi-bandgaps on ubwavelength scale. J Sound Vibr 423:322–339
17. Du Q et al (2018) H-fractal seismic metamaterial with broadband low-frequency bandgaps. J Phys D Appl Phys (in press) https://doi.org/10.1088/1361-6463/aaaac0
18. Fedotov V, Metamaterials. Springer handbook of electronic and photonic materials. https://doi.org/10.1007/978-3-319-48933-9_56
19. Ginis V, Tassin P, Veretennicoff I, Amplifying optical gradient forces with metamaterials. 978-1-4577-0733-9/12/$26.00 ©2012 IEEE
20. Liu AQ, Zhu WM, Tsai DP, Zheludev NI, Micromachined tunable metamaterials: a review. IOP Publ J Opt. https://doi.org/10.1088/2040-8978/14/11/114009
21. Felbacq D, Antezza M (2012) Quantum metamaterials: a brave new world, https://doi.org/10.1117/2.1201206.004296 2012 SPIE
22. Metamaterial wall amplifies sound: Phys Rev Lett 110:244302 (2013) JUNE 2013 | Vol 498 | Nature | 411
23. Larouche S, Rose A, Smith DR, A constitutive description of nonlinear metamaterials through electric, magnetic, and magnetoelectric nonlinearities. Springer Series in Materials Science 200. https://doi.org/10.1007/978-3-319-08386-5_1
24. Bogue R, Sensing with metamaterials: a review of recent developments. https://doi.org/10.1108/SR-12-2016-0281
25. Szabó Z, Park G-H, Hedge R, Li E-P (2010) A unique extraction of metamaterial parameters based on Kramers–Kronig relationship. IEEE Trans Microwave Theory Techn 58(10)
26. Arslanagić S, Hansen TV, Mortensen NA, Gregersen AH, Sigmund O, Ziolkowski RW, Breinbjerg O (2013) A review of the scattering-parameter extraction method with clarification of ambiguity issues in relation to metamaterial homogenization. IEEE Antennas Propag Mag 55(2):95
27. Chen X, Grzegorczyk TM, Wu B-I, Pacheco J, Kong JA (2004) Robust method to retrieve the constitutive effective parameters of metamaterials. The American Physical Society, Phys Rev E 70:016608

Anomaly Detection in Drones with Machine Learning Algorithms

Soumya Shaw, Kartik Joshi, Akhil Pathak, Abitha K. Thyagarajan, G. Vidya, Rahul Hemal Shah, V. Ram Kishan, and John Sahaya Rani Alex

Abstract Drones are expected to be widely used in near future in the areas of agriculture, surveillance and logistics. The behavior of the autonomous systems, especially the drones, needs to be robust, irrespective of the status of the sensors, environment and the collective conditions of the system. The dynamics of the autonomous system can go into unpredictable behavior because of the faultiness of the sensors, power failure or unexpected environment condition. This work detects whether the state of the drone is in order or not, using multiple streams of sensory data. An ensemble of machine learning models (supervised, unsupervised) is implemented to detect the anomaly in the drone. They are *k*-nearest neighbor (*k*-NN), *k*-means square, decision tree, support vector machine (SVM) and logistic regression. The models were developed and trained with individual sensor data as well as selected combinations of sensor data. Our experimental results demonstrate that detection of anomaly in drones with high accuracies close to 100%.

Keywords Anomaly detection · Drone · k-NN · SVM · Decision tree · Logistic regression

1 Introduction

Autonomous systems are deployed to aid the rapid development in the automation of the repeated processes, or tasks that are riskier for a human to do. Unmanned aerial vehicles (UAVs), also known as drones, are a subset of autonomous systems that find its applications in agriculture [1–4], logistics [5], archaeology [6], mining [7] and so on. Basic drones comprise a battery, computing unit, sensors and actuators. Depending on the application, the drone can be modified. The failure of a

S. Shaw · K. Joshi · A. Pathak · A. K. Thyagarajan · G. Vidya · R. Hemal Shah · V. Ram Kishan
School of Electronics Engineering, Vellore Institute of Technology, Chennai, India

J. S. R. Alex (✉)
Centre for Healthcare Advancement, Innovation and Research, Vellore Institute of Technology, Chennai, India
e-mail: jsranialex@vit.ac.in

A. Sivasubramanian et al. (eds.), *Futuristic Communication and Network Technologies*, Lecture Notes in Electrical Engineering 792,
https://doi.org/10.1007/978-981-16-4625-6_42

drone arises from either of its module malfunctioning or an unforeseen dynamic environment conditions. The early detection of such an anomaly can help the drone to have a safe landing. Detection of such sensors' real-time malfunctions may prove to be vital and is implemented with Kalman filter in connected automatic vehicles (CAVs) [8]. Dynamic programming distance measurements such as Mahalanobis distance is applied for anomaly detection in UAVs [9] Few of the isolated tasks such as motor anomaly detection and collision avoidance detection in UAVs are implemented with reinforcement learning [10, 10]. Using the multisensory data, the modified growing neural gas network is employed to predict the anomaly in the semi-autonomous systems with optimization done at the complexity of the algorithm [12]. Machine learning (ML) algorithms are widely employed in the classification task, prediction tasks. In particular, [13] explores Particle Swarm Optimization (PSO) with feed-forward network. Drone environments can be very diverse, and [14] presents MANFIS architecture (multiple adaptive neuro-fuzzy inference system), which has simulated multiple environments and presents navigation breakthroughs in different environments. In order to apply supervised learning to anomaly detection, labeled data is required; typically, supervised ML algorithms require large quantities of labeled data.

Anomaly detection is the detection of rare events or observations which raise suspicion by differing from the dataset. These sudden abrupt events also known as anomalous items will translate to some kind of catastrophic failure in some applications.

In this work, we revisit simple supervised ML algorithms for anomaly detection in the drone [15–17]. Section 2.1 talks about the datasets, data preparation and pre-processing of data. Section 2.3 gives a brief description of supervised ML algorithms applied in this work. We have implemented a few unsupervised algorithms also for comparative study which is mentioned in Sect. 2.4. In Sect. 3, our training results and test results are reported. The last section provides a summary of our contribution to this work.

2 Methodology and Implementation

2.1 Data Pre-processing and Preparation

The initial data comprises various parameters in each dataset file, which were not totally correlated to many of the important parameters that decided the outcome (anomalous or normal). Thus, data cleaning was done for all normal and abnormal dataset files. Only the required and affected parameters were chosen, and the string data was converted to numerical data logically for actual processing and modeling.

We extracted the individual sensor readings as CSV files from the provided bag files and extracted the image data as individual frames. We used this extracted data for all our training. The extracted data is then modeled for detecting anomalies in the drone environment.

Principal Component Analysis Principal component analysis was implemented with different values of the principal components. It was observed that two principal components were able to best describe the majority of variance in the extracted data.

2.2 *Training*

We trained the ML models, viz., k-NN, logistic regression, SVM (for three different kernels, i.e., default, sigmoid and polynomial) for all the sensor parameters. (The data for each was given in a.csv file.)

2.3 *Supervised Learning*

A supervised learning algorithm learns from annotated training data and produces a machine learning model. This learned trained model can be used for predicting the label for the new data, which is unseen by the model. This requires the supervised learning model to generalize from the training data to unseen situations with fewer false predictions. Thus, this approach helps in detecting the anomaly while neglecting false positives.

***k*-Nearest Neighbors (*k*-NN)** *k*-nearest neighbors are one of the traditional classification algorithms in machine learning. *k*-NN algorithm finds its application in regression as well as classification problem. It uses distance measurement to find the best nearest neighbors for the test data.

Support Vector Machine (SVM) The support vector machine is an ML algorithm which is capable of performing classification tasks using classifiers (by constructing hyperplanes) in a multidimensional space which separates the data into two different classes and can support both regression and classification tasks and can handle multiple continuous and categorical variables.

SVM algorithms may use various kernel functions such as linear, nonlinear, polynomial, radial basis function (RBF) and sigmoid to transform the input data to a required form. They are applied for binary classification problem. In this research work sigmoid and polynomial kernels were used.

Logistic regression Logistic regression is a statistical model which in its basic form uses a logistic function to model a binary dependent variable. In regression analysis, logistic regression is estimating the parameters of a logistic model (a form of binary regression). Here in this model it uses sigmoid function which converts a linear line to a curve to cover the binary points. This model is very useful to find probabilities of a certain event, and it can also be used to find whether a given data will fit into either of the binary values (i.e., 0 or 1).

$$y = b_0 + b_1 x \tag{1}$$

$$p = \frac{1}{1 + e^{-y}} \tag{2}$$

$$\ln\left(\frac{p}{1-p}\right) = b_0 + b_1 x \tag{3}$$

2.4 Unsupervised Learning

Unsupervised learning is a technique in which there is no need to supervise the model.

Instead, the model discovers the information while training. It majorly deals with the unlabeled data. It helps in finding all kinds of unknown patterns in data, features which can be useful for categorization. One of the major ways in which unsupervised learning can be achieved is by clustering the data with similar features.

Anomaly detection is the detection of rare events or observations which raise suspicion by differing from the dataset. These sudden abrupt events also known as anomalous items will translate to some kind of catastrophic failure in some applications.

***k*-Means Clustering** This is an iterative algorithm which tries to classify the dataset into k predefined subgroups (clusters) which are also non-overlapping, where each data point corresponds to only one group. It tries to keep each cluster different from one another making inter-cluster data points as similar as possible. Given N samples, divide them into m clusters in such a way the variance within the cluster is minimum. Initially, a random cluster centroid is assumed and the centroid is iteratively moved to minimize the variance within the cluster.

3 Results

The individual test results of the models are presented in Table 1 and Fig. 1. We observe that k-nearest neighbors, logistic regression, support vector machine (auto) and support vector machine (sigmoid) give 100% accuracy, and support vector machine (poly) and decision tree give an accuracy greater than 99%.

Test results for all the models for the combined parameters are presented in Table 2 and Figs. 2 and 3. It is clear from the figures that the detected outliers are indeed outliers considering the ground truth sensor values.

Test results for all the models for the combined parameters with dimensionality reduction by PCA with 2 components are presented in Table 3. Principal components were reduced to optimal value of 2, as the two components were able to successfully

Table 1 Test results for each model

Parameter	Accuracy					
	k-NN	Logistic regression	SVM (auto)	SVM (sigmoid)	SVM (poly)	Decision tree
Battery	100.0	100.0	100.0	100.0	100.0	100.0
Compass_hdg	100.0	100.0	100.0	100.0	100.0	100.0
Gps_vel	100.0	100.0	100.0	100.0	100.0	100.0
Imu_data	100.0	100.0	100.0	100.0	100.0	100.0
Imu_data_raw	100.0	100.0	100.0	100.0	99.398	99.096
Imu_mag	100.0	100.0	100.0	100.0	100.0	100.0
Imu_static_pressure	100.0	100.0	100.0	100.0	–	99.699
Imu_temperature	100.0	100.0	100.0	100.0	100.0	100.0
Local_position	100.0	100.0	100.0	100.0	100.0	100.0
Pose	100.0	100.0	100.0	100.0	100.0	100.0
Position_global	100.0	100.0	100.0	100.0	100.0	100.0
Position_local	100.0	100.0	100.0	100.0	100.0	100.0
Position_raw_fix	100.0	100.0	100.0	100.0	100.0	100.0
Rel_alt	100.0	100.0	100.0	100.0	100.0	100.0
Velocity_body	100.0	100.0	100.0	100.0	100.0	100.0
Velocity_local	100.0	100.0	100.0	100.0	100.0	100.0
Vfr_hud	100.0	100.0	100.0	100.0	100.0	100.0
Average	100.0	100.0	100.0	100.0	99.962	99.929

explain a majority of the variance in the dataset, when compared to 3 and 5. The feature space was reduced to just 2 of the principal components, and the resulting dataframe was passed on to models for further analysis. Higher-dimensional dataset can introduce a bias and the principal component analysis was performed to reduce the high-dimensional data into low-dimensional feature space for reducing the error due to the bias.

The test accuracy yielded by the *k*-NN (whose classification report is enumerated in Table 3) was 0.81 or 81%.

The results of unsupervised learning with PCA file principal.csv using *k*-means algorithm are presented in Figs. 4 and 5. Figure 4 represents the plot of principal components for the normal and abnormal dataset. Green marks depict the principal components for normal data and red for abnormal data. Figure 5 represents the plot of principal components with the clusters predicting which are principal components for normal data and which are for abnormal data.

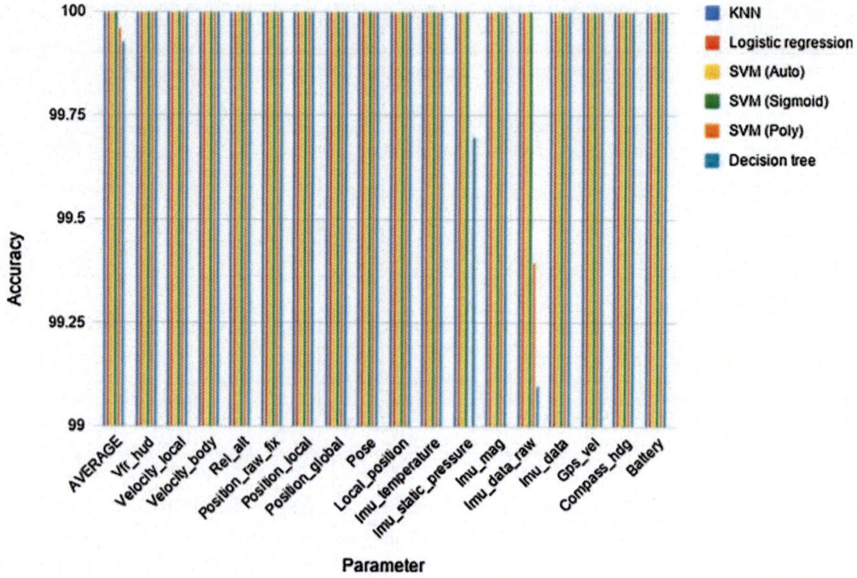

Fig. 1 Plot of test results for each model

Table 2 Supervised learning accuracy for combined.csv using logistic regression algorithm

Correctly predicted	75 out of 83
Wrongly predicted	8 out of 83
Accuracy	90.36%

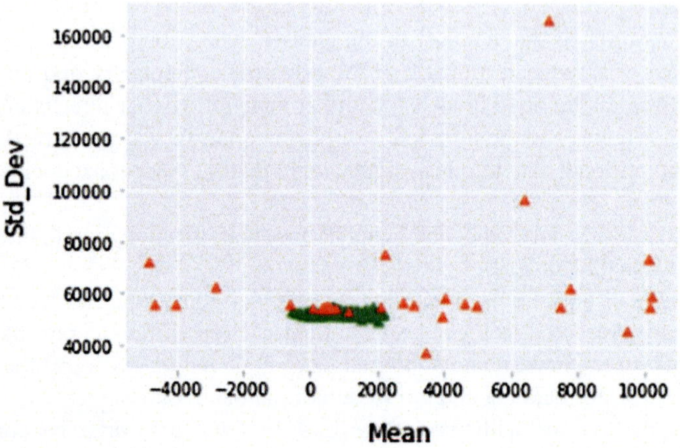

Fig. 2 Actual scatter plot of normal and abnormal dataset with parameters mean and stdev

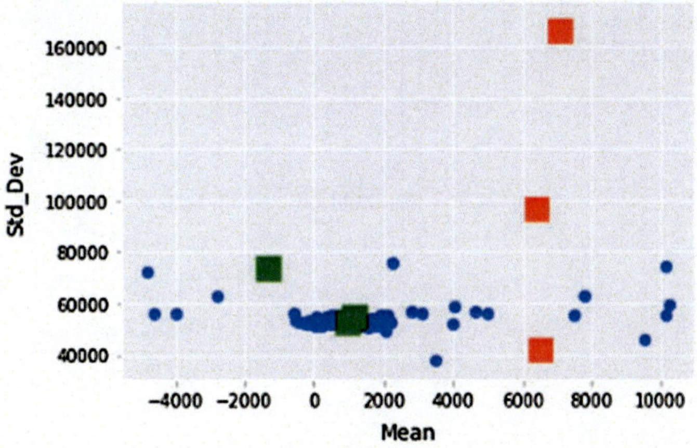

Fig. 3 Predicted clusters for normal and abnormal data after modeling

Table 3 Supervised learning with PCA file principal.csv using k-NN algorithm—classification report

	Precision	Recall	f_1-score	Support
0	1	0.09	0.17	11
1	0.81	1	0.89	42
Accuracy			0.81	53
Macro avg	0.9	0.55	0.53	53
Weighted avg	0.85	0.81	0.74	53

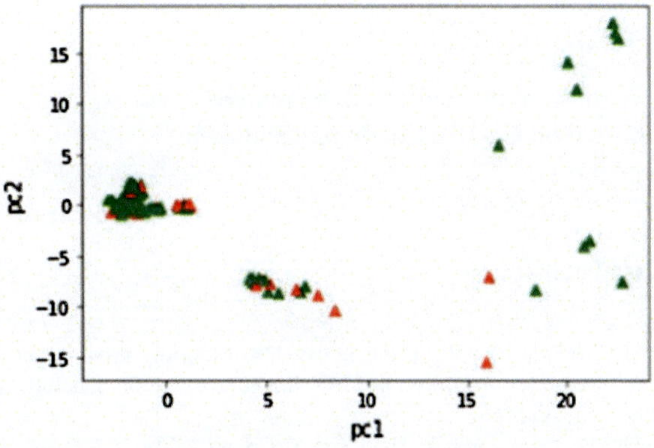

Fig. 4 Actual dataset plot for PC1 and PC2 before modeling

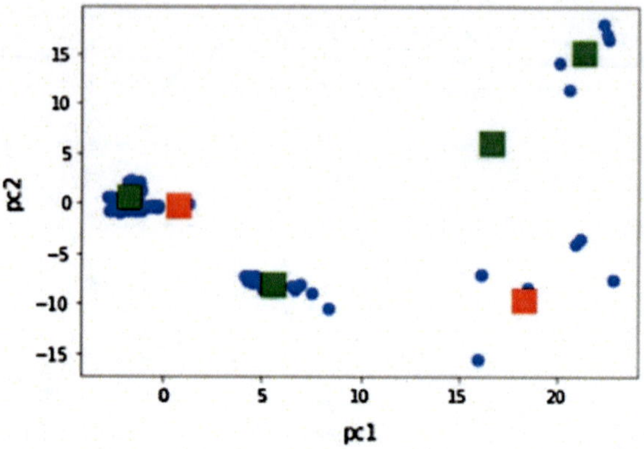

Fig. 5 Predicted clusters after modeling for PC1 and PC2

4 Conclusion

Machine learning algorithms are paving the way to implement an automatic learning behavior in autonomous systems. In this work, with the available data from the drone, we have implemented supervised and unsupervised learning models to detect the anomaly in the drone. Our experimental results demonstrate high accuracy with k-NN across all parameters.

Our study demonstrates good results with this particular scenario of detecting abnormal values among sensor readings taken by an autonomous vehicle. As this may not apply universally to various diverse scenarios, a focused empirical study would be recommended to test the viability of these methods in different situations calling for anomaly detection. In the future, this work may be extended with the use of recurrent neural networks.

Colab notebook

We have implemented all the models discussed in this report in these two notebooks hosted on Google Colab: *one* and *two*.

References

1. Meinen BU, Robinson DT (2020) Mapping erosion and deposition in an agricultural landscape: optimization of UAV image acquisition schemes for sfm-mvs. Remote Sens Environ 239:111666
2. Zheng H, Zhou X, He J, Yao X, Cheng T, Zhu Y, Cao W, Tian Y (2020) Early season detection of rice plants using rgb, nir-gb and multispectral images from unmanned aerial vehicle (UAV). Comput Electron Agric 169:105223
3. Radoglou-Grammatikis P, Sarigiannidis P, Lagkas T, Moscholios I (2020) A compilation of UAV applications for precision agriculture. Comput Netw 172:107148

4. Meng Y, Su J, Song J, Chen WH, Lan Y (202) Experimental evaluation of UAV spraying for peach trees of different shapes: effects of operational parameters on droplet distribution. Comput Electron Agric 170:105282
5. Kitjacharoenchai P, Lee S (2019) Vehicle routing problem with drones for last mile delivery. Procedia Manuf 39:314–324
6. Wiseman AL, Bezombes F, Moore AJ, De Groote I (2020) Non-invasive methods: the applicability of unmanned aerial vehicle (UAV) technology for recording fossilised footprints. Digit Appl Archaeol Cult Heritage 16:e00137
7. Kansake BA, Kaba FA, Dumakor-Dupey NK, Arthur CK (2019) The future of mining in Ghana: are stakeholders prepared for the adoption of autonomous mining systems? Resourc Policy 63:101411
8. van Wyk F, Wang Y, Khojandi A, Masoud N (2019) Real-time sensor anomaly detection and identification in automated vehicles. IEEE Trans Intell Transp Syst
9. Khalastchi E, Kaminka GA, Kalech M, Lin R (2011) Online anomaly detection in unmanned vehicles. In: The 10th international conference on autonomous agents and multiagent systems-vol. International foundation for autonomous agents and multiagent systems, pp 115–122
10. Lu H, Li Y, Mu S, Wang D, Kim H, Serikawa S (2017) Motor anomaly detection for unmanned aerial vehicles using reinforcement learning. IEEE Internet Things J 5(4):2315–2322
11. Shin SY, Kang YW, Kim YG (2020) Reward-driven U-Net training for obstacle avoidance drone. Expert Syst Appl 143:113064
12. Iqbal H, Campo D, Baydoun M, Marcenaro L, Gomez DM, Regazzoni C (2019) Clustering optimization for abnormality detection in semi-autonomous systems. In: 1st international workshop on multimodal understanding and learning for embodied applications, pp 33–41
13. Pandey DA, Panwar V, Hasan M, Parhi D (2020) V-REP-based navigation of automated wheeled robot between obstacles using PSO-tuned feedforward neural network 7, 1–8. https://doi.org/10.1093/jcde/qwaa035
14. Pandey DA, Kashyap AK, Parhi D, Patle B (2019) Autonomous mobile robot navigation between static and dynamic obstacles using multiple ANFIS architecture. World J Eng. https://doi.org/10.1108/WJE-03-2018-0092
15. Gu W, Mittu R, Marble J, Taylor G, Sibley C, Coyne J, Lawless WF (2014) Towards modeling the behavior of autonomous systems and humans for trusted operations. In: 2014 AAAI spring symposium series
16. Khalastchi E (2010) Anomaly detection in unmanned vehicles: thesis. The Maverick Group, Department of Computer Science, Bar Ilan University, Ramat Gan, Israel
17. Ahn H, Choi HL, Kang M, Moon S (2019) Learning-based anomaly detection and monitoring for swarm drone flights. Appl Sci 9(24):5477

Deep Learning-Based Multi-class 3D Objects Classification Using Digital Holographic Complex Images

R. N. Uma Mahesh, B. Lokesh Reddy, and Anith Nelleri

Abstract Deep learning for three-dimensional (3D) object classification is a promising task in digital holography. In this paper, we propose a convolutional neural network (CNN) that performs the multi-class classification of 3D objects information retrieved using off-axis Fresnel digital holography. The performance of the proposed method is evaluated for five different sets of 3D objects using loss and accuracy curves to demonstrate the proof of the concept.

Keywords Deep learning · Convolutional neural network · 3-D objects classification · Digital holography

1 Introduction

Deep learning is one of the sub-branches of artificial intelligence which also includes machine learning [1]. Recent developments in deep learning have stimulated the growth of new paradigms in digital holographic tasks like autofocusing, phase reconstruction, and hologram segmentation [2–5]. The traditional machine learning technique will do the manual feature extraction and then classifies the images. Therefore, deep learning which performs both feature processing and classification through neural networks is used in the present study. The neural network has input and output layers along with several hidden layers. There are different kinds of deep neural networks like convolutional neural networks (CNN), auto-encoder, and generative adversarial networks which are used in various applications in the field of computer vision and image processing [6]. The CNN has been employed in various digital holographic applications like classification, super-resolution, and image reconstruction [7–9]. The convolutional neural network was used in the hologram classification of anthrax spores for the bio-defense issue for three sets and also for five sets [10]. Here, the convolutional neural network is used for classification of five sets of reconstructed 2-D digital holographic complex images which contains amplitude and phase

R. N. Uma Mahesh · B. Lokesh Reddy · A. Nelleri (✉)
School of Electronics Engineering, Vellore Institute of Technology (VIT), Chennai, Tamil Nadu 600127, India
e-mail: anith.nelleri@vit.ac.in

A. Sivasubramanian et al. (eds.), *Futuristic Communication and Network Technologies*, Lecture Notes in Electrical Engineering 792,
https://doi.org/10.1007/978-981-16-4625-6_43

information. The CNN is a well-known neural network that contains two parts (a) feature extraction layer (b) classification layer. The convolutional layers and pooling layers are present in the feature extraction layer which will process raw data from the images. The convolutional layers employ convolutional filters, and the number and size of the filter are selected according to the application problem at hand. The output of feature extraction layer is given to the classification layer for further processing. Dense layers and output layer are there in the classification layer. The number of neurons and the selection of the activation function in the output layer are decided based on the task.

The present paper demonstrates the multi-class categorization of digital holographically reconstructed complex images that contains 3-D information of the object field using deep CNN. Five sets of digital holographic data for 3-D object classification problem are considered in the study.

2 Methodology

The architecture of CNN is shown in Fig. 1. The input to the network is a 2-D complex digital image which is numerically reconstructed from a digital hologram that contains 3-D information. The intensity and phase images are separated and given to the network for further processing. The network selects 150 × 150 pixels from the input image size of 1029 × 1029 pixels using image data generator. The first convolutional layer incorporates 64 filters, and each filter is of size 3 × 3. All convolutional layers have rectified linear unit activation (ReLU) function. There are pooling layer and convolutional layer successively for four stages. The pooling technique used is Max-Pooling which has a size of 2 × 2. The second, third, and fourth convolutional layers contain 128, 256, 256 filters, respectively. The output of the final pooling layer is converted into vector form and is given to the dense layer which contains 4096 neurons. Finally, the output of this dense layer is given to the output layer which has 5 neurons. Here, softmax activation function is used in the output layer for multi-class classification task to get the output. The main difference between binary classification and multi-class classification is that instead of considering the single neuron in the output layer, multiple neurons are considered.

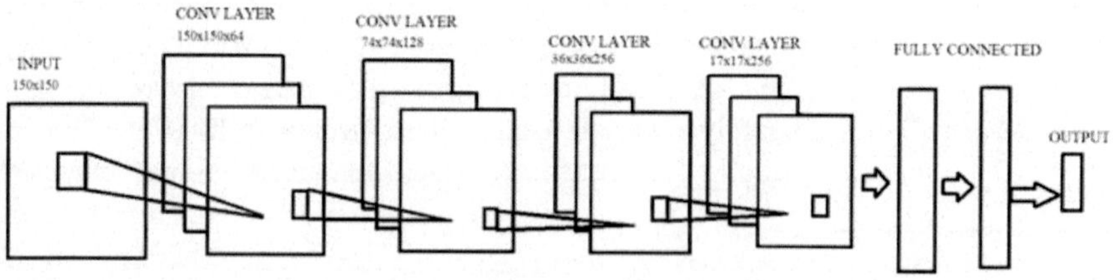

Fig. 1 Deep convolutional neural network for multi-class classification

3 Experimental Results and Discussion

To demonstrate the proposed multi-class CNN, we have experimentally considered five different 3-D objects such as circle–triangle, pentagon–square, rectangle–square, square–pentagon, and square–rectangle as shown in Fig. 2. For example, the features circle and triangle on different planes of the 3-D object named 'circle–triangle' are separated by a distance $d_o = 0.8$cm. The same separation was given between the features of the other 3-D objects chosen in the experiment. The 3-D object information was sensed using CMOS image sensor and off-axis digital holographic setup. The 3-D object information is in the form of 2-D digital complex image which was retrieved from the sensed off-axis digital hologram using computational techniques.

To record a digital hologram of each object, an off-axis scheme was built based on Mach–Zehnder interferometer geometry. A He–Ne laser of wavelength 632.8 nm was used as the light source. Each object was placed in the object plane at different distances $d = 18, 20, 30$cm from the recording plane. A CMOS camera with square pixel pitch format of 6 μm was used to acquire the digital holograms of size 1029 $\times$ 1029 pixels. The numerical reconstruction was performed on each recorded digital hologram using the complex wave retrieval algorithm [11] and inverse Fresnel transform to form the dataset of 2D digital complex images which contains 3D object information. The numerical reconstruction enables extraction of complex-valued information, i.e. intensity and phase information of the 3-D objects. The absolute part of complex information gives intensity image and arctan part of complex information gives the phase image. The reconstructed dataset was prepared for 525 intensity and phase images into five sets with the ratio of the training set (75%), validation set (15%), and test set (10%), respectively.

The network training was performed for 100 epoch using adaptive moment estimation (Adam) optimizer where learning rate of 0.0005 was used and categorical

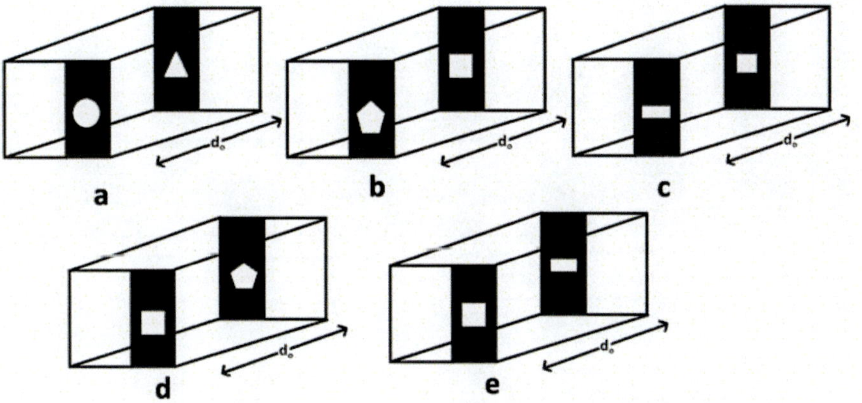

Fig. 2 Three-dimensional (3-D) objects used in the off-axis digital holographic experimental setup **a** circle–triangle, **b** pentagon–square, **c** rectangle–square, **d** square–pentagon, **e** square–rectangle

cross-entropy was used as a loss function. Each epoch was trained for 395 images from the training set and 80 images from the validation set with each batch consisting of 79 images and 16 images from training and validation set, respectively. Each epoch was iterated for 5 steps on training and validation set. The graph of loss and accuracy curves on the training set and validation set for intensity and phase images are shown in Fig. 3.

Figure 3a shows the loss and accuracy plots for intensity dataset. Once the training is completed, the training loss obtained is 3.3308×10^{-6} which is lesser than the

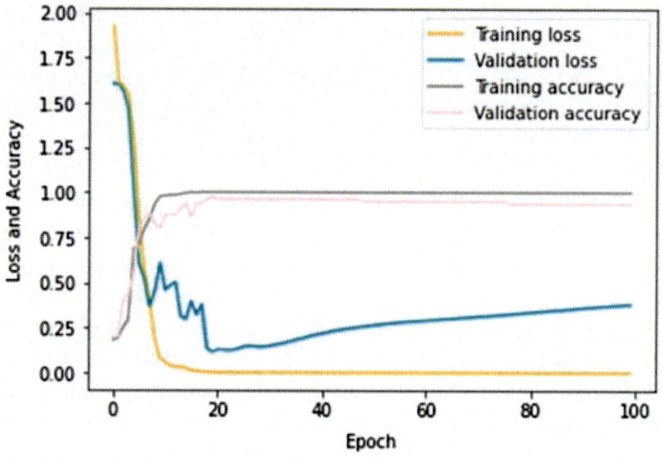

(a)

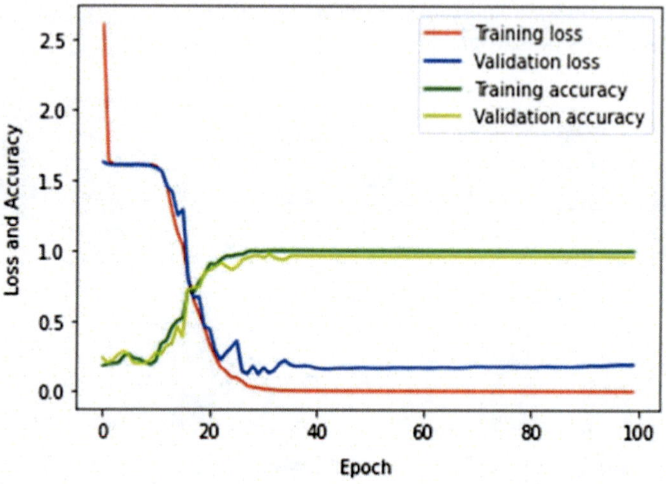

(b)

Fig. 3 Loss and accuracy curves of training set and validation set **a** intensity dataset and **b** phase dataset validation loss 0.3837 and training accuracy obtained is 1.00 which is higher than the validation accuracy 0.9375. Figure 3a shows that difference between the accuracy on the training data and validation data is not very high. Therefore, the proposed CNN is correctly fitted.

Figure 3b shows the loss and accuracy plots for phase dataset. From Fig. 3b, the training loss, validation loss is decreasing and training accuracy, validation accuracy is increasing, and after the training is completed, the training loss is 8.1361×10^{-5} and validation loss is 0.1974.The accuracy values on training set and validation set after the completion of the training are 1.00 and 0.9625. The margin between the training accuracy and validation accuracy is less. Therefore, the proposed CNN is not overfitting.

4 Conclusion

The proposed convolutional neural network is able to perform multi-class classification of 3-D objects for digital holographic reconstructed complex-valued images. The loss and accuracy curves are shown for the justification of the work. The network has minimal loss and maximum accuracy on the training set and validation set for both intensity and phase images. The validation accuracy and training loss for phase images are greater than that of the intensity images. Therefore, deep learning can be considered as a modern technique for the classification of 3-D objects task using digital holography.

Acknowledgements This work was supported by the Science and Engineering Research Board (SERB), Department of Science and Technology (DST), Government of India, under the grant no. CRG/2018/003906.

References

1. LeCun Y, Bengio Y, Hinton G, Deep learning. https://doi.org/10.1038/nature14539
2. Pitkäaho T, Manninen A, Naughton TJ (2019) Digital hologram reconstruction segmentation using a convolutional neural network. In: Digital holography and three-dimensional imaging, OSA technical digest (Optical Society of America, 2019), https://doi.org/10.1364/DH.2019.Th3A.1
3. Wang H, Lyu M, Situ G (2018) eHoloNet: a learning-based end-to-end approach for in-line digital holographic reconstruction. Opt Expr 26(18):22603–22614
4. Ren Z, Xu Z, Lam EY (2018) Learning-based nonparametric autofocusing for digital holography. Optica 5:337–344
5. Pitkäaho T, Manninen A, Naughton TJ (2019) Focus prediction in digital holographic microscopy using deep convolutional neural networks. Appl Opt 58(5):A202–A208
6. Wang H (2019) Deep learning enables cross-modality super-resolution in fluorescence microscopy (2019). https://doi.org/10.1038/s41592-018-0239-0

7. Shimobaba T, Kuwata N, Homma M, Takahashi T, Nagahama Y, Sano M, Hasegawa S, Hirayama R, Kakue T, Shiraki A, Takada N, Ito T (2017) Convolutional neural network-based data page classification for holographic memory. Appl Opt 56:7327–7330
8. Loy DCC, He K, Tang X (2016) Image super-resolution using deep convolutional networks. IEEE Trans Pattern Anal Mach Intell 38(2):295–307
9. Mousavi, Baraniuk RG (2017) Learning to invert: signal recovery via deep convolutional networks. In: International conference on acoustics, speech and signal processing
10. Jo Y, Park S, Jung J, Yoon J, Joo H, Kim M-H, Kang SJ, Choi MC, Lee SY, Park Y (2017) Holographic deep learning for rapid optical screening of anthrax spores. Sci Adv 3:e1700606 https://doi.org/10.1126/sciadv.1700606
11. Liebling M, Blu T, Unser M (2004) Complex-wave retrieval from a single off-axis hologram. J Opt Soc Am A 21:367–377

An Efficient Hybrid De-Noising Method for Remote Sensing Images

R. Avudaiammal, S. Devipriya, A. Swarnalatha, and R. Vijayarajan

Abstract In recent years, analysis of remote sensing imagery has become significant due to its good representation of data in region of interest (ROI) in the world. However, the noise-corrupted satellite image data may not be capable to recognize and analyze it accurately. It necessitates the usage of suitable image de-noising that aids in improving the clarity, sharpness of image for further processing like feature extraction and enhanced analysis of the image data. In this paper, an efficient hybrid bilateral-guided (HBG) filter has been proposed that can filter noise faithfully from the satellite digital image. The attractive feature of the HBG filter is its focus on preserving edges of the original image while removing noises, thereby supporting edge detection and other image processing in a better way. Weiner filtering, median filtering, bilateral filtering, guided filtering and HBG are implemented using MATLAB R2016a on Intel core i3 system. These filtering techniques are carried out on Gaussian noise, salt-and-pepper noise and speckle noise-corrupted satellite images. The performance measures used are mean square error (MSE) and peak signal-to-noise ratio (PSNR). PSNR of HBG is 83%, 76% and 81% more than that of conventional bilateral filter, and MSE of HBG is 49%, 50% and 49% lesser than that of conventional guided filter for 50% salt-and-pepper noise, speckle noise and Gaussian noise-corrupted image, respectively. It is also proved that PSNR of proposed HBG is 74, 71 and 69% more than that of existing hybrid median Wiener filter (HMW) and MSE of HBG is 85%, 95% and 84% lesser than that of hybrid median Wiener filter (HMW) for 50% Gaussian noise, salt-and-pepper noise and speckle noise-corrupted image, respectively. Thus, the simulation outcomes show that the HBG filtering is suitable for removal of all types of noise from satellite images

Keywords Hybrid bilateral-guided (HBG) filter · Hybrid median Wiener filter (HMW) · Weiner filter · Median filter · Bilateral filter · Guided filter

---

R. Avudaiammal · S. Devipriya (✉) · A. Swarnalatha
Department of ECE, St. Joseph's College of Engineering, OMR, Chennai 600119, India
e-mail: devipriyas@stjosephs.ac.in

R. Vijayarajan
Division of Healthcare Advancement, Innovation and Research, VIT University, Chennai 600127, India

A. Sivasubramanian et al. (eds.), *Futuristic Communication and Network Technologies*, Lecture Notes in Electrical Engineering 792,
https://doi.org/10.1007/978-981-16-4625-6_44

1 Introduction

Remote sensing is suitable to analyze earth regions [1] since it gives a good representation of providing the condition of surface ground at every point in the world by taking data only at certain points around the world. Remote sensing images are suitable in numerous applications such as earth resources tracking, geographical mapping, urban development, crops prediction, weather prediction, flood assessment and forest fire assessment.

Satellite images [2] affected by noise adversely degrade the image quality and its content. Prime factors accountable for creating noise in an image are the medium through which image is taken (climatic and atmospheric factors like pressure and temperature), the accuracy of the instrument used to take the image (for instance camera) and the quantization of data used to store the image. The noise types are Gaussian noise, salt-and-pepper noise and speckle noise. Gaussian noise [3] is an additive noise, produced by amplifier or detector or image sensor. Gaussian noise is a statistical noise. The pixels and its intensity are independent; their variation affects the mean of an image, i.e., probability density function (PDF) is identical to Gaussian distribution. Salt-and-pepper noise is produced by analog-to-digital converter, faulty memory cell and sensors of cameras. Salt-and-pepper noise [4] affected image black pixels present in bright sections and vice versa. Speckle noise [5] is a multiplicative-type noise, and granular pattern of gray level may present in the image.

Satellite image data comprising the noise may not be capable to recognize and to analyze it correctly. Therefore, suitable image de-noising is required to improve the clarity, sharpness of image for further processing like feature extraction and enhanced analysis of the image data. Research has shown that there are a number of parameters, which determine the quality of an image, most of which cannot be resolved manually without the support of a computer.

The rest of the paper is outlined as follows. Section 2 discusses related works. Section 3 details about proposed filtering techniques. Section 4 presents the performance analysis of filtering methods. Section 5 concludes the work.

2 Related Work

Image de-noising is the inevitable, vital pre-processing step in many image processing applications to reduce noise for guaranteed identification of image information and for attaining better results from further image processing. Image can be de-noised through numerous filtering procedures, but the choice of the procedure depends on the behavior and the image kind. The big challenge in filtering is to retain the details of the image during the noise removal from the image. Basically, two procedures are used to remove noise named as linear and nonlinear filtering [6]. In linear filtering, the random noise that is not relevant to the image is filtered by analyzing statistics of the noise samples in spatial domain. The main drawback of linear filtering is the requirement of priory statistical knowledge. Linear [7] methods are fast as compared to nonlinear methods, but linear methods are not able to preserve the details/information of the image in comparison with nonlinear methods. Nonlinear filtering [5] is powerful in removing additive noise and is used in medicine, remote sensing and so on. Guided, median and bilateral filters are nonlinear filters.

Wiener filter [8] is one of the linear filters that exploit spectral properties of image and noise to perform filtering. Median filtering [9] is one of the nonlinear filters that replace each pixel using median value computed from its surrounding neighborhood pixels.

Guided filter [4] is edge preserving smoothing filter and has better behavior near edges; currently, it is one of the fast edge preserving filter. The filtering is carried out by using guidance image content. Its time complexity is O (N) where N is the number of pixels independent of pixel size.

Bilateral filter [10] is a nonlinear edge preserving and smoothing filter for images. It works on two parameters, i.e., geometric closeness and their photometric similarity. In bilateral filter each pixel is replaced by the weighted average intensity computed from nearby pixels. This filtering may blur an image while respecting strong edges.

Hybrid filter [11] is the cascade of median filter and Wiener filter. The impulse noise is removed by median filter stage, and then, additive white noise is removed by the Wiener filter stage. The adaptive median filter and the Wiener filter are used for reducing Gaussian noise. The median filter is suitable in the low-frequency sections, whereas Wiener filter is suitable in sections with high Gaussian noise variances. But, the PSNR is not good at low levels of Gaussian noise variation.

Discrete wavelet transform [12] uses sparse representation, i.e., it contains many coefficients, which are close to zero. Earlier for the image de-noising the Fourier transform was used. In Fourier transform, the signal was present only in frequency domain, but in wavelet transform the signal is present in both frequency and time domain. The noisy image is applied with the wavelet transform, and then, hard and soft threshold values are applied.

An insight of some major work in image de-noising, robust discrete wavelet transform (DWT) and discrete cosine transform (DCT) approach is offered in [11]. This work is additionally enhanced by using the exposure-based sub-image histogram equalization. This framework gives better visual quality and better performance parameters. This framework mainly focuses on the picture recovery from its noisy counterpart. It gives higher PSNR & low MSE and can be further extended to video framework in future.

A modified decision-based median filter (DBMF) is proposed in [13] which eliminates the impulse noise from gray images. DBMF replaces noisy pixels alone with the median value, and another benefit is the window size that can be increased as per the requirement. The standard median filter types were effective only when the noisy density is less than 20% and when the noise density is greater than 20%, the edges are preserved and the window size was increased dynamically which increases the computation time.

A combination of wavelet transform (WT) with ICA (independent component analysis) is introduced in [14], and these two algorithms have different advantages but are not able to remove the noise ideally. The WT-ICA de-noising methods retain texture & edge details very well. The drawback is that this filter is suitable only for Gaussian noise.

A morphological mean filter technique is proposed in [15] to recover high-density noise-corrupted image. It first finds the number and the position of the noise-free pixels. Further the oblation process of these noise-free pixels is carried out based on morphological image processing, and this is done iteratively in order to replace the neighborhood noise pixels. This filter consumes only moderate time of execution and processes the highest PSNR values.

Threshold modulation of Gaussian filter is explored in [16] to preserve texture information. It uses the standard deviation to discriminate the flat portion and texture portion. The low value of standard deviation in the flat portion makes the filter to function like an average filter, and the high value of standard deviation in the texture portion makes the filter to function like normal distribution filter. Texture information is preserved to a larger extent after applying the Gaussian filter with threshold modulation.

A two-stage filter based on fuzzy rule is proposed in [17] for noise removal. In the first stage, fuzzy derivative values for all eight directions with the central pixel as reference are used to calculate the noisy pixels. In the second stage fuzzy smoothing is carried out by using the neighborhood pixels. This fuzzy logic is used iteratively applied on the noised image until the anticipated PSNR value comes. This algorithm works well on the high-density salt-and-pepper noise and on high variance Gaussian noise. The time complexity is lesser than Weiner filter.

A double bilateral filter is introduced in [18]. Generally, the bilateral filter smoothes the image by preserving edges by nonlinear combination of the adjacent pixels. Their approach combines the range filtering with domain filtering for smoothing and edge preserving. Their technique focused only on the Gaussian noise, but it is also found that it is good at handling the impulse noise. It provides better results in returning the affected image with mixture of both Gaussian noise and impulse noise.

Multi-resolution bilateral filter is proposed in [19] where bilateral filter has been combined with wavelet thresholding filter. The method decomposes the image into low- and high-frequency components, and the bilateral filtering is used in the approximation BW and the wavelet thresholding is used in the detailed BW. The main advantage is that, once if the parameter is decided, there is no further change depending on the image. The multi-resolution helped in eliminating the coarse-grain noise in the image.

A comparative study is done [20] between the Gaussian-based filters extended Kalman filter (EKF), the unscented Kalman filter (UKF) and the Gaussian particle filter. They have noted the GPF provided an improved approximation to subsequent distribution when compared to the EKF and UKF, but its computational complexity is higher.

From the literature it is inferred that an ultimate image with 100% accuracy is not possible with any type of de-noising. The main objective of the image de-noising is to get the best promising image with least noise.

3 Proposed HBG Filtering

In this work, a hybrid bilateral-guided (HBG) filter has been proposed which incorporates the best features of bilateral and guided filtering techniques to recover the satellite image from the noise. HBG is a two-stage filter that supports edge detection in a better way by preserving edges while removing noises. In the first stage, the intensity value of all the pixels in an image is changed by a weighted average intensity of adjacent pixels. In order to preserve sharp edges, the weight is calculated by considering both the spatial kernel and range kernel. Spatial kernel is computed by Euclidean distance of pixels for smoothing differences in coordinates, whereas range kernel is computed as color intensity, for smoothing differences in intensities. These two kernels follow Gaussian distribution function. Since regional characteristics are not considered, the output image of first stage introduced blurring in image while concerning strong edges and also causing in smoothed textured regions.

In the second stage HBG removes the blurredness and also preserves edges of an image by considering regional characteristics of a guidance image. It considers the statistics characteristics of a matching region in the guidance image for computing weight of each pixel. The possibilities of choosing guidance image are a dissimilar version of the image or a totally dissimilar image. In HBG, the original image is chosen as the guidance image. HBG algorithm is given below.

Algorithm

Algorithm

1. Read the input RGB Image I (m,n)
2. Convert RGB image into gray level image
3. Calculate weight of each pixel by using spatial closeness and intensity differences

$$W_B = \sum_{x \in C} K_r(\|(I(x_i) - I(x)\|)\, K_s(\|(x_i) - x\|) \quad (1)$$

where

I -Noisy Image

x - Coordinates of the pixel to be filtered

C - Window centered at x

K_r - Range kernel

K_s - Spatial kernel

4. Obtain output of I stage $F_B(m,n)$ by replacing each pixel of I by the weighted average of intensity values from its nearby pixels.

5. Assign $F_B(m,n)$ as a guidance image G.

6. Compute the weight of each pixel using statistical characteristics of a matching region in the guidance image G. Assign the regularization parameter values such as the local window radius r=5 &blur degree of the filter ε =.01.

6. Compute Mean and Variance of $F_B(m,n)$, Mean of G(m,n), Average cross product of $F_B(m,n)$ and G(m,n).

7. Compute the value of linear coefficients.

$$a = F_B(m,n)G(m,n) - Meun(F_B(m,n)).Mean(G(m,n)). / (\mathrm{var}(F_B(m,n)+\varepsilon))$$

$$b = Mean(G(m,n)) - a.{*}Mean(F_B(m,n)) \quad (2)$$

8. Compute the Mean(a,b)

9. Obtain denoised output image F(m,n) using Mean(a,b),

$$F_B(m,n) = Mean(a.{*}F_B(m,n)) + mean(b) \quad (3)$$

4 Results and Discussion

In this section, the performance metrics of proposed HBG is compared with various conventional de-noising methods such as Weiner filtering, median filtering, bilateral filtering, guided filtering and Gaussian filtering. Filtering is simulated using MATLAB R2016a on Intel core i3 system.

Simulation Environment: The multispectral satellite image of Madurai city shown in Fig. 1 acquired on March 17, 2011, by WORLDVIEW 2 satellite is used for analysis. The spatial resolution of the image is 1.84 m. The image size is 256 × 256.

Performance Analysis: The performance measures used for analysis are mean square errors (MSE) and peak signal-to-noise ratio (PSNR). MSE (1) characterizes collective squared error among the noisy image $I(m, n)$ and the filtered image $F(m, n)$ [21].

$$\text{MSE} = \frac{\sum [I(m,n) - F(m,n)]^2}{(m * n)} \tag{1}$$

where m and n indicate the number of rows and columns of the images, respectively. Peak signal-to-noise ratio (PSNR) (2) between two images is ratio between the maximum possible power of the original image and a noisy image [22].

$$\text{PSNR} = 10 \log_{10}\left(\frac{R^2}{\text{MSE}}\right) \tag{2}$$

where R refers the extreme intensity in the input image data and R is 255 for grayscale image. De-noising effect of various filters is tested on Gaussian noise, salt-and-pepper

Fig. 1 Satellite image of study area

noise and speckle noise-corrupted satellite images with noise probabilities from 25 to 75%.

Performance measures of various filtering on Gaussian noise, salt & pepper noise and speckle noise with various noise densities from 25 to 75% are given in Tables 1, 2 and 3, respectively.

It is very evident from Tables 1, 2 and 3 bilateral filtering provides higher PSNR and guided filter provides lesser MSE for Gaussian noise, salt-and-pepper noise and speckle noise-corrupted image. The obtained results gave the motivation of proposing a hybrid filter that provides good performance measures for all the three types of noise. Hybrid filter is devised by incorporating the features of guided and bilateral filter.

In this paper, de-noising effect using hybrid filter is also tested on Gaussian, salt-and-pepper and speckle noise-affected satellite images with 50% noise density and it is compared with existing hybrid median Wiener (HMW) filter. Performance of the proposed HBG filter is presented in Table 4. It is evident that HBG provides higher PSNR and lesser MSE than HMW for all three types of noise. PSNR of HBG is 83, 76 and 81% more than that of conventional bilateral filter, and MSE of HBG is 49, 50 and 49% lesser than that of conventional guided filter for salt-and-pepper noise, speckle noise and Gaussian noise-affected image, respectively.

Figure 2a shows 50% Gaussian noise-corrupted image. Figure 2b shows HMW effect and Fig. 2c shows HBG effect on Gaussian noise-corrupted image, respectively. Figure 3a shows 50% salt-and-pepper noise-affected image. Figure 3b shows HMW

Table 1 Performance measures of filtering on Gaussian noise

Filter	25%		50%		75%	
	PSNR	MSE	PSNR	MSE	PSNR	MSE
Gaussian	18.02	1.02	15.47	1.85	14.52	2.30
Weiner	18.16	2.84	15.85	4.78	14.81	6.16
Median	14.43	3.67	12.7	3.38	11.03	4.32
Guided	7.17	0.21	6.83	0.23	8.57	3.63
Bilateral	28.78	2.38	27.89	3.65	27.66	1.24

Table 2 Performance measures of filtering on salt-and-pepper

Filter	25%		50%		75%	
	PSNR	MSE	PSNR	MSE	PSNR	MSE
Gaussian	11.19	0.49	11.18	0.49	11.19	0.49
Weiner	11.41	2.22	9.98	3.69	9.34	4.22
Median	9.39	2.67	8.02	3.84	7.35	2.19
Guided	5.32	0.24	4.72	2.42	4.42	0.28
Bilateral	27.43	3.31	27.01	4.26	26.96	3.79

Table 3 Performance measures of filtering on speckle noise

Filter	25%		50%		75%	
	PSNR	MSE	PSNR	MSE	PSNR	MSE
Gaussian	14.25	2.44	10.94	5.24	8.86	8.46
Weiner	13.97	1.06	10.29	1.86	8.35	2.78
Median	11.01	0.34	7.99	2.73	6.04	2.15
Guided	7.94	0.21	4.74	0.29	3.82	0.20
Bilateral	28.47	2.96	27.07	2.39	27.02	4.25

Table 4 Performance analysis of proposed filter (HBG)

Filter/noise	Gaussian noise		Salt-and-pepper noise		Speckle noise	
	PSNR	MSE	PSNR	MSE	PSNR	MSE
Bilateral	27.01	4.26	27.07	2.39	27.89	3.65
Guided	4.72	2.42	4.74	0.29	6.83	0.23
HMW	14.03	2.59	15.24	1.96	17.35	1.21
HBG	52.35	0.38	52.35	0.08	55.21	0.19

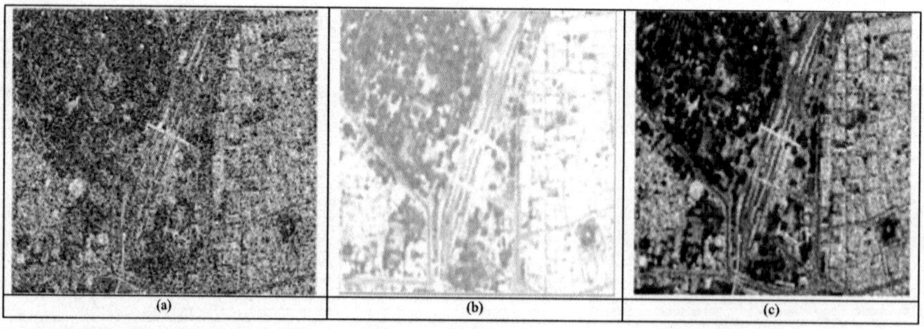

Fig. 2 **a** Gaussian noise-corrupted image, **b** HMW filtered image, **c** HBG filtered image

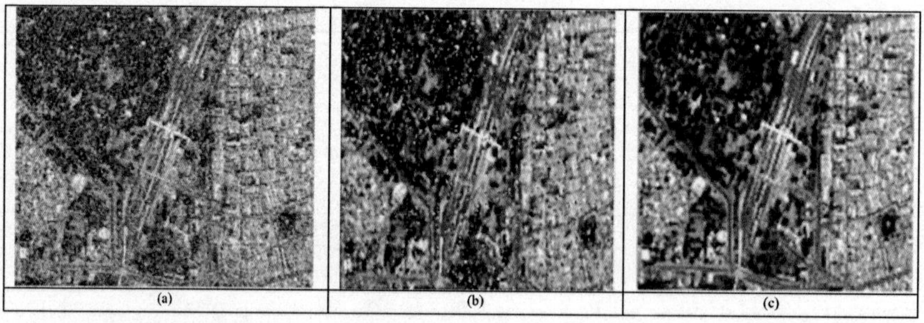

Fig. 3 **a** 50% Salt & Pepper noise-corrupted image, **b** HMW filtered image, **c** HBG filtered image

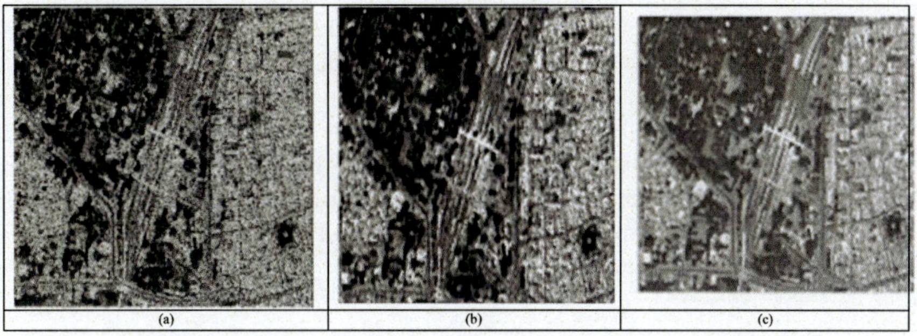

Fig. 4 **a** 50% speckle noise-corrupted image, **b** HMW filtered image, **c** HBG filtered image

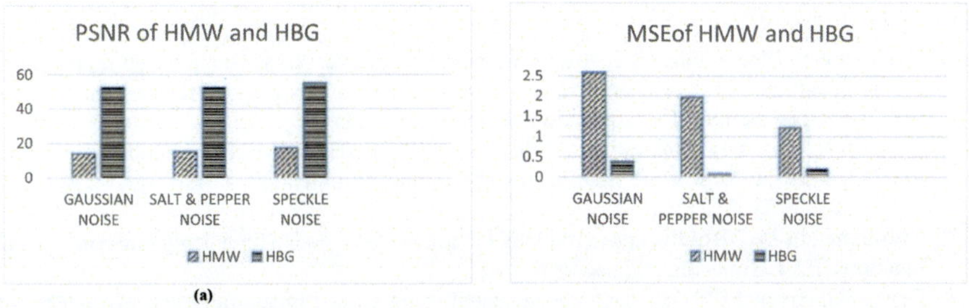

Fig. 5 **a** PSNR of HMW and HBG **b** MSE of HMW and HBG

effect and Fig. 3c shows HBG effect on salt-and-pepper noise-affected image, respectively. Figure 4a shows 50% speckle noise-affected image. Figure 4b shows HMW effect and Fig. 4c shows HBG effect on speckle noise-affected image, respectively.

Figure 5a shows PSNR analysis of HMW with proposed HBG. It is evident from Fig. 5a that PSNR of proposed HBG is 74, 71 and 69% more than that of existing hybrid median Wiener filter (HMW). The higher value of PSNR indicates that quality of the reconstructed image is better. Figure 5b shows MSE analysis of HMW with HBG. It is clear from Fig. 5b that MSE of HBG is 85, 95 and 84% lesser than that of HMW for 50% Gaussian noise, salt-and-pepper noise and speckle noise-affected image, respectively. The lower value of MSE indicates that the error is minimum.

5 Conclusion

In this work, a hybrid bilateral-guided (HBG) filter has been proposed which is composite of bilateral filter and Guided filter to remove the noise faithfully from the satellite digital image. It supports edge detection in a better way by preserving edges while removing noises. The various de-noising methods such as Weiner filtering, median filtering, bilateral filtering, guided filtering and Gaussian filtering are simulated and tested on Gaussian, salt-and-pepper and speckle noise-corrupted satellite images, using the performance measures such as mean square errors (MSE) and peak signal-to-noise ratio (PSNR). The results display that the proposed HBG method is suitable for removal of noise from satellite because its PSNR is 74, 71 and 69% more than that of HMW filter for 50% Gaussian noise, salt-and-pepper noise and speckle noise-corrupted image, respectively. MSE is 85, 95 and 84% lesser than that of HMW filter for 50% Gaussian noise, salt-and-pepper noise and speckle noise-corrupted image, respectively.

References

1. Vakalopoulou M, Karantzalos K, Komodakis N, Paragios N (2016) Graph-based registration, change detection, and classification in very high resolution multi-temporal remote sensing data. IEEE J Sel Top Appl Earth Observations Remote Sens 9(7)
2. Chandra N, Ghosh JK, Sharma A (2016) A cognitive based approach for building detection from high resolution satellite images. IEEE Explore Document.https://doi.org/10.1109/ICACCA.2016.7578878
3. Charalampidis D (2016) Recursive implementation of the Gaussian filter using truncated cosine functions. IEEE Trans Signal Process 64(14)
4. Devi S, Cheriyan J (2012) Image enhancement using guided image filter and wavelet based edge detection. Int J Modern Eng Rese 3(3)
5. Ganesan P, Rajini V (2012) Comparative study of denoising methods for satellite image restoration using Matlab. Int J Adv Res Comput Sci 4(5)
6. Al-amri SS, Kalyankar NV, Khamitkar SD (2010) A comparative study removal noise from remote sensing image. Int J Comput Sci Issues (IJCSI) 7(1)
7. Babu PA (2015) Comprehensive analysis of image filtering techniques. Int J Eng Res Sci Technol 4(4). ISSN 2319-5991
8. Janalipour M, Mohammadzadeh A (2016) Building damage detection using object-based image analysis and ANFIS from high-resolution image (case study: BAM Earthquake, Iran). IEEE J Sel Top Appl Earth Observations Remote Sens 9(5)
9. Pande S, Bhadouria VS (2012) A study on edge marking scheme of various standard edge detectors. Int J Comput Appl 44(9)
10. Wang J, Wu J, Wu Z, Jeon G, Jeong J (2017) Bilateral filtering and directional differentiation for Bayer Demosaicking. IEEE Sens J 17(3)
11. Singh S, Bhadauria S (2017) Wavelet based image de-noising using weighted high pass filtering coefficient & exposure based sub—image histogram equalization enhancement technique. In: 5th international conference on emerging trends in engineering, technology, science and management ICETETSM-17, vol 6, no 8, pp 1094–1104
12. Koranga P, Singh G, Verma D, Chaube S, Kumar A, Pant S (2018) Image denoising on wavelet transform using visu thresholding technique . Int J Math Eng Manag Serv. ISSN:2455-7749
13. Kunsoth R, Biswas M (2016) Modified decision based median filter for impulse noise removal. In: 2016 international conference on wireless communications, signal processing and networking (WiSPNET) pp 1316–1319. https://doi.org/10.1109/WiSPNET.2016.7566350
14. Li W, JiaY (2015) A new image de-noising method by combining WT with ICA. Hindawi Publishing Corporation Mathematical Problems in Engineering, Article ID 582640
15. Lin P, Chen B, Cheng F, Huang S (2016) A morphological mean filter for impulse noise removal. J Disp Technol 12(4):344–350. https://doi.org/10.1109/JDT.2015.2487559

16. Jain A, Gupta R (2015) Gaussian filter threshold modulation for filtering flat and texture area of an image. In: 2015 international conference on advances in computer engineering and applications, pp 760–763. https://doi.org/10.1109/ICACEA.2015.7164804
17. Rai CS, Vidhi MK, Varun V (2015) Removal of high density gaussian and salt and pepper noise in images with fuzzy rule based filtering using MATLAB. In: 2015 IEEE international conference on computational intelligence & communication technology, pp 166–172. https://doi.org/10.1109/CICT.2015.75
18. Chang H, Chu WC (2009) Double bilateral filtering for image noise removal. In: 2009 WRI world congress on computer science and information engineering, pp 451–455. https://doi.org/10.1109/CSIE.2009.414
19. Zhang M, Gunturk BK (2008) Multi-resolution Bilateral Filtering for image de-noising" IEEE Trans Image Processing 17(12)
20. Vermula M, Bugallo MF, Djuric PM (2007) Performance comparison of Gaussian based filters using information measures. IEEE Signal Processing Lett (12)
21. Senthilnath J, Kulkarni S, Benediktsson JA (2016) A novel approach for multispectral satellite image classification based on the bat algorithm. IEEE Geosci Remote Sensing Lett 13(4)
22. Patidar P, Srivastava S (2016) Image de-noising by various filters for different noise. Int J Comput Appl 9(4)

A Dual Band Metamaterial Loaded CPW-Fed Antenna for GSM, WLAN, LTE and RFID Applications

K. A. Ansal, Akhila Ann Kuruvilla, S. Keerthana, Keerthi Mariam Harries, Varsha Susan Johns, and Anu Raj

Abstract A dual band metamaterial loaded CPW-fed antenna for GSM, WLAN, LTE and RFID applications is presented in this paper. Metamaterials are new artificial materials which are used to achieve the physical properties that do not exist in natural materials. Here, the metamaterial provided gain enhancement for the proposed antenna. The overall size of the proposed antenna is 114×114 mm^2. The proposed antenna is imprinted on FR4 substrate with a relative permittivity 4.5, loss tangent 0.02 and thickness 0.8 mm. For the gain improvement and bandwidth enhancement, antenna is modified using a complimentary circular split ring resonator (CSRR). The measured dual band operating frequency is in the range of 0.92–2.76 GHz and 5.02 GHz–7.11 GHz. The circular unit cells are utilized to resonate the antenna at 1.8 GHz and 5.7 GHz, respectively. Thus, the proposed antenna is suitable for GSM, WLAN, LTE and RFID applications.

Keywords Metamaterial · GSM · WLAN · LTE · RFID · CSRR

1 Introduction

In today's fast developing world, the most unavoidable thing in our life is mobile phone. As its need is increasing, the requirement of wireless technologies such as GSM, LTE, and WLAN [1] is also increasing. The rapid growth of mobile phones is also leading to a threat of harmful radiations from these mobile phones and the subsequent health problems. Since no one can think about avoiding the phones, the only possible solution is to reduce the radiation emitted by the mobile phones [2]. Certain methodologies have to be implemented while designing these mobile phones. When compared with the conventional antennas, the metamaterial antenna decreases the number of the antenna elements and enhances antenna performance. Metamaterial-supported antennas will have gain improvement, [3] bandwidth enhancement, high efficiency, etc. The highly efficient antennas are in a huge demand especially to work

K. A. Ansal (✉) · A. A. Kuruvilla · S. Keerthana · K. M. Harries · V. S. Johns · A. Raj
Department of Electronics and Communication Engineering, SAINTGITS College of Engineering, Pathamuttom 686532, India
e-mail: ansal.ka@saintgits.org

A. Sivasubramanian et al. (eds.), *Futuristic Communication and Network Technologies*, Lecture Notes in Electrical Engineering 792,
https://doi.org/10.1007/978-981-16-4625-6_45

in service bands like global system for mobile communication (GSM-900, GSM-1800), [4] wireless LAN(WLAN) at 2.4 GHz [5] and 5 GHz, long-term evolution (LTE) [6] cover three bands (698–966 MHz, 1.427–2.69 GHz, 3.4–3.8 GHz), etc. [7]. The enhanced performance of metamaterial antennas makes them highly potential structures in wireless communication systems.

2 Metamaterials

The word metamaterial is derived from Greek word 'meta' and Latin word 'materia' meaning beyond and matter, respectively [8]. Metamaterials are artificial materials on which a particular property is being induced which the material does not possess naturally. These property may be due to its shape, structure, size or the geometry which it has, leading to changes in the electromagnetic property of the material [9]. So these properties are not derived from the base materials but from the internal microstructure [10]. The core objective of usage of metamaterial is that these materials help us to achieve the desired properties [11]. These desired properties are achieved, when necessary changes are made in the geometry which results in change in refractive index [12], either to a positive value, or a negative value resulting in achieving the property.

3 Split Ring Resonator

Split ring resonators (SRR) are structures which consist of two metallic rings that are etched on the dielectric material, and they also consist of slit which are etched at opposite sides [13]. These are structures that are mainly used by metamaterials in order to get the derived property [14]. It is mainly used for producing desirable magnetic susceptibility. Split ring resonators are having strong magnetic coupling and they are much more than ferromagnetic materials. Several kinds of split ring resonators are present and one such is complementary split ring resonator (CSRR) (Fig. 1).

4 Antenna Design

Antenna 1 in Fig. 5a is the initial antenna with a CPW L-shaped fed structure. Apparently on the dielectric substrate of the CPW structure, a lump with two diminished slits ground electrodes lied alongside and parallel to the strip on the same surface. It is fabricated using FR4 substrate which is having a relative permittivity of 4.5, loss tangent of 0.02 and thickness of 0.8 mm. The material is capable of retaining

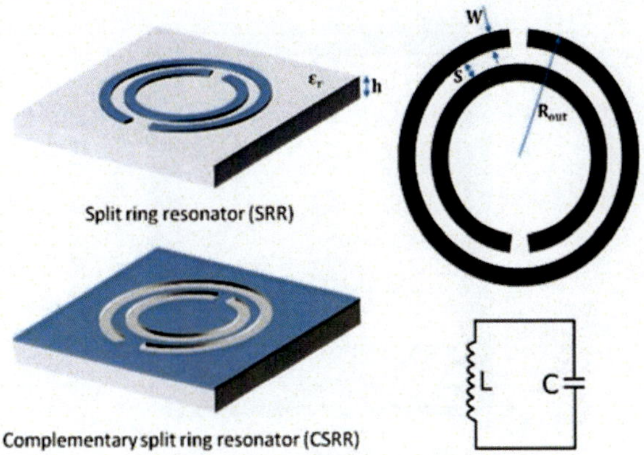

Fig. 1 Split ring and complementary split ring resonator

its mechanical characteristics and electrical insulation in dry as well as wet conditions. Along with good manufacturing, it can be utilized for various mechanical and electrical applications. All the dimensions of the antennas are in millimetres (mm). The substrate is in a square shape with an edge G1, and the overall size was 114 × 114 mm.

The geometry has L-shaped microstrip line acting as the patch. Here, dual band is obtained, but at lower frequencies the antenna gain and bandwidth are less. Usually for the gain improvement and bandwidth enhancement metamaterials are used, as they are utilized for achieving the desired properties. The antenna is modified by using a complementary circular split ring (CSRR)-based metamaterial as shown in Fig. 2b. The circular unit cell used in the antenna structure is shown in Fig. 2c. The complementary split ring is placed at the bottom of the CPW and is utilized to resonate at 1.8 GHz and 5.7 GHz, respectively. We obtained a band of 0.92–2.76 GHz and 5.02–7.1 GHz where 1.8 GHz and 5.7 GHz are included in these bands so that the proposed antenna is suitable for GSM, WLAN, LTE and RFID applications.

5 Results and Discussion

5.1 Return Loss

The return loss plot, VSWR, radiation patterns and 3D polar plots are shown in Figs. 3, 4 and 5.

Return loss values of − 18.9, − 11.2 dB are obtained at 1.8 and 5.7 GHz for antenna without metamaterial and − 24db, − 20.5db are obtained at 1.8 GHz and 5.7 GHz for antenna with metamaterial which is useful for GSM, WLAN, LTE and RFID applications.

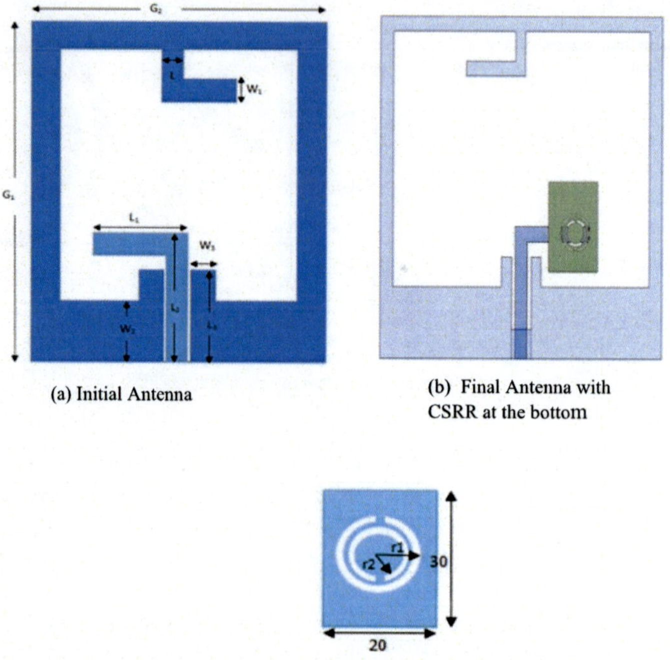

(a) Initial Antenna

(b) Final Antenna with CSRR at the bottom

(c) Complementary circular split ring with r1=6mm, r2=4.5mm

Fig. 2 Structure and dimensions of initial and proposed final antenna: $G_1 = G_2 = 114$ mm, $L = 4$ mm, $L_2 = 40$ mm, $W_1 = W_3 = 5$ mm, $L_1 = 30$ mm, $W_2 = 24$ mm, $r1 = 6$ mm and $r2 = 4.5$ mm

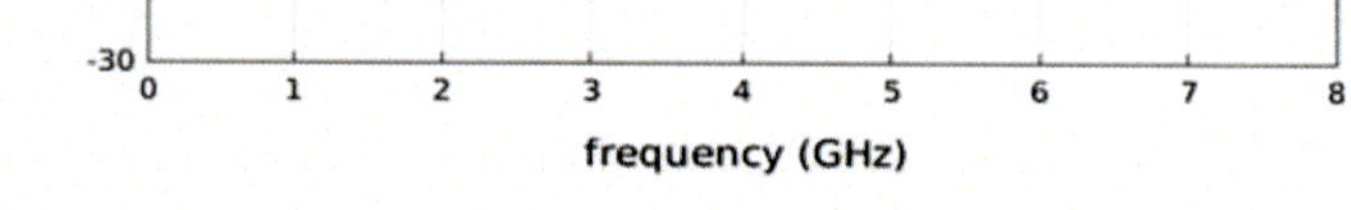

Fig. 3 Return loss plot for the antennas shown in Fig. 3

5.2 *VSWR*

The values of VSWR are 0.93 and 1.54 at 1.4 and 5.2 GHz for antenna without metamaterial. The values of VSWR are 1.05 and 1.7 at 1.8 GHz and 5.7 GHz for

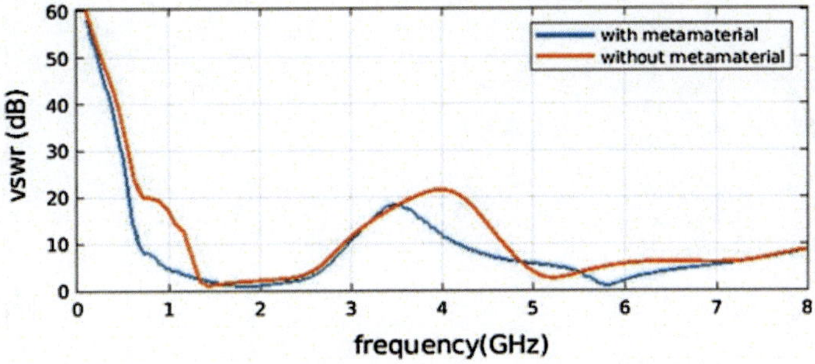

Fig. 4 VSWR for the antennas shown in Fig. 1

antenna with metamaterial. This shows that the proposed design is well matched and antenna is well suited for GSM, WLAN, LTE and RFID applications (Fig. 4).

5.3 Radiation Pattern

Figure 5a, b depicts the radiation pattern of antenna 1 in Fig. 2. Figure 5c, d depicts the radiation pattern of antenna 2 in Fig. 2. The variation of the power radiated by an antenna as a function of the direction away from the antenna is depicted by the radiation pattern of an antenna [5]. Figure 5a shows the radiation pattern centred at 1.4 GHz which is slightly deviated. Figure 5b shows the radiation pattern at 5.2 GHz which is more deviated than the pattern at 1.4 GHz. Figure 5b shows the radiation pattern centred at 1.8 GHz and Fig. 5c shows the radiation pattern at 5.7 GHz. It shows that at higher frequencies the radiation pattern undergoes distortion.

5.4 Gain

Figure 6a, b represents the 3D polar plot of the antenna 1 shown in Fig. 5. Figure 6c, d represents the 3D polar plot of the antenna 2 shown in Fig. 2. It can be observed from the plots that the gain at 1.4 GHz and 5.2 GHz is 2.5 dB and 5 dB. The gain at 1.8 GHz and 5.7 GHz is 4.8 dB and 6.2 dB, respectively. Hence, it shows that the gain is increased when antenna structure is modified using metamaterial.

This project mainly emphasizes on wireless bands, which mainly includes global system for mobile communication (GSM), wireless local area network (WLAN), long-term evolution (LTE) and radio frequency identification (RFID) applications. We could see that the designed antenna could be effectively used in these wireless band frequencies.

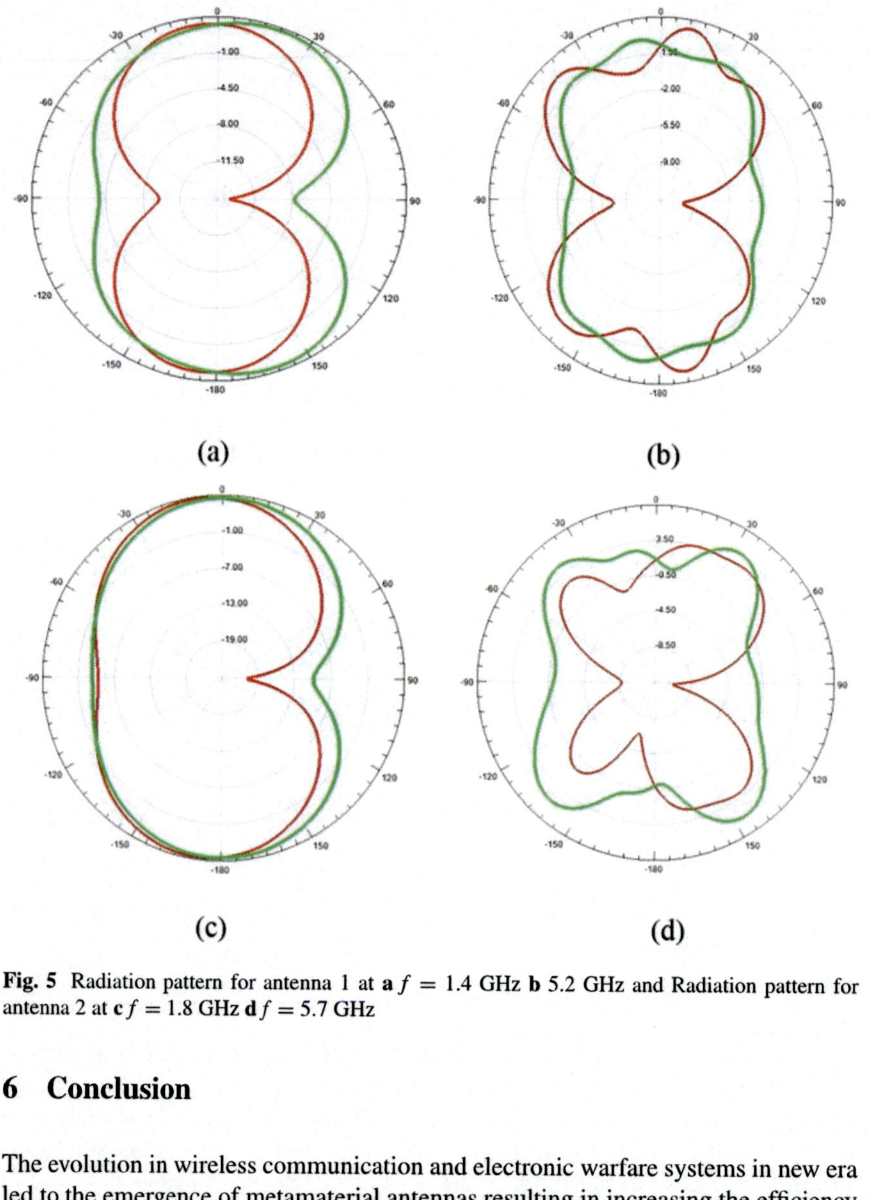

Fig. 5 Radiation pattern for antenna 1 at **a** $f = 1.4$ GHz **b** 5.2 GHz and Radiation pattern for antenna 2 at **c** $f = 1.8$ GHz **d** $f = 5.7$ GHz

6 Conclusion

The evolution in wireless communication and electronic warfare systems in new era led to the emergence of metamaterial antennas resulting in increasing the efficiency or performance of the antenna to a greater extent. In order to adapt with the wireless communication systems in this modern era, 'A dual band metamaterial loaded CPW-fed antenna for GSM, WLAN, LTE and RFID applications' is introduced. The antenna is designed in two stages by introducing and without the complimentary split ring resonators. It was found that by using complimentary split ring resonators the

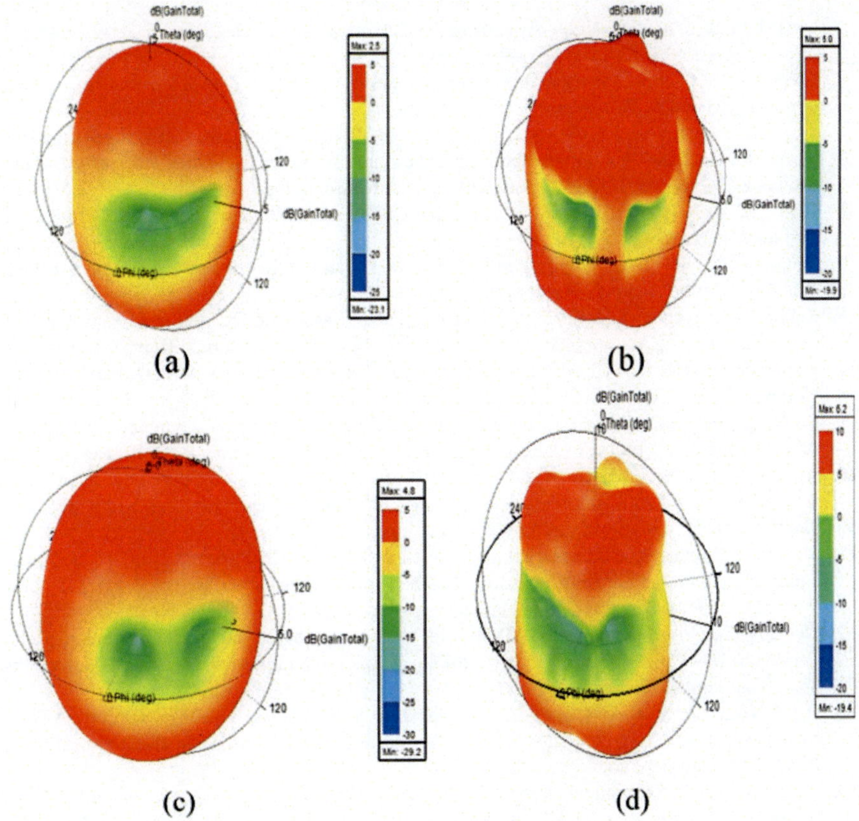

Fig. 6 3D polar plot for antenna 1 at **a** $f = 1.4$ GHz **b** $f = 5.2$ GHz and antenna 2 at **c** $f = 1.8$ GHz **d** $f = 5.7$ GHz

antenna is able to provide high gain and bandwidth. The results obtained after simulation ensure that the antenna functions well in the lower (0.92–2.76 GHz) as well as in upper (5.02–7.11 GHz) frequency range. Since it covers the frequency band of GSM, WLAN, LTE and RFID, it is possible to use this antenna in these applications. The fabrication of antenna is much easier as well as it is simple, economical and also appropriate for small portable devices.

References

1. Krzysztofik WJ, Cao TN (2018, November 5) Metamaterials in application to improve antenna parameters. Metamaterials Metasurfaces Josep Canet-Ferrer, IntechOpen. https://doi.org/10.5772/intechopen.80636
2. Aznar F, Gil M, Siso G, Bonache J, Martin F (2009) SRR- and CSRR-based metamaterial transmission Lines: modeling and comparison. In: 2009 IEEE MTT-S international microwave

workshop series on signal integrity and high-speed interconnects, Guadalajara, 2009, pp 49–52. https://doi.org/10.1109/IMWS.2009.4814907
3. Caloz C, Itoh T (2006) Electromagnetic metamaterials: transmission line theory and microwave applications. Wiley, New Jersey
4. Iyer AK, Eleftheriades GV (2002) Negative refractive index metamaterials supporting 2-D waves. In: IEEE-MTT international microwave symposium, vol 2, Seattle, WA, pp 412–415
5. Siwiak K, McKeown D (2004) Ultra wideband radio technology
6. Park S-Y, Oh S-J, Park J-K (2010) A simple CPW Fed UWB antenna design. J Koren Inst Electromagn Eng Sci
7. Antonino-Daviu E, Cabedo Fabres M, Ferrando Bataller M, Valero-Nogueira A (2003) Wideband double-fed planar monopole antennas. Electron Lett
8. Subbarao A, Raghavan S (2014) Compact coplanar waveguide-fed planar antenna for ultra wideband and WLAN applications, wireless personal communications. Kumar A, Shanmuganantham T (2014) A CPW fed octagonal patch UWB antenna with WiMAX band notched. Commun Embed Syst (ICICES2014)
9. Lu G, von der Mark S, Korisch I, Greenstein LJ, Spasojevic P (2004) Diamond and rounded diamond antenna for ultra wideband communications. IEEE Antennas Wirel Propag Lett
10. Abdalla MA, El-Sobky NA, El-Gabry MN (2016) Metamaterials inspired dual-wide band CPW-fed antenna using split ring resonator structure. In: 10th international congress on advanced electromagnetic materials in microwaves and optics—metamaterials
11. Kukreja J, Kumar Choudhary D, Kumar Chaudhary R (2017) CPW fed miniaturized dual-band short-ended metamaterial antenna using modified split-ring resonator for wireless application. Wiley Int J RF Microwave Comput Aided Eng
12. Denceli D, Yogeshwaran A (2016) Split ring resonator metamaterial antenna for WiMax/WLAN applications. In: 2016 online international conference on green engineering and technologies (IC-GET)
13. Pandeeswari R, Samson Daniel R, Deivalakshmi S, Raghavan S (2017, November) Non-bianisotropic split ring resonator based CPW-fed dual band antenna. In: Proceedings of the 2017 IEEE region 10 conference (TENCON)
14. Ashok Kumar S, Shanmuganantham T, Dileepan D (2017) Design and development of CPW fed monopole antenna at 2.45 GHz and 5.5 GHz for wireless applications. Alexandria Eng J 56:231–234 (2017)

Autonomous Wheeled Bot with Monocular Camera

M. Aswath, Abitha K. Thyagarajan, S. Melvin Noel, Nikshith Narayan Ramesh, and P. Reena Monica

Abstract Recognition and mapping is the key innovation to rising technology, for example, intelligent robotic systems and autonomous driving. Our goal is to provide a robot with systems that will aid its recognition and mapping of its environment for better locomotion and navigation. In the case of navigation, we use homography on a semantic segmentation to build an occupancy grid. For localization of the camera, the ORBSLAM2 is used, pose estimation is used to improve the robustness of the tracking over the period of time and also results in a better environment reconstruction. Monocular camera-based occupancy grid construction was developed, and our trial results have indicated great execution of our algorithms under testing it in a real-world environment.

Keywords Monocular robot navigation · ROS · Arduino · Computer vision · ORBSLAM2

1 Introduction

With our growing dependency on automation in this world, it is quite evident that the use of robots for simple or mundane tasks in our lives is imminent. One such application is last mile delivery in e-commerce and shipping services—i.e., using a car to deliver an item to a building or neighborhood, and then a different means of conveyance to deliver the item to its final destination. To meet the need for delivering on this "last mile" of the journey, smaller robots can be used, which have the ability to move around places without the explicit help of humans. These robots must be able to analyze their surroundings and maneuver themselves through their paths to the ultimate destination. Feature extraction from camera images can be used to localize the robot more precisely; also, as an added benefit, the surroundings can be mapped to facilitate obstacle avoidance. This can be achieved using computer vision techniques

M. Aswath (✉) · A. K. Thyagarajan · S. M. Noel · N. N. Ramesh · P. R. Monica
School of Electronics Engineering, Vellore Institute of Technology, Chennai, India

A. Sivasubramanian et al. (eds.), *Futuristic Communication and Network Technologies*, Lecture Notes in Electrical Engineering 792,
https://doi.org/10.1007/978-981-16-4625-6_46

such as semantic segmentation, simultaneous localization and mapping (SLAM), and many more models. In this paper, we have proposed a smart autonomous robot that can do exactly this.

2 Literature Review

André M. Santana et al. have proposed using planar information to build a 2D occupancy grid map from monocular vision [1]. In this paper, segmentation was done by sampling a cropped image of the ground and then applying the Mean Shift algorithm. This approach cannot handle anomalies if the ground plane contains textures or a tiled pattern—even if the walls and the ground have the same color or texture, it cannot distinguish between them.

Chenyang Lu et al. presented a compilation of various cases for Convolutional Neural Network (CNN)-based semantic segmentation [2]. This paper gave us insights into using this approach with Neural Networks. One benefit of using such an approach is that the image is processed as a global perspective instead of local sampling and thresholding. With the global understanding of the environment, CNNs are able to yield better segmentation results.

Raul Mur-Artal, Juan D. Tardos came up with SLAM based on ORB features [3]. ORB-SLAM2 supports monocular, stereo, and RGB-D cameras. Because stereo and RGB-D cameras are expensive, we wanted to try a more economical approach with an ubiquitously available monocular camera. The focus of interest in our paper is the Monocular-based ORB-SLAM2.

Dylan Jennings conducted a study which comprises various types of Autonomous Delivery Bots in the development [4]. The study includes parameters like time of travel, on-road, and off-road analysis, study of driving on pedestrian, and sidewalks, government rules and regulations in this area.

For homography or birds eye view conversion, M. Venkatesh and P. Vijayakumar provided insights on the mathematical formulations and pipeline for processing the camera feed and then presented the results of their physical implementation on embedded hardware [5]. LinBo Luo et al. provided yet another approach and implementation for the bird's eye view homography with deeper pipelines to process the image for robust conversion [6].

3 Methodology and Implementation

There are two main components, software and hardware implementations. The hardware architecture shows the interconnection between the components of the system. The software architecture shows the flow of data and the pipelines for the complete autonomous navigation stack.

Hardware architecture, as summarized in Fig. 1, consists of a Raspberry Pi as

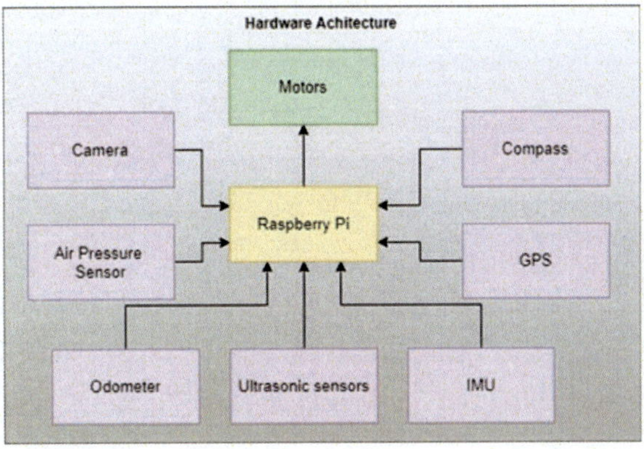

Fig. 1 Hardware architecture of the wheeled bot

the primary controller interconnecting the following seven sensors and actuators. An air pressure sensor is used to measure altitude and thereby to determine which floor the bot is on. A compass and GPS are used for global positioning and navigation. An Inertial Measurement Unit (IMU) sensor and an odometer are used for precise local state estimation. An ultrasonic sensor is used for crude obstacle avoidance. A monocular camera is used for our computer vision applications, mainly to plan the occupancy grid, determine the position of obstacles, and for executing SLAM, as elaborated further. An odometer is used to precisely estimate the distance traversed over short periods of time; the drift is accounted for with visual feedback from SLAM. The actuators used are motors.

Software architecture, as summarized in Fig. 2, comprises three major parts: localization, motion planning, and drive control. Localization determines the current position of the robot with respect to its immediate surroundings (walls, obstacles, etc.); its state estimate functionality is further elaborated in Sect. 3.1.

The next step after localization is to determine which parts of the environment are free to move in, and based on that, to generate the best trajectory to proceed on. This is handled by the motion planning stack, as elaborated in Sect. 4.2. Finally, the drive control takes care of the actual motion of the robot, by providing commands to the actuator based on the calculated trajectory. By using a pure pursuit controller, the robot can follow the given trajectory. Our assembled bot, incorporating the above hardware and software architecture, is shown in Fig. 3.

3.1 Localization with Kalman Filter

Our localization stack consists of a Kalman filter that uses various sensors to perform state estimation, which includes position, velocity, and heading angle. The estimate state is provided by sensors with high refresh rate, and the correction is performed

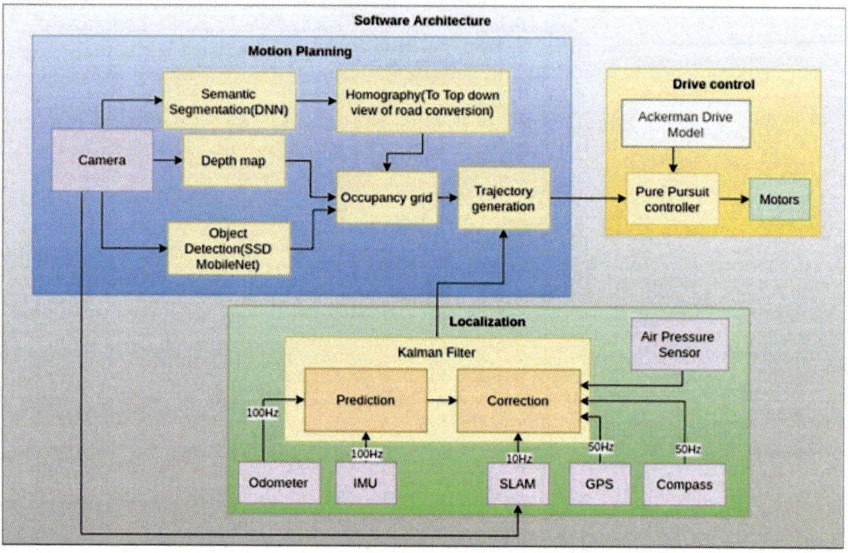

Fig. 2 Software architecture of the wheeled bot

Fig. 3 Implementation of the autonomous wheeled bot

by SLAM and GPS which run at a low refresh rate. Both are fused to get the best estimate of the ground truth state.

The initial state estimate mean $\hat{x}_0$ and covariance P_0 are predefined by the user at the start of the state estimation procedure. The Kalman gain K_k is computed based on the previous covariance P_k', the sensor measurement matrix or selection matrix

H, and the Gaussian noise model R. The current state estimate mean $\hat{x}_k$ is calculated based on the previous state estimate, the Kalman gain, and the current sensor readings $\hat{y}_k$. Finally, the current covariance P_k is calculated based on the Kalman gain and the previous state covariance matrix. These steps are defined in Eqs. 1–4.

$$\hat{x}_0 = E[x],\ P_0 = E\big(x - \hat{x}_0\big)\big(x - \hat{x}_0\big)^T \tag{1}$$

$$K_k = P_k' H^T \big(H P_k' H^T + R\big)^{-1} \tag{2}$$

$$\hat{x}_k = \hat{x}_{k-1} + K_k H\big(y_k - H_k \hat{x}_{k-1}\big) \tag{3}$$

$$P_k = (I - K_k H_k) P_{k-1} \tag{4}$$

4 Results and Discussion

The implementation of the navigation software stack is presented in the following subsections. The details about the concept and algorithms are written, and the results obtained in each stage are also displayed in the figures, respectively.

4.1 Localization with Visual SLAM

SLAM is used for localizing the bot. There are many types of SLAM, each for use with different sensor types (such as LIDAR or camera). For our use case, we are exploring a camera-based approach in order to be cost-efficient. We are using mono-camera-based ORB-SLAM as proposed by Mur-Artal and Tardós [3]. The camera frame is localized in its surrounding as shown in Fig. 4. Since this is a monocular camera, the motion of camera frame is in pixel space. To convert it to real-world coordinates, a scale factor needs to be determined. In the start of the program, the bot is allowed to move for a predefined distance to calculate the scale factor. The scale factor is determined by taking the magnitude of the predefined real-world translation to the magnitude of the camera pose translation. Subsequently, the same scale factor is multiplied with the camera pose translation to get the real-world pose at any given time of the robot's movement.

ORB Feature Detection ORB, also known as Oriented FAST and rotated BRIEF, is a fast-robust local feature detector in images. It is used in tasks such as object recognition, 3D construction, and other computer vision applications. Our bot needs to construct a map of its environment (which is unknown to it), as well as map the path to where it is supposed to go. This path has to be used to guide the bot around

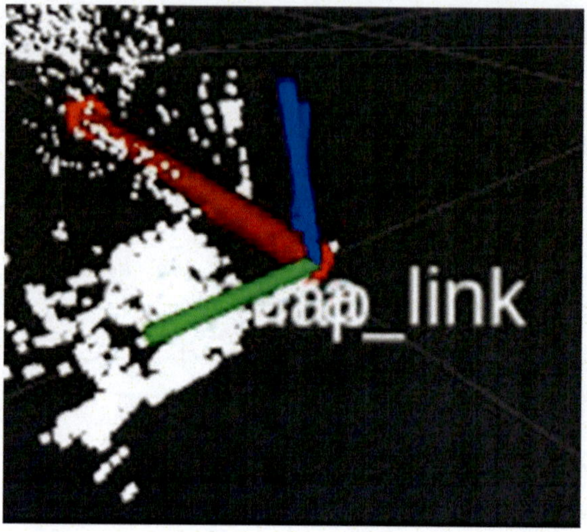

Fig. 4 Camera frame localization with ORB-SLAM2

obstacles. Moreover, the location of the bot has to be kept track of at all times. Hence, the robot is our agent. Thus, SLAM is ideal for our requirements. It provides a foundation for the rest of the processes, namely path planning, path following, and so on. This is where ORB feature detection comes into play—it forms an integral part of our project.

RANSAC Random sample consensus (RANSAC) is an algorithm used to remove any outlier points that may add noise when estimating the camera poses. This method is more reliable than SVD approach in practical implementation. If a really large outlier is present, then the while optimal solution deviates from the desired solution. In RANSAC, the issue of this large outlier problem is handled as follows. A number of image feature points (greater than eight, required for fundamental matrix) are taken from the previous view to find the fundamental matrix. Then, all the remaining image feature points are transformed with the fundamental matrix to get its position in the current time-step. Taking any two points reference, the line connecting the two chosen points is considered and two other parallel lines on either sides are calculated. The inliers are calculated as the number of points present within the two lines. The algorithm has found the solution when the ratio between the inliers to the total number of sample points is above the threshold. Finally, the fundamental matrix that gives the largest number of inlier points is chosen as the best estimate.

Bundle Adjust When a set of images is rendered, describing a set of 3D points from several viewpoints, bundle adjust (BA) is said to be a condition of simultaneous refinement of 3D coordinates depicting the scene geometry and the thresholds of relative motion, including the optical parameters, extrinsic parameters in our case. Optimization is done based on the projection of world points onto the image plane. This ensures to have a proper estimate of the six degrees of freedom (6DOF) for the camera trajectory, as well as 3D map (3D point cloud) directing from the input feature tracks. Bundle adjustments are mainly of three types, namely Full, Local, and Motional. ORB-SLAM uses a local bundle adjustment in its pipeline.

4.2 Motion Planning

A vital part of our bot's operation is to understand its surroundings. This is achieved through a perception stack comprising two separate pipelines, localization and motion planning. For motion planning, semantic segmentation is used to classify the pixels belonging to the ground plane. Then, a homography operation is performed on the segmented image to give a virtual top down view of the road. The obstacles detected are used to build an occupancy grid of the environment, which is then used to generate a safe trajectory map for the robot to follow. There are many path planning algorithms; we have implemented A* path planner, a heuristic algorithm as shown in Fig. 5. This planner takes into account the turning radius of the wheeled bot. The same concept can be extended for online path planning as well. For ground plane detection, seed-based threshold and segmentation results are shown. The floor

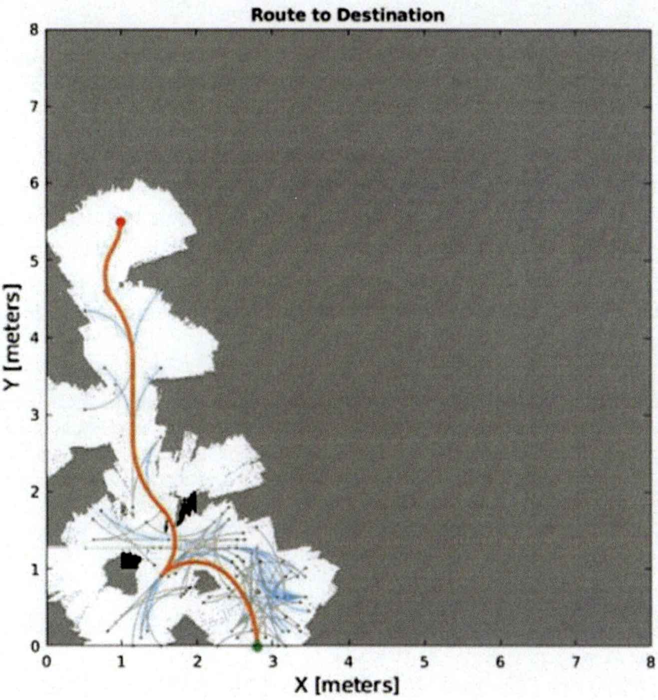

Fig. 5 Path planning with A* path planner

Fig. 6 Segmentation and applied to in-door environment

is reflective, and hence, only the left part of the ground is segmented properly as seen in Fig. 6.

5 Conclusion

The localization of monocamera was done successfully, and it proved to be robust. The algorithms were tested successfully, proved to be efficient and useful. The sensor fusion has to be integrated for more precise state estimation. The segmentation as shown in the results section does not cover the whole ground, so to improve this, a Gaussian blur can be done for the whole image to even out the reflection before segmenting it.

Acknowledgements The authors would like to express their sincere thanks and deep sense of gratitude to their project guide, Dr. Reena Monica, Associate Professor, School of Electronics Engineering, VIT Chennai, for her consistent encouragement and valuable guidance offered to us throughout the course of the project work.

References

1. Santana AM, Aires KRT, Veras RMS, Medeiros ADA (2011) An approach for 2D visual occupancy grid map using monocular vision. Electron Notes Theor Comput Sci 281:175–191
2. Lu C, Molengraft M, Dubbelman G (2019) Monocular semantic occupancy grid mapping with convolutional variational encoder–decoder networks. IEEE Rob Autom Lett 4(2):445–452

3. Mur-Artal R, Tardós JD (2017) ORB-SLAM2: an open-source slam system for monocular, stereo, and RGB-D cameras. IEEE Trans Rob 33(5):1255–1262. https://doi.org/10.1109/TRO.2017.2705103
4. Jennings D, Figliozzi M (2020) Study of road autonomous delivery robots and their potential effects on freight efficiency and travel. Trans Res Rec 2674(9):1019–1029. https://doi.org/10.1177/0361198120933633
5. Venkatesh M, Vijayakumar P (2012) Transformation Technique. Int J Sci Eng Res 3(5)
6. Luo L, Koh I, Park S, Ahn R, Chong J (2009) A software-hardware cooperative implementation of bird's-eye view system for camera-on-vehicle. In: 2009 IEEE international conference on network infrastructure and digital content, pp 963–967. https://doi.org/10.1109/ICNIDC.2009.5360920

Real-Time Pothole Detection with Onboard Sensors and Camera on Vehicles

M. Aswath, S. Jeevak Raj, and K. Mohanaprasad

Abstract Road conditions play an important role in our everyday commute. With the proliferating number of vehicles on the road each year, it has become necessary to access the road conditions very frequently, and this would ensure that the traffic also flows smoothly. Even the smallest crack in the road could be easily be chipped into a large pothole due to changing surface temperatures of the road and from the force of vehicles riding over it. In this paper, we have addressed how we could better identify these potholes in real time with the help of onboard sensors in vehicles so that the data could be useful for analysis and better management of potholes on a large scale. For the implementation, we used an SVM classifier to detect potholes, and we achieved 98.1% accuracy based on data collected from a local road for about 2 km which had 26 potholes distributed along the road.

Keywords Support vector machine · Computer vision

1 Introduction

Real-time pothole detection is needed to report any potholes that are taking root in the roadways. Informing authorities of the potholes' location and its effect on vehicles, by reporting to the authorities, it becomes much easier to plan, manage and fix the potholes. This ensures safety for drivers, prevents accidents that occur due to driving over potholes, vehicle damage in wheel rims which require the owners to change the whole alloy rim which is expensive and also non-economical. Furthermore, the collected information about the potholes can be used for alerting drivers of upcoming potholes in their route. The digital platform for this system has easier manageability and precise outlook. Massive data collection is possible if implemented on large scale. Based on the data, it is also easy to estimate the cost required for fixing the potholes, the typical cost to repair potholes is approximately \$35 to \$50 per pothole.

M. Aswath · S. J. Raj · K. Mohanaprasad (✉)
School of Electronics Engineering, Vellore Institute of Technology, Chennai, India
e-mail: kmohanaprasad@vit.ac.in

A. Sivasubramanian et al. (eds.), *Futuristic Communication and Network Technologies*, Lecture Notes in Electrical Engineering 792,
https://doi.org/10.1007/978-981-16-4625-6_47

There may be an initial mobilization cost of about $100 to $150 [1] to bring trucks and crew out to the repair site.

2 Literature Review

Wang [2] used a combination of acceleration thresholds, Z-Thresh, Z-DEV, STDEV Z, for detecting potholes. If readings cross the threshold, a pothole is detected. Disadvantages of this method can be false triggers or failure to detect different types of potholes while driving at different speeds and different terrain. This gave us an overview of the existing algorithms' performances.

Bhatt et al. [3] used SVM model to classify the sensor data. They used a total of 26 features formed with raw data collected from sensors. This might require more computational time for the CPU. We try to address this by experimenting with a lesser number of features.

Mednis et al. [4] evaluated various methods like Z-THRESH, Z-DIFF, STD-DEV(Z), G-ZERO algorithms on a mobile running Android operating system. Each detection method is evaluated with a range of values of threshold points in terms of ratio of G-force. The percentage of correctly detected potholes to the original pothole counts are shown for each methods. Ability of correctly detecting a various types of potholes such as large, small, pothole clusters, gaps and drain pits was also evaluated.

Singh and Shekhar [5] used Nvidia GPU to detect potholes, and this provided us with the insight to explore such options, and their choice of the neural net was also explained.

Chen et al. [6] explored a different approach to localize and detect a pothole on the road instead of searching for it in the whole image, but rather localizing certain regions and then continue the search.

3 Methodology and Implementation

We explore three modes of pothole detection as follows:

1. Offline method: the user captures the image and uploads it to the server.
2. Online method: the user starts a live stream to detect pothole in real time.
3. Online method: using accelerometer and gyroscope to detect potholes in real time.

The computer vision algorithm can detect the presence of pothole from the image both at day and night time and can extract its intrinsic parameters such as dimension (length, width), area and the effect of riding. If the pothole is detected, the app acquires the latitude and longitude position of the device at that instant, and captures the image of the pothole and sends it to the SQL database server, which is running on

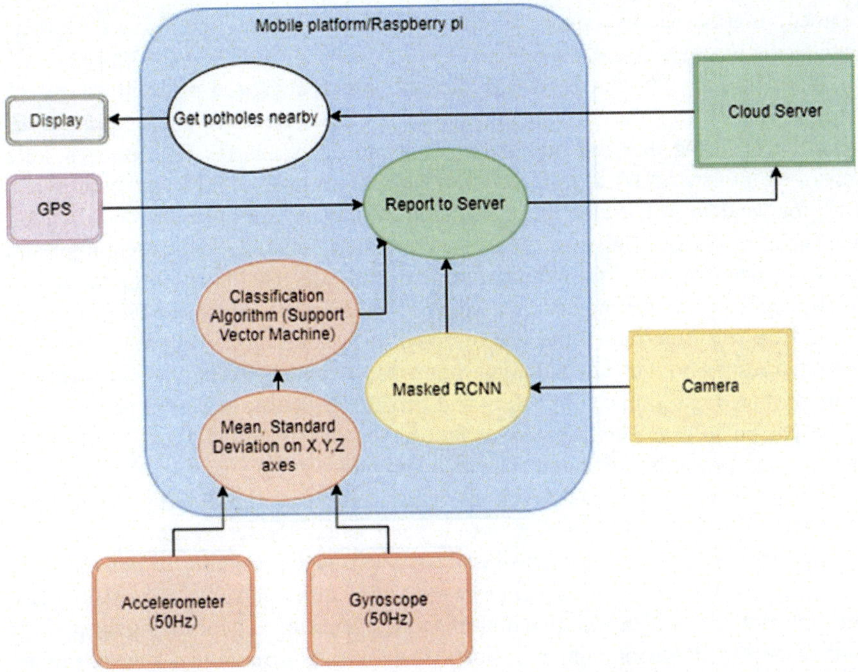

Fig. 1 System architecture

Node-RED platform. Once the pothole is detected and sent to the server, the image of the pothole can be viewed in the maps of Node-RED server. If the user feels that taking photos of the pothole is difficult or if there are many potholes present, then the live stream option can be used. The potholes detected from the live stream are uploaded to the server automatically. During night time or in heavy traffic, taking photos is difficult and here comes the use of the third option, with the help of IMU (Inertial Measurement Unit).

For inertial motion sensors, in case of roads with heavy traffic, the camera cannot see the road because of dynamic obstacles like other vehicles on the road. This is where the accelerometer and gyroscope sensors is a useful substitute. The overall system architecture is shown in Fig. 1.

3.1 Image Processing from the Captured Image

The aim is to detect the pothole exactly and apply a mask to it. To do this, we are using the algorithm called Mask-RCNN which was developed by FAIR which uses the concept of instance segmentation and object detection. The object detection technique is used to classify the object and label it with a confidence score. Instance segmentation is used to separate things into groups, isolating the pixels associated with each object.

Training First, the dataset was collected, which consisted of 1500 images of potholes each with different camera perspectives and under different lighting conditions, even during day and night time, wet and dry scenarios. The images were separated into two classes, train and validation, in a ratio of 90% for training and 10% for validation. For labelling the region of interest, the open-source VIA tool, developed by Oxford University, was used to draw a polygon around the potholes to label it. Then, the JSON file with the ground truth was used for training the model for 160 epochs, and the model was tested to provide an accuracy of 81%.

Testing Once the model has been trained, it was tested in real time. Good and expected results both at day and night time were obtained. To improve the visibility at night time, logarithmic contrasting and stretching algorithm was employed. As a result, the potholes are highlighted, and the area is also obtained. Based on the area, the effect of riding is calculated. The area and effect parameters are added into the list along with the latitude and longitude values, and we used free stack API to create the URL of the image and added that to list and sent to the server through socket.

Measures to prevent duplicate entries There is a possibility of detecting the same pothole multiple times on different days or by different vehicles. In such cases, the same potholes will be reported multiple times. Marking and localizing the pothole with respect to the surrounding may not be feasible, i.e. surroundings keep changing. So we devised an algorithm that receives nearby potholes data and stores it on the edge device; next, it compares the length, breadth, or area of detected potholes with the images obtained from the cloud; if the difference is significant, then it uploads to the server as a record of newly detected pothole.

3.2 Pothole Detection in Live Stream

Detection through live stream needs a faster algorithm to capture the details instantaneously; so we preferred the canny edge algorithm which is quite faster than the Mask-RCNN so that the pothole is detected along with its count and area values and updated to the server.

Canny Edge detection It is an edge detection technique used to detect edges in an image; we tuned the edge detector so that the edges of the potholes are highlighted and the value of the threshold came around 0.6; next, we dilated the image into a binary image taking a kernel of ones.

Finding the potholes and its area The Mask-RCNN gives an image with highlighted pixel mask for region with potholes. The area of the potholes is calculated as the ratio of number pixels in the highlighted region to the total number of pixels in the image.

Uniformity Suppose the potholes are detected from a distant view, then the area would be small and when compared to detecting it from a closer position. So to avoid this discrepancy and maintain uniformity, a horizontal threshold line in the image was taken, and only when the pothole overlaps with the virtual line, the detected pothole is registered. As an alternative solution, homography can also be used to get a birds eye view of the dashboard camera feed, to obtain the correct area, irrespective of the distance.

3.3 *Accelerometer and Gyroscope*

For the classification of sensor data as potholes or plain road, a machine learning classifier called SVM (support vector machines) was used. For real-time implementation, a smartphone centric mechanism was explored. Two apps were built—one for data acquisition and another for real-time prediction. The data acquisition app was built with the MIT App Inventor V2. Then, the SVM machine learning classifier was trained with the collected data. For the SVM classifier, six features were used for the model, that is, mean and standard deviation of accelerometer and gyroscope. The support vectors are then used to classify data as either potholes or plain road. The advantages of this approach are that it is more accurate and reliable than thresholding.

Data Acquisition Data collection was decided to be done with the help of smartphone sensors. Input data was formatted into seven fields as follows: Time-stamp, Accelerometer X, Accelerometer y, Accelerometer Z, Gyroscope X, Gyroscope y, Gyroscope z, Pothole-detected flag. These are collected at 50 Hz. The user opens the app as shown in Fig. 2 and starts logging accelerometer and gyroscope sensor data. The user presses the 'pothole' button every time the vehicle drives over a pothole. This

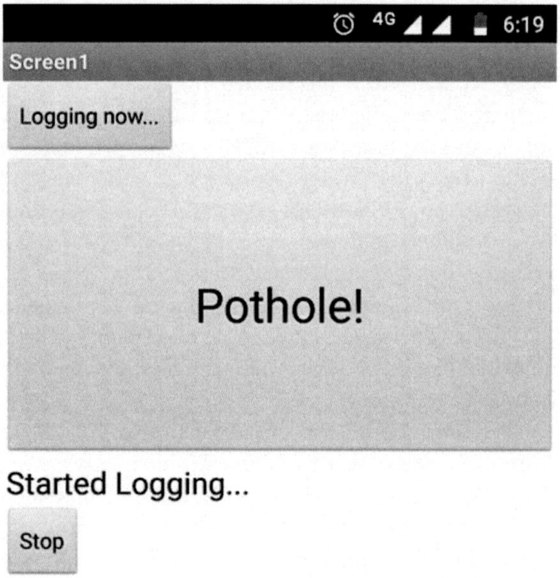

Fig. 2 Data acquisition app

Fig. 3 Detection app

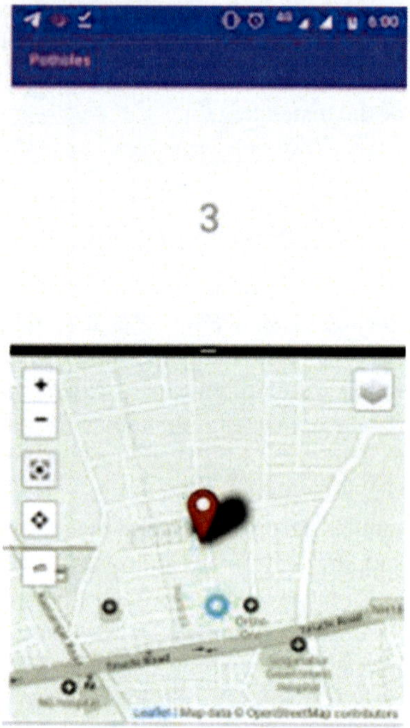

associates the acceleration reading for the subsequent 1 s period as a pothole data. The phone was fixed to the floorboard of the vehicle, and gently the 'Pothole' button was pressed when the driver rides the vehicle over the potholes. Around 300 Kb of data was collected for training (Fig. 3).

Data classifier creation Classification of potholes from the images is done with Mask-RCNN, and it was trained on Google Colab. As a quick alternative prototype method, Canny edge detection was also tested in MATLAB; however, there were a lot of false triggers. Mask-RCNN makes pixel classification possible; this way, the exact area of the pothole is calculated. The Mask-RCNN model is also robust to various lighting conditions, and it has been tested for both day and night detection as well. Problems addressed with the video feed from the camera are area calculation, detection during both day and night, and the effect of riding on a pothole.

The accelerometer and gyroscope data obtained are visualized in a 3D graph plot for a better understanding and to identify which classifier should be used. As shown in Fig. 4 and Fig. 5, a linear classifier can be used to classify the pothole and non-pothole data.

The SVM model was prototyped in MATLAB with linear kernel for two-class learning. A test bench was created which calls the entry-point functions and passes the features as parameters. After testing and validation, the entry-point function was exported to C++ code for the ARM Cortex-A processor using MATLAB Coder. The generated C++ code is copied into the Android Studio Projects folder under the JNI directory. Then, JNI interface was used to call C++ SVM predict entry-point function

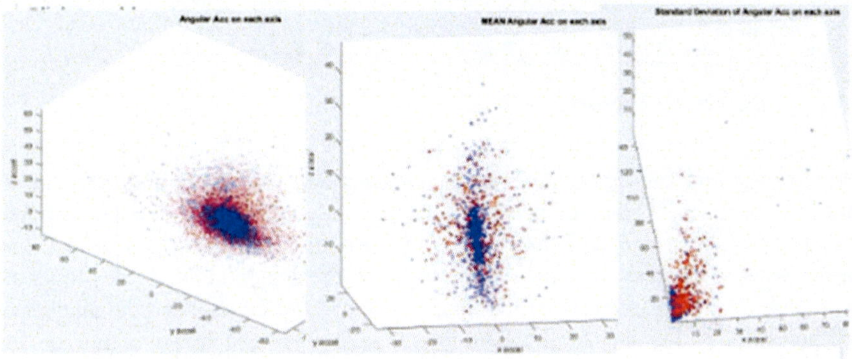

Fig. 4 Accelerometer data points

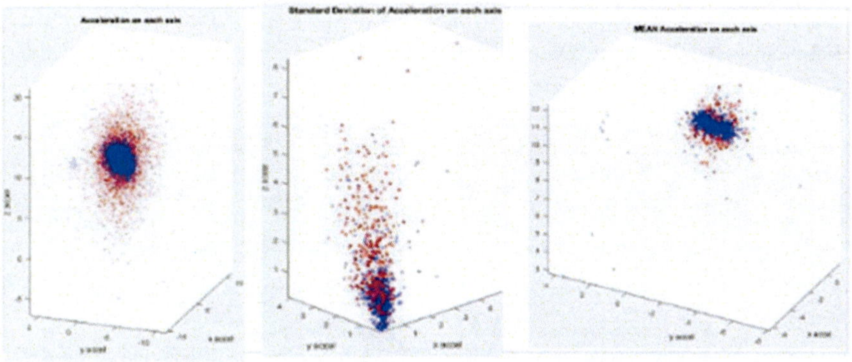

Fig. 5 Gyroscope data points

by passing the six features as input. On the Android device, the minimum, maximum, mean, standard deviation of the accelerometer and the gyroscope separately were calculated, totalling up to six features, and feeding it as input to the SVM classifier. If a pothole is detected, 1 s duration was given to refresh the sensor data values stored in the array which is used to calculate the mean and variance; during this period, the values are not passed into the SVM Classifier.

This is to prevent multiple triggers of the same pothole, as new data points obtained just after the detection would also be most likely classified as a pothole. Problems addressed in this section is the detection of a pothole when the camera is not available or the road is not in view, possibly due to heavy traffic. In these cases, the location of the pothole and the effect of riding on it is reported.

The code for the detection App is available here and the code for SVM modelling in MATLAB here.

4 Results

4.1 Computer Vision

So now the app can detect potholes at both day and night time and calculate its intrinsic parameters like length, width, area, the effect of riding both in the captured image and live stream, and then we tested the pothole detection app in real time and it gave us impressive results. Figure 6 shows the detection during day time, the pothole's area is nearly 18% of the image (1040 × 780 pixels), and the number of potholes detected is one. As shown in Fig. 7, the image was taken at night time, the pothole's area is around 32% of the image (1040 × 780 pixels), and the number of potholes detected is one.

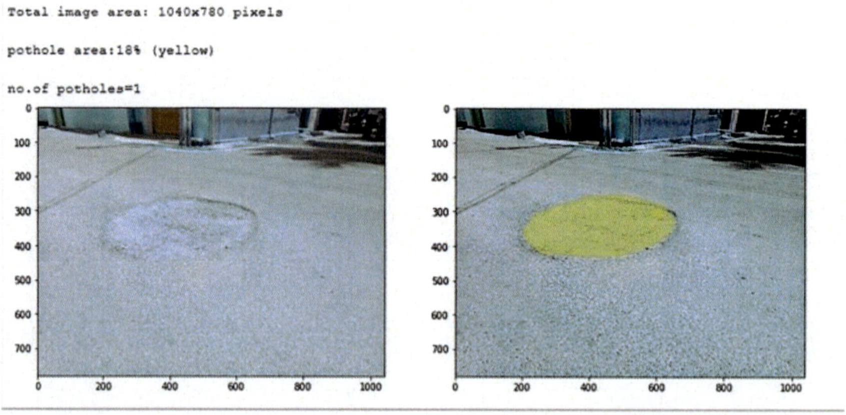

Fig. 6 Day time detection

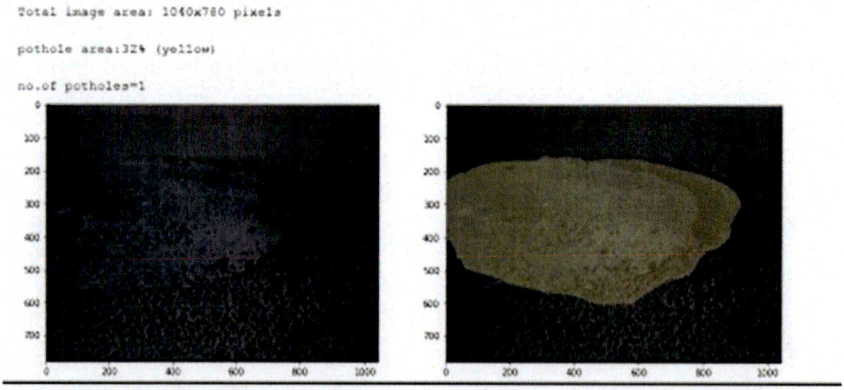

Fig. 7 Night time detection

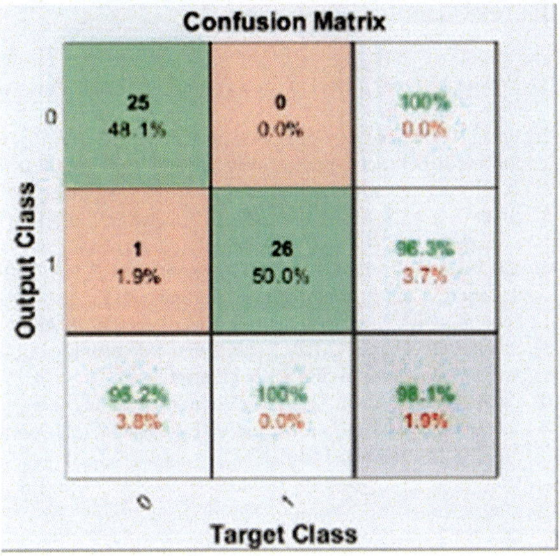

Fig. 8 Confusion matrix

4.2 Accelerometer and Gyroscope

The confusion matrix was generated, as shown in Fig. 8, for a set of 26 data points obtained while driving on potholes and 26 data points while driving on a non-pothole area. The failure rate of 1.9% was present in the system, 1 time out of 102 times, a pothole road was classified as a plain road.

This is reasonable for detection without any visual feed. Legend: Class 1 is Plain road and Class 0 is Pothole.

5 Future Work

Integration of computer vision and the accelerometer prediction can be fused with Kalman filter. Idea and codes can then be easily exported to embedded platforms for dedicated detection mechanism. SVM model can be improved by adding more features such as velocity, min–max of sensor values, classify road as good or bad.

Acknowledgements The authors would like to express their sincere thanks and deep sense of gratitude to their faculties, Dr. R. Menaka and Dr. S. Muthulakshmi, who organized this competition provided us with the opportunity to present this idea for the Ericssion Designathon competition. The Ericssion team also gave their insights and valuable feedback for us throughout the course of the project work.

References

1. How much Does it Cost to Fix a Pothole in Asphalt Pavement. https://sealmaster.net/faq/much-cost-fix-pothole/
2. Wang HW, Chen CH, Cheng DY, Lin CH, Lo CC (2015) A real-time pothole detection approach for intelligent transportation system. Math Prob Eng 2015:869627 (Aug 2015).https://doi.org/10.1155/2015/869627,10.1155/2015/869627. Publisher: Hindawi Publishing Corporation
3. Bhatt U, Mani S, Xi E, Kolter JZ (2017, October) Intelligent pothole detection and road condition assessment. arXiv:1710.02595 [cs]. http://arxiv.org/abs/1710.02595. arXiv: 1710.02595
4. Mednis A, Strazdins G, Zviedris R, Kanonirs G, Selavo L (2011) Real time pothole detection using android smartphones with accelerometers. In: 2011 international conference on distributed computing in sensor systems and workshops (DCOSS), pp 1–6
5. Singh J, Shekhar S (2018) Road damage detection and classification in smartphone captured images using mask r-cnn. arXiv preprint arXiv:1811.04535
6. Chen H, Yao M, Gu Q (2020) Pothole detection using location-aware convolutional neural networks. Int J Mach Learn Cybern **11**(4):899–911. https://doi.org/10.1007/s13042-020-01078-7,10.1007/s13042-020-01078-7

Brain Tumour Segmentation Methods Based on DWT

S. Mary Cynthia and L. M. Merlin Livingston

Abstract Image processing plays a vital role in the medical field. For the successful implementation of medical image processing accuracy of the segmented result is very important. The purpose of the proposed technique is to determine the best brain tumour segmentation method by comparing the segmented output of different methods. Early detection of brain tumour is very important because early treatment can help humans with brain tumour to live long. But the detection of brain tumour with the help of an expert is the time-consuming work. Computer-aided automatic detection methods are used to overcome the drawbacks faced by manual prediction. Because of varying nature of pixel distributions in MRI image, the segmentation process becomes very difficult. The selection of the best segmentation method is needed to extract the desired feature. In this paper, four magnetic resonance brain image segmentation techniques were used to extract the tumour, Otsu thresholding, level set method (LSM), fuzzy C-means level set method (FCMLSM) and discrete wavelet transform (DWT) with morphological processing. High-frequency noise is removed using anisodiffusion filter. Also these segmentation algorithms were compared by various quality metrics like accuracy, precision, sensitivity, specificity, etc. These quality parameters were evaluated with the help of true positive (TP), true negative (TN), false positive (FP) and false negative (FN) values.

Keywords LSM · DWT · FCMLSM · Otsu thresholding · Anisodiffusion filter

1 Introduction

Brain tumour MRI segmentation algorithms are classified into three types. They are manual, semi-automatic and automatic. The first one is the manual method, in which the tumour is identified by an expert with the help of aiding software. Experts will have different opinions, and it leads to produce variations among the results. Also this method takes more time to detect tumour. The second method is semi-automatic,

S. Mary Cynthia (✉) · L. M. Merlin Livingston
Department of Electronics and Communication Engineering, Jeppiaar Institute of Technology, Sriperumbudur, India

A. Sivasubramanian et al. (eds.), *Futuristic Communication and Network Technologies*, Lecture Notes in Electrical Engineering 792,
https://doi.org/10.1007/978-981-16-4625-6_48

which is the combination of automatic and manual algorithm. In this method, only at the initial stage human observer is needed.

[1] Described the segmentation technique, to extract the brain tumour in MR images using the parameters run-time complexity and accuracy. Various segmentation methods based on edge based, colour based, line based and pixel based and the quality metrics in terms of MSE, PSNR and sensitivity are described in [2, 3, 6]. Segmentation technique for the identification of brain tumour using fast fuzzy C-means algorithm [4, 5, 7] reduces the noise and timing problems, where a number of clusters are not automatically calculated. A combination of various techniques like K-means clustering, wavelet transform, fuzzy C-means and region growing techniques was used for the segmentation of Brain tumour from MR images [8, 10, 11]. Problems associated with 2-D medical image segmentation was analysed [12], and the computed values of the Jaccard and Dice coefficient were described in [9, 12].

2 Implementation

The anisotropic diffusion filter is applied to brain tumour magnetic resonance image to remove noise. Then, four different segmentation methods are used to detect tumour, also its resultant image is validated using quality metrics and best segmentation method is identified.

Figure 1 shows the work flow of the proposed work. After the pre-processing of the input image to remove the noise, various segmentation methods are applied to extract the tumour from the MRI image.

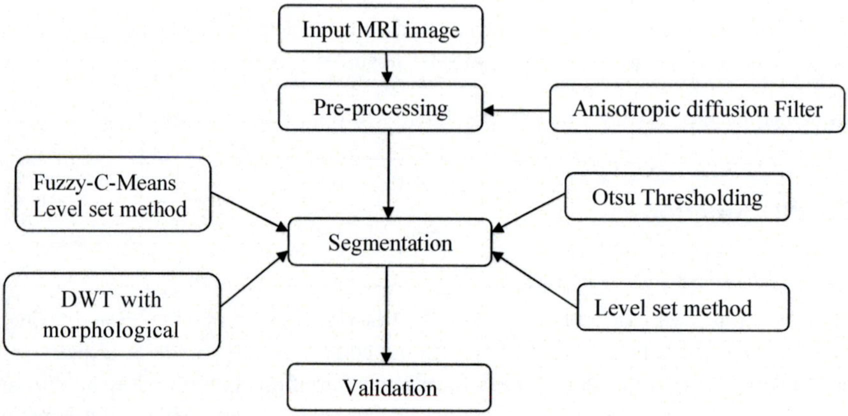

Fig. 1 Work flow of the proposed work

2.1 Anisotropic diffusion Filter

In image segmentation, image filtering is essential to remove noise. Due to this filtering process, colour of the image is changed because it changes pixel value without altering pixel positions. Anisotropic diffusion filter is used to remove high-frequency noise present in the magnetic resonance image. This filter is used to remove noise adaptively with the maintenance of image edges. Also it needs initial filtering for the purpose of clustering. Its diffusion function is given by,

$$\frac{\partial I}{\partial t} = \mathrm{div}(c(x, y, t)\nabla I) = \nabla c \cdot \nabla I + c(x, y, t)\Delta I \tag{1}$$

where the diffusion constant c controls the diffusion rate.

2.2 Image Segmentation

For the successful diagnosis and treatment of brain tumours, medical image analysis is very essential. Image segmentation process is used to extract features of particular regions of the image. This technique is used to aid brain tumour detection, and the effectiveness depends on the resolution of the segmented image. In this paper four different segmentation methods applied for brain magnetic resonance image and their quality metrics were compared.

2.3 Otsu Thresholding

It is the simplest threshold-based segmentation method. It automatically selects threshold value based on the interclass variance maximization. It depends only on the grey value of the image. For that it requires computation of grey level histogram before selecting threshold value. In this method centroid of the tumour is calculated and the binary image which contains pixel values less than the threshold is removed.

2.4 Level Set Method

In the level set method, a level set function is used to represent contours or surfaces. By using this level set representation, the process of image segmentation is obtained by well-defined mathematical formulations like partial differential equation and calculus of variations. The usage of fixed Cartesian grid in the numerical computations involved in the process of making segmentation is more favourable one.

Also complex topology can be used to represent contours/surfaces. In this method, topology can be changed in a natural way. But this method can only detect objects with edges defined by gradient.

2.5 FCMLSM

This clustering method groups the data into two or more clusters. This algorithm uses membership function which is determined by finding the distance between the cluster centre and each data point. The data point nearer to the cluster has more chance to become the member of that particular cluster.

This level set method needs more time compared with other segmentation algorithms, because of iteration process. Therefore for brain tumour segmentation, to increase the accuracy of tumour size detection and identify the stage of tumour level, set method is used with fuzzy C-means algorithm. This proposed method is efficiently reduce the computation time. Because of adding single-point pixel distribution with the energy function in each iteration the step-size becomes increased which makes better improvement in the efficiency of the algorithm. The quality metrics obtained for this method clearly shows that this segmentation process provides very good results compared with the level set segmentation method.

2.6 Discrete Wavelet Transform with Morphological Processing

In this segmentation technique, the first step is performing the fourth-level discrete wavelet method, then morphological operations like contrast enhancement and histogram thresholding are applied. DWT plays a vital role for representing image into different domains. The DWT decomposed the image into approximate, vertical, horizontal and diagonal coefficient matrices. Low-frequency components of the image are represented by approximate coefficient. The vertical, horizontal and diagonal coefficient matrices are called detailed coefficient matrices which are used to represent high-frequency components of an image. In reconstruction, only the magnitude values of negative coefficient is considered. This segmentation method was applied to test the input MRI image, and the result indicates that wavelet-based segmentation method produces higher resolution compared with other methods.

3 Experimental Results

The results of various segmentation methods are shown in Fig. 2.

3.1 Evaluation and Validation

In medical image analysis the validation of segmented brain tumour MRI image is very essential because surgical planning depends on the segmentation result. This

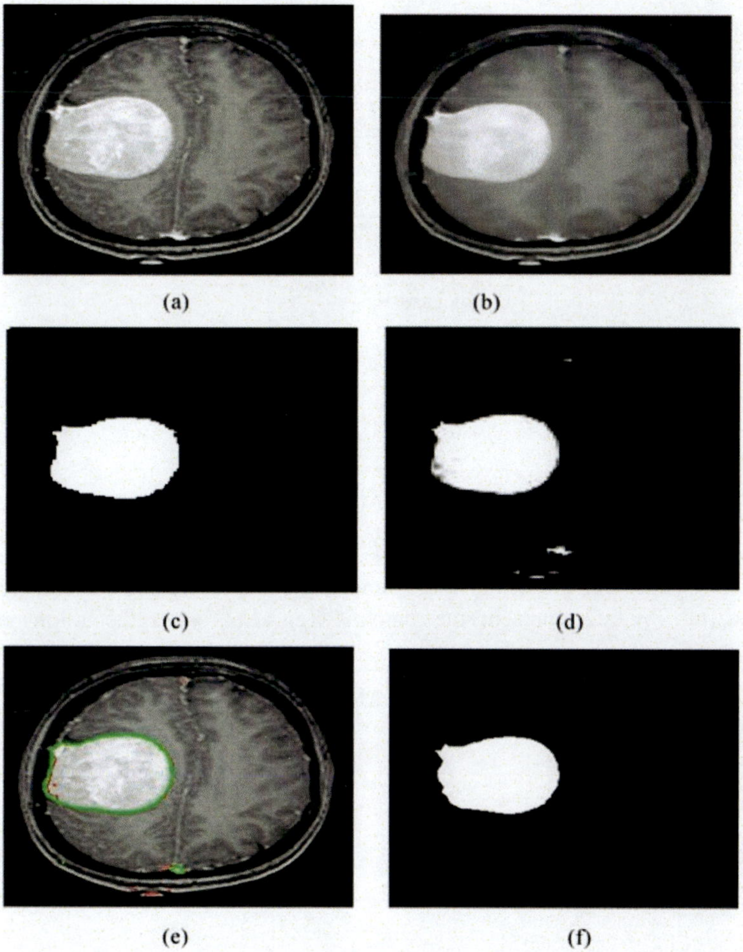

Fig. 2 **a** Input MRI image. **b** Filtered image. **c** Otsu segmentation. **d** LSM segmentation. **e** FCMLSM segmentation. **f** DWT with morphological-based segmentation

segmented image is quantitatively evaluated by using the overlap image with the ground truth. Image fusing process is used to obtain overlapped image. The quality metrics are calculated by using the following formulas.

$$\text{Accuracy} = \frac{\text{TP} + \text{TN}}{\text{FN} + \text{FP} + \text{TP} + \text{TN}} \tag{2}$$

$$\text{Sensitivity} = \frac{\text{TP}}{\text{TP} + \text{FN}} \tag{3}$$

$$\text{Precision} = \frac{\text{TP}}{\text{TP} + \text{FP}} \tag{4}$$

$$F\text{measure} = \frac{2 \times \text{TP}}{(2 \times \text{TP}) + \text{FP} + \text{FN}} \tag{5}$$

$$\text{MCC} = \frac{(\text{TP} \times \text{TN}) - (\text{FP} \times \text{FN})}{\sqrt{(\text{TP} + \text{FP})(\text{TP} + \text{FN})(\text{TN} + \text{FP})(\text{TN} + \text{FN})}} \tag{6}$$

$$\text{Dice} = \frac{2 \times \text{TN}}{(2 \times \text{TN}) + \text{FP} + \text{FN}} \tag{7}$$

$$\text{Jaccard} = \frac{\text{Dice}}{2 - \text{Dice}} \tag{8}$$

$$\text{Specificity} = \frac{\text{TN}}{\text{TN} + \text{FP}} \tag{9}$$

where

MCC— Matthews's correlation coefficient,

TP—true positive, FP—false positive,

TN—true negative, FN—false negative.

These quantitative metrics give the similarity between segmented result and the ground truth. Figure 3 represents the comparison curve of four segmentation methods

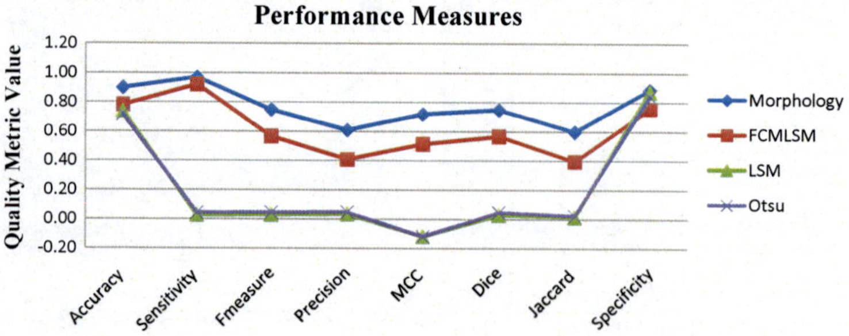

Fig. 3 Comparison curve of quality metrics of four segmentation methods

Table 1 Quality metrics for segmented MRI image

Segmentation method	Accuracy	Sensitivity	*F* measure	Precision	MCC	Dice	Jaccard	Specificity
DWT morphology	0.96	0.97	0.75	0.61	0.72	0.75	0.60	0.89
FCMLSM	0.78	0.92	0.57	0.41	0.51	0.57	0.40	0.76
LSM	0.74	0.03	0.03	0.03	-0.12	0.03	0.01	0.87
Otsu	0.72	0.04	0.05	0.05	-0.12	0.05	0.02	0.84

used in this proposed technique.

Table 1 presents the various quality metrics to measure the quality of the segmented output. The proposed method using DWT-morphological technique produces good results when compared with other segmentation techniques.

4 Conclusion

The applications of image segmentation are very large. In all applications, the quality of segmented output is very important which is identified by using the quality metrics. In this paper, three segmentation methods namely FCMLSM, LSM and OTSU thresholding are compared with the proposed DWT-based morphological processing. The experimental results show that the DWT-based morphological processing provides better segmented output when compared with other segmentation methods. The combination of discrete wavelet transform with morphological technique gives more than 95 percentage accurate results to extract the tumour from the MRI image. Also anisotropic diffusion filtered image shows the removal of high-frequency noise from the input image. This enhancement method helps to give better segmented output.

References

1. Gupta KK, Dhanda N, Kumar U (2020) Depth analysis of different medical image segmentation techniques for brain tumour detection. In: Advances in bioinformatics, multimedia, and electronics circuits and signals. Springer, Singapore, pp 197–214
2. Jacily SJ, Therese AB (200) Selection of suitable segmentation technique based on image quality metrics. Imaging Sci J 67(8):475–480
3. Merlin Livingston LM, Mary Cynthia S (2020) Region of interest prediction using segmentation. Int J Eng Adv Technol 9(5)
4. Busa S, Vangala NS, Grandhe P, Balaji V (2019) Automatic brain tumour detection using fast fuzzy C-means algorithm. Springer, Berlin
5. Bilenia A et al (2019) Brain tumour segmentation with skull stripping and modified fuzzy C-means. Springer, Berlin
6. Mary Cynthia S, Merlin Livingston LM (2019) Automatic detection and classification of brinjal leaf diseases. Int J Innov Technol Explor Eng 8(12)

7. Lei T et al (2017) Significantly fast and robust fuzzy C-means clustering algorithm based on morphological reconstruction and membership filtering. IEEE Trans Fuzzy Syst
8. Mukambika PS, Uma Rani K (2017) Segmentation and classification of MRI brain tumour. Int Res J Eng Technol (IRJET) 4(7):683–688
9. Jemila JS, Therese AB (2020) Artificial intelligence based myelinated white matter segmentation for a pediatric brain a challenging task. Recent Adv Comput Sci Commun
10. Kaur H et al (2016) Segmentation of tumour region from brain MRI images using FCM clustering and seeded region growing. IOSR J Comput Eng 18:5
11. Hooda H et al (2014) Brain tumour segmentation: a performance analysis using k-means, fuzzy C-means and region growing algorithm. In: IEEE international conference on advanced communication control and computing technologies (ICACCCT)
12. Vijayalakshmi S, Dahiya S (2017) A review and proposal of an innovative approach for medical image segmentation. In: International conference on computing, communication and automation (ICCCA2017). IEEE, p 1167–1172

Prediction of Jowar Crop Yield Using K-Nearest Neighbor and Support Vector Machine Algorithms

S. Pavani and P. Augusta Sophy Beulet

Abstract This paper describes the machine learning strategies to the crop yield prediction in agriculture. Agriculture is the important occupation in India. The demand for crop production increases more for future generations, so prediction of crop yield becomes more important. The research was conducted on Telangana state to predict the crop yield. The main purpose of the study is to apply K-nearest neighbor and support vector machine algorithms for prediction of jowar yield. These machine learning (ML) algorithms are implemented in MATLAB and the results are verified. The parameters used in the dataset are soil moisture, humidity, temperature, and rainfall. From the comparison of algorithms, the best accurate model for prediction of crop yield is suggested. Support vector machine model gave the better accuracy comparatively.

Keywords KNN · SVM · Crop yield prediction · Accuracy

1 Introduction

Precise and fast choice is currently a modern method, as the technology is growing rapidly. The evolution of technology became the reason for considerable improvement of agriculture [1]. With the emerging population day by day, the demand for crop yield is increasing. The erratic and unpredictable trends due to global warming leads to difficult situations for the farmers [2]. Precision agriculture is an executive approach that uses information technology to get data from different resources to bear on assessments correlated with crop production [3]. For implementing precision agriculture, it requires agriculture machinery and is time consuming. The farmer should also be able to apply it.

S. Pavani · P. Augusta Sophy Beulet (✉)
School of Electronics Engineering, Vellore Institute of Technology, Chennai, India
e-mail: augustasophyt.p@vit.ac.in

S. Pavani
e-mail: sunkara.pavani2017@vitstudent.ac.in

A. Sivasubramanian et al. (eds.), *Futuristic Communication and Network Technologies*, Lecture Notes in Electrical Engineering 792,
https://doi.org/10.1007/978-981-16-4625-6_49

Machine learning is a dawn for the assessment of these kinds of complicated issues. The change in data producing devices is popular in agriculture which enables farmers to select smart farming, which is nothing, but data led decision making. A review of different techniques used in smart farming is described [4]. Crop yield prediction using machine learning techniques is an advantage for the farmers due to early forecast. Agriculture applications with support of data mining can help to take fast decisions with more advantages [5]. A comparative analysis between supervised machine learning algorithms is made to determine which algorithm is more precise and accurate in predicting the crop yield, which will help the farmers [6].

In this work, Jowar crop yield prediction is considered. Jowar is a tropical crop and blooms between the temperatures around 25–32 °C [7]. Telangana state crop yield data and other input parameters from 2008 to 2018 are collected from Telangana state development planning society. The data is preprocessed before applying as an input for the supervised machine learning algorithms used in the analysis. After the collected data is preprocessed, a model is built from the training cases, and the classification model finds correlation between interpreters and objectives and thus summarizes the model. In testing stage, the model is tested on a sample, which class labels are known but not used for training. Finally, the model is used for classification on the new data whose class labels are not known. The input parameters considered are rainfall, maximum and minimum temperatures, humidity in the morning and evening, and soil moisture of everyday.

This paper emphasizes on the application of supervised machine learning algorithms. The work discussed in this paper has uneven data which is preprocessed to get an unswerving dataset. The paper is drafted as Sect. 2 which explains the methodology and preprocessing of data. Section 3 discusses the results and Sect. 4 is the conclusion.

2 Methodology

2.1 Data Attainment and Processing

The data required is collected from Telangana state development planning society (TSDPS). It is preprocessed according to the need of the input parameters applied to the machine learning algorithms. The main important parameters for a crop like rainfall, maximum and minimum temperatures, humidity, and soil moisture for all the mandals of Telangana state for everyday are collected for both the seasons of kharif and rabi. The data is averaged for every season of all the mandals for every district for both the seasons of every year from 2008 to 2018. From the data maximum of it is used for training the model and one-year data is used for testing purposes. KNN and SVM algorithms are used for predicting the crop yield and the accuracies are compared to analyze the best performance in prediction of the jowar crop yield. The algorithms are applied to the model as shown in Fig. 1.

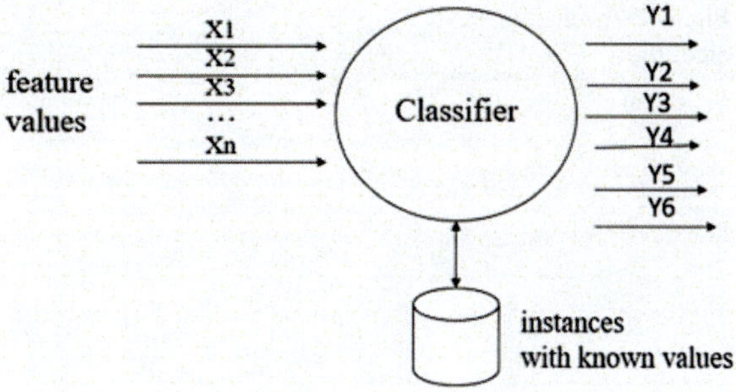

Fig. 1 Classifier model

Table 1 Classification of output labels

$Y1$	Extremely poor crop yield
$Y2$	Poor crop yield
$Y3$	Average crop yield
$Y4$	Good crop yield
$Y5$	Very good crop yield
$Y6$	Excellent crop yield

The input feature values X_1 to Xn are maximum temperature, minimum temperature, humidity in the morning and evening, rainfall, and soil moisture that are collected every day at a particular time. Value of n in X_n varies from 1 to 300 for every season. The crop yield is categorized into six classes from poor to excellent crop yield for the respective input parameter ranges. The output labels are classified depending on the amount of crop obtained as shown in Table 1.

2.2 *Prediction of Crop Yield Using KNN*

KNN depends on feature closeness. It is a supervised algorithm which is used to predict the unknown value from the known values which are near, by calculating the Euclidean distance between them [8]. It is how individuals judge by monitoring our friends or colleagues. We move with individuals of similar traits, similarly the data. If $p = (p1, p2)$ and $q = (q1, q2)$, the Euclidean distance is calculated by using the below equation.

$$d(p, q) = \sqrt{(q1 - p1)^2 + (q2 - p2)^2} \tag{1}$$

How KNN works: KNN flowchart is shown in the Fig. 2.

KNN algorithm is used to predict the jowar crop yield from the known parameters, i.e., the trained data by calculating the Euclidean distance between test data and

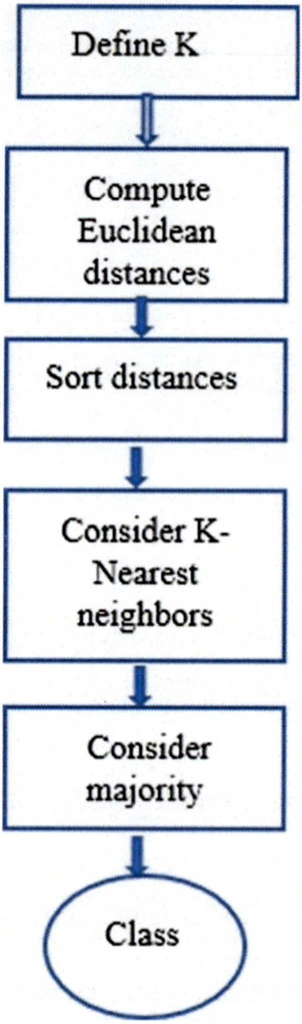

Fig. 2 Flowchart of KNN algorithm

known values of nearest neighbors. The results of accuracy are shown in the below graphs Figs. 3 and 4 for both the seasons.

2.3 Support Vector Machine (SVM)

Support vector machine is used for both classification and regression analysis. It is useful for linearly separable data that is the data can be classified into two classes

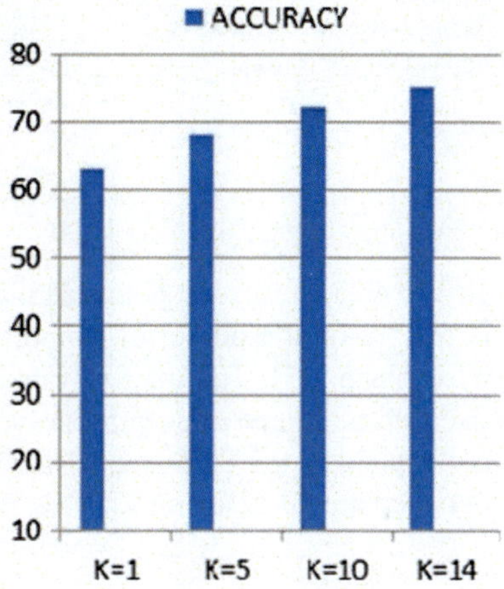

Fig. 3 Accuracy for different values of "*K*" (kharif season) for jowar crop

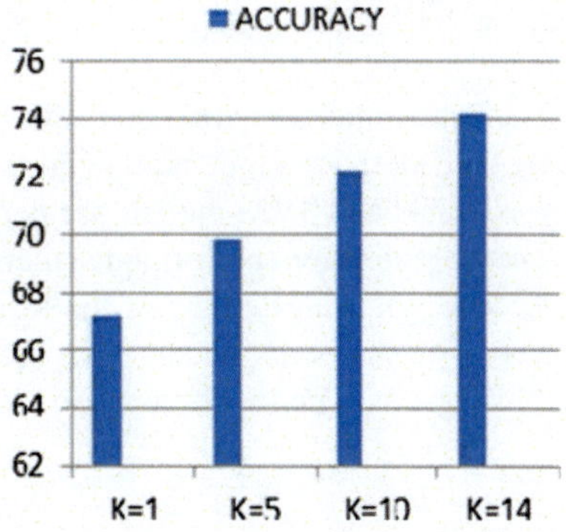

Fig. 4 Accuracy for different values of "*K*" (rabi season) for jowar crop

using a hyperplane [9]. The coordinates are called as support vectors. The best hyperplane is defined as the one which is having the maximum distance between the closest point and the hyperplane of the two classes. They are more popular for their capability to handle numerous continuous and definite variables [10]. The margin classifier is shown in Fig. 5.

Fig. 5 Margin classifier

Table 2 Results of models applied on the dataset

Classifier	Accuracy % (kharif season) [%]	Accuracy % (rabi season) [%]
KNN	75.23	74.16
SVM	93.12	91.28

The area, the closest points define around the hyperplane is called as margin. That is the reason the border of support vector machine is called as maximum margin hyperplane [11]. The main advantage of SVM is its accuracy. Nine years data of the crop is trained and with one-year data the model is tested. When the datasets are applied to the SVM model for both the kharif and rabi seasons the accuracies obtained are 93.12% and 91.28%, respectively.

3 Results

The main important parameters like rainfall, temperature, soil moisture, and humidity are considered as input parameters where the input parameters can be varied for analysis. KNN and SVM models are performed in MATLAB for calculating accuracies. There are different performance metrics and accuracy is calculated in this paper [12]. Accuracy is well-defined as the form of being precise or appropriate. SVM model has given the better accuracy in prediction of jowar crop yield for both kharif and rabi seasons. The results are provided in the below Table 2.

4 Conclusion

Crop prediction makes farmers to take better decisions for their farms. Already there is high reimbursement in using machine learning techniques when the nature of the data and its rapport is known. In the above analysis stated, we have used the supervised machine algorithms in agriculture field, such as KNN and SVM models for predicting the jowar crop yield. SVM model gave better accuracy in predicting the crop yield compared to KNN algorithm. Other input parameters and more data can be considered for more accurate results.

References

1. Ding Y, Wang L, Li Y, Li D (2018) Model predictive control and its application in agriculture: a review. Comput Electron Agricul 151:104–117
2. Baral S, Kumar Tripathy A, Bijayasingh P (2011) Yield prediction using artificial neural networks. In: Das VV, Stephen J, Chaba Y (eds) Computer networks and information technologies. CNC 2011. Communications in Computer and Information Science, vol 142. Springer, Berlin, Heidelberg. https://doi.org/10.1007/978-3-642-19542-6_57
3. Tohidyan Far S, Rezaei-Moghaddam K (2017) Determinants of Iranian agricultural consultants' intentions toward precision agriculture: integrating innovativeness to the technology acceptance model. J Saudi Soc Agric Sci 16(3).https://doi.org/10.1016/j.jssas.2015.09.003. ISSN 1658-077X
4. Kamilaris A, Kartakoullis A, Prenafeta-Boldú FX (2017) A review on the practice of big data analysis in agriculture. Comput Electron Agric 143:23–37
5. McQueen RJ, Gamer SR, Nevill-Manning CG, Witten IH (1995) Applying machine learning to agricultural data. Comput Electron Agricul 12:275–293
6. Shakoor MT, Rahman K, Rayta SN, Chakrabarty A (2017) Agricultural production output prediction using supervised machine learning techniques. In: 2017 1st international conference on next generation computing applications (NextComp), pp. 182–187. https://doi.org/10.1109/NEXTCOMP.2017.8016196
7. Statistical yearbook 2017, Directorate of Economics and Statistics, Government of Telangana
8. Gonzalez-Sanchez A, Frausto-Solis J, Ojeda-Bustamante W (2014) Predictive ability of machine learning methods for massive crop yield prediction. Spanish J Agricul Res. https://doi.org/10.5424/sjar/2014122-4439
9. Vanitha CN, Archana N, Sowmiya R (2019) Agriculture analysis using data mining and machine learning techniques. In: 2019 5th international conference on advanced computing & communication systems (ICACCS), pp. 984–990. https://doi.org/10.1109/ICACCS.2019.8728382
10. Patel AA, Kathiriya DR (2017) Data mining trends in agriculture: a review
11. Elavarasan D, Vincent DR, Sharma V, Zomaya AY, Srinivasan K (2018) Forecasting yield by integrating agrarian factors and machine learning models: a survey. Comput Electron Agricul 155:257–282
12. Espejo-Garcia B, Martinez-Guanter J, Pérez-Ruizb M, Lopez-Pellicer FJ, Zarazaga-Soria FJ (2018) Machine learning for automatic rule classification of agricultural regulations: a case study in Spain. Comput Electron Agriculture 150(2018):343–352

E-Biometric Voting Machine

B. Vignesh, P. P. Sricharan, S. Shankrith Chokkalingam, J. Bhuvana, and B. Bharathi

Abstract The voting process plays a key role in electing the decision-makers in democratic countries like India. Nowadays the electronic voting machine is widely used, which has extensively reduced counting times as compared to the older election process that used ballots. This is the motivation to propose a work that automates the process of voting using biometrics, the Internet of Things (IoT), and cloud storage. This would reduce the time, expenses, and human effort. The proposed system prevents fake votes using biometric validation, making it more credible. This mechanism is more secure as it would use a two-step verification process implemented using OTP sent through SMS. The proposed system is to be made on Raspberry Pi and contains an LCD touch screen, enabling a smooth interface for voting and a fingerprint scanner for biometric verification. Results will be instantly available on storage which reduce counting time. We are confident that introducing this system in the general elections would bring a remarkable change in the election process.

Keywords Biometric authentication · Secure elections · Internet of Things (IoT) · Encryption

1 Introduction

Voting in elections to choose leaders is the fundamental right of citizens in a democratic country like India. How elections are conducted has changed over the years. Paper-based methods use paper ballots on which voters mark their preferences. It also includes postal ballots to allow people from remote locations to vote. In contrast to secret ballots, an open ballot takes place in public and is commonly done by a show of hands. With the industrial revolution, there came the machine-based voting systems. These include lever-style voting machine and punch card voting machine.

B. Vignesh · P. P. Sricharan · J. Bhuvana · B. Bharathi
Department of Computer Science and Engineering, SSN College of Engineering, Chennai, India

S. Shankrith Chokkalingam (✉)
Department of Electronics and Communications Engineering, SSN College of Engineering, Chennai, India

A. Sivasubramanian et al. (eds.), *Futuristic Communication and Network Technologies*, Lecture Notes in Electrical Engineering 792,
https://doi.org/10.1007/978-981-16-4625-6_50

With modernization, there came the electronic voting machines like direct-recording electronic voting machines which are currently being used in India [1].

In the direct-recording electronic voting system, individuals after reaching the age of 18 must register themselves with the Election Commission of India, and after which, they are issued with their voter IDs. On the day of voting, the person must head to the nearest polling booth and get themselves verified with the officials and get their nail inked to prevent a person from casting multiple votes. Then, they head to a cabin where an electronic voting machine is placed. The official in charge initializes the machine's control unit after every person casts their vote. In the ballot unit, the candidates' names are listed alongside the voting buttons. A maximum of 2000 votes can be recorded in a single EVM with at most 384 candidates in it. An LCD screen displays the candidate name the voter has chosen. The vote has to be saved each time, and the system is re-initialized. The process is repeated for the next voter when he arrives. After the poll, the control unit is shifted to a secured place to save it from tampering. On the day of counting, the results are known by using the "Result" button which can be opened only after breaking the seal which prevents it from tampering. The main limitation is that a candidate can know the count from a polling station who voted in favor of him which may lead to the candidate showing favoritism or holding a grudge in that area. There has been rising demand to go back to ballots as these machines are not transparent enough [2].

There is a dire necessity to create a voting machine that would provide a solution to the problems in the current voting system. Manual verification is not trustworthy. The manual verification by polling officers takes lots of time which makes people wait in long queues. This wait has resulted in decreased turnout on polling day. There are lots of complaints about individual names missing from the voter roll. The voter cannot poll from any other stations assigned to them. The current election process requires lots of employees and officials. The safe storage of control units requires extensive security arrangements. The counting process is also cumbersome and time-consuming. In our work, the proposed biometric voting system (BVS), an efficient, secure, robust, and fool-proof system, is designed. BVS enables voters to cast votes from any polling stations. The verification is carried out using a fingerprint scanner. The system is lightweight and can be transported easily. It is made fool-proof by using live SMS alerts to prevent faking of votes. The results are instantaneously generated. The details of the proposed system are explained in further sections.

The further sections of the paper are structured as follows. The survey of existing works related to this system is presented in section II. The proposed system along with the motivation and problem statement is presented in section III. Further, the implementation details of the system are elaborated in section IV. Section V discussed the performance of our proposed system and the results. Section VI provides a conclusion.

2 Survey of Existing Works

A paper ballot system is used for recording and tabulating the votes that are casted on paper cards. Records of using ballot systems date back to 130 BCE. Usually, voters write down their candidate of choice in a sheet of paper that is then stored in a secure box. [1]

In punch card voting systems, voters are required to punch holes in specialized punch cards with a ballot marking device (BMD). A ballot marking device (BMD), as known as a vote recorder, allows voters to record votes on ballots which are then stored/tabulated elsewhere. These devices generally carry a label that is mapped to a certain candidate with each punching position present on the card [1].

Voter verifiable paper audit trail (VVPAT) or verifiable paper record (VPR) is a method of providing feedback to voters using a ballot-less voting system. A VVPAT is intended as an independent verification system for voting machines designed to allow voters to verify that their vote was cast correctly, to detect possible election fraud or malfunction, and to provide a means to audit the stored electronic results. A ballot marking device (BMD) or vote recorder is a type of voting machine used by voters to record votes on physical ballots. In general, ballot marking devices neither store nor tabulate ballots, but only allow the voter to record votes on ballots that are then stored and tabulated elsewhere [1].

In a ballot-less voting system, voters are not provided with any mechanism to check if their vote has been recorded accurately. These systems only allow a count/recount of the stored votes. There is no guarantee that if the system was rigged, these votes represent the voter's actual intent. Voter verifiable paper audit trail (VVPAT) functions as a verification method for such ballot-less voting systems [1]. It allows voters to check if their vote was casted correctly. This decreases the chances of electoral fraud or system malfunctions.

In direct-recording electronic (DRE) voting systems, voters are required to cast votes on electronic ballots. The vote data is transmitted from the polling booth to a central database through a secure network. Vote data may be transmitted as individual votes, periodically as batches, or as one full batch after a day of voting [1].

This paper proposes a voting system where biometrics is used as the mode of authentication. Fingerprints collected from voters are compared with the ones in the database, and if authentication is successful, voters are allowed to cast their votes, after which, they receive an acknowledgment slip. The final vote count is sent to authorities via a GSM modem [3, 3]. The system's increased security due to the use of biometrics is an advantage. However, the usage of local databases to store voter information that is devoid of encryption instead of cloud storage is a drawback since maintaining several local storage devices is more expensive and less secure.

This model uses biometric authentication and a central database server for an efficient and faster election process. Upon successful biometric authentication, voters are allowed to cast their votes. Voters receive a confirmation message, and the votes are simultaneously transferred online to the central server [5]. The advantages of this model include increased security during authentication and faster results. However, since the vote data transferred to the central server is not encrypted, it is vulnerable to cyber-attacks and data manipulation problems.

In this model, a special algorithm—Bio-hash [6]—is used for authentication. The 13-digit key that is generated from a voter's fingerprint is used during the Election Day for verification. Upon successful authentication, the voters are allowed to cast their votes [7]. This data is stored in a local database. Even though this model is relatively more secure, the use of an extra algorithm makes it slower and increases the storage needs.

3 Proposed System

India is the largest independent democratic country in the world. One of the crucial features of our democratic government is the elections. However, conventional methods of voting have many disadvantages. The amount of time, money, and human resources spent on elections is huge, and in this technological century, it is quite unnecessary. Moreover, people have lost their faith in these systems for they have multiple loopholes that can be manipulated for an individual's gain. We felt that if a voting system that could guarantee the free will of a country's citizens existed, it would bring hope to the people and make them believe in our democracy.

The conventional voting methods are slow because the votes are counted manually after elections. The process also requires a lot of human resources from people in booths to check voter IDs of the people counting the votes. The identification method used in India has many loopholes since it requires only certain documents from individuals and time, and again, such documents have been prepared illegally and used for fake votes. There are many more methods using which people can cast false votes. This deprives the citizens of their fundamental right to vote, a fundamental right that was the base on which our government was built. This problem has caused the citizens of our country to completely lose faith in the system. We thought if we could automate the booth for voter identification and also the vote counting, it would save a lot of time and human resources that could very easily be put to better work to develop the country. With this motivation, we propose to develop a biometrically secure voting system that is connected to the cloud for real-time vote counting (see Fig. 1).

Our objective is to find a more reliable and trustworthy voting system (see Fig. 1.) where much of the voting processes were automated to save time and human resources. After much consideration and after reviewing the existing systems, we concluded that using biometric data for voter identification would be the best approach to develop a reliable voting system.

We chose fingerprint data as our mode of identification as no two people have the same fingerprints, and it has been tested a lot. To automate the counting process, we realized having the data on a centralized cloud could help us relay the election results in real time. To ensure there were no loopholes, we also wanted to implement a two-step verification system apart from the fingerprint. The verification system

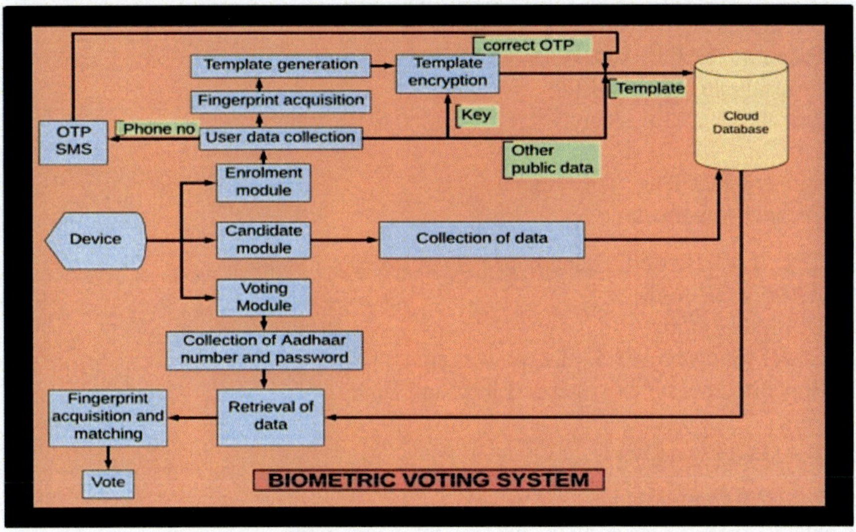

Fig. 1 Proposed system architecture

included a user password and an OTP sent to the registered mobile number. To safely store the fingerprint data of the voters, we encrypted the data using the user password known only to them. All of these features ensure that the system would be fool-proof and very reliable yet remaining a very fast and efficient method of voting.

4 Implementation of System

4.1 Hardware Components

Fingerprint Sensor. The system uses a fingerprint sensor module R307. The sensor can be interfaced either via TTL UART or via USB2.0. The default baud rate of the sensor is 57600, which can be altered if needed. It can be used with microcontrollers directly.

Technical parameters:

- Power supply: DC 4.2–6.0V
- Normal current: 50mA (typical); Peak current: 80mA
- Fingerprint image input time: <0.3 seconds
- Window area: 14x18 mm
- Matching method: Comparison method (1: 1)
- Search method (1: N)
- Characteristic file: 256 bytes
- Template file: 512 bytes

- Storage capacity: 1000 templates
- Fake rate (FAR): <0.001%
- Refusal rate (FRR): <1.0%
- Search time: <1.0 seconds (1:1000 hours, mean value)

Raspberry Pi 3 B+.
Technical parameters:

- CPU: 1.4 GHz 64-bit quad-core ARM Cortex-A53 CPU
- RAM: 1 GB RAM
- USB 2.0: 4 ports
- Power supply: 5 V/2.5A DC power input
- Supported operating systems: Linux and Unix.

USB—UART Adapter.

- USB 2.0 complaisant
- Standard USB type A male and TTL 6pin connector.
- 6 pins for 3.3V, RST, TXD, RXD, GND and 5V.
- Baud rates: 300 bps–1.5 Mbps

7-inch Touch Display.

- 800×480 high resolution.
- Capacitive touch control.
- HDMI interface for displaying and USB interface for touch control
- Backlight control to lower power consumption.

4.2 Hardware Implementation

The fingerprint sensor is connected to the Raspberry Pi using the USB–UART adapter in one of the four USB inputs on the Pi (see Fig. 4.) The touch display is connected using an HDMI cable for display and a micro USB cable for touch inputs (see Fig. 2.). The whole system is driven by a 5V/2.5A power supply connected to the mains. The fingerprint sensor reads the fingerprint, processes it, and extracts the features onto a template of size 512 bytes. Therefore, we need not process the fingerprint images separately for feature extraction. The Raspberry Pi has an on-board Wi-Fi adapter that allows it to connect to the cloud database. A camera is connected to the Raspberry Pi for image acquisition (see Fig. 3.)

Fig. 2 Raspberry Pi and touch screen connection

Fig. 3 Raspberry Pi camera connection

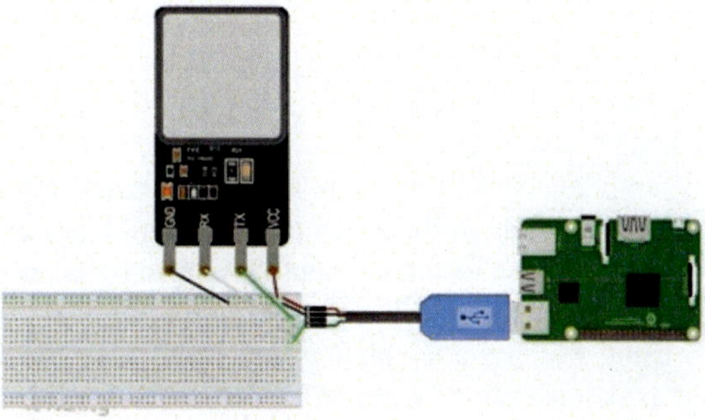

Fig. 4 Raspberry Pi fingerprint sensor connection

4.3 Software Components

Our proposed system has the following functional modules namely,

- Voter enrollment module

- Candidate enrollment module
- Authentication module

We have implemented the proposed voting system using Linux, Firebase, and Python. This system uses Python libraries including,

- SHA256 Python library
- AES Python library
- Way2sms Python library
- Pyrebase Python library
- Tkinter Python library
- Fingerprint Python library
- OpenCV Python library

4.4 Software Implementation

To interface the fingerprint sensor, a library called *pyFingerprint* is used. It has methods like e*nroll, share, and delete. Tkinter* Python library is used to develop a graphical user interface. *Pyrebase* library helps to connect to the Firebase database and services. It also has methods to add or remove items to or from the database. SHA25 Python library helps in hashing a string using the SHA256 algorithm. AES Python library helps in encrypting the fingerprint templates. Way2sms Python library helps to connect to the way2sms API to send an SMS to a given number.

Voter Enrollment Module. The module was developed using the Python *Tkinter* library. It reads the voter details, namely Aadhar number, name, gender, date of birth, and phone number. After the voter details have been entered, the next step is to read the fingerprint. A dialog box opens and prompts the voter to place his left index on the fingerprint sensor. Once the user places his finger, the sensor verifies if the image pertains to that of a fingerprint or not. If it does, it tries to extract the features out of the image. The process happens three times, and the average of the three is converted to a template. Next, a user password is obtained and used to encrypt the fingerprint template as it has to be stored on the cloud. Considering that the fingerprint of an individual is used to secure their phone and bank accounts, it needs to be encrypted when stored on a cloud. The password is hashed using SHA256 to give an encryption key using which the template is encrypted using AES encryption. Then, an OTP is sent the voter's phone number for verification. After the correct OTP is entered, the details are uploaded to the database and can be retrieved whenever necessary. This module is now refreshed to accept the next voter's details. While in this module, the voter cannot exit the app without the help of the system admin. The details have to be entered via a virtual keyboard, and the close button of the module does not work unless the voter details are completely filled.

Candidate Enrollment Module. This module is designed to get the details of the candidates of the election. The party symbol has to be uploaded to the system either via means of the internet or via physical means such as USB. This module asks for a candidate name and party name and allows to choose the party symbol from the system files. Since this is an admin procedure, no focus was given to the security aspect. These details are uploaded to the database.

Authentication Module. On the day of the election, this module will be initially run by the admin. Once entered, the module cannot be terminated without admin privileges. The voter is asked for the Aadhar number and password they entered at the time of registration. The system retrieves the details of the given Aadhar number. This module displays the details and asks the user to confirm if the details are correct. If the details are wrong, then the voter is allowed to enter the Aadhaar number again. If the details are correct, then the voter is asked to place his finger on the sensor for biometric verification. The password entered is hashed again to get the decryption key for the fingerprint template on the database. The template on the sensor and the decrypted template are matched. If the match fails, then an SMS is sent to the registered phone number intimating them of potential fraud. The system gets locked until the admin restarts the module. If the fingerprint matches, the second layer of protection is present in terms of OTP via SMS. Only if both layers are successfully passed, the voter can cast his vote. Once the voter casts his vote, an SMS will be sent to him containing the details of where the vote was casted and at what time.

5 Performance Metrics

The performance measures that are used to evaluate our proposed system are:

- **Vote casting time (VCT)**
- **SMS generating time (SGT)**
- **Result generating time (RGT)**
- **Multiple voting rate (MVR)**
- **Response time (RT)**
- **Fingerprint capture time (FCT)**
- **Encryption type—symmetric (AES)**
- **Key size—256 bits**

6 Results

We have compared our proposed system with conventional voting in practice nowadays. Our database consists of 1500 pre-enrolled voters.

Table 1 Summary of attacks against encryption algorithm

Attack	Rounds	Data	Workload	Memory
Rectangle—RK	10	2^{114}	2^{173}	–
Subkey diff	10	2^{48}	2^{49}	2^{33}
Differential—RK[a]	14	2^{131}	2^{131}	2^{65}
Boomerang—RK	14	$2^{99.5}$	$2^{99.5}$	2^{78}

[a]signifies that the attack works for a weak key class, and the workload includes the effort to find related keys from the class

Table 2 Model performance against several metrics

Metrics	Performance
VCT	3–5 min
SGT	4 s
RGT	< 2 min
MVR	0%
RT	1 s
FCT	1 s

6.1 Vote Casting Time (VCT)

VCT is the average time taken to cast a vote from when the voter enters the booth till he/she successfully casts his/her vote. Although VCT of this system is 3–5 min (see Table 2), actual time taken to vote, i.e., time taken by the system to authenticate a voter's biometrics and time taken by the voter to cast his/her vote is **less than 1 min**. This is much faster than the time spent to cast a vote using the traditional voting method which is approximately 15 min per person. Moreover, this can further be brought down if dedicated servers are used (Table 1).

6.2 SMS Generating Time (SGT)

Average time for receiving an SMS during enrollment, authentication, and voting for verification, our system achieved it by 4 s (see Table 2).

6.3 Result Generating Time (RGT)

Result generation takes less than 2 min (see Table 2) in our model compared to a minimum of 2–3 days with the existing system. Although this is the case when a single system is used, during nationwide elections, all the voting booths need not necessarily finish their voting at the same time. This may lead to an increase in the time taken for result generation.

6.4 Multiple Voting Rate (MVR)

MVR refers to the possibility that the system will fail to identify a voter who is trying to vote more than once. MVR is 0 for our model (see Table 2).

6.5 Response Time (RT)

RT is the time taken by this system to produce a match for the given fingerprint with one in the database. RT for this system is 1 s/unit (see Table 2).

6.6 Fingerprint Capture Time (FCT)

This refers to the time taken for the designed system to successfully capture the fingerprint of a voter which is one second (see Table 2).

6.7 Attack Against Encryption

A summary of the effectiveness of the encryption algorithm (AES-256) against common attack methods [8] (see Table 1).

Given below is a comparison of our model with existing paper-based voting systems and a proposed voting system using crypt-watermark methodology [9] (see Table 3).

Table 3 Comparison of proposed system against existing systems

Metrics	Paper-based systems	E-voting system using crypt-watermark methodology	E biometric voting system
Time taken to vote	> 15 min	30 s	< 1 min
RGT	2–3 days	1 min	< 2 min
Possibility of MVR	Yes	No	No
RT	–	1.2 s	1 s

7 Conclusion

In this paper, we have presented a biometric voting machine which overcomes the shortcomings in the current voting systems. The authentication data is stored securely in the cloud preventing any hacking attacks. The voters are alerted using SMS after they casted their vote which makes our system trustworthy. The system is compact and affordable which reduces election expenses by a large margin.

The voting process can be further made simple if we could get access to the Aadhar card database. The registration process can then be carried out online instead of in-person registration. Thus, in practical implementation, the voters need not come to the booth to get themselves registered.

Through this work, we have made it possible to conduct fair elections efficiently and robustly. We strongly believe that our system would bring a revolution in the way elections are conducted in India.

Acknowledgements We are grateful to the management of SSN College of Engineering, Chennai, for encouraging us to conduct our research in areas of interests by supporting us financially.

References

1. Voting and Elections Homepage. http://homepage.divms.uiowa.edu/~jones/voting/. Last accessed 2020/10/17
2. Michael D, Laurence M (2018) Overview of software security issues in direct-recording electronic voting machines. In: ICCWS 2018 13th international conference on cyber warfare and security. Int J Inf Comput Secur 182–190 (Washington)
3. Anandaraj S, Anish R, Devakumar PV (2015) Secured electronic voting machine using biometric". In: International conference on innovations in information, embedded and communication systems (ICIIECS). IEEE, Coimbatore, pp 1–5
4. Manjulatha CH, Sridhar S (2016) Electronic voting machine using fingerprint. In: IJPRES VII(4):274–277
5. Deepika J, Kalaiselvi S, Mahalakshmi S, Shifani SA (2017) Smart electronic voting system based on biometric identification-survey. In: Third international conference on science technology engineering & management (ICONSTEM). IEEE, Chennai, pp 939–942
6. Andrew J, David N, Alwin G (2004) Biohashing: two-factor authentication featuring fingerprint data and tokenized random number pattern recognition. In: Pattern recognition, vol 37, issue no 11, pp 2245–2255
7. Arooj A, Riaz M (2016) Electronic voting with biometric verification offline and hybrid EVMs solution. In: Sixth international conference on innovative computing technology (INTECH). Dublin, pp 332–337
8. Alex B, Johann G (2011) Cryptanalysis of the full AES using GPU-like special-purpose hardware. In: IACR cryptology archive 2011/710
9. Mohamed Z, Nakirya K (2018) A secure e-voting system using biometric fingerprint and crypt-watermark methodology. In: ASCENT international conference proceedings—information systems and engineering. Kuala Lumpur

Compressive Single Shot Off-Axis Digital Holography for High-Resolution Image Reconstruction

B. Lokesh Reddy and Anith Nelleri

Abstract In this paper, we demonstrate a high-resolution intensity and phase image reconstruction from a single shot off-axis digital Fresnel hologram from a fewer hologram pixels detection using compressive sensing (CS) algorithm. The computational results reveal that the proposed CS method is feasible and has enhanced the resolution of the complex object wave reconstructed compared with direct complex wave retrieval method. The CS implementation is achieved by solving an l_1-norm optimization problem.

Keywords Digital holography · Compressive sensing · Complex image reconstruction · Resolution

1 Introduction

Digital holography is a three-dimensional imaging system, and the hologram is optically produced by interference between object and reference waves. The digital hologram is acquired using a digital sensor array, and it is then transferred to a computer for numerical reconstruction of complex object wavefield which contains both quantitative phase and intensity information. Compressive sensing (CS) [1–4] is a signal reconstruction technique which is widely used in various digital holographic applications [5–17] like optical information processing and security, holographic displays, interferometry, metrology, reconstruction of noise-free images, etc.

CS is a novel mathematical framework utilized to recover the original signal of size M from a far fewer measurements N than conventional Nyquist sampling method, where $N \ll M$. CS is the best suitable method to recover the original object wavefield from the detected digital holograms to remove the noisy measurements in the reconstruction process. This can be realized by utilizing the sparsity of the original object wave. The CS has been applied in various inline digital holographic schemes [5–10] such as CS-based single exposure inline holography (CS-SEOL),

B. Lokesh Reddy · A. Nelleri (✉)
School of Electronics Engineering, Vellore Institute of Technology (VIT), Chennai, Tamil Nadu 600127, India
e-mail: anith.nelleri@vit.ac.in

A. Sivasubramanian et al. (eds.), *Futuristic Communication and Network Technologies*, Lecture Notes in Electrical Engineering 792,
https://doi.org/10.1007/978-981-16-4625-6_51

CS-based phase shift digital holography (CS-PSDH) [9] and CS-based parallel phase shift digital holography (CS-PPSDH) [10]. The disadvantage of CS-PSDH is that the recording procedure of hologram requires multiple exposures, and it is not suitable for imaging dynamic specimens. The advantage of CS-PPSDH recording procedure is that it requires only a single shot hologram with reference wave phase shifts addressed on pixels in an interleaved manner. Recently, the CS has been applied in off-axis digital holography [11, 12] as it involves only single shot hologram recording procedure and suitable for dynamic specimens imaging. The experimental procedure of off-axis digital holography is simple and useful for 3D macroscopic or microscopic phase-contrast imaging of both static and dynamic specimens. An off-axis compressive digital holography [11] system was proposed to image the object wavefield in low-light conditions from the far fewer measurements. In this method, two consecutive holograms were recorded with and without object wavefield to eliminate DC term and also to neglect the nonlinear term involved in the holographic formulation. This method was implemented based on a total variation (TV) regularized problem in compressive holography for intensity image reconstruction.

The present paper demonstrates, the reconstruction of high-resolution images, i.e. both intensity and phase from the compressed off-axis digital Fresnel hologram. We present a mathematical model of compressive off-axis digital holography, and the numerical simulation results are performed to verify the proposed CS theory.

2 Methodology

Let us assume a complex object wavefield $T(x, y)$ of size M and its complex Fresnel field $U(x, y)$ is propagated by a distance d mm from the detector plane is related using a linear operator $\mathcal{F}_d[..]$. The linear operator $\mathcal{F}_d$ is a Fresnel transform of the object wavefield, i.e. $U(x, y) = \mathcal{F}_d[T(x, y)]$, and therefore, it is direct implementation to employ CS to retrieve the high-resolution object wavefield from noisy and incomplete observations of $U(x, y)$. The off-axis digital Fresnel hologram $I(x, y)$ is obtained at recoding plane by interfering the Fresnel field $U(x, y)$ and reference wave $R(x, y)$. Let us consider only N samples of the measured hologram $I(x, y)$ and $N \ll M$. The CS algorithm is applied on N measurements to recover the input object wavefield $T(x, y)$ by solving an l_1-norm optimisation problem [18] shown in Eq. (1).

$$\underbrace{min}_{T(x,y)} \frac{1}{2}\|U(x,y) - \tilde{\mathcal{F}}_d T(x,y)\|_2^2 + \tau\|T(x,y)\|_1 \tag{1}$$

Here, $U(x, y)$ indicates incomplete Fresnel measurements of complex object wave $T(x, y)$, and τ is a non-zero positive constant. The main requirement of CS framework is linear measurement model. Therefore, the holographic reconstruction procedure of complex wave retrieval method [19] is linear and suitable for the CS model. The numerical model of complex wave retrieval method for computing the Fresnel field [19] at the recording plane from a single digital Fresnel hologram is denoted as $\tilde{\mathcal{F}}_d$.

The retrieved Fresnel field at the recording plane is inverse Fresnel propagated to retrieve the original object wavefield. The proposed CS method intends to search for original object wavefield reconstruction in an effective manner.

3 Computational Results and Discussion

In the numerical simulation, a binary image 'E' of size 256 × 256 was taken as input complex object wavefield with phase function as e^{1j}. Figure 1 shows the input object wavefield used in the simulation. The parameters used in this simulation were wavelength $\lambda = 532$nm, pixel pitch of the sensor is $\Delta x = \Delta y = 6\mu m$, and propagation distance from object plane to recording plane is $d = 80mm$. The digital Fresnel hologram was simulated with an off-axis angle $\theta = 1.16^{\circ}$ on the hologram plane is shown in Fig. 2.

The object wavefield reconstructed with complex wave retrieval method (direct method) using 60% of the hologram pixels is shown in Fig. 3. It can be noticed in Fig. 3 that the resolution of retrieved object wavefield was highly degraded in the presence of noise. The retrieved complex object wavefield using proposed CS algorithm is shown in Fig. 4. By using proposed CS algorithm, it can be noticed from Fig. 4 that the resolution of retrieved object wavefield is effectively enhanced in the CS method than that of the direct method.

Figures 5 and 6 show the retrieved original input object wavefield by using direct method and proposed CS method, respectively, when only 30% of the hologram pixels were used for reconstruction. In Fig. 4 and Fig. 6, the estimated object wavefield obtained by solving Eq. (1) using CS is comparable to the original complex object wavefield. Thus, CS has the capability to achieve single shot high-resolution imaging in off-axis digital Fresnel holography.

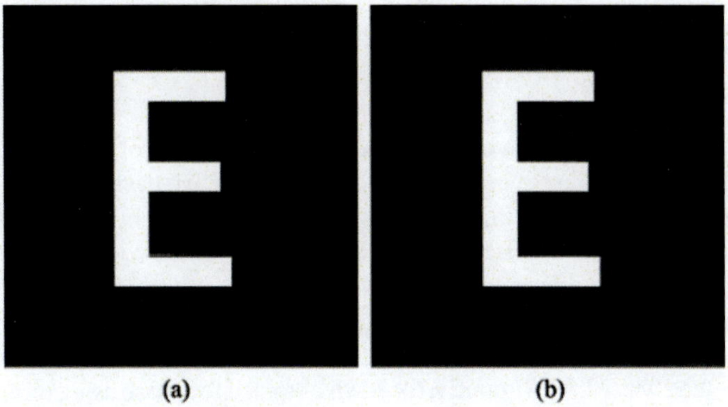

Fig. 1 Original input object wavefield **a** intensity and **b** phase

Fig. 2 Numerically computed digital hologram using 60% of the detected hologram pixels

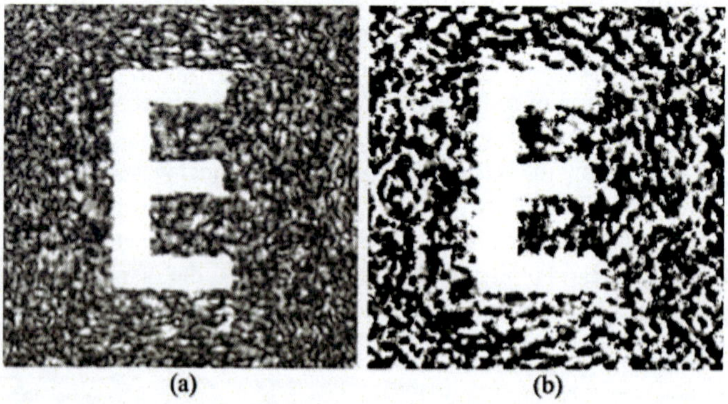

Fig. 3 Retrieved object wavefield using direct method using 60% of the detected hologram pixels **a** intensity and **b** phase

In the process of hologram compression, the reconstructed object wavefield resolution will deteriorate due to pixels loss and noise. Here, we have analysed the mean square error (MSE) of reconstructed object wave image quality by varying percentage of pixel loss. The MSE of retrieved complex object wavefield with different percentages of detected hologram pixels is shown in Table 1. With only 30% of detected hologram pixels, the recovered object wavefield for both methods is shown in Figs. 5 and 6, respectively. The MSE obtained for intensity using direct method was 57.30, and CS method was 27.53. Similarly, the MSE obtained for phase using direct method was 150.63, and CS method was 13.08. It can be inferred from Table 1 that the MSE

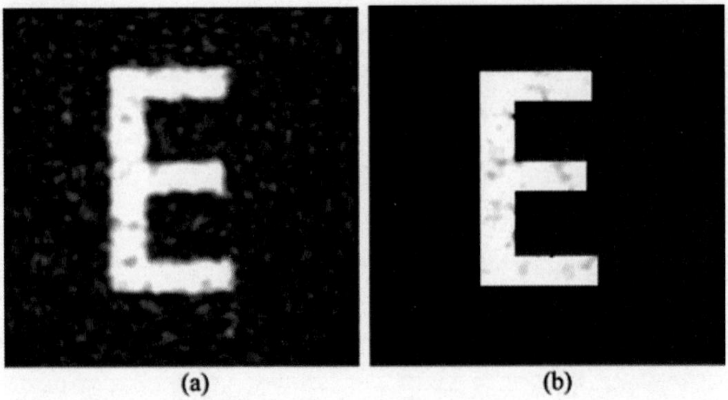

Fig. 4 Retrieved object wavefield using proposed CS method using 60% of the detected hologram pixels **a** intensity and **b** phase

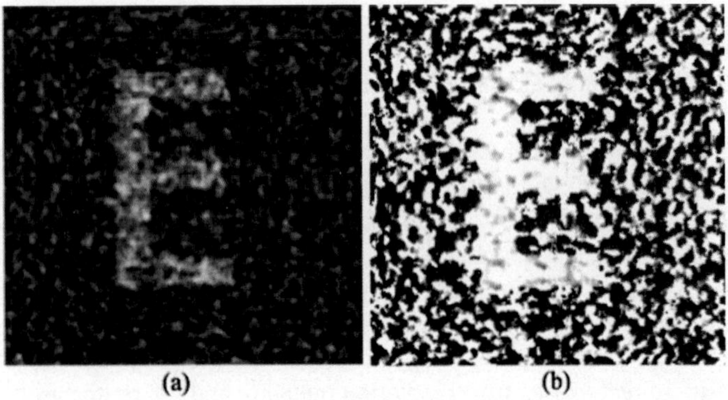

Fig. 5 Retrieved object wavefield using direct method using 30% of the detected hologram pixels **a** intensity and **b** phase

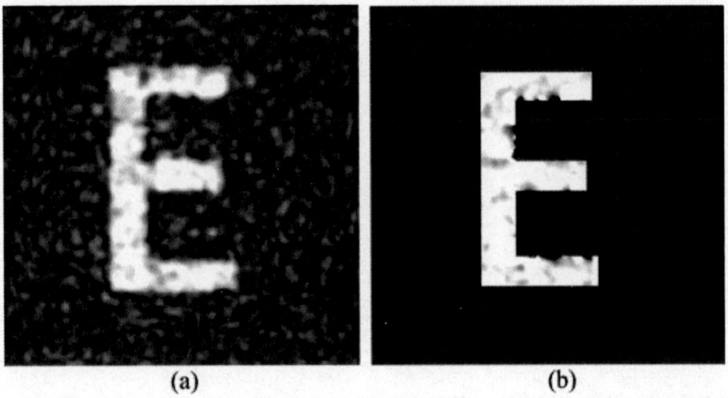

Fig. 6 Retrieved object wavefield using proposed CS method using 30% of the detected hologram pixels **a** intensity and **b** phase

Table. 1 The MSE of retrieved complex object wavefield by using percentages of hologram pixel detection

Percentage of hologram pixel detection (%)	Intensity MSE		Phase MSE	
	Direct method	Proposed CS method	Direct method	Proposed CS method
1	115.12	56.98	154.62	23.21
10	110.23	50.87	152.08	19.13
30	57.30	27.53	150.63	13.08
60	45.57	25.21	145.61	7.84
90	33.05	25.02	145.02	7.24
100	32.10	24.17	144.87	7.01

error in the proposed CS scheme is very less compared with direct method. Therefore, the object wavefield can be reconstructed with better quality using proposed CS scheme.

Figure 7 shows the MSE plot of the recovered intensity and phase of the object wave with varying percentage of hologram pixels used for reconstruction in direct method and proposed CS method. From Fig. 7, it can be deduced that even by retaining 30–40% of hologram pixels, the reconstruction quality of both intensity and phase is stable with least error in the case of proposed CS method. In direct method, the MSE increases rapidly by decreasing the percentage of hologram pixels used for reconstruction. It is significant to observe that the reconstruction quality of the phase information is good even after discarding the 70–80% of the detected hologram pixels. Therefore, it can be concluded that the proposed CS scheme has the capability to reconstruct high resolution intensity and phase images.

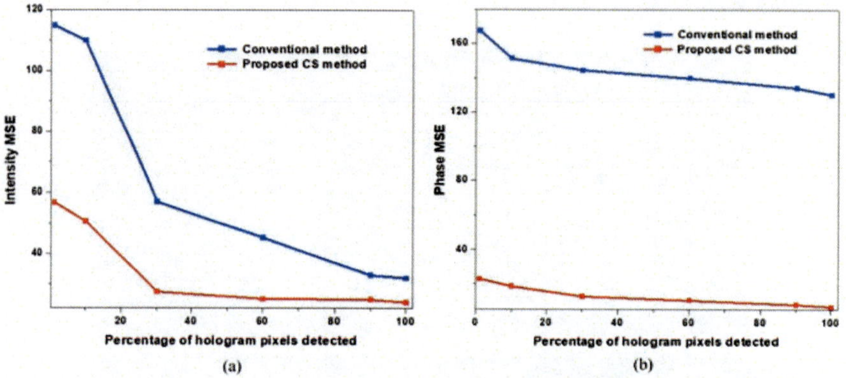

Fig. 7 Comparison of MSE vs Percentage of hologram pixels detected in the proposed CS method and direct method **a** intensity reconstruction and **b** phase reconstruction

4 Conclusions

In this paper, a CS-based complex wave retrieval method is presented to improve the resolution of the reconstructed object wavefield from off-axis digital Fresnel hologram. The linearity requirement of the proposed CS algorithm is achieved by using a linear model of the complex wave retrieval method. The numerical simulation results show that the proposed CS method is viable and has enhanced the resolution of the object wavefield than the direct method. The MSE of the intensity and phase computed from the original and reconstructed object waves is very less in the case of CS method compared to direct complex wave retrieval method.

Acknowledgements This work was supported by the Science and Engineering Research Board (SERB), Department of Science and Technology (DST), Government of India under the grant no. CRG/2018/003906.

References

1. Boyd S, Vandenberghe L (2004) Convex optimization. Cambridge University Press
2. Donoho D (2006) Compressed sensing. IEEE Trans Inf Theory 52:1289–1306
3. Brady DJ, Choi K, Marks DL, Horisaki R, Lim S (2009) Compressive holography. Opt Express 17(15):13040–13049
4. Ke J, Ashok A, Neifeld MA (2011) Block-wise motion detection using compressive imaging system. Opt Commun 284(5):1170–1180
5. Rivenson Y, Stern A, Javidi B (2013) Overview of compressive sensing techniques applied in holography. Appl Opt 52(1):A423–A432
6. Rivenson Y, Stern A, Javidi B (2010) Compressive Fresnel Holography. IEEE/OSA J Disp Technol 6(10):506–512
7. Rivenson Y, Stern A (2011) Conditions for practicing compressive Fresnel holography. Opt Lett 36(17):3365–3367
8. Rivenson Y, Stern A (2011) Conditions for practicing compressive Fresnel holography. Opt Lett 36:3365
9. Ramachandran P, Alex ZC, Nelleri A (2015) Compressive Fresnel digital holography using Fresnelet based sparse representation. Opt Commun 340:110
10. Ramachandran P, Alex ZC, Nelleri A (2018) Phase reconstruction using compressive two-step parallel phase-shifting digital holography. Opt Eng 57(4):043105
11. Marim MM, Atlan M, Angelini E, Olivo Marin JC (2010) Compressed sensing with off-axis frequency-shifting holography. Opt Lett 35:871–873
12. Reddy BL, Ramachandran P, Nelleri A (2020) Compressive complex wave retrieval from a single off-axis digital Fresnel hologram for quantitative phase imaging and microlens characterization. Opt Comm 126371
13. De Souza JC, Freire BR Jr, Dos Santos PAM (2019) Compressive holography with resolution improvement and lensless adjustable magnification. Opt Comm 437:337–341
14. Kumar S, Manjunatha M, Pranab Kumar D (2020) Compressive holography from poisson noise plagued holograms using expectation-maximization. IEEE Trans Comput Imaging
15. Brodoline A, Rawat N, Alexandre D, Cubedo N, Gross M (2019) 4D compressive sensing holographic imaging of small moving objects with multiple illuminations. Appl Opt 58:G127–G134

16. Brodoline A, Rawat N, Donnarumma D, Alexandre D, Cubedo N, Gross M (2019) Compressive sensing holographic microscopy for imaging of sparse moving objects in 3D. In: Digital holography and three-dimensional imaging 2019, OSA Technical Digest, pp Th2B-7
17. Reddy BL, Pandurangi U, Nelleri A (2019) Compressive digital fresnel holographic encryption using circular harmonic key. In: Frontiers in Optics + Laser Science APS/DLS, OSA Technical Digest, pp JTu4A-104
18. Figueiredo MAT, Nowak RD, Wright SJ (2007) Gradient projection for sparse reconstruction. IEEE J Sel Top Signal Process 1:586
19. Liebling M, Thierry B, Michael U (2004) Complex-wave retrieval from a single off-axis hologram. J Opt Soc Am A 21:367–377

An Analytical Model of Propagation Channels for Terahertz Wireless Communication System

Abdelmounim Hmamou, Mohammed EL Ghzaoui, Sudipta Das, and Jaouad Foshi

Abstract Multipath channels cause large differences between transmitted and received signals. These variations depend on time, frequency and space. Since we cannot act on the transmission channel, it becomes a drawback for signal transmission. So, the interest in characterizing and modeling the transmission channel takes on its full meaning. Modeling a channel enables us to design, validate and test a technology, as well as to establish an effective communication algorithm, which could optimize the transfer of information. Simulations of CDF (cumulative distribution function) have been carried out in order to validate the proposed model.

Keywords MIMO · THz · Transfer function · Channel modeling · Wireless communication system

1 Introduction

Over the past 25 years, indoor channel modeling has been investigated for a frequency of less than 6 GHz [1]. For frequencies beyond 100 GHz, many researches have been done on channel characterization; however, a full model has not been carried out so far. A time domain channel model for LOS configuration with spreading loss and molecular absorption has been proposed in [2]. A comparison between the derived time domain response and experimental data was carried out. It has been demonstrated that in the THz band, the deep notches are produced by multipath phenomena caused by THz channel as well as by molecular absorption. Already demonstrated in [3] that molecular absorption is a deterministic phenomenon in THz channel. a comparison between Path loss and azimuth at 28 GHz and 140 GHz in a shopping mall scenario can be found in [4]. Also, in [5], a measurement comparison between

A. Hmamou · J. Foshi
FST Errachidia, EEIMP, BP 512, Boutalamine, Meknes, Morocco

M. EL Ghzaoui
Faculty of Sciences, Sidi Mohamed Ben Abdellah University, Fez, Morocco

S. Das (✉)
Department of ECE, IMPS College of Engineering and Technology, Jadupur, India

A. Sivasubramanian et al. (eds.), *Futuristic Communication and Network Technologies*, Lecture Notes in Electrical Engineering 792,
https://doi.org/10.1007/978-981-16-4625-6_52

Path loss and DS in the outdoor environment for the band [0.8–73] GHz is carried out. In the above works, authors have studied in detail the path loss and molecular absorption in the Terahertz (THz) band (0.1–120.0 THz). Nowadays, THz technology has been proposed in order to meet the market need in the telecommunications field [6]. In [7] a simultaneous measurement at 6.75 GHz, 30 GHz, and 60 GHz in a LOS configuration in two different locations (Ilmenau and Tokyo) was investigated. The authors have analyzed the impact of the architecture on the LSPs (large-scale parameters). The aim of this work is to identify the presence of deterministic scatterers and showing their influence on the performance of the THz system. In [8], a THz channel model in scattered-rich environments is carried out by measurements. In order to study the THz channel, authors have to take into account the propagation loss and its variation as well as the multipath behavior in the THz channel. The measurement results confirmed that the proposed models can be used in millimeter-wave 5G applications. The paper [9] addressed the small communication range, high path loss, and high dynamics of the electromagnetic (EM) wave problems under penetrating through soil, sand, and water and through the very specific crude oil medium. A novel communication channel model was presented taking into account the propagation properties of terahertz EM waves. It was demonstrated that the Terahertz propagation channel depends not only on transmission distance but also on the molecular composition of the transmission channel. Authors conclude that the high channel capacity of the Terahertz band offers data rate as well as enables new information encoding and modulation techniques to be used in such an environment. In [10], a time domain channel model with a minimum phase in the THz band has been studied. Authors have mainly focused on short-range wireless communications for distances less than 100 cm were examined. In [11], a measurement of the transfer function in the frequency band 126–156 GHz was carried out by considering three indoor scenarios for Tx–Rx distances up to 10.6 m: a conference room, a laboratory and an office. The measurement results confirmed that the delay spread increases with the distance. In [12], several measurements were investigated for a frequency band up to 86 GHz for two different scenarios: an office and in-street. Based on the measurements, the delay spread does not increase with frequency. Related measurements in V2V scenarios have shown also similar results in [13]. Indeed, in the academic community, several researches have been done to discover the property of the THz channels by measurements, while only a few scientific publications on the channel modeling for THz bands have been published [1]. The conclusion is that several measurements in different situations and scenarios can be found in the literature; however, the THz channel modeling has been poorly investigated. Consequently, a full model for THz bands has not been provided until now. This reason motivated us to work on modeling the THz channel. In this work, we propose a novel analytical channel model for THz band, based on spatial correlation at the transmitter. The proposed model considers the propagation channel as a random process characterized by a set of statistical laws. The advantage of this model is that it is flexible. The main objective of this work is to drive a mathematical expression defining the THz MIMO channel model based on correlation. To analyze the spatial covariances at the transmitter C_{Tx} and at the receiver C_{Rx}, we will use the Kronecker model. In this paper, we particularly focused on knowledge and modeling of the THz MIMO propagation channel in order to develop suitable tools when analyzing the performance of a MIMO transmission chain.

2 Analytical Channel Modeling

Analytical models also called "non-physical" models, consider the propagation channel as a random process characterized by a set of statistical laws. The advantage of these models is that they are flexible. However, the identification of statistical laws governing the behavior of characteristic parameters of the channel remains difficult to achieve. In practice, the choice of laws most suited to the environments under consideration is made with the help of large-scale measurement campaigns or simulations. Three types of analytical models exist in the literature. These are correlation-based analytical models, models inspired by propagation parameters, and models including polarization diversity. In this work, we were interested in the analytical model by correlation.

2.1 Matrices of Complex Gaussian Random Variables

Many narrowband models have matrix coefficients described from complex multidimensional Gaussian distributions. In this work, we will start from the distribution of Rice since we consider LOS configuration.

The law of probability of rice distribution is given by:

$$p\left(\frac{m}{d_r}, \mu\right) = \frac{m}{\mu^2}\exp\left(-\frac{m^2 + d_r^2}{2\mu^2}\right)J_0\left(\frac{md_r}{\mu^2}\right) \tag{1}$$

where d_r and μ are the distribution parameters of Rice and J_0 is the modified Bessel function of order 0. This distribution is reduced to the distribution of Rayleigh if $d_r = 0$. The channel power is given by $2\mu^2 + d_r^2$. From the ratio between the coherent power d_r^2 and the non-coherent power μ^2, the Rice factor k is given by:

$$k = \frac{d_r^2}{2\mu^2} \tag{2}$$

The channel is characterized by its coherent H_c and non-coherent H_{nc} contributions using the following expression:

$$H = \sqrt{\frac{K}{1+K}} H_c + \sqrt{\frac{1}{1+K}} H_{nc} \quad (3)$$

The modeling of the coherent contribution of the H_c channel consists of a matrix whose amplitudes are identical but whose phases are different. For the sake of clarity, we will be interested in modeling the covariant matrix C_H considering the channel only non-coherent, i.e., $K = 0$. To analyze the spatial covariances at the C_{Tx} transmitter at the C_{Rx} receiver, we will use the Kronecker model.

2.2 Kronecker Model

The Kronecker model assumes that the spatial correlation at the transmitter as well as the spatial correlation at the receiver is independent. Besides, the channel correlation matrix can be expressed as the Kronecker product of the transmitting correlation matrix and the receiving correlation matrix as described by Eq. (4).

$$H_{\text{corr}} = C_{Tx} \otimes C_{rx} \quad (4)$$

$$C_{rx} = E\left(H_{\text{Kron}}^{H} H_{\text{kron}}\right)\text{: correlation matrix at the transmitter}$$

$$C_{rx} = E\left(H_{\text{Kron}}^{H} H_{\text{kron}}\right)\text{: correlation matrix at the receiver}$$

$\otimes$: represents Kronecker's product.

It is recalled that the Kronecker product of two matrices M and M' is defined by:

$$M \otimes M' = \begin{pmatrix} M_{11}M' & M_{12}M' & \cdots \\ M_{21}M' & M_{22}M' & \cdots \\ \cdots & \cdots & \cdots \end{pmatrix}$$

According to the Kronecker model, the channel matrix is expressed by:

$$H_{\text{Kron}} = C_{Rx}^{\frac{1}{2}} H_{\varpi} \left(C_{Tx}^{\frac{1}{2}}\right)^{t} \quad (6)$$

H_{ϖ}: is a Gaussian identical independent distribution matrix.

As described in [14], the MIMO channel can be given by

$$h\left(r_{Tx}^{j}, r_{Rx}^{i}, \tau\right) = \begin{pmatrix} h_k^{11}\left(r_{Tx}^{j}, r_{Rx}^{i}, \tau\right) & h_k^{12}\left(r_{Tx}^{j}, r_{Rx}^{i}, \tau\right) \\ h_k^{21}\left(r_{Tx}^{j}, r_{Rx}^{i}, \tau\right) & h_k^{22}\left(r_{Tx}^{j}, r_{Rx}^{i}, \tau\right) \end{pmatrix} \tag{7}$$

where h_k^{ij} corresponds to the impulse response of the channel that associates antenna i with antenna j.

At this stage, it becomes simple to calculate the coefficients of the matrix H for a 2×2 MIMO Channel. For SISO case, transfer function of each path can be given by:

$$H^{ij}(f, d) = \frac{c}{4\pi f d} \exp\left(-\frac{1}{2}\gamma(f, T_k, p)d\right) \exp(-j2\pi f \tau) \tag{8}$$

where d is the distance between the transmitter and receiver; $\gamma(f, T_k, p)$ is the molecular absorption coefficient.

The impulse response h^{ij} can be calculated easily by the Inverse discrete-time Fourier transform (DTFT) of H^{ij}.

3 Simulation Results

Figure 1 gives CDF as a function of delay spread for a number of paths equal to 3, 6 and 10. Based on this figure, we can conclude that delay spread increases for an increasing number of trips. We note that, regardless of the number of trips, we have a large gap between the curves for 10 trips and the curves for 3 and 6 trips. This gap results in a large error of accuracy. On the other hand, between the curves for 3 and

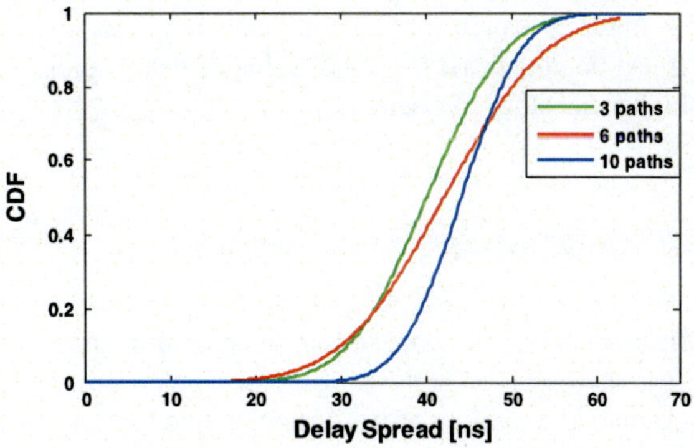

Fig. 1 CDF versus delay spread for 3, 6 and 10 paths

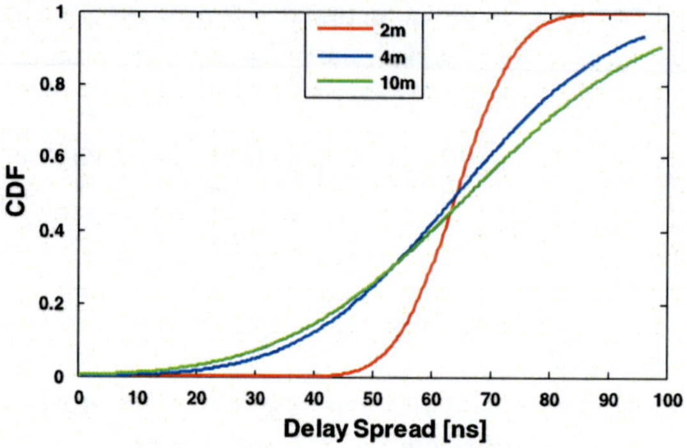

Fig. 2 CDF versus delay spread for 2,4, 10 m

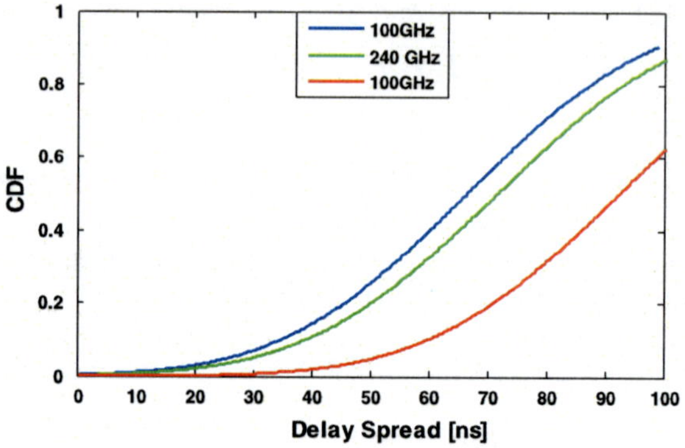

Fig. 3 CDF versus delay spread for 100 GHz, 240 GHz, 300 GHz

6 trips, we have a small deviation associated with a minor error. The CDF is well separated with an increasing order of paths, which shows that the number of paths is detrimental to the total power of a signal emitted on this kind of channel. The same can be concluded in Fig. 2. The CDF is well separated with an increasing number of paths. So, the longer the distance increases the longer the delay spread also increases. From Fig. 1 and Fig. 2, it can be noted that the simulated delay spread matches the normal distribution very well.

Figure 3 depicts the CDF of the delay spread for 100 GHz, 240 GHz and 300 GHz. It can be noted that the small value of delay spread can be found for 300 GHz and higher for other frequencies.

4 Conclusion

In this paper, we proposed an analytical channel model for THz bands by using spatial correlation at the transmitter. The proposed model considers the propagation channel as a random process characterized by a set of statistical laws. The flexibility of this model is one of its advantage. It has been validated by simulation that the proposed model can be used to determine the performance of the THz system.

References

1. Majed MB, Rahman TA, Aziz OA, Hindia MN, Hanafi E (2018) Channel characterization and path loss modeling in indoor environment at 4.5, 28, and 38 GHz for 5G cellular networks. Int J Antennas Propag 1–14
2. Tsujimura K, Umebayashi K, Kokkoniemi J, Lethomäki J (2016) A study on channel model for THz band. In: Proceedings of International symposium on antennas and propagation, Okinawa, Japan, 24–28 October, pp 872–873
3. Paine S (2019) The am atmospheric model. Smithsonian Astrophys Observ 152
4. Nguyen SLH, Jarvelainen J, Karttunen A, Haneda K, Putkone J (2017) Comparing radio propagation channels between 28 and 140 GHz bands in a shopping mall. arXiv:1712.09438
5. Raimundo X, El-Faitori S, Salous S (2018) Multi-band outdoor measurements in a residential environment for 5G networks. In: 12th european conference on antennas and propagation (EuCAP 2018). Institution of Engineering and Technology, London, UK, vol 368, 4 pp
6. El Ghzaoui M, Das S (2020) Data transmission with terahertz communication systems. In: Biswas A, Banerjee A, Acharyya A, Inokawa H, Roy JN (eds) Emerging trends in terahertz solid-state physics and devices: sources, detectors, advanced materials, and light-matter interactions. Springer, Singapore, pp 121–141
7. Dupleich D, Mueller R, Landmann M, Shinwasusin EA, Saito K, Takada JI, Luo J, Thoma R, Galdo GD (2019) Multi-band propagation and radio channel characterization in street canyon scenarios for 5G and beyond. IEEE Access 7:160385–160396
8. Huang Z, Rodriguez-Pineiro J, Yin X, Lv Y, Wang H (2020) Measurement-based characterization of 73 GHz propagation channels in scatterer-rich environments. In: 2020 IEEE wireless communications and networking conference (WCNC). IEEE, Seoul, Korea (South), pp 1–5
9. Akkaş MA, Sokullu R (2015) Wireless underground sensor networks: channel modeling and operation analysis in the terahertz band. Int J Antennas Propag 1–12
10. Tsujimura K, Umebayashi K, Kokkoniemi J, Lehtomaki J, Suzuki Y (2018) A causal channel model for the terahertz band. IEEE Trans Terahertz Sci Technol 8:52–62 (2018).
11. Pometcu L, D'Errico R (2018) Characterization of sub-THz and mmwave propagation channel for indoor scenarios. In: 12th European conference on antennas and propagation (EuCAP 2018). Institution of Engineering and Technology, London, UK, vol 632, 4 pp
12. Nguyen SLH, Medbo J, Peter M, Karttunen A, Haneda K, Bamba A, D'Errico R, Iqbal N, Diakhate C, Conrat J-M (2017) On the frequency dependency of radio channel's delay spread: analyses and findings from mmMAGIC multi-frequency channel sounding. arXiv:1712.09435
13. Boban M, Dupleich D, Iqbal N, Luo J, Schneider C, Muller R, Yu Z, Steer D, Jamsa T, Li J, Thoma RS (2019) Multi-band vehicle-to-vehicle channel characterization in the presence of vehicle blockage. IEEE Access 7:9724–9735
14. El Ghzaoui M, Hmamou A, Foshi J, Mestoui J (2020) Compensation of non-linear distortion effects in MIMO-OFDM systems using constant envelope OFDM for 5G Applications. J Circ Syst Comput 2050257

MIMO Channels Modeling for Terahertz Radio Communications Systems

Bilal Aghoutane, Hanan El Faylali, Mohammed EL Ghzaoui, Sudipta Das, and Ravi Prakash Dwivedi

Abstract THz channel is a multipath channel that causes large differences between transmitted and received signals. To do transmission over this environment, the channel model must be realized. In fact, modeling the THz channel will allow researchers to develop the THz communication system, as well as to establish an effective communication algorithm, which could improve the data transmission. In this work, we derived mathematical expressions defining an analytical model of the THz MIMO channel based on correlation. Moreover, we focused on the knowledge and modeling of the THz MIMO propagation channel in order to develop suitable tools when analyzing the performance of a MIMO transmission chain. Simulations of the transfer function and impulse response have been carried out in order to validate the proposed model.

Keywords MIMO · THz · Transfer function · Channel modeling · Wireless communication system

1 Introduction

Thanks to, among other things, the success of mobile telephony and the Internet, wireless communications have become ubiquitous in everyday life. In fact, the need for applications and services is increasing to such an extent that wireless communications cover a multitude of scenarios with diverse requirements such as high data rate and range of consumption reliability or connectivity. For that, THz technology has been proposed in order to meet the market need in the telecommunications field

B. Aghoutane · H. E. Faylali
ISO Laboratory, Department of Computer Science, IbnTofail University, Kenitra, Morocco

M. EL Ghzaoui (✉)
Faculty of Sciences, Sidi Mohamed Ben Abdellah University, Fez, Morocco

S. Das
Department of ECE, IMPS College of Engineering and Technology, Jadupur, India

R. P. Dwivedi
SENSE, VIT University Chennai Campus, Chennai, Tamil Nadu, India

A. Sivasubramanian et al. (eds.), *Futuristic Communication and Network Technologies*, Lecture Notes in Electrical Engineering 792,
https://doi.org/10.1007/978-981-16-4625-6_53

[1, 2]. On the other hand, MIMO systems have been used to increase throughput and reduce the harmful effects of the propagation channel [3].

Indeed, multiple-input multiple-output (MIMO) systems have been the subject of several types of researches since the last decade of the twentieth century [4, 5]. The first works were published in 1984 by Jack Winters of Bell Labs [6] and presented a study of fundamental limits of the bit rates of multiple antenna systems in a multi-path environment with Rayleigh fading. The progress of MIMOs systems led to the appearance of the first commercial MIMO system which was developed in 2001 by Iospan Wireless Inc. [7]. From 2006, several companies such as Broadcom and Intel have adopted new communication techniques using MIMO technology application to LANs (local area network). The new standard is called IEEE 802.11n. MIMO technology has achieved channel capacity and high performance for radio communications. As a result, MIMO technology is nowadays presented in standards for advanced systems such as those of the fourth-generation 4G [8]. The range of MIMO applications is wide. Currently, MIMO system is one of the techniques suitable for 5G applications [9]. Several areas of research have been developed around multi-antenna systems. These axes cover multiple access MIMOs systems, ad hoc networks, as well as modeling of the wireless transmission channel [10].

Compared to SISO (single-input single-output) techniques, MIMO systems techniques exploit the spatial dimension of the propagation channel characterized by spatial correlation. In this case, special modeling of the THz channel is very important to design and analyze the THz MIMO communication systems. Thus, by simulating the spatial characteristics of THz channel, we can study the performance of a MIMO system as a function of the spatial correlation.

In fact, it seems that the association of MIMO and THz systems has been poorly investigated in the literature. Building this architecture will provide rich performance perspectives and wide fields of application. The object of this work is then to establish a new model of the THz MIMO channel. The aim of this work is to approach the modeling of the THz MIMO channel in order to study the transmission chain based on the MIMO system in a THz context. For this, mathematical expressions defining an analytical model of the THZ MIMO channel based on correlation have been given. A simulation part has been presented.

2 MIMO System

Unlike conventional communication systems including one transmitting antenna and one receiving antenna, MIMO systems consist of multiple antennas both transmitting and receiving. The MIMO propagation channel is then represented by its impulse responses associated with the pairs of transmitting–receiving antennas. Thus, the impulse response of the MIMO channel is represented by the matrix $h(t, \tau)$:

$$h(t,\tau) = \begin{pmatrix} h^{11}(t,\tau) & \cdots & h^{1N_R}(t,\tau) \\ \vdots & \ddots & \vdots \\ h^{N_R1}(t,\tau) & \cdots & h^{N_RN_T}(t,\tau) \end{pmatrix} \quad (1)$$

where $h^{ij}(t,\tau)$ corresponds to the impulse response of the channel that associates the receiver *i* with the transmitter *j*. Note that this representation takes into account the effects of the antennas (diagram, number, geometry of the network and coupling) as well as the width of the bandwidth of the system.

In the literature, different classes of models are distinguished for MIMOs systems [11]. These models are mainly listed in two categories: physical models and analytical models. Physical models cover both deterministic and geometry-based models. Analytical models include correlation-based models, stochastic diffuser models, and propagation-based models.

The mathematical representation of the propagation channel, with a pair of transmitter–receiver antennas, can be expressed by a linear transformation between the input and the output. Assuming that the propagation medium is stationary and linear, the channel can be modeled by a filter. This filter includes the effect of the antennas of the transmitter and receiver and the propagation medium. For further simplification, we will not take into account the antenna array of the transmitter and receiver.

The propagation channel then depends only on the propagation medium and the geometric, temporal, and frequency configuration of the link.

The realistic description of the propagation channel in a complex environment as in the case of the THz system requires the use of adapted models. In this work, we were interested in the analytical approach to characterize the THz channel. In this case, the impulse response is said to be doubly directional and results from the contribution of all the paths propagating between the transmitter (Tx) and receiver (Rx). Each path is characterized by its complex amplitude, delay and departure and arrival directions.

3 MIMO Channel Response

In the case of a MIMO system with a double-directional impulse response, the impulse response of each elementary sub-channel is expressed according to the following relation:

$$h^{ij}(t,\tau) = \iiint h_k\left(r^j_{Tx}, r^i_{Rx}, t, \tau, \varphi_d, \varphi_a\right) D_{T_x}(\varphi_d) D_{R_x}(\varphi_a) g(t-\tau) \mathrm{d}\tau \mathrm{d}\varphi_d \mathrm{d}\varphi_a \quad (2)$$

where τ is the delay, φ_d is the direction of departure, φ_a is the direction of arrival, r^j_{Tx} is the position of the transmitter j, r^i_{Rx} is the position of receiver i. Also, $D_{T_x}(\varphi_d)$, $D_{R_x}(\varphi_a)$, respectively, represent the antenna diagram at the transmitter and at the receiver. $g(\tau)$ is the pulse response of the *Tx* and *Rx* antennas.

In the absence of cross-polarization of the antennas (i.e., only vertical or horizontal polarization is considered), the effects of delay spreading as well as angular spreading are described by the following equation:

$$h\left(r_{Tx}^{j}, r_{Rx}^{i}, t, \tau, \varphi_d, \varphi_a\right) = \sum_{k=1}^{N} h_k\left(r_{Tx}^{j}, r_{Rx}^{i}, t, \tau, \varphi_d, \varphi_a\right) \tag{3}$$

where N is the number of paths between the transmitter and receiver and h_k is the contribution of the path k. For plane waves, the contribution of each path is expressed as:

$$h_k\left(r_{Tx}^{j}, r_{Rx}^{i}, t, \tau, \varphi_d, \varphi_a\right) = a_k \delta(\tau - \tau_k)\delta\left(\varphi_d - \varphi_{a_k}\right)\delta(\varphi_a - \varphi_{dk}) \tag{4}$$

where a_k represent the complex amplitude, τ_k is the delay spread of path k, φ_{dk} is the direction of the departure of path k, and φ_{a_k} is the direction of the arrival path k.

In the general case (i.e., vertical and horizontal polarizations are considered), the double-directional impulse response is written as:

$$h\left(r_{Tx}^{j}, r_{Rx}^{i}, t, \tau, \varphi_d, \varphi_a\right) = \begin{pmatrix} h_k^{11}\left(r_{Tx}^{j}, r_{Rx}^{i}, t, \tau, \varphi_d, \varphi_a\right) & h_k^{12}\left(r_{Tx}^{j}, r_{Rx}^{i}, t, \tau, \varphi_d, \varphi_a\right) \\ h_k^{21}\left(r_{Tx}^{j}, r_{Rx}^{i}, t, \tau, \varphi_d, \varphi_a\right) & h_k^{22}\left(r_{Tx}^{j}, r_{Rx}^{i}, t, \tau, \varphi_d, \varphi_a\right). \end{pmatrix} \tag{5}$$

For the wave planes, the contribution of each path $h_k\left(r_{Tx}^{j}, r_{Rx}^{i}, t, \tau, \varphi_d, \varphi_a\right)$ is given by:

$$h\left(r_{Tx}^{j}, r_{Rx}^{i}, t, \tau, \varphi_d, \varphi_a\right) = \begin{pmatrix} a_k^{11} & a_k^{12} \\ a_k^{21} & a_k^{22} \end{pmatrix} \delta(\tau - \tau_k)\delta\left(\varphi_d - \varphi_{a_k}\right)\delta(\varphi_a - \varphi_{dk}). \tag{6}$$

Here, the complex amplitude is a polarimetric matrix that takes into account the nature of the diffusers. In addition, Eq. (6) shows that the double-directional impulse response describes only the propagation channel and is therefore completely independent of the antenna type and configuration, the bandwidth of the system, or the shape of the emitted wave.

4 Simulation Results

In order to validate the proposed model, it is crucial to study the THz channel characteristics. A 2 × 2 MIMO PLC transfer function is shown in Fig. 1. It is clear from this figure that the THz channel is a selective channel with strong destructive interference

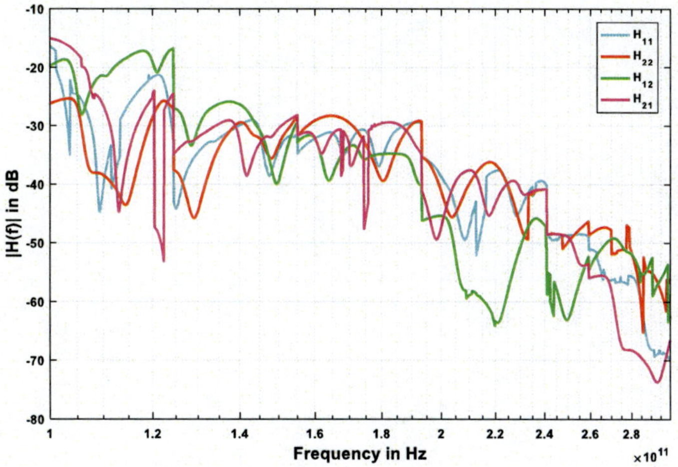

Fig. 1 Transfer function of the proposed 2 × 2 MIMO channel model

and deep notch. Also, we can notice that for four channels, there is a high variation of the amplitude of each channel caused by the presence of the multipath effect. Another important remark is that the channels (H12, H21) present deeper notches than the channels (H11, H22).

The amplitude of the impulsive response $h(t)$ can be calculated from the absolute value and phase of transfer function simulated in Fig. 1. Figure 2 depicts the amplitudes of the impulse responses of the channels $h_{11}(t)$, $h_{12}(t)$, $h_{21}(t)$ and $h_{22}(t)$. We can notice from Fig. 2 the same peaks for the four cases which confirm the multipath characteristics of THz channels. However, the maximum value of the amplitude is different, which means there is a small difference between those channels.

5 Conclusion

In this work, we derived mathematical expressions for THz MIMO channel model. In order to develop the MIMO transmission chain, the study was focused on the knowledge and modeling of the THz MIMO propagation channel. Simulations of the transfer function and impulse response have been carried out to validate the proposed model for THz radio communication systems.

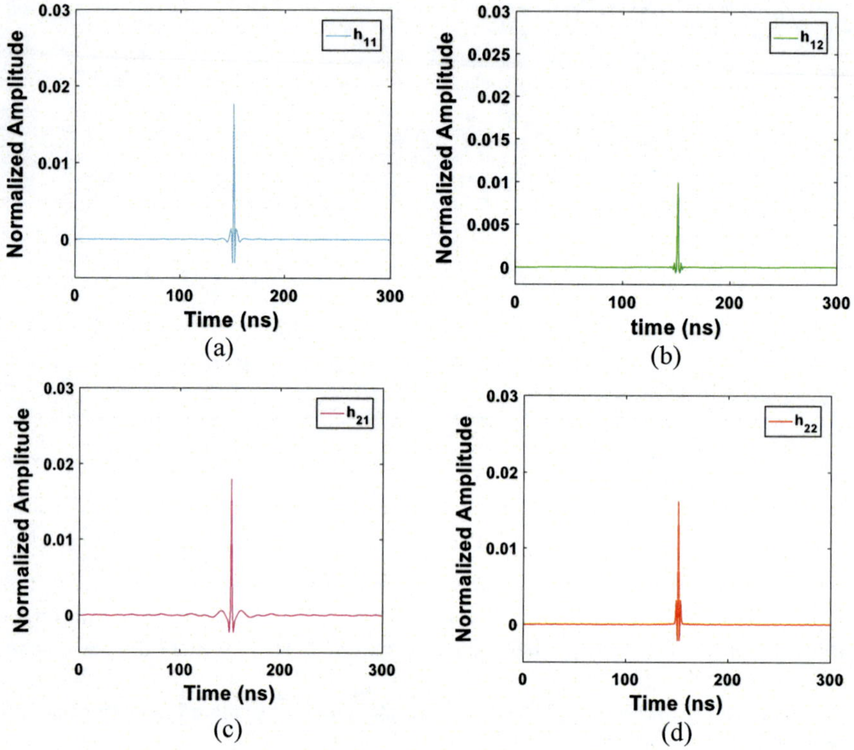

Fig. 2 Impulse responses of the channels: **a** $h_{11}(t)$, **b** $h_{12}(t)$, **c** $h_{21}(t)$ and **d** $h_{22}(t)$

References

1. El Ghzaoui M, Das S (2020) Data transmission with terahertz communication systems. In: Biswas A, Banerjee A, Acharyya A, Inokawa H, Roy J (eds) Emerging trends in terahertz solid-state physics and devices, Springer, Berlin
2. Tekbiyik K, Ekti AR, Kurt GK, Görçin GA (2019) Terahertz band communication systems: challenges, novelties and standardization efforts. Phys Commun 35:100700. https://doi.org/10.1016/j.phycom.2019.04.014
3. Mestoui J, El Ghzaoui M, Fattah M et al (2020) Performance analysis of CE-OFDM-CPM Modulation using MIMO system over wireless channel. J Ambient Intell Human Comput 11:3937–3945
4. Singh J, Kedia D (2020) Performance Improvement in Large-Scale MU-MIMO System with Multiple Antennas on User Side in a Single-Cell Downlink System. Arab J SciEng 45:6769–6789
5. Abu Hilal H (2019) Error rate analysis of ZF and MMSE decoders for massive multi cell MIMO systems in impulsive noise channels. Int J Wireless Inf Networks 26:80–89
6. Winters J (1987) On the capacity of radio communication systems with diversity in a Rayleigh fading environment. IEEE J Sel Areas Commun 5(5):871–878
7. Piazza D, Kirsch NJ, Forenza A, Heath RW, Dandekar KR (2008) Design and evaluation of a reconfigurable antenna array for MIMO systems. IEEE Trans Antennas Propag 51(3):869–881

8. Chataut R, Akl R (2020) Massive MIMO systems for 5G and beyond networks-overview, recent trends, challenges, and future research direction. Sens (Basel Switz) 20(10):2753
9. El Ghzaoui M, Hmamou A, Foshi J, Mestoui J (2020) Compensation of non-linear distortion effects in MIMO-OFDM systems using constant envelope OFDM for 5G applications. J Circ Syst Comput 29(16):2050257. https://doi.org/10.1142/S0218126620502576
10. Shuguang C, Goldsmith AJ, Bahai A (2004) Energy-efficiency of MIMO and cooperative MIMO techniques in sensor networks. IEEE J Sel Areas Commun 22(6):1089–1098
11. Poutanen J, Haneda K, Liu L, Oestges C, Tufvesson F, Vainikainen P (2011) Parameterization of the cost 2100 MIMO channel model in indoor scenarios. In: Proceedings of 5th European conference on antennas and propagation (EUCAP), pp 3606–3610

MIMO-OFDM Technique Over Terahertz Radio Channels

Jamal Mestoui, Mohammed EL Ghzaoui, Sudipta Das, Bilal Aghoutane, Hanan El Faylali, and Ravi Prakash Dwivedi

Abstract Wireless communications require more and more communication speed. One solution to meet this growing demand is the use of the millimeter band. In fact, it has been already shown that a system operating in millimeter bands offers communication speeds of several gigabits. This drew the attention of researchers to work on the THz band. This made this band a candidate for future 5G communications systems. However, the major drawback remains a severe attenuation due to the absorption of oxygen. Certain studies carried out to characterize the THz channel propose the use of directional antennas to compensate for the attenuation of the propagation effect. In this work, we will propose the MIMO-OFDM technology, which allows converting the THz channel into several flat fading subchannels and at the same time increase the transmission rate. In the present work, we present the performance of the MIMO-OFDM communication system operating in the THz band. The performance of the MIMO-OFDM system has been studied in terms of BER. In order to evaluate the performance of our system, the THz channel has been modeled.

Keywords MIMO · THz · 5G

1 Introduction

A good alternative to wired communications would be the use of wireless communications systems. In order to combat the two major problems of the THz channel, namely multipath propagation and frequency selectivity [1–3], in addition to targeting high bit rates, we are offering MIMO-OFDM technology in the context of millimeter

J. Mestoui · M. EL Ghzaoui
Faculty of Sciences, Sidi Mohamed Ben Abdellah University, Fez, Morocco

S. Das (✉)
Department of ECE, IMPS College of Engineering and Technology, Malda, West Bengal, India

B. Aghoutane · H. E. Faylali
ISO Laboratory, Department of Computer Science, IbnTofail University, Kenitra, Morocco

R. P. Dwivedi
SENSE, VIT University Chennai Campus, Chennai, Tamilnadu, India

A. Sivasubramanian et al. (eds.), *Futuristic Communication and Network Technologies*, Lecture Notes in Electrical Engineering 792,
https://doi.org/10.1007/978-981-16-4625-6_54

bands. MIMO-OFDM is a technology that has already been adopted by several fourth-generation (4G) standards [4]. We can cite, for example, 3GPP-LTE, WiMAX, IMT Advanced as well as WLAN (IEEE 802.11a, IEEE 802.11n). This technology is the combination of two technologies: The first is MIMO (Multiple Input Multiple Output) technologies [5]. This type of system consists of the use of several antennas for transmission and reception. Such a system makes it possible to strengthen the radio link by adopting diversity techniques. It also makes it possible to increase the throughput of the communication with multiplexing techniques. It has been shown that the capacity of a MIMO system, at NT transmitting antennas and NR receiving antennas, increases by a factor equal to min (NT, NR). The second technique is OFDM (Orthogonal Frequency Division Multiplexing). It allows several data beams to be sent in parallel on orthogonal subcarriers [6]. This modulation technique is easy to implement with a simple FFT (Fast Fourier Transform). In addition to this, it allows the subdivision of a wideband channel into several narrowband subchannels. OFDM is also interesting because it allows the efficient use of the frequency resource which is an increasingly scarce resource [7, 8]. Several studies have been carried out in order to assess the performance of MIMO systems in a Terahertz environment [9–12]. In [9], a performance comparison is made between a MIMO system with omnidirectional antennas and MIMO with directional antennas for 2.4 GHz communications, in a THz channel. Another work has been presented in [10] where directional patch antennas are suggested for 60 GHz communications. On the other hand, a study of the performance of OFDM for 5G is presented in [5]. In [11], OFDM has been proposed to enhance the performance of the THz channel. The performance of the THz system has been studied in terms of BER and data rate. In [12], the performance of the THz communication system has been enhanced by using the MIMO technique. It has been proved that the data rate can be improved by using the MIMO system.

In the present work, we carry out a study of the performance of MIMO-OFDM in a THz channel for the wireless communications systems. The present work, therefore, consists of a study of the performance of the MIMO-OFDM communication system operating in millimeter bands. To do that, we first give the channel model of the THz system. It has been demonstrated by simulation that by combining these two techniques, the performance of the overall system improves considerably in terms of data rate and Bite error rate simultaneously.

2 OFDM System

During the OFDM block period, the signal is written in the form:

$$s(t) = \sum_{i}\left[\sum_{k=0}^{N-1} I_{i,k} e^{j2\pi f_k t}\right] g(t - iT_B), 0 \leq t \leq T_B \tag{1}$$

With $g(t)$ is a rectangular filter of duration T_B given by the following relation:

$$g(t) = \begin{cases} 1, 0 \leq t < T_B \\ 0, \text{otherwise} \end{cases} \quad (2)$$

where $I_{i,k}$ are the symbols of the data, the exponential term represents the subcarriers, N is the total number of subcarriers, $f_k = \frac{k}{T_B}$ is the center frequency of the kth subcarrier, and T_B is the period of the block.

The signal at the receiver can be expressed as:

$$r[i] = \sum_{m=0}^{N_c-1} h[m]s[i-m] + n[i], i = -N_g, 0, N_B - 1 \quad (3)$$

where $h[m]$ is the impulse response of the channel. I order to evaluate the performance of the THz the channel response must be calculated.

3 MIMO Channel Modeling

3.1 Spatial Correlation

A MIMO terahertz channel can be given by:

$$h\left(r_{Tx}^{j}, r_{Rx}^{i}, t, \tau, \varphi_d, \varphi_a\right) = \begin{pmatrix} a_k^{11}\delta(\tau-\tau_k)\delta\left(\varphi_d-\varphi_{a_k}\right)\delta(\varphi_a-\varphi_{dk}) & a_k^{12}\delta(\tau-\tau_k)\delta\left(\varphi_d-\varphi_{a_k}\right)\delta(\varphi_a-\varphi_{dk}) \\ a_k^{21}\delta(\tau-\tau_k)\delta\left(\varphi_d-\varphi_{a_k}\right)\delta(\varphi_a-\varphi_{dk}) & a_k^{22}\delta(\tau-\tau_k)\delta\left(\varphi_d-\varphi_{a_k}\right)\delta(\varphi_a-\varphi_{dk}) \end{pmatrix} \quad (4)$$

If the static channel and the plane waves propagate only in the azimuthal plane, then the last expression becomes:

$$h\left(r_{Tx}^{j}, r_{Rx}^{i}, \tau\right) = \begin{pmatrix} a_k^{11} & a_k^{12} \\ a_k^{21} & a_k^{22} \end{pmatrix} a_k\delta(\tau-\tau_k) \quad (5)$$

The spatial correlation R(f) of the transfer function deduced from the last equation by Fourier Transform is given by:

$$R(f) = E\left(H\left(r_{Tx}^{j}, r_{Rx}^{i}, f\right)H\left(r_{Tx}^{j}, r_{Rx}^{i}, f\right)\right) \quad (6)$$

$$R(f) = \frac{1}{2\pi}\iint E\left(a_k\left(r_{Tx}^{j}, r_{Rx}^{i}, f\right)a_k\left(r_{Tx}^{j}, r_{Rx}^{i}, f\right)\right)\exp(j\nu(\varphi_d)r_0)\exp(-j\nu(\varphi_a)r_0)\mathrm{d}\varphi_d s\varphi_a \quad (7)$$

This relationship reveals the intimate connection between spatial correlation and the angular spectrum of the local area.

3.2 Generation of Channel Impulse Response

In this section, we give a general explanation for the channel impulse response generation procedure. At this stage, it becomes simple to calculate the coefficients of the matrix H for a 2×2 MIMO channel [13]:

$$h\left(r_{Tx}^{j}, r_{Rx}^{i}, \tau\right) = \begin{pmatrix} h_k^{11}\left(r_{Tx}^{j}, r_{Rx}^{i}, \tau\right) & h_k^{12}\left(r_{Tx}^{j}, r_{Rx}^{i}, \tau\right) \\ h_k^{21}\left(r_{Tx}^{j}, r_{Rx}^{i}, \tau\right) & h_k^{22}\left(r_{Tx}^{j}, r_{Rx}^{i}, \tau\right) \end{pmatrix} \tag{8}$$

where h_k^{11} corresponds to the impulse response of the channel that associates antenna i with antenna j.

The generation of Channel coefficients for each SISO subchannel is obtained from a sequence of steps.

First step: generating channel parameters.

These are the location of delays ρ_{τ}, the location of departure angles ρ_{φ_d}, the location of arrival angles ρ_{φ_a}, fading related to multipath and finally the rice K-factor (" Ricean K-factor ").

2th Stage: choice of propagation conditions.

This step allows setting the propagation conditions, i.e., whether the antennas are in direct visibility of each other (LOS/NLOS).

3th Step: calculation of mean attenuation and multipath fading.

The ratio between the power emitted P_{Tx} and the power received P_{Tx} is generally presented as the parameter PL as a function of distance d and frequency f between the Tx and Rx.

$$PL(d, f) = \frac{P_{Tx}(f)}{E(P_{Rx}(d, f))} \tag{9}$$

Which can be expressed by:

$$PL_{\text{dB}}(d, f) = PL_{\text{dB}}(d_0, f_0) + 10n_f \log_{10}\left(\frac{f}{f_0}\right) + 10n_d \log_{10}\left(\frac{d}{d_0}\right) \tag{10}$$

4th Step: Calculation of correlation.

This step generates the correlation between the broadband parameters (i.e., delay spread, angular spread, Ricean K-factor, and fast multipath fading). We will not describe this step, which is detailed [14].

5tsh Step: Generation of departure and arrived angles.

This step generates the angles of each of the paths in each cluster. The start described here remains the same for the calculation of the four angles, i.e., the starting and ending angles in azimuth and elevation. Angles are determined by applying the inverse of the following Gaussian function:

$$\psi_k = \frac{2\chi\sqrt{-\ln(P_k/\max(P_k)}}{F} \tag{11}$$

where χ corresponds to the standard deviation of the starting (or finishing) angles and F is a "scale factor" which is given by [14]:

$$F^{\mathrm{LOS}} = C\left(1.1035 - 0.028K - 0.002K^2 + 0.0001K^2\right) \tag{12}$$

where C is a scaling factor related to the total number of clusters and is given in [13]. And P_k the power of each cluster which is calculated using the formula (13):

$$P_k = \frac{P_k^{'}}{\sum_{k=1}^{N} P_k^{'}} \tag{13}$$

where $P_k^{'}$ is defined by the following relation:

$$P_k^{'} = \exp\left(\frac{-\tau_k}{\mu_\tau}\right) \times 10^{\frac{-L_k}{10}} \tag{14}$$

In the last relation, μ_τ represents the spread of delays, $-\tau_k$ corresponds to the delay calculated in the previous step and L_k corresponds to a log-normal law (LN) that reproduces the behavior of fading related to multipaths.

ψ_k is an angle that is always positive. In order to assign a sign to it, the following equation should be used:

$$\psi_k^{'} = \mathrm{X}_k\psi_k + Y_k + \psi_{\mathrm{LOS}}^{'} \tag{15}$$

where X_k corresponds to a random variable that follows a uniform distribution defined over the interval [−1, 1], X_k is a component that follows a normal distribution and $\psi_{\mathrm{LOS}}^{'}$ corresponds to the direction of the direct path.

6thStep: generating channel coefficients.

- Random choose of the phase

To calculate the channel impulse response between antennas we should consider the displacement of antennas. The channel response at point (x, y) near to (x_0, y_0),will be calculated based φ which is the azimuth, we will consider these two equations:

$$\begin{cases} x - x_0 = r\cos(\alpha) \\ y - y_0 = r\sin(\alpha) \end{cases}. \tag{16}$$

$$\Delta l = r\cos(\varphi - \alpha) \tag{17}$$

This step involves performing a random choice of the initial phase $(\varphi^{11}, \varphi^{12}, \varphi^{21}, \varphi^{22})$ of each path. The initial phases follow a normal distribution defined on the interval [−180, 180] degree.

- Generation of H coefficients $h\left(r_{Tx}^{j}, r_{Rx}^{i}, t, \tau, \varphi_d, \varphi_a\right)$.

This step generates the coefficients of the channel $h\left(r_{Tx}^{j}, r_{Rx}^{i}, t, \tau, \varphi_d, \varphi_a\right)$.

$$h\left(r_{Tx}^{j}, r_{Rx}^{i}, t, \tau, \varphi_d, \varphi_a\right) = \begin{pmatrix} a_k^{11}\delta(\tau - \tau_k)\delta(\varphi_d - \varphi_{a_k})\delta(\varphi_a - \varphi_{dk}) & a_k^{12}\delta(\tau - \tau_k)\delta(\varphi_d - \varphi_{a_k})\delta(\varphi_a - \varphi_{dk}) \\ a_k^{21}\delta(\tau - \tau_k)\delta(\varphi_d - \varphi_{a_k})\delta(\varphi_a - \varphi_{dk}) & a_k^{22}\delta(\tau - \tau_k)\delta(\varphi_d - \varphi_{a_k})\delta(\varphi_a - \varphi_{dk}) \end{pmatrix}. \tag{18}$$

the final step is to add to each channel coefficients $h_k\left(r_{Tx}^{j}, r_{Rx}^{i}, t, \tau, \varphi_d, \varphi_a\right)$ t attenuation according to the transmitter–receiver distance and the square root of the standard deviation of the fading related to multipath.

4 Simulation Results

In this section, we analyze the performance of the MIMO-OFDM modulation on a THz channel. Figure 1shows the transfer function of the THz channel which is a frequency selective channel. The proposed model is a multipath model taking into account attenuation and reflection. It can be seeing that all channel coefficients (s_{11}, s_{22}, s_{12}, s_{21}) are flat fading channels as observed in Fig. 1. This channel will be used to analyze the performance of the THz system. On the other hand, Fig. 2 showsBER simulation results for the CE-OFDM system in a THz channel for deferent values of modulation order: 16, 64, 256 and 1024. There is an improvement in the spectral

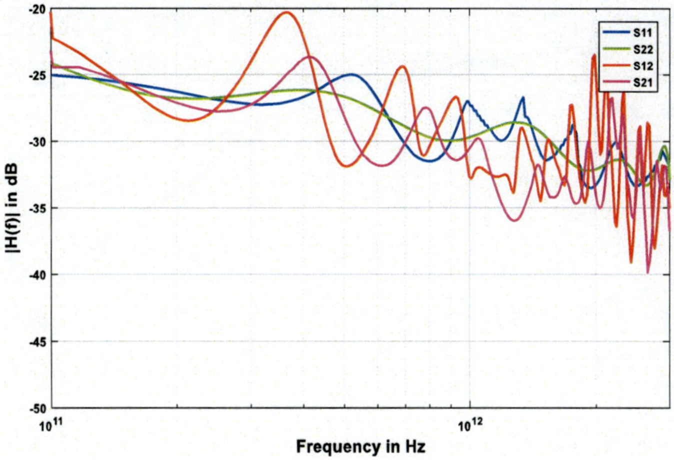

Fig. 1 MIMO channel transfer functions

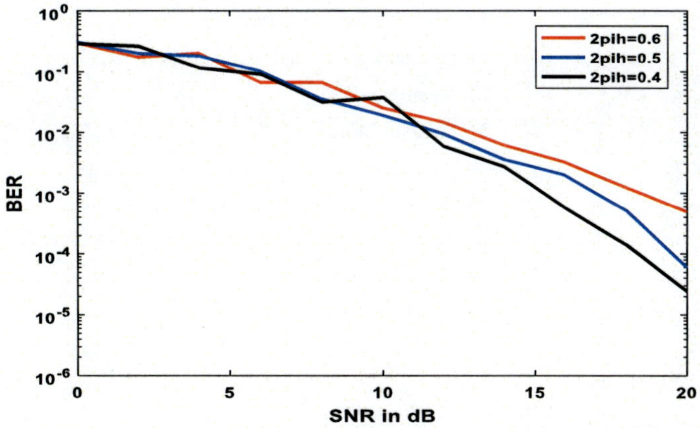

Fig. 2 BER of CE-OFDM over THz channel, $2\pi h = 0.6$

efficiency of CE-OFDM, with the increase of the order of modulation at the cost of performance degradation in terms of BER. Figure 3 shows the simulated BER at the output of the CE-OFDM modem for a THz channel. Based on the results, it appears that the modulation index significantly affects the performance of the CE-OFDM system. These results of BER shown that the performance of CE-OFDM is more sensitive to the variation of modulation index. In contrast, the value of the index plays a crucial role. The simulation parameters are summarized in Table 1.

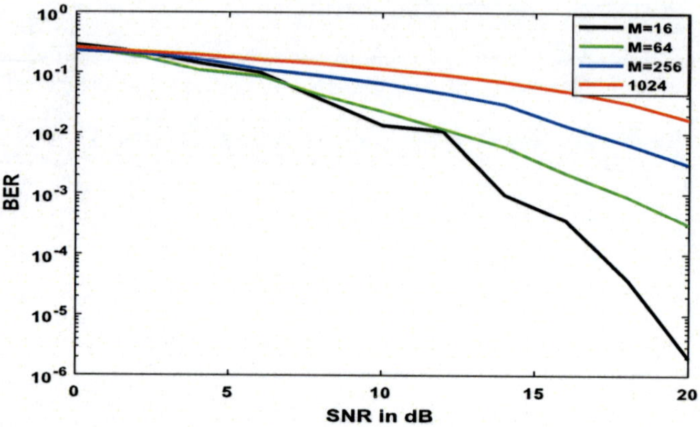

Fig. 3 BER of CE-OFDM over THz channel, $M = 64$

Table 1 Simulation parameters

Number of antennas at transmitter	2
Number of antennas at receiver	2
Maximum delay spread of channel (s)	$4e^{-6}$
Modulation order	16, 64, 256, 1024
Modulation index	$0.4/2\pi$, $0.5/2\pi$, $0.6/2\pi$
Number of subcarriers	2048
Blocks/channel realization (L)	8
Block time	TB = 32e−6
Oversampling factor	$J = 4$
Guard time (Tg)	6e−6
DFT size (for equalizer)	4096

5 Conclusion

In this work, we studied the performance of the MIMO-OFDM system in a THz context. the modulated channel is used to simulate MIMO-OFDM communications through a THz channel. A good characterization of the channel allowed us to determine the optimal parameters of OFDM that are necessary to combat the inter-symbol interference and the frequency selectivity of the THz channel. The 2 × 2 MIMO-OFDM system improved the performance of the THz system. we can say that the integration of MIMO-OFDM in THz wireless communications, will open the door to a wide range of applications that need a high data rate.

References

1. Song H, Nagatsuma T (2011) Present andssss future of terahertz communications. IEEE Trans Terahertz Sci Technol 1(1):256–263
2. El Ghzaoui M, Das S (2020) Data transmission with terahertz communication systems. In: Biswas A, Banerjee A, Acharyya A, Inokawa H, Roy J (eds) Emerging trends in terahertz solid-state physics and devices, pp 121–141
3. Federici J, Moeller L (2010) Review of terahertz and subterahertz wireless communications. J Appl Phys 107(11):111101
4. Aarab MN, Chakkor O (2020) MIMO-OFDM for wireless systems: an overview. In: Ezziyyani M (ed) Advanced intelligent systems for sustainable development (AI2SD'2019). AI2SD 2019, LNNS vol 92, pp 185–196
5. El Ghzaoui M, Hmamou A, Foshi J, Mestoui J (2020) Compensation of non-linear distortion effects in MIMO-OFDM systems using constant envelope OFDM for 5G applications. J Circ Syst Comput 29(16):2050257 (21 pp)
6. Mestoui J, ELGhzaoui M, Hmamou A, Foshi J (2019) BER performance improvement in CE-OFDM-CPM system using equalization techniques over frequency-selective channel. Procedia Comput Sci 151:1016–1021
7. Patra JP, Singh P (2019) A novel LMMSE-EM channel estimator for high mobility STBC-OFDM system. J Circ Syst Comput 28(13)
8. Konguvel E, Kannan M (2018) A survey on FFT/IFFT processors for next generation telecommunication systems. J Circ Syst Comput 27(3)
9. Mabrouk B, Talbi L, Nedil M (2012) Performance evaluation of a MIMO system in underground mine gallery. IEEE Antennas Wirel Propag Lett 11:830–833
10. Mabrouk B, Hautcoeur J, Talbi L, Nedil M, Hettak K (2013) Feasibility of a millimeter-wave MIMO system for short-range wireless communications in an underground gold mine. IEEE Tran Antennas Propag 61(8):4296–4305
11. Bharathi M, Amsaveni A, Sasikala S (2019) Orthogonal frequency division multiplexing for improving the performance of terahertz channel. J Infrared Millimetere Waves 38(3):63–269
12. Bharathi M, Sasikala S, Vanmathi J (2017) Performance analysis of terahertz channel with multiple input multiple output techniques. J Infrared Millimeter Waves 36(6):641–645
13. Mestoui J, El Ghzaoui M, Fattah M et al (2020) Performance analysis of CE-OFDM-CPM Modulation using MIMO system over wireless channels. J Ambient Intell Human Comput 11:3937–3945
14. Kyösti P, Meinilä J, Hentilä L, Zhao X, Jämsä T, Schneider C, Narandzić M, Milojević M, Hong A, Ylitalo J, Holappa V-M, Alatossava M, Bultitude R, de Jong Y, Rautiainen T (2008) IST-4-027756 WINNER II D1.1.2 v1.2 WINNER II channel models. Inf Soc Technol 11

Classification of Retinal Lesions in Fundus Images Using Atrous Convolutional Neural Network

Radha, Suchetha, Rajiv Raman, Madhumitha, Sorna Meena, Sruthi, and Nada Philip

Abstract The long-lasting continuation of diabetes can cause damage to the nerves in the retina and may threaten the vision of the eye. These nerve lesions are ranked into micro aneurysms, hard exudates, hemorrhages, and cotton wool spots. Early-stage detection of lesions in the retina improves the betterment of successful treatments. Automatic detection of retinal lesions makes it easier for the ophthalmologists to analyze them without spending much time on manual segmentation. Image classification can be achieved by using atrous convolution method to improve the view of the lesions depending on the various lesions of retinopathy. The present technique includes the study of fundus images, normalization of shape, and normalization of size, segmentation and image classification. Atrous convolutional neural network is proposed to extract the features and classify the retinal lesions in fundus images. We achieved a classification accuracy of 0.944 which indicates the suitability of the proposed method in retinal lesion classification.

Keywords Retinal lesions · Fundus images · Shape normalization · Size normalization · Segmentation · Atrous convolutional neural network (ACNN)

Radha (✉) · Suchetha · Madhumitha · S. Meena · Sruthi
Division of Healthcare Advancement, Innovation and Research, VIT, Chennai, India

Suchetha
e-mail: suchetha.m@vit.ac.in

Sruthi
e-mail: sruthi.rangarajan2016@vitstudent.ac.in

R. Raman
Shri Bhagwan Mahavir Vitreo-Retinal Services, Sankara Nethralaya, Chennai, India

N. Philip
School of Computer Science and Mathematics, Kingston University, Kingston, UK
e-mail: N.Philip@kingston.ac.uk

A. Sivasubramanian et al. (eds.), *Futuristic Communication and Network Technologies*, Lecture Notes in Electrical Engineering 792,
https://doi.org/10.1007/978-981-16-4625-6_55

1 Introduction

Diabetic retinopathy is the origin of vision loss in adults. Diabetes influences around 422 million people worldwide. In recent years it is extending as a notable problem [1]. The risk factor for the development of diabetic retinopathy is the duration of diabetes. If the person having diabetes for a long time which results in vision loss. This will be categorized into macular edema, micro aneurysms, and hemorrhage's and cotton wool spots. Disruption of blood-retinal barrier results in diabetic macular edema. It will occur at any point in the progression of diabetic retinopathy. The survey taken by the World Health Organization says that the rapid change in diabetes patients from 153 million to 347 million [2]. The diabetes problem is exponentially increasing in both developed as well as developing countries. The study reveals that diabetic retinopathy patients in developing countries are almost 75%. Diabetic retinopathy is identified by the person only when the vision of the gets disturbed. Duration of the disease and age decides the prevalence of diabetic retinopathy [2]. The diabetes population is presumed to rise from 2.8 to 4.8% globally within 30 years. Uncontrolled diabetes can result in vision loss and even blindness because the retinal blood vessels get affected starts to swell and the oxygen content in the blood vessels gets reduced [2].

Micro aneurysm is the first noticeable sign of retinal damage. Micro aneurysm is the small dark red spot, we cannot decide the shape of the red spot this will happens because there is a chance of opening blood vessels and drives the blood into the retinal tissue encompassing it [4]. Because the shape will get differ when the disease starts progress [3]. If the blood sugar increases continuously for a long period of time, blood vessels will get swell and leak lipoproteins and some other form of proteins. It appears as a white or yellowish white pigment deposited on the retinal layer. In the retinal outer layer, hard exudates are located and it looks like a circinate ring. At the edges of the retinal edema hard exudates are present in irregular shape.

Axoplasmic flow provides communication between the eye and the brain, the uneven blood flow in the retina affects the axon flow and stacked up across the retinal layer and looks like fluffy cotton-like lesions. Reduced oxygen content in the blood flow for a long period of time weakens the blood vessels in the retina. Hemorrhages also looks like a red spot which is smaller than the micro aneurysms with the dimension of 125micro meter [3]. These lesions can be identified from the fundus images. Figure 1 shows different types of lesions in fundus images.

Fluorescence angiography can be used to get a more precise diagnosis. We can see a clear vessel structure using fluorescence angiography. Fluorescein dyes are not available in small hospitals and it creates allergy to some patients. Nowadays fundus photography is the most popular method for screening diabetic retinopathy patients. Fundus images have sufficient visibility of lesions and it is convenient for acquisition.

Access of fundus photography equipment is easy and the population of diabetes patients is high. The prevalence of diabetes patients in India from 2015 to 2019 is 16.9% while sight threatening diabetic retinopathy prevalence is 3.6% [16]. The first

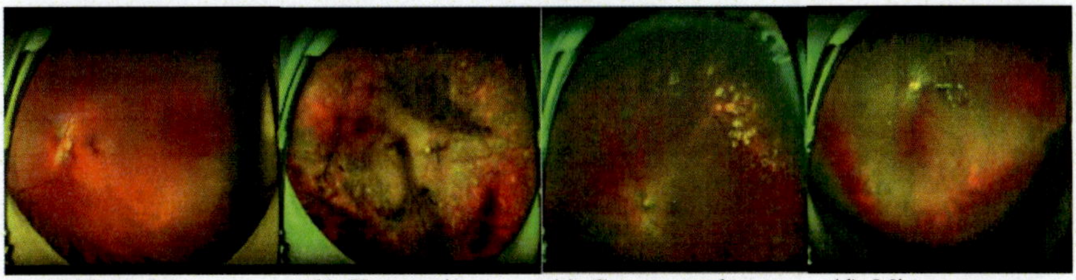

(a) Normal eye (b) Haemorrhage (c) Cotton wool spot (d) Micro aneurysms

Fig. 1 Different types of lesions in fundus images

National diabetes and Diabetic Retinopathy survey taken by Union Health Ministry from 2015 to 2019 indicates that the uneven distribution of ophthalmologist is less among the country. According to the survey done from 2005 to 2020 estimated number of ophthalmologists across the India is 25000 [5]. This necessitates the need for automatic diagnosis of diabetic retinopathy using computer. The key process in the automatic diagnosis is the feature extraction. After the feature extraction the extracted features are fed to the classifier to classify which type of lesion. The handcrafted feature-based methods might be less sufficient compared to the automatic diagnosis [5].

In recent years, many deep neural network algorithms achieved better results. A much-attracted method for diagnosis of diabetic retinopathy is DNN [17]. In this paper, we proposed atrous convolutional neural network to get better results in classification of lesions in the retina.

2 Literature Review

There are many existing algorithms to detect diabetic retinopathy. Nowadays CAD-based diabetic retinopathy diagnosing and grading systems are implemented to help the ophthalmologist to detect the DR and also used as virtual second opinion device in the medical world. Recently Discovery Randomics concept has been a more promising one for the study and the treatment of tumors. Discovery Randomics which leverage the architecture of the deep neural network to extract and learn the features. In the deep neural network, there is one limitation which is a visualization of how the features are extracted and the decision is made. To overcome this approach de-convolutional layer is attached parallel to the convolutional layer to visualize how the features are extracted and the decision was made. After creating the DNN layer, CLEAR—DR maps are created to make the decision which is influences the level of DR. It computes a set of attentive individual responses which are types of grades in the diabetic retinopathy [9].

Segmenting the blood vessels in the retinal is one of the accepted methods to diagnose retinal disease [10]. The automatic diagnosis of diabetic retinopathy can be achieved by two levels, first is detecting the lesion and the second is grading the lesion. The severity of DR can be stated by their presence and the numbers to classify it as mild, moderate, or serious. Hemorrhages are described as blood leaking deeper in the retinal layer. We cannot define the shape of the hemorrhages and cannot match the single template with all possible sizes and shapes. High contrast green channel is first extracted to extract hemorrhages. The brute force approach extracts a regional minimum which is a group of connected pixels with constant intensity and this method is highly sensitive to noise. To overcome this dynamic transformation is applied it groups the pixel according to the local contrast. To estimate the noise intensity local standard deviation is calculated. To differentiate lesion and non-lesion region random forest classifier is used. It is suitable for noisy and high dimensional data [11].

Optic disk and the position of vessel origin are the main anatomical feature of the retinal fundus photography. In bright optic disk region extraction circular structuring element with half diameter of the estimated optic disk region works more effectively. The Gaussian mixture model is used to extract the possible OD regions and bright non-OD regions. OD region is extracted by two categorizing method. The first is based on the shape, structure, and color. The second one is based on the retinal vasculature properties [12]. The two-stage deep convolution network to detect lesions as well as finding the severity. Input images are preprocessed and divided into patches, these patches are categorized into different lesion types using a local network. For each image map, a weighted lesion map is generated with labels. Lesion path with higher probability has higher weight whereas lower probability will get low weight. A global network is implemented to separate the input data into different stages which are severity [13]. Here the two classifiers are used for blood vessel segmentation CNN and Random Forest classifier. The convolution layer works as a feature extractor that extracts the same feature map at different positions. The subsampling layer provides local averaging. A fully connected layer performs linear multiplication. It multiplies the input with the weight matrix. Random forest classifier uses tree predictors. Here training set is used to build the tree. Then the tree is split with different variables which are labels to predict the severity [14, 15].

An automated hessian-based candidate selection algorithm is one of the methods to detect the retinal nerve lesion. For each image pixel, the second-order derivative is calculated in such a way the hessian matrix is formed. This will calculate the intensity differences and selects the local intensity image pixels. The fundus image with a green channel is given as input and hessian matrix extracts circular dark regions. Finally, the regions are extracted by applying thresholding. In diabetic retinopathy detection, optic disk removal is a very important part because the color of the optic disk is very similar to hard exudates and cotton wool spots. A-IFS representation deals with hesitancy in deciding the pixel intensity. Thresholding and region merging techniques are used to segment the optic disk. Fundus image with green component is used as input to extract exudates and hemorrhages because these two are the highest contrast components. Threshold of 0.35 is used for extraction in binary image. Absolute threshold of 0.1 is used for final extraction. For blood vessel extraction 2D gabor matched filter approach is used [18, 19].

3 Materials and Method

The proposed method takes the color fundus image as input data. It outputs the different types of lesions that are classified into four types. Ad hoc, expert-designed feature extractors and classifiers are the conventional methods of medical image segmentation. Semantic image segmentation, object detection and image classification are the vision-based tasks and using Deep Convolutional Neural Networks provide better results than other conventional methods. Training more number of data sets will extract the feature automatically and classify it into different types in Deep CNN. In DCNN each output node is connected to a section of an input node whereas in a fully connected traditional neural network each output is connected to all the input nodes. Cascaded arrangement of more convolutional layers and pooling layers provides the depth of view for the receptive field. Feature map also down-sampled in this type of Deep Convolutional Neural Network. This can quite unfriendly in segmentation process. In medical images lesions are very small in size, so the down sampling process may discard some important features. For this small dimension of features many techniques have been proposed. One of the techniques to retrieve the down sampled feature in fully connected neural network is up sampling or incorporating DE-convolutional layer in the network.

To increase the depth of view in traditional Deep Convolutional Neural Network we can use atrous convolution as an alternative to the DE-convolution or up sampling layer. Atrous convolution inserts zero between the nonzero filters. Inserting zero does not affect the spatial dimension of the feature with less computational complexity and provides the depth view of the feature [6]. The object or a feature map present at multiple scales to deal with this issue, traditional way in DCNN is clustering the feature maps. This approach increases the system performance but it increase the computational cost. In Deep convolutional neural network layers computing feature responses in multiple scales for each layer has to be calculated. Atrous convolution is computationally efficient method for resampling because the feature layer is resampled with the multiple filter size at each layer. Here multiple parallel atrous convolutional layers are used with different sampling rate at each layer [7].

3.1 Data Collection

In this proposed method we used a fundus image dataset with four categories of diabetic retinopathy. In the dataset each one of the image represents any one of the four categories of DR. The dataset is collected from Sankara Nethralaya hospital and Institution for ophthalmic care. The fundus images in the dataset are labelled with one of these four categories of diabetic retinopathy: micro aneurysm, cotton wool spots, hemorrhages and normal with the help of ophthalmologists.

3.2 Size Normalization

The input image should have uniform size before it is given to the input to an Atrous Convolutional Neural Network. Images are resized under selected threshold by summing up the images vertically and horizontally and the borders are removed. Finally the images are resized to the required dimension.

3.3 Augmentation

To train the neural network successfully we should have a sufficient training data. To utilize the available data fully some data augmentation techniques have to be done [8]. The following steps are carried out with the image to augment the data.

- Horizontal flip
- Vertical flip
- Zoom in
- Zoom out
- Random rotation

To augment the each images in the data set all of the above steps are carried out.

3.4 Network Architecture

For dense prediction tasks traditional deep convolutional networks are employed. The repeated combination of convolutional layer and max pooling layer reduces the spatial dimension of the feature because of down sampling. To avoid this one of the remedy is to use up sampling or DE convolution. If up sampling or DE convolution techniques are used means, it increases the computational complexity. So we go for Atrous Convolutional Neural Network. The Network architecture contains multiple convolutional layers with different filter size at each layers, i.e. inserting zero elements between the nonzero filter elements. [6]

Consider the output $Y[n]$ with the input signal $x[n]$ and the filter $w[i]$ with the atrous rate a.

$$Y[n] = \sum x[n + a \cdot k]w[i]$$

Input signal is sampled with the atrous rate a. For a standard deep convolution method atrous rate $a = 1$ for all layers in the network. Feature extraction with normal convolution operation is explained in Fig. 2. Resolution is reduced by the factor of two in the down sampling operation. At last we will get only a quarter of

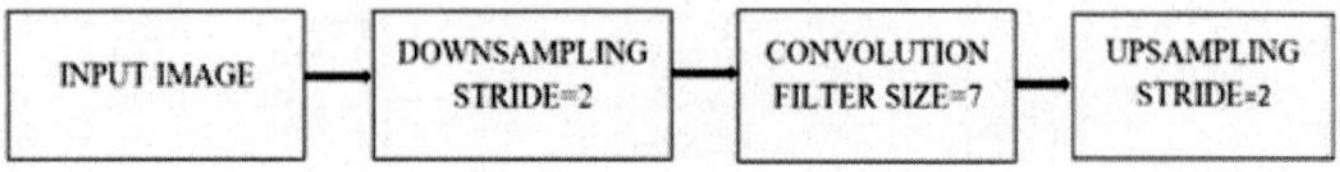

Fig. 2 Feature extraction with a standard deep convolution

the image position response at the output. Up sampling or DE convolution operation is performed to obtain the dense feature this will increase the computation cost.

Atrous rate "a" introduces $a - 1$ holes between the serial filter values. By enlarging the filter size of $k \times k$ to $K0$

$$K_0 = k + (k - 1)(r - 1)$$

One way to perform the atrous convolution without increasing the computation complexity is to insert holes inside the filter for up sampling. For smooth noisy segmentation conditional random fields is employed. In this model neighboring nodes are coupled to favor the same label assignment. The primary purpose of the conditional random fields to clean the weak classifiers superior prediction. This restriction can be overcome by implementing fully connected conditional random field. This fully connected conditional random field employs the energy function,

$$\frac{\text{Iteration } N}{\text{image}} = \text{Iteration}_{N,1} + \text{Iteration}_{N,2} + \cdots + \text{Iteration}_{N,n} / \text{Train}_{NO,1} + \text{Train}_{NO,2} + \ldots \text{Train}_{NO,m}$$

where x is the label assignment for pixel.

$$\theta_i(x_i) = -\log p(x_i)$$

Here $p(xi)$ is the probability of label assignment at pixel i.

To connect the all pairs of pixel in the image below equation is used.

$$\theta_i(x_i, x_j) = \mu(x_i, x_j)\left[w1\exp\left(-\frac{\|p_i - p_j\|}{2\sigma_\alpha^2}\right) - \left(-\frac{\|I_i - I_j\|}{2\sigma_\beta^2}\right) + w2\exp\left(-\frac{\|p_i - p_j\|}{2\sigma_\gamma^2}\right)\right]$$

When $xi \neq xj$ means $\mu(xi, xj) = 1$ and otherwise zero. In the above equation first kernel w1 has the pixel position p and the RGB color denoted as I and the second kernel w2 depends only on the pixel position. Scale of the Gaussian kernel is controlled by 0α, $\sigma\beta$ and $\sigma\gamma$. To have the similar labels pixel with similar color and the position were forced by the first kernel w1. Second kernel takes care only the pixels that are close together are grouped when considering smoothness. The

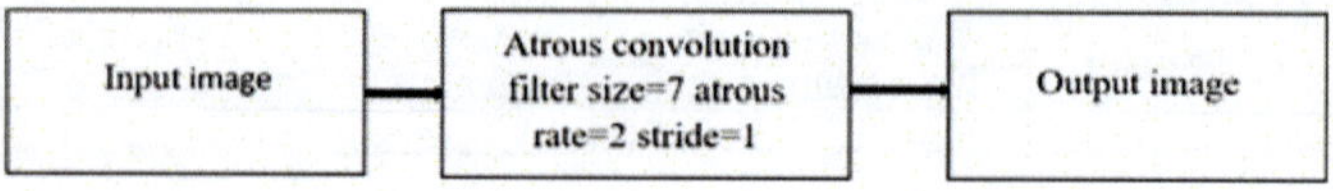

Fig. 3 Feature extraction with atrous convolution

mean field approximation can be expressed as a Gaussian convolution by using the following equation,

$$B(x) = \prod_i b_i(x_i)$$

Feature extraction with atrous convolution is explained in Fig. 3. To get all image position responses perform the convolution operation with the filter along with holes inserted between the nonzero values of the filter. The nonzero filter values are taken into attention and also the filter size is increased and we get constant operations per position and number of filter parameters.

4 Results and Discussion

Nowadays diabetic retinopathy is the most serious health problem for people having diabetes for a long period of time. Detecting diabetic retinopathy lesions in early stages is difficult and time consuming. So this system will take fundus image as input and process on them using various techniques like size normalization, augmentation, feature extraction and classification using atrous convolution neural network algorithm. Image classification is very important part of digital image analysis which is computational procedure that sort images into groups according to the similarities. Major techniques and analysis have been studied for classification of nerve lesions in the retinal fundus image (Fig. 4).

The proposed method includes study of fundus images to find the lesions in the retina. ACNN is performed to provide depth view of lesions in the fundus image data set. The images are first loaded for extraction which is functioned with the different convolutional layer with the different atrous rate for each layer. The functions used here is convolution layer with atrous rate, pooling layer and fully connected layer. The images are now trained and classified into high level features. After the training the images are tested with new dataset for further classification.

Figure 5 shows the clustered pixels. Figure. 6 indicates unmasked feature. The classified output with cotton wool spot, hemorrhages and micro aneurysms is shown in Fig. 7.

In order to compare the performance of the proposed method we have calculated the accuracy for different number of Images for both ACNN and traditional CNN. Table 1 shows the performance comparison. In all cases the performance of ACNN exceeds than CNN. We achieved a classification accuracy of 0.944 for 100 images

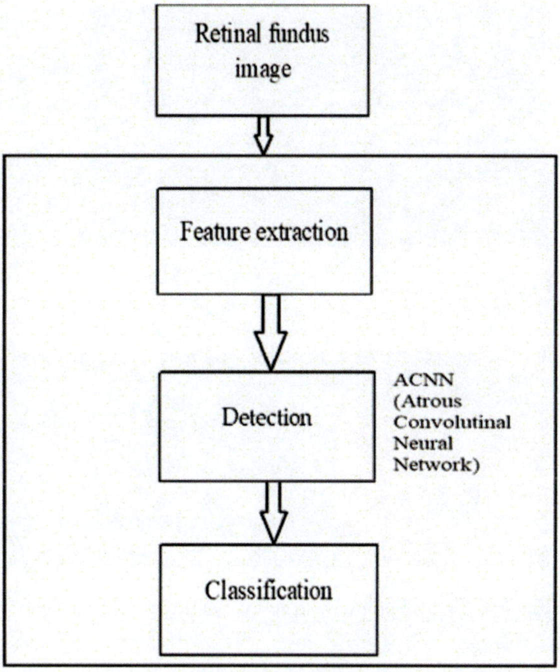

Fig. 4 Block diagram

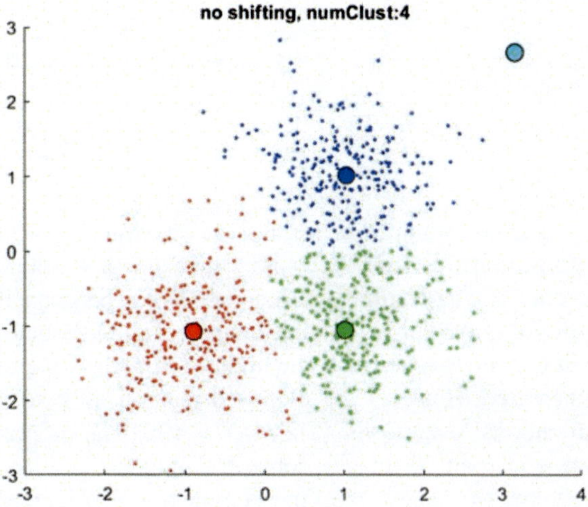

Fig. 5 Clustered pixels

compared to 0.922 by CNN. The high accuracy achieved indicates the applicability of ACNN for diagnostic applications using nerve lesion classification.

The proposed method was evaluated using performance metrics specificity (*SP*), sensitivity (*SE*) and accuracy (*AC*).

Fig. 6 For mask

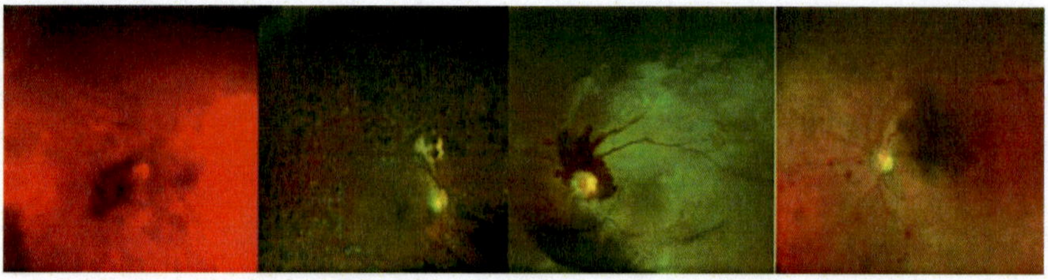

Fig. 7 Classified output with cotton wool spot, hemorrhages and micro aneurysms

$$\text{Specificity}(S_p) = \frac{\text{TP}}{\text{TP} + \text{FN}}$$

$$\text{Sensitivity}(S_E) = \frac{\text{TN}}{\text{TN} + \text{FP}}$$

$$\text{Accuracy}(S_E) = \frac{\text{TN}}{\text{TN} + \text{FP} + \text{TP} + \text{FN}}$$

The performance of this system was evaluated with the traditional method CNN [20] and also performance was evaluated with Kaggle dataset and the dataset collected from Sankara Nethralaya Hospital. The experimental results were evaluated using different number of training images (Train_{NO}) such as 10, 20, 25, 60 and 100. In all the above schemes if the number of images increases number of iterations also increased. However the proposed method took 40 iterations which less than the traditional method for (Train_{NO}) = 100. The proposed method consumes 10 min 48 s to train 20 images. Since the number iteration increases with respect to the number of images, the time complexity also increases. The number of iterations needed to train one image is calculated by,

Table 1 Comparison of number of training images, iterations, time complexity and accuracy

Algorithm	No of training images for ($Train_{NO}$)	No of iterations $iteration_N$	Time complexity $train_{time}$	Accuracy (%)
ACNN	10	5	5.24	73.3
	20	10	10.48	82.1
	25	13	13.1	83
	35	18	18.34	83.7
	45	23	23.58	84
	50	25	26.2	85.5
	60	30	31.4	86.2
	70	35	36.68	88.3
	80	40	41.9	89.8
	90	45	47.16	91.5
	100	51	52.4	94.4
CNN	10	7	7.5	72.1
	20	14	15	80.1
	25	18	19	80.4
	35	25	26.2	81.2
	45	32	33.7	83.3
	50	36	37.5	79.8
	60	43	45	81.1
	70	50	52.5	83
	80	58	60	84.5
	90	65	67.5	86.6
	100	72	75	92.2

$$\frac{\text{Iteration}_N}{\text{image}} = \text{Iteration}_{N,1} + \text{Iteration}_{N,2} + \cdots + \text{Iteration}_{N,n} / \text{Train}_{NO,1} + \text{Train}_{NO,2} + \ldots \text{Train}_{NO,m}$$

For the proposed method the number of iterations required to train the image is $\frac{\text{Iteration}_N}{\text{image}} = 0.5059$, while the traditional CNN method requires 0.7205. Similarly the time required to train the one image is calculated by,

$$\frac{\text{Iteration}_{\text{time}}}{\text{image}} = \text{Iteration}_{\text{time},1} + \text{Iteration}_{\text{time},2} + \cdots + \text{Iteration}_{\text{time},n} / \text{Train}_{\text{NO},1} + \text{Train}_{\text{NO},2} + \ldots \text{Train}_{\text{NO},m}$$

For the proposed method the time taken to complete the training for one image is 1.048s, while the traditional CNN method takes 1.312s to complete. Figure 8a–c

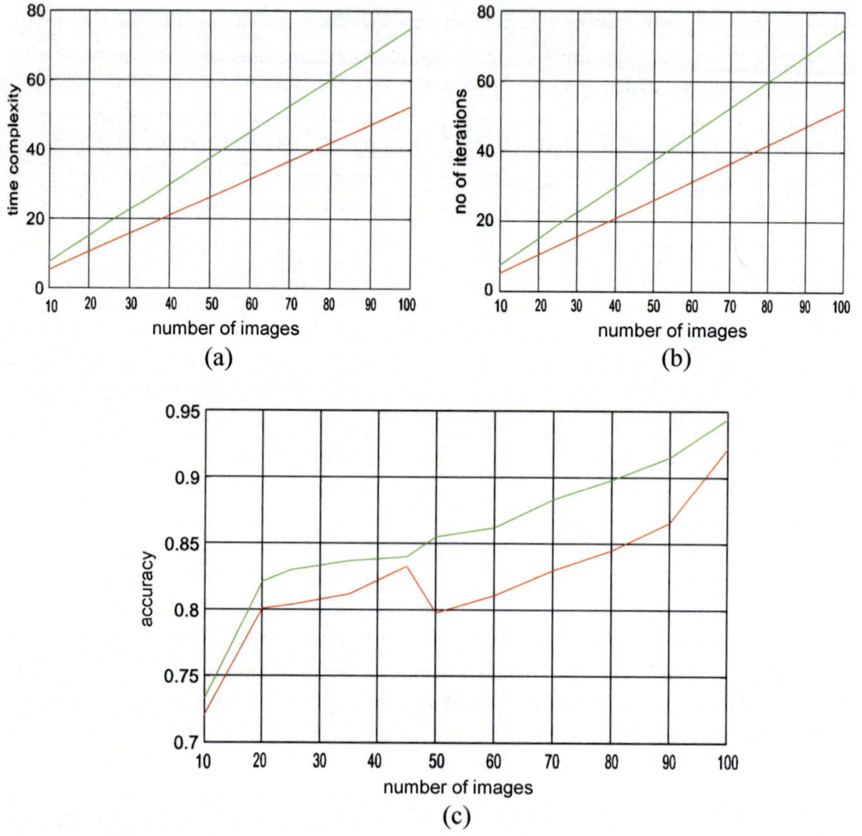

Fig. 8 Graphical comparison of the number of iteration, time complexity, and the accuracy with the number of images

shows the graphical comparison of number of iteration, time complexity and the accuracy with number of images. As the graph shows that accuracy increases if number of images increases.

5 Conclusion

Diabetic retinopathy is one of the most fatal diseases in the world that will cause vision loss to the people affected with diabetes. Detecting the lesions in the retina at the early stages is difficult and it requires dense prediction algorithms to find it at the early stages. The best approach is to detect and classify the various types of retinal lesions in the fundus images is deep learning. The proposed model is Atrous Convolutional Neural Network with fundus images as an input. We achieved a high classification accuracy of 0.94. The high accuracy and the reduction in memory usage makes ACNN suitable for using in diagnostic application.

References

1. Kar S, Maity S (2018) automatic detection of retinal lesions for screening of diabetic retinopathy. IEEE Trans Biomed Eng 65(3):608–618
2. Mansour R (2017) Evolutionary computing enriched computer-aided diagnosis system for diabetic retinopathy: a survey. IEEE Rev Biomed Eng 10:334–349
3. Antal B, Hajdu A (2012) An ensemble-based system for microaneurysm detection and diabetic retinopathy grading. IEEE Trans Biomed Eng 59(6):1720–1726
4. Dai L, Fang R, Li H, Hou X, Sheng B, Wu Q, Jia W (2018) Clinical report guided retinal microaneurysm detection with multi-sieving deep learning. IEEE Trans Med Imaging 37(5):1149–1161
5. Gao Z, Li J, Guo J, Chen Y, Yi Z, Zhong J (2019) Diagnosis of diabetic retinopathy using deep neural networks. IEEE Access 7:3360–3370
6. Vijayalakshmi S, Dahiya S (2017) Medical image segmentation using various techniques: a survey. Int J Recent Trends Eng Res 3(2):120–130. https://doi.org/10.23883/ijrter.2017.3014.wni4f
7. Chen L, Papandreou G, Kokkinos I, Murphy K, Yuille A (2018) DeepLab: semantic image segmentation with deep convolutional nets, Atrous convolution, and fully connected CRFs. IEEE Trans Pattern Anal Mach Intell 40(4):834–848
8. Wisaeng K, Sa-Ngiamvibool W (2019) Exudates detection using morphology mean shift algorithm in retinal images. IEEE Access 7:11946–11958
9. Kumar D, Taylor G, Wong A (2019) Discovery radiomics with CLEAR-DR: interpretable computer aided diagnosis of diabetic retinopathy. IEEE Access 7:25891–25896
10. Imran A, Li J, Pei Y, Yang J, Wang Q (2019) Comparative analysis of vessel segmentation techniques in retinal images. IEEE Access 7:114862–114887
11. Seoud L, Hurtut T, Chelbi J, Cheriet F, Langlois J (2016) Red lesion detection using dynamic shape features for diabetic retinopathy screening. IEEE Trans Med Imaging 35(4):1116–1126
12. Imran A, et al (2019) Comparative analysis of vessel segmentation techniques in retinal images. IEEE Access 7:114862–114887. https://doi.org/10.1109/access.2019.2935912
13. Ieeexplore.ieee.org. 2020. Optic disc boundary and vessel origin segmentation of fundus images. IEEE J Mag [online]. Available at https://ieeexplore.ieee.org/document/7225107/. Accessed 13 June 2020@@@
14. Yang Y, Li T, Li W, Wu H, Fan W, Zhang W (2017) Lesion detection and grading of diabetic retinopathy via two-stages deep convolutional neural networks. Medical Image Comput Comput Assisted Intervention MICCAI 2017 Lecture Notes in Computer Science, pp 533–540
15. Wang S, Yin Y, Cao G, Wei B, Zheng Y, Yang G (2017) Corrigendum to 'hierarchical retinal blood vessel segmentation based on feature and ensemble learning'. Neurocomputing 149(2015):708–717. Neurocomputing 226:270–272
16. Rubini SS, Kunthavai A (2015) Diabetic retinopathy detection based on eigenvalues of the Hessian matrix. Procedia Comput Sci 47:311–318
17. Gadkari S, Maskati Q, Nayak B (2016) Prevalence of diabetic retinopathy in India: the all India ophthalmological society diabetic retinopathy eye screening study 2014. Indian J Ophthalmology 64(1):38
18. Quellec G, Charrière K, Boudi Y, Cochener B, Lamard M (2017) Deep image mining for diabetic retinopathy screening. Med Image Anal 39:178–193

19. Mookiah M, Acharya UR, Martis RJ, Chua CK, Lim C, Ng E, Laude A (2013) Evolutionary algorithm based classifier parameter tuning for automatic diabetic retinopathy grading: a hybrid feature extraction approach. Knowl Based Syst 39:9–22
20. Liang G, Hong H, Xie W, Zheng L (2018) Combining convolutional neural network with recursive neural network for blood cell image classification. In: IEEE access, vol 6, pp 36188–36197. https://doi.org/10.1109/ACCESS.2018.2846685

Frequency Reconfigurable Antennas—A Contemporary Review

K. Jayamani, V. Vinodhini, and Rahul Krishnan

Abstract Nowadays, wireless technologies are required to be vigorous in their utilities to enhance the prosecution which may be realized by a single reconfigurable antenna where the characteristics of antenna can be modified depending on the user requirement. The commonly used antenna in the communication is the reconfigurable frequency antenna. Numerous techniques used to attain frequency reconfiguration are reviewed, and discussions are made on the essential characteristics and particularities of frequency reconfigurable antennas. The comparative study of various switching practices and their utilization in different frequency bands is suggested.

Keywords Reconfigurable antenna · Switching methods · Features of frequency reconfiguration

1 Introduction

In the last several decades, wireless communication innovation and its utilization like Bluetooth, WiFi, GPS, WLAN, and numerous others have approved phenomenal development [1, 3, 4]. To make use of the accessible electromagnetic spectrum by means of less snooping, diminished multifunctional antenna is needed. For this scenario, a reconfiguration characteristic of antenna is helped.

Since the implication of the frequency reconfigurable antennas is on the rise every day, the latest explorations accomplished in this field are reviewed and summarized in this document. Various tactics used in favor of attaining frequency reconfiguration in antenna geometry, applications of those antennas as well as analogy of their performances are given. Finally, conclusion is presented.

K. Jayamani (✉) · R. Krishnan
Department of ECE, Rajalakshmi Institute of Technology, Chennai 600124, India

V. Vinodhini
Department of ECE, University College of Engineering, Ariyalur Campus, Kavanur 621704, India

A. Sivasubramanian et al. (eds.), *Futuristic Communication and Network Technologies*, Lecture Notes in Electrical Engineering 792,
https://doi.org/10.1007/978-981-16-4625-6_56

2 Electrical Reconfiguration

The electronic knobs like PIN diodes, varactors or MEMs are helped to realize electrical reconfiguration [7–10]. Antenna parts are attached or detached via these knobs [1]. Because of this modification in the antenna structure, the distribution of antenna current is varied, and the modification of available frequency is realized [12, 22, 23]. By assimilating these knobs in antenna design itself, design intricacy is abridged. Even though these antennas are set up through tremendous segregation of knobs and little destruction environment, the imprecise property of knobs, presence of biasing stripes and also the deprivations connected with the biasing stripes affect these antennas efficiency. Different methods also with illustrations of electrically reconfigurable antennas to attain the reconfigurability features are presented.

2.1 Frequency Reconfiguration by Using PIN Diode

With the use of PIN diode, the size of patch or ground is altered to achieve the frequency reconfigurability [20, 24]. PIN diode switching speed lies between 1 and 100 nsec. Usage of PIN diode requires appropriate forward bias. PIN diode has the advantages of reliability and extremely low cost [17]. The main disadvantages of using PIN diodes are high power deprivation, little linearity, small quality factor, massive insertion deprivation.

PIN diode to change slot length. A circular monopole antenna has been preferred as an elementary layout since it is used to achieve desired performance over widespread bandwidth. FR4 substratum size of 40 × 40 mm^2 provided a value of $\varepsilon_r = 4.4$, and a thickness of 1.6 mm is used to construct the antenna [5]. Microstrip feeding approach is applied to feed the circular patch [13]. The defected ground has a slot with 4 PIN diodes. Figure 1a, b shows the outlook of the antenna. Via turning on/off the PIN diodes, the dimension of the slot is varied. Because of this change in slot length reconfigurability is achieved. To build the RF linking of the PIN diode and to segregate the RF signal from the dc, for every PIN diode, the slit was fitted with a 100-pF dc blocking capacitor. For multi-radio wireless applications like WiMAX, WiFi, UWB and cognitive radio, this antenna is used.

PIN diode to change patch size. Reformed biconical unipole antenna is proposed for broadband (VHF/UHF/L-bands) applications [11]. The antenna of size 100 × 216 mm^2 is mounted on the substratum which has the density of 8 mm, and the ε_r value of 4.4 which is represented in Fig. 2. Ground plane is gasket type with the diameter of 1 M. PIN diode is selected as an electrical knob, positioned among the termination of aerial and the track. When the diode is on, the antenna acts like traveling antenna, and when it is off, the antenna works as unipole antenna. It also can be utilized as superior potential antenna designed to resist as much as 18 W for aircraft applications. It is bounded in a streamlined housing which then may be set up in mobile rostrums.

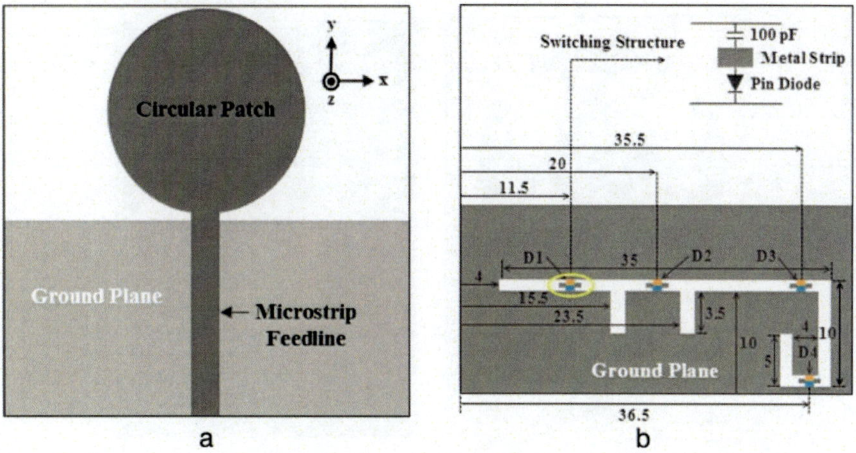

Fig. 1 **a** Antenna front view [5]. **b** Antenna back view [5]

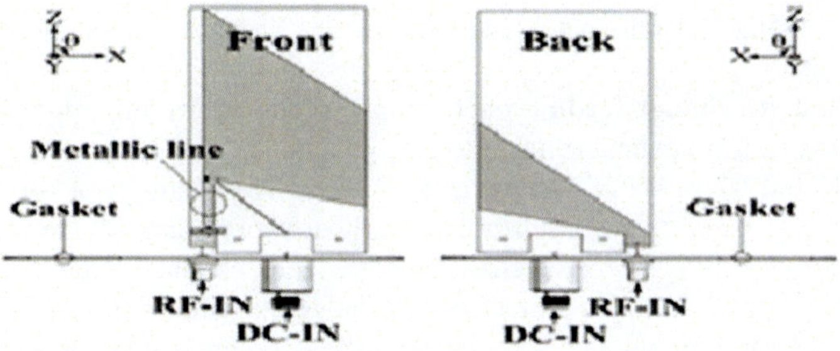

Fig. 2 Antenna structure front and back [11]

PIN diode to change operating mode (patch/slot). For WiMAX and WLAN application, reconfigurable antenna panoply is suggested in the paper [15–17]. The unit cell consists of an abridged spherical radiator, C-fashioned parasitic rudiments as well as disconcerted rhombic slot component with two orthogonal feed lines made on a 1.6-mm-thick FR4 substratum featuring dielectric constant of 4.3 plus loss tangent as 0.023 which is shown in Fig. 3. The planned antenna functions in dual modes as patch and slot mode. The PIN diode that is mounted on the feed structure is used to stimulate an abridged spherical patch radiator plus the slit in the antenna front and rearmost flank. In principle method, on the backside, a disconcerted rhombic slot is stimulated. To activate the slot, dual feed lines having 90° phase variance are coupled to a standard 50Ω SMA connector. In another mode, a solo feed line excites the spherical patch with parasitic rudiments attached to the external surface. Using 5 PIN diodes, all the modes are obtained. To accomplish frequency, radiation pattern and polarization reconfiguration in the operating bands, six distinct practical conditions are used. The orthogonal feeding mechanism is helped to provide distinct

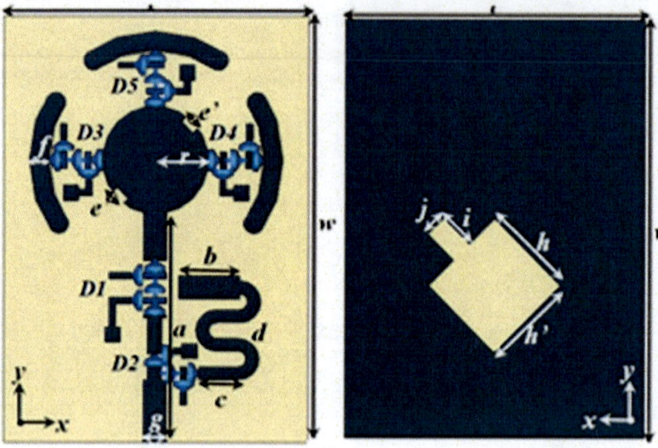

Fig. 3 Antenna prototype top view and bottom view [17]

swapping among slot plus patch and removes the use of different slot and patch antenna.

PIN diode to change feeding port. Graphene conductive ink imprinted fabric-centered microstrip antenna suitable to change among "TM 10 mode of S-band at 3.03 GHz, TM 02 mode of C-band at 5.17 GHz, TM 20 mode at 6.13 GHz" is proposed [14]. A multifaceted substratum methodology is employed to enhance the emanation functioning of the antenna since graphene conductive ink has the low-conductivity issue. Multiple port technique is helped to achieve frequency reconfigurability. RF knob assimilated transmission line is premeditated and employed to facilitate the feeding ports, ensuing in frequency reconfigurability. The PIN diode RF knob consists of dual λ_g /2 microstrip line, a PIN diode, RF block inductor of 1 mH and DC block capacitor of 1uF. The supply voltage of 1 V is connected to the PIN diode to switch on the RF knob. If no bias is applied, the RF knob is deactivated. SMA connectors attach the premeditated RF knob to Port-2 of the double fodder antenna, and then, straight fodder is given to Port-1. Port-2 must have an energetic state where no bias is added to the PIN diode so that the antenna gets activated in tripartite band mode. The port-2 is disabled when the bias voltage is delivered to a PIN diode, after which it can run on a double band (via port-1). In this way, frequency reconfigurability is achieved. The complete configuration is given in Fig. 4a, b.

2.2 *Frequency Reconfiguration by Using a Varactor Diode*

An antenna arrangement consisting of double arc rings in the shape of triangular consists of Ring1 and Ring2 on the topmost layer, and an edge-truncated ground plane with two short shreds on the bottom most layer of substratum of 0.8 mm thickness

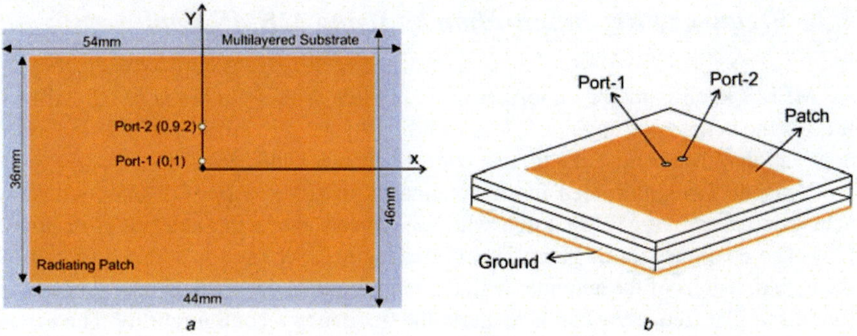

Fig. 4 **a, b** Geometry of multilayer antenna with multiple port and multilayer structure [14]

($\varepsilon_r = 4.4$, $\tan\delta = 0.02$) is suggested in the paper [21]. The defected ground is used to realize impedance equivalence and also to improve frequency reciprocation. Varactor diodes employed on the ground plane are helped to enhance the beam scanning. On top layer, another set of varactor diode is placed to divide the excitation signal and to control the feeding. To achieve the frequency reconfigurability, any one of the top layer varactor diode is tuned, whereas ground plane varactor diode should be kept fixed at the lowest value. Figure 5 shows the proposed antenna structure.

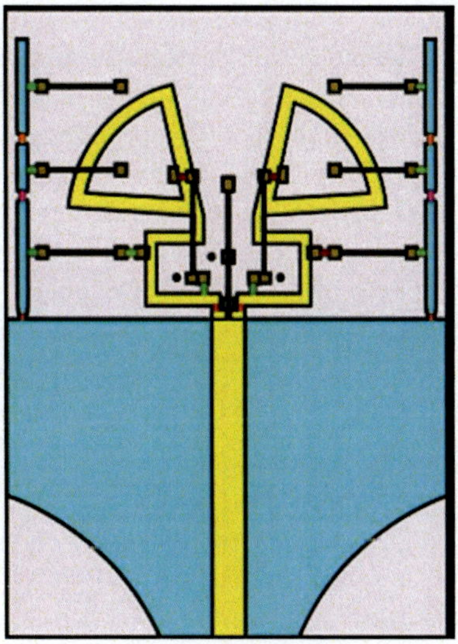

Fig. 5 Proposed antenna design [21]

2.3 Frequency Reconfiguration by Using MEMS Knob

RF MEMS knob is major important basic module within the arena of RF MEMS. It takes the benefits of tiny mass, less weight, little energy utilization, less inclusion deprivation, great seclusion and also wide frequency band [25–27].

Coplanar waveguide feed mode antenna of size10 × 10 × 0.5 mm^3 with low actuation capacitive MEMS knob which is placed above the antenna is designed [25]. The configuration is given in Fig. 6. This MEMS knob is adopted to control the both divisions of the antenna, with the intention that the antenna may function in four dissimilar frequency bands to attain the frequency reconfigurability. The extent of the MEMS knob is small, and the density of the beam is 1μm. To diminish the actuation voltage of the knob, selection of knob beam material plays major role. Since gold has the moderately great elastic modulus in addition with less residual stress and also has the property of quick reformation when the voltage is canceled, it is selected as knob beam material. Also low skin depth property of gold has low effect on signal communication besides the actuation voltage is as small as 14 V. Down-state plus up-state mechanism of the both knobs is agreed to accomplish the frequency reconfigurability.

3 Optical Reconfiguration

In the paper, [28] designs an optically operated frequency-reconfigurable antenna constructed on a multi-band slotted-waveguide antenna array with dual photoconductive knobs that have been integrated to achieve 28 and 38 GHz frequency reconfigurability. These two photoconductive knobs are liable to accomplish the electrical slot length by properly hindering some portion of the energy that has to be given out to the air. Antenna array has an aluminum waveguide pulverized through slots. With the help of denser bits of copper tape to irradiate the knobs from the top, slightly angled optical fibers are mounted. A laser beam irradiates the minute the silicon die,

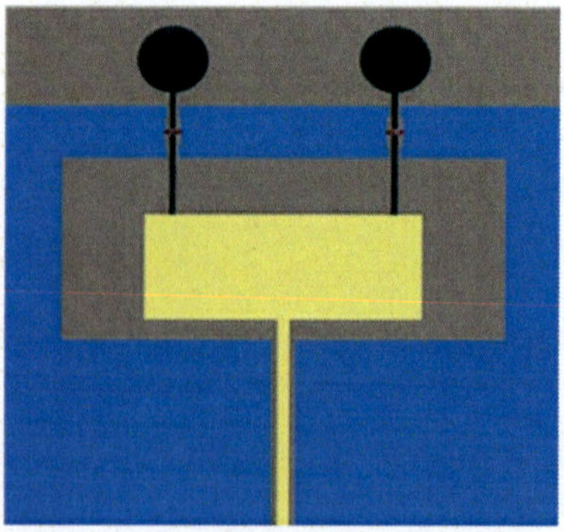

Fig. 6 Configuration of CPW feed antenna with MEMS knob [25]

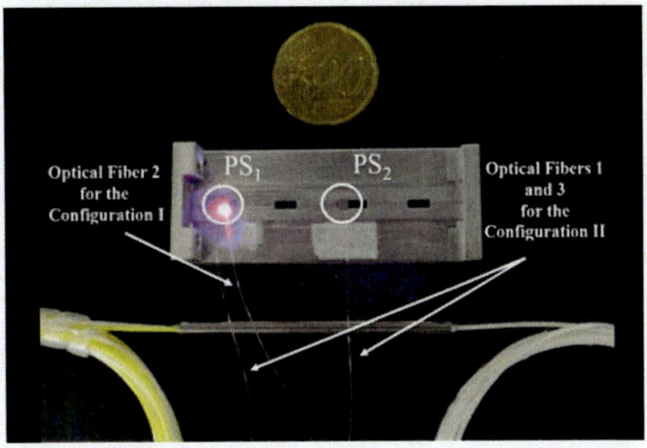

Fig. 7 Prototype of OCAA [28]

and it is known as on state; otherwise, it is off state. The OCRAA operates from 27.6 to 30.8 GHz in the 28 GHz frequency range, while the PS1 is on. The antenna array bandwidth is reconfigured at 38 GHz band, from 36.8 to 38.4 GHz, while both the PS1 and PS2 nodes are on. The prototype is represented in Fig. 7. This OCAA is used for indoor mm-wave system built on optical wireless networks where optical fibers can be used to transmit data and control the electromagnetic properties of the antenna remotely. Since the photoconductive approach does not need any bias line, the structure is modest and can lie in the antenna's main board to change the radiation structure.

Once the knobs are removed, antenna impedance matching is vanished and the antenna gain is reduced. When the laser is turned off, antenna principle lobe is piloted to alternative direction and the misalignment is even more critical.

4 Electronic Reconfiguration

The input signal from microcontroller or FPGA or processor is helped to attain the reconfigurability. Reconfigurable reflect array (RRA) element with a sole PIN diode is introduced to work at double frequencies by 1-bit phase resolution [18]. RRA encompasses 40 × 40 elements lit up thru a wide band conical horn. For easy fabrication and assembling, 40 × 40 array is distributed into 5 identical 8 × 40 sub-arrays. Every single element is regulated independently through an analogous transputer incorporated at the rear side of the aperture. The incorporation does not disturb the element depiction because the digital circuits are organized beneath the RF ground. An FPGA-based control board transmits the direction indications for the entire 1600-element array to every single component by FFC (fixable flat circuit) connection. The RRA setup is given in Fig. 8. The present array could be protracted to a more immense aperture proportions thru connecting additional sub-arrays. The component can work at X and Ku bands concurrently (11.1 GHz, and14.3 GHz).

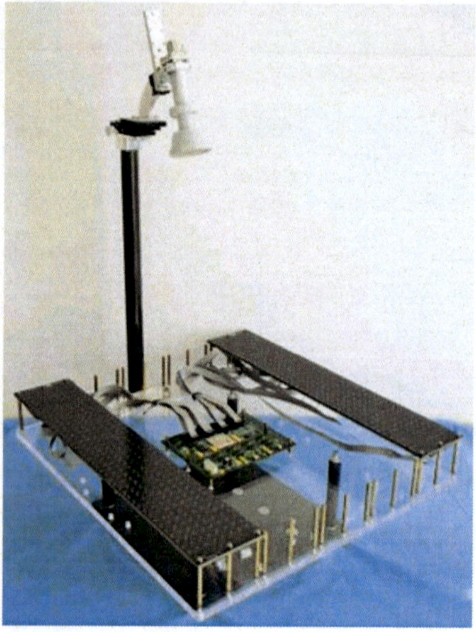

Fig. 8 RRA setup [18]

5 Material Change Reconfiguration

Reconfiguration using the smart materials is new growing research field. By pumping the liquid fluid in the hollow space of the antenna structure, varying the substratum dielectric property is done. This is helped to achieve the change the operating frequency. Even though this technique has the advantages of minor loss, added power handling and better tuning range, challenges like reliability and efficiency are yet to be examined thoroughly.

5.1 Using Liquid Crystal Polymer (LCP)

Rogers RT 5870 substratum of size 65 × 50 mm^2 with permittivity (ε_r) of 2.33, density of 0.787 mm and tangential loss (tanδ) of 0.0012 is taken to design the antenna; also the substratum is covered by 35 μm copper layer on each side [7]. Two circular patches with the slit or separation (0.5 mm wide) were then constructed. The cavity or separation (0.5 mm wide) among each of the patches is occupied with liquid crystal polymer (LCP), a density of 35 μm. In the end, 0.1-mm-thick and 1-mm-wide layer of mercury is located on the top of LCP. Figure 9 represents the top view. Liquid crystal polymer (LCP) has numerous benefits comprising lesser value of loss tangent (tanδ) (i.e., from 0.002 to 0.0045), great tractability, and little cost, robust chemical inertness, awfully less density plus very slight moisture absorption ratio. In the nonappearance of LCP and mercury, the both circular patches turn out to be

Fig. 9 Top view of the antenna with LCP [7]

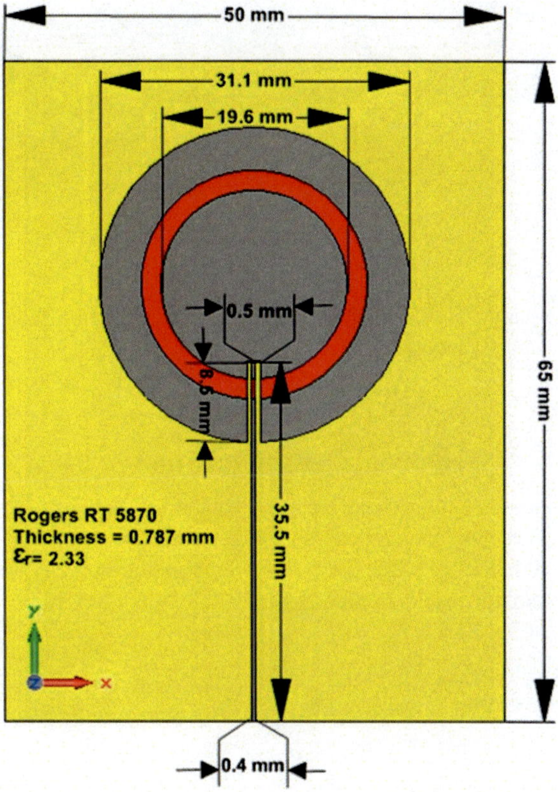

inaccessible from each other. In this case, interior patch is radiated at the frequency of 5 GHz. If the gap is filled with LCP, both patches are combined and then, diameter of the patch became 31.1 mm and is radiated at 3.6 GHz. The antenna is used for WLAN applications.

5.2 *Using Liquid Metal*

Reconfigurable frequency patch antenna with switchable slots (PASS) based on the deployment of liquid metal in a microfluidic channel printed in 3D is proposed [32]. Ceaseless transportation of the liquid metal droplet above the channel concealing the slots helps achieve frequency reconfiguration over a wide frequency range. For prototyping and measurements, EGaln carried by mineral oil has been taken. By integrating the 3D printed microfluidic module with the patch antenna on PCB through patterned adhesive layer, the model is fabricated. This production process is shown valuable in fabricating liquid metal antenna functioning EGaln thru minimum skin

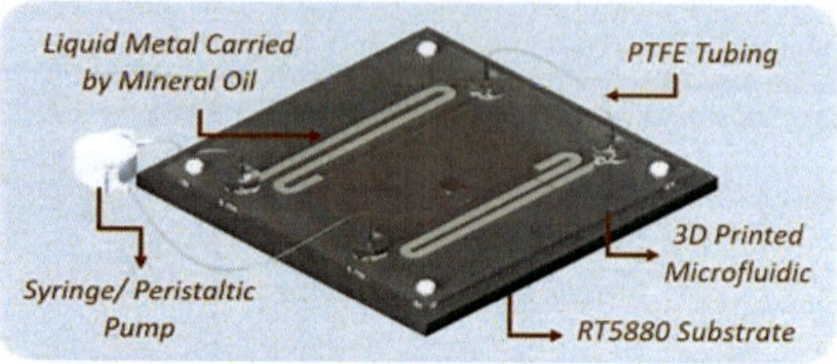

Fig. 10 PASS structure [32]

residue. By fixing this pattern on a 3D printed bracket, the location of syringes and tubes is secured (Fig. 10).

6 Mechanical Reconfiguration

This type of antenna can deliver extensive scope of reconfiguration. Even though added setup raises the cost, weight and complexity, mechanical shift of a portion of radiator results in reconfigurable antenna [2, 4, 6].

6.1 Reconfiguration Using Stepper Motor

An asymmetric coplanar strip (ACS)-fed monopole reconfigurable antenna is designed on a FR4-epoxy substratum that has a dielectric constant of 4.4, and the dimensions of $14 \times 16 \times 1$ mm^3 [33]. Rectangle patch with folded shape slit with stub which acts as repudiating device is used for the UWB antenna. An arc-shaped stub and an inverted L-shaped stub are used for the mobile antenna function. Reconfigurability is accomplished by allowing rounded movement of the patch. A modest clock stepper motor is placed behind the antenna to regulate the rotational patch part and stimulate the reverberating constituents. By revolving the circular substratum segment plus patch, the projected radiator may function in the triple-mobile-band (2.4–2.9, 3.6–4.4, and 5–6 GHz) and the UWB frequency range with stop bands of 3.3–4.2 GHz and 5–6 GHz. Antenna configuration with different modes of operation is given in Fig. 11.

6.2 Reconfiguration Using Fillers

Reconfigurability is achieved by shifting the dielectric constant of the substratum physically. A rectangular microstrip antenna is placed on FR4 substratum of density 2.4 mm, $\varepsilon_r = 4.4$ and loss tangent 0.02 [31]. Figure 12 represents the physical

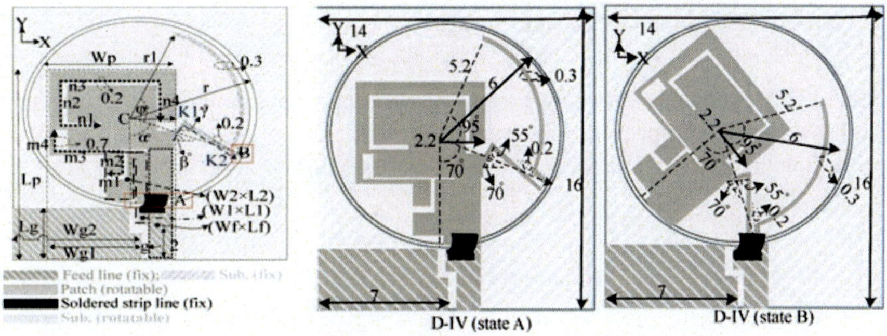

Fig. 11 Antenna configuration with different modes of operation [33]

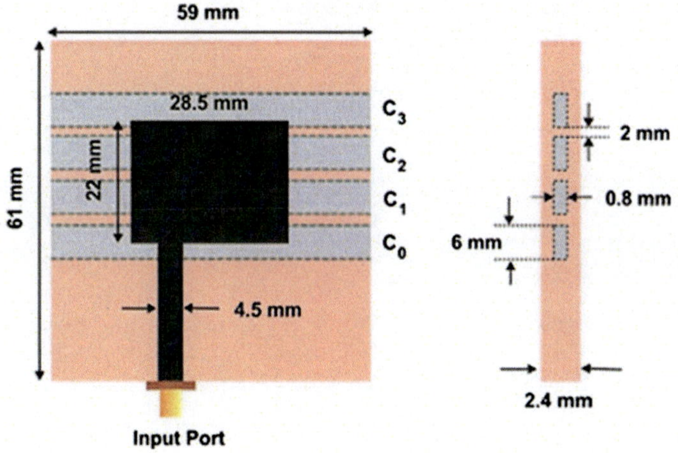

Fig. 12 Physical reconfigurable antenna [31]

configuration. As there are no electric knobs, spurious ohmic losses are eliminated. Quadruple rectangular back-to-back craters route through the width substratum. Every single crater is deliberated like a bit, and its grade can be stated as 0 (if it is empty) or 1 (if it is loaded by certain matter excluding air). Adding certain material inside the craters varies the dielectric constant of the whole substratum structure; therefore, the operating frequency of the antenna system is varied. Due to quadruple craters, overall 2^4 or 16 diverse situations (0000–1111) with unique resonant frequency are achieved. If the dielectric constant of the filler is higher than the substratum, the resonant frequency variation ranges possibly vaster as related to when the substratum substance itself is applied to fill the cavities. This antenna is used as mechanically reconfigurable chipless RFID tag.

7 Summary of Literature Review

Table 1 comprises the comparative study of several antennas deliberated in this paper. This table provides specifics of the switching mechanism used, the size and nature of the antenna substratum, and the frequency of reconfiguration accomplished with its operation.

Comparative summary of various antennas

Ref. No	Switching mechanism	Substratum material	Antenna type	Antenna size	Reconfiguration Frequency (GHz)	Application
[5]	Electrical-4 PIN diodes	FR4 of 1.6 mm thick	Circular monopole	40×40 mm^2	2.3–2.4, 2.5–2.7, 3.3–3.8, 5.15–5.85, 2.4–2.48, 5.15–5.85 3.1–10.6	Multiradio wireless applications and cognitive radio
[7]	Material-LCP	0.787 mm thick Rogers RT 5870	Circular patches	65×50 mm^2	3.6 5	WLAN
[11]	Electrical-1 PIN diode	FR4 of 8 mm thick	Modified biconical unipole	100×216 mm^2	VHF/UHF/L-bands	Broadband Airborne Applications
[14]	Electrical-1 PIN diode	Multilayered	Microstrip	44×36 mm^2	3.03 5.17 6.13	Satellite communication
[17]	Electrical-5 PIN diode	1.6 mm thick FR4	Truncated circular radiator	40×50 mm^2	3.6 5.8	WiMAX and WLAN
[18]	Electronic-FPGA signal	Taconic TLX-8 of 1.58 mm thick	Patch	5 identical 8×40 sub-arrays	11.1 14.3	X and Ku bands
[21]	Electrical-4 varactor diodes	0.8 mm thick FR4	Triangular-arc rings	$0.27\lambda_0 \times 0.34\lambda_0 \times 0.005\lambda_0$	2.21–2.55 2.77 to 2.82	Wireless applications
[25]	Electrical-2 MEMS knob	0.5 mm thick FR4	Coplanar waveguide feed mode	$10 \times 10 \times 0.5$ mm^3	10 30	Satellite communication
[28]	Optical-2 photo conductive knobs	Optical fiber	Slotted-waveguide array		28 38	5G

(continued)

(continued)

Ref. No	Switching mechanism	Substratum material	Antenna type	Antenna size	Reconfiguration Frequency (GHz)	Application
[31]	Physical-cavity filler	2.4 mm thick FR4	Micro-strip	61×59 mm^2	2.400–2.480 2.496–2.690	WSN, WLAN, LTE based IoT applications, chipless RFID tag
[32]	Material-Liquid metal	125 mil thick RT5880	Patch antenna with switchable slots	100×100mm^2	2.1–3.8	Wireless applications
[33]	Mechanical -stepper Motor	1 mm thick FR4	ACS fed monopole	14×16 mm^2	2.4–2.9, 3.6–4.4, and 5–6 3.2–11 GHz with two stop bands frequencies	WLAN

8 Comparison of Different Reconfiguration Techniques

Due to the soothing implementation and low cost, the electrical reconfiguration technique is the extensively employed method to achieve reconfiguration. Reconfiguration is done within a few nanoseconds if the PIN diode or varactor diode is used. Through using these knobs, fast switching and relatively high current handling are realized. At the same time, PIN diode has the disadvantages of considerable tuning speed, huge dc bias in on-state, and excellent power handling ability. Varactor diode provides small current flow with uninterrupted tuning, and it is easy to integrate. But when varactor diode is used, the antenna performance is affected by its nonlinear property.

MEMS knobs can be used without bias lines and provides high quality factor with low loss. But the response is slow in nature. Fabrication of antennas with MEMS requires expensive equipment, and fabrication time and cost are higher than RF knobs.

When optical reconfiguration technique is used, bias lines are not required and no intermodulation distortion is achieved. Due to its high transition speed, fast switching is achieved. At the same time, optical knobs establish additional losses and also require complex triggering system [28–30].

No active elements and no biasing lines are needed for physical reconfiguration technique. When this type is implemented, the cost and antenna complexity are increased. For this type of antennas, more space is needed to accommodate all the components. Extra power source also required to support mechanical tuning [19].

Low profile, less weight reconfigurable antenna is designed by using material change reconfiguration technique. On the contrary, they are least effective at microwave frequencies because of this drawback; the application of this type of reconfigurable antenna is limited.

Nowadays, reconfiguration is also done with the help of electronic components like microcontrollers and FPGAs. These devices are used to give inputs to control the RF knobs. CPU integrated system is needed to implement this technique.

9 Conclusion

A pervasive study of various techniques used to realize frequency reconfiguration is presented. The examined reconfiguration methods are categorized into electrical, optical, mechanical, physical and smart-material-loaded reconfigurable structures. Few applications of frequency reconfigurable antenna are also mentioned. Numerous patterns of reconfigurable antennas for diverse solicitations are deliberated, and their features are reported. Prognostication of the type of best reconfiguration mechanism well ahead is very challenging. If the design satisfies the purpose for which it will be entirely used without any deterioration, the reconfiguration mechanism is regarded as the best one at the period. This comprehensive investigation of reconfigurable antennas shall assist the academics in accomplishing eventual exploration in the field of reconfiguration.

References

1. Jayamani K, Rahul krishnan S, Nachiya RA, Raj A, Atchaya C (2020) a survey on frequency reconfigurable antenna for wireless applications. Int J Psychosoc Rehabil 24(5):5734–5744
2. Yang S-LS, Kishk AA, Lee K-F(2008) Frequency reconfigurable U-slot microstrip patch antenna. IEEE Antennas Wirel Propag Lett 7:127–130
3. Anagnostou DE, Gheethan AA (2009) A coplanar reconfigurable folded slot antenna without bias network for WLAN applications. IEEE Antennas Wireless Propag Lett 8:1057–1060
4. Sharma BL, Parmar G, Kumar M (2015) Design of frequency reconfigurable Mictrostrip patch antenna for S-band and C-band applications. In: 4th international conference on reliability, Infocom Technologies and Optimization: trends and future directions (ICRITO), pp 1–4
5. Boudaghi H, Azarmanesh M, Mehranpour M (2012) A frequency-reconfigurable monopole antenna using switchable slotted ground structure. IEEE Antennas Wirel Propag Lett 11:655–659
6. Cleetus RMC, Bala GJ (2017) Frequency reconfigurable antennas: a review. In: International conference on signal processing and communication (ICSPC'17), pp 160–164
7. Farooq W, Ur-Rehman M, Abbasi QH, Qaraqe K (2015) A circular patch frequency reconfigurable antenna for wearable applications. In: 2015 the first international workshop on advances in body-centric wireless communications and networks and their applications, pp 103–107
8. Mohanta HC, Kouzani AZ, Mandal SK (2019) Reconfigurable antennas and their applications. Universal J Electr Electron Eng 6(4):239–258
9. Selvam YP, Elumalai L, Alsath MGN, Kanagasabai M, Subbaraj S, Kingsly S (2017) A novel frequency and pattern reconfigurable rhombic patch antenna with switchable polarization. IEEE Antennas Wirel Propag Lett 16:1639–1642
10. Ren J, Yang Xi, Yin J, Yin Y (2015) A Novel antenna with reconfigurable patterns using H-shaped structures. IEEE Antennas Propag Lett 14:915–918
11. Rhee CY, Kim JH, Jung WJ, Park T, Lee B, Jung CW (2014) Frequency-reconfigurable antenna for broadband airborne applications. IEEE Antennas Wirel Propag Lett 13:189–193
12. Ban Y-L, Sun S-C, Li P-P, Li JL-W, Kang K (2014) Compact eight-band frequency reconfigurable antenna for LTE/WWAN tablet computer applications. IEEE Trans Antennas Propag 62(1):471–475
13. Rahul krishnan, Vanaja S, Chinnammal C, Jayamani K, Karunakaran A (2020) Design of circular patch antenna with coplanar waveguide feed for LTE application. Int J Innov Technol Exploring Eng 9(4):1620–1622
14. Kumar J, Basu B, Talukdar FA, Nandi A (2018) Graphene-based multimode inspired frequency reconfigurable user terminal antenna for satellite communication. IET Commun Spec Sect Satell Commun Antennas 12(1):67–74
15. Li PK, Shao ZH, Wang Q, Cheng YJ (2014) Frequency and pattern reconfigurable antenna for multi-standard wireless applications. IEEE Antennas Wirel Propag Lett 14:333–336
16. Majid HA, Rahim MKA, Hamid MA, Ismail MF (2014) Frequency and pattern reconfigurable slot antenna. IEEE Trans Antennas Propaga 62(10):5339–5343
17. Selvam YP, Alsath MGN, Kanagasabai M, Elumalai L, Palaniswamy SK, Subbaraj, Kingsly S, Kulandhaisamy I (2018) A patch-slot antenna array with compound reconfiguration. IEEE Antennas Wirel Propag Lett 17(3):525–528
18. Yang H, Yang F, Cao X, Shenheng Xu, Gao J, Chen X, Li M, Li T (2017) A 1600-element dual-frequency electronically reconfigurable reflectarray at X/Ku bands. IEEE Trans Antennas Propag 65(6):3024–3032

19. Yang X, Shenheng Xu, Yang F, Li M, Hou Y, Jiang S, Liu L (2017) A Broadband high-efficiency reconfigurable reflectarray antenna using mechanically rotational elements. IEEE Trans Antennas Propag 65(8):3959–3966
20. Boukarkar A, Lin XQ, Jiang Y, Yang XF (2018) A compact frequency-reconfigurable 36-states patch antenna for wireless applications. IEEE Antennas Wirel Propag Lett 17(7):1349–1353
21. Alam M, Abbosh A (2019) A compact reconfigurable antenna with wide tunable frequency and 360° beam scanning. IEEE Antennas Wirel Propag Lett 18(1):4–8
22. Zainarry SNM, Nguyen-Trong N, Fumeaux C (2018) A frequency- and pattern-reconfigurable two-element array antenna. IEEE Antennas Wirel Propag Lett 17(4):617–620
23. Nghia N-T, Piotrowski A, Fumeaux C (2017) A frequency-reconfigurable dual-band low-profile monopolar antenna. IEEE Trans Antennas Propag 65(7):3336–3343
24. Khan T, Rahman MuhibUr, Akram A, Amin Y, Tenhunen H (2019) A low-cost CPW-fed multiband frequency reconfigurable antenna for wireless applications. Electronics 8:1–17
25. Xu Y, Tian Y, Zhang B, Duan J, Yan L (2018) A novel RF MEMS switch on frequency reconfigurable antenna application. Microsyst Technol 24(9):3833–3841
26. Lakshmi P, Manohar P, Naga Sayanu P (2017) Optimization of structures of DC RF MEMS series switches for low actuation. Microsyst Technol 23(7):2371–2379
27. Deng Z, Guo X, Wei H, Gan J, Wang Y (2016) Design, analysis, and verification of Ka-Band pattern reconfigurable patch antenna using RF MEMS knobes. Micromachines 7(8):144
28. da Costa IF, Arismar Cerqueira S Jr, Spadoti DH, da Silva LG, Ribeiro JAJ, Barbin SE (2017) Optically controlled reconfigurable antenna array for mm-wave applications. IEEE Antennas Wirel Propag Lett 16:2142–2145
29. Sodré Junior AC, Feliciano Da Costa I, Manera LT, Diniz JA (2014) Optically controlled reconfigurable antenna array based on E-shaped elements. Int J Antennas Propag 1–8
30. Tawk Y, Costantine J, Hemmady S, Balakrishnan G, Avery K, Christodoulou CG (2012) Demonstration of a cognitive radio front end using an optically pumped reconfigurable antenna system (OPRAS). IEEE Trans Antennas Propag 60:1075–1083
31. Mathur P, Madanan G, Raman S (2020) Mechanically frequency reconfigurable antenna for WSN, WLAN, and LTE 2500 based internet of things applications. Int J RF Microw Comput Aided Eng 31(7):1–12
32. Song L, Chui WGO, Rahmat-Samii Y (2019) Wide-band frequency reconfigurable patch antenna with switchable slots (PASS) based on Liquid Metal and 3D printed microfluidics. IEEE Trans Antennas Propag 67(5):2886–2895
33. Lotfi P, Azarmanesh M, Soltani S (2013) Rotatable dual band-Notched UWB/Triple-band WLAN reconfigurable antenna. IEEE Antennas Wirel Propag Lett 12:104–107

Single-Precision Floating-Point Addition Under HUB

N. S. Sathyavathi and P. Augusta Sophy Beulet

Abstract Floating-point arithmetic implementation follows basic operations like addition, subtraction, multiplication and division. In scientific computation, digital signal processing rounding the number to true value leads to error in representing the original number. This paper includes a single-precision half unit biased floating-point format representation and provides the implementation technique of HUB addition and compared with standard floating-point addition. Implementation of HUB format in adding two floating-point numbers reduces the area by 20.6%, reduces the power by 16.7%. It also reduces the error due to rounding.

Keywords Floating-point addition · Round to nearest · Rounding · HUB

1 Introduction

The rounding operation is performed in various floating-point arithmetic operations to get exact result. There are many ways to implement the rounding process, rounding-to-nearest has the better characteristic among other methods. It gives the closest possible number to the exact true value, but if the correct value is precisely in the center between two numbers, then anyone of the numbers is selected. However, implementing the rounding mode is complex, which increases the area and power leading to increase in the critical path. This leads to decreased performance of the circuits. Various rounding methods have been proposed to reduce the delay, by combining the different rounding methods and removing the rounding circuit from the critical path. Consider a Floating-point adder, where the inputs for addition may be the result obtained from another operation. The addition result may be the input to the next operation. To reduce errors, the number has to be similar to a real value. To perform the normal floating-point rounding guard bit, round bit, stick bit are used,

N. S. Sathyavathi · P. Augusta Sophy Beulet (✉)
School of Electronics Engineering, Vellore Institute of Technology, Chennai, India
e-mail: augustasophyt.p@vit.ac.in

N. S. Sathyavathi
e-mail: sathyavathi.2018@vitstudent.ac.in

A. Sivasubramanian et al. (eds.), *Futuristic Communication and Network Technologies*, Lecture Notes in Electrical Engineering 792,
https://doi.org/10.1007/978-981-16-4625-6_57

this bits are used in different algorithms for rounding [1]. To simplify the implementation of round to nearest, a different approach is used by encoding the real-number. Round to nearest representations with half unit biased (HUB) format is proposed in this paper.

2 Floating-Point Format and HUB Format

The three fields in floating-point numbers are Sign (S), Mantissa (M) and Exponent (E). S is a sign bit. S represents one for negative, zero for a positive value. E is an exponent field or fractional field. The Exponent field is represented by 8 bits, a value biased with + 127 for single-precision floating-point numbers. The radix for the number is β. Twenty-three bits are available for the mantissa representation [2]. It turns out that if floating-point numbers are stored in their normalized form, then the leading bit or most significant bit or the bit on the left is always one. This bit is implicit most significant bit this implicit most significant bit is a hidden bit. It does not need any memory but involved in the calculation [3]. Figure 1 shows a format diagram of a 32-bit Floating-point number (single precision).

2.1 Floating-Point Round to Nearest

While adding two floating-point numbers with different exponents, rounding is needed for the exponent to align to the true value. In case of rounding in regular floating-point numbers, guard (g), round (r) and stick (s) bits are computed based on round to nearest even algorithm. The least significant bit is denoted as 't' [4].

The below Table 1 is formulated into the below equation as

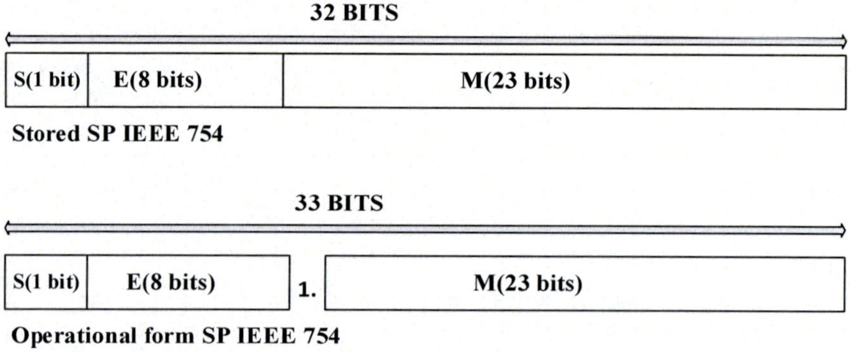

Fig. 1 The stored and operational format of the single-precision floating-point number

Table 1 Round to nearest using t, r, g and s

t	g	r	s	Action
x	0	x	x	Truncate
0	1	0	0	Truncate
1	1	0	0	add 1 to t
x	1	1	x	add 1 to t
x	1	0	1	add 1 to t

$$\text{RN} = \begin{cases} g = 0, & \text{truncate} \\ g = 1\&(t(\text{or})r(\text{or})s) = 1 & \text{add 1 to } t \end{cases}$$

The Error range for the round to nearest is,

$$-\frac{1}{2}\beta^{-p} < \text{Error} < -\frac{1}{2}\beta^{-p}$$

p is the position of the least significant bit.

2.2 *Floating-Point HUB*

A floating-point format using Half Unit Biased involves adding one more mantissa bit at the end of the twenty-third bit of mantissa. The addition of one more bit at the end of the mantissa is called as an implicit least significant bit (ILSB) [5]. Finally, the HUB format for single-precision has two implicit bits, the 23-bit mantissa is changed to 25-bit mantissa mentioned in below Fig. 2. The position of the two implicit bits is at the first bit and the last bit of mantissa. The IEEE floating-point representation has only one implicit bit. In both cases, the stored bits are 23 bits. On comparing the accuracy of HUB and Round to nearest rounding, both have the same accuracy [6].

The Error range for the round to nearest is,

$$-\frac{1}{2}\beta^{-p} < \text{Error} < -\frac{1}{2}\beta^{-p}$$

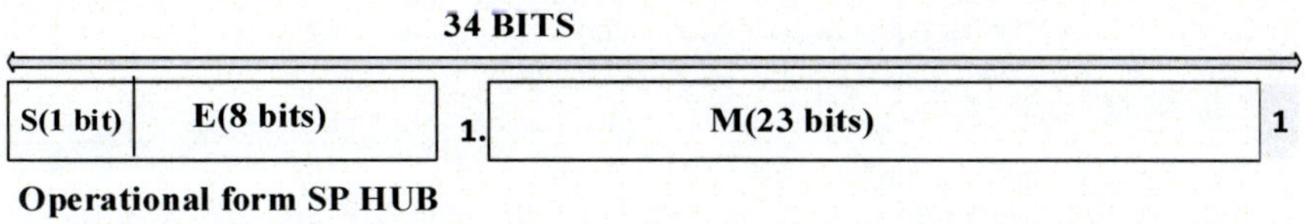

Fig. 2 The stored and operational format of the HUB floating-point number

Thus, the error range of HUB is similar to round to nearest [7]. For converting the number to its 2's complement value in case of standard floating-point format, the number has to be inverted and add 1 bit is added at the LSB but for the Half Unit Biased it is implemented simply by an inversion of bits and addition of one to LSB is not needed. HUB rounding is similar to truncation. If extra bits to be removed are zero, then implicit LSB has to be changed to 1 [8]. If the digit is EVEN, and the least significant bit is zero, nothing changes. If the digit is ODD, and the least significant bit is one, subtraction is performed by making the least significant bit is zero.

2.3 Illustration

Let us consider *A*, *B* as the single-precision Floating-point numbers.

```
     S  E        M
A =  0 01111101 00000000000000011100000

B =  0 10000101 10010000000000000000000
```

The exponents of the two floating-point numbers must be made equal while performing addition. When the mantissa part is shifts toward left by 1 or 2 bits, it results in decreased exponent by 1 or 2 bits, respectively [9]. When the mantissa part is shifted toward right by 1 or 2 bit, it results in increase of an exponent by 1 or 2 bits, respectively [10].

```
shifting by 10000101
           -01111101
           ---------
            00001000   shifting to 8 places.
```

```
0 01111101 1.00000000000000011100000 (original value)
0 01111110 0.10000000000000001110000 (shifted 1 place)
                                    t g r s
0 01111111 0.01000000000000000111000 0 0 0
0 10000000 0.00100000000000000011100 0 0 0
0 10000001 0.00010000000000000001110 0 0 0
0 10000010 0.00001000000000000000111 0 0 0
0 10000011 0.00000100000000000000011 1 0 0
0 10000100 0.00000010000000000000001 1 1 0
0 10000101 0.00000001000000000000000 1 1 1
```

Here increase it's exponent to make exponent equal.

In the algorithm Round to nearest, based on LSB, guard, round and stick bits as mention in above Table 1, one should be added to t which results in 0 10,000,101 0.00000001000000000000001 but in case of HUB format the value is 0 10,000,101 0.00000001000000000000001. On comparing the values, the HUB format and round to nearest, HUB format has the value nearest to the true value. The use of round to nearest format need adders but in case of HUB one implicit bit get stored as 24th bit which does not require computation like round to nearest. Thus HUB provides better accuracy.

3 Addition Using HUB and Floating-Point Number

The addition of 32-bit floating-point numbers $S = X + Y$ is more complicated than the multiplication or division of floating-point numbers because of shifting to make exponents equal [12].

Floating-point addition progresses in 4 steps as follows:

- The exponents are made equal by aligning the mantissa of X and Y.
- Addition/subtraction of the aligned mantissa.
- Normalization of the sum S.
- The resulting sum is rounded.

The alignment step keeps a "guard bit," "round bit" and "stick bit" for the rounding. In case of HUB rounding done by truncation because of having one implicit least significant bit [11].

4 Result and Discussion

Two 32-bit floating-point numbers are added in HUB floating-point format which has given more accurate results. The results using the HUB addition and standard floating-point addition are compared below. Comparing the number of gates for Round to nearest and HUB proves round to nearest requires 36% more gates. The number of gates are increased due to the computation of guard, round and stick bits for rounding.

The addition of floating-point numbers has been described in Verilog hardware description language and implemented. The standard IEEE floating-point and HUB format addition have been implemented using the cadence tool. The number of gates of both approaches is compared. The below Figs. 3, 4 and 5, clearly shows that the use of HUB provides the least amount of gates which results in a reduced area and power.

The addition of floating-point numbers using a Half Unit Biased format always requires significantly less area and power.

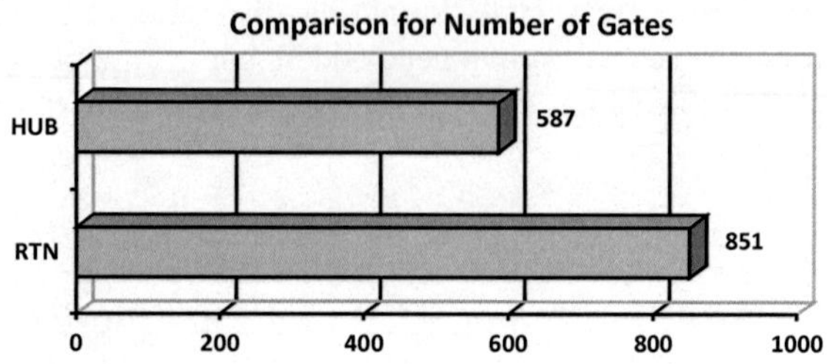

Fig. 3 The number of gates required for standard RTN and the HUB-based adders for single precision

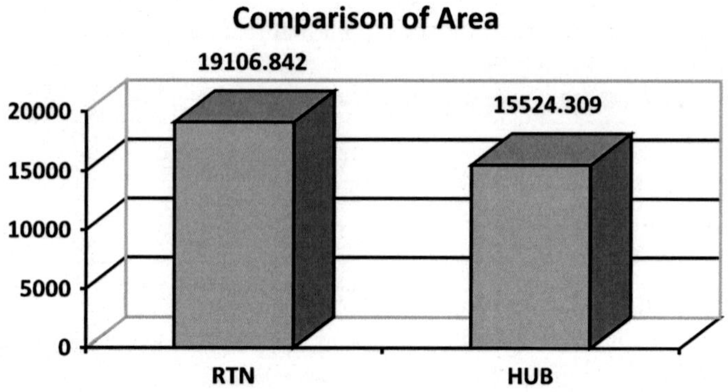

Fig. 4 The comparison of area for standard round to nearest and the HUB-based addition

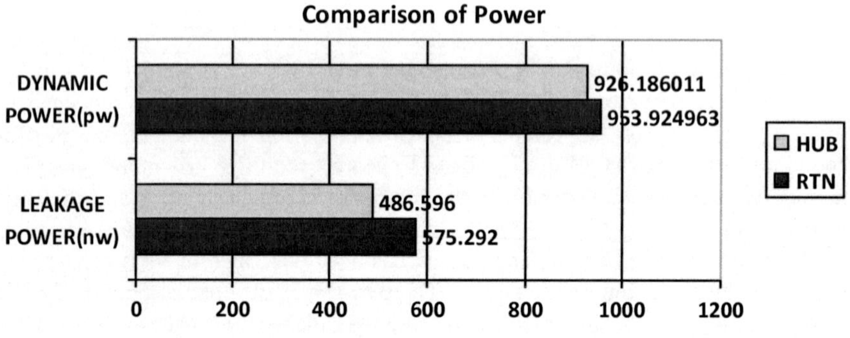

Fig. 5 The comparison of area standard round to nearest and the HUB-based addition

5 Conclusion

HUB format explains the most comfortable way of the Floating-point systems. When the rounding mode is Round to nearest, then implementing an arithmetic function using HUB provides better performance. It has been demonstrated using the cadence tool. Both round to nearest and HUB addition have the same error range, also have the same requirement of storage. This HUB format later has to be further implemented in subtraction, multiplication and division in future work.

References

1. Hormigo J, Villalba J (2014) Optimizing DSP circuits by a new family of arithmetic operators. In: 2014 Asilomar conference on signals systems, and computers, pp 871–875
2. IEEE Task P754, IEEE 754–2008, Standard for Floating-Point Arithmetic, Aug. 2008. [Online]. Available: http://ieeexplore.ieee.org/servlet/opacpunumber=4610933
3. Villalba-Moreno J (2016) Digit recurrence floating-point division under hub format. In: 2016 IEEE 23nd symposium on computer arithmetic (ARITH), pp 79–86
4. Quach NT, Takagi N, Flynn MJ (2004) Systematic IEEE rounding method for high-speed floating-point multipliers. IEEE Trans Very Large Scale Integr (VLSI) Syst 12(5):511–521
5. Hormigo J, Villalba J (2015) New formats for computing with real numbers under round-to-nearest. IEEE Transactions on Computers 99
6. Munoz SD, Hormigo J (2015) Improving fixed-point implementation of Q.R. decomposition by rounding-to-nearest. In: 19th IEEE international symposium on consumer electronics (ISCE 2015), pp 1–2
7. Beuchat J-L, Muller J-M (2005) Multiplication algorithms for radix-2 RN-codings and two's complement numbers, technical report, IRIA. http://www.inria.fr/rrrt/rr-5511.html
8. Hormigo J, Villalba J (2016) Measuring improvement when using hub formats to implement floating-point systems under round-to-nearest. IEEE Trans Very Large Scale Integr (VLSI) Syst 24(6):2369–2377
9. Santoro MR, Bewick G, Horowitz MA (1989) Rounding algorithms for IEEE multipliers. In: Proceedings of 9th IEEE symposium on computer arithmetic, Santa Monica, CA, pp 176–183
10. Ercegovac MD, Lang T (2004) Digital arithmetic. CA, USA Morgan Kaufmann, San Francisco
11. Ercegovac MD, Lang T (1994) Division and square root: digit recurrence algorithms and implementations. Kluwer Academic Publishers, Boston
12. Reiser JF, Knuth DE (1975) Evading the drift in floating-point addition. Inform Process Lett 3(3):84–87

Smart Energy Amassment: Using Piezosensors and Photovoltaic Cell with Two-Sided Mirror Technique

Aman Pandey, Sarsij Mishra, and O. Jeba Shiney

Abstract To be able to harvest energy from the usual day-to-day human activities like walking and running is only a very noble way of fighting environmental crisis and the least of a contribution from the society; this project focuses precisely on the same. Piezosensor is attached to a doormat/tile-like structure to generate current when pressure is applied externally. The system is capable of producing voltage of 80–100 V which could be stored with the help of capacitors or supercapacitors or rechargeable battery. Also, the photovoltaic cells complement in power generation using sun rays with the added two-sided mirror technique, increasing device efficiency with better output. Inclining more toward resources like "solar energy," which is present in abundance, must be the primary agenda of sustainable development. This project proposes an intelligible and judicious way of escalating the statistics of power generation of India and provides an eco-friendly and cost-effective method of helping the large population of the country or even world out of the darkness.

Keywords Piezosensors · Photovoltaic cells · Two-sided mirror technique · Footsteps · Electrified India · Energy harvesting

1 Introduction

The idea of over-consumption and perennial shortage of energy resources in accordance with human needs has become synonymous to the reality of contemporary lifestyle. When the tragedy of over-population stares directly at our faces, even the slightest exhibition of ignorance on our part is only very irresponsible and morally unacceptable. Therefore, a rethinking of our day-to-day lifestyle in this mechanical era is vital. Since every scientific experiment is essentially premised on the singular aim of social amelioration, likewise this project aims to address immediate social concerns by putting forward methods of sustainable development and harnessing of

A. Pandey · S. Mishra · O. J. Shiney (✉)
School of Electrical, Electronics and Communication Engineering, NCR, Galgotias University, Greater Noida, India

A. Sivasubramanian et al. (eds.), *Futuristic Communication and Network Technologies*, Lecture Notes in Electrical Engineering 792,
https://doi.org/10.1007/978-981-16-4625-6_58

energy. This project focuses on generating energy—this energy to be stored in capacitors/supercapacitors or rechargeable battery(s) which can be used in variety of applications. Even we could convert DC to AC with the help of an inverter circuit for its use in AC appliances. The energy produced can be used in low-power-consumption devices like emergency lamps, portable speakers/device charging station, etc. To bring to notice is the fact that the project does not use any conventional energy resource; rather energy from fundamental human activities like walking or running can be channeled to generate electricity using the device/setup. Propounding this idea, this project uses "piezoelectric sensors" to generate electricity. An external pressure on the piezosensor is converted into electricity. This does not require use of any other equipment/material. The increase (or decrease) of the current and voltage required is regulated through parallel and series connections of the piezo, respectively. A mat/tile structure, for example, could be designed with 12 piezoplates in a matrix of 4×3 or with 20 such plates in 4×5 matrix, wherein four piezos would be in series with five other such connections in parallel. The range is in microampere of current generated by one piezosensor. Since the current generated by piezosensors is feeble enough to not cater to the desired range, "solar panels" are used to accentuate the current generation. Voltage generated is in the range of 7–12 V/piezosensor. Photovoltaic cells are used to increase the device efficiency at lower costs. An approach to increase the efficiency of it has been done. Two mirrors are placed at such an angle along the breadth of the panel so as to concentrate even more solar rays onto the panels, thus enabling an increased output than the normal mark. These mats/tiles can be placed in shopping malls, hospitals, universities, railway stations, airports, and many such public places where heavy commuters are expected. The constant pressure on the mats/tiles will enable current generation in continuity and more power could be conserved. Not only as mats, but also the setup which is easy-to-handle and quite portable can be installed in public footpaths, crossover bridges, even in dance academies where excessive physical movement and footwork is a usual thing. The concept of this project is outlined with the need to target the collective conscience of the society, unfortunately becoming apathetic to the alarming environmental situation. This aims to make the society at large, conscious of its role in not only judiciously conserving energy but also creating it. The existence of human species is intricately interlaced and greatly dependent on nature and its resources. As experts agree, the global trend of accelerated depletion of non-renewable resources like coal for example has created an alarming situation. Zooming into the national circumstances, an estimate of 319.02 billion tonnes of geological resource of coal has been recorded as of April 2018. Comparing it to the essential factors, like national population, which is palpably exploding, the many industries predominantly based on coal as their raw material only give us a rather unnerving picture of threat out society faces. Also, to inform the citizen of its responsibility to adopt judicious ways of consuming energy resources over exploitative manners of overuse and thus protect the human species from the dangers of shrinking options of resources. Therefore, a rethinking of developmental processes on global level is necessary. Scientific endeavors along with governmental policies have to focus on bringing non-conventional energy resources to the center stage; to identify the wealth that could be generated from resources discarded away as wastes and put them into use is the need of the hour. This energy harnessed can be used to light up houses of those eighty million people still having to live in darkness in our country. This eco-friendly method of harvesting energy can help in making India a hundred percent electrified nation.

2 Literature Review

Kamboj et al. [3] have presented a design in which they have used wooden board as base to generate power using piezo. According to their table, the output so produced is very feeble. Hard foam could have been a better base to provide more stress for the piezoplates to generate more power as the external force is directly proportional to power produced in the case of piezosensors. Rubber has been used to protect plates, which actually absorbs shocks hence providing lesser stress over the piezoplates and consequently lesser amount of power is generated. Also, no storage circuitry like capacitor has been discussed to shift its non-static nature of output to static. More parallel connections would add up to more amount of current generation for wider use. Bridge rectifier would have been a better replacement than normal diode for appropriate DC output [4]. In [5], the authors have considered a surface to provide appropriate amount of stress for piezo to produce better output but clearly have missed a factor that the surface above it should allow uniform pressure over the plates to generate effective results. This could be done by applying a hot glue over the center of every piezoplate as the efficiency is maximum at the center and place a hard surface above it for uniform pressure at each and every plate. The output produced with the help of the proposed model will provide feeble energy generation. A hard foam base is required to provide more stress for the piezoplates in order to generate more power. Also, full-wave bridge rectifier using Schottky diode would provide better DC output with low voltage drop than simple diode. A capacitor in parallel to the rectified will help in smoothening of the signal as well as provide better static output [6]. A storage circuitry is not mentioned which is essential to provide a static output in order to be further utilized by device like LTC3588 [7]. Without it, it will not be possible to provide a static input to the buck converter, resulting in no output. The current produced by piezosensor is in the range of micromilli amperes and also is only for a fraction of second, momentarily. A storage unit with capacitor or battery is required before giving the output out of diode directly to any load or module for further enhancement of the power produced.

3 Proposed System

See Fig. 1.

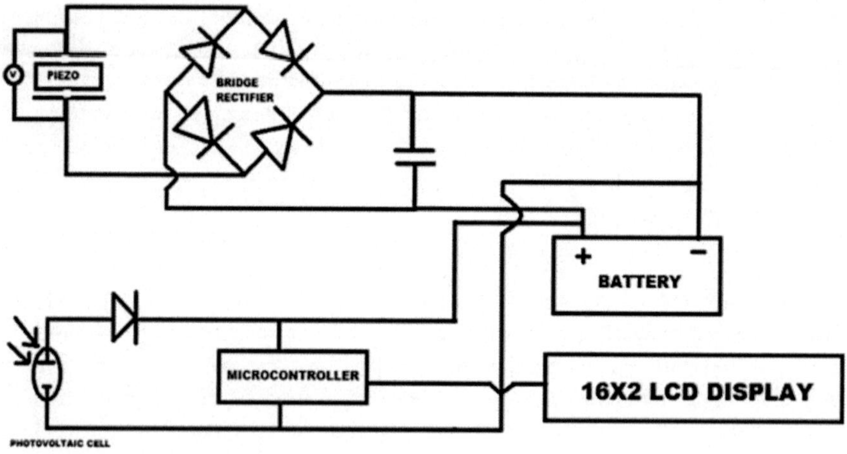

Fig. 1 Proposed block diagram

4 Methodology

This project of harvesting energy is based on two systems which is piezosensor and solar energy-based system. The more focused source of conserving energy in this project is piezosensor which when supplied with external pressure generates AC signal. This AC signal is then converted into pulsating DC signal by passing it through a full-wave bridge rectifier. This is made using four Schottky diode because it provides low voltage drop. The DC signal produced after having it pass through the bridge rectifier is, however, for a fraction of second and not enough to influence even a reasonable energy generation. Thus, this output is fed into a capacitor. A capacitor not only smoothens the resultant DC signal, but is also conducive to generate a current production in continuity.

This project includes the usage of two capacitors of different values of capacitance, for a comparative analysis; one with a capacitance of 470 μF and the other of 1000 μF. The comparison of the respective readings of the two capacitors creates a space for comparative performance analysis. A piezosensor operates through external pressure application so the type of material that is used as the base of the sensor is decisive of the efficiency output of the whole device. It is through a series of experimentation that "hard foam" as the most appropriate material for the base of the piezo has been included in this project.

With wooden board as the base was recorded to generate output of a rather lower reading than desired. To check for any difference in the reading of current generation, the wooden base was then replaced with rubber. But since the insulation of a material like rubber is high enough to absorb all the shock/suspension, it did not create the stress between the plates and the surface to generate required output. Hence, a material like "hard foam" which absorbs the external pressure in the right amount, enabling a stress between the surface and piezo and consequently producing desired

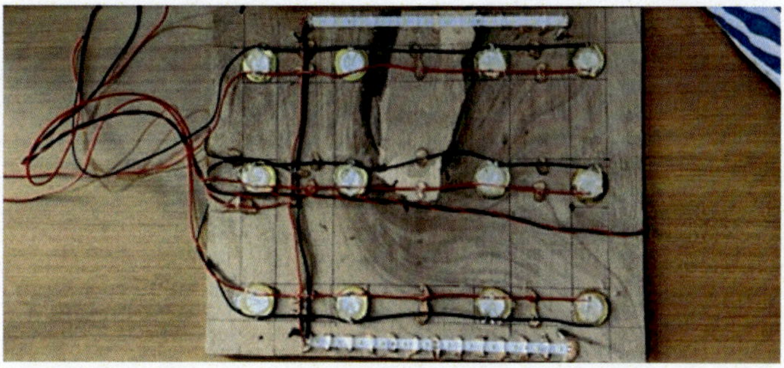

Fig. 2 Piezoplates matrix setup over a wooden base

output, is used. This reading when juxtaposed with the readings of the previous two cases—wooden board and rubber—is higher (Fig. 2).

The piezo itself acts as a parallel plate capacitor and has an internal resistance which plays against the motive of maximum current generation. To counter this internal resistance, the piezo has been put together in a matrix of more parallel connections. To enable more current generation, the matrix has more parallel connections than series connections because series connection contributes to more voltage generation and not current production. The piezo design that this project is based on is put into a matrix of 6×9. This model of it being only a prototype and caught within the limitation of academic exhibition has the potential to be further developed into a larger system with a larger matrix, which can even touch the mark of unit ampere. To extract the maximum efficiency from the piezosensor, hot glue is fixed at the center of every piezo which is producing maximum output when stimulated with external pressure than the peripheral regions of the sensor. However, a factor that poses challenge to the idea of consumer-friendly model is the cost of the piezosensor. Developing this prototype into a larger system with a larger matrix, for example, 6×100 or 6×200, produces higher reading of current and would also be very expensive and inaccessible to the larger mass. Therefore, a "solar energy"-based system as a more consumer-friendly source of energy has been incorporated into this project. Not only does this complement in sourcing current generation but also takes forward the imperative notion of sustainable energy harvesting. In times of shrinking options of energy production due to the overuse of the limited conventional resources, solar energy is the solution to proceed with a sustainable scientific development; solar is the future. Including it as one of the sources to compensate with the current generation reduces the production cost of the system to a palpable extent. However, losses in the energy production of solar panels are bound to happen. Wherein the efficiency of solar panel in laboratory is recorded around 40%, it can barely touch the range of 25% in normal environmental conditions (usually between 10 and 25%) [2]. But there are various techniques to curb this loss of energy, like solar tracker, solar concentrator, panel orientation, etc. The technique of solar concentrator conventionally uses "Fresnel lens," which again are expensive. Deriving from this very technique but using mirrors instead of Fresnel lens, the loss of energy in solar panels is compensated for, here. Based on "double sun technique," the effective area or the area exposed to the sun is increased. This is done by placing two mirrors along the sides of the solar panel, thus enabling more solar rays to be concentrated on the larger area. It is calculated by the ratio between the effective area (the area seen by the sun) and the active cell area as defined by the following equation:

$$C = A_{\text{eff}}/A_{\text{module}};$$

where A_{eff} is the effective area exposed to sun

$$= (A_{\text{mirrors}} + A_{\text{module}})/A_{\text{module}}$$

where A_{mirrors} is the area of the mirrors and A_{module} is the total area of the module.

$$C = \mathbf{1} + (\mathbf{A_{mirrors}}/\mathbf{A_{module}}) \tag{1}$$

This increases the "geometric concentration factor" and pushes the efficiency of the solar panel by 10–15% than usual.

5 Case Study and Performance Analysis

5.1 Power Generated at Different Voltage–Current Level Reached

At different stages in this project, variable voltage–current generation took place ranging from 20 to 100 V and 0.05–0.25 mA, respectively. Taking five different readings into account, power generated at different stages has been calculated using the formula (Figs. 3 and 4):

$$\mathbf{P} = \mathbf{V} \cdot \mathbf{I} \tag{2}$$

5.2 Time Required to Charge a Super Capacitor with the Help of this System

It took 4 steps (approx.) to fully charge a 470 μF capacitor. The result shows that it will take 40,000 steps (approx.) to fully charge a 4.7 F supercapacitor. The other case study showed that it took 20 steps approx. to fully charge a 1000 μF capacitor. Studying both the results we can say that on steps count in the range of 20–50 K, the

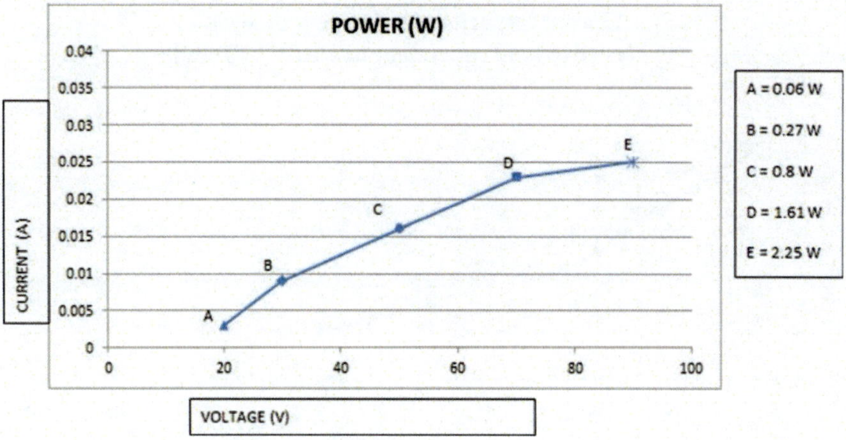

Fig. 3 Power generated at different levels

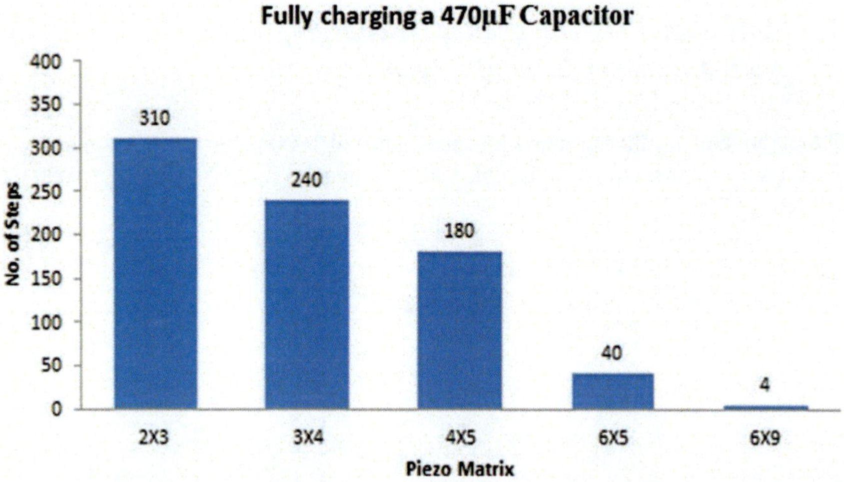

Fig. 4 Complete charging of a 470 μF capacitor

piezosensors will be able to successfully charge a supercapacitor of 1–5 F value. For the first case study let us consider that,

$$1\ \text{s} = 2\ \text{Steps}$$

therefore, $40{,}000\, steps = 20{,}000\ \text{s}$

This time in hours will be 5.5 h. Approximating the hours taken to 6–7 h, we can say that, the time required to charge a supercapacitor of 1–5 F is 6–7 h.

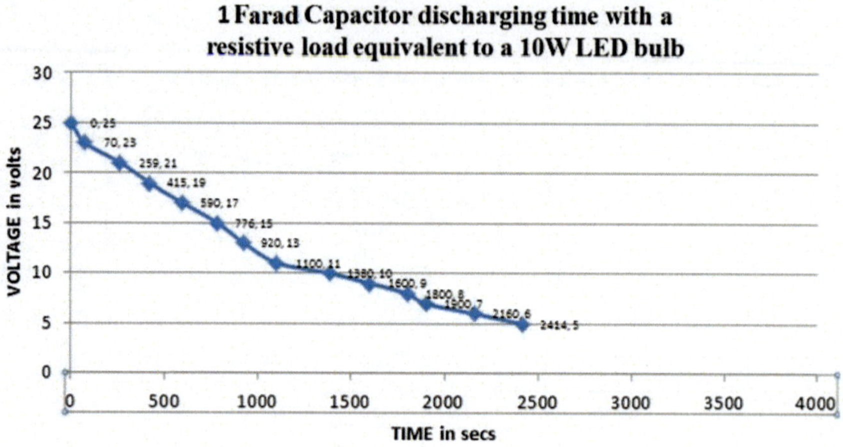

Fig. 5 Unit Farad capacitor discharging time with a resistive lead equivalent to a 10 W LED bulb

5.3 Unit Farad Supercapacitor Discharging Time with a Resistive Load of a 10 W LED Bulb

It took 50–60 s for this system with capacitance of 1000 μF to reach 5 V mark.

if 1 s = 2 steps, therefore, it took 100–120 steps to reach 5 V mark in 1000 μF capacitor (Fig. 5).

The initial voltage considered is 25 V, and the safety threshold voltage is 5 V. The resistive load of a 10 W LED bulb is equivalent to 1500 Ω.

Formula used for calculation of this graph is:

$$V_C = V_0 \, e^{-t/RC} \tag{3}$$

where

V_c = voltage across capacitor, V_0 = initial voltage, t = time elapsed, R = resistance, C = capacitance.

The time taken by the load to discharge capacitor to threshold voltage is 2414 s which is approximately 40 min. Hence, we can conclude this study by saying that within 6–7 h of placing this system in active mode one can conserve the amount of energy which is sufficient to light a 10 W LED bulb for good 40 min.

5.4 Piezoelectric Energy Calculation

Considering 1000 μF capacitor into account with the system working for 1 min, it will be able to reach up to the mark of 5 V (as per previous case study).

So, piezoelectric energy of the system would be calculated using the same formula as that of capacitance energy in low frequency condition, due to its working in such conditions is similar to that of a parallel-plate capacitor;

$$U = \frac{1}{2} C V^2 \tag{4}$$

where C = capacitance, V = voltage, U = piezoelectric energy.

Using this formula, we get:

$$U = 12.5\,\text{mJ} \tag{5}$$

5.5 *Solar Energy Calculation with Double Sun Technique and Comparison with Flat Solar Panels*

As explained in and using Eq. (1) from above we can calculate the geometric concentration factor (C) (Fig. 6):

$$A_{\text{mirrors}} = 2X(10 * 20.5)\,\text{sq. inch} = 410\,\text{sq. inch}$$
$$A_{\text{module}} = (12.5 * 18.5)\,\text{sq. inch} = 231.25\,\text{sq. inch}$$
$$C = 410/231.25 = 1.77x$$
$$\mathbf{C \approx 2x}\ \text{(approx.)}$$

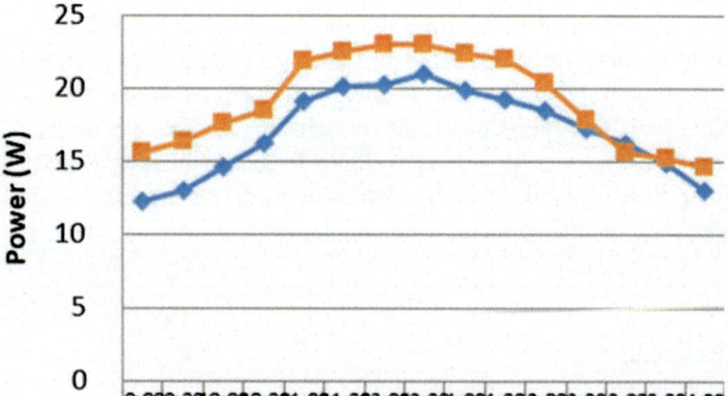

	9:00 AM	9:30 AM	10:00 AM	10:30 AM	11:00 AM	11:30 AM	12:00 PM	12:30 PM	1:00 PM	1:30 PM	2:00 PM	2:30 PM	3:00 PM	3:30 PM	4:00 PM
Flat Solar Panel	12.2	13	14.6	16.2	19.1	20.1	20.2	20.9	19.8	19.2	18.4	17.3	16.2	14.8	13
Double Sun Technique	15.6	16.3	17.5	18.4	21.8	22.5	23	23	22.3	22	20.3	17.8	15.6	15.2	14.6

Fig. 6 Power versus daytime scatter-line plot of flat solar and double sun solar panel

$$\text{Avg. Power (Flat Solar Panel)} = 16.99\,\text{W}$$

$$\text{Avg. Power(Double Sun Technique)} = 19.05\,\text{W}$$

We can clearly see from the graph that double sun technique provides 12.14% more power as compared to the base average power from flat solar panel. Furthermore, this can be increased by installing a setup prepared by small concave mirror on a panel to concentrate more sun rays on solar panels, increasing the efficiency by more percentage, but that would increase the cost too.

6 Conclusion

This paper positions itself around the notion of development through non-conventional and renewable energy resources. Piezoelectric sensor provides an intelligible alternative to the declining reserves of non-renewable resources like coal. Piezo has the potential to furnish a better energy storage module by amassing energy from regular human activities like walking and running. The installation of piezosensors in heavy commuting spaces like railway stations, bus stands, gyms, malls, shopping complexes will create an energy reservoir, much needed to meet the ever-growing demands. As proposed, developing the prototypical piezo model with increased number of parallel connections will lessen the internal resistance of the device and make it more widely accessible. The piezoelectric system paired with photovoltaic cell which harvests solar energy using the technique of solar concentration is a better substitute for pollution-inducing coal and challenges its popularity as a cheap energy resource. This piezoelectric system apart from being a cleaner, noiseless, and eco-friendly means of energy production can be made even more economical than coal. An official news release states that India's electricity consumption would increase up to 4 trillion units by 2030. To secure the future and to be able to meet the enormous demands, adoption of non-conventional methods of energy production, like piezoelectric system, is essential.

References

1. Donev JMKC et al (2018) Energy education—solar cell efficiency [Online]. Available https://energyeducation.ca/encyclopedia/Solar_cell_efficiency. Accessed 24 Apr 2020
2. Kamboj AH, Kumar A, Sharma VK, Kumar A (2017) Design of footstep power generator using piezoelectric sensors. In: 2017 international conference on innovations in information, embedded and communication systems (ICIIECS), Coimbatore, pp 1–3

3. Panthi N, Gupta A, Baby E, Santhosh N (2018) Footstep energy harvester using piezoelectric transducer. Int J Latest Res Sci Technol 3:54–57
4. Pisharody HG (2011) An optimal design for piezoelectric energy harvesting system. In: ISGT2011-India, Kollam, Kerala, pp 244–247
5. Dubey CK et al (2016) Power generation using piezoelectric transducer. Int J Eng Res Appl 6(5):71–73. ISSN: 2248-9622
6. Patil A, Jadhav M, Joshi S, Britto E, Vasaikar A (2015) Energy harvesting using piezoelectricity. In: 2015 international conference on energy systems and applications, pp 517–521. IEEE
7. Carazo AV (2000) Novel piezoelectric transducers for high voltage measurements. Universitat Politècnica de Catalunya, p 242

Substrate-Integrated Waveguide filters—A Review

R. Boopathi Rani, P. Reshma, and S. Ramkumar

Abstract Substrate-integrated waveguide refers to a planar transmission line with parallel metallic vias to inherit the advantages of waveguide. It compliments the merits of both planar and waveguide technologies to some extent. Manufacturing of waveguide is difficult and it is bulky. Microstrip devices are lossy and it has less power handling capability. Small size, less weight, better power handling capability and easy manufacturing are the salient features of SIW technology. In this paper, a review of SIW filters is presented which were designed using SIW technology.

Keywords Substrate-integrated waveguide · SIW · Filter

1 Introduction

A novel concept of integrated circuits that can be used at high frequencies was proposed by Wu in the year 2003 which is known as the substrate-integrated waveguide (SIW) technology. SIW technology has features like high Q-factor, low cost, low loss, simple structure, and mass producible, which make it more popular over standard waveguide technology. SIW structure is similar to that of a planar dielectric substrate with two arrays of metallic vias in a parallel manner connecting the layers of the substrate as shown in Fig. 1. The two walls of via-fence connect the two layers of the substrate. The SIW parameters are as following: W_{eff}: SIW effective width, h: Substrate height, ϵ: Substrate dielectric constant, d: Metallic via diameter, P: Space between two consecutive vias.

R. Boopathi Rani (✉) · P. Reshma · S. Ramkumar
National Institute of Technology Puducherry, Karaikal, India
e-mail: rbrani@nitpy.ac.in

A. Sivasubramanian et al. (eds.), *Futuristic Communication and Network Technologies*, Lecture Notes in Electrical Engineering 792,
https://doi.org/10.1007/978-981-16-4625-6_59

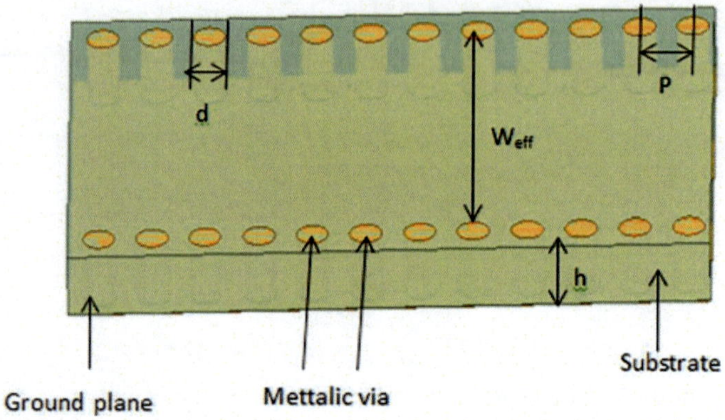

Fig. 1 Structure of SIW

While calculating the parameters listed above, we need to consider some constraints. SIW filter basic design includes parameter calculation such as the calculation of width, which depends on the SIW cut-off frequency. Calculations of the parameters W_{eff}, d, p, h can be chosen possibly low for designing the SIW as the waveguide cut-off frequency, solely dependent on the width of the guide. Relationship between the width and cut-off frequency for TE_{10} mode as the dominant mode is given by

$$f_{10} = \frac{c}{2w\sqrt{\varepsilon_r \mu_r}} \tag{1}$$

The effective width and width are related as

$$W_{\text{eff}} = w - 1.08 * \left(\frac{d^2}{p}\right) - 0.1 * \left(\frac{d^2}{w}\right) \tag{2}$$

SIW structure is different from rectangular waveguide as the modes are not identical in both cases; the only modes that are present in SIW structure are TE_{m0} modes as TM modes couldn't be extracted from the surface current distribution, so TM modes don't exist in the SIW. The feeding mechanism is another essential factor to be considered for the SIW filter design, and feeding techniques are developed accordingly. Conductor losses, dielectric loss, and leakage loss are the few losses present in SIW.

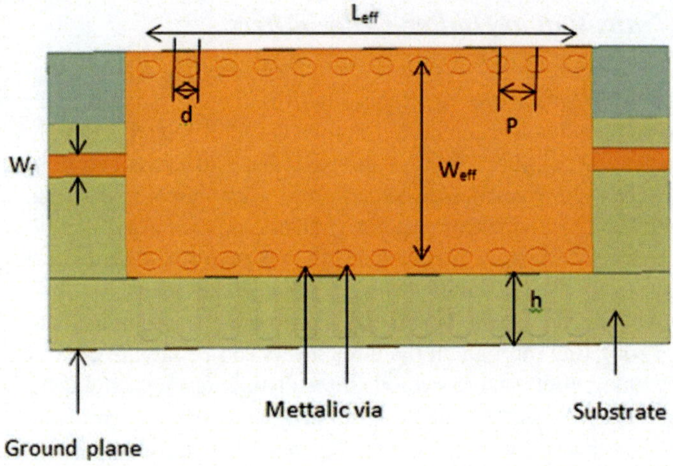

Fig. 2 Basic model of SIW filter

2 Design Process of SIW Filter

The filter design model includes calculation of parameters listed as in equations. From Fig. 2, the different geometrical parameters to be calculated are as follows: W_f = microstrip feed line width, d = via diameter, P = via spacing, L_{eff} = SIW length, W = SIW width, W_{eff} = SIW effective width. Modified taper or tapered via is used for transition from MSL (microstrip line) to SIW.

3 SIW Filter Design Enhancement and Usage

A recent study tells about the new filter designing using SIW (substrate-integrated waveguide) technology, as they exhibit low cost and ease of fabrication. Though the stopband characteristics are better in single-layer SIW bandpass filters, they are not considered as they require a rectangular cavity, and it increases the size because of this reason as larger the size it results in the square cavity for the identical fundamental frequency. Still, we opt this SIW technology to construct miniaturized filters, we need to fabricate small size, and with large bandwidth. Band accommodation is also required. For this, a lot of size reduction techniques are employed by including folded structure [1].

3.1 A Transition to Higher-Order Filters

The author used SIW cavities to present a fourth-order cross-coupled bandpass filter with wide-stopband designed and to improve the filter selectivity, realize the transmission zeros (TZs) at both sides of the passband with cross-coupling topology is adopted [2]. The authors explained how a fourth-order filter obtained by including two planar resonators loaded with capacitors (transmission zeros are produced by signal interference from resonators) are induced in a two-cavity SIW structure that results in quasi-elliptic response [3]. Filter design with three bands using an ultra-miniaturized SIW cavity operated at evanescent mode is proposed by author, a circular patch that is identical trisecting through all the three layers can be seen in the design, and the host cavity along with patches exhibits three distinct resonant frequencies to create the filter's three passband frequencies, six passband-edge TZs are created for good selectivity. Resonance frequencies are independently controlled [4]. To suppress the modes in higher order of the SIW cavity, author proposed two methods, one is that the narrow slots to be etched on the metallic plate of cavities, and the other is to replace single iris with double coupling irises between two cavities [5]. Without an increase in size, filters of higher order with fan-shaped miniaturized half-mode SIW resonator is designed, with a SIW cavity in a semicircular shape and along its circumference-metalized vias are distributed and including one or more distinct length radial slots. Independently, we can control the resonant frequencies of half-modes, enable second- or third-order filter design with maintainable bandwidth and wide stopband [6]. Using the coplanar-to-SIW connection, a SIW Chebyshev bandpass filter is designed as two different stages of impedance connection is involved, and the transition of low impedance at the SIW end yields the filter Q_{ext} easily [7]. Using electric and magnetic couplings technology, the SIW bandpass filter with order 6 is designed. SIW cavity filter metal plane is properly etched with slots, and resonators formed, electric coupling created by slots, metallic vias create magnetic coupling. Besides the coupling achieve two TZ points, lower and upper transmission zeros are influenced by magnetic and electric coupling [8]. Odd-order response SIW filter design is made here with combining stripline and SIW technologies filter consists of one stripline resonator and two SIW cavity resonators to form passband and suppressing the rest frequencies to enhance the selectivity, two extra filters with cross-couplings are included, which generates transmission zeroes (TZs) [9]. The dielectric substrate is perforated so to change the SIW filter topology, and from this principle, the author modified the SIW structure in some of the regions to be perforating the substrate such that the region in between the perforated and the performed region behaves as resonant cavities and impedance inverters, respectively [10]. A combination of SIW technology along with a linear phase filter has been implemented using the cross-couplings, as it can be used for millimeter and microwave integrated circuits, respectively [11].

3.2 Frequency Enhancement

Pelluri et al. proposed a dual-band filter structure with an E-shaped DGS (defected ground structure) that is patched on cavity as a top layer that results in a low-frequency passband range with radiation loss and roll-off rate as low and high values, respectively. TE_{101} and TE_{102} are cavity's lower modes, which achieve high-frequency passband by mode shifting with slot perturbation, this gives a compact and broadly separated passbands filter designs [12]. Ismail et al. studied about a SIW bandpass filter (BPF) with wideband which was incorporated with two complimentary-split-ring-resonators (CSRRs) arranged reversely. Now, the surface of SIW is integrated with CSSRs with the higher frequency region due to extended passband area, yielding the wideband response of filter [13].

Lee et al. implemented a practical method for SIW filter, which is post-loaded, produce a response that is asymmetric with a pair of transmission zeros at the near band, practical coupling structure which is frequency-dependent for asymmetric response post-loaded SIW resonators was designed [14]. By replacing a tunable capacitor with a negative impedance converter (NIC), a new non-Foster filter is developed through which a wideband, compact, SIW filter with half-mode using internal non-Foster element was designed [15]. Dual-band dual-mode SIW filter with suppressing the spurious passband (it can be achieved through coupling lines, as they are loaded on feeders at input and output) is designed by Sun et al., using rectangular SIW by tuning its size according to the requirement, through that we can determine the passband [16].

Through E-shaped slot lines, a dual bandpass filter with SIW cavities etched edge sides with resonant modes that are coupled with magnetic fields where the cavity operates below its fundamental resonant frequency has been designed [17]. Arranging the coupling region and feeding ports, integration of bandpass responses was done along with isolation between different channels. Through this, the design of SIW filter with multi-channel crossovers with control in the frequencies, and filter channels can be extended to produce intersecting signals two sets of each easily, extension of channel number is similar to single-channel bandpass filter [18].

3.3 Implementing with Different Modes

Huang et al. proposed a SIW technology using quarter-mode packaged bandpass filter with four TZs, attenuation of stopband filter was increased using the design methodology of transmission zeros (TZs). High-mode suppression of SIW resonators was observed properly by setting their feed points [19]. Using a wide-resonance SIW resonator structure, a compact SIW DBPF was designed [20]. A fully low-pass prototype with extracted pole synthesis where zeros and poles are equal and the left, there exists one reflection zero at infinity. Using microstrip open stubs, Macchiarella et al. obtained the extracted poles through this a half-mode SIW filter with a single-sided extracted pole that is designed [21]. A comparison between half-mode and folded SIW filters was studied to design a compact cavity half-mode SIW filter at 4.5 GHz [22]. Filter with loaded varactors that is reconfigurable using S-CSRR was designed that has invariable fractional bandwidth and feed problem of capacitance diodes can be solved using non-metallic vias in the center of S-CSRR, and by changing the dc bias applied to the varactors, bandpass filters center frequency can be tuned electronically [23].

3.4 Harmonic Mode Suppression

Zhaosheng He et al. proposed a hairpin-loaded compact SIW bandpass filter (hairpin structure increases effective capacitance) defected ground structure (DGS). Thus, the proposed system design owns a larger Q factor, which gives a better band-rejection sharpness and deepness performance. Combined band-rejection characteristics of DGS, and the high-pass characteristic of SIW, the performance of the filter was good in both stopband and passband [24]. We can achieve a passband which is wider in range and bandwidth with a large rejection ratio through the integration of the SIW filter with pre-filter, thereby suppressing the spurious almost thrice the passband frequency considered and designed [25].

3.5 SIW Filter Design for Different Applications

The transition of microstrip to SIW bandpass filter design and fabrication was explained by Güvenli et al. and can be used in X-band and Ku band applications [26]. To generate modes of higher-order cruciform slot, the top layer was etched properly to cut off the surface current, which was used to design SIW bandpass filter (BPF) with two structures. We can achieve an increase in resonance frequency and a decrease in the size of the filter structure through slot and vias, this design can be used in mm-wave (millimeter wave) applications [27]. Introducing planar resonators within the cavity can increase the bandwidth of filter, and resonators that are embedded create a single-cavity filter with two-stage effect, through this 3D design of integrated system on package (SoP) low-temperature Co-fired c (LTCC) two-stage SIW single-cavity filters with planar resonators embedded can be used in 5G applications and beyond [28]. SIW filter to work in Q-band was fabricated using LTCC with 8 layers, i.e., multilayer technique, designed in a manner that is feasible for mm-wave (millimeter wave) in filtering stage used in Q-band millimeter wave communications [29]. The transition of microstrip to air-filled SIW filter as it reduces effectively the insertion loss is observed and balun bandpass filter using the SIW filter is designed, which can be used for X-band RADAR applications and communication systems [30]. For wireless communications, the system designs are presented in [13, 23]. Development of integrated circuits system design can be seen

Table 1 Comparison of SIW filters with different design techniques

Ref./filter type	F (GHz)	IL (dB)	RL (dB)	FBW (%)	Size	Techniques/approaches
[2]/BPF	10	2.1	>10	3.5	–	Cross-coupling
[3]/BPF	2.35	1.65	<–20	5	1517mm^2	Capacitor-loaded planar resonator
[4]/LPF	2.1	2.31	–	–	$0.22 \times 0.22\ \lambda_g{}^2$	Multiple SIW cavities
[7]/BPF	1.684	1.3	>15	4	–	Chebyshev
[12]/BPF	5.37	1.35	20	6.9	–	E-shaped DGS
[21]/BPF	12	<1	>40	–	6.5×21.7 mm^2	Half-mode SIW
[26]/BPF	11.016	<3	>10	13.9	20×30 mm^2	Microstrip transition
[28]/BPF	28.21, 28.12	0.82, 0.53	>10, >10	13, 15	33.5 mm^2	Embedded planar resonators in LTCC package, feed is given in center post and middle of the resonator
[29]/BPF	39	1.954	>10	–	4.382 mm^2	LTCC technology with 8 layers
[30]/BPF	11	3.9	>4	8.6	–	TE_{20} mode air-filled substrate-integrated waveguide (AFSIW)-based balun bandpass filter

F—Frequency; IL—Insertion Loss; RL—Return Loss; FBW—Fractional Band Width

in paper [17]. The comparison of SIW filters, its frequency of operation, Insertion Loss, Return Loss, Fractional Bandwidth and their design techniques are presented in Table 1.

4 Advantages and Disadvantages

Advantages of SIW technology are as follows which helps in designing and fabrication of SIW filters effectively

- Power handling capability is high.
- Radiation losses are minimal.
- Reduction in fabrication cost.
- Conduction losses are low as metal vias are used.

Disadvantages of SIW technology are

- In the case of mm-wave (millimeter wave) applications, one needs to be cautious while designing as it is dependent on frequency.
- Dielectric losses arise as the dielectric is used in its structure.
- SIW exhibits lower values of cut-off frequency as its structure is of the waveguide.
- One should take care while designing the separation of via holes that are set on the top planes, as they induce the leakage losses.

5 Conclusion

SIW filters, the way they can be used and designed as per today's requirement of compact size and properties, are discussed briefly. The key features of SIW technology are easy integration, low cost, and high Q factor. Besides this, the coaxial SIW technology induces flexibility, tunability, and compactness. With this technology, filters, resonators with simple and advanced structures have been designed. By including some additional building blocks, we can implement dual-mode filters. The use of cavity resonators suppresses the higher-order modes. The capacitive part of the resonator helps to increase the tunability of filter by using tuning elements like trimmers, varactors, and disks. Currently, SIW is a promising technology which made tremendous contribution to the characteristics of microwave devices including filters. Finally, the paper concludes that SIW filters have shown better characteristics and the lot more improvements are yet to explore.

References

1. Martinez JD, Sirci S, Boria VE, Sanchez-Soriano MA (2020) When compactness meets flexibility: basic coaxial SIW filter topology for device miniaturization, design flexibility, advanced filtering responses, and implementation of tunable filters. IEEE Microw Mag 21(6):58–78
2. Azad AR, Mohan A (2020) Substrate integrated waveguide cross-coupled bandpass filter with wide-stopband. In: URSI regional conference on radio science (URSI-RCRS), pp 1–4
3. Dhwaj K, Hadi R, Yang T, Jiang LJ, Kovitz J, Itoh T (2018) Planar resonator embedded substrate integrated waveguide (SIW) cavity filter. In: 48th European microwave conference, pp 761–764
4. Ho M-H, Tang K-H (2020) Miniaturized SIW cavity tri-band filter design. IEEE Microw Wirel Compon Lett 30(6):589–592
5. Ma L, Li J, Zhou Y, Sun X, Honghao Zhu N (2018) Substrate integrated waveguide filter with higher order modes suppression. In: International conference on microwave and millimeter wave technology, pp 1–3
6. Xie H-Y, Wu B, Xia L, Chen J-Z, Su T (2020) Miniaturized half-mode fan-shaped SIW filter with extensible order and wide stopband. IEEE Microw Wirel Compon Lett 30(8):749–752
7. Nwajana AO, Yeo KSK, Dainkeh A (2016) Low cost SIW Chebyshev bandpass filter with new input/output connection. In: 16th Mediterranean microwave symposium, pp 1–4
8. Dong M, Shen D, Ma C, Zhang X (2017) A cascaded six order bandpass SIW filter using electric and magnetic couplings technology. In: Sixth Asia-Pacific conference on antennas and propagation, pp 1–3

9. Zhu Y, Dong Y (2020) Stripline resonator loaded compact SIW filters with wide suppression and flexible response. IEEE Microw Wirel Compon Lett 30(5):465–468
10. Silvestri L, Massoni E, Tomassoni C, Coves A, Bozzi M, Perregrini L (2016) Modeling and implementation of perforated SIW filters. In: 2016 IEEE MTT-S international conference on numerical electromagnetic and multiphysics modeling and optimization, pp 1–2
11. Chen X, Hong W, Cui T, Chen J, Wu K (2005) Substrate integrated waveguide (SIW) linear phase filter. IEEE Microw Wirel Compon Lett 15(11):787–789
12. Pelluri S, Fasil M, Mondal D, Kartikeyan MV (2020) Dual-band bandpass filter using SIW cavity with E-shaped DGS. In: URSI regional conference on radio science, pp 1–3
13. Ismail N, Siregar RA, Nusantara H, Munir A (2018) Wideband substrate integrated waveguide BPF incorporated with complementary split ring resonators. In: Progress in electromagnetics research symposium, pp 1134–1137
14. Lee S, Nam S, Lee J (2020) Practical design and implementation method for asymmetric-response post-loaded substrate-integrated waveguide filters. IEEE Access 8:101528–101538
15. Shi T, Tang M-C, Ziolkowski RW (2018) The design of a compact, wide bandwidth, non-foster-based substrate integrated waveguide filter. In: IEEE Asia-Pacific conference on antennas and propagation, pp 54–56
16. Sun J, Zhang D, Liu Q, Wang X (2018) Dual-mode dual-band substrate integrated waveguide bandpass filter with improving the spurious passband. In: International conference on microwave and millimeter wave technology, pp 1–3
17. Zhang H, Kang W, Wu W (2018) Miniaturized dual-band SIW filters using E-shaped slot lines with controllable center frequencies. IEEE Microw Wirel Compon Lett 28(4):311–313
18. Zhan W-L, Xu J-X, Zhao X-L, Hu B-J, Zhang XY (2020) Substrate integrated waveguide multi-channel filtering crossover with extended channel number and controllable frequencies. IEEE Trans Circ Syst II: Express Briefs
19. Huang X-L, Zhou L, Völkel M, Hagelauer A, Mao J-F, Weigel R (2018) Design of a novel quarter-mode substrate integrated waveguide filter with multiple transmission zeros and higher mode suppressions. IEEE Trans Microw Theory Tech 6(12):5573–5584
20. Umma Habiba H, Prashanth TV, Keerthipriya S, Naveed Ahamed Sayeed L, Sandhya R (2017) A compact full mode SIW UWB Bandpass filter using novel input/output transmission-line-structure. In: IEEE international conference on computer, communication, and signal processing, pp 1–3
21. Macchiarella G, Tomassoni C, Massoni E, Bozzi M, Perregrini L (2017) A novel class of half-mode SIW filters with extracted poles. In: 47th European microwave conference, pp 807–810
22. Moscato S, Delmonte N, Silvestri L, Bozzi M, Perregrini L (2015) Half-mode versus folded SIW filters: modeling and design. In: IEEE MTT-S international conference on numerical electromagnetic and multiphysics modeling and optimization, pp 1–3
23. Yue H, Zhang X, Wei F, Song J, Shi X (2018) A half mode substrate integrated waveguide reconfigurable bandpass filter based on S-CSRR. In: International conference on microwave and millimeter wave technology, pp 1–3
24. He Z, Zheng L, Chen W (2018) A compact substrate integrated waveguide bandpass filter with harmonic suppression. In: International conference on microwave and millimeter wave technology, pp 1–3
25. Tomassoni C, Delmonte N, Macchiarella G, Bozzi M, Perregrini L (2018) A technique for spurious suppression in substrate integrated waveguide filters. In: 48th European microwave conference, pp 753–756
26. Güvenli K, Yenikaya S, Seçmen M (2018) Design and implementation of substrate integrated waveguide filter to work on X-Band and Ku-Band. In: 9th International conference on ultrawideband and ultrashort impulse signals, pp 198–200
27. Peng W, Xiao Q, Zhang J (2018) Design of compact SIW bandpass filter for millimeter-wave applications. In: International conference on microwave and millimeter wave technology, pp 1–2
28. Showail J, Lahti M, Kari K, Arabi E, Rantakari P, Huhtinen I, Vähä-Heikkilä T, Shamim A (2018) SIW cavity filters with embedded planar resonators in LTCC package for 5G applications. In: 48th European microwave conference, pp 757–760

29. Hizamel M. Hizan, Zulkifli Ambak, Azmi Ibrahim, Mohd Zulfadli Mohamed Yusoff.: Q-Band Millimeter-Wave SIW Filter Using LTCC Technology. IEEE Asia-Pacific Conference on Applied Electromagnetics, pp. 199–202 (2014).
30. Adhikary M, Sarkar A, Sharma A, Biswas A, Akhtar MJ (2018) TE20 mode air filled SIW based Balun Bandpass filter. In: IEEE international symposium on antennas and propagation, pp 1–2

Frequency Selective Structures and Its Applications

R. Boopathi Rani, T. N. Sreedhar Kumar, and S. Padmavathi

Abstract Frequency Selective Structures (FSS) have made revolutionary changes in the size and characteristics of many of the microwave devices. This paper presents the short review on FSS and its widespread applications. Initially, FSS was used in the place of reflector in parabolic antenna. Then, significance of FSS in antenna was explored much by changing the structures in it and by making it in 3-dimensional and changing its materials such as permittivity of the substrate, etc. The FSS is prominently used in Satellite Communications where the antenna number and size reduction play a vital contribution. The FSS can also increase gain, reduce size and also plays a key role in communication mainly 5G antennas and also for terahertz region. The number of researches on FSS is keep on increasing. It is also used in sensor applications like detection of strain/crack on the material. Hence, this paper reports the review on FSS.

Keywords Frequency selective structures · Antennas · Applications

1 Introduction

Frequency Selective Structures in short form famously called as FSS. In 1919, Marconi and Franklin got patents for the invention of FSS. In 1957, FSS was first used in antenna to reduce the radar cross section. First FSS was a 2-dimensional FSS which can be printed on a dielectric surface or etched out of a conductor surface [1] as shown in Fig. 1.

A 3-dimensional FSS can be formed with two 2-dimensional FSS of multimode cavity made by a substrate integrated waveguide (SIW) structure [1] which is shown in Fig. 2.

R. Boopathi Rani (✉) · T. N. Sreedhar Kumar · S. Padmavathi
National Institute of Technology Puducherry, Karaikal, India
e-mail: rbrani@nitpy.ac.in

A. Sivasubramanian et al. (eds.), *Futuristic Communication and Network Technologies*, Lecture Notes in Electrical Engineering 792,
https://doi.org/10.1007/978-981-16-4625-6_60

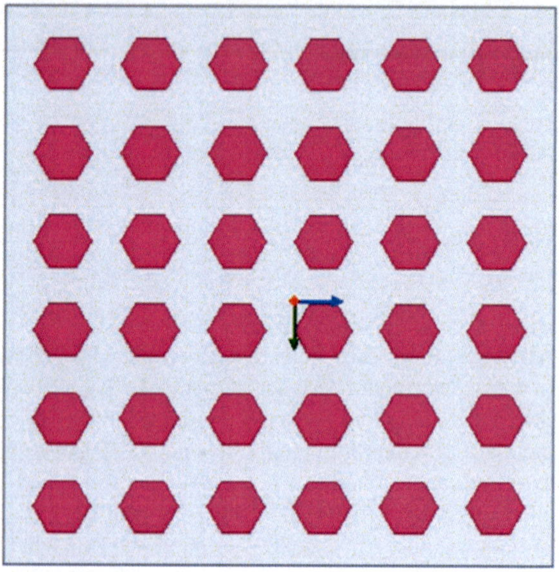

Fig. 1 Model of 2-dimensional FSS

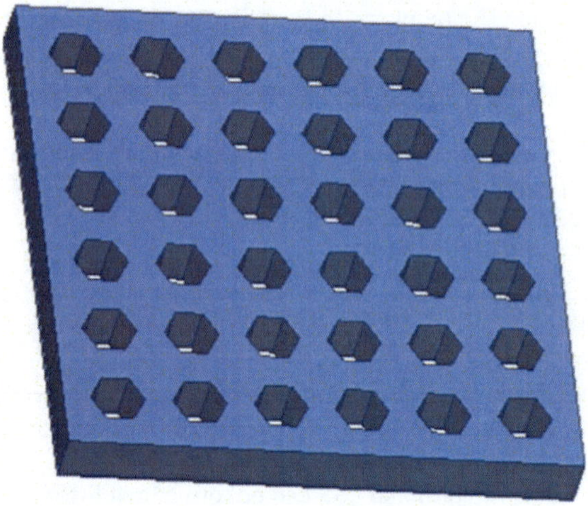

Fig. 2 Model of 3-dimensional FSS with cavity [1]

In subsequent section, we can see that the FSS is used in different areas mainly for reducing the size of antennas and improve the performance of antennas [2–30].

2 Applications of FSS

In [1], the authors proposed that the FSS can be used to reduce the radar cross section, and in paper [2], the authors explained that they want to replace the main reflector or sub-reflector or both in reflector antenna with the FSS reflector which is formed by arranging the metallic cross dipoles on dielectric substrate [2]. FSS is mainly used in antennas to reduce the size of antenna, increase the gain of the antenna [3–10], etc.

By using FSS in satellite communication, we can design an antenna which can reflect and transmit instead of separate antennas by which size will be decreased and can be done by arranging an antenna array on FSS as 9 × 9 elements and excited through horn antenna [3]. If we use coplanar wave guide fed ultrawide band strawberry artistic shaped printed monopole with FSS then it showed gain improvement from 1.65 to 7.87 dB in the lower band and 6.3 to 9.68 dB in upper band and also dimension of the antenna reduced to 61 mm × 61 mm × 1.6 mm size which is smaller than previous antenna for ultrawide band application and ground penetrating radar application [4]. Antenna with FSS inclusion having overall size of 52 mm × 62.5 mm × 23.2 mm, improved the gain up to 9.4 dB [8]. Single layer FSS was proposed in ultrawide band for band stop response from 3.1 to 13.3 GHz [5], and ultraminiaturized FSS was proposed to improve stop band frequency in ultrahigh frequency region [7]. Transmission performance of the band pass filter can also be improved using double layer FSS [6]. By making FSS on RO3010 substrate, the size of the antenna can be reduced to 0.15 of wavelength in the center of frequency which is less than 90% of conventional multi array antenna and has a bandwidth of 9 GHz [9]. In paper [10], the author proposed three FSS with two dielectric substrates so that we can get band pass frequency with wide stop rejection ranging from 1 to 40 GHz and pass band ranging at 10 GHz with 12% fractional bandwidth.

FSS is used in communication sectors [11–16] to improve efficient usage of bandwidth, less power consumption, low cost, high gain, etc., Active frequency selective structures (AFSS) are used for wireless applications at 2.4 and 2.9 GHz in which FSS has four circles, and one slotted cross dipole with circles on top is 1.618 times greater than bottom [11]. Dual band miniaturized FSS are used at 3.4 and 4.9 GHz with bandwidth of 950 MHz and 850 MHz, respectively, and a band ratio of 1.44 with unit dimension of 0.069 $\lambda \times 0.069\lambda$, where λ is the first resonant wavelength in free space [12]. FSS is used in 5G cellular and 5G Wi-Fi applications by incorporating it as in 3D structure with multiple antennas for MIMO system [13]. Payloads for satellite communication are increasing due to increase in application, so we want to decrease the size of antennas and also want to get same or increased gain and clarity by using FSS that we can make a 20 GHz uplink and 30 GHz downlink for Ku-band and also 12 GHz downlink and reflective at 20 and 30 GHz for Ka-band [14]. FSS can be used in millimeter wave applications due to which gain is enhanced of about 11 dB [15]. We can reduce the grating lobes in Ka-band by using multilayer FSS on a dual polarized aperture antenna array [16]. Like this FSS can also be used in different applications [17–30].

I-shaped structure inside two rectangular structures as FSS was used as reflector, absorber, transmitter and wave guide filter. I-shaped structure is placed in two rectangular structures by which it provides negative characteristics, gap in rectangular structures act as inductor and gap between them act as capacitor [17].

Three-dimensional FSS was proposed to act as transmission window by making an aperture antenna and absorption window by using lossless resonators on the antenna [18]. By using Graphene-based FSS, we can also make a terahertz absorber [19], and we can reduce the multi-coupling in nano antenna arrays of multiband ultra-massive MIMO systems [20]. Cylindrical slot FSS consists of 12 columns with 8 slots each dividing the cylinder by 30° each which can be used for beam switching operation conveniently [21]. FSS which is in the form of convoluted ring can also be used for X-band signals which is used in wireless communication devices [22]. FSS can be made as RF/EM shield [23, 24]. It can be made as RF shield by dual band FSS through glass windows [23] and can be made as EM shield by using miniaturized and flexible FSS which is used for conformal applications [24]. Self-complementary FSS can be used to make excellent multiband polarization separation for particular frequency, and polarization separation can be fixed for particular frequency for different application [25]. Wireless security can be achieved using slow phase frequency switching and which can be made using FSS in which BER is also low [26]. In [27–30], it is extended to usage as a medical body area networks [27], we can use for indoor localization without a chip [28], and it can also be used as sensor which can sense strain [29] and also can sense crack in materials [30].

3 Conclusion

This paper has presented the review on Frequency Selective Structures and its applications. Even though it was invented much early and got patent in the dawn of twentieth century, the usage of FSS was started from 1957 as the replacement for reflector. FSS can be used for both transmit and receive antennas. It reduces the size of the antenna which in turn reduces the space occupied by the transmitter module in the satellite. By this way, FSS proves to improve the payload in satellite communication. We can also use FSS for increasing the gain, reducing the cost, making absorbers, etc. We can achieve all these things by using the appropriate FSS by varying its structure, material and dimension. Now, the research on FSS is extended to sensing applications like finding strain or crack on materials and also detection of objects. Hence, it is concluded that the FSS has made revolutionary changes in microwave devices and hope to continue in future. Table 1 presents the comparison of different types antenna with number of FSS unit cells and the achieved gain.

Table 1 Types of antenna and their respective gain

Reference	Type of antenna	Number of unit cell	Gain
[3]	Reconfigurable array antenna	9 × 9	18.2–20.9 dBi
[4]	Ultrawide band strawberry artist printed monopole antenna	10 × 10	7.87–9.68 dB
[5]	Single layer compact FSS	30 × 30	10 dB
[8]	Compact ultrawide band antenna	25 × 25	9.4 dBi
[9]	Convoluted Frequency Selective Surface	29 × 41	15 dBi
[12]	Dual band miniaturized FSS	80 × 80	30 dBi
[13]	Multi band MIMO systems	70 × 50	3–4 dBi
[14]	Reflect array backed by FSS	36 × 36	28.5 dBi
[15]	High dense dielectric patch antenna	8 × 8	17.78 dBi
[16]	Dual polarized aperture array antenna	8 × 8	26 dBi

References

1. Rashid AK, Li B, Shen Z (2014) An overview of three-dimensional frequency selective structures. IEEE Antennas Propag Mag 56(3):43–67
2. Lin F, Yingzheng R (1996) Application of FSS to reduction of antennas RCS. IEEE J Syst Eng Electron 7(1):71–74
3. Li W, Wang Y, Sun S, Shi X (2020) An FSS—Backed Reflection/transmission reconfigurable array antenna. IEEE Access 8:23904–23911
4. Al-Gburi AJA, Ibrahim IBM, Zeain MY, Zakaria Z (2020) Compact size and high gain of CPW-fed UWB strawberry artistic shaped printed monopole antennas using FSS single layer reflector. IEEE Access 8:92697–92707
5. Sampath SS, Sivasamy R (2020) A single-layer UWB frequency selective surface with band—stop response. IEEE Trans Electromagn 62(1):276–279
6. Huang M, Meng Z (2015) Transmission upper limit of band-pass double layer FSS and method of transmission performance improvement. IEEE J Syst Eng Electron 26(2):224–231
7. Zhao L, Liang X, Chen Z-M, Li Y, Zhang S, Ma H, Shen X (2020) An ultraminiaturized dual—stopband frequency selective surface for ultra high frequency. IEEE Access 8:44830–44835
8. Mondal R, Soni Reddy P, Sarkar DC, Sarkar PP (2019) Compact ultra-wideband antenna: improvement of gain and FBR across the entire bandwidth using FSS. IET Microw Antennas Propag 14(1):66–74
9. Habib S, Kiani GI, Butt MFU (2019) A convoluted frequency selective surface for wideband communication applications. IEEE Access 7:65075–65082
10. Liu N, Sheng X, Zhang C, Guo D (2019) Design and synthesis of band-pass frequency selective surface with wideband rejection and fast roll-off characteristics for radome applications. IEEE Trans Antennas Propag 68(4):2975–2983
11. Paiva SB, Silva Neto VP, Assuncao AGD (2020) A new compact, stable and dual-band active frequency selective surface with closely spaced resonances for wireless applications at 2.4 and 2.9 GHz. IEEE Trans Electromag Compat 62(3):691–697
12. Wang H, Qu S, Wang J, Yan M, Zheng L (2020) Dual-band miniaturized FSS with stable resonance frequencies of 3.4/4.9 GHz for 5G communication systems applications. IET Microw Antennas Propag 14(1):1–6
13. Saleem R, Bilal M, Chattha HT, Ur Rehman S, Mushtaq A, Shafique MF (2019) An FSS based multiband MIMO system incorporating 3D antennas for WLAN/WiMAX/5G cellular and 5G Wi-Fi applications. IEEE Access 7:144732–144740

14. Abdollahvand M, Forooraghi K, Encinar A, Atlasbaf Z, Martinez-de-Rioja E (2020) A 20/30 GHz reflectarray backed by FSS for shared aperture Ku/Ka-band satellite communication antennas. IEEE Antennas Wirel Propag Lett 19(4):566–570
15. Asaadi M, Afifi I, Sebak A-R (2018) High gain and wideband high dense dielectric patch antenna using FSS superstrate for millimeter-wave applications. IEEE Access 6:38243–38250
16. Sanchez-Escuderos D, Ferrando-Rocher M, Herranz-Herruzo JI, Valero-Nogueira A (2020) Grating lobes reduction using a multilayer frequency selective surface on a dual-polarized aperture array antenna in Ka-Band. IEEE Access 8:104977–104984
17. Sahai J, Tiwari G (2020) A metamaterial based I-shaped frequency selective surface for band pass and band stop applications. In: IEEE international conference on computer communication and informatics, pp 1–3
18. Zhang B, Jin C, Kechen, Shen Z (2020) Aperture antenna embedded notched parallel plate waveguide and its application to dual-polarized 3-D absorptive frequency selective transmission structure. IEEE Access 8:94833–94841
19. Mishra R, Sahu A, Panwar R (2019) Cascaded graphene frequency selective surface integrated tunable broadband Terahertz metamaterial absorber. IEEE Photonics J 11(2):1–10
20. Zhang B, Jornet JM, Akyildiz IF, Wu ZP (2019) Mutual coupling reduction for ultra-dense multi-band plasmonic nano-antenna arrays using graphene-based frequency selective surface. IEEE Access 7:33214–33225
21. Liang B, Sanz-Izquierdo B, Parker EA, Batchelor JC (2014) Cylindrical slot FSS configuration for beam-switching applications. IEEE Trans Antennas Propag 63(1):166–173
22. Yong WY, Rahim SKA, Seman FC, Himdi M, Suong DL, Elmobarak HA, Ramli MR (2018) Flexible convoluted ring shaped FSS for X-band screening application. IEEE Access 6:11657–11665
23. Farooq U, Shafique MF, Mugual MJ (2020) Polarization insensitive dual band frequency selective surface for RF shielding through glass windows. IEEE Trans Electromag Compat 62(1):93–100
24. Bilalm M, Saleem R, Abbasi QH, Kasi B, Shafique MF (2020) Miniaturized and flexible FSS-based EM shields for conformal applications. IEEE Trans Electromag Compat 62(5):1703–1710
25. Wang H, Yan M, Qu S, Zheng L, Wang J (2019) Design of a self-complementary frequency selective surface with multi-band polarization separation characteristic. IEEE Access 7:36788–36799
26. Roberts J, Ford KL, Rigelsford JM (2016) Secure electromagnetic buildings using slow phase-switching frequency selective structures. IEEE Trans Antennas Propag 64(1):251–261
27. Ashyap AYL, Abidin ZZ, Dahlan SH, Majid HA, Kamarudin MR, Alomainy A, Adb-Alhameed RA, Kosha JS, Noras JM (2018) Highly efficient wearable CPW antennas enabled by EBG-FSS structure for medical body area network applications. IEEE Access 6:77529–77541
28. Jimenez-Saez A, Schubler M, El-Absi M, Abbas AA, Solbach K, Kaiser T, Jakoby R (2020) Frequency selective surface coded retroreflectors for chipless indoor localization tag landmarks. IEEE Antennas Wirel Propag Lett 19(5):726–730
29. Soltani S, Taylor PS, Parker EA, Batchelor JC (2020) Popup tunable frequency selective surfaces for strain sensing. IEEE Sens Lett 4(4):1–4
30. Mahmoodi M, Vanzant L, Donnell KM (2020) An aperture efficiency approach for optimization of FSS-based sensor resolution. IEEE Trans Instrum Measur 69(10):7837–7845

Effective Utilization of Channel Spacing in Elastic Optical Network for 400 Gb/s Transmission

M. Adarsha and S. Malathi

Abstract This work focusses on flexi grid elastic optical network (EON) and demonstrates high bit rate transmission. Traditional wavelength division multiplexing (WDM) technology uses fixed grid spectral spacing. Underutilization channels are avoided in flexi grid and caters to traffic demand at a high bit rate (1 Tb/s). To overcome this problem, flexi grid channel spacing is used and transmitted 400 Gb/s data rate in EON. In the proposed EON architecture, 12.5 GHz channel spacing granularity is used for channel allocation. Finally, EON channel spacing is compared with a fixed grid network.

Keywords WDM · DWDM · EON

1 Introduction

System components required to establish an optical communication as shown in Fig. 1, that consists of an optical transmitter, a communication channel, and an optical receiver.

1.1 Optical Transmitter

The optical transmitter section mainly consists of the information source, data encoder, laser, and modulator. The information source is encoded before modulation as per the requirement of the modulator used in the transmitter section. Optical modulators are classified into two groups depending upon the material property.

Absorptive Modulators: In absorption, coefficient of the material is changed in this sort of modulator; it can be achieved by manipulating Franz–Keldysh effect, the quantum-confined Stark effect, excitonic absorption, changes of Fermi level, or

M. Adarsha · S. Malathi (✉)
M. S. Ramaiah University of Applied Sciences, Bangalore, India

A. Sivasubramanian et al. (eds.), *Futuristic Communication and Network Technologies*, Lecture Notes in Electrical Engineering 792,
https://doi.org/10.1007/978-981-16-4625-6_61

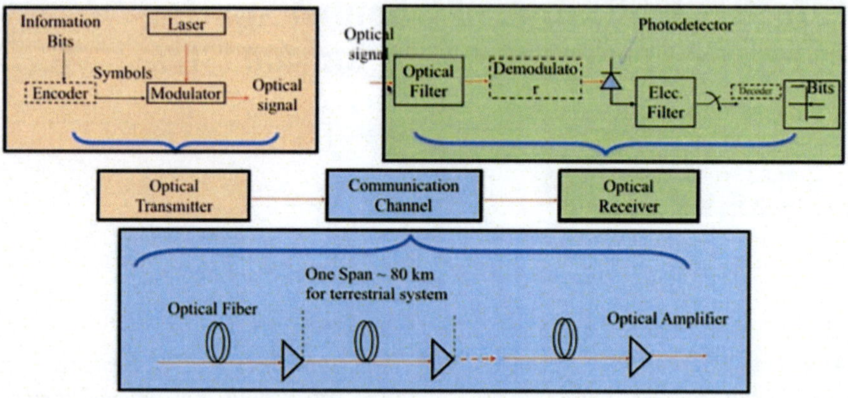

Fig. 1 General block diagram of optical communication system

changes of free carrier concentration. If a couple of such effects appear together, the modulator is called an electro-absorptive modulator [1].

Refractive Modulators: The refractive index of the material is changed mostly electro-optic effect is used for modulation. Some modulation takes advantage of polarization changes in liquid crystals or uses an acousto-optic effect or magneto-optic effect. This type of modulator is named depending upon the effect used, namely electro-optic modulators, acousto-optic modulators [2]. Modulated signals are multiplexed and transmitted with the help of an optical cable.

1.2 Communication Channel

The transmission medium of optical fiber will help to establish communication between two different points. Optical fiber is classified depending upon: Refractive index profile and the number of modes transmitted through the fiber. When an optical signal travels for a long distance, then the signal strength will start attenuating. Optical losses are due to various fiber characteristics such as absorption, scattering, and fiber bend losses. To overcome this issue, advanced fibers are used in addition to that Erbium-doped optical amplifier (EDFA), and dispersion compensating fibers (DCF) is used.

1.3 Optical Receiver

In receiver section, optical signal is filtered and de-multiplexed to convert optical power into electrical current that will be achieved with the help of a photodiode.

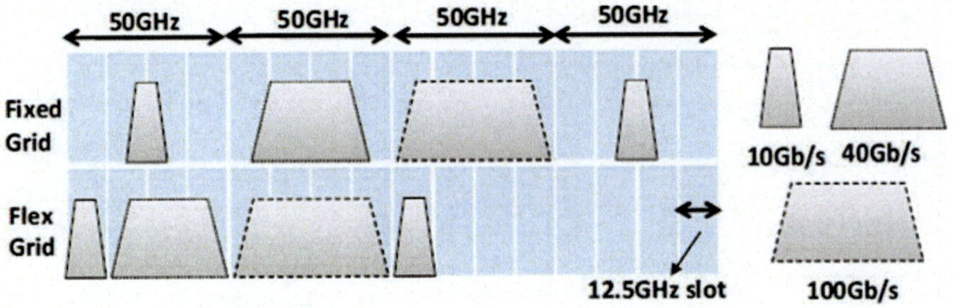

Fig. 2 Fixed grid versus flexi grid [6]

After that, electrical domain filters and decoders are used to receive the transmitted information.

2 Fixed Grid and Flexi Grid

Wavelength division multiplexing (WDM) technology uses fixed grid spectral spacing [3]. WDM technology is categorized into coarse WDM and dense WDM depending upon the channel spacing. CWDM and DWDM define in ITU-T G.671 [4] and ITU-T G.694.1 [5], respectively. DWDM network has a fixed grid (channel spacing) of 50 GHz. Figure 2 shows the fixed and Flexi grid channel spacing. In the fixed grid, spectrum wastage occurs at a low bit rate and it is impossible to manage the traffic at a high bit rate (1 Tb/s). ITU-T G.694.1 frequency grids support a variety of channel spacing ranging from 12.5 to 100 GHz and wider. DWDM can be made cost-effective by incorporating Bandwidth Variable Transponder (BVT), super-channel, and flexible grid technology. To overcome the spectral efficiency problem in the fixed grid can be handled with the help flexible grid elastic optical network (EON).

3 Elastic Optical Network (EON) Design

To design EON, it requires a few additional submodules in addition with DWDM design. The requirement of advance control planes is needed at the network level. In the same way, flex ROADMs, SBVT, super-channels, and high power amplifiers are required in the hardware level [7]. All this makes EON a little expensive, but it provides a large capacity and also helps to overcome the network traffic issues. Figure 3 shows several features and will help us define some important terms.

Elastic Optical Networking (EON): The elastic or flexible optical network refers to a system with variable bit and flexible channel spacing [7].

Bandwidth Variable Transceivers (BVTs): It handles the physical layer parameters like transmitted bit rate, transparent reach, and spectral efficiency. Multi-flow

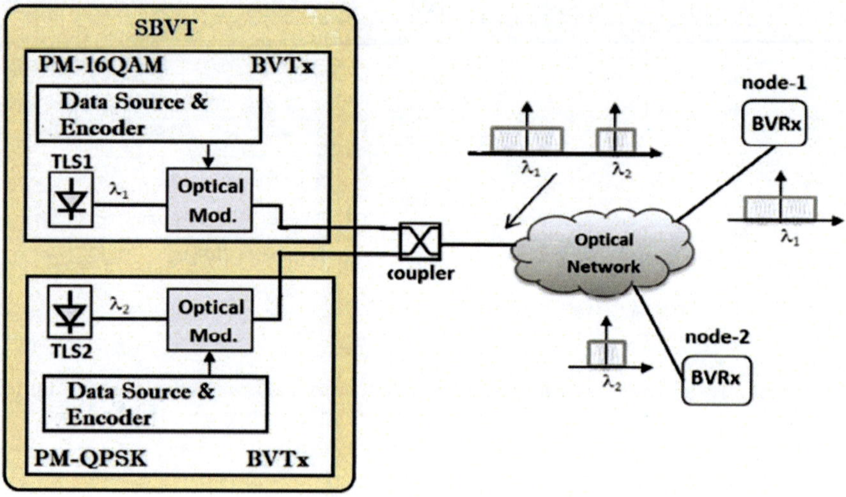

Fig. 3 Elastic optical network

transponder assists various optical streams with various information rates running from 10 Gb/s to 1 Tb/s which flows multiple directions at a time.

Sliceable Bandwidth Variable Transponder: SBVT is a collection of "virtual" lower capacity bandwidth variable transceivers (BVTs) and it is a very important element in the EON to make network flexible and efficient [9]. Dynamic adaption of variable modulation format and different distance is achieved by using software control [8].

Super-Channel: It is an advancement in dense wavelength division multiplexing (DWDM) in which numerous coherent optical carriers are joined to make a bound together channel of a higher information rate and which runs in a single operation cycle [7]. DWDM channel allocation is driven by dividing optical bandwidth into equal slots of bandwidth (Δf). In EON, minimum frequency spacing is used for the channel allocation, which is given by:

$$\Delta f_c = n \times \Delta f \tag{1}$$

A lower bound to the total optical bandwidth for the unequal channel spacing is mentioned below.

$$B_{\text{un}} \geq \left(1 + \frac{\frac{N}{2} - 1}{n}\right) B_{\text{eq}} \tag{2}$$

where $B_{\text{eq}} = (N - 1) \times \Delta f$; N is number of channels (Small); $n =$ minimum number of slots [9].

Modulation and Demodulation Schemes: To achieve a higher bit rate, advanced polarized modulating formats are used. Depending upon the optical reach, different modulation schemes are allocated, for example, PM-16QAM for small, PM-QPSK for medium, and PM-BPSK for large optical reach [10].

Bandwidth Variable Receiver: In the receiver side, a polarization demultiplexer is used to separate the optical signals depending upon the modulation scheme used. This block consists of different demodulators section. In this proposed work, it will segregate the two different demodulators circuits, namely PM-16QAM and PM-QPSK. After extracting the proper demodulated signal, it will drive to the digital signal processing unit for signal condition before entering to end-user.

4 EON Simulation and Result

The proposed work is realized with the help of OptiSystem™, and Fig. 4 gives clear details about the processing flow of simulation. The client interface will help to set the data rates which can be 10 Gb/s to 1 Tb/s. In our proposed elastic optical network, 400 Gb/s data is considered to transmit, and two different modulation schemes are used, namely PM-QPSK and PM-16QAM. For the realization, we consider four 100 Gb/s interface instead of one 400 Gb/s. It is hard to implement one 400 Gb/s interface practically because it requires a baud rate of nearly 100 Gbaud. That is the

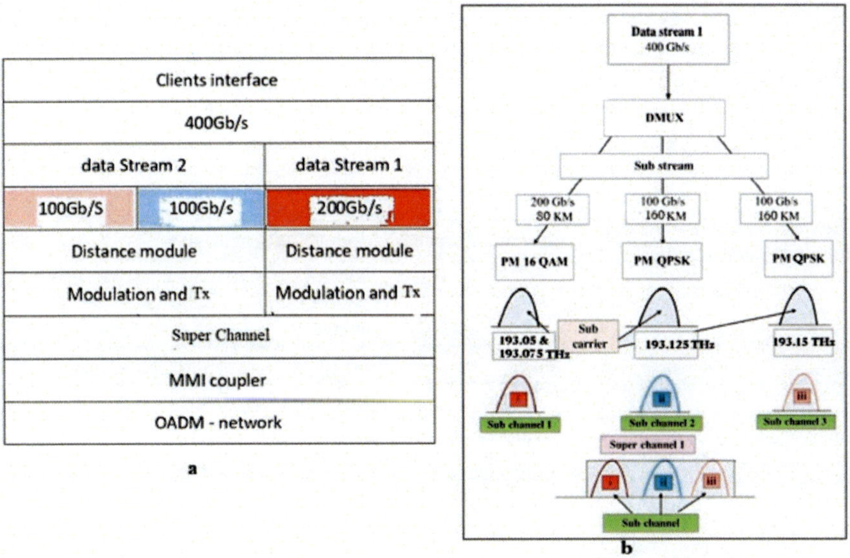

Fig. 4 **a** Processing flow of transmission section. **b** Processing flow of distance module and receiver

reason we considered 25 Gbaud with 10 GHz Guard bandwidth. That is given by channel bandwidth = spectral bandwidth + guard bandwidth (GHz).

Two different cases are considered:

(A) 200 Gb/s PM-16QAM modulation: In order to achieve this, two 100 Gb/s PM-16QAM modulation is used with the center frequency of 193.05 and 193.075 THz. This user demand request came from 80 km away so that PM-QAM modulation is chosen and it considers the first super-channel.
(B) 100 Gb/s PM-QPSK two different users: Two different requests arrived at 160 km from the source. That is the reason two different PM-QPSK modulation scheme is chosen and both combined with the central frequencies of 193.15 and 193.175 THz.

Two different super-channels are coupled and transmitted to the optical network. Table 1 gives the clear details of Dual Polarization 16QAM and Dual Polarization BPSK modulation schemes. It also gives the bit rate, sample rate, and a number of samples information.

Decision-Making Block: This particular block is used to figure out an appropriate modulation technique for the received optical signal, and it will be directed to into a suitable demodulator circuit for demodulation. Demodulator block consists of two different demodulators (PM-16QAM, PM-QPSK) circuits.

Advanced Digital Signal Processing: Low-pass filter and front-end corrections are done in this block, and in addition to that, chromic dispersion is also addressed here. At the end polarization, de-multiplexing will segregate the signal for the phase estimation and equalization and finally down-sample the signal at the receiver end.

The data rate of 200 Gb/s signal is modulated with the help of DP-16QAM, and also, taken 12.5 GHz channel spacing slots are considered. 50 GHz channel spacing is required to transmit 200 Gb/s demand. Figure 5 shows the constellation diagram of the DP-16QAM signal; it gives the clear vision of both X and Y polarization of

Table 1 System parameters for modulation schemes

System parameter	Modulation schemes	
	Value of PM-16QAM scheme	Value of PM-QPSK scheme
Bit rate	200 Gbps	100 Gbps-2 user
Sample rate	400 GHz	400 GHz
Sequence length	65,336 bits	65,536 bits
Sample per bit	4	4
Number of samples	262,144	262,144
Length of fiber	80 km	160 km
Super-channel reference wavelength	193.05 and 193.075	193.150

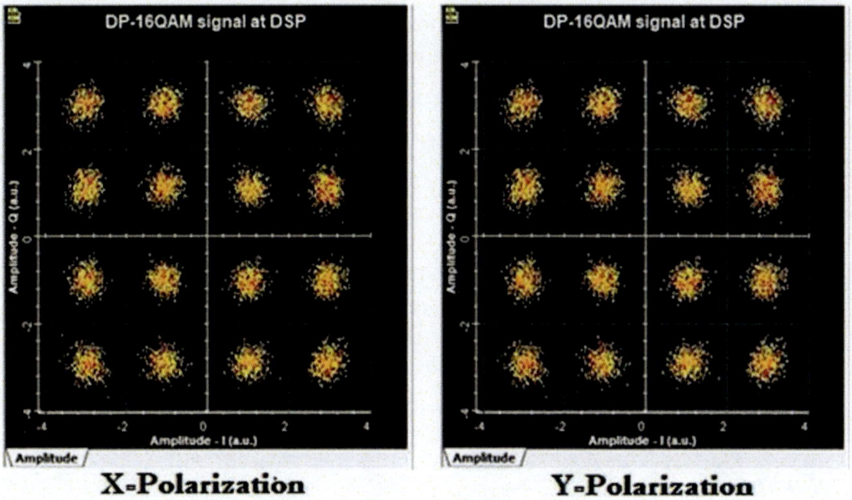

Fig. 5 Constellation diagrams for a 200 Gbit/s DP 16-QAM signal

the modulated signal. Ones it is reached to the receiver block, specific demodulated signal is extracted and passed to advance DSP for signal conditioning. Same way 100 Gb/s DP-QPSK signal is demodulated at the receiver end.

The simulating setup in OptiSystem™, optical spectra for both fixed and flexi channel spacing is observed. Figure 6 describes channel utilization of the fixed grid

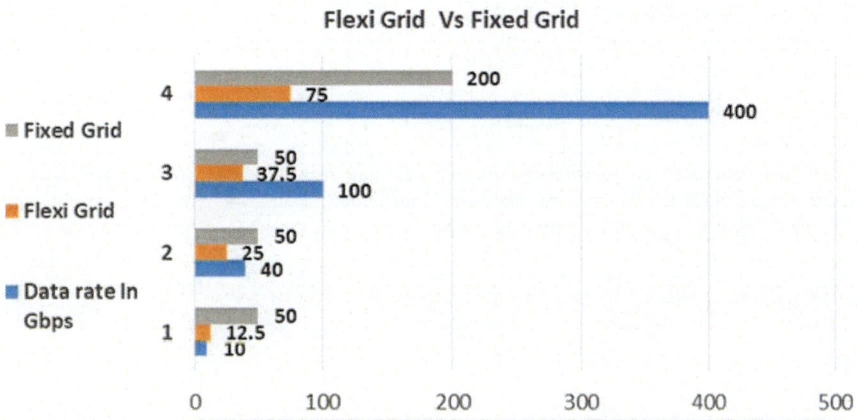

	1	2	3	4
Fixed Grid	50	50	50	200
Flexi Grid	12.5	25	37.5	75
Data rate In Gbps	10	40	100	400

Fig. 6 Flexi grid versus fixed grid

and flexible grid. It clearly shows that flexi grid consumes less spectrum slots for both lower and higher bit data transmissions.

5 Conclusions

The elastic optical network has been studied in detail. The super-channel technique is used to transmit a 400 Gb/s data rate. To realize this network, flexi grid channel spacing is utilized; in addition to that, advanced digital signal processing block is used for the signal conditioning. Elastic optical network gives better performance, and it uses minimum channel spacing.

References

1. Gupta S et al (2015) 50 GHz Ge waveguide electro-absorption modulator integrated in a 220 nm SOI photonics platform, vol 1, no c, p. Tu2A.4
2. Olusegun A et al (2012) We are IntechOpen, the world's leading publisher of Open Access books Built by scientists, for scientists TOP 1%. Intech, vol i, no. tourism, p 38
3. Li H (2016) Network topology design
4. ITU (2002) ITU G.694.1 (06/2002): spectral grids for WDM applications: DWDM frequency grid
5. ITU-T (2012) Spectral grids for WDM applications: DWDM frequency grid. Itu-T
6. Imran M, Anandarajah PM, Kaszubowska-Anandarajah A, Sambo N, Poti L (2018) A survey of optical carrier generation techniques for terabit capacity elastic optical networks. IEEE Commun Surv Tutorials 20(1):211–263
7. Gerstel O, Jinno M, Lord A, Yoo SJB (2012) Elastic optical networking: a new dawn for the optical layer? IEEE Commun Mag 50(2)
8. Fischer JK et al (2014) Bandwidth-variable transceivers based on four-dimensional modulation formats. J Light Technol 32(16):2886–2895
9. Maharjan R, Lavrinovica I, Supe A, Porins J (2016) Minimization of FWM effect in nonlinear optical fiber using variable channel spacing technique. Adv Wirel Opt Commun 3(3):1–4
10. Thangaraj J (2019) Review and analysis of elastic optical network and sliceable bandwidth variable transponder architecture, vol 57, no 11

Analysis of Unit Cell with and Without Splits for Understanding Metamaterial Property

S. Rajasri and R. Boopathi Rani

Abstract Unit cells are the building blocks for creating the metamaterial periodic surface. Generally, the metamaterial unit cell exhibits the negative value of permittivity and permeability. Determination of metamaterial scattering parameters is the important step involved in the design of periodic structures. Here, a square-shaped unit cell having two loops is analyzed by placing on the FR4 substrate. The overall dimension of the substrate is 7 mm × 7 mm × 0.25 mm. Step-by-step procedure is done to optimize the unit cell which is suitable to extract the metamaterial properties are proposed. Unit cell is simulated using high-frequency structured simulator (HFSS) version HFSS 2020 R2, and the results are exported for the parameter retrieval. Scattering parameter retrieval method is used to analyze the unit cell with proper dimension and boundaries. Permittivity and permeability of the double-loop unit cell with and without split rings were determined using MATLAB R2020a program from the exported data of the simulations done. Extracted parameters of all the different combinations are compared and the metamaterial properties are determined for each unit cells.

Keywords Double-loop unit cell · Metamaterial · Negative permittivity · Negative permeability

1 Introduction

Meta means 'altered, changed,' or 'higher, beyond.' Metamaterial is defined that the material is not found in nature and it is artificially obtained material having the negative values of permittivity and permeability. These materials are also called as left handed (LH), double-negative (DNG), negative refractive index (NRI), backward-wave (BW), Veselago medium. In 1968, Veselago introduced the material which exhibits negative values of permittivity and permeability [1]. In 1999, John Pendry introduced the material with periodic structures [2]. In 2000, David Smith introduced the composite medium with single unit cell having negative values of permittivity

S. Rajasri (✉) · R. Boopathi Rani
National Institute of Technology Puducherry, Puducherry, India

A. Sivasubramanian et al. (eds.), *Futuristic Communication and Network Technologies*, Lecture Notes in Electrical Engineering 792,
https://doi.org/10.1007/978-981-16-4625-6_62

and permeability [3]. Veselago presented the equivalent circuit for the metamaterial unit cell which is making the periodic structures [4]. There are researches which present the equivalent circuit for the artificial magnetic inclusion in the metamaterial unit cell [5–7].

In order to confirm the unit cell which exhibits the metamaterial property, parameter extraction is done on the unit cells. Different techniques have been used to determine the metamaterial properties with some important parameters. A. M. Nicolson and G. F. Ross, in 1970, introduced the first method for determining the complex values of permittivity and permeability [8]. Smith et al. presented scattering parameter method to calculate the permeability and permittivity of the proposed unit cell having negative values [9]. Xudong Chen et al. proposed the robust method to retrieve the effective parameters for metamaterials [10]. In the earlier stage, square-shaped unit cell and circular-shaped unit cell were designed for the negative values of permittivity and permeability. Later on, there are many different shapes were introduced for the metamaterial. Tukmen et al. introduced the U-ring resonator to provide multi-band operation, and it is geometrically simple unit cell [11]. Yudurshan et al. introduced the hybrid-shaped single-loop resonator which is the combination of square and triangular-shaped unit cell [12]. There are similar papers came up with these different shapes of unit cell and its equivalent circuits with metamaterial properties. In recent years, antennas designed with these unit cells called 'metamaterial inspired antennas.' Boopathi Rani et al. proposed an antenna with complementary split ring resonator (CSRR) for notch in UWB antenna [13]. Pandeeswari presented a CPW fed triple band antenna inspired with split ring resonator (SRR) and non-bianisotropic complementary split ring resonator (NBCSRR) [14]. Samson Daniel et al. presented an antenna with ELC metamaterial resonator for WIMAX and WLAN applications [15]. In metamaterial, a unit cell with negative values of permittivity and permeability plays the most important role. In order to find the metamaterial properties, a unique technique is needed for the proper extraction of parameters.

2 Design and Construction of Unit Cell

The basic unit cell consists of metal conductor on the top of the substrate. It is a square-shaped structure and the material used is FR4 with dielectric constant 4.4 and loss tangent 0.02. The height of the substrate is 0.25 mm. Six steps are done to analyze the entire unit cell. Length and width of the substrate are 7 mm × 7 mm. The outer length and width of the metal conductor are 6 mm and the inner conductor having the length and width as 4 mm. Thickness of the conductor is 0.7 mm and the split gap is 0.3 mm. And the gap between the two metal conductors is also 0.3 mm (Fig. 1).

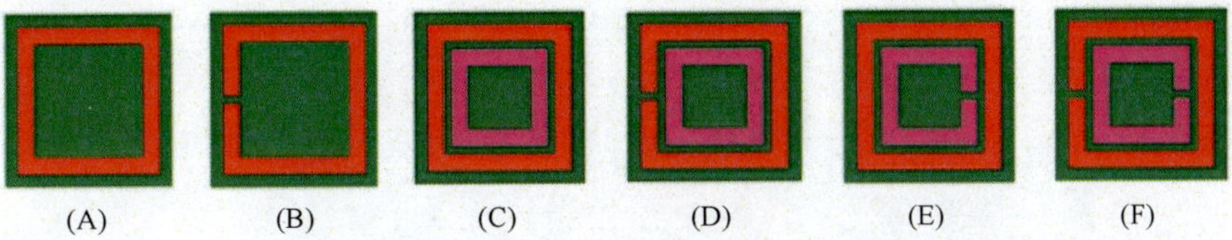

Fig. 1 **a** Single-loop unit cell without split, **b** Single-loop unit cell with split, **c** double-loop unit cell without split, **d** double-loop unit cell with split on the outer conductor, **e** double-loop unit cell with split on the inner conductor and **f** double-loop unit cell with double split

3 Method of Analysis

This unit cell is analyzed using the scattering parameter retrieval method [9] using PEC and PMC boundaries. Two wave ports were assigned at the sides to analyze the unit cell. Conductors must be assigned as perfect boundary. This is also called as waveguide setup to analyze the unit cell. It helps to understand the band pass and band stop characteristics of the designed unit cell.

In order to achieve the effective values of permittivity and permeability, [5, 16], Permittivity, $\varepsilon = n/z$, Permeability, $\mu = nz$.

Let us consider the metamaterial of thickness '*d*,' refractive index '*n*,' and impedance '*z*.' The reflection and transmission parameters are calculated by the formulae given in Eqs. (1) and (2), respectively,

$$S_{11} = \frac{R\left(1 - \mathrm{e}^{i2nk_0 d}\right)}{1 - Re^{i2nk_0 d}} \tag{1}$$

$$S_{21} = \frac{(1 - R)e^{ink_0 d}}{1 - R^2 e^{i2nk_0 d}} \tag{2}$$

where reflection at single interface, $R = \frac{z-1}{z+1}$, wave number $k_0 = \frac{2\pi}{\lambda_0}$ and λ_0 is the wavelength.

Using the scattering parameters, impedance and refractive index can be found which are given in Eqs. (3)–(5) (Fig. 2)

$$Z = \pm\frac{\sqrt{(1 + S_{11})^2 - S_{21}^2}}{\sqrt{(1 - S_{11})^2 - S_{21}^2}} \tag{3}$$

$$\mathrm{e}^{ink_0 d} = \frac{S_{21}}{1 - S_{11}\left(\frac{z-1}{z+1}\right)} \tag{4}$$

$$n = \frac{1}{k_0 d}\left(-i \ln \mathrm{e}^{ink_0 d} + 2m\pi\right) \tag{5}$$

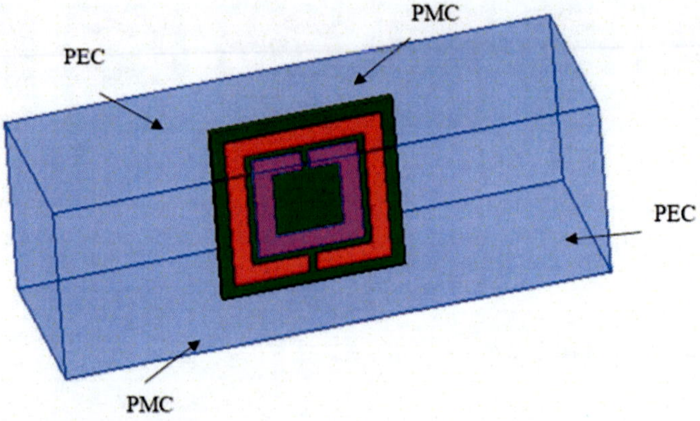

Fig. 2 Scattering parameter retrieval method boundary setup

Using these mentioned equations, parameters can be retrieved. Initially, the reflection (S_{11}) and transmission (S_{21}) parameters are determined from the simulation using HFSS. Export the values into the MATLAB R2020a software to find the values of permittivity and permeability.

4 Results and Discussion

4.1 Single-Loop Unit Cell Without Split

Figure 3 shows that reflection and transmission characteristics of single-loop square-shaped unit cell without split. It gives the band stop characteristics at 6.6 and 32.79 GHz. It is not giving the proper values of permittivity and permeability. Hence, it would not be considered as the metamaterial unit cell.

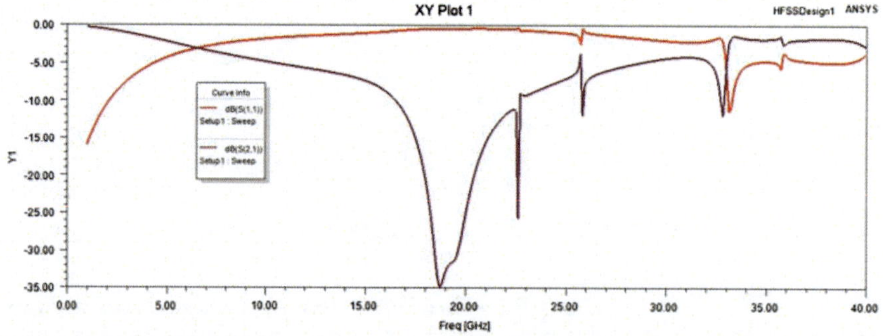

Fig. 3 S_{11} and S_{21} characteristics of the single-loop unit cell without split

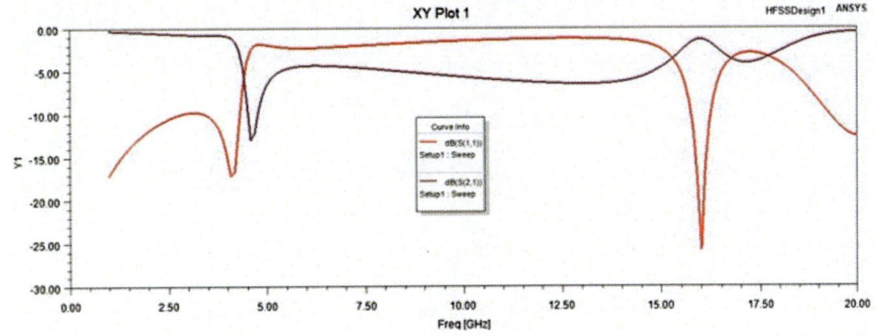

Fig. 4 S_{11} and S_{21} characteristics for the single-loop unit cell with split

4.2 Single-Loop Unit Cell with Split

Figure 4 shows that the S_{11} and S_{21} characteristics of the single-loop unit cell with split. It gives the band pass characteristics at 4.14 and 16 GHz. It exhibits the negative values of permeability. Hence, it is suitable for the metamaterial.

4.3 Double-Loop Without Split

Figure 5 shows that the S_{11} and S_{21} characteristics of double-loop unit cell without splits on the conductors. From the graph, it is clearly shown that it gives the good band stop characteristics from 12.7 to 18.21 GHz. But the values of permittivity and permeability are not negative. Hence, it is not given proper response for the metamaterial property.

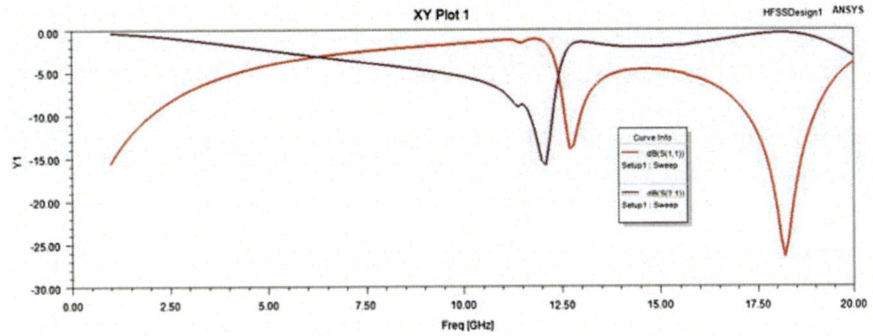

Fig. 5 S_{11} and S_{21} characteristics of the double-loop unit cell without splits

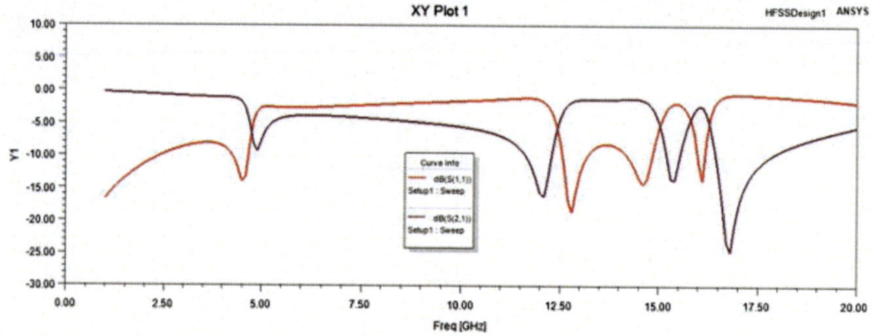

Fig. 6 S_{11} and S_{21} characteristics of the double-loop unit cell with split on the outer conductor

4.4 Double-Loop with Single Split on the Outer Conductor

Figure 6 shows the S_{11} and S_{21} characteristics of double-loop unit cell with split on the outer conductor. It was observed that it provides the strong band pass characteristics from 4.9 to 12.08 GHz. So it can able to provide the negative value of permeability exactly at 4.9 GHz. From the observation, it is clearly shown that the split on the outer conductor on the unit cell exhibits the metamaterial property.

4.5 Double-Loop Unit Cell with Split on the Inner Conductor

Figure 7 shows the S_{11} and S_{21} characteristics of the double-loop unit cell with split on the inner conductor. The results showed that it gives the band stop characteristics from 12.74 to 18.62 GHz. This is also not a metamaterial unit cell.

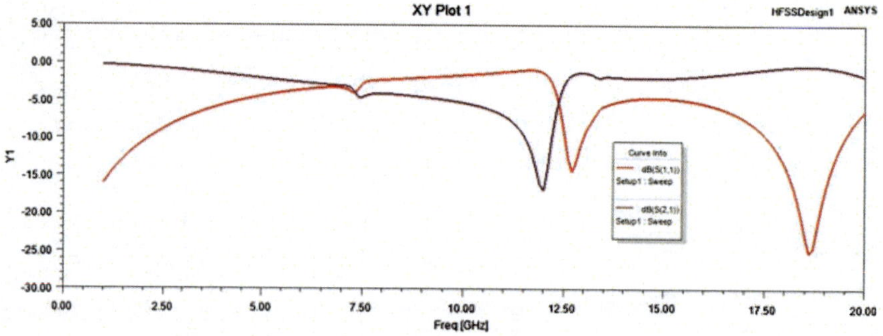

Fig. 7 S_{11} and S_{21} characteristics of the double-loop unit cell with split on the inner conductor

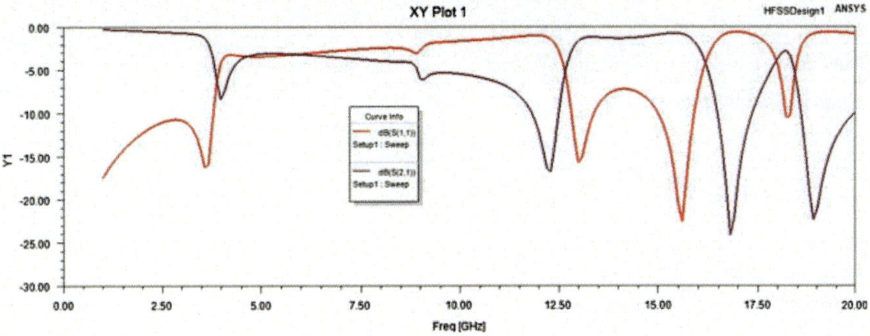

Fig. 8 S_{11} and S_{21} characteristics of the double-loop unit cell with double splits

4.6 Double-Loop Unit Cell with Double Splits

Figure 8 shows the S_{11} and S_{21} characteristics of the double-loop unit cell with double splits on the conductors. It gives the good band pass and band stop characteristics. Based on the applications, one can decide the characteristics. It gives the good negative value of permeability compared to all the different combinations. Hence, this unit cell is considered as the metamaterial unit cell.

Table 1 gives the overall views about the analyzed unit cell with and without splits. Permittivity and permeability have taken to analyze the unit cell. The designed unit cell was simulated in the HFSS 2019 R2 and gets the values of S_{11} and S_{21}. The results are exported to the MATLAB R2020a. Using the real and imaginary parts of the S_{11} and S_{21}, the values of permittivity and permeability were calculated. single-loop unit cell without split, double-loop unit cell without split, and double-loop unit cell with split on the inner conductor were not given the values of negative permeability. Remaining single-loop with split, double-loop unit cell with split on the outer conductor, and double-loop unit cell with double splits were given the good results. Hence, these three types of unit cell give the negative value of permeability. These unit cells are suitable for the metamaterial property. Based on the metamaterial properties, the unit cell discussed above can be used for the K_u, K, and K_a band application in the super high-frequency range.

5 Conclusion

Square-shaped unit cell was analyzed in different combinations such as with split and without split. At one scenario, the split is on the inner conductor and the results were analyzed. At other scenario, the split is on the outer conductor and the results were analyzed. Both the results were compared. Band pass and band stop characteristics were observed using the HFSS software. The exported results from the HFSS 2019 R2 were used for the parameter retrieval using scattering parameter retrieval method. The

Table 1 Comparison of permittivity and permeability of the unit cell with and without splits

Unit cell	Permittivity	Permeability
Single-loop unit cell without split	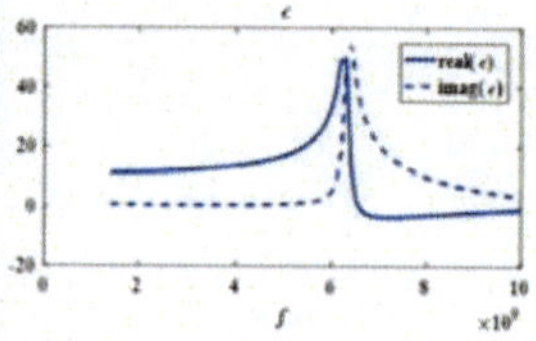	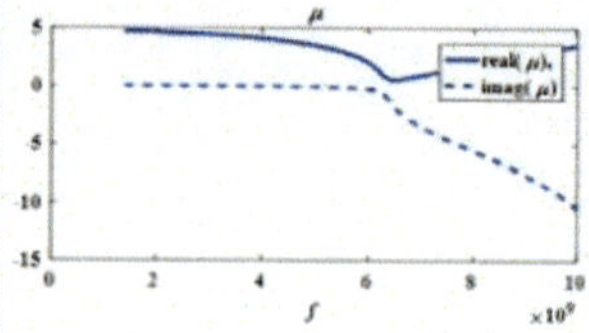
Single-loop unit cell with split	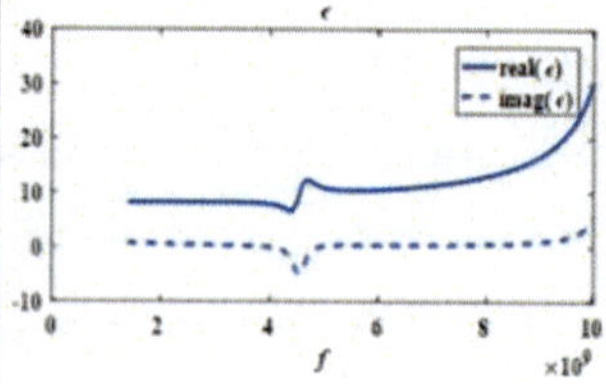	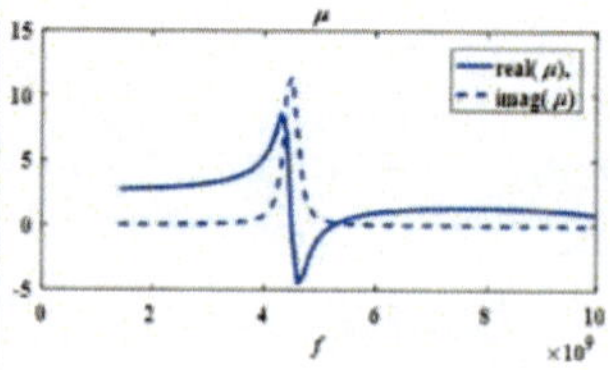
Double-loop unit cell without split	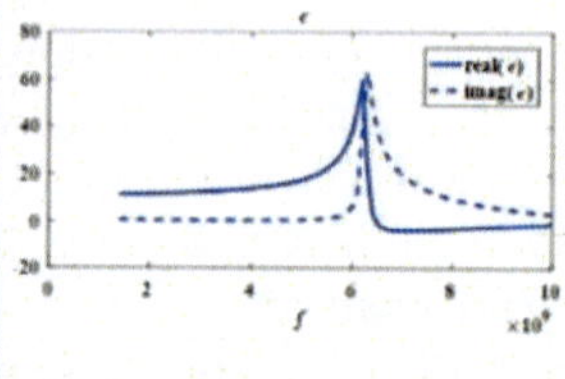	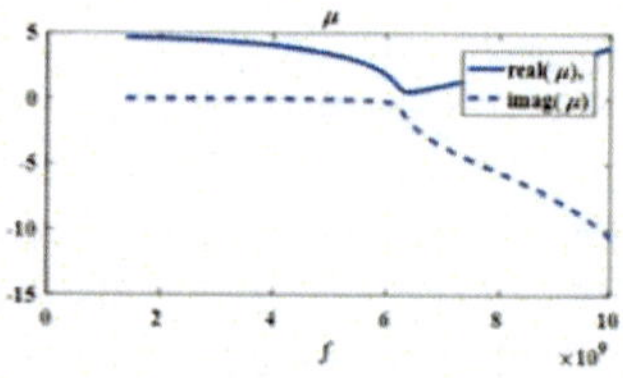
Double-loop unit cell with split on the outer conductor	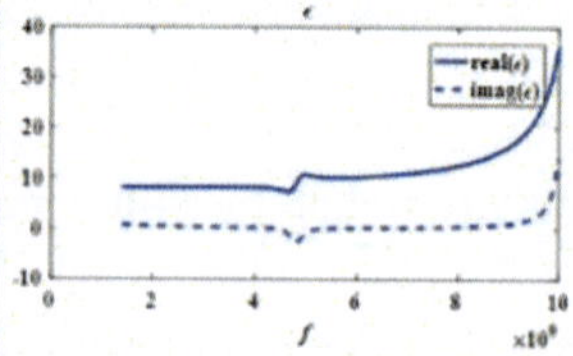	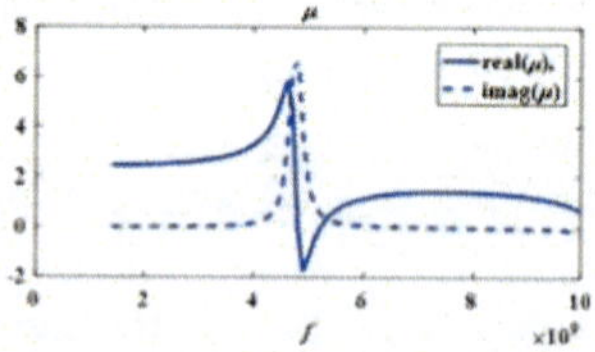
Double-loop unit cell with split on the inner conductor	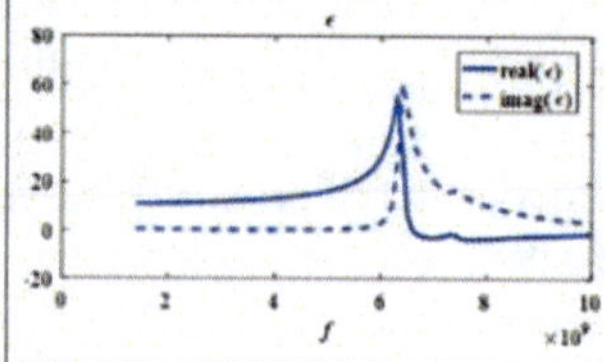	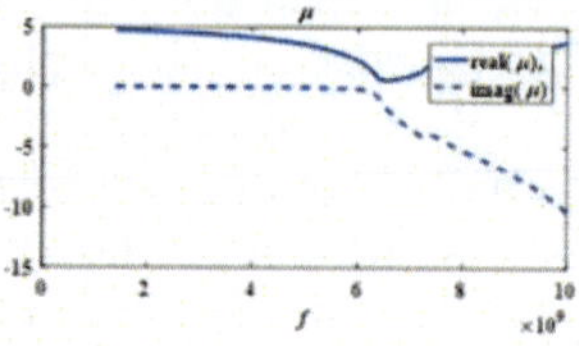
Double-loop with double split	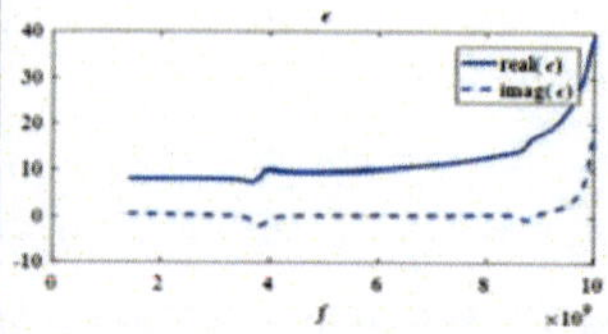	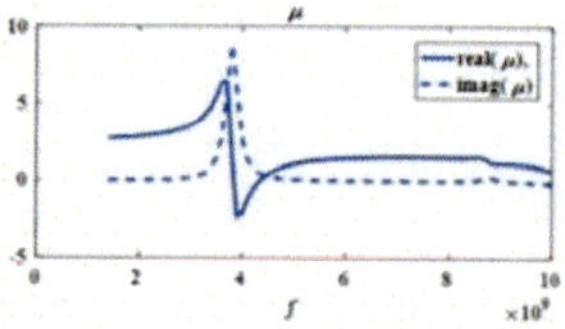

values of permittivity and permeability were compared for all different combinations, and the metamaterial properties were extracted using MATLAB R2020a.

References

1. Veselago VG (1968) The electrodynamics of substances with simultaneously negative values of ϵ and μ. Phys Usp 10(4):509–514
2. Pendry JB, Holden AJ, Robbins DJ, Stewart WJ (1999) Magnetism from conductors and enhanced nonlinear phenomena. IEEE Trans Microw Theory Tech 47(11):2075–2084
3. Smith DR, Padilla WJ, Vier DC, Nemat-Nasser SC, Schultz S (2000) Composite medium with simultaneously negative permeability and permittivity. Phys Rev Lett 84(18):4184
4. Veselago V, Braginsky L, Shklover V, Hafner C (2006) Negative refractive index materials. J Comput Theor Nanosci 3(2):189–218
5. Baena JD, Marqués R, Medina F, Martel J (2004) Artificial magnetic metamaterial design by using spiral resonators. Phys Rev B 69(1):014402
6. Bilotti F, Toscano A, Vegni L, Aydin K, Alici KB, Ozbay E (2007) Equivalent-circuit models for the design of metamaterials based on artificial magnetic inclusions. IEEE Trans Microw Theory Tech 55(12):2865–2873
7. Baena JD, Bonache J, Martin F, Sillero RM, Falcone F, Lopetegi T, Sorolla M (2005) Equivalent-circuit models for split-ring resonators and complementary split-ring resonators coupled to planar transmission lines. IEEE Trans Microw Theory Tech 53(4):1451–1461
8. Nicolson AM, Ross GF (1970) Measurement of the intrinsic properties of materials by time-domain techniques. IEEE Trans Instrum Meas 19(4):377–382
9. Smith DR, Vier DC, Kroll N, Schultz S (2000) Direct calculation of permeability and permittivity for a left-handed metamaterial. Appl Phys Lett 77(14):2246–2248
10. Chen X, Grzegorczyk TM, Wu BI, Pacheco Jr J, Kong JA (2004) Robust method to retrieve the constitutive effective parameters of metamaterials. Phys Rev E 70(1):016608
11. Turkmen O, Ekmekci E, Turhan-Sayan G (2012) Nested U-ring resonators: a novel multi-band metamaterial design in microwave region. IET Microw Antennas Propag 6(10):1102–1108
12. Yurduseven O, Yilmaz AE, Turhan-Sayan G (2011) Hybrid-shaped single-loop resonator: a four-band metamaterial structure. Electron Lett 47(25):1381–1382
13. Boopathi Rani R, Pandey SK (2017) A CPW-fed circular patch antenna inspired by reduced ground plane and CSRR slot for UWB applications with notch band. Microw Opt Technol Lett 59(4):745–749
14. Pandeeswari R (2018) SRR and NBCSRR Inspired CPW fed triple band antenna with modified ground plane. Progr Electromag Res 80:111–118
15. Daniel RS, Pandeeswari R, Raghavan S (2018) Dual-band monopole antenna loaded with ELC metamaterial resonator for WiMAX and WLAN applications. Appl Phys A 124(8):570
16. Smith DR, Vier DC, Koschny T, Soukoulis CM (2005) Electromagnetic parameter retrieval from inhomogeneous metamaterials. Phys Rev E 71(3):036617

Design of Compact Sierpinski Gasket Fractal Multi-band Antenna for Wireless Applications

Rakesh Narayanan, P. Santhana Krishnan, S. Shivashankar, and V. Sasikala

Abstract Traditional patch antennas are designed to operate in a single frequency. This factor limits the use of the traditional antenna in devices, which has multiple applications at multiple resonant frequencies. This paper proposes a rectangular shaped patch antenna with iterative hexagonal slots (Sierpinski Gasket Fractal) is designed. The proposed structure has a dimension of $50 \times 38 \times 1.67$ mm^3. It operates on S, C and X bands (2–4 GHz), (4–8 GHz) and (8–12 GHz) frequencies, respectively. Microstrip line feeding is the feeding technique used. The antenna is fabricated with copper metal with thickness 0.035 μm on a FR4 dielectric substrate having $\varepsilon_r = 4.4$ which is mounted on copper material ground. The fabricated antenna is tested using Vectored Network Analyzer. The antenna parameters such as Voltage Standing wave Ratio (VSWR) and reflection coefficient (S_{11}) for the resonant frequencies. Six different resonant frequencies appeared at 2.49, 4.3, 5.8x, 7.4, 8.1 and 10.85 GHz with Reflection coefficient below −10 dB for all frequencies during simulation. The resonant frequencies that appeared during testing using VNA were 2.43, 4.19, 5.7, 7.17, 8.052 and 10.85 GHz. CST microwave studio 2019 software tool was used to design and simulate the prospective antenna design. Applications of this antenna include Bluetooth 802.15 and GPS (2.4–2.48 GHz), WLANs 802.11, WiMAX, Radio altimeters, RADARs, Aviation services and UWB applications (3.1–10.6 GHz).

Keywords Multiband antenna · Slot antenna · Sierpinski Gasket fractal

1 Introduction

Patch antenna plays an essential role in the field of wireless communication. Advancements in the modern electronic equipment and communication technologies are reducing the size of the device smaller day by day. A traditional antenna is designed to operate at a single frequency and also its size is larger, which makes it difficult to

R. Narayanan · P. Santhana Krishnan (✉) · S. Shivashankar · V. Sasikala
Department of ECE, Sri Sairam Engineering College, Chennai, Tamil Nadu, India

V. Sasikala
e-mail: sasikala.ece@sairam.edu.in

A. Sivasubramanian et al. (eds.), *Futuristic Communication and Network Technologies*, Lecture Notes in Electrical Engineering 792,
https://doi.org/10.1007/978-981-16-4625-6_63

be integrated in today's systems. Thus, a compact miniaturized patch antenna with multiple iterative hexagonal slots operating at multiple frequencies is investigated, as they are small in size, less in weight and have better performance that are vital for various applications in the field of wireless communication. Microstrip antennas (MSAs) are low-profile miniaturized antennas and can be incorporated in a device easily. Various types of multiband and ultra-wide band slot antennas were studied from various literature references. The design of low-profile compact size wideband and multiband antenna is the main objective in the wireless communication system [7]. Several designs of multi-band slot antennas have been designed in [1–6].

The proposed antenna consists of three iterations of hexagonal Sierpinski Gasket fractals which contains five hexagonal slots on the inner surface and four fractals at the corners on the microstrip patch which undergoes multiple resonance at six different frequencies 2.49, 4.19, 5.7, 7.17, 8.052 and 10.85 GHz with low reflection coefficient and considerable gain which can be used for S, C and X bands of 2–4 GHz, 4–8 GHz and 8–12 GHz respective applications. Section 2 of this paper contains the design of the antenna. The simulation result is presented in Sect. 3 of this paper. The conclusion and applications of the proposed design is presented in Sect. 4 of this paper.

2 Design of the Proposed Structure

The prospective antenna is made of a copper patch in rectangular shape with three iterations of hexagonal Sierpinski Gasket fractals which contains five hexagonal slots on the inner surface and four fractals at the corners. The antenna is designed on FR-4 substrate with thickness of 1.6 mm and relative permittivity of 4.4 which is mounted on the ground plate as shown in Fig. 1 [7]. CST Microwave Studio software is used for the design and simulation of the antenna. The conventional rectangular microstrip patch antenna was designed initially for 2.49 GHz which is shown in Fig. 2.

The width and length of the antenna are calculated from the following equations [7].

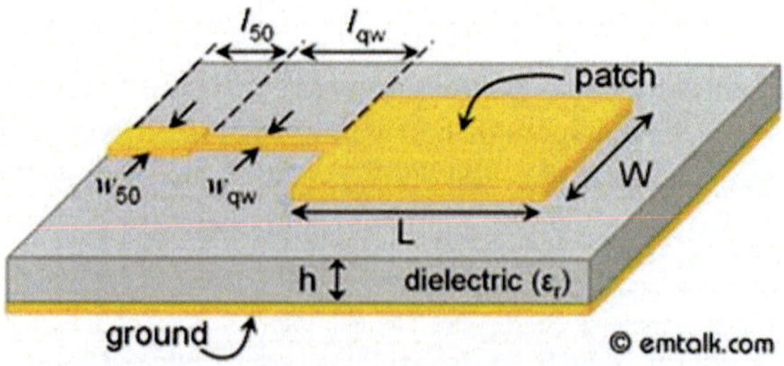

Fig. 1 Typical structure of patch antenna [7]

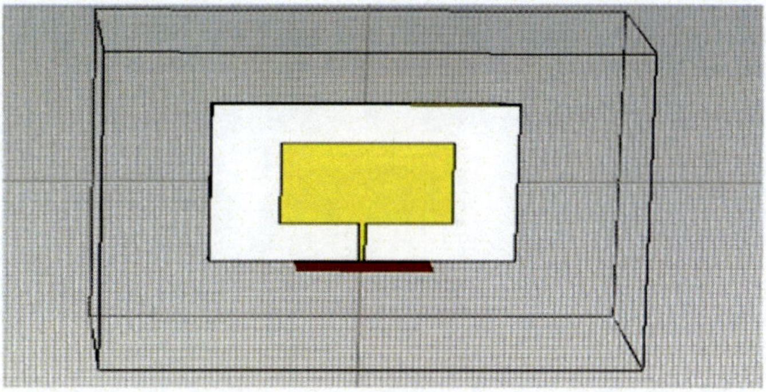

Fig. 2 Rectangular patch antenna designed for 2.49 GHz

$$L = \frac{C_0}{2f_r\sqrt{\varepsilon_{\text{reff}}}} = -2\Delta L$$

$$W = \frac{C_0}{2f_r}\sqrt{\frac{2}{\varepsilon_r + 1}}$$

$$\frac{\Delta L}{h} = 0.412\frac{(\varepsilon_{\text{reff}} + 0.3)\left(\frac{W}{h} + 0.264\right)}{(\varepsilon_{\text{reff}} - 0.258)\left(\frac{W}{h} + 0.8\right)}$$

$$\varepsilon_{\text{reff}} = \frac{\varepsilon_r + 1}{2} + \frac{\varepsilon_r - 1}{2}\left[1 + 12\frac{h}{W}\right]^{-\frac{1}{2}},\ W/h > 1$$

$$L_g = 6h + L$$
$$W_g = 6h + W$$

where L is the length, W is the width of the rectangular copper patch, respectively; L_g is the length and W_g is the width of ground plane. Design specifications for the initial design of 2.4 GHz antenna are as shown in Table 1.

The antenna structure shown in Fig. 2 is designed for the dimensions specified in Table 1. The initial triangular antenna is the basic patch antenna, with no slots.

Table 1 Design specifications for 2.49 GHz rectangular patch antenna

Antenna design parameters	Dimensions (mm)
Patch length (L)	28
Patch width (W)	32
Ground/substrate length (L_g)	38
Width of ground/substrate (W_g)	50
Length of feedline (L_f)	10
Width of feedline (W_f)	1

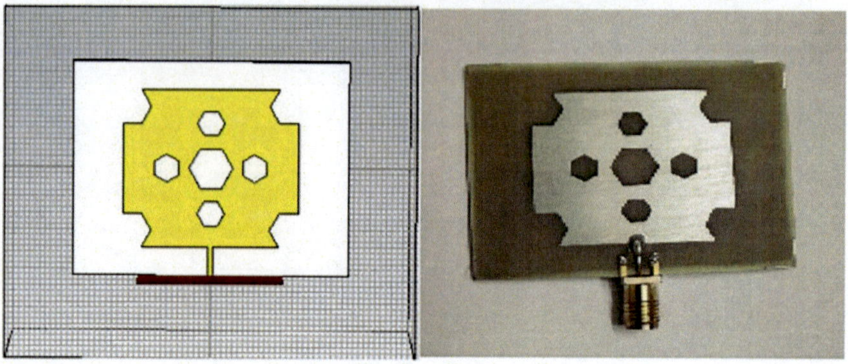

Fig. 3 Proposed and fabricated antenna structure

The antenna is excited using microstrip line feeding technique and is matched to an input impedance of 50 Ω. Two pairs of hexagonal slots with radius 3.5 mm are added to the slot to possess multiband and wideband characteristics. The proposed antenna structure physically compact with the dimensions $50 \times 50 \times 1.6$ mm^3 is shown in Fig. 3. The hexagonal slot starts to radiate at multiple resonant frequencies where the electrical and magnetic field intensity values are maximum. The obtained six multiple frequencies are found to be useful in S, C, and X band applications.

The proposed antenna structure is then fabricated on the FR4 substrate using the same design constraints (shown in Fig. 3). 1 mm SMA (SubMiniature version A) connector is soldered to the microstrip feed line, for testing the fabricated antenna. The fabricated antenna is then tested using a vectored network analyzer.

3 Results and Discussion

The proposed antenna structure in Fig. 4 exhibits frequency multiband characteristics at 2.49 GHz, 4.3 GHz, 5.9672 GHz, 7.402 GHz, 8.194 GHz, 10.856 GHz with reflection coefficients −13.065 dB, −41.574 dB, −21.641 dB, −17.883 dB, −12.946 dB, −17.707 dB, respectively, during simulation. The graphical plot of *S*-parameter S_{11} value measured at various generated resonance frequencies is shown in Fig. 5. The VSWR at various operating frequencies can be determined from the VSWR plot shown in Fig. 6. The fabricated antenna is then tested using vectored network analyzer. The reflection coefficient and VSWR were measured. It was observed that there was a slight shift in measured frequencies from the simulated output frequencies. This shift in frequencies may be due to the SMA connector, the soldering joints, used to solder the SMA connector to the feeding line, and other electro-magnetic interference signals in the atmosphere. The frequencies and their respective reflection coefficients and VSWR obtained during testing are tabulated in Table 2 (Fig. 7).

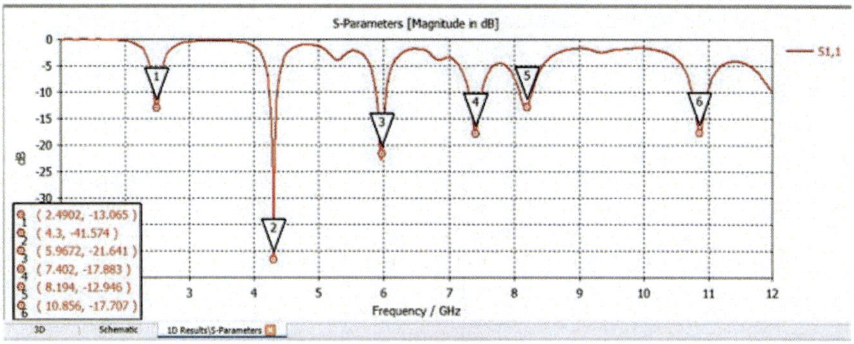

Fig. 4 Simulated return loss of proposed antenna structure

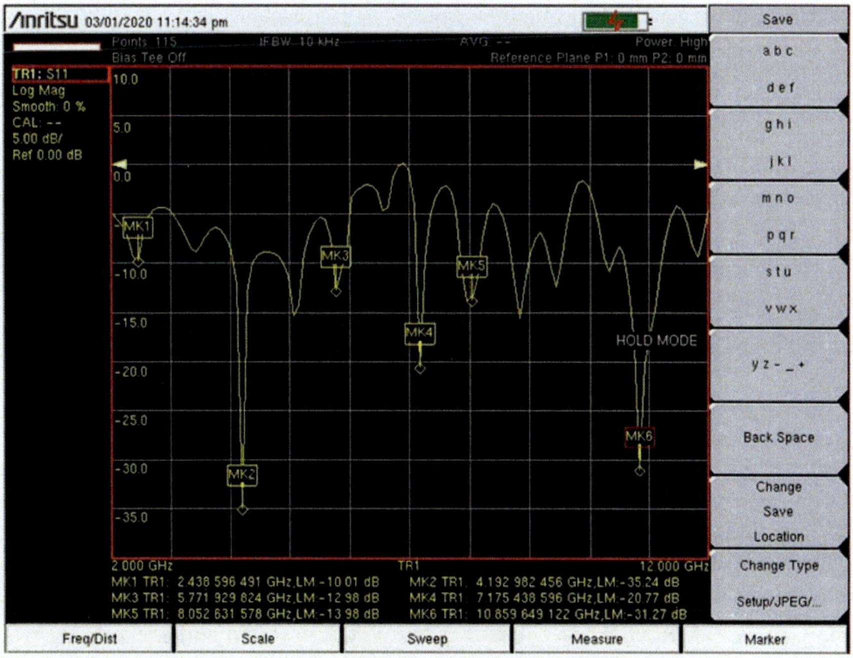

Fig. 5 Measured return loss of proposed antenna structure

4 Conclusion

The design, simulation and testing of a Sierpinski Gasket fractal multiband antenna using CST Microwave Studio software is presented in this paper. The multiband characteristics have been achieved with a considerable gain and low return loss by employing four iterations of hexagonal gaskets in a rectangular patch antenna. The

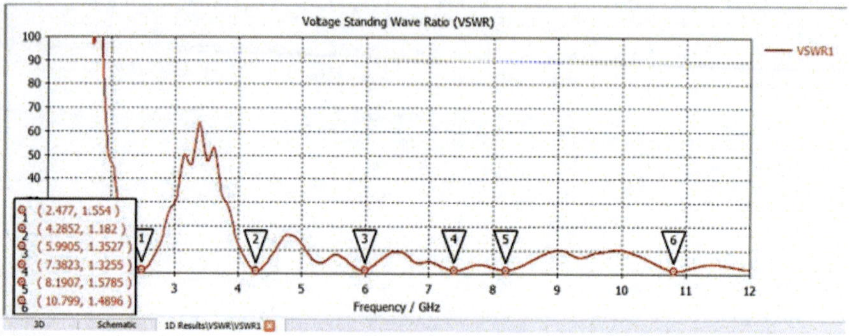

Fig. 6 Simulated result of VSWR plot of the antenna structure

Table 2 Simulated and tested performance results of the antenna structure

Simulated results			Tested results		
Frequency (GHz)	Reflection coefficient (dB)	VSWR (no unit)	Frequency (GHz)	Reflection coefficient (dB)	VSWR (no units)
2.49	−13.065	1.554	2.438	−10.01	1.92
4.3	−41.574	1.182	4.192	−35.24	1.03
5.96	−21.641	1.3527	5.771	−12.98	1.58
7.4	−17.883	1.3255	7.175	−20.77	1.20
8.194	−12.946	1.5785	8.052	−13.98	1.50
10.85	−17.707	1.4896	10.859	−31.27	1.06

designed antenna resonates at six individual resonant frequencies at 2.438, 4.192, 5.771, 7.175, 8.025, 10.859 GHz and all frequencies had a return loss less than −10 dB. Harmonics and slight deviation in from the simulated results were observed in between the desired frequencies were identified to be interference with the testing device and other electrical and electronic equipment in the test site. Applications of this antenna includes S-band, C-band and X-band applications like Bluetooth 802.15 and GPS (2.4–2.48 GHz), WLANs 802.11, WiMAX, Radio altimeters, RADARs, Aviation services and UWB applications (3.1–10.6 GHz).

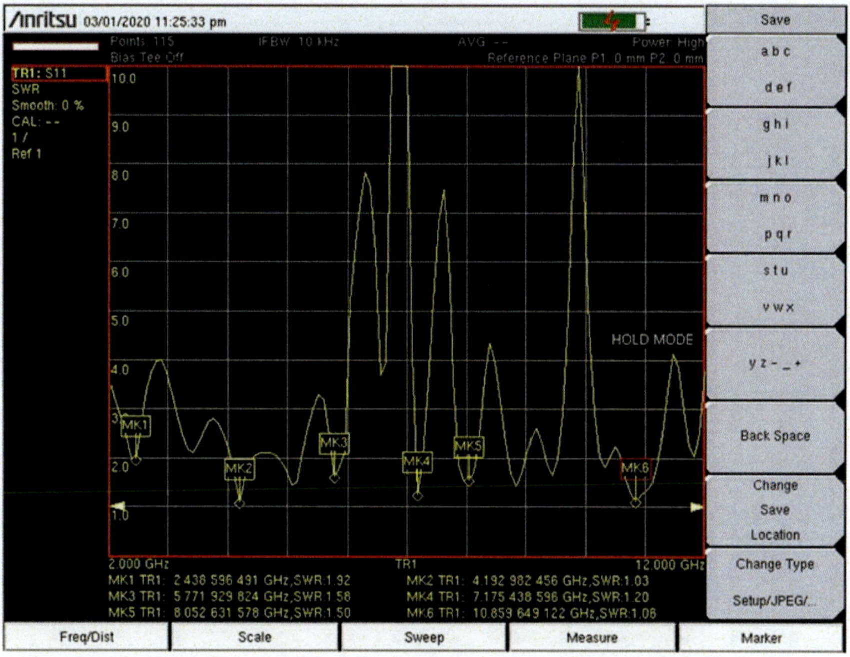

Fig. 7 Measured VSWR plot of the antenna structure

References

1. Chaouche YB, Nedil M, Hammache B, Belazzoug M (2019) Design of modified Sierpinski Gasket fractal antenna for tri-band applications. In:2019 IEEE international symposium on antennas and propagation and USNC-URSI radio science meeting, Atlanta, GA, USA, pp 889–890
2. Sharma N, Sharma V, Bhatia SS (2018) A novel hybrid fractal antenna for wireless applications. Progr Electromag Res M 73:25–35
3. Abdullah N, Mohamud Shire A, Ali MA, Mohd E (2015) Design and analysis of Minkowski fractal antenna. In: International conference on electrical and electronic engineering 2015 (IC3E 2015), 10–11 August 2015, Melaka, Malaysia
4. Abd Kadir MF, Ja'afar AS, Abd Aziz MZA (2007) Sierpinski carpet fractal antenna. In: 2007 Asia-Pacific conference on applied electromagnetics, Melaka, pp 1–4
5. Kaur A, Singh G, Kaur M (2017) Miniaturized multiband slotted microstrip antenna for wireless applications. Wirel Pers Commun 96:441–453
6. Hwang KC (2007) A modified Sierpinski fractal antenna for multiband application. IEEE Antennas Wirel Propag Lett 6:357–360
7. Balanis CA, Antenna theory, 2nd edn
8. https://www.emtalk.com/mpacalc.pp
9. https://www.emtalk.com/mscalc.php

Triple Band Cylindrical Dielectric Resonator Antenna with a Ring-Shaped Microstrip Line for Wi-Fi/5G Applications

B. Manikandan and D. Thiripurasundari

Abstract A compact multiband cylindrical dielectric resonator antenna (CDRA) fed by ring-shaped microstrip line along with two stubs is presented in this paper. The proposed antenna consists of cylindrical resonator with relative permittivity of 9.8 and FR4 substrate with dielectric constant of 4.4. Two bands are radiated through DRA by orthogonally connecting the L-shaped stub on the microstrip. The proposed antenna offers three separate bands ranging from 2.12 GHz to 2.6 GHz, 3.5 GHz to 3.89 GHz and 8.1 GHz to 8.6 GHz with the fractional bandwidth of 20.33%, 10.55% and 5.98%, respectively. The proposed design is suitable for Wi-Fi, 5G and X-band application.

Keywords CDRA · Microstrip fed · Multiband · Ring

1 Introduction

In modern communication, researchers mainly focus on lightweight and multifunctional performance antenna. This ensures that the antenna can operate efficiently in various wireless communication applications. Microstrip patch antenna can meet their needs but it suffers from low gain and narrow bandwidth due to high conductor loss at high frequency [1]. To overcome this problem, DRA was presented in 1980 by S. A. Long. The dielectric resonator antenna has gained more popularity due to various advantages: low profile, broad bandwidth, high gain, high radiation efficiency, high power handling capacity and low conductor loss on mm wave frequencies [2]. In addition, DRA has different degrees of design freedom, such as rectangular, cylindrical, hemispherical and triangular geometries [3]. For more than one frequency band service, different techniques have been introduced: Microstrip line along with

B. Manikandan (✉)
Department of Electronics and Communication Engineering, Nadar Saraswathi College of Engineering and Technology, Theni, Tamil Nadu, India

D. Thiripurasundari
Department of Electronics and Communication Engineering, VIT Chennai, Tamil Nadu, India
e-mail: dthiripurasundari@vit.ac.in

A. Sivasubramanian et al. (eds.), *Futuristic Communication and Network Technologies*, Lecture Notes in Electrical Engineering 792,
https://doi.org/10.1007/978-981-16-4625-6_64

two ring slots of conical DRA and various feed techniques such as coaxial probe, CPW, aperture coupled with DRA, microstrip fed with L-shaped and C-shaped DRA in which the two radiating resonators are closely coupled and radiate at different frequencies [4–6]. Another method for multiband generation such as multi-segment DRA stated by Bemani which gives good level of isolation but geometry of antenna becomes bulky [7]. Similarly, several techniques like parasitic element loading, band notched monopole, hybrid structure DRA and higher order mode generation are used for to achieve multiband characteristics in DRA [8–10].

In this letter, a novel feeding mechanism of ring-shaped microstrip line with cylindrical DRA is presented. The two L-shaped rectangular-shaped stubs are introduced at opposite side of the feed line in bifurcated fashion, which in turn acts as a dual band-pass filter. A circular slot has been mounted on the ground plane in order to reduce the slow wave effect and make antenna as a compact structure. The proposed antenna exhibits three different bands at reasonable bandwidth. Sections 2 and 3 address the antenna geometry and parametric analysis, respectively. Experimental results and conclusion are described in Sects. 4 and 5.

2 Antenna Geometry

The design of proposed antenna with ring-shaped microstrip line feed is illustrated in Fig. 1. The geometrical dimension of it is shown in Table 1. A ring-shaped microstrip line using FR4 substrate with relative permittivity ($\varepsilon_r = 4.4$), thickness $h = 1.6$ mm and loss tangent tan $\delta = 0.009$ support the 50 Ω characteristic impedance with width of strip line $W_f = 2.7$ mm and thickness of the strip $t = 0.035$ mm. The DR is made up of alumina dielectric material of dielectric constant ($\varepsilon_r = 9.8$), and loss tangent

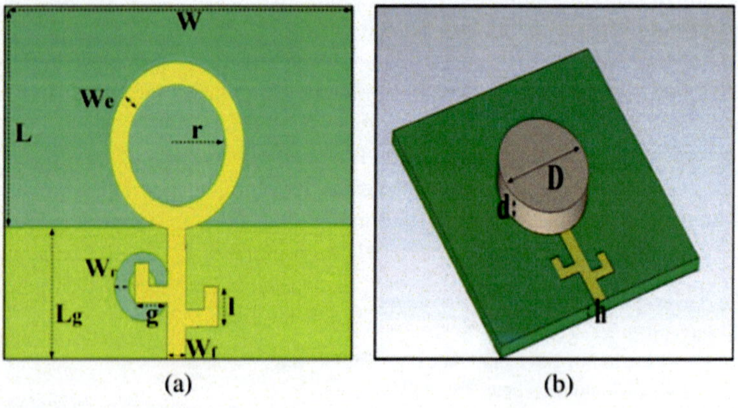

Fig. 1 Schematic diagram of the proposed antenna **a** feeding mechanism **b** isometric view of CDRA

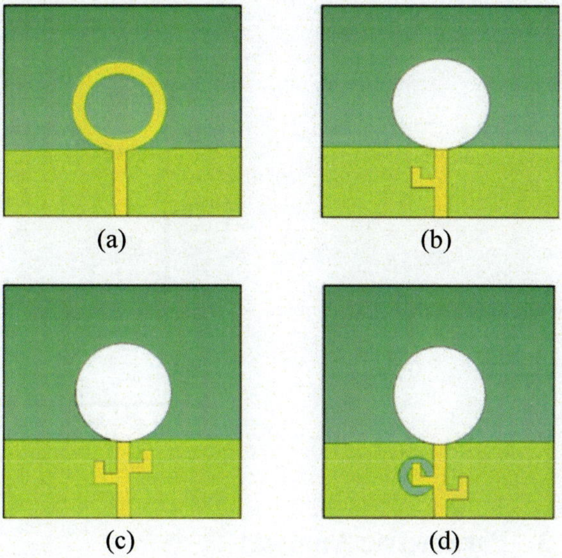

Fig. 2 Steps of antenna evolution **a** Ring antenna (Ant 1) **b** Single L strip feed (Ant 2) **c** Dual L strip feed (Ant 3) **d** Dual L feed with DRA (Ant 4)

Table 1 Geometrical dimension

Parameter	Dimension (mm)
L	50
W	48
W_f	2.7
l_f	2.7
W_g	2
f	9.3
l	4.6
g	4.6
D	10
r	7
d	3
h	1.6

$\tan\delta = 0.002$ is placed on the circular ring patch. The bandwidth performance can be controlled by choosing the dielectric material of height ($d = 3$ and radius ($a = 2$).

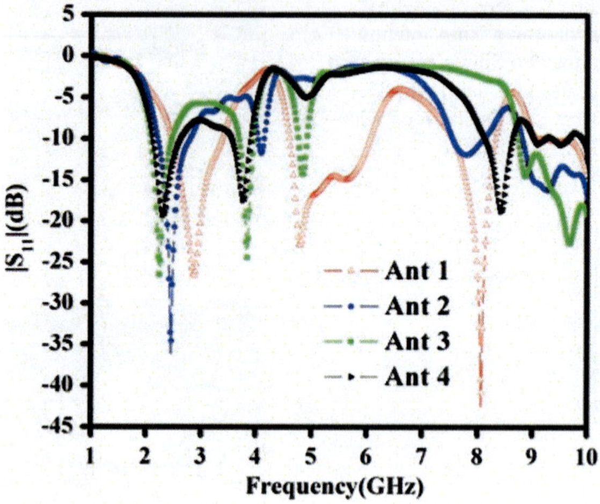

Fig. 3 Return loss of antennas prototype

3 Parametric Analysis

Hierarchical development of the antenna is shown in Fig. 2, and the corresponding return loss characteristic is shown in Fig. 3. Antenna architecture starts with a ring-shaped microstrip line with three separate bands at 2.8, 5.3 and 8 GHz. For frequency tuning, two L-shaped stubs are introduced in the design. Slight transformation occurred in return loss, while inserting the L-shaped stub on the left side of the microstrip line. This made the antenna to resonate at a single frequency resonance of 2.4 GHz in Ant 2. One more stub was added to the right side of the microstrip line resulting in another 3.7 GHz frequency response. At this stage, the two stubs in the prototype design serve as a band-pass filter. A circular-shaped slot has been inserted in the ground plane to make antenna small, which makes shift the resonant peak to lower frequency. As a result, the third operating frequency is generated, as shown in Fig. 2d. In the proposed structure, when circular slot is increased, the resonant peak has been shifted to lower frequency resulting in slow wave effects are generated and effective permittivity increased. Thus, the circular slot in the ground plane makes antenna compact and low profile.

4 Result and Discussion

The antenna S_{11} shown in Fig. 4 indicates the impedance bandwidth of three different bands ranging from 2.12 to 2.6 GHz, 3.5 to 3.89 GHz and 8.1 to 8.6 GHz is used for various applications such as Wi-Fi, 5G mid-range and X-band frequency applications. A VSWR plot shown in Fig. 5 indicates the antenna matching. The antenna gain shown on Fig. 6 shows that the maximum gain is 4 dBi. Figures 7, 8, and 9 indicate

Fig. 4 S_{11} of proposed antenna

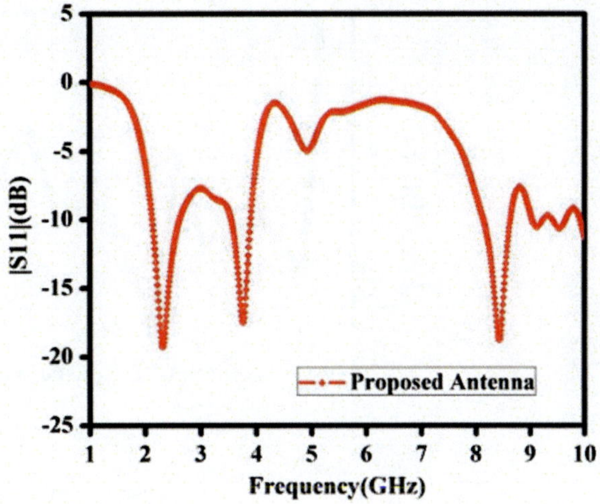

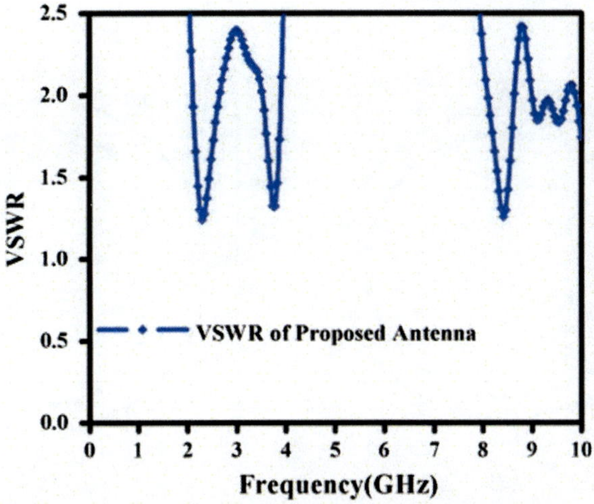

Fig. 5 VSWR of proposed antenna

the radiation pattern at three different operating frequencies such as 2.4, 3.7 and 8.4 GHz (Table 2).

5 Conclusion

A ring-shaped microstrip line with the cylindrical dielectric resonator antenna was investigated in this letter. A two opposite L-shaped stub was inserted into the microstrip line resulting in the two different frequency bands operating at 2.4 GHz

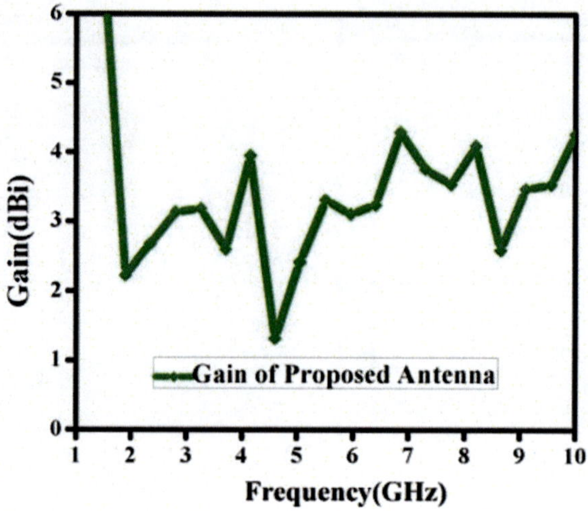

Fig. 6 Gain of the proposed antenna

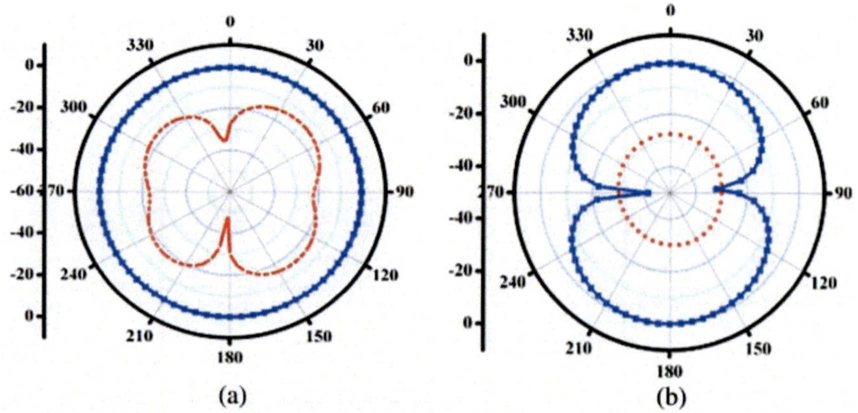

Fig. 7 Radiation pattern at 2.4 GHz **a** XZ-plane **b** YZ-plane

and 3.7 GHz, respectively. A circular slot was inserted on the ground plane to make antenna compact which radiates the third band of operating frequency at 8.4 GHz. By use of the circular slot, resonant frequency shifts from 9.7 to 8.4 GHz. One significant aspect of the total volume of the DR is the height of the DR, as it then raises the problem of modifying the prototype within the communication system. The proposed approach reduces the structure's total volume, which makes the antenna lightweight. The proposed antenna radiates triple separate frequency band, thereby antenna covers the Wi-Fi, 5G mid-range and X-band applications.

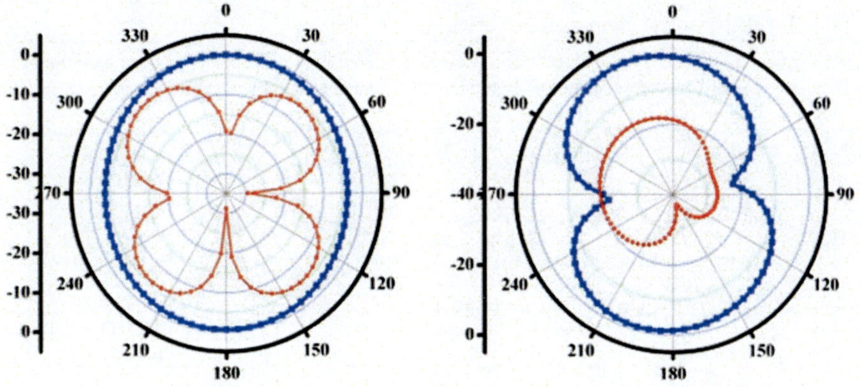

(a) (b)

Fig. 8 Radiation pattern at 3.7 GHz, **a** XZ-plane **b** YZ-plane

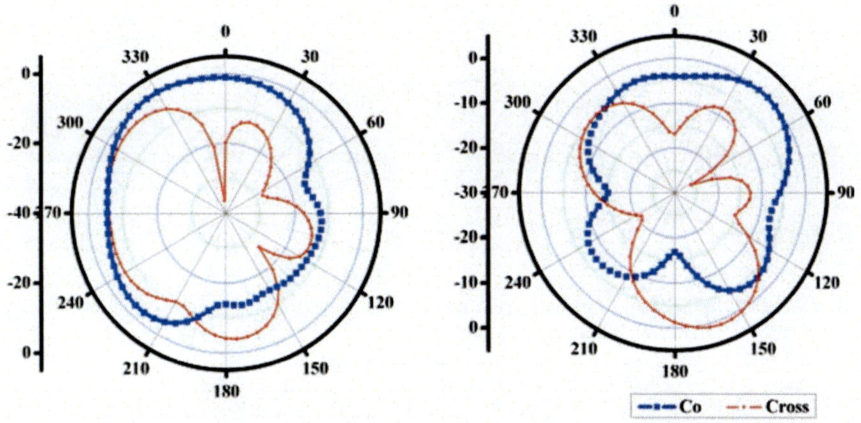

(a) (b)

Fig. 9 Radiation pattern at 8.4 GHz, **a** XZ-plane **b** YZ-plane

Table 2 Comparison of performance of antenna

Ref triple band antenna	Dimension (mm × mm × mm)	Height of the DRA (mm)	Lower band (%)	Middle band (%)	Upper band (%)	Feed techniques
[11]	$0.790\lambda_g \times 0.790\lambda_g \times 0.023\lambda_g$	$0.190\lambda_g$	18.34	2.6	7.47	Psi-shaped microstrip
[12]	$0.728\lambda_g \times 0.728\lambda_g \times 0.023\lambda_g$	$0.139\lambda_g$	8.33	17.14	39.93	Two ring-shaped microstrip
[13]	$1.096\lambda_g \times 0.6944_g \times 0.022\lambda_g$	$0.067\lambda_g$	6.06	26.4	35.7	MIMO + folded microstrip
[14]	$0.819\lambda_g \times 0.819\lambda_g \times 0.026\lambda_g$	$0.213\lambda_g$	18.31	23.81	–	Pentagon-shaped aperture
[15]	$0.759\lambda_g \times 0.759\lambda_g \times 0.184\lambda_g$	$0.189\lambda_g$	21.42	25.35	10.32	Annular ring + U-shaped print line-shaped microstrip
Proposed antenna	$0.731\lambda_g \times 0.701\lambda_g \times 0.023\lambda_g$	$0.043\lambda_g$	20.33	10.55	5.98	Ring-shaped microstrip + two rectangular stub

References

1. Petosa A, Ittipiboon A (1998) Recent advances in dielectric-resonator antenna technology. IEEE Antennas Propag Mag 40(3):35–48
2. Luk KM, Leung KW (2003) Dielectric resonator antenna. Res Stud
3. Petosa A (2007) Dielectric resonator antenna handbook. Artech House, USA
4. Hamsakutty V, Bindu G, Mathew KT (2005) A multi frequency coaxial-fed metal coated dielectric resonator antenna. Microw Opt Technol Lett 47(6):573–575
5. Huitema L, Koubeissi M (2011) Compact and multiband dielectric resonator antenna with pattern diversity for multistandard mobile handheld devices. IEEE Trans Antennas Propag 59(11):4201–4208
6. Denidni TA, Rao Q, Sebak AR (2005) Broadband L-shaped dielectric resonator antenna. IEEE Antennas Wirel Propag Lett 4:453–454
7. Bemani M, Nikmehr S (2011) A novel small triple band rectangular dielectric resonator antenna for WLAN and Wimax applications. J Electromag Waves Appl 25(11):1688–1698
8. Chen H-M, Wang Y-K, Lin Y-F, Lin S-C (2009) A compact dual-band dielectric resonator antenna using a parasitic slot. IEEE Antennas Wirel Propag Lett 8:173–176
9. Kim J-Y, Kim N, Lee S (2013) Triple band-notched UWB monopole with two resonator structures. Microw Opt Technol Lett 55(1):4–6
10. Sharma A, Gangwar RK (2015) Compact triband cylindrical dielectric resonator antenna with circular slots for wireless application. J Electromag Waves Appl 30(16):331–340
11. Sharma A, Das G, Ranjan P (2016) Novel feeding mechanism to stimulate triple radiating modes in cylindrical dielectric resonator antenna. IEEE Access 4:9987–9992
12. Sharma A, Gangwar RK (2015) Triple band two-segment cylindrical dielectric resonator antenna with a novel micro strip feed for WLAN/WiMAX applications. Microw Opt Technol Lett 57(11)

13. Sharma A, Das G (2016) Dual polarized triple band hybrid MIMO cylindrical dielectric resonator antenna for LTE2500/WLAN/WiMAX applications. Int J RF Microw Comput Aided Eng 26(9):763–772
14. Sharma A, Das G, Gangwar RK (2017) Dual-band dual polarized hybrid aperture-cylindrical dielectric resonator antenna for wireless applications. Int J RF Microw Comput Aided Eng 27(5)
15. Sharma A, Gangwar RK (2017) Tri-band composite cylindrical dielectric resonator antenna with hybrid mode excitation and cross-polarization suppression. Int J RF Microw Comput Aided Eng 27(8)

Failure Detection Using Artificial Neural Networks

B. R. Menaghapriya and R. G. Sangeetha

Abstract Failures in optical networks are of two categories: soft and hard failures. In this paper, the failure detection is monitored by a machine learning (ML) algorithm called artificial neural networks (ANN). The implementation of ANN algorithm for failure management has achieved an accuracy of 98%, when compared with other ML algorithms such as K-nearest neighbor, decision tree and SVM.

Keywords ANN—artificial neural network · QoT—quality of transmission · FM—failure management

1 Introduction

Failure in optical network leads to severe service disruption and loss of data transmission, which degrade the quality of transmission. Machine learning techniques have provided effective solutions to address challenges which are difficult to tackle manually in optical network. ML-based algorithms are popular for efficient performance in optical communication network. In this new era of artificial intelligence, the role of human operators in the network management systems can be replaced by the availability of accurate data monitoring systems [1]. The failure management in optical network communication is very important than any network issues as failures can induce service loss and disturbance. Due to its importance, failure management systems are always in need of complex and time-consuming manual intervention which can be avoided by increasing automatic failure recovery algorithms [2]. The utilization of ML algorithms is very promising to automate the network failure monitoring and management operations.

The failures in optical network communication are of two types: soft and hard failures. The hard failures are those which immediately lead to service loss and can

B. R. Menaghapriya · R. G. Sangeetha (✉)
School of Electronics Engineering, VIT University, Chennai, India
e-mail: sangeetha.rg@vit.ac.in

B. R. Menaghapriya
e-mail: menaghapriyab.r2019@vitstudent.ac.in

A. Sivasubramanian et al. (eds.), *Futuristic Communication and Network Technologies*, Lecture Notes in Electrical Engineering 792,
https://doi.org/10.1007/978-981-16-4625-6_65

be easily identified, e.g., fiber bends/cuts. The soft failures differ from above as it slowly leads to degradation of signal quality and reduce the QoT. Soft failures are mainly caused by signal overlapping, laser drift, filter shift, tight filtering, nonlinear interference, noise, etc. [3]. Optical performance monitoring (OPM) is the process of estimating physical layer parameters of the optical signals. OPM functionalities are reliable and improve the network efficiency. OPM is a helpful technology for software-defined networks (SDNs). With the help of OPM, the SDNs can be aware of real-time working network conditions and can adjust parameters such as modulation formats, power and spectrum assignment. ML algorithms are very much cost-effective for monitoring impairments in optical communication networks [4].

ML algorithms are widely used to identify and estimate soft failures. A method based on KNN and SVM classifier is used to monitor the BER as input and identify the filtering faults induced by wisdom systems is explained in [5–7]. In this paper, we implement ANN to detect failures caused by filtering, NLI interference and ASE noise by analyzing the BER as input. The approach is based on a set of classifiers that make predictions using features extracted from the optical signal spectrum. These approaches are designed with the ANN-based classifier approach, in which the failures are detected, identified and estimated. We also compare the accuracy of proposed ANN-based approach with decision tree (DT), support vector machines (SVM) and K-nearest neighbor algorithms.

2 Artificial Neural Network

Artificial neural network is inspired model from human brain architecture. The neural network with single neuron is known as perceptron, as shown in (Fig. 1).

A **Perceptron** is a combination of one input and one neuron. The input layer receives the input from the dataset. The number of nodes depends upon the number

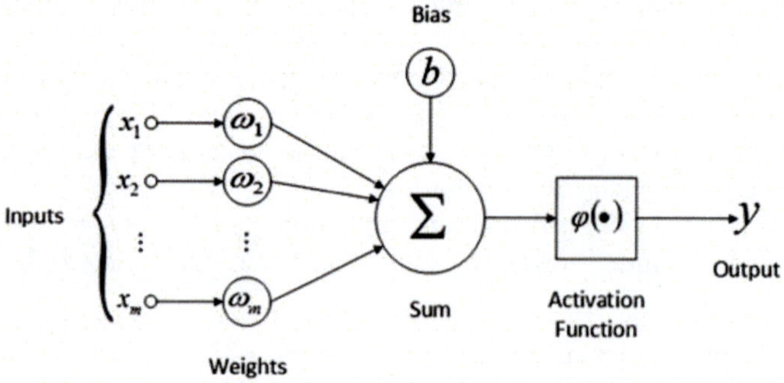

Fig. 1 Perceptron [8]

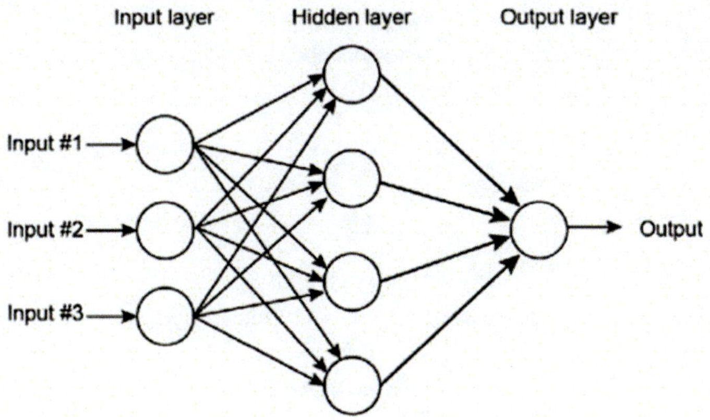

Fig. 2 ANN architecture [9]

of input features of a dataset. The weights are mostly initialized with random values. The input is multiplied by the weights and added together. The sum is passed into the activation function (which is similar to human neural system's neurons) to obtain the desired output.

The combination of N number of single perceptrons together is known as artificial neural network (or) **multilayer perceptron** (Fig. 2). The perceptron only has the ability to problems which are linearly separable with n number of input features. So the efficiency of a perceptron decreases in the case of nonlinearly separable dataset. To overcome the above disadvantage the multiple layers of perceptrons called MLP is used. The MLPs are a combination of multiple numbers of neurons connected together as a network. The major three layers in an MLP are an input layer, one (or) more hidden layer and an output layer. Feed-forward and back-propagation are the two phases executed in the MLP (Fig. 3).

In **feed-forward phase**, the input layer receives the values and multiplies with the weights. To avoid null value, a bias is added to the sum of inputs. The added value from input layer is passed into the first hidden layer based on their weights and bias. The activation function in the hidden neuron layers operates on the values received from input layer. Some of the activation functions are s sigmoid function, step function, tanh and relu function. Mostly the really and sigmoid functions were used for hidden and output layers. The output of the first hidden layer is passed to the second hidden layer and then to the next layer. The process of adding hidden layer is continued till reaching the outer layer. The calculated values at the outer layer are real outputs of the MLP algorithm. To improve the accuracy of predictions, the back-propagation phase is executed.

Back-propagation: the difference between predicted and desired output value is called error. The function which is used to calculate error is called 'loss function.' The minimization of error is done with partial derivation the error. The partial derivative of the error function is derived with respect to the biases and weights and it is known as gradient descent. The slope of the error function is estimated from derivatives. For

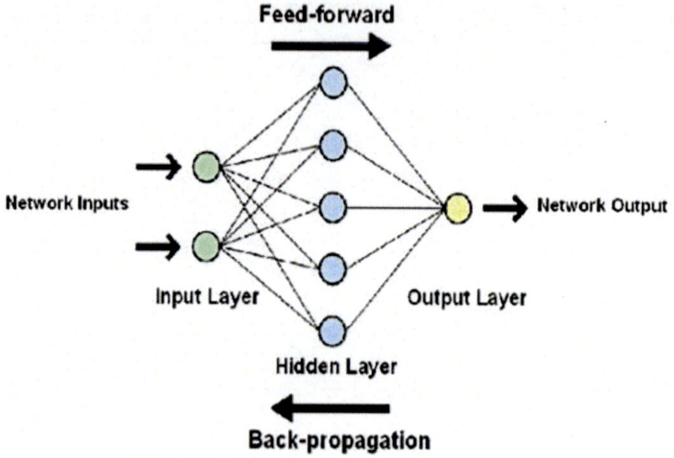

Fig. 3 Two phases of ANN algorithm [10]

positive slope, the weight value is reduced and for negative slope the weight value is increased, so that overall error gets reduced automatically. The process of reducing error calls optimization function. AN epoch is a combination of one feed-forward and back-propagation and it is continued until the accuracy is achieved.

3 Proposed Failure Management Approach

The proposed work involves in three functions: failure detection, failure identification and failure magnitude estimation. The algorithm is designed in following flow: data preprocessing, training and testing of dataset (Fig. 4). In the preprocessing stage, the dataset is collected and BER values are set as initial parameter to determine failure.

The input feature is the optical signal from which the BER can be classified. The dataset consists of x-input features such as maximum BER, minimum BER, mean

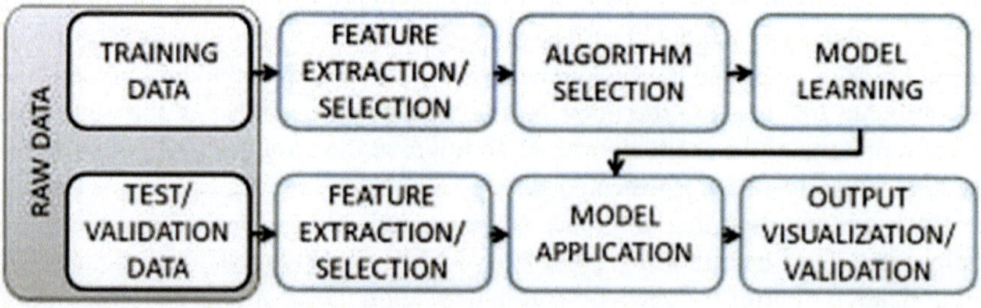

Fig. 4 ML algorithm train and test model flow [2]

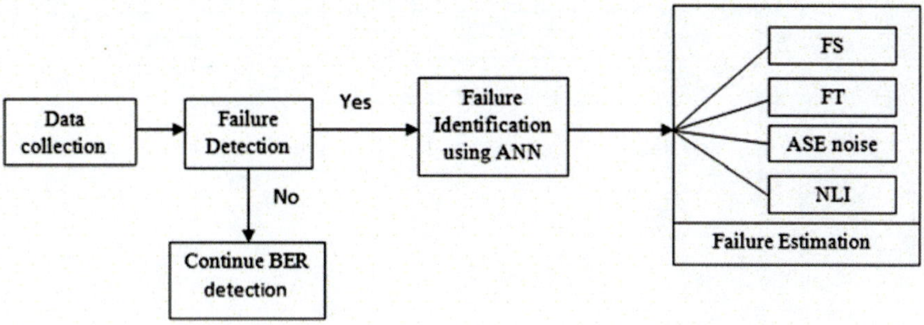

Fig. 5 Proposed ML algorithm design

BER values and output *y* attributes were failure detection, failure type and failure severity [2]. The collected input data '*x*' is fed into the ANN and detected output '*y*' is determined. The entire dataset is divided into two parts: 75% used for training and 25% data is used for testing. When the output value $y = 0$, then there is no failure and the OPMs can continue dataset collection, if $y = 1$, the failure is detected. The failure detection involves in classification of dataset two classes (normal = 0, failure = 1). The failure management model is designed as below (Fig. 5).

The failure detection feed the input $y = 0$ into the failure identification stage, where the type of failures are classified as filtering, ASE noise, nonlinear interference. The identification is done by assigning values to all the parameters, when there is change in normal level the failures are separated into different classes using ANN. For the filter failure, the central frequency offset of WSS is 14 to 18 GHz with the step size of 0.1 GHz. For the ASE noise, the noise at the receiver is set to with an OSNR of 9GHz and a step size of 0.08 dB. The launch power is set to 7–8 dBm with a step size of 0.008 dB for the NLI, where as the optimum launch power value is 0 dB. The threshold value is assumed to be 0.03 for FEC_BER. Among the dataset, 80% is spilt for training and 20% for testing phases. Further, the failure magnitude estimation is done to find out the severity of the failures in network. In the failure magnitude estimation stage, the impact of failures and their types and failure accuracy using ANN is predicted.

4 Results and Discussions

The above experiment is done using python software. The loss values on the training and testing during predictions are plotted in Fig. 6. During the training and testing process, the loss values on both sets are closer, which avoids over fitting. The heat map represents the performance of ANN over the test dataset. The confusion matrix provides information about failure identification and failure root cause in rows and columns. In order to further improve and validate the performance of the ANN, we train the model extra 10–50 times, and each time, the testing and training data are

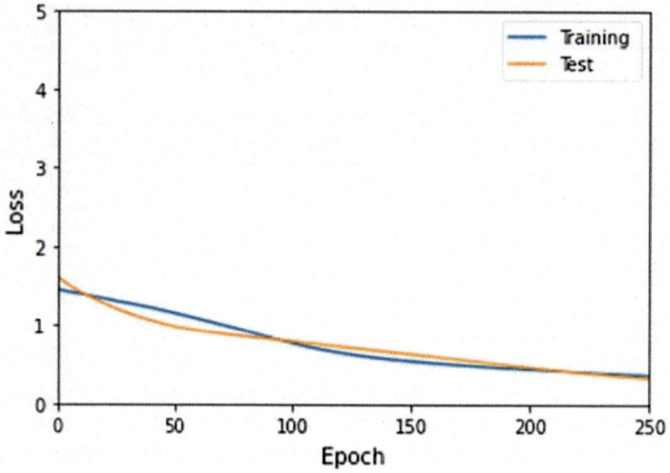

Fig. 6 ANN performance

randomly chosen from dataset. Finally, the accuracy of 98% is achieved for the given data.

The same dataset is fed into the algorithms such as SVM (support vector machine), DT (decision tree) and K-nearest neighbor (KNN) to evaluate the performance of proposed ANN algorithm (Fig. 7). The performance scores of each algorithm were compared with their respective F1 score and accuracy values. The F1 score and accuracy of SVM are 90 and 92%, for DT output, they are 91 and 93%, whereas for KNN output, they are 95 and 96%. As observed, ANN-based classifiers significantly outperform with an accuracy of 98% and F1 score value of 99 than all other algorithms

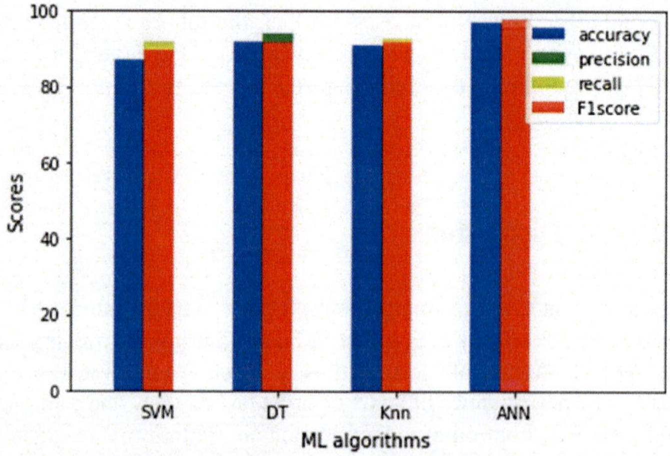

Fig. 7 Performance comparison with different ML algorithms

in failure detection and management. As a result, ANN-based classifiers can be the preferred option for failure management approaches.

5 Conclusion

In this work, the most efficient soft failure identifier based on ANN-MLP is proposed. Its performance and fast computations are demonstrated in systems with soft failures such as filtering, ASE noise and nonlinear interference. In future work, the implementation of proposed algorithm in real-time dataset is to be developed.

References

1. Vela AP, Ruiz M, Velasco L (2018) Applying data visualization for failure localization. In: Optical fiber communication conference, OSA Technical Digest (online). Optical Society of America, paper W1D.5
2. Musumeci F, Rottondi C, Corani G, Shahkarami S, Cugini F, Tornatore M (2019) A tutorial on machine learning for failure management in optical networks. J Lightwave Technol 37(16):4125–4139. https://doi.org/10.1109/JLT.2019.2922586
3. Velasco L, Shariati B, Vela AP, Comellas J, Ruiz M (2018) Learning from the optical spectrum: soft-failure identification and localization [Invited]. In: 2018 optical fiber communications conference and exposition (OFC), San Diego, CA, pp1–3
4. Khan FN, Fan Q, Lu C, Lau APT (2018) Machine learning-assisted optical performance monitoring in fiber-optic networks. In:2018 IEEE photonics society summer topical meeting series (SUM), Waikoloa Village, HI, pp 53–54. https://doi.org/10.1109/PHOSST.2018.8456700
5. Vela AP et al (2018) Soft failure localization during commissioning testing and lightpath operation. IEEE/OSA J Opt Commun Netw 10(1):A27–A36. https://doi.org/10.1364/JOCN.10.000A27
6. Allogba S, Tremblay C (2018) K-Nearest Neighbors classifier for field bit error rate data. In: 2018 Asia communications and photonics conference (ACP), Hangzhou, pp 1–3. https://doi.org/10.1109/ACP.2018.8596133
7. Vela AP, Shariati B, Ruiz M, Comellas J, Velasco L (2018) Soft failure localization in elastic optical networks. In: 2018 20th international conference on transparent optical networks (ICTON), Bucharest, pp 1–4. https://doi.org/10.1109/ICTON.2018.8473716
8. De Oliveira R, Araújo F, A system based on artificial neural networks for automatic classification of hydro-generator stator windings partial discharges. J Microw Optoelectron Electromag Appl 16:628–645. https://doi.org/10.1590/2179-10742017v16i3854
9. Manning T, Sleator R, Walsh P (2013) Biologically inspired intelligent decision making. Bioengineered 5. https://doi.org/10.4161/bioe.26997
10. Chan PhooiM'ng J, Mehralizadeh M (2016) Forecasting East Asian indices futures via a novel hybrid of wavelet-PCA denoising and artificial neural network models. PLOS ONE 11:e0156338. https://doi.org/10.1371/journal.pone.0156338

A Smart System for Monitoring Flow in Drip Bottles in Healthcare

Siddharth Yadav, Murali Manohar, O. Jeba Shiney, B. Priestly Shan, and Gnana Jeba Das

Abstract In this paper a system which is useful in monitoring the fluid level in intravenous drip bottles that can be used in intensive care units (ICUs) and post-surgical healthcare wards has been proposed. With the help of the proposed system the accidents and associated loss of lives occurring due to negligence of attendant/nurse while monitoring the fluid level can be avoided. A weight sensor to detect the level of fluid, Arduino Uno microcontroller to read the data, GSM module to send short message, SparkFun Blynk Board to communicate over IoT for real-time monitoring, an LCD to display percentage of fluid left, a buzzer to alert when the bottle is about to be empty and a solenoid valve to close the outflow of liquid from bottle to the patient's body when bottle is completely empty are used in the system design.

Keywords Arduino Uno microcontroller · GSM · SparkFun Blynk Board · IoT · LCD · Solenoid valve

1 Introduction

A number of automated devices for patient health monitoring are on the verge of development these days, to reduce stress among the patients as well as the healthcare professionals [6]. In a healthcare environment, safety of the patient is the most essential component to be given utmost importance while considering the design of the system. The use of Internet of Things can be a healthier counterpart in achieving this. Using IoT the 24*7 automated observation of patient is helpful in better treatment [7]. IoT system is related to sensors, digital and computing devices which are spread athwart the globe over the internet. These devices share and transfer information among themselves using unique ID for every device [8]. The task of IoT in healthcare is to extend a personalized and efficient healthcare to the users [9]. A number of advanced devices are used for monitoring patient's health during the course of therapeutic treatments. But this comes with the cost of this advanced equipment which

S. Yadav · M. Manohar · O. Jeba Shiney (✉) · B. Priestly Shan · G. J. Das
School of Electrical, Electronics and Communication Engineering, Galgotias University, Greater Noida, India

A. Sivasubramanian et al. (eds.), *Futuristic Communication and Network Technologies*, Lecture Notes in Electrical Engineering 792,
https://doi.org/10.1007/978-981-16-4625-6_66

could not be afforded by common people. In this work we have picked one such common case study of automating the intravenous drug delivery system. In most of the cases a nurse/attendant is responsible for monitoring the level of liquid inside the IV bottle. At times due to heavy workload often they forget to stop the outflow from the bottle when it turns empty due to which reverse flow of blood to the patient's body occurs which may cause pulmonary embolism (clotting of blood vessels in the lungs). To avoid this several automated systems were developed which uses android smart phone, Zigbee, IoT [10–13].

In our proposed system we have used GSM-based messaging system to alert the staff when the bottle is about to be empty. The system uses a load cell to measure the weight of the liquid, and a microcontroller reads the data and sends it to GSM module [14]. Also an LCD board displays the percentage of liquid in the saline bottle; when the liquid level is below the threshold value (100 ml) it sends an SMS to the attendant and the alert will be issued by a buzzer sound. Additionally the real-time fluid level can be monitored by a team of healthcare professionals in their workplace using the IoT module. In case if the staff/attendant is unable to reach the patient upon alarming, the outflow of liquid is automatically closed by the solenoid valve.

2 Literature Review

The work presented by Chiang [1] demonstrates a CMOS liquid level to frequency converter with calibration circuits for detecting liquid level of intravenous drip. An additive innovation to this idea is that the outputs are directly digitized so that their transmission could be possible directly over radio, IR sensors, ultrasonic sensors, PSN, etc. Also a calibration techniques used here, before performing calibration the maximum linear error is 46.44%, and post calibration, it is reduced to 2.04%. The chip which is proposed is suitable for detecting liquid levels of intravenous liquids.

In this paper [2] the system used power line communication device and telemetry intravenous drip monitoring system, an additional LAN network is not required for data communication, and using PLC the costs were reduced by using already formed power grids. Additionally, a wireless capacitive sensor was used to detect the remaining amount of IV set, and for transmission of result by RF communication microcontroller was used. This article [3] displays a microwave reflectometry-based framework utilized for constant observing and programmed control of stream of fluid in intravenous clinical implantations. The framework is planned for dodging some regular downsides getting from the optical detecting, in this manner permitting an effective elective answer for consequently observing the momentary progression of intravenous clinical arrangements. To this reason, the proposed framework consolidates microwave time area reflectometry (TDR) estimations with a non-intrusive detecting component (i.e., strip anodes straightforwardly connected to the outer surface of the imbuement bottle). The results accomplished show that through minimal effort convenient TDR gadgets stream control can be accomplished. Extravasation (spillage of fluid out of a holder) is a typical peril in intravenous (IV) treatment; babies (0–2 years matured) are at high hazard. In any case, in clinical work on, nursing staff are liable for checking the infusion status consistently [4], which prompts the danger of postponed treatment of extravasation of IV treatment in youngsters. Sensor combination makes up for the confinement of utilizing a solitary sensor, expands the estimation run, and is progressively impervious to natural impedance.

The article [5] investigates a RFID-based admonition arrangement of coming up short on infusion liquid for attendants; a RFID tag is planned and joined on a sack of intravenous dribble to show the advantages in the current framework. The fundamental thought of this framework is that tag is debilitated when the pack is not vacant in light of the EM stacking because of the fluid contained. The sack can be any sort in the present market and be with no electronic connection or alteration. LAN (local area network) is additionally applied as a piece of this framework for information transmission.

3 Methodology

Existing systems for the problem use IR sensor, ultrasonic sensor, load sensor [15, 16]. Our system consists of a load cell, load cell amplifier, microcontroller, GSM module, Wi-Fi module, electronic solenoid valve, LCD display, buzzer, IoT platform. A load cell is a weight sensor/transducer which converts weight into electrical signal; the electrical signal thus obtained goes to the HX711 amplifier which amplifies it and passes it on to the microcontroller. The signal is read by Arduino Uno (microcontroller), and the LCD display connected to the microcontroller displays the percentage of liquid left inside the bottle. In this developed prototype 500 ml is taken as 100% which can be changed as per requirement. Also a GSM900a module is also connected to the controller which sends SMS to the attendant/nurse when the liquid inside the bottle is below the threshold value (this is 100 ml in our case); additionally, a buzzer also sounds along with this message. Upon alerting if the outflow of liquid is stopped, the task is completed, but if the attendants not able to close it manually a solenoid valve is used to automatically close the outflow when the bottle is completely empty to avoid any misfortune. The solenoid valve is connected to the Arduino Uno which is programmed to close the valve when required (Fig. 1).

The Wi-Fi module used is SparkFun Blynk Board which is used to send real-time data to the staff. Using Blynk app the staff can monitor the real-time level of the liquid inside the bottle. The Blynk app is a platform over which with the use of IoT different electronic devices can be controlled and monitored. It is set up by providing it with a user name and password, and the same are mentioned in the SparkFun Blynk Board's code so as to restrict its access. After the complete setup is done, SparkFun Blynk Board is powered along with the microcontroller; a hotspot point is created and the bottle can be monitored by logging in to the Blynk app by using the same user name and password as mentioned above (Fig. 2).

Fig. 1 Automated drip monitoring system

4 Results

The individual components of the system were tested and verified successfully before executing it as a whole. As stated, the real-time level of liquid in the bottle is shown on the android app (the figure of the same is given at the end), and the interface used is Blynk app. This helps to incorporate a monitor which displays the real-time data in a central monitoring room where one can keep track of the system which adds an additional back up to the monitoring process even if the staff associated fails. The LCD displays the corresponding percentages of liquid left in the bottle, and when the level is below 100 ml, it shows warning and sends an SMS repeatedly with a time period of 10 s until the system is shut down along with it a buzzer also sounds the pictorial proofs of the same results are attached at the end of the file. If the attendant manually stops the outflow, the job is done; otherwise, the solenoid valve comes into play which closes the outflow automatically once the bottle is complete (Figs. 3 and 4).

5 Conclusion

In this paper, the smart bottle provides the real-time solution of observing the fluid level in the bottle with reliability. This system can be adopted in the general wards and post-ICU wards of the hospital to help the patients and nurses to understand its performance and utilization. It alerts nurses when the fluid level goes below 100 ml in the bottle and also displayed on LCD. When the bottle is completely empty the outflow is stopped using solenoid valve. The system provides an efficient approach to address a crucial problem in the healthcare environment which can be deployed at a minimum cost and hence utilized by a larger population.

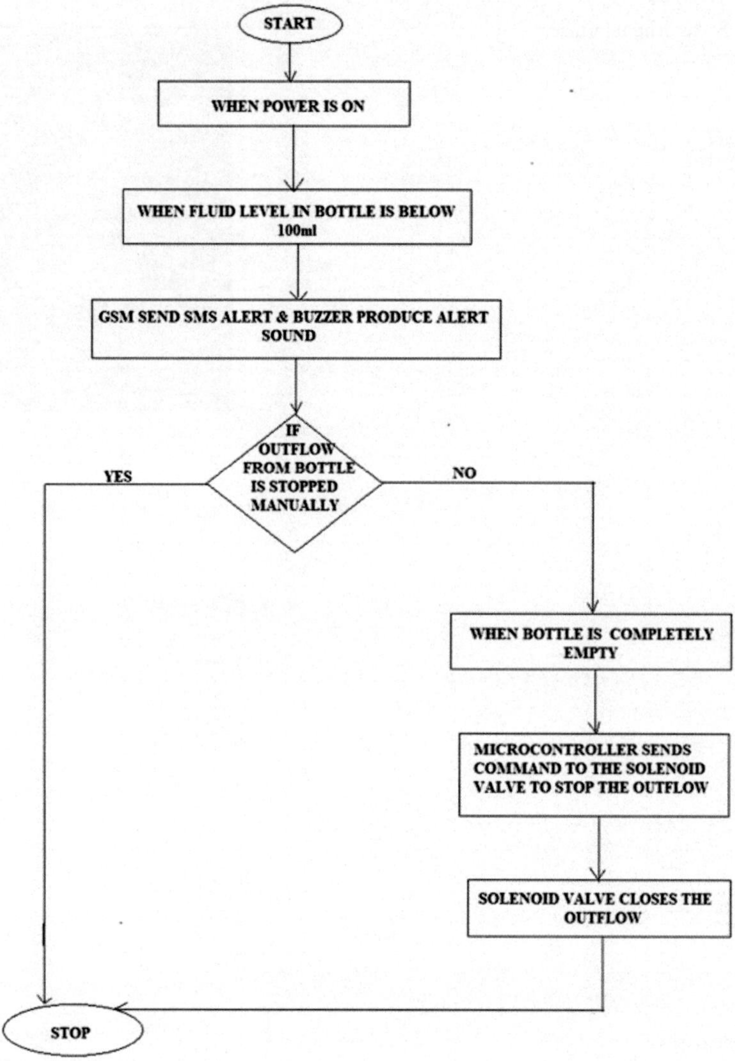

Fig. 2 Flowchart

Fig. 3 Blynk App interface

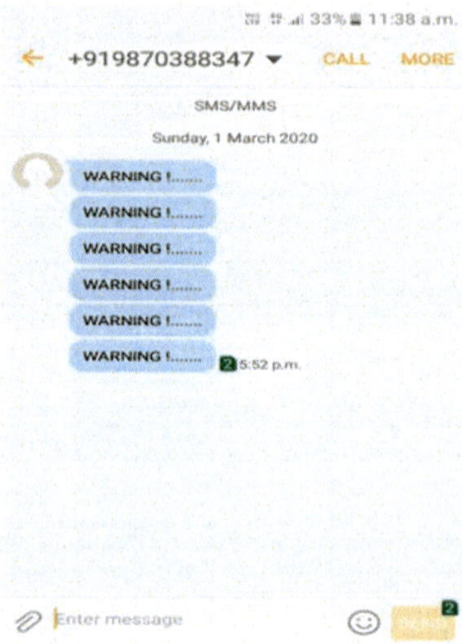

Fig. 4 Warning messages

References

1. Chiang C, Tsai P (2014) A CMOS liquid level to frequency converter with calibration circuits for detecting liquid level of intravenous drip. In: 2014 IEEE international conference on mechatronics and automation, Tianjin, pp 342–346
2. Kim DJ et al (2012) Design a wireless capacitive sensor detection system with power line communication for liquid volume of intravenous drip measurement. In: Proceedings of 2012 IEEE-EMBS international conference on biomedical and health informatics, Hong Kong, pp 269–272
3. Cannazza CG, Giaquinto N, Trotta A, Andria G (2011) Development of a remote system for real-time control of intravenous drip infusions. In:2011 IEEE international symposium on medical measurements and applications, Bari, pp 234–237

4. Lin J, Kuo C, Huang W, Lee H (2019) Demo abstract: a sensor-fusion approach system for detecting early extravasation of infant intravenous infusion. In: 2019 18th ACM/IEEE international conference on information processing in sensor networks (IPSN), Montreal, QC, Canada, pp 344–345
5. Huang C, Lin J (2011) A warning system based on the RFID technology for running-out of injection fluid. In:2011 annual international conference of the IEEE engineering in medicine and biology society, Boston, MA, pp 2212–2215
6. Priyadharshini R, Mithuna S, Kumar UV, Devi SK, Automatic intravenous fluid level indication system for hospitals. Int J Res Appl Sci Eng Technol (IJRASET) 3(VIII):427–432
7. Kishore KH, Nath KVS, Krishna KVNH, Kumar DP, Manikanta V, Basha FN (2019) IOT based smart health monitoring alert device. Int J Innovative Technol Exploring Eng (IJITEE) 8(6S)
8. Gupta K, Johari R (2019) IOT based electrical device surveillance and control system. In: 2019 4th international conference on internet of things: smart innovation and usages (IoT-SIU), Ghaziabad, India, pp 1–5
9. Chatterjee P, Cymberknop LJ, Armentano RL (2017) IoT-based decision support system for intelligent healthcare—applied to cardiovascular diseases. In: 2017 7th international conference on communication systems and network technologies (CSNT), Nagpur, pp 362–366
10. Singh J, Tiwari M, Shelar M (2012) ZigBee based patient monitoring system. Int J Comput Appl 51(22):17–20
11. Belgali PP (2015) Implementation of GSM based real time patient monitoring system. Int J Innovative Res Sci Eng Technol 4(8):6875–6878
12. Anbumani S, Kavin R, Saranya A, Dineshkumar T, Premkumar R (2015) An intelligent patient tele-monitoring system using android technology. Int J Res Eng Technol 04(02):477–482
13. Ghosh D, Agrawal A, Prakash N, Goyal P (2018) Smart saline level monitoring system using ESP32 and MQTT-S. In: 2018 IEEE 20th international conference on e-health networking, applications and services (healthcom), Ostrava, pp 1–5
14. Gayathri S, Sundar Ganesh CS (2017) Automatic indication system of glucose level in glucose trip bottle. Int J Multidisciplinary Res Mod Educ 3(1), 148–151
15. Chen F, Wang J, Chen S, Tu S, Chen K (2015) A hang-and-play intravenous infusion monitoring system. In: 2015 3rd international conference on applied computing and information technology/2nd international conference on computational science and intelligence, Okayama, pp 278–281
16. Bustamante P, Bilbao U, Solas G, Guarretxena N (2007) A new wireless sensor for intravenous dripping detection. In: 2007 international conference on sensor technologies and applications (SENSORCOMM 2007), Valencia, pp 30–35

A Review on: Machine Learning Techniques to Mitigate Security Risks in IoT Framework State of the Art

Pradeep Singh Thakur, Usha Kiran, and Om Prakash Sahu

Abstract The network of devices that are connected to communicate, can sense the data from the environment, and respond intelligently, is in the massive demand nowadays and being used in billions. Despite such an enormous demand, IoT devices suffer from various security issues as the protocol designed for the IoT environment is not capable enough to prevent the system from those attacks. Therefore, the security issues related to IoT devices have become a prominent research area. The major security attacks faced by IoT devices include fake device identity attack, denial-of-service (DoS) attack, distributed denial-of-service (DDoS) attack, malfunctioning, and attacks related to user data privacy. In this report, the comprehensive review of vulnerabilities related to IoT systems and security solutions based on machine learning techniques to mitigate security attacks on IoT devices has been discussed. Comparative analysis has also been shown to clearly understand the existing approaches, their limitations, and future scope.

Keywords Machine learning · IoT security · Intrusion detection system

1 Introduction

The IoT devices are used in many services, consisting of earthquake monitoring systems, building health systems, identification of landslides, energy management (smart parking and lighting), public building automation, and noise monitoring system air quality [1]. The perception, network, and application layers are the main

P. S. Thakur (✉)
National Institute of Technology, Raipur, India
e-mail: psthakur.bme@nitrr.ac.in

U. Kiran
Indian Institute of Technology, Bhilai, India
e-mail: ushak@iitbhilai.ac.in

O. P. Sahu
SENSE, Vellore Institute of Technology, Vellore, India
e-mail: omprakash.sahu@vit.ac.in

A. Sivasubramanian et al. (eds.), *Futuristic Communication and Network Technologies*, Lecture Notes in Electrical Engineering 792,
https://doi.org/10.1007/978-981-16-4625-6_67

basic layers in the three-layer architecture in the IoT framework. IoT devices are vulnerable to various security risks. As it follows, security constitutes an essential part of the IoT-based applications environment. The conventional IoT security architecture and IoT standards are just not enough. There major security issues related to IoT devices are phishing, malware, and X scripts [2], integrity, confidentiality, denial of services (DoS), eavesdropping, a middle-man attack, RFID intrusion, WSN node jamming, fake node, malicious code injection, sensor data protection, side-channel attack [3], etc.

The security solutions that require a lot of computation and memory capacity are generally avoided in IoT devices as the devices are resource constraint in nature in terms of CPU power, storage capacity, transmission bandwidth, and battery power. Therefore, we need security solutions that not only mitigate the security attacks but also are compatible with IoT devices [4]. Machine learning solutions are becoming more prominent for protecting IoT devices from attacks [5]. With the advancement and emerging compatibility of machine learning approaches with IoT devices, the selection of security mechanisms and parameters that influenced more with attacks has become easy. While designing the security protocols, we always have to keep in mind the limitation, heterogeneity, and dynamic nature of the networks [6].

This paper has reviewed the machine learning approaches used in IDS to be applied in the IoT environment. The primary focus of this review is to discuss the security issues related to the IoT system, how machine learning algorithms are becoming promising and reliable techniques to prevent the system from anomalies, and what are the current challenges related to machine learning algorithms design and deployment in the IoT environment.

2 Attacks Related to IoT Devices and Machine Learning Approaches Against Those Attacks

In this section, we discuss the various security issues [7] related to IoT devices and network and the machine learning solutions to mitigate those attacks.

2.1 Denial-of-Service (DoS)/Distributed Denial-of-Service (DDoS) Attack

The denial-of-service (DoS) attack prevents the IoT devices from using the network resources by consuming the resources in an unauthorized way [8]. The distributed denial-of service (DDoS) attack does the same thing as the DoS attack but in a distributed manner. The DDoS attack makes use of thousands of IP addresses and makes the identification of legitimate devices among attacking devices more difficult. The lightweight devices suffer a lot from DDoS attacks because of their limited capabilities. There are some machine learning-based techniques against DoS/DDoS attacks that we discuss in this section.

2.2 Machine Learning Techniques to Mitigate Against DoS/DDoS Attack

In [9], the author has presented a technique of how neural network techniques can be applied in a wireless sensor network for security purposes. The proposed technique supports the media access control (MAC) protocol. The multilayer perceptron model (MLP) has been developed to detect DoS attacks against CSMA-based wireless networks. The MLP constantly monitors the parameters that get affected by attacks and hence increases the security of a WSN. The technique is efficient because as soon as any suspicious activity is sensed, it promptly shuts off the sensor nodes working in the MAC layer and physical layer. The backpropagation and particle swarm optimization algorithms have been used here for the training of the neural network model. Results got from the simulation have shown the proposed technique is successful in increasing the lifetime of WSN. The false-positive alarms may affect the effectiveness of the network and cause the nodes to shut down even in the normal situation.

The DoS attack detection system using multivariate correlation analysis (MCA) has been introduced in [10]. The technique accurately characterizes network traffic by extracting geometric correlations using anomaly-based detection. Learning of legitimate network traffic patterns is sufficient to detect DoS attack, and support of triangle-area-based technique improves the detection. KDD cup99 dataset has been used here for the testing of the model. Both non-normalized data and standardized data have been used for the performance analysis. The detection accuracy is good compared to previous techniques. However, it also has false-positive alarms which degrade the network performance. The technique is not able to detect land, Neptune, and teardrop attack records. It performs well with normalized data, but not with original data.

Similar to the signal-to-noise ratio, the SINR represents the ratio of interest signal and the sum of total signal and background noise. This SINR has been used in [11] to detect DoS attack. After analyzing the two-player game with multiple power levels, the game framework using Markov has been implemented to interactively represent the decision-making process. The game is an interactive decision-making process that differentiates between the current state and the gathered state of information. A modified Nash Q-learning algorithm has been applied to solve the associated optimality (Bellman) equations to obtain the optimal solutions. The estimation performance depends on the initial estimations of the state.

2.3 Jamming Attack

The jamming attack is the attack on bandwidth, energy, central processing units (CPUs), and memory resources. It is initiated by attackers by transmitting fake signals to interrupt the ongoing radio transmissions of IoT devices and further decimate the resources [12]. The following techniques have been applied to detect jamming attacks.

2.4 Machine Learning Techniques Against Jamming Attack

The reinforcement learning approach has been proposed in [13] to detect the jamming attack. The multi-agent environment has been assumed here that operates wideband autonomous cognitive radios (WACRs). To handle a jamming attack, all the sensors try to prevent the transmissions of other WACRs as well as avoid the jammer signal that moves through the spectrum band. This technique helps to avoid both the jammer signal and any other interference. The computational complexity has been proven to be low by the results. If the WACR faces any interference in the network, it switches to an available sub-band to continue uninterrupted transmission. In case of more interference, switching among sub-band may itself become time-consuming. Therefore, this technique cannot be applied in low-cost applications and systems where the delay is not a big concern.

In the cognitive radio network, to deal with the jamming attack, a two-dimensional communication scheme has been implemented in [12]. The spread spectrum and mobility of users have been exploited in this technique to detect the jamming attack by the secondary user (SU). The jamming area of SU is determined using deep Q-network. Hopping frequency patterns have been analyzed here to defeat smart jammers. Prior knowledge of the jamming and radio channel model is not needed. The use of a deep neural convolution network with a large number of frequency channels speeds up the learning process. This technique is efficient to detect a cooperative jamming attack but increases the signal-to-interference-plus-noise ratio. The DQN-based technique requires high computational cost, and the resource requirement concludes that the technique is not suitable for resource-constraint devices.

The detection technique based on Dyna-Q-based reinforcement learning algorithm has been proposed in [14] for the software-defined network (SDN) to combat software-defined radio. Dual controllers to meet collective decision making have been used here. It includes the SDN controller and the UAV flight controller. It is built on top of the SDN, and the entire network's state knowledge converges rapidly. The UAV preparation into a particular space for action and power allocation has been consolidated. The result got from the simulation proves the proposed technique is efficient to detect smart jamming attacks. This technique is comparatively more rapid in learning than the existing techniques. The deployment of Dyna-Q-based environmental model is not easy and, therefore, often overlooked.

2.5 *Spoofing Attack*

In a spoofing attack [15], the attacker node pretends to be a legal user with the identity of any legal user. The identity that an attacker may use in spoofing attacks is the medium access control (MAC) address, IP address, and RFID tag to gain illegal access to the IoT system. The spoofing attack may be used to launch further attacks such as DoS/DDoS and man-in-the-middle attacks [16]. The techniques have been discussed in the following to mitigate and prevent spoofing attacks.

2.6 *Machine Learning Techniques Against Spoofing Attack*

The authentication has been applied in the physical layer [17] to detect spoofing attack. The technique is based on the game zero-sum. The receiver selects the test threshold in the hypothesis test to maximize its usefulness in spoofing detection. With the use of the Q-learning- and Dyna-Q-based spoofing detection schemes, they have achieved the optimal test threshold via reinforcement learning. The technique has been deployed over universal device radio peripherals, and for the testing, a laboratory environment has been used. Results obtained from experiments confirmed the efficacy of the methods proposed. The authentication technique to detect spoofing attack can be applied in MAC and network layer as well.

The technique proposed in [18] exploits any prior information available related to the system and to resolve limitations imposed from the distributed problem structure. Supervised and semi-supervised learning has been used to model the problem of attack detection with the fusion of decision and feature levels. Learning algorithms have been analyzed with statistical learning methods to detect non-observable attacks. The algorithms have been analyzed over different IEEE test systems. Results after experiments show that the performance of the proposed technique is higher than existing machine learning-based detection techniques. The large amount of data are required to train the machine. As the network grows, the false-positive cases get increased too.

The paper [19] suggested a system in which sensors capture specific aspects of human physiology and their everyday actions. A deep learning-based user authentication scheme has been applied to the extracted features to accurately identify each device. The logistic regression analysis along with the Frank–Wolfe algorithm [20] has been used for estimating the logistic regression model parameters. A distributed PHY layer authentication minimizes the overhead contact between the landmarks and the security agencies. The proposed construction provides higher accuracy and lower overhead computation online authentication. The system is feasible for most of the wireless networks.

2.7 Intrusion in IoT Network

Intrusion is any kind of malicious activity performed by the attacker to disturb the normal functioning of the network or system [21]. The intruder may be an insider or an outsider. To prevent or detect such activity, the system is designed called intrusion prevention or detection system. This intrusion detection system can be used to protect the whole network or a single system, and it means IDS can be network based or host based [22]. The following are the machine learning-based approaches in support of the intrusion detection system.

2.8 Machine Learning Techniques for Intrusion Detection

In this paper [23], the author has addressed the problem related to unsupervised outliers detection in wireless sensor networks. The technique helps to minimize resource consumption like energy and bandwidth. The communication is managed using only a single hop. The technique allows efficient detection of node failure, carrier sense facility, and deftly manage the complex data. This method is well suited for applications in which either an adjustment of the sliding window size or the number of neighbors used in a distance-based outlier detection technique can measure the confidence of an outlier rating. The technique is not suitable for resource-constraint devices as communication is costly and requires the most accurate data, including outliers.

The author has implemented the deep learning approach for the intrusion detection in [24]. In this technique, they have designed a multi-convolution neural network (multi-CNN) fusion framework. The feature data are split into four parts and are then converted into a grayscale graph, depending on the correlation. The experimental results show that the multi-CNN fusion model is highly appropriate for the provision of high-precision and low-complexity grading system on the NSL-KDD dataset. Furthermore, the performance of this system is superior to traditional machine learning methods. This work may help to protect IoT devices used in industrial applications. Deep learning fusion and online learning for the issue of network intrusion detection, which can intelligently secure industrial IoT data, can be implemented in future.

2.9 Software Attacks

The software attacks include attacks like worms, viruses, malware, and trojan [25]. A few of the techniques have been listed in the following that help to detect malware with the support of machine learning approaches.

2.10 Machine Learning Techniques to Detect Malware

In [26], the author discusses malware detection based on cloud, where mobile devices download traces of their application via simple stations or in complex networks to secure servers. The Q-learning has been used here to detect malware that offers the optimum rate of offloading. The detection efficiency is increased using Dyna architecture and reinforcement learning. The results of the simulation show that the proposed method increases the accuracy of detection, decreases detection time, and makes a mobile device more useful. The highest detection accuracy is yet to be achieved.

In this paper [27], the author has addressed the security issues related to mobile devices. The proposed technique detects malware using the anomaly-based approach. Two datasets have been used in the assessment: private and public (i.e., MalGenome). Based on the results of the assessment, the Bayes network and random forest classifications have been given more precise readings. The proposed approach gives a true-positive rate (TPR) of 99.97 percent with the MalGenome dataset, as opposed to the multilayer perception with only 93.03%. Processing time for the detection of malware, precision, recall, and f-measure value increases concerning large size databases.

In [28], the author examined the ability of deep learning using recurrent neural network (RNN) in detecting IoT malware. This method utilized RNN to evaluate the execution operation codes (OpCodes) of ARM-based IoT applications. An IoT framework dataset composed of 281 malware and 270 benign ware was used to train the models. The trained model then has been used to test 100 new IoT malware samples with three different long–short duration memory configurations (LSTMs). The model is not previously known to the model. Because of the use of tenfold cross-validation and two-layer neural network have achieved 98.18% accuracy in detecting new samples of malware. Also, a comparative analysis of the LSTM approach with other classifications indicated that it offers the best possible outcome. In this technique, two-layer neurons provide the highest accuracy (98.18%). Deep learning techniques need to be exercised to increase the accuracy, speed, and scalability of detection techniques.

3 Conclusion and Challenges

The vulnerabilities related to IoT devices have been explored in detail in this paper. After that, machine learning-based methods have been addressed to defend the device from the various security attacks. It is clear from the literature that machine learning approaches help to improve IoT devices and network efficiently. However, detailed analysis and comprehensive study on approaches to machine learning are still to be done. The resource-constraint nature of devices, data analysis, and heterogeneity of devices are major challenges in IoT environment. While implementing any protocol, we must take into account all the limitations of IoT devices. In IoT-connected devices, communication losses often occur and reliable signals are needed to ensure data reach the other end of time. Training the machines effectively with accurate data is a big challenge. In the case of machine learning approaches, we need to gather and analyze a large amount of data. The data that come from different sources and in different formats need to be processed. The real-time processing of these data often makes it difficult because the data are comprehensive, so that training the machines effectively is a big challenge. A variety of devices working over different protocols are connected to the IoT network. The application built for such an environment should be compatible with all available devices and protocols.

References

1. Hancke GP, Hancke GP Jr (2013) The role of advanced sensing in smart cities. Sensor 1:393–425
2. Meneghello F, Calore M, Zucchetto D, Polese M, Zanella A (2019) IoT: Internet of threats? A survey of practical security vulnerabilities in real IoT devices. IEEE Internet Things J 6(5)
3. Adat V, Gupta BB (2018) Security in Internet of Things: issues, challenges, taxonomy, and architecture. Telecommun Syst 67(3):423–441
4. Hussain F, Hussain R, Hassan SA, Hossain E (2020) Machine learning in IoT security: current solutions and future challenges. IEEE Commun Surv Tutorials
5. Singla A, Sharma A (2019) Physical access system security of IoT devices using machine learning techniques. In: Proceedings of international conference on sustainable computing in science, technology and management (SUSCOM). Amity University Rajasthan, Jaipur–India
6. Restuccia F, D'Oro S, Melodia T (2018) Securing the Internet of Things in the age of machine learning and software-defined networking. IEEE Internet Things J 5(6):4829–4842
7. Chaabouni N, Mosbah M, Zemmari A, Sauvignac C, Faruki P (2019) Network intrusion detection for IoT security based on learning techniques. IEEE Commun Surv Tutorials 21(3):2671–2701
8. Kasinathan P, Pastrone C, Spirito MA, Vinkovits M (2013) Denial-of-Service detection in 6LoWPAN based Internet of Things. In: 2013 IEEE 9th international conference on wireless and mobile computing, networking and communications (WiMob), pp 600–607
9. Kulkarni RV, Venayagamoorthy GK (2009) Neural network based secure media access control protocol for wireless sensor networks. In: 2009 international joint conference on neural networks, pp 1680–1687
10. Tan Z, Jamdagni A, He X, Nanda P, Liu RP (2013) A system for denial-of-service attack detection based on multivariate correlation analysis. IEEE Trans Parallel Distrib Syst 25(2):447–456
11. Li Y, Quevedo DE, Dey S, Shi L (2016) SINR-based DoS attack on remote state estimation: a game-theoretic approach. IEEE Trans Control Netw Syst 4(3):632–642
12. Han G, Xiao L, Poor HV (2017) Two-dimensional anti-jamming communication based on deep reinforcement learning. In: 2017 IEEE international conference on acoustics, speech and signal processing (ICASSP), pp 2087–2091
13. Aref MA, Jayaweera SK, Machuzak S (2017) Multi-agent reinforcement learning based cognitive anti-jamming. In: 2017 IEEE wireless communications and networking conference (WCNC), pp 1–6
14. Li Z, Lu Y, Shi Y, Wang Z, Qiao W, Liu Y (2019) A Dyna-Q-based solution for UAV networks against smart jamming attacks. Symmetry 11(5)
15. Nawir M, Amir A, Yaakob N, Lynn OB (2016) Internet of Things (IoT): taxonomy of security attacks. In: 2016 3rd international conference on electronic design (ICED), pp 321–326

16. Cekerevac Z, Dvorak Z, Prigoda L, Cekerevac P (2017) Internet of Things and the man-in-the-middle attacks–security and economic risks. MEST J 5(2):15–25
17. Xiao L, Li Y, Han G, Liu G, Zhuang W (2016) PHY-layer spoofing detection with reinforcement learning in wireless networks. IEEE Trans Veh Technol 65(12):10037–10047
18. Ozay M, Esnaola I, Vural FT, Kulkarni SR, Poor HV (2015) Machine learning methods for attack detection in the smart grid. IEEE Trans Neural Netw Learn Syst 27(8):1773–1786
19. Xiao L, Wan X, Han Z (2017) PHY-layer authentication with multiple landmarks with reduced overhead. IEEE Trans Wirel Commun 17(3):1676–1687
20. Clarkson KL (2010) Coresets, sparse greedy approximation, and the Frank-Wolfe algorithm. ACM Trans Algor (TALG) 6(4):1–30
21. Mishra A, Nadkarni K, Patcha A (2004) Intrusion detection in wireless ad hoc networks. IEEE Wirel Commun 11(1):48–60
22. Singh AP, Singh MD (2014) Analysis of host-based and network-based intrusion detection system. IJ Comput Netw Inf Secur 8:41–47
23. Branch JW, Giannella C, Szymanski B, Wolff R, Kargupta H (2013) In-network outlier detection in wireless sensor networks. Knowl Inf Syst 34(1):23–54
24. Li Y, Xu Y, Liu Z, Hou H, Zheng Y, Xin Y, Zhao Y, Cui L (2020) Robust detection for network intrusion of industrial IoT based on multi-CNN fusion. Measurement
25. Xiao L, Wan X, Lu X, Zhang Y, Wu D (2018) IoT security techniques based on machine learning: how do IoT devices use AI to enhance security? IEEE Sig Process Mag 35(5):41–49
26. Xiao L, Li Y, Huang X, Du X (2017) Cloud-based malware detection game for mobile devices with offloading. IEEE Trans Mob Comput 16(10):2742–2750
27. Narudin FA, Feizollah A, Anuar NB, Gani A (2016) Evaluation of machine learning classifiers for mobile malware detection. Soft Comput 20(1):343–357
28. Haddad Pajouh H, Dehghantanha A, Khayami R, Choo KK (2018) A deep recurrent neural network based approach for internet of things malware threat hunting. Futur Gener Comput Syst 85:88–96

A Comparative Analysis of PAPR Diminishment Techniques for Next Generation Network: LTE-MIMO to 5G

Shailesh S. Hajare and D. Vydeki

Abstract In the wireless communication system frameworks, OFDM is an efficient and capable multicarrier modulation approach for far-off transmission of data like audio and video via radio waves. The rising requirement for high-speed and high-quality data led to the evolution of 5G mobile communication. The 5G requirements such as information rate, high spectral efficiency, low latency and complexity could be achieved using multicarrier modulation which is being researched as a piece of 5G physical layer arrangements. This paper presents the comparative analysis of PAPR diminishment techniques used in next generation networks. A simulation result shows that the performance of adjacent techniques reduces the PAPR. The accomplishment PAPR diminishment techniques is also analyzed by changing the number of transmitted symbols, oversampling factor over various OFDM sub-block lengths, number of subcarriers for different modulation techniques. The Bit error rate analysis is performed to compare performance of various systems.

Keywords Orthogonal frequency division multiplexing (OFDM) · Selective mapping method (SLM) · Bit error rate (BER) · Peak to average power ratio (PAPR)

1 Introduction

Third era (3G) versatile communication technologies actualized almost around the globe are neither ready to meet the prerequisites of the high data rates nor throughputs. Besides, voice communication in 3G relies on circuit switching technology, the same as in second-generation (2G) communication systems, rather than pure Internet protocol (IP) approach. Researchers have been trying for the next evolutionary fourth generation (4G) communication systems to give a far reaching and secure IP arrangement where voice, information and media can be offered to clients

S. S. Hajare (✉)
Vellore Institute of Technology, Chennai, India

D. Vydeki
SENSE, Vellore Institute of Technology, Chennai, India

A. Sivasubramanian et al. (eds.), *Futuristic Communication and Network Technologies*, Lecture Notes in Electrical Engineering 792,
https://doi.org/10.1007/978-981-16-4625-6_68

at "Anytime, anyplace" with higher information rates than past ages. Orthogonal frequency division multiplexing (OFDM) is the multicarrier modulation technique used within LTE, 4G signal format [1]. OFDM, orthogonal frequency division multiplex, is the simple layout used, and that is changed to offer the more than one get entry to scheme: OFDMA, orthogonal frequency division more than one get entry to with inside the downlink and SC-FDMA, single channel orthogonal frequency division more than one get entry to with inside the uplink. It has significantly progressed the overall performance as compared to preceding solutions. However, with inside the previous couple of years, the evolution of necessities and use instances posed a venture to next generation systems such as 4G and 5G which ought to be able, amongst others, to cope with the massive variety of customers and offer better information rates [2].

The current paper is organized as follows: Section 1 gives prerequisite of wireless communication standard, Sect. 2 gives literature survey of PAPR techniques and its impact on system performance. In Sect. 3, definition of PAPR and current challenges are mentioned along with reduction techniques. Section 4 presents comparative analysis of PAPR diminishment methods in SISO OFDM and MIMO models. Simulated results and its discussion are provided in Sect. 5. The conclusion is presented as Sect. 6.

2 Literature Survey

There is considerable research carried out in PAPR reduction. A complete overview of literature on MIMO OFDM for PAPR diminishment methods has been presented here.

Ben Mabrouk et al. [1] their paper has addressed a necessary condition to achieve better PAPR performance than OFDM as tone reservation and active cancellation technique.

Yi and Wang et al. [2] in their paper have addressed a technique for PAPR discount through random constellation mapping. Using random mapping methods, signals are converted to data symbols. Mapping is performed across each symbol.

Then, the candidate symbol with least PAPR is dispatched out for further communication in conjunction with its candidacy number.

Sghaier et al. [3] describe a MIMO OFDM system with blind and reduced complexity methods. High-frequency-selective channels are considered.

Boccardi et al. [5] addressed PAPR techniques with Fourier-based and Wavelet-based multicarrier communication techniques. Based on results, they categorized multicarrier modulation as high PAPR, low PAPR and moderate PAPR systems.

3 Peak to Average Power Ratio

In OFDM, the original information which is to be transmitted consists of high power and very narrow bandwidth. This signal is divided into N parallel blocks using IFFT operation. After IFFT operation, parallel data is converted into serial with last sample and cyclic prefix (CF). Therefore, OFDM structures are acknowledged to have an excessive PAPR, as compared with single-carrier structures. In fact, the excessive PAPR is one of the maximum destructive elements within side the OFDM system because it decreases the signal-to-quantization noise ratio (SQNR) of analog-to-digital converter (ADC) and digital-to-analog converter (DAC) even as degrading the performance of the power amplifier within side the transmitter. The PAPR hassle is greater essential within the uplink because the performance of the power amplifier is essential because of the confined battery energy in a cellular terminal. High PAPR is one of the massive drawbacks of multicarrier modulations. PAPR is described as the ratio among the maximum power of the signal and its average power. The PAPR expressed as given in Eq. (1)

$$\text{PAPR} = \frac{P_{\text{peak}}}{P_{\text{average}}} = 10\log 10 = \frac{\max\left[\{|X_n|\}^2\right]}{E\left[\{|Xn|\}^2\right]} \tag{1}$$

As per Eq. (1), peak energy traits also can be defined in phrases in their magnitudes (now no longer energy) with the aid of using defining the crest factor (CF) as shown in Eq. (2);

$$\text{Pass band Condition CF} = \sqrt{\text{PAPR}} \tag{2}$$

3.1 *Challenges to Mitigate PAPR Requirements*

PAPR approach in OFDM has till now confronted substantial challenges related to power amplifiers used at transmitter and receiver.

As per power distribution defined by radio technologies, higher power is consumed due to high-power amplifier as shown in Fig. 1. One of the primary parameters that have an effect on all cell UE devices is their battery life. It is consequently essential to make certain financial and green energy use within the transmission and reception of signals. The complexity of the PAPR perspectives unexpectedly will increase while many digital items combat for display screen area to mark bodily entities within the dynamic perspectives of more than one fast moving roaming user.

- ***High PAPR*** and inter-carrier interference (ICI) are the two foremost issues within the implementation of an OFDM system. OFDM systems require tight frequency synchronization in contrast to single carrier systems, due to the fact that in OFDM,

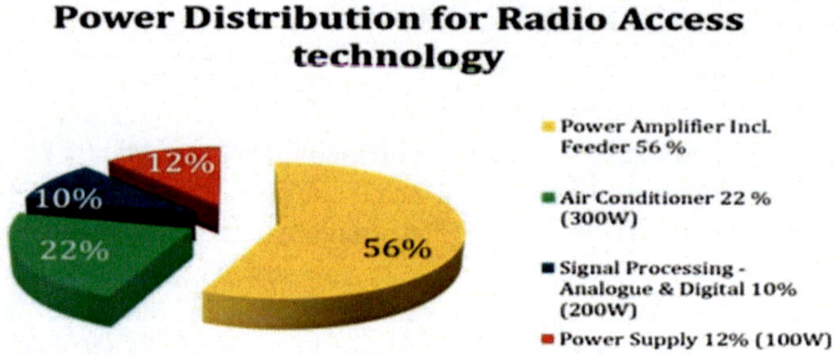

Fig. 1 Power distribution for radio access technology

the subcarriers are narrowband. Therefore, it is far touchy to small frequency offset among the transmitted and the obtained signal, which may also rise up because of Doppler Effect within side the channel or because of mismatch among transmitter and receiver nearby oscillator frequencies.

- ***Inter-service interference (ICI)*** is a main assignment inside the error-loose reception and detection of OFDM symbols.
- ***Increase within the spectral efficiency*** of wireless communication systems is one of the finest demanding situations confronted with the aid of using wireless communication engineers.
- ***Bandwidth efficiency***: The available bandwidth is scarce and costly, whereas there can be a massive call for data price created through developing the amount of subscribers and growth in multimedia applications which require big bandwidth.

3.2 PAPR Diminishment Techniques

Multicarrier modulation structures are affected by excessive PAPR. If the high-power digital equipment (HPA) is operated in this similar linear region, then the excessive PAPR does not influence the standard of the sign on the transmission. But, this example finally ends up in an excessive value in phrases of strength efficiency, normally for cellular applications and Ad hoc networks. To address the problem of high power amplifiers (HPA), this is operating in its saturation region; several strategies are familiar with cut back PAPR in OFDM structures [1, 2]. Various PAPR diminishment methods such as filtering and clipping were used initially. Furthermore, as per need of high BER achievements, selective mapping (SLM) and partial transmit sequence (PTS), tone reservation and hybrid methods are considered for PAPR diminishment. Similarly, in next generation wireless networks LTE, 5G the equipment of strategies of lowering PAPR were confined in filter bank multicarrier techniques [13].

PAPR diminishment strategies are labeled into the various approaches: clipping approach, coding approach, probabilistic (scrambling) approach, adaptive predistortion approach and DFT-spreading approach proven in Fig. 2

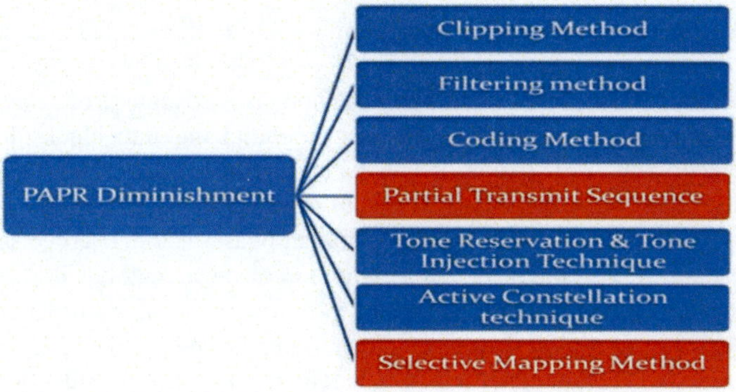

Fig. 2 PAPR diminishment methods

- *Clipping Method*: The clipping approach of PAPR diminishment is made possible within the nonlinear saturation region of a high-power amplifier across the peak envelope of the signal. The implementation of PAPR diminishment method in LTE and MIMO OFDM framework found easy, but it has in band and out of band signaling. This in-band signaling distortion results in lower BER. In the clipping method, out band distortion is reduced by using filtering techniques on another side Aliasing effect will occur in the sampling process. In the current scenario repeated, reconstructed and interactive clipping with filtering approaches are used [12].
- *Coding Method*: In this method, it employs selection of different code words that will reduce PAPR in the MIMO OFDM network. Due to use of code words reduction in in-band and out band distortion is made possible. The complexity of this method is responsible due to large loops and program size for the encoding and decoding purpose [13].
- *The Active Cancellation Technique*: In proposed method we are using clippling methods whereas high amplitude peaks are going to removed. The limitations of this method such as high conversion time and mean square error (MSE). This limitations are overcome by using proper broadcasting method and with the aid of designing suitable schooling signals [3].
- *Selective Mapping Method (SLM)*: In selective mapping method of PAPR diminishment, generation of number of OFDM symbols with lowest PAPR is achieved by representing each data blocks and transmits over a time [9]. Higher complexity and data rates in this method are increased due to use of number of IFFT blocks. Errors are increased due to loss of data blocks of data blocks in SLM method. The phase constant and input data blocks are multiplied so that alternate input symbols are generated [13].
- *Companding Technique (CT)*: A satisfactory achievement in diminishment of PAPR is possible due to companding method. This method compresses the signal at the input so it expands at the receiver so as to maintain the amplitude beneath

level as low as to achieve the maximum amplitude throughout communication [13].

- *Partial Transmit Sequence (PTS)*: This method gives advantage of achievement of PAPR diminishment of PAPR by dividing sub-block into data blocks. Rotation in phase factors achieved by rotational factor and sub-blocks. The sub-block consists of multiples of non-overlapping blocks of information. A rotation mechanism is employed in such a way that all the rotational phase factors become statistically independent of each other. A major limitation of these methods is poor spectral efficiency due to higher order complexity and filters. Similarly to the reality that it impacts the spectral performance of the system [3, 4].
- *Tone reservation method (TR)*: It is also known as peak tone reduction method. The word "Tone" refers to a set of information. In TR method, isolation of energy made for cancellation of higher peaks with a predefined set of reserved tones. The TR method gives advantage of less complexity, simple receiver design and absence of side information due to this higher spectral performance obtained [4, 9].

3.3 MIMO-OFDM System

In rapid-velocity next generation wireless communication, OFDM may be implemented to convert frequency-selective MIMO channels into parallel flat MIMO channels in multipath fading surroundings; additionally, excessive statistics price sturdy transmission may be achieved via way of means of lowering the hassle of the receiver [2]. A maximum transmitting antenna such as (1 × 1, 2 × 2, 4 × 4) antenna systems is used [5, 10]. An enter facts bit circulate is provided into space-time coding, then modulated via way of means of OFDM and eventually fed to antennas for transmission of information signal. At receiver side, early detection of information signal is made by using detectors blocks so that exact signal is recovered. Time space encoding and pace time decoding used at transmitter and receiver side, respectively [3]. Figure 3 shows the basic configuration of a MIMO-OFDM system.

PAPR is the most immediate power normalized via means of the average power among all possible signal patterns. Hence, we define $\text{PAPR}_{\text{MIMO}}$ is the maximum of all PAPER related to N_t MIMO path; hence, we have

$$\text{PAPR}_{\text{MIMO}} = \text{PAPR}_i \tag{3}$$

An excessive dynamic variety HPA has low energy efficiency. The energy may want to store with the aid of using lowering PAPR as shown in Eq. (3). This energy saving is carried out in the sort of manner to offer an instantaneous correlation with the favored common output energy.

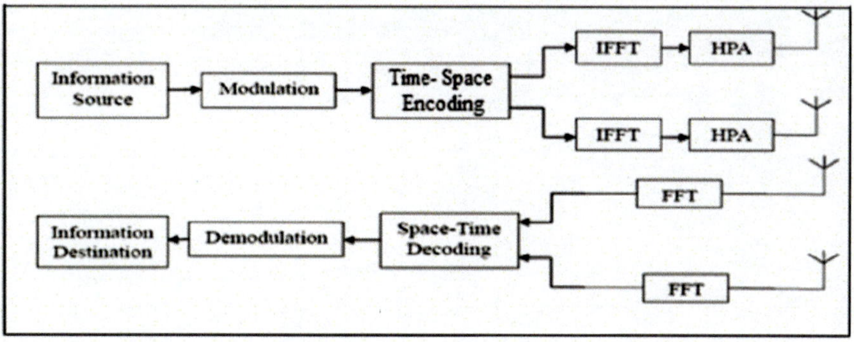

Fig. 3 Configuration of MIMO OFDM system

4 Comparative Analysis and Simulation Results of PAPR Diminishment in MIMO OFDM

In proposed method of research we are considering following parameters of OFDM such as shown in Table 1.

The said system uses Rayleigh selective fading channel with the parameters shown in Table 1. In this work, spatial multiplexing with N_t independent streams of data symbol on each antenna is considered; hence, the waveforms on different antennas are not couples in any way to avoid ISI.

The PAPR is calculated by using Eq. 1 and complimentary cumulative distribution function CCDF. The BER performance of MIMO OFDM for different transmitted symbols such as 64,128 7 1024 over QPSK, 16QAM, BPSK modulation techniques [represented in Figs. 4, 5 and 6]. The detection performance of all PAPR diminishment techniques, i.e., effects on BER is irrelevant as the various PAPR diminishment methods mentioned.

Hence, the waveform generated at the transmitter side is simulated to measure the PAPR. The comparative analysis is thoroughly done for single antenna SISO (single input single output) and MIMO antenna, including comparison with methods. BER performance of SISO OFDM for different transmitted symbols such as 64, 128 and 1024 over QPSK, 16QAM and BPSK modulation scheme is shown in Figs. 4, 5 and 6.

Table 1 LTE parameter

Frequency range	2.4–5 GHz
Bandwidth	1.25 MHz
FFT size	128,256
Number of transmit symbol	64,128,1024
Modulation type	QPSK
Spacing frequency	30 MHz

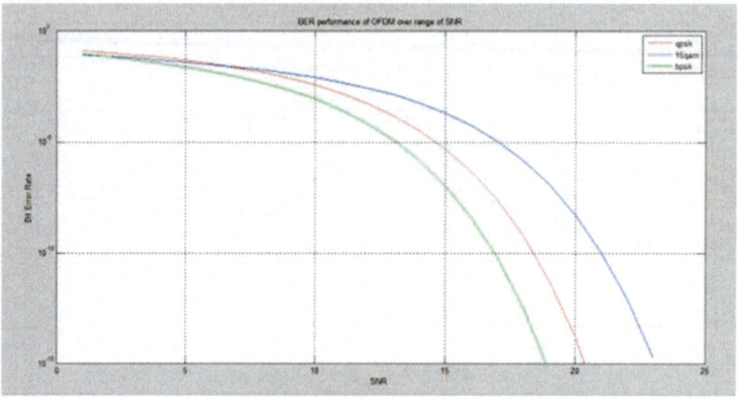

Fig. 4 BER performance for transmitted symbols = 64

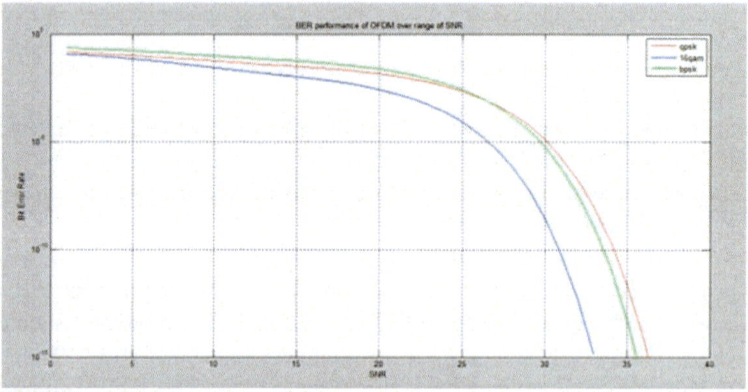

Fig. 5 BER performance for transmitted symbols = 128

Figure 4 shows BER performance of MIMO OFDM over 64 symbols with QPSK modulation techniques. The SNR at BER 10^{-15} is equal to 23.25 dB with QPSK modulation.

Figure 5 gives SNR at BER 10^{-15} is equal to 36 dB with QPSK modulation and highest SNR is achieved by using proposed system across all existing methods.

Figure 6 shows BER performance of MIMO OFDM over 64 symbols with QPSK modulation techniques. The SNR at BER 10^{-15} is equal to 70 dB with QPSK modulation.

After constructing the MATLAB-simulated version for above-mentioned PAPR diminishment methods, overall performance of the system is evaluated with distinctive strategies motioned in this work. The MATLAB simulation performed with modified PTS with respect to original OFDM and old PTS method.

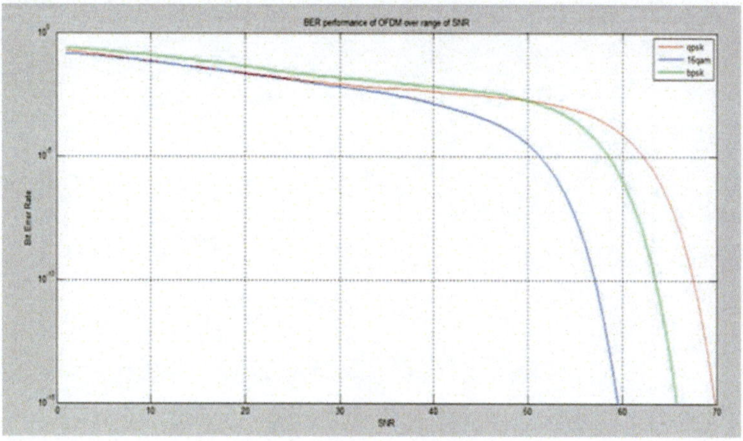

Fig. 6 BER performance for Transmitted symbols = 1024

The CCDF curve for the case of single antenna OFDM with $N = 64$, QPSK and 16QAM modulation orders are shown in Fig. 7. The CCDF of modified PTS here is improved and PAPR of 5.1, whereas normal OFDM has high PAPR 10.8 as shown in Fig. 7. Since it is approximated that with modified PTS 50%, reduction in PAPR is possible.

As mentioned, in proposed work, OFDM symbol generation for LTE network with frequency range 2.4–5 GHz is shown in Fig. 8.

Figure 9 shows band-limited spectrum with frequency spacing 30 MHz for MIMO

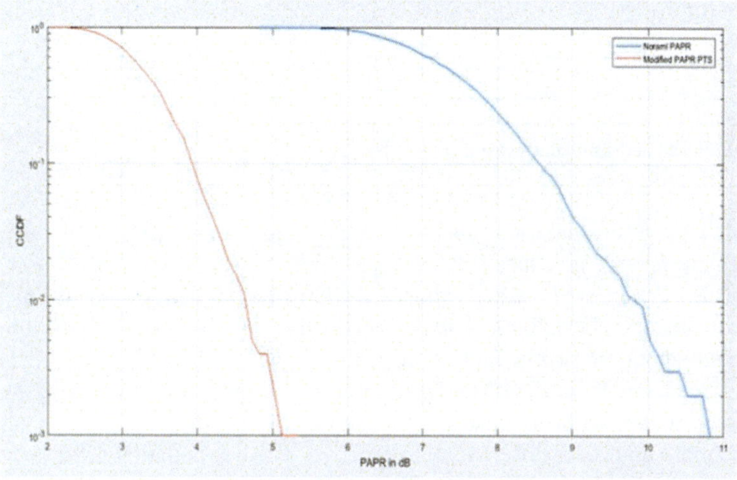

Fig. 7 CCDF functions of modified PTS scheme

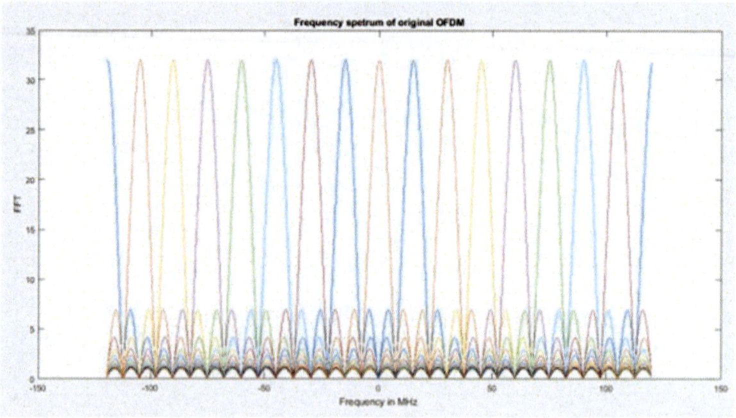

Fig. 8 OFDM frequency spectrum

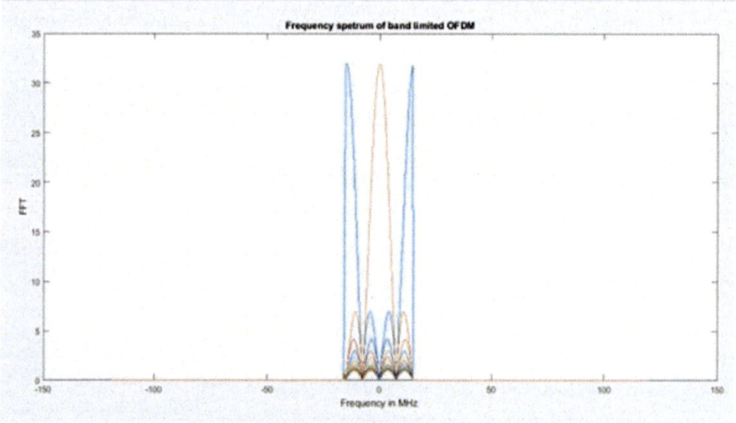

Fig. 9 OFDM band-limited spectrum at 30 MHz

OFDM. This means that each subcarriers are spaced with 30 MHz; due to this, inter symbol interference is reduced.

The sub-carriers are together orthogonal within the frequency domain which alleviates the results of ISI. Table 2 shows results and comparison of PAPR diminishment methods. As shown in Table 2, oversampling factor $L = 1$ is considered due to fast-fading channels in MIMO OFDM.

It is clear that with modified PTS technique, PAPR diminishment is achieved over normal OFDM for different alphabet sizes. The constellation order size of the alphabet considered is 4, 16 and 32. Simulation were performed on MATLAB with an LTE toolbox over 1000 OFDM symbols.

Figure 10 shows graphical analysis of PAPR techniques over normal OFDM

Table 2 Results and comparison with PAPR diminishment methods

Parameters	Input to MATLAB simulation model					
L factor (1–1.5)	1	1	1	1	1	1
Transmitted symbol (Power of 2)	16	32	64	32	64	128
Size of alphabet	4	4	4	16	16	32
PAPR with normal OFDM—dB	18.32	16.04	18.77	17.13	21.15	20.04
PAPR with SLM technique—dB	9.9	11.09	10.37	10.21	13.59	15.53
Efficiency of SLM	45%	30.85	32	40.34	35.73	22.49

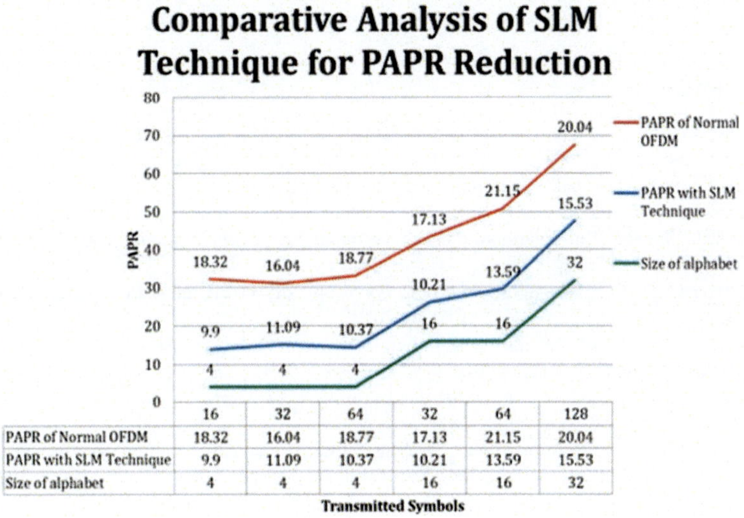

Fig. 10 Graphical analysis of SLM technique

symbolic methods and with SLM techniques. In this analysis, we used different alphabet symbols over which performance is measured. Using SLM method, PAPR of 15.53 is achieved, whereas normal OFDM methods gives high PAPR 20.04.

5 Results and Discussion

In above-presented literature survey and PAPR diminishment method, we examine numerous or many current studies in phrases of PAPR diminishment in OFDM. Using methods of Selective Mapping Method (SLM), Pratial Tranmit Sequence (PTS) and another different method used for PAPR reduction in MIMO OFDM network, this offers regular verbal exchange and conscious from the error. In the PAPR surroundings, each method maximizes and reduces statistics associated with every alphabet symbol.

In this paper, we have analyzed different PAPR diminishment techniques, and additionally, in comparison, those are strategies for distinctive parameters in next generation wireless networks. We look at that SLM is the only approach to mitigate PAPR to a remarkable quantity and additionally enhance BER overall performance of the system. The factors influencing the PAPR were studied; this also helps to understand the LTE and 5G network. In LTE network new method incorporated to uplink multiple-access scheme of SC-FDMA has been used which is known as Single Carrier FDMA (SC-FDMA).

References

1. Ben Mabrouk M et al (2017) Low-PAPR condition for 5G-candidate waveforms. In: 32nd URSI GASS, Montreal, 19–26 August 2017©2017. IEEE
2. Yi W, Wang Y, et al (2017) Efficient PAPR reduction through random constellation mapping. In: 2017 3rd IEEE international conference on computer and communications, 978-1-5090-6352-9/17/ ©2017. IEEE
3. Sghaier M et al (2016) An efficient reduced complexity PAPR reduction approach for 3GPP LTE system. In: IEEE, WCNC, pp 1558–2612
4. Tosato F, Sandell M, Tanahashi M (2016) Tone reservation for PAPR reduction: an optimal approach through sphere encoding
5. Boccardi F, et al (2017) Low-PAPR condition for 5G-candidate waveforms. In: 32nd URSI GASS, Montreal, 19–26 August 2017
6. Jebbar H, Hassani SE, Abbassi AE (2017) Performance study of 5G multicarrier waveforms. In: 2017 international conference on wireless networks and mobile communications (WINCOM)
7. Rodriguez J (2015) Fundamentals of 5G mobile networks. Wiley © 2015
8. Ben Mabrouk M, Chafii M, Louet Y, Bader F (2017) A precoding based PAPR reduction technique for UF-OFDM and filtered-OFDM modulations in 5G systems. In: European Wireless 2017; 23th European Wireless Conference, Dresden, Germany, pp 1–6.F
9. Rost P et al (2014) Cloud technologies for flexible 5G radio access networks. IEEE Commun Mag 52(5):68–76
10. Osseiran A et al (2014) Scenarios for 5G mobile and wireless communications: the vision of the METIS project. IEEE Commun Mag 52(5):26–35
11. Ibraheem ZT, Mijanur Rahman M, Yaakob SN, Razalli MS, Salman F, Ahmed KK (2014) PTS method with combined partitioning schemes for improved PAPR reduction in OFDM system. Telkomnika Indonesian J Electr Eng 12(11):7845–7853
12. la Sorte N, Justin Barnes W, Rafai HH (2008) The history of orthogonal frequency division multiplexing. In: Global telecommunication conference. IEEE, pp 1–5
13. Jebbar H, El Hassani S, El Abbassi A (2018) PAPR reduction for 5G waveforms. In: 2018 6th international conference on wireless networks and mobile communications (WINCOM)

An Optically Transparent Antenna Based on Transparent Conductive Oxides for Tera-Hertz Applications

S. Syed Feroze Hussain and D. Thiripurasundari

Abstract A U-shaped transparent patch antenna made up of indium tin oxide (ITO) and fluorine doped tin oxide (FTO) conductive material is designed on polyimide substrate proposed to resonate at 750 GHz. The substrate material has dielectric constant and loss tangent of 3.5 and 0.008, respectively. The ground plane is made out of conductive oxide material such as ITO and FTO. The thickness of patch is 0.4 μm. The gain, impedance bandwidth, return loss and VSWR are analysed, and results are plotted. The return loss of −30.30 dB, −30.84.50 dB and −16.65 dB are obtained for the U-shaped ITO, FTO and copper-based antenna, respectively. An impedance bandwidth of 9%, 14% and 5.23% and gain of 2.772 dB, 1.965 dB and 4.915 dB are obtained for ITO, FTO and copper-based antenna, respectively. The transparency of ITO and FTO are greater than 80%. The low density portions are identified and removed to form the shape of rake. This structure enhances the transparency of the antenna with very little compromise in performance. The rake-shaped antenna has gain of 2.117 dB and 1.1707 dB along with impedance bandwidth of 9.3% and 12.62% is obtained for indium tin oxide and fluorine doped tin oxide, respectively. The rake-based antenna has return loss of −30 dB and −37 dB for ITO and FTO, respectively. There is an increment in bandwidth over 0.3% for indium tin oxide, and increase in return loss over 10 dB for fluorine tin oxide antenna is obtained with rake structure. The co-cross polarization ratio for both the U-shaped transparent antenna as well as the rake-shaped antenna structure are more than 30 dB. These type of transparent antenna result in larger bandwidth, high data rates and are suitable for microscopic scanning, military applications and inter-satellite communications.

Keywords Transparent antenna · Indium tin oxide · Fluorine tin oxide

S. Syed Feroze Hussain (✉) · D. Thiripurasundari
SENSE, VIT University, Chennai, India
e-mail: syedferoze.hussain2019@vitstudent.ac.in

D. Thiripurasundari
e-mail: dthiripurasundari@vit.ac.in

A. Sivasubramanian et al. (eds.), *Futuristic Communication and Network Technologies*, Lecture Notes in Electrical Engineering 792,
https://doi.org/10.1007/978-981-16-4625-6_69

1 Introduction

The invention of micro strip antennas by Sir Deschamps in 1953 has found its applications in mobile technologies because of its light weight, small structure and compactness. The mounting of antenna on objects such as glass panels on buildings, automobile windows, monitors etc. without deviating much from its transparency has provided path to the development of transparent antennas. The applications integrated with optical transparent antenna are optical LED [1], liquid crystal display [2, 3] solar cells [4, 5], energy harvesting, light, display devices such as monitors and wearable applications [6]. Various conductive films such as variant of tin oxide (ITO/FTO) [7–10], aluminium-doped zinc oxide (AZO) [11–13], fluorine-based tin oxide (FTO), gallium-doped zinc oxide (GZO) [13] and silver-coated polymer (AgHT) [14–16] films are used in design of transparent antenna to produce reasonable performance and transparency. The methods used for designing transparent antenna are transparent conductive oxide (TCO), mesh type, micro metal mesh conductive (MMMC), meta-material loading, rectangular and circular patch antenna which can be used to implement dual band, multiband and broadband characteristics. The terahertz antenna creates an easier and faster data transfer, high capacity, higher data rate and low latency, which will be demanded by end-user in future generation. A wide variety of transparent substrates can be opted to design a transparent micro strip antenna such as glass, polyimide, etc. These thin films are deposited using various deposition methods such as magnetron-based sputtering, spray-based techniques, sol–gel process and pulsed deposition technique [17–29]. The increase in film thickness causes conductivity of film to increase while the optical transparency decreases. The radiation efficiency increases as the conductivity increases, thereby resulting in compromise between optical transmission and radiation performances. The factors such as low resistivity, higher transparency, stability, growth temperature, sheet resistance, heat resistant and film thickness are considered for optimal radiation performances. By using thick substrates, both impedance bandwidth and antenna size increases. The radiation performance is reduced because of higher permittivity therefore results in reduction in size of antenna. However, transparent conductors have lesser conductivity and better optical transmittance than non-transparent conductors.

2 Antenna Design

A U-shaped transparent patch antenna made up of indium tin oxide (ITO) and fluorine tin oxide (FTO) conductive material is designed on polyimide substrate proposed to resonate at 750 GHz. The substrate material has a dielectric constant of 3.5 and loss tangent of 0.008. The ground plane is made out of conductive oxide material such as ITO and FTO. The antenna has a patch thickness of 0.4 μm, while the antenna made up of ITO and FTO have a finite conductivity [30, 31]. These transparent antenna

Table 1 Optimized dimension of transparent and copper based antenna

Parameter	ITO dimensions (μm)	FTO dimensions (μm)	Copper dimensions (μm)
Patch length (L × L1)	88.98 × 83.08	88.98 × 86.71	88.98 × 80
Patch width (*W* × W1 × W2)	133.2 × 39.5 × 45.3	133.2 × 46 × 42	133.2 × 35 × 38
Thickness of patch (*t*)	0.4	0.4	0.4
Substrate (*L* sub × *W* sub)	208.98 × 433.2	208.98 × 433.2	208.98 × 433.2
Substrate thickness (*h*)	20	20	20
Feed line length and width (L2 × W3)	60 × 13.9	60 × 15	60 × 12
Ground plane (L g × W g)	208.98 × 433.2	208.98 × 433.2	208.98 × 433.2

are compared with non-transparent material made of copper having conductivity of 5×10^6 S/m.

The overall size of the antenna is 208.98 × 433.2 μm. Table 1 illustrates the dimension of antenna. The design of antenna is suitable for applications such as microscopic imaging, inter-satellite communication and military applications.

The parameters L1, W1, W2 and feed strip width (W2) are adjusted to provide impedance matching of antenna. The L1 and W1 causes the changes in resonance frequency. *L* is the length of the patch, *W* is the width of the patch, W1 is the width of first arm, W2 is the width of second arm, L1 is the length of inner arm, W3 is the feed line width and L2 is the length of feed line. The resonant frequency and length and width dimensions are calculated using following equations:

$$fr = \frac{c}{2(L + 2\Delta L)\sqrt{\varepsilon_{\text{eff}}}} \quad (1)$$

where *fr* is frequency of operation, *L* is patch length, *c* is velocity of light propagation, ε_{eff} is effective dielectric constant, ΔL is patch extension length.

$$W = \frac{c}{2fr}\frac{1}{((\varepsilon_r + 1)/2)} \quad (2)$$

$$L = \frac{c}{2fr\sqrt{\varepsilon_{\text{eff}}}} - 2\Delta L \quad (3)$$

$$\Delta L = 0.412h \frac{(\varepsilon_{\text{eff}} + 0.3)\left(\left(\frac{w}{h}\right) + 0.264\right)}{(\varepsilon_{\text{eff}} - 0.258)\left(\left(\frac{w}{h}\right) + 0.8\right)} \tag{4}$$

$$\varepsilon_{\text{eff}} = \frac{(\varepsilon_r + 1)}{2} \frac{(\varepsilon_r - 1)}{2} \frac{1}{1 + 12\left(\frac{h}{w}\right)^{0.5}} \tag{5}$$

where ε_r, is dielectric constant of substrate material, h is the height of substrate and w is width of the patch.

2.1 *RAKE Shaped Antenna*

The antenna structure is modified to form RAKE-shaped structure in which the low-current regions are identified, and those regions are removed without changes in the resonant frequency of 750 GHz. The modified structure resembles the rake-shaped structure. This structure will enhance the transparency of the antenna and with very little compromise in performance. The gain, return loss, impedance bandwidth and VSWR are analysed, and results are plotted.

3 Results and Discussion

A U-shaped transparent patch antenna made up of ITO and FTO conductive material is designed on polyimide substrate proposed to resonate at 750 GHz as shown in Fig. 1. The transparent antenna is compared with copper-based antenna. Fig. 2 shows the −10db bandwidth for ITO ranges from 718 to 786 GHz with return loss of −30.30 dB. Figure 3 shows the simulated VSWR value of ITO is 1.063 which is in an acceptable limit (VSWR < 2). The antenna has a gain of 2.772 dB at 750 GHz. Figure 4a shows radiation pattern of ITO having co-cross polarization ratio more than 30 dB. Figure 5a shows the surface current density of antenna which

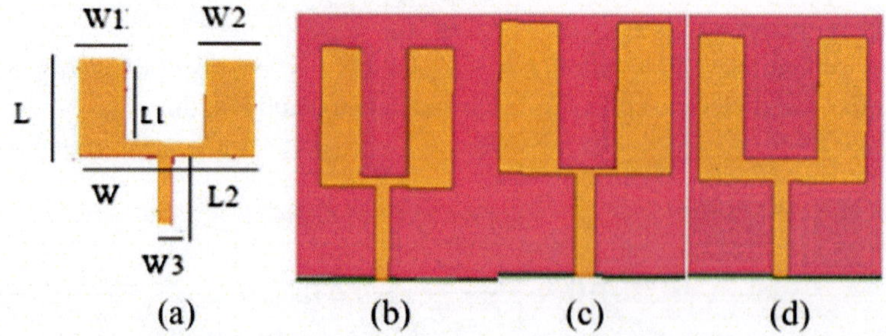

Fig. 1 Structure of transparent antenna **a** Antenna physical parameter **b** ITO **c** FTO **d** Copper

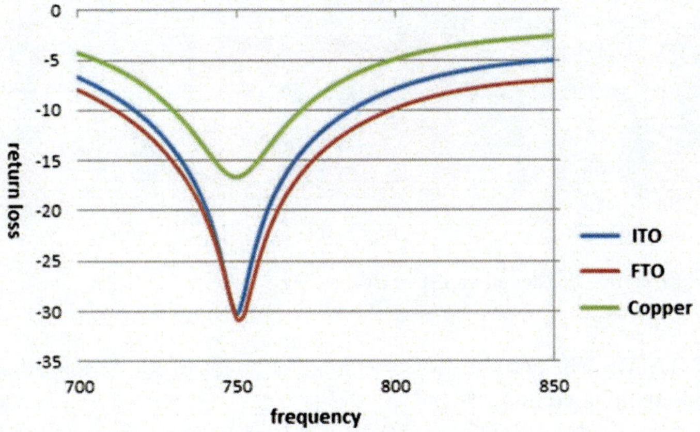

Fig. 2 Return loss versus frequency plot **a** ITO **b** FTO **c** Copper

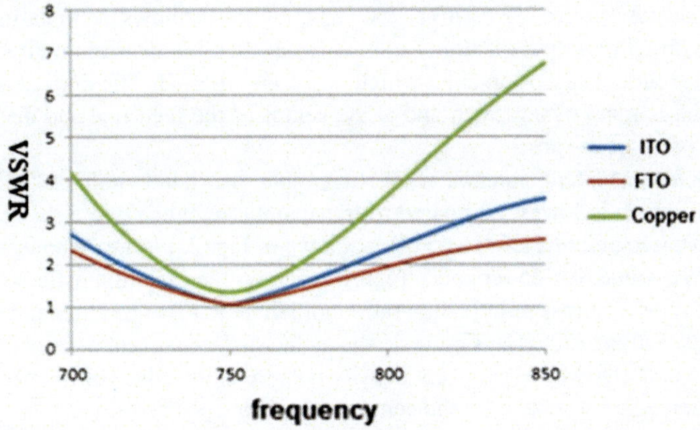

Fig. 3 VSWR versus frequency **a** ITO **b** FTO **c** Copper

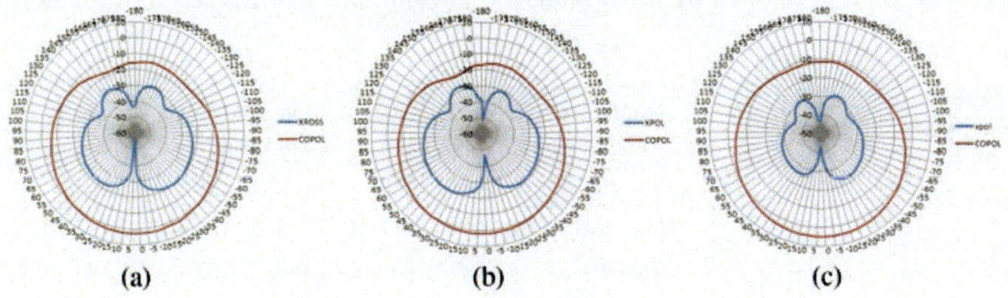

Fig. 4 Antenna radiation pattern **a** ITO **b** FTO **c** Copper

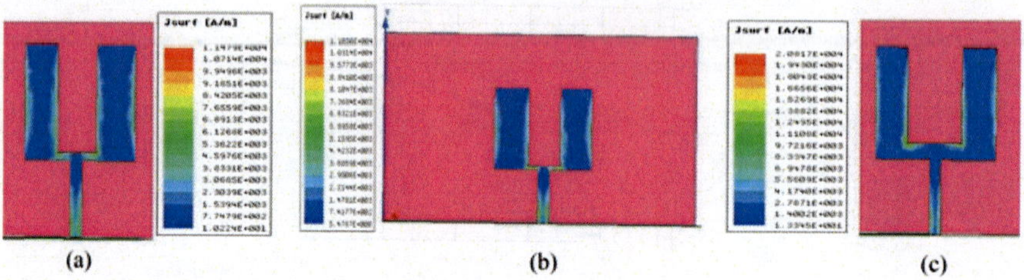

Fig. 5 Surface current density plot **a** ITO **b** FTO **c** Copper

is 1.141e^4[A/M]. The current density is more at edges of the U-shaped arm patch and also along the feed line.

The antenna made of FTO has −10 dB bandwidth ranges from 712 to 820 GHz which is highest among the other materials. The return loss of FTO is –30.84 dB as shown in Fig. 2. Figure 3 shows the simulated VSWR value of FTO which is 1.3449. The antenna has a gain of 1.965 dB at 750 GHz. Figure 4b shows the radiation pattern of FTO with co-cross polarization ratio is greater than 30 dB. Figure 5b shows the surface current density of antenna which is 1.105e^4[A/M]. The current density is more at the borders of the patch and at the centre of the feed line and decreases at the inner patch of the arms.

The non-transparent antenna made of copper has bandwidth of 5.3% (730–770 GHz) which is lowest compared to transparent antenna. While the return loss of copper-based antenna is −16.65 dB as shown in Fig. 2. Figure 3 shows the simulated VSWR value of 1.35 (copper). Figure 4c shows the radiation pattern of copper antenna having co-cross polarization ratio more than 30 dB. The antenna has a peak gain of 4.915 dB at 750 GHz. Figure 5c shows the surface current density of antenna which is 2.08e^4[A/M]. The current density is more at the edges of the patch and at the feed line while it reduces at the centre of the U-shaped arms.

Table 2 shows the comparative analysis of the radiation performances between transparent oxide-based antenna and copper-based antenna. The RAKE-shaped structure has been developed by identifying the low-current regions, and those regions

Table 2 Performance comparison of transparent and copper-based antenna

Parameters	Indium Tin Oxide	Fluorine doped Tin Oxide	Copper
Return loss	−30.30 dB	−30.84 dB	−16.64 dB
Operation band	718–786 GHz	712–820	730–770
Impedance bandwidth	9%	14%	5.3%
VSWR	1.063	1.3449	1.35
Gain	2.772 dB	1.965 dB	4.915 dB
Surface current density	1.141e^4	1.105e^4	2.08e^4

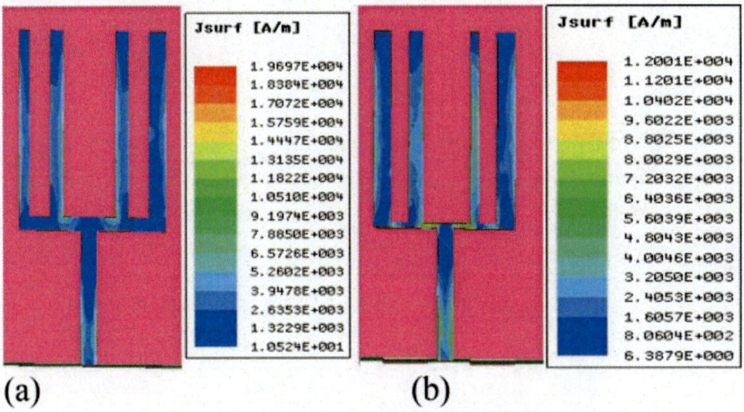

Fig. 6 Surface charge density plot **a** RAKE shaped ITO, **b** RAKE-shaped FTO

are removed without changes in the resonant frequency of 750 GHz. Figure 6a and b shows that the current density is more at the border of the inner rakes and at the bottom of the feed line for ITO. In FTO it is more at the edges of the inner rakes, borders of outer rakes, top portion and the central part of feed line. Figure 7 shows that the reflection coefficient of −30 dB and −37 dB is obtained for indium and fluorine oxide, respectively. Figure 8 shows the VSWR plot of rake-shaped structure which is 1.069 and 1.029 for ITO and FTO-based antenna, respectively. Figure 9a and b shows the radiation pattern for rake-shaped antenna with co-cross polarization ratio is more than 30 dB. Table 3 illustrates the gain of the antenna for ITO and FTO-based antenna are 2.117 dB and 1.1707 dB, respectively. The impedance bandwidth for rake-shaped transparent antenna is 9.3 and 12.62% with surface charge density 1.969e^4[A/M] and 1.200e^4 [A/M] is obtained for indium and fluorine-based antenna. This structure will enhance the transparency of the antenna and with very little compromise in

Fig. 7 Return loss versus frequency plot

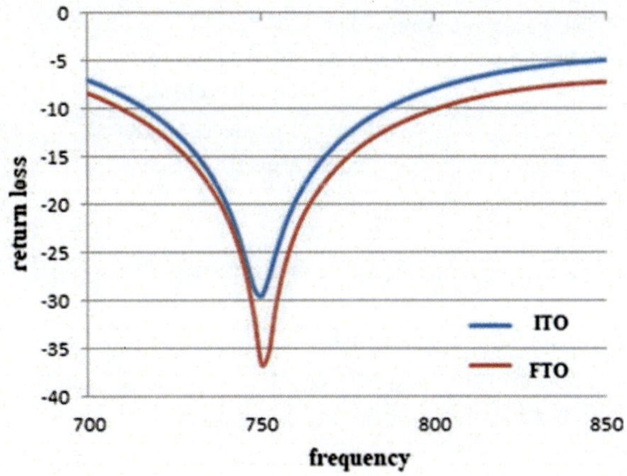

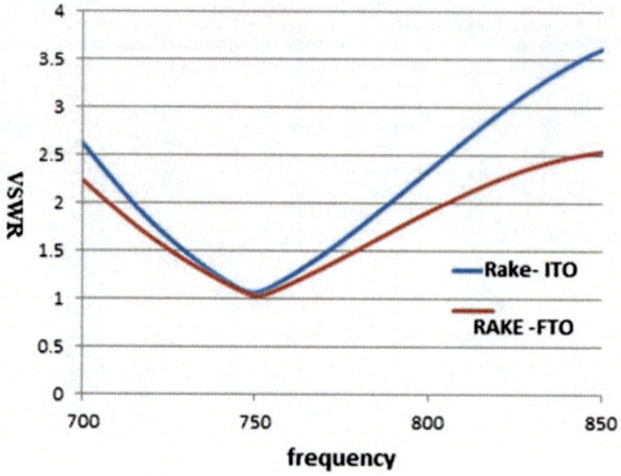

Fig. 8 VSWR versus frequency plot

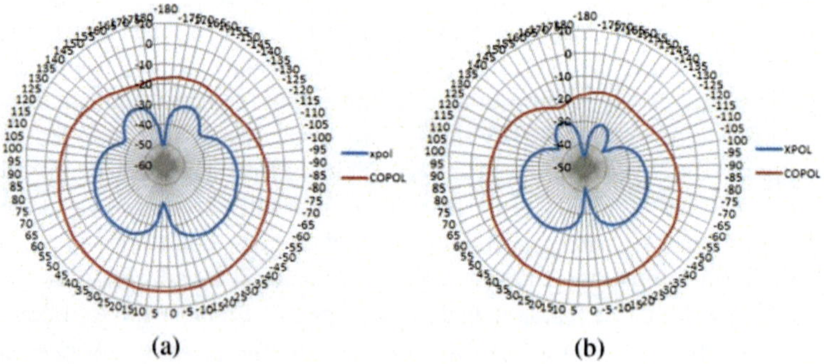

Fig. 9 Radiation plot **a** RAKE shaped ITO **b** RAKE-shaped FTO

Table 3 Comparison of RAKE-based transparent antenna

Parameters	ITO	FTO
Return loss	−30 dB	−37 dB
Operation band	716–786 GHz	705–800 GHz
Impedance bandwidth	9.3%	12.62%
VSWR	1.069	1.029
Gain	2.117 dB	1.1707 dB
Surface current density	1.969e^4	1.200e^4

radiation performance. The antenna overcomes the drawbacks of narrow impedance bandwidth (less than 5%) and low gain (less than 2 dB). Hence, the antenna is more suitable for high-frequency applications.

4 Conclusion

A U-shaped transparent patch antenna made up of ITO and FTO conductive material is designed on polyimide substrate proposed to resonate at 750 GHz. The antenna is compared to copper-based antenna which is non-transparent. The return loss of −30.30 dB, −30.84 dB and −16.65 dB are obtained for ITO, FTO and copper-based antenna, respectively. An impedance bandwidth of 9%, 14% and 5.23% and gain of 2.772 dB, 1.965 dB and 4.915 dB are obtained for ITO, FTO and copper-based antenna, respectively. The transparency of ITO and FTO are greater than 80%. The low-density portion are identified and removed without changing in resonance resulting in shape of rake structure. This structure enhances the transparency of the antenna with little compromise in radiation performance. The rake structure has reflection coefficient of −30 dB and −37 dB along with impedance bandwidth of 9.3% and 12.62% is obtained for indium and fluorine oxide-based antenna, respectively. The RAKE-based antenna has a gain of 2.117 dB and 1.1707 dB for indium tin oxide and fluorine tin oxide antenna, respectively. The antenna has broad impedance bandwidth and reasonable gain. The difference between co and cross polarization ratio is more than 30 dB for both U-shaped transparent and rake-based antenna. There is an increment in bandwidth over 0.3% for indium tin oxide and increase in return loss over 10 dB for fluorine tin oxide-based antenna. This antenna can be suitable for high data rate and secured communications due to its wider bandwidth. The antenna can be incorporated on glasses in buildings, windows of automobile, laptop screen and transparent surface to get rid of excess spacing condition and can be used along with integration with solar cell structure for photovoltaic applications.

References

1. Kim J, Granström M, Friend R, Johansson N, Salaneck W, Daik R, Feast W, Cacialli F (1998) Indium–tin oxide treatments for single-and double-layer polymeric light-emitting diodes: the relation between the anode physical, chemical, and morphological properties and the device performance. J Appl Phys 84(12)
2. Oh B, Jeong M, Moon T, Lee W, Myoung J, Hwang J, Seo D (2006) Transparent conductive Al-doped ZnO films for liquid crystal displays. J Appl Phys 99(12):124505
3. Leterrier Y, Medico L, Månson JA, Betz U, Escola MF, Kharrazi Olsson M, Atamny F (2004) Mechanical integrity of transparent conductive oxide films for flexible polymer-based displays. Thin Solid Films 460
4. Anand S, Darak MS, Sriram Kumar D (2014) In advances in signal processing and intelligent recognition systems. Springer International Publishing, Switzerland, pp 195
5. Mohd Ali N, Misran N, Mansor M, Jamlos M (2017) Transparent solar antenna of 28 GHz using transparent conductive oxides (TCO) thin film. J Phys Conf Ser 852:012036. https://doi.org/10.1088/1742-6596/852/1/012036
6. Raad H, White C, Schmitzer H, Tierney D, Issac A, Hammoodi A (2017) A 2.45 GHz transparent antenna for wearable smart glasses. In: 2017 progress in electromagnetics research symposium—fall (PIERS-FALL). Available https://doi.org/10.1109/piers-fall.2017.8293117
7. Rashed A, Sharshar H (2013) Optical micro strip patch antennas design and analysis. Optik 124(20):4331–4335. Available https://doi.org/10.1016/j.ijleo.2013.01.003

8. Paul L, Pramanik R, Ur Rashid M, Sarker S, Mahmud M, Tariqul Islam M (2019) An ITO based high gain optically transparent wide band micro strip antenna for K band satellite communication. In: 2019 international conference on robotics, electrical and signal processing techniques (ICREST). Available https://doi.org/10.1109/icrest.2019.8644203
9. Sheikh S, Shokooh-Saremi M, Bagheri-Mohagheghi M (2015) Transparent microstrip patch antenna based on fluorine-doped tin oxide deposited by spray pyrolysis technique. IET Microw Antennas Propag 9(11):1221–1229. Available https://doi.org/10.1049/iet-map.2015.0048
10. Maged MA, Elhefnawi F, Akah HM, El-Hennawy HM (2018) C-band transparent antenna design for intersatellites communication. Int J Sci Eng Res 9(3):248–252
11. Awalludin M, Ali M, Mamat M (2015) Transparent antenna using aluminum doped zinc oxide for wireless application. In: 2015 IEEE symposium on computer applications & industrial electronics (ISCAIE). Available https://doi.org/10.1109/iscaie.2015.7298323
12. Zamudio M, Busani T, Tawk Y, Costantine J, Christodoulou C (2016) Design of AZO film for optically transparent antennas. In: 2016 IEEE international symposium on antennas and propagation (APSURSI). Available https://doi.org/10.1109/aps.2016.7695772
13. Green R et al (2017) An alternative material for transparent antennas for commercial and medical applications. Microw Opt Technol Lett 59(4):773–777. Available https://doi.org/10.1002/mop.30404
14. Hakimi S, Rahim S, Abedian M, Noghabaei S, Khalily M (2014) CPW-fed transparent antenna for extended ultra wide band applications. IEEE Antennas Wirel Propag Lett 13:1251–1254. Available https://doi.org/10.1109/lawp.2014.2333091
15. Desai A, Upadhyaya T, Patel J, Patel R, Palandoken M (2020) Flexible CPW fed transparent antenna for WLAN and sub-6 GHz 5Gapplications. Microw Opt Technol Lett 62(5):2090–2103. Available https://doi.org/10.1002/mop.32287
16. Desai A, Upadhyaya T, Patel R (2018) Compact wideband transparent antenna for 5G communication systems. Microw Opt Technol Lett 61(3):781–786. Available https://doi.org/10.1002/mop.31601
17. Agura H, Suzuki A, Matsushita T, Aoki T, Okuda M (2003) Low resistivity transparent conducting Al-doped ZnO films prepared by pulsed laser deposition. Thin Solid Films 445(2):263–267
18. Minami T (1999) Transparent and conductive multicomponent oxide films prepared by magnetron sputtering. J Vacuum Sci Technol A: Vacuum, Surfaces, Films 17(4):1765–1772
19. Colombel F, Motta Cruz E, Himdi M, Legeay G, Castel X, Vigneron S (2009) Ultrathin metal layer, ITO film and ITO/Cu/ITO multilayer towards transparent antenna. IET Sci, Measure Technol 3(3):229–234
20. Moholkar A, Pawar S, Rajpure K, Bhosale C (2007) Effect of solvent ratio on the properties of highly oriented sprayed fluorine-doped tin oxide thin films. Mater Lett 61(14–15):3030–3036
21. Chopra KL, Major S, Pandya DK (1983) Transparent conductors—a status review. Thin Solid Films 102(1)
22. Kim D, Park M, Lee H, Lee G (2006) Thickness dependence of electrical properties of ITO film deposited on a plastic substrate by RF magnetron sputtering. Appl Surface Sci 253(2):409–411
23. Alam M, Cameron D (2000) Optical and electrical properties of transparent conductive ITO thin films deposited by sol–gel process. Thin Solid Films
24. Elamurugu E, Ramamurthi K (2005) Studies on micro-structural and electrical properties of spray-deposited fluorine-doped tinoxide thin films from low-cost precursor. Thin Solid Films 476
25. Bisht H, Eun H, Mehrtens A, Aegerter M (1999) Comparison of spray pyrolyzed FTO, ATO and ITO coatings for flat and bent glass substrates. Thin Solid Films 351(1–2)
26. Sarhaddi R, Shahtahmasebi N, Rokn-Abadi MR, Bagheri-Mohagheghi MM (2010) Effect of post-annealing temperature on nanostructure and energy band gap of indium tin oxide (ITO) Nano-particles synthesized by polymerizing—complexing sol–gel method. Physica E43: Low Dimensional Syst Nanostr 43(1)
27. Kim H, Horwitz J, Kushto G, Kafafi Z, Chrisey D (2001) Indium tin oxide thin films grown on flexible plastic substrates bypulsed-laser deposition for organic light-emitting diodes. Appl Phys Lett 79(3):284–286

28. Bender M, Seelig W, Daube C, Frankenberger H, Ocker B, Stollenwerk J (1998) Dependence of film composition andthicknesses on optical and electrical properties of ITO–metal–ITO multilayers. Thin Solid Films 326(1–2):67–71
29. Gorjanc T, Leong D, Py C, Roth D (2002) Room temperature deposition of ITO using r.f. magnetron sputtering. Thin Solid Films 413(1–2):181–185
30. Porch A, Morgan DV, Perks RM, Jones MO, Edwards PP (2004) Electromagnetic absorption in transparent conducting films. J Appl Phys 95(4734)
31. Fukano T (2004) Low-temperature growth of highly crystallized transparent conductive fluorine-doped tin oxide films byintermittent spray pyrolysis deposition. Solar Energy Mater Solar Cells

Wideband Planar 2 × 2 MIMO Antenna Design for 5G NR Wireless Devices

Maruti Tamrakar, K. Usha Kiran, Anjali Singh, and P. Archana

Abstract In this paper, wide-band, multiple-input-multiple-output (MIMO) antenna system is proposed for mobile applications. The MIMO is an emerging field which aims to improve the signal-to-noise ratio in communication channels. The MIMO antenna system consists of two compact monopole antenna elements for 5G standard covering bands (n48, n77, n78) between 3.3 and 4.2 GHz. The dimensions of the board are equal to 100 × 60 × 1.6 mm, which is identical to typical smartphone's back panel. The size of antenna is only 10 × 7.8 mm. The length of the antenna trace is for about $\lambda g/4$ which has been L-shaped folded to make it resonate at desired frequency. Both antennas give better than −10 dB return loss and 80% efficiency. The antennas are placed at the PCB corner and the center edge of the board, with both positions supporting optimum MIMO performance. This compact size of antenna allows the designer the freedom to allocate the other electronic components. The proposed antennas are well suitable for small form factor wireless devices.

Keywords 5G · NR · MIMO · Monopole antenna · Planar · Mobile device

1 Introduction

The growth in demand of higher data rate in wireless communication requires higher bandwidth, and to meet channel capacity with increased number of users, the multiple-input-multiple-output (MIMO) technology has to be adapted in any wireless system. The new radio (NR) wireless technology, i.e., 5G can help to meet these (bandwidth and channel capacity) requirements. The 5G technology is not only evolution of 1G-2G-3G-4G technologies but also provide universal wireless connectivity for multiple applications like automobile, healthcare, industries, smart home and buildings. The 5G is supported by MIMO technology to improve the communication channel and enhance the user experience. The MIMO is an emerging field

M. Tamrakar · K. Usha Kiran (✉) · A. Singh · P. Archana
School of Electronics Engineering, Vellore Institute of Technology, Chennai, Tamil Nadu 600127, India
e-mail: usha.kiran@vit.ac.in

A. Sivasubramanian et al. (eds.), *Futuristic Communication and Network Technologies*, Lecture Notes in Electrical Engineering 792,
https://doi.org/10.1007/978-981-16-4625-6_70

which aims in reduction of error bit rates and increase the signal-to-noise ratio of carrier signal in communication channels.

These 5G and MIMO technologies are adapted in this work to develop an antenna. The NR 5G bands n48, n77 and n78 are considered for antenna design. These bands cover the frequency from 3.3 to 4.2 GHz. This frequency band is allocated for Citizens Broadband Radio Service (CBRSS) and C-band service with time division duplexing (TDD) mode.

The six MIMO antennas are presented in [1] for mobile phone-size devices, where four antennas for sub-6 GHz band and two antennas for mm-wave bands. The decoupling technique is given in [2] by canceling the higher order modes for MIMO platform, where all antennas shares the common ground plane. The coupled line resonator is used in [3] to get isolation between two antennas. The similar MIMO antenna design is presented in [4] for 3.4–3.6 GHz frequency. The [5] presents the MIMO antenna with five antennas in as system, but all these antennas [1–5] have narrow band response.

The two compact 5G printed circuit board (PCB) wideband monopole antennas are proposed in this paper for mobile wireless devices to support 2 × 2 MIMO application. These antennas are placed at the PCB corner and the center edge of the board, with both positions supporting optimum MIMO performance. The antenna size is very compact, which is suitable for small form factor of the wireless devices. This compact size of antenna allows the designer the freedom to allocate the other electronic components. The new antenna is an ideal component for flexibility embedding 5G into any wireless device and an easy way to upgrade any product to 5G.

2 Antenna Design

The planar monopole type antenna is designed is this work. The monopole antenna has quarter wavelength, which is compact and developed in FR4 laminate/board (double side copper) of size 100 × 60 mm. The FR4 PCB permittivity is 4.4 and thickness is 1.6 mm. The higher dielectric substrate helps to reduce the physical dimension of antenna. To achieve miniaturization, the antenna is developed on both sides (top and bottom layer) of PCB and antenna traces are bended in inverted L shape. The antenna design, simulation and analysis are done using CST Microwave Studio tool. Two monopole antennas are designed and developed to support MIMO operation. The geometry of the proposed compact antennas is shown in Fig. 1. Due to different placement (board corner and edge center), both antenna dimensions are different. The feed points and live trace are given on top layer of PCB as shown in Fig. 1a and coupled shorted trace is printed on bottom layer of PCB presented in Fig. 1b. Each of antenna elements are an inverted-L monopole and feed using the 50 Ω microstrip feeding line. Due to the close space, a strong mutual coupling exists between the two inverted-L traces. The top layer trace (live trace) is capacitively coupled to bottom layer trace, which is shorted with ground plane. These monopole traces (top and

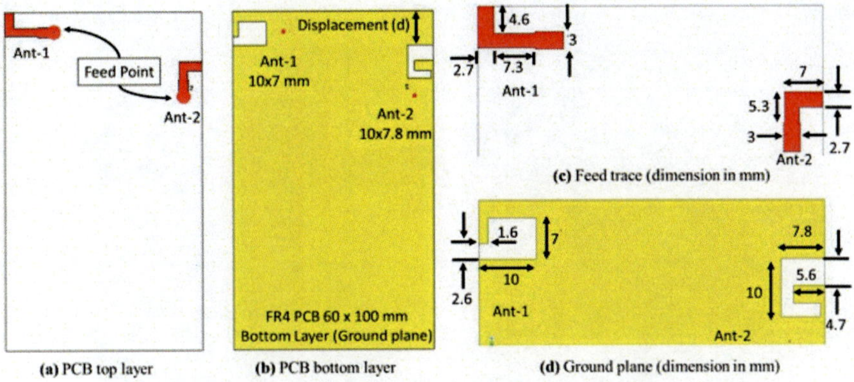

Fig. 1 **a** PCB top layer **b** PCB bottom layer **c** antenna feed trace in top layer (dimension in mm) **d** ground plane and antenna coupled trace in bottom layer (dimension in mm)

bottom layer trace) and capacitive coupling together provides the wideband bandwidth to the antenna. Both antenna designs are different due to different placement option. One antenna is placed on corner of the board and another antenna is placed on the edge center of the board.

The microstrip feed line width is 3 mm and length is 5 mm in the design. Antenna-01 placement is on the right top corner of the board and total dimensions is 10 × 7 mm. Antenna-01 top layer trace length is 14.6 mm, and bottom layer trace length is 14.4 mm. Antenna-02 is placed on the board edge center and has total dimensions 10 × 7.8 mm. Antenna-02 top layer trace length is 12.3 mm, and bottom layer trace length is 14.6 mm. The detail dimension of antennas is given in Fig. 1c and d. The antenna-02 is displaced form board corner position to achieve required isolation between two antennas. There is current coupling between these antennas through common ground plane (in PCB bottom layer). The isolation between antenna-01 and antenna-02 can be controlled by displacement the antenna-02 position. The displacement, $d = 15$ mm, is selected for antenna prototype and testing as it gives out-of-phase coupling between antenna and better isolation. The fabricated antenna prototype is shown in Fig. 2.

3 Result and Discussion

The antenna reflection coefficient parameters $|S_{11}|$ and $|S_{21}|$ are measured using portable Keysight VNA (vector network analyzer). The measurement setup and VNA screen images are shown in Fig. 3. The S-parameter measurement is done for frequency range 1–5 GHz. Figure 4 presents return loss and isolation of the both proposed monopole antennas. It is observed that S_{11} and S_{22} parameters possess good return loss (below −10 dB) for operating frequency band 3.3–4.2 GHz (fractional bandwidth of 24% and more), which indicates that the reflected power is less than

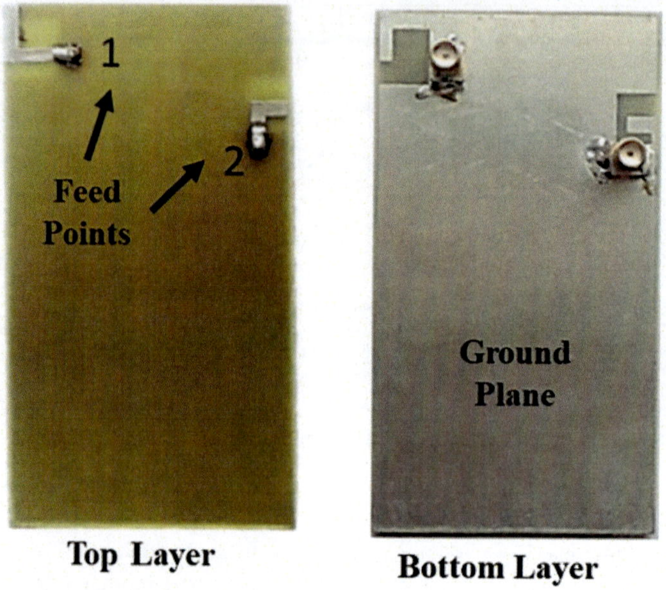

Fig. 2 Fabricated prototype (FR4 PCB) of proposed antennas

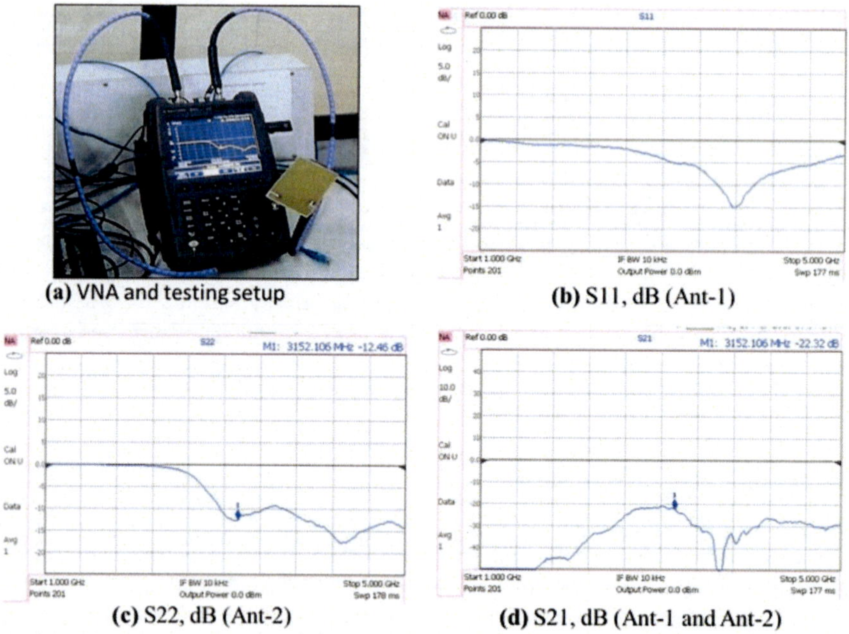

Fig. 3 S-parameter testing using VNA

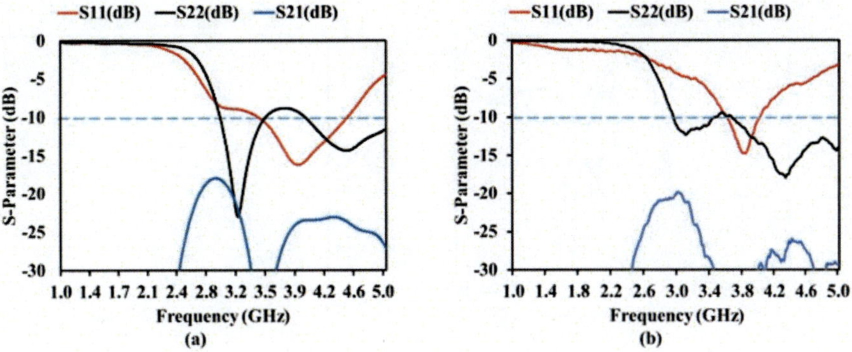

Fig. 4 *S*-parameter of proposed antenna **a** Simulated **b** Measured

10% from these antennas due to impedance mismatch. The antenna is having good isolation ($S_{21} < -15$ dB) between antenna-01 and antenna-02 for frequency range 3.3–4.2 GHz. Any MIMO system requires an isolation of −15 dB for its operating frequency bands. The isolation is required to avoid interference among the multiple wireless in a device and to achieve good MIMO performance. The simulated efficiency is above 80% and peak gain is between 4 and 6 dBi for both proposed antennas. The antenna efficiency and gain plot are shown in Fig. 5.

The isolation parameter has been improved antenna-02 displacement from board edge corner. Both the antennas have current coupling through common ground plane, and to achieve good isolation between them, the antennas should be out-of-phase coupled. The parametric sweep for displacement parameter (d) is varied from 10 to 25 mm and the isolation curve for different value of (d) is shown in Fig. 6a. The isolation is better for $d = 15$ mm for operating frequency band (3.3–4.2 GHz).

MIMO performance can be described and evaluated by the correlating coefficient. The correlation coefficient can be computed from far-field pattern using CST Microwave studio tool post-processing template [6]. The correlation coefficient graph

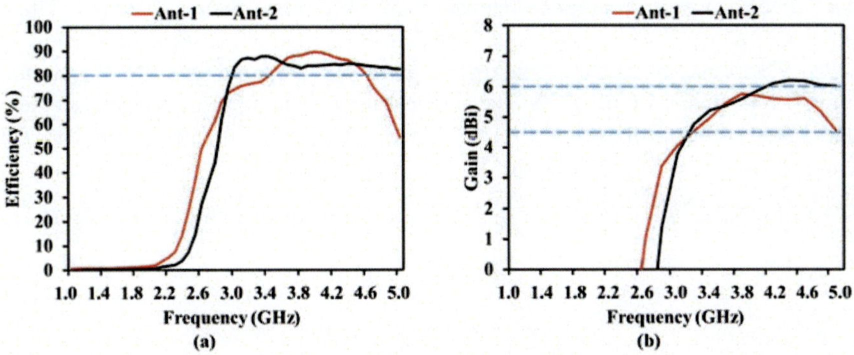

Fig. 5 Simulated **a** efficiency (%) **b** peak gain (dBi) of proposed antenna

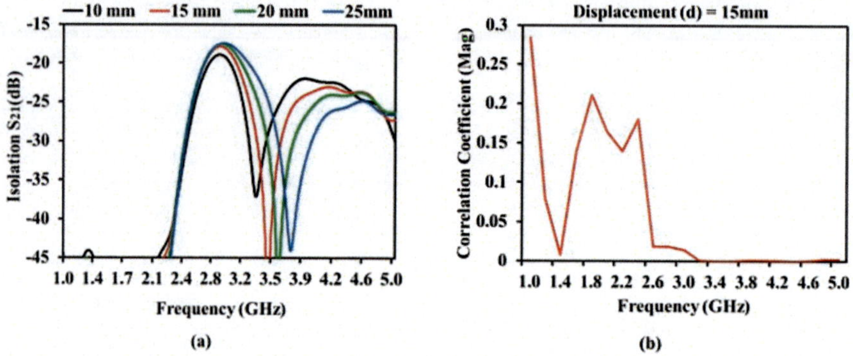

Fig. 6 **a** Simulated isolation S_{21} (dB) between antenna with Ant-2 displacement (d) from PCB edge **b** MIMO correlation coefficient (mag) for $d = 15$ mm

is shown in Fig. 6b for $d = 15$ mm, and it is less than 0.05 for proposed antennas and placement.

The simulated radiation pattern of both proposed antennas is shown in Fig. 7. In Fig. 7, the E-plane is xy-plane (theta = 90°), and H-plane is xz-plane (phi = 0°). It can be observed that the antennas have omnidirectional pattern. The radiation pattern of both antennas has different directional of maxima and minima due to antenna placement in the board. The cross-polarization level is less than 15 dB.

The author contribution (this work) is compared with references and tabulated in Table 1. Table 1 shows that the proposed design is compact and having higher bandwidth, while efficiency and gain parameters are good.

4 Conclusion

In this paper, two inverted L-coupled monopole wideband antennas are proposed for 5G NR bands. Both proposed antennas have return loss below −10 dB and total efficiency above 80% for frequency band 3.3–4.2 GHz (fractional bandwidth of 24%). The proposed antenna solution and placement are good for MIMO operation and 5G NR bands n48, n77, n78. The antennas are very compact, which makes them suitable for small form factor wireless devices. The proposed antennas are an ideal component for flexibly embedding 5G into any wireless device and an easy way to upgrade any product to 5G.

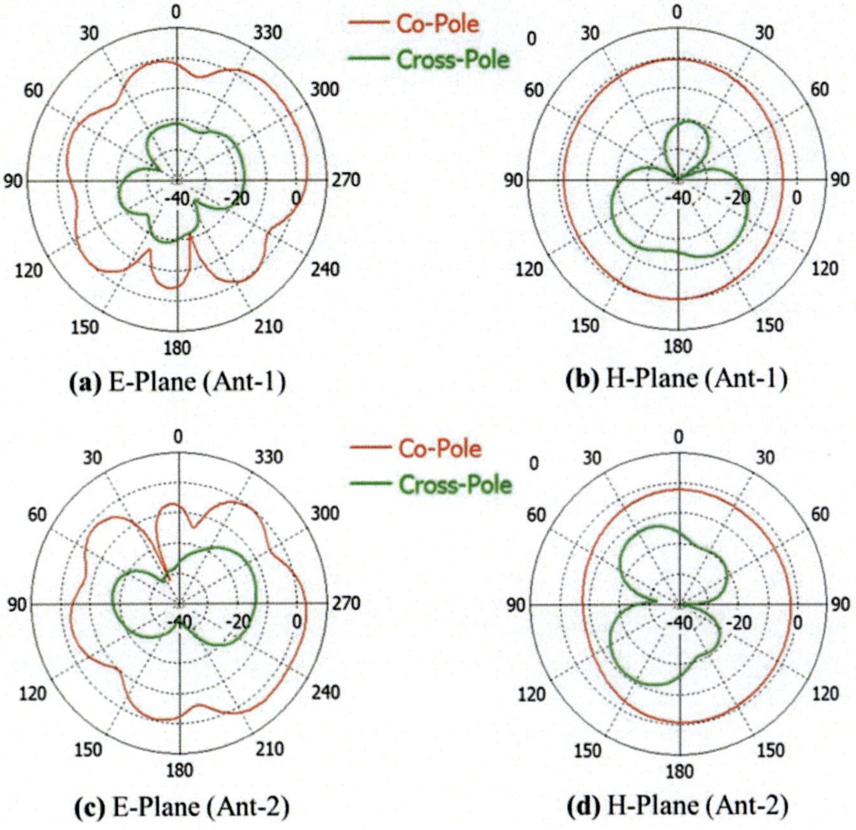

Fig. 7 Simulated radiation pattern at 3.8 GHz **a** E-plane of Ant-1 **b** H-plane Ant-1 **c** E-plane of Ant-2 **d** H-plane Ant-2

Table 1 Design compared

Ref.	Dimension	Frequency band	Efficiency	Peak gain
[1]	11 × 14.5 mm	2.0–3.0 GHz	–	5 dBi
[3]	7.0 × 10.5 mm	3.4–3.6 GHz	–	–
[4]	15 × 15 mm	3.4–3.6 GHz	–	5.1 dBi
[5]	6.0 × 27 mm	2.1–2.3 GHz	74%	3.4 dBi
This work	7.8 × 10 mm	3.3–4.2 GHz	80%	5.9 dBi

References

1. Ikram M, Sharawi MS, Attia H (2017) A compact dual standard MIMO antenna system for mobile applications. In: IEEE 28th annual international symposium on personal, indoor, and mobile radio communications (PIMRC), pp 1–3

2. Zhao X, Yeo SP, Ong LC (2018) Decoupling of inverted-F antennas with high-order modes of ground plane for 5G mobile MIMO platform. IEEE Trans Antennas Propag 66(9)
3. Zhang XY, Xue C-d, Cao Y, Ding C-F (2017) Compact MIMO antenna with embedded decoupling network. In: IEEE international conference on computational electromagnetics (ICCEM), pp 64–66
4. Abdullah M, Ban Y-L, Kang K, Li M-Y, Amin M (2017) Compact four-port MIMO antenna system at 3.5 GHz. In: International applied computational electromagnetics society symposium—Italy (ACES), pp 1–2
5. Ikram M, Hussain R, Sharawi MS (2017) 4G/5G antenna system with dual function planar connected array. IET Microw Antennas Propag 11(12):1760–1764
6. Tian R, Lau BK, Ying Z (2011) Multiplexing efficiency of MIMO antennas. IEEE Antennas Wirel Propag Lett 10:183–186

Component Reliability Analysis of a 4 × 4 Symmetric 2D Torus Optical Interconnection Network Node Architecture

Abhilasha Sharma and R. G. Sangeetha

Abstract Reliability analysis is one of the crucial performance parameters which can affect the performance of any network architecture. This paper focuses on evaluating the reliability of the basic building block of a two-dimensional 4 × 4 symmetric torus optical interconnection network. In this paper, we analyze the component reliability for a single node of the network using the reliability block diagram method.

Keywords Component reliability analysis · Torus optical interconnection network · Reliability block diagram

1 Introduction

In this scenario of the COVID 19 pandemic, there has been a surge in the traffic due to facilitating various services (like offering work from home services to the employees, online shopping, etc.) to maintain the social distancing and home isolation. These Internet and cloud services are accessed through the optical interconnection networks providing high-speed end to an end operation. To provide highly reliable connectivity at the user end, reliability analysis of the network and the node is a crucial concern for any optical interconnection network. Network reliability can be analyzed as terminal and broadcast reliability from the type of connectivity and usage point of view which provides a probabilistic overview of network failure assuming a constant node failure rate. To achieve a more accurate analysis, it is necessary to analyze the reliability of a single network node considering all the components within the architecture. This analysis is termed as component reliability analysis. This analysis is important as it provides a probabilistic overview about maintaining a single link inside a node in between the input and the output ports of the node [2, 3, 10].

In [4], we proposed a novel routing algorithm for the two-dimensional torus optical interconnection network that improved the latency and throughput significantly as compared to the conventional XY routing algorithm. This showed that

A. Sharma · R. G. Sangeetha (✉)
VIT University, Chennai, India
e-mail: sangeetha.rg@vit.ac.in

A. Sivasubramanian et al. (eds.), *Futuristic Communication and Network Technologies*, Lecture Notes in Electrical Engineering 792,
https://doi.org/10.1007/978-981-16-4625-6_71

a two-dimensional symmetric torus optical interconnection network is a suitable candidate for the data center networks. In our earlier work, we have analyzed time-dependent [6] and time-independent [5] network reliability of the torus optical interconnection network using the binary optical routing algorithm and the XY routing algorithm. This work also showed that the torus optical interconnect provides high reliability using a binary optical routing algorithm as compared to the XY routing algorithm. Similarly, in [8], we analyzed the two different types of reliability (such as terminal and broadcast) for the torus optical interconnect at various network sizes, implementing the binary optical routing algorithm and the XY routing algorithm. The analysis showed improved reliability using the proposed routing algorithm. Still, component reliability analysis remains a major concern as a single faulty component connected in series can cause the failure of node and links connected to it. In [3], the authors performed the component reliability analysis for the 4 × 4 optical data vortex switch fabric using the reliability block diagram method. This work realized the reliability value for every single component present within the node architecture, contributing to the node reliability based on the series/parallel interconnectivity in between the components. Using a similar approach now, this paper analyzes the component reliability of a two-dimensional torus node architecture. The hardware model of the torus node architecture is explained in Sect. 2. Section 3 describes the analysis of component reliability and Sect. 4 concludes the paper.

2 The 4 × 4 Torus Optical Node Architecture

The two-dimensional symmetric 4 × 4 torus optical interconnection network using binary optical routing algorithm [7] is composed of a total of 16 nodes as shown in Fig. 1. Every single node in this architecture has two inputs and two outputs and it provides bi-directional connectivity. In this architecture, each node processes a single packet in a one-time slot. The hardware model of the torus node architecture is shown in Fig. 2 along with all the components. When the packet enters the node through any of the port, the power tap extracts 20% of the optical power for header processing and 80% of the signal power is used to route the payload at the appropriate output. The 20% extracted optical power is further passed through the coupler to obtain the source and destination node addresses with the help of filters. The optical source/destination addresses are passed through photodetectors to obtain the electrical signal. This electrical signal is fed to the electronic signal processing unit which provides a control signal to trigger the SOA at the output port.

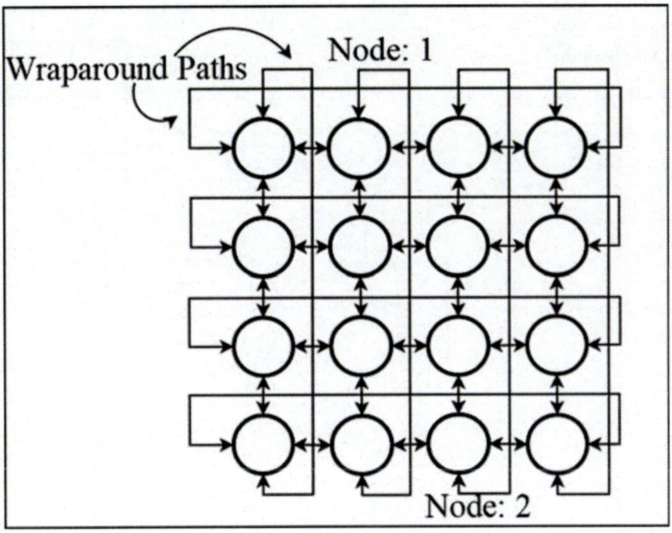

Fig. 1 A 4 × 4 two-dimensional torus optical interconnection network

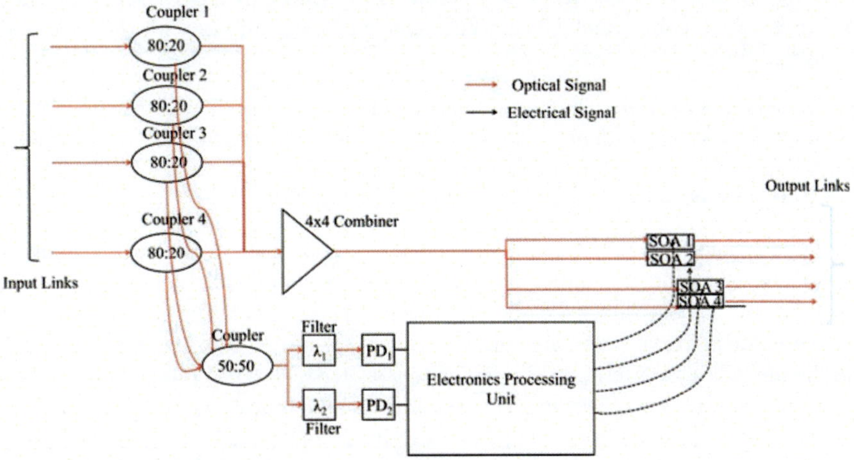

Fig. 2 Hardware model of torus node

3 Analysis of Component Reliability of 4 × 4 Torus Optical Node

The component reliability analysis is carried out considering both node component failures along with the fiber links using reliability block diagram method. Each node architecture has two optical couplers, two optical filters, two photodetectors, four power taps, and four SOAs. All of the above-mentioned components are connected in

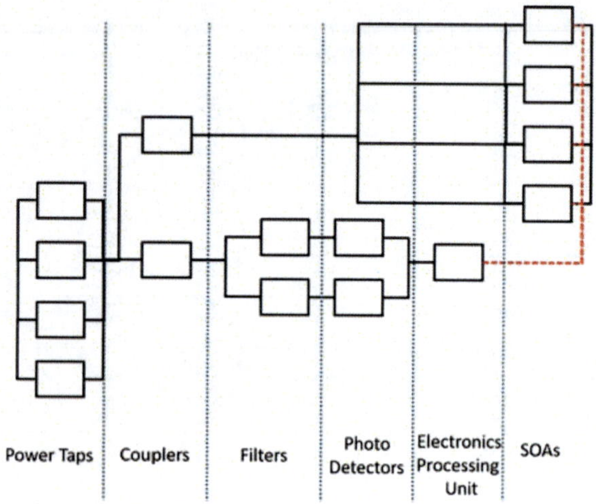

Fig. 3 Reliability block diagram of a torus node

a combination of series-parallel. The analysis is carried out using the reliability block diagram method. The component reliability block diagram for the torus node architecture shown in Fig. 3 is developed based on the interconnectivity (series/parallel) in between the components. The entire analysis is carried out with the following considerations:

- The component failure event is independent.
- If any component with the node architecture connected in series fails, causes node failure.
- Multiple node failures cause the entire network failures.

For the analysis the components are considered to be fault-free before the network is put into operation. The entire analysis is based on the values of failure rates of the components present within the node architecture [1–3, 9, 10]. The node architecture is composed of various components. These components are connected in series with each other while the same components are connected in parallel. For example, power taps and couplers are connected in series while all the four power taps present at all the ports are connected in parallel. The unreliability of the components also depends on the types of arrangements or connectivity (series/parallel) between them. Based on the type of arrangement between the similar type of components the reliability is evaluated. The reliability block diagram shown in Fig. 3 is developed from the hardware model of the node architecture (Fig. 2) following the aforementioned approach. Along with the failure rate, it is necessary to evaluate the component reliability over a specific period. Here, for the analysis, we have extended the analysis for 10, 20, and 30 years as shown in Table 1. The reliability of all the components is evaluated using Eqs. (1–5) formulated using Fig. 3 based on the series/parallel connectivity in between them.

$$R_{\text{Power Tap}} = 1 - \left(1 - e^{-4\lambda t}\right)^4 \quad (1)$$

Table 1 Analysis of component reliability of a 4 × 4 torus node

Components	Quantity	Failure rate (λ) ($\times 10^{-9}$)	Arrangement	Reliability evaluated for 10, 20 and 30 years		
				10 years	20 years	30 years
Coupler	2	20	Parallel	0.996502	0.993016	0.989543
Filter	2	50	Parallel	0.991278	0.982632	0.974062
Photodetector (PD)	2	70	Parallel	0.993016	0.986082	0.979195
Power Tap	4	50	Parallel	0.999999	0.999998	0.999993
SOA	4	1000	Parallel	0.992365	0.935572	0.820963

$$R_{\text{Coupler}} = 1 - \left(1 - e^{-2\lambda t}\right) \quad (2)$$

$$R_{\text{SOA}} = 1 - \left(1 - e^{-4\lambda t}\right)^4 \quad (3)$$

$$R_{\text{Filter}} = 1 - \left(1 - e^{-2\lambda t}\right) \quad (4)$$

$$R_{\text{PD}} = 1 - \left(1 - e^{-2\lambda t}\right) \quad (5)$$

The failure rate (λ) of the components considered for the analysis based on practical implementations is based on [2, 3, 10]. As depicted from the analysis, each component shows high reliability which in turn increases the reliability of a single node.

4 Conclusion

This paper analyses the component reliability of the torus node architecture using the binary optical routing algorithm. This analysis shows that all the components are highly reliable and this contributes to the availability of fault-free path through that node.

References

1. Bhagavath VK, Sobol H (1992) Reliability evaluation and protection schemes for dense WDM network architectures. In: [Conference Record] SUPERCOMM/ICC'92 discovering a new world of communications, pp 174–178. IEEE
2. Kuo W, Prasad VR, Tillman FA, Hwang CL (2001) Optimal reliability design: fundamentals and applications. Cambridge University Press

3. Sangeetha RG, Chandra V, Chadha D, 4 × 4 optical data vortex switch fabric: component reliability analysis
4. Sharma A, Sangeetha RG (2018) Performance analysis of high speed low-latency torus optical network. In: 2018 10th international conference on communication systems & networks (COMSNETS), pp 488–491. IEEE
5. Sharma A, Sangeetha RG (2018) Reliability analysis of data center network. In: Optical and microwave technologies, pp 71–80. Springer
6. Sharma A, Sangeetha RG (2018) Time dependent network reliability analysis of optical data center network. In: 2018 international conference on fiber optics and photonics (PHOTONICS). OSA
7. Sharma A, Sangeetha RG (2020) Performance analysis of torus optical interconnect with data center traffic. ETRI J
8. Sharma A, Sangeetha RG (2020) Terminal and broadcast reliability analysis of direct 2-d symmetric torus network. J Supercomput 1
9. Wosinska L, Thylen L (1998) Comparative study of reliability performance of optical cross-connect switches. In: Optical networks and their applications. Optical Society of America, p WP2
10. Zhang C, Hu W (2008) Design and analysis of a multicast-capable optical cross-connect. In: Optical transmission, switching, and subsystems VI, vol 7136. International Society for Optics and Photonics, p 71364H (2008)

A Neoteric Image Encryption System Using Nonlinear Chaotic Strange Attractors

Suchindran Srinivasan, Varun Subramaniam, V. S. Ramya Lakshmi, and N. R. Raajan

Abstract The science of securing the message by altering the information into an uninterruptable form so that it cannot be recognized is called cryptography. Currently, a wide range of algorithms have been employed for image encryption, but chaos blended image encryption techniques are preferred due to their excellent security and speed. This scheme primarily consists of two steps: chaos-based diffusion and scrambling of the pixel locations. This paper provides a concise introduction to chaotic image encryption and then an analysis of the properties of the chaotic attractors used. The inherent properties of the attractors which are relevant to cryptography are the salient reasons for their utilization in information security applications. Here, we introduce an innovative technique to encrypt a standard RGB image using the pseudorandom numbers generated by the 3D chaotic attractors which has an advanced response to a broad range of control parameters. The unpredictability of the attractors and its sensitivity to initial conditions has enhanced the security of the image immensely. This encryption scheme involves the chaotic bit streams and the image parameters being combined over a simple logical operation, thus reducing the possibility of retrieving the original data by intruders. From the results of the variety of statistical tests conducted, we come to an inference that encryption algorithm proposed is up to the mark and can be deployed for safe transmission of information.

Keywords Cryptography · Encryption · Chaotic attractors · Diffusion

1 Introduction

A large amount of data is being relayed over the Internet and this not only includes text but also image, audio and multimedia files. With the accelerated advancement in technology, transmission of information through a safe and secure network and its storage has become a matter of great importance. It is imperative to shield the cyber-related information from illegal retrieval, copying and distribution by attackers [1].

S. Srinivasan (✉) · V. Subramaniam · V. S. Ramya Lakshmi · N. R. Raajan
SASTRA Deemed University, Thanjavur 613401, India
e-mail: 122101008@sastra.ac.in

A. Sivasubramanian et al. (eds.), *Futuristic Communication and Network Technologies*, Lecture Notes in Electrical Engineering 792,
https://doi.org/10.1007/978-981-16-4625-6_72

Images have a lot of extensive applications ranging from telemedicine, multimedia systems to military communication [2]. The privacy of people's data is put in jeopardy due to the leakage of data while transmitting through wireless and wired media such as Wi-Fi and Ethernet [3].

Hence, it is imperative to formulate a data encryption system that is resilient and provides data integrity, authentication and confidentiality [4]. The obsolete encryption algorithms like DES, SHA, AES, RSA and elliptic curve-based cryptography are not effective for images due to its speed and its inability crack the correlation between the adjacent pixels [5–8]. Another reason why these algorithms and its variations are ineffective is because they have small keyspace and use elementary encryption operations thus, making it susceptible to attacks such as the brute force attack [6]. It is not advisable to employ these techniques in real time because of the high degree of data redundancy. Therefore, it is pivotal to design a cryptosystem that can provide excellent data security, speed and retrieval of decrypted image without loss in quality. In this context, chaos theory has turned out to be one of the most budding research fields for information security for the past few decades [9]. The robust encryption schemes that involve chaos have superseded the above-mentioned traditional algorithms and play an indispensable role in the surveillance of IT communication. Cryptography and chaos theory are closely connected and play a paramount role in the development of contemporary information security [10]. Research scholars are trying to incorporate the properties of the nonlinear chaotic attractors in secure communication. Ergodicity, unstable behavior, no periodicity, high sensitivity to system constraints and initial conditions are a few outstanding features of these chaotic systems that make it suitable for data encryption [9]. One-dimensional chaotic maps yield only single elementary chaotic paths and have average keyspace [11]. As a result, it becomes easy for the intruder to obtain the key necessary for decryption. Hence, we opt for three-dimensional chaotic attractors for securing data as incrementing the dimensions increases the nonlinearity, provides enhanced security and has greater numbers of positive exponents of Lyapunov [12, 13]. The entropy generated by the chaotic attractor produces three sets of pseudorandom sequences that can be utilized in carrying out the scrambling and diffusion process [14, 15]. In this paper, the standard 256 × 256 Pepper test image is encrypted using the Genesio-Tesi attractor, Lorenz attractor and the Modified Chua attractor [16–18]. The results of the proposed scheme and a detailed analysis of the various statistical tests conducted are provided toward the end of the paper.

2 Methodology

2.1 Realization of the Nonlinear Strange Attractors

The comprehensive analysis and the realization of the nonlinear chaotic attractors used to encrypt the RGB image is shown below.

Lorenz Attractor: The Lorenz nonlinear chaotic attractor is a continuous system that is represented by a family of differential equations which is notable for having chaotic solutions for a certain range of parameter values and system conditions. This chaotic attractor is an example of a strange attractor and is a deterministic system as the future values can be predicted as it changes with time when the exact starting values are known. The Lorenz system of equations is given below

$$\frac{dx}{dt} = m_1(y - x)$$
$$\frac{dy}{dt} = m_2x - y - xz$$
$$\frac{dz}{dt} = xy - m_3z \quad (1)$$

In this system of equations, m_1, m_2, m_3 are the system constraints and the left-hand side of the family of differential equations represent the state of the system. The terms xy and xz contribute to the nonlinearity of the chaotic system. Normally, the values of m_1, m_2, m_3 are taken to be positive. Here, in this case, we will designate the values of the constraints as $m_1 = 10$, $m_2 = 8/3$, $m_3 = 28$ and the preliminary constraints of the chaos-based system as $(x_0, y_0, z_0) = (0, 1, 0)$. The Lorenz system displays chaotic behavior for the aforementioned values and proximate values as well. The 3D view of the Lorenz attractor is displayed in Fig. 1a.

Genesio-Tesi Attractor: The Genesio-Tesi chaotic attractor is also another continuous system that is represented by a set of nonlinear differential equations shown below

$$\frac{dy}{dx} = y$$
$$\frac{dy}{dx} = z \quad (2)$$
$$\frac{dz}{dt} = -p_3x - p_2y - p_1z + x^2$$

The state variables are represented by x, y, z and p_1, p_2, p_3 are positive real system parameters such that the parameters satisfy $p_1.p_2 < p_3$. The values for the initial conditions and the system constraints are designated as $(x_0 = 0.1, y_0 = 0.1, z_0 = 0.1)$ and $p_1 = 1, p_2 = 3.03, p_3 = 5.55$ so that chaotic system displays the anticipated chaotic behavior. The plot of the Genesio-Tesi attractor is shown in Fig. 1b.

Modified Chua Attractor: The Modified Chua chaotic attractor system is a multi-scroll attractor with system parameters $b_1 = 10.82$, $b_2 = 0.11$, $b_3 = 14.286$, $e = 1.3$, $d = 0$ and initial conditions $(x_0 = 1, y_0 = 1, z_0 = 0)$. The system of equations representing the Modified Chua attractor is shown below

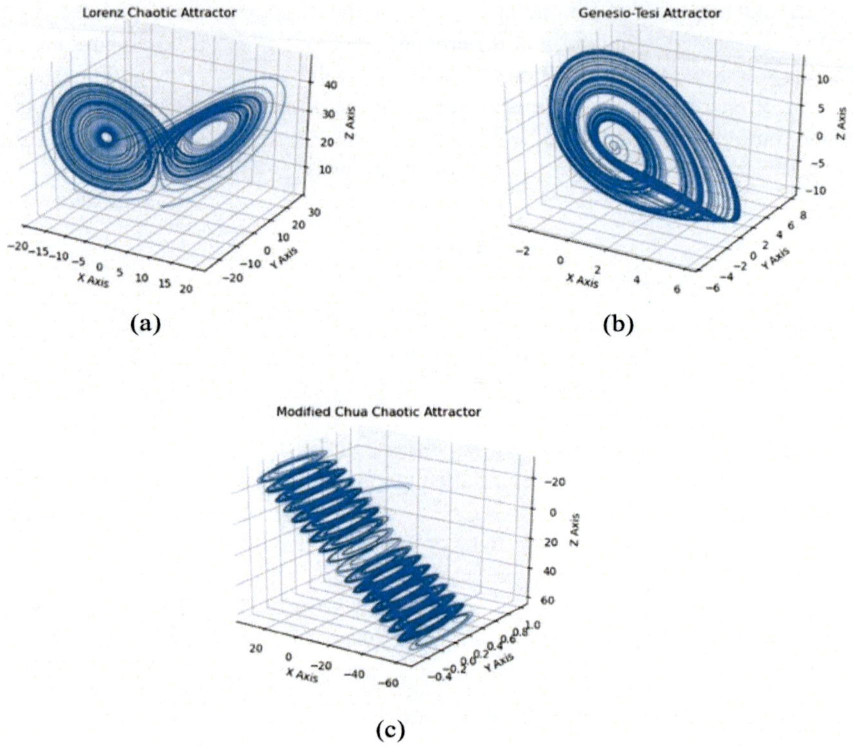

Fig. 1 **a** Lorenz chaotic attractor, **b** Genesio-Tesi chaotic attractor, **c** Modified Chua chaotic attractor

$$\begin{aligned}
\frac{\mathrm{d}x}{\mathrm{d}t} &= b_1(y + b_2(\sin(\frac{\pi x}{2e} + d))) \\
\frac{\mathrm{d}y}{\mathrm{d}t} &= x - y + z \\
\frac{\mathrm{d}z}{\mathrm{d}t} &= -b_3 y
\end{aligned} \tag{3}$$

The 3D graph of the Modified Chua attractor is shown in Fig. 1c.

2.2 Implementation of the Proposed Encryption Scheme

Discretization of the continuous Lorenz, Genesio-Tesi and Modified Chua system based on Runge–Kutta technique: A broad range of techniques can be harnessed to acquire the numerical results of the system of nonlinear differential equations. Difference equations are obtained after the differential equations undergo discretization. In this case, the function is performed with the aid of Runge–Kutta

4th numerical technique. An approximate value of x is estimated for a known value of t. The limitation of Runge–Kutta technique is that it is applicable only differential equations of the first order. The formula used for the mentioned operation is shown below

$$\begin{aligned} m_1 &= \text{if}(t_n, x_n) \\ m_2 &= \text{if}\left(t_n + \frac{i}{2}, x_n + \frac{m_1}{2}\right) \\ m_3 &= \text{if}\left(t_n + \frac{i}{2}, x_n + \frac{m_2}{2}\right) \\ m_4 &= if(t_n + i, x_n + m_3) \\ x_{n+1} &= x_n + \frac{m_1}{6} + \frac{m_2}{3} + \frac{m_3}{3} + \frac{m_4}{6} \end{aligned} \tag{4}$$

The formula given above is used to determine the next value of x, i.e., x_{n+1} from x_n. The variable i is the increment size and $t_{n+1} = t_n + i$. The increment size is set to a very small value for greater accuracy. We fix the increment value to $i = 0.001$ and the initial conditions and subsequently, the three sets of pseudorandom numbers namely X, Y, Z discrete series of the three attractors are obtained.

Proposed Encryption Scheme: The standard 256 × 256 Pepper RGB image is taken and processed. The image is transformed into a two-dimensional matrix of 3 different layers, i.e., R, G, B planes. The planes are further converted into one-dimensional arrays. The diffusion process is implemented by performing bit-wise XOR operation between the X, Y, Z pseudorandom sequences and the one dimensional R, G, B plane 8-bit values respectively.

This is followed by a scrambling process where the same random number sequences are used to shuffle the location of the matrix elements. This is used to produce indiscernible 1D array to increase the ambiguity of the enciphered image and the complexity of the encryption key. The encrypted information is transformed back into a 2D matrix, thus producing the encrypted image. This algorithm involves three layers of scrambling and diffusion to boost the security of the data. The entire process of encryption is shown as a block diagram in Fig. 2. The original test image can be obtained by performing the inverse operation of encryption, i.e., decryption. The diffused image, final encrypted image along with the decrypted image are shown in Fig. 3. From Fig. 3c, we can clearly observe that the process of securing the given test image is efficient and successful as the details of the plain test image cannot be retrieved from the enciphered image by an unintended recipient without the key.

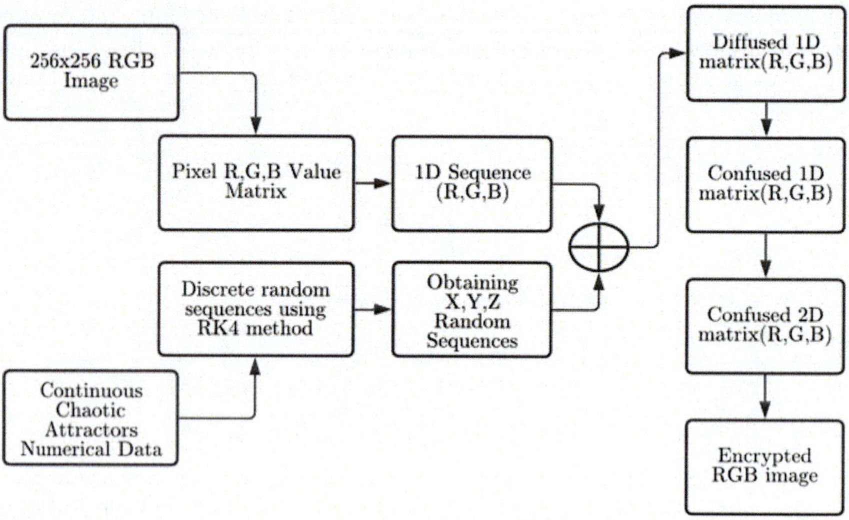

Fig. 2 Proposed encryption scheme

Fig. 3 **a** 256 × 256 Pepper test image, **b** Diffused image, **c** Final encrypted image, **d** Decrypted image

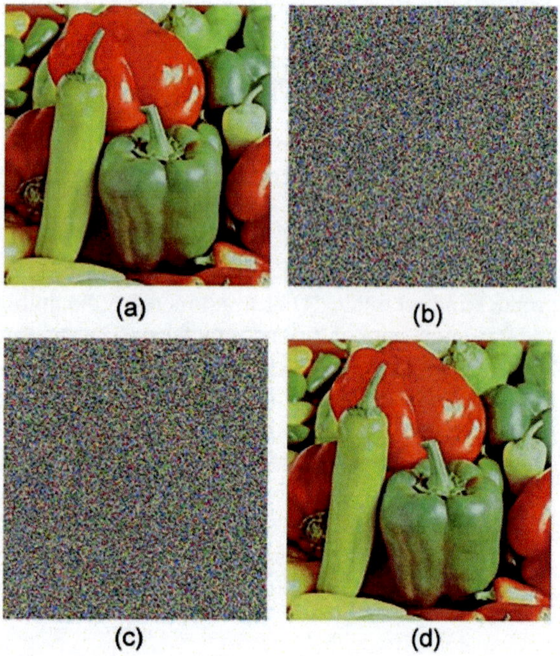

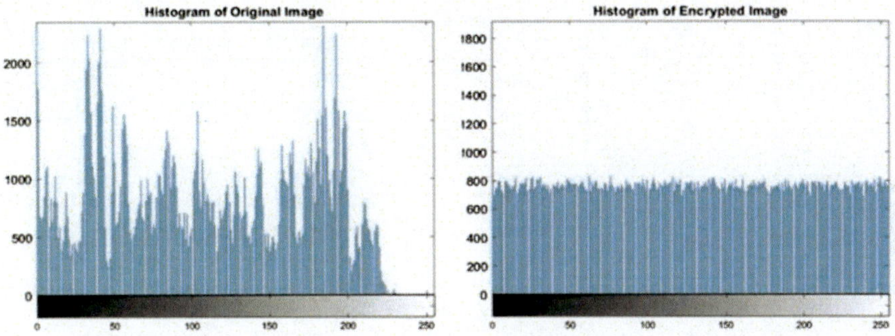

(a) (b)

Fig. 4 **a** Histogram plot of test image, **b** Histogram plot of ciphered data

3 Results and Discussion

3.1 Histogram Plot Analysis

The quantity of pixels in an image for different intensity values present in an image is given by the histogram plot. It is indicative of how the tones are distributed in an image. The histogram plot of the original test image along with the encrypted image are displayed in Fig. 4a and b. From the figure, it is observed that the histogram plot of the original data comprises of irregular rise and falls whereas the cipher data has a fairly consistent tonal distribution. The histogram plot of the encrypted data substantiates that the suggested cryptosystem is extremely effective against statistical attacks.

3.2 Entropy Analysis

Entropy analysis is one of the most primary metrics to estimate the degree of randomness in the image and evaluate the effectiveness of the scrambling and diffusion technique. The entropy of the digital image increases with the increase in the distribution of the R, G, B values in the image. The most suitable value of the entropy for an 8-bit encrypted image is 8.000. The entropy values for the enciphered image are clearly shown in Table 1.

Table 1 Comparison between the entropy of the test image and the enciphered image

Image	R	G	B	Total
Original pepper	7.1730	7.2192	7.0244	7.6271
Encrypted pepper	7.9977	7.9975	7.9976	7.9991

Table 2 Analysis of the correlation coefficients of the encrypted image and original image

Image	Horizontal	Vertical	Diagonal
Plain pepper	0.9615	0.9659	0.9435
Enciphered pepper	−0.0000	−0.0019	−0.0019

Table 3 Comparison of the UACI and NPCR values

Image		NPCR	UACI
Pepper	R	99.6735	28.5168
	G	99.6735	28.5168
	B	99.6735	28.5168

3.3 *Correlation Coefficient Analysis*

The relationship between two proximate pixels in a digital image is given by the coefficient of correlation. It is the measure of resemblance between two pixels.

The value of the coefficient varies from -1 to $+1$. For a cipher image, the correlation coefficient values should be very low and should be high for plain image. With the aim of testing the correlation of the proximate pixels of the enciphered image, 3000 pixels are randomly selected from the image and their correlation coefficients are estimated. The horizontal, diagonal and vertical coefficients of correlation are listed in Table 2.

3.4 *NPCR and UACI Analysis*

The NPCR value is indicative of the rate at which the number of pixels changes in the enciphered image, when a certain pixel value of the test image is altered and the UACI value is indicatory of the mean intensity of differences between the test data and the enciphered data. These values are directly proportional to the ability of the system to withstand differential invasions. The NPCR and UACI values for the suggested cryptosystem are listed in Table 3.

3.5 *PSNR Analysis*

The quantitative mathematical relationship between peak strength of a signal and the strength of noise that degrades the characteristics of the signal rendition is given by the peak signal-to-noise ratio (PSNR) [19, 20]. The PSNR is a direct indicator of the features of the image obtained. It is proportional to the quality of the image [21, 22]. PSNR and MSE is a very beneficial performance evaluation test to assess the quality of reconstruction of encryption codes. The average of errors raised to the second

Table 4 PSNR and MSE values of the image

Image		PSNR	MSE
Pepper	R	9.2636	7.7041e + 03
	G	7.6610	1.1142e + 04
	B	7.5478	1.1437e + 04

Table 5 Comparative study of the statistical tests

Images	NPCR			UACI			Entropy
	R	G	B	R	G	B	
Ref. [3]	99.64	99.64	99.64	33.49	33.56	33.50	7.9992
Ref. [23]	99.45	99.43	99.41	29.58	29.65	29.43	7.7036
Ref. [24]	99.6216	99.6307	99.6017	34.2633	34.0188	34.1718	7.9993
Ref. [25]	99.6357	99.6158	99.6247	33.4570	33.4705	33.4423	7.9991
Our algorithm	99.6735	99.6735	99.6735	28.5168	28.5168	28.5168	7.9991

degree between the plain test image and the image with noise is given by the MSE. The error is denotative of the amount by which the values of the plain test image differ from the depreciated image. The PSNR values obtained using the suggested algorithm is shown below in Table 4.

3.6 Comparative Study of the Encryption Technique

With a plethora of research papers being published on information security throughout the year, it is essential to compare the results of the encryption standard put forth with the already existing novel cryptosystems. The comparative study of various statistical tests conducted for the encryption scheme proposed in this paper and the reference papers are shown clearly in Table 5.

4 Conclusion

In this paper, an innovative and robust approach has been implemented in securing digital images by making use of nonlinear chaotic iterative attractors. The proposed scheme is efficient due to the use of three-dimensional chaotic attractors as they have more keyspace and it becomes very hard for the unintended recipient to obtain the original data from the cipher image. It is corroborated that the scheme proposed masks the original data effectively and the decrypted image obtained has no loss in quality. This efficiency of the cryptosystem is validated by the results of the various statistical tests conducted. This algorithm can be extended to encrypt other forms of data as well such as audio and video. This cryptosystem has been executed using software without any major obstacles and forms a solid groundwork for hardware implementation. The encryption method validated by this paper can be used in real-time communication and chaos theory has a splendid future in cryptography and information security.

References

1. Abinaya Kumari, Akshaya B, Umamaheswari B, Thenmozhi K, Rengarajan Amirtharajan, Padmapriya Praveenkumar (2018) 3D Lorenz map governs DNA rule in encrypting DICOM images. Biomed Pharmacol J 11(2):897–906
2. AlRababah A (2017) Digital image encryption implementations based on AES algorithm. VAWKUM Trans Comput Sci 13(1):1–9
3. Khan M, Masood F, Alghafis A, Amin M, Naqvi SIB (2019) A novel image encryption technique using hybrid method of discrete dynamical chaotic maps and Brownian motion. PLoS ONE 14(12):e0225031
4. Al-Haj A (2015) Providing integrity, authenticity, and confidentiality for header and pixel data of DICOM images. J Digit Imaging 28(2):179–187
5. Sheela SJ, Suresh KV, Tandur D (2017) A novel audio cryptosystem using chaotic maps and DNA encoding. J Comput Netw Commun 2017
6. Kumari M, Gupta S, Sardana P (2017) A survey of image encryption algorithms. 3D Res 8(4):37
7. Patil P, Narayankar P, Narayan DG, Md Meena S (2016) A comprehensive evaluation of cryptographic algorithms: DES, 3DES, AES, RSA and Blowfish. Procedia Comput Sci 78(1):617–624
8. Na Su, Zhang Y, Mingyue Li (2019) Research on data encryption standard based on aes algorithm in internet of things environment. In: 2019 IEEE 3rd information technology, networking, electronic and automation control conference (ITNEC). IEEE, pp 2071–2075
9. Alvarez G, Li S (2003) Cryptographic requirements for chaotic secure communications. arXiv preprint arXiv:nlin/0311039
10. Mainardi F, Pagnini G, Gorenflo R (2007) Some aspects of frac- tional diffusion equations of single and distributed order. Appl Math Comput 187(1):295–305
11. Xiao S, Yu Z, Deng Y (2020) Design and analysis of a novel chaos-based image encryption algorithm via switch control mechanism. Secur Commun Netw 2020
12. Xu L, Li Z, Li J, Hua W (2016) A novel bit-level image encryption algorithm based on chaotic maps. Opt Lasers Eng 78:17–25
13. Essaid M, Akharraz I, Saaidi A, Mouhib A (2019) A novel image encryption scheme based on permutation/diffusion process using an improved 2D chaotic system. In: 2019 International conference on wireless technologies, embedded and intelligent systems (WITS). IEEE, pp 1–6
14. Jeong Y-S, Oh K, Cho C-K, Choi H-J (2018) Pseudo random number generation using LSTMs and irrational numbers. In 2018 IEEE international conference on big data and smart computing (BigComp). IEEE, pp 541–544
15. Wang X, Su Y, Luo C, Wang C (2020) A novel image encryption algorithm based on fractional order 5D cellular neural network and Fisher-Yates scrambling. PLoS ONE 15(7):e0236015
16. Sambas A, Vaidyanathan S, Mamat M, Sanjaya WSM, Prastio RP (2016) Design, analysis of the Genesio-Tesi chaotic system and its electronic experimental implementation. Int J Control Theory Appl 9(1):141–149
17. Arpacı B, Kurt E, Çelik K (2020) A new algorithm for the colored image encryption via the modified Chua's circuit. Eng Sci Technol Int J 23(3):595–604

18. Patel S, Muthu RK (2020) Image encryption decryption using chaotic logistic mapping and dna encoding. arXiv preprint arXiv:2003.06616
19. Choudhary R, Arun JB (2014) Secure image transmission and evaluation of image encryption. Int J Innov Sci Eng Technol 1(2):65–69
20. Goyal M, Lather Y, Lather V (2015) Analytical relation comparison of PSNR and SSIM on babbon image and human eye perception using matlab. Int J Adv Res Eng Appl Sci 4(5):108–119
21. Patil VP, Gohatre UB, Sonawane RB (2017) An enhancing PSNR, payload capacity and security of image using bits difference base on most significant bit techniques. Int J Adv Electron Commun Syst
22. Hore A, Ziou D (2010) Image quality metrics: PSNR vs. SSIM. In: 2010 20th international conference on pattern recognition. IEEE, pp 2366–2369
23. Suryanto Y, Ramli K (2017) A new image encryption using color scrambling based on chaotic permutation multiple circular shrinking and expanding. Multimed Tools Appl 76(15):16831–16854
24. Guesmi R, Farah MAB, Kachouri A, Samet M (2016) A novel chaos-based image encryption using DNA sequence operation and secure Hash algorithm SHA-2. Nonlinear Dyn 83(3):1123–1136
25. Niyat AY, Moattar MH, Tor-shiz MN (2017) Color image encryption based on hybrid hyper-chaotic system and cellular automata. Opt Lasers Eng 90:225–237

Design of CPW-Fed Dodecagon Monopole Antenna for Ultra-Wideband Applications with Hilbert Curve Fractal

C. Muthu Ramya and R. Boopathi Rani

Abstract This paper deals with the design of CPW-fed Dodecagon Monopole antenna for Ultra-wideband applications that exhibits the bandwidth ranges from 2.97 to 15.58 GHz which corresponds to the fractional bandwidth of 135.96%. By introducing a second-order Hilbert curve slot in the patch nearby feed line, the rejection of band occurred around 3.5 GHz, to eliminate the Worldwide interoperability for Microwave Access (WiMAX) band was achieved. The peak antenna gain of 3.2 dBi is detected. The proposed antenna was designed on FR-4 substrate with a dimension of $40 \times 40\ mm^2$.

Keywords Dodecagon monopole · Hilbert fractal · Ultra-wideband · CPW feed · Bandwidth enhancement · Fractal antenna · Band notch

1 Introduction

The Ultra-wideband antenna (UWB) is mostly preferred for High speed and short range communications to meet the requirements of recent technological advancements in mobile communication. The UWB antenna with band rejections is desirable to mitigate the problem of interference with the conventional wireless communication systems. It is possible to obtain band notch characteristics by using any one of the techniques such as fractal structure [1], Slots [2], Parasitic stub [3], Electromagnetic Band Gap structures [4]. In this paper, the band notch characteristics were created by inserting a Hilbert curve slot in the Dodecagonal monopole radiating patch.

By incorporating a Hilbert curve slot, the miniaturization as well as the frequency rejection characteristics is obtained. The third iteration order Hilbert curve is used to make the antenna compact and the band notch creation [5]. An antenna that uses Hilbert like curve for multiple band operation was proposed [6]. A hexagonal boundary Sierpinski carpet fractal antenna by modifying hexagonal monopole

C. Muthu Ramya (✉) · R. Boopathi Rani
National Institute of Technology Puducherry, Karaikal, India

R. Boopathi Rani
e-mail: rbrani@nitpy.ac.in

A. Sivasubramanian et al. (eds.), *Futuristic Communication and Network Technologies*, Lecture Notes in Electrical Engineering 792,
https://doi.org/10.1007/978-981-16-4625-6_73

antenna for UWB application was proposed in [7]. By introducing a 'Y' like slot, the antenna can reject the band from 5.15 to 5.825 GHz.

Designing an antenna to serve for the modern-day wireless communication necessitates broadband, multiband and high gain which is being the biggest challenge today. The incorporation of fractal is one of the techniques utilized to satisfy such needs. In this proposed work a dodecagon monopole has been presented in which a Hilbert fractal was introduced to get the band rejection characteristics. This monopole antenna produced the wideband resonance from 2.97 to 15.58 GHz. A Hilbert curve slot is etched from the patch to create a notch band around 3.5 GHz through which the disturbances from WiMAX communication system will be eliminated.

This paper is arranged as follows: Sects. 2 and 3 present the antenna design and simulated results and discussions respectively. Section 4 reports the conclusion about the work.

2 Antenna Design

A dodecagon monopole has been proposed for UWB antenna. The dodecagon-shaped patch radiator is shown in Fig. 1. There are research papers proposed hexagonal or octagonal patch for bandwidth enhancement [8, 9]. In line with that, it is aimed to find the dodecagon as radiating element. Hence the construction starts with a dodecagonal of radius 10 mm. The CPW feed of width 3.2 mm provides a good amount of impedance matching over a wide frequency range also offers a less dispersion at higher frequencies [10].

Fig. 1 Basic dodecagonal shape patch

The expression given in Balanis [11] is utilized to calculate the dimension of the proposed patch antenna since the circular shape patch and the dodecagonal shape patch are closely related to each other. The mathematical expression for resonant frequency corresponding to a circular radiating patch is as follows:

$$F_r = \frac{X_{\text{nm}} C}{2\pi a_e \sqrt{\varepsilon_r}} \tag{1}$$

where the value of $'X'_{\text{nm}} = 1.8412$ for TM_{11} mode. 'C' is free space light velocity, $'\varepsilon'_r$ is the dielectric constant of the substrate, and $'a'_e$ is the effective radius of circular patch. The $'a'_e$ in terms of physical radius is given as,

$$a_e = a\sqrt{\left[1 + \frac{2h}{\pi a \varepsilon_r}\left(\ln\frac{\pi a}{2h} + 1.7726\right)\right]} \tag{2}$$

where 'a' is the physical radius of the same patch, the thickness of the substrate is represented as 'h'. The resonant frequency of the proposed dodecagonal patch is calculated by equating the area of circular patch with that of the dodecagon patch.

$$\pi (a_e)^2 = 11.196 r^2 \tag{3}$$

where the side length of dodecagon patch is denoted as 'r'.

The radius and side length of the dodecagon patch are 10 mm and 5.176 mm respectively, and by using the Eqs. (1)–(3), the resonant frequency obtained is 4.28 GHz.

The German mathematician David Hilbert introduced an algorithm for Hilbert curve in 1891, [12]. The Hilbert curve is built with the iterative steps as shown in Fig. 2. Each step is built with four duplicates of the previous structure. The total length S of whole line portions is given as (4) [5],

$$S = \left(2^{2n} - 1\right)d = \left(2^{2n} + 1\right)L \tag{4}$$

$$d = \frac{L}{\left(2^{2n} - 1\right)} \tag{5}$$

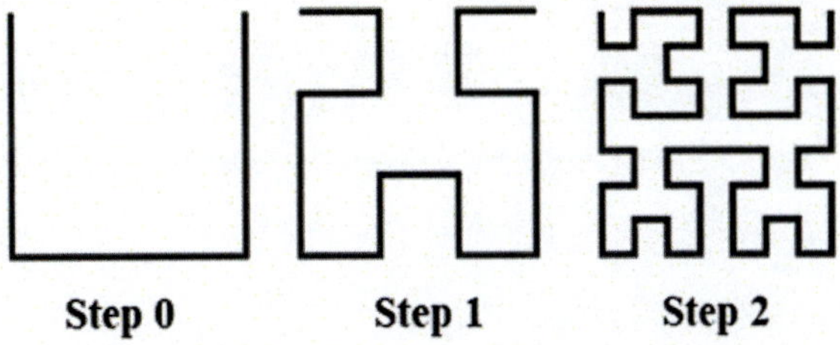

Fig. 2 Building process of Hilbert curve [5]

Here, the Hilbert curve's side dimension is denoted as '*L*', the line segment's length is '*d*' and the iteration order as '*n*'. Hilbert curve of second order given in step 1 was utilized as a band notch segment in this proposed antenna by etching it from the main radiating patch.

The proposed antenna shown in Fig. 4a was designed on FR4 material of 1.6 mm height with dimensions of 40 × 40 mm^2, which has relative dielectric constant and loss tangent are 4.4 and 0.02 respectively. The space between the dodecagon and the ground is 1.25 mm. The width of each ground plane is 17.11 mm. The suitable width of the Hilbert curve in the proposed antenna is 0.52 mm.

3 Results and Discussions

The design and analysis of the proposed antenna is carried out by using commercial electromagnetic simulator ANSYS HFSS. It uses Finite Element Method to analyze antenna. Figure 3 shows the dodecagon-shaped monopole antenna and its return loss output. It produces the bandwidth ranges from 2.95 GHz to 15.95 GHz which corresponds to the fractional bandwidth of 137.57%.

Figure 4a and b shows the proposed antenna consists of second-order Hilbert curve slot on the dodecagonal patch and its return loss results respectively.

Figure 5a and b shows the E plane and H plane radiation patterns of the final antenna at 4.28 GHz and 4.78 GHz frequencies.

The return loss results plot states that the proposed dodecagon monopole antenna with Hilbert curve fractal slot can be used for UWB applications. The radiation pattern expresses the radiation characteristic of the antenna is suitable for the proposed application. The peak antenna gain detected is 3.2 dBi.

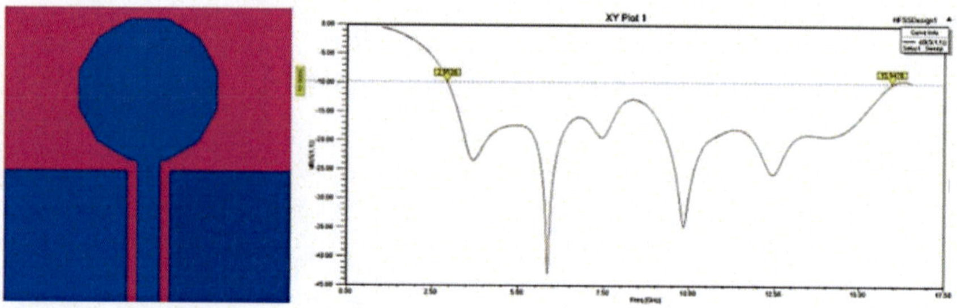

(a) (b)

Fig. 3 **a** Dodecagonal shaped monopole, **b** return loss characteristics

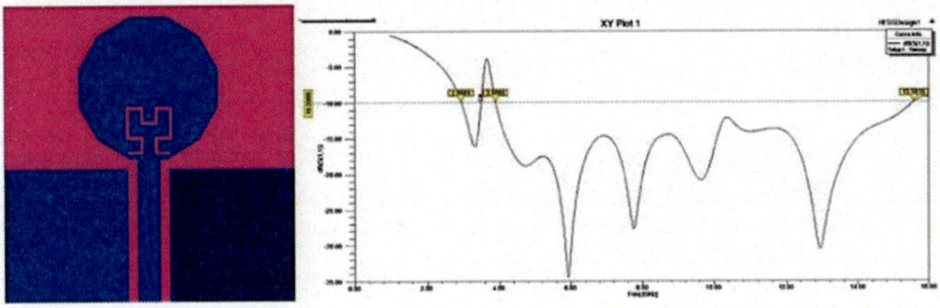

(a) (b)

Fig. 4 **a** Proposed antenna, **b** return loss results of proposed antenna

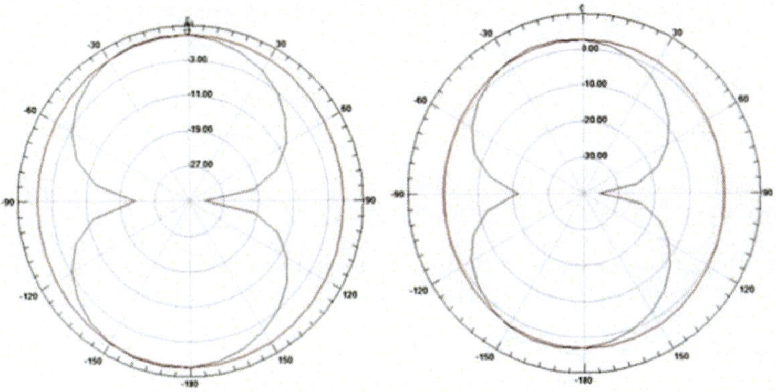

(a) (b)

Fig. 5 **a** Radiation pattern at 4.28 GHz, **b** radiation pattern at 4.78 GHz

4 Conclusion

This paper proposes and analyzes a dodecagonal monopole antenna with CPW feed for Ultra-wideband communications. The antenna operates in the frequency range of 2.97–15.58 GHz for S_{11} better than -10 dB. The notch band is created around 3.5 GHz by means of incorporating the Hilbert curve slot on the radiating patch. The return loss and gain are determined to analyze the performance of the given antenna. This antenna performs well for UWB communication and X-band satellite communication systems.

References

1. Banerjee J, Gorai A, Ghatak R (2020) Design and analysis of a compact UWB MIMO antenna incorporating fractal inspired isolation improvement and band rejection structures. AEU Int J Electron Commun 153274
2. Boopathi Rani R, Pandey SK (2016) A parasitic hexagonal patch antenna surrounded by same shaped slot for WLAN, UWB applications with notch at vanet frequency band. Microw Opt Technol Lett 58(12):2996–3000
3. Ojaroudi Parchin N, Basherlou HJ, Abd-Alhameed RA (2020) UWB microstrip-fed slot antenna with improved bandwidth and dual notched bands using protruded parasitic strips. Prog Electromagn Res 101:261–273
4. Sanmugasundaram R, Natarajan S, Rajkumar R (2020) Ultrawideband notch antenna with EBG structures for WiMAX and satellite application. Prog Electromagn Res 91:25–32
5. Kim DO, Kim CY, Park JK, Jo NI (2011) Compact band notched ultra-wideband antenna using the Hilbert-curve slot. Microw Opt Technol Lett 53(11):2642–2648
6. Kaka AO, Toycan M, Bashiry V, Walker SD (2012) Modified Hilbert fractal geometry, multi-service, miniaturized patch antenna for UWB wireless communication. COMPEL Int J Comput Math Electr Electron Eng
7. Ghatak R, Karmakar A, Poddar DR (2013) Hexagonal boundary Sierpinski carpet fractal shaped compact ultrawideband antenna with band rejection functionality. AEU-Int J Electron Commun 67(3):250–255
8. Sawant KK, Kumar CS (2015) CPW fed hexagonal micro strip fractal antenna for UWB wireless communications. AEU-Int J Electron Commun 69(1):31–38
9. Singhal S, Singh P, Kumar Singh A (2016) Asymmetrically CPW-fed octagonal Sierpinski UWB fractal antenna. Microw Opt Technol Lett 58(7):1738–1745
10. Kumar R, Malathi P, Sawant K (2011) On the design of wheel-shaped fractal antenna. Microw Opt Technol Lett 53(1):155–158
11. Balanis CA (2016) Antenna theory: analysis and design. Wiley
12. Sagan H (2012) Space-filling curves. Springer Science & Business Media

Performance Enhancement of FSO Communication Under Complex Weather Environment

Ramika Chakhaiyar, Risha Singh, and Mangal Singh

Abstract Free space optical communication is a well-known line of sight technology with the use of lasers for optical bandwidth connections. Bad weather conditions like dust storms in the Middle-East affect the performance in free space. These weather conditions lead to an increase in the bit error rate (BER) and decrease the Q-factor to low levels in FSO channel. To improve the performance of the system in such conditions a dual FSO channel has been proposed in which each channel includes optical amplifiers (specifically, an Erbium-doped fiber amplifier (EDFA)). This system shows performance improvement in terms of both BER and Q-factor. Additional communicating distance can be realized by exploiting 1550 nm by multiple channels and EDFA.

Keywords Free space optical communication (FSO) · Q-factor · Bit error rate (BER) · Atmospheric attenuation · Laser · Dust attenuation · Erbium-doped fiber enhancer (EDFA) · Optisystem

1 Introduction

Free Space Optical Communication (FSO) can be referred as an optical communication revolution that operates on light propagation in free space in order to transmit information for broadcast wireless communications. This appears differently in relation to wired medium, for example, optical fiber links. This free space optical research is of great importance when the physical connections are seen to be unreasonable because of significant expenses or different considerations [1].

FSO has become popular because of its ability to carry high bandwidth data for long distances (up to 8 km) with low BER. Weather conditions must be considered while using FSO channels because FSO channels get disturbed by carrier medium and dust storms result in suspended particles in air. Complex weather conditions like fog, dust storm, snow, rain, haze, etc. have a degrading effect on FSO channel. Absorption and scattering have a reciprocation effect on laser photons. Numerous

R. Chakhaiyar · R. Singh · M. Singh (✉)
Institute of Technology, Nirma University, Ahmedabad, India

A. Sivasubramanian et al. (eds.), *Futuristic Communication and Network Technologies*, Lecture Notes in Electrical Engineering 792,
https://doi.org/10.1007/978-981-16-4625-6_74

methods have been proposed to overcome the setbacks of bad weather conditions, Optical Amplifying, Wavelength Division Multiplexing (WDM), etc. WDM has been observed to overcome the problem of beam divergence [2–7].

In this paper, FSO models have been executed using an optical amplifier (Erbium-doped fiber amplifier (EDFA)) at the transmitter and receiver end to obtain performance comparison in terms of BER and Q-factor values along with a distance up to 1 km. The weather conditions have been simulated using Kim's Model aimed at laser beam propagation. Dual-channel FSO system and then three-channel FSO system have been implemented. The comparison of the parameters for the different systems proposed above has been summarized [2].

The paper is organized as follows: Sect. 2 discusses the attenuation by dust particles for FSO link. Next, the effect of turbulence is studied. System models in order to optimize the output have been proposed and implemented. Later, the results obtained from the proposed models have been compared and analyzed in Sect. 3. Finally, appropriate conclusions have been obtained in Sect. 4.

2 Effect of Dust Storms in FSO Link Communication

2.1 Attenuation Condition

Attenuation refers to weakening of signal strength during propagation. It occurs due to absorption, scattering and scintillation of laser photons. Water particles and carbon dioxide are responsible for the absorption of optical signals. Beer's Law can be utilized in order to realize the relation which is present between the powers of the transmitted and the received signals, when atmospheric attenuation is taken into account [3]. This Law can be given as expressed in (1):

$$\tau = P_L = P_0 \times \text{EXP}(-\alpha \times L) \tag{1}$$

where,

τ = Atmospheric attenuation.
P_L = Laser Power at length L.
P_0 = Laser Power at the source.
L = Distance.
α = Atmospheric Attenuation Constant.

The atmospheric attenuation will be calculated (in dB) by using (2)

$$\tau_R = 4.3429 \times (\alpha \times L) \tag{2}$$

For the calculation of the Atmospheric Attenuation Coefficient of (2) as per Kim's formula, the value of α will be

Table 1 Attenuation by FSO link (Kim model) [3]

Visibility (km)	dB/km (785 nm)	dB/km (1550 nm)	Weather
0.05	340	340	Heavy fog
0.07	242	242	Heavy dust
0.2	85	85	Moderate
0.5	34	34	Dust-fog
1	14	10	Rain
2	7	4	Haze
4	3	2	
10	1	0.4	Clear
23	0.5	0.2	

$$\alpha = \frac{3.91}{V}\left(\frac{550\,\text{nm}}{\lambda}\right)^{-q} \tag{3}$$

where,

$\lambda =$ Wavelength in nm.
$q =$ Size distribution of scattering particles.
$V =$ Visibility.

Scattering is a phenomenon under the condition of the optical signal in collision with a scatterer in the medium. As a result, there is an occurrence of a directional redistribution of energy [8]. This results in the loss of beam power in the direction of propagation which is highly dependent on the relative size of the scatters in the medium with reference to the optical signal wavelength [3].

For the constant wavelength, attenuation is only a factor of visibility. In this paper, the comparison of two different wavelengths has been done and the attenuation effect on each one of them has been analyzed. The atmospheric attenuation coefficient has been summed up as shown in following Table 1.

2.2 *Turbulence Condition*

The signal loss caused due to Turbulence can be described with two parameters of intensity fluctuation [9]: (1) Correlation length, (2) Correlation time.

The performance of any FSO link gets detoriated due to atmospheric turbulence, which leads to the temporal and spatial fluctuation of light intensity. The Space Diversity Reception Technique (SDRT) as well as the Advanced Modulation formats help in order to remove the impairments of the atmospheric turbulence [10].

When the aperture of the receiver is larger than the length of correlation then loss can be reduced significantly by using aperture averaging technique. But this might not always be a practical solution. In such cases, there can two approaches [4]:

1. Temporal domain technique—In this technique, only one receiver is used. If the receiver has knowledge of marginal fading distribution but has no knowledge of the temporal fading correlation and instantaneous fading state then the technique employed is Maximum Likelihood Detection (MLD). But if the receiver has knowledge of joint temporal fading distribution and has no knowledge of instantaneous fading state then Maximum Likelihood Sequence Detection (MLSD) is employed.
2. Spatial domain technique- In this technique, a minimum of two receivers are used at different locations or at different spatial angles. For improving the gain of reception, these two receivers should be kept at large distance.

Temporal pulse broadening results in an aliasing effect if the duration of temporal pulse is more than that of bit time. This causes loss of signal and hence results in higher BER [5]. The receiver SNR can be expressed as given in (4):

$$\mathrm{SNR}(z) = \frac{(P_r(z)R_d)^2}{\sigma_n^2} \tag{4}$$

where,
R_d = Receiver responsivity.
σ_n^2 = Dark Noise, thermal noise and shot noise.
$P_r(z)$ = Received power under pulse broadening.
z = Distance at which receiver is kept.

The power received at distance z under pulse broadening is given by (5):

$$P_r(z) = P(0)\frac{\sigma_0}{\sigma(z)}\int \exp\left(-\frac{t^2}{4\sigma^2(z)}\right)\mathrm{d}t \tag{5}$$

One of the techniques that can be employed to reduce the effect of Attenuation and Turbulence is Co-operative diversity. In this technique, virtual arrays are created for transmission and reception of the signal [11]. Relay-assisted communication technique can also be applied which receives a signal, passes it through an amplifier (optical) and finally this signal is passed down to the consecutive node [6].

2.3 *Literature Survey on Existing Systems*

In FSO communication technology high bit rate of laser beams are transmitted. Since light in a laser follows a straight line, anything obstructing its path would cause loss of signal [7]. Therefore, this becomes a limitation of this system. Obstruction caused by rain, haze, fog, temperature, pressure, etc. lower the range of such devices to only a few kilometers.

Some parameters have to be highlighted to optimize the model. The best priority order for various parameters is as illustrated in Fig. 1 [11]. Under hazy weather

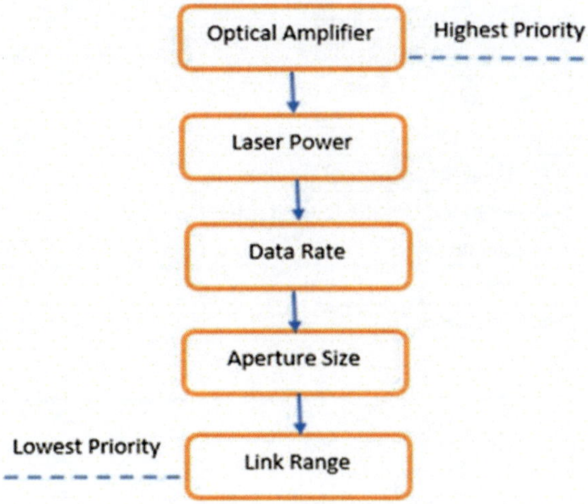

Fig. 1 Priority order for FSO performance

conditions (visibility > 2 km), 1550 nm and 785 nm light differ in their value of atmospheric attenuation, but in foggy weather, the laser light attenuation is not dependent on the wavelength. All the wavelengths get attenuated equally by fog. Mie scattering calculations results show this condition [12].

2.4 System Model

The system has been modeled with the optical simulator software "Optisystem-17". This software allows the users in order to plan, test, as well as simulate optical links in the optical networks.

In this model, the transmitter has a pseudo-random bit generator, a non-return-to-zero pulse generator, Mach–Zehnder modulator and continuous wave laser diodes. This generates wavelengths 750 nm and 1550 nm. In order to reduce the jitter timing by 40%, an avalanche photodiode (APD) and a low-pass Gaussian filter are used at the receiver [3]. The parameters as specified in Table 2 had been taken into account.

Figure 2 shows the modeling of two-channel FSO system without any amplifier. In Fig. 3, two optical amplifiers EDFAs have been enforced in two FSO channels. This allows the benefit of the constructive combination of the light power [3].

The gain of the amplifiers is 20 dB. In Fig. 4, a three-channel FSO having amplifiers has been represented. The parameters are kept same in all the models.

Table 2 FSO link parameters

Channel configurations	Parameter
Dynamic Range	1000 m
Attenuation for dust storms	25 dB/km–200 dB/km
Diameter of the aperture of the receiver	20 cm
Diameter of the aperture of the transmitter	2.5 cm
Transmission power	160 mV
Wavelength	785 nm, 1550 nm
Beam divergence	1 mrad

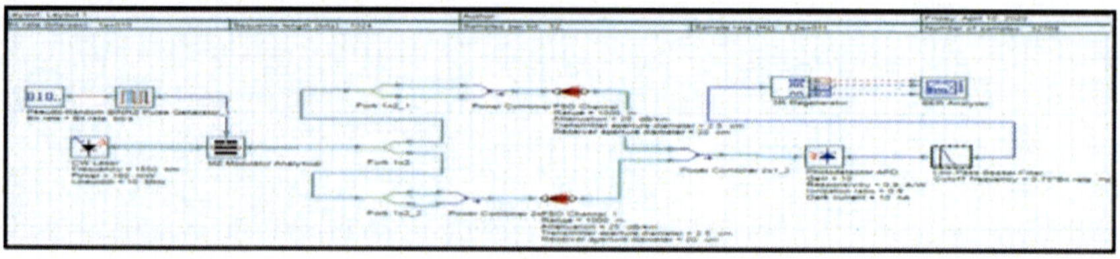

Fig. 2 Two-channel FSO system

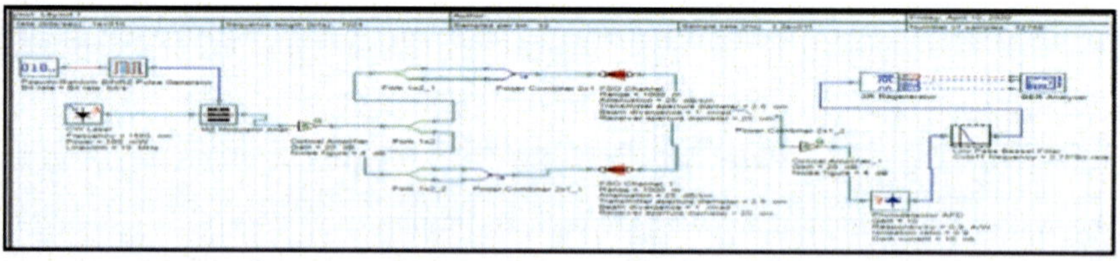

Fig. 3 Two-channel FSO system with amplifier

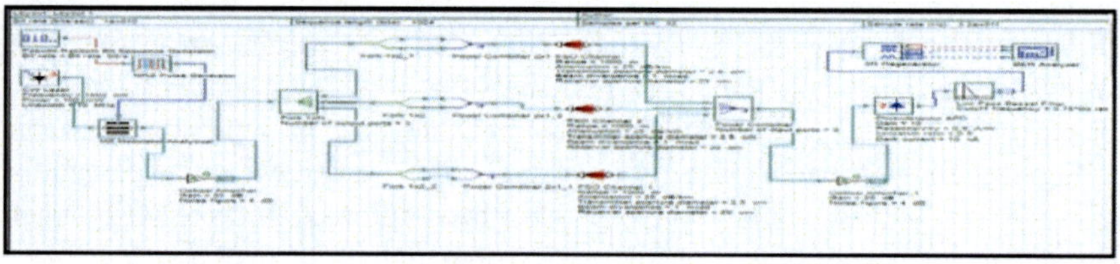

Fig. 4 Three-channel FSO system with amplifier

3 Results and Discussions

The parameters are set as per Table 2 and the following results for different wavelengths and different systems have been obtained.

Table 3 Q-factor and BER for various models-785 nm wavelength of laser beam

Model	Q-factor (max)	BER (min)
Two-channel without amplifier	16.22	1.13e−0.59
Two-channel with amplifier	3131.11	0
Three-channel with amplifier	3899.89	0

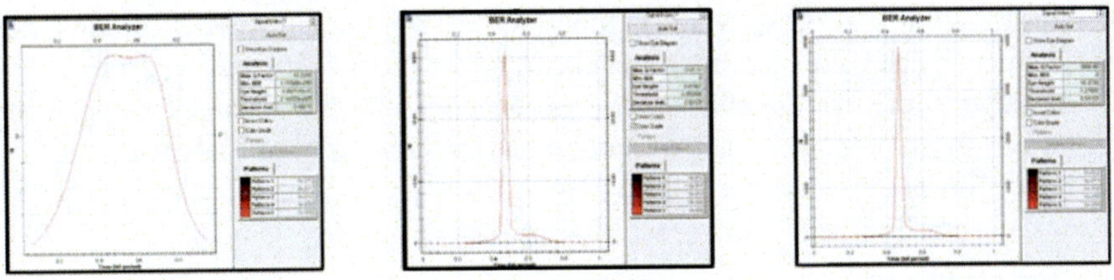

(a) (b) (c)

Fig. 5 **a** Two-channel-no amplifier; **b** Two channels with amplifier; **c** Three channels with amplifier

3.1 For 785 nm Wavelength Laser Beam

The Q-factor versus time graph of the systems after setting the wavelength as 785 nm has been shown. Table 3 shows the Q-factor (max) and BER (min) obtained.

Figure 5a shows the resultant BER curve two-channel FSO system without amplifier. The result was having a Q-factor (max) of 16.22 and the BER's minimum value as 1.133e−0.59.

For further improvement of the result, it was implemented with the addition of amplifiers. This upgraded the result, as evident in Fig. 5b. The maximum value of Q-factor is 3131.11 and the BER value is 0. It can be seen that performance has been improved for three-channel FSO (Fig. 5c). The result is having Q-factor 3899.89 (higher than the latter ones), and BER is 0 as desired.

3.2 For 1550 nm Wavelength Laser Beam

Table 4 shows the results that were obtained by different proposed systems by setting the wavelength to 1550 nm.

Table 4 Q-factor and BER for various models-1550 nm wavelength of laser beam

Model	Q-factor (max)	BER (min)
Two-channel without amplifier	17.74	6.102e−0.71
Two-channel with amplifier	1137.37	0
Three-channel with amplifier	1344.04	0

In Fig. 6a, the methodology of finding this BER Analyzer is similar to that which is done in Fig. 5a, but the wavelength is 1550 nm. The result in Fig. 6a can be seen to be as 17.74 as Q-factor along with a BER value of 6.102e−0.71.

In Fig. 6b, Q-factor of the two-channel FSO with amplifier model comes out to be 1137.37 and the BER is 0. Therefore, the Q-factor of a two-channel FSO with amplifier is more than that of without amplifier. Also, the Q-factor of two-channel FSO with amplifier model of 1550 nm is lesser as compared to that of 785 nm. As demonstrated in Fig. 6c, Q-factor is found to be 1344.04 and BER is 0. Thus, the three-channel FSO with amplifier model can be seen to give the better results.

In Table 5, the eye diagram of the two-channel and three-channel model with amplifiers in two conditions at wavelength of 1550 nm has been analyzed.

Figure 7a demonstrates the eye diagram of a two-channel model when exposed to

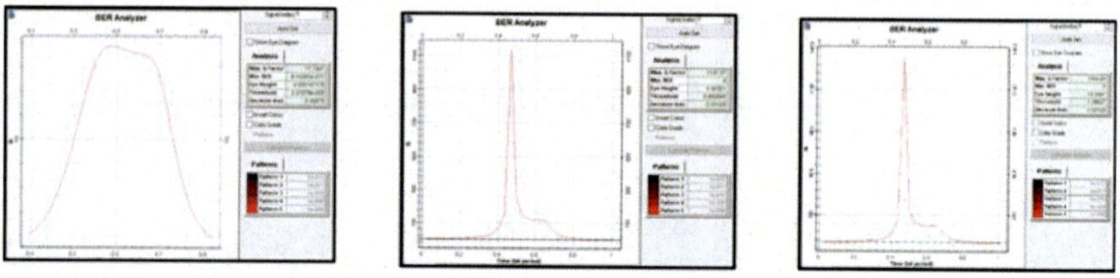

(a) (b) (c)

Fig. 6 a Two-channel-no amplifier; b Two channels with amplifier; c Three channels with amplifier

Table 5 Q-factor and BER for various models-1550 nm wavelength of laser beam

Model	Q-factor (max)	BER (min)
Heavy dust 200 dB/km of two channels at 380 m	6.317	1.103e−0.10
Moderate dust 32 dB/km of two channels at 1.7 km	20.159	7.352e−0.91
Heavy dust 200 dB/km of three channels at 380 m	7.692	6.126e−0.15
Moderate dust 32 dB/km of three channels at 1.7 km	25.266	2.437e−141

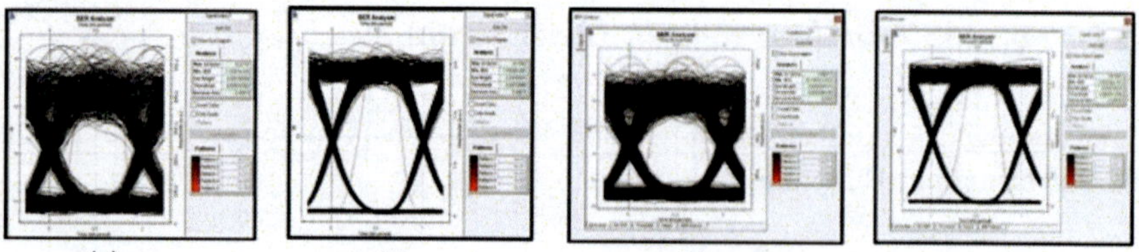

(a) (b) (c) (d)

Fig. 7 **a** Eye diagram depicts Q-factor on heavy dust 200 dB/km of two channels and amplifier system at 380 m; **b** Eye diagram depicts Q-factor on moderate dust 32 dB/km of two channels and amplifier system at 1.7 km; **c** Eye diagram depicts Q-factor on heavy dust 200 dB/km of three channels and amplifier system at 380 m; **d** Eye diagram depicts Q-factor on moderate dust 32 dB/km of three channels and amplifier system at 1.7 km heavy dust and high attenuation of 200 dB/km. The maximum Q-factor obtained in such a case is 6.317 and the minimum value of BER is found to be 1.103e−010. As per Fig. 7b, shows the eye diagram of a two-channel model in moderate dust. The maximum Q- factor obtained is 20.159 and the minimum BER is 7.352e−0.91.

After implementation Fig. 7d in three channels, the Q-factor (maximum value) is 25.266 and the minimum value of BER is 2.437e−141. As can be seen from Fig. 7a and c, the two-channel model has low Q-factor as compared to the three-channel model in similar situations. Same can be seen with Fig. 7b and d. The eye diagram also suggests that the model gives better results when the attenuation is less, as seen by comparing Fig. 7a with b and Fig. 7c with d. The three-channel system gives the optimum output as compared to any other proposed model. The Q-factor obtained when the wavelength is 1550 nm is lower than that obtained when the wavelength is 785 nm in all models. This is valid as the dynamic range is kept as 1000 m.

4 Conclusion

The FSO system has a Q-factor performance which varies over diverse climatic conditions. As per the results, the FSO links can be made more effective for increased attenuations with the help of several channels which have EDFA amplifiers. The Q-factor performance can be increased with the help of the growing quantity of transmitting links added to the influence of amplifiers. The higher the transmission links and amplifiers, the better is the performance amplifiers.

The idea of multiple channels and the power of amplifiers to increase Q-factor has been proposed in this paper for the same. Extending the idea may lead to even better results, directing to better FSO performance even in adverse climatic conditions.

References

1. Kumar P (2018) Free space optics—a review. Int J Res Appl Sci Eng Technol 6(2). https://doi.org/10.22214/ijraset.2018.2087
2. Mohammed NA, El-Wakeel AS, Aly MH (2012) Performance evaluation of FSO link under NRZ-RZ line codes, different weather conditions and receiver types in the presence of pointing errors. Open Electr Electron Eng J 6(1):28–35. https://doi.org/10.2174/1874129001206010028
3. Alnajjar SH, Noori AA, Moosa AA (2017) Enhancement of FSO communications links under complex environment. Photonic Sens 7(2):113–122. https://doi.org/10.1007/s13320-017-0336-1
4. Mikolajczyk J, Bielecki Z, Szabra D (2018) Free-space laser communications. Prz Elektrotechniczny 94(8). https://doi.org/10.15199/48.2018.08.11
5. Malik A, Singh P (2015) Free space optics: current applications and future challenges. Int J Opt 2015. https://doi.org/10.1155/2015/945483
6. Burdah S, Alamtaha R, Samijayani ON, Rahmatia S, Syahriar A (2019) Performance analysis of Q factor optical communication in free space optics and single mode fiber. Univ J Electr Electron Eng 6(3). https://doi.org/10.13189/ujeee.2019.060311

7. Mikołajczyk J et al (2017) Analysis of free-space optics development. Metrol Meas Syst 24(4). https://doi.org/10.1515/mms-2017-0060
8. Raja AR, Kagalwala QJ, Landolsi T, El-Tarhuni M (2007) Free-space optics channel characterization under UAE weather conditions. In: ICSPC 2007 Proceedings of 2007 IEEE International conference on signal processing and communication, pp 856–859. https://doi.org/10.1109/ICSPC.2007.4728454
9. Zhu X, Kahn JM (2002) Free-space optical communication through atmospheric turbulence channels. IEEE Trans Commun 50(8):1293–1300. https://doi.org/10.1109/TCOMM.2002.800829
10. Wang Z, Zhong W, Fu S, Lin C (2009) Performance comparison of different modulation formats over free-space optical (FSO) turbulence links with space diversity reception technique. IEEE Photonics J 1(6):277–285. https://doi.org/10.1109/JPHOT.2009.2039015
11. Karimi M, Nasiri-Kenari M (2011) Free space optical communications via optical amplify-and-forward relaying. J Lightwave Technol 29(2):242–248. https://doi.org/10.1109/JLT.2010.2102003
12. Bouhadda M, Abbou FM, Serhani M, Boutoulout A (2019) FSO communication capacity in turbulent and dispersive channel with AWGN noise. In: 2019 International conference on wireless technologies, embedded and intelligent systems (WITS 2019), pp 1–4. https://doi.org/10.1109/WITS.2019.8723853

Voltage and Other Parameters Monitoring System Using Smart E-Glass

B. Rajalakshmi, V. Aparna, G. Divyaswathy, and R. Pooja

Abstract In this paper, have proposed smart e-glasses display for transformer for easy voltage monitoring while working. When testing and troubleshooting the electrical works/PCB, the person face issues while placing probes on 2 points on the electric lines and simultaneously noting the value in multi-meter. The conventional model consumes more time as well as it also leads to faulty/improper measurements and also causes accidents when in contact with live wire. To overcome the issue, have proposed to integrate voltage display through the user glasses for a virtual voltage display while troubleshooting/testing the system. An Arduino is used in the circuit for processing and displaying output. The glass frame used is made of advanced organic led display. The frame is constructed to fit a display along with circuitry and the display lens used for desired reflection on the glass frame which is done by the organic led using augmented reality technology. The voltage and current measured, system is smartly designed to fit easily on a person ears. WeMos Arduino display is integrated with e-glass for efficient monitoring of current, voltage and other parameters.

Keywords Voltage monitoring · Troubleshooting/testing · Virtual organic display · Augmented reality technology

1 Introduction

Today's world requires safety and smart technology. This project is to overcome the accidents in the livewire and the transformer line burns. For the past years, there have been nearly 200 people are reported as victims. This has been increasing day by day. These systems allow supervisors, technician, managers to control and monitor the performance of power line or transmission line over the long distance. The monitoring system is very important when working in the field of three-phase systems like transformers, some companies use smart monitoring software programs.

B. Rajalakshmi (✉) · V. Aparna · G. Divyaswathy · R. Pooja
Sri Sai Ram Engineering College, Chennai, India
e-mail: rajalakshmib.ece@sairam.edu.in

A. Sivasubramanian et al. (eds.), *Futuristic Communication and Network Technologies*, Lecture Notes in Electrical Engineering 792,
https://doi.org/10.1007/978-981-16-4625-6_75

Fig. 1 Workers find faults in the transmission lines by climbing the transformer

The employers install the program or company install program to check for error. The main objective of this paper is to create a smart monitoring system using smart technology. The proposed system is a smart e-glass to monitor voltage and current [1]. measuring. The system design consists of two parts namely: the first is the control system uses Wemos Arduino to calculate and read current and voltage from sensors. The WeMos D1 is an ESP8266 operated with voltage of 3.6 V, also a WiFi-based board that uses the Arduino layout, this is an open source platform that is a low cost, very flexible, and has data processing capabilities in the cloud. This paper helps the electric workers to monitor the faults in the transformer without the direct intact with the electric lines. (Fig. 1) The workers can monitor and view the values from the ground without taking effort of climbing the transformer and having the direct contact with the electrical transmission lines through the organic LED.

The ZMPT101B current transformer is used as a voltage sensor unit (Inter-plus Industry Co. Ltd., Shenzhen, China) and the amplification of signal using a LM358 IC (Texas Instruments, Dallas, TX, USA). The current sensor unit is based on ACS712 chip (Allegro Microsystems, Worcester, MA, USA). Temperature unit, voltage and current units are isolated, low cost and flexible to use. The cloud computing is the final part of the unit placed between the control system and the end user (monitoring system) [2].

Cloud Computing is combination of the hardware and software to deliver a desired service over the network (typically the Internet). Users can access files and application using cloud computing. Its services includes the easy monitoring for servers, high storage, analyzing of database, network, analysis and intelligence through Internet ("the cloud") to offer latest innovation, flexible resources and economics of the scale [3].

Cloud computing is special computing which relies on shared resources better than having the local servers or personal devices to handle applications. The cloud can access any remote data for any related work. The cloud computes on a corporate level as well house hold operation.

The cloud storage is digital data is just stored in logical pools. The physical environment has data storage spans multiple servers (sometimes may be in multiple locations) is typically used in monitoring and managed by a hosting companies.

Cloud computing provides the top-of-the-security-line to protect user's data. Hence the Cloud computing is completely safe for the data storage.

The ThingSpeak is one of the communication library by the Arduino such as the *ESP8266* and ESP32. *This* is the open platform for the Internet of Things with the MATLAB analytics and visualization. This system enables the Arduino or some of other flexible hardware to either read or write the data to ThingSpeak or from the ThingSpeak [4, 5].

The second part of this paper is the monitoring system or monitoring application displaying using the organic LED [6]. The data received from transmission line or power line passes through microcontroller. An open source platform uses a new application designed by Google uses virtual display is implemented in organic LED display.

2 Literature Review

This paper is divided into two parts, the first part is for the Arduino microcontroller [2, 3] to read the data from the voltage and current sensors [7] and send the data are shared by advanced cloud computing technology over Bluetooth to the end user [4, 5]. The second part of the system is monitoring using organic display, where the received data using the microcontroller, i.e., Arduino Nano [6, 8] (Table 1).

Many methods of controlling and monitoring a three-phase system circuit [11] used either for controlling or displaying the values. Organic LED where used in paper instead of traditional smartphones and LCD displays or an analog method for monitoring and displaying the results [9]. The control unit includes voltage sensor, current sensor, WeMos Arduino and cloud [10, 13, 14], while the monitoring of voltage and current done by virtual organic display [12], Google has provided with android project application (Fig. 2).

The existing system has only a small range of transmission over the sensors with the electric lines through the android application as the output display [7, 15]. This allows only a few distance to be operated for monitoring the electric lines by direct intact with the electric lines [12, 13, 16].

3 Proposed System

The Proposed system is the basic monitoring method by taking the input data from the sensors placed in the transformers. This helps the workers to view and monitor the faults and troubleshoot in the transmission lines without climbing the transformer (Fig. 3). This makes the work so easier by viewing the faults from the ground surface through the organic LED display.

This organic LED is also safe for the eyes as it does not cause any harm to the vision. This organic LED can send the data within the period of 20 ns for every data it

Table 1 Literature review related to monitoring system

Related work (Reference)	Year	Application	Communication system	Types of sensors	Microcontroller	Monitoring system
Gill et al. [9]	2012	Smart power monitoring	Zigbee	Voltage and current	PC	PC
Aurilio et al. [2]	2014	Smart meters	—	Voltage and current	Arduino Shield	Ordinary
Tamkittikhun et al. [10]	2015	Power meter design	Ethernet shield	Voltage and current	Arduino Mega 2560	PC
Salamone et al. [11]	2016	Smart lamp	Bluetooth	Temperature and humidity	Arduino Mega 2560	Smartphone
Sung et al. [18]	2013	Smart LED	WiFi + ZigBee	Light sensor	XP-8000	Smartphone
El-Latif Mowad et al. [7]	2014	Monitoring system	ZigBee	pH Probe	Raspberry Pi	PC
Calderon et al. [12]	2016	Monitoring system	Cable	Temperature	Arduino Mega and PC	PC
Alsibai et al. [4]	2015	Monitoring system	WiFi	Webcam	Embedded linux board and Arduino	Smartphone

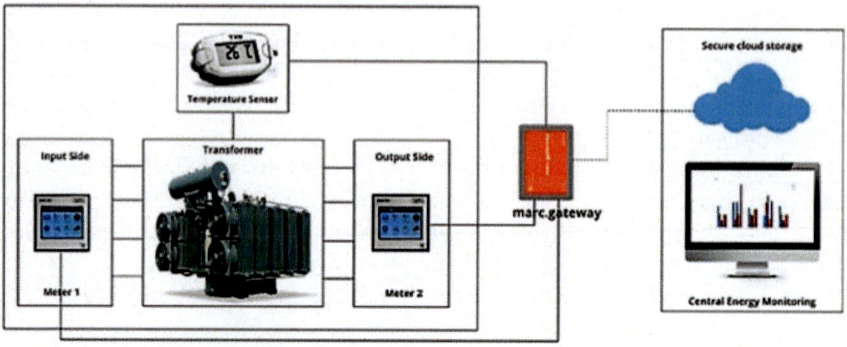

Fig. 2 Process of receiving data from the sensors

is received. The organic LED operates under the basis of the augmented technology. The principle for this display is the reflection. The data are displayed by the basis of reflection is viewed by the user to monitor the fault from the range of distance from the ground [17, 18] (Fig. 4).

Fig. 3 Structural view of the transmission lines of the transformer

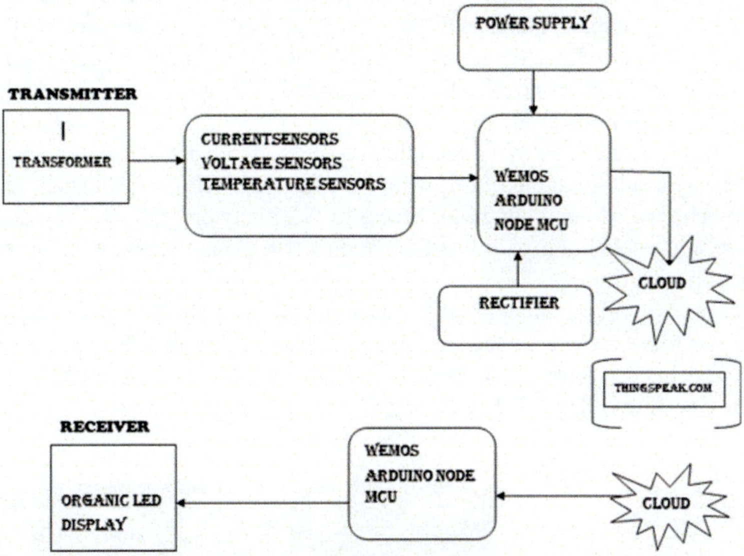

Fig. 4 Block diagram of smart e-glass voltage and other parameters monitoring system

The range of the WeMos Arduino working is around 3.6 V. The WeMos Arduino WiFi module is an indoor point to multipoint arrangement using 802.11n and the stock antenna has the range of 50–55 m nearly 162 ft or less.

The WeMos D1 is similar to Arduino UNO [10] but with the WiFi development based on the ESP8266 where it is a wireless module with the full-fledged development board circuit system. The program is done by using the Arduino IDE [11, 19].

The cloud storage involves the function of stashing data on circuits in hardware on a Internet to access remote physical locations. Clients instead of storing the data in hardware drives now send files to a data server maintained by a cloud provider. The data is sent to ThingSpeak using Wemos (Fig. 5) which requires either network

Fig. 5 WeMos ESP8266 WiFi module

connectivity through shield or on board.

ThingSpeak creates a user account and the channel for communication. A channel is for sending and receiving data. Every channel requires up to 8 data fields, location fields and the data fields. The received by ThingSpeak is 15 s for complex application up to every minute. Since ThingSpeak has the efficiency higher as more than one value can be sent. This contains a library which provides various organized by board type to help to start the program.

The WeMos ESP8266 WiFi Module [8, 20] is an advanced technology and is equipped with self-contained SOC which supports TCP/IP protocol stack that can support any type of microcontroller access to WiFi network system. The ESP8266 can host any application or offload all WiFi networking functions from another application (Fig. 6).

Voltages over approximately 50 V can be usually very dangerous for the human health. The transformer is framed to step down at the range of 12 V. The current and voltage sensors and temperature sensors and also the ground connectivity is placed over the transformer [8, 21] (Fig. 7).

Fig. 6 Smart system architectural design

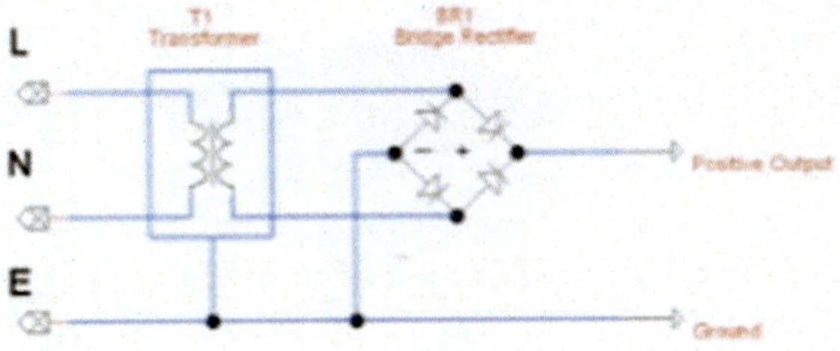

Fig. 7 Design of transformer

4 Design Methodologies

The WeMos D1 has 4 Mb of flash in it and uses 3 Mb for file data as the size of the File system depends on the flash chip size. The ESP8266 is low of cost WiFi microchip with full of microcontroller and also TCP/IP stack in it. It is manufactured by the Espressif Systems. This is the 32-bit microcontroller with 80 MHz to 160 MHz and it has nearly 80 Kb user data and 16GPIO pins (Fig. 8).

The basics of the OLED structure are simple—an organic emitter placed between the electrodes as to create efficiency. This OLED uses several intermediate layers like the electron transport and blocking layers in it. The whole structure is deposited on the substrate and the display is driver electronics. This uses the principle of the augmented reality (Figs. 9 and 10).

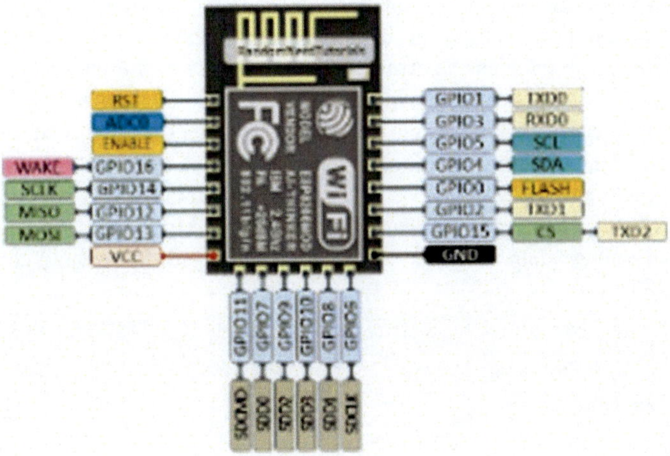

Fig. 8 Pin configuration of ESP8266 WiFi module

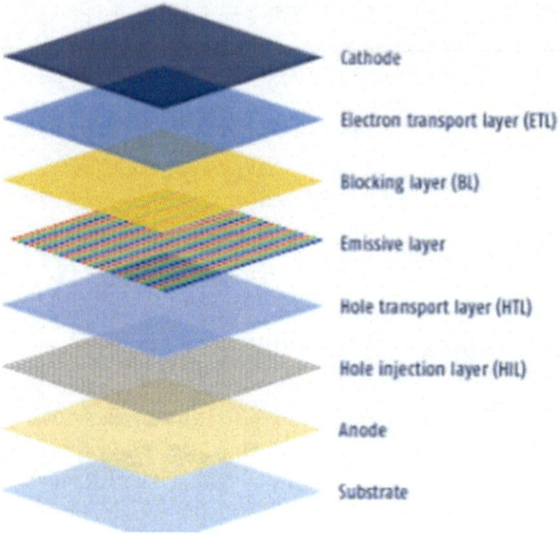

Fig. 9 Inner structure of organic LED

5 Result

The output sensed from the ACS712 is used as current sensor, ZMPT101B is used as voltage sensors, flame sensor. MQ-2 is the gas sensor is used to detect CO from the transformer. WeMos Arduino node MCU collects the data from the sensors. The cloud storage is a model of the computer storage in which the digital data is stored in logical pools. The ThingSpeak is the communication library for the Arduino for ESP8266. This enables the Arduino to write or read the data to or from the ThingSpeak. This organic LED can send the data within the period of 20 ns for every data it is received. The organic LED operates under the basis of the augmented technology. (Fig. 11) from the range of nearly 50 m or up to 160 feet range. The monitoring of the transmission is very safe for the workers without direct intact with the electric lines. The display is viewed from the ground to monitor the temperature, current and voltage data from the transmission lines with the ground connectivity (Tables 2, 3 and 4).

6 Future Scope

In future, the system can be modeled to find the fault from their current location through the wireless communication system by machine language. This technology can be used to find the faults and troubleshoots of the transmission lines.

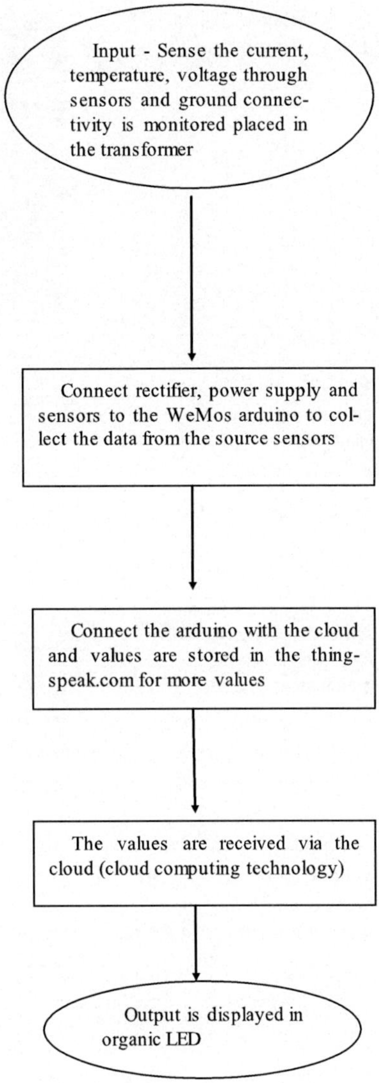

Fig. 10 Flowchart of the process in the smart voltage and other parameters monitoring system

7 Conclusion

In this paper, how the voltage and other parameters can be monitored through the smart e-glass has been discussed. The unique property can be used for safety purpose of the electric workers. Looking at the result of different methodologies this proposal will be useful for the monitoring of the parameters over the range of distance without the direct contact with the electric wires. This work can be further extended by using better alternative technique.

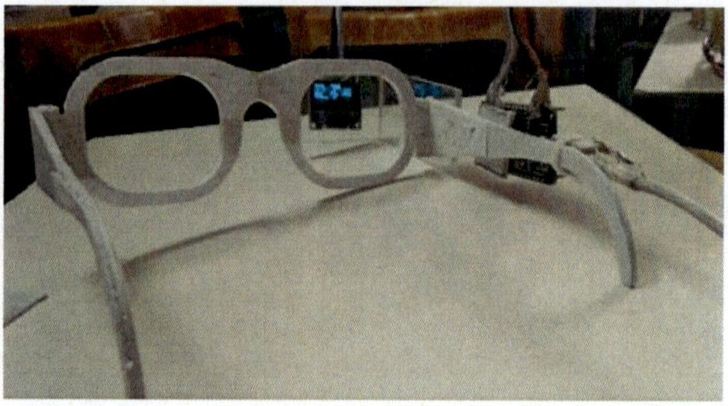

Fig. 11 Organic LED display outer structure

Table 2 Analysis of smart monitoring system for transformer of 12 V

Sensors	Values
Voltage	11.09 V
Current	820 mA
Flame	25 C
Gas	Normal

Table 3 Analysis of smart monitoring system for transformer of 5 V

Sensors	Values
Voltage	4.259 V
Current	520 mA
Flame	25 C
Gas	Normal

Table 4 Analysis of smart monitoring system for transformer of 12 V abnormal

Sensors	Values
Voltage	11.09 V
Current	820 mA
Flame	45 C
Gas	Normal

References

1. Kor A, Yanovsky M, Pattinson C (2016) Smart-item: IoT-enabled smart living. In: Proceedings of the 2016 Future technologies conference. San Francisco, CA, USA, pp 739–749, 6–7 December 2016
2. Siang HM, Alsibaai MH (2015) A smart driver monitoring system using android application and embedded system. In: IEEE International conference on control system, computing and engineering. Penang, Malaysia, 27–29 November 2015
3. Chisab RF, Van den Bossche A, Mnati MJ (2017) A smart voltage and current monitoring system for the three phase inverters using the android smartphone application. Sensors

4. Alsibai MH, Siang HM, A smart driver monitoring system using android application and embedded system. In: Proceedings of the 5th IEEE International conference on control systems, computing and engineering. ICCSCE 2015. Penang, Malaysia, pp 242–247, 27–29 November 2015
5. Salamone F, Belussi L, Danza L, Ghellere M, Meroni I (2016) An open source "smart lamp" for the optimization of plant systems and thermal comfort of offices. Sensors 16. https://doi.org/10.3390/s16030338
6. Tamkittikhun N, Tantidham T, Intakot P (2015) AC power meter design based on Arduino: multichannel single-phase approach. In: Proceedings of the 19th International computer science and engineering conference (ICSEC): hybrid cloud computing: a new approach for big data era. Chiang Mai, Thailand, 23–26 November 2015
7. El-Latif Mowad MA, Fathy A, Hafez A (2014) Smart home automated control system using android application and microcontroller. Int J Sci Eng Res 5:935–939
8. Players B, Theaters H, Sensors G, Magnet P, Motor S, Scales W, Ranges WS, Drain LS (2015) LMx58, LMx58x, LM2904, LM2904V dual operational amplifiers package IN + IN −; Texas Instruments: Dallas. TX, USA
9. González FC, Osiris O, Villegas V, Torres Ramírez DE, Guadalupe V, Sánchez C, Ochoa Domínguez H (2014) Smart multi-level tool for remote patient monitoring based on a wireless sensor network and mobile augmented reality. Sensors 14:17212–17234
10. Arduino NANO Pinout Diagram Available online: https://forum.arduino.cc/index.php?topic=147582.0. Accessed 13 March 2017
11. David N, Anozie FN, Ebuka FO, Nzenweaku SA (2016) Design of an Arduino based wireless. Int J Sci Eng Res 466–469
12. Gill SPS (2012) Smart power monitoring utility system using wireless sensor networks. In Proceedings of the 2012 Sixth international conference on sensing technology. Kolkata, India, pp. 5–114, 18–21 December 2012
13. Aurilio G, Gallo D, Landi C, Luiso M, Cigolotti V, Graditi G (2014) Low cost combined voltage and current transducer for smart meters. In: Proceedings of the 2014 IEEE International instrumentation and measurement technology conference. Montevideo, Uruguay, pp 1459–1464, 12–15 May 2014
14. .Online website: http://electronicsforyou.com
15. Al-qaseemi SA, Almulhim MF (2016) IoT architecture challenges and issues: lack of standardization. In: Proceedings of the 2016 future technologies conference. San Francisco, CA, USA, pp 731–738, 6–7 December 2016
16. Rahman M, Alfaki A, Shafiullah GM, Shoeb A, Jamal T (2016) Demand response opportunities in residential sector incorporated with smart load monitoring system. In: Proceedings of the 2016 IEEE innovative smart grid technologies. Asia, Melbourne, Australia, 28 November–1 December 2016
17. Online websites: smart e-glass monitoring system: http://nevonprojects.com.
18. Sung WT, Lin JS (2013) Design and implementation of a smart LED lighting system using a self adaptive weighted data fusion algorithm. Sensors 13:16915–16939
19. De Santis D, Giampetruzzi DA, Abbatantuono G, La M, Fellow S, Bari P (2016) Smart metering for low voltage electrical distribution system using Arduino due. In: Proceedings of the 2016 IEEE workshop on environmental, energy, and structural monitoring systems. Trento, Italy, pp. 1–6, 13–14 June 2016
20. Kebir ST, Bouhedda M, Mekaoui S, Guesmi M, Douakh A (2016) Gesture control of mobile robot based Arduino. In: Proceedings of the 2016 8th International conference on modelling, identification and control. Algiers, Algeria, pp 1081–1085, 15–17 November 2016
21. Integrated F, Linear HE Sensor C ACS712; Texas Instruments datasheet. Dallas, TX, USA, pp 1–15

Gain Enhancement for Metamaterial Loaded UWB Antenna Using TSCCR AMC

K. M. Neeshu and Anjini Kumar Tiwary

Abstract In this paper, an artificial magnetic conductor (AMC) with metamaterial-based planner monopole antenna has been proposed and examined. To realize unidirectional radiation, 5 × 5 AMC surface is used as a back cavity. It consists of four triangular shape capacitive coupled resonator (TSCCR) as a unit cell which is placed below the antenna at height of $\lambda_0/12$ mm. The proposed AMC unit cell exhibits ±90° wideband response from 2.8 to 8.6 GHz (94%). The highest gain of 8 dBi is achieved in both lower (3.1–8.3 GHz) and upper-frequency band (8.4–14 GHz). The antenna's positive gain range is 4–8 dBi for its operating band. The proposed antenna gives impedance matching of -10 dB for ultrawide frequency band 2.88–14 GHz. The UWB and high-gain properties of the proposed antenna make it suitable for wideband wireless application such as antenna for base station and narrow beam tracking like RFID.

Keywords Antenna · Efficiency · Epsilon negative (ENG) · Metamaterial · Gain · UWB

1 Introduction

Metamaterials are artificially engineered structure which gives unusual electromagnetic properties which cannot be found naturally. [1, 2]. These metamaterials are composed of periodically arranged unit cell. One of the important applications of metamaterial is artificial magnetic conductor (AMC) [3–5] which has become a hot topic in antenna designing because, it gives 0° phase reflection to the incident wave which helps to design the low-profile antenna. Conventionally, to increase the gain of the antenna, the perfect electric conductor (PEC) is positioned as a back cavity [6] and there should be at least λ/4 distance between PEC and the antenna. In the AMC surface due to 0° reflection phase to the incident wave, it is possible to design

K. M. Neeshu (✉) · A. K. Tiwary
Birla Institute of Technology, Mesra, Ranchi, India

A. K. Tiwary
e-mail: aktiwary@bitmesra.ac.in

A. Sivasubramanian et al. (eds.), *Futuristic Communication and Network Technologies*, Lecture Notes in Electrical Engineering 792,
https://doi.org/10.1007/978-981-16-4625-6_76

a low-profile antenna with high gain simultaneously. However, the main restriction of using AMC is the narrow bandwidth of the unit cell although the AMC surface can exhibit wideband when it is placed at an optimum distance from the antenna. By using AMC as back cavity, many methods [7–12] have been proposed to improve the radiation performance of the antenna. The radiation performance of the bidirectional circularly polarized aperture antenna was made unidirectional by applying the square-AMC reflector in ref. [7]. In ref. [8], high permeability material has been integrated as AMC with an antenna to increase antenna bandwidth and efficiency. A wideband textile antenna with rectangular ring AMC and circularly polarized crossed dipole for WLAN application has been discussed in refs. [9, 10]. In ref. [12], a wideband antenna (1.1–3.03 GHz) bow-tie antenna which gain was improved using a fractal-shaped AMC unit cell has been proposed. However, the original bandwidth of the antenna deteriorates slightly by using AMC at the distance of λ/8. In ref. [13], parasitic arms were used in the AMC cell to control its surface current, it gives bandwidth improvement of 41.65% and increased gain of about 4 dBi. In this paper, the first time we proposed AMC with a planer monopole antenna based on metamaterial to improve the radiation performance of an antenna. The AMC unit cell possesses bandwidth of 3.1–8.6 GHz. The composite response of antenna and AMC results in nearly the same impedance bandwidth with improved gain. The operating band gain is improved by ~4 dBi in comparison with antenna without AMC. The simulation software used is CST Microwave Studio which is based on the finite integration technique (FIT) method.

2 Proposed Design

2.1 AMC Unit Cell Design

To design unit cell, an approach is followed here which ensure that the negative value of permittivity of the unit cell, appears after the desired frequency band. The electromagnetic properties of the AMC could be tracked by its equivalent LC network. This equivalent LC network of AMC surface gives parallel resonance hence gives high impedance. The impedance of AMC surface can be calculated

$$Z_S = \frac{j\omega L}{1 - \omega^2 LC}$$

The AMC is placed below the antenna at height of h_2 as shown in Fig. 1c. The propagating wave in AMC is combination of incident wave and reflected wave. Thus, to have constructive interference, total phase difference should be zero. The AMC gives total phase difference which is given by

$$\varphi_T = \varphi_2 + \varphi_1\varphi_2 = \varphi_1 - \frac{2\pi}{\lambda}2h_2 - \theta$$

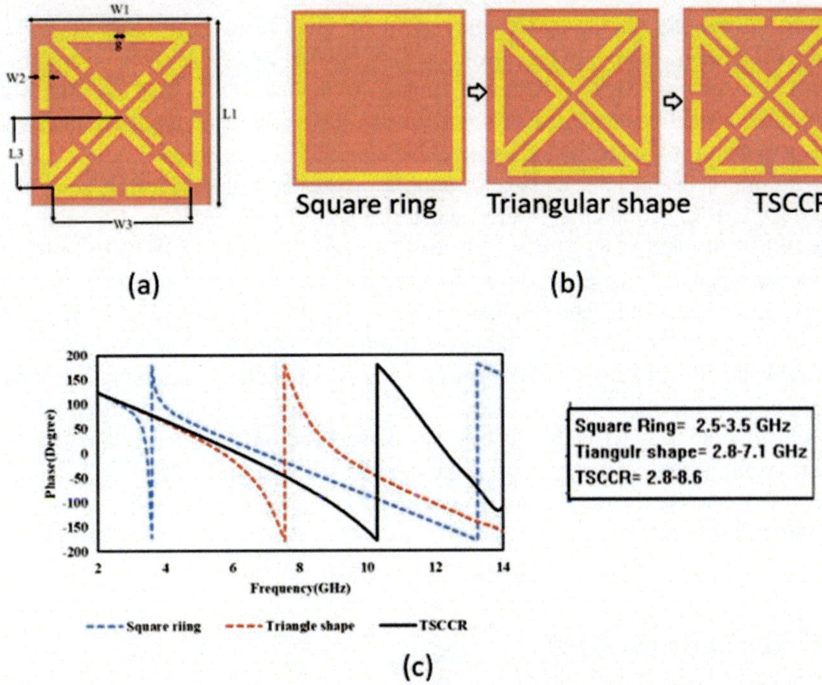

Fig. 1 Geometry of proposed unit cell: **a** Unit cell (TSCCR), **b** Unit cell designing steps, **c** Reflection coefficient phase (°)

where

φ_T = total phase difference
φ_1 = Phase of incident wave
φ_2 = Phase of Reflected wave
h_2 = Height of AMC below antenna
θ = Phase shift.

The AMC unit cell can be analyzed in terms of its equivalent LC circuit of frequency

$$f = \frac{1}{2\pi\sqrt{LC}}$$

where L and C are equivalent inductance and capacitance.

The bandwidth of AMC unit cell is defined as a band from −90° to +90° of the reflection coefficient phase, which is proportional to L/C [13, 14]. The dimensions of the proposed unit cell are shown in Fig. 1a where $W1 = 10$ mm, $L1 = 10$ mm, $W2 = 0.20$ mm, $W3 = 2.40$ mm, $L3 = 5.30$ mm, $g = 0.50$ mm. The design steps of the proposed AMC unit cell are shown in Fig. 1b. Firstly, a square ring has been chosen which gives narrow band from 2.5 to 3.5 GHz. Then, four triangles are taken to form LC coupling in the AMC unit cell which gives band from 2.8 to 7.1 GHz. The triangle shape is chosen because it utilizes the inner space of the conventional split-ring resonator (SRR) efficiently. At last, triangular shape with a center slit in all three arms is designed with three slots in each arm. Two types of coupling exist in this unit cell which is intra-capacitive coupling between four TSCCR within a single unit cell and inter capacitive coupling between immediate unit cell. The material used for simulation is FR4 substrate, with thickness $h1 = 1.6$ mm, the relative permittivity of $\varepsilon_r = 4.4$, and loss tangent of tan $\delta = 0.025$. To observe the electromagnetic characteristic of the proposed metamaterials unit cell, perfect electric and perfect magnetic boundary with waveguide port is applied in x, y and z-direction simultaneously. When square ring unit cell is simulated as AMC unit cell, it gives bandwidth of 2.5–3.5 GHz which is very low. To enhance coupling effect for bandwidth enhancement further, triangular-shaped unit cell has been chosen which gives bandwidth 2.8–7.1 GHz. At last, slit between all arms of triangle is etched which results in induced capacitive coupling in proposed unit cell TSCCR which gives the widest bandwidth from 2.8 to 8.6 GHz as depicted in the reflection phase diagram in Fig. 1c.

2.2 Antenna with AMC

The design of the proposed monopole antenna loaded with metamaterial and planar AMC surface as back cavity is shown in Fig. 2. A planer monopole antenna based on metamaterial loading is shown in Fig. 2a. The proposed antenna of size 31.8 $\times$ 27.6 mm is simulated using the FR4 substrate of relative permittivity $\varepsilon_r = 4.4$, loss tangent 0.025, and thickness (h) 1.6 mm. This antenna comprises the loading of triangular-shaped loaded unit cell and cross slot on the patch and ground plane, respectively. The partial ground plane of length 7.90 mm below the antenna to excite the port. The rest of the ground part is covered with 7 $\times$ 6 cross slot unit cell to improve the bandwidth. Top patch of antenna is loaded with 4 $\times$ 3 triangular-shaped unit cells and a window is slotted on the backside of patch antenna which improve the radiation efficiency. The optimized dimension of antenna is $W = 27.6$ mm, $L = 31.8$ mm, $W5 = 12$ mm, $L4 = 16$ mm, $W6 = 3$ mm, $L6 = 7.90$ mm, $W7 = 20.40$ mm and $L5 = 8.20$ mm (Length $L5$ is the sum of the partial ground plane and gap) $W8 = 3.60$ mm, $L8 = 3.60$ mm, $W9 = 0.20$ mm, $L9 = 3.20$ mm, $W3 = 2.40$ mm, $L3 = 1.70$ mm, $g = 0.50$ mm, $W4 = 3.75$ mm and $L4 = 2.65$ mm. The effects of the AMC height h2 on reflection coefficients and antenna gain are shown in Fig. 3a and b. The approach based on the parasitic is used to give the wideband response in the AMC unit cell. Due to the etching of different slots into the AMC unit cell, additional inductance and capacitance occur which give multiple resonating frequencies close to each other thus wideband achieved. Therefore, by combining electrical and geometrical characteristics of AMC, wideband and gain enhancement of antenna can be achieved. The optimum dimensions of the slots are calculated by parametric studies.

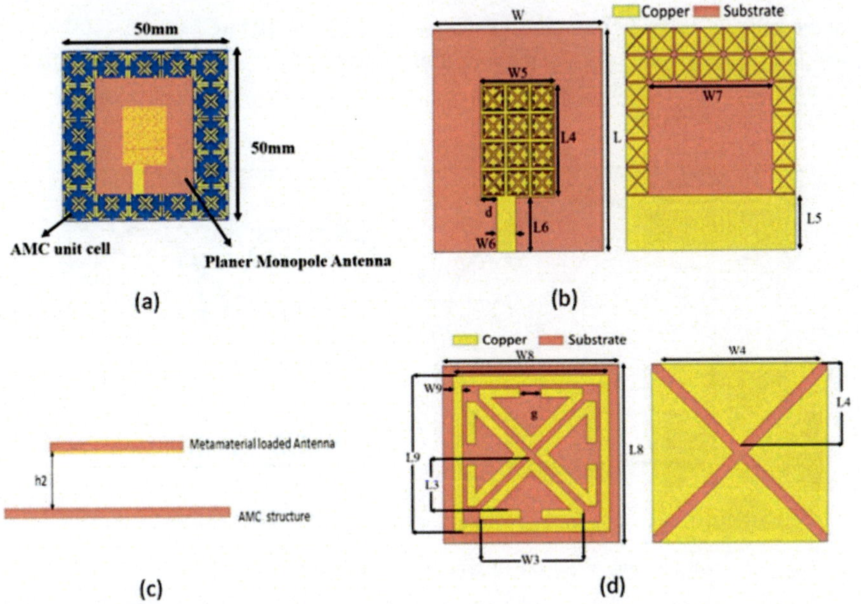

Fig. 2 Geometry of proposed antenna: **a** Front view of antenna with AMC, **b** Top and bottom view, **c** Side view of antenna with AMC, **d** Metamaterial unit cell loaded in planer monopole antenna

3 Results and Discussion

3.1 Reflection Coefficient and Gain

Simulated reflection coefficients and gain characteristics of the proposed design with and without AMC are depicted in Fig. 3. In Fig. 3a, the antenna with AMC gives nearly the same bandwidth of 3.1–13.6 GHz (121%), as the antenna without AMC gives bandwidth of 2.88–14 GHz (131%).The impedance bandwidth of proposed antenna at the AMC height $h_2 = 8, 10, 12$ mm are, respectively, 3.1–13.6 GHz (121%), 2.9–13.8 GHz (123%), 2.8–13.7 GHz (123.0%). It is near −5dB impedance matching from 3.1 to 4 GHz and below −10 dB from 4 to 14 GHz this may be because of high losses around 3 GHz. The proposed antenna shows the good impedance matching of 51 dB in the operating band. Figure 3b demonstrates antenna gain improved sharply over the frequency of 2.8–14 GHz when $h_2 = 8$ mm. The antenna shows gain of 4–8 dBi in the operating band, which is higher than the antenna without AMC by 4 dBi. To summarize, the optimum height h_2 of AMC is taken as 8mm (0.08λ at the resonating frequency of 3 GHz) for gain enhancement of the proposed antenna. Conventionally when PEC is used as reflector, it gives 180° phase reflection to the incident wave. Due to 180° phase difference between incident and reflecting waves, the energy of the incident wave gets weaken. Therefore, the distance of 25 mm (*λ/4* at 3 GHz) is required when PEC reflector is used in antenna which gives 3 times larger antenna profile compared to the antenna with AMC as a back cavity. Thus, using AMC as a back cavity in the proposed antenna, high gain and broad bandwidth are achieved in compact antenna size.

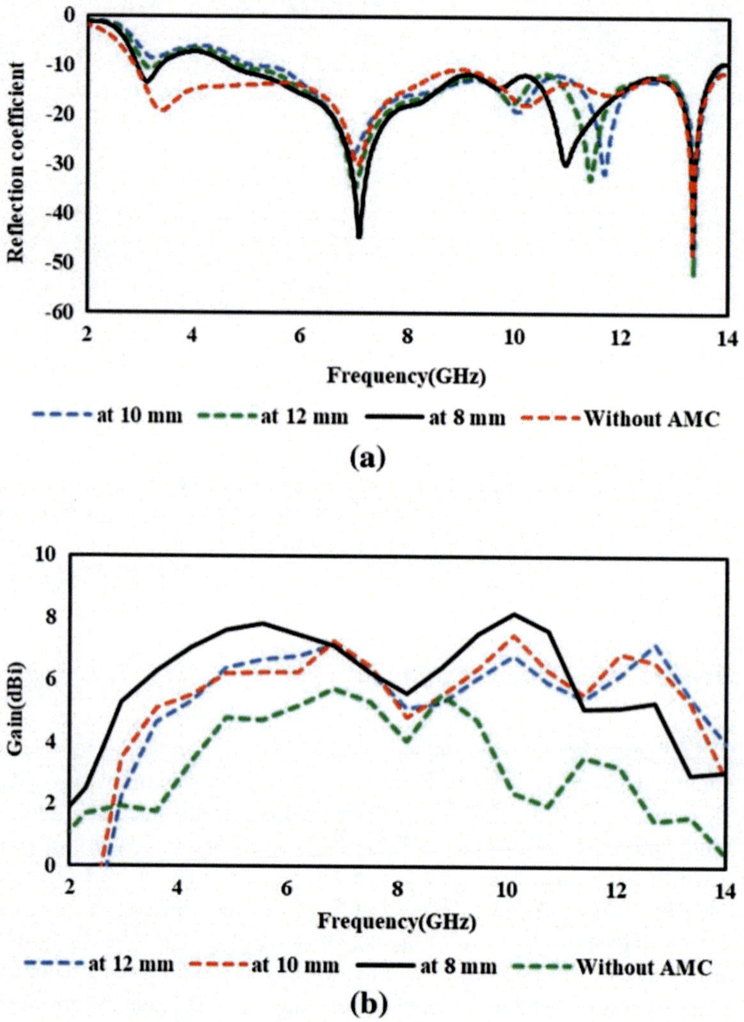

Fig. 3 a Simulated reflection coefficient curve, **b** Gain

3.2 Radiation Pattern

The radiation pattern of the proposed antenna is analyzed at the frequencies of 4, 8, and 12 GHz. At 4 GHz antenna gives an omnidirectional radiation pattern as shown in Fig. 4a. The radiation pattern at 8 and 12 GHz are presented in Fig. 4b and Fig. 4c, respectively. The maximum radiation is observed in opposition to the ground plane. The radiation pattern changes from omnidirectional to directional due to the AMC reflector used as a back cavity. Polar plot for radiation pattern for given frequency is shown in Fig. 5. it is clearly observed at higher frequency radiation pattern changes and multilobes generated due to higher modes.

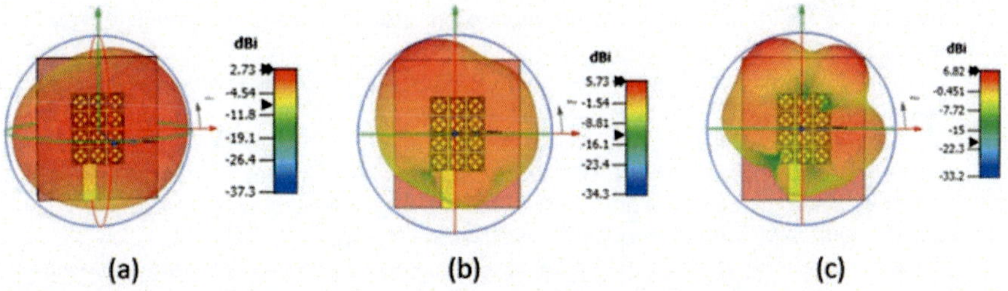

Fig. 4 3D radiation pattern at **a** 4 GHz, **b** 8 GHz, **c** 12 GHz

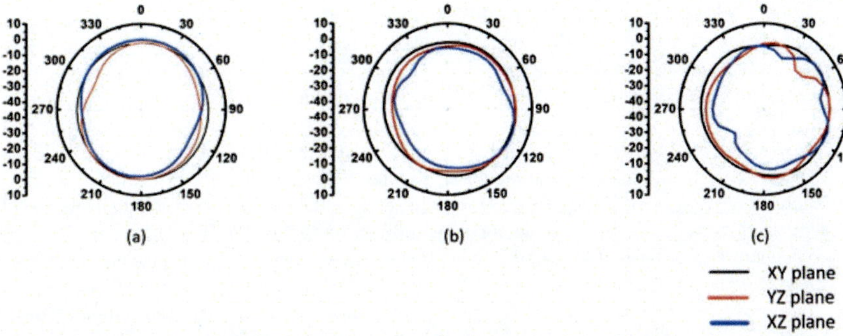

Fig. 5 2D radiation pattern at **a** 4 GHz, **b** 8 GHz, **c** 12 GHz

4 Conclusion

In this paper, a novel wideband AMC of bandwidth 2.8–8.6 GHz is proposed and applied in a planer monopole antenna for gain improvement maintaining antenna size compact. The proposed antenna with the AMC surface as a back cavity exhibits good radiation characteristics and miniaturization. The proposed antenna exhibits bandwidth from 2.88 to 14 GHz (131%). The antenna gives unidirectional radiation pattern with gain improvement of ~4 dB in the operating bandwidth. The proposed design can be used as low-profile unidirectional antenna with high-gain and wideband application.

References

1. Fonseca NJG, Mangenot C (2016) High-performance electrically thin dual-bandpolarizing reflective surface for broadband satellite applications. IEEE Trans Antennas Propag 64(2):640–649
2. Agarwal K, Nasimuddin N, Alphones A (2013) Wideband circularly polarized AMC reflector backed aperture antenna. IEEE Trans Antennas Propag 61(3):1456–1461
3. Xu WW, Wang JH, Chen M et al (2015) A novel microstrip antenna with composite patch structure for reduction of in-band RCS. IEEE Antennas Wirel Propag Lett 14:139–142
4. Joubert J, Vardaxoglou JC, Whittow WG et al (2012) CPW-fed cavity-backed slot radiator loaded with an AMC reflector. IEEE Trans Antennas Propag 60(2):735–742
5. Ghosh A, Mandal T, Das S (2016) Design of triple band slot-patch antenna with improved gain using triple band artificial magnetic conductor. Radioengineering 25(3):442–448
6. Almutawa AT, Mumcu G (2013) Small artificial magnetic conductor backed log-periodic mcrostrip patch antenna. IET Microw Antennas Propag 7(14):1137–1144
7. Agarwal K, Nasimuddin N, Alphones A (2013) Unidirectional wideband circularly polarised aperture antennas backed with artificial magnetic conductor reflectors. IET Microw Antennas Propag 7(5):338–346
8. Lago H, Jamlos MF, Soh PJ et al (2017) Bandwidth enhancement of an array antenna using slotted artificial magnetic conductors. Appl Phys A 123(1):53
9. Aun NFM, Soh PJ, Jamlos MF et al (2017) A wideband rectangular-ring textile antenna integrated with corner-notched artificial magnetic conductor (AMC) plane. Appl Phys A 123(1):19
10. Ta SX, Park I, Ziolkowski RW (2013) Circularly polarized crossed dipole on an HIS for 2.4/5.2/5.8-GHz WLAN applications. IEEE Antennas Wirel Propag Lett 12(3):1464–1467
11. Zhong YW, Yang GM, Zhong LR (2015) Gain enhancement of bow-tie antenna using fractal wideband artificial magnetic conductor ground. Electron Lett 51(4):315–317
12. Malekpoor H, Jam S (2016) Improved radiation performance of low profile printed slot antenna using wideband planar AMC surface. IEEE Trans Antennas Propag 64(11):4626–4638
13. Wu J, Yang S, Chen Y et al (2017) A low profile dual-polarized wideband omnidirectional antenna based on AMC reflector. IEEE Trans Antennas Propag 65(1):368–374
14. Li K, Li L, Cai YM et al (2015) A novel design of low-profile dual-band circularly polarized antenna with meta-surface. IEEE Antennas Wirel Propag Lett 14:1650–1653

Virtual Image Representation and Adaptive Weighted Score Level Fusion for Genetic Face Recognition

S. Deepa, A. Bhagyalakshmi, V. Vijaya Chamundeeswari, and S. Godfrey Winster

Abstract Face recognition is the most interesting and wide area of research over the past few decades. This research work proposes the effective virtual image representation and adaptive weighted score level fusion face recognition-based algorithm to classify the genetic faces and non-genetic faces. The algorithm integrates the virtual image and original image, which is a nonlinear transformation of original image. This virtual representation improves moderate intensities pixels and changes the high or low intensities pixels. In face images, various pixels have distinctive significance. Subsequently, it is sensible to fix various weights to individual pixels. Then score level fusion is performed for recognizing the faces, which overcomes illumination due to lighting in the image. Genetic face recognition accuracy gets impacted by the variations in face due to age. The proposed score level fusion scheme gives the best optimal solution for genetic face recognition with respect to age variation. Using statistical measure, distance is calculated between the virtual image and original image for both test and training set of images. The performance results reveal that the proposed score level fusion is robust to different test conditions and performs well compared to existing algorithms.

Keywords Genetic face recognition · Score level fusion · Virtual image representation

Present Address:
S. Deepa (✉)
SRM Institute of Science and Technology, Ramapuram Campus, Chennai, India
e-mail: deepas1@srmist.edu.in

A. Bhagyalakshmi · V. V. Chamundeeswari
KGISL Institute of Technology, Coimbatore, India

S. G. Winster
SRM Institute of Science and Technology, Chennai, India

A. Sivasubramanian et al. (eds.), *Futuristic Communication and Network Technologies*, Lecture Notes in Electrical Engineering 792,
https://doi.org/10.1007/978-981-16-4625-6_77

1 Introduction

Face recognition is an interesting specialization of pattern recognition, and at the same time, it is a complex issue considered in the field of computer vision. There are many challenges in the face recognition process. Significant efforts have been given to the development of exact and robust face recognition algorithms. It is yet, one of the greatest challenges to recognize the faces genetically irrespective of enormous appearance changes because of variations on illumination, expression, aging and poses. Genetic face recognition is identifying the person belonging to the same gene based on similar facial characteristics. Accuracy of genetic face recognition gets impacted by the variations in face due to age. Researchers have tried in the last few decades to develop a face recognition system to recognize the faces with different modalities, unfortunately, no approach present till now for recognizing the same faces with age variation. To overcome this, an algorithm has been developed using nonlinear transformation and adaptive weighted score level fusion method for genetic face recognition, to identify the faces irrespective of age variation with limitations. In spite of the variations, the algorithm can still extract particular highlights of the original face which improves the efficiency of the genetic face recognition and to find the group of similar faces with respect to age variations. In this method, adaptive weighted score level fusion technique is implemented for original image and virtual image to recognize the faces efficiently and to overcome the variations due to illumination, pose and expression. The proposed score level fusion algorithm recognizes the faces with respect to age difference for similar person in different background.

In paper [1], the algorithm has been proposed that joins weighted virtual images to understand a higher face recognition exactness. Image fusion incorporates highlights from numerous sources and can improve performance in the fields of face recognition. When all is said in done, fusion is implemented at three levels, decision level, feature level, and score level. In paper [2], the fusion of score level approach is broadly attributable to the great achievements and it has been arranged into four fusion rules: max rule, min rule, product rule, and sum rule. The paper [3] proposes another age assessment system which exploits different-stage highlights from a feature extractor of generic system, a proposed convolutional neural organization (CNN), and decisively consolidated these features with a determination of handcrafted feature with age-related. The papers [4–8] propose an another fusion scheme, implemented by feature and score level fusion to improve the accuracy of recognition which is referred to as the matcher performance-based (MPb) fusion scheme. In paper [9], a novel adaptive fusion and category-level dictionary learning approach (AFCDL) overcomes the limitations such as Hugh variations in background, speed of the object, motion in video surveillance which is implemented for Multiview action recognition. In paper [10], multi-biometric framework of human recognition is implemented by consolidating biometric hints from various sources (algorithms, numerous sensors, and modalities) at various levels. The scores acquired from various uni-biometric frameworks can be resolved by utilizing score standardization before fusion and afterward proposes an effective score fusion procedure dependent on Dezert-Smarandache theory (DSmT). This paper [11] will propose cutting edge fusion of multimodal biometric procedures to increase acknowledgment execution of delicate biometrics. The key commitment of this paper is the investigation of the impact of separation on delicate biometric qualities and an investigation of the power of acknowledgment utilizing fusion at different distances. This paper [12] intends to examine the impact of fusion of multimodal biometric frameworks at fusion of feature and score level for gender orientation characterization. These papers [13, 14] used genetic algorithm for Monaural speech separation for better results, likewise it can be used for genetic face recognition for better performance.

In this paper, an algorithm is proposed to recognize the similar faces with respect to age, with the help of virtual image representation and adaptive weighted score level fusion. The paper work is described as follows. Proposed score level fusion genetic face recognition algorithm is explained in Sect. 2. In Sect. 3, Implementation and results are carried out for different databases emphasizing that the algorithm works satisfactorily. Conclusion and References are discussed in Sect. 4.

2 Proposed Genetic Face Recognition System

Figure 1 demonstrates the architecture diagram of the proposed approach for score level fusion genetic face recognition. The original training and test face image from the dataset are pre-processed and converted into virtual image representation which is nonlinear transformation. The first distance is calculated between the pre-processed image from training and test database, and then second distance is calculated between virtual image from test and training database. Euclidean distance measure is used to calculate the distances between the two pre-processed and virtual images. These two distances are normalized to calculate the weight using the distance values. Then score level fusion is applied for calculated weight and distance values. Using this fusion values, genetic face images are classified with respect to age. Face images are taken from different perspectives and different age to prove the classification of genetic and non-genetic faces and with respect to age difference. Statistical measures of the classification results demonstrate that the proposed algorithm is very productive in recognizing the faces genetically and group similar faces with different age.

The following subsections briefly describe the stages of the proposed score level fusion face recognition system.

2.1 Pre-processing of Images

The aim of pre-processing is an improvement of the image data that enhances some image features or suppresses unwanted distortions important for further processing.

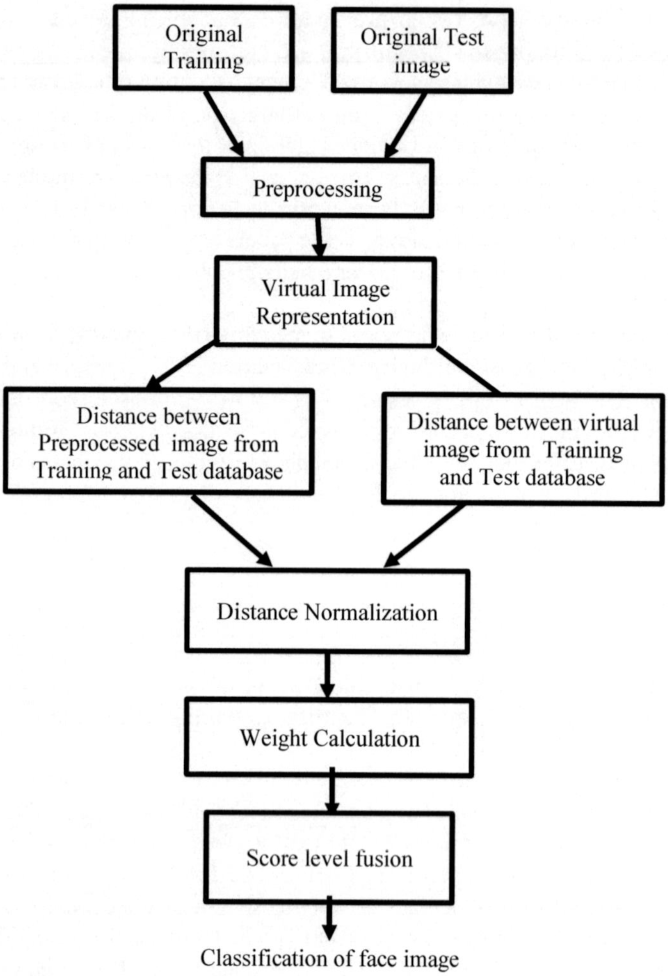

Fig. 1 Architecture diagram of the proposed system

Histogram equalization is used for pre-processing. Histogram equalization is a technique to enhance the contrast by adjusting image intensities. Figure 2 shows the original and pre-processed image.

Let p be an image given as a m_r by m_c pixel matrix of intensities from 0 to M-1. M is the intensity value. Let f denote the histogram normalization of p. So

$$f_n = \frac{\text{number of pixels with intensity } n}{\text{total number of pixels}} \quad n = 0, 1, \ldots, M-1. \tag{1}$$

The image of the histogram g is defined by

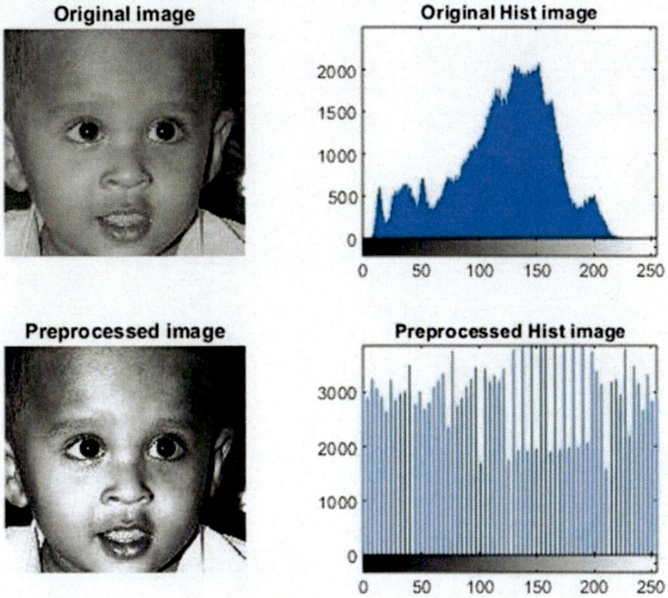

Fig. 2 Pre-processed image

$$g_{i,j} = \text{floor}\left((M-1)\sum_{n=0}^{p_{i,j}} f_n\right) \quad (2)$$

where floor () rounds to the closest integer. This is equal to intensity of pixel transformation, k of p by the equation

$$T(k) = \text{floor}\left((M-1)\sum_{n=0}^{k} f_n\right). \quad (3)$$

2.2 *Non-linear Transformation*

In order to recognize face images successfully, however, much valuable data as could be expected are required. The original image pixel with high intensity relates to virtual image of low intensity pixels in nonlinear transformation, and the other way around. To direct face recognition, the inverse of the image is used as a supplement of the source image. Figure 3 shows source image and its relating virtual image. The virtual face images have a converse portrayal, yet look like human appearances. Also, in addition the pixels with high intensity speak to significant facial highlights in virtual

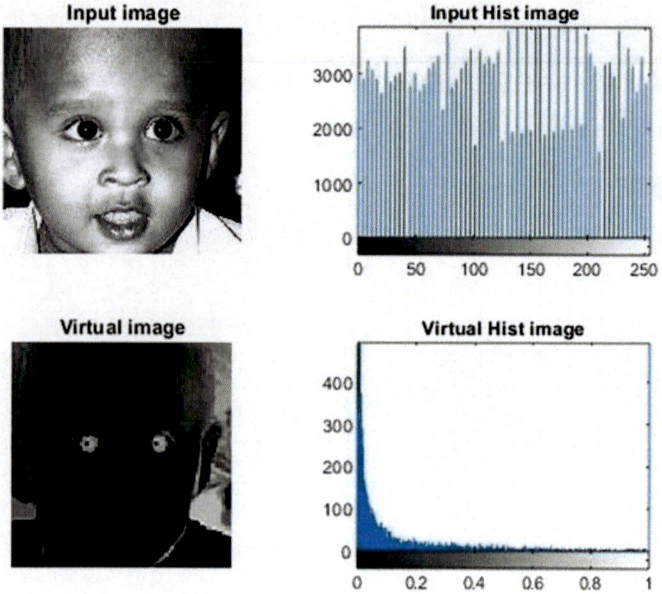

Fig. 3 Nonlinear transformation

images. The histogram of the original image and relating virtual image is represented in Fig. 3. The intensity of pixels average value in the source image is generally high. In actuality, intensity of the pixels of the virtual image is generally low. Additionally, the virtual image average gray level has been altogether diminished.

This virtual representation upgrades moderate intensities pixels and diminishes high- or low-intensity pixels. This virtual images show the particular highlights of the source face. The nonlinear transformation proposed in Eq. (4) can fulfill our prerequisite in light of the nonlinear characteristic. Subsequently, combination of integral virtual and source images gets best execution for recognition of faces.

$$V = \frac{255}{1 + e^{aX+b}} \tag{4}$$

where V is the virtual image, X is the original image, a and b are constants.

At the point when the data correlation gets stronger, the precision improvement will be decreased. Correlation coefficient is used to conduct the quantitative analysis between source images and virtual images. The meaning of coefficient of correlation is defined in Eq. (5)

$$\text{correlation} = \frac{\sum(x - \overline{x})(y - \overline{y})}{\sqrt{\sum(x - \overline{x})(y - \overline{y})}} \tag{5}$$

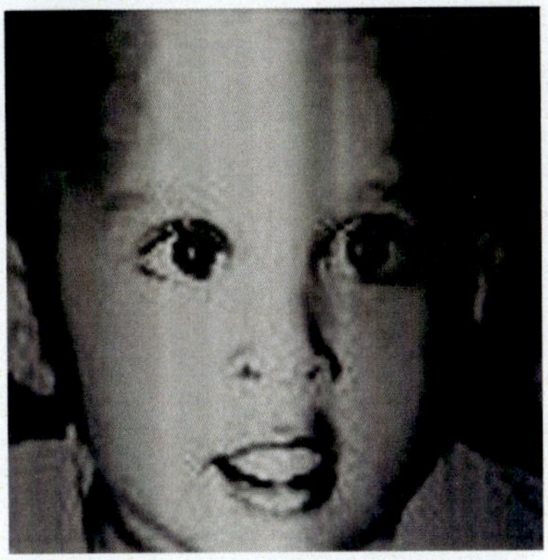

Fig. 4 Re-constructed image

The correlation coefficient should be determined to x and y which represents the two image vectors. The correlation factor varies for all the images, and there is a correlation of negative value between the source and the virtual data. The coefficient of correlation scope ranges from -1 to 1. On the off chance that the absolute estimation of the correlation coefficient is extremely near 1, at that point the virtual information to be sure doesn't give extra helpful data to fusion. Consequently, the algorithm can produce sensible additional information that also looks somewhat like faces. The moderately low relationship between source and virtual images demonstrates that they are integral to one another for face acknowledgment. It is also possible to re-construct the original image using the correlation factor. Figure 4 shows the re-constructed image by the following Eq. (6)

$$R = (V * C) + \text{Mean}(I) \tag{6}$$

where

V—Virtual Image.
C—Correlation factor.
I—Original Image.
R—Re-constructed image.

2.3 Euclidean Distance Calculation and Normalization

The database test image is pre-processed similar to that of training image and undergoes nonlinear transformation to get virtual images. The row vector of each original and virtual image class is shown in Eqs. (7) and (8)

$$X_i = \left[x_{(i-1)n+1}, \ldots, x_{in}\right] i = 1, 2, \ldots, Cl. \tag{7}$$

$$V_i = \left[v_{(i-1)n+1}, \ldots, v_{in}\right] i = 1, 2, \ldots, Cl. \tag{8}$$

Cl is the different classes of n training samples. The i-th class of training samples of n-th source and virtual data are denoted as x_{in} and v_{in}.

A linear function for each source image and virtual image from each class is defined by Eqs. (9) and (10)

$$\alpha_i = \left(X_i^T X_i + \lambda E\right) X_i y \quad i = 1, 2, \ldots, Cl. \tag{9}$$

$$\gamma_i = \left(V_i^T X_i + \lambda E\right) V_i y_v \quad i = 1, 2, \ldots, Cl. \tag{10}$$

Constant λ, identity matrix E, test image y, and virtual test image y_v is obtained by Eq. (4). Euclidean distance measure is used to calculate the distance $d1$ between test images and training images, and the distance $d2$ between the corresponding virtual test image and virtual training image. The closeness between the training image and the test image are calculated using the Euclidean distance.

$$d_1^i = ||y - X_i \alpha_i|| \quad i = 1, 2, \ldots, C. \tag{11}$$

$$d_2^i = ||y_v - V_i \gamma_i|| \quad i = 1, 2, \ldots, C. \tag{12}$$

The distance shows the commitment of a training image in order to distinguish a test image. A littler distance shows a higher recognition. Equation (13) is used to normalize the distance between the values 0 and 1

$$d_j^{'} = \frac{d_j^{\max} - d_j^{i}}{d_j^{\max} - d_j^{\min}}, (i = 1, 2, \ldots, Cl), (j = 1, 2). \tag{13}$$

where $d_j^{\max}$ and $d_j^{\min}$ are the maximum and minimum of d_j^{i}, respectively.

2.4 Weight Calculation

The main advantage of this proposed system is calculating the weight automatically. The distances $d1$ and $d2$ are sorted in ascending order, such that $P1, P2, P3 \ldots Pn$ are the sorted values of $d1$ and $V1, V2, V3 \ldots Vn$ are the sorted values of $d2$. The weights $w1$ and $w2$ for this proposed system are calculated by Eqs. (14) and (15)

$$w1 = \frac{P_2 - P_1}{(P_2 - P_1) + (V_2 - V_1)} \tag{14}$$

$$w2 = \frac{V_2 - V_1}{(P_2 - P_1) + (V_2 - V_1)} \tag{15}$$

The contrast value between the first score value and the second score value can decide a dependable weight for data fusion. The more noteworthy the contrast, the more certain the classification result.

2.5 Score Level Fusion

This work proposes an algorithm that joins virtual images weighted fusion to understand a face recognition exactness. Image fusion incorporates highlights from various sources and can improve performance in the fields of recognition of face, multi-biometrics, and video retrieval. The fusion of score value approach is generally used attributable to its great performance, and it very well may be sorted into four combination rules: max rule, min rule, product rule, and sum rule. Most of past research studies prove that the best result arrives from the sum rule. The fusion result f_i is expressed in Eq. (16).

$$f_i = w_1 d_1 + w_2 d_2 \tag{16}$$

Despite the fact that score fusion has focal points over other approaches, deciding the best possible weights so as to realize optimization is a difficult assignment.

3 Implementation and Results

The score level fusion face recognition system shown in Fig. 1 is evaluated with face image databases with respect to genetic faces and non-genetic faces, faces of same person in different ages. Totally 200 face images of 20 persons with different ages are used for this algorithm. The proposed algorithm has been used to classify the genetic and non-genetic faces and similar person with respect to different ages such that images of different ages will be in different illumination, background and poses. So, this algorithm has overcome the complexities involved in face recognition. There are four database of sample face images are used in different analysis. Genetic and non-genetic faces of different families are gathered for databases. In the person 1 database, there are 13 images of same person with different ages and persons from same family descent and different family with different background setup and different illumination.

Figure 5 shows the database of different faces in which 7 faces belongs to the same person of different age (1–6 yrs, 2–5 yrs, 3–7 yrs, 4–2 yrs, 11–1 yr, 12–3 yrs, 13–4 yrs) and 5th image belong to same family and faces 6–10 belong to different family members. Figure 6 shows the pre-processed image of original face database of person 1. Pre-processing using histogram equalization enhances the contrast of the image which will be very useful for old images taken long back ago with less contrast and brightness. Figure 7 shows the nonlinear transformation of Person 1 database.

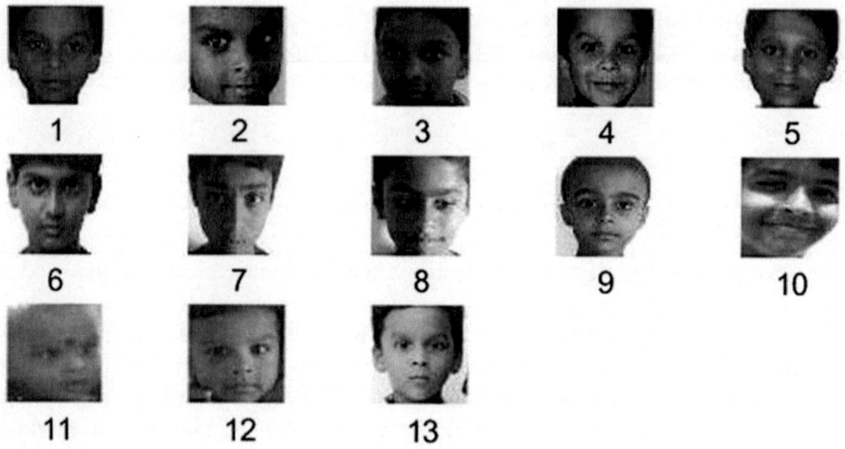

Fig. 5 Person 1 database

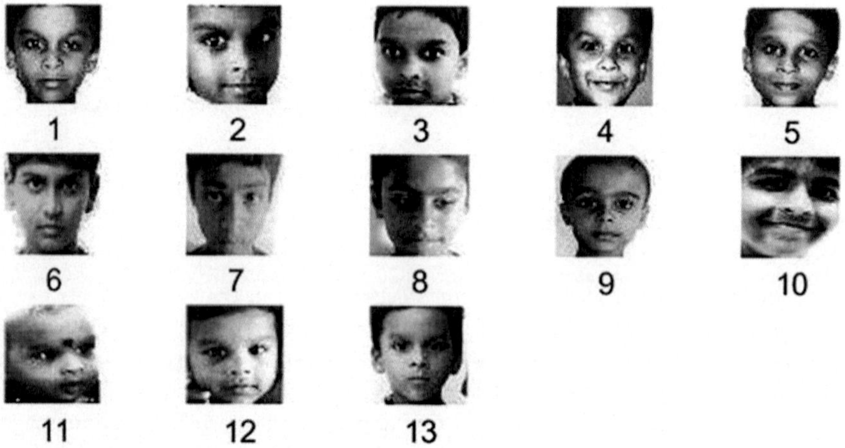

Fig. 6 Pre-processed image of Person 1 database

The virtual image representation, which is the complement of original image, can still show particular highlights of the original face which improves the efficiency of the genetic face recognition and finds the group of similar faces with respect to age variations.

From this output image, it is concluded that the main features have been extracted using these proposed algorithm and classified. For the classification purpose, distance values are calculated between the input image and database image for comparison. In this paper, Euclidean distance measure is used for calculating the difference error values for all the images. In Person 1 Database, 3rd face image is the input image

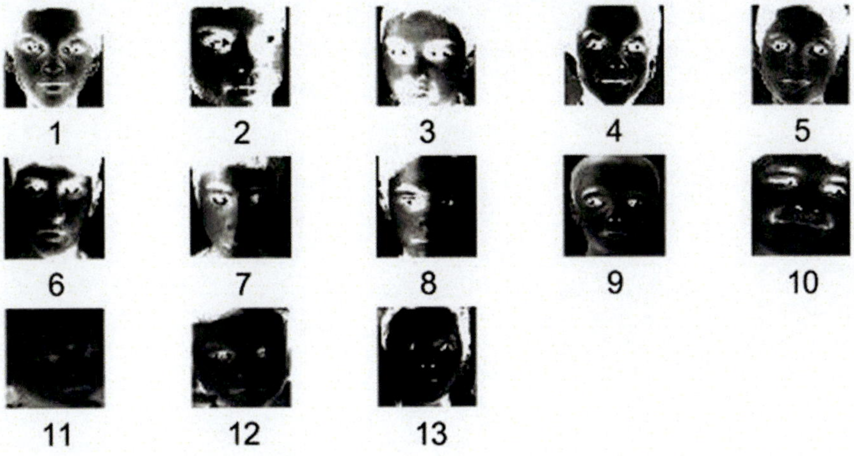

Fig. 7 Virtual image of Person 1 database

which is compared with all the remaining face images. The classification of similar, genetic and non-genetic faces for Person 1 Database is shown in Fig. 8.

From Table 1, it is proved that the difference error value of the Images 1–5, 11, 12, 13 (genetic faces and similar faces) is less than the error value of the Images 6–10 (non-genetic faces). Specifically, the analysis of similar person with respect to age is proved. First three image from Table 1 is having less value compared to the 4th image.

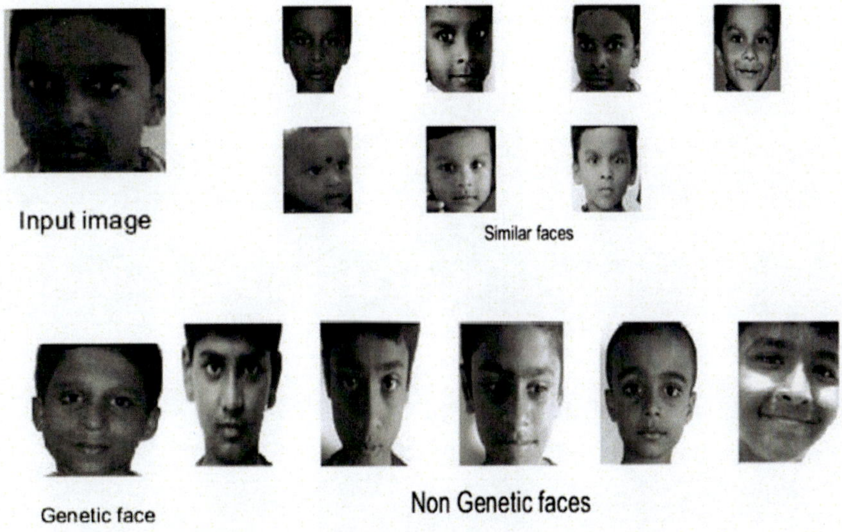

Fig. 8 Classification of similar, genetic and non-genetic faces

Table 1 Score-based fusion values for Person 1 database

	Face database 1												
Proposed algorithm	Image 1	Image 2	Image 3	Image 4	Image 5	Image 6	Image 7	Image 8	Image 9	Image 10	Image 11	Image 12	Image 13
Score level fusion	123.71	139.1	0	208	170.28	275.3	284.03	288.89	356.78	234.35	187.96	224.87	194.21

The classification graph has been plotted using these values, and the classification has been done for similar and genetic faces with non-genetic faces. From Fig. 9, it is proved that the genetic and similar faces are very clearly classified compared to non-genetic faces. This experimental results dependent on statistical measure acquired for genetically similar and non-similar faces lead to a road of genetic and age-wise similarity-based face recognition. Inverse transform score level fusion algorithm is the proposed approach, and it has demonstrated as a successful tool in removing the remarkable features belonging to the genetic resemblance and similar faces with respect to ages.

Table 2 shows the distance values of persons with respect to their ages. From these values, it is proved that there are more or less similar distance values for the same age of all the persons.

Figure 10 shows the graph for the age group similarity values of all the persons.

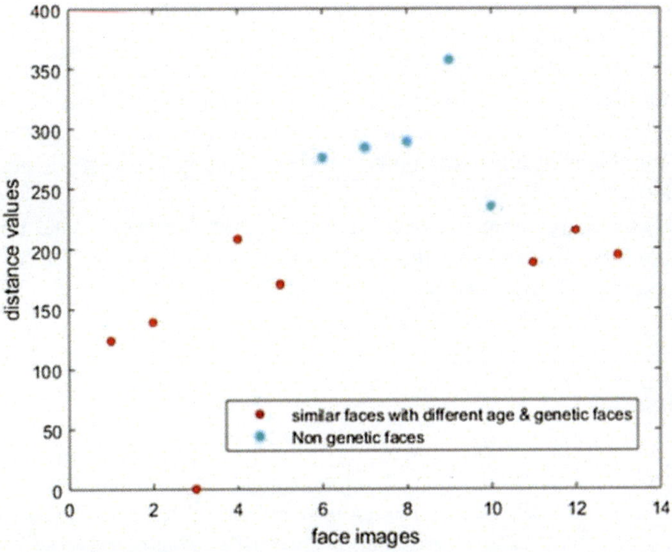

Fig. 9 Classification graph of Person 1 database

Table 2 Distance values for all the persons related to ages

Ages	1	2	3	4	5	6	7	8	9	10
Person 1	188.0	224.9	208	194.2	139.1	123.7	136.2	172.5	185.1	230
Person 2	262.3	202.2	199.6	191.3	151.1	178.5	126.2	153.4	223.5	220.2
Person 3	250.6	186.8	226.8	183.6	170.6	135.9	147.7	164.8	174.6	246.3
Person 4	208.2	178.5	194.2	188.0	172.5	174.6	145.1	151.1	214.6	237.9
Person 5	196.4	195.6	218.5	179.8	146.8	165.9	139.6	163.6	204.7	225.1
Person 6	218.3	222.5	219.3	201.8	163.9	128.6	123.8	153.3	195.2	263.6

Fig. 10 Graph of distance values (w.r.t reference images) versus age group for person 1–6

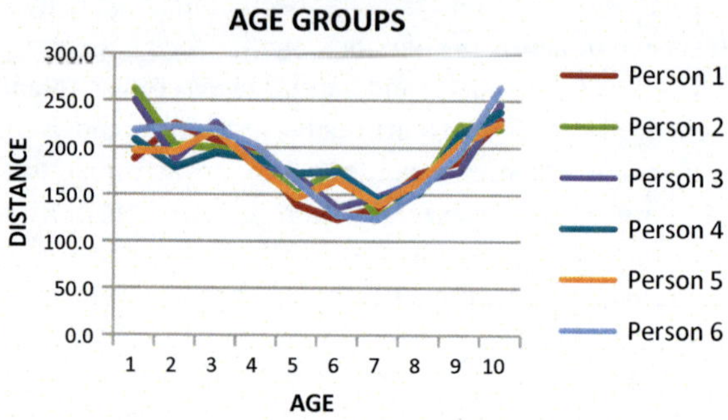

This analysis has been experimentally done based on statistical measure obtained by using the proposed algorithm. From this analysis, it is proved that inverse transform and score level fusion is a successful tool for extracting the significant features for classifying genetically similar and non-similar faces.

The proposed work implementation is done in MATLAB. Face image databases are gathered with respect to genetic faces and non-genetic faces, faces of the same person in different ages, faces of the same person with the same age at different locations and faces of different family members. Totally 100 face images are used for this algorithm.

The proposed algorithm has been used to classify the genetic and non-genetic faces. Sample faces have been illustrated in this paper. There are four database of images used in different analysis.

4 Conclusion

This research work represents the virtual image representation and adaptive weighted score level fusion for genetic face recognition. For recognizing the similar faces with respect to age, virtual image representation and adaptive weighted score level fusion is proposed in this research work. The algorithm integrates the original image and virtual image, which is a nonlinear transformation of original image. This virtual representation upgrades pixels with moderate intensities and changes the pixels with high or low intensities. Using statistical measure, distance is determined between the original image and virtual image for both test and training set of images. Additionally, various pixels have distinctive significance in representing face images. Subsequently, it is sensible to set various weights to individual pixels. Then the weight is calculated by the two least distance values, so the adaptive weight varies for all the images. Then score level fusion is performed for recognizing the faces, which overcomes illumination due to lighting in the image. According to the experimental results, the proposed algorithm performs well compared to the previous work in recognizing accuracy. The proposed score level fusion scheme gives the best optimal solution for genetic face recognition with respect to age variation.

References

1. Qian R (2018) Inverse transformation based weighted fusion for face recognition. Multimedia Tools Appl 77:28441–28456
2. Vukotić V, Raymond C, Gravier G (2018) A crossmodal approach to multimodal fusion in video hyperlinking. IEEE MultiMedia, 25(2):11–23
3. Taheri S, Toygar Ö (2019) Multi-stage age estimation using two level fusions of handcrafted and learned features on facial images. IET Biometrics 8(2):124–133
4. Kabir W, Omair Ahmad M, Swamy MNS (2018) Normalization and weighting techniques based on genuine-impostor score fusion in multi-biometric systems. IEEE Trans Inf Forensics Secur 13(8):1989–2000
5. Kabir W, Omair Ahmad M, Swamy MNS (2019) A multi-biometric system based on feature and score level fusions. IEEE Access 7:59437–59450
6. Huang Z-H, Li W-J, Wang J, Zhang T (2015) Face recognition based on pixel-level and feature-level fusion of the top-level's wavelet sub-bands. Inf Fusion 22:95–104
7. Cheniti M, Boukezzoula N-E, Akhtar Z (2018) Symmetric sum-based biometric score fusion. IET Biometrics 7(5):391–395
8. Aboshosha A, El Dahshan KA (2015) Score level fusion for fingerprint, iris and face biometrics. Int J Comput Appl 111(4)
9. Gao Z, Xuan H-Z, Zhang H, Wan S, Raymond Choo K-W (2015) Adaptive fusion and category-level dictionary learning model for multi-view human action recognition. IEEE Internet Things J 6(6):9280–9293
10. Sharma R, Das S, Joshi P (2018) Score-level fusion using generalized extreme value distribution and DSmT, for multi-biometric systems. IET Biometrics 7(5):474–481
11. Guo BH, Nixon MS, Carter JN (2019) Soft biometric fusion for subject recognition at a distance. IEEE Trans Biometrics Behav Identity Sci 1(4):292–301
12. Eskandari M, Sharifi O (2019) Effect of face and ocular multimodal biometric systems on gender classification. IET Biometrics 8(4):243–248
13. Shoba S, Rajavel R (2020) A new genetic algorithm based fusion scheme in monaural CASA system to improve the performance of the speech. J Ambient Intell Humaniz Comput 11(1):433–446
14. Sivapatham S, Ramadoss R, Kar A, Majhi B (2020) Monaural speech separation using GA-DNN integration scheme. Appl Acoust 160:107140

Anomaly Detection in Drone-Captured Images Using Machine Learning Techniques and Deep Learning Architectures

Kartik Joshi, G. Vidya, Soumya Shaw, Abitha K. Thyagarajan, Akhil Pathak, Rahul Hemal Shah, V. Ram Kishan, and John Sahaya Rani Alex

Abstract Drones have found application in many fields ranging from agriculture to defense. With their on-board sensors, they provide useful insights in all the domains. It becomes important to study the abnormal behaviors observed due to the various sensors which hamper the drones' smooth functioning. From temporally coherent streams of footage obtained, we aim to analyze the image frames, both quantitatively and qualitatively. The pixel level intensity variation was studied to find the patterns in them. Based on the intensity fluctuation distribution, the frames were classified into normal or anomalous. The images were evaluated using machine learning techniques. The best result was yielded by naïve Bayes classification with an accuracy of 92.079%. From the images, semantic inference was established through deep learning architectures through the application of a convolutional neural network. The best result was obtained when using a one-block VGG, yielding a validation accuracy of 92.17% and a test accuracy of 90.435%.

Keywords Anomaly detection · Drone · CNN · Softmax · Support vector classification · Decision tree · Random forest classification · Naïve Bayes

1 Introduction

With the world moving toward more data-driven approaches enabled by the advent of big data, there is intense effort in data collection and data analytics. Sensors and cameras capturing our data are now ubiquitous. Such data may contain anomalies, and detecting these anomalies can save computing power as well as money. It may also yield meaningful insights to many problems.

K. Joshi · G. Vidya · S. Shaw · A. K. Thyagarajan · A. Pathak · R. H. Shah · V. R. Kishan
School of Electronics Engineering, Vellore Institute of Technology, Chennai, India

J. S. R. Alex (✉)
Centre for Healthcare Advancement, Innovation and Research, Vellore Institute of Technology, Chennai, India
e-mail: jsranialex@vit.ac.in

A. Sivasubramanian et al. (eds.), *Futuristic Communication and Network Technologies*, Lecture Notes in Electrical Engineering 792,
https://doi.org/10.1007/978-981-16-4625-6_78

Anomaly detection is the detection of rare events or observations which raise suspicion by differing from the dataset. These sudden abrupt events also known as anomalous items will translate to some kind of catastrophic failure in some applications.

In recent times, vast quantities of data are collected by autonomous devices such as drones. Drones are rapidly becoming ubiquitous, finding application in fields ranging from in agriculture [1–4], to logistics [5], to archaeology [6], to mining [7]. Drones are often equipped with on-board sensors, actuators, and cameras, besides the basic components of a drone which include a battery, computing unit, and motors. As drones are typically airborne, they are prone to failure due to unforeseen environmental factors and circumstances; if these are identified quickly, the drone may be able to avoid its failure. By analyzing the data collected by a drone's on-board sensors and camera feed, such situations can be rapidly recognized by identifying anomalous data.

In this paper, we have worked on image data provided as part of the 2020 Signal Processing Cup [8] organized by IEEE Signal Processing Society and ICASSP. To detect anomalies, we have used supervised learning methods as well as deep learning architectures, as detailed in Sect. 3. We have achieved promising results with these methods as outlined in Sect. 4. The implications of these results and some suggestions for future improvement are made in Sect. 5.

2 Literature Review

Uruma et al. [9] propose an algorithm for colorization-based image coding, in which a color image is compressed by encoding its luminance by a standard coding method such as JPEG and by storing several color pixels called representative pixels (RPs). Additionally, the RPs compression algorithm using the graph Fourier transform is suggested to improve image coding performance.

An anomaly detection algorithm for road map images is proposed by Kamitani et al. [10]. The images processed in this paper were collected through hand-held cameras (whereas we use drone-captured images). Experimental results presented in this paper show that the use of local regions significantly improves detection performance compared with that of whole images.

Another approach using deep neural networks (DNNs) is proposed by Lile and Yiqun [11]. An automatic method is devised to detect anomalies in thermal images by training a DNN model to learn statistical regularities in normal images and thereby detect anomalies based on pixel-wise comparisons using the learned reference temperatures as a guideline.

A systematic study on deep learning-based approaches for Drones is done by Wang et al. [12] traditional methods are compared with deep learning-based solutions and are found to be better in terms of deducing the abnormality. The study takes into account real-world scenario and establishes a hypothesis on usefulness of Deep Learning-based approach.

Chalapathy and Chawla [13] present a detailed survey on various deep learning techniques that are available for the task of anomaly detection. This comparison helps in studying the state-of-the-art models available and puts forward the potential challenges involved in the task of anomaly detection. This thesis represents a holistic view of anomaly detection with variety of applications spread across all the domains.

A specific use case of drones in defense is explored by Yong et al. [14]. Here, the values of embedded sensors are analysed for detecting abnormality. This is also supported by statistical inferences. The KPCA algorithm proposed is found useful with AUC score of 0.982 for pulse anomaly detection.

Anomaly Detection is further studied for different types of faults that are responsible. Keipour et al. [15] introduce ALFA dataset specifically for UAV Anomaly and Fault Detection. Their detailed collection also includes 13 min of post anomaly detection. This gives enough room to analyse the actual reason for the anomaly in a precise way.

In this paper, Liu and Zheng [16], an anomaly detection algorithm is proposed based on jpeg images. This algorithm directly works on the jpeg image data and eliminate the jpeg image decoding process. To determine whether the anomaly or abnormal target appeared, this algorithm compares two frames of the extracted DC parameters of the jpeg image. This proposed algorithm costs less storage space, access time and computation, which makes it perfectly suitable for Embedded Systems with limited resources.

Researchers at Google [17] have proposed a novel framework for the task of anomaly detection in drones. This comes with the dynamic anomaly detection with data trained on many real-time flight scenarios.

Automated Surface inspections are very important and widely used by many manufacturing industries to produce high quality product, which often use machine learning-based solutions. As supervised learning algorithms can prove to be bit challenging, in Ref. [18] author proposes an unsupervised learning algorithm on fault-free samples by training a deep convolutional neural network to complete images whose center regions are cut out. Results of this methodology demonstrate that it is suitable for detection of visible anomalies which were performed on the surface of decorated plastic parts.

3 Methodology

Drones can capture video footage through on-board cameras. The analysis of this video footage can reveal valuable information related to the area and objects monitored by the drone. A large volume of data can be captured in this manner, which also leads to better analysis with increase in the dataset size. However, this also begets the need to identify and scrub anomalous data, which is increasingly difficult to do manually for large datasets. Hence, we can apply artificial intelligence techniques to analyze the data and identify anomalous image frames.

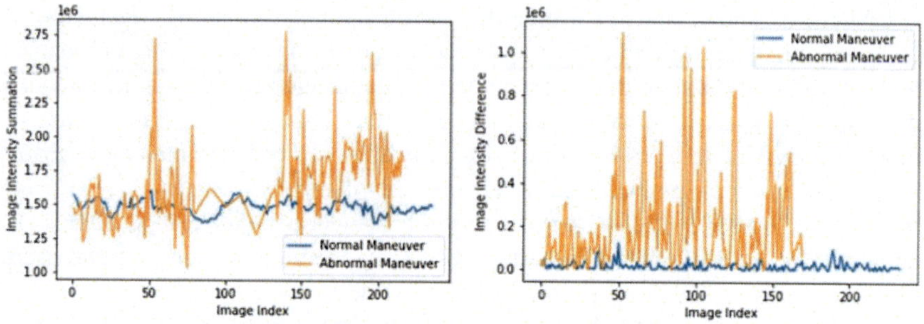

(a) Intensity fluctuation comparison (b) Intensity difference comparison between subsequent frames

Fig. 1 Analysis of the intensity of the image frames

To apply this approach to our drone-captured video data, we first split the videos into their constituent video frames. Then, in order to identify a pattern in the images, we compared the normalized pixel intensity of each frame for normal and abnormal labeled datasets. The sum of normalized pixel intensity is plotted in Fig. 1a. The intensity ranges of each frame largely overlap with one other, and hence intensity is not a good measure to differentiate; however, the variance is sparse. Thus we used the difference of consecutive frame intensity fluctuation as the measure for classification.

The difference of the intensity summation of subsequent frames was determined and plotted, as seen in Fig. 1b. The graph suggests that the classes may be disjoint. Then, we use artificial intelligence algorithms to classify the drone motion into normal or abnormal based on intensity fluctuation between consecutive frames.

3.1 Data Pre-processing and Preparation

The initial data was provided as a ROSBAG file, comprising sensor readings as well as images captured from drones. We extracted the image data as individual frames, and used this extracted data for all our training. As the data was labeled, we performed supervised learning.

As our essential approach was exploring classification possibilities of the image data captured through aerial maneuvers of drones, our workflow was divided into two major paths. One was the use of deep learning architectures, namely a convolutional neural network (CNN), and the other was the use of various machine learning techniques. Hence, it was our attempt to fine-tune the algorithms using the intensity of each frame as an empirical parameter. This approach yielded promising results.

3.2 Machine Learning Techniques

The quantitative methods that we have implemented are logistic regression, softmax, support vector classification, k-nearest neighbors (k-NN), decision tree, random forest classification, and naïve Bayes classification. The former two models, logistic regression and softmax, did not yield satisfactory results due to overfitting and so have been excluded.

We have used the typical parameters for the models. For k-NN, a value of $k = 3$ was taken. The support vector classifier used a Gaussian radial basis function and performed non-linear classification. The naïve Bayes classifier used the *maximum* a posteriori decision rule.

3.3 Deep Learning Architectures

A convolutional neural network (CNN) is a deep neural network that comprises a regularized version of multilayer perceptrons, which consist of fully connected layers of neurons. These networks perform well on image recognition tasks and are favored for their good performance as well as their ease of implementation and modular nature [19]. For our purpose, a baseline convolutional neural network model was developed first, following visual inspection of the dataset images; normal and abnormal images are shown in Fig. 2. Then, a CNN model with 1 and 2 VGG blocks was implemented. The architecture is shown in Fig. 3.

4 Results and Discussion

As described in our methodology, we employed several machine learning techniques and models on the drone-captured images. The results of the quantitative analysis using the machine learning techniques is enumerated in Sect. 4.1. For qualitative analysis, we employed a deep convolutional network with some variants as enumerated in Sect. 4.2.

4.1 Machine Learning Techniques

Overall, the most accurate model for identifying anomalous image frames was naïve Bayes, having an accuracy of 92.079%. Of all the machine learning techniques we implemented, the best performance overall was given by naïve Bayes, having the

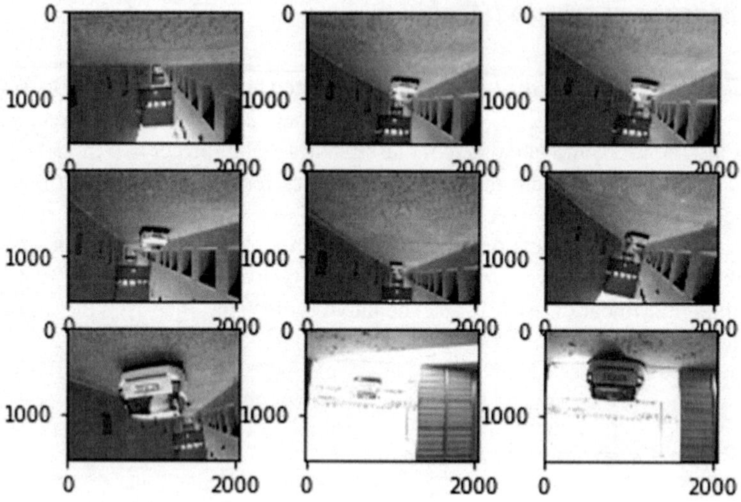

(a) Normal

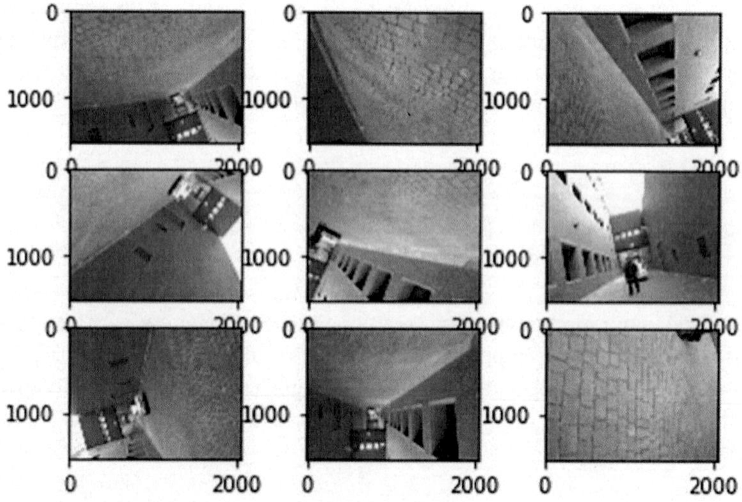

(b) Abnormal

Fig. 2 Images from the video footage captured by the drone

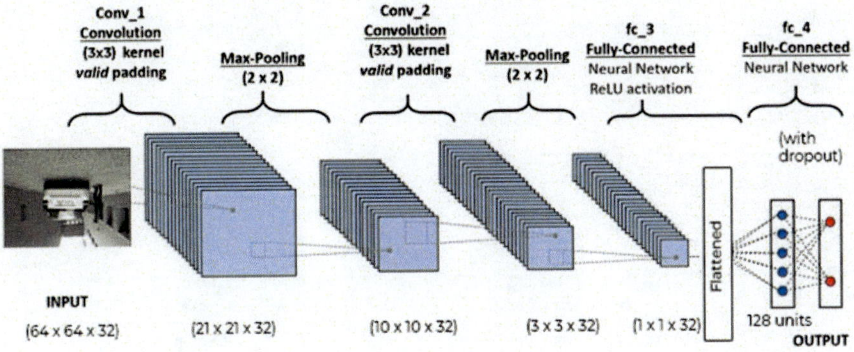

Fig. 3 CNN architecture

Table 1 Performance metrics for machine learning techniques

Parameter	SVC	k-NN	Decision tree	Random forest	Naïve Bayes
f_1 score	0.857	0.86	0.82	0.84	**0.9**
Accuracy (%)	86.139	86.139	85.148	83.168	**92.079**
Sensitivity (%)	97.826	95.652	85.484	86.956	**98.333**
Specificity (%)	76.364	78.182	**84.615**	80.0	82.927
Precision	77.586%	**78.571%**	–	78.431%	–
AUC & ROC (%)	87.095	86.917	85.049	83.478	**90.630**
MCC (%)	74.723	73.980	69.245	66.693	**83.857**

highest f_1 score, accuracy, sensitivity, AUC & ROC, and MCC. The highest specificity was yielded by the decision tree, while the highest precision was exhibited by *k*-nearest neighbors algorithm (Table 1).

4.2 Deep Learning Architectures

With CNN, a validation accuracy of 82.5% and a test accuracy of 85.21% was observed. Its accuracy and loss are shown in Fig. 4. Better results were obtained with VGG-1 and VGG-2; for the former, the validation and test accuracies were 92.17 and 90.435%, and for the latter, they were 91.3% and 91.304% respectively. The accuracy and loss for VGG are shown in Fig. 5. We can observe that best results are obtained with one-block VGG.

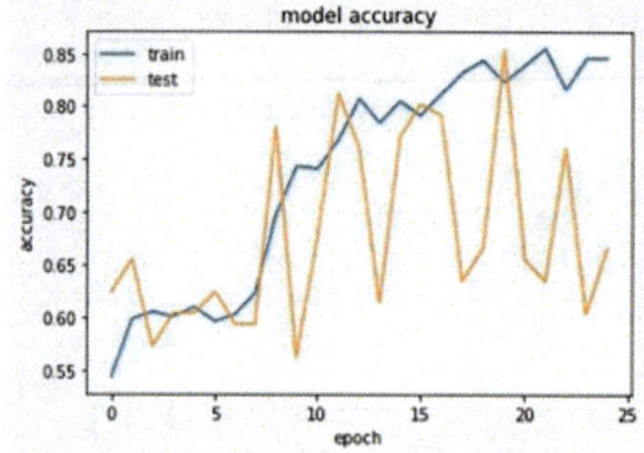

(a) Accuracy for CNN

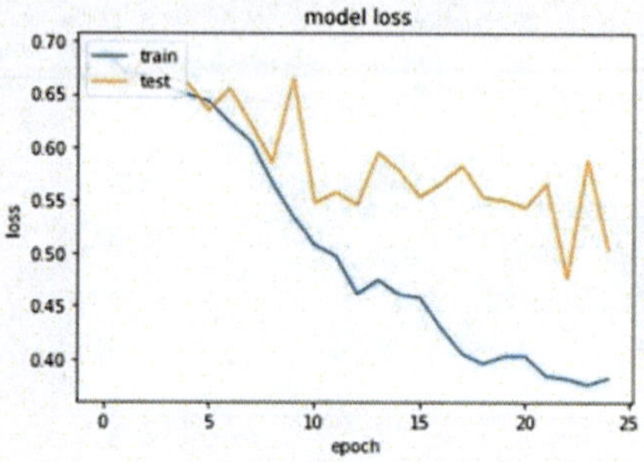

(b) Loss

Fig. 4 CNN accuracy and loss of the intensity of the image frames

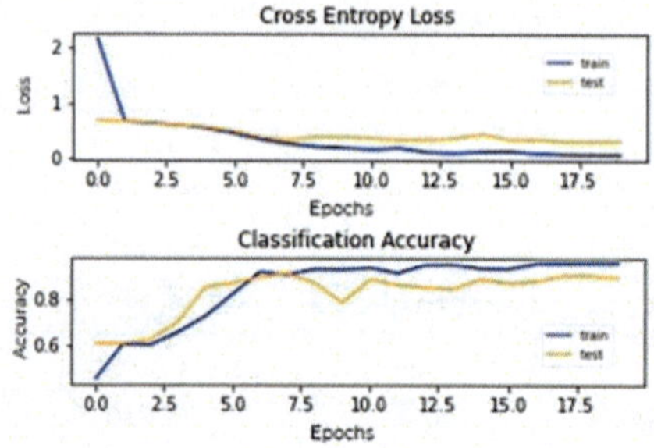

(a) 1 block VGG

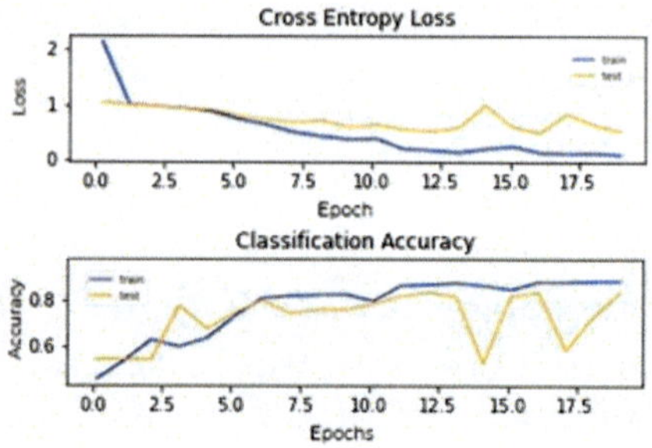

(b) 2 block VGG

Fig. 5 Results with 1 and 2 VGG blocks

5 Conclusion

In this paper, we have presented two methods for anomaly detection in the drone-captured images. Firstly, using regression and classification models. Secondly, by using the concept of convolution neural networks. We observe that best results are obtained with softmax and one-block VGG with CNN.

For future work, further machine learning and deep learning techniques can be considered. More recently, Self Organizing Maps have been impressive with learning the semantic information over less data. Furthermore, the Growing Neural Gas has also shown promising results with the random Gaussian distribution pattern observed during training.

References

1. Meinen BU, Robinson DT (2020) Mapping erosion and deposition in an agricultural landscape: optimization of UAV image acquisition schemes for sfm-mvs. Remote Sens Environ 239:111666
2. Zheng H, Zhou X, He J, Yao X, Cheng T, Zhu Y, Cao W, Tian Y (2020) Early season detection of rice plants using rgb, nir-gb and multispectral images from unmanned aerial vehicle (UAV). Comput Electron Agric 169:105223
3. Radoglou-Grammatikis P, Sarigiannidis P, Lagkas T, Moscholios I (2020) A compilation of UAV applications for precision agriculture. Comput Netw 172:107148
4. Meng Y, Su J, Song J, Chen WH, Lan Y (2020) Experimental evaluation of UAV spraying for peach trees of different shapes: effects of operational parameters on droplet distribution. Comput Electron Agric 170:105282
5. Kitjacharoenchai P, Lee S (2019) Vehicle routing problem with drones for last mile delivery. Procedia Manuf 39:314–324
6. Wiseman AL, Bezombes F, Moore AJ, De Groote I (2020) Non-invasive methods: the applicability of unmanned aerial vehicle (UAV) technology for recording fossilised footprints. Digit Appl Archaeol Cult Heritage 16:e00137
7. Kansake BA, Kaba FA, Dumakor-Dupey NK, Arthur CK (2019) The future of mining in Ghana: are stakeholders prepared for the adoption of autonomous mining systems? Resour Policy 63:101411
8. SP Cup (2019) https://2020.ieeeicassp.org/program/sp-cup/. Accessed 15 Sept 2020
9. Uruma K, Saito K, Takahashi T, Konishi K, Furukawa T (2017) Representative pixels compression algorithm using graph signal processing for colorization-based image coding. In: 2017 IEEE International conference on image processing (ICIP). IEEE, pp 3255–3259
10. Kamitani T, Fujimoto S, Yoshimura H, Nishiyama M, Iwai Y (2018) Anomaly detection using local regions in road images acquired from a hand-held camera. In: 2018 IEEE 7th Global conference on consumer electronics (GCCE). IEEE, pp. 375–378
11. Lile C, Yiqun L (2017) Anomaly detection in thermal images using deep neural networks. In: 2017 IEEE International conference on image processing (ICIP). IEEE, pp 2299–2303
12. Wang R, Nie K, Wang T, Yang Y, Long B (2020) Deep learning for anomaly detection. In: Proceedings of the 13th International conference on web search and data mining. WSDM '20. Association for Computing Machinery, New York, NY, USA, pp 894–896. https://doi.org/10.1145/3336191.3371876
13. Chalapathy R, Chawla S (2019) Deep learning for anomaly detection: a survey
14. Yong D, Yuanpeng Z, Yaqing X, Yu P, Datong L (2017) Unmanned aerial vehicle sensor data anomaly detection using kernel principle component analysis. In: 2017 13th IEEE international conference on electronic measurement instruments (ICEMI), pp. 241–246
15. Keipour A, Mousaei M, Scherer S (2019) Alfa: a dataset for UAV fault and anomaly detection
16. Liu G, Zheng Z (2011) Anomaly target detection algorithm based on jpeg images. In: 2011 International conference on multimedia technology, pp 2952–2955
17. Sindhwani V, Sidahmed H, Choromanski K, Jones B (2020) Unsupervised anomaly detection for self-flying delivery drones. In: International conference on robotics and automation
18. Haselmann M, Gruber DP, Tabatabai P (2018) Anomaly detection using deep learning based image completion. In: 2018 17th IEEE international conference on machine learning and applications (ICMLA), pp 1237–1242
19. Simonyan K, Zisserman A (2014) Very deep convolutional networks for large-scale image recognition

Antenna Selection in Cognitive Radio for MIMO System

Shelesh Krishna Saraswat

Abstract Cognitive radio is the most important and emerging technologies of today, as it solves spectral scarcity in highly traffic channels. The main aim of CR is to reduce the interference as low as possible of the secondary user at primary user. Multiple input–output systems are implemented with CR to enhance diversity gain. However, all the antenna used at the same time in the system is wastage of transmitter power and also increases the interference. The better way to deal with this problem is to select a few groups of antennas which optimize the system performance in a dynamically varying channel environment. Here, this issue is considered as an optimization issue where the main objective function is to maximize data transmission rate, and constraints are to minimize total transmission power and interference.

Keywords Antenna selection · CR system · Multiple input–output system · Optimization

1 Introduction

1.1 Cognitive Radio Systems (CRS)

Having three key abilities, for example, cognitive capability, learning ability and reconfiguring capacity CRS have enough potential of being a detrimental force within the spectrum management. Utilizing SDR for actualization of reconfigurable radios, CRS gives extra adaptability and offers better efficiency for the use of overall spectrum. It might also utilize the white spaces or spectrum holes of licensed spectrum bands given that it doesn't create unsafe impedance for any of main subscribers to upgrade efficiency of the range of frequency usage. The principal issue confronting future remote frameworks is the place to discover appropriate range groups to satisfy the need for future services. While basically the part of radio range is distributed to different applications, and customers, observation gives proof, the use of range

S. K. Saraswat (✉)
GLA University, Mathura, India
e-mail: Shelesh.saraswat@gla.ac.in

A. Sivasubramanian et al. (eds.), *Futuristic Communication and Network Technologies*, Lecture Notes in Electrical Engineering 792,
https://doi.org/10.1007/978-981-16-4625-6_79

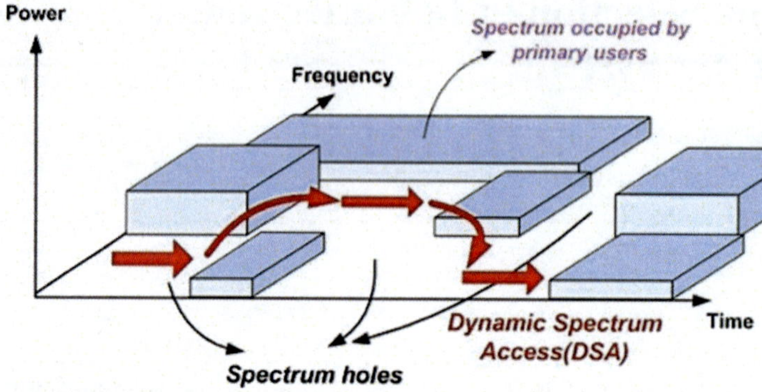

Fig. 1 Spectrum hole

is low. So as to conquer this issue and to improve the spectrum use, CR ideas have been presented. The CR target is to utilize scarce and to restrict normal resources proficiently without effecting uncontrolled interference to any of main licensed subscribers. CR will recognize and understand its spectrum environment, associate its temporarily idle spectrum, gain from its conduct and transmit adaptively [1]. Currently used radio systems aren't sensible about their radio spectrum context as it was schemed in such a way that it could be operated in defined band by using frequency range access system. Some inquires for range of frequency utilization specifies that the spectrum is not effectively utilized most often. All in all spectrum utilization could be improved considerably after permitting secondary unlicensed users to vigorous approach toward spectrum holes as it is not temporarily occupied by the main subscriber in the geographical part of notice as shown in Fig. 1.

The spectrum space is the range of frequencies allocated to the main subscribers, at a specific time and different region; the band isn't used by that subscribers. The space idea may be summed up as transmission open doors in radio white hole which is a hypothetical space involved by signals that have location dimensions, time, angle of reach, energy, frequency and others [2]. Radio developed is able to perceive and feel its near spectrum atmosphere, to recognize space in radio spectrum hole able to craft decision by their own about how it accesses spectrum. Such CR utilizing dynamic spectrum can possibly altogether upgrade the spectrum efficiency utilization bringing about simple and adaptable spectrum access for present or future purpose to provide wireless services.

CR is a present-day concept that is first introduced by Joseph Mitola and Gerald Maguire [3] which introduced CR as conservatory of software-defined radio which was able to enhance the litheness of wireless help with radio domain-based model interpretation by a unique language called RKRL. CR's structural design is an corporate agent for software-based radio of private technology at the crossing point of intersection, and computational intelligence was additionally evolved in Mitola's doctoral thesis [4].

The concept behind CR has emerged with the need to use the radio spectrum productively, and to have the option to keep up the most proficient type of correspondence for certain conditions like overarching and using the level of dealing out that are available currently, it is conceivable to build up a radio that is capable to see the spectrum, distinguish the frequency and it can be easily identified, and afterward execute the supreme type of communication for the needed conditions. So, in this way CR innovation can choose the frequency band and power levels that generally fit the necessities, the type of modulation, prevailing conditions and geographic dogmatic requirements. CR may be defined as an intellectual system that knows about general condition, gains by the atmosphere and adjusts inside states to measurable quantity in the further frequency stimuli by making changes accordingly in assured parameter in practical time, with two main ideas: reliable communications when- and wherever important as well as proficient deployment of radio spectrum [5].

In general, the CR might be relied upon to take a gander at parameters, for example: free channels, inhabitancy, the kind of transmitted data and the modulation types that are used. It should likewise take a gander at the intended requirements. For certain conditions, it may be requisite to use a software-defined radio and might be reconfigured itself to meet and accomplish the optimal transmission technology for a specified arrangement of parameters. Respectively, CR technology and software-defined radio are regularly firmly linked. In request to accomplish these targets, CR is obligatory to rapid change the properties and to get the spectrum without creating worthless interference to the main subscribers. Steps followed by the cognitive cycle are: frequency range sensing, frequency range decision, sharing and mobility [6].

Spectrum observing: It is dynamic range understanding process where CR screens geographical surroundings and radio atmosphere as well as distinguishes between utilization measurements of certain users (primary as well as secondary) and decides to calculate spectrum holes. Spectrum observing can be possible by many CR terminals or by unknown observing networks exchange data in a way that will improve on the whole accuracy.

Spectrum decision: By seeing above data, CR chooses when to start its event and operates the frequency and its parameters used. CR's main target is to send useful data and to fulfill the quality of service. At the same time, CR may utilize information from the regulatory and policy database in order to better its outage operation.

Spectrum sharing: As large number of unlicensed subscribers taking interest in the use of present spectrum space, CR has to get a balance between its aim of sending data in an efficient manner and considerate aims to share the present sources with other subscribers either cognitive or non-cognitive subscribers. But this can be done with rules deciding that CR conducts in specific conditions.

Spectrum mobility: Let the main subscriber start to work, then CR has to halt the activity or renounce right now the utilized spectrum and changed frequency. So, in order to evade intrusion for the main subscribers, it function has to be implemented in practical scenario; therefore, CR has to regularly explore all the feasible substitute spectrum holes.

Radio Environment: Purpose behind Radio Environment Maps (REMs) designs is to make a decision to store the information in a sorted manner and the way it is able to access different–different radios (CR or otherwise). It could cover the multi-domain ecological data like spectral regulations, geographical importance or location of various entity to notice (reflectors, obstacles, radios) plus radio equipment capability profiles, pertinent policies and many more. Its information is often modernized with respect to the interpretations from CR nodes and promulgates all through CR networks.

1.2 Multiple Input–Output Systems

These wireless systems are the systems whose both ends have multiple antenna elements, i.e., transmitter and receiver [7]. Computer simulations were the first one to present this system in the 1980s [8], and then, papers reconnoiter analyzed [9, 10]. From then to till now, interest in the above system has been fulminate that is used for WCDMA as well as discussed for best-performance mode of the booming IEEE 802.11 paradigm for LAN. Topics which are related to Multiple Input–Output Systems occupy main component of communications. In the above system, many antennas are used in two ways: The first is the foundation of efficient antenna diversity system, whereas the second one is for the transmission of different parallel data streams using the multiple antennas to enhance the capacity of the system.

Digital communications using MIMO systems have come out together of the important research field in communications and also figures on the chart of technologies that can have a chance to sort out the bottlenecks of traffic in the coming high-speed rate broadband Internet access system.

The random system given a link to the transmitting end as the receiving end is provided with many antenna elements, as given in Fig. 2. The idea of this system is that the signals on the TX antennas at one end and RX antennas on the former end are "mutual" in approach that the eminence (Bit Error Rate) or the data rate (Bit/s) of the communication will be better. Space–time processing method needed by the above

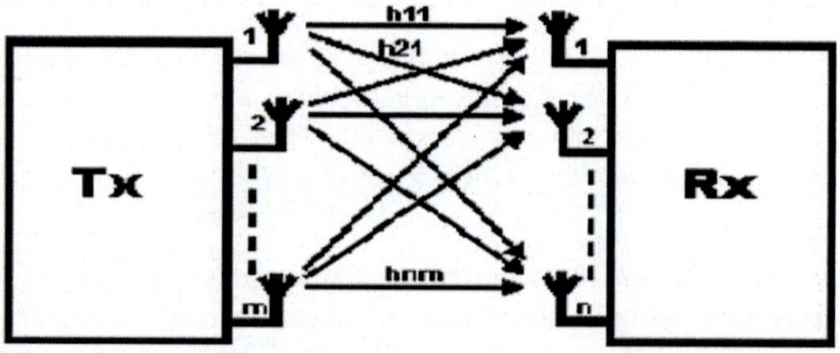

Fig. 2 General outline of MIMO system

system in which the time dimension is finished with the spatial dimension brought by different antennas. These networks are observed as an addition of the alleged "smart antennas," a method, first invented in the 1970s, for enhancing wireless transmission [11].

A channel could also be suffering from dwindling, and this may affect the SNR that affects the error rate, by making assumptions of the transmission of digital data. The idea of diversity is to give many versions of an equivalent signal on the receiver. And this is impact in many ways from signal path; the chances that they are going to exaggerated at an equivalent total time are decreased. In similar way, diversity is used to stabilize a link and increase performance and error rate. Different diversity modes provide a large number of benefits:

- *Time diversity*: In this, data signals are transmitted at different times, e.g., via dissimilar time slots and channel coding.
- *Frequency diversity*: This uses different frequencies; it involves transmission of same information signals of different carriers such as spread spectrum.
- *Space diversity*: Multiple antennas are located at spatial positions or locations which results in different received signals.
- MIMO network is a radio method as multiple antennas are used at the TX and RX to permit a different signal paths to hold data, by selecting distinct paths for every antenna to let and use multiple signal paths.
- Common idea behind MIMO systems is wireless systems space signal processing in which time goes together with the dimension integrated within the use of many spatially scattered antennas. Accordingly, MIMO wireless systems are considered as main idea for the smart antennas that are in used for many years to increase communication system performance.

It is obtained among TX and RX; the signal can acquire many paths. And also, by moving the antennas a distance, paths may change the variability of paths given which happens as result of large number of elements appearing to side between the TX and RX at the same time the multiple paths used are just to introduce interference. By using Multiple Input–Output System, these additional paths usually don't have any advantage; they grant supplementary strength to the link from recuperating SNR. Two specific formats for Multiple Input–Output System are mentioned below:

- *Spatial diversity*: It is used in the narrower way often present for TX and RX diversity which are used to offer improvements within SNR and identified from ameliorating the dependability of the network w.r.t. the fading.
- *Spatial multiplexing*: MIMO network is generally used to supply auxiliary information capacity by availing the many paths to hold added traffic, which directly increases the info throughput potential.

This results in the utilization of many antennas, and this method enhanced the range of the particular channel while following Shannon's law. So, to enhance the value of RX and TX antennas, there is a chance to enhance the throughput with every pair of antennas given to the network. It makes Multiple Input–Output System technology, the most important wireless technique used in the current period. For radio communications systems, spectral bandwidth is going to become the most estimable commodity and these methods are needed to use the present bandwidth more and MIMO network method is one of these techniques.

A learning article [12] is given an association among SISO and MIMO wireless. They need to compare Shannon egrodic ability of the SISO link and its diverse realization with the MIMO network channel. It is shown wireless information rate augments linearly with enhancement of many antennas both at TX and RX. MIMO network wireless connects capacities once TX and RX each have high information of channel matrix. The sole TX has the matrix awareness and when the sole RX has the awareness is given. MIMO System channel decomposition using eigen amount and singular values decomposition technique is proposed. The generic issue can be specified as the joint selection of TX-RX antennas in a MIMO CR network. The idea is to enhance the attainable rate as much as possible of the CR during satisfying any intrusion constraint due to the PU RX (either capable with an antenna or many antennas) functioning in locality. We take the describe task by performing an exhaustive search all over combinations of antenna parts and optimize over the TX CM of the CR subscribers. After having simulation results, other optimization methods are also applied on it.

1.3 Organization of Paper

The present paper is organized as: Sect. 2 discusses the network model, Sect. 3 discusses framework, and Sect. 4 gives performance details analysis. Section 5 concludes the given network.

2 System Model

The model is shown in Fig. 3 CR transmitter and receiver equipped with many antennas. Channel matrix CH_{CR} is formed in between the CR transmitter and CR recover. Here, assumed channel is Rayleigh channel with quasi-stationary type, which remains stationary for particular time. In the cognitive system, interference of secondary users at primary users plays a major role. Interference channel matrix CH_{INT} is formed between CR transmitter and PU receiver. Here there is the linear relationship of interference with the number of PU receivers; if the amount of PU receivers increases, then interference also increases. So our main constraint of the algorithm is to minimize the interference between CR users and PU. Another interference of PU transmitters at CR receivers can be formed. Channel matrix CH_{PC} shown in figure gives idea about interference of primary user transmitters at CR receivers. Same linear relationship of interference has also been applied here.

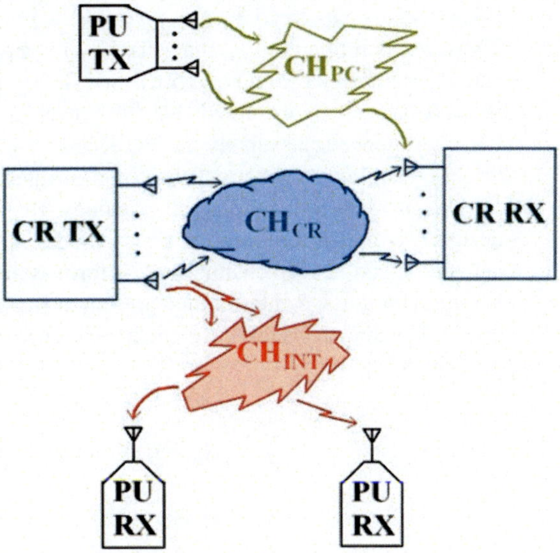

Fig. 3 System model

At the initial stage, we have considered that only one PU transmitter and one PU receiver are present in the MIMO cognitive system. The problem can be generalized by taking many numbers of primary users. As more as our system model is concerned, there are three-channel matrices $\mathrm{CH_{CR}}$, $\mathrm{CH_{INT}}$ and $\mathrm{CH_{PC}}$, where each of them formulated by group of antennas used at TX and RX. $\mathrm{CH_{CR}}$ matrix has T_{CR} and R_{CR} number of columns, $\mathrm{CH_{INT}}$ matrix has T_{CR} number of rows and $\mathrm{R_{PU}}$, and $\mathrm{CH_{PC}}$ matrix has T_{PU} rows and R_{CR} columns.

3 Analytical Framework

The network mode which is proposed here as shown in Fig. 3. Let us assume that transmitter and receiver have with their corresponding T_{CR} and R_{CR} antennas. PU consists of two antennas, i.e., T_{PU} TX antennas and R_{PU} RX antennas. It is assumed that there is more than one single receiving antenna when PUs operation in CR environs. These situations are considered as single and multiuser (SU, MU), respectively. The signal on CR receiver is as given:

$$f_{\mathrm{CR}}(n) = \mathrm{CH_{CR}}a(n) + b(n) + g(n) \tag{1}$$

where

1. $\mathbf{CH_{CR}} \in \mathbb{C}^{\mathrm{RCR} \times \mathrm{TCR}}$ is the channel gain matrix, whereas $g(n)$ is Gaussian noise. $a(n)$ and $f_{\mathbf{CR}}(n)$ are the send and the RX signal vectors correspondingly; $b(n)$ is the interference. In addition to this, the transmit covariance matrix of the CR

subscribers is expressed $\mathbf{V_{CR}} = [\boldsymbol{a}(n)\boldsymbol{a}(n)^{\dagger}]$. Let consider the total average CR TX power is limited to P_{CR} that is$\mathrm{Tr}(\mathbf{V_{CR}}) \le \mathbf{P_{CR}}$. The CM of the interference and noise is distinct from $\mathbf{D} = [\mathbf{b}(n)\mathbf{b}(n)\dagger + \mathbf{g}(n)\mathbf{g}(n)^{\dagger}] = \mathbf{I_{RCR}} + \mathbf{CH_{PC}CH_{PC}}^{\dagger}$. In multi-user, it is presumed that the power decays exponentially as well as CR interference transversely the PU receivers [13]. Various simulation results might be applied to construct the chain of interfering links which is based on a series of random locations, other values, etc. Consequently, to simplify our design it is useful to apply the decay model. CR transmitter and receiver for antenna selection keep channel state information ($\mathbf{CH_{CR}}$, $\mathbf{CH_{INT}}$ and $\mathbf{CH_{PC}}$). It is important that CR interference present at PU–RX must be below the defined threshold value. For a test case, the interference constraint for a subscriber may be written as:

$$\begin{aligned} &\sum_{i=1}^{R_{PU}} \mathrm{CH_{INT}}(i,:) V_{\mathrm{CR}} \mathrm{CH}_{\mathrm{INT}}^{\dagger}(i,:) \le \Theta \\ &\Rightarrow T_r\left(\mathrm{CH_{INT}} V_{\mathrm{CR}} \mathrm{CH}_{\mathrm{CR}}^{\dagger}\right) \le \Theta \end{aligned} \tag{2}$$

where $\mathbf{CH_{INT}}$ (i,:) $\in \mathbb{C}^{1\times TCR}$represents channel from the CR transmitter to the ith receive antenna of the PU receiver and Θ is the adequate total power at the PU receiver.

In addition to idea of doing constrained joint TX-RX antenna selection at the CR, here it usually begins with the familiar idea that the achievable rates of the CR network using total present antennas are [14]

$$R(\mathrm{CH_{CR}}, V_{\mathrm{CR}}) = \det\left(I_{\mathrm{RCR}} + \mathrm{CH_{CR}} V_{\mathrm{CR}} \mathrm{CH}_{\mathrm{CR}}^{\dagger} D^{-1}\right) \tag{3}$$

Similar to the approach of [15], it shows the diagonal selected matrices $E1$, $E2$ of dimension $R_{\mathrm{CR}} \times R_{\mathrm{CR}}$ and $T_{\mathrm{CR}} \times T_{\mathrm{CR}}$ with binary diagonal entries, respectively. Mathematically, elements are defined as

$$\begin{aligned} &(Ei)_{kk} = 1 \quad \text{if the } k\text{th antenna is choosed} \\ &0 \quad \text{otherwise} \end{aligned} \tag{4}$$

where I is equal to 1, 2. Diagonal elements of $E1$, $E2$ are the elements of the antennas at the CR receiver and CR transmitter. Then, the expression (3) reduces to

$$R(\underline{\mathrm{CH}}_{\mathrm{CR}}, \underline{V}_{\mathrm{CR}}) = \det\left(\underline{I}_{\mathrm{RCR}} + \underline{\mathrm{CH}}_{\mathrm{CR}} \underline{V}_{\mathrm{CR}} \underline{\mathrm{CH}}_{\mathrm{CR}}^{\dagger}\right) \tag{5}$$

where $\underline{\mathrm{CH}}_{\mathrm{CR}} = \underline{D}^{(-1/2)} E_1 \mathrm{CH_{CR}} E_2$. By selecting RX antennas, we get a noise $\mathbf{CMD}_{re}$ (dimension $m_{cr} \times m_{crd}$) and a new interference and the matrix is increased to Form $\underline{D}$, an $R_{\mathrm{CR}} \times R_{\mathrm{CR}}$ matrix, by placing columns and rows of zeros to the non-certain RX antennas. Similarly, with the obtained TX antennas, compact V_{red}

matrix ($t_{cr} \times t_{cr}$) is inflated to $\underline{V}$, ($T_{CR} \times T_{CR}$) by placing columns and rows of zeros with respect to the non-selected TX antennas. So, the problem of combining TX-RX antenna selection with CR power allowance may mathematically cast in the SU case as the constrained optimization issue $P1$ as.

$P1$: maximization

$E1$, $E2$, VCR

$$\det\left(I_{\text{RCR}} + \underline{D}^{(-1/2)} E_1 \text{CH}_{\text{CR}} E_2 V_{\text{CR}} E_2^{\dagger} \text{CH}_{\text{CR}}^{\dagger} E_1^{\dagger} \underline{D}^{(-1/2)}\right)$$

Subject to

$$\{1, 0\}, j = 1, \ldots, R_{CR} if i = 1 and j = 1, \ldots, T_{CR} if i = 2$$

$$T_r\left(CH_{INT} E_2 V_{CR} E_2^{\dagger} CH_{INT}^{\dagger}\right) \leq \Theta, V_{CR} \geq 0$$

$$Tr(\text{VCR})\text{PCR}, \ Tr(E1) = \text{rcr}, Tr(E2) = \text{tcr} \tag{6}$$

Assume that.

Here P1 is an objective function for maximizing data rates between selected antennas E_1 and E_2 at receiver and transmitter, respectively. Constraints: covariance matrix V_{CR} must be positive and interference is less or equal to threshold Θ. It is assumed that power is less or equal to addition of diagonal parts of CM V_{CR} and sum of diagonal elements of E equals to the number of antennas selected at each side.

4 Simulation Results

In this system, BPSK data are considered for data communication in cognitive network. Here, MIMO channel matrix $\mathbf{CH_{CR}}$ is 2×2 configured, and also one primary user transmitter $\mathbf{T_{PU}}$ & one primary user receiver $\mathbf{R_{PU}}$ are present in the CR environment. By performing number of iterations for calculating interference, it can be summarized that CR interference observed at PU RX value should not be more than the threshold value $\Theta = 3.9373$ which is already defined. It is obvious that transmitting and receiving power is minimized in a selected pair of antennas compared to using all antennas simultaneously. So assuming that minimum power requirement, data rates for two pairs of $\mathbf{T_{CR}}$-$\mathbf{R_{CR}}$ have been calculated with interference threshold as a main constraint.

Figure 4 shows that the throughputs (data rates) of antennas chosen by our algorithm are quite higher than using the fixed configured antennas. The BPSK scheme has two values for symbols are -1 and 1, which can be generated randomly for communication in our system model.

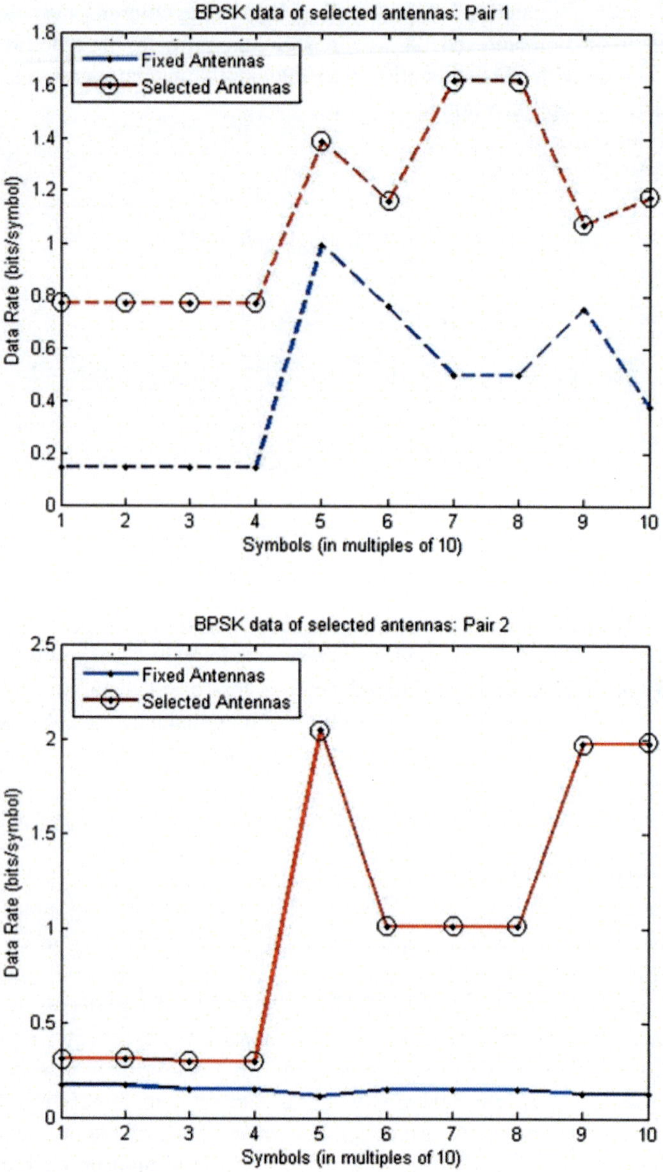

Fig. 4 Simulation results for BPSK

5 Concluding Remarks and Future Work

It has been observed that antenna selection schemes outperform conventional fixed antenna schemes with interference and saving energy as constraints at same time. Here it is considered that the channel is Rayleigh fading distribution during simulation. So our antenna selection system gives higher and better data rates as compared to fixed arbitrary antenna selection configuration. In this scheme, simulation is performed only for both antennas which are at both transmitter and receiver end, where they are combined with a single antenna. Here only BPSK technique is used for data communication; at later stage the scope of system is elaborated for defining algorithm and could be applicable for one or more number of antennas at both secondary and primary side, and better optimization method is selected to reduce complexity at transmitter and receiver side with individual modulation schemes. Nowadays in communication, multi-input and multi-output are the fundamental key parts of wireless communication in terms of 5G technological development. In this light, this study will be of immense nature in the field of wireless communication. Through this paper, different dimensions of antenna have identified its prudent role in this modern technological era of development and it will also serve the dual purpose for technical professional and stakeholders so that they can get maximum advantage to fulfill the competitive demand of data communication and simulation energy. This research has also presented the required modulations and prevailing conditions to develop an intellectual system for the practical and reliable communications.

References

1. Ding Z et al (2017) A survey on non-orthogonal multiple access for 5Gnetworks: Research challenges and future trends. IEEE J Sel Areas Commun 35(10):2181–2195
2. Mitola J (2000) CR: an integrated agent architecture for software defined radio. Ph.D. Thesis, KTH Royal Institute of Technology, Stockholm, Sweden
3. Akyildiz IF, Lee W-Y, Chowdhury KR (2009) CRAHNs: CR ad hoc networks. Ad Hoc Netw 7(5):810–836
4. Gesbert D, Shafi M, Shiu DS, Smith PJ, Naguib A (2003) From theory to practice: An overview of MIMO space-time coded wireless systems. IEEE J Sel Areas Commun 21:281–302
5. Mitola J, Maguire GQ (1999) CRs: Making software radios more personal. IEEE Pers Commun 6(4):13–18
6. Winters JH (1987) On thc capacity of radio communications systems with diversity in Rayleigh fading environments. IEEE J Sel Areas Commun 5:871–878
7. Foschini GJ, Gans MJ (1998) On limits of wireless communications in a fading environment when using multiple antennas. Wireless Personal Commun 6:311–335
8. Gesbert D, Akhtar J Breaking the barriers of Shannon's capacity: an overview of MIMO wireless systems. Telenor's J Telektronikk
9. Khan A, Vesilo R A tutorial on SISO and MIMO channel capacities
10. Gao H, Smith PJ, Clark MV (1998) Theoretical reliability of MMSE linear diversity combining in Rayleigh-fading additive interference channels. IEEE Trans Commun 46(5):666–672
11. Telatar E (1999) Capacity of multi-antenna Gaussian channels. Europ Trans Telecommun 10(6):585–595

12. Dua A, Medepalli K, Paulraj AJ (2006) Receive antenna selection in MIMO systems using convex optimization. IEEE Trans Wireless Commun 5(9):2353–2357
13. Scutari G, Palomar D, Barbarossa S (2008) Cognitive MIMO radio. IEEE Signal Process Mag 25(6):46–59
14. Zhang R, Liang YC (2008) Exploiting multi-antennas for opportunistic spectrum sharing in CR networks. IEEE J Sel Topics Signal Process 2(1):88–102
15. Zhang L, Xin Y, Liang YC (2008) Weighted sum rate optimization for CR MIMO broadcast channels. In: Proceedings of ICC, pp 3679–3683
16. Sridharan S, Vishwanath S (2008) On the capacity of a class of MIMO CRs. IEEE J Sel Topics Signal Process 2(1):103–117
17. Molisch AF, Win MZ (2004) MIMO systems with antenna selection. IEEE Microw Mag 5(1):46–56
18. Sanayei S, Nosratinia A (2004) Antenna selection in MIMO systems. IEEE Commun Mag 42(10):68–73
19. Smith PJ, King TW, Garth LM, Dohler M (2008) A power scaling analysis of norm-based antenna selection techniques. IEEE Trans Wireless Commun 7(8):3140–3149
20. Zhang R, Liang Y-C, Cui S (2010) Dynamic resource allocation in CR networks. IEEE Signal Process Mag 27(3):102–114
21. Zhou J, Thompson J, Krikidis I (2009) Multiple antennas selection for linear precoding MISO CR. In: Proceedings of WCNC, pp 1–6
22. Islam SMR, Zeng M, Dobre OA (2017) NOMA in 5G systems: exciting possibilities for enhancing spectral efficiency. IEEE 5G Tech Focus 1(2):1–6

Wide-Angle, Polarization Insensitive, Ultrathin Tuneable Metamaterial Absorber for C-Band Application

Toji Thomas, G. Boopalan, and C. K. Subramaniam

Abstract Proposed in this paper is a tuneable metamaterial absorber that demonstrates single band absorption in the C-band. The unit cell structure is a wire cross with square patches on the corners of the cell. The simulated structure is analysed and demonstrated to show an absorption greater than 95% at 4.8 GHz. The proposed work shows polarization independence and also a wide angle insensitivity of up to 70° for transverse electric (TE) mode and up to 40° for transverse magnetic (TM) modes on the structure that maintains more than 90% absorption near 4.8 GHz. The variation of length and width of the wire cross shows the structure has the potential to be tuned to both higher sides and lower sides of the resonant frequency. The entire thickness of the optimized structure is found to be 0.0086 λ. Such an ultrathin single band absorber could find potential application in selective filtering of radiations and noises.

Keywords Angle insensitive · C-band · Ultrathin · Metamaterial · Absorber

1 Introduction

The exploration for manipulation of the EM waves with materials that are arranged in arrays started from nineteenth century onwards. The idea of negative refractive index was not a digestible theory, until in 1968, Vesselago first theorized about the negative index of refraction (NIR) materials [1]. A practical understanding of the NIR metamaterial started to ignite the research interests when John Pendry showed this material can be used to make perfect lens [2]. The ability to focus energy on the desired region by negative permeable materials and their use for energy transfer applications are discussed in [3, 4]. Recently metamaterials find practical applications in a wide range, such as in defence, communication, solar energy harvesting, antennas, acoustics and biomedical, and metamaterials prove to be the next phase of

T. Thomas (✉) · C. K. Subramaniam
School of Advanced Sciences, Vellore Institute of Technology, Vellore, India

G. Boopalan
School of Electronics Engineering, TIFAC CORE, Vellore Institute of Technology, Vellore, India

A. Sivasubramanian et al. (eds.), *Futuristic Communication and Network Technologies*, Lecture Notes in Electrical Engineering 792,
https://doi.org/10.1007/978-981-16-4625-6_80

advancement in technology. Many practical difficulties and sophistications are to be considered in designing a metamaterial. Most of the critical applications start with an absorber, whether it is a sensor or an energy harvester. There have been many types of metamaterial absorber (MMA) reported throughout the world that operates in microwave regime, infrared regime, optical regime and terahertz regime [5, 6]. There have been reports of metamaterial absorbers that demonstrate single band, dual band, multiband and broadband absorbers [7–9]. Broadband absorbers are the most explored for the current world scenarios. The practical metamaterials absorbers [10] that demonstrate perfect absorption by impedance matching and wide angle polarization independent structures are demonstrated [11] which finds high usage in sensing applications. Advancing world requirements uses a surfeit of absorbers in improving the performance of a wide range of applications such as in photodetectors, gas sensors, solar and thermal photovoltaic, spatial light modulators, antennas [12, 13], etc. The lower band side of the microwave spectrum such as the S-band, C-band, X-band [14–17] is also explored as they find extensive applications in satellite communications, Wi-Fi, etc.

There have been many reports on C-band applications in industrial and medical field [18, 19]. A highly efficient single band, tuneable absorber was simulated and studied for 4.8 GHz. The fourfold symmetry of the structure makes it polarization insensitive. The structure was desirably made slender to be used in many potential applications. The tuneability of the structure is studied based on the variation in geometrical parameters. This facilitates in targeting individual frequencies for filtering or band stop applications in the C-band. The impedance matching [20] and interference theory [21] provide the most established theories on absorption. The maximum resonance happens when the free space impedance matches with the effective impedance [20] of the structure so that the reflection and transmission become minimum. The EM waves that penetrate into the structure, reflecting off from the metallic backplate of the metamaterial, destructively interfere cancelling the reflection and in effect improving the absorption [21].

2 Design and Simulation

The unit cell structure proposed is a metal-dielectric-metal composite that is simulated and studied using the CST Microwave Studio Suite. Frequency domain solver was used for the analysis. The cheaply available copper was used as the metal for the top geometry as well as the bottom metallic layer. The demonstrated FSS design is obtained by removing a smaller cross portion of provided geometrical parameters from a complete metallic cross structure. The proposed structure was based on a modified loop-type element discussed in Munk [22]. The same methodology discussed in it was adopted for the initial stage. Later the top dielectric was removed and a metal backplate was introduced to obtain the metamaterial structure which was optimized by improving the dielectric constant and other FSS geometrical parameters. Later, a small square patch was added at the corners of the proposed unit cell.

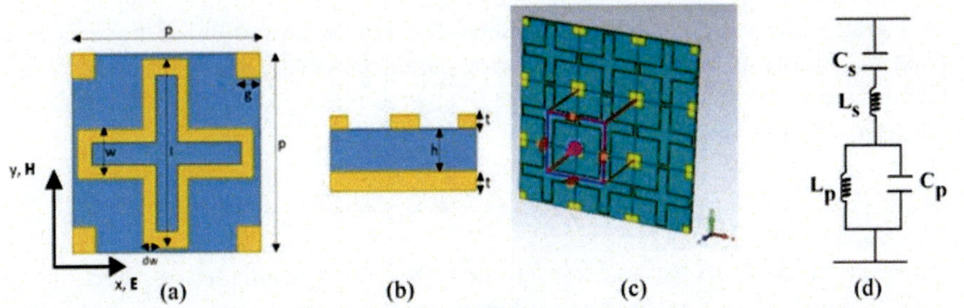

Fig. 1 The unit cell (wire cross with square patches on corners). **a** Top view, **b** side view, **c** the 3 × 3 array of the unit cells with unit cell boundaries and **d** the equivalent circuit for the FSS

The thickness of the metal is selected based on how much the radiation can penetrate through it. Rather than a particular dielectric material, the dielectric constant is given much attention. The proposed structure is seen in Fig. 1.

The parameters such as the length of the wire cross (l), width of the cross (w), the small width of the cross wire (dw), the side length of the square patches (g) on the corners of the cell and the thickness of the dielectric substrate (h) are varied and studied for the improvement and tuneability of the structure. The conductivity of the copper of thickness (t) 0.017 mm was set to be $\sigma = 5.96 \times 10^7$ S/m. The optimized dielectric constant used throughout the simulation is $\varepsilon_r = 12$ which corresponds to TMM13i. The structure is constructed on the x–y plane with the direction of propagation of the EM wave in the -z direction. Unit cell boundary conditions are applied in both x and y direction and an open add space boundary condition in the propagation direction. The propagating distance between the source and the surface of the structure is kept half the period (p) of the unit cell. The optimized parameters of the structure are provided in Table 1. The equivalent circuit for the structure was extracted from [23], where the L_p and C_p correspond to the parallelly connected inductor and capacitor and L_s and C_s are the inductor and capacitor of the series LC circuit.

Table 1 Optimized design parameters

Parameters	Optimized value (mm)
Length of the wire cross (l)	6.25
Width of the wire cross (w)	1.25
Smaller width of the wire (dw)	0.171
Thickness of the wire cross (t)	0.017
Side length of the square patch (g)	0.575
Dielectric thickness (h)	0.508
Period of unit cell (p)	6.609

The structure is tested between 4 and 6 GHz, and the absorption of the resonance peak was evaluated based on the scattering parameters [16]. The absorption characteristics would be based on the percentage of absorption possessed by the absorber. Absorptivity is evaluated as:

$$A(\omega) = 1 - R(\omega) - T(\omega) \quad (1)$$

where reflectivity ($R(\omega)$) and transmission ($T(\omega)$) are expressed as functions of frequency ω. They are easily evaluated based on the scattering parameters [16]. That is, $(R(\omega)) = |S_{11}(\omega)|^2$ and $(T(\omega)) = |S_{21}(\omega)|^2$. The scattering parameters are obtained from the simulations. From the absorptivity equation, in order to have maximum absorption, both the reflection and transmission have to be minimum. For a unity absorption, both have to be zero. The thickness of copper is selected such that it is much greater than the skin depth making all radiation that irradiates on it to get reflect back, and hence, zero transmission is expected. Therefore, the (1) reduces to

$$A(\omega) = 1 - R(\omega) = 1 - |S_{11}(\omega)|^2 \quad (2)$$

The normal incidence frequency response of the structure is provided in Fig. 2, which shows the reflection and absorption response of the structure at resonance.

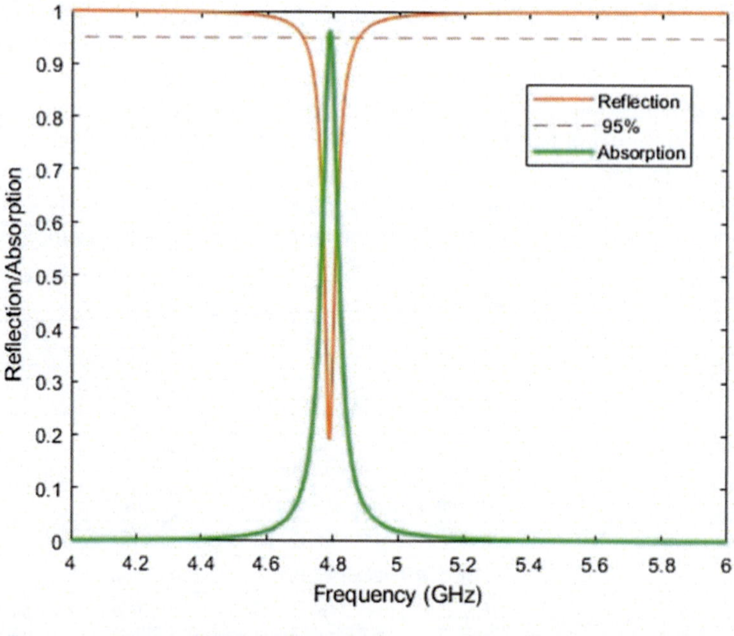

Fig. 2 Frequency response of the optimized structure at resonance. It demonstrates a single band absorption greater than 95% (96.27%) at 4.8 GHz

The frequency response demonstrated an absorption of 96.27% at 4.8 GHz. The horizontal dotted line in Fig. 2 is the 95% absorption.

3 Results and Discussion

The optimized structure is simulated, and the resonance for normal incidence for transverse electric and transverse magnetic polarization was found to be at 4.8 GHz. The absorption was at 96.27%, as demonstrated in Fig. 2. The dielectric constant plays an important role in the tuneability of the structure. For an improved value of absorption, the permittivity of the dielectric was set to be 12. Each of the parameters was varied by keeping rest of them as constant, and the tuneability of the structure was explored. Figure 3 demonstrates the structure's response to variation to the wire cross's length. The length was varied between 6.5 and 5.75 mm in a step of 0.25 mm. As the length reduced, the frequency response shifted towards the higher side of the C-band, maintaining the constant absorption above 95%. It clearly shows the contribution of the structures length in tuning the single band frequency in C-band. Similarly, the contribution of width of the wire cross was also studied and is reflected in Fig. 4. The width was varied from 3.25 to 1.25 mm in a step of 1 mm. The response shifted towards the lower side of the testing frequency, with near unity absorption. Reduction in width, more than the demonstrated, will show response in

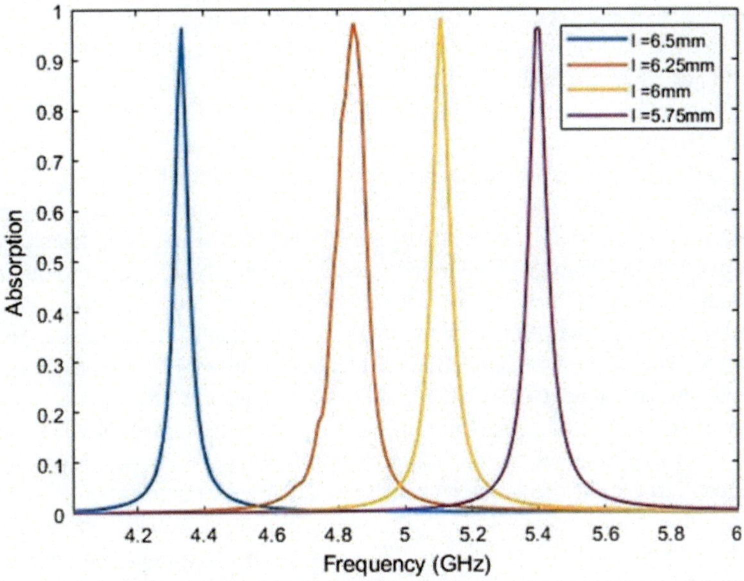

Fig. 3 Resonance variation to change in the length of the wire cross. Resonance shift towards higher side as the length is reduced

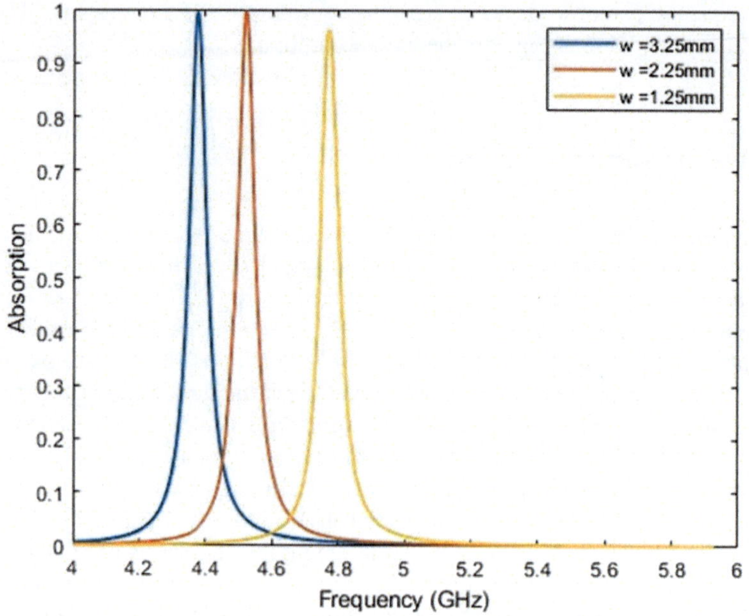

Fig. 4 Resonance shift to the variation of width of the wire cross. The absorption peak shift towards higher edge as the width is reduced

higher-frequency side. The absorption response to variation in the patch size at the corner (g), the thickness of the dielectric (h) and the wire width of the wire cross (dw) did not affect the results.

The fourfold symmetry exhibited by the structure makes it insensitive towards different polarization angles when light falls on the structure normally. The polarization independence is tested at different oblique angles of incidence for transverse electric (TE) and transverse magnetic (TM) polarization modes. The result is discussed in Fig. 5a and 5b, where the angle of incidence (θ) is varied from 0° to 80° in a step of 10° for TE and TM polarized waves, respectively. The results demonstrate a wide-angle insensitivity up to 70° which produces more than 90% absorption near 4.79 GHz for TE polarization. More than 95% absorption is maintained up to 60°.

For TM modes, the wide-angle independence was up to 40° that produced more than 90% absorption near 4.79 GHz and up to 50° the absorption is greater than 80%. Further increase showed a reduction in the absorption percentage for both modes. Figure 6 demonstrates the surface current distribution at resonance. More currents are concentrated at the tips of the wire cross indicating the contribution of width and length in tuning the frequency. The direction of current distribution on top surface is opposite to that on the bottom plane. This curling in current distribution produces the magnetic response in the structure. The lumped components of the equivalent circuits can be analysed from the direction of the surface currents on the FSS. The side patches show negligible surface currents, showing a weak contribution as a lumped

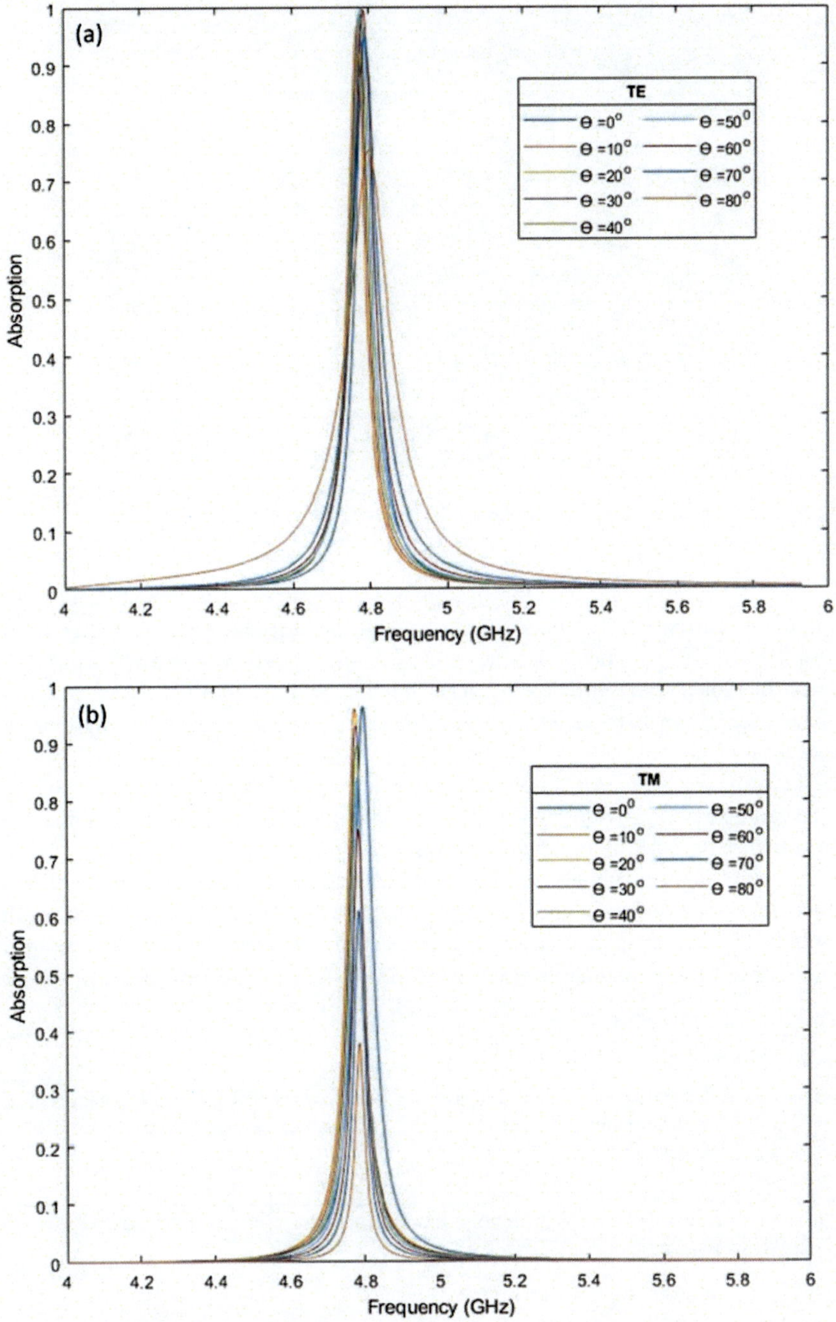

Fig. 5 The absorption response to different polarizations at different angles of incidence, **a** transverse electric (TE) mode, **b** transverse magnetic (TM) mode

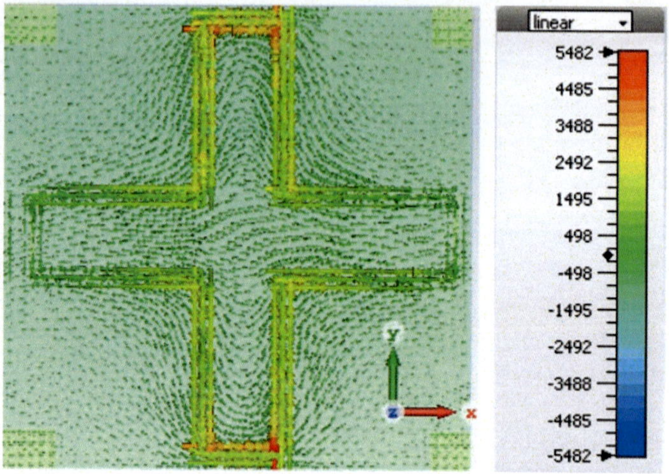

Fig. 6 Surface current distribution at 4.8 GHz

element. The flow of currents through the arms contributes to the inductance, and the flow of current in the opposite direction of the adjacent arms gives rise to the capacitance. The parameter dependence of L and C can be evaluated from [24], where the inductance is affected by the arm length and the width of the wire and the capacitance by the distance between adjacent arms and the facing side length of two adjacent loops.

The proposed structure is compared with few of the existing literature in Table 2. The comparison provides an insight into the unique features that are exhibited by the structure in terms of angle independence and thickness.

The work demonstrates a much thin MMA that has an improved angle of incidence, insensitivity for TE and TM polarized waves. The structure proposed results in minimum thickness as compared to other structures available in the literature. Being much thinner and angle insensitive MMA, it finds great potential in sensors

Table 2. Comparison with available literature

Reference	Maximum absorption in C-band (%)	Unit cell size (mm)	Angle independency (absorption greater than 80%)		Thickness (λ)
			TE (°)	TM (°)	
[8]	93	20 × 20 × 1.5	40	40	0.0112
[9]	95.49	10 × 10 × 1	60	60	0.0563
[10]	99.64	11 × 11 × 1	60	60	0.014
[11]	98.8	6 × 6 × 0.8	60	60	0.015
Proposed	96.27	6.609 × 6.609 × 0.542	70	50	0.0086

and antenna applications. Owing to the tuneability it demonstrated based on length and width variation, it can be used for selective absorption in C-band and even in X-band. Especially, 4.8 GHz is explored in wireless implantable body area network applications [25].

4 Conclusion

The work successfully demonstrated a metamaterial absorber unit cell in the shape of a wire cross with side patches at the corners that gives a single peak resonance in the C-band. 96.27% of absorption was reported at 4.8 GHz which demonstrated a greater wide-angle independence for transverse electric and transverse magnetic modes as well as polarization independence. The thickness of the entire unit cell was up to 0.0086 λ, making it an ultrathin structure. The variation in length and width of the wire showed tuneability of the resonance peak within the C-band. The proposed structure finds good use in filtering, noise cancellation and antenna applications.

References

1. Veselago VG (1968) Reviews of topical problems: the electrodynamics of substances with simultaneously negative values of epsilon and mu. Soviet Physics Uspekhi 10:R04
2. Pendry JB (2000) Negative refraction makes a perfect lens. Phys Rev Lett 85(18):3966
3. Boopalan G, Subramaniam CK (2017) Influence of a negative refractive index material on the magnetic flux profile of a circular loop. Int J Microwave Opt Technol 12:434–439
4. Boopalan G, Subramaniam CK (2014) Frequency dependence of magnetic flux profile in the presence of metamaterials for wireless power transfer. In: IEEE International symposium on circuits and systems (ISCAS). Melbourne VIC, Australia, pp 2437–2440
5. Bilal RMH, Saeed MA, Choudhury PK, Baqir MA, Kamal W, Ali MM, Rahim AA (2020) Elliptical metallic rings-shaped fractal metamaterial absorber in the visible regime. Sci Rep 10(1):1–12
6. Ghosh SK, Yadav VS, Das S, Bhattacharyya S (2019) Tunable graphene-based metasurface for polarization-independent broadband absorption in lower mid-infrared (MIR) range. IEEE Trans Electromagn Compat 62(2):346–354
7. Ishii Y, Takida Y, Kanamori Y, Minamide H, Hane K (2017) Fabrication of metamaterial absorbers in THz region and evaluation of the absorption characteristics. Electron Commun Japan 100(4):15–24
8. Jain P, Singh AK, Pandey JK, Garg S, Bansal S, Agarwal M, Kumar S, Sardana N, Gupta N, Singh AK (2020) Ultra-thin metamaterial perfect absorbers for single-/dual-/multi-band microwave applications. IET Microwaves Antennas Propag 14(5):390–396
9. Cong LL, Cao XY, Song T, Gao J, Lan JX (2018) Angular-and polarization-insensitive ultrathin double-layered metamaterial absorber for ultra-wideband application. Sci Rep 8(1):1–12
10. Han Y, Gong S, Wang J, Li Y, Qu S, Zhang J (2019) Reducing RCS of patch antennas via dispersion engineering of metamaterial absorbers. IEEE Trans Antennas Propag 68(3):1419–1425
11. Hoa NTQ, Tuan TS, Hieu LT, Giang BL (2019) Facile design of an ultra-thin broadband metamaterial absorber for C-band applications. Sci Rep 9(1):1–9

12. Liu Z, Liu X, Wang Y, Liu G, Tang C (2020) Silicon antennas metasurface based light absorber with quantitatively adjustable operating frequency and intensity. IEEE Journal of Selected Topics in Quantum Electron
13. Alkurt FO, Altintas O, Ozakturk M, Karaaslan M, Akgol O, Unal E, Sabah C (2020) Enhancement of image quality by using metamaterial inspired energy harvester. Phys Lett A 384(1):126041
14. Edries M, Mohamed HA, Hekal SS, El-Morsy MA, Mansour HA (2020) A new compact quad-band metamaterial absorber using interlaced I/square resonators: design, fabrication, and characterization. IEEE Access 8:143723–143733
15. Garg P, Jain P (2020) Novel ultrathin penta-band metamaterial absorber. AEU-Int J Electron Commun 116:153063
16. Mishra N, Chaudhary RK (2019) Design and development of an ultrathin triple band microwave absorber using miniaturized metamaterial structure for near-unity absorption characteristics. Prog Electromagn Res 94:89–101
17. Kaur M, Singh HS (2019) Simulation study of ultra compact polarization independent dual-band metamaterial absorber. In: 2019 IEEE Indian conference on antennas and propogation (InCAP). IEEE, pp 1–4
18. Schmidt M, Trömel S, Ryzhkov AV, Simmer C (2018) Severe hail detection: hydrometeor classification for polarimetric C-band radars using fuzzy-logic and T-matrix scattering simulations. In: 2018 19th International radar symposium (IRS). IEEE, pp 1–7
19. Elwi TA (2019) Printed microwave metamaterial-antenna circuitries on nickel oxide polymerized palm fiber substrates. Sci Rep 9(1):1–14
20. Smith DR, Vier DC, Koschny T, Soukoulis CM (2005) Electromagnetic parameter retrieval from inhomogeneous metamaterials. Phys Rev E 71(3):036617
21. Chen HT (2012) Interference theory of metamaterial perfect absorbers. Opt Express 20(7):7165–7172
22. Munk BA (2005) Frequency selective surfaces: theory and design. Wiley, New York
23. Costa F, Monorchio A, Manara G (2012) Efficient analysis of frequency-selective surfaces by a simple equivalent-circuit model. IEEE Antennas Propag Mag 54(4):35–48
24. Tong XC (2018) Functional metamaterials and metadevices. Springer, Bolingbrook, IL
25. Ramli NH, Ayob MF (2018) Design of a chip printed antenna at 4.8 GHz for wireless implantable body area network applications. J Telecommun Electron Comput Eng (JTEC) 10(1–17):27–31

Interpretation of Lung Sounds Using Spectrogram-Based Statistical Features

G. Shanthakumari and E. Priya

Abstract Lung sounds convey useful information related to pulmonary pathology. Auscultation is the widely used clinical practice. An auscultation is considered as an inferior diagnostic tool in differentiating adventitious lung sounds from normal since it relies on physician's hearing ability and experience. World Health Organization has declared that approximately 3 million people die each year because of chronic lung diseases. Computerized auscultation along with signal processing methods has enhanced the diagnostic abilities of lung sound. Anatomical changes in lung reflect in time and frequency domain attribute of lung sound. Spectrogram is a graphical representation of the signal in time frequency domain. This work is an attempt to categorize the lung sound as normal, crackle and wheeze using the spectrogram of the lung sounds. The texture and statistical analysis of the spectrogram is computed for the demarcation purpose. The statistical parameters such as mean, variance, skew, kurtosis and entropy of spectrogram images are computed and plotted. Results demonstrate that the features variance and kurtosis extracted from the spectrogram demarcate the normal lung sounds from crackle and wheeze with better efficiency.

Keywords Lung sounds · Spectrogram · Texture · First-order statistics

1 Introduction

Lung sounds are generated due to the flow of air in respiratory system. Lung sounds include normal breath sounds and adventitious sounds. Normal and adventitious lung sounds have specific characteristics typical to certain pulmonary pathologies. Respiratory disorders can be clinically assessed by examining acoustical pattern of

G. Shanthakumari (✉) · E. Priya
Department of Electronics and Communication Engineering, Sri Sai Ram Engineering College, West Tambaram, Chennai, Tamil Nadu 600044, India
e-mail: shanthakumari.ece@sairam.edu.in

E. Priya
e-mail: priya.ece@sairam.edu.in

A. Sivasubramanian et al. (eds.), *Futuristic Communication and Network Technologies*, Lecture Notes in Electrical Engineering 792,
https://doi.org/10.1007/978-981-16-4625-6_81

the lung sounds [1]. Adventitious lung sounds are categorized as continuous adventitious sound (CAS) and discontinuous adventitious sound (DAS). CAS includes wheezes, rhonchi, squawks and stridor where DAS includes crackles and friction rub. The associated pulmonary pathology for CAS is constriction in the airways and DAS is development of fluid in airways [2]. The bandwidth of lung sound varies from 20 to 2000 Hz. Wheeze is a periodic sound with dominant frequency above 100 Hz and time period of 100 ms whereas the period of crackle is less than 20 ms and occupies a broader frequency range [3]. Auscultation is the important diagnostic tool for respiratory diseases. It relies on the hearing capacity and experience of the physician and hence an instinctive process. The major downside of auscultation is quantitative measurement and permanent record of the measurement is not possible [4]. The process involved in computerized lung sound analysis is digital recording of lung sound followed by preprocessing, feature extraction and classification. Recorded respiratory sounds include heart sounds, muscle sounds and other moving artifacts [5]. Hence, preprocessing of recorded respiratory sound plays a vital role.

Lung sounds are nonlinear and non-stationary signals. Hence, it is difficult to study the characteristics of lung sound in time domain or frequency domain alone. Traditional systems use either time or frequency domain study [6]. The spectrum of lung sounds affords diagnostic information. It is also evident that the spectral features of lung sound could be analyzed using spectrogram. Crackles can be distinguished from normal and wheeze signal because of its wide frequency range. Normal lung sound has a narrow frequency range [7]. The short-time Fourier transform (STFT) of a signal results in spectrogram of the signal. Spectrogram displays the frequency components at different time points. Time frequency representation of lung sounds using spectrogram provides rich visual representation enabling to demarcate the signal visually [8]. Spectrogram is considered as an image with distinct texture and hence image processing techniques can be exploited to extract the statistical features. Spectrogram of the signal projects power distribution of different frequency components at different instants of time [9].

Texture is a recurrent pattern of information. One of the significant image processing technique is texture feature extraction. Texture features are extracted using methods such as statistical, structural, model-based and transform information. Statistical methods characterize the texture indirectly with respect to the nondeterministic properties of the image [10].

In this work, spectrogram is obtained for the preprocessed normal sound, wheeze and crackle lung sounds. The spectrograms are considered as image from which parameters such as mean, variance, skewness and kurtosis are computed and plotted. Along with first-order features, entropy is calculated. The graphical representation of the features aids in analyzing and demarcating the adventitious lung sound wheeze and crackle from the normal lung sound.

2 Methodology

The lung sounds acquired from the International Conference on Biomedical Health Informatics Challenge—2017 (ICBHI) database are utilized in this work. The recordings are collected utilizing heterogenous hardware with duration ranges from 10 to 90 s. The respiratory sounds are recorded by AKG C417L Microphone, 3 M Littmann Classic II SE Stethoscope. In addition, the electronic stethoscopes such as 3 M Littmann 3200 and Welch Allyn Meditron Master Elite are used. The lung sounds are recorded from various chest areas that incorporate trachea, left and right side of anterior, posterior and lateral chest regions [11]. These recorded signals are to be preprocessed before it is examined further. Sixty lung sounds including twenty normal, twenty wheezes and twenty crackles are utilized in this work. Spectrogram of the lung sounds is obtained by applying short-time Fourier transform (STFT).

2.1 Short-Time Fourier Transform

Fourier transform (FT) converts a time domain signal to a meaningful frequency domain information. But comprehensive information about non-stationary signals cannot be extracted from FT. FT resolves the frequency components of the signal alone but not its position in time. STFT overcomes this drawback [12]. Representation of non-stationary signals in time frequency domain is possible with STFT which is obtained by windowing the input signal and applying discrete Fourier transform. For a given signal x_n and window w_n, STFT is expressed by

$$X(t, f) = \int_{\infty}^{\infty} x(\tau)w(t - \tau)\mathrm{e}^{-j2\pi mf\tau}d\tau \tag{1}$$

where $w(t - \tau)$ is a window function and $x(\tau)$ is the signal to be decomposed.

2.2 Spectrogram

STFTs are typically imagined employing the log of the spectra, $20\log_{10}(X(h, k))$. The two-dimensional log-spectra can be picturized with a temperature map and is recognized as a spectrogram. A visual interpretation of the range of frequencies in change with time results in a spectrogram [13]. The spectrograms of normal, wheeze and crackle lung sounds are calculated for 1024 FFT points with 75% hamming window overlap. The computation results in a matrix that generates a three-dimensional image of the input signal with frequency, time and amplitude.

The graph is represented by time and frequency on x and y axis, respectively. The amplitude is represented in terms of color [14].

2.3 Texture Analysis of Spectrogram Based on Statistics

Image textures are intricate graphical patterns composed of objects or sections with sub-patterns with the characteristics of brightness, color, shape, size, etc. As a source of extracting information, texture analysis characterizes spatial variation within the image [15]. These statistical approaches involve in the study of spatial dispersal of gray values that are obtained by manipulating the local features at all points in the image. First-order texture measures are statistics computed from the original image [16]. Histogram-based texture method relies on the intensity value concentrations on the image represented as histogram. First-order statistical features extracted from the texture are mean, variance, skewness, kurtosis and entropy [17]. Mean indicated the general brightness of the image and is given by

$$\mu = \sum nI_n.p(I_n) \tag{2}$$

where I_n and $p(I_n)$ are the image pixel values and probability distribution of I_n, respectively. Variance is the square of standard deviation. The measure of the frequency distribution of a pixel value of an image is known as standard deviation. It can be expressed as

$$\sigma^2 = \frac{1}{n-1}\sum_{i=1}^{n}(I_n - \mu)^2. \tag{3}$$

The statistical measure of randomness that characterizes the texture of an image is called entropy. It gives the measure of information content. It is measured using

$$H = -\sum np(I_n).\log_2 p(I_n). \tag{4}$$

Skewness and kurtosis can be referred by the histogram technique. The symmetry of distribution of the pixel intensity value is termed as skewness. Skewness of a normal distribution is zero and is expressed as

$$\alpha_3 = \sum_n \frac{E(p(I_n) - \mu)^3}{\sigma_3}. \tag{5}$$

In a normal distribution, the range of data distributed near the mean is measured using kurtosis. Distinct peak and rapid decline near the mean are an indication of a group of data with high kurtosis whereas low kurtosis results in flat top near the mean. It can be measured using the formula

$$\alpha_4 = \sum_n \frac{E(p(I_n) - \mu)^4}{\sigma_4}. \tag{6}$$

3 Results

An illustration of normal, wheeze and crackle lung sound is shown in Fig. 1a, b and c, respectively. It is evident from the presentation that lung sound is a complex signal comprising of different frequency components. It is pointless to analyze the frequency contents of the signal in a time domain representation. Since lung sounds are non-stationary in nature, the occurrence of instants of different frequency components is insignificant. Hence, time frequency domain analysis of the lung sound is attempted in distinguishing the details present in wheeze and crackle from normal subjects. The signal in time frequency domain is attained using the spectrogram technique.

The spectrogram for a typical wheeze, crackle and normal lung sound is represented in Fig. 2a to c. It is evident from Fig. 2a that the frequency range of normal lung sound is approximately up to 100 Hz. The inspiration cycle is of short duration compared to the expiratory cycle. Figure 2b presents the spectrogram of wheeze

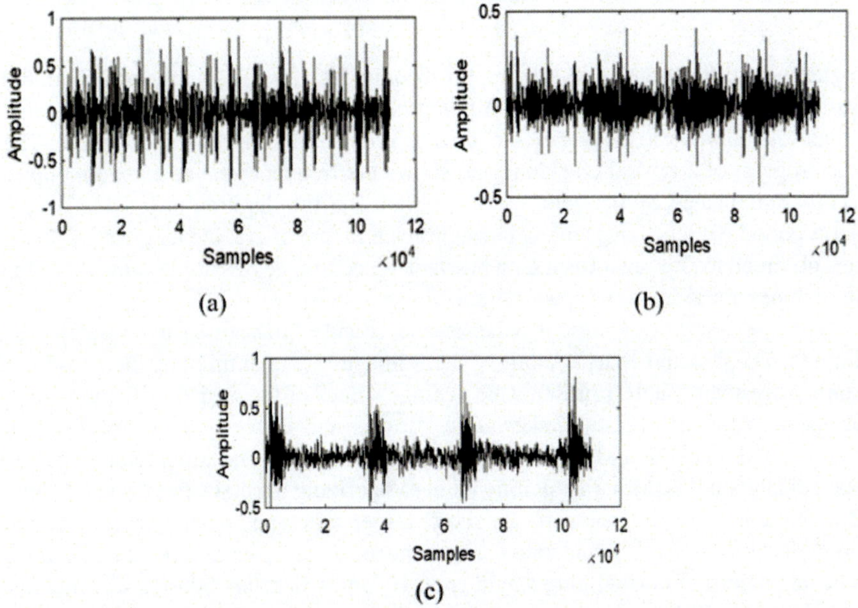

Fig. 1 A representative lung sound of **a** normal, **b** wheeze and **c** crackle

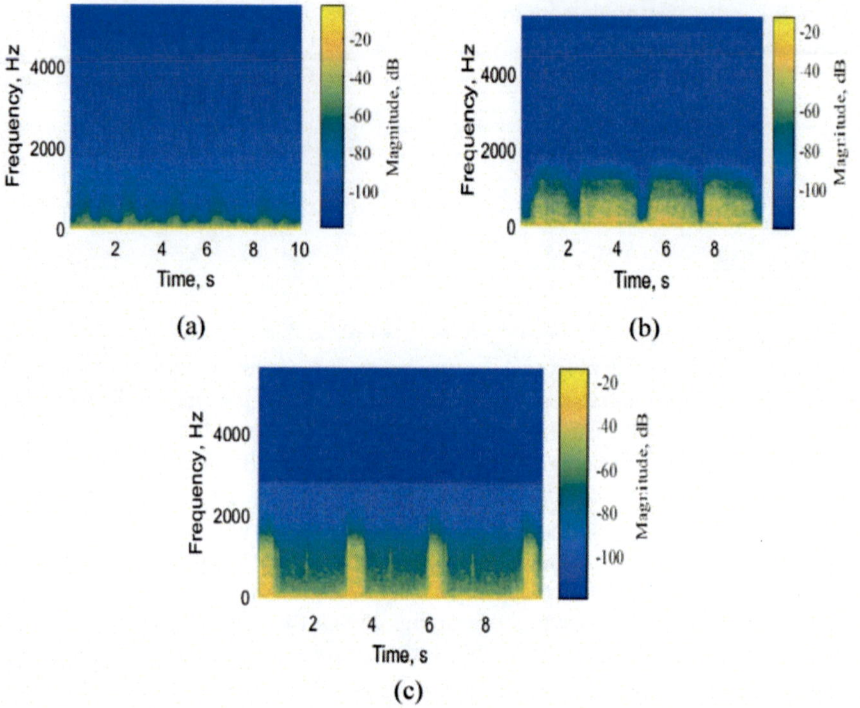

Fig. 2 Spectrogram of **a** normal, **b** wheeze and **c** crackle lung sound

signal where the frequency extends up to only 500 Hz. They follow a rhythmic pattern exhibiting continuous adventitious sound. It is understood from Fig. 2c that crackles exhibit wide frequency range when compared to normal and wheeze. The spectrogram of a typical crackle signal shows the frequency range extending up to 1000 Hz indicated by the yellow color region. But the duration of crackle is very short period, which is one among the significant nature of crackle lung sound. Thus, it is observed to explore information relevant to occurrence of sample values in terms of texture pattern from the spectrogram.

The spectrogram of normal, wheeze and crackle lung sound is considered as images. The fine and coarse details of these images are quantified by the first-order statistical values. Table 1 presents the average values of the statistical features such as mean, standard deviation, skewness, kurtosis and entropy of the spectrogram image for the crackle, wheeze and normal lung sounds. The interpretation of the statistical feature is primarily significant in categorizing normal wheeze and crackle. The numerical value of the variance shows larger difference when compared to the numerical difference of other statistical parameters among the different lung sounds. The bandwidth of normal lung sound is 60 Hz and wheeze is 500 Hz. Crackle has a wider bandwidth as they are not periodic and exists for short time duration. The lungs sounds are low-frequency signals and hence the entropy value is almost zero.

Table 1 Average statistical features of spectrogram

Class of lung sound	Statistical features				
	Mean	Variance	Skewness	Kurtosis	Entropy
Normal	−103.17	131.7885	3.68881	21.14288	0
Wheeze	−88.83	293.6804	1.98483	6.78117	0.002488
Crackle	−82.21	169.8777	1.89528	7.88005	0.002705

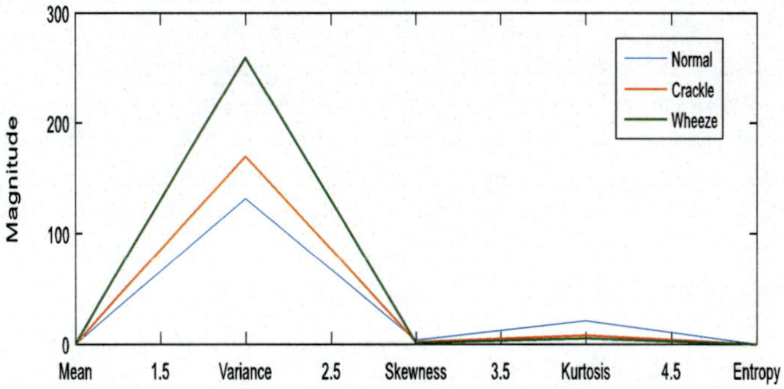

Fig. 3 First-order statistical features of normal, wheeze and crackle lung sound

The graphical representation of first-order statistical characteristics of normal, wheeze and crackle lung sound is shown in Fig. 3. Among the five statistical features, the highest value is obtained for the variance. The mean, skewness and entropy of all the three lung sounds are almost equal to zero. The contribution of each pixel intensity for the complete image is mean; therefore, the mean of normal, wheeze and crackle lung sound is nearly equal and low. The distinction of a pixel from its adjacent pixel is obtained from the variance and hence it is used to demarcate dissimilar regions. It is observed that the variance value shows a better demarcation of normal, wheeze and crackle classes. There exists significant difference between the variance value of wheeze and crackle. The wheeze sound is periodic and its duration is longer compared to crackle and hence the numerical value of variance is high in wheeze in comparison with normal and crackle lung sound. The skewness of a normal distribution is equal to zero. The values of skewness of the three classes of lung sounds are close to zero. Hence, it can be concluded that the lung sounds follow normal distribution. The numerical value of kurtosis reflects a small difference between crackle and wheeze and is well distinct from normal lung sounds. The bandwidth of normal lung sound is very much less compared to the adventitious lung sounds and hence the intensity variation is less in the spectrogram of normal lung sound. Since the variation in intensity is less, the histogram of normal lung sound results in sharp-peaked normal distribution curve and thus kurtosis value is high. The methodology used in this work aids to categorize the normal from wheeze and crackles using the statistical features obtained from the spectrogram of the lung sounds.

4 Conclusion

The automated auscultation procedure along with signal processing methods has enhanced the ability in diagnosing lung sound. The anatomical changes in lung are reflected in time and frequency domain characteristics of lung sound. In this work, the spectrogram of the normal, wheeze and crackle lung sounds is derived from the respective signals and analyzed. The spectrogram distinguishes the frequency components and reflects the identity of the three lung sound classes. Spectrogram provides only qualitative information regarding the characteristic of the signal. In order to perform a quantitative analysis, statistical features such as mean, variance, skewness, kurtosis and entropy are obtained from the spectrogram image. The numerical values of features obtained are tabulated and plotted. Results demonstrate that the first-order statistical parameter variance shows significant difference for normal, wheeze and crackle lung sound. The statistical features adopted in this methodology are a unique pipeline in the workflow. Thus, this methodology seems to be useful as a decision support for the physician in mass screening of respiratory disorders.

References

1. Qiu Y, Whittaker AR, Lucas M, Anderson K (2005) Automatic wheeze detection based on auditory modelling. Proc Inst Mech Eng [H] 219(3):219–227
2. Sovijarvi ARA, Vanderschoot J, Earis JE (2000) Standardization of computerized respiratory sound analysis. Eur Respir Rev 10(77):585–585
3. Gurung A, Scrafford CG, Tielsch JM, Levine OS, Checkley W (2011) Computerized lung sound analysis as diagnostic aid for the detection of abnormal lung sounds: a systematic review and meta-analysis. Respir Med 105(9):1396–1403
4. Alsmadi S, Kahya YP (2008) Design of a DSP-based instrument for real-time classification of pulmonary sounds. Comput Biol Med 38(1):53–61
5. Tocchetto MA, Bazanella AS, Guimaraes L, Fragoso JL, Parraga A (2014) An embedded classifier of lung sounds based on the wavelet packet transform and ANN. IFAC Proceedings Volumes 47(3):2975–2980
6. Rizal A, Hidayat R, Nugroho HA (2016) Lung sounds classification using spectrogram's first order statistics features. In: 2016 6th International annual engineering seminar (InAES). IEEE, pp 96–100
7. Saatci E, Akan A (2005) Heart sound reduction in lung sounds by spectrogram. IFMBE Proc 11(1):1727–1983
8. Aviles-Solis JC, Storvoll I, Vanbelle S, Melbye H (2020) the use of spectrograms improves the classification of wheezes and crackles in an educational setting. Sci Rep 10(1):1–8
9. Ravichandran D, Nimmatoori R, Dhivakar MA (2016) A study on image statistics and image features on coding performance of medical images. Int J Adv Comput Eng Commun Technol (IJACECT) 5(1):1–6
10. Hory C, Martin N, Chehikian A (2002) Spectrogram segmentation by means of statistical features for non-stationary signal interpretation. IEEE Trans Signal Process 50(12):2915–2925
11. Rocha BM, Filos D, Mendes L, Vogiatzis I, Perantoni E, Kaimakamis E, Paiva RP (2017) A respiratory sound database for the development of automated classification. In: International conference on biomedical and health informatics. Springer, Singapore, pp. 33–37
12. Haider NS, Joseph J (2017) An investigation on the statistical significance of spectral signatures of lung sounds

13. Parkhi A, Pawar M (2011) Analysis of deformities in lung using short time Fourier transform spectrogram analysis on lung sound. In: 2011 International conference on computational intelligence and communication networks. IEEE, pp 177–181
14. Hadjileontiadis LJ (2008) Lung sounds: an advanced signal processing perspective. Synth Lect Biomed Eng 3(1):1–100
15. Sengupta N, Sahidullah M, Saha G (2016) Lung sound classification using cepstral-based statistical features. Comput Biol Med 75:118–129
16. Palaniappan R, Sundaraj K, Sundaraj S, Huliraj N, Revadi SS (2018) Classification of pulmonary pathology from breath sounds using the wavelet packet transform and an extreme learning machine. Biomed Eng (Biomedizinische Technik) 63(4):383–394
17. Kochetov K, Putin E, Azizov S, Skorobogatov I, Filchenkov A (2017) Wheeze detection using convolutional neural networks. In: EPIA conference on artificial intelligence. Springer, Cham, pp 162–173

Index for Assessment of EEG Signal in Ischemic Stroke Patients

R. Geetha and E. Priya

Abstract Stroke is an acute condition of sudden compromise in vascular perfusion of brain and manifestation of neurological deficit. Worldwide, stroke is the second leading cause of death and also the third leading cause of morbidity and disability. Electroencephalogram (EEG) is a noninvasive method that captures the electrical activity of brain as a signal from the scalp. Detection and reduction of artifacts play an important task to acquire clean EEG signals so as to examine and detect brain activities. In this work, EEG signals from normal and subjects with acute ischemic stroke (AIS) are acquired under standard signal acquisition protocol from public database. The quality of the signal is improved by the techniques. An attempt has been made to detach artifacts by independent component analysis. The preprocessed EEG signal is decomposed by discrete wavelet transform method into wavelet coefficients to reduce the signal dimension. The decomposed signal is categorized as the sub-waves namely alpha, beta, delta, theta and gamma. The index such as delta–alpha ratio (DAR), delta–theta to alpha–beta ratio (DTABR), brain symmetry index (BSI) obtained by Welch's method helps to distinguish AIS from controlled subject. Also, the performance of the procedure is evaluated by statistical measures such as skewness, kurtosis, entropy, mean and variance. It is observed from the results that AIS patients have a high DAR, DTABR and BSI. Results also demonstrate that the extracted statistical metrics are high for AIS compared to that of normal individuals. Thus, the index and statistical metrics used in this work are significant in classifying AIS from normal subjects.

Keywords Acute ischemic stroke · Discrete wavelet transform · Delta–alpha ratio · Delta–theta to alpha–beta ratio · Brain symmetry index

---

R. Geetha (✉) · E. Priya
Department of Electronics and Communication Engineering, Sri Sairam Engineering College, West Tambaram, Chennai, Tamil Nadu 600044, India

E. Priya
e-mail: priya.ece@sairam.edu.in

A. Sivasubramanian et al. (eds.), *Futuristic Communication and Network Technologies*, Lecture Notes in Electrical Engineering 792,
https://doi.org/10.1007/978-981-16-4625-6_82

1 Introduction

Globally, brain stroke is the significant and foremost cause of mortality as well as long-lasting disability. Ischemic stroke is the most common type and has a global prevalence of approximately 11.6 million new cases per year. Among various etiologies, acute ischemic stroke (AIS) is one of the significant causes of morbidity. Around 87% of all strokes are ischemic stroke that is due to the blockage of oxygen-rich blood flow through the brain vessels [1]. Brain function is monitored by electroencephalography (EEG) signals particularly in intensive care unit. Brain ischemia is detected with respect to the characteristics changes that take place in EEG. The feature changes in EEG are correlated with cerebral blood flow and brain metabolism [2, 3].

The feature extraction of EEG signal is done in wavelet transformed domain and feed-forward neural network with extreme learning machine algorithm to demarcate normal and ischemic stroke signals [4]. The characteristics of brain wave are analyzed by different categories of stroke level using relative power ratio [5]. Augmentation in delta power (1–4 Hz) is one of the main features found to be related to ischemic stroke [6]. Delta to alpha power ratio (DAR) demonstrates maximal accuracy in differentiating AIS patients from controls [7]. The prediction of stroke outcome is identified by using neurophysiological markers such as delta–theta to alpha–beta ratio (DTABR) and alpha relative power. The functional outcome of stroke is also determined by measuring the symmetry index, absolute and relative frequency band indices and dynamic time change captured from quantitative electroencephalography (QEEG) [8]. Acute ischemic hemispheric stroke is monitored by continuous bedside monitoring of quantified EEG to measure brain symmetry index (BSI). This measure is consequently correlated with National Institute of Health Stroke Scale (NIHSS) [9, 10]. Among various EEG-derived parameters, BSI is one of the most accepted parameters which is used in the research field for the prediction of stroke prognosis [11].

In this work, an attempt is made to analyze EEG signals automatically by extracting features using wavelet transformation. The ischemic stroke patients from normal subjects are thus distinguished based on indices such as DAR, DTABR and BSI. The performance evaluation of the preprocessing procedure is done by statistical measures such as skewness, kurtosis, entropy, mean and variance.

2 Methodology

2.1 EEG Dataset

The EEG signals are obtained from public open-source repository for open data (RepOD), BNCI Horizon 2020 and the Temple University Hospital EEG Corpus (TUH-EEG) datasets. In these datasets, the EEG signal is recorded for 10 min from each patient using the standard 10–20 EEG electrode placement system (Fig. 1).

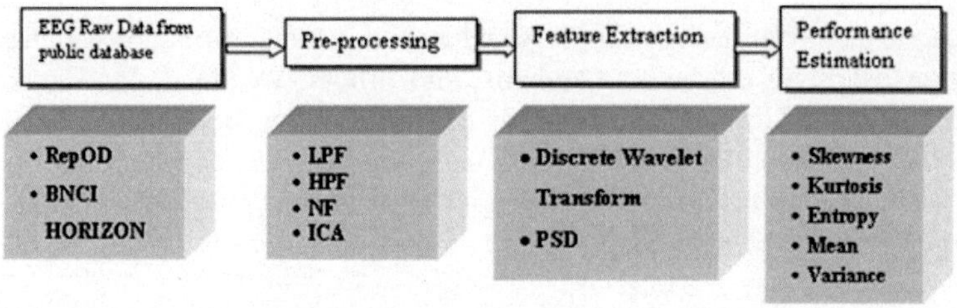

Fig. 1 Block diagram of the proposed approach

The raw ischemic stroke EEG signals from 16 channels comprise all prominent regions of human brain. The acquired signal is sampled at a rate of 250 Hz. These signals are preprocessed, to acquire useful information. Features from preprocessed signals are extracted and compared with normal EEG signals.

2.2 *Preprocessing*

The spatial information will get lost due to noise and artifacts while recoding the EEG signals from the scalp. The brain signals are filtered by using 30 Hz low-pass filter (LPF), 70 Hz high-pass filter (HPF) and 50 Hz Notch filter. The artifacts are aimed to be removed by independent component analysis (ICA) using EEGLAB.

2.3 *Feature Extraction*

The preprocessed EEG signal data from the EEGLAB in European data format (.edf) are converted into ASCII data. Then, using Matlab script, the wavelet transformation is done. Wavelet transform estimates spectral information which is expressed as an infinite series of wavelets. The signal is decomposed to a set of coefficients called wavelet coefficients. Adequate numbers of coefficients are computed to reconstruct the signal back. The wavelet transform of a signal is defined as

$$W_{\psi}x(s,\tau) = \frac{1}{\sqrt{|s|}} \int_{-\infty}^{\infty} x(t)\psi * \left(\frac{t-\tau}{s}\right) dt \tag{1}$$

where $\psi(t)$ is a mother wavelet, s and τ are the scaling parameters, s denotes the oscillatory frequency representing the length of the wavelet, τ indicates its shifting position, and the asterisk denotes complex conjugate.

Discrete wavelet transform (DWT) provides significant information which enables the signal to be analyzed and to decompose into different frequency bands by consecutive high-pass and low-pass filtering of the signal. DWT utilizes scaling and wavelet functions. The smoothing feature of the Daubechies wavelet of order 8 (db8) is used to detect changes in EEG signals. The signal dimension is reduced using db8. DWT decomposition can be expressed as

$$M_{(a)}(b) = x(a) * \varphi_{a,b}(c) \tag{2}$$

$$N_{(a)}(b) = x(a) * \psi_{a,b}(c) \tag{3}$$

where $M_{(a)}(b)$ and $N_{(a)}(b)$ are the approximate coefficients and detailed coefficients resolution of a with c is the time sequence.

The normalized wavelet and scale basis function of $\varphi_{a,b}(c)$, $\psi_{a,b}(c)$ are represented as

$$\varphi_{a,b}(c) = 2^{a/2} g_a\left(c - 2^a b\right) \tag{4}$$

$$\psi_{a,b}(c) = 2^{a/2} h_a\left(c - 2^a b\right) \tag{5}$$

where $2^{a/2}$ specifies the inner product normalization, a and b are the scaling and translation parameter [11–13].

One of the most common features to analyze the EEG signal is power spectral density. The spectral power density of each channel is computed from the decomposed EEG signal into four sub-waves based on the signal frequency range of alpha (8–12 Hz), beta (14–20 Hz), theta (4–7 Hz) and delta (1–4 Hz) [6]. The calculation of power spectral ratios such as DAR and DTABR is defined as

$$\text{DAR} = \frac{\text{Relative Power Ratio Delta}}{\text{Relative Power Ratio Alpha}} \tag{6}$$

$$\text{DTABR} = \frac{\text{Relative power Ratio Delta} + \text{Theta}}{\text{Relative Power Ratio Alpha} + \text{Beta}}. \tag{7}$$

The degree of symmetry between left and right brain activities is measured using BSI, which is defined as the arithmetic average of absolute value of difference in mean hemispheric power. The BSI determines the amount of ischemic damage in brain. The value of BSI is nearer to 0 which represents perfectly symmetrical and it has higher value closer to 1 that symbolizes the maximal asymmetrical wave. It estimates the overall asymmetry within the identical channels and computes the average of frequency ranging from 1 to 25 Hz. It is defined as

$$BSI = \frac{1}{PO}\sum_{l=1}^{O}\left\|\sum_{k=1}^{P}\frac{RIH_{kl} - LEH_{kl}}{RIH_{kl} + LEH_{kl}}\right\| \tag{8}$$

where RIH_{kl} and LEH_{kl} represent the power spectral density of the left and right hemisphere, O refers to the overall number of Fourier coefficients and P is the total number of electrode pairs [9].

2.4 Performance Estimation

The performance of the procedure is evaluated by statistical measures such as skewness, kurtosis, entropy, mean and variance. These metrics are helpful in distinguishing ischemic stroke patients from normal [14].

3 Results and Discussion

Typical representation of Fig. 2 shows the extracted beta, alpha, theta and delta waveform of EEG spectra for normal individuals and AIS patients. The frequency variation in the EEG spectra noticeable from the figure represents brain function too.

Decomposition of EEG signal using DWT provides the relative power ratio of different frequency segments such that of beta, alpha, theta and delta waves. Relative power indices for all these frequency bands are derived by using absolute value of power from each of the bands as a percent of absolute power summed over the five frequency bands. The comparison of relative power for all frequency bands of AIS patient and normal individual is presented in Fig. 3a–d.

In this scattered graph, the even number of electrodes determines the right hemisphere while the odd number of electrodes represents the left hemisphere. The delta activities of both cerebral hemispheres are high in AIS patient than normal individual as shown in Fig. 3a. The changes in delta activity demonstrate the strong correlation with cerebral blood flow (CBF) and electrical activity of cortical neurons during ischemia. The activities of theta on right and left hemisphere are highly elevated in AIS patients which represents reduction in the brain metabolism than normal individual as shown in Fig. 3b. The relative band power of beta and alpha as shown in Fig. 3c and d tends to be lower when compared with relative band power of delta and theta as shown in Fig. 3a and b. So it is inferred that the beta and alpha components of EEG do not contribute much on assessing the AIS patient condition.

The comparison of normalized average values of band power ratio for AIS patient and normal individual is shown in Fig. 4a. Similarly, the normalized average values of BSI for AIS patient and normal individual are shown in Fig. 4b.

The overlay graph as shown in Fig. 4a represents the average values of band power ratio for AIS patient and normal individual. EEG parameters such as DAR and DTABR value of AIS patient are found to be higher than normal individual. The higher value of DAR and DTABR implies a significant reduction in CBF in AIS patient. BSI is a measure used to find out the symmetry of brain waves between left and right hemisphere. The average values of BSI of AIS patient tend to be higher compared to that of normal individual as shown in Fig. 4b. An average value of BSI nearer to 1 signifies the maximal asymmetry that occurs due to cerebral hypo-perfusion in AIS patients.

The overlay graph as shown in Fig. 5 represents the variations in normalized average values for AIS patient and normal individual. The performance evaluation of the statistical features extracted from the EEG signals such as skewness, kurtosis, entropy, mean and variance of AIS is high compared to that of normal individual. The extracted EEG signals are precised by means of higher-order moments such as skewness and kurtosis. The lack of symmetry or the asymmetry of intrinsic brain activity is deliberated by skewness. A higher-level statistic kurtosis provides the measure of spikiness of EEG signals and is increased relatively to a smaller extent. An increment in the entropy reflects the slowing in the brain activity following

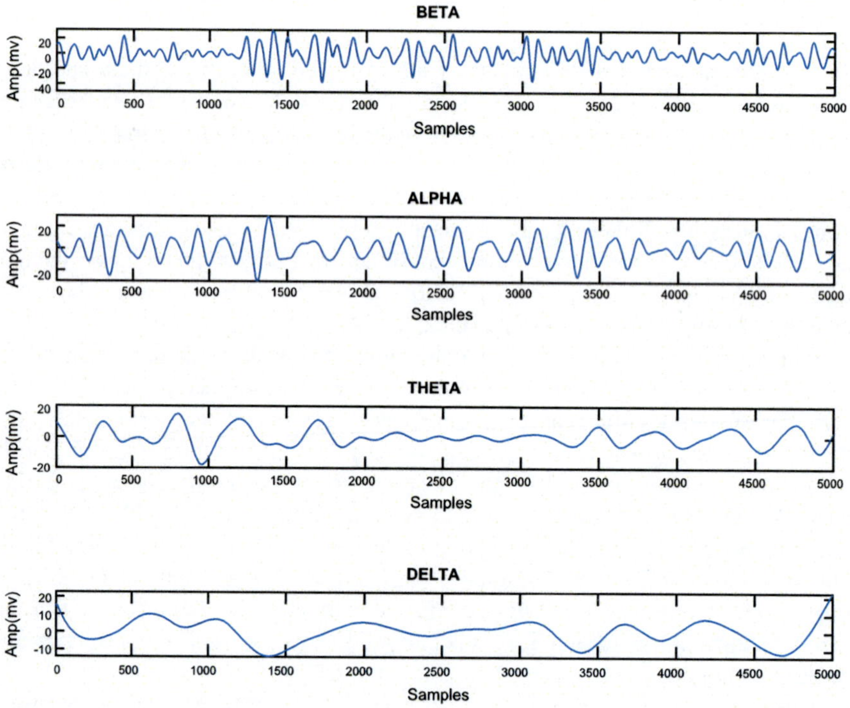

(a)

Fig. 2 Representation of beta, alpha, theta, delta wave of EEG signals for **a** normal individual and **b** AIS patient

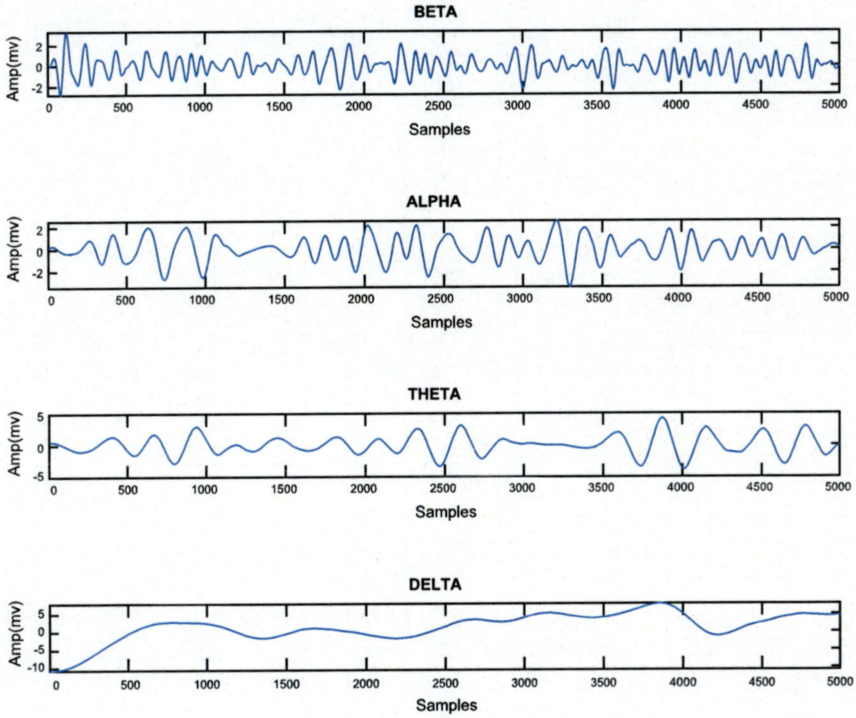

(b)

Fig. 2 (continued)

cerebral stroke. The mean and the variance were selected to measure the changes in EEG waves and are used to reveal the cortical involvement in the injured cerebral hemisphere.

4 Conclusion

Assessment of EEG signal helps in distinguishing acute ischemic stroke from the controlled subjects. In this work, appropriate indices and metrics are used as a promising tool to detect acute ischemic stroke lesion. The EEG signal obtained from the public database is preprocessed using a combination of low-pass, high-pass filter and a notch filter to eliminate the noise and artifacts. The decomposition of EEG signals is done by using DWT with Daubechies wavelet level 8 as a mother wavelet. The level 8 is categorized into four sub-waves namely beta, alpha, theta and delta. The feature index such as DAR and DTABR is extracted from each sub-waves. The extracted features are further correlated with normal individual. It is observed from

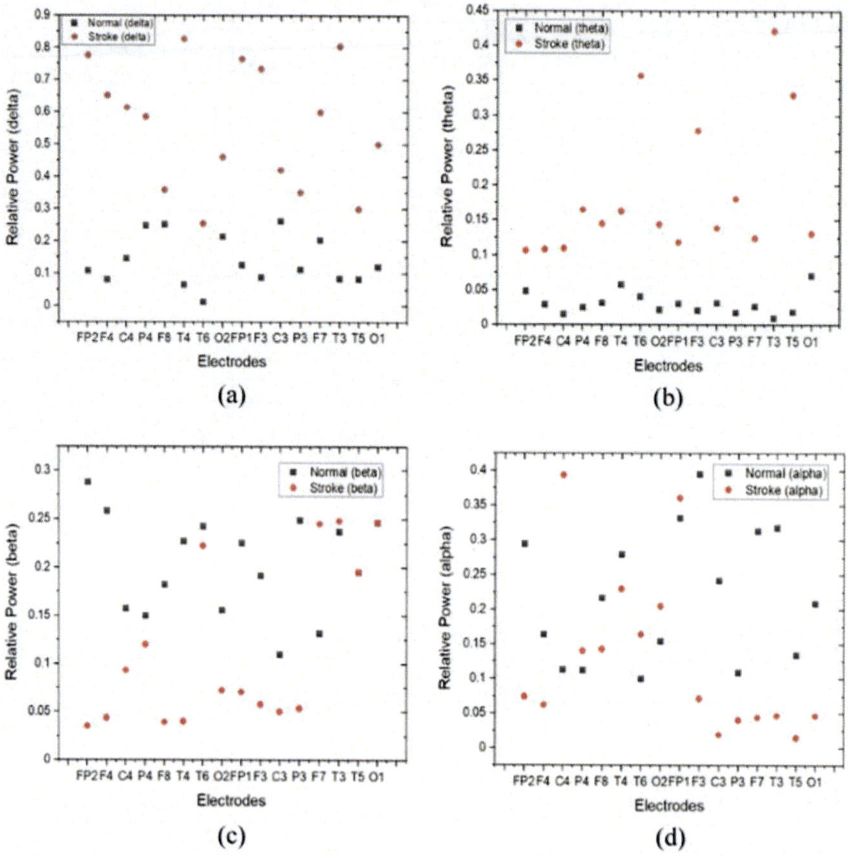

Fig. 3 Comparison of relative power of **a** delta, **b** theta, **c** beta and **d** alpha between normal individual and AIS patient

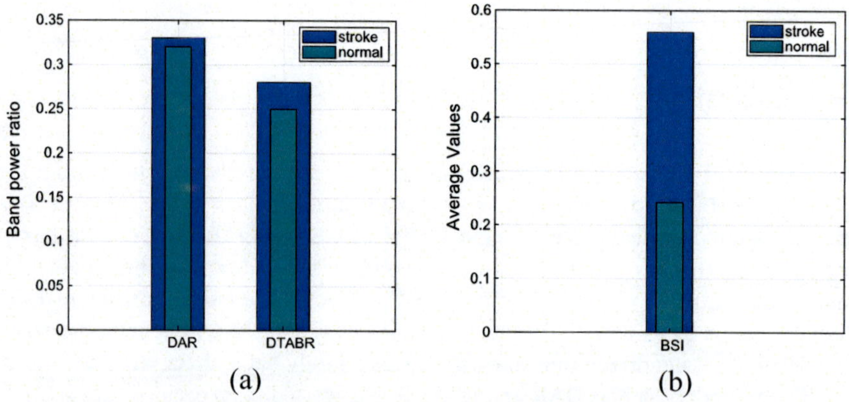

Fig. 4 Average values of **a** band power ratio and b BSI for stroke patient and normal individual

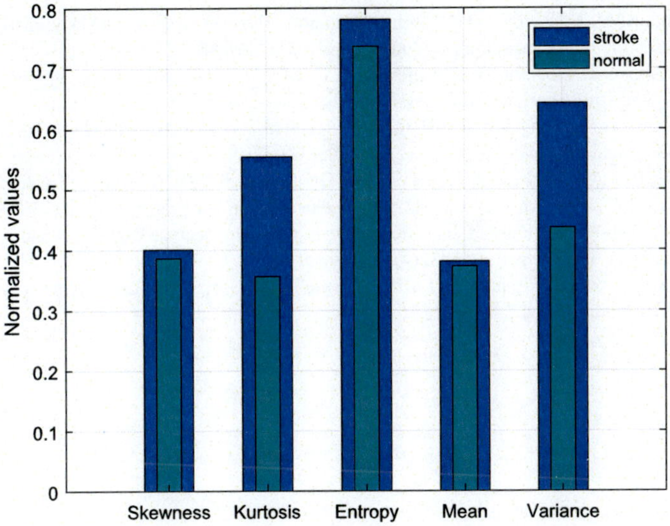

Fig. 5 Variations in normalized average values between normal and stroke patient

the results that AIS patients have high DAR, DTABR, BSI and statistical metrics. This higher value of the extracted features signifies severe cerebral infarction which helps in demarcating AIS from controlled subjects.

References

1. Kumar JS, Bhuvaneswari P (2012) Analysis of Electroencephalography (EEG) signals and its categorization–a study. Procedia Engi 38:2525–2536
2. Foreman B, Claassen J (2012) Quantitative EEG for the detection of brain ischemia. In: Annual Update in Intensive Care and Emergency Medicine 2012. Springer, Berlin, Heidelberg, pp. 746–758
3. Rahma ON, Wijaya SK, Prawito B (2016) Electroencephalogram analysis with wavelet transform and neural Network as a Tool for acute ischemic stroke identification. In: Proceedings IASTEM International Conference, Bali, Indonesia, Vol 9, January
4. Omar WRW, Jailani R, Taib MN, Isa RM, Sharif Z (2013) Assessment of acute ischemic stroke brainwave using Relative Power Ratio. In: 2013 IEEE 9th International Colloquium on Signal Processing and its Applications, pp. 310–313. IEEE
5. Tolonen U, Sulg IA (1981) Comparison of quantitative EEG parameters from four different analysis techniques in evaluation of relationships between EEG and CBF in brain infarction. Electroencephalogr Clin Neurophysiol 51(2):177–185
6. Finnigan S, Wong A, Read S (2016) Defining abnormal slow EEG activity in acute ischaemic stroke: Delta/alpha ratio as an optimal QEEG index. Clin Neurophysiol 127(2):1452–1459
7. Bentes C, Peralta AR, Viana P, Martins H, Morgado C, Casimiro C, Melo TP (2018) Quantitative EEG and functional outcome following acute ischemic stroke. Clin Neurophysiol 129(8):1680–1687
8. Finnigan S, van Putten MJ (2013) EEG in ischemic stroke: quantitative EEG can uniquely inform (sub-) acute prognoses and clinical management. Clin Neurophysiol 124(1):10–19

9. Van Putten MJ, Tavy DL (2004) Continuous quantitative EEG monitoring in hemispheric stroke patients using the brain symmetry index. Stroke 35(11):2489–2492
10. Agius Anastasi A, Falzon O, Camilleri K, Vella M, Muscat R (2017) Brain symmetry index in healthy and stroke patients for assessment and prognosis. Stroke Research and Treatment
11. Ji N, Ma L, Dong H, Zhang X (2019) EEG signals feature extraction based on DWT and EMD combined with approximate entropy. Brain Sci 9(8):201
12. Omerhodzic I, Avdakovic S, Nuhanovic A, Dizdarevic K (2013) Energy distribution of EEG signals: EEG signal wavelet-neural network classifier. arXiv preprint. arXiv:1307.7897
13. Subasi A (2007) EEG signal classification using wavelet feature extraction and a mixture of expert model. Expert Syst Appl 32(4):1084–1093
14. Vimala V, Ramar K, Ettappan M (2019) An intelligent sleep apnea classification system based on EEG signals. J Med Syst 43(2):36

A Preliminary Step to Realize Digital Filter for Electrocardiogram Signal Denoising

R. Chitra and E. Priya

Abstract The electrocardiogram (ECG) is one of the most commonly used technique in the assessment of cardiovascular disease. Cardiovascular disease is one of the major cause of death in many countries so it is important for interpretation of heart beat. The ECG signal during its acquistion gets commonly affected by baseline drift that hinders physician in the correct diagnosis of heart condition. In this work, ECG signal is preprocessed to alleviate noise. Digital filters such as Infinite Impulse Response (IIR) and Finite Impulse Response (FIR) filter are designed for the purpose of removal of artifacts from the signal. The IIR Butterworth low pass filter and least square linear phase low pass FIR filter are designed to reduce the baseline drift artifact in ECG signal. The preprocessing block is designed in MATLAB filter design and analysis tool. The corresponding model is realized in simulink environment. The performance measures of IIR and FIR filter are analyzed. The corresponding HDL code is generated and is targeted to a field programmable gate array board to observe the simulation results.

Keywords Cardiovascular disease · Electrocardiograph · Baseline drift · Infinite impulse response

1 Introduction

Biomedical signal processing helps in extracting the required signal in a proper way which helps the physician to diagnosis the condition of a human accurately. The necessity of extracting the required information from the measured ECG signal is to derive at a decision about the physiological condition of heart by the physician. The vital role of signal processing is to remove noise [1, 2]. Denoising is a process

R. Chitra (✉) · E. Priya
Department of Electronics and Communication Engineering, Sri Sai Ram Engineering College, West Tambaram, Chennai, Tamil Nadu 600044, India
e-mail: chitra.ece@sairam.edu.in

E. Priya
e-mail: priya.ece@sairam.edu.in

A. Sivasubramanian et al. (eds.), *Futuristic Communication and Network Technologies*, Lecture Notes in Electrical Engineering 792,
https://doi.org/10.1007/978-981-16-4625-6_83

to extract the original information from a noisy signal. ECG measures the activity of heart which is a continuous time varying signal. The ECG signal is acquired with help of electrodes, its frequency and voltage range lies between 0.05–100 Hz and 0.5–4 mV [3, 4].

ECG signal is contaminated by different noises such as power line interference, baseline drift, electrode contact noise, electromyography signal interference. Baseline drift may be due to breathing of the patient while measuring ECG. The frequency range of baseline wander signal is 0.5–0.6 Hz [5–7]. Many authors have worked on denoising the ECG signal by removing the baseline wander [8–11]. The authors have attempted denoising of ECG signal either in MATLAB environment or in FPGA including various optimization techniques. This work aims at incurring the advantages of both, denoising filter in MATLAB environment and FPGA implementation. The advantage includes complex design and processes large data.

Compared with analog filters, digital filters are more accurate [12]. Digital filters such as Infinite Impulse response (IIR) and Finite Impulse Response (FIR) filter process the input signal to extract the information required to analyze the condition of the heart [13]. IIR filter is unique in its feedback mechanism. The filter uses the current and past output whereas FIR filter uses no feedback so the system in inherently stable. Filter Design and Analysis Tool (FDATool) is one of the design tools in MATLAB environment to design and simulate digital filter such that the design specification are met [14]. FDATool helps in rapid design of the process using the specifications of digital filter. The performance measures Mean Absolute Error (MAE), Peak Signal to Noise Ratio (PSNR), Peak magnitude Root Mean Square (PRMS) and Fisher's ratio are measured to find the best of FIR and IIR filter. A model is realized from the FDATool which is simulated in MATLAB simulink environment. Compiling the filter design leads to a code in Hardware Descriptional Language (HDL). The generated code is programmed in to a Field Programmable Gate Array (FPGA) to analyze the performance of a filter [15].

In this work, the digital filter is designed in FDATool for the specifications that reduce the baseline wandering in ECG signal. The error measures of FIR and IIR filter are analyzed. A model is realized in a MATLAB simulink environment that generates a HDL code. Simulating the code in integrated software environment helps in providing an optimal design.

2 Methodology

The ECG signal is obtained from the standard MIT-BIH database. The signal is recorded in the laboratory from different people. The signal is digitized at 360 samples per second [16].

Digital filters possess linear characteristics which process input signal to produce an output signal of desired frequency range. Impulse response of FIR is finite whereas impulse of IIR is infinite [17]. FIR filter is designed by adding a delay element for input signal. The algorithm involves multiplying each sample with corresponding coefficients and thereby the results are accumulated. FIR filter is expressed as

$$y[n] = \sum_{k=0}^{N-1} h[k] \cdot x[n-k] \quad (1)$$

where $x[n]$ is the input, $y[n]$ is the output and $h[n]$ is the impulse response.The symmetry between the coefficients are not required unless to reduce the hardware complexity [18].

The algorithm in IIR filter involves, a recursive impulse response such that the signal is fedback to the filter.

It is expressed as

$$H(Z) = k\Pi_{K=1}^{n} \frac{1 + b_k z^{-1}}{1 + a_k z^{-1}} \Pi_{i=1}^{m} \frac{1 + d_{i1} z^{-1} + d_{i2}}{1 + c_{i1} z^{-1} + c_{i2}} \quad (2)$$

where k is the gain of the filter, a_k and b_k for $k = 1, 2, \ldots$ n are the first order coefficients and $c_{i1}, c_{i2}, c_{i3}, d_{i1}$ and d_{i2} for are the second order coefficients [19].

The FDATool is a user interface for designing and analyzing the digital filters in a faster way. It helps in analyzing the filters, such as magnitude response, phase response, pole zero plot, filter coefficients, filter information, magnitude response estimate and round off noise power spectrum [20]. The MAE, PSNR, PRMS and Fisher's ratio are computed and compared. MAE is the error measure and PSNR signifies the peak level of signal to the noise present in the signal. The peak magnitude to RMS ratio is the ratio of the largest absolute value to root mean square value. Fisher's ratio implicitly reveals correlation between two signals. Simulink model is realized from the design for the corresponding digital filter and simulated in the simulink environment.The Hardware Description Language (HDL) code i.e., Verilog code is generated using generate HDL option. The filter code is simulated using Isim simulator for device xc3s500e.The synthesis report summarize the synthesis option summary, HDL compilation, design hierarchy analysis, HDL analysis, HDL synthesis, partition report, device utilization of the design.

3 Results and Discussion

In this work, the 100.dat ECG signal obtained from MIT-BIH Arrhythmia database is combined with BW.dat from MIT- BIH noise stress database for analysis. The digital filter is designed to reduce the baseline drift from the contaminated ECG signal. The low pass FIR filter is designed in FDATool with the following specifications of order, cutoff frequency as 10 Hz and sampling frequency as 360 Hz. The filter order is chosen as 4 because as the order of the filter is increased it leads to instability. It

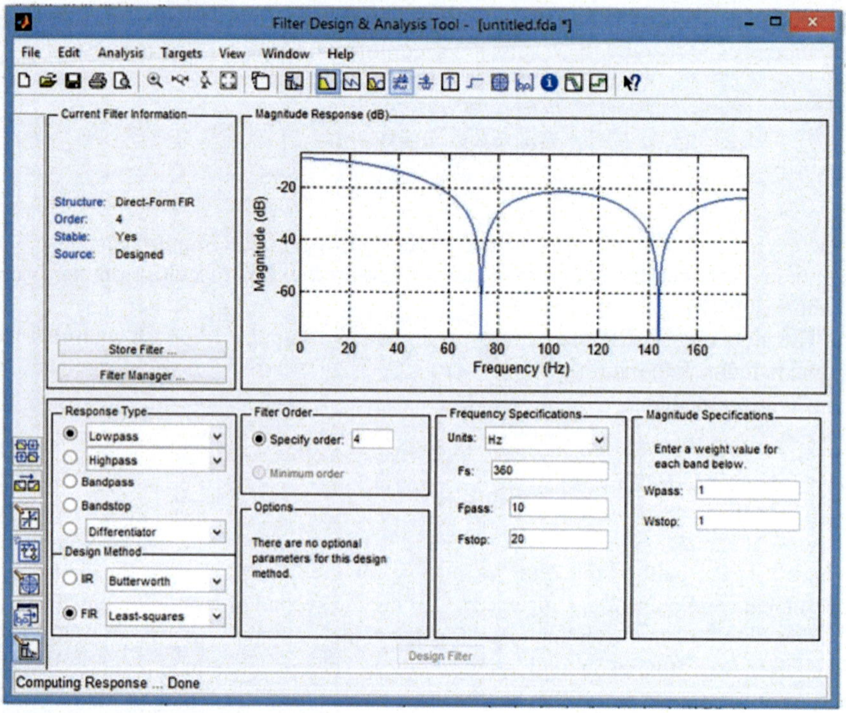

Fig. 1 A low pass FIR filter design in FDATool

is difficult to reconstruct the signal for lower order filters. The least square low pass FIR filter design is shown in Fig. 1.

The corresponding model of a low pass FIR filter is realized from FIR filter design by using realize model option. The realized model simulated in MATLAB simulink environment for the contaminated ECG signal obtained from database is presented in Fig. 2. The filtered signal received as output from the realized model is viewed with a scope. The corresponding input output signal of the filter is shown in Fig. 3.

In a same way IIR Butterworth filter is designed in FDATool with same specification as FIR. Corresponding low pass IIR filter design using FDATool is shown in Fig. 4.

The corresponding model of a low pass IIR filter is realized from the IIR filter design by using the realize model option. The realized model simulated in MATLAB simulink environment for the contaminated ECG signal obtained from database is presented in Fig. 5. The output of the realized model is viewed with a scope. The input output signal of the realized model is shown in Fig. 6.

It is observed from Table 1 that FIR filter has low MAE than IIR filter. The PSNR is high for FIR filter and this indicates the signal value being high after denoising when compared with IIR filter. Increase in peak magnitude to RMS ratio increase distortion in the signal but it is observed that the value of peak to RMS ratio is less

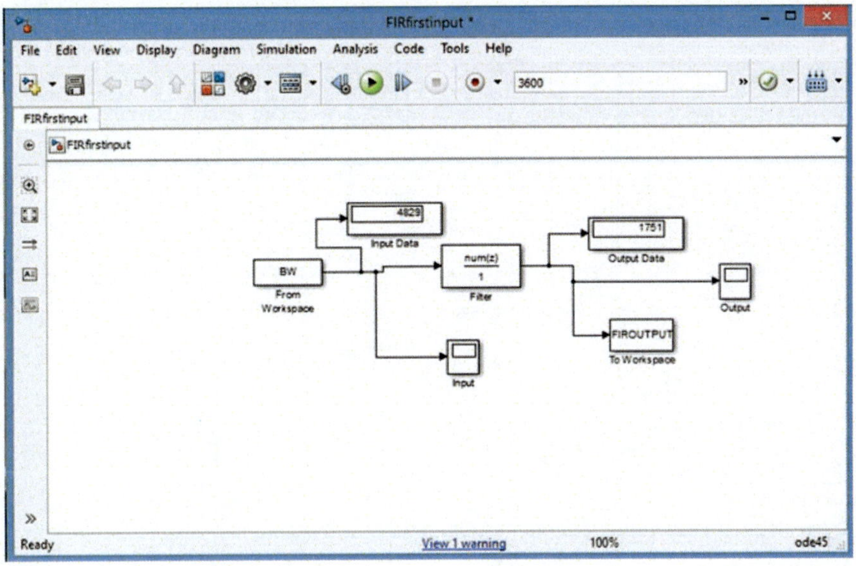

Fig. 2 A low pass FIR filter model

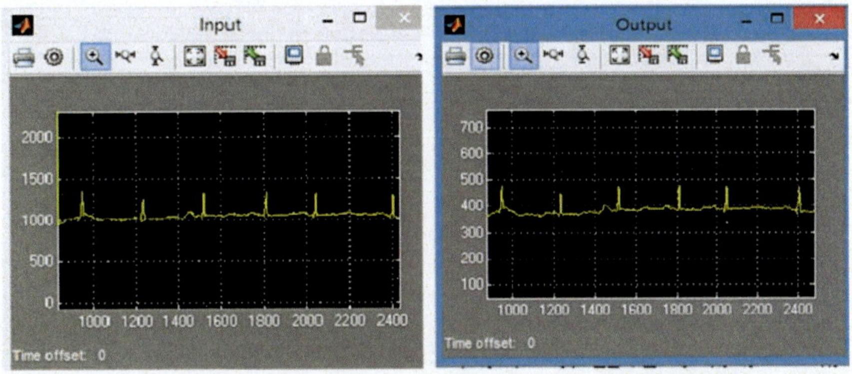

(a) Input (b) Output

Fig. 3 Input–output waveform captured in the scope for the realized FIR filter model

in FIR than IIR filter. Lower Fisher's value indicates FIR filter having reduced noise better than IIR filter.

The filter design HDL coder user interface generates the HDL code for the filter design and test bench.The filter design and testbench code is generated in VHDL. The 100.dat ECG signal obtained from the database is given as input stimulus of 3600 samples length. The compilation of generated code is done in ISE Design suite 14.5. The Isim graphical user interface provides the waveform of input output signals of FIR and IIR filter. The respective waveforms are presented in Fig. 7 and Fig. 8.

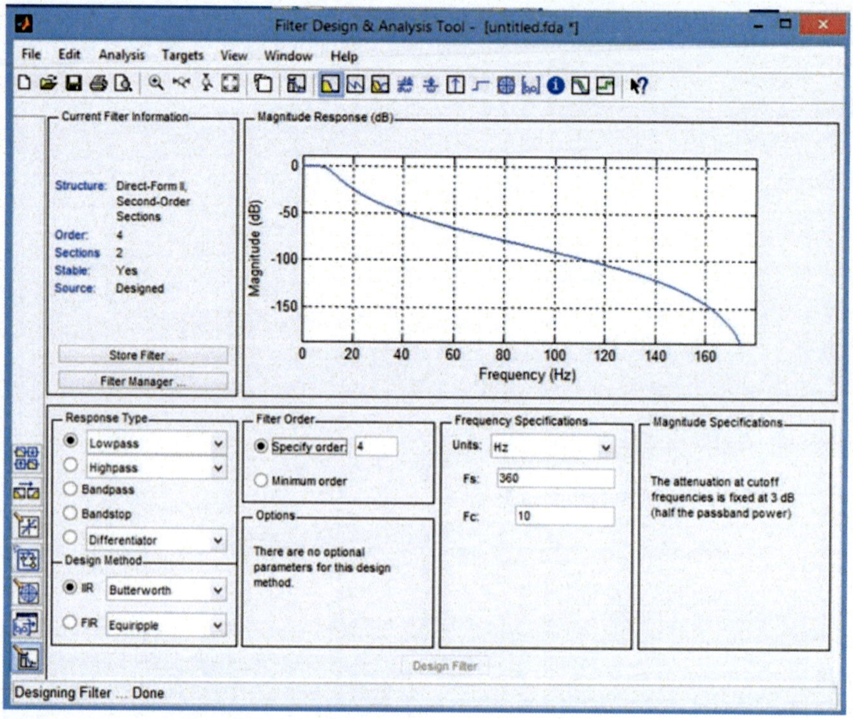

Fig. 4 A low pass IIR filter design in FDATool

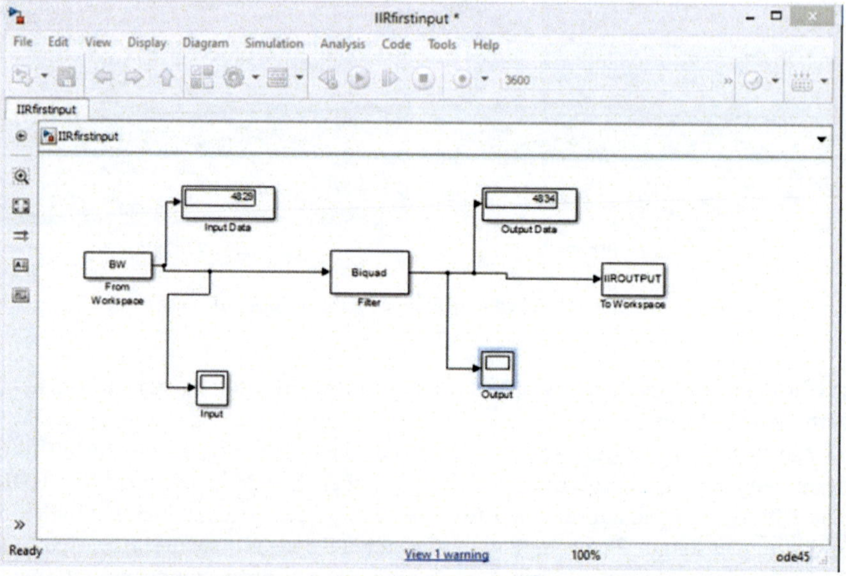

Fig. 5 A low pass IIR filter model

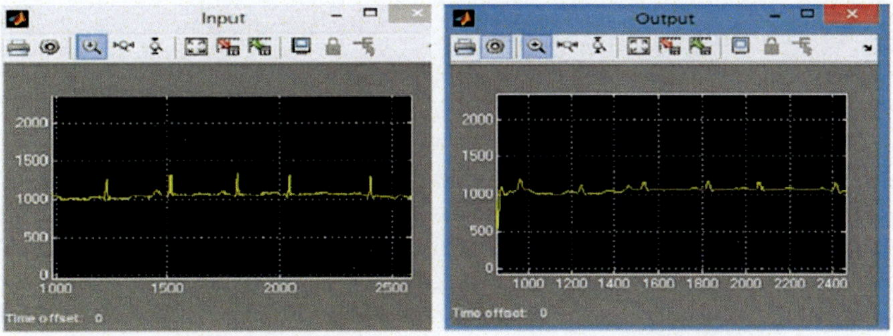

(a) Input (b) Output

Fig. 6 Input – Output waveform captured in the scope for the realized IIR filter model

Table 1 Performance measures of IIR and FIR filter

Filter	MAE	PSNR	Peak to RMS ratio	Fishers ratio
IIR	2893.21	3.7081	13.7933	2.1266
FIR	34.2609	25.3692	1.5241	9.3792×10^{-7}

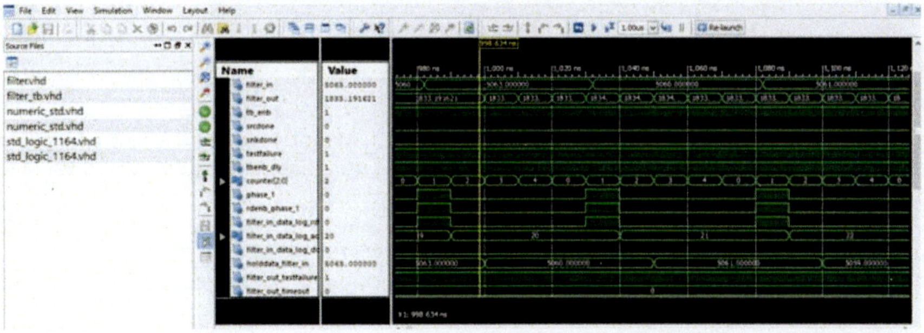

Fig. 7 A low pass FIR filter in ISim

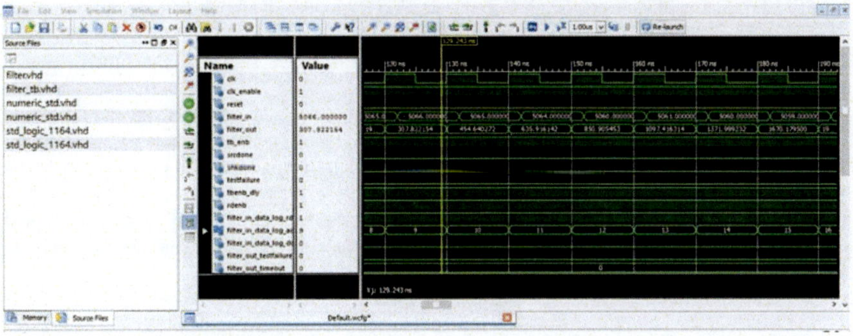

Fig. 8 A low pass IIR filter in ISim

4 Conclusion

ECG signal is the prime signal among all bioelectrical signals. It characterizes the electrical activity of the human heart. As ECG comprises of low frequency components with less amplitude, it is prone to different internal and external noises [21]. In this work, it is attempted to filter the baseline wander noise from the ECG signal obtained from the standard MIT-BIH database. The FIR with least squares method and IIR Butterworth low pass filter designed with appropriate specifications are designed to remove the baseline wander noise. The FIR and IIR filter is designed by FDATool. It is observed from the performance measures that FIR filter is better than IIR filter in removing the baseline wandering. The model for corresponding FIR and IIR filter are realized in the MATLAB simulink environment. The simulated waveform of IIR and FIR filter in Isim validates the output obtained from the realized model of MATLAB simulink environment. The simulation result helps in synthesis of design to implement in field programmable device. The FPGA device work in parallel manner when compared to microprocessor and digital signal processor and well suited for image and digital signal processing application.

References

1. Lastre-Dominguez C, Shmaliy YS, Ibarra-Manzano O, Munoz-Minjares J, Morales-Mendoza LJ (2019) ECG signal denoising and features extraction using unbiased FIR smoothing. BioMed Research International
2. Chen SW, Chen YH (2015) Hardware design and implementation of a wavelet de-noising procedure for medical signal preprocessing. Sensors 15(10):26396–26414
3. Joshi N, Jain P ECG based heart rate monitoring system implementation using FPGA for low power devices and applications. IJRTE
4. Kirti SH, Jain S (2019) FPGA implementation of Power-Efficient ECG pre-processing block. IJRTE 8(1):2899–2904
5. Joshi SL, Vatti RA, Tornekar RV (2013) A survey on ECG signal denoising techniques. In: 2013 International Conference on Communication Systems and Network Technologies, April, pp 60–64. IEEE
6. Sharma H, Sharma KK (2015) Baseline wander removal of ECG signals using Hilbert vibration decomposition. Electron Lett 51(6):447–449
7. Kher R (2019) Signal processing techniques for removing noise from ECG signals. J. Biomed. Eng. Res 3:1–9
8. Chandrakar C, Kowar MK (2012) Denoising ECG signals using adaptive filter algorithm. IJSCE 2(1):120–123
9. Belkacem S, Messaoudi N, Dibi Z (2018) Artifact removal from electrocardiogram signal: A comparative study. In: 2018 International Conference on Signal, Image, Vision and their Applications (SIVA) , November, pp 1–5. IEEE
10. Kabir MA, Shahnaz C (2012) Denoising of ECG signals based on noise reduction algorithms in EMD and wavelet domains. Biomed Signal Process Control 7(5):481–489
11. Blanco-Velasco M, Weng B, Barner KE (2008) ECG signal denoising and baseline wander correction based on the empirical mode decomposition. Comput Biol Med 38(1):1–13
12. Kadam G, Bhaskar PC (2012) Reduction of power line interference in ECG signal using FIR filter. Int J Comp Eng Res 2(2):314–319

13. Gholam-Hosseini H, Nazeran H, Reynolds KJ (1998) ECG noise cancellation using digital filters. In: Proceedings of the 2nd International Conference on Bioelectromagnetism (Cat. No. 98TH8269), February, pp 151–152. IEEE
14. Jiang X, Bao Y (2010) FIR filter design based on FPGA. In: 2010 International Conference on Computer Application and System Modeling (ICCASM 2010), Vol 13, pp V13–621, October. IEEE
15. Vázquez-Sedano A, Pérez-Suárez ST, Travieso-González CM, Hernández JBA. (2012). Cardiac pathologies detection over FPGA using electrocardiogram. In: BIOSIGNALS, February, pp 360–364
16. Egila MG, El-Moursy MA, El-Hennawy AE, El-Simary HA, Zaki A (2016) FPGA-based electrocardiography (ECG) signal analysis system using least-square linear phase finite impulse response (FIR) filter. J Electrical Syst Info Tech 3(3):513–526
17. Chandrakar B, Yadav OP, Chandra VK (2013) A survey of noise removal techniques for ECG signals. Int J Adv Res Comp Comm Eng 2(3):1354–1357
18. Meidani M, Mashoufi B (2016) Introducing new algorithms for realising an FIR filter with less hardware in order to eliminate power line interference from the ECG signal. IET Signal Proc 10(7):709–716
19. Singh N, Ayub S, Saini JP (2013) Design of digital IIR filter for noise reduction in ECG signal. In: 2013 5th International Conference and Computational Intelligence and Communication Networks, September, pp 171–176. IEEE
20. Li C (2010) Design and realization of FIR digital filters based on MATLAB. In: 2010 International Conference on Anti-Counterfeiting, Security and Identification, July, pp 101–104. IEEE
21. Khiter A, Adamou-Mitiche AB, Mitiche L (2020) Denoising Electrocardiogram Signal from Electromyogram Noise Using Adaptive Filter Combination. Revue d'Intelligence Artificielle 34(1):67–74

A Preprocessing Techniques for Seismocardiogram Signals in Removing Artifacts

V. A. Velvizhi and E. Priya

Abstract Seismocardiogram (SCG) is a non-invasive technique for cardiomechanical assessment by analyzing local vibrations on chest surface. SCG signals have numerous clinical and health awareness applications. The SCG signals utilized in this work are obtained from public domain database which are acquired by standard signal acquisition protocol. The acquired SCG signal includes artifacts such as base line wander, random noise, and predictive power line interference. The artifact-free signal helps significantly in analyzing them either in time scale or in frequency domain. In this work, an attempt has been made to remove motion-induced noises and the low frequency noise including respiratory sounds. The artifact removal methods include moving average and median filter and finite impulse response-based smoothing filter named Savitzky Golay filter. The performance of all these methods is compared using the denoising metrics such as mean square error, mean absolute error, signal to noise ratio, peak signal to noise ratio. Results demonstrate that median filter along with Butterworth filter performs better in removing the motion-induced artifact and low frequency and respiratory sounds. The methodology used in this work is helpful further for further annotation of the signals.

Keywords Seismocardiogram · Motion artifact · Smoothing filter · Median filter · Moving average filter · Savitzky Golay filter · Denoise metrics

1 Introduction

The cardiovascular system provides blood flow to all parts of human body. Any abnormalities in the functioning of the system may lead to mortality. The world

V. A. Velvizhi (✉) · E. Priya
Department of Electronics and Communication Engineering, Sri Sai Ram Engineering College, West Tambaram, Chennai, Tamil Nadu 600044, India
e-mail: velvizhi.ece@sairam.edu.in

E. Priya
e-mail: priya.ece@sairam.edu.in

A. Sivasubramanian et al. (eds.), *Futuristic Communication and Network Technologies*, Lecture Notes in Electrical Engineering 792,
https://doi.org/10.1007/978-981-16-4625-6_84

health organization reports that 17.7 million people lose their lives due to cardiovascular disease which is about 31% of the global deaths [1]. Early detection of heart diseases not only reduces the cost of treatment, it also simplifies the treatment process and aid in effective treatment. For early diagnosis, heart activity should be tracked in long-term and monitored daily to get important information. The device should also be light weight, long-lasting, ease of use and wearable. Such devices should be simple and reliable to provide instant information. Electrocardiography (ECG) and seismocardiography (SCG) are simple, non-invasive and inexpensive as well as provide much useful and reliable information about the heart [2].

Though ECG is considered to be the golden standard to access heart problems, ECG cannot show congenital heart defects including defects in heart valves. These structural abnormalities in the heart valve do not alter the electrical depolarization of myocardium and hence cannot be reflected in ECG. But these abnormalities are reflected in SCG. SCG is more sensitive to identify anatomical and physiological diseases. Seimocardiography is a non-invasive technique to assess the mechanical activity of the heart. The vibrations on chest surface are acquired from MEMS-based accelerometers. SCG signal provides reliable information that can be used for healthcare and clinical applications. During acquisition, the data are prone to power line or external interference. Hence, noises are common. Accurate measurements of fiducial points are very important to extract the features of the SCG signal. Special software is used to make decision on heart diseases.

Many authors have reported in their work to remove the noises and artifacts from the SCG signal. The literature reveals the usage of filters being used to remove the noises and artifacts from SCG signal. The methods are digital filters like FIR and IIR, wavelet-based denoising methods and empirical mode decomposition [3–7].

In this work, an attempt has been made to remove motion-induced noise present in the SCG using smoothing filters. The smoothing filters used are moving average, median and Savitzky Golay filter. They eliminate motion-induced artifact present in the SCG signal. The performance measures aid in identifying the best suitable denoising method. In addition, Butterworth filter is used in eliminating the low frequency respiratory sounds.

2 Methodology

2.1 SCG Signal Database

The SCG signals are obtained from CEBS publicly available database in PhysioNet archive. The signals are affected by motion-induced artifacts, trending effect, baseline drift, sudden variations in systolic and diastolic profiles and irregular rhythms. The instrument used to record the signal is BIOPAC MP36 acquisition system (Santa Barbara, CA, USA). The sampling rate is fixed as 5000 Hz. The signals are processed for 10 ms and are downsampled to 1000 Hz.

2.2 Denoising Filters

The literature reveal that adaptive filters also called as moving average filters eliminate motion-induced noise. This filter uses a sliding window to compute the average value of the data within the window. Moving average filter is defined by the difference equation:

$$y(n) = \left[\frac{1}{\text{Window Size}}\right][x(n) + x(n-1) + \ldots + x(n - (\text{Window Size}))] \quad (1)$$

where $y(n)$ is the present output data, $x(n)$ is the present input data, and $x(n-1)$ is the past input [8].

It helps to remove sporadic noise but not suitable to remove sudden spikes and high amplitude noise. Higher order filter may deteriorate the signal quality. Moving average FIR filter is apt to smooth the envelope of the SCG signal [9, 10].

Median filters are non-linear digital filters that are widely used to eliminate noise from any signal. The filter operates along the signal from sample to sample such that each sample is replaced by the median value of the neighboring sample values. Window corresponds to the set of neighbors and determines the frequency and fitting of a signal [11].

For any non-linear discrete dynamic FIR system, in which the linear combinations are replaced by non-linear function of adequate number of variables,

$$y(n) = F[x(n), x(n-1), \ldots, x(n-M)] \quad (2)$$

where M is an arbitrary positive integer, $y(n)$ and $x(n)$ are the present output and input sample, $x(n-1)$ and $x(n-M)$ are the previous input samples, etc [12].The median filter is derived from vector of samples,

$$x(n) = \{x(n), x(n-1), \ldots, x(n-M)\}^T \quad (3)$$

Savitzky Golay filter is a polynomial FIR filter that can retain moments even beyond third order around the transition points. This is not possible for a digital FIR filter [13]. The output data set of points is expressed as,

$$Y_j = \frac{1}{N}\sum_{i=0}^{N-1} C_i y_{j+1}; \left(\frac{m+1}{2}\right) \le j \le n - \left(\frac{m-1}{2}\right) \quad (4)$$

where $j = 1, 2, 3, \ldots, n$ and y is an observed value, C_i is the set of N convolution coefficients of the filter, and m is the window size.

In order to remove the low frequency respiratory sounds embedded in the SCG signal, IIR filters are attempted. IIR filters have non-linear phase. Also, filtering a signal by IIR introduces delay that leads to a shift in time in the output of the filter. This delay introduces distortion in the SCG signal. To overcome this delay, a zero-phase delay compensation FIR filter is used to remove the low frequency respiratory sounds [14]. The complete data sequence is obtained from the public database, and hence, it is non causal. This characteristic makes IIR filter a zero-phase filter where non-linear distortions are removed [15]. Hence, a fifth order Butterworth filter is used extract the desired frequency [16].

2.3 *Performance Analysis*

The metrics used to analyze the performance of various types of smoothing filters are mean squared error, mean absolute error, signal-to-noise ratio and peak signal-to-noise ratio [17]. Mean squared error is given by

$$\mathrm{MSE} = \frac{\sum_{n=1}^{N} [x(n) - x'(n)]^2}{N} \tag{5}$$

Accuracy of the filter is improved when the MSE is minimum. Accuracy of the denoising is evaluated using mean absolute error. Mean absolute error is defined as:

$$\mathrm{MAE} = \frac{\sum_{n=1}^{N} \left|[x(n) - x'(n)]\right|}{N} \tag{6}$$

MAE should be minimum to ensure the accuracy of denoising. Performance of denoising method is commonly examined from signal-to-noise ratio. SNR is inversely proportional to logarithmic of MSE and is given by:

$$\mathrm{SNR} = 10 \log_{10} \frac{\sum_{n=1}^{N} x(n)^2}{\sum_{n=1}^{N} [x(n) - x'(n)]^2} \tag{7}$$

Peak signal-to-noise ratio is defined as:

$$\mathrm{PSNR} = 20 \log_{10} \frac{\max[x(n)]}{\mathrm{RMSE}} \tag{8}$$

For accurate denoising, PSNR should be larger. Performance of the SCG signal can further be improved by using FIR and IIR filters.

3 Results and Discussion

A typical representation of the raw SCG signal is presented in Fig. 1 (a). Savitzky Golay, median and moving average filters are attempted to eliminate the motion artifacts. Figure 1b–d shows the filtered results of the respective filters. Qualitative analysis shows that Savitzky Golay filter smoothens the signal better than median and moving average filter. This is because the Savitzky Golay filter approximates a function to a polynomial of specified order. Best match can be obtained for a given function using this filter in terms of least square sense. This filter also preserves higher-order moments than other types of FIR filters.

The performance analysis of these filters over the SCG signal is presented in Fig. 2. It is observed from the bar graph that the median filter provides better SNR and PSNR than the other two filters.

The average values of the performance metrics are tabulated in Table 1. It is observed that the Savitzky Golay and moving average filter are inferior to median filter.

The reason behind the improvement in SNR is that the median filter is less sensitive to outliers as it discards the sample value that differ much from the other samples

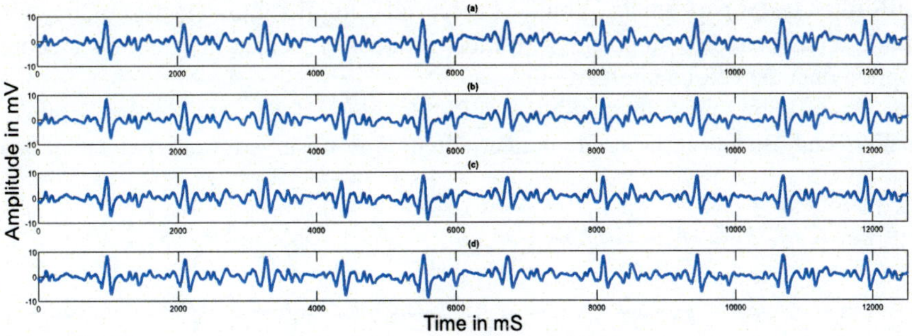

Fig. 1 SCG **a** raw signal **b** Savitzky Golay **c** median **d** moving average filter output

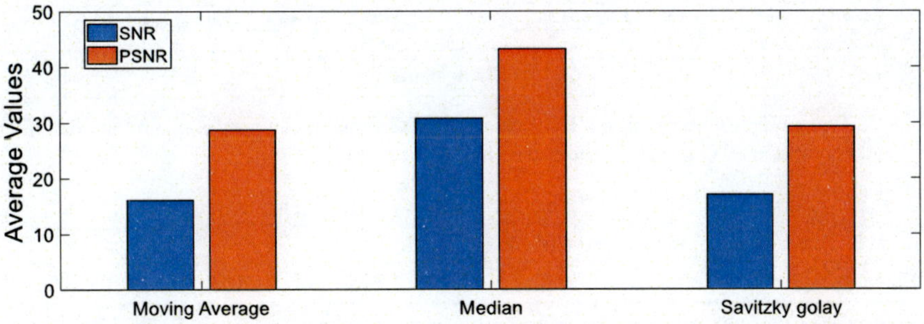

Fig. 2 SNR, PSNR of smoothing filters

Table 1 Average value of the performance metrics

Filter Type	MSE	MAE
Savitzky Golay	0.399272	0.326398
Moving Average	0.327960	0.300748
Median	0.013627	0.063677

present in the window. The spike-like samples are the erroneous component present in the raw signal. The moving average filter considers all sample valves within the window. Thus, the addition of noisy spikes degrades the performance of moving average filter.

A qualitative analysis of the median filter over the SCG signals is presented in Fig. 3. Typical representative of the SCG signal that is less and highly corrupted by noise is presented in Fig. 3a and c. Figure 3b and d shows the filtered output of the respective signals. It is observed from the figures that the noisy spike-like component is smoothened by the median filters and thus eliminates the motion artifacts present in the signal.

The SCG signal is infrasonic, and hence, noise (low frequency respiratory sounds) present in the signals is attempted to be removed by zero-phase delay compensation filter. This significantly improves the SNR and PSNR of the signal. Though it is a FIR filter, better performance could be achieved using IIR filter which is monotonic both in passband and stopband. A Butterworth-based median filter is thus a better option than the FIR counterpart.

Table 2 shows presents the SNR and PSNR values of zero-phase delay compensation and Butterworth-based median filter. It is observed from the table that

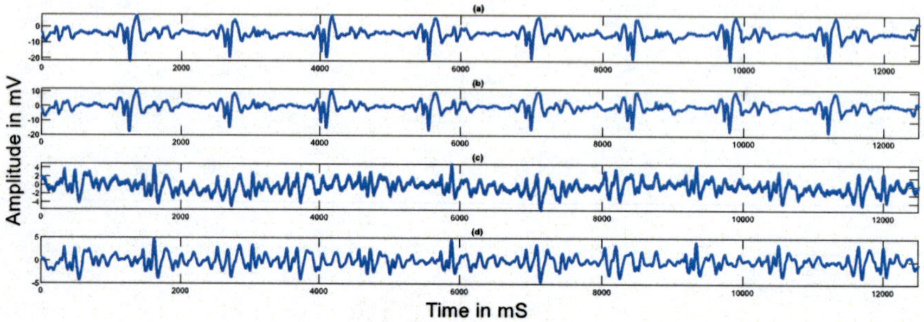

Fig. 3 Typical representative of a **a** less corrupted SCG signal **b** denoised signal by median filter **c** highly corrupted SCG signal **d** denoised signal by median filter

Table 2 Performance of the IIR and FIR filters

Filter type	SNR	PSNR
Zero-phase delay compensation-based median	46.934167	59.787270
Butterworth-based median	53.786589	66.740566

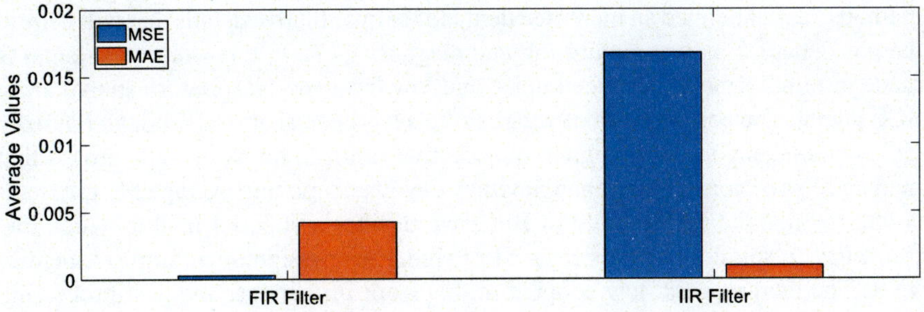

Fig. 4 Average MSE and MAE of FIR and IIR filter

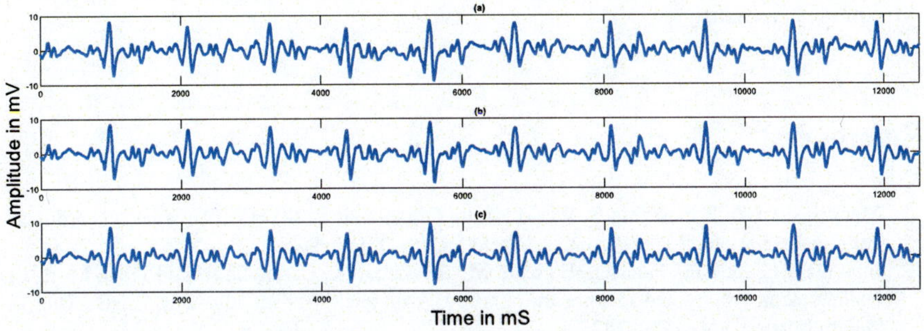

Fig. 5 **a** Median filtered SCG signal, **b** filtered output of zero-phase delay compensation median filter and **c** filtered output of Butterworth-based median filter

Butterworth-based median filter performs better. This is because zero-phase delay compensation filter is an FIR filter. But IIR filter has steep transition region and hence results in better performance than FIR filter.

Figure 4 shows the bar plot variation of MSE and MAE between the FIR and IIR filter. It is observed that MSE in the zero-phase delay compensated FIR filter is much larger than that of the Butterworth filter. Butterworth filter is able to remove most of the noise present in the signal as it is an infinite response filter.

Figure 5a represents median filtered SCG signal which is given as input to the zero-phase delay compensation filter and Butterworth filter. Both qualitative and quantitative analyses depict that Butterworth-based median filter performs better in removing the motion artifact and the low frequency respiratory noise from the SCG signal.

4 Conclusion

SCG measures cardiac-persuaded mechanical variations at the surface of the chest and is well below the human hearing. Recent advancement in signal processing methods has challenged an increased demand for quantitative details that will helpful the physician for patient monitoring and diagnosis [17]. In this work, an attempt is made to remove motion-induced noise and low frequency respiratory sounds from SCG signal. The performance of various filters has been analyzed. Though Savitzky Golay smoothens the SCG signal, quantitative result in terms of SNR proves that median filter performs better than Savitzky Golay and moving average filter. Also, it is inferred that the performance of Butterworth filter is superior in elimination low frequency‘ respiratory sounds from SCG signal with a significant improvement in SNR. The unique work flow adopted in this work further ease the post-processing stage in distinguishing the abnormal heart vibrations from the normal. This work involves denoising of SCG signals in time domain. As a future work, significant improvement can be achieved by analyzing the signal in frequency domain and time – frequency domain.

References

1. Mendis S, Puska P, Norrving B, World Health Organization (2011) Global atlas on cardiovascular disease prevention and control. World Health Organization
2. Inan OT, Migeotte PF, Park KS, Etemadi M, Tavakolian K, Casanella R, Di Rienzo M (2014) Ballistocardiography and seismocardiography: A review of recent advances. IEEE J Biomed Health Inform 19(4):1414–1427
3. Jane R, Laguna P, Thakor NV, Caminal P (1992) Adaptive baseline wander removal in the ECG: Comparative analysis with cubic spline technique. In: Proceedings of Computers in Cardiology, pp. 143–146. IEEE
4. Poornachandra S, Kumaravel N (2008) A novel method for the elimination of power line frequency in ECG signal using hyper shrinkage function. Digital Sig Process 18(2):116–126
5. Levkov C, Mihov G, Ivanov R, Daskalov I, Christov I, Dotsinsky I (2005) Removal of power-line interference from the ECG: a review of the subtraction procedure. Biomed Eng Online 4(1):50
6. Vaseghi SV (2013) Advanced signal processing and digital noise reduction. SpringerVerlag
7. Chang KM, Liu SH (2011) Gaussian noise filtering from ECG by Wiener filter and ensemble empirical mode decomposition. J Sig Process Sys 64(2):249–264
8. Oppenheim AV (1999) Discrete-time signal processing. Pearson Education India.
9. Lee JW, Lee GK (2005) Design of an adaptive filter with a dynamic structure for ECG signal processing. Int J Control Autom Syst 3(1):137–142
10. Choudhary T, Sharma LN, Bhuyan MK (2018) Heart sound extraction from sternal seismocardiographic signal. IEEE Signal Process Lett 25(4):482–486
11. Cao X, Li Y, Zhan Y, Chen X, Kang F, Wang J, Liang J (2016) Removing noises induced by gamma radiation in cerenkov luminescence imaging using a temporal median filter. BioMed Res Int
12. Miček J, Kapitulík J (2003) Median filter. J Info Control Management Syst 1:51–56
13 . Pandia K, Ravindran S, Cole R, Kovacs G, Giovangrandi L (2010) Motion artifact cancellation to obtain heart sounds from a single chest-worn accelerometer. In: 2010 IEEE International Conference on Acoustics, Speech and Signal Processing, pp590–593. IEEE
14. Castiglioni P, Meriggi P, Rizzo F, Vaini E, Faini A, Parati G, Di Rienzo M (2011). Cardiac sounds from a wearable device for sternal seismocardiography. In: 2011 annual international conference of the IEEE engineering in medicine and biology society, pp. 4283–4286. IEEE

15. Lee H, Lee H, Whang M (2018) An enhanced method to estimate heart rate from seismocardiography via ensemble averaging of body movements at six degrees of freedom. Sensors 18(1):238
16. Yang C, Tavassolian N (2018) An independent component analysis approach to motion noise cancelation of cardio-mechanical signals. IEEE Trans Biomed Eng 66(3):784–793
17. Sundar A, Pahwa V, Das C, Deshmukh M, Robinson N (2016) A comprehensive assessment of the performance of modern algorithms for enhancement of digital volume pulse signals. Int J Pharma Medicine Biological Sciences 5(1):91

Exploring Efficient Preprocessing Techniques for Breast Cancer Diagnosis

Y. K. Anupama, S. Amutha, and D. R. Ramesh Babu

Abstract Preprocessing is a significant data preparation step in data mining or machine learning techniques, to improve the performance of the classification models and to obtain better results. Exploring efficient preprocessing technique plays a vital role. In this study, principle component analysis (PCA) preprocessing technique compared with the proposed Preprocess_Integration(PI) preprocessing technique. The PCA is one of a preprocessing technique that reduces the dimension of the original data. The algorithm proposed in this paper integration of the PCA, correlation matrix, covariance matrix, and Chi-square test. Both the PCA and proposed methodology are applied on the breast cancer dataset. Results obtained with PCA are better than the PI technique.

Keywords Principal component analysis · Correlation matrix · Covariance matrix · Chi-square test

1 Introduction

Extracting useful relevant information and hidden knowledge from the dataset helps in developing an efficient algorithm to determine and analyze trends. This phenomenon refers to data mining. Real-world data are incomplete and can contain many errors. Data preprocessing technique transforms raw data into an understandable format. In machine learning process, data preprocessing is a very important step as it is responsible for converting the data into a format that can be understood by the machine learning (ML) algorithm. Sensors are used to collect the data for any ML application. However, there can be errors generated due to instrumental failures, human errors, noise, redundant data, and incorrect information being collected by these sensors. This erroneous information can influence the performance of the model.

Y. K. Anupama (✉) · S. Amutha · D. R. Ramesh Babu
Department of Computer Science and Engineering, Dayananda Sagar College of Engineering, VTU University, Bangalore, Karnataka, India

A. Sivasubramanian et al. (eds.), *Futuristic Communication and Network Technologies*, Lecture Notes in Electrical Engineering 792,
https://doi.org/10.1007/978-981-16-4625-6_85

Different preprocessing techniques can be used to remove these anomalies from the collected data. Principal component analysis (PCA) [1] is one such technique used to analyze the high-dimensional data. The main idea of PCA is to reduce the dimensionality of a highly correlated dataset and extract important information from the collected data. This is achieved by converting the largely correlated variables into a small set of variables, known as principal components.

Aim of the paper is to explore an efficient preprocessing technique for breast cancer data. So, the performance of the machine learning model can be improved. The rest of the paper is categorized as follows. Section 2 is about literature survey. Section 3 introduces related work. Section 4 explains about experimental results. Conclusion and future work is in Sect. 5.

2 Literature Survey

Shruti Sehgal et al. [1] proposed a method to separate two different objects having different combinations for four sensors and identified similar sensors or different sensors using the association between variables. The main objective is to consider the PCA for identifying significant related information and ignoring insignificant information. Asma Saleem et al. [2] proposed strategies to accommodate the values of mislaid variables and to identify an instance in a set of results that is very much vast or minor than the nearby existing instances. 2-Sigma (2S) and box-plot (BP) techniques are used to find exceptional instances. The weighted average strategy is used to accommodate the values for mislaid variables. As a future scope, they suggest to extend this method to general applications.

Trupti A. Aneyrao and R. A. Fadnavis [3] explored a strategy to eliminate bias and anticipation from data, mainly focused on direct bias elimination from the dataset. Lin Rui et al. [4] explored an ensemble strategy for data preprocessing of the digital information. The ensemble data preprocessing technique is a fusion of PCA and WA (weighted average) algorithm. Accordingly, 73% prediction accuracy was obtained by applying PCA.

Raksha Upadhyay et al. [5] proposed a methodology to reduce the high dimensionality of databases. In this paper, PCA with covariance matrix is used to check the correlated input dimensions. This method is used for analyzing the PC in the simulated RSS pattern. R. R. Janghel et al. [6] proposed a novel methodology for the prediction of breast cancer. PCA is used to reduce the features of the dataset, and the results are predicted using different neural network architectures. Best results are obtained for the SANE (symbiotic, adaptive neuro-evolution) strategy after applying PCA to the dataset.

Duygu Kaya and MustafaTurk 7] proposed a methodology for breast cancer diagnosis. PCA-support vector machine (SVM) model classifier implemented in parallel using graphical language programming LABVIEW. Muduli et al. [8] explored a methodology to classify breast mass of a mammogram image as normal or abnormal. Moth flame-efficient optimization technique with EMI is used for classification.

Table 1 Real-valued features of each cell nucleus

Radius	Mean of distances from center to points on the perimeter
Texture	Standard deviation of gray scale values
Perimeter	Length of the continuous line forming the boundary of the nucleus
Area	Measurement of the surface of nucleus
Smoothness	Local variation in radius length
Compactness	Perimeter^2 /area -1.0
Concavitiy	Severity of concave portions of the contour
Concave Points	Number of concave portions of the contour
Symmetry	The orientation of nucleus symmetric axis present in the nucleus
Fractal Dimension	Coastline approximation -1

Classification accuracy results are obtained as 97% and 98% for datasets MIAS and DDSM, respectively.

3 Related Work

3.1 Dataset

The Wisconsin Diagnostic Breast Cancer dataset (WDBC) [9] is used. A total of 569 instances are present in the dataset out of which 357 are Benign and 212 are malignant.Ten real-valued features are computed for each cell nucleus from a digitized image of a fine needle aspirate (FNA) of a breast mass as shown in Table 1 are used as predictors. Based on these feature values, target variable diagnosis is classified as 'B'(Benign) or 'M'(Malignant). The dataset is in.csv(comma separated values) text file format.

3.2 Existing Methodology

Principal component analysis is a dimensionality reduction preprocessing technique. It analyzes a large amount of data and tries to convert its high dimensionality to low dimensionality so that it becomes easier to understand. It converts the correlated data into uncorrelated variables and then stores the most important information as principal components and is as shown in Fig. 1.

The PCA algorithm works as follows:

Step1: Creating/collecting a dataset and deciding the predictor and the response variables.

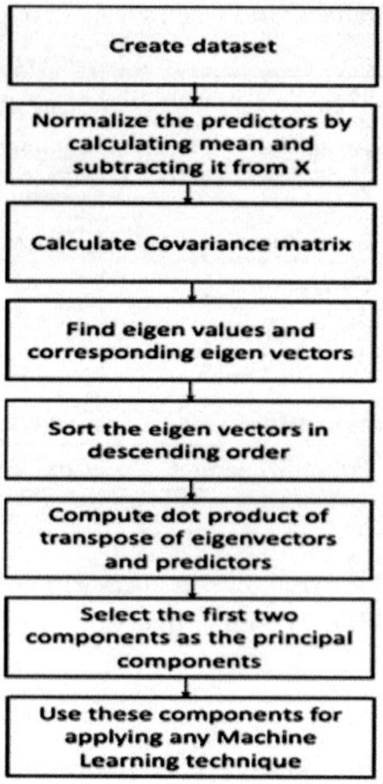

Fig. 1 Principal component analysis

Step2: Normalizing the predictors by calculating the mean of the data and subtracting it from itself.

$$X = X - \text{mean}(X) \tag{1}$$

Step3: Calculating the covariance matrix for the data to determine the covariance present in the dataset.

Step4: Finding eigen values and the associated eigen vectors of the matrix returned from step 3.

Step5: Sorting the eigen vectors in descending order to obtain important information present in the initial components.

Step6: Determining the score matrix by calculating the dot product of eigenvectors and predictors obtained from step 2.

Step7: Selecting the first two components, known as the principal components and further apply any machine learning algorithm required.

In this way, the dimensionality of the dataset is reduced from n to 2, where n is the number of features present in the dataset. It is a very useful technique as it not only decreases the dimensionality of the dataset but also tends to improve the accuracy of the predictions made.

Proposed Methodology

Proposed methodology preprocess integration algorithm has as two paths.

First path consists of following steps:

Step1: Creating/collecting a dataset and deciding the predictor and the response variables.

Step2: Computing contingency table(X').

$$(X')^2 = \Sigma(m - N)^2/N \tag{2}$$

where, m = observed value and N = expected value

Step3: Normalizing the data present in the contingency table

Step4: Determining the covariance matrix and different types of correlation matrix (Spearman, Kendall, Pearson)

Step5: Find eigenvalues and the associated eigenvectors of the matrix returned from steps 4 and 6.

Step6: Sort the eigenvectors in descending order to obtain important information present in the initial components

Step7: Determining the score matrix by calculating the dot product of eigenvectors and predictors obtained from step 2 and step 5

Step8: Selecting the first two components, known as the principal components and further apply any machine learning algorithm required.

Second path consists of following steps:

Step1: Creating/collecting a dataset and deciding the predictor and the response variables

Step2: Normalizing the original dataset consisting of predictors

Step3: Calculate different types of correlation matrix (Spearman, Kendall, Pearson)

Step4: Find eigenvalues and the associated eigenvectors of the matrix returned from steps 4 and 6.

Step5: Sort the eigenvectors in descending order to obtain important information present in the initial components

Step6: Determining the score matrix by calculating the dot product of eigenvectors and predictors obtained from step 2 and step 5

Step7: Selecting the first two components, known as the principal components and further apply any machine learning algorithm required.

Proposed methodology block diagram is as shown in Fig. 2

4 Experiment Results

Experiment conducted using different types of correlation matrix. Logistic regression is used as a classifier and the performance compared using accuracy and confusion matrix.

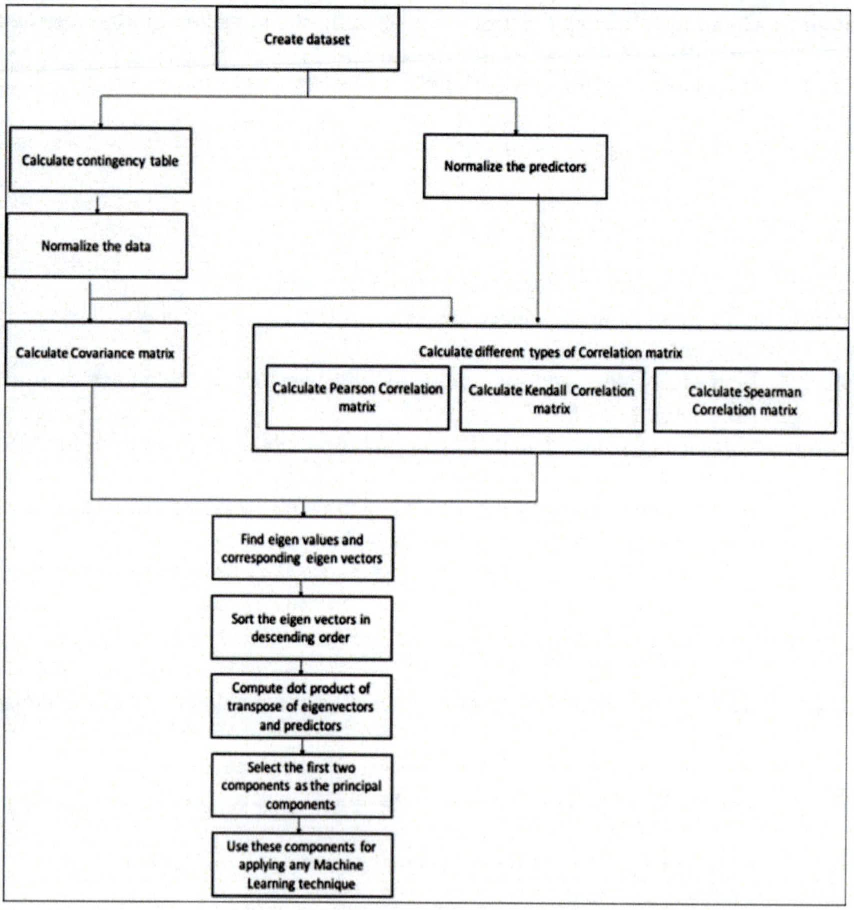

Fig. 2 Proposed methodology: Preprocess_Integration algorithm

4.1 Types of Correlation Matrix

The correlation matrix [10] is a matrix where a cell depicts the correlation between two variables. Three types of correlation matrix are considered:

1. Kendall correlation matrix [10]: Determines strength of dependence between two variables denoted by σ.

$$\sigma = \frac{A - B}{\frac{1}{2}M(M-1)} \tag{3}$$

where $A=$ Number of concordant variables, $B=$ Number of discordant variables and M = Number of observations.

2. Pearson correlation matrix [10]: Determines the relationship between linearly related variables.

$$R_{ab} = \frac{N\sum a_i b_i - \sum a_i \sum b_i}{\sqrt{N\sum a_i^2 - \left(\sum a_i\right)^2}\sqrt{N\sum b_i^2 - \left(\sum b_i\right)^2}} \tag{4}$$

where, R_{ab} = Pearson R correlation coefficient between a and b, N = number of observations, a_i = value of a (for ith observation) and b_i = value of b (for ith observation).

3. Spearman correlation matrix [10]: Determines the degree of a monotonic relationship between variables.

$$\beta = 1 - \frac{6\sum x_i^2}{N\left(N^2 - 1\right)} \tag{5}$$

where, β = Spearman rank correlation, x_i = the difference between the ranks of corresponding variables and N = number of observations.

Chi-square test [11] is used to check whether there is independence between the categorical variables. Contingency table used to check whether a variable depends on the other variable or not.

4.2 Machine Learning Technique

The WDBC dataset been split into 25:75 as the test data and training data, respectively. Logistic regression is used as a linear classification machine learning technique to classify the cancer into benign or malignant.

The sigmoid function used by logistic regression [12] is:

$$\mathbf{s(m)} = \frac{1}{1 + \mathbf{e}^{-\mathbf{m}}} \tag{6}$$

4.2.1 Accuracy and Confusion Matrix

Accuracy is the result obtained by dividing the total number of correct predictions made by the model and total number of predictions made my the model. It can also be defined as the result obtained by dividing the sum of true negatives and true positives by sum of all the predictions made.

$$\mathbf{Accuracy} = \frac{\mathbf{A + C}}{\mathbf{A + B + C + D}} \tag{7}$$

Table 2 Confusion matrix

	Predicted as A	Predicted as B
Observed as A	A	D
Observed as B	B	C

A = True positives, **B** = False positives, **C** = True negatives and **D** = False negatives.

Confusion matrix represents the summary of all the predictions made by the model is as in Table 2.

Visualization of Results

Categorical data, i.e., breast cancer dataset which determines whether the cancer is malignant or benign based on various predictors. The accuracy and confusion matrix are obtained as in Table 3 and results are visualized for accuracy vs different cases as in Fig. 3.

5 Conclusion and Future Work

In the proposed methodology, PCA is used in combination with Chi-square, Kendall, Pearson, and Spearman correlation matrix. Obtained accuracy of 91%. Using PCA as a data preprocessing technique for breast cancer dataset obtained accuracy of 94% higher than the other methods. So, PCA is a better preprocessing technique with respect to other techniques comparatively.Work can be extended to experiment PCA in combination with different discriminant analysis techniques.

Table 3 Accuracy and confusion matrix

Categories	PCA which uses covariance matrix	Kendall correlation matrix	Spearman correlation matrix	Pearson correlation matrix	Chi-square test + Kendall correlation matrix	Chi-square test + Spearman correlation matrix	Chi-square test + Pearson correlation matrix	Chi-square test + covariance matrix
Accuracy (%)	94.4	93.0	93.0	93.7	91.6	91.6	91.6	91.6
Confusion matrix	[87 3] [5 48]	[83 7] [3 50]	[83 7] [3 50]	[83 7] [2 51]	[86 4] [8 45]	[86 4] [8 45]	[86 4] [8 45]	[86 4] [8 45]

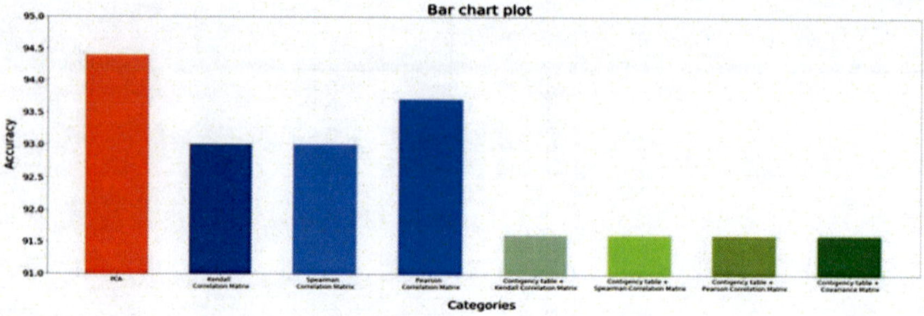

Fig. 3 Accuracy versus categories (cases)

Acknowledgements We wanted to thank Nishtha Goel, Engineering student studying in CSE department of Dayananda Sagar college of Engineering at Bangalore, has contributed in publishing this paper.

References

1. Sehgal S, Singh H, Mohit Agarwal V, Bhasker S, Data analysis using principal component analysis. In: 2014 International Conference on Medical Imaging, m-Health and Emerging Communication Systems (MedCom), IEEE, pp 45–48
2. Aleem AS, Ali A, Asif KH, Awan SM, Pre-processing methods of data mining. In: 2014 IEEE/ACM 7th International Conference on Utility and Cloud Computing, IEEE, pp 451–456
3. Aneyrao TA, Fadnavis RA, Analysis for data preprocessing to prevent direct discrimination in data mining. In: 2016 World Conference on Futuristic Trends in Research and Innovation for Social Welfare (WCFTR'16)
4. Rui L, Jiamei W, Bingze L, Changsen Y, Digital ınformation mining technology based on principal component analysis. In: 2017 IEEE 3rd International Conference on Control Science and Systems Engineering, IEEE
5. Upadhyay R, Panse P, Soni A, Rathore Bhatt U (2019) Principal component analysis as a dimensionality reduction and data preprocessing technique. In: Proceedings of Recent Advances in Interdisciplinary Trends in Engineering & Applications (RAITEA)
6. Janghel RR, Kala R, Shukla A (2012) Breast cancer data prediction by dimensionality reduction using PCA and adaptive neuro evolution. Int J Info Syst Social Change, 1–9. https://doi.org/10.4018/jissc.2012010101
7. Kaya D, Turk M (2020) Labvıew based robust cascade predictive model for evaluating cancer prognosis. In: Physica A: Statistical Mechanics and Its Applications
8. Muduli D, Dash R, Majhi B (2020) Automated breast cancer detection in digital mammograms: A moth flame optimization based ELM approach. In: Biomedical Signal Processing and Control, 59
9. Lichman M (2013) UCI machine learning repository [Online]. Available from: https://archive.ics.uci.edu/ml/datasets/Breast+Cancer+Wisconsin
10. https://easystats.github.io/correlation/articles/types.html
11. https://www.statisticshowto.com/probability-and-statistics/chi-square/
12. https://developers.google.com/machine-learning/crash-course/logistic-regression/model-training

Smart Logistics for Quality Food During Pandemic Crisis

T. Aghil, B. Nagajayanthi, S. Buvan Kumaar, and G. Sanjay

Abstract The conventional method of checking the safety and freshness of goods (fruits and vegetables) is done manually by a person during the loading and unloading process. In this real-time monitoring system, the goods are monitored using sensors. Temperature, photoresistor and ultrasonic sensors are used to gather information about the freshness of the product. The ESP8266 module in the model sends data given by the sensors to the Arduino board and then sends that to a phone/computer for remote monitoring in transit. This system prevents the goods from being damaged or getting rotten during transportation. There were several instances in which stale food was delivered to the restaurants and several people got infected. This system could be monitored by the health department to avoid decayed food consumption by the consumers.

Keywords Smart Logistics · Internet of Things · NodeMCU · Light dependent resistor · If This Then That Protocol (IFTTT)

1 Introduction

Our country's diverse climatic conditions have ensured the availability of all kinds of fresh fruits and vegetables. While China stands first, India stands second in the world in the production of fruits and vegetables. It has been estimated that around ninety million metric tons and one seventy million metric tons of fruits and vegetables had

---

T. Aghil · S. B. Kumaar · G. Sanjay
Vellore Institute of Technology, Chennai, India
e-mail: t.aghil2018@vitstudent.ac.in

S. B. Kumaar
e-mail: sbuvan.kumaar2018@vitstudent.ac.in

G. Sanjay
e-mail: g.sanjay2018@vitstudent.ac.in

B. Nagajayanthi (✉)
Associate Professor, Vellore Institute of Technology, Chennai, India
e-mail: nagajayanthi.b@vit.ac.in

A. Sivasubramanian et al. (eds.), *Futuristic Communication and Network Technologies*, Lecture Notes in Electrical Engineering 792,
https://doi.org/10.1007/978-981-16-4625-6_86

been produced by India, respectively, in the year 2015. In India, around six million hectares of land area is used for the cultivation fruits and ten million hectares of land area is used for the cultivation of vegetables [1].

India ranks first in the production of okra and ginger, and it is the second-largest producer of brinjal, cabbage, cauliflower and onion among vegetables. Among fruits, India stands second in the world in terms of producing papayas, mangoes and bananas.

Smart Logistics System has been designed in such a way that the sensing of the goods' condition is done continuously, and the data are simultaneously sent to the cloud server from where the owner will be able to access the data anytime for quality check. We have made use of different sensors like DHT11 sensor to measure the temperature, LDR (Light-Dependent Resistor) sensor to measure the intensity of light and ultrasonic sensor to measure the distance in our proposed system. We have discussed about the system's architecture in Sect. 2, its implementation in Sect. 3, Sect. 4 covers the obtained results, and the paper is concluded in Sect. 5 [2, 3].

2 Smart Logistics Architecture

Figure 1 depicts the inputs from three sensors comprising of ultrasonic sensor, DHT11 temperature sensor and LDR sensors. The gathered entities from the sensors are sent to the cloud through ESP8266 (NodeMCU Wi-Fi module), and then if the values exceed the given threshold limit, then it activates the buzzer to alert the driver. This is simultaneously sent through an IFTTT (If This Then That) protocol to send an alert message/mail to smartphone/computer which is monitored by the sender.

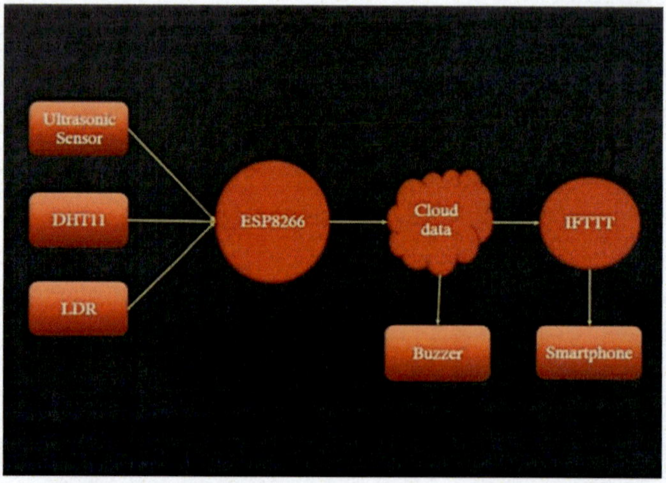

Fig. 1 Architecture of the proposed system

3 Implementation

The model takes input from three sensors, temperature sensor to measure the surrounding temperature, ultrasonic sensor to measure the distance and LDR to measure the intensity of light. All the three sensors take input in analog format first, and then the analog to digital converter formats it into digital signal and this is used to transmit to cloud through NodeMCU. This project was tested using Tinkercad and Thingspeak. The same project is simultaneously tested using hardware equipment in real time.

As shown in Fig. 2, the Arduino board gets the input from the sensors, analyzes the data and sends it to ESP8266 which sends the data to the cloud. This is achieved through coding which is done with Arduino IDE in C language.

For cloud data, we are using Thingspeak, which is an open-source platform where data are viewed by authenticated users in the form of graphs and chart.

The data, which are received, are processed, transmitted and shown in real time as a graphics tool in Thingspeak. Arduino Uno board is based on real-time emulation and also supports embedded fast memory. It is easy to use in places where size matters for access control because of its low-power consumption and small size. It is best used for protocol convertors and gateways in communication and hence gives large processing power and high buffer size [4].

ESP 8266 is an economical Wi-Fi module that has a Wi-Fi chip which includes TCP (Transmission Control Protocol)/IP(Internet Protocol) protocols. It starts directly from external flash when it hosts the applications. Due to its integrated cache, the system performance increases, and the memory requirement decreases. When the Wi-Fi module acts as a Wi-Fi adapter, wireless Internet access can be achieved for any microcontroller-based design using the UART interface [5].

The Wi-Fi module makes use of serial pins for communication of data. For establishing communication between a Wi-Fi module and a microcontroller, it needs two wires. It sends Wi-Fi tasks to the board therefore allowing Arduino board code to be quite simple. ESP8266 is called through SPI (Serial Peripheral Interface) and UART (Universal Asynchronous Receiver/Transmitter) protocols and hence makes

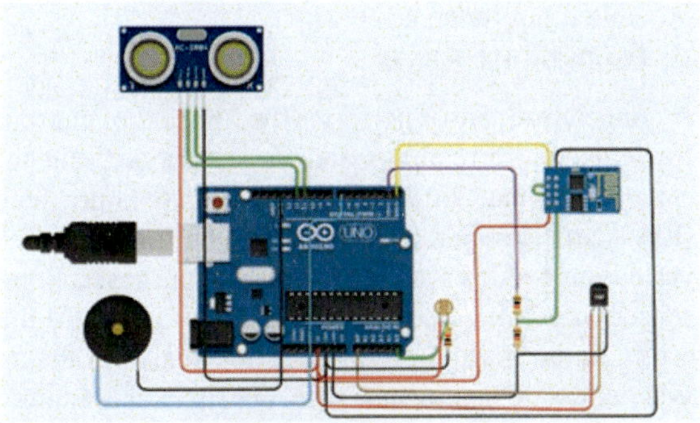

Fig. 2 Software implementation using Tinkercad and ThingSpeak

Fig. 3 Hardware implementation of smart logistics

it effortless to build an IoT application. We do not need to have TCP/IP running in our own board because we can use AT (Attention) commands to connect to networks. We can send data to the Internet, i.e., to the central server by simply connecting the board to ESP8266 directly [6].

Figure 3 shows the hardware implementation which was performed with the given circuit diagram when plugged into a power source. Data transfer will begin and data will be logged in real time. Data alerts will be sent as per the parameters monitored using the appropriate sensors as described below.

3.1 Sensors

A sensor is a semiconductor device that performs the sensing of changes in its surroundings continuously and the data are simultaneously sent to the computer in the form of electrical signals. This decreases the error in reading data, minimizes the workload and reduces the time taken to read the data. Depending on its functions, it is used in various embedded system devices. In fact, IoT would not have been possible if not for sensors [7].

Temperature sensor

A temperature sensor performs the function of continuously sensing the surrounding for temperature readings and these data are sent to the computer in the form of electrical signal. We have different temperature sensors like thermocouples, RTD (Resistance Temperature Detector), DHT11, etc. Temperature sensors are used in various embedded systems for various purposes. Some key areas where temperature sensors are being used are in air conditioners to maintain the temperature of the room, in the transportation of fish to maintain the appropriate temperature. In our system, we are making use of DHT11 sensor to monitor the surrounding temperature of fruits and vegetables. Since it ranges from (0–50) °C, this can be used to monitor the temperature for all varieties of fruits and vegetables.

Ultrasonic Sensor

An ultrasonic sensor is used to measure the distance. It works on principle of transmitting and receiving sound waves. First, sound waves are transmitted from the sensor. Once the waves hit on any object, it is reflected back to the sensor. Depending on the total time taken by the sensor to transmit and receive the sound waves, the distance is calculated using the distance formula. It also comes with few drawbacks. If there is no object in the direction of wave emission, then there will be any reflected wave which results in infinite distance. Also, if the object does not have an even surface, then the emitted waves will be reflected in a different direction, which again results in loss of reflected wave and infinite distance. In our proposed system, ultrasonic sensor is used to keep a check on the fruits and vegetables from being displaced. In case, while being transported in an uneven road, there is a possibility of the goods to fall down and if left like that it'll be damaged. In case of theft, this procedure comes handy. So, to make sure nothing happens to the goods in transit, ultrasonic sensor is used to monitor the goods.

LDR Sensor

An LDR is a photoresistor that is used in an embedded system device to measure the intensity of light. Photoconductivity is the principle that is followed by LDR. Smart streetlight is a device that works using an LDR. Each variety of fruits and vegetables requires a particular lighting condition to keep it fresh, if exposed to excessive light, it may get spoiled in shorter span of time. In our proposed system, we have made use of LDR sensor to ensure sufficient lighting conditions for respective fruits and vegetables. This ensures fresh transportation of fruits and vegetables.

3.2 Working on Thingspeak

- Sign in to your account on Thingspeak.
- Create a new channel with required parameters.
- Give the field names: Temp, Light intensity, Distance changed.
- Click on the Save Channel option and the API key will be generated.
- Copy the read and write API keys in the body of the program.
- Modify the program so that it writes to separate fields.
- Select charts to create multiple charts with the received data.
- Data which are sent and received both happens in real time.

3.3 Working with IFTTT

- Login to the IFTTT (If this then that) account using google/mail.
- Create a wen hooks Applet which will be available in applet window.
- Give an input option.
- Select webhooks services to get Http Request.
- Select to send notification and then create the action.

3.4 Quality of the food

Quality of the food is ensured from factors like heat, light and oxygen that can cause food to taste bad and become malnutritional. The data that we collect display the temperature and the light intensity in which the food is being transported which will be helpful in ensuring the quality of the food. This is sent as an alert message if it exceeds the threshold value [8].

3.5 Security of the system

The first way of enhancing security of data is by providing an authentication to it no matter even if a user wants to update a little information.

The next step will be encrypting the data which are being sent or received as it must be protected during its transition from the device across the cloud. It is best to use industry-standard, peer-reviewed cryptographic functions for encrypting the data.

A secure update mechanism is provided as updating the system is crucial and it gives new firmware to the system as most of the times data attacks happen to system while updating as it becomes vulnerable, secure mechanism is provided.

To enhance security in cloud, AWS cloud can be used, and by using IOT core in AWS, we can use MQTT Publish Subscribe model.

As open-source software's are used, chance of vulnerability attacks is more compared to licensed software. So, an open-source update plan is generated, and an individual is assigned to watch for published vulnerabilities, and when it arises, software is updated, integrity of the data is checked to reduce the user's data risk [9, 10]

4 Results

Approved users can log in to Thingspeak to see the results by entering the username and password as shown in Figs. 4 and 5.The results are displayed for DHT11 sensor,

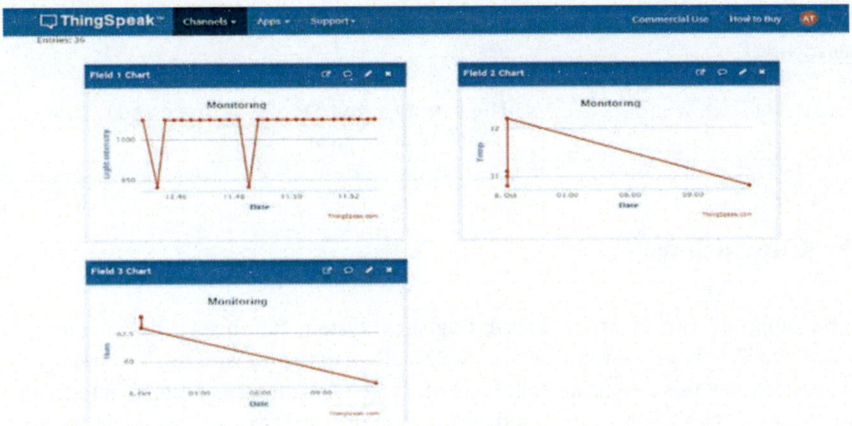

Fig. 4 Data logged in Thingspeak

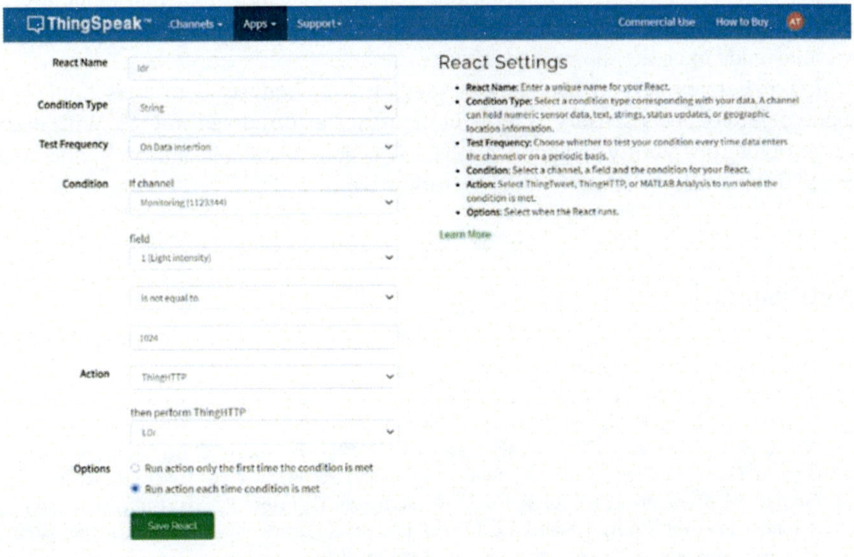

Fig. 5 Creating react for IFTTT in Thingspeak

ultrasonic sensor and LDR sensor, respectively, in real time. While using IFTTT, we can see the received notification through the smartphone as shown in Fig. 6.

From Figs. 4 and 5, we can see that whenever there is temperature drop or change to the given maximum temperature which was set, a notification/Alert message is sent to the desired user as shown in Fig. 6 which will help the user to find the error in real time and fix it at that moment.

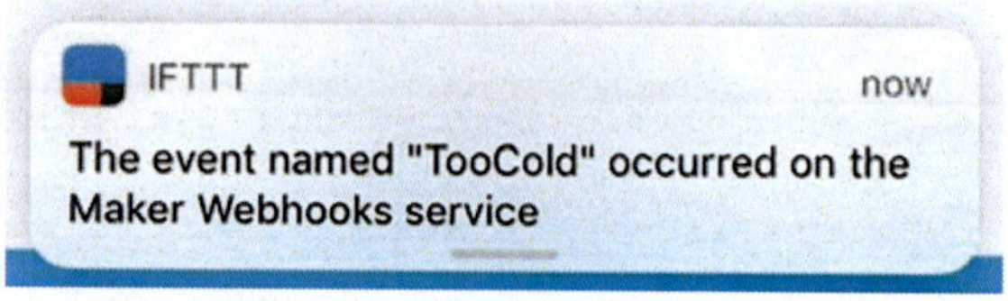

Fig. 6 Notification on smartphone

5 Conclusions

The proposed Smart and Secured Logistics system has been implemented and checked. This sends warning messages and alerts to the transporter. The transporter can implement this system and take care of the goods without any human interference to check the condition of the goods. This system is user-friendly and can be easily installed and monitored by less-skilled individuals.

As many IoT devices have security issues, we have implemented key security measures to overcome those issues as the data will be safe and private to the users. IoT is becoming a part of our day-to-day life, and it is important to implement in various fields to create the best ecosystem.

IoT integrates several applications ranging from industrial to household. For example, it connects sensors to tools in industries as discussed in [11]. With wide range of interoperability and applications, IoT becomes scalable and this application could be extended to other trending sectors as well.

References

1. 'Fresh Fruits and Vegetables', apeda.gov.in
2. Vermesan O, Friess P (2014) Internet of Things–From research and innovation to market deployment. River Publishers' series in communications
3. Vermesan O, Friess P, Guillemin P, Gusmeroli S et al (2011) Internet of Things strategic research agenda, Internet of Things -Global Technological and Societal Trends. River Publishers
4. Serrano M, Barnaghi P, Carrez F, Cousin P, Vermesan O, Friess P (2015) Internet of things semantic interoperability: research challenges, best practices, recommendations and next steps. European Research Cluster on the internet of things, IERC
5. Sebudde S, Kabagambe R, Geofrey M, (2014) Hygiene and sanitation in public eating places in one municipal health system of Uganda, pp 1–8
6. Lee K (2015) The Internet of Things (IoT): Applications, investments, and challenges for enterprises, business horizons. ScienceDirect 58(4):431–440
7. Sundmaeker H, Guillemin P, Friess P, Woelfflé S (2010) Vision and Challenges for Realizing the Internet of Things. Cluster of European Research Projects on the Internet of Things, European Commission. https://doi.org/10.2759/26127
8. https://oxfordbiomed.com/tech-resources/resources/food-quality-and-freshness
9. Atzori L, Iera A, Morabito G (2010) The Internet of Things: A survey. Comput Netw 54(1):52787–52805
10. Gartner. (2014), Gartner says the Internet of Things will transform the data center. Retrieved from https://www.gartner.com/newsroom/id/2684616

11. Vermesan O, Friess P (2013) Internet of things: Converging technologies for smart environments and integrated ecosystems. River Publishers' Series in Communications

A Survey: Beyond 5G Toward 6G—Network Security Issues, Thrust Areas and Challenges

K. S. Lavanya and B. Nagajayanthi

Abstract Fifth generation (5G) network is being rapidly deployed in real time. The exploratory research mind drives toward 6G networks. For further investigation and research toward 6G technologies, foundational research is carried out in a 6G network security, and privacy is presented as a survey in this paper. The survey begins with the key areas to be considered beyond 5G network toward 6G network such as intelligent radio, distributed artificial intelligence, real-time intelligent edge and 3D intercom. The survey also states the key issues of 5G network such as data security and authentication along with the countermeasures to overcome these issues in 6G network. Finally, the survey concludes with a new design idea based on the deep learning method (CNN-LSTM) for implementation in a 6G network for authentication by detecting the attack in network traffic.

Keywords 6G · Network security · Artificial intelligence · Deep learning

1 Introduction

Even though 5G network is not deployed fully, certain inherent limitations tend to move research focus toward 6G network. Researchers worldwide have started working on 6G communication and its network security challenging facts. There is no solid differentiation with respect to standards or specifications between 5 and 6G networks. Research focus has evolved from 5 to 6G to improvise the network and spectrum as depicted in Fig. 1 and also due to the features enlisted in Table 2. 6G network is expected to be faster than 5G Network.

K. S. Lavanya
Vellore Institute of Technology, Chennai, India
e-mail: lavanya.ks2020@vitstudent.ac.in

B. Nagajayanthi (✉)
Associate Professor, Vellore Institute of Technology, Chennai, India
e-mail: nagajayanthi.b@vit.ac.in

A. Sivasubramanian et al. (eds.), *Futuristic Communication and Network Technologies*, Lecture Notes in Electrical Engineering 792,
https://doi.org/10.1007/978-981-16-4625-6_87

The 6G network is expected to work with higher artificial intelligence (AI) technology. It should be "6G-AI-empowered." The 5G networks also use AI technology, but here, the future limitations show that 6G network should use AI tools in deep integration manner. Over the years, security and privacy of communication networks become the important factor. Beyond 5G, the network security and privacy concern with four major components stated (Fig. 2) as intelligent radio subsided with distributed AI and real-time intelligent edge.

These three areas are interconnected as AI exists with intersection factor and 3D intercoms. These are the areas of focus for AI-empowered 6G researchers. The main security issues here in 6G network are related to data transmission, encryption, access control and authentication. However, the issues are somehow particularly sensitive in some nature of communication. The communication intended in 6G are visible light communication (VLC), molecular communication, block chain technology-based communication and quantum communication.

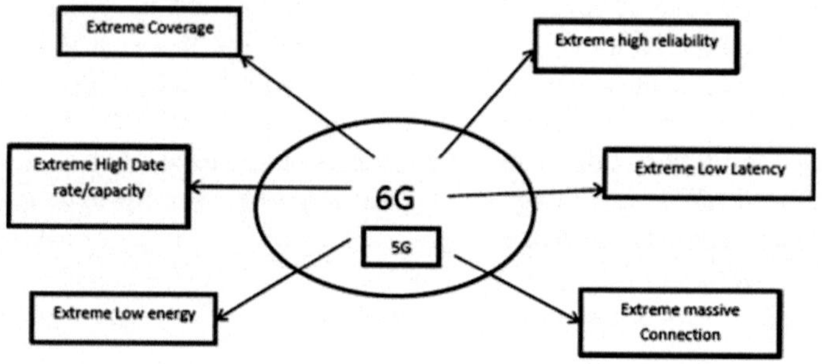

Fig. 1 Beyond 5G toward 6G

Fig. 2 Key areas of 6G network

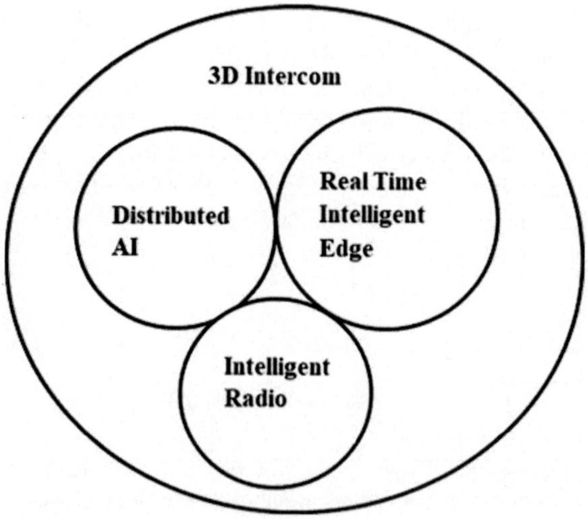

VLC along with real-time intelligent edge is weak against malicious behavioral attacks. Molecular communication has issues toward authentication and encryption while it processes under THz technology. Both quantum and block chain technology face the issue due to malicious attack.

2 6G Network: Key Areas and Security Issues

The 5G coverage is deployed only at ground level since it has risk factors to be dealt with for work in space and undersea communication at 3D intercom levels. Four thrust areas in 6G networks with added network security are highlighted as follows:

2.1 Distributed Artificial Intelligent Edge

Fifth generation is deployed in Internet of Things (IoT), whereas 6G is deployable in Internet of Everything (IoE). 6G network could take intelligent decision at a different level (i.e., decentralized system). To have the decentralized system, the dataset is trained unevenly over edge devices, with the edge devices capable of accessing and controlling the data part. Data is computed and stored independently by reducing the dimension. It also cleans and makes abstract out of the data [1]. Here, the shared data is processed as the training set irrespective of the state of personal information. Using this process, security and privacy of the network are enhanced. Machine learning plays a vital role to solve data integrity issues. Block chain methods can solve issues raised in distributed edge technology due to authentication.

2.2 Intelligent Radio

Over the years in network generation, the transceiver and device algorithm are designed together. In recent research and for 6G networks, the possible ways are measured to separate the transceiver algorithm from hardware by making the development in antenna and circuit boards. Intelligence-based radio can operate as a framework in a unified manner where the algorithm gets configured dynamically, and from the hardware, information is updated automatically.

Huang et al. [2] noted that the design of operating system and interfacing layer is based on the information available from the hardware, and also with AI, the transceiver algorithm gets configured. They also researched that design constraints should make the frequency band to adhere to hardware as well as to the environment. AI-enabled spectrum sharing is deployed. Further, Jiang et al. [3] stated that data security issues arise from signal jamming and wideband interference.

2.3 3D Intercom

Sixth generation networks need to improve certain factors relating to optimization capabilities, planning and analyzing of the network from previous network generation. Apparently, beyond 5G, 6G networks should support 3D space communication; i.e., it should be able to conform to both undersea communication and satellite communication. The 3D intercom could serve a service-based communication in such aspects.

The main question of 3D intercom is whether the 6G network would be able to operate in undersea is a controversial fact because of its complex nature. Grimmett [4] discussed the security issues involved in the transmission process in undersea communication network. From this, it can be inferred that undersea communication network will be the primary aspect of 6G network.

2.4 Real Time-based Intelligent Edge

The 5G networks still face the issue in the implementation of unmanned aerial vehicle (UAV) network because of its lack in controlling the network due to its high latency. UAV needs low latency and real-time intelligence where the current technologies are limited to this [5]. The 6G network should be able to handle these functions to support all power services interactively. The key areas concerning 6G network technologies are shown (see Table 1) and the key comparison between 5G and 6G is shown (see Table 2).

Table 1 Key areas in relation with 6G

Key areas	6G relation	Features
Distributed artificial intelligence	Capable of decision-making	Intelligent decision-making
Intelligent radio	Communication responsibility	Self-adaptive nature
3D intercom	Responsible for coverage	3D coverage
Real-time intelligent edge level heading	Capacity control	Real time response

Table 2 Fifth generation versus sixth generation

Key areas	5G	6G
Performance	Speed: 0.1 Gbps 1 ms—low latency 2D positioning Improved performance	Speed: 1–10 Gbps 0.1 ms—very low latency 3D positioning Guaranteed performance
Spectrum	Approx. 100 GHz Use in unlicensed spectrum High bands: limited no of national Licenses	Up to 10 THz Open in shared spectrum bands High bands: large no of local licenses
Network	Massive machine type communication Area traffic: 10Mbps/s/m^2	Machine support: broadband Area traffic: 1 Gbps/s/m^2
Network characteristics	Cloudization Softwarization Virtualization Slicing	Intelligentization Cloudization Softwarization Virtualization Slicing
Other aspects	Limited AI	Massive AI

3 Network Security: New Paradigm Shift

3.1 Security Issues

The 6G core key is to achieve customized secure services in human-centric mobile communication. The potential privacy and security issues faced in 5G networks are due to the following aspects mentioned below. The countermeasures for the issues in 5G and the reasons for the evolution from 5 to 6G are also highlighted below.

AI: Security Issue: Though it is always stated that 6G is AI-empowered, AI also raises the security issue like malicious utilization of AI technology. To train the AI model, the service provider collects user information (sensitive information) such as location, trajectory and identity of the user [6]. There is a high probability of the data getting leaked during transmission of data and during processing of data. The AI model could be attacked via white-box attacks though the attacker does not have any data about the architecture and parameter of the AI model. This approach could be made by checking the input and output. For dynamic protection of AI, a model needs to be framed.

IoT Network: Security Issue: For IoT deployment, an efficient authentication mechanism is required. Key management and resources for computing require encryption and decryption. IoT device security is limited by constrained storage space and battery energy [7]. Therefore, to overcome these limitations, a lightweight security algorithm is designed to support low power IoT devices with energy efficiency.

UAV Networks: Security Issue: Like terrestrial-based station, UAV could not support complex nature of cryptographic algorithm due to power and weight requirements. As a result, UAV is prone to network security vulnerabilities. To overcome this, mechanisms are required to counteract the internal attack.

3.2 Security Issue: Possible Countermeasures

To overcome the above stated issues, advancement in security and privacy technologies need to be framed.

Method based on Physical Layer: This method provides defense layer besides the upper cryptographic layer in 6G network. It also provided secure transmission in wireless signals. The combination of secret key and authentication key of the physical layer can give secure protection with a lightweight algorithm for the air interface. Most robust and efficient physical layer security (PLS) technology should be framed to satisfy throughput, latency and overhead requirements [8]. PLS provides lightweight mechanism to provoke against the security threat. Also, to achieve intelligence-based network security, distributed PHY protocol could be designed.

Method based on Traditional Cryptography with lightweight algorithm: In network security and data privacy, traditional encryption, authentication, authorization, signature and privacy are still a safeguarding procedure. Anonymous and group-based authentication gives lightweight and flexible solutions for development and research direction in cryptographic-based security and privacy. This method could be incorporated into both IoT network and UAV network security issue. Both the networks have the need to work under lightweight mechanism to overcome threat issue. This lightweight cryptography method or protocol under elliptic curve cryptographic (ECC) method provides reliable data security and authentication.

Method based on Quantum Technique: More lightweight and higher processing capability with absolute randomness is obtained by quantum-based method (quantum communication and quantum computing) when compared with the binary-based technique. The pre- and post-quantum cryptography is a stronger solution for data security, and privacy in quantum key distribution (QKD) gives unconditional guarantee of a secure key. By the above-stated process, the key gets generated in a manner irrespective of certainty and productivity of quantum state. The key is implemented randomly [2]. The evolution of 5G toward 6G is more of an AI network where multiple data are to be processed, so QKD will play the major role for high secured key generation to provoke against AI network issue. This method could be designed with continuous variable-quantum key distribution (CV-QKD) for both heterodyne and homodyne detections.

Method based on Block Chain: Block chain-based methods are decentralized methods. By this, it has the anti-corruption ability (stronger) and recovering ability which tend to have a stronger factor for authentication. This block chain-based algorithm could be designed for strong authentication and secured data sharing, whereas this method would be a wall to the authentication process in AI Network and IoT network. This block chain-based method could be implemented with the interplanetary file system (IPFS) for high data privacy and authentication.

Method based on Artificial Intelligence: The 6G network has the highest peak rate where it promotes the integration of AI in 6G network security mainly for anomalies detection and authentication. This AI-based method is used especially in machine learning-based auto-coder work for the detection of traffic date abnormality [9]. ML-based physical layer authentication is a much needed technique to achieve robust and reliable authentication. As we move forward, in intelligence-based automation, ML algorithm will provide security against abnormal attack in AI network. Convolutional neural network with long short-term memory (CNN–LSTM) technique can be used for abnormality detection in AI network platform since it gives instant prediction factor with known results of the previous state.

4 New Research Design for Security Issue

Beyond 5G toward 6G network, many researches have been deployed in various aspects. As mentioned above, 6G is AI driven where the main part deals with security and privacy. Deep learning technique helps to build a better networking platform prone to security attacks [10, 11].

4.1 Deep Learning-based Design

When the future networks get attacked, the certain factor is to find the attack with its category and information. By combining convolutional neural network with long short-term memory (CNN-LSTM), the spatial and temporal characteristics of the network are known in order to know the abnormality in network traffic. The design can be framed with three parts. The flow model of CNN-LSTM model is stated in Fig. 3 where network traffic states the temporal data.

1. Spatial feature of CNN can be implemented to process the network traffic data (to form spatial classification network model).
2. CNN-processed transmission vectors are nest processed in training set. Using LSTM, time characteristics are learned to form model which works for recognition.
3. By combining spatial and temporal model, the network traffic classification model is designed, and it acts as a prone to the network attacks.

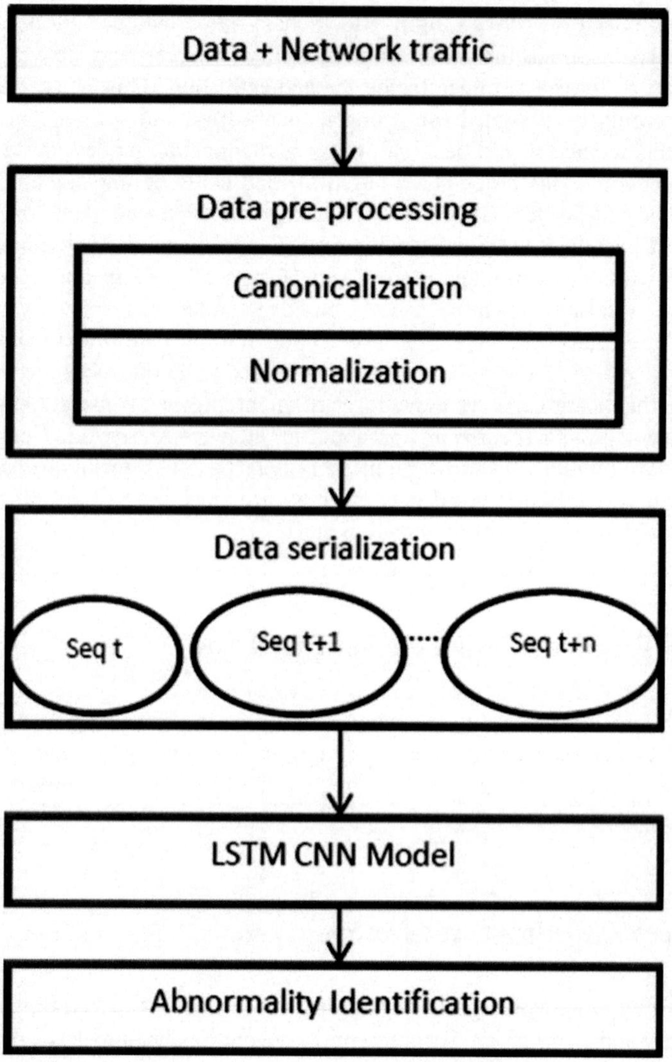

Fig. 3 CNN-LSTM model for abnormality detection (authentication)

5 Conclusion

The 6G network is the new agenda for many researchers. From the survey, it is inferred that 6G network gives high-level service compared to the previous generations. In this survey, analysis of 6G is based on security and privacy perspectives. The new research idea is stated under deep learning for implementation. The survey provides an insight into the technologies relating to 6G network.

References

1. McMahan HB, Moore E, Ramage D, Hampson S, et al (2017) Communication-efficient learning of deep networks from decentralized data. In: Proceedings of the 20th International Conference on Artificial Intelligence and Statistics (AISTATS), JMLR: W&CP, Vol 54
2. Huang W, Stokes JW (2016) MtNet: a multi-task neural network for dynamic malware classification", Detection of Intrusions and Malware, and Vulnerability Assessment. Springer, Berlin, Germany, pp 399–418
3. Jiang C, Zhang H, Ren Y, Han Z, Chen K-C, Hanzo L (2016) Machine learning paradigms for next-generation wireless networks. IEEE Wireless Communications 24(2):98–105
4. Grimmett DJ (2007) Message routing criteria for undersea acoustic communication networks. in: OCEANS, IEEE, Europe, pp 1–6
5. Katz M, Pirinen P, Posti H (2019) Towards 6G: Getting ready for the next decade. In: 16th International Symposium on Wireless Communication Systems (ISWCS), pp 714–718, IEEE
6. Buczak AL, Guven E (2016) A survey of data mining and machine learning methods for cyber security intrusion detection. IEEE Communications Surveys & Tutorials 18(2):1153–1176
7. Zhang Y H, Deng R, Bertino E, et al (2020) Robust and universal seamless handover authentication in 5G HetNets. In: IEEE Trans Dependable Secure Computer
8. Liang YC, Larsson EG, Niyato D, et al (2020) 6G mobile networks: Emerging technologies and applications
9. Sharafaldin I, Lashkari AH, Ghorbani AA (2018) Toward generating a new intrusion detection dataset and intrusion traffic characterization. In: Proceedings of the 4th International Conference on Information Systems Security and Privacy (ICISSP), Madeira, Portugal
10. Rawat DB, Reddy SR (2017) Software defined networking architecture, security and energy efficiency: a survey. IEEE Communications Surveys & Tutorials 19(1):325–346
11. Ren J, Hussain A, Zhao H et al (2020) Advances in brain inspired cognitive systems. In: International Conference on Brain Inspired Cognitive Systems, Vol. 11691 of Lecture Notes in Computer Science, Springer, Berlin, Germany

Enhancement of Productivity Using Chatbots

B. Balatamoghna and B. Nagajayanthi

Abstract Chatbots automate and personalize conversations. Similar queries are grouped and resolved without human trafficking. This user-friendly keyword-based texting conversational framework chats with the client, takes the request, and responds to the query. WhatsApp encryption maintains the integrity of the conversation. WhatsApp bot attracts attention and keeps the groups active. This is especially useful during pandemic situation where high social media presence is abundant. Educational institutions can take advantage of automated responses to satisfy an ever-growing demand caused by high levels of social media usage. Manual resources are scarce in pandemic crisis. This provides an economical replacement for manual information transfer and infrastructure. WhatsApp is deployed and is popular in social media for entertainment purposes. Smartbot used in WhatsApp is used to reduce quality chatting time. By using customized keyword, smartbot provides accurate information. Smartbot implementation in WhatsApp with monitored chat and automated response is the novelty of this implementation. This paper proposes a solution to streamline academic queries during pandemic crisis and keeps them connected.

Keywords Smartbot · Automated response · WhatsApp · Social media · Analysis

1 Introduction

WhatsApp is a conversational app that encourages free flow of conversation rather than a formal setting and rigid text structure. WhatsApp has grown in popularity among youngsters as well as professionals due to its simplicity and user-friendly features. This has well-established itself as a secure instant messaging app. This feature has made WhatsApp a preferred mode of communication and holds potential to create closeness between customers and businesses [1].

B. Balatamoghna
Student, Vellore Institute of Technology, Chennai, India

B. Nagajayanthi (✉)
Associate Professor, Vellore Institute of Technology, Chennai, India
e-mail: nagajayanthi.b@vit.ac.in

A. Sivasubramanian et al. (eds.), *Futuristic Communication and Network Technologies*, Lecture Notes in Electrical Engineering 792,
https://doi.org/10.1007/978-981-16-4625-6_88

The objective of this paper is to provide a solid proof of the positive effects on productivity for enthusiasts as well as official groups using automated bots to take care of providing spontaneous communication along with specialized target demographics. This could be visualized and experienced by creating a chatbot which responds immediately to the clients [2].

During this pandemic situation, it is essential to have proper management as well as quick reply assistance to cover for the loss in manpower mishandlings. By having a bot set to make announcements over multiple groups as well as ban users, the information could be centrally monitored and securely transmitted. By having a quick response chatbot, customers earn trust and could be better assisted. This gains better relationship between the client and the service provider. Researcher Oanh et al. [3] has analyzed on the user utterances in chatbots to better communicate to the customers.

1.1 Motivation

The bot was made to alleviate the stress of lack of human resources during the pandemic situation. WhatsApp is a major social media messaging platform and is well integrated within official groups which require laborious and intensive management to keep productivity high.

People prefer chatbots over graphical representations [4], and so, the demand for a chatbot over a popular instant messenger is an essential motivation factor.

With the use of the bot, features such as announcing, banning, anti-spamming, and having custom messages are achieved. Thus, group management and clear customized communication could be achieved with less work and less time.

2 Methods

2.1 Participants and Procedure

The chatbot was developed and introduced to one specific group that continually tested and used the bot in order to better integrate it seamlessly into chats. Data was collected through the chat log of the group. The group consists mainly of college going individuals (N) ($N = 35$, aged 18–22) who are well acquainted to technology and heavily use WhatsApp and other forms of social media.

A message is defined as a single sent/received data from the any source [5], be it user or the bot. The total message count reached 39,878 within the group comprising of 36 members within a short span of ~2 months.

The factors that were taken into consideration for the data analysis were:

(1) Timestamps of when the message was sent, (2) number of messages sent by each user, (3) text contained in the messages, and (4) user's phone number.

The bot was used mostly as a way to improve messaging clarity during conversations. This can be directly related to productivity since message clarity is an important factor to improve productivity generally [4].

2.2 Limitation

The bot is existent in only one group and data on the group has not been recorded anywhere other than on WhatsApp's chat log in view of privacy and security ideology. The data cannot conclusively be extrapolated to project other inferences without extensive time durations.

3 Data Analysis

WhatsApp bot is programmed to manage groups as well as provide chatting enhancements. The overall chat's health and activity raises tremendously. This group was formed as a casual chatting group for people across India to converse and exchange information. The graph depicts the lack of messages before August 2020, which was just before the bot was introduced. Chatbots functionality depends on the algorithms used [6].

This group was formed on July 2020. Initially, it had around 2.5 k messages. After the bot's introduction, the messages per month peaked to 12.5 k messages. Group chat was exported and analyzed using a chat analyser as shown in Fig. 1.

The resulting data shows the tremendous traffic and attention generated by the bot.

Data in Fig. 2 provides another interesting observation. There are no messages between 12 a.m. and 7:30 a.m. The bot was shut down during these hours. This proves customized efficient maintenance. The data here shows the importance of the bot in regular conversation.

During weekends, the bot was used much more than on weekdays as shown in Fig. 3. During weekdays, due to online classes, the usage is less. New features are mostly introduced and developed on Wednesdays, reason being time allocated for online classes on Wednesdays is much shorter. The bot has sent a total of 3971 messages to the group as shown in Fig. 4.

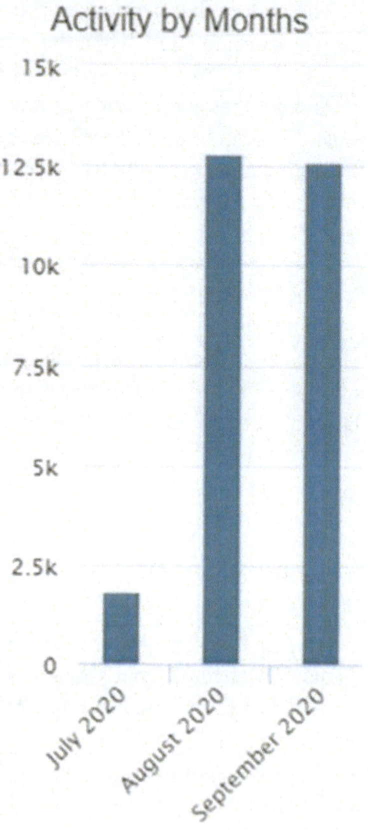

Fig. 1 Chat analyser report based on group chat

4 Implementation of Smartbot

The bot was added to a small group of 12 members. Initially, the group was used as a place to casually converse. After the bot was added, the current total of the members in the group has reached 36 people in total.

5 Salient Features of Smartbot

The bot currently can do many functions that aim for quality of life (QoL) improvements as well as entertainment features. Along with these features, the bot also can ban and announce to multiple groups to keep chat health high which in turn keeps groups secure and gain high attention retention.

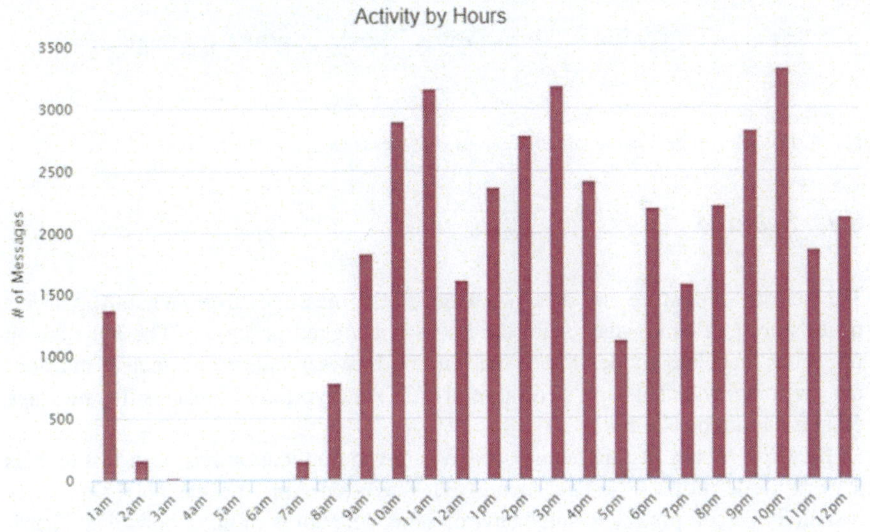

Fig. 2 Shows hourly usage. Data shown here is from chat analyser, where the group chat was exported

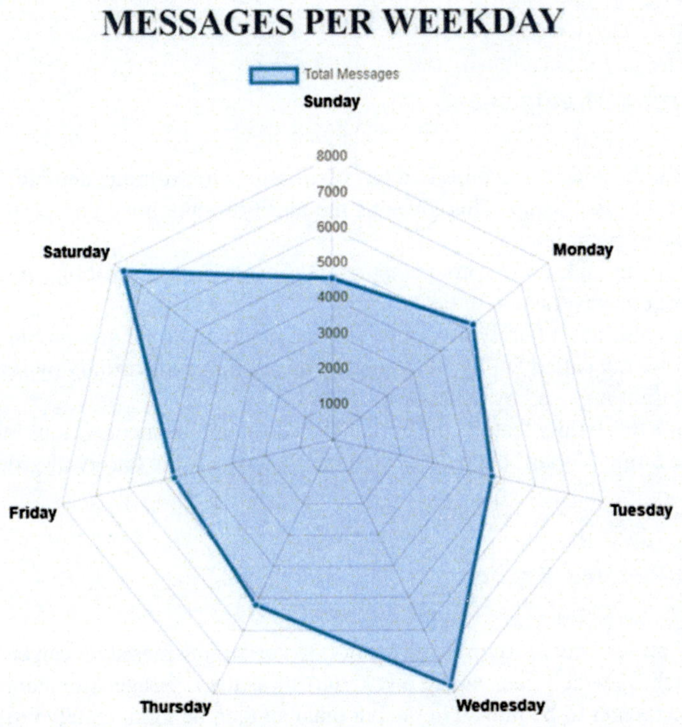

Fig. 3 Shows weekly usage. Data shown here is from chat analyser, where the group text was exported

Name	Messages	Words per Message	Sent Pics	Sent Audio	Most Used Words
Bot	3971	11.7			your - you - to -

Fig. 4 Data shown here is the total number of chat messages

5.1 Secured

The bot does not store any data; whatsoever, the messages from WhatsApp app is never stored. All messages from the bot is not logged or shown. The bot runs on puppeteer headless mode, where the chrome browser runs as a process and does not give display. WhatsApp's end-to-end (E2E) encryption [7] ensures that message objects cannot be stored.

Building on top of this feature, the WhatsApp application has been installed as a work-mode profile [8]. This means that the WhatsApp app has been separately installed in a completely isolated environment. The bot is running on isolated hardware as well, namely on a Raspberry Pi 4(R-Pi). The R-Pi is also running with no display connection and is only accessible within the same network through virtual network computing (VNC).

5.2 Group Management

The bot can be added to groups and has permissions to promote, demote, kick, and add users into the group. This gives us the ability to automate and perform basic group management features.

The bot can kick users who are spamming inside groups, enabling friendly and normal-paced conversation to take over.

The bot also has a family of groups feature, where multiple groups can be added in a single entity called family. This enables huge groups to partially bypass the 256 members limit imposed by WhatsApp.

Through this family feature, the bot can then do announcements to multiple groups inside the family. Smartbot could as well ban feature for the family of groups.

5.3 Automated Replies

The bot features a custom command creator, where simple messages can be recorded into a database which then can be invoked to show the messages immediately. This enables messages to be pinned in the bot that can then be immediately retrieved.

5.4 Entertainment

The bot features a simple toss game that keeps users continually playing and keeping the group active. Along with the bot's responses, there is a secluded Web site for the bot where users can continue to play even between 12 a.m. and 7:30 a.m.

5.5 Educational

The bot has a simple API attachment that can be used by members to search Japanese and Kanji definitions. Currently, there are other use cases for the bot, but the use case for this existing feature is niche and would require a more formal setting which the bot is not provided with. WhatsApp is used for chatting purposes. In this novel approach, smartbots are used to reduce quality time in getting the required information. Chatbots are implemented in Web sites and Facebook. Since WhatsApp is a user-friendly messaging app [9], smartbot is implemented in WhatsApp to provide automated responses. Students have gained the required information with less messaging time.

6 Conclusion

The WhatsApp bot shows that WhatsApp is a good platform for bots, leaving management tasks to be automated. The bot helps users to keep their conversation focused on productivity and explicit communication.

The bot provides a secured dynamic group management with monitored features to keep groups secure and converse with high quality, thus improving the chat activity and attention retention greatly.

If this system of management and attention retention is implemented officially, educational institutions [10] as well as corporate will have a greater and more productive environment as it decreases workload on humans and revolutionizes the way the information is handled in WhatsApp. In the current era, WhatsApp is used for spreading false messages and is misused. This bot is tested and analyzed for educational purposes.

Disclaimer
The bot was made solely for research and data analysis and has no commercialization or monetary purposes. The data shown here was an exported chat from a different account in the same group.

Conflict of Interest
The authors declare that they have no conflict of interest.

References

1. Maina T (2013) Instant messaging an effective way of communication in workplace. Computers and Society
2. Nadella (2016) "Bots are the new apps". USA Today. https://www.usatoday.com/story/tech/news/2016/03/30/microsof-ceo-nadella-bots-new-apps/82431672/
3. Tran O, Luong TC (2020) Understanding what the users say in chatbots: A case study for the Vietnamese language. Engineering Applications of Artificial Intelligence. Elsevier, Vol 87
4. Folstad A, Brandtzæg P (2017) Chatbots and the new world of HCI. www.interactions.acm.org
5. Hill J, Ford WR, Farreras IG (2015) Real conversations with artificial intelligence: A Comparison between human-human online conversations and human-chatbot conversations. Comp Human Behav 49:245–250
6. Vijayaraghavan V, Cooper JB, Leevinson RJ (2020) Algorithm inspection for chatbot performance evaluation. In: Third International Conference on Computing and Network Communications (CoCoNet'19), ScienceDirect, Vol 171, pp 2267–2274
7. Herzberg A, Leibowitz H (2016) Can Johnny finally encrypt?: evaluating E2E encryption in popular IM applications. In: Proceedings of the 6th Workshop on Socio-Technical Aspects in Security and Trust STAST '16. https://doi.org/10.1145/3046055.3046059
8. Google support: "What is a work profile?" https://support.google.com/work/android/answer/6191949?hl=en
9. Zumstein D, Hundertmark S (2017) Chatbots an interactive technology for personalized communication, transactions and services. IADIS Int J WWW/Internet 15(1):96–109
10. Smutny P, Schreiberova P (2020) Chatbots for learning: A review of educational chatbots for the Facebook Messenger. Computers and Education. Elsevier, Vol 151

Medical Image Fusion Based on SWT and Cross Bilateral Filter

Shaik Gowhar, Maruturi Haribabu, and Velmathi Guruviah

Abstract Image fusion is a process of combining relevant information from a different set of medical images into a single image. Multimodal medical image fusion techniques are used to improve the quality of a fused image. The cross bilateral filter (CBF) is an edge-preserving filter to highlight the edge's information in a fused image used for better analysis. In this paper, SWT and cross bilateral filter-based medical image fusion is proposed. Firstly, the input images are decomposed into a frequency coefficient using stationary wavelet transform. Secondly, the low coefficients are fused by CBF and high coefficients are fused with maximum fusion rules. Thirdly, the fused image is obtained by reconstruction process. This proposed method is superior to the conventional CBF method in terms of visualization and quality. The proposed method is evaluated by quality parameters such as mean (M), standard deviation (STD), and entropy (E).

Keywords SWT · CBF · Medical images · Image fusion · Quality parameters

1 Introduction

Image fusion is the process of merging the multiple source images into a single resultant image, which is used for desired medical diagnosis in a better way. The resultant fused image is more informative than the input images. Image fusion has been used in many application areas like multisensory image fusion, which is used to achieve high spatial and spectral resolution, and multi-focus image fusion is suitable for better visualization. In medical field, image fusion is a prominent domain

S. Gowhar
ECE Department, QISCET, Ongole, Andra Pradesh, India

M. Haribabu (✉)
QISCET and Research Scholar in SENSE, Vellore Institute of Technology, Chennai, Tamilnadu, India
e-mail: maruturi.haribabu2018@vitstudent.ac.in

V. Guruviah
School of Electronics Engineering, Vellore Institute of Technology, Chennai, Tamilnadu, India

A. Sivasubramanian et al. (eds.), *Futuristic Communication and Network Technologies*, Lecture Notes in Electrical Engineering 792,
https://doi.org/10.1007/978-981-16-4625-6_89

because single medical image does not provide required information and which is quite difficult to the physicians for diagnosis such as magnetic resonance imaging (MRI) image reveals delicate tissue information but lack of bone information, and computed tomography (CT) image gives bones information but lack of tissue. So, medical image fusion is wider and suitable for real-time applications and used to diagnosis of medical-related problems. The pair of MRI and CT images provides soft and hard tissue information, and T1-MR and MRA images provide soft and lesion locations, respectively, and these combinations are suitable for better diagnostic and interpretation of diseases. The image fusion methods comprise the modest technique [1–3]. CBF is one such algorithm that can be applied for multimodal imagining applications. The remaining paper will be arranged as follows: Description of SWT and CBF is in sect. 2. Sects. 3 and 4 are proposed method and experimental results. Lastly, the conclusion part is discussed in Sect. 5.

2 Preliminary

From past decades, many researchers proposed various algorithms related to medical image fusion for better diagnosis such as M. B Abdulkareem, et al. [4] proposed DWT-based MIF. Initially, the input images are preprocessed by filters and then decomposed using DWT to get approximation and detailed coefficients. The coefficients are fused by average fusion rule. Finally, the fused image is created by reconstruction process with good quality but lack of shift-invariant due to downsampling. B. M. Reddy [5] proposed DWT and PSO-based image fusion. The input images are fused by DWT and then weights are created by PSO. This method produces fused image with artifacts and is not suitable for diagnosis. Rabia Bashir et al. [6] suggested MIF using SWT and PCA. Firstly, the source images are spilt into low and high bands using SWT and then fused by PCA fusion rule with help of principle components. Secondly, the fused image is obtained by inverse DWT. This method is superior to SWT and PCA techniques in terms of visibility. P. S. Akshara et al. [7] proposed CBF-based denoising and image fusion. This method produces fused image and the superior to Gaussian and bilateral filter in terms of visibility and quality.

Image fusion has been performed in two different domains like spatial and spectral. Spatial domain techniques such as Intensity-Hue-Saturation (HIS), principal component analysis (PCA), and Brovey provide fused image with high spectral distortions and limited complementary information. The spectral domain techniques are Discrete wavelet transform (DWT), stationary wavelet transform, and contourlet transform, and these are useful to overcome the above mentioned drawbacks. The DWT technique provides good localization in both time and frequency, but shift-invariant is the major limitation. To overcome that problem, stationary wavelet transform is preferable technique and is used to fill the gap caused by decimation in the DWT. In SWT, the input image is decomposed into four coefficients such as LL (approximation), LH (horizontal), HL (vertical), and HH (diagonal). The diagrammatic representation of SWT is shown in Fig. 1.

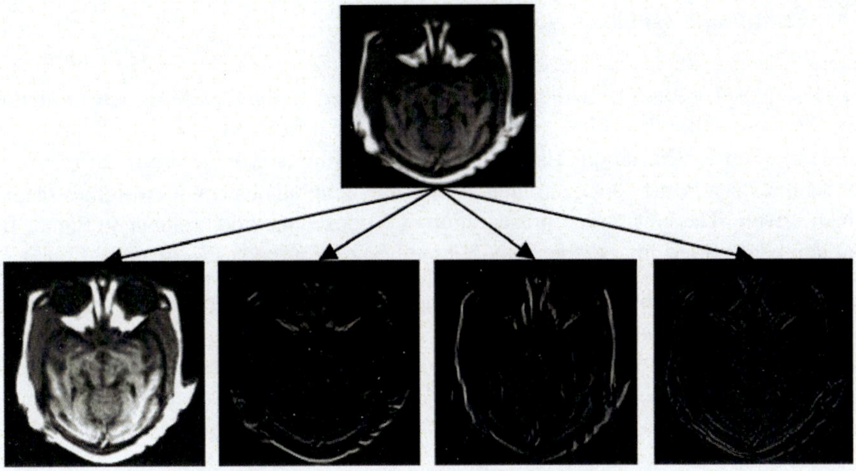

Fig. 1 Diagrammatic representation of SWT-based decomposition levels

Cross bilateral filter is a non-iterative and linear technique to enhance the edges information in an image, which includes low pass filter and edge function. In CBF, the calculation of edge strength of every pixel is in both vertically and horizontally. The merits of this filter are smoothness and edges information, which are obtained using neighboring pixels. The CBF is performed between input image and guided image to get base and detailed layers, respectively. However, the construction of weight map is not appropriate and unable to assign optimal weights to fuse source images. This method may introduce artifacts into the fused image. To rectify that drawbacks in conventional CBF, we propose cross bilateral filter using efficient fusion rule. The proposed method can able to address multimodal image fusion applications [8–10].

The procedure for CBF-based MIF is as follows:

1. Firstly, consider the input images and apply CBF to get approximation layer or base layer.
2. Secondly, the detailed layers are obtained by based layers, i.e., subtraction between input and base layers.
3. Thirdly, detailed-layer edge strengths are calculated in both horizontally and vertically.
4. Fourthly, weighted map creation.

Merits:

1. The CBF method is suitable for various image fusion domains such as multimodal, remote sensing, and multi-focus.
2. Fused has good edges information.

Demerits:

1. Fused image quality is low due to artifacts.
2. CBF method is expensive in terms of computations.

3 Proposed Method

The proposed method is medical image fusion based on the maximum selection rule in SWT and CBF. The MRI gives only soft tissue information, while CT provides only fractional information. The CBF is used to highlight the edges information using weighted maps, which are generated by base and detailed layers with weighted maps, respectively. The flowchart representation of proposed method is shown in Fig. 2. To obtain better fused image, the proposed method is arranged as follows:

1. Consider two different source images (*I1 and I2*). Both images are having the same size.
2. Apply SWT on both images to get for coefficient values like LL, LH, HL, and HH.
3. Generate base and detailed layers of two-input low coefficients by using CBF and described as follows.

$$\begin{aligned} \mathrm{BA1} &= \mathrm{CBF}\ (\mathrm{I1},\ \mathrm{I2}), \\ \mathrm{BA2} &= \mathrm{CBF}\ (\mathrm{I2},\ \mathrm{I1}). \end{aligned} \tag{1}$$

$$\begin{aligned} \mathrm{DE1} &= \mathrm{LL1} - \mathrm{BA1} \\ \mathrm{DE2} &= \mathrm{LL2} - \mathrm{BA2} \end{aligned} \tag{2}$$

4. Generate saliency information and generalized by

$$\mathrm{S1} = |\mathrm{DE1}|,\ \&\, \mathrm{S2} = |\mathrm{DE2}|$$

$$\alpha 1 = \mathrm{S1}/(\mathrm{S1} + \mathrm{S2}) \quad \text{and} \quad \alpha 2 = \mathrm{S2}/(\mathrm{S1} + \mathrm{S2}) \tag{3}$$

5. The detailed layers are fused using the following equation:

$$FdE = \begin{cases} DE1 & if \quad \alpha 1 > \alpha 2 \\ DE2 & if \quad \alpha 1 < \alpha 2 \\ (DE1 + DE2)/2\ . & otherwise \end{cases} \tag{4}$$

6. The base layers are fused using the following equation:

$$\mathrm{Fb} = \max\ (\mathrm{BA1},\ \mathrm{BA2}) \tag{5}$$

7. The final fused LL coefficients are obtained by averaging.

$$\mathrm{FLL}\ = \mathrm{FbA} + \mathrm{FdE} \tag{6}$$

8. The maximum selection rule is used to get new detailed fused coefficient values and then final fused image is obtained by inverse SWT.

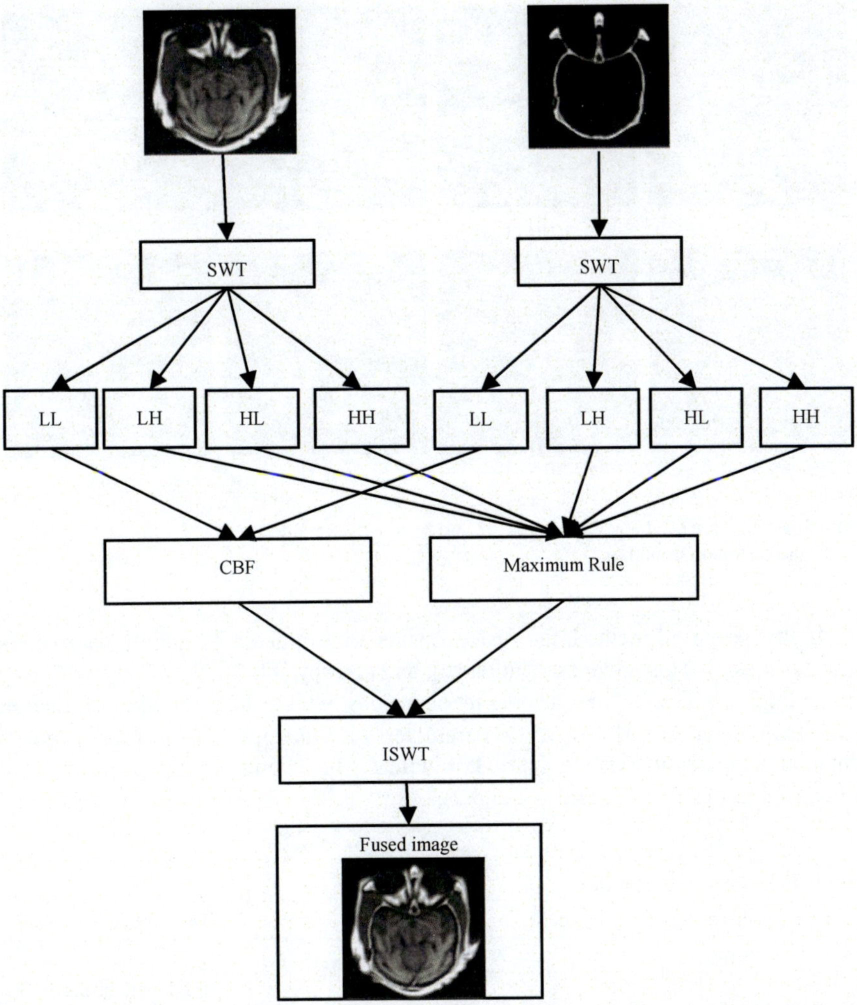

Fig. 2 Proposed method flowchart

4 Experimental Results

The simulation of fusion results is done by MATLAB, and existing fusion methods are low fusion performance than the proposed method in terms of visual and quantitative representations. This proposed method was tested by two different sets of multimodal medical image. The results are performed over the two modalities of MRI and CT images. The maximum content of the information for identification of disease is obtained by fusion process of bones and soft tissue information of CT and MRI images. Fig. 3 shows input images, existed method results, and proposed method results.

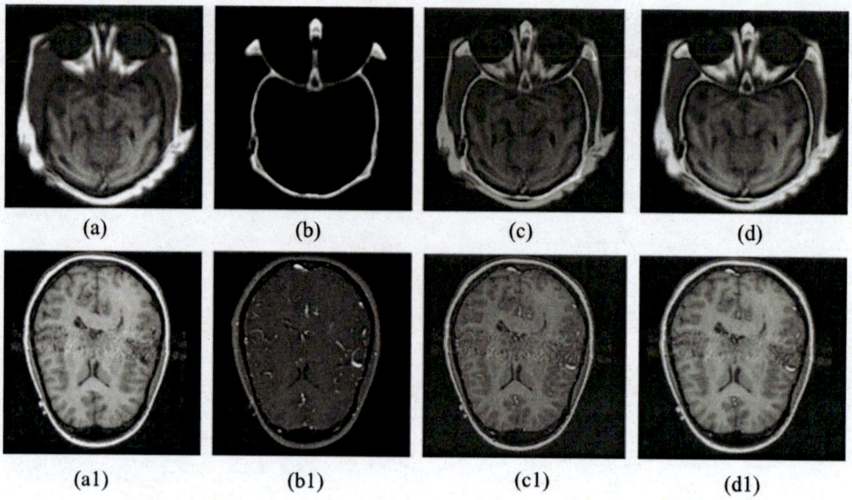

Fig. 3 **a** and **b** are MRI and CT images, **a1** and **b1** are T1-weighted MR-MRA images, **c** Prasad's [10], and **d** Proposed method

In this proposed method, the fusion results are evaluated by quality parameters such as mean [11], standard deviation [6], and entropy [6].

In Fig. 3, a–a1 and b–b1 are the input images, which show limited information of human organ. The *M*, SD, and *E* parameters give the highest values for proposed method compared to existing method; it indicates fused image has good quality and enhanced in Fig. 3d–d1 (Tables 1 and 2).

Table 1 List of quality metrics

S.no	Metrics	Formula	Description
1	Mean	$M = \frac{1}{D\times F}\sum_{d=1}^{D}\sum_{f=1}^{F}[\text{Fused}(d, f)]$	It represents brightness of the image
2	Standard deviation	$SD = \sqrt{\frac{1}{D\times F}\sum_{d=1}^{D}\sum_{f=1}^{F}[\text{Fused}(d, f)] - M]^2}$	It shows contrast of the image
3	Entropy	$E = -\sum_{i-1}^{l} p \log p$	This metric indicated high amount of information is present in a fused image

Table 2 Comparison of descriptive statics

Datasets	Schemes	Fusion parameters		
		Mean	STD	Entropy
MRI-CT	Prasad [10]	32.9105	38.2887	4.0876
	Proposed method	**60.0295**	**61.1446**	**4.7323**
T1_MR-MRA	Prasad [10]	45.7504	48.3068	4.0444
	Proposed method	**67.3236**	**68.4670**	**4.6601**

5 Conclusion

In this paper, the proposed method is image fusion based on the stationary wavelet transform domain and CBF. After applying the SWT, the LL's coefficient of base and detailed layers is fused based on CBF, and LH, HL, and HH coefficient values are fused based on the maximum selection rule. Finally, inverse SWT to get the final fused image. The proposed method is superior to the conventional CBF, and also, the quality parameters prove better results. The proposed method is more suitable for real-time scenarios.

References

1. Jameel A et al (2014) Improved guided image fusion for magnetic resonance and computed tomography imaging. The Scientific World J, 1–8
2. Giesel FL et al (2009) Image fusion using CT, MRI and PET for treatment planning, navigation and follow up in percutaneous RFA. Exp Oncol 31(2):106–114
3. Haribabu M et al (2014) Image fusion with Biorthogonal Wavelet Transform based on maximum selection and region energy. ICCCI, 1–6
4. Abdulkareem MB et al (2018) Design and development of multimodal medical image fusion using discrete wavelet transform. IEEE-ICICCT, 1629–1633
5. Reddy BM et al (2018) Image fusion using stationary wavelet transform and particle swarm optimization. IJETA
6. Bashir R et al (2018) SWT and PCA image fusion methods for multi-modal imagery. Multimed Tools Appl, 1–29
7. Akshara PS et al (2018) Medical image denoising and fusion using cross bilateral filter. IRJET, 4538–4543
8. Sultana T et.al (2016) Analysis on SWT based Image Fusion Techniques using Intuitionistic Fuzzy Set Operations. IJTEEE, 16–19
9. Pradnya M et al (2013) Image fusion based on stationary wavelet transform. Int J Adv Eng Res Studies, 99–101
10. Prasad D et al (2018) A new image and video fusion based on cross bilateral filter. ISIF, 1511–1518
11. Manu VT et al (2012) A novel statistical fusion rule for image fusion and its comparison in non subsampled contourlet transform domain and wavelet domain. Int J Multimedia Appl, 69–87

Assistive Device for Visually Impaired People

N. Sriram, Anirudh Hosur, Akash Reshan, and P. Vetrivelan

Abstract In this project, we intend to create an assistive system for the blind, which will help them identify the person in front of them, read text written on signboards, etc., and assist them while crossing a road based on the traffic signal color. All the output will be in the form of audio instruction that will be played via a headphone that is connected to the system. The Raspberry Pi controller will be used to integrate the various components and control the processing of the images and also to train and store the model, using OpenCV and Tesseract; for the audio output, we will be using Google-text-to-Speech (gTTS). This system will assist the blind and partially sighted people in the known and unknown environment without much help from other people. This device can improve mobility as well as the safety for the visually impaired people.

Keywords gTTS · TTS · Recognition system · Assistive system · OpenCV.

1 Introduction

Moving through an unknown environment is a real challenge for the visually impaired or blind people, although they have to rely on their other senses. According to the World Health Organization, there are roughly about quarter billion visually impaired people around the world, out of which 39 million are blind [1]. The World Health Organization has estimated the number of blind people will double by 2022. They usually use a white-colored cane or Seeing Eye dog for their assistance.

However, these methods are not effective because they do not help in saving blind people from risks [2].

Assistance systems that were designed previously for visually impaired people consist of ultrasonic canes, voice-assisted navigation canes, laser-based walkers,

N. Sriram · A. Hosur · A. Reshan · P. Vetrivelan (✉)
School of Electronics Engineering, Vellore Institute of Technology, Chennai, India
e-mail: vetrivelan.p@vit.ac.in

A. Hosur
e-mail: anirudh.hosur2017@vitstudent.ac.in

A. Sivasubramanian et al. (eds.), *Futuristic Communication and Network Technologies*, Lecture Notes in Electrical Engineering 792,
https://doi.org/10.1007/978-981-16-4625-6_90

infrared canes etc. [3]. There are some benefits of these systems, but they also have some disadvantages. For instance, infrared cane assistance infrared has a short-range for detecting obstacles, produces not so favorable results in the dark. When an obstacle is detected, the ultrasonic sensors which are present on top of the cane start vibrating signaling obstacles nearby or in that surrounding. As the cane can only scan the nearby surroundings compared to a person, it is not so effective in its function [4]. In case of the laser-based walking assistance, obstacles are detected by scanning the surrounding with lasers.

A laser is fixed and another is rotated by a DC motor. There is one belt of five motors that give feedback to users when an obstacle is detected [5]. In this assistance system, lasers are used which may harm other people if it strikes them in their eyes or skin.

An assistance system is suggested by the team in this research work which is aimed at assisting blind people, which will be entirely safe for people and the environment as it uses a camera and a headphone or a speaker to alert the user if there is an obstacle in front of him and to notify him to change his/her direction thus eliminating the use of the cane and laser-based projects as it can harm people vigorously and can cause untreatable damage, whereas our project is very cost-effective, efficient and safe to use and user friendly.

2 Design and Methodology of Blind Assistive Device

The idea implemented in this project is a voice assistive system for visually impaired people. It will have multiple functionalities that help the blind in different situations.

The text-to-speech (TTS) module is used to generate audio output so that the person can hear the instructions. To implement this, we use the Google text-to-speech (gTTS) module; it creates an mp3 file of the audio based on the string that is input, and this can be played using the default audio player on your device.

The face recognition module is used to identify the name of the person that is in front of you. For this, we will use Python OpenCV to process the image. We create a dataset by adding images of a person's face. The features alone are stored in the dataset along with the name of the person. Now, when a new image is an input, using Haar Cascade, we try to identify the name of the person by matching the features of the images. This name is then sent to the TTS module for audio.

The traffic signal detection system is used to help the blind while crossing roads, etc.; it uses Python and NumPy to convert and store the image into arrays. We initially set the RGB values for the colors in the traffic signal. So, based on the RGB values in the array, we can identify the signal color and hence generate an instruction string that is sent to the TTS module for audio output.

The sign-board text recognition module is used to read the signboard and provide audio output. For this, we use the PyTesseract module to extract the text present in the image and convert it into a string and send it to the TTS.

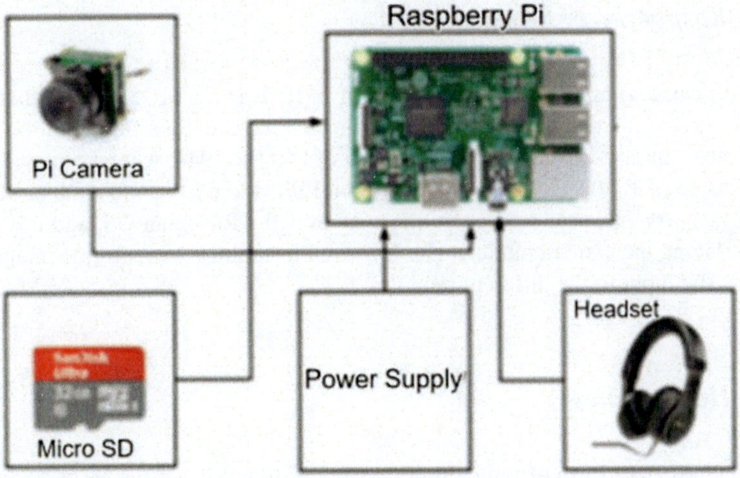

Fig. 1 Block diagram of the assistive device for the visually impaired

2.1 Block Diagram

Figure 1 depicts the schematic block diagram of the proposed model of the assistive device for the visually impaired. Various components involved in this project diagram are explained in the following sections. The interface between user and system will be controlled by using buttons.

2.2 Raspberry Pi 3B

The third generation of the Raspberry Pi is called a Raspberry Pi 3B. It is a compact single-board computer that is widely used in various fields. It has a Broadcom system on a chip and processing unit which is central (CPU) and a graphics processing unit (GPU) which is on chip which are ARM-compatible. Its subsequent features:

- Broadcom BCM2387
- 802.11bgn Wireless LAN
- Bluctooth 4.1
- 1.2 GHz Quad-Core ARM
- 4xUSB ports
- 1 GB RAM, 64 Bit CPU [6].

The purpose of Raspberry Pi in this project is to interface all the components, process the image for the various modules, and to provide the audio output through the 3 mm audio port.

2.3 Raspberry Pi Camera

The RPi camera module has a resolution of 5MP. It is a CMOS camera that has a fixed focus lens that helps in capturing still images and also for high-definition video. By default, the still images have a resolution of 2592 × 1944, whereas the video has a resolution of 1080p at 30 FPS or 720p at 60 FPS, and 640 × 480 at 60 or 90 FPS. The Raspberry Pi's preferred operating system is the Raspbian OS, and it helps us in interfacing the camera module. [7]. It is used to capture the real-time images and provide the input to the different modules.

2.4 Headphones

Headphones are a form of miniature loudspeaker-like drivers that are worn over or inside a user's ears/head, and thus enables them to listen to audio in a more private environment. They are a type of electro-acoustic transducers that mainly convert electrical signals to their respective sound waves. It is used to provide an audio result of the various modules; we can use any mobile headphones. A better headphone produces a better sound quality.

2.5 Power Supply

An external power supply such as a power bank is required to power the Raspberry Pi and hence power the entire system.

3 Algorithm

This section discusses the platform and major modules used to implement the programming of the algorithm used to test the images for the various functionalities of the project.

3.1 OpenCV

OpenCV is a collection of functions that enables users to apply real-time computer vision in their code [8]. Since our project involves vision, this module is of utmost importance. It has ways of feature extraction, model training, etc.

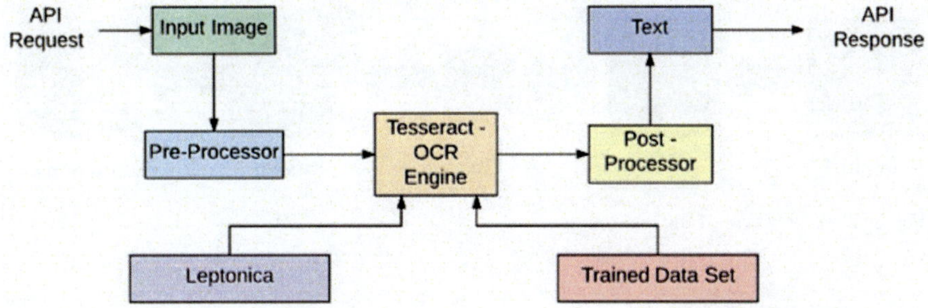

Fig. 2 Text recognition using Tesseract

Digital image processing is the use of computer algorithms to process, communicates and display digital images. Digital image processing algorithms are often used to: improve resolution and eliminate noise and other artifacts.

3.2 *Tesseract*

Tesseract engine in Python contains a vast set of functions for optical character recognition. Optical character recognition refers to the process of detecting text in images. Tesseract is provided to us using apache 2. This can be used directly or through a set of functions/APIs so that it can be used in languages like Python [9]. However, it is a line-based command and does not have a GUI.

Using wrappers (classes that allow you to use certain functions), we can use Tesseract in a huge set of languages. This has two ways of implementation. It can be used directly to detect text in a document, or it can be combined with an image detector to recognize text from an image. This is the application that we use in our project.

Tesseract uses a convolution neural network that recognizes the image with the character. Inside this, there is another neural network which is set to be a text line detector. However, the problem of detecting the length of the text is solved using LSTM which is a type of recurrent neural network (Fig. 2) [10].

3.3 *Image Pre-Processing*

Preprocessing is used to eliminate unwanted distortions and the correct image is used Next is that the process. Preprocessing steps within the image include increasing the contrast, resizing the RGB to form changes (Fig. 3).

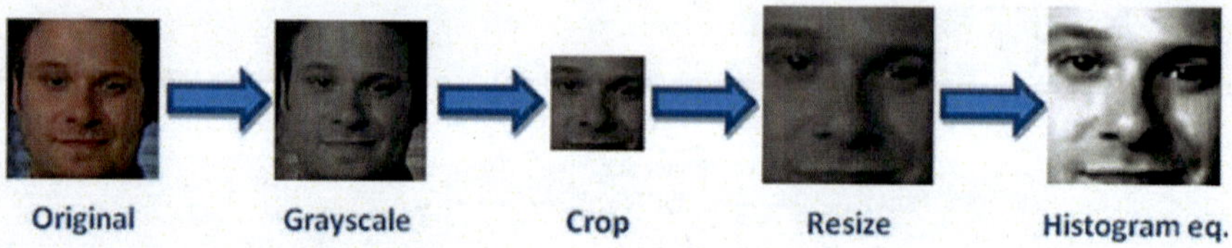

Fig. 3 Preprocessing of the test image

3.4 Segmentation

This segment is usually supported by image pixel structures. These edges can define locations. Other methods separate the image from areas based on color.

Color image segmentation supported the color characteristic of the image pixels assumes that similar colors within the image imply different objects. Hence, object classifications happen inside an image. This basically means that each group has a category of pixels that have similar color characteristics.

In this project, we use color image separation to extract the properties of the test image. Partitioning stages include morphological operations on RGB to HSV (color, saturation and value), binary conversion, binary black and white image and clear boundaries of the infected area.

3.5 Feature Extraction

Feature extraction describes the process of obtaining information about the shape that we are interested in. This process helps it easier for the recognition function to recognize the pattern [11]. In image processing, pattern recognition primarily occurs in the form of dimension reduction. Many algorithms make this possible.

If the data fed to the algorithm is too big, then we convert it to a reduced set called the feature vector. The features that make it here go through a feature selection process, and the features obtained after this selection process are the ones that have the useful information from the input data. We can use this set in the algorithm as it will be faster and easier as it is a smaller set to process.

In our project, we use Hear Cascades. Haar Cascades is a machine learning technique in which the model is trained using several positive images that help in improving the model and negative images which help in removing errors [12]. This process is used to find out facial features.

Figure 4 describes the various features of the face that are extracted and compared in the face recognition module. The most prominent features of the face are extracted; the eyes, nose and mouth determine a theoretical model of the face. This information is then carried out towards the output modules.

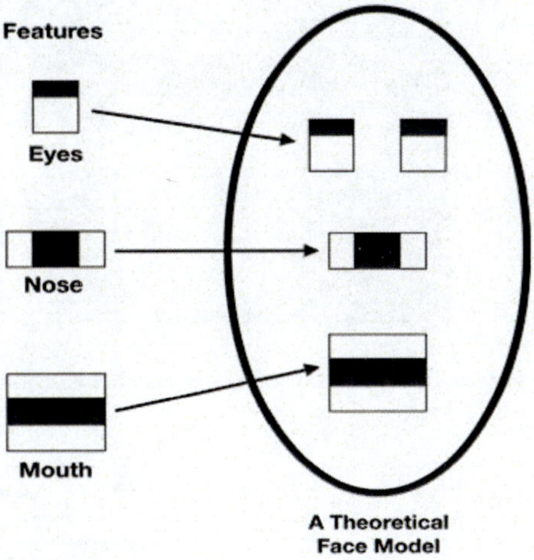

Fig. 4 Extracted features of the face

3.6 *NumPy*

NumPy is the primary extension in Python that is used for advanced mathematical operations as well as simple processing of multidimensional arrays [13]. Since most images are multi-dimensional arrays, this is of prime importance.

In our project, by storing the images read as a NumPy array, various image processing can be performed using NumPy functions. NumPy allows acquisition and rewriting of pixel values, trimming by the slice, concatenating, etc.

In our traffic light detection module, we store the pixel values of the image as a NumPy array, and using this, we compare the values with the standard values set to identify the region with the required colors.

3.7 *Bilateral Filter*

A bilateral filter is a very common filter used in image processing. The main reason for this is that it keeps the edges sharp while removing noise effectively [14]. Another option we had was the Gaussian filter. The advantage of the Gaussian filter is that it is faster. However, the Gaussian filter algorithm of finding the weighted averages of nearby pixels is a flaw for our project as our focus is the features that may get ignored while using the Gaussian filter. The Gaussian filter also blurs the edges which can cause problems for the model to recognize [15].

This is used in our signboard text detection, the edges of the letters can be clearly identified, while the other regions of the images are blurred (Figs. 5 and 6).

As you can see in the above images, the edges are preserved, and this helps to identify the image text in our project.

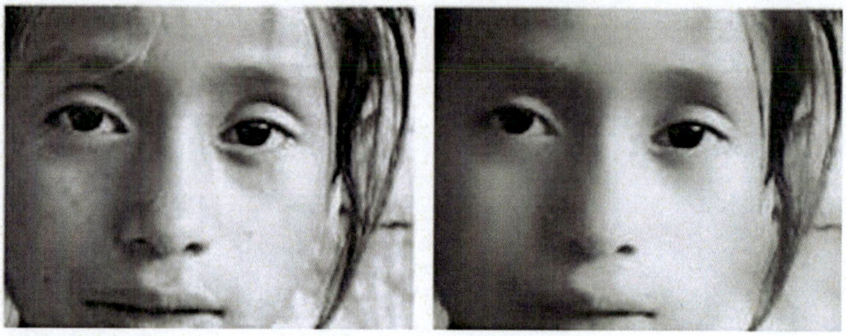

Fig. 5 The application of a bilateral filter is seen in the above pair of images

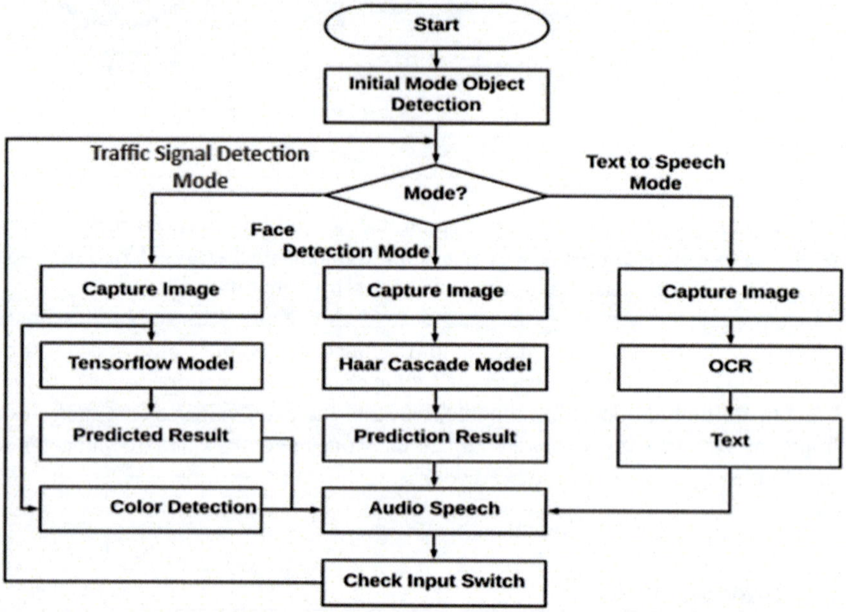

Fig. 6 Flow chart of our system

4 Results and Discussion

The traffic light detection module identifies the color of the signal, and if it is red, we instruct the user to cross as all the vehicles would have come to a halt; similarly, if it is yellow or green, we ask them to wait through the audio output.

Here are a few sample images where the colors are identified (Figs. 7 and 8).

The face recognition system stores the sample images of the face and then compares the input image with the available dataset to classify the person. Some of the output images of the module are provided below in Fig. 9. A classification

Fig. 7 Snapshot of the detection of red signal color

Fig. 8 Snapshot of the detection of green signal color

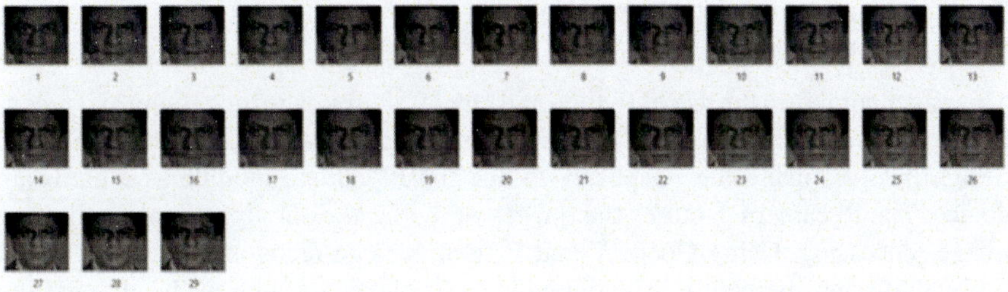

Fig. 9 Snapshot of the sample dataset

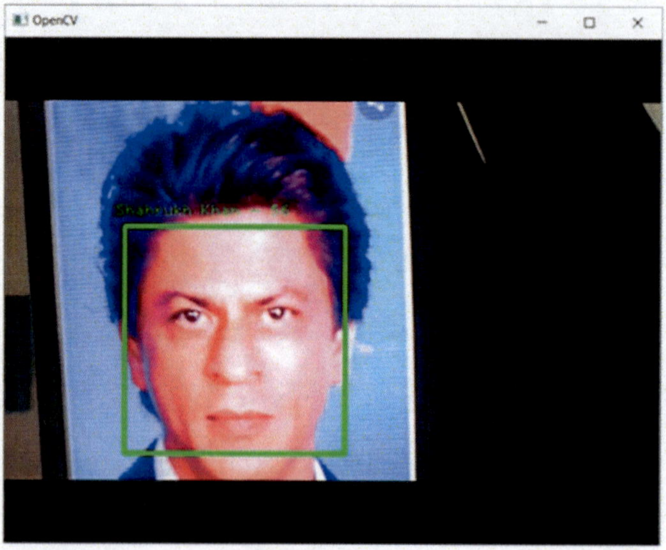

Fig. 10 Snapshot of the command window during runtime

algorithm then classifies the image based on the training data (Fig. 10).

The signboard text detection module is capable of reading short phrases and very small sentences, like those on the signboards. The text-to-speech is capable of converting any English string to an mp3 file that can be played out loud using any of the default audio player present in the device. The interface between the user and the system is done using buttons to choose which module is to be used.

5 Conclusion and Future Work

This paper combines the use of Python and OpenCV for image processing and recognition. This paper proposes an efficient and cost-effective system for the blind. System design involves using the Raspberry Pi and the Pi camera module as a hardware device. The Pi camera captures the images of the face/traffic light/signboard, using image processing. Using OpenCV and Python, we can recognize faces and detect traffic lights and signboards. The output audio is played through the headphone and provides necessary instruction/information to the user.

The system allows the blind to act independently to a certain extent. The user can choose the mode of operation using buttons, depending upon the assistance that is required. He/she has three modes to choose from, which are the 3 modules in our project.

The future work on this system can be as follows:

- We can improve the signboard recognition module to help read books, etc., and hence serve as a book reader.
- We can enhance the model by retraining the model on a cloud GPU and use it to recognize a wide number of faces, signs, etc.
- We could add a GPS module to our device which will send the information about the current location of the user to their care-taker, etc.
- The information could be received on the care-takers mobile as a text message or also we could develop an app using Android Studio, which could show us the real-time location of the blind person (like Google maps).

References

1. Pascolini D, Mariotti SP (2011) Global estimates of visual impairment: 2010. British J Ophthalmology 96(5):614–618
2. Dakopoulos D, Bourbakis N (2010) Wearable obstacle avoidance electronic travel aids for blind: A survey. IEEE Trans Syst, Man, Cybern Part C (Appl Rev) 40(1):25–35
3. Varpe KM, Wankhade MP (2013) Visually impaired assistive system. Int J Comput Appl 77(16):5–10
4. Elmannai W, Elleithy K (2017) Sensor-based assistive devices for visually-impaired people: current status, challenges, and future directions. Sensors 17(3):565
5. Agarwal AWP, Zink M, Adame MR, Möller K, Burgard W (2016) Navigating blind people with walking impairments using a smart walker. Auton Robot 41(3):555–573
6. Gay WW (2014) The Raspberry Pi. In: Mastering the Raspberry Pi. Apress, pp 5–8.
7. Gay WW (2018) Pi camera. In: Advanced Raspberry Pi. Apress, pp 493–499.
8. Pulli K, Baksheev A, Kornyakov K, Eruhimov V (2012) Real-time computer vision with {OpenC}. Queue 10(4):40–56
9. Geetha C (2019) Optical character recognition with tesseract. J Mech Contin Math Sci 1(2)
10. Google Open Source Google, [Online]. Available: https://opensource.google/projects/tesseract
11. Setiono R, Liu H (1998) Feature extraction via Neural networks. In: Feature extraction, construction, and selection. Springer, pp 191–204
12. Automatic gate using Face recognition technique using HAAR cascade algorithm. Int J Eng Adv Technol 9(3):1302–1305.
13. Harris CR, Millman KJ (2020) Array programming with NumPy. Nature 585:357–362
14. Tomasi C, Manduchi R, Bilateral filtering for gray and color images. In: Sixth International Conference on Computer Vision
15. Sagar V (2015) Gaussian blur filter effect for CAPTCHA security. Int J Comput Appl 124(3):29–31

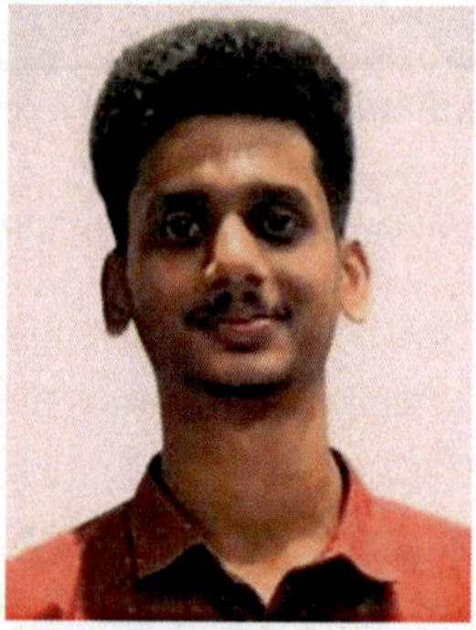

N. Sriram is a final year student at Vellore Institute of Technology pursuing his B.Tech in Electronics and Computer Engineering. His research work is concentrated in the fields of computer vision and machine learning. He has a strong sense of problem-solving and strives to solve real world problems using state-of-the-art technology.

Anirudh Hosur is a final year student at Vellore Institute of Technology pursuing his B.Tech in Electronics and Computers. He's research work is concentrated in the fields of deep learning and computer vision. He has a strong sense of problem-solving and a data-driven enthusiast.

Akash Reshan a final year student at Vellore institute of technology pursuing his B.Tech in Electronics and Communication. His research work is concentrated in the fields of deep learning and computer vision. He has a strong sense of problem-solving and wants to improve the current systems using newer technology.

Vetrivelan Pandu is an Associate Professor Senior and Head of the Department (HoD) for Bachelor of Technology (Electronics and Communication Engineering) in School of Electronics Engineering at Vellore Institute of Technology (VIT), Chennai, India. He has completed Bachelor of Engineering from the Madras University, Chennai and both Master of Engineering in Embedded Systems Technologies and Doctor of Philosophy in Information and Communication Engineering from Anna University, Chennai. He has around 17 years of teaching experience altogether in CSE and ECE Departments in both private Engineering Colleges in Chennai (affiliated to Anna University, Chennai) and Private Engineering University in Chennai respectively. He is the author of five book chapters and one proceeding in lecture notes published by reputed springer publisher, and has authored 30+ Scopus indexed Journal papers and few other papers published in reputed international conferences. He was member in Board of Studies, doctoral committee, doctoral thesis Examiner, doctoral oral Examiner in both private and government Universities. He worked as a reviewer for reputed International Journals and International Conferences. His research interests include Wireless Networks, Adhoc, VANETs, Embedded Systems and Internet of Things (IoT) with Machine Learning.

Loan Origination System in Housing Sector Using Blockchain

Utkarsh, Hemant Pamnani, Puneet Saran, and P. Vetrivelan

Abstract This paper tends to analyze the pattern of automated loan origination, particularly in housing sector, with a clear demarcation of "potential" customers via an adequate blend of blockchain technology to smooth-out the very procedure unlike the tedious traditional fashion of loan sanctioning. It's basically a decentralized setup that stores debt history: Debt repayment of debt default. The proposed model can be of great aid to the SMEs (Small and Medium Enterprises) as they would be able to attain access to loans with no collateral in place, thus, with low risk and high-quality appetite, they would certainly be able to display their credibility and key-performance indicators through adequate distribution of information. The entrance of blockchain approach in our proposed model (inspired by Peer to Peer lending) ensures an accelerated and a quite transparent procedure of accessing loan funds without getting into the tiresome loop of conventional transactional dealings. This paper is superficially tending toward defining a feasible business model for Peer to Peer lending from a technical and business standpoint.

Keywords Blockchain · Lending · Peer to Peer · SMEs

1 Introduction

P2P lending is basically, a lending system that is free from the involvement of any collaterals [1]. The market segment targeted is essentially those with negligible or no credit history. Typically, this form of lending in risky as compared to regular

Utkarsh · H. Pamnani · P. Saran · P. Vetrivelan (✉)
School of Electronics Engineering, Vellore Institute of Technology, Chennai, India
e-mail: vetrivelan.p@vit.ac.in

Utkarsh
e-mail: utkarsh.2018a@vitstudent.ac.in

H. Pamnani
e-mail: hemant.pamnani2018@vitstudent.ac.in

P. Saran
e-mail: puneet.saran2018@vitstudent.ac.in

A. Sivasubramanian et al. (eds.), *Futuristic Communication and Network Technologies*, Lecture Notes in Electrical Engineering 792,
https://doi.org/10.1007/978-981-16-4625-6_91

lending methods as it involves no collateral or security against the loan offer. P2P lending is facilitated by the process of bidding, profiling, matching of offers, loan amount transfers and loan collections or recovery [2]. This entire process is performed digitally, and hence is bound to include certain extra costs like processing charges, maintenance charges, etc. Operational charges, however, would be reduced because it is an online process and the intervention of a human being would be much lesser as compared to a traditional process. This method of P2P lending helps with faster on-boarding processes and financial inclusion [3]. This paper explores the usage and the power of blockchain technology in the Peer to Peer lending space.

1.1 Objectives

The followings are the objectives of this project:

Return on sales (ROS) maximization, extended Peer to Peer lending, implementation of optimization algorithms, entire turn around time (TAT) reduction, self-instantiation and futuristic modeling, operationalize on real-time constraints, employment of validation checkers as well as robust machine learning algorithms to ensure smooth/transparent decentralization, encryption as well as decryption of relevant meta-data with the aim of intensive parallelization (as followed in Hadoop Distributed File System).

1.2 Background and Literature Survey

1.2.1 History of P2P Lending and Its Growth

Earlier known models of P2P lending were mainly dealing regarding lending money to friends, family and close ones—This is where a factor of trust began to show importance as the lending bandwidth grew in terms of people and ticket sizes. A problem of this kind gave rise to a lending system based on blockchain that facilitates Peer to Peer lending without involvement of an intermediary in the entire process. In India, P2P lending is primarily regulated by the RBI (Reserve Bank of India). The RBI typically grants licenses to market players to operate under the P2P lending space [4]. Companies with licenses can borrow money from central lenders and lend it further to their customers but cannot end it out from their own books. The limit on lenders is 10 Lakhs, and the limit on borrowers is also the same at 10 Lakhs [1, 5].

1.2.2 Blockchain

The role of blockchain in reducing costs, time of processing and the elimination of intermediary, in turn reducing operational costs have been covered in the previous sections. Blockchain facilitates decentralization of data—this helps to eliminate the dependency of a single source for data storage and retrieval. Data immutability helps make the data encrypted cryptographically. A hashtag considered as an identification point for such a situation. To enhance and to add another layer of security, computers can act as validators for the data—this improves transparency and adds to the accountability of the process. Elimination of sub processes and/or simplification of the same using blockchain technology helps reduce costs as compared to traditional lending procedures. Blockchain technology as per (Fig. 2) holds the potential for being applied in financial processes like Government cash management, transaction processing, clearing and settlement of financial assets, bank ledgers, etc. Blockchain also offers solutions to issues like:

Efficiency, Lag in transactions, Fraud, Risks in operations.

Cryptocurrencies are implementations of blockchain that fit in the architecture of P2P lending. Flow of funds into the system would be faster and would take place without any additional bank processing charges or delays. This platform also has the capability of being scaled to various geographies globally and would limit the role of bankers to such implementations. This system could possibly eliminate the need of a third-party credit rating system. This is because the blockchain system is robust to keep track and would also be able to maintain the credit score of borrowers [6].

2 Blockchain Futuristic Model

The entrance of blockchain approach in our proposed model (inspired by P2P lending) ensures an accelerated and a quite transparent procedure of accessing loan funds without getting into the tiresome loop [7] of conventional transactional dealings.

2.1 Methodology

This paper is inspired by P2P lending mechanism and the proposed model in this paper aspires to thrive on two major dependencies/ criticalities:

Self-instantiation of the built model, modern system integration, (optional) default cases computation and segregation of the "potentials."

Some of the robust algorithms as part of "AI-powered blockchain-involvement" around which the analysis is supremely aspired to work upon:

Validation Algorithms: For ensuring sufficient iterations of check, we could employ K-fold cross validation, Random bootstrapping, sub-sampling, etc.

Search Algorithms: For keeping demographic records of the customer-base, it's recommended to go with the virtual wrappers, being trained on informed search, breadth first search (BFS), depth first search (DFS), iterative DFS, etc.

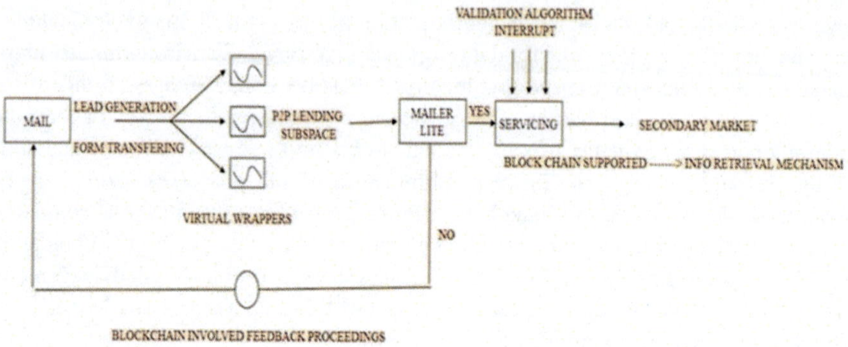

Fig. 1 Block diagram of proposed model

Predictive Algorithms: For the clear demarcation of the "potentials" as well as "weak learners," this could have been done with ease by decision trees as well as logistic regression, but our primed concern here is to work upon the strengthening of the ensemble model (blend of random bootstrapping as well as bagging). Also, boosting can be followed to make the ensemble more predictive.

Intensive Parallelization approach: Like HDFS where files are distributed and not copied or transferred, the similar approach is appreciated in loan origination system (LOS) blockchain model to attain a believable model with no dependence on collaterals (with constraints).

2.2 *Implementation of Proposed Methodology*

2.2.1 Origination

In origination, the borrower applies for a new loan and then application is then sent to a lender for processing. This is the standard procedure among all loans. With the help of blockchain technology, the above initial step can be improved with accurate record keeping [6]. This will give lenders access to borrowing information and data from their lead generation procedure as shown in (Fig. 1).

2.2.2 Fulfillment

Once the loan is fulfilled, the lender gives estimates of the loan amount and issues the loan itself. Blockchain can speed up this process by providing estimates within 3 days and also help with the details on the time stamped receival of documents.

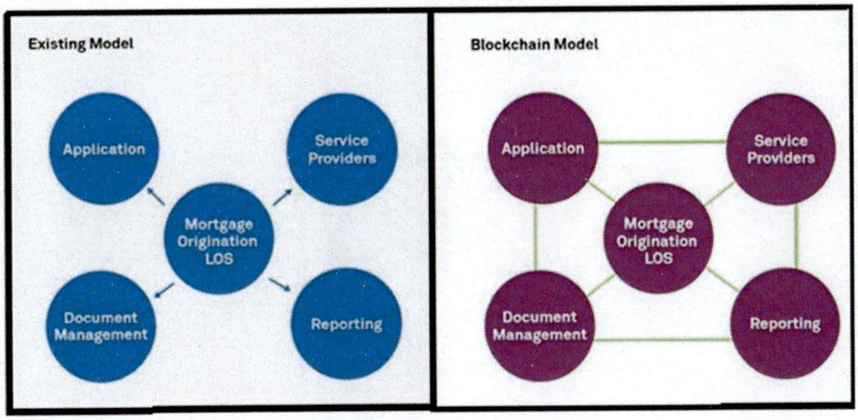

Fig. 2 Existing model in contrast with blockchain Model

2.2.3 Settlement

During a loan closing, many documents needed to be presented by the borrower's side, thereby adding stress to this complex process. The closing of a loan can be accelerated with precision with smart contracts via blockchain.

2.2.4 Servicing

In this step, the lender collects interest, principal and escrow from the borrowing. This ends up being a complicated exchange of information and money. With the aid of blockchain, servicing can be improved with heightened tracking of payments, increased data accuracy within loan transfers and ease of asset movement to servicers. This is due to blockchain's validation requirement.

2.2.5 Secondary Market

The secondary market provides a space for loans to be bought and sold by various lenders and investors. Blockchain offers extreme transparency of ownership of assets and streamlines the transaction processes like smart contracts can be implemented in master services agreements, pooling and service agreements and investor contracts which cannot be easily manipulated and can be viewed within the chain. Also, blockchain's automatic, immediate update of information too ensures the safety of secondary market deals as shown in (Fig. 2).

Compiling all the discussed formulations, the model would head toward as: Primarily, the external borrower approaches P2P lending sub-space ("allowed" space for P2P lending by RBI [8]), major stakeholders being SMEs, potential guarantors,

Fig. 3 Breakdown of Mortgages with blockchain

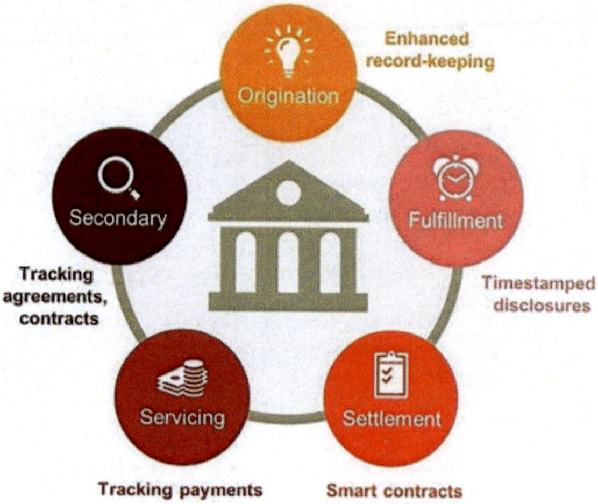

etc. via "lead generation" form filling as shown in, followed by lending process fulfillment (AI-powered blockchain involved: blend of predictive algorithms and optimization classifiers). Thirdly, we go for settlement section which involves the blockchain-involved platform-independent dealings: MailerLite (email automation, landing page activation, etc.).

Followed by this, the concluding procedures include servicing as shown in (Fig. 3) using robust search algorithms and secondary market (Blockchain updation features as well as information-retrieval methodologies) [9].

2.3 Design Approach

The very purpose of introducing the concept of blockchain in proposing the design of the futuristic model is to ensure the fact that the crucial clouds embedded with decisive information regarding the customer-base remain privately secured from any random alteration, hack, or cheat by externals.

The concept of blockchain employed here, being superficially a digital ledger of transactions (DLT), is duplicated and distributed across the entire section of the proposed wireframe. Each block in the chain contains a number of transactions and relevant ensures in the form of discussed active algorithms, and every time a new feed occurs on the blockchain, and a record/clone of that transaction is added to every participant's ledger.

In short, the discussed concept is thriving on an immutable cryptographic signature termed as "hash."

A BPMN (Business Process Model and Notation) is ideally configured which is inline with the thought-process.

It starts from interface interactivity (when borrower steps into the P2P lending sub-space), followed by "Lead generation" automation through email. Then enters the procedure of loan assessment as well as fulfillment with the aid of optimization algorithms, followed by settlement and servicing with the aid of validation algorithms being applied on. Finally, stepping into the secondary market, where blockchain-supported information-retrieval approaches are employed to obtain the lowest level of abstraction (being more specific).

3 Conclusion

Decentralization at each section of the proposed model with the aid of blockchain technology (DLT as shown in (Fig. 4) along with the employment of vivid ML algorithms such as recommendation, search, validation, information-retriever, parser, locator, etc.) smoothens the tedious conventional procedure and the very process is tending to mirror-out the P2P lending with "no" collaterals [3], however, virtual parties are there to ensure veracity.

Furthermore, on-boarding procedure is taking less time due to the adequate digital customization with blockchain-involvement. Cost reduction due to the absence of any sort of physical intermediaries [4] except virtual wrappers and databases to carry out the smooth online transactional processing (OLTP).

Also, it leads to total TAT reduction, that is, turn around time decrementation due to the additional virtual controls.

Smarter interest rate generation following "Attribute Subset Selection" approach.

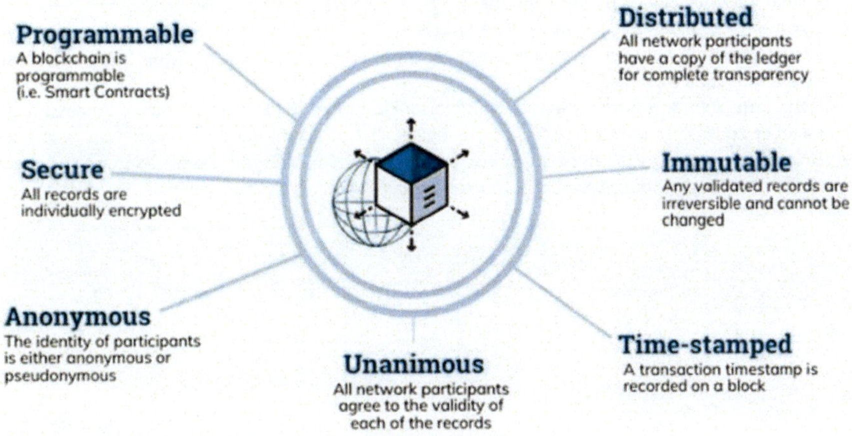

Fig. 4 Properties of distributed ledger technology (DLT)

4 Future Directions

Throwing light on the model feasibility, since the procedure is digitally adhered (automated), it almost resembles P2P lending (with soft allowance or liberty of "discussed constraints"). Loan origination system's, primary dependence for its future thriving, is on 3 criticalities:

Self-instantiation, modern system integration and default cases (cannot be accurately predicted).

Thus, the area of interest and scope of research/ analysis are more in the accurate prediction of default cases corresponding to the number of potential customers with an adequate blend of blockchain technology.

References

1. Degong MF, Yancun CS, Wei YT (2009) Evaluation of the financial derivative risk from the probability of default angle. In: IEEE 3rd UKSim European Symposium on Computer Modeling and Simulation, pp 293–298
2. Abd Elminaam DSF (2016) Smart kitchen: automated cooker technique using IoT. Int J Electronics Info Eng 1:1–10; 2(5):99–110
3. Adriana D, Dhewantoa W (2018) Regulating P2P lending in Indonesia: Lessons learned from the case of China and India. J Internet Banking Commerce 23(1)
4. Allen F, Santomero AM (2001) What do financial intermediaries do? J Bank Finance 25(2):271–294
5. Fan PF, Liu YS, Zhu JT, Fan XF, Wen LF (1999) Identity management security authentication based on blockchain technologies. Int JNetwork Security 6:912–917
6. Ali F, Azad S, Smith T, David F (2019) Blockchain and mortgage lending process: A study of people, process, and technology involved. J Applied Knowledge Management 7(1):53–66
7. Larrimore F, Jiang S, Larrimore T, Markowitz F, Gorski F (2011) Peer to Peer lending: The relationship between language features, trustworthiness, and persuasion success. J Appl Communication Res, pp 19–37
8. RBI, https://www.rbi.org.in/Scripts/NotificationUser.aspx?Id=11764&Mode=0. Accessed 22 October 2020
9. Sridhar Sathyanarayan, https://www.wipro.com/en-IN/banking/blockchain-solutions-formortgage-loan-origination/. Accessed 21 October 2020

Blockchain-Based Secure IoT Telemedicine System

Dhruv Piyush Parikh, Aayush Dhanotiya, and P. Vetrivelan

Abstract We can extensively define the usage of telecommunication systems to provide clinical facts and alerts which can be cross-validated with human intelligence which in-turn gives better accuracy in the health industry. The diagnosis of the disease is a significant and tedious task in health care. The health-care industry gathers large quantities of disorder databases that, unfortunately, are not "mined" to determine concealed records for effective decision-making by health-care practitioners. Heart disease is a term also used interchangeably for cardiovascular disease that describes conditions involving blocked or narrowed blood vessels which can lead to a heart attack, stroke, or chest pain (angina). Myocardial infarction and cardiovascular disease are a few categories of heart diseases. In this paper, we will research cardiovascular diseases, heart attack, breast cancer, and providing a network for better medical solutions. The data classification is based on supervised machine learning algorithms which engendered accuracy, time taken to build the algorithm. The data evaluation is done using fivefold cross-validation, and the end results are compared to get higher accuracy. We have used bolt IoT modules for predicting risk index for patient's health, and the unique alert system is used to circumspect medical practitioners.

Keywords Random forest classifier · Telemedicine network · Bolt IOT · Twilio

1 Introduction

A significant endeavor confronting medical services associations (clinics, medical facilities) is the arrangement of value administrations at lower estimated costs.

D. P. Parikh · A. Dhanotiya · P. Vetrivelan (✉)
School of Electronics Engineering, Vellore Institute of Technology, Chennai, India
e-mail: vetrivelan.p@vit.ac.in

D. P. Parikh
e-mail: dhruv.parikh2018@vitstudent.ac.in

A. Dhanotiya
e-mail: aayush.dhanotiya2018@vitstudent.ac.in

A. Sivasubramanian et al. (eds.), *Futuristic Communication and Network Technologies*, Lecture Notes in Electrical Engineering 792,
https://doi.org/10.1007/978-981-16-4625-6_92

Quality carrier infers diagnosing patients adequately and directing effective medicines. We are living in the era where innovation is a need, telemedicine gives patient-focused methodologies which help in improving the idealness of care and health. Patients can address medical services issues rapidly with ongoing dire consideration meetings and find out about therapy alternatives inside a short time period, this is the primary explanation telemedicine is acquiring prevalence. Poor logical decisions can prompt lamentable results which are accordingly inadmissible. Emergency clinics ought to likewise limit the cost of clinical tests. They can gain these outcomes via utilizing reasonable innovation-based data as well as choice help structures. These days, most medical clinics utilize a type of facility insights to deal with their medical services or individual information. These frameworks commonly produce huge amounts of data which appear as numbers, literary substance, graphs and pictures. Tragically, this information is not frequently used to manage clinical decision-making. There is an abundance of data in those insights that are generally undiscovered [1]. This brings up a basic issue: "How might we transform information into helpful data that can empower medical care professionals to settle on astute clinical choices?" This is the foremost inspiration for this task can be comprehend from Fig. 1.

Major health-care organization's use informative devices to aid patient invoice, medical commodities management and developing simple insights. A few medical clinics use decision supportive networks, yet they are to a great extent restricted to metropolitan cities. They can answer basic questions like "What is the normal time of victims who have coronary illness?", "The number of medical procedures had brought about wellbeing office remains longer than 10 days?", "Recognize the female victims who're single, over 30 years collectible, and who have been

Fig. 1 Depiction of IoT telemedicine system

managed for cancer" [2]. However, they can't answer complex inquiries like "Distinguish the essential preoperative indicators that expansion the time of wellbeing office stay", "Given patient records on malignant growth, should therapy incorporate chemotherapy alone, radiation alone, or both chemotherapy and radiation?", and "Given patient insights, anticipate the likelihood of victims getting a coronary heart disease." Clinical decisions are regularly made dependent on clinical specialists' instinct and experience as opposed to on the information rich information covered up in the database [3]. This activity prompts undesirable predispositions, blunders and inordinate clinical charges which influences the nature of administration gave to patients. We propose that reconciliation of clinical decision guides with innovation-based patient measurements could diminish clinical slip-ups, enliven influenced individual security, decline undesirable practice variety, and improve patient eventual outcomes. This idea is promising as data demonstrating and examination instruments, e.g., information mining, have the capacity to produce an understanding-rich climate which can serve to significantly improve the assistance of clinical treatments [4].

The main reason for going forward with this idea came around 2018, the average prevalence of the cardiovascular disorder in India was evaluated to be 54.5 million. From every four deaths on was caused due to cardiovascular disorder in India. We are choosing our system to be working on these disorders because CVDs with ischemic heart disease and stroke are the main factors for over 80% of the burden [2]. Breast cancer is the most common cancer in India. It reports for 25–32% of female cancers in all cities across India. One woman is diagnosed with breast cancer every four minutes and dejectedly one woman dies of breast cancer, every 14 min in India.

2 Design and Methodology of Telemedicine System

The idea implemented in this project is a Blockchain-based IoT Telemedicine system. It will have multiple functionalities that help people as well as health practitioners in different situations.

The Bolt-Python module is used to generate digital output so that the person can view his health risk index. To implement this, we use the bolt chip that creates a connection between hardware and software.

The Random Forest module is used to identify the health risk of the person that is having these specified reports. For this, we will use python Sklearn to process our report data and predicting the outcome with respect to the dataset, we create a model by scikit learn. The features alone are stored in the dataset along with the name of the person, id and contact number of the person. Now, when a new report data is input, using sklearn. Ensemble we try to identify the health risk index of the person by matching the features of the dataset. This output is then sent to the local database module for generating a static database with all vital information [5].

The concept of blockchain applied here is superficially a Digital Ledger of Transaction, is duplicated and distributed across the entire section of proposed wireframe.

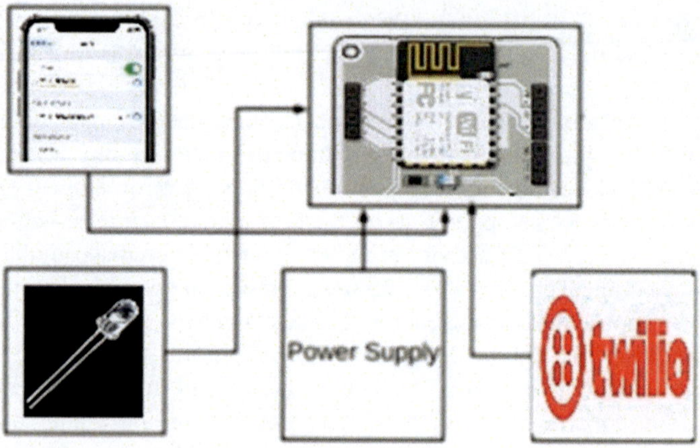

Fig. 2 Block diagram of the telemedicine system

Each block in the chain contains a number of transaction and relevant insurers in form of discussed active algorithm. Every time, the new feed occurs on the blockchain a clone of that transaction is added to every participant's ledger. This concept of resulting in an immutable cryptographic signature is called hash.

The Twilio system is used to help the health practitioner while getting an alert SMS, led high output, etc.; it uses python and bolt-SMS modules to convert and store the response. We initially set the bolt response to low values for the risk in the patients' health. So, based on the predicted response values in the array, we can identify the health risk to the patient and hence generate an instruction string which is sent to the hospital database for health-care service.

2.1 *Block Diagram*

The systematic work flow of the proposed assistive device for the health-care practitioners is shown in Fig. 2. Various components involved in this block diagram are explained in the following sections. The interface between user and system will be controlled by using the python programming interface.

2.2 *Bolt IOT Module*

Bolt is an Internet of Things Platform for the most rapid development of IoT products and services. This powerful platform comes with a Wi-Fi chip to interface your sensors and devices. We get free access to the Bolt Cloud to collect, store and visualize your data. The API's let you import or integrate your own personalized tools for your IoT projects.

We should use bolt because it has an amazing machine learning interface that helps to reduce over 75 percentage of our code as well as is its deployment time is 10 times faster. Bolt works with Over Integromat as well as works with SMS, E-Mail, iOS and Android Notifications, Facebook, Twitter, Instagram and hundreds of other services.

2.3 Led

A light-emitting diode is a semiconductor light-emitting entity that radiates light while current passes through device. It is used here to depict high or low risk indexes.

2.4 Twilio

Twilio Inc. represents a US organization leading in the CPaaS segment. CPaaS stands for Communication Platform as a Service. It's easy to get Twilio solutions setup into any project by means of the API. Twilio is for those who need communication services to manage their business. I'm talking about voice calls, video conferences, business messaging and chatting, etc.

Twilio is a kind of telephone exchange for offices, and it simplifies the whole process of handling multiple incoming calls. We prefer Twilio due to it premium parameters it provides such as a secure authorization process, call queuing and API to manage those queues. It helps in conversion of texts into speech and opposite transformations. They provide tools to build own scripts with PBX possibilities. Experts claim Twilio has made a real communication breakthrough.

2.5 Salient Dataset Features

The intricate part of any such system is the salient features of the dataset we choose. The machine learning algorithms are quintessential of the future, they can be specious at times thus we must choose features in such a manner that it precipitates to desired outcome. We have taken care of all salient features such as cholesterol, blood glucose, oxygen level, previous history engendered to current situation, rest ECG value, etc. [6]. Considering all these salient features, our model will provide a stimulating response.

2.6 Power Supply

An external power supply such as a power bank is required to power the Bolt IoT, and hence power the entire system.

3 Algorithm

This section discusses the platform and major modules used to implement the programming of the algorithm used to test the health disorders for the various user reports of the project.

3.1 Bolt IoT

Bolt IoT platform gives us the ability to command our system and gather data from connected IoT devices safely and securely. We get practical perception by deploying deep learning algorithms with a reliable correspondence to detect peculiarity as well as predict values.

Bolt IoT provides a confederate to quickly connect our system to IoT instances over general purpose input output ports, asynchronous receiver transmitter and analog to digital converter with power modulation. Chip is equipped with standard protocols to ensure a reliable and quick communication of our device data with the instances. Bolt IoT systems are often used to: improve functionality of machine learning algorithms, eliminate signal interference and other communication issues.

3.2 Random Forest Classifier

The random forest is a model consisting of two subdivisions known as Regression and classifier. It is made up of multiple decision trees from tree libraries. Instead of just averaging the prediction values of decision trees, this forest model applies two ideology that gives it the name random. Its capable of taking random samples of data points while building trees and splitting data into random subsets of key labels (features) while considering nodes.

While we are training the model, every tree in the random forest takes insights from a random sample of the data. The samples are taken with a proxy, called bootstrapping, which means that some samples will be used numerous times in a single decision tree. The main reason behind this is that by training each tree on non-identical samples, despite the fact that each tree might have high variation regarding

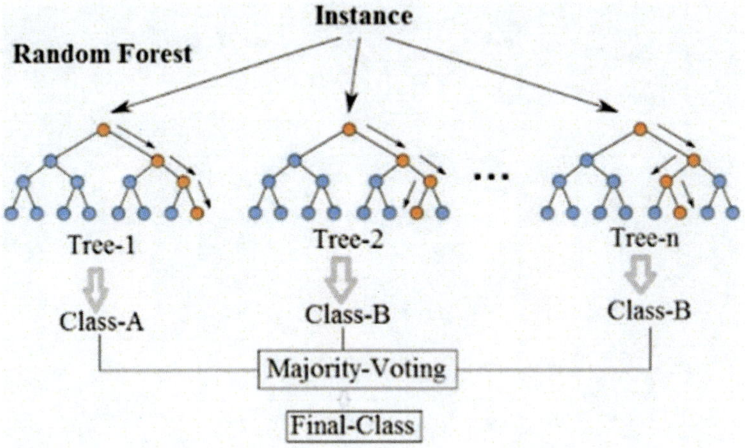

Fig. 3 Random forest classifier algorithm

a particular set of the training data. Generally, the entire forest will produce lower variance but not at the cost of higher bias.

The Random Forest Classifier is the key supervised algorithm that we have used in our project. This algorithm uses a set of multiple decision trees which randomly selects subsets of training dataset can be visualized in Fig. 3. It is compiled of votes from different decision trees to resolve the final class of the test variable or object.

Random forest is part of an ensemble library, this algorithm works efficiently even without hyper-parameter tuning, with a precise result most of the time. It is also one of the most used algorithms, because of its intelligibility and heterogeneity.

The main cause of choosing a random forest classifier for this system was that it provides better results as compared to SVM as well as logistic regression.

Secondly, the ameliorated reason behind choosing random forest classifier algorithm is that it deals with missing parameters and maintains accuracy of a large proportion of data. Even if there are more trees, it won't allow overfitting trees in the model. We have successfully achieved 78.32% accuracy, 74.41% accuracy in justifying cardiovascular disorder and whooping 93.41% accuracy in account for breast cancer can be visualized in Fig. 4 [7]. The threshold value of risk index is kept as 74% with a steady error of 5%. We have tried to generalize this for our complete system so that even if we have a mild chance, we won't take any risk with human health.

In this project, we have used this algorithm to predict that a person with a specified report has the possibility of corresponding disorder/health risk. The Accuracy of our model is the decision boundary for the health workers to keep a track on such cases and provide essentials.

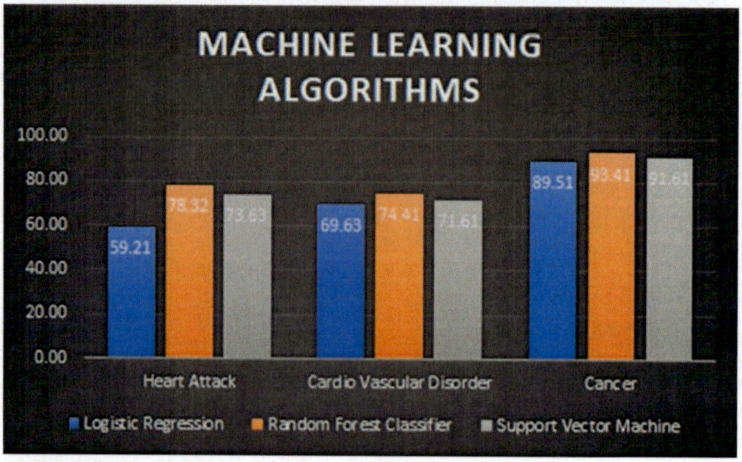

Fig. 4 Analysis of three different types of machine learning algorithm

3.3 *Model Framework*

A model is made robust by using the train test split method. In this technique, we evaluate the performance of our algorithm by creating multiple partitions and training our machine learning algorithm on them. After splitting data into a test set and training set, we can apply any classification model and can also apply any sort of supervised learning algorithm for the given dataset. For achieving higher accuracy, we prefer performing odd cross-validation sets and stratifying the output label for greater accuracy and precise output values.

We use this to verify our model accuracy and examine the loose ends of our algorithm. Figure 5 depicts the splitting of dataset into training, validation and testing.

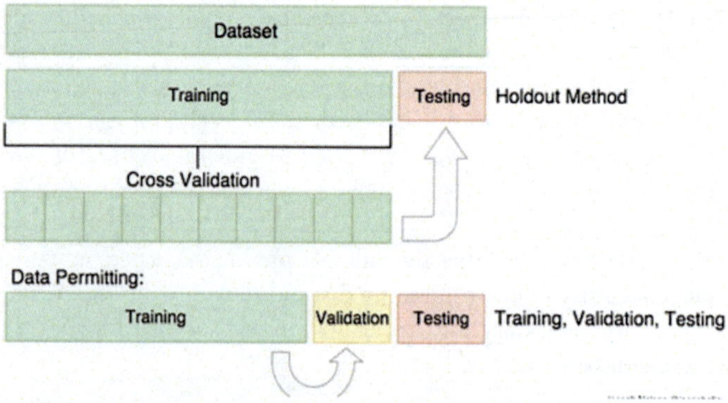

Fig. 5 Train test split on our dataset

3.4 Alert System Module

We have used the rest API from Twilio for this purpose, Twilio platform is very safe and secure for Ajax calls. We are generating alerts on the respective health-care device so that health regulatory bodies can perform their needful task as well as create availability for the higher risk patients. This might be supportive for patient's recovery [8].

In this project, we use powerful APIs to create interactive alerts that keep our users close, while making our infrastructure robust. Our alert systems infrastructure is built for exuberant live calls and low-visibility so we can scale quickly while maintaining accuracy. Thus, the most secure way to transfer response of the patient to the health worker and hospitals.

3.5 Local Database

Database is the chief constituent for Backend as Service of our model. After taking vital information from the user our system stores, it in a local database with comma separated format. This database is then connected to the cloud server for the hospital or the health firm of the specified locality [9].

When the input data to an algorithm is distributed under a model framework and the outcome is stored into the database in the last column, similar changes are passed on the hospital server through the cloud. If the algorithm states that the patient's health risk index is high the database will connect the vital information to the alert system and mark the details for further investigation and checkups.

In our system, we use a CSV base for storing the key information of the users and keep a local set for future aspects and creating a self-training artificial intelligence model. The online Database is helpful in creating a server which is accessible to hospitals, NGO's, Health Workers, etc. for a robust system.

Figure 6 shows the database architecture of the project that is formed by performing all above steps of the algorithm.

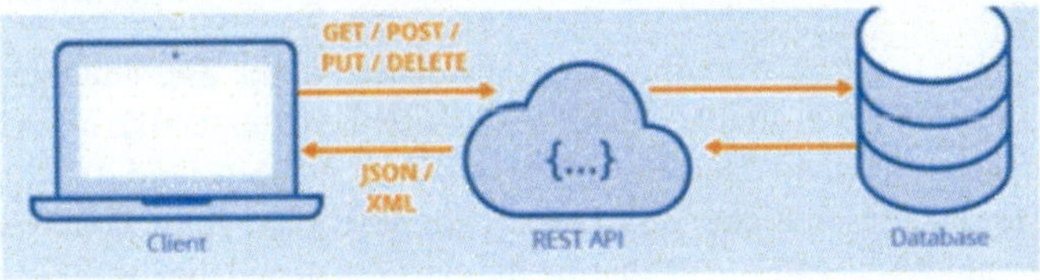

Fig. 6 Local database of our algorithm

Fig. 7 Picture of IoT system

3.6 *IoT System*

We can define the Internet of Things in many different ways. One way to describe IoT would be this: IoT is a system of interrelated devices, machines, objects that are provided with unique identifiers and can transfer data over a network without the need of human to computer interaction.

In our project, we have connected the bolt IoT system depicted in Fig. 7 with a power source and led that acts as an indicator for the machine learning algorithm we have used to scrutinize the person's medical report for an analysis on their risk of disorder. The system will give a high output (LED ON) if the patient is at higher risk index and using Twilio it sends an alert signal to the designated health service number of the locality.

4 Results and Discussion

The Bolt IoT module integrated with our algorithm identifies the risk of the patient. If led is on as depicted in Fig. 8, we alert the user to take necessary preventive measures and send the response to the health facility for the emergency service. If risk index is low for the patient, we just turn off the led or keep it in steady state as depicted in Fig. 9. Here are few sample images where the system is identifying risk to health for the patient.

The machine learning algorithm stores the dataset of the user and then compares the input data with the available dataset to classify the patients risk index. Figure 10 depicts the newly formed dataset; Fig. 11 justifies the alert module which will be generated for patients at higher risk index [csv—comma separated values].

The alert module is capable of sharing the alert signal and the details of the patient to the respective health service number. Our system will help the health practitioners to easily secure the patients at higher risk index. The interface between the user and the system is created using a python programming interface to help them select the report to be verified.

Fig. 8 Snapshot of the detection of high risk

Fig. 9 Snapshot of the detection of low risk

5 Conclusion and Future Work

Our system provides an accurate estimation on the health risk of the patient. The bolt IoT cloud is robust and quick with response time to ease the speed of the database connection with the hospital system. We have taken major diseases that have high risk in India so our main ideology behind this system was to address them. Our assiduous

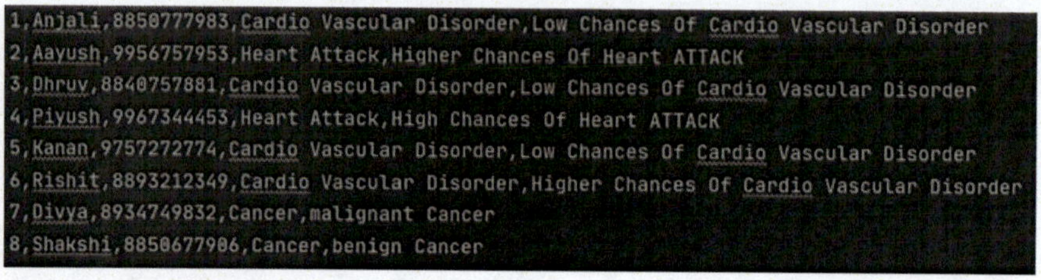

1,Anjali,8850777983,Cardio Vascular Disorder,Low Chances Of Cardio Vascular Disorder
2,Aayush,9956757953,Heart Attack,Higher Chances Of Heart ATTACK
3,Dhruv,8840757881,Cardio Vascular Disorder,Low Chances Of Cardio Vascular Disorder
4,Piyush,9967344453,Heart Attack,High Chances Of Heart ATTACK
5,Kanan,9757272774,Cardio Vascular Disorder,Low Chances Of Cardio Vascular Disorder
6,Rishit,8893212349,Cardio Vascular Disorder,Higher Chances Of Cardio Vascular Disorder
7,Divya,8934749832,Cancer,malignant Cancer
8,Shakshi,8850677906,Cancer,benign Cancer

Fig. 10 Snapshot of a CSV database

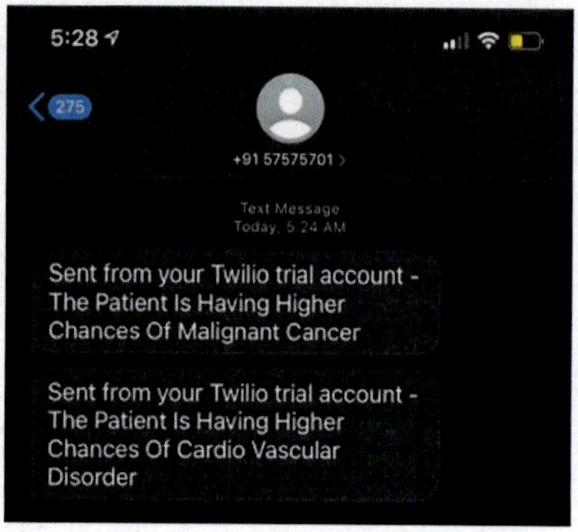

Fig. 11 Snapshot for the alert message to the health service

efforts have unfettered a new horizon in the medical industry. This paper combines the use of Python and Scikit learn for making integration of medical decisions supported with server-based patient records. This paper proposes an efficient and cost-effective system for the health-care industry. System design involves using the Bolt IoT and the led as a hardware device. The Random Forest Algorithm gives a decision boundary for the patients' health risk index, using Supervised Learning from Sklearn and Python, we are able to recognize risk in the patients' health. The output is stored in the database and provides necessary instruction / information to the user, if they are at high risk as well as sends an alert message to the health-care personnel [10].

The system allows the hospitals and health-care firms to act independently to a certain extent by helping general patients with required consultation within a timeframe. The user can choose the health affliction of operation using python interface, depending upon the assistance that is required. He/she has 3 consultation reports to choose from, which are the three major disease algorithms in our project.

In future, we want to add some commensurate functionalities to make an autonomous system for scheduling appointments with doctors and monitoring their consultation based on meeting id. We can also connect our platform with third parties for providing necessary supplements and providing a virtual environment for ease of our users. Using global positioning system, we can even share accurate coordinates of patients to the health-care team for efficient services in emergency situation.

References

1. Alajmi D, Almansour S, Househ M (2013) Recommendations for implementing telemedicine in the developing world. Stud Health Technol Inform 190:118–120. https://pubmed.ncbi.nlm.nih.gov/
2. Kruse CS, Soma M, Pulluri D, Nemali NT, Brooks M (2017) The effectiveness of telemedicine in the management of chronic heart disease—A systematic review. JRSM Open 8. http://dx.doi.org/10.1177/2054270416681747
3. Wason D, (2014) Design of telehealth trials—Introducing adaptive approaches. Int J Med Inform 83:870–880. http://dx.doi.org/10.1177/193229681300700617
4. Sood S, Mbarika V, Jugoo S, Prakash N, Merrell RC (2017) "What is telemedicine?" A collection of 104 peer-reviewed perspectives and theoretical underpinnings. Telemed. e-Health 13:573–590. http://dx.doi.org/10.1089/tmj.2006.0073
5. Choi WS, Choi JH, Oh J, Shin I.-S, Yang J-S (2019) Effects of remote monitoring of blood pressure in management of urban hypertensive patients: A systematic review and meta-analysis. Telemedicine. e-Health. http://dx.doi.org/10.1089/tmj.2019.0028
6. Otto L, Harst L, Schlieter H, Wollschlaeger B, Richter P, Timpel P (2018) Towards a unified understanding of eHealth and related terms—Proposal of a consolidated terminological basis. In: Health Informatics Europe; HEALTHINF. Madeira, Portugal
7. Hanlon P, Daines L, Campbell C, McKinstry BH (2017) Telehealth interventions to support self-management of long-term conditions: a systematic metareview of diabetes, heart failure, asthma, chronic obstructive pulmonary disease, and cancer. J Med Internet Res, 19:e172. http://dx.doi.org/10.2196/jmir.6688
8. Timpel P, Harst L (2019) Research implications for future telemedicine studies and innovations. Eur. J. Public Health 29(Suppl. 4). http://dx.doi.org/10.1093/eurpub/ckz185.098
9. Kidholm K, Ekeland AG, Jensen LK, Rasmussen J, Pedersen CD, Bowes A, Flottorp S.A, Bech M (2012) A model for assessment of telemedicine applications: Mast. Int J Technol Assess Health Care 28:44–51. http://dx.doi.org/10.1017/S0266462311000638
10. Dookhy R, Doarn CR (2005) Benefits and drawbacks of telemedicine. 11(2):60–70,Issue published: Hjelm NM (2005) Department of Chemical Pathology, March 1, St George's Hospital, London, UK

Item-Based Collaborative Filtering Blockchain for Secure Movie Recommendation System

Shikhar Kumar Padhy, Ashutosh Kumar Singh, and P. Vetrivelan

Abstract A movie or web show or song recommendation plays a very important role in day to day daily life in giving a common person enhanced entertainment. Such a kind of technology has the ability to recommend a group of snaps to user according their likes and dislikes, or how much a movie or films is popular. But for an efficient, accurate and precise recommender system, large volumes of data along with high quality are required. As a result role of customer's privacy comes into play. Many companies try to accommodate these concerns with various tactics but are unable to provide high decipherment security. So we are proposing a system that tends to recommend films or movies to as a replacement user using the block-chain technology that will reduce misuse of privacy concerns and will in fact entice them to share more data. It finds and search movie databases available already to cluster all the important information about the preferences and movie, such as, popularity and attractiveness, required for recommendation. It generates films flocks not only comfortable for movie makers or producer to plan a replacement movie but also useful for movie recommendations for the user of a platform.

Keywords Recommendation · Block chain replacement · Entertainment · Databases · Efficiency

1 Introduction

A recommendation system may be defined as a subclass of data filtering system that seeks to predict ratings based on user's preferences for any item primarily utilized for commercial applications. The basic idea about a recommendation system is mostly

---

S. K. Padhy · A. K. Singh · P. Vetrivelan (✉)
School of Electronics Engineering, Vellore Institute of Technology, Chennai, India
e-mail: vetrivelan.p@vit.ac.in

S. K. Padhy
e-mail: shikharkumar.padhy2018@vitstudent.ac.in

A. K. Singh
e-mail: ashutoshkumarsingh.2018@vitstudent.ac.in

A. Sivasubramanian et al. (eds.), *Futuristic Communication and Network Technologies*, Lecture Notes in Electrical Engineering 792,
https://doi.org/10.1007/978-981-16-4625-6_93

software or a program which has the ability to produce a list of videos, movies, films, songs, etc. [1]. Many companies are using this technology for their products such as YouTube for recommending a video to its user or a ecommerce company like Flipkart or Amazon to suggest a particular kind of product to its user. Some of the commonly used recommendation or suggestion systems are material-based and collaborative filtering recommender systems.

1.1 Collaborative Filtering

In collaborative based filtering, the recommendation is made upon the preferences of other users. For instance, person A liked a movie that movie will be recommended to other person B because of the reason that a person which has some connections with person B has liked it [2]. The simple and effective nature of memory-based technique makes it optimum to use from the latter. They are divided into two:

User-defined collaborative based filtering: In this model, suppose user A and user B liked a movie and a new movie comes out which is liked by person A then that movie is recommended to user B.
Item-based collaborative filtering: In this model, suppose user A gives 5 star ratings to two movies X and Y then another user B when watches movies X then movie Y is also recommended to him/her based on the similar ratings criteria.

Model-based methods are made and improved using data available and digging deep into it this process is also termed as data mining, algorithms of machine learning such as clustering and classification to predict users' rating [1, 3]. For this type of system, for better accuracy, dimensionality reduction techniques from Principal Component Analysis are used. Examples of such type of systems include Latent factor, Bayesian Model, Rule-based Model.

Content-based systems use metadata like genre, director, actor, singer to suggest things like films and songs. Such a suggestion would be for instance suggesting Inception that was performed by Leonardo Di Caprio because anyone saw and rated it high. The same thing happens in the case of music industry for certain artists such as Charlie Puth because you liked his Attention cover. The basic difference between the two filtering systems is explained in Fig. 1.

1.2 Block-Chain Implementation

Today, the importance of recommender systems has increased progressively as it assists people in examining decisions in a broader perspective. Especially in the areas of e-learning, e-library, e-government and e-business services it reduces transaction costs of finding and selecting items in an online and computerized environment.

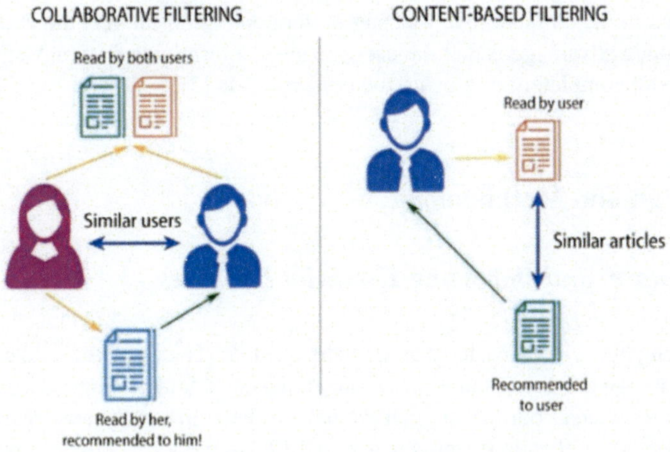

Fig. 1 Difference between collaborative and content-based Filtering

According to various marketing strategies, mass personalization helps in revenue augmentation as well as increasing customer gratification. As a result, companies tend to create a better customer interaction or behavior. Even if they know that the quality of recommendation will be increased, customers show reluctance toward sharing personal data because they fear fraud and misuse of their data. So, here we are using block-chain approach in order to store all the digital information in the form of blocks and the databases in the form of chain. Anchor aging block-chain technology enables the customers to be trustworthy of the recommender system and encourages them to fully provide more data to build it more efficient and improve its performance [4]. The contracts in the form of blocks are connected through databases in the form of chains is shown in Fig. 2.

The working is pretty simple. First the contract regarding how and what user's data are taken into account and are stored in the block-chain system by the company. Then the user can request for that contract, inspect it and allow for further sharing of his/her personal data. Finally the company can utilize that data as several input

Fig. 2 Block-chain technology

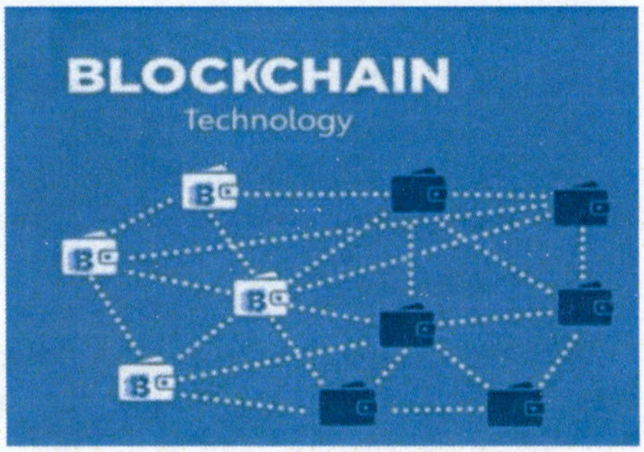

parameters to the collaborative filtering system and build the recommender engine [4]. The main advantage is that the user can stop or terminate from any stage at any time without completely disclosing the personal data [5].

2 Design and Methodology

2.1 Dataset and Importing Essential Libraries

For building our system, we have used the Movie Lens Dataset. We could use several datasets but we are using movies.csv and ratings.csv dataset from Kaggle. ***Here***'s the link to download our dataset that we have utilized in our System Project which consists of 105,339 ratings applied over 10,329 movies [1]. For this we need to install and load four packages or libraries—'reshape2', 'data.table', 'ggplot2' and 'reshape2'.

2.2 Data Retrieval

First we try to fetch our data points directly from movies.csv into movie data, data frame and ratings.csv into rating data and can make use of str, head(), summary() to display information about the movie data, data frame, view summary of the movies and first six lines of movie data, respectively. Similarly we can try it for the 'rating data' data frame [6].

2.3 Data Pre-processing

After giving head function for movie data the result we obtained is as shown in Fig. 3.

The table clearly describes about the integral values from user Id and movie Id column. Henceforth, it's important to change the genre present in the movie data data frame into a format that is more understandable by the users [6]. As there are limited number of genres so we will convert the string representing the genre into a numeric value using the one hot encoding method and we will create a matrix of it.

Now the task is to select the movies as easily as possible to do this genres that are already in the data list we will create a 'search matrix.' There are movies that have several genres, for example, Titanic, which is an American romantic film also falls under the genres of Thriller, Tragedy, Epic and Historical Drama [1].

Now our task is to pull the most different rating by creating a table and then we have to look for the movies that have been watched by most number of people in the database we have. Our first task is to count the number of views in a film

```
head(rating_data)
```

```
##    userId movieId rating  timestamp
## 1       1      16    4.0 1217897793
## 2       1      24    1.5 1217895807
## 3       1      32    4.0 1217896246
## 4       1      47    4.0 1217896556
## 5       1      50    4.0 1217896523
## 6       1     110    4.0 1217896150
```

Fig. 3 Top 6 movieId with their ratings

and then organize them in a data frame which will cluster them in descending order. Visualization of total number of views of top films could be carried out using ggplot2 function and finally visualized a heat map of the movie ratings that consisted of first 25 rows and 25 columns.

2.4 Data Preparation

Now our task is to prepare data in the given three steps.

2.5 Data Selection

From the dataset we tried to set the threshold for the minimum number of users who rated the film to 50. So it actually acts as a minimum number of views for a certain film resulting in filtering a list of watched films from the least-watched ones. And finally we will select those lists of films whose rating value is greater than the threshold value [3].

2.6 Data Normalization

There can be a case where some movies will have very high rating from the user and on the other hand some will have very low rating so in that case the movies with very high rating compared to others will have more effect on the mode so for that we have to normalize out data in a range so that a particular type of data does not have very high influence on the model compared to other data points [6, 3].

After normalizing the data points we then plot a heat map that delineates our normalized ratings which is visualized as given in Fig. 4.

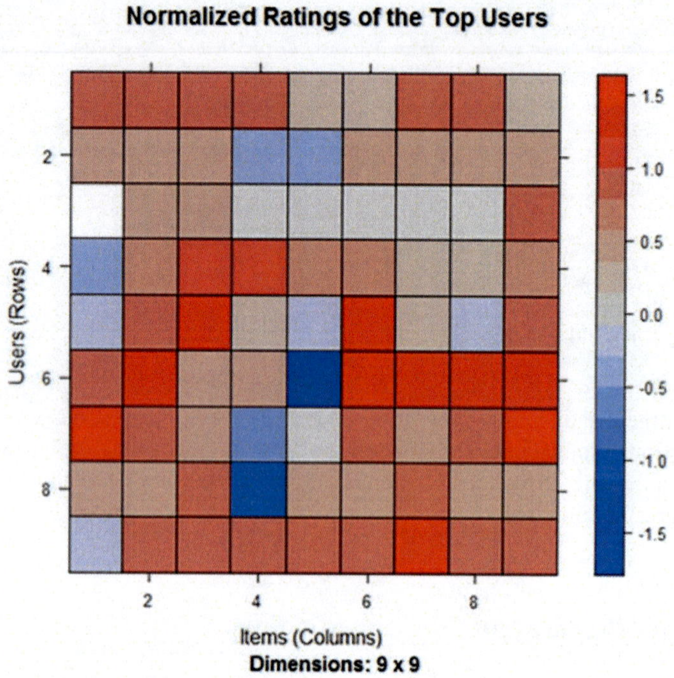

Fig. 4 Heat map normalized ratings of top users

2.7 Data Binarization

Now we will carry out with the data binarization process which basically means to convert the data in the form of discrete values 1 and 0 that will make the recommendation system more efficient. We will define a threshold value like 3 and then based on it will create a matrix that consists of 1 if the rating is above 3 otherwise it is 0.

2.8 Collaborative Filtering System

Here comes our most important part of this project in developing the recommendation engine i.e., Item-based Collaborative Filtering System. The definition of Item-Based Collaborative Filtering System is explained with the help of Fig. 5.

Basically what it does is it finds similarities between items based on people's ratings. Initially it creates a record of the customers who purchased similar items and then caters into the recommendation system. For instance, say, customer A purchases a product x and also product y[1, 2] Create a record or table of those items purchased

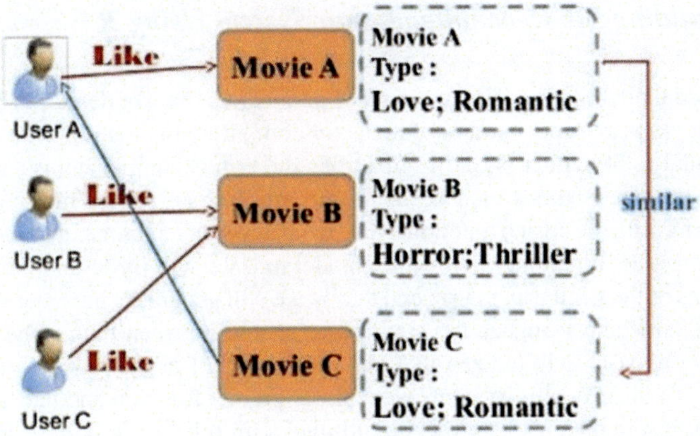

Fig. 5 Item-based collaborative filtering system

by that individual and then calculate the similarity. Figure 6 describes about the similarity between two movies watched by person A according to the Item-based collaborative filtering system.

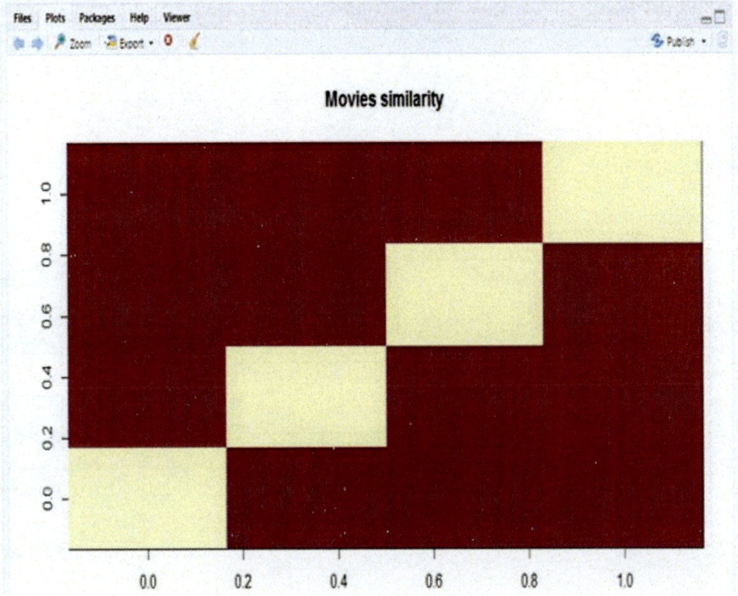

Fig. 6 Similarities between movies among different users

2.9 Building the Recommendation System Using R

Item-Based Collaborative Filter consists of parameters that are default in nature. So initially, let k denotes the number of items for computing their similarities. Here, we took k equal to 30 which primarily identifies the k most similar items and store it. There are various methods such as cosine, jaccard, pearson for the filtering process; we are going with the cosine function. Now, to retrieve the recommend model, we will use the getModel function. Then within model info we will try to find the class and dimensions of the similarity matrix and finally we will generate a heat map consisting of top 20 items and visualize the similarities shared between them. The next step is to create a variable initialized to 10 that keeps count of the movies watched by each user. To identify similar items we will use predict function and then rank them accordingly. And then each rating is multiplied with related similarities and finally adds them [1, 7].

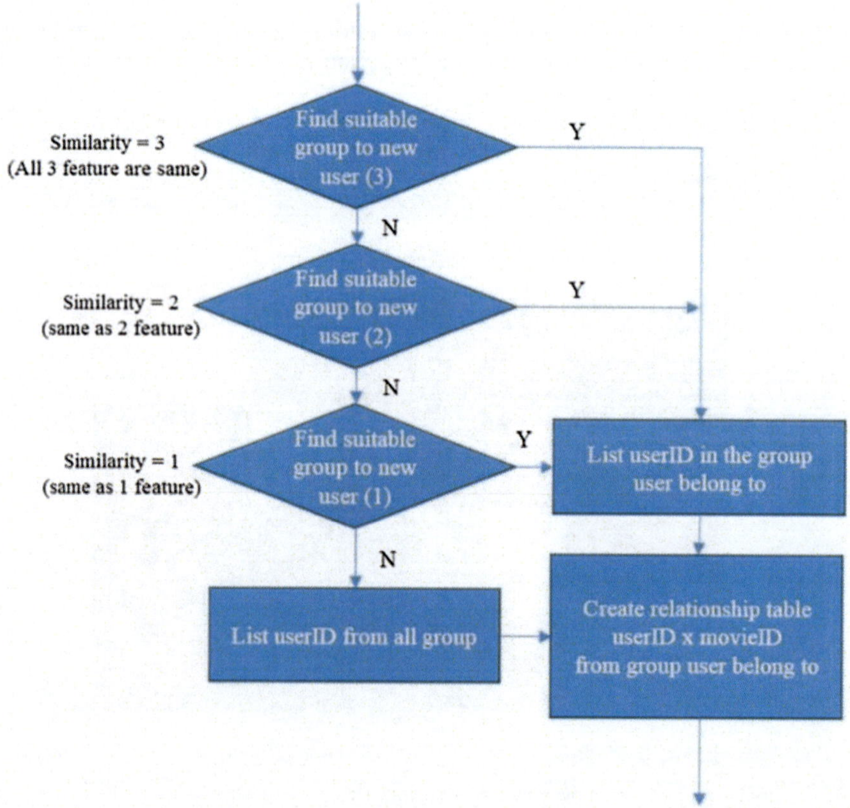

Fig. 7 Block diagram of the relationship table

Figure 7 describes the block diagram of the proposed K-NN algorithm used to create relationship table between user id and movie id from n clusters of data.

3 Steps and Implementation

This section discusses the platform and major modules used to implement the programming of the algorithm used to build the movie recommendation engine.

Processes

Step-1: Import the dataset.

Step-2: Data Exploration and Cleaning

Step-3: Merging the 2 csv files

Step-4: Working with genres column

Step-5: Working with the Cast column

Step-6: Working with directors column

Step-7: Working with Keywords column

Step-8: Similarity between movies

Step-9: Building the Recommendation engine

3.1 *Algorithm for Item-Based Collaborative Filtering System*

For user u, get items set N(u) that this user liked before. Recommend items which are similar to many items in N(u) to user u [2]. The necessary steps to be followed are as shown in Fig. 8.

4 Conclusion

Recommendation Engines are the foremost widespread sort of ML applications that is employed in various sectors. The evolution of these recommendation systems has drastically improved over time and has in fact assimilated several advanced ML techniques to provide the users with the content that they need [5]. In today's life, especially during this pandemic these engines have efficiently helped people to binge-watch movies according to their choices and preferences keeping them home quarantine.

However this system also has some disadvantages which obviously can be improved or upgraded by building a more advanced Memory-based cooperative filtering process where we basically divide the information database into a coaching set and a check set and then use techniques like circular function similarity to reckon

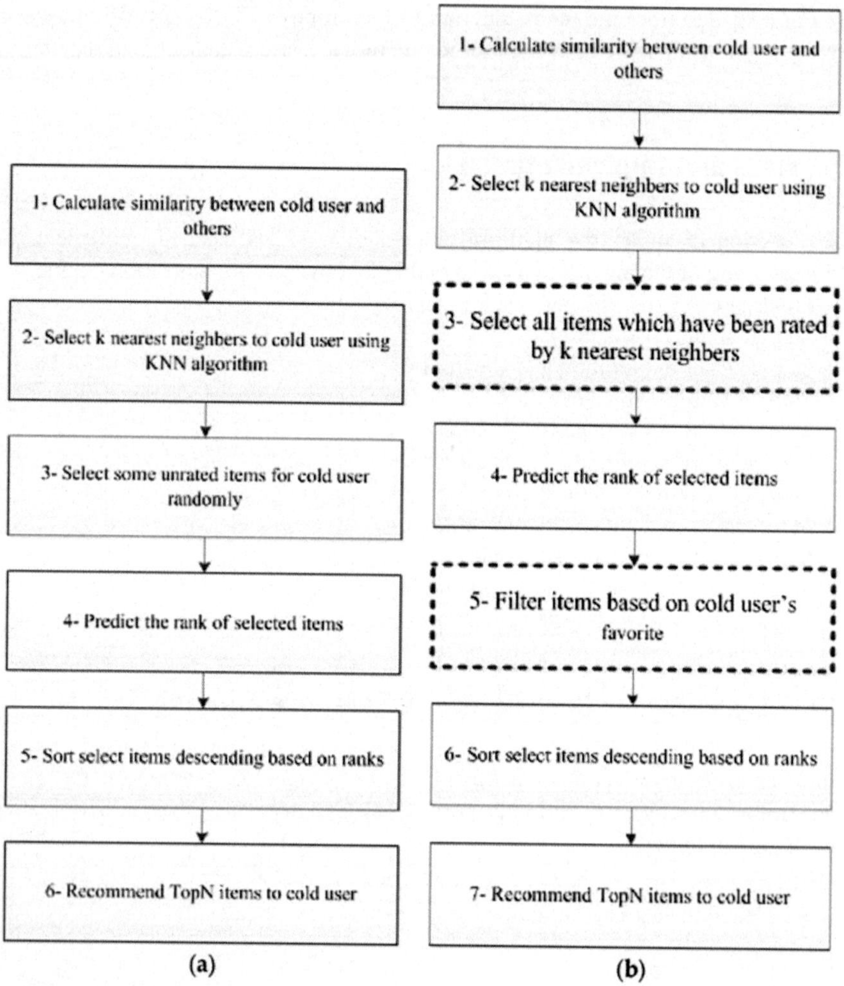

Fig. 8 Block diagram of selecting K-NN clusters of data

the similarity between the films. An alternate is to create a Model-based cooperative Filtering system. Matrix factoring is sweet at coping with quantifiability and scantiness than the previous [5].

We will then valuate your model mistreatment techniques like Root Mean square Error (RMSE).

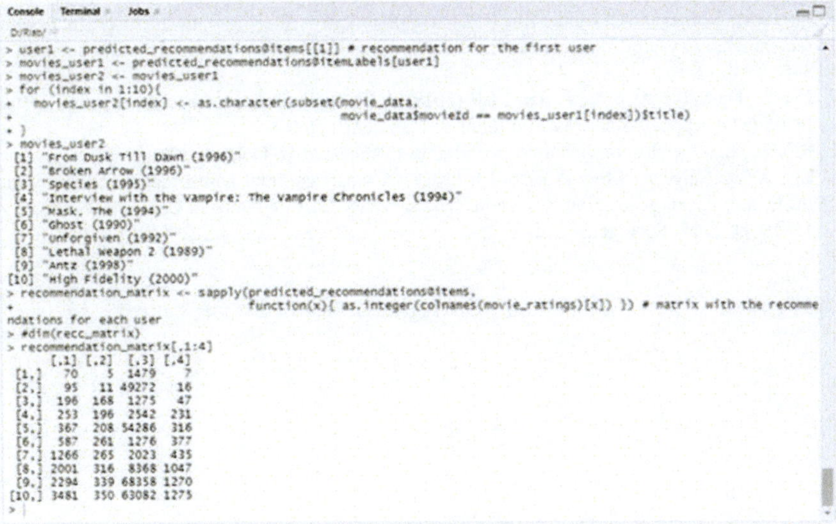

Fig. 9 Snapshot of the final results of data analysis

5 Results and Discussion

The algorithmic program of collaborative filtering system used for filtering the films based on choices first builds a similar-items table of the purchasers who have bought them into a mixture of comparable things and the latter can then be provided to advice system[2, 7]. The similarity between single merchandise and connected merchandise will be determined with the subsequent algorithmic program–

- Every item x present in the product catalog, purchased by customer A.
- Also each item y purchased by the customer A.
- Create a record of the items x and y purchased by C.
- Calculate the similarity between i1 and i2.

Figure 9 shows our final result of Movie Recommendation Engine.

References

1. Data Flair Blogs, https://data-flair.training/blogs/data-science-r-movie-recommendation/Last. Accessed 13 September 2020
2. Yunkyoung Lee F (2005) Recommendation system using collaborative filtering. San Jose State University, pp 390–409
3. Andreas Vogl, R-Bloggers, https://www.r-bloggers.com/2020/04/movie-recommendation-with-recommenderlab/. Accessed 16 October 2020

4. Frey RM, Worner D, Alexander Illic F (2016) Collaborative filtering on the block chain: A secure recommender system for e-commerce. In: Americas Conference On Information Systems, pp 3–6
5. Lisi A, De Salve A, Mori P, Ricci LF (2019) A block chain based recommendation system. Institute Of Applied Sciences And Intelligent Systems, pp 7–9
6. RPubs, https://rpubs.com/jeknov/movieRec/last. Accessed 16 October 2020
7. Lisi A, De Salve A, Mori P, Ricci LF (2019) A smart contract based recommender system. In: Economics of grids, clouds, systems, and services. Lecture Notes in Computer Science, vol 11819, pp 1–13. Springer, Cham

A 3-D Printed Trapezoidal Antenna for X-Band Wireless Communications Applications

Ashraf E. Ahmed, Wael A. E. Ali, Apurba Chowdhury, and Sudipta Das

Abstract By reason of a fast growth in the 3-D printing technology, 3-DP trapezoidal antenna is proposed, designed and then simulated. This paper that introduces a new 3-D printed trapezoidal antenna operates accurately in X-band for wireless communication applications. It is designed by a low-cost polylactic acid (PLA) substrate with relative permittivity of 3.1 and having compact size of $13.11 \times 21.36 \times 1$ mm^3. The proposed antenna has S_{11} of –21.67 dB at 10.1 GHz. The suggested antenna has gain of 5dBi as individual element. Modeling and simulation of the suggested antenna are performed by Ansoft High Frequency Structure Simulator (HFSS 13).

Keywords Antennas · 3-D printing · Trapezoidal patch · X-band applications

1 Introduction

Recently, 3-D printing (3-DP) is an advanced technique that attracts the most attention from the researchers and industrialists, and there is another name for this technology which is called additive manufacturing (AM) technology. The 3-DP improves very quickly and undertakes significant advance in the near future [1, 2]. The three-dimensional objects can be built using the 3-D printing technique after the implementation of computer-aided design (CAD) prototype, and this is successfully done by adding the desired material layer over layer. 3-D printing method has several

A. E. Ahmed
Electronics and Communication Engineering Department, College of Engineering, Higher Institute of Engineering and Technology, King Marriott, Alexandria, Egypt

W. A. E. Ali
Electronics and Communication Engineering Department, College of Engineering, Arab Academy for Science, Technology and Maritime Transport (AASTMT), Alexandria, Egypt
e-mail: wael.ali@aast.edu

A. Chowdhury (✉)
Mallabhum Institute of Polytechnic, Bishnupur, Bankura, India

S. Das
Department of ECE, IMPS College of Engineering and Technology, Malda, West Bengal, India

A. Sivasubramanian et al. (eds.), *Futuristic Communication and Network Technologies*, Lecture Notes in Electrical Engineering 792,
https://doi.org/10.1007/978-981-16-4625-6_94

advantages over conventional fabrication methods such as low profile, less weight, low cost and the ability to produce very complex shapes [3].

The AM technology can prolong many advantages such as the decrease in cost, time and complex shapes industrialization. The 3-DP can be utilized in the microwave engineering in the design of complex 3-D antenna structure; also it can be used to design a patch antenna with wire mesh as reported in [4]. Furthermore, a low-cost 3-D printed and inkjet techniques are applied to fabricate a circularly polarized patch antenna as reported in [5]. A log-periodic dipole antenna (LPDA) is 3-D printed in [6] to achieve the desired performance.

In this paper, a trapezoidal antenna using 3-DP technique is introduced. The prototype of the antenna is fabricated using 3-D printing technology to carry out the substrate material using PLA (polylactic acid) material as a high-strength thermoplastic material, and then, the radiator and the ground plane parts are created by coating a copper material on the 3-D printed layer. The proposed antenna operates at X-band application accurately which is used for marine radar [7], military communications satellites and wireless computer networks since the higher frequencies in the X-band permit for target identification and recognition using high-resolution imagery radar systems. The proposed antenna design is able to work at 10.1 GHz and covers X-band applications. The antenna model is fixed with substrate and ground by two screw nails from the same PLA substrate material. The characteristics of the proposed structure of 3-D printed trapezoidal antenna are obtained and then simulated using HFSS simulator to validate the desired performance.

2 3-D Printed Trapezoidal Antenna Design

For the proposed antenna design, resonant frequency (f_r) is chosen at 10.1 GHz, PLA with dielectric constant $\varepsilon_r = 3.1$, and thickness 1 mm is used as a substrate material. The trapezoidal patch is designed as a combination of a rectangular patch with an equilateral triangular patch, and the effective side length (a_e) of the triangular structure is calculated in terms of the rectangular width (W). The used rectangular patch sides dimensions are $L = 10.47$ mm, $W = 7.7$ mm and after using the following equation, the effective side length of the trapezoidal structure (a_e) is found to be equal to 8.89 mm [8].

$$a_e \times \cos 30^o = W \tag{1}$$

The radiator and the ground plane of the suggested antenna are modeled as a copper sheet of 1 mm thickness for each layer of the antenna. Figure 1 shows the proposed 3-DP trapezoidal antenna design with all detailed dimensions (in mm). The proposed antenna is fed by a transmission line (TL) of length (λ/4) and two stubs of length (λ/8) are used to satisfy the optimum matching and tuning. The stubs are united with the patch and TL to achieve accurate matching at the desired operating frequency. The

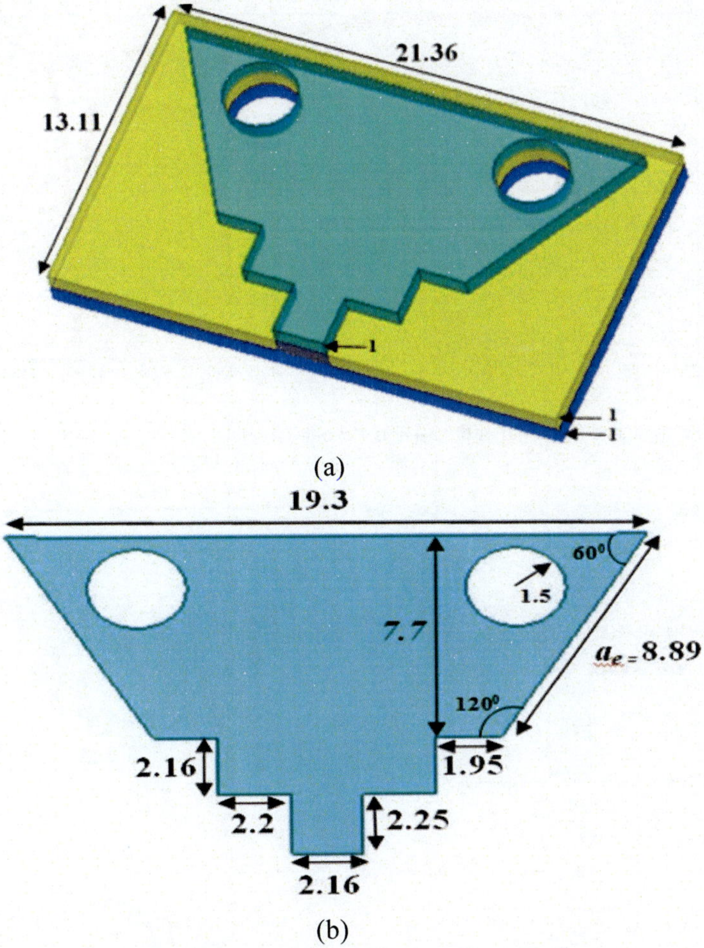

Fig. 1 The suggested 3-DP trapezoidal antenna. **a** 3-D view. **b** Front view

simulation result of S_{11} is depicted in Fig. 2, and the antenna succeeded to resonate at 10.1 GHz.

3 The Methodology of the Suggested Antenna Design

The design steps of the proposed trapezoidal antenna are demonstrated in Fig. 3. In the beginning, the antenna is constructed by using a sheet of copper (thickness = 1 mm)for the radiator and the ground plane, and the layout is cut by a laser machine. The dielectric material is made of PLA with a dielectric constant of 3.1, 0.001 loss

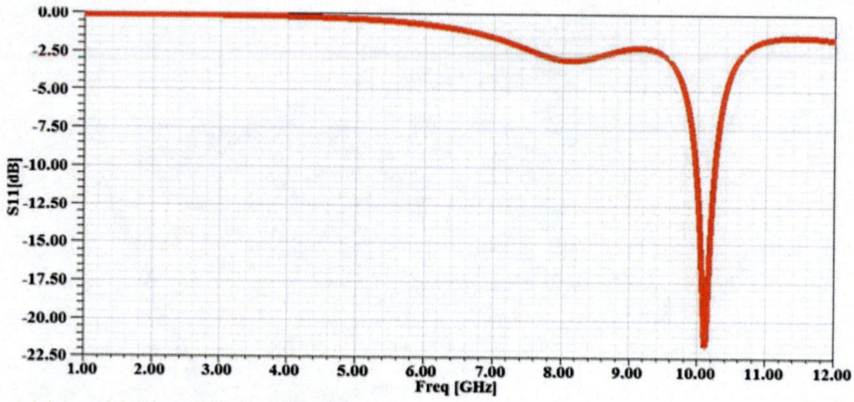

Fig. 2 The simulated S_{11} result of the 3-DP trapezoidal antenna

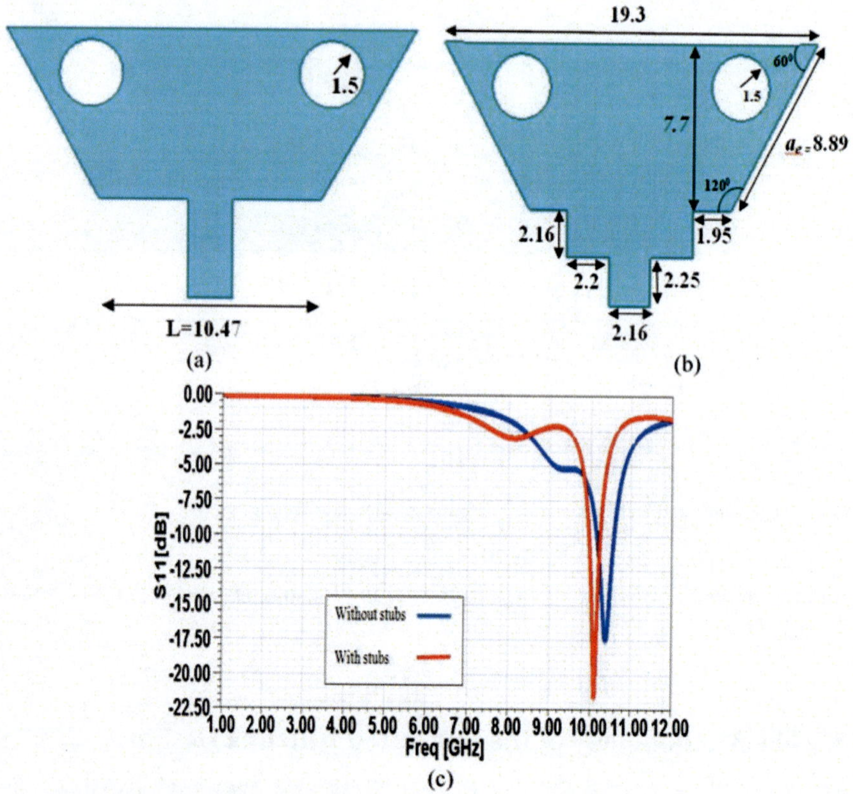

Fig. 3 The 3-DP trapezoidal antenna. **a** Patch without stubs. **b** Patch with stubs. **c** The simulated S_{11} results

tangent and 1 mm thickness. The substrate is 3-D printed which is different from the normal substrates used in many studies [9–14] to achieve accurate performance. The two sides of the antenna were fixed by using screw nails with a radius of 1.5 mm made from the same material of the substrate.

The ultimate proposed trapezoidal antenna provides impressive performance in terms of S_{11}. Trapezoidal shape will be printed on the PLA substrate by 3-D printing machine. The layouts and the results of the two antennas are depicted in Fig. 3 since it shows the top view, and the simulated S_{11} results of the 3-DP trapezoidal antenna with/without stubs. The existence of stubs improves the results since the achieved value of reflection coefficient S_{11} is –21.67 dB at 10.1 GHz which is deeper than the achieved result when the stubs are not used; also the resonance frequency of the red curve coincides with the desired frequency for X-band applications as depicted in Fig. 3c.

4 Current Distribution Analysis

In order to show the effectiveness of the suggested trapezoidal antenna, the current distribution is carried out in this section at the achieved results of the model with the two stubs. As noticed from Fig. 4, the surface current density distribution increases from the center of the antenna to the upper edges due to the use of stubs and the trapezoidal geometry shape which in turn causes an improvement in the fringing field effect. Furthermore, it enhances the impedance matching performance of the antenna as previously demonstrated in Fig. 3c.

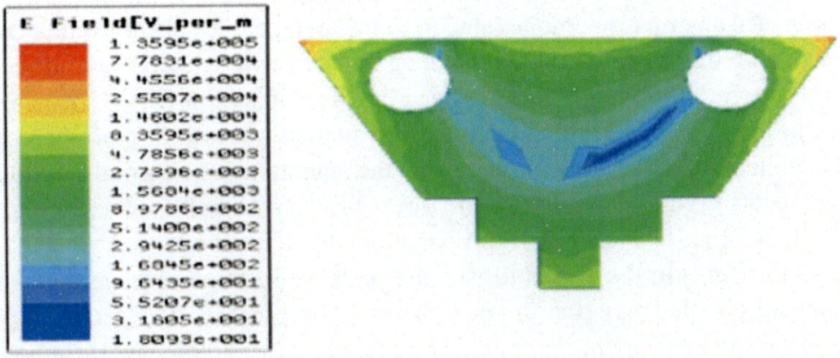

Fig. 4 Current distribution on the trapezoidal antenna

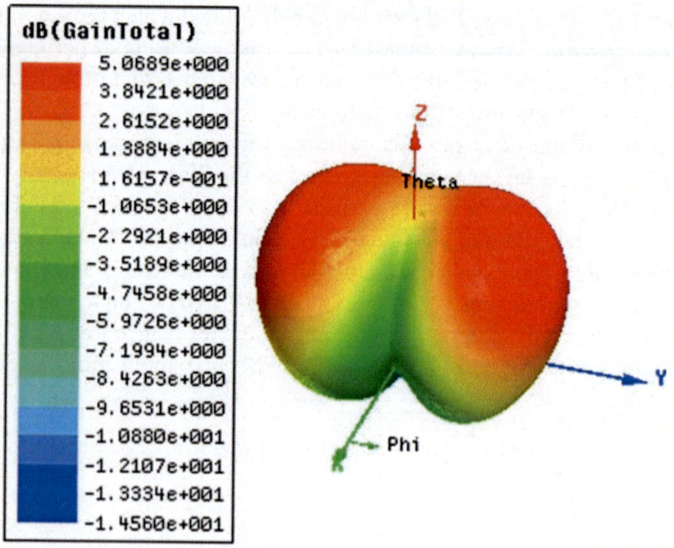

Fig. 5 The simulated gain of the 3-DP trapezoidal antenna

5 Results and Discussions

The structures of the proposed 3-DP trapezoidal antenna were designed using low-cost PLA substrate. Two copper sheets for patch and ground plane are stacked on the both sides of PLA substrate, then fixed by screw nails and nuts to construct the whole antenna. In order to confirm the validity of the suggested 3-DP antenna, the gain of 3-DP proposed antenna is demonstrated in Fig. 5 and it can be observed that the gain reaches to 5.06 dBi as an individual element antenna with a nearly omnidirectional radiation pattern. The simulation S_{11} curve gives a resonance at 10.1 GHz with return loss reaches to –21.67 dB which confirms the distinct transmission of the proposed antenna at the X-band accurately; therefore, the antenna can be a good candidate for various aspects such as satellite comm., direct broadcast satellite TV, point to point link (PTP) and line-of-sight (LOS) propagation characteristics.

In order to confirm the availability of the suggested antenna for various wireless communications, the radiation characteristics are presented. The required performance of the antennas operating for different wireless technologies is the omnidirectional with stable radiation pattern, and this requirement is fulfilled with our antenna as shown in Fig. 6. The patterns are nearly omnidirectional in each plane (E-plane and H-plane) at the operating frequency of the final design. It is clearly obvious from the previously displayed results that the 3-DP substrate trapezoidal antenna is capable of achieving the optimal impedance and radiation characteristics for X-band wireless applications.

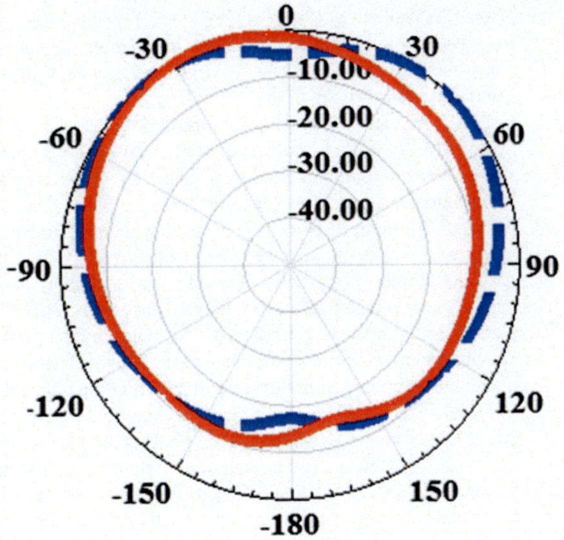

Fig. 6 Simulated radiated electric field in E-plane (solid) and H-plane (dashed) at 10.1 GHz

6 Conclusion

In this paper, an innovative 3-DP trapezoidal antenna has been designed with two stubs matched at 10.1 GHz for X-band wireless application. The antenna substrate is 3-DP using PLA material with the aid of advanced and inexpensive 3-D printing technique which expands an excellent performance in terms of renovation and simplicity. Moreover, a trapezoidal-shaped patch with two stubs is utilized to enhance the impedance matching of the proposed antenna. The trapezoidal antenna has an omnidirectional radiation pattern in both planes with a distinguished gain value, which makes it suitable to be used for modern marine radar, satellites comm. and all X-band applications.

References

1. Lipson H, Kurman M (2013) Fabricated: the new world of 3D printing. Wiley, New York, NY, USA
2. 3-D printing steps into the spotlight (2013). Upstate Bus J. Retrieved 20 Dec 2019
3. Sanz-Izquierdo B Parker E (2014) 3-D printing of elements in frequency selective arrays. IEEE Trans Antennas Propagat 62(12):6060–6066
4. Liang M, Shemelya C, Macdonald E (2015) 3-D printed microwave patch antenna via fused deposition method and ultrasonic wire mesh embedding technique. IEEE Antenna Wirel Propag Lett 14:1346–1349
5. Jun S, Sanz-Izquierdo B, Heirons J, Mao C, Gao S, Bird D, McClelland A (2017) Circular polarised antenna fabricated with low-cost 3D and inkjet printing equipment. Electron Lett 53:370–371

6. Nassar IT, Weller TM, Tsang H (2014) A 3-D printed miniaturized log-periodic dipole antenna. In: Antennas and propagation society international symposium (APSURSI). IEEE, pp 11–12
7. Fang-Yao K, Hwang RB (2014) High-isolation X-band marine radar antenna design. IEEE Trans Antennas Propag 62(5):2331–2337
8. Garg R. Microstrip antenna design handbook. Prakash Bhartia, InderBahl
9. Salamin MA, Ali WAE, Zugari A (2019) A novel UWB antenna using capacitively-loaded fork-shaped resonator and modified fork-shaped DMS for interference mitigation with WiMAX and WLAN applications. J Instrum 14(03)
10. Ali WAE, Moniem RMA (2017) Frequency reconfigurable triple band-notched ultra-wideband antenna with compact size. Prog Electromagn Res C73(37)
11. Ali WAE, Ibrahim AA (2018) Tunable band-notched UWB antenna from WLAN to WiMAX with open loop resonators using lumped capacitors. ACES J 33(6):603–609
12. Hamad EKI, Ali WAE, Hamdalla MZM, BassiunyMA (2018) High gain triple band microstrip antenna based on metamaterial super lens for wireless communication applications. In: International conference on innovative trends in computer engineering (ITCE), Aswan, 2018, pp 197–204
13. Ali WA, Mansour AM, Mohamed DA (2016) Compact UWB wearable planar antenna mounted on different phantoms and human body. Microw Opt Technol Lett 58:2531–2536
14. Salamin MA, Ali WAE, Das S, Zugari A (2019) Design and investigation of a multi-functional antenna with variable wide- band/notched UWB behavior for WLAN/X-band/UWB and Ku-band applications. AEU-Int J Electron Commun 111:152895

An Inverted L-Shaped ACS-fed Antenna with SRR for Bandwidth Enhancement Application

Rajkumar Rengasamy, Sathyamoorthy Sellapillai, and Dileepan Dhanasekaran

Abstract In this paper, an inverted L-shaped asymmetric coplanar strip (ACS)-fed antenna with split-ring resonator (SRR) structure is proposed for bandwidth enhancement. Antenna design has an inverted L-shaped monopole structure and split-ring resonator (SRR) structure. Antenna monopole is responsible for obtaining a lower frequency band (2.4 GHz), and SRR gives the response of 5 GHz band. The bandwidth and impedance of the proposed antenna are improved by embedding the metamaterial-inspired structure. Antenna design has a compact size of $12 \times 22 \times 1.6$ mm^3 and dual-band resonance frequencies at 2.45 GHz (2.05–2.65 GHz) and 5.9 GHz (5.35–8.10 GHz) with a bandwidth of 600 MHz and 2750 MHz, respectively, which covers 2.4/5.8 GHz WLAN, 5.5 GHz WiMAX, 5.9 WAVE and 7.25–7.75 GHz X-band downlink frequency. An impact of SRR structure is analysed and compared with the proposed antenna results. The antenna has better radiation characteristics at the required frequency bands.

Keywords Asymmetric coplanar strip · WLAN · WiMAX · WAVE · Split-ring resonator

1 Introduction

Today's technological era has replaced maximum wired communication into wireless communications; towards this improvement, antenna plays a vital role. Antennas used for wireless communication need to be compact, low profile and also should accommodate a greater number of applications/standards. As a single, multiband antenna will resonate for various frequencies which in turn reduces the size and cost of the communication devices. Multiband resonance is achieved by introducing slots and metamaterial concepts, where four bands resonance is achieved for GPS, WLAN and WiMAX applications [1]. Inverted-F antenna is presented in [2]; it covers WiMAX, WLAN and Bluetooth applications. In [3], S-shaped fractal antenna is

R. Rengasamy (✉) · S. Sellapillai · D. Dhanasekaran
Vel Tech Rangarajan Dr. Sagunthala R&D Institute of Science and Technology, Chennai 600062, India
e-mail: drrajkumarr@veltech.edu.in

A. Sivasubramanian et al. (eds.), *Futuristic Communication and Network Technologies*, Lecture Notes in Electrical Engineering 792,
https://doi.org/10.1007/978-981-16-4625-6_95

proposed which covers the application of [2]; also, in addition, it covers X-band. Antenna design [4] presents a multiband fractal antenna designed by using modified Sierpinski Carpet Fractal. Koch curve and Minkowski and Moore and Gosper curves [5] combined hybrid fractal antenna proposed for multiband. Dual-band covering WiMAX and WLAN are proposed in [6] using a 'C'-shaped strip and an 'L'-shaped strip. A triple-band dual-sense circularly polaraized antenna is discussed [7]. Fractal-structured monopole antenna is proposed for dual-band application [8], split-ring resonators is proposed and introduced in a planar slot antenna in [9], and split-ring resonator (SRR) and metallic strips created in a slot antenna have been proposed in [10].

Microstrip feeding technique has a radiating strip on one side, and ground plane is present on another side which is less complicated to fabrication but provides narrow resonance bandwidth, and MMIC integration is also hard [1, 4, 5, 8]. CPW feeding has both radiating and ground on the same surface, and it supports easy integration [2, 3, 9]. The compactness of an antenna can be achieved by using asymmetric coplanar strip (ACS), and it consumes half of CPW design size. ACS feed dual-band monopole antenna with semicircular structure with reduced ground size has been proposed for LTE and WLAN/WiMAX applications [11]. In [12], multiband metaatom-based ACS antenna is proposed. SRR loaded, mirrored L-shaped conductor with the truncated ground has been proposed in [13] and covers WLAN, WiMAX and ITU applications. In [14], a seven-shaped monopole antenna is discussed which covers WLAN and WiMAX applications. In [15], portable system applications proposed through compact asymmetric coplanar strip with an extended rectangular strip with tri-band resonance. Metamaterial was artificially constructed materials created to introduce negative permeability and permittivity which are discussed in [16–19].

The proposed antenna structure is the asymmetric coplanar strip with rectangular split-ring resonator proposed for dual-band operation. The bandwidth and impedance enhancement performance are enhanced, and it is confirmed analysis of with and without SRR structure. Further, the current distribution analysis is done to confirm the impact of the SRR structure.

2 Proposed Antenna Design

The proposed antenna structure is designed on FR4 substrate which has a compact size ($W \times L \times H$) of $12 \times 22 \times 1.6\ \text{mm}^3$ as shown in Fig. 1. The feed width, ground width and length are having a size of 3 mm, 6.7 mm and 6 mm, respectively, and the gap between feed line and ground is 0.25 mm. Rectangular split-ring resonator is introduced to realize the improved bandwidth and enhanced impedance performance at a higher frequency. The rectangular split-ring resonator has a thickness of 0.5 mm, split gap of 0.4 mm and embedded at the height of 6 mm between the ground and upper conductor of the antenna (Table 1).

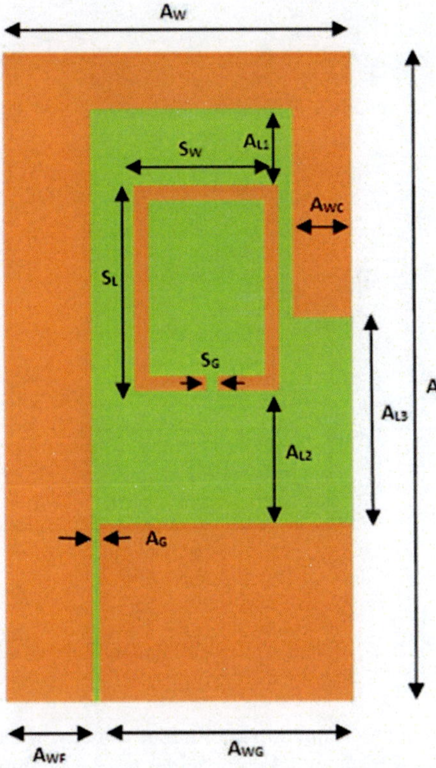

Fig. 1 Schematic diagram of the proposed antenna

Table 1 Antenna parameters measurement

Parameters	Size (mm)	Parameters	Size (mm)
A_L	22	A_{WC}	2
A_W	12	S_L	5
A_{WF}	3	S_W	3
A_{WG}	6.7	S_G	0.4
A_{Lg}	6	A_{L2}	6
A_G	0.25	A_{L1}	2.6

3 Results and Discussion

The proposed antenna is designed on a low-cost FR4 substrate. The structure occupies a smaller volume of 442.4 mm^3 in dimension which is very compact in nature. In the proposed antenna design, without metamaterial structure, antenna has two frequencies at 2.4 GHz and 8 GHz with reflection coefficient value of −14.5 and −12 dB as illustrated in Fig. 2. It has a bandwidth of 550 MHz and 1250 MHz at lower and higher frequencies, respectively. The above design does cover WLAN (2.4 GHz)

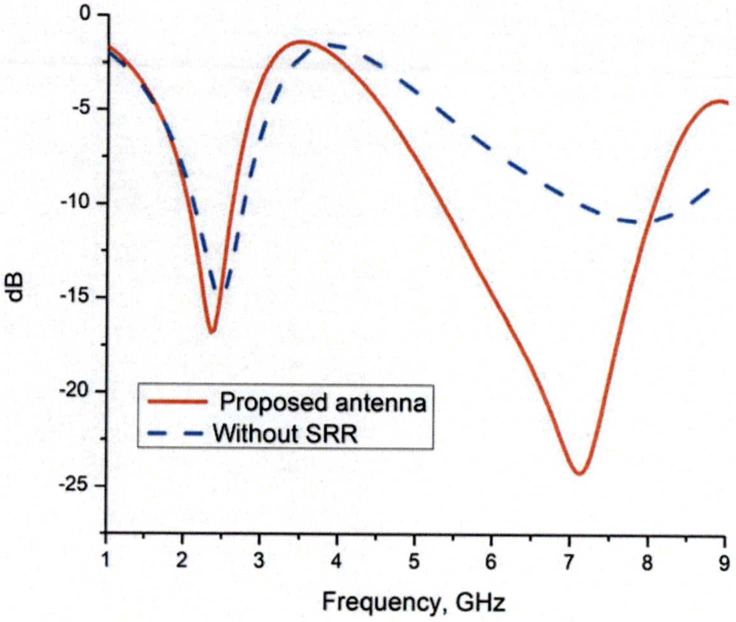

Fig. 2 S_{11} comparison of the proposed antenna with and without SRR structure

with less reflection coefficient value, while at higher frequency, it does not cover any other frequency bands, and it has less bandwidth.

In order to improve the bandwidth and impedance performance in existing antenna design, metamaterial-inspired split-ring resonator can be used. The metamaterial is man-made material; it does have special properties which can be used to improve the various antenna performance. In the proposed design, SRR is embedded along with a monopole antenna. Bandwidth and impedance performance are improved due to the process of mutual coupling between monopole and SRR structure.

The proposed antenna has −10 dB resonance frequencies centred at 2.45 and 7 GHz with refection coefficient value of −16 and −24 dB which covers 2.05–2.65 GHz and 5.35–8.10 GHz with a bandwidth of 600 MHz and 2750 MHz, respectively, as shown in Fig. 2. It has wide application coverage such as WLAN (2.4/5.8 GHz), WiMAX (5.5 GHz) and WAVE (5.9 GHz) applications. The proposed rectangular SRR structure is effectively used for improving antenna impedance and bandwidth performances, and its parameters are tabulated in Table 2.

Voltage standing wave ratio (VSWR) response of the proposed antenna is illustrated in Fig. 3. It is observed that the antenna has a VSWR good results at 2.45 and 5.8 GHz frequencies. VSWR values range between 1 and 2 in the VSWR scale in the required frequencies. Similarly, antenna impedance responses are illustrated in Fig. 4. It is seen that antenna's real and imaginary values are plotted from 1 to

Table 2 Antenna performance comparison

Frequency band	First resonance frequency (GHz)	Second resonance frequency (GHz)	Bandwidth (MHz)	Reflection coefficient (dB)
Without SRR	2.2–2.75	7.25–8.50	550 1250	−14.5 −12
Proposed antenna	2.05–2.65	5.35–8.10	600 2750	−16 −24

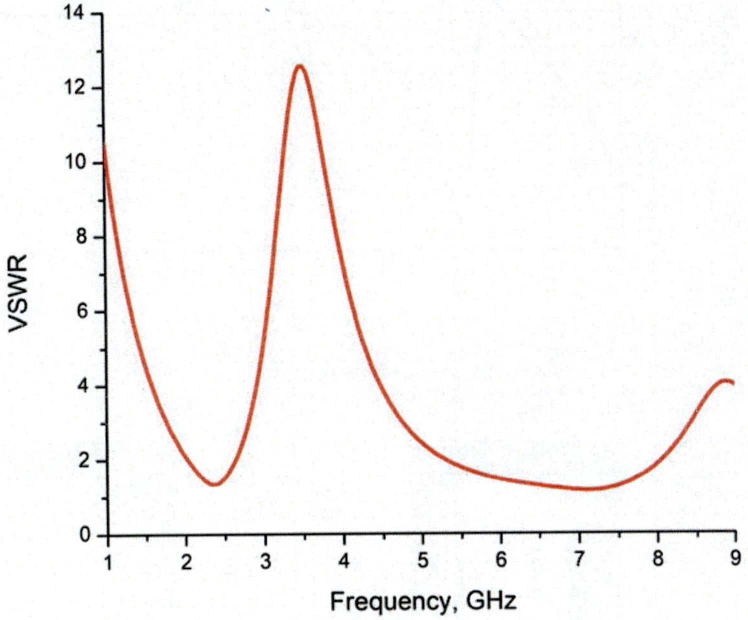

Fig. 3 VSWR performance of the proposed antenna

9 GHz, and around 50-Ω and 0-Ω impedances are observed for both real and imaginary values, respectively, in the (lower and upper) frequency axis. From this analysis, it confirmed that antennas are designed effectively for particular applications.

An impact of a split-ring resonator of the proposed antenna can be observed from the surface-current distribution of the proposed antenna. Figure 5a represents the monopole antenna without SRR structure, and maximum current is concentrated on the inverted antenna, while no current is concentrated on the SRR structure. The proposed antenna current concentration is shown in Fig. 5b for 5.80 GHz. It is observed that maximum current is concentrated on the SRR structure and parallel side of the monopole antenna. From this, it is confirmed that due to mutual coupling between ring and monopole is responsible for generating higher frequency band.

The radiation pattern of the proposed structure (E and H plane) for the frequencies of 2.45 and 5.80 GHz are shown in Fig. 6a, b. The antenna has dumb-bell-shaped

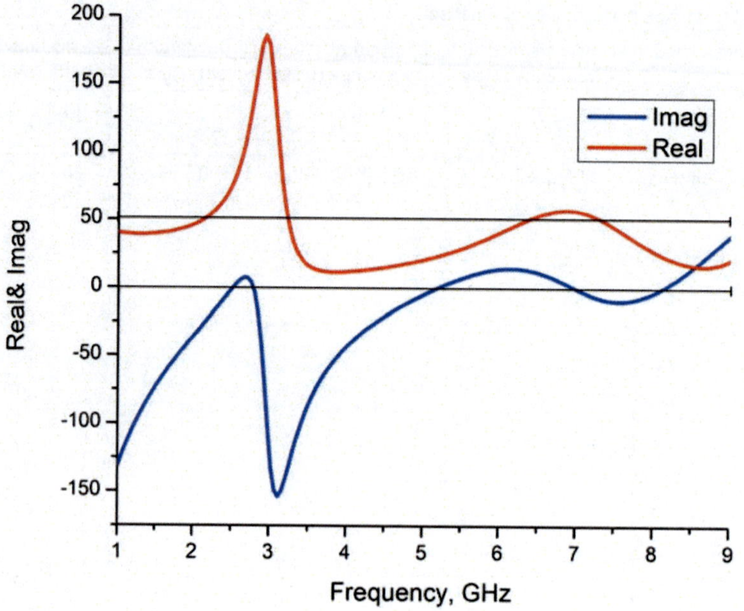

Fig. 4 Impedance performance of the proposed antenna

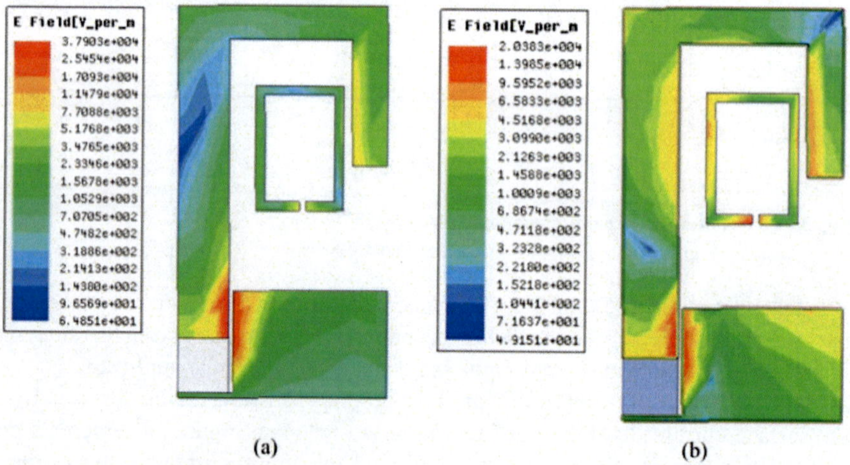

Fig. 5 Surface-current distribution of the proposed antenna. **a** 2.45 GHz and **b** 5.80 GHz

E-plane and omnidirectional-shaped H-plane in the desired frequencies. The antenna has a gain of 0.58 and 1.49 dBi at 2.45 and 5.8 GHz, respectively.

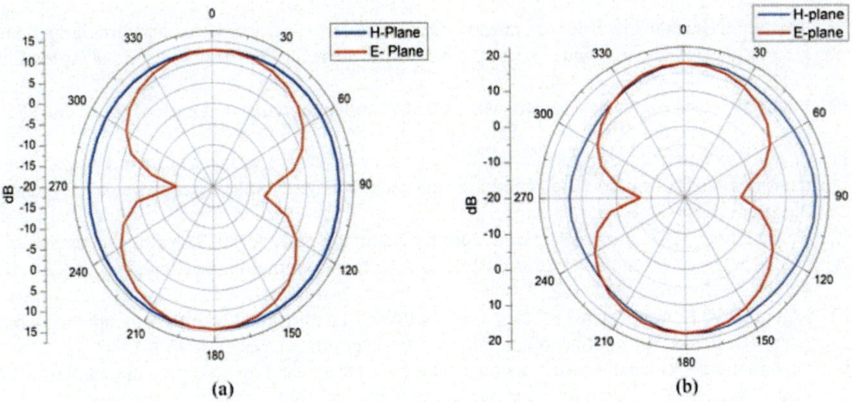

Fig. 6 Antenna radiation pattern at **a** 2.45 GHz and **b** 5.8 GHz

4 Conclusion

In this paper, an inverted L-shaped asymmetric coplanar strip (ACS)-fed antenna with split-ring resonator (SRR) structure has been proposed and discussed for dual-band applications. The proposed structure shows good impedance bandwidth of 600 and 2750 MHz. It has been constructed on simple basics, and it is compact in size. The impact of SRR is observed using current distribution analysis. The proposed antenna covers 2.4/5.8 GHz WLAN, 5.5 GHz WiMAX and WAVE applications which is viable for use in various communication systems.

References

1. Ali T, Khaleeq MM, Pathan S, Biradar RC (2017) A multiband antenna loaded with metamaterial and slots for GPS/WLAN/WiMAX applications. Microw Opt Technol Lett 60:79–85
2. Al-Khaldi M (2017) A highly compact multiband antenna for Bluetooth/WLAN, WiMAX, And Wi-Fi applications. Microw Opt Technol Lett 59:77–80
3. Cheong HR, Yeap KH, Lai KC, The PC, Nisar H (2017) A compact CPW-fed antenna with Fractal s-shaped patches for Multiband applications. Microwave Opt Technol Lett 59:541–546
4. Sivia JS, Kaur G, Sarao AK (2017) A modified Sierpinski carpet fractal antenna for multiband applications. Wireless Pers Commun 95:4269–4279
5. Kaur K, Sivia JS (2017) A compact hybrid multiband antenna for wireless applications. Wireless Pers Commun 97:5917–5927
6. Tan M, Wang B (2016) A dual-band circularly polarized planar monopole antenna for WLAN/Wi-Fi applications. IEEE Antennas Wirel Propag Lett 15:670–673
7. Midya M, Bhattacharjee S, Mitra M (2019) Triple-band dual-sense circularly polarised planar monopole antenna. IET Microwaves Antennas Propag 13(12):2020–2025
8. Sahoo R, Vakula D (2019) Compact metamaterial inspired conformal dual-band antenna loaded with meander lines and fractal shaped inductor for Wi-Fi and WiMAX applications. IET Microw Antennas Propag 13(13):2349–2359

9. Tharehalli Rajanna PK, Rudramuni K, Kandasamy K (2019) Compact triband circularly polarized planar slot antenna loaded with split ring resonators. Int J RF Microw Comput Aided Eng e21953
10. Paul PM, Kandasamy K, Sharawi MS (2018) A triband circularly polarized strip and SRR-loaded slot antenna. IEEE Trans Antennas Propag 66(10):5569–5573
11. Naidu PV, Kumar A, Rengasamy R (2018) Design, analysis and fabrication of compact dual band uniplanar meandered ACS fed antenna for 2.5/5 GHz applications. Microsyst Technol 25(1): 97–104
12. Rajeshkumar V, Rengasamy R, Vummadisetty Naidu P, Kumar A (2019) A compact meta-atom loaded asymmetric coplanar strip-fed monopole antenna for multiband operation. AEUE—Int J Electron Commun 98:241–247
13. Rajkumar R, Kommuri UK (2016) A Compact ACS-fed mirrored L-shaped monopole antenna with SRR loaded for multiband operation. Prog Electromagn Res C 64:159–167
14. Gunamony SL, Gnanadhas JB, Lawrence D (2020) Design and investigation of a miniaturized single-layer ACS-fed dual band antenna for LTE and 5G applications. J Electromagn Eng Sci. 20(3):213–220
15. Naidu PV, Kumar A, Rengasamy R (2019) Uniplanar ACS fed multiband high gain antenna with extended rectangular strips for portable system applications. Int J RF Microw Comput Aided Eng 29(10):e21870
16. Baena JD et al (2005) Equivalent-circuit models for split-ring resonators and complementary split-ring resonators coupled to planar transmission lines. IEEE Trans Microw Theory Tech 53(4):1451–1461
17. Pendry JB, Holden AJ, Robbins DJ, Stewart WJ (1999) Magnetism from conductors and enhanced nonlinear phenomena. IEEE Trans Microw Theory Tech 47(11):2075–2084
18. Smith DR, Padilla WJ, Vier DC, Nemat-Nasser SC, Schultz S Composite medium with simultaneously negative permeability and permittivity. Phys Rev Lett 84:4184–4187
19. Smith DR, Pendry JB, Wiltshire MCK (2004) Metamaterials and negative refractive index. Science 305:788

Photonic Crystal Based Devices and the Methods Implemented—A Review

S. Preethi Manjou and K. Chitra

Abstract In this paper, the finite-difference time-domain, pulse wave expansion method are the major methods used to study characteristics of photonic band gaps in photonic crystals. The 2D structure of photonic crystals comprises of nonlinear rods with different types of geometrical shapes with lattices on air or plasma background. The embedded dielectric rod containing Kerr effect which is a nonlinear property has relative permittivity which is a function of the external light intensity, but this demands a high amount of optical intensities. The simulation results show the transmission efficiency, and the optical field distributions for both TE and TM modes can be obtained.

Keywords Polarization splitter · Photonic crystals · Kerr effect

1 Introduction

The different types of dimensional structures called as photonic crystals (PCs), a tedious pioneering works are started [1, 2], as effective control of the propagation of the waves in the similar way with semiconducting materials influencing the conduction of electrons. The advantage of photonic crystals (PCs) [3–5] is that they provide the photonic band gap structures which provides the frequency at which propagation takes place. Both optical modes and spontaneous emission do not occur inside the band gaps. Many applications require photonic band gap structures (PBG) [6, 7]. The 3D PCs controls the propagation of light in means of spatial directions. Hence, more attention is put on the two-dimensional PCs. The 2D crystal lattices consists of dielectric rods that are arranged in air in a dielectric background. These are polarization sensitive so considerable conditions of the band gap structures are almost different. For TE polarization mode, it is parallel to the periodic plane in these structures and for the transverse magnetic (TM) mode, the magnetic field is parallel

S. Preethi Manjou (✉) · K. Chitra
Vellore Institute of Technology, Chennai, India
e-mail: preethimanjou.s2019@vitstudent.ac.in

A. Sivasubramanian et al. (eds.), *Futuristic Communication and Network Technologies*, Lecture Notes in Electrical Engineering 792,
https://doi.org/10.1007/978-981-16-4625-6_96

[8]. A PBG for a two-dimensional photonic crystal for which both the TE and TM polarization are obtained overlapping in the alike frequency region [9].

In real structures, as a result of the intrinsic imperfections that is found in the fabrication, crystal coupling is possible. The mode balance property is vital for real-time applications [10]. A PBG can be used in both TE and TM polarizations of light, and the waveguides that are dependent on defect modes which can guide light as well. PCs including dispersive materials are used because of their specific properties in comparison with conversional type of PCs. Plasma being a dispersive medium whose refractive index is connected to the frequency of the incident EM wave [11]. Recently Zang et al. [12] proposed a 2D plasma photonic crystal where a PBG is full observed [13, 14]. Nowadays, there is an enormous demand for the PBG in the normal photonic crystal for the designing of dynamical or switchable devices [15].

2 Related Work

In this section, various works that have been carried out in photonic crystals. The advantages of each designed structures are put forth. The methods used for constructing the optical devices using photonic crystals are mentioned clearly with their future scope. The software's such as RSOFT and COMSOL can be used to design the photonic crystal structures.

In [16], the finite-difference time-domain method is used to study the complete photonic band gap structure for the photonic crystals with nonlinear Kerr effect. The 2D structure is made of nonlinear rods with square and honeycomb lattices on a plasma background. The embedded dielectric rods in the crystal structures contains Kerr nonlinearity which is a function of the external light intensity. The simulation results obtained from the structure illustrate that a nonlinear structure gives a photonic band gap whose parameters can be changed by the alteration of light intensity.

In [17], a two-dimensional photonic crystal structure comprising of the dielectric rods in air background gives a photonic band gap structure from which the operating wavelength is obtained. This structure is obtained by utilizing the finite difference time domain method based on Yee's algorithm in an orthogonal coordinate system.

In [18], for the calculation of the band structure of the crystal structure, plane wave expansion method is used. This algorithm solves the Eigen value. An effective-medium tensor is applied to improve the convergence and get improved efficiency.

In [19], the photonic crystals have gathered attention because of its ability to be used as optical components. The photonic crystals that are fabricated are inverse opal for photonic crystals where colloidal crystal templates are used. Materials used in the fabrication the structure are composed of complex oxides comprising more number of metal ions. The materials, structure for photonic crystals, fabrication procedures, and evaluation properties are discussed.

In [20], optical logic devices are future for all the optical computing and data processing. An optical half-subtractor that acts as all optical devices is designed using photonic crystal structure. This structure is realized using threshold switching and beam interference. The designed structure is analysed using the FDTD. The contrast ratio of the device is obtained from the simulation result for the *D* and *B* ports. The maximum rise time and fall time are noted; similarly, the minimum is also noted.

In [21], the nonlinear optics is more significantly used to overcome the obstacle faced by lasers at particular frequencies were unable to meet. Recent researches in photonic crystal are frequency conversion in nonlinear crystal which is ferroelectric with different lattice structures. In this paper, they have discussed the fabrication methods of photonic crystals with nonlinear properties. Certain other methods are introduced, they are quasi-phase matching and the nonlinear diffraction harmonic which have been used in the development of the photonic crystals. Nonlinear photonic crystal has great scope in practical applications such as detection, communication, imaging and so on.

In [22], a four-channel wavelength division demultiplexer is designed using 2D-photonic crystal. The plane wave expansion method and FDTD methods are used for the calculation of the operating frequency from the photonic band gap and the transmission spectrum of the demultiplexer. The certain functional parameters for the designed structure are analysed from the simulation results. The transmission efficiency of this structure is 93% which can be in any mode (TE or TM), and the *Q* factor of this designed device 781. The size of the demultiplexer is around 681 μm^2 designed suitable for the integrated optics that is the future scope for all optical network system.

In [23], an optical one-bit half-adder is realized and designed based on the 2D photonic crystals. A hexagonal lattice of 25 × 20 with silicon rods air medium is the proposed structure. The wavelength at which the structure propagates is 1550 nm. Simple structure and integral size are the advantages of this proposed structure that made this recognised to be optical logic gate design. A distinct gap between the states of outputs for a long time is the major advantage of this proposed structure. These advantages are making it easy to cut-off the errors in identifying the logical states at output end.

In [24], all optical half-subtractor based on the photonic crystal is proposed and designed. The light control is proceeded by constructing the input and output waveguides in the proposed structure. An all optical subtraction can be fulfilled by introducing a single-rod defect and an extra waveguide. The structure does not poses any nonlinear properties. The wavelength at which the device propagates is 1550 nm from which the lattice constant, and the radius of the rods are obtained. The radii of the rods with air holes are 214 nm, and the lattice constant is 414. The device is faster compared to other designed half-subtractor so far as the have the minimum rise time and fall time.

In [25], an optical logic gate is designed using a 2D photonic crystals. The structure is designed by air holes in silicon on a square lattice. It consists of three input lines and one output line. The centre input line which is an additional one is considered as the reference port. A square-shaped lattice with a resonator design is implemented for the input ports. The simulation results of structure exhibits logic gate functions of certain gates. In the launch field, a phase shift is introduced in order to improve the contrast ratio; as a result, some logic functions provide improved power at the output ports. The footprint value for the structure is obtained. Plane wave expansion and FDTD method are used in this proposed structure.

In [26], an optical full-adder-based photonic crystal is the idea proposed and designed. The lattice structure for the designed structure is 15 × 22 with dielectric rods of radius 109nm. The structure is built on square lattice with a lattice constant of 409nm. The designed structure is simple and compact with a surface of 72 μm^2. The wavelength at which the proposed structure propagates is 1550 nm. To control the light behaviour in the waveguide two defect are used either line defect or rod defect. The average delay time observed for this proposed device is less than 2 ps. The proposed structure has a compact size, quick response, and time gap is more between logic level 0 and 1, compared to existing structures.

In [27], a splitter device is designed using photonic crystal and the photonic band gap structures are obtained. Self-collimation method is used which is an advantage of this designed structure. The splitter is built on a hexagonal lattice structure with number of rods placed in air background. The material used in rods is GaAs. The rod radius is taken as 100 nm with a 500 nm lattice constant which is used for obtaining the band structure. The RSOFT software is used for simulation of this proposed splitter device. The simulation results show that the splitter device gives optical field distribution in both the TE and TM modes. A better transmission spectral is also obtained for this structure. Power dissipation is reduced. A transmission efficiency of 90% is obtained for the proposed model.

In [28], a 4 to 2 encoder is designed using photonic crystal which serves as an all optical device. The idea of this encoder is to design an encoder coupling the ring resonator with high coupling efficiency. A ring resonator designed using a photonic crystal has been placed in between the both input and output ports which gives improved coupling efficiencies. By doing so, the cross-reflection between the ports is reduced. The central wavelength obtained for this structure of encoder is 1555 nm. The wider bandwidth in the structure provides an extensive range of use. The coupling efficiency has been increased in this design. The advantage of this design structure is non-usage of nonlinear properties in the material in designing this structure. As a result, high amount of optical intensities is not required.

3 Proposed Work

Encoders usually have 2*N* input lines and *N* output lines. *N* is considered to be 2, so here we are designing a 4 to 2 encoder with a fundamental photonic crystal structure. The proposed photonic crystal ring resonator is designed on a square lattice where the photonic crystals are made of dielectric rods. The material of the rod has a refractive index of 3.46. The radius of the rods is considered to be 116.2 nm. The lattice constant of the proposed structure is 518 nm. The operating wavelength of the structure lies between 1400 and 2000 nm. The ring resonators are coupled by T-shaped waveguides. The coupling rod radius is 110 nm. Each ring resonator has different rod radius. The resonant ring has a lattice structure of 4 * 4 square-shaped lattices. The resonant mode of the resonator varies for each ring resonator, approximately 1555 nm is considered as the centre wavelength [28].

3.1 All-Optical Encoder

An optical encoder is composed of a 35 * 67 on a square lattice based on photonic crystal. The dielectric rods are placed in air medium. The radius and lattice constant values are similar to that of the existing design. The material is same as that of existing design, so the refractive index remains the same. The proposed design has four input ports and two output ports. Additional output port is added considered as the reference port. The input lines are I1, I2, I3 and I4. The output ports are O1 and O2, and the additional output port is V (Fig. 1).

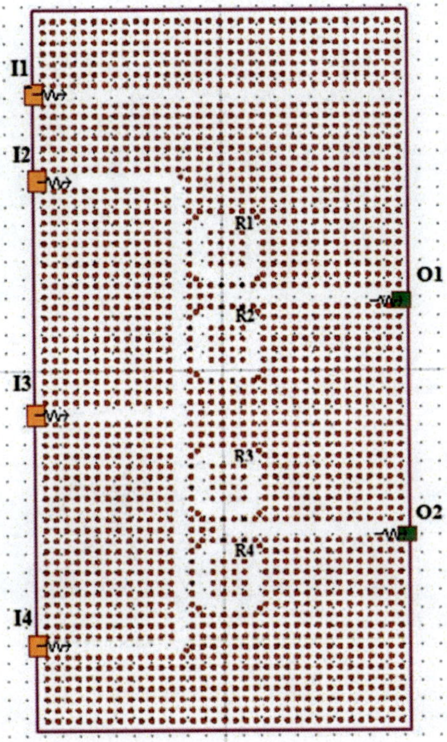

Fig. 1 Proposed structure of 4 to 2 encoder [28]

4 Simulation Results

In this section, the simulation results and functions of each port of the proposed structure are presented. A Gaussian wave pulse with wavelength of 1555 nm which is the operating frequency of the device is propagated in the waveguides to obtain the simulation result. This structure consists of four input ports in the active status, but only one input line can be ON at a time. When all the input lines are OFF which is denoted as the active case of the structure, all the output ports will be OFF. This issue is not discussed in most of the previously designed structures (Fig. 2).

This shows when I3 is ON, the waves propagate towards resonator R2 and R3 that drops them into resonator and sends them to O1 and O2 port. In this case, the output ports O1 and O2 are ON, but the port V is OFF [28] (Fig. 3).

The normalized intensities at the ports V, the output ports O1 and O2 are 0, 75 and 75%. In this case, the time delay for the output port O1 is about 4.6 ps.

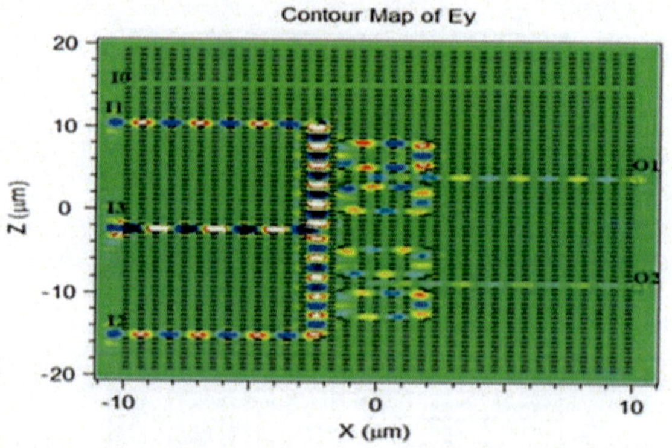

Fig. 2 Optical field distribution when I3 is ON

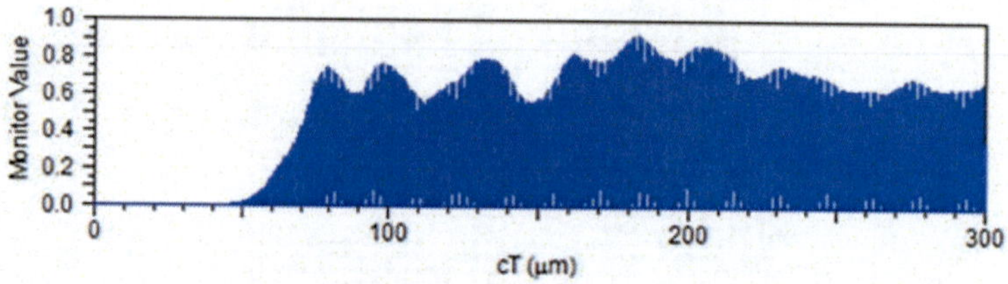

Fig. 3 The normalized optical intensities at V, O1 and O2

Table 1 The working states of the proposed 4 to 2 encoder [28]

Input ports				Output ports					
				Normalized intensity (%)			Logic gates		
I1	I2	I3	I4	V	O2	O1	V	O2	O1
0	0	0	0	0	O	0	0	0	0
1	0	0	0	100	0	0	1	0	0
0	1	0	0	0	5	95	0	0	1
0	0	1	0	0	95	5	0	1	0
0	0	0	1	0	(i) 75 (ii) 25	25 75	0	1	1

5 Working Principle

The truth table designed for the structure is put up in Table 1. In this structure, the maximum delay time is observed to be about 4.8 ps. The major advantage is its higher coupling efficiency at the output. The resonant ring resonator is able to couple 90% of optical waves from input into output ports. The cross-reflection obtained for the structure in between input lines is very low.

The PhCRRs are combined using the T-shaped optical waveguides, and development of all-optical 4 to 2 encoder with coupling efficiencies in the waveguides is done. The additional output port namely V which is the reference port used for distinguishing the active and inactive status when the input port I0 is ON. The designed structure has four different working conditions when it is active. The transmission and time response of the structure are obtained. The coupling frequencies or efficiency in the input port and output waveguides obtained through the proposed structure is more compared to that of the existing structure.

6 Conclusion

In this paper, the overall designs and concepts of photonic crystals are discussed. The methods used and their advantages have been put forth. Fabrication of photonic crystal is possible. Hence, this survey is useful for design and fabrication of photonic crystal devices such as polarization splitter, polarization convertor devices and so on for which practical applications are possible.

References

1. Yablonovitch E (1987) Inhibited spontaneous emission in solid-state physics and electronics. Phys Rev Lett 58:2059

2. John S (1987) Strong localization of photons in certain disordered dielectric superlattices. Phys Rev Lett 58:2486
3. Anderson CM, Giapis KP (1996) Larger two-dimensional photonic band gaps. Phys Rev Lett 77:2949
4. Florescu M, Torquato S, Steinhardt PJ (2009) Complete band gaps in two-dimensional photonic quasicrystals. Phys Rev B 80:155112
5. Rezaei B, Khalkhali TF, Soltani Vala A, Kalafi M (2009) Enlargement of absolute photonic band gap in modified 2D anisotropic annular photonic crystals. Opt Commun 284(13):3315–3322
6. Kurt H, Citrin DS (2007) Reconfigurable multimode photonic-crystal waveguides. IEEE J Quant Electron 43:78
7. Zhou WD, Sabarinathan J, Bhattacharya P, Kochman B, Berg E, Yu PC, Pang S (2001) Properties of PC filters in one-dimensional photonic crystals containing defects. IEEE J Quant Electron 37:1153
8. Alipour-Beyraghi M, Khalkhali TF, Bananej A, Mostafavi-Amjad J (2018) Polarization-independent and tunable flat lenses based on graded index two-dimensional photoniccrystals. Opt Commun 420:133
9. Takayama S, Kitagawa H, Tanaka Y, Asano T, Noda S (2005) Experimental demonstration of complete photonic band gap in two-dimensional photonic crystal slabs. Appl Phys Lett 87:061107
10. Khalkhali TF, Bananej A (2016) Complete photonic band gap characteristics of two-dimensional Kerr nonlinear plasma photonic crystals. Opt Commun 369:79
11. Hojo, Mase A (2004) Dispersion relation of electromagnetic waves in one-dimensional plasma photonic crystals. Plasma Fusion Res 80:89
12. Zhang HF, Ding GW, Li HM, Liu SB (2015) The properties of two-dimensional fractal plasma photonic crystals with Thue-Morse sequence. Phys Plasmas 24:042116
13. Khalkhali TF, Bananej A (2016) Tunable complete photonic band gap in anisotropic photonic crystal slabs with non-circular air holes using liquid crystals. Phys Lett A 380:4092
14. Khalkhali TF, Bananej A (2017) Full photonic band gap properties of plasma photonic crystals with triangular structure. J Mod Opt 64:830
15. Qi L, Yang Z, Lan F, Gao X, Shi Z (2010) Properties of obliquely incident electromagnetic wave in one-dimensional magnetized plasma photonic crystals. Phys Plasmas 17:042501
16. Khalkhali TF, Shiri R, Shahrokhabadi H, Bananej A (2019) Complete photonic band gap characteristics of two-dimensional Kerr nonlinear plasma photonic crystal. Indian J Phys
17. Rafiee Dastjerdi S, Ghanaatshoar M (2013) Finite difference time domain method for calculating the band structure of a 2D photonic crystal and simulating the lensing effect. J Phys: Conf Ser 454
18. Shi S, Chen C, Prather DW (2004) Plane-wave expansion method for calculating band structure of photonic crystal slabs with perfectly matched layers. J Opt Soc Am A 21(9)
19. Kuwabara M (2018) Photonic crystals fabricated by sol-gel process. Springer
20. Askarian A, Akbarizadeh G, Fartash M (2019) A novel proposal for all optical half-subtractor based on photonic crystals. Opt Quant Electron. Springer
21. Huangjia LI, Boqin MA (2019) Research development on fabrication and optical properties of nonlinear photonic crystals. Optoelectron
22. Venkatachalam K, Robinson S, Sriram Kumar D (2017) Design and analysis of dual ring resonator based 2D-photonic crystal WDDM. Springer
23. Abdollahi1 M,·Parandin F (2019) A novel structure for realization of an all-optical, one-bit half-adder based on 2D photonic crystals. J Comput Electron. Springer
24. Talebzadeh R, Beiranvand R, Mir A (2020) Very simple all-optical half-subtractor based on two-dimensional photonic crystals. Optical Quant Electron. Springer
25. Saranya D, Anbazhagan R (2020) Design and analysis of optical logic gates based on trifurcation structured 2D photonic crystals. Opt Quant Electron. Springer
26. Vali-Nasab A-M, Mir A, Talebzadeh R (2020) Design and simulation of an all optical full-adder based on photonic crystals. Quant Electron. Springer

27. Manjou P, Sivasindhu SM, Samundiswary P (2019) Design of GaAs-photonic crystal based polarization splitter. In: IEEE Conference, NIT Trichy
28. Manjou P, Sivasindhu SM, Samundiswary P. Design of 4 to 2 encoder based on photonic crystal ring resonator. IET J (communicated)

IoT Based Gesture-Controlled Surveillance Rover

Nikhil Gulati, Anmol Oberoi, Pranshu Dhyani, and V. Berlin Hency

Abstract The design, assembly, and working of the proposed model, "IoT based gesture-controlled surveillance rover" aims to provide a better result and communication between the operator and the communication in either of the aimed use; aid in post-disaster and surveillance. The cameras are mounted inside the chassis of the rover, giving out live stream at a specific link while being connected to the Internet to receive movement commands from the operator. Furthermore, implementation of such system would make the rescue process much accurate and quick. The rescue team would know the position of any living being under the debris through the thermal imaging from the rover. The rover is controlled by the gestures of our hand, and we get live thermal imaging at our place by using IoT. Results show that the model is working on multiple terrains with a constraint like Internet connectivity and speed is directly proportional to the livestream and control of the rover and high-power consumption.

Keywords IoT · Hand gesture recognition · Surveillance · Thermal imaging · Disaster management

1 Introduction

During cases of disaster, living beings can get trapped under the demolished buildings. The removal of debris using heavy machinery in India rescues these lives. Considering the number of people who died because of such disaster and the population of highly populated countries, heavy machinery cannot be accessed and mobilized to every location, due to which rescue party has to save lives by removal of rubble manually. Determining the location of the body trapped under debris can help save time, effort, and resources needed to rescue the body [1]. The proposed

N. Gulati (✉) · A. Oberoi · P. Dhyani · V. B. Hency
School of Electronics Engineering (SENSE), Vellore Institute of Technology, Chennai (VIT), Chennai, Tamil Nadu 600127, India

V. B. Hency
e-mail: berlinhency.victor@vit.ac.in

A. Sivasubramanian et al. (eds.), *Futuristic Communication and Network Technologies*, Lecture Notes in Electrical Engineering 792,
https://doi.org/10.1007/978-981-16-4625-6_97

model of "IoT based gesture-controlled surveillance rover" aims to develop a prototype to detect the presence of a living being by locating heat signatures. To detect any presence, thermal sensors have been used [2]. When the rover is sent inside the debris, the sensors scan the area for any heat signatures. It is connected to the Internet, enabling the operator to control the rover from any part of the world. The chassis of the rover has been mounted with two cameras: a standard camera and a thermal camera, which are used to provide a live stream to the operator [3]. This feature can be used by the operator to avoid any false heat signatures. Once detected, quick action can be taken by the rescue party to save lives. The tank like chassis enables its operation on multiple terrains [4]. Its robustness makes it suitable for aid in post-disaster management and surveillance [5].

2 Literature Survey

Smart bands and other devices have established themselves as a better option for human–machine interaction; the readiness and availability of such devices enables the user to continuously link between many devices in order to control that particular device from one's smart device [6–8]. Similarly, with the emergence of autonomous vehicles, safety of pedestrians has become an upmost priority issue. This can be done with the use of proper algorithm to detect and decide the presence of a thermal body using unidirectional thermal camera [2]. The incorporation of organic light-emitting diodes in higher-end devices, namely smart phones and TVs, alongside the inclusion of cameras embedded in day-to-day use gadgets, provides the chance to establish optical camera communication systems in light of OLED emitters for Internet of Things [3]. Assessing any situation during a disaster is a must, and for that, a thermal monitoring system for microelectronics dependent on the vision is applied to examine through pictures using an infrared camera [9]. The system which is being used to assess the situation has to maintain a proper use of power in order to function efficiently, and for longer duration, camera is one such power extracting unit. For this, an always-on low power camera has been proposed, but on the contrary, if the stream is not manually monitored, the system would require a very strong processing power in order to make the right decision based on all the data it is collecting [10]. As every situation has to be worked around in a particular manner, similarly, a simple rover cannot function on all platforms. The rover design has to incorporate the feasibility to work even on terrains which cannot be accessible normally [4, 11]. The working and movements of the rover has to be studied, predicted, and planned accordingly for further control; thus, proper embodiment of the kinematics is necessary [12]. The locomotion inside the debris and avoidance of the obstacle can be adapted from the standard snake rover which can operate in even small spaces [13, 14].

3 Proposed System

The idea implemented in this proposed model is to create a remotely accessed and controlled rover, which installed with a thermal camera will be able to detect heat signatures in areas where human traversal is not possible thus aiding the rescue process.

For demonstration of our proposed model, we assembled an all-terrain tank chassis installed with a Raspberry pi 3.0 as its controller which is connected to the motor drivers that in turn moves the whole tank chassis. The controller part of our proposed model, i.e., the glove uses an accelerometer, which determines the extent of movement of the rover.

The accelerometer readings are relayed to the Arduino board which is connected to Internet via ESP8266. The Arduino then uploads the accelerometer values on a designated cloud. The accelerometer readings on the cloud is then remotely accessed by the Raspberry pi module installed in the rover which then moves the rover in the specified direction with the help of a motor driver. A Raspberry pi cam is also connected to the Internet via pi module providing live stream of the rover.

The glove part is connected to a 220 V AC supply which is then stepped down and converted into DC. An LCD is also used which displays the accelerometer values. Our

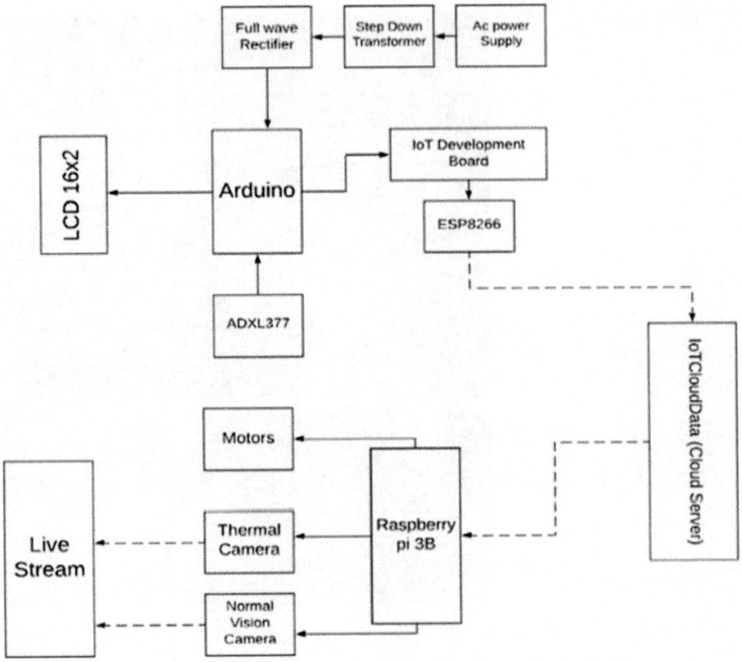

Flow Chart 1 Physical block diagram

proposed model can be mainly divided into three fundamental parts (Flow Chart 1, Figs. 1 and 2).

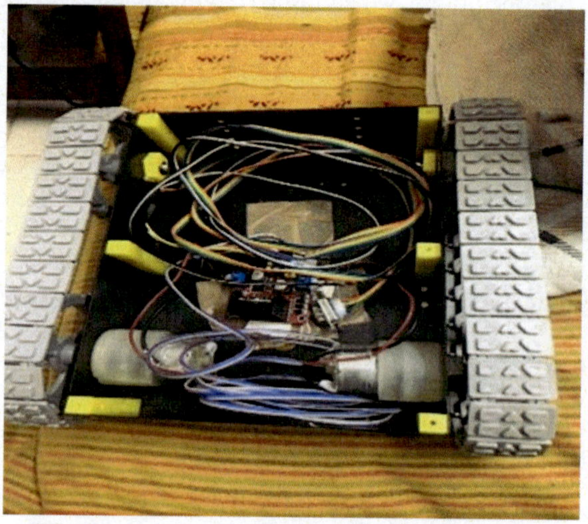

Fig. 1 Rover component

Fig. 2 Glove component

4 Implementation

Our proposed model can be divided into three fundamental parts.

1. Data transfer between the Arduino and the server

- ADXL accelerometer contains three axes, the data is collected when the sensor is moved or tilted. This data is getting stored in the Arduino's memory under three different variables for each axis, named "SensorValuet," "SensorValuep," and "SensorValuez."
- The raw data is then read and processed in the Arduino to get the proper value of the required classification of the axis's coordinates.
- To check whether the system is dead or working, we are also printing the coordinates on an LCD screen.
- The processed data is then compared to get binary 4-bit output associated with four basic types of the movement of the rover, i.e., forward, backwards, right, and left. Each of the 4-bits represents a type of movement. For example, "1000" represents forward, "0100" represents backward, "0010" represents left, and "0001" represents backwards.
- The raw sensor values are converted to the above represented bits if they come under the specific range.
  - If the value calculated at "SensorValuet" is greater than the desired value of 200, the transmitted data to the Wi-Fi module is "1000" that means the command is to move forward else "0100" is transmitted to move backwards.
  - If the value calculated at "SensorValuep" is greater than the desired value of 190, the transmitted data to the Wi-Fi module is "0010" that means the command is to move left else "0010" is transmitted to move backwards.
  - While this data calculation and serial communication is happening, the Wi-Fi module connects to the Internet and establishes connection with the server with the help of IoT development board.
- While the Arduino determines the value, which has to be transmitted, the Wi-Fi module established the connection with the cloud server and serially prints if the connection was successful or not.
- The 4-bit code is transmitted 1-bit at a time to the cloud server "http://www.iotclouddata.com/wilog/" and stored in the table at the cloud which is being transmitted at a regular interval of 5 ms.

Flow chart 2 represents the above procedure in a flow diagram.

In order to access the table with all the inputs from the glove, we are first greeted with the login page on the Cloud Server. Upon logging in, we gain the access to the DataLogs table where every single transmitted input is stored. The Data Logs can be seen in Fig. 3 where the table consists of LogID, Data, LogDate, and LogTime.

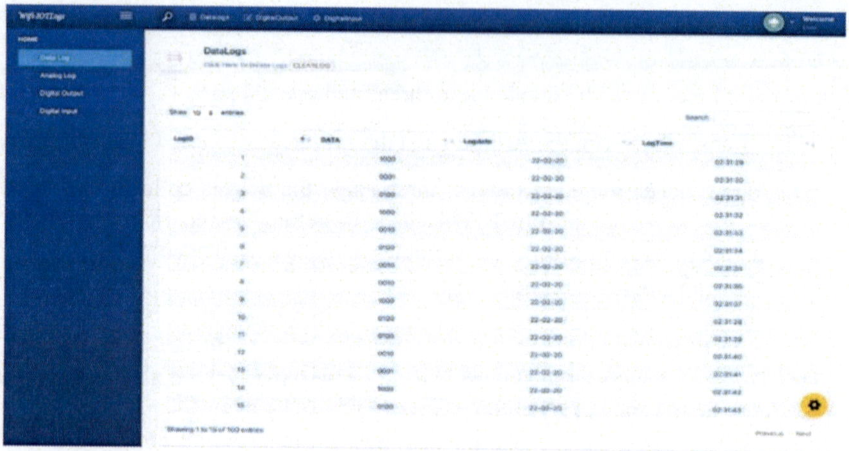

Fig. 3 Database page

2. Data transfer between rover and the cloud

   - If the received bits are "0," "1," "0," "0," then the rover will move backward.
   - If the received bits are "0," "0," "1," "0," then the rover will move right.
   - If the received bits are "0," "0," "0," "1," then the rover will move left.
   - The rover receives data at the interval of every 5 ms.

   Flow chart 3 represents the above procedure in a flow diagram.

3. Live stream of the cameras

   - The first step in order to stream the data from the camera is to set up the Web page using HTML.
   - The normal vision camera used is a 5 mega-pixel camera with a 60-degree field of view. The camera has the specification to go over 1920 × 1080 resolution over 30fps, i.e., 30 frames per second with 1080p resolution.
   - To make the stream more fluid and faster, we have set the resolution of the stream to be at 640 × 480 with 90fps, i.e., 480p with streaming over 90 frames per second.
   - Format of the output is set to MPEG. We chose to go with MPEG over H-264 due to some reasons:
     a. Setup of a H-264/265 video output system is much complex as compared to MPEG or MPEG4.
     b. H-264/265 does provide a clear video, but it fails to provide constant stability of the video quality which MPEG does.
   - The Web page truncates the existing data and continuously streams the data from the camera module in the middle of the page.

The set link for the camera stream is the IP address though which Raspberry pi is connected to the Internet and at a port, 8080 (Figs. 4 and 5).

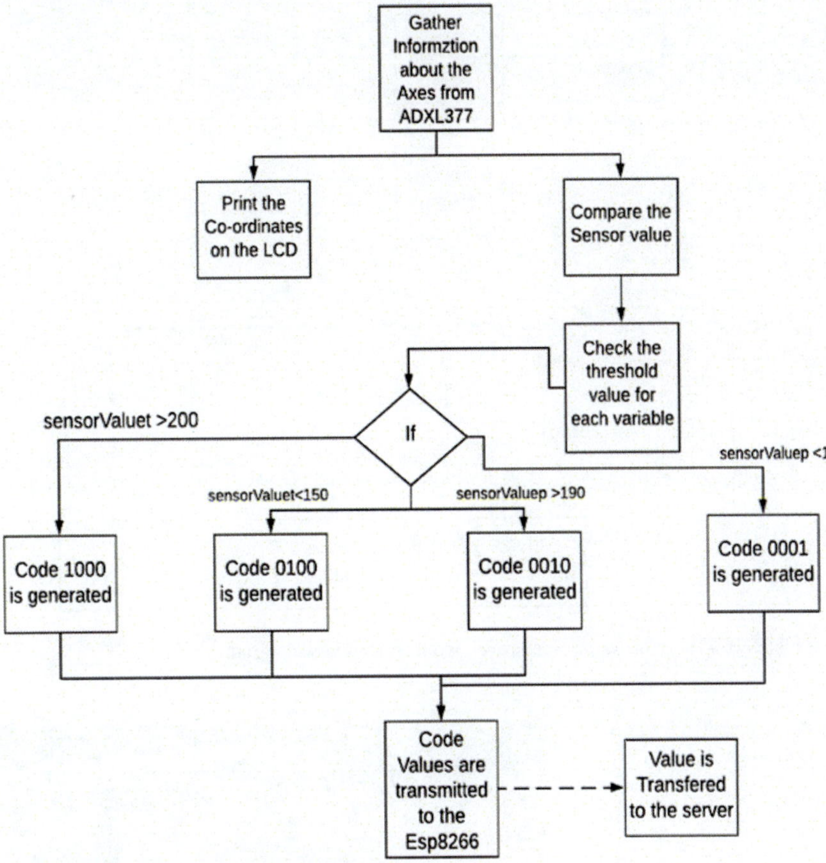

Flow Chart 2 Data transfer between Arduino and server flow diagram

5 Results

5.1 Movement of the Rover

The rover code process is initiated on the hardware manually to start reading the values from Table 1 located at the cloud. The rover receives the value and starts movement of the wheels accordingly as described in the previous sections. The hardware for glove when mounted on a plate is shown as in Fig. 2.

The limitation faced is the relation between Internet connection speed and the reaction speed of the rover. If the Internet connection provided to the Raspberry pi is slow, then despite the high-speed connection of the Glove hardware, the rover will react to the commands with a significant delay.

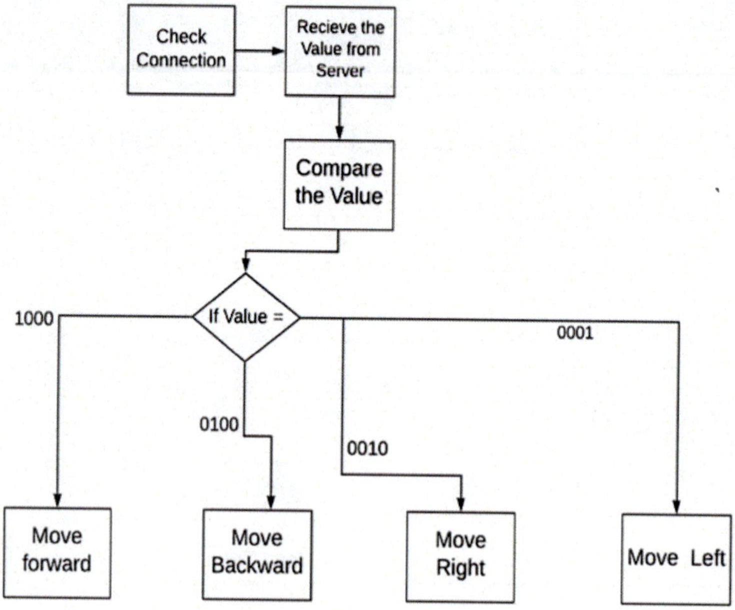

Flow Chart 3 Data transfer between rover and the cloud flow diagram

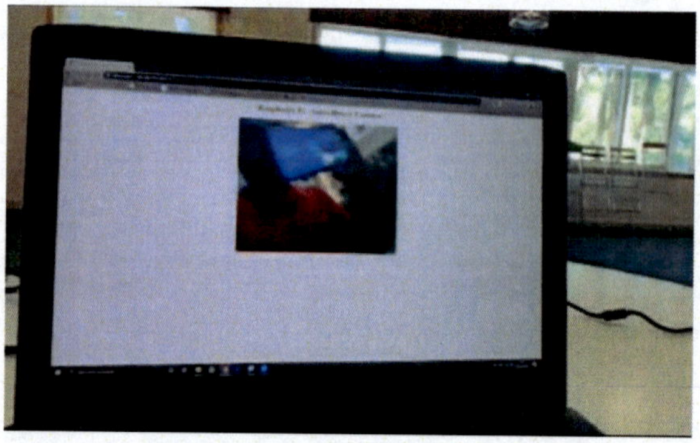

Fig. 4 Camera live feed on laptop browser

Furthermore, due to the multi-terrain class of the rover, the motor used for the wheels is a 12 V Dc motor with 100 RPM. This motor requires excess power. For that, additional power is being given to the rover. The rover draws power from two different ports of a 20,000 mAH power-bank, one port if connected to give power to the Raspberry pi, while the other gives additional power to motor driver. The rover hardware is shown as in Fig. 1 upon sending in the rover, one can close the top lid of the rover to prevent damage to the sensors and hardware of the rover. The metallic tank like chassis enable it to be robust and operate on multi-terrain.

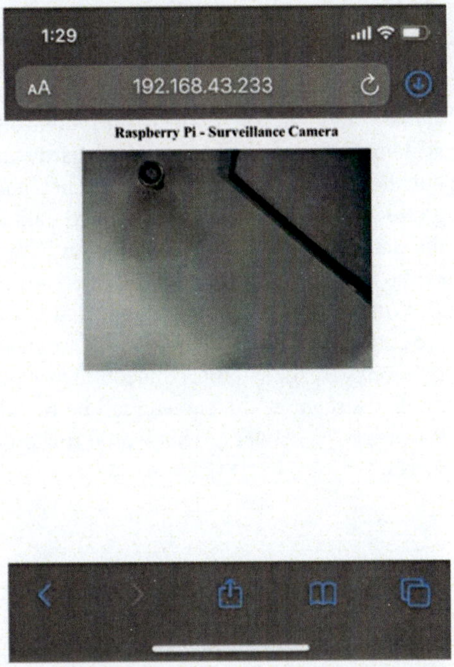

Fig. 5 Camera live feed on mobile browser

Table 1 Processing of commands in rover

Gesture	Transmitted Bits	Rover's Response
Forward	"1," "0," "0," "0"	Forward
Backward	"0," "1," "0," "0"	Backward
Left	"0," "0," "1," "0"	Left
Right	"0," "0," "0," "1"	Right

5.2 *Live Stream of the Cameras*

The camera stream is also started when the code is run manually. The camera streams within the network of the Raspberry pi. The Web page is kept to be a simple blank page with a media player situated in the middle of it. The resolution used is 640 × 480 pixels or 480p as it not only provides significant quality to know the presence of someone but also provides us with the smoothness of 90 frames per second. The livestream platform can be viewed on any device as depicted as per Figs. 4 and 5.

6 Conclusion and Future Work

In this proposed model, we control a multi-terrain rover with two different types of cameras mounted on it. The rover is controlled using the gestures sent from the glove system of the hardware. The glove hardware can be situated at any part of the world but should be connected to the Internet, and it will be able to control the rover. The glove will be sending in the received data to the cloud, and the rover would receive the data from the cloud. The relation between the Internet speed and reaction time of the rover is also observed.

With more capital investment and resources, the ideas to upgrade the system are endless. When being used in the rescue operation, the Internet connection can be replaced with the satellite connection to avoid the delay in reaction time. Furthermore, the live stream of the camera can be transferred on a live cloud rather than on the Raspberry pi network. This would make the rover accessible from any part of the world.

References

1. Cheng K, Khokhar MS, Liu1 Q, Tahir R, Li M (Member, IEEE) (2019) Data-driven logical topology inference for managing safety and re-identification of patients through multi-cameras IoT. Received 17 Oct 2019, accepted 28 Oct 2019, date of publication 4 Nov 2019, date of current version 13 Nov 2019, pp 1594666–159478
2. Chen Z, Huang X, Senior Member, IEEE (2019) Pedestrian detection for autonomous vehicle using multi-spectral cameras. IEEE Trans Int Veh 4(2):211–219
3. Chavez-Burbano P, Vitek S, Teli SR, Guerra V, Rabadan J, Perez-Jimenez R, Zvanovec S (2019) Optical camera communication system for Internet of Things based on organic light emitting diodes. Electron Let 55(6):334–336
4. Liu C, Chau KT, Lin F (2015) A new hybrid-structure machine with multimode fault-tolerant operation for Mars Rover. IEEE Trans Magn 51(11):103893–103917
5. Kalaivani A, Perumal P, Ali MAM, Yahya ZH (2019) Fire fighter robot with night vision camera. In: 2019 IEEE 15th international colloquium on signal processing and its applications (CSPA 2019), 8–9 Mar 2019, Penang, Malaysia, pp 270–275
6. Benli E, Student Member, IEEE, Motai Y, Senior Member, IEEE, Rogers J, Senior Member, IEEE (2019) Human behavior-based target tracking with an omni-directional thermal camera. IEEE Trans Cogn Dev Syst 11(1):36–50
7. Tai T-M, Jhang Y-J, Liao Z-W, Teng K-C, Hwang W-J (2018) Sensor-based continuous hand gesture recognition by longshort-term memory. IEEE Sensors Letters 2(3):66000704
8. Kalantarian H, Alshurafa N, Sarrafzadeh M (2016) Detection of gestures associated with medication adherence using smartwatch-based inertial sensors. IEEE Sens J 16(4):1054–1061
9. Lakhssassi A, Member, IEEE, Palenychka R, Member, IEEE, Savaria Y, Fellow, IEEE, Saydé M, Zaremba M, Senior Member, IEEE (2016) Monitoring thermal stress in wafer-scale integrated circuits by the attentive vision method using an infrared camera. IEEE Trans Circuits Syst Video Technol 26(2):412–424
10. Desai SJ, Shoaib M, Member, IEEE, Chowdhury AR, Senior Member, IEEE (2015) An ultra-low power, "Always-On" camera front-end for posture detection in body worn cameras using restricted Boltzman machines. IEEE Trans Multi-Scale Comput Syst 1(4):187–194
11. Weiren Wu, Ning X, Liu L (2013) New celestial assisted INS initial alignment method for lunar explorer. J Syst Eng Electron 24(1):108–117

12. Pickering CA (2005) Gesture recognition driver controls. IEE Comput Control Eng
13. Liljebäck P, Member, IEEE, Pettersen KY, Senior Member, IEEE, Stavdahl Ø, Member, IEEE, Gravdahl JT, Senior Member, IEEE (2011) Controllability and stability analysis of planar snake robot locomotion. IEEE Trans Autom Control 56(6):1365–1380
14. Transeth AA, Member, IEEE, Leine RI, Glocker C, Pettersen KY, Senior Member, IEEE, Liljeb ˚ack¨ P, Student Member, IEEE (2008) Snake robot obstacle-aided locomotion: modeling, simulations, and experiments. IEEE Trans Robot 24(1):88–104

Inverse Intermodulation-Based Transition Encoder Scheme with Minimal Power Consumption

V. Sasikala, K. Chitra, and A. Sivasubramanian

Abstract In the current scenario, optical computing is the highly demanded technology needed in high-speed signal processing applications. As this demand is increasing day by day, the focus is emphasized more on reliability, switching speed and minimal power consumption. In this paper, we proposed the experimental setup for a secured transition encoder based on the Hamming distance algorithm. It is used to encrypt the data using a control bit and also minimize the power consumption by reducing the number of transitions between the successive elements. The major components of this setup are designed using Semiconductor optical amplifiers and are simulated by selecting frequencies in inverse intermodulation products. The dependence of the probe power, channel spacing are examined for nonlinear elements such as fiber and semiconductor optical amplifiers. The obtained observations confirmed that the suggested setup is capable of operating at minimal power with a good extinction ratio and quality factor.

Keywords Nonlinear optics · All optical computing · Semiconductor optical amplifier · Minimal power consumption · Transition encoder

1 Introduction

Optical signal processing is the most demanded technology in the current scenario for making reliable and flexible high-speed photonic networks. The few limitations in improving the performance of conventional computing systems are the size, speed, bandwidth and networking demands. The challenges in conventional computing are overcome by the fast growing all optical technology. The basic conventional electronics functions are realized by all optical gates such that it satisfies the information carrying signals to remain as photons and produce the ultra-compact high-speed

V. Sasikala (✉) · K. Chitra · A. Sivasubramanian
Sri Sairam Engineering College, Chennai, India
e-mail: sasikala.v2014phd1152@vit.ac.in; sasikala.ece@sairam.edu.in

A. Sivasubramanian
School of Electronics Engineering, VIT University, Chennai Campus, Chennai, India

A. Sivasubramanian et al. (eds.), *Futuristic Communication and Network Technologies*, Lecture Notes in Electrical Engineering 792,
https://doi.org/10.1007/978-981-16-4625-6_98

systems. Switching operation in all optical signal processing plays the important role in making reliable and flexible high-speed photonic networks [1–3].

Optical signal processing is the most demanded technology in the current scenario for making reliable and flexible high-speed photonic networks. The few limitations in improving the performance of conventional computing systems are the size, speed, bandwidth and networking demands. The challenges in conventional computing are overcome by the fast growing all optical technology. The basic conventional electronics functions are realized by all optical gates such that it satisfies the information carrying signals to remain as photons and produce the ultra-compact high-speed systems. Switching operation in all optical signal processing plays the important role in making reliable and flexible high-speed photonic networks [1–3].

This paper presented the experimental setup for an all optical transition encoder using inverse intermodulation nonlinear effect. The introduction and the challenges in conventional computing are discussed in Sect. 1. Section 2 discusses the performance shortcomings in optical switching. Nonlinear effects in all optical computing and the inverse inter-modulation effect are briefly described in Sect. 3. We proposed the experimental setup for all optical transition encoders to minimize the power consumption in Sect. 4. The simulation output of the XOR gate and optimization techniques based on inverse inter-modulation nonlinear effects in SOA are given in Sect. 5 and finally the conclusion is made in Sect. 6.

2 Optical Switching-Performance Shortcomings

Currently, the computers are consuming more energy of up to 250 watts in performance mode. An important point to be analyzed is that most of the power consumption is not used effectively; instead it is consumed as idle power due to internal friction in the ICs. Friction between the elementary particles releases a high amount of heat and is to be overcome by an alternate technology. All optical signal processing uses light as an information carrier and perform the signal processing at very high speed. However, it uses the nonlinear properties of the fiber or other optical components to realize most of the conventional computing operations. A relatively weak optical nonlinearity needs strong switching fields to implement the required operation. Such a strong field will induce many performance degradations [4]. Several encoding algorithms are being developed by the researchers to improve the switching speed and compensate for the losses. This paper proposed a multi-level coding technique that enhances the efficiency of all optical switching transition activities. It also minimizes the dynamic power dissipation [5] using the principle of bit inversion encoding.

3 Nonlinear Effects in All Optical Computing

The linear systems are characterized by the system function irrespective of inputs. Any light signal coupled into the optical fiber is shaped by dispersion (phase) and attenuation (amplitude). The linear system does not create any new frequency where only the existing frequencies are modified. The origin of nonlinear response is the effect of the intense electromagnetic field from any dielectric to the light. When we send an intense signal it is no longer sustained in linear because of changes in the refractive index with intensity and is referred to as the Kerr effect.

In the beginning era of fiber optic telecommunications, nonlinear effects were the fundamental limitation and it decides the amount of data that can be sent on a single optical fiber. Various methods have been proposed using different optical filters like infinite impulse response (IIR), and Bessel's filter [6, 7] to mitigate the nonlinear effects such as parametric amplification. Another technique also focused on unequal channel spacing demonstration to mitigate the effects in FWM [8, 9]. All these techniques lack by complexity and cost. After finding the solutions to overcome the basic linear hurdles like attenuation and dispersion, the nonlinear optics also became important in many applications. It was very important to consider the nonlinear optics when designing the ultra-modern fiber optic systems.

When a high-intensity optical signal impinges on the optical field, it causes changes in the refractive index. The total refractive index n is given by Eq. 1 as,

$$n = n_{\mathrm{L}} + n_{\mathrm{nL}} \tag{1}$$

When n_{L} is Linear refractive index, n_{nL} is nonlinear refractive index.

The linear function of the electric field is represented by the electric polarization P and is given as Eq. 2,

$$P = \varepsilon_0 X E \tag{2}$$

where X is linear susceptibility, ε_0 is permittivity and E is the electric field intensity. At high intense field, the medium becomes nonlinear and the polarization is as shown in Eq. 3,

$$P = \varepsilon_0\left(XE + X^{(2)}E^{(2)} + X^{(3)}E^{(3)} + \cdots\right) \tag{3}$$

where $X^{(2)}$ and $X^{(3)}$ are second and third harmonic generations, responsible for parametric interactions and intensity-dependent refractive index respectively.

The optical inter-modulation effect due to nonlinear optics is called FWM. The interception due to third-order harmonics generates signals at new wavelengths. It is also called a parametric nonlinear process in which the energy of the incoming photon is conserved and the efficiency of the new wavelength is depending upon the phase-matching condition of the original and the generated signals [10, 11].

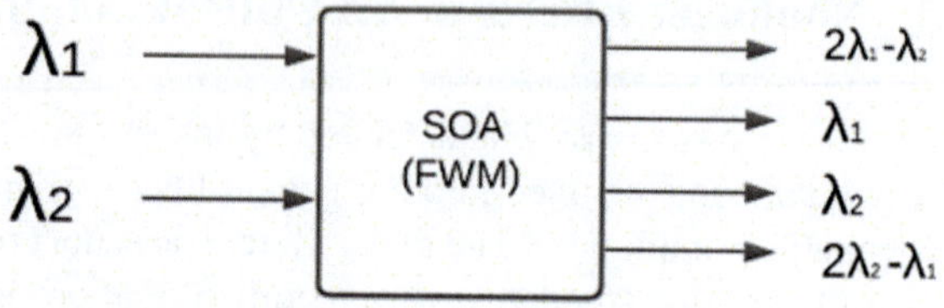

Fig. 1 Two signals intermodulation effect

When two input signals at different wavelengths impinge on a nonlinear medium, it generates additional frequencies as shown in Fig. 1.

The nonlinear polarization P_{nl} for a single signal is given by Eq. 4.

$$P_{\text{nl}} = \varepsilon_0 X^3 E^3 \tag{4}$$

The reason for FWM is the phase mismatch between different signals traveling at different group velocities in any nonlinear medium. In general, for n numbers of signals launched into the fiber, the total numbers of generated signals are given as M and are given in the following Eq. 5.

$$M = \frac{n^2}{2}(n - 1) \tag{5}$$

There are two categories of newly generated frequencies. The sum frequency is always higher than the input frequency whereas the difference frequency is typically the sum of two and minus the third. The sum and difference frequencies are generated due to the exploitation of the incoming photons at the point of resonance in nonlinear medium. The sum frequency is generated closer to the resonance however it changes the refractive index and generates additional co-linear vectors. Thus the mixing path length will have phase mismatching components of the additional vectors. This phase mismatching causes diverse effects especially during sum frequency but it is more easily compensated for different frequency generations. The FWM generates additional frequencies which is usually widening the frequency band due to intermodulation effects. By properly tuning to different frequencies, the diverse effects are compromised and the spectrum also becomes narrowed. This conflicting outcome to FWM reaction is called the inverse intermodulation effect which is also beneficial in many optical signal processing applications.

4 Experimental Setup for Efficient Optical Switching

In optical signal processing, the transitions between successive bits are to be reduced to minimize the power consumption due to switching. Otherwise, these power penalties will limit the scalability. If it exceeds the ample transceiver power budget, optical amplifiers become necessary in between the elements. This in turn raises both power consumption and interconnects cost between the components. Hamming distance algorithm is used to determine the difference between the code vectors and reduce the transitions between the successive data.

Consider n words of w bit size, the hamming distances are assumed as $\{h_1, h_2 \ldots h_n\}$. The hamming distance of ith word is given in Eq. 6.

$$h_i = \sum_{j=1}^{w} S_{(i-1)} \oplus S_{ij} \tag{6}$$

The number of transitions between successive elements in the given data is represented in Eq. 7.

$$\sum_{i=1}^{n} h_i \tag{7}$$

The experimental setup of the Transition encoder which encodes the data using Hamming distance is shown in Fig. 2. The encoder output is the continuous stream of 1 s. However, the control bit is necessary to decode and bring the original data at the following chip in the network. This setup reduces the number of transitions in the encoded information compared to the number of transitions in the original data.

Let the Previous data = 0 0 1 1; and present data = 1 1 0 1.

Hamming distance between the successive element X and Y is defined by $d(X; Y) = 3$. Here only one bit is the same for each other. Hence $d = 3$. After encoding the transitions, difference between the two data is reduced to one.

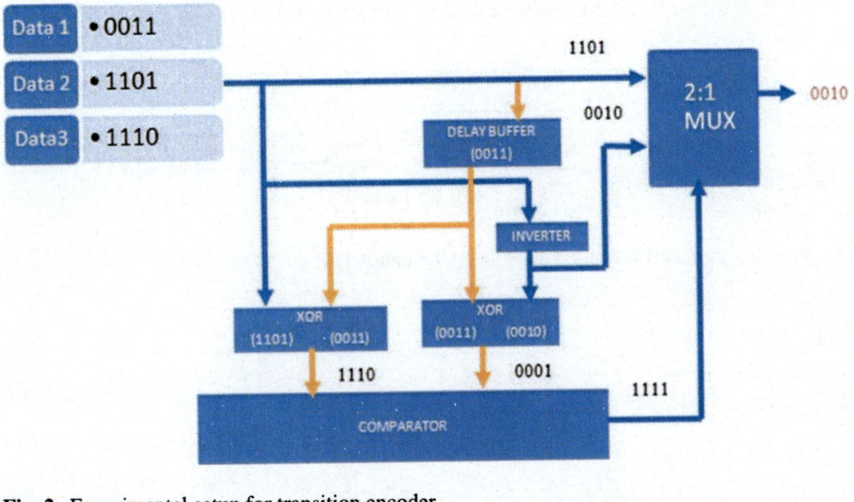

Fig. 2 Experimental setup for transition encoder

5 Results and Discussions

In the proposed encoder design, the XOR gate is the major component to perform the encryption. The simulation setup shown in Fig. 3 represents the XOR operation using intermodulation effect. The data block represents two pump signals of wavelength 1556.75 and 1562.15 nm. The wavelength of 1549.3 nm is given as a low power probe signal. These signals are coupled and given to the SOA having a pump gain of 0.350 A and wavelength gain resonance at 1548.3 nm.

The results obtained for a single XOR gate are shown in the figure for the custom data of 4 bits. The XOR gate can be efficiently designed using XPM or FWM nonlinear effects. The design using Mach–Zehnder interferometer needs two SOA whereas the design using the FWM effect needs careful detuning to select a particular wavelength [7].

The simulation results of XOR gate are presented in Fig. 4. The analysis is also made on the channel spacing, input power of probe signal and phase and resonant conditions. Figures 5 and 6 show the relative power vs. Extinction ratio with respect to time for fiber and SOA-based intermodulation products respectively. It is shown that the fiber-based FWM effect produces resonance only for the perfect phase-matching condition, whereas the SOA-based products can be adjusted with polarization components and generate resonant signals even when the phases are orthogonal to each other. Figure 7 shows the spectroscopy view of input and output of two-channel FWM signals. It is clearly shown that the original component and the generated sum and difference component are satisfying the FWM equation. When more number of inputs impinged on the SOA, there may be additional effects due to Raman effects and stimulated scattering. By properly detuning to a particular wavelength, the respective effect may be analyzed and utilized for further research.

Figure 8 shows the graph between extinction ratio and the probe power. The probe

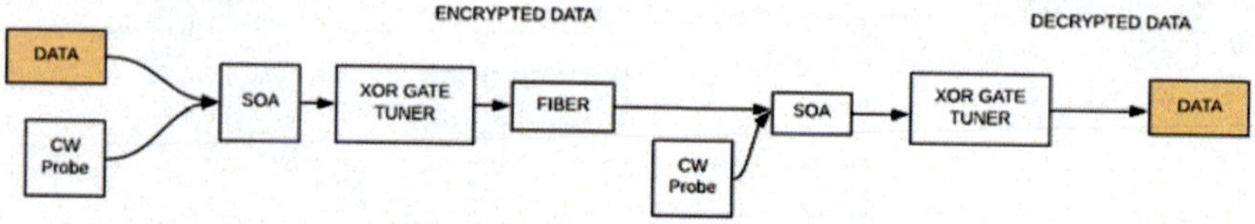

Fig. 3 Encryption and decryption using intermodulation effect in SOA

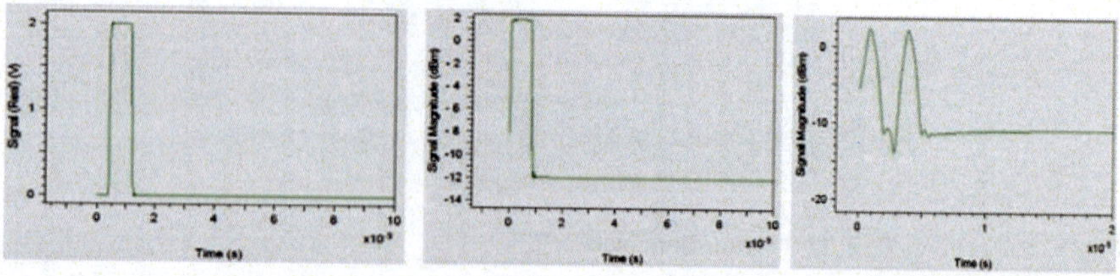

Fig. 4 Input data 1at 1557.75 nm, input data 2 at 1563.15 nm, XOR gate output

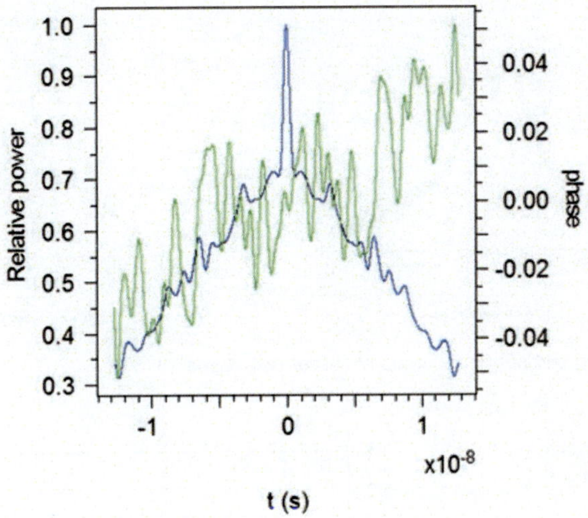

Fig. 5 Power versus phase for fiber-based FWM

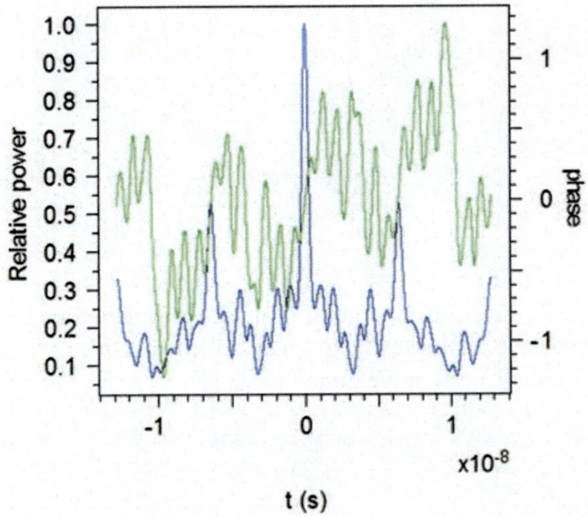

Fig. 6 Power versus phase for SOA-based FWM

power is to be kept at a small level to avoid any unwanted gain modulation. The FWM signal preserves the phase and amplitude information and gives a good extinction ratio by maintaining proper input power and channel spacing. In considering the existing encoding scheme designs based on Silicon Mach–Zehnder interferometer, it consumes 51–200 mW for 10Gbps corresponding to 5 pJ/bit of energy and 100 μm long [12–14]. In comparison with this, the proposed inversion scheme consumes only 11.78 mW power by properly choosing the down frequency range in SOA

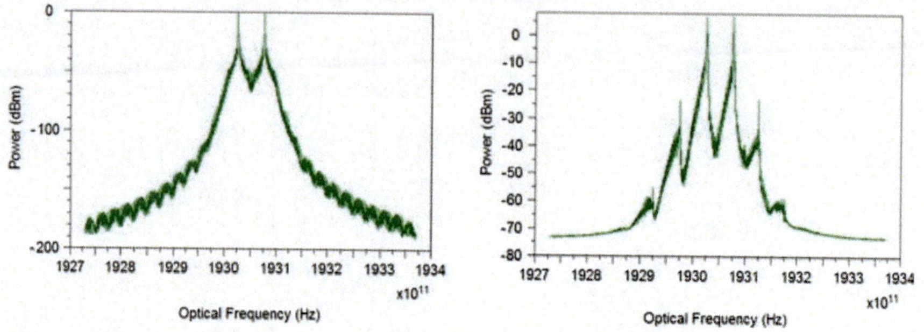

Fig. 7 Input (left) and output spectrum (right) of two-channel FWM

Fig. 8 Probe power versus extinction ratio of FWM-SOA

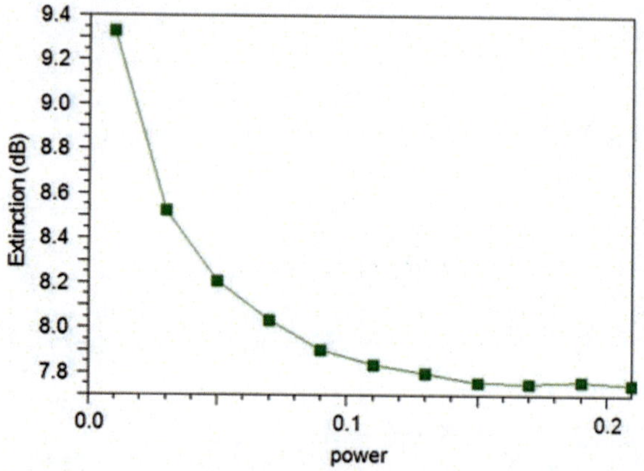

FWM effect. The highly efficient control of output wavelength is made possible by FWM process and is being the key factor to avoid any distortions in adjacent signals of the generated wavelengths at the output [15].

6 Conclusion

All optical signal processing is a fast growing technology replacing the conventional electronics by all optical gates. The nonlinear effects play a major role in computing-based applications. SOAs are one of the useful building blocks for optical computing. The XOR gate using SOA non linearity is the fundamental and major component in realizing many operations. The simulation results of XOR based on SOA inverse inter-modulation effect are presented. The design is extended to the application as a dual-purpose transition encoder which encrypts the data using a control bit and also reduces the number of transitions between the successive elements. The analysis is also made on the channel spacing, input power of probe signal and phase and resonant conditions. The FWM signal preserves the phase and amplitude information and gives a good extinction ratio by maintaining proper input power and channel spacing. All optical computing is an inspiring field in research where we can bring much innovation in different optical circuits and hence increase switching performance at ultra-high-speed applications.

References

1. Sasikala V, Chitra K (2018) All optical switching and associated technologies: a review. J Opt 47(3):307–317
2. Dimitriadou E, Zoiros KE (2013) All-optical XOR gate using single quantum-dot SOA and optical filter. J Lightwave Technol 31(23):3813–3821
3. Sasikala V, Chitra K, Sivasubramanian A (2016) Design of efficient all optical switching encoder using XOR gate based on SOA non linearity. In: International conference on fibre optics and photonics. Optical Society of America
4. Kotb A (2016) Simulation of high quality factor all-optical logic gates based on quantum-dot semiconductor optical amplifier at 1 Tb/s. Optik 127(1):320–325
5. Jacobs BC et al (2012) All-optical computing using the Zeno effect. Johns Hopkins APL Tech Digest 30(4):346–360
6. McKinstrie CJ, Karlsson M (2017) Effects of polarization-mode dispersion on degenerate four-wave mixing. J Lightwave Technol 35(19):4210–4218
7. Kavitha K, Vishnu Priya R, Vel Rathina Devi M (2017) Mixed fiber optical parametric amplifiers for broadband optical communication systems with reduced nonlinear effects. In: 2017 International conference on communication and signal processing (ICCSP). IEEE
8. Khalid R, Zafrullah M (2017) Analysis of the Q factor of the external modulation with different optical filters to overcome the FWM non-linearity in the fiber networks. In: 2017 International symposium on wireless systems and networks (ISWSN). IEEE
9. Paz JL et al (2018) Three levels of propagation of the Four-wave mixing signal. Results Phys. 11:414–421
10. Azzam ES, Aly MH (2018) FWM in DCF based four-input all-optical NOR/AND gate. Opt Quant Electron 50(4):166
11. El-Saeed EM et al (2016) Optical logic gates based on semiconductor optical amplifier Mach–Zehnder interferometer: design and simulation. Opt Eng 55(2):025104
12. Barwicz T et al (2007) Silicon photonics for compact, energy-efficient interconnects. J Opt Netw 6(1):63–73
13. Gan F et al (2007) Compact, low-power, high-speed silicon electro-optic modulator. In: Conference on lasers and electro-optics. Optical Society of America
14. Green WMJ et al (2007) Ultra-compact, low RF power, 10 Gb/s silicon Mach-Zehnder modulator. Optics Exp 15(25):17106–17113
15. Anjum OF et al (2019) Bandwidth enhancement of inter-modal four wave mixing Bragg scattering by means of dispersion engineering. APL Photon 4(2):022902

A Comparative Analysis of Standard Classifiers with CHDTC to Detect Credit Card Fraudulent Transactions

C. M. NalayiniI and T. Gayathri

Abstract Nowadays, credit card frauds grow rapidly due to the technological advancements. Fraudsters are well trained with the technology factors, so it is quite tough enough to avoid such frauds in real-life scenario. Even though many fraud detection systems are introduced to reduce the fraud transactions, still some loop holes will be the cause for such frequent fraud transactions. It is necessary to build a model to detect fraud transactions at an earlier stage to prevent the further loss. We proposed a chained hashing decision tree classifier to detect the fraud transactions at a faster rate and also to reduce false positives. We did data normalization first and then applied feature correlation algorithm to select the appropriate attributes for training and testing purpose. After our algorithm is trained and tested with different subdatasets to improve accuracy, it is also compared with other existing models such as NB and RF to prove its high performance through various parameters such as false alarm rate and sensitivity in association of confusion matrix. Our CHDTC algorithm is divided into two phases like pattern analysis and pattern detection phase; during the pattern analysis phase, the rules are stringent to identify the behavioral pattern, and during the pattern detection phase, pattern is detected based on the decision like either fraud or legitimate. With the use of chained hashing along with decision tree classifier, it is fast enough to predict the class either fraud or legitimate for any input transactions.

Keywords CHDTC—Chained hashing decision tree classifier · NB—Naïve Bayes · RF—Random forest · Feature correlation

C. M. NalayiniI (✉) · T. Gayathri
Information Technology, Velammal Engineering College, Chennai, India

T. Gayathri
e-mail: gaya3thangadurai@gmail.com

A. Sivasubramanian et al. (eds.), *Futuristic Communication and Network Technologies*, Lecture Notes in Electrical Engineering 792,
https://doi.org/10.1007/978-981-16-4625-6_99

1 Introduction

In today's scenario, everything is digitized, customers are seeking for digitized kind of transaction, and they feel quite comfortable with current advancements and technology with the ease of use and convenient transaction. Credit card transactions have both positives and negatives. Fraudsters are skilled enough to perform fraud transactions frequently. It is most essential to find a best fraud detection system where it can be implemented with best machine learning model to detect the fraud transaction at a faster rate. Various machine learning algorithms such as Naïve Bayes, random forest, SVM, decision tree and many including both supervised and unsupervised are intended to aim for increasing the detection rate and decreasing the false alarm rate [1].

1.1 Traditional Versus Machine Learning Approach

Nowadays, most of the industries started using machine learning approaches rather than the traditional rule-based approach.

Traditional Rule-Based Approach in Credit Card Fraud Detection System
In this approach, rules are written by sophisticated programmers. If any changes need to be done to find a new fraud case, then it is very difficult to do changes in the existing rules, a set of new rules to be created again. It has its own drawbacks such as high time consumption, high cost, unable to identify the hidden patterns due to its strict rules and showing high false positives. This in turn leads to lose the genuine customers. Therefore, an optimized system is required to analyze the patterns and also to perform predictions of fraud cases at the earliest.

Machine Learning Approach
To overcome the drawbacks of the traditional rule-based system, machine learning techniques are the suitable solution to detect credit card fraud at a fast rate. It does not require a sophisticated analyst to write rules pertain to changes. It self-learns the data and provides better solution by experience. It also reduces the false positives by its self-learning behavior and also identifies the hidden patterns which help to detect a credit card fraud the one which is not yet recognized previously. The following factors make ML more popular for the detection of fraud cases such the speed (fast computation), efficiency (reduces false positives) and scalability (producing accuracy for huge dataset).

1.2 Detection Techniques

Machine learning plays a vital role in providing solutions to the real world applications such as health-care system, banking sector, trading systems, insurance and many other applications.

Logistic regression is a statistical model; with the use of logistic regression formula, computation is quite easy to detect credit card fraud. When compared with other models, its performance is proved low. It is a pure binary-dependent variable. It is used to predict the categorical value [2].

Linear regression is a statistical model that provides relationship between dependent and independent variables under linear type. It is used to predict continuous value and also quite sensitive outliers [2].

Naïve Bayes is a kind of classification algorithm and depends on two conditional probabilities like fraud and legitimate [3]. It uses binary classification rules to predict the class. Sometimes due to over fitting, there are chances of retraining the new fraud types.

Random forest algorithm is suitable for both regression and classification. It works well on missing values and also suitable for both categorical and numerical features. It is a kind of stable algorithm.

Decision tree [4] is applicable for both regression and classification. Here, the data are continuously divided into n nodes until reaching a certain parameter. It involves both decision-making nodes and leaves. Final outcome will be obtained from the respective leaf based on the decision. Information gain and entropy are baseline to make the respective decision for the given dataset. It has the over fitting issue but that can be addressed by random forest algorithm.

Support vector machine (SVM) [5] is most widely used for classification problems also suitable for regression. It involves the hyperplane concept to be implemented in a N-dimensional space which uniquely classifies the data points into respective groups.

Neural network is one of the most widely used machine learning algorithm to detect fraud transactions. It is not suitable for medium-scaled real-time operations due to high computational power. There are chances of retraining for a few fraud cases [2].

1.3 Our Objective

There are several challenges such as lack of high accuracy due to highly imbalanced dataset and increase of false positives. Even though we have come across many machines learning model, each model has its own merits and demerits. Therefore, our objective is to develop a fraud detection system which uses a best algorithm say CHDTC to detect the fraud transaction earlier when compared to the existing machine learning models and also to reduce the false positive at the maximum extend. Our main strategy is to apply hashing data structure on a decision-making system such a way that it will perform faster on any input dataset and provides efficient detection of fraud transactions. Hashing speeds up the computation process and reduces the time factor.

2 Related Work

Kumar and Iqbal [1] tried both supervised and unsupervised learning in order to find the best model which produces high accuracy. For highly imbalanced data set, there exist some loop holes still.

Ms. Amruta et al. [6] have done survey on various detection techniques, but every technique has its own merits and demerits. They have shown all possible performance metrics to be used in machine learning models.

Patel and Singh [7] proposed a system to detect the fraud transactions with the use of genetic algorithm. It produces good accuracy, but it is not suitable for highly imbalanced dataset.

S. Panigrahi et al. proposed a credit card fraud detection system [8]; it involves many components such as transactional DB, Bayesian learner, rule-based filter and Dumpster Shafer. With the help of dumpster theory, they created an initial belief by considering various evidences which in turn classifies three cases, normal, abnormal and suspicious and proved 98% true positive rate.

D. Sánchez et al. created a set of association rules [9] for extracting the behavioral patterns of fraud transactions. They used fuzzy logic and data mining tool to simplify the work and bring better results. Their focus on retail companies and accordingly used the data samples.

E. Kirkos et al. used decision tree classifier and Bayesian belief networks to identify the fraud transactions. They proved BNN had produced 90.3% accuracy and decision tree produced 73.6% [10]. They identified 38 financial fraud with the help of their techniques.

3 Proposed Framework

In our proposed framework, we followed data preprocessing, feature selection and training and testing of our proposed model with appropriate performance metrics by comparing standard classifiers such as Naïve Bayes and random forest. In data preprocessing stage, we used data normalization [11] to improve the training time of all data with same scale [11]. Next, we extracted required features by feature correlation algorithm and then trained our CHDTC algorithm with different subdatasets and tested finally with the appropriate test data set.

A chained hashing decision tree classifier algorithm is proposed to predict the class either the fraud case or legitimate case by self-learning the predefined rules inferred from the selected features. Our CHDTC is divided into two phases like pattern analysis and pattern detection. Here, cardholder details are recorded and stored in a chained hashing methodology for every transaction based on identity, time, location and amount. Fast computation is the base of our proposed algorithm. It predicts fraud or legitimate case at a faster rate and reduces the false positives to the maximum extent. It checks how many times the transaction was done accordingly count variable will be incremented and it is checked with threshold level every time a new transaction come into picture, and also it continuously maintains the time, location for tracking the fraudulent scenario. Considering threshold and other features such as time, location and amount, decision is made by checking with attempt parameter. If attempt is said to be 1, then the decision to be made is FRAUD; otherwise, it is decided as legitimate case and stored the information into the transaction database.

3.1 Pattern Analysis Algorithm

Input: Transaction Database.

Output: Transaction pattern analysis for fraud or legitimate.

Each transaction is assigned a unique transaction ID, whenever a transaction has been processed, its entire details such as customer id, date, time, amount and other details are recorded as a timestamp and stored in the DB.

Each transaction is allowed based on the threshold; thereby, suitable decision can be taken into account.

Set threshold (TH) to 5, where ATTEMPT is one of the fields maintained in every node of the chained to track frauds.

For I = 1 to N cardholder.

Record the following attributes into the chained hash entry of the respective cardholder.

```
L=H->Tran[HASH(Last_3digit(CID), TSIZE)]
                    Fetch(H->Tran[L].CID, H->Tran[L].TH, H->Tran[L].TID, H-
                    >Tran[L].TIME, H->Tran[L].AMT,H->Tran[L].ATTEMPT, H-
                    >Tran[L].LOC_COUNT, H->Tran[L].COUNT)

For I -1 to N Cardholder
        Call L=H->Tran[HASH(Last_3digit(CID), TSIZE)] to get into the Chain
        entry of the respective Card holder
                For J=1 to N Transaction of each Cardholder
                        if((H->Tran[L] .COUNT > TH) && ((H-
                        >Tran[L].LOC_COUNT>4 )&&
                        ((H->Tran[L].TIME) > MIN_TIME)) && (H-
                        >Tran[L].AMT>50000))
                          Set H->Tran[L]. ATTEMPT =1
```

3.2 Pattern Detection Algorithm

```
Input: Transaction Database
Output : Transaction attempted for fraud or legitimate
For I =1 to N Cardholder
        L=H->Tran[HASH(Last_3digit(CID), TSIZE)]
                if (H->Tran[L].ATTEMPT )== 0
                        Decision is legitimate
                else
                        Decision is Fraud
```

4 Experimental Setup

Our proposed chained hashing decision tree classifier algorithm uses different groups of datasets from the original dataset to obtain better accuracy. The model depicted in Fig. 1 shows the data preprocessing and feature selection. We used financial credit card dataset obtained from "Kaggle Data Repository," which is a publicly available one with proof of fraudulent cases. It has 284,808 entries and 31 attributes with class labels [11]. Out of 284,808 entries, we considered 200,000 records as input dataset and performed data normalization and then applied feature correlation algorithm as shown in Fig. 1. Since highly imbalanced dataset is taken as the input, it is essential to remove the certain irrelevant attributes in order to improve the accuracy. Feature selection is one the best approach to remove redundant and irrelevant data. Here, we have used the feature correlation algorithm to remove the highly correlated attributes based on rule-based comparison. In this algorithm, FF-Correlation (feature_feature) and FC-Correlation (feature_class) are the two correlation measures applied on every individual attribute as shown in Fig. 1. If the FC-Correlation value is higher than the FF-Correlation value, then the respective feature is identified as the required feature of the training and testing dataset. Based on this rule, we obtained 23 features required for training and testing the model. Further, the dataset is divided into three groups of

Fig. 1 Data preprocessing and feature correlation algorithm

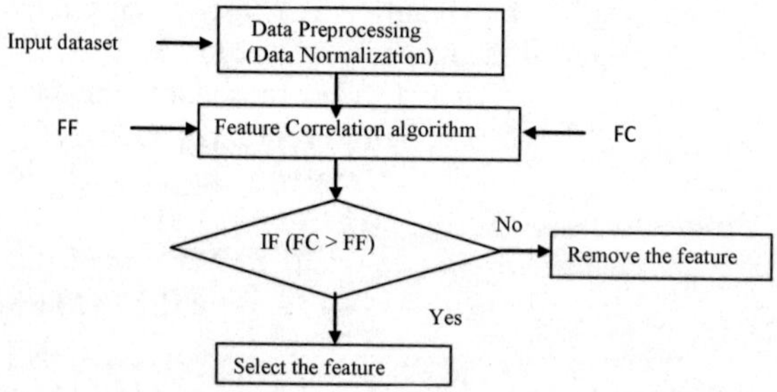

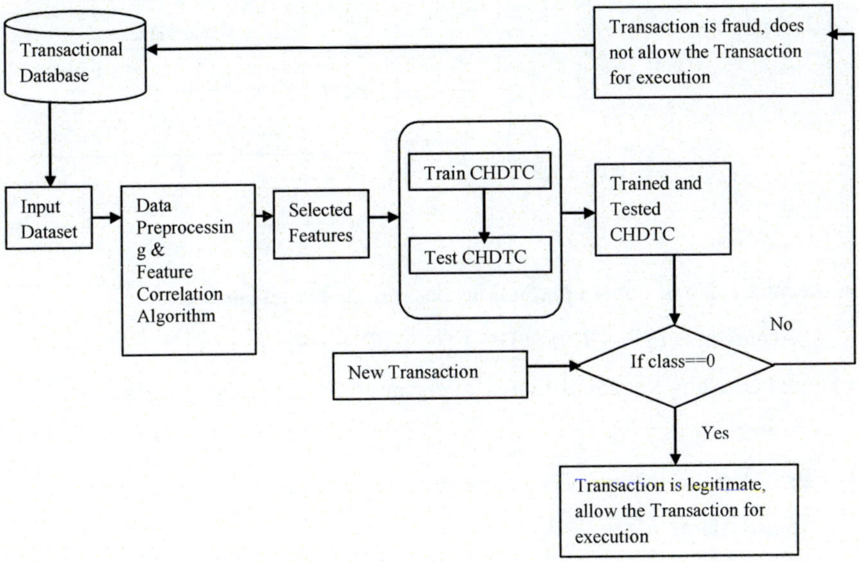

Fig. 2 Fraud detection systems

subdatasets, say D1, D2 and D3 with the fraud rate in each subdataset approximately being 15%, 10% and 5%, respectively [12]. Among the selected attributes, some of the main attributes of credit card transactions such as transaction id, cardholder id or customer id, amount, time and date, transaction location, frequency of the transaction and class or label where the value 0 (zero) is associated with legitimate transaction and the value 1 is associated with fraud transaction are involved for making final decisions in our CHDTC as shown in Fig. 2. The accuracy of model depends on more the data better the model. We used python for completing our proposed work.

4.1 Performance Metrics

Performance is evaluated based on the following classification metrics.

In our work, fraud transaction is associated with the class labeled 1 or positive class and legitimate transaction is associated with the class labeled 0 or negative class. The terms used in the confusion matrix which is shown in Fig. 3 are TP, TN, FP and FN.

Positive (P): No of fraud transactions.
Negative (N): No of legitimate transactions.
True Positive (TP): No of fraud transactions predicted as fraud.
True Negative (TN): No of legitimate transactions predicted as legitimate.
False Positive (FP): No of legitimate transactions predicted as fraud.
False Negative (FN): No of fraud transactions predicted as legitimate.

Fig. 3 Confusion matrix

		Actual Values	
		Positive(1)	Negative(0)
Predicted Values	Positive(1)	TP	FP
	Negative(0)	FN	TN

- Accuracy: No of correct predictions decides the model correct
  - Accuracy = (TP + TN) / (TP + TN + FP + FN)
- Fraud Catching Rate/Sensitivity: It indicates the fraud detection rate
  - Sensitivity = TP/P
- False Alarm rate: Negatives are predicted as positives
  - False Alarm rate = FP/N

5 Results

Our proposed algorithm involves two phases like pattern analysis and pattern detection which predicts the accuracy during training phase as 79.8%, testing phase as 85.7 and finally during validation phase as 93.45 which is shown in Fig. 4. If CHDTC is periodically applied for large volume of dataset, then its accuracy will go higher and higher which in turn reduces false positive rate automatically.

We compared our CHDTC algorithm with standard classifiers such as NB and RF [1] and proved that our algorithm CHDTC has high performance pertaining to metrics which uses the parameters of confusion matrix which is shown in Fig. 3 in order to evaluate fraud catching rate or sensitivity and false alarm rate as shown in Fig. 5. This in turn reduces the false positives to the maximum extent. The real-life dataset used is highly imbalanced, and hence, it is divided into three subsets say D1, D2 and D3 as shown in Table 1. After comparing with above existing classifiers, the fraud catching rate or sensitivity is high and false alarm rate is very low for our CHDTC. Classifiers

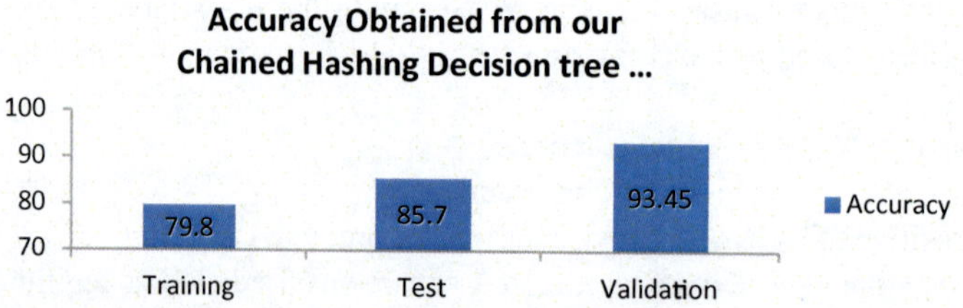

Fig. 4 Accuracy of CHDTC algorithm

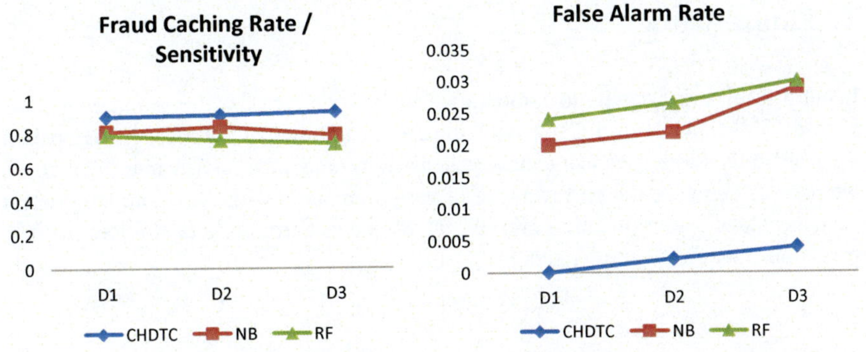

Fig. 5 Performance analysis of the classifiers CHDTC, NB and RF using highly imbalanced dataset

Table 1 Datasets used for performance evaluation

Group of datasets	Legal	Fraud	Total	Fraud rate (%)
D1	19,000	2900	21,900	15
D2	29,400	3100	32,500	10
D3	50,690	4500	55,190	5

NB (Naïve Bayes) and RF (random forest) have poor performance, especially for the Dataset D3 due to its highly imbalanced factor, but our proposed model works better and better for any large dataset which is highly imbalanced.

6 Conclusion

We examined the performance of standard classifiers such as NB and RF with our proposed CHDTC algorithm and proved that our proposed model works better and better for any real-life highly imbalanced dataset and produces high accuracy at all situations. Another advantage of using our model is that we can reduce the false positives with the use of fast computation hash technology and framed rules. Our proposed model also detects the fraud transactions very faster which in turn improves computational speed and reduces the time component factor of the credit card fraudulent domain. This kind of hashing-based methodology can used in business reacted domains such as telecom and insurance to address the fraud transactions effectively.

7 Future Work

In our future work, we will do comparative analysis on various other standard models such as SVM, ANN and CNN with maximum number of metrics. In our paper, we used only three metrics accuracy, sensitivity and false alarm rate. Therefore, we need to compare the performance of various models with few more parameters such as balanced classification rate (BCR), Mathews correlation coefficient (MCC), precision, specificity and F1 score.

References

1. Kumar P, Iqbal F. Credit card fraud identification using machine learning approaches. In: 2019 1st International conference on innovations in information and communication technology (ICIICT).
2. Randhawa K, Loo CK, Seera M, Lim CP, Nandi AK (2018) Credit card fraud detection using AdaBoost and majority voting. IEEE Access 6. https://doi.org/10.1109/ACCESS.2018.2806420
3. Adetunmbi AO, Awoyemi JO, Oluwadare SA (2017) Credit card fraud detection using machine learning techniques: a comparative analysis. In: 2017 IEEE
4. Kamboj M, Gupta S (2016) Credit card fraud detection and false alarms reduction using support vector machines. Int J of Adv Res Ideas Innov Technol 2(4). ISSN 2454-132X
5. Kokate S, Dr. Sheela Rani CM (2020) Fraudulent detection in credit card transactions using radial basis function kernel method based support vector machine. Int J Adv Sci Technol 29(12s):2557–2565
6. Pawarl AD, Prof. Kalavadekar PN, Ms. Tambe SN (2014) A survey on outlier detection techniques for credit card fraud detection. IOSR J Comput Eng (IOSR-JCE) 16(2), Ver. VI. ISSN 2278 8727
7. Patel RD, Singh DK (2013) Credit card fraud detection & prevention of fraud using genetic algorithm. Int J Soft Computing & Eng (IJSCE), 2(6)
8. Panigrahi S, Kundu A, Sural S, Majumdar AK (2009) Credit card fraud detection: a fusion approach using Dempster-Shafer theory and Bayesian learning. Inf Fusion 10(4):354–363
9. Sánchez D, Vila MA, Cerda L, Serrano JM (2009) Association rules applied to credit card fraud detection. Expert Syst Appl 36(2):3630–3640
10. Kirkos E, Spathis C, Manolopoulos Y (2007) Data mining techniques for the detection of fraudulent financial statements. Expert Syst Appl 32(4):995–1003
11. Nayak SC, Misra BB, Behera HS (2014) Impact of data normalization on stock index forecasting. Int J Comput Inf Syst Ind Manage Appl 6:257–269
12. Zareapoor M, Shamsolmoali P (2015) Application of credit card fraud detection: based on bagging ensemble classifier. In: International conference on intelligent computing, communication and convergence (ICCC-2015)

Design and Implementation of MEMS-Based Gesticulation Control for Disabled People

S. Nivash, E. N. Ganesh, and Racha Purna Chander

Abstract The invention in this paper aims to depict an adaptable, cost-cogent and conveyable MEMS positioned subjection ideology, for the physically disabled people with practicability and high accuracy. People do not compass time to look after their old and needy ones because of their hassle work hours. This invention helps them be absolute and recede a frontage of unessential laceration. The transmitter and receiver modules data can be cogent to keep track of instant action that can be taken. The features embedded in designing a scalable, reliable, cost-effective, and portable wireless MEMS-based device module for the physically disabled with high precision and in better functionality. The venture centers on finger movement recognition acknowledgment through displacement in various ways. Numerous methodologies utilizing camera modules and artificial intelligence to interpret sign language were in existence. Gesture recognition with sensor modules is a way for machines in depth in understanding human body language, thus building a high interface bridge and connectivity between humans and machines. Roughly 6 million people in the world face the issue of disability due to paralysis of various degrees and our aim to conquer the problems faced by them in their daily routine life.

Keywords MEMS · Disabled · Sensor · Wireless · Arduino · Zigbee · Accelerometer

1 Introduction

Developing new technologies makes life easier those who are suffering from disabled have less muscular activity than normal people. Stroke is known as the reduction of athletic movements. CVA (cerebra vascular accident) belongs to the category of brain blood nerves collision resulting from paralysis. These occur in old age for the people

S. Nivash (✉) · E. N. Ganesh
Department of ECE, Vels Institute of Science, Technology and Advanced Studies, Chennai, India

R. P. Chander
Department of Electronics and Communication Engineering, SRMIST, Kattankulathur, Chennai, India

A. Sivasubramanian et al. (eds.), *Futuristic Communication and Network Technologies*, Lecture Notes in Electrical Engineering 792,
https://doi.org/10.1007/978-981-16-4625-6_100

due to genetic disorders or injuries earlier in their lives. These activities of CVA cause the patients to get paralyzed and can't work without any human assistance in their daily routine. With the increase in technology, sensible tactile nerds have been initiated with embedded systems for these people's emergency need. With the increase in CVA, the pressure to convince the demand to forget their needs has been deployed. As we know, the rapid growth of these category people has made many equipped developments like a wheelchair and the use of lane sticks to support and humanitarian assistance. But with the lack of opposing opinions, the reduced service and stress have been considered for these people's benefits in embedded systems in the interfacing terminology and soft corner implementation with reliable support. Hence, the paralyzed people can work with these designs in the system's form, only handling functions. Thus, the technology has been advantageously taken into considerations and to stabilize the consequences without hazard carving methodologies.

2 Literature Review

A brief review of the existing research outcomes is discussed in this section.

The project [1] demonstrates an adaptable robotic staging refers to a type of structure that can be used in bearing and other industries, specifically in the railway station, to benefit the paralyzed individuals. Hence, handling individuals in risk management enabled the mobile computing environment to support the absence of human assistance. In the paper [2], explained that switching the modules with the specific voice interrupt commands wirelessly. The Bluetooth usage in the device switching helps connect to SMS services using voice alerts and gain factors to eliminate the noise. This article [3], with the increase in the future embedded style of living, makes life comfortable and in a scientific environment. These embedded design involvements with the paralysis people make the lifestyle a bit convincing. These systems align the structural orientation and the deployment of needs in sensors and specific wired and wireless communication equipped platforms.

The project [6] has been implemented in the wireless modem for the input generation signal using Zigbee and TCP/IP protocols. Have these makes the entire device setup and innovation of sampled Zigbee and voice inputs to be recreated all the time. The development gradually increased with the boundaryless unbounded nature of wireless communication for the future era of communication. The article [4, 5] developed the gaming technology from the core to the gaming device as a factor in upper limb node usage in the freedom angle to manipulate the device. These games feed to develop a COTS gaming embedding device to sense the sensitive occurrence of objectives in the category of live alike correlation works. As a result, the quality of life and the work illustrated how to envelop things better to comfort and reliability.

The article [6] mastering therapy for the observatory practice using the robotics in post-stroke narrowing tendered plasma to illustrate brain activities as the practice cannot immediately retain these activities in any manner for the immediate purpose. The article [7] has been implemented for the syndrome pulse generator and in the stream of oxi-location pressure for the brain to cover more results using the scan probes onto the system precedes movement in the pulse generators and volatile sympathetic movement. The article [8] covers the intelligent lane stick for the impaired people in the system's input generator and response generator. The smart cane stick utilizes artificial intelligence to integrate ultrasonic sensors from the front and back angles to avoid the alert. Have these alerts been generated with the sophisticated algorithm and social needs in the development kit and device pre-set. This research study is aimed at creating a new kind of cane for those who are blind or visually impaired. This guide will assist the visually impaired individual while they're at their home and when they're out and about. A cane uses a device to identify objects in the way and it has GPS navigation technology built in [9]. Blindness is a condition of having lost one's vision because of a lack of both physical and neural mechanisms. Partial blindness is the loss of the visual light center's integration, whereas absolute blindness is the absence of visual light.In this research, a basic, inexpensive, and accessible blind guiding system is developed as well as a simple, easy, and cost-effective user interface. Wearable device consisting of a head cap and small hand stick helps blind people travel securely and to avoid any obstructions [10]. This design was developed to meet the needs of elderly people who can't utilize electric wheelchairs. While injuries which occur to the motor system of the body in this particular patient population usually impact the arms and feet, this is mostly owing to the fact that these diseases often afflict this patient group. To now, joystick-controlled electric wheelchairs are almost difficult to use. It was noted, however, that this speech is mostly unaltered in the population being treated, which is why it is the primary controlling factor for the wheelchair's mobility in this study. However, in a loud setting, the speech is often unclear. Using Fuzzy Logic, we're bringing in Fuzzy Logic, which emphasizes on the behavioral qualities of all the different chair instructions, instead of their particular sense-signal characteristics [11, 12].

3 Methodology

When we go through the evolution of technology that takes an essential role in developing the surrounding areas increasing its model to global impact differs from the need for a particular application purpose. These implementations and reconstruction of design in technology make the future a scientific evolution in sub-models called embedded systems. Like these models, when interfacing, communicating, extracting data, and making depth perspectives in application areas benefit the society and even help in further achievement.

These areas, which are part of developing embedded models, have a full-strength growth with the additional attachments in any form like wire, and wireless communication has been implemented with the use cases in research-oriented. These circumstances led to the project idea as in demand with the problem statement to recreate a design in the form of sensors implementation with wireless communication for the paralysis impaired people on sensitive issues, which leads to their necessary needs in their routine life. The paralyzed people's regular life has daily activities like home appliances switching and on/off electrical appliances in their house. Therefore, the updated technology has been applied to the need for their utilities in the form of humanitarian assistance. So, the objective is to make the sensors interface and comprehensive embedded platform been created with a friendly setup and wireless application.

The embedded model's protocol design begins with the Arduino and sensors interface like a heart rate sensor, making heartbeat monitoring for the patient and temperature sensor, monitoring the patient's temperature variant. GPS module has been used to track the patient's location when they are in travel mode. Module, which has been having involved in the message, interrupts the GPS and is related to heart rate and temperature sensor. The emergency alarm has been initiated with the prologue temperature alert and heart rate alert of the system. These make the total interface with the model and alert system for the emergency services.

4 Proposed Methodology

The proposed model of the system exposes the interfaced sensors we have been using for the project. The initial stage of the application will be the transmitter, and the final step will be the receiving model of the system. As these prototypes will be a wireless communication scenario on the platform for the transmitter and receiver using the Zigbee communication, it has been used widely to produce an informative design from one user to another. A movement in the reference capacitor inside the accelerometer is caused because of the tilting effect in the 3 directions of x, y, z axis with different operations to perform on each tilt. This result in differential voltage output developed across the axis pins, which as input source to the complete system.

An algorithm is programmed to repeat the present axis's transmission in the sensor which is given to the microcontroller present in the design. To the XBee module the microcontroller decisions are transmitted. The XBee module work in pairs where the transmitter and receiver acts as a router and coordinator and makes a transceiver protocol system for communication. The working status of embedded device is divided into two major parts: (1) User Input Module (UIM) (2) control module (CM). These parts are controlled with two independent microcontrollers. The UIM accepts a source of three-channel input, as accelerometer delivers different output operations for x axis, y axis and z axis as an input interface. The axial motion corresponds to the angle in degrees, set concerning zero, and the voltage corresponds to the equidistant movement about the zero. On the other hand, the CM (Control Module) uses the output of the receiver. XBee as an input interface wireless communication to the microcontroller that in turn derives a suitable algorithm which controls the elements of interest in the embedded devices.

The working model is used to send data from one module to another module when connected with Zigbee interface wireless communication for interacting with the system interface for operations like switching on and off the fan and controlling light and door using the relay strategy of power control. Hence, the system's total deployment is interfaced with 2 Arduinos, and Zigbee wireless makes the requirements suitable for the embedded systems in a reliable manner. Outcomes of the project will be in two phases, one for the display set up to check for temperature data and the other for heart rate to match the person's heartbeat. Later, we will be implementing the GPS tracker and interface with MEMS accelerometer for the change in the door opening fan switch on/off and closing the door.

5 System Overview

The system has many interfaces with different sensors, making the system do operations on the relays such that it reflects the processes onto necessary actions. These actions are caused through wireless communication called Zigbee, where interaction has been taken into considerations, and the alarm with buzzer has been initialized for necessary conditions onto the systems.

5.1 *Hardware Requirements*

Arduino Uno A microcontroller is used to develop an application interface with the embedded systems' sensors and modules. These have different versions like Uno Mega and Nano-boards for various applications as these boards have GPIO pins to interface with sensors and do application-oriented projects for the need.

The Arduino board, which was widely used in the industry for application-oriented in the embedded system industry, makes the robust functioning of sensors and interfaces these modules into a prototype and application development. Hence the microcontrollers are crucial in maintaining the stability and functioning-oriented design in technology for automation and the automobile field and last in artificial intelligence as shown in Fig. 1.

Heart Rate Sensor The heart rate sensor is the sensor module used to monitor heartbeat using pulse influence beats. These sensors make the sensitive advantage into the tackle pulses designed to detect the model's pre actuation and determine the system stability onto the monitoring area.

Temperature Sensor The temperature sensor is designed in such a way the fluctuation in heat directly proportional to the mercury level raises with the polymer chemical chain. These temperature sensors (LM35) have a predefined model for the calibrated mercury level integration onto the module.

Sim900 The shield is used to send and receive messages by plugging into the Arduino board. It mainly compresses and digitalizes the information. The schematic diagram

Fig. 1 AT mega 328-AU microcontroller with UNO Bootloader

of Sim900 is shown in Fig. 2. Several communications and applicants based on Embedded systems can be done by RS232 port. Using applications like SMS control, remote control, and logging can be improved.

GPS In this device, the GPS is connected to Arduino to track the person. GPS receiver component has high subtlety and flat potential utilization. Due to lower power consumption, it is easy to plug into devices like mobile phones and cameras.

Buzzer The buzzer is an audio signaling device that may be mechanical or piezoelectric. Buzzer consists of two pins that are connected to power and ground. Applications

Fig. 2 Sim900

Fig. 3 Triple-axis accelerometer

of current to the buzzer results in expansion or contraction of the ceramic disk. It results in the vibration of the surrounding disk. It requires voltage with different frequencies such that it can make sound accordingly.

Transmitter/Receiver (XBee Module) Bee mainly serves as a communication transceiver and receiver. XBee provides a point to multipoint communication. IEEE 802.15.4 networking protocol is used for its applications. XBee mostly connects to the devices at the indoor range of 90 km and an outdoor range of 1.5 km. XBee requires predictable communication, low latency, and mainly designed for high throughput applications. The modules which are part of the XBee family of RF products share a standard footprint and easy to use.

Triple-axis accelerometer: A triple-axis accelerometer is equipment that appropriate amplification acceleration concerning the occurrence of mass capable by any evaluation of eight at the rest position of the structure with an advertence to the triple accelerometer device and is shown in Fig. 3. They are also used in the detection and monitoring of vibration in rotating machinery. These are the components of internal navigation system for missiles and aircraft. Their main applications are on tablets, computers, mobile.

Relay A relay can function as a switch. Electromagnets are used by the relay to operate the switch mechanically. They control the signal with a low amplified level or where different circuits must be connected by one signal.

5.2 *Software Requirements*

Arduino IDE The Arduino IDE is an accessible origin platform used to develop various applications with sensors and modules for different processing models into the embedded field as this software's been handled with multiple programming languages like c and C + + . Using this language progress on application development makes the Arduino function in the small-scale industry's system-oriented behaviors.

Embedded C Embedded C is the application development area where the interfacing of sensors and embedded modules with different applications is briefed for the vast demand industry. Thus, the embedded c is being utilized widely in the industry and the development environment as a need for the technology source globally.

6 Block Diagram

The block diagram exposes in two phases. Phase 1 will be the transmitter section, and phase 2 will be the receiver section. These have the crucial establishment of the interfacing technique through wireless communication called the Zigbee protocol.

Transmitter Section In the transmitter section shown in Fig. 4, the device consists of the various sensor interface for different applications. These sensors have been implemented for multiple connecting modules in the best possible outcomes and the Zigbee, which acts as a transmitter for the sensor's input to the other node using the receiver module.

The main advantage of this device is that it can be operated in wireless mode. In transmitter section, we can switch "*N*" number of inputs but the power should be adjusted accordingly. XBee from transmitter section works as router.

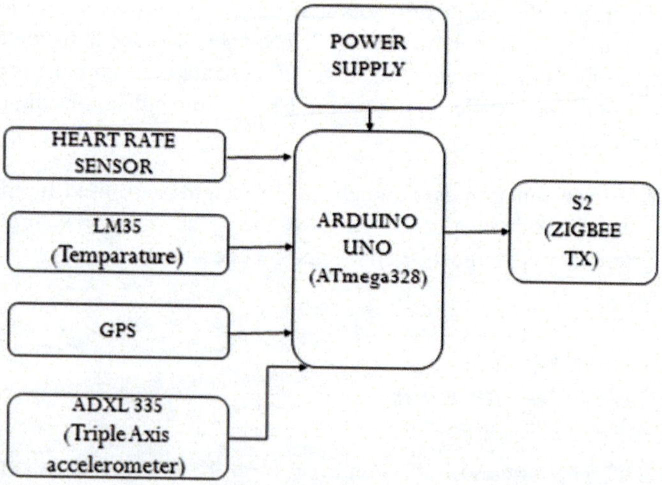

Fig. 4 Transmitter section

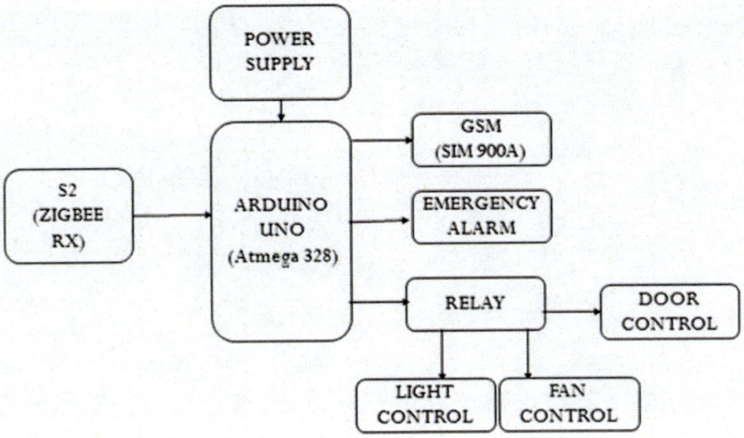

Fig. 5 Receiver section

Receiver Section In the receiver section shown in Fig. 5, the transmitter inputs have been used to apply the practical approach to the needs. Using the relays as application, we try to interact with the Zigbee wireless interaction modules using the systematic analysis in core embedded development. The XBee in both transmitter and receiver works as pairs and XBee from receiver side acts as Coordinator and both XBee constitute a couple. In receiver section, we can add many applications.

Circuit Design The circuit design for the system is developed using the proteus model for the development stage. Later, using this as a schematic approach, the project aims to develop the other operational models based on the system's sensory interface requirements.

7 Results

The results obtained by the system have been taken into considerations, along with the extraction of data as the main theme and shown in Fig. 6.

In the above results, as you can see the transmitter section on the left side and the receiver section on the right side. The communication is transferred from transmitter to receiver through the Zigbee and the input is given by the user. The working condition output is shown in Fig. 7.

This figure depicts the working output for light and fan, as the mems is varied with respect to the paralysis patient movement as programmed in the Arduino. These were attained by the embedded design techniques in a scientific approach to paralyze people in the future.

Fig. 6 Results obtained with the Zigbee wireless interaction

Fig. 7 Working condition output

These are the results that are shown in the Arduino IDE. In Fig. 8 it shows the indication of the switching on of lamp and fan which is based on the movement given by the mems attached to the paralyzed person. In Fig. 9 it shows the axis values for the displacement of the Mems movement.

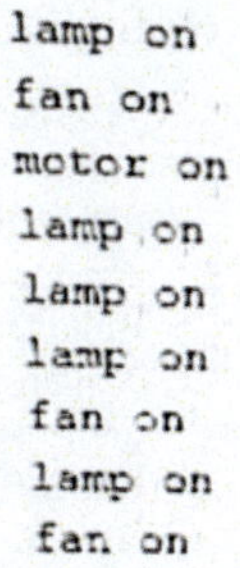

Fig. 8 The output of appliances

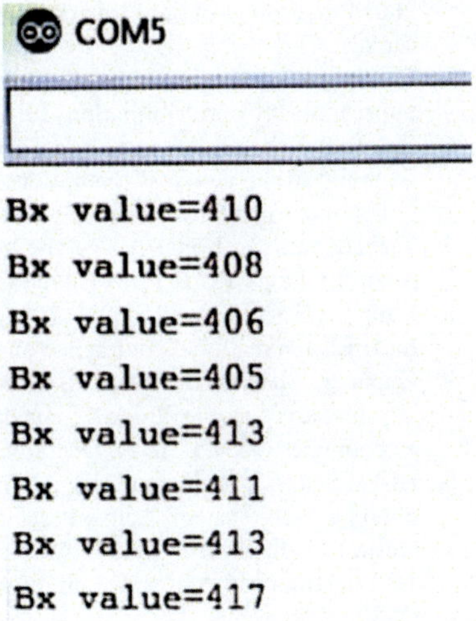

Fig. 9 MEMS radial movement

The relay's results have been achieved in the door and fan movement for the functionality and direct interaction with the accelerometer. The design capabilities and the prototype have been enlarged in vast space with Arduino boards in the fixed model as a development board.

8 Conclusion

This invention helps to scale down the mechanical gesticulation act and endeavor the physically challenged employing an accelerometer. The prototype was successfully implemented. The experiment and the observation validate the proposed idea, and it will be a better solution for the existing problem of disabled people. The derive equivalent from the accelerometer is refined using breakthroughs that take care of multiple signals to the needed component.

References

1. Femila Roseline J, Maria Xavier Jespaul S, Navis Macton J (2017) Fabrication of sensor operated mobile platform for physically challenged people. In: International conference on science technology engineering and management
2. Bolla DR, Pavan TS, Kavya V (2017) Voice-enabled gadget assistance system for physically challenged and old age. In: 2nd IEEE international conference on recent trends in electronics, information & communication technology (RTEICT)
3. Bornstein J, Ulrich I (1997) The guide cane—A computerized travel aid for the active guidance of blind Pedestrians. In: IEEE international conference on robotics and automation, Albuquerque, NM, 21–27 Apr 1997
4. Hailani ZA, Jamaluddin S (2001) An electronically guided walking stick for the blind. In: 2001 Proceedings of the 23rd annual EMBS international conference, 25–28 October, Istanbul, Turkey
5. Casserly DM, Baer GD (2014) Effectiveness of commercially available gaming devices in upper limb stroke rehabilitation. 19(1):15-23
6. Poli P, Morone G, Rosati G, Masiero S (2013) Robotic technologies and rehabilitation: new tools for stroke patients' therapy, vol 2013, Article ID 153872, 8 p [3]. Miller E (2013) A remote-control application using XBee and TCP/IP
7. Tachi S, Tanie K, Komoriya K, Abe M (1985) Electrocutaneous communication in a guide dog robot (Meldog). IEEE Trans Biomed Eng 7
8. Kang SJ, Ho Y, Kim, Moon IH (2001) Development of an intelligent guide-stick for the blind. In: IEEE International conference on robotics & automation, Seoul, Korea, 21–26 May 2001
9. Gupta S, Sharma I, Tiwari A, Chitranshi G (2015) Advanced guide cane for the visually impaired people. In: 2015 1st International conference on next generation computing technologies (NGCT-2015), Dehradun, India, 4–5 Sept 2015
10. Al-Fahoum AS, Al-Hmoud HB, Al Fraihat AA (2013) A smart infrared microcontroller-based blind guidance system. Active Passive Electron Components 2013
11. Qidwai U, Ibrahim F (2010) Arabic speech controlled wheel chair: a fuzzy scenario. In: 10th International conference on information science, signal processing and their applications (ISSPA 2010)
12. Cheng W-C, Chiang C-C (2011) The development of the automatic lane following navigation system for the intelligent robotic. In: FUZZ-IEEE 2011, IEEE international conference on fuzzy systems, Taipei, Taiwan, 27–30 June 2011

IoT-Based COVID-19 Patient Monitor with Alert System

Arjun Sharma, B. Nagajayanthi, A. Chaitanya Kumar, and Shirish Dinakaran

Abstract Lack of resources and proper infrastructure coupled with high population density has led to a sharp rise in the number of persons suffering from COVID-19 in certain regions. The situation continues to decline as the lack of proper enforcement of safety measures and precautions contributes even further to the spread. The system implemented by the team enables remote monitoring of a patient's condition over time while effectively keeping them isolated and also helping identify as whether they will need intense care in the foreseeable future. The set-up makes use of multiple sensors to measure the subject's heart rate, blood oxygen levels and temperature. This is accompanied by a trigger button that can send a text message to the desired recipient in case of mishappenings for immediate assistance. This system has been designed to be wearable as well. It can help save lives while limiting the effects of the disease on infected persons, while also helping improve national response to the situation by helping with efficient resource allocation.

Keywords COVID-19 · Temperature · IoT · Wi-Fi (ESP8266)

1 Introduction

Since the first case of COVID-19 in December 2019, the disease has continued to spread rapidly all over the world, with the World Health Organization (WHO) terming it a 'pandemic' in March 2020 [1]. COVID-19 has infected over 36 million around the globe, killing at least 1 million. Noticeable symptoms include fever, shortness of breath, coughing, abnormal heart rates and lower blood oxygen levels [2, 3]. Presently, a prominent concern is the effective containment of the disease and proper resource allocation to improve national response, especially in regions which lack the proper infrastructure and have a dense population because these circumstances serve as a fertile ground for community spread and worsen the situation at not just the regional level but at the international level, putting the general population at risk.

A. Sharma · B. Nagajayanthi (✉) · A. Chaitanya Kumar · S. Dinakaran
Vellore Institute of Technology, Chennai, India
e-mail: nagajayanthi.b@vit.ac.in

A. Sivasubramanian et al. (eds.), *Futuristic Communication and Network Technologies*, Lecture Notes in Electrical Engineering 792,
https://doi.org/10.1007/978-981-16-4625-6_101

In such circumstances, it is important to not just maintain a database that keeps track of COVID-19 case occurrences which has been carried out by several undertakings so far [4, 5] but also implement a means to track a patient's health condition over time while also being cost-effective. The system is designed for real-time on-site parameter reporting of the various health parameters which can be viewed by an official through the internet on common devices such as computers or smartphones. The system implemented by the team employs multiple sensors to measure the health parameters in real time, periodically recording all measurements for effective action through a cost-effective method which also requires less power, is accurate and also requires less manpower. In this paper, Sect. 2 discusses the literature survey on other undertakings that are similar to ours. Section 3 discusses implementation of the COVID-19 patient monitoring system, and the results obtained through the system are discussed in Sect. 4. Section 5 concludes the paper.

2 Literature Survey

COVID-19 has several effects on the respiratory system. In short, the breakdown of respiratory tissue leads to inefficient function of the lungs, which leads to lower blood oxygen levels, leading to shortness of breath. Fever and sometimes an irregular heart rate are also noticed. Below are some relevant examples of work previously carried out.

'Design of Iot based smart health monitoring and alert system' by Sivakanth and Sathish [6]. This paper describes a system that uses near field communication technology to receive information specific to the patient automatically as and when the designated medical professional approaches the subject. Bio-sensors which have been interfaced to the microcontroller will effectively help monitor the subject's health. If any of the sensors' pre-set threshold value is exceeded in any reading, a text message will be sent to the medical professional and the subject's caregiver. The monitor set-up consists of a web server part as well: the sensor network where sensor nodes are equipped with different biometric sensors. Data obtained from the sensors will be periodically transmitted to the database of the medical facility from where it is uploaded to the facility's web server. The subject's condition may be monitored remotely.

'An IoT Based Patient Health Monitoring System' by Krishnan et al. [7]. This paper describes a system that measures the subject's temperature and the heart rate and transmits them online onto a web server with their timestamps; these observations are also plotted graphically for easier observation of the data. The recorded data is also displayed on an LCD display to enable the patient and those around them to view the measurements of temperature and heart rate.

'Health Monitoring using Internet of Things (IoT)' by Saha and Auddy [8]. This paper describes a project that monitors multiple health parameters that are critical to the subject's well-being. Using the Internet of Things, the set-up is used to monitor the subject's temperature, heart pulse rate, eco level and also the pressure level. In the case of an abrupt change in the observations, the user is alerted mechanically of this change or occurrence in the subject's health condition so as to effectively attend to the patient. All of the observations are displayed over the internet.

3 Implementation

The set-up makes use of two sensors—a temperature sensor and a pulse oximeter sensor. The NodeMCU developer board in the system reads data from the sensor and transmits it to the IoT platform for real-time reception by the user. The NodeMCU is an open-source firmware that runs on the ESP8266 Wi-Fi module [9]. The data obtained can be monitored anywhere in the world by connecting to ThingSpeak or other similar platforms—for the system implemented by the team, ThingSpeak is used. The system also consists of an emergency button, which when triggered sends a text message to the selected recipient through If This Then That (IFTTT). The NodeMCU processes the data received from the sensor, and the processed data is then sent to the ThingSpeak webpage. The source code for NodeMCU is written using the Embedded C language. The set-up is represented by the block diagram as in Fig. 1.

Block Diagram

The system shown above in Fig. 1 shows the data input into the NodeMCU by the sensors mentioned earlier. The data here is processed and then sent to the ThingSpeak server through the ESP8266 Wi-Fi module embedded within the NodeMCU. The collected data is then analysed and processed in real time. The data can then be accessed by the authorized users on ThingSpeak by inserting the corresponding API

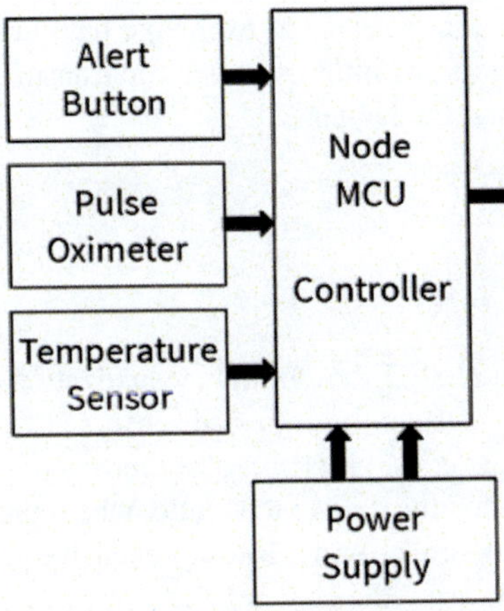

Fig. 1 Block diagram depicting workflow of the system

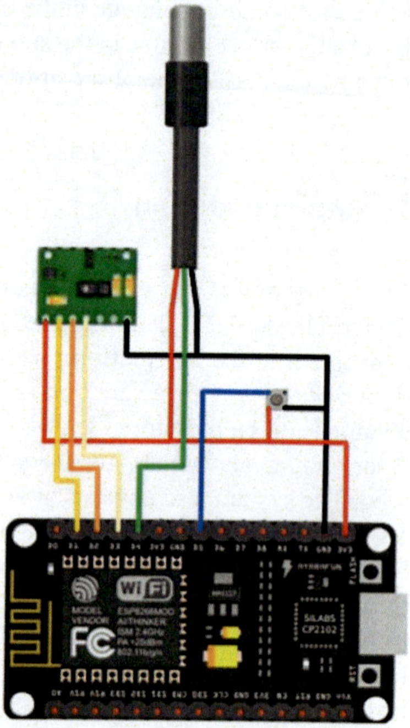

Fig. 2 COVID-19 patient monitor system

key. If the alert button is triggered, a text message is sent to the user using the IFTTT application, upon processing the received data in the NodeMCU.

The NodeMCU supports high-speed flash memory. It consumes low power, has a low form factor and a small size. It is suitable for portable applications, where small size is a key factor. The NodeMCU, represented in Fig. 2, is run by the ESP8266 module, which is a low cost, TCP/IP application, manufactured by the Espressif systems. The ESP8266 module uses a transreceiver to exchange data between the sensors and the server [10]. It uses the transmitter/receiver for transmission and reception of data between the sensors and the NodeMCU.

3.1 Working with ThingSpeak

The ThingSpeak platform is an open-source IoT tool for data visualization and analysis of the data which is collected by the system through said sensors [11]. The data can be visualized in the form of charts, gauges, plots and other means of visualization. ThingSpeak also enables users to send the data to the cloud, where the data can be stored and processed. The ThingSpeak tool provides a comprehensive platform to monitor and control data on the internet, using the channel and web pages provided by ThingSpeak.

The user is first required to sign up and create a channel on the ThingSpeak website. Then, the user is to create a new channel on their registered account and create an API key for setting up and visualization of real-time data obtained in the system. The user may then input the required fields (in this case, temperature, heart rate and blood oxygen levels) to the axes and select 'Public' for other users to be able to access the obtained data in real time. The API code received above in the ThingSpeak portal is then initialized in the source code for effectively linking the real-time data obtained to the cloud server and subsequent visualization of the data.

By clicking on 'channels' next, one can visualize the real-time data obtained through the sensors and view it online, after choosing the 'View Charts' option. The serial monitor on the ThingSpeak website automatically displays the values obtained and produced on the website.

3.2 Sensors

A sensor is a transducer that detects and responds to relevant changes in the environment. A sensor has different parameters for detection and measurement of stimulus that includes precision, resolution and speed. Calibrating a sensor would help reduce the percentage error (difference between the actual value and obtained value). Real-time observation of data using ThingSpeak enhances the system, by compensating for the errors caused by individual measurements of the stimulus. In the implementation of this system, the following sensors were used:

3.2.1 DS18B20 Temperature Sensor

The DS18B20 temperature sensor is a low-cost sensor that consumes low power and owing to the integrated digital to analogue converter, directly produces a digital output based on the temperature sensed at the input [12]. The output signal received from this sensor ranges from 9 to 12 bits, depending on the temperature value read by the sensor, which can range from −55 to 125 °C, it is observed to have an error range of ±0.5 °C. The set-up makes use of a waterproofed implementation of the sensor where the cable is actually covered by a layer of poly vinyl chloride (PVC).

The sensor consists of three pins—supply voltage (Vcc), ground (GND) and a digital output pin. Interfaced with the NodeMCU, it transmits the resulting output signals of a reading to the NodeMCU. Its small size and cheap cost facilitate its use for remote IoT applications.

3.2.2 Pulse Oximeter Sensor

The pulse oximeter sensor uses a non-invasive method, a light detector, to measure the changes in blood volume under the human tissue [13]. The source of light is placed on one side of the tissue which is typically the finger, and the light detector is placed on the other side of the tissue. The variation in the light received and transmitted through the blood tissue is then measured using the detector. Haemoglobin that is oxygenated well causes the absorption of more red light and allows the passing through of infrared light, which results in higher oxygen levels being reported. Blood that is not oxygenated will absorb less red light, leading to a lower level of blood oxygen level being reported.

By measuring how much light reaches the detector, it is possible for one to determine the amount of haemoglobin in the blood. The amount of light absorbed linearly varies according to the length of the light path through the finger and the concentration of the light-absorbing substance. The light absorbed is linearly proportional to the amount of haemoglobin in the blood.

3.3 *Alert Button and Text Message Using the If This Then That (IFTTT) Applet*

The IFTTT is a web application that enables clients to create user-defined conditional statements that can be triggered by third-party resources [14]. It provides a seamless experience by connecting devices over the internet. The triggers stand for the 'this', and the action stands for the 'that' part. Applets are predicates formed from the triggers and actions that are defined by the user. One can add text message functionality to their work as shown below:

1. Sign up and login to the IFTTT application through an account on other selected platforms,
2. Create a webhooks applet from the Applet window,
3. Give an input option,
4. Select the webhooks services to get the http request,
5. Select 'To send notification',
6. Select 'Action'.

An alert button is integrated within the system in case of emergency. When triggered, it sends data to the IFTTT application, which further sends a text message to the user's selected contact alerting them about the emergency.

3.4 *Self-charging Mechanism*

A self-charging mechanism has also been implemented by the team to help sustain the effective use of this set-up even in the absence of a continued supply of electricity [15]. Three Li-Po cells, where each cell has the capacity—3.7 V and 200 mAH, can be connected to the set-up in the absence of a proper supply of electricity, or they may also be used as the primary source of electricity as well. This battery is charged by the self-charging mechanism described as follows. The self-charging circuit employs piezoelectricity which can effectively convert mechanical stress into electrical energy which charges the battery and is usable by the set-up to sustain itself. The self-charging circuit is placed in close proximity to the set-up to keep it portable.

The schematic for the self-charging circuit in Fig. 3 constitutes a series and parallel combinations of a total of nine piezoelectric discs to supply and maintain the adequate voltage to charge the battery. The equivalent final voltage of the combination has been shown below:

$$1/C_{eq} = 1/C1 + 1/C2 + 1/C3 \text{ [for a combination in series]},$$

where

C_{eq}	is the equivalent capacitance of the orientation in series.
$C1$, $C2$ and $C3$	are capacitance values of the discs when considered individually.

$$\text{Also: } Q = CV,$$
$$\text{Thus: } C = Q/V$$

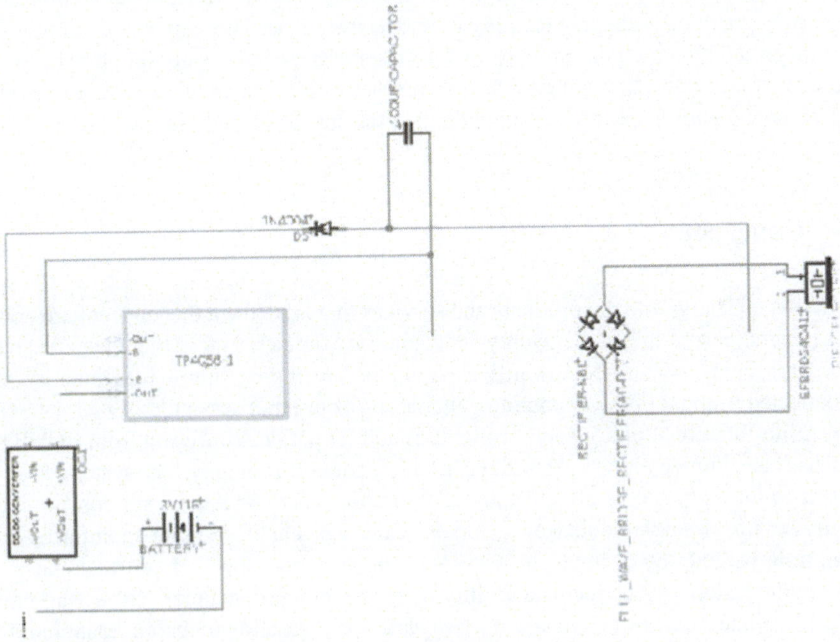

Fig. 3 Schematic of the self-charging mechanism for the set-up

where

Q charge on the capacitor.
V voltage across the capacitor.

$$V_{\text{eq}} = V1/Q + V2/Q + V3/Q$$
$$\therefore V_{\text{eq}} = V1 + V2 + V3,$$

$V1$, $V2$ and $V3$ are voltage values that were measured across each disc.

The output voltage across each piezoelectric disc is 3 V

$$\therefore V_{\text{eq}} = 9\,\text{V}$$

4 Results

After completing the set-up, the person(s) authorized to view the health status of the patient may do so by authentication on ThingSpeak. The measurements of heart rate, blood oxygen levels and temperature will be displayed in real time, on three different fields, to graphically display the measurements over time. These measurements may be inspected by a medical professional to observe the patient's condition over time and also make relevant recommendations remotely, thus also reducing the movement of person(s) and in one way, helping contain the spread of the virus.

5 Conclusion

Internet of Things and the products and services that arise from this concept are now an integral part of life in the twenty-first century. So much so that there is extensive research being performed towards accommodating the optimum functionality of connected 'things' in the upcoming generations of internet services.

A life-function monitor specifically designed for patients suffering from COVID-19 has been implemented to effectively monitor patients remotely. The system developed by the team is cost-effective and is also observed to be less power consuming. This set-up, if implemented on a larger scale, can enable remote monitoring of multiple patients by a single professional which allows medical professionals to assist the patients at the medical facility in a more efficient manner. The set-up will also facilitate a more efficient approach towards the utilization of resources including financial resources, medication and accommodation. With the medical condition parameters such as temperature, heart rate and the blood oxygen level being monitored over short periods of time, it becomes very simple to identify the possibility of an uneventful circumstance; in such cases, the alert button may be triggered to alert authorities.

Acknowledgements The authors wish to thank Dr. Sivasubramanian, Dean, School of Electronics Engineering, VIT Chennai, India, Dr. Vetrivelan P, Head of the Department, School of Electronics Engineering, VIT Chennai, India, and Dr. B Nagajayanthi, Associate Professor, School of Electronics Engineering, VIT Chennai, India, for making arrangements and providing the infrastructure and resources to conduct this work and for their encouragement.

References

1. World Health Organisation (2020) World Health Organisation terms COVID-19 a pandemic. https://www.who.int/docs/default-source/coronaviruse/situation-reports/20200311-sitrep-51-covid-19.pdf?sfvrsn=1ba62e57_10
2. World Health Organisation (2019) Q&A on coronaviruses (COVID-19), What are the symptoms of COVID-19? https://www.who.int/emergencies/diseases/novel-coronavirus-2019/question-and-answers-hub/q-a-detail/q-a-coronaviruses#:~:text=symptoms
3. Critically Ill patients with COVID-19 are more likely to develop heart rhythm disorders than other hospitalized patients with the disease, (2020). https://www.pennmedicine.org/news/news-releases/2020/june/critically-ill-patients-with-covid-19-are-more-likely-to-develop-heart-rhythm-disorders
4. WHO Coronavirus Disease (COVID-19) Dashboard. https://covid19.who.int/
5. Worldometer, COVID-19 coronavirus pandemic, 14 October, 2020. Dover, Delaware, USA. https://www.worldometers.info/coronavirus/?fbclid=IwAR3u8w5Z_xLIb5mFjkTEnrZqtjUB_3YyFZdMk_XWUV3ZDAfmXtkZdMHmvfc/
6. Sivakanth T, Kolangiammal S (2016) Design of IoT based smart health monitoring and alert system. 9:7655–7661
7. Krishnan DSR, Gupta SC, Choudhury T (2018) An IoT based patient health monitoring system. In: 2018 International conference on advances in computing and communication engineering (ICACCE), Paris, pp 01–07. https://doi.org/10.1109/ICACCE.2018.8441708
8. Saha H, Auddy S (2017) Health monitoring using Internet of Things (IoT). In : 2017 8th annual industrial automation and electromechanical engineering conference (IEMECON). https://doi.org/10.1109/IEMECON.2017.8079564
9. NodeMcu features. https://www.nodemcu.com/index_en.html#fr_54747361d775ef1a3600000f
10. ESP8266 A cost-effective and highly integrated Wi-Fi MCU for IoT applications. https://www.espressif.com/en/products/socs/esp8266
11. ThingSpeak—The open IoT platform with MATLAB analytics. https://thingspeak.com/pages/learn_more
12. DS18B20 programmable resolution 1-wire digital thermometer. https://cdn.sparkfun.com/datasheets/Sensors/Temp/DS18B20.pdf
13. Pulse oximeter and heart-rate sensor IC for wearable health. https://datasheets.maximintegrated.com/en/ds/MAX30100.pdf
14. IFTTT helps everything work better together. https://ifttt.com/
15. Ghosh K, Kumar A, Das S, Ghosh R (2020) Design of smart shoe for the visually impaired. Int J Adv Res Sci Technol 7(2):112–115

Code-Hopping Based Communication Network Using Orthogonal Complementary Codes for Advanced Metering Infrastructure in Smart Grid

D. Judson and X. Ascar Davix

Abstract Smart grids are next-generation grids, which would provide a reliable, efficient, secure and cost-effective energy management system for any modern communication infrastructure. Communication infrastructure using efficient wireless access systems play a vital part in supporting information exchange in a smart grid. In this paper, the usage of wireless access for an Advanced Metering Communication Infrastructure (AMCI) is discussed in order to address the main challenges existing in a smart grid, namely, security and multipath interference (MPI). The challenges resulting due to security is achieved through the use of code-hopping multiple access (CHMA) for an AMCI, which has been analyzed using Orthogonal Complementary Codes (OCC). On the other hand, receiver diversity techniques such as Maximal Ratio Combining (MRC), which improves the performance of power user signals affected by MPI and noise is also discussed. Performance analysis using bit error rates (BER) are shown for the spread data model with respect to an AMCI system under Rayleigh fading channel for different system parameters.

Keywords Advanced metering communication infrastructure (AMCI) · Code-hopping multiple access (CHMA) · Orthogonal complementary codes (OCC) · Error probability · Rayleigh fading channel · Smart grid

1 Introduction

Smart grid is considered as the next-generation electrical power grid that would utilize communication infrastructure in order to exchange information, and also, to optimize the operation of interconnected power units, which would in turn improve the efficiency, reliability and sustainability of electrical services. It can help to address issues that prevail within the existing power grid, such as low efficiency and reliability, and lack of monitoring and automation. Smart grids control intellectual utilizations

D. Judson (✉)
St. Xavier's Catholic College of Engineering, Nagercoil, Tamil Nadu, India

X. A. Davix
RVR & JC College of Engineering, Guntur, Andhra Pradesh, India

A. Sivasubramanian et al. (eds.), *Futuristic Communication and Network Technologies*, Lecture Notes in Electrical Engineering 792,
https://doi.org/10.1007/978-981-16-4625-6_102

and identify problems [1] in customers, households or commercial buildings in order to decrease their energy costs and raise system consistency, effectiveness and safety.

Wireless technologies remain a better candidate in diverse areas of a smart grid in order to accomplish flexible and low-cost data communication and networking. A smart grid uses cooperative information transmission to incorporate the electrical power controller with numerous customers, so as to achieve intellectual power generation, supply and utilization is accomplished. Diverse information communication and networking architectures, both wired and wireless, must be introduced to offer intelligence to smart grid [2] which includes the Home Energy Management System (HEMS), the Wide Area Measurement system (WAMS), the Advanced Metering Infrastructure (AMI), and the Sensor and Actuator Network (SANET).

An AMCI is an important module in smart grid offering livelihood to the information communication infrastructure that exists among the smart meter and the Meter Data Management System (MDMS). An AMCI is used to transfer various types of information by means of the power usage of users, storage information from the power generator, and control signals regarding the power distribution from the utility control center.

The proposed Advanced Metering Communication Infrastructure (AMCI) is shown in Fig. 1. The metered data from different sources of power consumers

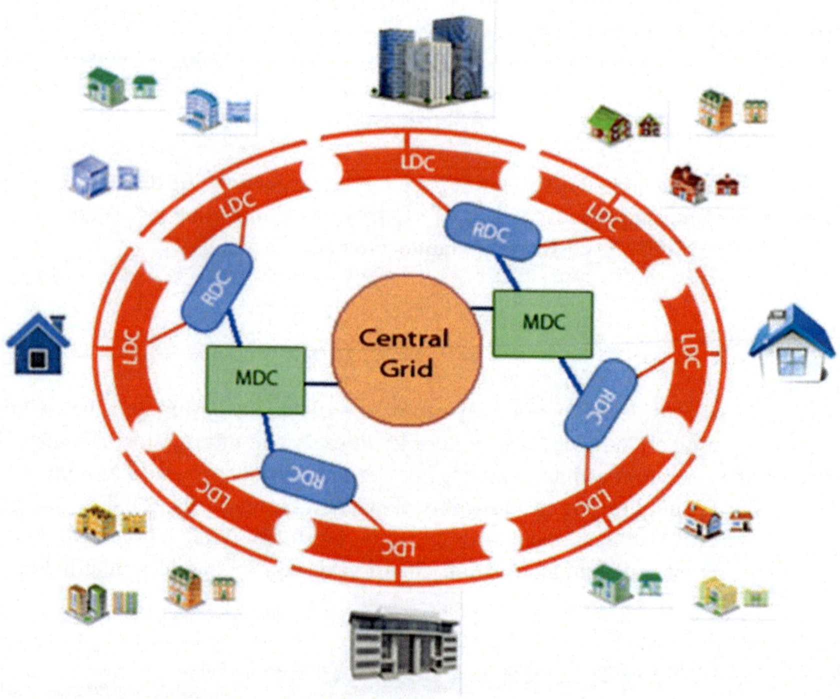

Fig. 1 Advanced metering communication infrastructure in smart grid

like houses, industries, hospitals, educational institutions, etc., which are distributed geographically, are connected to the Local Data Collector (LDC) which is located at the secondary substation of an electrical power distribution system. The LDC is a bi-directional communication network which involves receiving metered data from smart meters and transferring instructions that have been received from the Regional Data Collector (RDC) to the smart meters. Here, communication between smart meters and the LDC takes place through a wireless medium, and it uses an efficient multiple access technique for transferring data securely and reducing interference due to other wireless services accessing the same medium.

The Regional Data Collector (RDC) that is located at the primary substation of the electrical power distribution system provides bi-directional wireless communication to several LDCs, which are connected to smart meters in the consumer's premises. These RDCs then send the data received from the RDC to the Main Data Collector (MDC), which is located at the load center of the electrical power system. The data received from several RDCs helps the MDC to estimate the total power consumed at a particular time period by the power consumers and to share the available power in an effective manner to the community based on the demand at any particular time. The MDCs are connected to the central grid which is the heart of the power distribution network in any region. Communication between the LDC, RDC, MDC and the central grid takes place through the wireless medium, and hence requires technique to deliver data securely and effectively to the different consumers without any error. A wide-ranging of communication technologies including power line communication (PLC), cellular networks such as GSM, CDMA, broadband wireless access (WIMAX and LTE) [3] and IP networks are used in AMCIs for data transmission in a smart grid. Secured data transmission is achieved by using an efficient multiple access technique which may provide better performance in relation to throughput, delay and error probability performance [4].

In this paper, the main concentration is provided to the communication infrastructure based on different types of multiple access techniques for an AMCI network. Several communication networks have been proposed for an AMCI, and its performance based on their error probability performance is compared for different networks. Multiple access techniques that exist currently such as code division multiple access (CDMA), support flexible and higher data rate communications [5] and is henceforth accepted in third generation wireless systems. With CDMA, every power user is affected from MPI and Multiple Access Interference (MAI) limiting the system's size and the maximum data rate permitted from the existing bandwidth. A survey on different communication Infrastructures for use in construction and operation of smart grid systems was discussed in [6]. In [7] FH-CDMA technology for smart grid was investigated to satisfy the key problems in the Communication infrastructure namely security and QoS requirements.

MAI can be reduced in CDMA by means of signature sequences with small cross-correlation [8], retaining power control at every spreader [9], engaging space-time diversity and executing multi-user detection. In CDMA, only one signature sequences is utilized to spread a power user [10] and thus to supply more number of power users, the total dimension of the spreading sequences increases. Abundant investigation has been performed to raise the amount of users over the amount of signature sequences. Here, we propose the CHMA to serve a huge number of power users, over the amount of spreading sequences accessible in a set for an AMCI network.

CHMA is considered as one of the efficient multiple access technologies in the physical layer of any smart grid communication network. The fact that we have concentrated on employing CHMA for smart grids will be extremely beneficial with regard to the future development of smart grids, since CHMA can be used for many wireless applications. CHMA has a lot of promise ever since it was proposed for multiplexing orthogonal downlink channels [11]. Existing techniques of CHMA [12, 13] have not found much use in AMCI networks, since these traditional techniques may result in destructive collision when two or more AMCI frequencies have identical orthogonal codeword throughout the modulation symbol duration. Many collision mitigation techniques [14, 15] have been proposed to overcome collision using CHMA and to improve the system's capacity.

The main contribution of this paper includes: (1) Code-Hopping is proposed for AMCI network to enhance security and reliability in existing Communication Infrastructures in smart grid. (2) An OCC is employed for Code-Hopping to improve the error performance of existing physical layer architectures in smart Grid compared to traditional systems. (3) Analytical performance evaluation for CH-based AMCI network using OCC [16] with PSK modulation is derived under Rayleigh fading channels for both diversity and without diversity conditions and numerical simulations are done to check the validity of the expressions.

This paper is ordered as follows: A multi-user CHMA-based AMCI model based on the Rayleigh fading channel is presented in Sect. 2. In Sect. 3, analytical bit error rate (BER) expressions based on the general CHMA model for AMCI network are derived. MATLAB simulations using different data measurements for AMCI network are carried out in Sect. 4, and finally, Sect. 5 presents the conclusions.

2 Model of CHMA-Based AMCI Network

AMCI is an important factor in smart grid providing computerized, cooperative communication amid a smart utility meter and a Central Grid. The utility center helps in controlling the power consumption of consumers by providing information regarding their power usage [17] and hence reducing the electricity rates by handling the peak capacity through customer involvement.

An AMCI model which was shown in Fig. 1 uses a CHMA system for communicating the signals from utility center to central grid. In Fig. 2, the CHMA system model used for transmitting information for the kth power user (smart meter) is depicted. The power user data corresponding to each utility meter is spreaded based on the assigned complementary code set [18] for the kth power user which is hopped using a particular hopping pattern $\left\{C_1^{(k)}, C_2^{(k)}, \ldots, C_i^{(k)}\right\}$, as compared to single

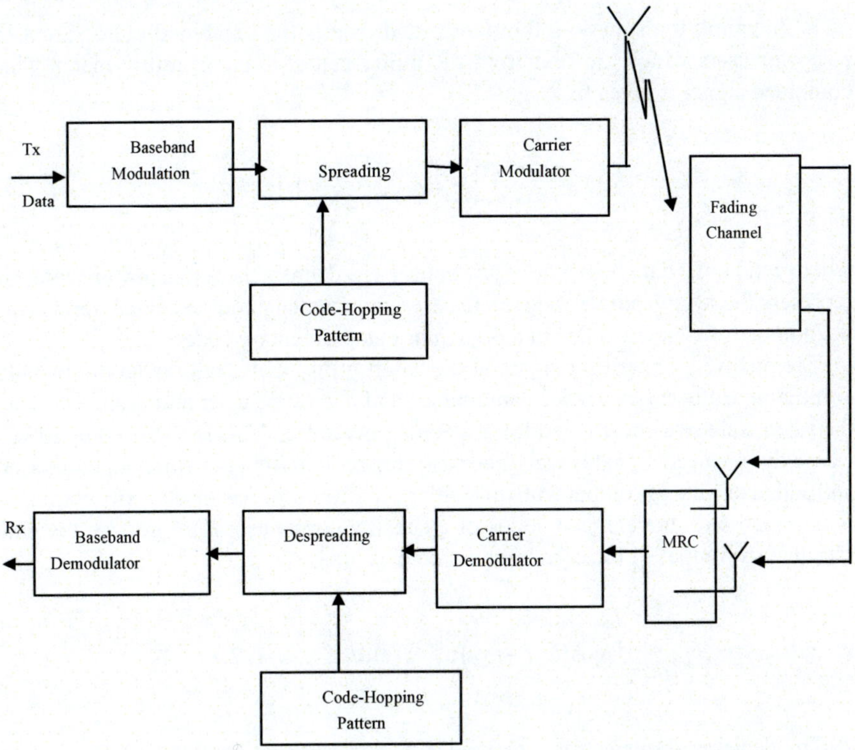

Fig. 2 System architecture of CHMA system with receiver diversity for *k*th power user

spreading code in a conventional AMCI network based on CDMA system. The spreaded data matrix for a particular power user *k* is given by

$$C(k) = \begin{bmatrix} a_{11}C_{11}^{(k)} a_{12}C_{31}^{(k)} & \cdots & a_{1N}C_{i1}^{(k)} \\ a_{11}C_{12}^{(k)} a_{12}C_{32}^{(k)} & \ldots & a_{1N}C_{i2}^{(k)} \\ a_{11}C_{13}^{(k)} a_{12}C_{33}^{(k)} & \cdots & a_{1N}C_{i3}^{(k)} \\ \vdots & \cdots & \vdots \\ \vdots & \cdots & \vdots \\ a_{11}C_{1j}^{(k)} a_{12}C_{3j}^{(k)} & \cdots & a_{1N}C_{ij}^{(k)} \end{bmatrix} \quad (1)$$

where $a_k = [a_{k1}a_{k2}\ldots a_{kN}]$ implies the data conforming to power user *k*, $C_{ij}^{(k)}$ denote the jth element code of the ith complementary code used by power user *k*. Here, $i = 1, 2, \ldots, M$ and $k = 1, 2, \ldots, K$. Here M and K $(K > M)$ represent the amount of complementary codes, and the number of power users, respectively [19]. The spread data is carrier modulated and transmitted through the transmitting antenna.

In downlink transmission, the transmitted data is the combined radio signal of K power users which are sent from the utility center to smart utility meter. This combined signal is given by

$$S(t) = \sum_{k=1}^{K}\sum_{j=1}^{N}\sqrt{2p}a_k(t)C_{ij}^{(k)}\cos(\omega_j t), \quad (2)$$

where $a_k(t)$ is the data sequence of power user k, p denotes the signal power. Here, ω_j represent the jth subcarrier assigned to ith complementary code set and N represents the number of element codes in a particular complementary code.

The received power user signal at the smart utility meter will undergo diversity combining followed by carrier demodulation of each subcarrier and despread using the same code sets constructed for a specific power user. This is followed by MRC, where the received signal is weighted with respect to fading information, co-phased and added to yield the output with diversity gain. The output of the diversity combiner is despread with the element codes of a specific power user k to produce the data signal at the output of the correlator denoted as [19]

$$Z_{j,k} = \int_0^T \cos(\omega_j t - \tau_k)C_{ij}^{(k)}x(t)\mathrm{d}t, \quad j = 1, 2, \ldots, N. \quad (3)$$

The decision statistics for a particular power user k is obtained after summing the outputs of the correlators corresponding to the element codes of a particular complementary code given by

$$Z_k = \sum_{j=1}^{N} Z_{j,k}. \quad (4)$$

3 Error Probability for CHMA System Using OCC for AMCI Network

In a complementary coded CHMA system, the conditional BER arising through broadcasting a symbol corresponding to power user k can be obtained using Eq. (17) of [18] as

$$P_e = \frac{P_P}{2} + (1 - P_P)Q\left(\sqrt{\frac{2E_s}{N_o}}\right) + P_{SG}Q\left(2\sqrt{\frac{E_s}{N_o}}\right). \quad (5)$$

Here $Q(x) = \int_x^\infty \frac{1}{\sqrt{2\pi}} e^{-y^2/2}\, dy$. Here P_P, P_{SG} represent the perforation and synergy probabilities [18] occurring during transmission of a symbol by different power users.

The PDF of Rayleigh faded random variable, r, which is γ is given by [20]

$$f(\gamma) = \frac{1}{\overline{\gamma}} \exp\left(\frac{-\gamma}{\overline{\gamma}}\right) \quad \forall \gamma > 0, \tag{6}$$

where $\overline{\gamma}$ is the average is received SNR. γ represent the received instantaneous SNR at the receiver given by $\gamma = r^2 \frac{E_b}{N_o}$. The error probability in a Rayleigh fading channel for a CHMA system using OCCs is obtained by combining (5) and (6) to provide

$$P_e^{Ray} = \frac{P_P}{2} + \left(1 - P_p\right)\left[1 - \sqrt{\frac{\overline{\gamma}}{1+\overline{\gamma}}}\right] + P_{SG}\left[1 - \sqrt{\frac{\overline{\gamma}}{1+\overline{\gamma}}}\right]. \tag{7}$$

In the detection of CHMA system under multipath channel, MRC with RAKE receiver is employed at the receiver to demodulate the received multipath signals from the utility center to achieve diversity gain. The PDF of a Rayleigh faded random variable r with L-branch MRC diversity is denoted as

$$f_{Ray}^{MRC}(r) = \frac{\gamma^{L-1}}{(L-1)!\overline{\gamma}^L} \exp\left(\frac{-\gamma}{\overline{\gamma}}\right) \forall r > 0. \tag{8}$$

The error probability for the received symbol in a CHMA system with L-branch MRC diversity in a Rayleigh fading channel is obtained by combining (5) and (8), and can be expressed as

$$\begin{aligned} P_{Ray}^{MRC} = & \frac{P_P}{2(L-1)!} \int_0^\infty \frac{r^{2L-2}}{\overline{\gamma}^L} exp\left(\frac{-r^2}{\overline{\gamma}}\right) dr \\ & + \frac{\left(1 - P_p\right)}{2} \int_0^\infty \frac{r^{2L-2}}{(L-1)!\overline{\gamma}^L} \exp\left(\frac{-r^2}{\overline{\gamma}}\right) dr \\ & - \frac{\left(1 - P_p\right)}{2} \int_0^\infty \operatorname{erf}\left(\frac{\gamma_1 . r}{\sqrt{2}}\right)\left(\frac{r^{2L-2}}{(L-1)!\overline{\gamma}^L}\right) \exp\left(\frac{-r^2}{\overline{\gamma}}\right) dr \\ & + \frac{P_{SG}}{2} \int_0^\infty \frac{r^{2L-2}}{(L-1)!\overline{\gamma}^L} \exp\left(\frac{-r^2}{\overline{\gamma}}\right) dr \\ & - \frac{P_{SG}}{2} \int_0^\infty \operatorname{erf}\left(\frac{\gamma_1 . r}{\sqrt{2}}\right)\left(\frac{r^{2L-2}}{(L-1)!\overline{\gamma}^L}\right) \exp\left(\frac{-r^2}{\overline{\gamma}}\right) dr \end{aligned} \tag{9}$$

Here, erf(.) is the error function in probability theory obtained from [20] as

$$Q(x) = \frac{1}{2}\left(1 - \text{erf}\left(\frac{x}{\sqrt{2}}\right)\right). \tag{10}$$

Using the infinite series expansion for erf(x) from [21], we have

$$\text{erf}(x) = \frac{2}{\sqrt{\pi}} e^{-x^2} \sum_{k=0}^{\infty} \frac{2^k x^{2k+1}}{(2k+1)!}. \tag{11}$$

Using infinite series expansion from [21], we obtain the error probability for CHMA using MRC under a Rayleigh fading channel as

$$\begin{aligned} P_{\text{Ray}}^{\text{MRC}} = {} & \frac{P_{\text{P}}}{4(L-1)!} \frac{\Gamma\left(L - \frac{1}{2}\right)}{\sqrt{\overline{\gamma}}} \\ & + \frac{(1 - P_{\text{P}})}{4\sqrt{\overline{\gamma}}(L-1)!} \Gamma\left(L - \frac{1}{2}\right) - \frac{(1 - P_{\text{P}})}{2\overline{\gamma}^L \sqrt{\pi}(L-1)!} \\ & + \frac{P_{\text{SG}}}{4\sqrt{\overline{\gamma}}(L-1)!} \Gamma\left(L - \frac{1}{2}\right) \\ & - \frac{P_{SG}}{2\overline{\gamma}^L \sqrt{\pi}(L-1)!} \sum_{k=0}^{\infty} \frac{2^k \left(\frac{\gamma_2}{\sqrt{2}}\right)^{2k+1}}{(2k+1)!} \cdot \frac{\Gamma(k+L)}{\left(\frac{\gamma_2^2}{2} + \frac{1}{\overline{\gamma}}\right)^{k+L}} \end{aligned} \tag{12}$$

where $\gamma_1 = \sqrt{\frac{2E_s}{N_o}}$ and $\gamma_2 = 2\sqrt{\frac{E_s}{N_o}}$ represent the bit energy to noise spectral density.

The infinite series expansion in (12) converges to a finite value because of the following criterion

$$\lim_{n \to \infty} \left| \frac{\sum_{n=0}^{\infty} \frac{2^{n+1}(\gamma_1/\sqrt{2})^{2(n+1)+1}}{(2(n+1)+1)!}}{\sum_{n=0}^{\infty} \frac{2^n(\gamma_1/\sqrt{2})^{2n+1}}{(2n+1)!}} \right| = 0 < 1 \tag{13}$$

4 Performance Analysis

Here, statistical and simulation results are carried out to evaluate the performance of an AMCI system using CHMA employing OCC under Rayleigh fading channel. In the simulation results discussed in this section, the number of input bits mostly used is 10^5, and BPSK modulation scheme is assumed. The analytical plot of SISO-CHMA under Rayleigh fading channel using (7) is plotted in Fig. 3 for different values of perforation probabilities (P_{p}) in an AMCI network. The numerical parameters

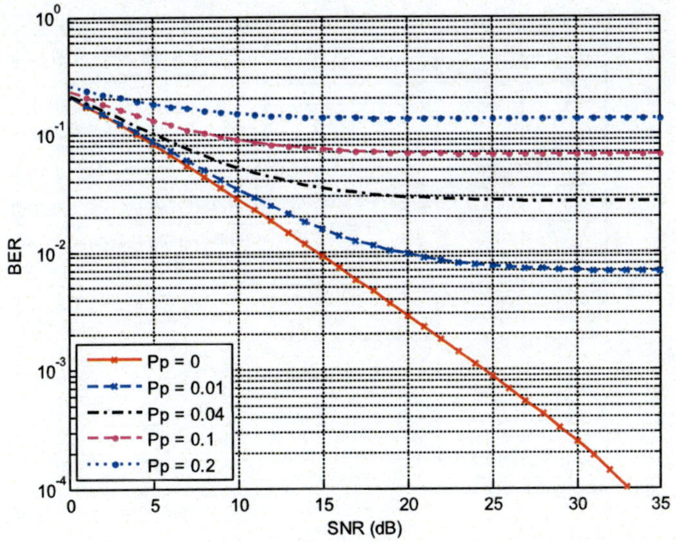

Fig. 3 Theoretical performance of CHMA system in Rayleigh fading channel for dissimilar values of perforation probability for AMCI network

assumed are $\overline{\gamma} = 2$, the code collision probability, $P_c = 0.01$. It is apparent that the bit error rate (BER) rises with variation in perforation probability. This occurs because of increase in destructive collision occurring between different power users using the same code in the specified time slot, resulting in poor BER. The performance of CHMA system can be improved by incorporating diversity techniques at the receiver end to obtain improved BER for specific values of perforation probability. The analytical expressions in (12) are in the form of infinite summation terms, but these expressions converge rapidly [using (13)] and hence can be used efficiently in performance analysis of CHMA-based AMCI network. It can also be observed from the analytical results of Fig. 4, based on (12), that the reduction in error probability decreases with increase in perforation probability for CHMA system having receiver diversity of order 2. The numerical parameters assumed are $\overline{\gamma} = 1$, the code collision probability, $P_c = 0.01$. The BER performance can be improved by increasing the no of receiver antennas based on specific values of perforation probability.

The analytical versus simulated performance comparison of CHMA for both diversity and without diversity for AMCI network is plotted using the superset of perfect complementary codes (PCC) [22] under Rayleigh fading channel is shown in Fig. 5. The comparison is made using OCC with its distinguishing factors, $M = 8$, $N = 8$, and element code length $L_c = 4$ having PG of 32 in a multipath fading channel. The number of power users used for evaluation is 8 for both systems. The analytical BER for CHMA system without diversity using OCC under Rayleigh fading channel is plotted using (7) for the case of $P_p = 0$. The numerical parameters assumed are $\overline{\gamma} = 2$ and $P_c = 0$. It can be observed that the analytical BER at 12 dB is 0.0150,

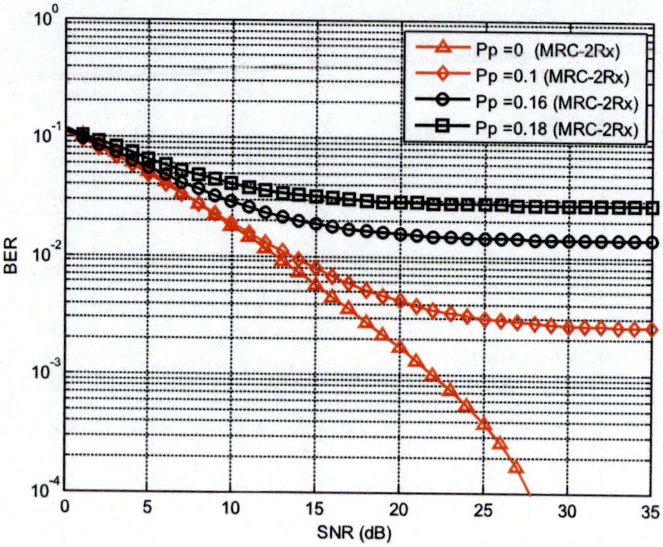

Fig. 4 Performance comparison CHMA system under Rayleigh fading channel using MRC in communicating over AMCI network

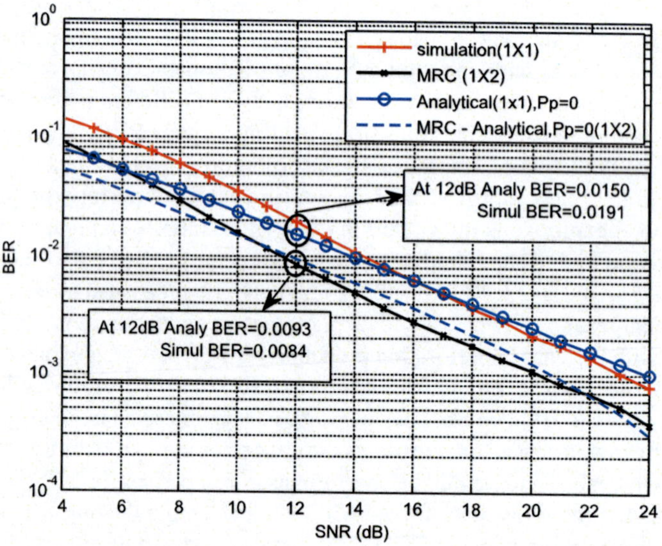

Fig. 5 Comparison of analytical and simulated BER for a CHMA-based AMCI network using OCC under multipath fading channel for system with and without diversity combining

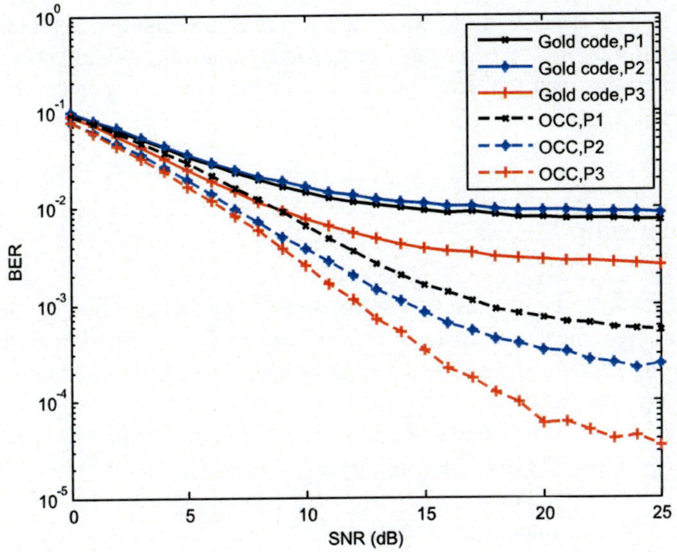

Fig. 6 Performance comparison of a CHMA-based AMCI system using OCC and Gold codes in a multipath fading channel under different multipath channel environments

which is approximately equal to the simulated BER of 0.0191 validating the desired expression for BER of SISO-CHMA using OCC for Rayleigh fading channel in (7).

To demonstrate the ability of CHMA-based AMCI system to remove both MAI and MPI existing between different power users operating in the same wireless channel, BER performance of CHMA using OCC is compared with Gold codes in Fig. 6. Here the number of power users was assumed to vary from 2 to 8. The PG of OCC is 32 and Gold code length is $L_c = 31$. The Gold codes are Quasi-orthogonal codes through its fine well-ordered 3-level cross-correlation function which show better performance compared to Walsh codes in multipath channel, where the orthogonality does not hold in channels with delay [20]. Here, the interpath delay is assumed to be 2 chips and a single correlator is used at the receiver for both the systems. Further the path gain coefficient vector used in Fig. 6. is [1 0.6 0.4].

In order to evaluate the performance of CHMA-based AMCI network using OCCs under different multipath fading scenarios, three multipath delay profiles have been measured for assessment in Fig. 6. Here, we have selected models with delay profiles [1, 0.6, 0.4], [1, 0.65, 0.2, 0.15], [1, 0.55, 0.2, 0.15.0.1], and denoted as P1, P2, and P3, respectively. It can be visualized that through a rise in the amount of multipath waves, the simulated BERs for OCC in an AMCI network decreases rapidly compared to Gold coded CHMA. At 15 dB SNR, the BER is found to be 9×10^{-2}, 1×10^{-3}, and 7×10^{-3} for P1, P2, and P3, respectively, which is much better compared to the error rate using Gold coded CHMA. It can be observed OCC-based AMCI system attains improved BER in a frequency selective fading channel using a single receiving antenna under multipath channel conditions. This feature shows the capability of the

AMCI system to achieve multipath diversity gain in addition to MAI and MPI free operation offering increased power user capacity compared to Gold coded CHMA in communicating over multipath fading channel from the utility center to the smart utility meter.

5 Conclusion

In this paper, we have proposed the CHMA technology for smart grid applications in order to progress the performance of present communication network architectures. AMCIs using OCC are investigated in order to progress security and minimize interference for power users under MAI and MPI scenarios. Receiver diversity structures are suggested for CHMA-based AMCIs using OCC to advance the performance of the current system and to offer improved performance over systems functioning without diversity. Simulation studies display that the suggested CHMA-based AMCI networks approve reduction in error rates and lower interference from new power users over fluctuating environments.

References

1. Huang J, Wang H, Qian Y, Wang C (2013) Priority-based traffic scheduling and utility optimization for cognitive radio communication infrastructure-based smart grid. IEEE Trans Smart Grid 4(1):78–86
2. Niyato D, Wang P (2012) Cooperative transmission for meter data collection in smart grid. IEEE Commun Mag, 90–97
3. Gungor VC (2011) Smart grid technologies: communication technologies and standards. IEEE Trans Ind Electron 7(4):529–539
4. Selvaraj K, Judson D, Ganesh Kumar P, Anandaraj M (2020) Low complexity linear detection for multiuser MIMO SC-FDMA uplink transmission system. Wirel Pers Commun 112:631–649
5. Judson D, Bhaskar V, Arun S (2019) Space Time Regularized zero forcing in downlink code division multiple access systems with complementary codes. Wirel Pers Commun 109:333–347
6. Yan Y, Qian Y, Sharif H, Tipper D (2013) A survey on smart grid communication Infrastructures: motivations, requirements and challenges. IEEE Commun Surveys Tutor 15(5)
7. Zeng Q, Li H (2012) Frequency-hopping based communication network with multi-level qoss in smart grid: code design and performance analysis. IEEE Trans Smart Grid 3(4):1841–1852
8. Chen Y, Suehiro (2001) A multi-carrier CDMA Architecture based on orthogonal complementary codes for new generations of wideband wireless communications. IEEE Commun Mag 39(10):126–135
9. Moshavi S (1996) Multi-user Detection for DS-CDMA communication. IEEE Commun Mag 34:124–136
10. Dian EH, Jabbari B (1998) Spreading codes for direct sequence CDMA and wideband CDMA cellular networks. IEEE Commun Mag 0163–6804 (1998)
11. Park S, Sung DK (2002) Orthogonal code hopping multiplexing. IEEE Commun Lett 6(12): 529–531
12. Judson D, Albert Raj A, Jonisha Miriam LR, Mourice Seraphin JJ (2013) Performance of multiuser CHMA systems over Rayleigh flat fading channels. In: Proceedings of IEEE SEISCON

13. Kwon JK, Park S, Sung DK, Lee H (2003) Adaptive code rate for orthogonal code hopping multiplexing (OCHM) in synchronous downlink. In: Proceedings of IEEE WCNC, vol 2, pp 855–859
14. Kwon JK, Park S, Sung DK (2003) Log likelihood ratio (LLR) conversion schemes in orthogonal code hopping multiplexing. IEEE Commun Lett 7(3):104–106
15. Kwon JK, Park S, Sung DK (2006) Collision mitigation by log likelihood ratio (LLR) conversion in orthogonal code hopping multiplexing. IEEE Trans Veh Technol 55(2):709–717
16. Judson D, Bhaskar V (2018) Interference cancellation in complementary coded CDMA in Rician fading channels. Wirel Pers Commun 101:897–914. https://doi.org/10.1007/s11277-018-5732-3
17. Backer D (2007) Power quality and asset management, the other two-thirds of AMI value. In: IEEE rural electric power conference, pp 6–8
18. Chen HH, Zhang HM, Huang ZK (2012) Code hopping multiple access based on orthogonal complementary codes. IEEE Trans Veh Technol 61(3):1074–1083
19. Judson D, Bhaskar (2018) Error rate analysis of SIMO-CDMA with complementary codes under multipath fading channels. Wirel Pers Commun 98:1663–1677
20. Proakis JG (2001) Digital communications, 4th edn. Mc Graw Hill Inc., NY
21. Gradshteyn IS, Ryzhik IM (2000) Table of integrals series and products, 6th edn. Academic Press, San Diego
22. Ascar Davix X, Judson D (2019) Successive interference cancellation in asynchronous CC-CDMA systems under Rician fading channels. Telecommun Syst 72:261–271
23. Chen HH (2007) The next generation CDMA technologies. Wiley, NY

Design and Analysis of a Vertical PINI Junction Phase Shifter in Silicon Mach–Zehnder Modulator

R. G. Jesuwanth Sugesh and A. Sivasubramanian

Abstract High-speed data transmission with high modulation efficiency is the essential condition for a high-performance modulator. The phase shifter plays a significant role in determining modulator performance. In this paper, a vertical PINI junction phase shifter is proposed to attain high modulation efficiency. The optical group index is matched to the RF effective index to have a strong coupling of optical and RF mode for high-speed operation. The circuit-level analysis is performed on the optimised phase shifter imported in a silicon Mach–Zehnder modulator. At 60Gbps, a maximum extinction ratio of 10.5 dB with a bit error rate of 5.45×10^{-6} is obtained at $V\pi L$ of 1.25 V cm for the phase shifter length of 2.5 mm. The energy per bit transmission for the proposed transmission is calculated to be 4pJ/bit. The proposed designs demonstrated to support high data rate transmission for commercial applications. The proposed device can also find other applications such as optical switches, delay lines, and free space optics.

Keywords Silicon photonics · Phase shifter · Silicon Mach–Zehnder modulator · Extinction ratio (ER) · Bit error rate (BER)

1 Introduction

The present pandemic situation has shown us the importance of data communication for our daily tasks. Recent growth in network demands has triggered a need for advanced optical communication. Silicon photonics has attracted the interest of researchers because it enables the cost-effective mass production of optical devices. Optical modulators play a vital role in a communication link by modulating the light beam propagating in the waveguide. Electro-optic modulators are preferred as they support high-speed data transmission. The modulation of the light beam is produced by charge carrier movement, i.e. carrier accumulation, depletion or injection, in an electro-optic modulator.

R. G. Jesuwanth Sugesh (✉) · A. Sivasubramanian
School of Electronics Engineering, Vellore Institute of Technology, Chennai 600127, India
e-mail: jesuwanthsugesh.rg2016@vitstudent.ac.in

A. Sivasubramanian et al. (eds.), *Futuristic Communication and Network Technologies*, Lecture Notes in Electrical Engineering 792,
https://doi.org/10.1007/978-981-16-4625-6_103

Silicon carrier depletion electro-optic modulators have been widely studied as they are compatible with standard fabrication technologies and support high-speed operation but are flawed by losses that limit commercial adoption, when compared with InP and $LiNbO_3$-based conventional modulators [1, 2]. Optical modulators, being a part of the transmitter circuit, should only consume a small fraction of the power of the circuit. The modulator drive voltage is responsible for obtaining the π-phase shift and plays a significant part in the energy consumed per bit (E_{bit}). In long haul applications, power consumption is of less importance. The modulator's loss includes the loss of all the constituent components, i.e. splitter, waveguide, phase shifter, electrodes, and combiner. The modulation frequency of a modulator is the maximum frequency at which a signal can be modulated by the modulator. The phase efficiency is measured as $V\pi L$ (V cm), where $V\pi$ is the drive voltage required to obtain a π-phase shift and L required length of the modulator. The modulation depth or extinction ratio (ER) is a measure of the output power between 1 and 0 states. The footprint is the physical dimensions of the modulator on the chip. The ring modulator (RM) is generally preferred over the Mach–Zehnder modulator (MZM) because of its compactness.

In contrast, the MZM is preferred over the RM as it is less temperature-sensitive. The BER plays a significant role in communication systems as it defines the number of bits in error among the total number of bits transmitted per unit time. Thus, to realise high performance, the modulator should operate at a low modulator drive voltage, have high modulation speed, and ER, low phase efficiency ($V\pi L$), and BER, decreased temperature sensitivity, and DCD value, and a small footprint.

A phase shifter plays a vital role in determining the performance of a modulator. Phase shifters based on different doping patterns, junction placements, and doping concentrations have been studied experimentally, such as those with a lateral, vertical, interleaved, and serpent PN junctions [3–14]. The doping pattern, intrinsic gap, junction placement, and doping concentration play a significant role in determining the performance of the phase shifter. When the doping concentration is increased, the number of carriers is increased. This increases the capacitance, and with drive voltage, a larger effective index change occurs, leading to an increased phase shift. Thus, even at low drive voltage and short phase shifter length, the π-phase shift can be obtained (low $V\pi L$). But, with increasing carriers, the absorption loss due to the carriers increases. By forming multiple PN junctions, low-drive voltage π-phase shift can be obtained at the expense of increased insertion loss and ease of fabrication. The vertical junction reduces the size of the doping region, but it necessitates higher concentration, and the placement of the vertical slabs determines the loss. The loss is reduced by the inclusion of an intrinsic region between the P-type and N-type regions of the PN junction in the light propagation path. Increasing the intrinsic gap reduces the loss but also reduces the phase shift, and thus necessitates a high phase efficiency ($V\pi L$). The $LiNbO_3$ modulator was designed to improve the modulation speed [5, 6]. It has been demonstrated that Si modulators are also capable of high-speed data transmission up to 56 Gbaud [7]. Vertical doped slabs of thickness 35 and 20 nm with high doping concentration were used in [9]. When the doping concentration is increased, the absorption loss due to free carrier absorption will increase. In order to improve optical overlap with the depletion region for improved modulation, a U-shaped PN junction was designed in [10]. Slot-like structures were used to reduce the loss and to increase the light-matter interaction in [11, 12].

Lumped-mode design in a phase shifter suffers from RC time constant; thus, travelling wave (TW) mode electrode design is desired. TW supports carrier depletion but requires a large footprint [13, 14]. The optical group index is matched to the RF effective index to have a strong coupling of optical and RF mode. This index matching reduces inter-symbol interference, thus improving the performance of the modulator. The impedance matching to 25 Ω or 50 Ω will reduce the return loss in TW [13].

Among modulator types, the MZM is employed for the work due to its low cost, thermal stability, and ease of fabrication [14]. Considering the objectives and the trade-off conditions, a phase shifter with PINI junction is proposed. The intrinsic region plays a major role in the performance of a phase shifter. The objective of the work is to design an MZM modulator with high ER and sufficient BER for data centre application. The designed structure and the simulation process flow are described in Sect. 2, and the results are discussed in Sect. 3.

2 Design Structure

Commercially available Lumerical tool is used for the simulation analysis of the proposed phase shifter's performance [15]. The TE_1 mode is preferred due to its better optical confinement. Thus, a rib waveguide is designed and simulated with Lumerical mode, accordingly. The dimensions of the rib waveguide are set to 500 nm × 220 nm. The simulated light propagation through the waveguide is shown in Fig. 1. Since holes have a larger index shift and lower absorption than electrons, [16] the concentrations of P (boron) and N (phosphorus) dopants are set to 9×10^{17} cm^{-3} and 7×10^{17} cm^{-3}, respectively. The thicknesses of the p-doped and n-doped layers

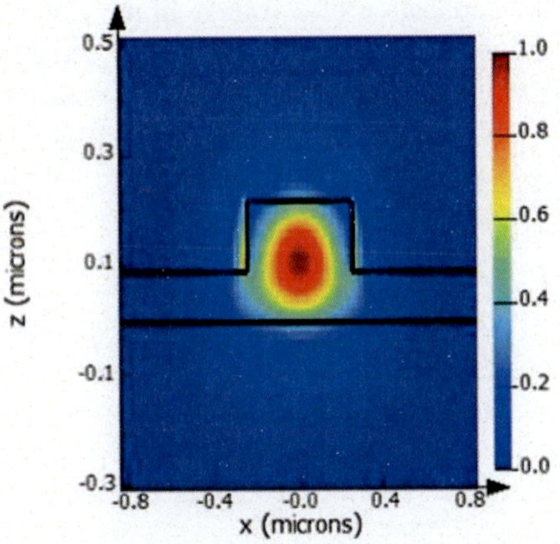

Fig. 1 Light propagation through waveguide (500 nm × 220 nm)

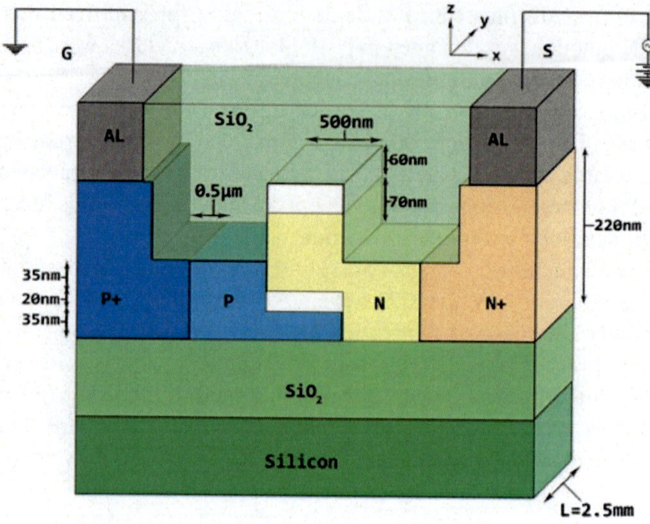

Fig. 2 Cross-sectional view of the proposed phase shifter

is set to 35 and 105 nm in the rib region. The heavily doped P and N wells, with carrier concentrations of 1×10^{19} cm^{-3}, are placed at the edges of the slab to reduce slab resistance. The P-slab is placed at the bottom of the structure as boron atoms have high diffusivity, and it is difficult to form a shallow high implanted doped well. An intrinsic layer of thickness 20 nm is between P- and N-doped regions. The intrinsic region in the top of the rib waveguide is of thickness 60 nm. The electrical contacts to the phase shifter are provided by aluminium pads on the sides on the slab. A cross-sectional view of the phase shifter model is presented in Fig. 2. The proposed phase shifter is designed over a silicon dioxide layer (SiO_2) of thickness 2 μm on a 220 nm thick silicon substrate. The optical mode coupling into the silicon substrate is prevented by the SiO_2 layer. The carriers in the junction is depleted with the application of a voltage in the cathode (reverse bias), and the other contact is grounded. In order to keep power consumption to a minimum, the driving voltage was limited to a maximum of 5 V. A surface oxide layer of SiO_2 is formed at the top of the device to protect the device from external impurities.

3 Result Analysis

In the device-level analysis, the electro-optic and RF characteristics of the proposed phase shifter (PS) are analysed using Lumerical mode and device-charge transport module. In the system-level analysis, the proposed phase shifter parameters are imported into silicon MZM (PS-MZM) in Lumerical interconnect module.

3.1 Device-level Analysis

The electro-optic performance of the proposed PS is analysed by varying the reverse bias voltage (0–5 V) to the PS. When the voltage is increased, the carriers are depleted from the centre of the waveguide. The decrease in free carriers reduces the capacitance in the junction, as shown in Fig. 3. Thus, the capacitance of the junction (C) is based on the electron (N_n) and hole (N_p) densities (1),

$$C = \frac{t_{\text{rib}} \times \sqrt{(q\varepsilon_0\varepsilon_r)}}{\sqrt{\left(2\left(N_p^{-1} + N_n^{-1}\right) \times (V_v - V)\right)}} \tag{1}$$

where q denotes the electric charge,ε_0 the dielectric constant, ε_r the relative permittivity, and V_v the diffusion potential.

The carrier depletion has a major role in the absorption coefficient ($\Delta\alpha$) and refractive index (Δn) of the PS. Their mathematical relationship [(2) and (3)] was expressed by Soref and Bennett in [16].

$$\Delta n = -8.8 \times 10^{-22} N_n - 8.5 \times 10^{-18} N_p^{0.8} \tag{2}$$

$$\Delta\alpha = 8.5 \times 10^{-18} N_n + 6 \times 10^{-18} N_p \tag{3}$$

The change in the effective index (5) with respect to voltage alters the optical property of the PS. Thus, phase shift (6) on the propagating wavelength along the PS length (L) is obtained.

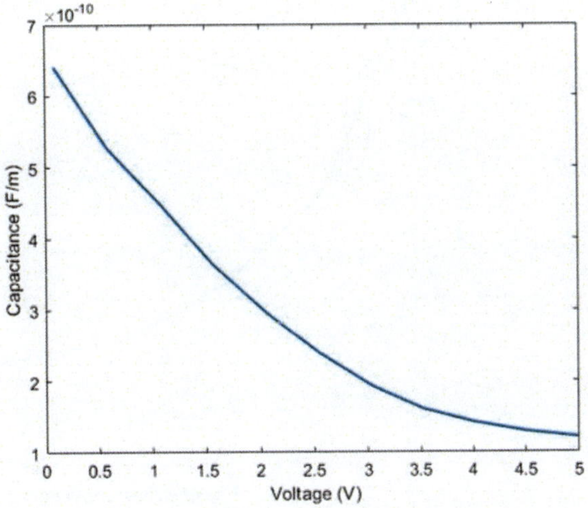

Fig. 3 Capacitance measured with respect to reverse bias voltage

$$n_{\text{eff}}(V) = n_{\text{eff,i}} + \int \Delta n(V)\text{d}V \tag{4}$$

where $n_{\text{eff,i}}$ denotes the effective index of the waveguide without doping.

$$\Delta n_{\text{eff}}(V) = n_{\text{eff}}(V) - n_{\text{eff}}(0) \tag{5}$$

$$\varphi(V) = \frac{2\pi\,\Delta n_{\text{eff}}(V)}{\lambda L} \tag{6}$$

The absorption coefficient α along the PS (*y*-axis) is obtained from (7).

$$\alpha(y) = \frac{\iint \Delta\alpha(V)\left|E(x, y, z)^2\right|\text{d}x\text{d}z}{\iint \left|E(x, y, z)^2\right|\text{d}x\text{d}z} \tag{7}$$

where *x* and *z* denote the waveguide dimension coordinates and y denotes the length coordinate of the PS, and $E(x, y, z)$ is the optical intensity distribution of the waveguide mode.

The interaction of carriers with the propagating wavelength reduces with the carrier depletion. The free carrier absorption of the photons contributes to a significant portion of the PS loss. The intrinsic gap between P and N junction reduces the absorption loss in the middle of the waveguide. The phase shift required for modulation (π) is obtained at 5 V for the proposed PS design of length 2.5 mm with absorption loss of 6 dB. Figure 4 shows the $V\pi$ and loss variation with respect to PS length. With the increase in length, the voltage required to obtain π phase shift is

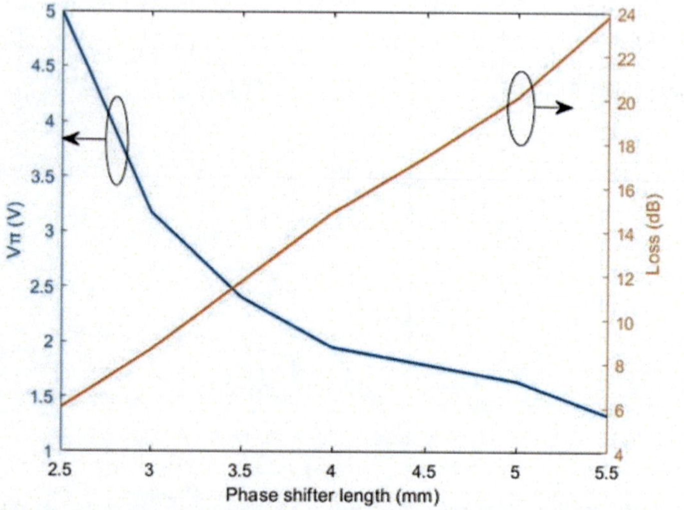

Fig. 4 $V\pi$ and loss variation with respect to the length of the proposed phase shifter design

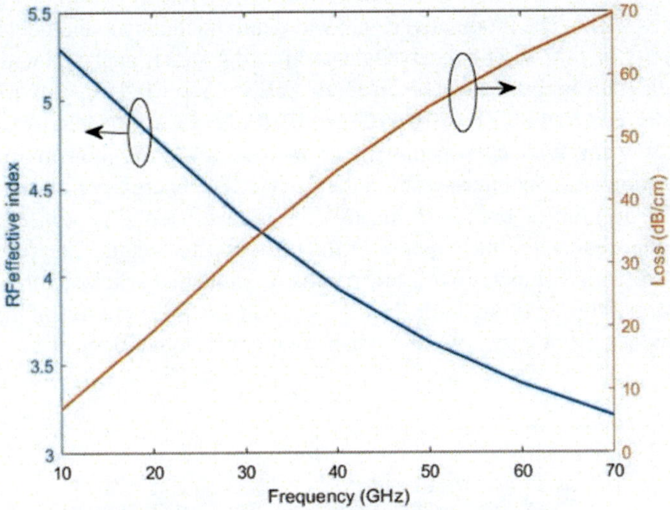

Fig. 5 RF effective index and loss acquired as a function of frequency

reduced due to the increase in exposure of the phase shifter. This increases the loss along the phase shifter length. This is the trade-off between the length, $V\pi$, and loss in the phase shifter. Loss is also contributed through imperfectness or roughness in the sidewalls of the waveguide during fabrication. This can be reduced by optimising the etching process.

In order to reduce the inter-symbol interference and have strong coupling between the RF and optical mode, the optical group index and the effective index should be matched. At 40 GHz, the index matching occurs, and loss is calculated to be 4.4 dB (Fig. 5). The obtained impedance for the proposed design is near 25 Ω, thus reducing the return loss. The 6 dB bandwidth of 14.6 GHz is obtained for the proposed design of length 2.5 mm.

3.2 *System-level Analysis*

The proposed PS design's parameter values are imported into an unbalanced MZM (PS-MZM), and the dynamic performance was analysed. A length variation of 100 μm is used to make the unbalanced PS-MZM, and a fixed bias voltage of 1 V was provided to the long arm. An alternating voltage swing (V_{pp}) with DC bias (V_{dc}) was applied to the shorter arm of PS-MZM. The message data was generated from a pseudo-random bit sequence (PRBS) generator at 60 Gbps. An electrical signal generator with NRZ line coding technique was used to convert the message data to the electrical message signal. The carrier optical signal of 1553.5 nm was produced from a continuous wave laser. The message signal was modulated onto the carrier signal in PS-MZM. The modulated signal was demodulated by a photodetector with a responsivity of 1 A/W. The received demodulated signal is analysed with the help of an eye diagram analyser and a bit error rate (BER) tester. At 5 V_{pp} and 0.75 V_{dc}, an eye diagram (Fig. 6) with ER of 10.5 dB and BER of 5.45×10^{-6} was obtained for a $V\pi L$ of 1.25 V.cm. Wide eye-opening and eye-crossing at 50% show minimum duty cycle distortion and low inter-symbol interference. The energy per bit transmission ($E_{bit} = (CV^2)/4$) for the proposed transmission is calculated to be 4 pJ/bit. The E_{bit} can be further reduced at the expense of the phase shifter length. The performance of the PS-MZM is compared with few results of published articles with depletion-type MZM in Table 1. From Table 1, it is clear that designed phase shifter in MZM provides higher ER at very low $V\pi L$ when compared to other devices.

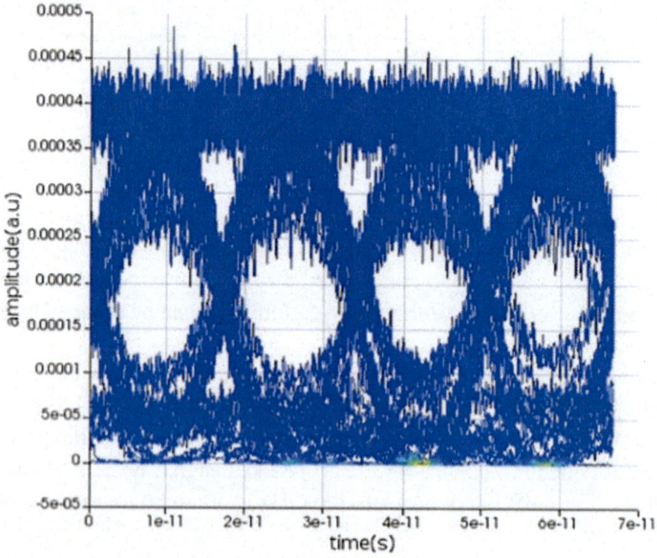

Fig. 6 Eye diagram for PS-MZM with $V\pi L = 1.25$ V cm

Table 1 Parameter comparison with published results

Refs.	L (mm)	Gbps	$V\pi L$ (V cm)	ER (dB)
[3]	0.75	40	1.5	7.01
[4]	3	10	1.08	11
[5]	8	–	3.1	18
[7]	1.5	112	2.3	–
[8]	5	100	2.5	5.5
[11]	4	70	2.1	4.9
This work	2.5	60	1.25	10.5

4 Conclusion

A vertical PINI junction phase shifter was proposed for high-speed optical modulation. The proposed design obtained high modulation efficiency. The proposed phase shifter was imported into a silicon Mach–Zehnder modulator, and dynamic performance analysis was performed. At 60 Gbps, an extinction ratio of 10.5 dB with a bit error rate of 5.45×10^{-6} was obtained at $V\pi L$ of 1.25 V cm for the length of the phase shifter of 2.5 mm. The energy per bit transmission is calculated to be 4 pJ/bit. The proposed device can also find other applications such as optical switches, delay lines, and free space optics.

References

1. Tu X, Liow TY, Song J, Luo X, Fang Q, Yu M, Lo GQ (2013) 50-Gb/s silicon optical modulator with traveling-wave electrodes. Opt Express 21(10):12776–12782
2. Reed GT, Mashanovich GZ, Gardes FY, Nedeljkovic M, Hu Y, Thomson DJ, Li K, Wilson PR, Chen SW, Hsu SS (2014) Recent breakthroughs in carrier depletion based silicon optical modulators. Nanophotonics 3(4–5):229–245
3. Xu H, Xiao X, Li X, Hu Y, Li Z, Chu T, Yu Y, Yu J (2012) High speed silicon Mach-Zehnder modulator based on interleaved PN junctions. Opt Express 20(14):15093–15099
4. Ogawa K, Goi K, Ishikura N, Ishihara H, Sakamoto S, Liow TY, Tu X, Lo GQ, Kwong DL, Lim ST, Sun MJ (2016) Silicon-based phase shifters for high figure of merit in optical modulation. In: Silicon photonics XI, vol 9752. SPIE, US, p 975202
5. Rao A, Patil A, Rabiei P, Honardoost A, DeSalvo R, Paolella A, Fathpour S (2016) High-performance and linear thin-film lithium niobate Mach-Zehnder modulators on silicon up to 50 GHz. Opt Lett 41(24):5700–5703
6. He M, Xu M, Ren Y, Jian J, Ruan Z, Xu Y, Gao S, Sun S, Wen X, Zhou L, Liu L (2019) High-performance hybrid silicon and lithium niobate Mach-Zehnder modulators for 100 Gbit s^{-1} and beyond. Nat Photonics 13(5):359–364
7. de Farias GB, Bustamante YR, de Andrade HA, Moura UC, Freitas AP, Motta DD (2018) Demonstration of >48 GHz single-drive push-pull silicon Mach-Zehnder modulator with low $V\pi L$. In: 2018 IEEE 15th international conference on group IV photonics (GFP). IEEE, Mexico, pp 1–2
8. Palmer R, Alloatti L, Korn D, Schindler PC, Schmogrow R, Heni W, Koenig S, Bolten J, Wahlbrink T, Waldow M, Yu H (2013) Silicon-organic hybrid MZI modulator generating OOK, BPSK and 8-ASK signals for up to 84 Gbit/s. IEEE Photonics J 5(2):6600907
9. Azadeh SS, Merget F, Romero-García S, Moscoso-Mártir A, von den Driesch N, Müller J, Mantl S, Buca D, Witzens J (2015) Low $V\pi$ silicon photonics modulators with highly linear epitaxially grown phase shifters. Opt Express 23(18):23526–23550
10. Yong Z, Sacher WD, Huang Y, Mikkelsen JC, Yang Y, Luo X, Dumais P, Goodwill D, Bahrami H, Lo PGQ, Bernier E (2018) U-shaped PN junctions for efficient silicon Mach-Zehnder and microring modulators in the O-band: erratum. Opt Express 26(25):32757
11. Jain S, Rajput S, Kaushik V, Kumar M (2019) High speed optical modulator based on silicon slotted-rib waveguide. Opt Commun 434:49–53
12. Jain S, Rajput S, Kaushik V, Kumar M (2020) Efficient optical modulation with high data-rate in silicon based laterally split vertical PN junction. IEEE J Quantum Electron 56(2):1–7
13. Baehr-Jones T, Ding R, Liu Y, Ayazi A, Pinguet T, Harris NC, Streshinsky M, Lee P, Zhang Y, Lim AEJ, Liow TY (2012) Ultralow drive voltage silicon traveling-wave modulator. Opt Express 20(11):12014–12020

14. Jesuwanth Sugesh RG, Sivasubramanian A (2018) Redesigning Mach-Zehnder modulator with ring resonators. In: Gnanagurunathan G, Sangeetha R, Kiran K (eds) Optical and microwave technologies, vol 468. Lecture notes in electrical engineering. Springer, Singapore, pp 185–191
15. Lumerical software. https://www.lumerical.com/products/. Last accessed 28 Sept 2020
16. Soref R, Bennett B (1987) Electrooptical effects in silicon. IEEE J Quantum Electron 1:123–129

A Machine Learning and Blockchain Platform for Operation Risk Management—An Application to Real-time Risk Awareness System Development

Michael Zipperle, Florian Gottwalt, Marius Becherer, Ke Wang, Yu Zhang, and Elizabeth Chang

Abstract This paper presents a machine learning and blockchain platform (the Platform) for real-time risk awareness, management and control for large enterprises in the area of big data analytics, and the ground truth of heterogeneous information powered by AI and blockchain. We evaluated the platform through a real-world application known as real-time risk awareness system (ReRAS) to address enterprise risks, and the big data analytics and the platform achieved 95.8% risk identification accuracy and a possibility of human effort reduction on the risk identification at 90%.

Keywords Risk management · Big data · Heterogenous information · Machine learning · Blockchain platform

1 Introduction

Big data analytics, heterogeneous information management, accountability, and security using AI and machine learning are the biggest challenges for many governments and large private enterprises [1, 2]. Real-time security and risk management

M. Zipperle (✉) · F. Gottwalt · M. Becherer · K. Wang · Y. Zhang · E. Chang
University of New South Wales, Canberra, Australia
e-mail: m.zipperle@adfa.edu.au

F. Gottwalt
e-mail: f.gottwalt@adfa.edu.au

M. Becherer
e-mail: m.becherer@adfa.edu.au

K. Wang
e-mail: ke.wang9@adfa.edu.au

Y. Zhang
e-mail: yu.zhang@adfa.edu.au

E. Chang
e-mail: e.chang@adfa.edu.au

A. Sivasubramanian et al. (eds.), *Futuristic Communication and Network Technologies*, Lecture Notes in Electrical Engineering 792,
https://doi.org/10.1007/978-981-16-4625-6_104

are the key issues and require a mandraulic human effort to investigate and time consuming to determine the risks and achieve greater accuracy. Currently, there is no ready used real-world tools or platforms available that can be deployed or plug into the existing inter- and intra-enterprise network environment. This is mostly due to the technical difficulties in transforming the large-scale unstructured data into structured data and particularly in real-time while data are continuously changing and evolving in the real-world environment [3].

In order to address the above issues and reduce the risks including big data security, single source of truth, the ground truth of transactions, data asset, and accountability for large government enterprises and or multinational corporations, we developed a practical machine learning and blockchain technical platform that can be used and deployed in the real-world setting. We validated the platform through the development a real-time risk awareness system, coined ReRAS, which employs AI techniques, such as text mining, machine learning, and blockchain and can handle large unstructured enterprise data in real time, extract, and perform interrogation and audits against enterprise data policies, processes, and predefined business rules without human intervention. We achieve this through three steps consist of (1) text mining to extract features for (2) the document classification, and consequently (3) document-type adjusted text mining is executed. During this process, the whole transaction history of all acquisitions is tracked in a blockchain to ensure transparency, data security, and accountability. Hereby, compliance-related data for prospective compliance checks are stored on the blockchain, whereas the actual evidence is stored on a distributed file system.

The structure of this paper is as follows: We conduct an overview the related work of our research in section 2. In Sect. 3, we present the overall framework of AI and blockchain platform and application to real-time risk awareness system. In Sect. 4, we evaluate the accuracy and the performance of the platform with ReRAS. Finally, we evaluate our platform through ReRAS in Sect. 5, and this is followed by the conclusion of strength and future work with the platform in Sect. 6.

2 Related Work

The automated enterprise risk identification in the area of data interrogation, auditing, and risk decision making was first presented in [4]. High-risk data and transactions should be analyzed more often than low-risk ones [5]. This leads to the requirement of using AI to carry out high-risk management not only continuously but in real time to proactively avoid the impact of such a risk. Real-time big data and heterogeneity in types of analytics require state-of-the-art automation technologies. Large institutions like IBM have proposed to use big data analytic techniques and implemented the concept [6]. However, achieving high efficiency, accuracy, and accountability while reducing costs, time, and effort, especially in the public sector, is not available in the marketplace. The main reasons quoted for that are governance complexities, cyber is everything (data protection, cybersecurity considerations, staff training), full transparency, reporting on steroids, and skillset shortages [7]. The authors in [8–10] have described requirements and problems for the introduction of continuous auditing in Nigeria, Malaysia, and the UK, but no ready used AI platform, or conceptual models, or concrete implementation or evaluations in the literature to date.

A core element of continuous auditing is the use of state-of-the-art text mining and machine learning techniques to automate manual processes of traditional auditing. These techniques have developed rapidly in recent years and offer great potential for the use in continuous auditing [11].

Furthermore, blockchain technology has not only established itself in cryptocurrencies but also offers various opportunities for the use in public sectors [12]. Through key characteristics of a private blockchain, being immutability, transparency, accountability, and security, the requirements of public sectors can be fulfilled, and thus, the introduction of continuous auditing in the public sector can be made possible. While related work in [13] have proposed a concept of how blockchain can be used for the trustful storage of machine learning data and authors in [14] have proven that blockchain can be used to track data accountability and provenance, to the best of our knowledge, there is no proposal for how blockchain can track the real-time auditing process.

In summary, continuous auditing is becoming increasingly important in the public sector, as the amount of data to be audited is growing rapidly. Only with an automated approach, public governance can be ensured in the future. To implement continuous auditing in the public sector, various challenges need to be addressed. Firstly, there is a lack of a concrete concept on how to replace the traditional auditing process with an automated continuous auditing process. Secondly, there is a lack of an approach that shows how state-of-the-art text mining and machine learning technologies can be used for process automation. Thirdly, there is a lack of an approach to how blockchain can be used to track auditing processes in real time.

3 An AI and Blockchain Platform for ReRAS

A machine learning and blockchain platform is developed in this study to automatically and timely analyzing the risks and its application to ReRAS is completed and consecutively accomplishes four tasks, including text extraction, document classification, compliance checking, and blockchain assurance.

The workflow of the platform performing the tasks is demonstrated in Fig. 1 in which each task is individually annotated. Specifically, when users upload a heterogeneous dataset following the risk policy provided by the enterprise risk division, the platform through its application to ReRAS automatically detects and conducts text mining on both the evidences and policy to extract the information required in the following tasks. The extracted information will be stored in a document-based database, based on which the evidences are classified by learning their textual features in order to determine the compliance level of the uploading process. After updating the database with the classification, the purchase entries specified in the checking

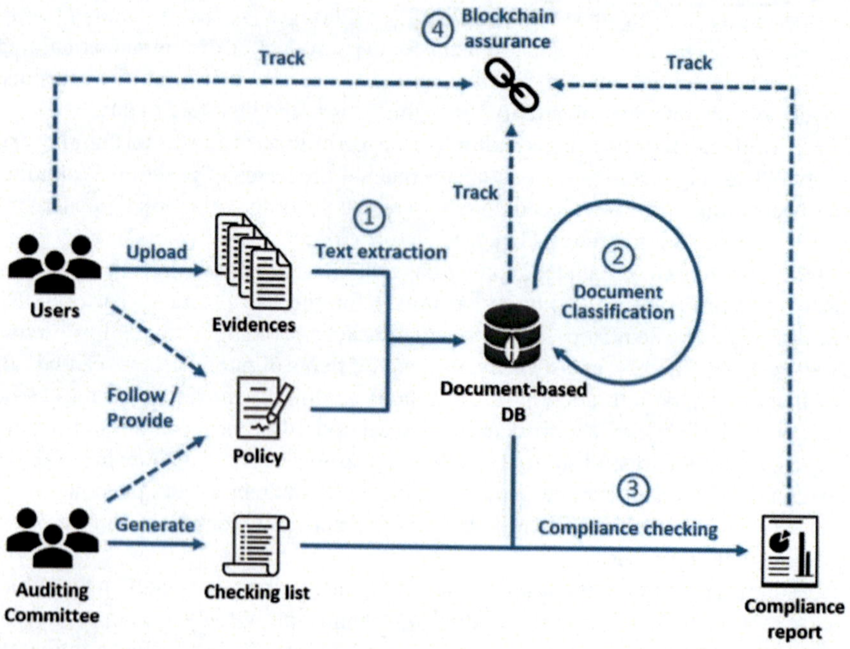

Fig. 1 ReRAS workflow

list from the auditing committee will be matched with the data extracted from the evidences for a numerical compliance audit. Combining both compliance results, a risk report is generated lastly. The heterogeneous data transactions through risk classification are hashed and stored on the blockchain to provide a full, immutable audit trail.

4 Evaluation Design

We prototypically implemented ReRAS and evaluated it on a real-world dataset provided by the procurement audit team of a large-scale public sector. This organization is currently suffering from the difficulties of traditional auditing methods for most of its business perspectives and endeavoring to improve this situation.

The major tasks of the auditing team include initiating annual procurement auditing process, requesting purchasing evidences from purchasing officers, comparing the unit price, quantity, total value, exchange rate, date, and description from the annual financial report form (checking list) against the information in the uploaded evidences, and generating a compliance report to procurement managers. We evaluate ReRAS by automatically accomplishing all these tasks, achieving auditing accuracy higher than satisfactory, as well as completing within a short period of time.

4.1 Dataset and Preprocessing

The organization provided 660 purchasing entries and corresponding evidences for evaluation. The entries are stored in an Excel sheet which contains columns of purchasing number, item name, unit price, quantity, total value, purchasing date, and exchange rate. This sheet is used as the checking list in our evaluation. In addition, the evidences for these 660 purchases are packed into file folders, and each folder is titled by the corresponding purchasing number. Different evidences are included in the folders, such as purchase orders, tax invoices, customer price list, communication emails, and screen captures, all of which can be considered as purchasing evidences according to the procurement policy of the organization. Various formats of files were detected in the uploads, such as .docs, .pdf, .xlsx, .png, .msg, and .zip. Without data preprocessing, there are 1480 files in total in the evidence folders.

4.2 Baseline

The true compliance level of the 660 purchasing entries is absent, and therefore, we need to adopt manual approaches to the obtain the truth as baseline for evaluation. To this end, four postgraduate students were hired to manually check the purchases against the evidences and figure out the real compliance situation of these purchases. They were required to record their working hours during the manual process, as well as the details of their manual auditing results such as insufficient evidences, incorrect uploads, and incorrect purchasing information.

4.3 Evaluation Procedure

The evaluation mainly focuses on three perspectives including auditing accuracy, processing time, and labor consumption.

Regarding evaluating the accuracy of ReRAS, we employ a classification approach to compare the auditing outcome predicted by the proposed ReRAS against the baseline. A confusion matrix is used to define the classes of the four situations, which are true positive (*TP*), false negative (*FN*), true negative (*TN*), and false positive (*FP*). Accordingly, we will measure three classification rates including *Accuracy*, false positive rate (*FPR*), and false negative rate (*FNR*), where the *Accuracy* gives an overall accuracy level of the compliance checking results generated by ReRAS, while the FPR and FNR show the Type I and Type II error rates, respectively, in our case. s As for the processing time, we will record the time that ReRAS spends in processing and analyzing the 660 purchases and the corresponding evidences. Then, this processing time will be compared to the hours spent by the four postgraduate students in obtaining the baseline.

With regard to the labor consumption, we will compare the number of staff usually involved in auditing 660 purchases to that operating the system and generating a compliance report.

5 Results and Discussion

5.1 Accuracy Evaluation Results

The results of the accuracy evaluation are given in Table 1 and insdicate that ReRAS is able to achieve an overall 95.8% accuracy in auditing the compliance level of the 660 purchases. It means that the system has 95.8% chance to predict the same results as the true situations, which is rather promising for a practical application.

In addition, the system achieved 0.007% false positive rate, which means the system almost did not make any Type I error in the evaluation. In other words, the system would not miss any in-compliant purchases in the checking list. Meanwhile, this leads to almost 100% precision rate indicating that the system had full confidence in finding all the compliant purchases without mistakes.

Moreover, the true positive rate reached 92.7%, which denotes the system mistakenly predicted 26 cases as in-compliant purchases while they were in fact compliant. These 26 mistakes belong to Type II error, and they would require extra human effort input in practice to confirm their true compliance situations. To figure out the cause of the mistakes, we manually explored the evidences corresponding to the 26 purchases. It turns out that among the 26 mistakes, 13 cases had the pictures with so low quality that the OCR could not properly extract any valid information from them; another 10 cases had the tax invoices in which the tabled values were not well aligned inside the table cells, thus leading to fault value extraction; and the last three cases were resulted from different round approaches being used by the auditing and purchasing officers. However, 26 out of 660 cases is acceptable according to the auditing officers of the organization as manual checking could also make such number of errors.

Table 1 Accuracy confusion matrix

		Baseline		
		Compliance	Incompliance	
Prediction	Compliance	TP: 347	FP: 2	Precision: 99.4%
	Incompliance	FN: 26	TN: 285	FOR: 8.4%
		TPR: 92.7%	FPR: **0.7%**	
		FNR: 7.3%	TNR: 99.4%	Accuracy: **95.8%**

Table 2 Comparison between ReRAS and manual audit

	Processing time	Labor consumption	Accuracy (%)	Type I and II Error rate (%)
ReRAS	5 min	1 operator & 1 auditing officer	95.8	0.007 and 7.3
Manual audit	120 h	8 auditing officers	≈ 90	≈ 5 and 5

5.2 *Other Evaluation Results*

In addition to accuracy evaluation, we also compared the time and labor consumption between ReRAS and traditional manual auditing approach, which are summarized alongside the results of accuracy and Type I and II error rates in Table 2.

ReRAS and the manual auditing approach are compared based on four indexes. Regarding processing time, ReRAS only took only 5 minutes to complete processing all the evidences of the 660 purchases, while 120 hours in total were spent by the four hired postgraduate students to audit the evidences. Therefore, the system substantially outperformed the manual audit in terms of processing time. In addition, ReRAS only needed one operator to run the system and another auditing officer to verify the final results, while eight auditing officers have been in charge of this manual auditing job in the organization costing more than 120 hours to accomplish it. In this case, ReRAS would save considerably amount of human workforce if ReRAS can be properly employed.

As for the indexes of auditing accuracy and Type I & II error rate, according to the auditing officers working for the organization, manual auditing usually makes 10% mistakes including both type errors due to fatigue, haste, typo, and other human caused fault. On the other hand, the ReRAS obtained 95.8% overall accuracy and 0.007% Type I error, which shows effectiveness and reliability of the system. Although there was 7.3% Type II error rate, these mistakenly picked purchases only cost half an hour of an auditing officer to inspect and confirm their true compliance situation, which was considered as acceptable by the organization.

5.3 *The Platform as a Tool for Risk Management*

The outstanding gains in the efficiency of the proposed continuous compliance auditing system have shown the great potential to reduce the workload of public servants. While this paper has addressed the particular risk of purchasing, there are many more risks in the asset and procurement life cycle which can be addressed with the proposed system. For future work, we are going to implement the auditing policies in a smart contract. This will allow a more transparent view of the policies for all entities involved in the auditing process. Further, we plan to extend our system to not only consider the purchasing risk but also other risks that occur in the entire asset lifecycle of public sectors.

6 Conclusion

Financial audits and risk management are becoming an increasingly important task in public sectors. Massive volumes of hard-copied evidence need to be audited regularly to ensure public funds that are spent responsibility according to policies and regulations. The lack of automated systems to check between computer records and physical (paper-based) records in large organizations leads to inaccurate financial data management and notable human effort in managing both computer and physical records. This is one of root causes to the enterprise risk, including poor asset management, poor productivity, and poor capability of operations. To address this problem, in this paper, we have proposed real-time risk awareness system (ReRAS), a system which enables continuous real-time risk management through continuous compliance auditing. The propsoed ReRAS demonstrates an auditing accuracy of 95.8% and save potential 90% human effort in auditing between the computer records and physical records.

References

1. Bhimani A, Willcocks L (2014) Digitisation,'big data' and the transformation of accounting information. Acc Bus Res 44(4):469–490
2. Dimyadi J, Amor R (2017) Automating conventional compliance audit processes. In: IFIP international conference on product lifecycle management. Springer, 324–334
3. Byrnes PE, Al-Awadhi A, Gullvist B, Brown-Liburd H, Teeter R, Warren JD Jr, Vasarhelyi M (2018) Evolution of auditing: from the traditional approach to the future audit. Continuous auditing: theory and application, pp 285–297
4. Groomer SM, Murthy US (1989) Continuous auditing of database accounting systems using embedded audit modules. J Inf Syst 3(1):53–69
5. Chan DY, Vasarhelyi MA (2011) Innovation and practice of continuous auditing. Int J Acc Inf Syst 12(2):152–160
6. Zhang J, Yang X, Appelbaum D (2015) Toward effective big data analysis in continuous auditing. Acc Horizons 29(2):469–476. http://orcid.org/10.2308/acch-51070
7. Antipova T (2019) Digital public sector auditing: a look into the future. Calitatea 20(S1):441
8. Orumwense JO (2017) Implementation of continuous auditing for the public sector in Nigeria. J Acc Bus Finance Res 1(1):19–23
9. Othman R, Aris NA, Mardziyah A, Zainan N, Amin NM (2015) Fraud detection and prevention methods in the Malaysian public sector: accountants' and internal auditors' perceptions. Procedia Econ Finance 28:59–67
10. Omoteso K, Patel A, Scott P (2008) An investigation into the application of continuous online auditing in the UK. Int J Digital Acc Res 8(14):23–44
11. Elshawi R, Maher M, Sakr S (2019) Automated machine learning: state-of-the-art and open challenges. arXiv:1906.02287

12. Olnes S, Ubacht J, Janssen M (2017) Blockchain in government: Benefits and implications of distributed ledger technology for information sharing, Government Information Quarterly 34(3):355–364. https://doi.org/10.1016/j.giq.2017.09.007
13. Wang T (2018) A unified analytical framework for trustable machine learning and automation running with blockchain. In: 2018 IEEE international conference on Big Data (Big Data). IEEE, pp 4974–4983
14. Neisse R, Steri G, Nai-Fovino I (2017) A blockchain-based approach for data accountability and provenance tracking. In: Proceedings of the 12th international conference on availability, reliability and security, pp 1–10